SPACE TECHNOLOGY AND APPLICATIONS INTERNATIONAL FORUM

SPACE TECHNOLOGY AND APPLICATIONS INTERNATIONAL FORUM (STAIF-2000)

"Bridging the Future-Space Station and Beyond"
January 30-February 3, 2000

Albuquerque Hyatt Hotel, NM

CONFERENCE ON INTERNATIONAL SPACE STATION UTILIZATION

CONFERENCE ON THERMOPHYSICS IN MICROGRAVITY

CONFERENCE ON ENABLING TECHNOLOGY AND REQUIRED SCIENTIFIC DEVELOPMENTS FOR INTERSTELLAR MISSIONS

CONFERENCE ON COMMERCIAL/CIVIL NEXT GENERATION SPACE TRANSPORTATION

17th SYMPOSIUM ON SPACE NUCLEAR POWER AND PROPULSION

Cosponsored by:

THE BOEING COMPANY
LOCKHEED MARTIN CORPORATION
NATIONAL AERONAUTICS AND SPACE
ADMINISTRATION
   Headquarters
   Field Centers

UNITED STATES DEPARTMENT OF ENERGY
   Headquarters
   Los Alamos National Laboratories
   Sandia National Laboratory

In cooperation with:

AMERICAN ASTRONAUTICAL SOCIETY
AMERICAN INSTITUTE OF AERONAUTICS AND
ASTRONAUTICS
   National & Local Sections
AMERICAN INSTITUTE OF CHEMICAL ENGINEERS
   Heat Transfer and Energy Conversion Division
AMERICAN NUCLEAR SOCIETY, *Trinity Section*

AMERICAN SOCIETY OF MECHANICAL ENGINEERS
   Nuclear Engineering Division & Heat Transfer Division
INSTITUTE OF ELECTRICAL AND ELECTRONICS
ENGINEERS, INC.
   Nuclear and Plasma Sciences Society
INTERNATIONAL ASTRONAUTICAL FEDERATION
NASA NATIONAL SPACE GRANT COLLEGE AND
FELLOWSHIP PROGRAM
   New Mexico Space Grant Consortium

Organized by:
INSTITUTE FOR SPACE AND NUCLEAR POWER STUDIES
 School of Engineering, University of New Mexico
Albuquerque, New Mexico
(505) 277-0446, http://www-chne.unm.edu/isnps

SPACE TECHNOLOGY AND APPLICATIONS INTERNATIONAL FORUM

Conference on International Space Station Utilization
Conference on Thermophysics in Microgravity
Conference on Enabling Technology and Required Scientific
   Developments for Interstellar Missions
Conference on Commercial/Civil Next Generation Space
   Transportation
17th Symposium on Space Nuclear Power and Propulsion

Albuquerque, NM   January 2000

PART TWO

EDITOR
Mohamed S. El-Genk

*Institute for Space and Nuclear Power Studies,
University of New Mexico*

AIP CONFERENCE PROCEEDINGS ■ 504   **Melville, New York**

Editor:

Mohamed S. El-Genk
Institute for Space and Nuclear Power Studies
The University of New Mexico
Farris Engineering Center, Room 239
Albuquerque, NM 87131-1341

E-mail: mgenk@unm.edu

L.C. Catalog Card No. 99-069016
ISSN 0094-243X

CD-ROM available: ISBN 1-56396-920-3

Casebound:
ISBN   1-56396-917-3   Part One
        1-56396-918-1   Part Two
        1-56396-919-X   Set

Printed in the United States of America

CONTENTS

PART ONE

CONFERENCE ON INTERNATIONAL
SPACE STATION UTILIZATION

OPENING SESSION: RESEARCH CAPABILITIES
OF THE INTERNATIONAL SPACE STATION

[A01] SPACE STATION INNOVATION DEVELOPED THROUGH THE
SMALL BUSINESS INNOVATION RESEARCH (SBIR) PROGRAM

[A02] PAYLOAD OPERATIONS ON THE INTERNATIONAL SPACE STATION

[A03] SPACE SCIENCES ON THE INTERNATIONAL SPACE STATION
SUB-SESSION I

[A04] LAUNCH SITE INTEGRATION OF PAYLOADS
FOR THE INTERNATIONAL SPACE STATION

[A05] BIOMEDICAL RESEARCH ON THE INTERNATIONAL SPACE STATION
SUB-SESSION I: HUMAN REACTIONS TO SPACEFLIGHT

[A14] GRAVITATIONAL BIOLOGY RESEARCH ON THE INTERNATIONAL SPACE STATION

[A15] COMBUSTION SCIENCE ON THE INTERNATIONAL SPACE STATION

[A16] EARTH SCIENCE AND REMOTE SENSING ON THE INTERNATIONAL SPACE STATION
SUB-SESSION II: EARTH SCIENCE ATTACHED PAYLOADS

[A17] ENGINEERING RESEARCH AND TECHNOLOGY DEVELOPMENT ON THE INTERNATIONAL SPACE STATION
SUB-SESSION I

[A18] PROTEIN CRYSTAL GROWTH RESEARCH ON THE INTERNATIONAL SPACE STATION

[A19] MATERIALS RESEARCH ON THE INTERNATIONAL SPACE STATION

[A20] EARTH SCIENCE AND REMOTE SENSING ON THE INTERNATIONAL SPACE STATION SUB-SESSION III: EARTH SCIENCE WINDOW OBSERVATIONAL RESEARCH FACILITY (WORF) PAYLOADS

[A21] ENGINEERING RESEARCH AND TECHNOLOGY DEVELOPMENT ON THE INTERNATIONAL SPACE STATION SUB-SESSION II

[A22] FUNDAMENTAL BIOTECHNOLOGY AND CELL TISSUE CULTURE RESEARCH ON THE INTERNATIONAL SPACE STATION

[A23] MICROGRAVITY MEASUREMENT AND DEVICES

[A24] INNOVATIVE APPROACHES TO COMMERCIAL ACTIVITIES ON THE INTERNATIONAL SPACE STATION: WHAT CAN THE PRIVATE SECTOR DO? SUB-SESSION I

[A28] COMMERCIAL COMMUNICATIONS SYSTEMS FOR THE INTERNATIONAL SPACE STATION
(NO PAPERS, PRESENTATIONS ONLY)

PART TWO

CONFERENCE ON THERMOPHYSICS IN MICROGRAVITY

[B01] FUNDAMENTALS OF TWO-PHASE FLOW AND HEAT TRANSFER IN MICROGRAVITY

[B02] EMERGING THERMAL CONTROL TECHNOLOGIES FOR FUTURE SPACECRAFT—I

[B03] EMERGING THERMAL CONTROL TECHNOLOGIES FOR FUTURE SPACECRAFT—II

[B04] EMERGING THERMAL CONTROL TECHNOLOGIES FOR FUTURE SPACECRAFT—III

[B05] FLUID PHYSICS RESEARCH IN MICROGRAVITY—I

[B06] FLUID PHYSICS RESEARCH IN MICROGRAVITY—II

CONFERENCE ON ENABLING TECHNOLOGY AND REQUIRED SCIENTIFIC DEVELOPMENTS FOR INTERSTELLAR MISSIONS

[C01] INTERSTELLAR MISSION CONCEPTS

[C02] PROPULSION TECHNOLOGIES FOR INTERSTELLAR PRECURSOR MISSIONS

[C03] PROPULSION TECHNOLOGIES FOR INTERSTELLAR FLYBY AND RENDEZVOUS MISSIONS

[C04] TOWARD BREAKTHROUGH PROPULSION: EXPERIMENTS WITH ELECTROMAGNETISM, SPACE, AND GRAVITY

[C05] TOWARD BREAKTHROUGH PROPULSION: QUANTUM VACUUM PHYSICS

[C06] TOWARD BREAKTHROUGH PROPULSION: VARIOUS ISSUES AND POSSIBILITIES

[C07] TOWARD BREAKTHROUGH PROPULSION: PHYSICS OF HYPERFAST TRAVEL

CONFERENCE ON COMMERCIAL/CIVIL NEXT GENERATION SPACE TRANSPORTATION

[D01] MARKETS—PRESENT AND FUTURE

[D02] CONTINUED EVOLUTION AND UP-GRADING OF TODAY'S FLEETS

[D03] NEW VEHICLE DEVELOPMENTS

[D04] TECHNOLOGY FOR COMMERCIAL AND CIVIL PROJECTS

[D05] FINANCING AND FUNDING THE ENTERPRISE
(NO PAPERS, PRESENTATIONS ONLY)

[D06] TECHNOLOGIES FOR FUTURE REUSABLE LAUNCH VEHICLES
(NO PAPERS, PRESENTATIONS ONLY)

[D07] SPACEPORT DEVELOPMENTS—I
(NO PAPERS, PRESENTATIONS ONLY)

[D08] SPACEPORT DEVELOPMENTS—II

17th SYMPOSIUM ON SPACE NUCLEAR POWER AND PROPULSION

[E01] POTENTIAL MANNED AND UNMANNED MISSIONS

[E02] NUCLEAR SYSTEM TESTING PROGRESS/RESULTS

[E03] ADVANCED PROPULSION CONCEPTS—I

[E04] STIRLING ENERGY CONVERSION

[E05] ADVANCED PROPULSION CONCEPTS—II

[E10] FUSION SPACE SYSTEMS APPLICATION

[E11] FUELS AND ADVANCED MATERIALS

[E12] ADVANCED RADIOISOTOPE POWER SYSTEMS

[E13] THERMOELECTRIC ENERGY CONVERSION AND APPLICATIONS

[E14] MULTI-MEGAWATT POWER SYSTEMS

INTRODUCTION

The technical program for the 2000 Space Technology and Applications International Forum (STAIF-2000) is one of the largest of its kind in terms of the number of hosted technical sessions and technical papers accepted for presentation at five conferences. The plenary as well as the technical sessions at STAIF-2000 emphasize this year's theme *"Bridging the Future-Space Station and Beyond."* The STAIF-2000 conferences cover a broad spectrum of topics in space science and technology. These topics span the range from basic research, such as fluid physics, biomedical research, and thermophysics in microgravity, to the most recent advances in space power and propulsion, next generation commercial/civil launch systems, and the current status, future utilization and planned research and technology development on board the International Space Station. Additional topics are related to interstellar and manned and unmanned planetary missions, advanced radioisotope space power systems, high energy electrical propulsion, high temperature materials, thermal management and thermal control of future spacecraft, and advances in energy conversion.

STAIF has been a very successful forum, since its inception in 1993, for information exchange and technical and professional interaction among members of academia, industry, government, program managers, and developers of space technologies in the US and abroad. STAIF actively promotes international participation and collaboration, which is reflected in the membership of the various committees, the conferences organizers, the invited speakers at the plenary sessions, and the many international papers in the technical program.

STAIF-2000 conferences, which share complementary and synergistic interests in many topical areas of planetary exploration, space science and technology, microgravity research, advanced space power and propulsion, space commercialization, and the international space station utilization, are:

Conference on International Space Station Utilization
Chairs: ***Tommy Holloway*** NASA Johnson Space Center, and
Brewster Shaw, The Boeing Company
Conference on Thermophysics in Microgravity
Chairs: ***Ad Delil***, National Aerospace Laboratory Space Division, The Netherlands, and
Ted Swanson, NASA Goddard Space Flight Center
Conference on Enabling Technology & Required Scientific Developments for Interstellar Missions
Chairs: ***Les Johnson***, NASA Marshall Space Flight Center, and
Marc Millis, NASA Glenn Research Center at Lewis Field
Conference on Commercial/Civil Next Generation Space Transportation
Chairs: ***Gary Payton***, NASA Headquarters, and
William A. Gaubatz, Universal Space Lines
17th Symposium on Space Nuclear Power and Propulsion
Chairs: ***Michael G. Houts***, Los Alamos National Laboratory/ NASA Marshall Space Flight Center, and
Robert Wiley, Booz-Allen & Hamilton, Inc.

The STAIF-2000 program features four plenary sessions. The two plenaries on Monday morning are *"Views from the Top"* and *"Programs and Technology"*. The plenary session on Tuesday morning is *"Human Physiology in Space"* and that on Wednesday morning is *"International Panel on Collaboration in the Use of the ISS"*. In addition, the program includes a special session, *"The Commercial Space Act – One Year Later"* on Wednesday evening. At these sessions, prominent national and international speakers from governmental agencies, industry and academia address timely topics related to this year's theme. We are grateful to the organizers, chairs, and the speakers at these sessions, and to the members of STAIF Steering and Advisory Committees for their effort in developing the meeting's theme and for organizing the plenary and special sessions.

We are very grateful to the program chairs and members of technical committees of the hosted conferences. Many thanks are due to the members of the executive technical program committee, and the chairs and co-chairs of the technical sessions for all their hard work and dedication in organizing this year's technical program. Special thanks to John-David Bartoe of NASA Johnson Space Flight Center and James Fountain of The Boeing Company for organizing and working out the details of the technical program of the International Space Utilization Conference. The contribution of these two individuals to the organization of the plenary and special sessions is appreciated.

We wish to express our appreciation to the authors and speakers as well for their contributions to the technical program and to the publications in the STAIF-2000 Proceedings.

The organizers of the **outreach program** of STAIF-2000 are celebrating the 13th anniversary of the Space Design Competition, an annual event which started in 1988 in conjunction with the 5th Symposium on Space Nuclear Power. This half a day program on Monday morning includes viewing and judging the entries of this year's space design competition *"Searching for Life on Europa,"* by Secondary School students from throughout the State of New Mexico. More than 150 students, teachers, and parents are expected to participate in this year's activities. The students of the best Space Design Competition projects, selected by members of the space science and technology community attending STAIF-2000, and their supervising teachers are recognized and receive their awards at the STAIF Luncheon. Another component of the outreach program is the high school special session which features a number of invited speakers on timely space science and technology topics of interest to the participating students, teachers, and parents. Special thanks are due to the organizers of these worthwhile events. The organizers are Irene El-Genk, West Mesa High School in Albuquerque, NM; Jeff King, University of New Mexico's Institute for Space and Nuclear Power Studies (ISNPS), and the members of the ISNPS's Education Outreach Advisory Board. This year's Outreach program is co-sponsored by *NASA's New Mexico Space Grant Consortium, the American Nuclear Society Trinity Section, and the University of New Mexico's Institute for Space and Nuclear Power Studies.*

The STAIF-2000 technical program offers 315 papers, shared among 72 topical sessions and a half day *"Hardware, Multimedia Display Session"* in which experimental and flight hardware and prototypes of components or actual systems are displayed. This unique session has more than ten entries on display for all conference attendees following the topical sessions on Monday.

I am pleased to introduce STAIF-2000's two-volume proceedings. This is the second year that STAIF proceedings are published in a searchable CD. Hard copies of this year's proceedings are also available at the American Institute of Physics (AIP) upon request. We are grateful to the authors for their cooperation and commitment to contribute to a clear and technically sound achievable publication. Special thanks are due to the ISNPS staff: Mary Bragg, STAIF Administrative Chair, and Carla Rogers and Yolanda Sanchez, Administrative Co-Chairs. Their help, dedication, hard work, and commitment in preparing and compiling the entries to the program as well as in editing and compiling the final manuscripts are greatly appreciated. Their effective communication with the session and conference organizers, authors, members of various committees, and the coordination of the many functions and events in the program are the key to the success of this annual event.

On behalf of the STAIF-2000 committees, I wish to express our thanks to the sponsoring organizations, national societies, and to the participating organizations from government, industry, national laboratories, and universities for their contribution to this year's program. Special thanks are also due to the many participating international organizations for their input and contributions to this year's program. We also wish to acknowledge the contributions of the exhibitors for their timely and informative displays on the latest in space technology and hardware, which are an integral part of this successful annual meeting.

We are grateful to the contributing organizations for helping to defray the cost of attending STAIF-2000, making it most affordable to professionals, academicians, and students. We are also grateful to the sponsors of this year's industry hospitality for their generosity of sponsoring one of the highlighted events of this annual meeting.

Without the commitment, dedication, and contribution of many individuals on the various STAIF committees, and of the sponsoring, contributing, and participating organizations, this year's meeting would not have been possible. My heartfelt thanks to the families of the members of the ISNPS staff and to my family for their understanding, patience, and continued encouragement and support through the demanding task of organizing this year's events. Special thanks are due to the University of New Mexico, its School of Engineering, and Department of Chemical and Nuclear Engineering for the continued support and encouragement.

This year's proceedings are dedicated to the memory of a highly recognized and appreciated colleague and friend of STAIF and the space commercialization community, *Ray Whitten, Sr.*, who passed away in 1999. The unexpected death of our young friend, Ray, is a tragic loss to our community. He is greatly missed.

Mohamed S. El-Genk
Regents' Professor and Director, ISNPS
STAIF Technical and Publication Chairman

DEDICATION

To The Memory of
Raymond P. Whitten, Sr., 1937-1999
Colleague and Friend

In 1984 NASA announced a new program to focus on the commercial development of space. The primary vehicle for this development was the Centers for the Commercial Development of Space (CCDS). Ray Whitten became a member of this new program because he believed in the great potential of space access for significant positive economic impact for the United States and eventually the world. He became the program's strongest advocate and its sharpest critic. He recognized that commercial results would come slowly but he insisted on progress. He personally exemplified the entrepreneurial spirit so essential to innovation and creativity. He constantly searched for new areas of promise for commercial initiatives. He was thoughtful, inquisitive, insightful and imaginative. He helped others to think clearly by his inquiring approach.

For many years he searched for a forum and means to display and record the activities and successes of the CCDS program in archival literature. In 1995, he initiated an association for the program with the Space Nuclear Power and Propulsion Symposium organized by the University of New Mexico's Institute for Space and Nuclear Power Studies, which resulted in the birth of an international forum for space technology. He was instrumental in the selection of the name and the founding of the Space Technology and Applications International Forum (STAIF), which hosts annually a number of conferences combined with a key focus on space technology, in addition to the annual Symposium of Space Nuclear Power and Propulsion. The establishment of STAIF, to the credit of Ray Whitten, has proven to be an effective platform and showcase of the research, development, and advances in space science and technology in many fields.

As the International Space Station approached its first element launch, Ray spearheaded the development of a conference focused on the ISS as a research platform for industry and integrated it with the annual presentation of results by the CCDS's. He was responsible for the introduction of STAIF to the International Space Station Utilization (ISSU) program at NASA. Since 1996, STAIF has become an effective international forum and a home for the annual ISSU conference.

Ray was never satisfied that he had done as much as he could. He always believed his efforts could have been better. By example, he set a very high standard of personal excellence for those around him. At the same time he was always a friend and often a mentor to this same group. He loved his work but he loved those around him more. And he loved life! To know Ray was to experience loyalty, diligence, and dedication in all he did, in his professional life and in his personal life. He deeply impacted the lives of many, many individuals. His early career contributions significantly enhanced pilot safety in aircraft. His work with young people as a Scoutmaster guided many to sound maturity and personal excellence. His collegial manner with his colleagues stimulated openness and unselfishness. He made a difference.

Ray will be missed but never forgotten for his legacy is in the minds and hearts of those with whom he came into contact, throughout his life. In memory of his most significant life, these Proceedings are dedicated to our friend and colleague.

ORGANIZING COMMITTEE

GENERAL CO-CHAIRS

Michael Griffin
Orbital Sciences Corporation
Dulles, VA

William D. Magwood IV
Department of Energy
Washington, DC

Arnauld Nicogossian
NASA Headquarters
Washington, DC

Brewster Shaw
The Boeing Company
Houston, TX

TECHNICAL AND PUBLICATION CHAIR
Mohamed S. El-Genk, The University of New Mexico (UNM)
Institute for Space & Nuclear Power Studies (ISNPS)

ADMINISTRATION

Mary J. Bragg
STAIF Administrative Chair
UNM-ISNPS

Carla Y. Rogers
STAIF Administrative Co-Chair
UNM-ISNPS

Yolanda M. Sanchez
STAIF Administrative Co-Chair
UNM-ISNPS

EDUCATION OUTREACH

Irene L. El-Genk, Chair
Secondary School Special Session
West Mesa High School

Jeff King, Chair
Space Design Competition
UNM-ISNPS

Carla Y. Rogers
UNM-ISNPS

CONFERENCE ON INTERNATIONAL SPACE STATION UTILIZATION
PROGRAM CO-CHAIR: **Tommy Holloway**, NASA Johnson Space Center, Houston, TX
PROGRAM CO-CHAIR: **Brewster Shaw**, The Boeing Company, Houston, TX

CONFERENCE ON THERMOPHYSICS IN MICROGRAVITY
PROGRAM CHAIR: **Ad Delil**, National Aerospace Laboratory Space Division, The Netherlands
PROGRAM CO-CHAIR: **Ted Swanson**, NASA Goddard Space Flight Center, Greenbelt, MD

CONFERENCE ON ENABLING TECHNOLOGY AND REQUIRED SCIENTIFIC DEVELOPMENTS FOR INTERSTELLAR MISSIONS
PROGRAM CHAIR: **Les Johnson**, NASA Marshall Space Flight Center, Huntsville, AL
PROGRAM CO-CHAIR: **Marc Millis**, NASA Glenn Research Center at Lewis Field, Cleveland, OH

CONFERENCE ON COMMERCIAL/CIVIL NEXT GENERATION SPACE TRANSPORTATION
PROGRAM CHAIR: **Gary Payton**, NASA Headquarters, Washington, DC
PROGRAM CO-CHAIR: **William Gaubatz**, Universal Space Lines, Newport Beach, CA

17th SYMPOSIUM ON SPACE NUCLEAR POWER AND PROPULSION
PROGRAM CHAIR: **Michael Houts**, LANL/NASA Marshall Space Flight Center, Huntsville, AL
PROGRAM CO-CHAIR: **Robert Wiley**, Booz-Allen & Hamilton, Arlington, VA

STEERING COMMITTEE

Michael Griffin, Chair
Executive Vice President and Chief Technical Officer
Orbital Sciences Corporation

Bonnie J. Dunbar
Associate Director for University
Resarch and Affairs
NASA Johnson Space Center
William C. Gordon
President
The University of New Mexico
Noel Hinners
Vice President-Flight Systems
Lockheed Martin Astronautics

Arnauld Nicogossian
Associate Administrator, Life and
Microgravity Sciences & Applications
NASA Headquarters
Shigeaki Nomura
Technical Special Advisory
Nat'l Space Dev. Agency of Japan
Ian Pryke
Head of European Space Agency,
Washington Office
European Space Agency

Michael J. Sander
Dir., Tech. & Applic. Programs
Jet Propulsion Laboratory
Brewster Shaw
Vice President
General Manager, International Space
Station, The Boeing Company
Earl Wahlquist
Deputy Associate Director
Office of Engineering and Tech. Dev.
U. S. Department of Energy,
Headquarters

ADVISORY COMMITTEE

Mohamed S. El-Genk, Chair
The University of New Mexico

John-David Bartoe
NASA Johnson Space Flight Center
Debra Bennett
Los Alamos National Laboratory
Dennis Berry
Sandia National Laboratories
Samit K. Bhattacharyya
Argonne National Laboratory
Stanley K. Borowski
NASA Glenn Research Center
David Boyle
Texas A&M University
Ad Delil
National Aerospace Laboratory
Space Division, The Netherlands
Jim Fountain
The Boeing Company
William Gaubatz
Universal Space Lines
Tim Gillespie
Lockheed Martin Astronautics
Richard Hemler
Martin Marietta Astro Space

Rodney Herring
Canadian Space Agency
Mark D. Hoover
Lovelace Respiratory Research Inst. ·
Michael Houts
LANL/NASA Marshall Space Flight Center
Tom Hunt
Advanced Modular Power Systems
Mary Kicza
NASA Goddard Space Flight Center
Gerald Kulcinski
University of Wisconsin
Clay Mayberry
Air Force Research Laboratory
John Metzger
State University of New York, Stony Brook
George H. Miley
University of Illinois
Jack Mondt
Jet Propulsion Laboratory
Mark Nall
NASA Marshall Space Flight Center

Gary Payton
NASA Headquarters
Ian Pryke
European Space Agency
Lyle Rutger
U. S. Department of Energy
Harrison Schmitt
Consultant
Joseph A. Sholtis, Jr.
Sholtis Engineering & Safety Consulting
R. Joseph Sovie
NASA Glenn Research Center
Ted Swanson
NASA Goddard Space Flight Center
Sadayuki Tsuchiya
Nat'l Space Dev. Agency of Japan
Giulio Varsi
Jet Propulsion Laboratory
Atsutaro Watanabe
Nat'l Space Dev. Agency of Japan
Bob Wiley
Booz-Allen & Hamilton, Inc.

EXECUTIVE TECHNICAL PROGRAM COMMITTEE

Mohamed S. El-Genk, Chair
The University of New Mexico

John-David Bartoe
NASA Johnson Space Flight Center
Frederick R. Best
Texas A&M University
Ad Delil
National Aerospace Laboratory
Space Division, Netherlands
Jim Fountain
The Boeing Company

William A. Gaubatz
Universal Space Lines
Tommy Holloway
NASA Johnson Space Center
Michael G. Houts
Los Alamos National Laboratory
Les Johnson
NASA Marshall Space Flight Center
Marc Millis
NASA Glenn Research Center

Gary Payton
NASA Headquarters
Brewster Shaw
The Boeing Company
Ted Swanson
NASA Goddard Space Flight Center
Robert Wiley
Booz-Allen & Hamilton

TECHNICAL PROGRAM COMMITTEES

CONFERENCE ON INTERNATIONAL SPACE STATION UTILIZATION

Tommy Holloway, Program Co-Chair
NASA Johnson Space Center

Brewster Shaw, Program Co-Chair
The Boeing Company

Iwan Alexander
Case Western Reserve
University
John Baras
University of Maryland
Shannon Bartell
NASA Kennedy Space Center
John-David Bartoe
NASA Johnson Space Center
Kul Bhasin
NASA Glenn Research Center
David Boyle
Texas A&M University
Frank Buzzard
NASA Johnson Space Center
Vita Cevenini
NASA Headquarters
Kathryn Clark
NASA Headquarters
Sharon Cobb
NASA Marshall Space Flight
Center
Richard DeLombard
NASA Glenn Research Center
Mark Deuser
Space Hardware Optimization
Technology

Jim Fountain
The Boeing Company
Jeffrey Irons
Teledyne Brown Engineering
Robert Jackson
NASA Ames Research Center
Gary Jahns
NASA Ames Research Center
Terry Johnson
BioServe Space Technologies
Michael Kearney
SPACEHAB, Inc.
John Kelley
NASA Headquarters
Feng-Chuan Liu
Jet Propulsion Laboratory
George May
ITD Space Remote Sensing Center
Giorgio Palumbo
University of Bologna, Italy
Betsy Park
NASA Goddard Space Flight Center
Neal Pellis
NASA Johnson Space Center
Ron Porter
NASA Marshall Space Flight Center
William Powell
NASA Marshall Space Flight Center

Howard Ross
NASA Glenn Research Center
Al Sacco
Northeastern University
Charles Sawin
NASA Johnson Space Center
Suzanne Schneider
NASA Johnson Space Center
Frank Schowengerdt
Colorado School of Mines
Kathy Schubert
NASA Glenn Research Center
David Seidel
Jet Propulsion Laboratory
Eun-Suk Seo
University of Maryland
Cathy Shields
The Boeing Company
Bhim Singh
NASA Glenn Research Center
Maynette Smith
NASA Kennedy Space Center
Helen Stinson
NASA Marshall Space Flight
Center
Mark Uhran
NASA Headquarters

CONFERENCE ON THERMOPHYSICS IN MICROGRAVITY

Ad Delil, Program Chair
National Aerospace Laboratory
NLR Space Division, Netherlands

Ted Swanson, Program Co-Chair
NASA Goddard Space Flight Center

Fred Best
Texas A&M University
Walt Bienert
Dynatherm
Dan Butler
NASA Goddard Space Flight Center
Martin Donabedian
Aerospace Corporation
Rodney Herring
Canadian Space Agency

Jean-Claude Legros
University of Brussels
Hans Rath
Center of Applied Space Technology
and Microgravity (ZARM)
Thomas Reinarts
United Technologies/USBI
Ziad Saghir
Consultant

CONFERENCE ON ENABLING TECHNOLOGY AND REQUIRED SCIENTIFIC DEVELOPMENTS FOR INTERSTELLAR MISSIONS

Les Johnson, Program Chair
NASA Marshall Space Flight Center

Marc Millis, Program Co-Chair
NASA Glenn Research Center

Leo Bitteker
NASA Marshall Space Flight Center
Sarah Gavit
Jet Propulsion Laboratory
Al Holt
NASA Johnson Space Center
James Ling
NASA Headquarters
Claudio Maccone
Alenia Spazio

G. Jordan Maclay
Quantum Fields, LLC
Gregory Matloff
New York University/New York City Technical College
F. Michael Serry
Digital Instruments, Inc.
Frank Mead, Jr.
Air Force Research Labs-AFRL/PRSP
Giovanni Vulpetti
Telespazio SpA

CONFERENCE ON COMMERCIAL/CIVIL NEXT GENERATION SPACE TRANSPORTATION

Gary Payton, Chair
NASA Headquarters

William Gaubatz, Program Co-Chair
Universal Space Lines

Herb Bachner
Federal Aviation Administration
Paul Birkeland
Kistler Aerospace
Hector Cueller
Banc of America Securities LLC
Bill Dettmer
New Mexico Space Commission

Lori Garver
NASA Headquarters
Karen Poniatowski
NASA Headquarters
Row Rogacki
NASA Marshall Space Flight Center
Jess Sponable
Universal Space Lines Inc.

17TH SYMPOSIUM ON SPACE NUCLEAR POWER AND PROPULSION

Michael G. Houts, Program Chair
Los Alamos National Laboratory
NASA Marshall Space Flight Center

Robert Wiley, Program Co-Chair
Booz-Allen & Hamilton

Samin Anghaie
University of Florida
Lester Begg
General Atomics
Deborah Bennett
Los Alamos National Laboratory
Samit K. Bhattacharyya
Argonne National Laboratory
Thierry Caillat
Jet Propulsion Laboratory
Robert Carpenter
Orbital Sciences Corporation
Mohamed S. El-Genk
The University of New Mexico
Bill Emrich
NASA Marshall Space Flight Center
Thomas Godfroy
NASA Marshall Space Flight Center
Lisa Herrera
U. S. Department of Energy

Ivana Hrbud
NASA Marshall Space Flight Center
Thomas K. Hunt
Advanced Modular Power Systems
Terry Kammash
University of Michigan
Roger X. Lenard
Sandia National Laboratories
Ron Lipinski
Sandia National Laboratories
George H. Miley
University of Illinois
Jiro Nagao
Hokkaido National Industrial Res. Inst.
Yuri Nikolaev
Scientific Industrial Assoc., "Lutch"
J. Boise Pearson
NASA Marshall Space Flight Center
James Polk
Jet Propulsion Laboratory

David I. Poston
Los Alamos National Laboratory
Lyle L. Rutger
U. S. Department of Energy
Amy Ryan
Jet Propulsion Laboratory
George Schmidt
NASA Marshall Space Flight Center
Al Schock
Orbital Sciences Corporation
Michael J. Schuller
Texas A&M University
Richard Shaltens
NASA Glenn Research Center
Joseph A. Sholtis, Jr.
Sholtis Engineering & Safety Consulting
Jean-Michel Tournier
The University of New Mexico-ISNPS
Melissa Van Dyke
NASA Marshall Space Flight Center

CONTRIBUTING ORGANIZATIONS

The Boeing Company
Lockheed Martin Corporation
NASA Headquarters and Field Centers
Sandia National Laboratories
U.S. Department of Energy

AWARDS AND OUTREACH

SCHREIBER-SPENCE ACHIEVEMENT AWARD

AWARD COMMITTEE: **Stan Borowski**, NASA Glenn Research Center, **Robert Cockfield**, Lockheed Martin; **Jim Fountain**, The Boeing Company; **Jess Sponable**, Universal Space Lines; and **Guilio Varsi**, Jet Propulsion Laboratory; **John Wheeler**, Department of Energy

OUTSTANDING PAPER AWARD

AWARD COMMITTEE: **Dave Poston** (Chair), Los Alamos National Laboratory; **Jim Fountain**, The Boeing Company; **George Miley**, University of Illinois; **Mel Montemerlo**, NASA Headquarters; **Nick Morley**, Air Force Research Laboratory; **Joseph A. Sholtis, Jr.**, Sholtis Engineering & Safety Consulting; and **Theodore Swanson**, NASA Goddard Space Flight Center

MANUEL LUJAN, JR. STUDENT PAPER AWARD

AWARD COMMITTEE: **Tom Reinarts** (Chair), United Technologies; **Les Begg**, General Atomics, **David Boyle**, Texas A&M University; **Ron Lipinski**, Sandia National Laboratories; **Greg Matloff**, New York University/ New York City Technical College; **Dennis Pelaccio**, Science Applications International Corporation; **Jean-Michel Tournier**, University of New Mexico/Institute for Space & Nuclear Power Studies; and **Jonathan Stabb**, NASA Kennedy Space Center

PUBLICATIONS

<u>Available from UNM's Institute for Space and Nuclear Power Studies</u> (Add $10 for shipping and handling within the U.S., $25 outside the U.S.)

Transactions of the 2nd - 5th Symposia (1985 - 1989) .... $10.00 (each)
Transactions of the 6th Symposium (1989) $15.00

<u>Available from the American Institute of Physics, c/o Springer-Verlag New York, Customer Service, 1-800-777-4643, or e-mail orders@springer-ny.com, or mail to Springer-Verlag, P. O. Box 2485, Secaucus, NJ 07096-2485, USA</u> (For North America, add $4.00 for shipping and handling for the first volume; plus $1.00 for each additional volume. For orders outside of North America, add $10.00 for first volume and $5.00 for each additional volume.)

Proceedings of the 8th Symposium (**1991**) (3-vol. hardcover set), ISBN 0-88318-838-4
    AIP Conference Proceedings 217 .... $175.00
Proceedings of the 9th Symposium (**1992**) (3-vol. hardcover set), ISBN 1-56396-027-3
    AIP Conference Proceedings 246 .... $225.00
Proceedings of the 10th Symposium (**1993**) (3-vol. hardcover set), ISBN 156396-137-7
    AIP Conference Proceedings 271 .... $275.00
A Critical Review of Space Nuclear Power and Propulsion (1984-1993) (Anniversary Issue)
    AIP Press, ISBN 1-56396-317-5 .... $ 75.00
Proc. 12th Symposium on Space Nuclear Power and Propulsion, Conf. on Alternative Power from Space,
    and Conf. on Accelerator-Driven Transmutation Technologies and Applications (**1995**) (2-vol. hardcover set)
    ISBN 1-56396-427-9 AIP Conference Proceedings 324 .... $225.00
Proc. 1st Conf. on NASA Centers for Commercial Development of Space (1-vol. hardcover book),
    ISBN 1-56396-431-7 AIP Conference Proceedings 325 .... $125.00
Proc. Space Technology and Applications International Forum (**STAIF-96**): 1st Conference on Commercial
    Development of Space; 1st Conference on Next Generation Launch Systems, 2nd Spacecraft Thermal Control
    Symposium, and 13th Symposium on Space Nuclear Power and Propulsion (1996) (3-vol. hardcover set),
    ISBN 1-56396-562-3 AIP Conference Proceedings 361 .... $275.00
Proc. Space Technology and Applications International Forum (**STAIF-97**): 1st Conference on Future Science and
    Earth Science Missions; 1st Conference on Synergistic Power and Propulsion Systems Technology; 1st
    Conference on Applications of Thermophysics in Microgravity; 2nd Conference on Commercial Development
    of Space; 2nd Conference on Next Generation Launch Systems; 14th Symposium on Space Nuclear Power and
    Propulsion (1997) (3-vol. Hardcover set), ISBN 1-56396-679-4 AIP Conference Proceedings 387 .... $295.00
Proc. Space Technology and Applications International Forum (**STAIF-98**): 1st Conference on Global Virtual
    Presence; 1st Conference on Orbital Transfer Vehicles; 2nd conference on Applications of Thermophysics in
    Microgravity; 3rd Conference on Commercial Development of Space; 3rd Conference on Next Generation
    Launch Systems; and 15th Symposium on Space Nuclear Power and Propulsion (1998) (3-vol. hardcover set),
    ISBN 1-56396-747-2 AIP Conference Proceedings 420 .... $320.00
Proc. Space Technology and Applications International Forum (**STAIF-99**): Conference on International Space
    Station Utilization; Conference on Global Virtual Presence; Conference on Applications of Thermophysics in
    Microgravity & Breakthrough Physics; Conference on Next Generation Launch Systems; 16th Symposium on
    Space Nuclear Power and Propulsion (1999), AIP Conference Proceedings 458, (2-vol. hardcover set),
    ISBN 1-56396-846-0 .... $300.00
    CD-ROM Version, ISBN 156396-879-7 .... $200.00
Proc. Space Technology and Applications International Forum (**STAIF-2000**): Conference on International Space
    Station Utilization; Conference on Thermophysics in Microgravity; Conference on Enabling Technology and
    Required Scientific Developments for Interstellar Missions; Conference on Commercial/Civil Next Generation
    Space Transportation; 17th Symposium on Space Nuclear Power and Propulsion (2000), AIP Conference
    Proceedings 504, (2-vol. hardcover set), ISBN 1-56396-919-X.. $300.00
    CD-ROM Version, ISBN 156396-920-3 .... $200.00

<u>Publications available from Orbit Book Company, P. O. Box 9542, Melbourne, FL 32902-9542,</u>
<u>Phone: (407) 724-9542</u>
Space Nuclear Power Systems (1984 - 1989) set $500.00

CONFERENCE ON THERMOPHYSICS IN MICROGRAVITY

Properties of Electrostatically-Driven Granular Medium: Phase Transitions and Charge Transfer

I.S. Aranson, D. Blair, V. A. Kalatsky, G.W. Crabtree, W.-K. Kwok, V.M. Vinokur and U. Welp

Materials Science Division, Argonne National Laboratory, 9700 South Cass Avenue, Argonne, IL 60439

phone: (630)-252-9725; e-mail: aronson@vortex.msd.anl.gov

Abstract. The experimental and theoretical study of electrostatically driven granular material are reported. It is shown that the charged granular medium undergoes a hysteretic first order phase transition from the immobile condensed state (granular solid) to a fluidized dilated state (granular gas) with a changing applied electric field. In addition a spontaneous precipitation of dense clusters from the gas phase and subsequent coarsening – coagulation of these clusters is observed. Molecular dynamics simulations shows qualitative agreement with experimental results.

INTRODUCTION

The dynamics of granular materials with their strong contact interactions and inelastic collisions between grains poses a challenge for physicists and engineers (Jaeger,1996; Kadanoff,1999; Melo,1995; Umbanhower,1996; Umbanhower,1998) . Peculiar properties due to new fascinating collective behaviors appear when small particles acquire an electric charge and respond to competing long-range electromagnetic and short range contact forces. Although of prime practical importance the properties of charged granular media and their response to the external field are still not fully understood.

The electrostatic excitation of granular medium offers unique new opportunities compared to traditional vibration techniques which have been developed to explore granular dynamics (Melo,1995; Umbanhower,1996; Umbanhower,1998; Hunt,1999; Aranson,1999; Boothryd,1971; Iinoya,1990, Tennakoon,1998). It enables one to deal with extremely fine powders under low gas pressure or zero gravity conditions. Fine particles are more sensitive to electrostatic forces which arise through particle friction or space charges (Fayed,1997). Their large surface to volume ratio amplifies the effect of water or other surfactants. These effects intervene in the dynamics causing agglomeration, charging, cohesion etc, making mechanical experiments uncontrollable. Electrostatic driving utilizes these *bulk* forces, and allows also for control of dynamic behaviors by long-range electric forces. Investigation of the collective dynamics of very small particles has an additional advantage of being attained to "thermodynamically large" ensembles of about 10^7 grains.

With this paper, we present our experimental and theoretical study of electrostatically driven granular material (Aranson,1999). We show that the charged granular medium undergoes a hysteretic first order phase transition from the immobile condensed state (granular solid) to a fluidized dilated state (granular gas) with a changing applied electric field. Charging of the particles results in electric current and possible increase of heat transfer. This current vanishes at the onset of particle motion at the critical field and grows gradually with the increase of the field. In addition we observe a spontaneous precipitation of dense clusters from the gas phase. These clusters then evolve via coarsening dynamics: small clusters disappear and large clusters grow. Eventually the large isolated clusters assume almost perfect circular form. In the process of coarsening the electric current also changes in time due to decrease of particles in the gas phase. We establish that the origin of coarsening dynamics and hysteresis is due to a screening of the electric field in dense particle clusters. Our results show surprisingly high sensitivity of the phase boundaries to surface coating of the grains due to humidity.

CP504, *Space Technology and Applications International Forum–2000*, edited by M. S. El-Genk

EXPERIMENTAL SET-UP

The experimental cell is shown in Figure 1 Conducting particles are placed between the plates of a large capacitor which is energized by a constant or alternating electric field. To provide optical access to the cell, the upper conducting plate is made transparent. We used 4×6 cm capacitor plates with a spacing of 1.5 mm and 35 μm copper powder. The field amplitude varied from 0 to 10 kV/cm and the frequencies on the interval of 0 to 250 Hz. The representative number of particles in the cell was about 10^7. The experiments were performed both in atmospheric pressure and in a high vacuum (5×10^{-5} Torr).

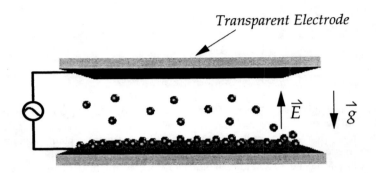

FIGURE 1. The electrostatic cell.

Let us first discuss the mechanism bringing particles into the motion. When the conducting particles come in contact with the capacitor plate they acquire a surface electric charge. As the magnitude of the electric field in the capacitor exceeds the critical value E_1 the resulting (upward) electric force overcomes the gravitational force mg (m is the mass of the particle, g is the acceleration due to gravity) and pushes the charged particles upward. When the grains hit the upper plate, they deposit their charge and fall back. Applying an alternating electric field $E = E_0 \sin(2\pi f t)$, and adjusting its frequency f one can control the particle elevation by effectively turning them back before they collide with the upper plate.

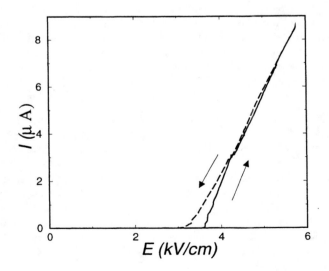

FIGURE 2. The electric current I across the cell vs applied electric field E. Solid line is obtained by increasing E from zero to E_{max}, dashed line is obtained by decreasing the field from E_{max} to zero.

Charging and discharging of the particles results in electric current across the cell. This current vanishes at the onset of particle motion at the first critical field E_1 and grows gradually with the increase of E, as it can be seen from Figure 2. For the value of E about $2E_1$ it achieves $2 - 3 \mu a$. The phase diagram for the experimental cell is shown in Figure 3. Isolated particles start to move at $E > E_1$. We have further found that at amplitudes of the electric field above a *second* threshold value, $E_2 > E_1$, the granular medium forms a uniform gas-like phase (granular gas). This second field E_2 is 50-70% larger then E_1 in nearly the whole range of the parameters used. This hysteresis can be seen from the charge transport measurements shown in Figure 2: the electric current corresponding increasing of the electric field from zero to E_{max} is smaller then the current for the decreasing field.

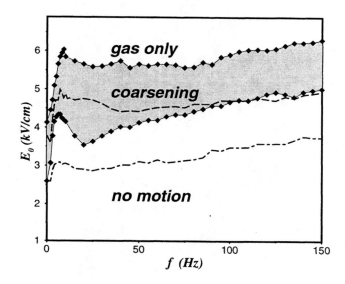

FIGURE 3. The phase diagram as a function of amplitude and frequency of the alternative field $E = E_0 \sin(2\pi f t)$. Solid lines with diamonds show $E_{1,2}$ in open cell. Dashed line shows E_2 and dash-dotted line shows E_1 for the evacuated cell correspondingly. In the gas only domain clusters do not form spontaneously, in the coarsening domain both clusters and gas coexist.

In the field interval $E_1 < E < E_2$, a remarkable phenomenon analogous to coalescence dynamics in systems exhibiting first order phase transitions is observed (Lifshtz,1981; Giron,1998; Aranson,1995). Upon decreasing the field below E_2, the gas phase looses its stability and small clusters of stationary particles, containing about 10^2 grains, surrounded by the granular gas spontaneously form. These clusters then evolve via coarsening dynamics: small clusters disappear and large clusters grow. Eventually the large isolated clusters assume almost perfect circular form (Figs 4 a-c). In the process of coarsening the electric current also changes in time due to decrease of particles in the gas phase. After a very large time ($t \approx 30000$ sec) only one large cluster containing about 10^6 grains survives. Even at the final stage, the coexistence of phases is preserved – not all the particles join the last cluster. On the contrary, a dynamic equilibrium between the granular solid in the cluster and the surrounding gas takes place.

For the cell under atmospheric pressure (open cell) we find that both fields E_1 and E_2 grow as the function of frequency for large f and show non-monotonous behavior for $f \approx 12$Hz. This indicates a characteristic time of the order 100 msec. We suggest that cohesion may be responsible for this relatively large time. Indeed, due to the humidity of the air a surface coating should exist, thus giving meaning to some characteristic time τ for the grain to detach from the capacitor plate. Although the cohesive forces are very small, they become comparable with the gravitational/electric forces. In order to reduce the cohesion we evacuated the cell to very low pressure. As demonstrated from Figure 3 , the frequency dependence is indeed substantially reduced and becomes almost flat. Small oscillations in the dependence for low frequency are probably due to a residual coating on the particles which does not completely evaporate in vacuum.

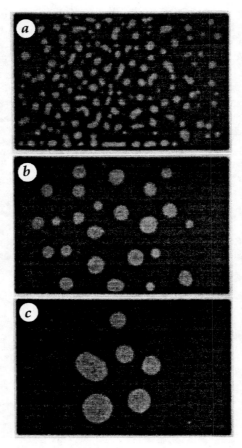

FIGURE 4. Top view of the experimental cell for the moments of time (a) $t = 1$ sec, (b) $t = 10$ sec, and (c) $t = 10910$ sec for $f = 40$ Hz and $E = 5.25$ kV/cm.

THEORETICAL MODEL

To clarify the mechanism of coarsening and the observed hysteresis we compare the forces exerted on an individual particle and on a particle which is held within a cluster at the same applied electric field.

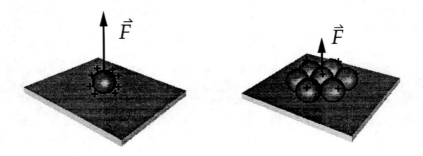

FIGURE 5. The forces acting on individual particle and the particle in the cluster. Small pluses show the surface charges.

The force between an isolated sphere and the capacitor plate in contact can be found as a limit of the problem of two conducting spheres when the radius of one sphere goes to infinity (Thompson,1853). Building on the technique of (Thompson,1853), we arrive at the force F_e in question:

$$F_e = 1.36a^2E^2 \tag{1}$$

where a is the radius of the sphere and E is the field in the capacitor. The constant 1.36 comes from summing of infinite series (Thompson,1853). The electric force F_e has to counterbalance the gravitational force $G = 4/3\pi\rho ga^3$, where ρ is the density of the material. Comparing F_e and G we find the first critical field E_1:

$$E_1 = \sqrt{\frac{4\pi\rho ga}{3 \cdot 1.36}} \tag{2}$$

Our theory indicates no frequency dependence of E_1. For the parameters of our experiments we obtain $E_1 = 2.05$ kV/cm, which is below than the experimental critical field $E_1 \approx 2.4$ kV/cm. This discrepancy seems reasonable since we did not take into account additional molecular and/or contact forces which will increase the critical field.

If the spheres are in contact with each other, the surface charge will redistribute: each sphere in the cluster acquires smaller charge than that of an individual sphere due to a screening of the field by its neighbors. The exact derivation of the force acting on the particle within the cluster is not available at present. The upper bound can be obtained by replacing the square lattice of spheres with radius a by a lattice of squares with area $4a^2$. Since the charge density of flat layer is $\sigma = E/4\pi$ and the total electric force on the square is $F_2 = 4a^2\sigma E/2$. Thus we obtain the ratio of the fields to lift individual particle E_1 and the particle in the square lattice E_2 (note again no frequency dependence):

$$E_2/E_1 = \sqrt{2\pi \cdot 1.36} = 2.92. \tag{3}$$

For closed-packed hexagonal lattice one obtains a slightly higher value $E_2/E_1 \approx 3.14$. The ratio of critical fields is independent of particle size and the density. For the constant applied field E, and, correspondingly, for the low frequency ac field, the ratio of the fields is about 3. This is consistent with the experimental findings, see Fig. 3, although exceeds experimental value by the factor of 2. More accurate account for the surface shape will reduce the ratio of the critical fields.

In order to study the details of coarsening process we performed molecular dynamics simulations. Several simplifications had been implemented: we neglected polarization effects of the particles and assumed that the charge is always located in the center of conducting sphere; we considered collisions between the particles and particle and conducting plate completely inelastic. The simulations show qualitative agreement with experiment: coarsening and first order phase transition between granular gas and granular solid. (Aranson,1999)

CONCLUSION

We have shown that conducting particles, when driven with electric fields show remarkable behavior similar to Ostwald ripening. This behavior, which can be understood as a first order phase transition, arises from a screening effect of the applied fields by a particles neighbors. An understanding of the competition between short range forces such as grain contact and collisions, and the long range electric forces will augment the existing understanding of granular materials. Also, using electrostatic driving allows for th careful investigation of large ensembles of fine powders, which are difficult to control mechanically.

ACKNOWLEDGMENTS

We thank Leo Kadanoff, Thomas Witten, Lorenz Kramer and Baruch Meerson for useful discussions. This research is supported by US DOE, grant W-31-109-ENG-38, and by NSF, STCS #DMR91-20000.

REFERENCES

Aranson, I.S., Blair, D., Kalatsky, V.A., Crabtree, G.W., Kwok, W.-K., Vinokur V.M., and Welp, U., "Electrostatically-Driven Granular Media: Phase Transitions and Coarsening," submitted to *Nature*, 1999.

Aranson, I.S., Blair, D., Kwok, W.-K., Karapetrov, G., Welp, U., Crabtree, G.W., Vinokur, V.M., and Tsimring, L.S., "Controlled Dynamics of Interfaces in a Vibrated Granular Layer," *Phys. Rev. Lett.* **82**, 731-734 (1999).

Aranson, I.S., Meerson, B., and Sasorov, P.M., " Front-Curvature Effects in the Dynamics of Confined Radiatively Bistable Plasmas: Perfect Patterns and Ostwald Ripening," *Phys. Rev. E* **52**, 948-971 (1995).

Boothroyd, R.G. *Flowing Gas-Solid Suspention*, Chapman & Hall, 1971;. Iinoya, K., Gotoh, K., and Higashitani, K., *Powder Technology Handbook*, N. Y. Marcel Dekker, 1990;

Fayed, M.E. and Otten, L. , ed., *Handbook of Powder Science & Technology*, second ed., Chapman & Hall, 1997.

Giron, B., Meerson, B. and Sasorov, P.M., "Weak Selection and Stability of Localized Distributions in Ostwald Ripening," *Phys. Rev. E* **58**, 4213-4116 (1998).

Hunt, M.L., Weathers, R.C., Lee, A.T., Brennen, C.E., Wassgren C.E, "Effects of horizontal vibration on hopper flows of granular materials," *Physics of Fluids* **11**, 68-75 (1999)

Iinoya, K., Gotoh, K., and Higashitani, K., *Powder Technology Handbook*, N. Y. Marcel Dekker, 1990;

Jaeger, H.M., Nagel, S.R., and Behringer, R.P., "Granular Solids, Liquids, and Gases," *Rev. Mod. Phys.* **68**, 1259-1273 (1996).

Jaeger, H.M., Nagel, S.R., and Behringer, R.P., "The Physics of Granular Materials," *Physics Today* **49**, 32-38 (1996);

Jeans, J. *The Mathematical Theory of Electricity and Magnetism*, Cambridge University Press, Vth Ed., 1948

Kadanoff, L., " Built Upon Sand: Theoretical Ideas Inspired by Granular Flows," *Rev. Mod. Phys.* **71**, 435-444 (1999)

Lifshtz, E.M., and Pitaevsky, L.P., *Physical Kinetics*, Pergamon, London, 1981.

Melo F., Umbanhowar, P.B., and Swinney, H.L., "Hexagons, Kinks, and Disorder in Oscillated Granular Layers," *Phys. Rev. Lett.* **75**, 3838-3841 (1995);

Mitsuya, T. and Hunt, M.L., "Toner Particle Packing in an Electrostatic Field and its Effect on Heating in E lectrophotography," *Powder Technology* **92**, 119-125 (1997).

Tennakoon, S.G.K., and Behringer, R.P, " Vertical and Horizontal Vibration of Granular Materials: Coulomb Friction and a Novel Switching State," *Phys. Rev. Lett.* **81**, 794-797 (1998)

Thomson, W., "On the Mutial Attraction or Repulsion between Two Electrified Spherical Conductors," *Phil. Mag.* s. 4, **5**, 287-297 (1853);

Umbanhowar, P.B., Melo F., and Swinney, H.L., "Localized Excitations in a Vertically Vibrated Granular Layer,"

Umbanhowar, P.B., Melo F., and Swinney, H.L., "Periodic, Aperiodic, and Transient Patterns in Vibrated Granular Layers", *Physica A* **249**, 1-9 (1998).

Air-Water Two-Phase Flow in a 3-mm Horizontal Tube

[1]Ing Youn Chen, [2]Yu-Juei Chang, and [3]Chi-Chung Wang

[1]*Department of Mechanical Engineering, National Yunlin University of Science and Technology, Taiwan*
[2,3]*Energy & Resources Laboratories Industrial Technology Research Institute, Hsinchu, Taiwan 310*

E-mail: [1]*Cheniy@flame.yuntech.edu.tw,*[3]*ccwang@.itri.org.tw*

Abstract. Two-phase flow pattern and friction characteristics for air-water flow in a 3.17 mm smooth tube are reported in this study. The range of air-water mass flux is between 50 to 700 $kg/m^2 \cdot s$ and gas quality is between 0.0001 to 0.9. The pressure drop data are analyzed using the concept of the two-phase frictional multipliers and the Martinelli parameter. Experimental data show that the two-phase friction multipliers are strongly related to the flow pattern. Taitel & Dukler flow regime map fails to predict the stratified flow pattern data. Their transition lines between annular-wavy and annular-intermittent give fair agreement with data. A modified correlation from Klimenko and Fyodoros criterion is able to distinguish the annular and stratified data. For two-phase flow in small tubes, the effect of surface tension force should be significantly present as compared to gravitational force. The tested empirical frictional correlations couldn't predict the pressure drop in small tubes for various working fluids. It is suggested to correlate a reliable frictional multiplier for small horizontal tubes from a large database of various working fluids, and to develop the flow pattern dependent models for the prediction of two-phase pressure drop in small tubes.

INTRODUCTION

The use of chlorofluorocarbon (HCFC) refrigerants had been restricted by the environmental regulations of the Montreal Protocal in 1996. Refrigerant 134a is the drop-in to replace Refrigerant-12 in automobiles for reducing the impact to the ozone layer. Refrigerant-22 will be banned by the year 2020 (or before). Presently, Refrigerants R-407C and R-410A are the most likely potential substitutes for R-22. During the past decade, the design of residential air conditioners has tended to downsize the refrigerant tube diameter in order to reduce air-side pressure drop and increase refrigerant and air heat transfer coefficients. The refrigerant tube inside diameter was greater than 10mm before 1980, and presently became 6.35~9.53 mm. Recently, investigations are toward to tube diameter < 5 mm. The refrigerant tube will soon be changed to around 4~5 mm after the year 2000 (Hsu, 1998). In addition, US NASA had utilized the two-phase flow concept to design the External Active Thermal Control System for Space Station Freedom (SSF). The tube diameters of the coldplates, heat exchangers, and the radiators in the SSF are small enough, i.e., 1.27 mm for the heat exchangers, 3.81 mm for the cold plates, and 1.7 mm for the radiator condensation tubes, (Chen, 1993).

Generally, there are several forces acting on the two-phase flow structure. Theses include the forces due to inertia, buoyancy, surface tension and turbulent eddies. The flow pattern seems to be determined by a delicate balance of these forces. There are four primary flow patterns of interest for two-phase flow in horizontal tubes. The flow patterns include stratified, slug, annular, and bubbly. These flow patterns are refereed as flow regimes. A particular flow pattern occurs only for limited ranges of gas and liquid flow. The gas/liquid flows, which mark the transition from one regime to another, are called regime transitions. For two-phase flow inside tubes, the pressure drop is strongly affected by the two-phase flow pattern because different two-phase structures yield different momentum and frictional characteristics. Knowledge of the two-phase flow regime would significantly improve the accuracy in predicting the two-phase pressure drop.

The majority of previous two-phase flow studies were for diameter on the order of 9.5~75 mm, tested at mass flux > 300 $kg/(m^2 \cdot s)$ (Wambsganss, 1991). For high performance compact plate fin heat exchangers, tube diameters are employed in the range of 1~10 mm and operated with the mass flux at 10~300 $kg/(m^2 \cdot s)$.

CP504, *Space Technology and Applications International Forum–2000*, edited by M. S. El-Genk
© 2000 American Institute of Physics 1-56396-919-X/00/$17.00

However, only a limited number of investigations are available for capillary tubes with diameters less than 4 mm. Barnea (1983) only presented flow pattern data in small diameter tubes (4-12 mm), an no frictional data were reported. Yang (1996) measured pressure drop for R-12 flowing in small hydraulic tubes used in automotive condensers and reported that the Chisholm equation (1967) did not give good predictions compared to their data.

Tests are conducted in horizontal tubes. Since the acceleration and gravitational pressure drops can be neglected in the adiabatic experiments, the pressure drop measured only came from the frictional loss. The pressure drop data are usually analyzed using the concept of the two-phase frictional multiplier:

$$\phi_L^2 = \frac{dP_f/dz}{dP_{f,L}/dz} \quad , \qquad \phi_{LO}^2 = \frac{dP_f/dz}{dP_{f,LO}/dz} , \tag{1}$$

where dP_f/dz is the measured two-phase frictional pressure gradient and $dP_{f,L}/dz$ is the frictional pressure gradient for liquid of the two-phase mixture flowing alone in the tube, and $dP_{f,LO}/dz$ is the frictional pressure gradient for total flow assumed liquid. The multiplier typically plotted versus the Martinelli parameter X.

$$X = \sqrt{\frac{\left(dP_{f,L}/dz\right)}{\left(dP_{f,G}/dz\right)}} . \tag{2}$$

Mishima (1996) had performed air-water frictional two-phase pressure drop measurements in vertical capillary tubes with diameters from 1 to 4 mm. One of Mishima's findings is a modification to the Chisholm equation in which the effect of tube diameter was included, i.e.

$$\phi_L^2 = 1 + \frac{C}{X} + \frac{1}{X^2} , \tag{3}$$

$$C = 21(1 - e^{-0.333d}) . \tag{4}$$

Chisholm (1983) assumed the C factor as a constant ranged from 5 to 20, depending on whether the liquid and vapor phases are laminar or turbulent. Where Φ_L^2 is the two-phase frictional multiplier based on pressure gradient for liquid flow alone, and d is the tube diameter in millimeters. In equation (4), Mishima (1996) correlated C factor from the diameters in which most of the diameter data were in the range of 0.7 to 7 mm. This equation may become questionable for extensive application because the C factor only contains a dimensional tube diameter and without any physical properties.

Wambsganss (1992) presented frictional pressure gradients of air-water mixtures in a rectangular geometry of 19.05 mm × 3.18 mm. They found that C is related to the mass flowrate and the Martinelli parameters as well. They modified the C factor in the Chisholm correlation as:

$$C = C(X, Re_{LO}) = aX^b, \tag{5}$$

where $a = -2.44 + 0.00939Re_{LO}$, $b = -0.938 + 0.000432Re_{LO}$, Re_{LO}: Reynolds number for liquid flow only.

Though there were some works related to the flow patterns and friction characteristics for small diameter tube. It seems that the two-phase characteristics in small diameter tube are still not fully understood. In view of this need, a program has been initiated to conduct a series of flow pattern visualization and measure the two-phase pressure drops in small tubes at Energy & Resources Laboratories of Industrial Technology Research Institute in Taiwan. Air-water and new refrigerants are utilized as the working fluids. This paper only describes the test results of air-water two-phase flow in a 3.17-mm diameter horizontal smooth tube at 25 °C. The objective of this study is to provide further test results and observations to this topic.

EXPERIMENTAL APPARATUS

The schematic of the experimental apparatus is depicted in Figure 1. The test rig is designed to allow adiabatic flow experiments with air-water mixtures. Air is supplied from an air-compressor and then stored in a compressed-air storage tank. The air flows through a pressure reducer and, depending on the mass flux range, is measured by a mass flowmeter (Aalborg® GFM 17 and GFM 47). The water flow loop consists of a variable speed gear pump, which delivers water to mix with airflow. The mixtures are near atmospheric pressure at 20 °C. The gear pump can provide refrigerant (water) mass fluxes ranging from 50 to 1000 kg/m²s. Three very accurate mass flowmeters with different flow range (Micromotion, model DS12S-100SU) are located at the downstream of the gear pump for measuring the adequate water flow rates. The accuracy of the mass flux flow meters is within ±0.2% of the test span. The pressure drop of the air-water mixtures is measured by a YOKOGOWA EJ110 differential pressure transducer, which has an adjustable span from 1300~13000 Pa and its accuracy is 0.3% of the measurements. The test tube is a round smooth copper tube having an inner diameter of 3.17 mm and the length of the test section is 995 mm. The test tube is well insulated by a 50-mm-thick rubber having a thermal conductivity of 0.032 W/m·K.

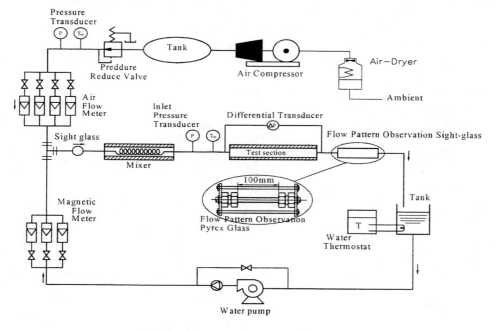

FIGURE 1. Schematics of test loop.

A pyex sight glass tube, having an identical inside diameter with the test section, is located at the down stream of the test section. The sight glass has a length of 100 mm. Two-phase flow patterns are obtained from directly visual observations made from a micro-camera (Nikon FM2) having extension tubes of 13 mm +21 mm +31 mm. For air-water mixture leaving the test section, air and water are separated by an open water tank in which the air is vented and the water is recirculated. The air and water temperatures were measured by resistance temperature device (Pt100Ω) having a calibrated accuracy of 0.05°C (calibrated by hp quartz thermometer probe with quartz thermometer, model 18111A and 2804A). The water inlet temperature to the pump is controlled by a thermostat in the water tank. To verify the accuracy of the instrumentation, single-phase pressure drops for air and water flow alone were measured. The result indicated that the single-phase data agree well with the exact laminar friction factor and the Blasius turbulent friction factor. The good agreement indicates the validity of the instrumentation and the experimental apparatus.

RESULTS AND DISCUSSION

Tests are conducted at the mixture mass flux (G) of 50, 100, 200, and 500 kg/m²·s, and the gas quality is in the range of 0.0001 to 0.9. The flow patterns for G = 50 and 100 kg/m²·s have observed the intermittent, stratified and wavy-stratified flow patterns. The annular flow pattern is not clearly seen for G < 100 kg/m²·s even at high quality region because the gas inertia force is not high enough. For G = 50 kg/m²·s and x > 0.2, one can see the occurrence of the liquid entrainment. The liquid entrainment may deposit on the tube wall to form liquid streak or film around periphery in the very high quality region. This phenomenon may become more pronounced for G = 100 and 200 kg/m²·s. At low mass flux region, gravity is the dominant force, the mechanism of transition from stratified to annular flow pattern is due to the deposition of liquid entrainment on the tube wall to form into the liquid film around the wall. Once the film of the annular flow is formed, one can see that the liquid film around the periphery is so thin that caused the atomization to be difficult. On the contrary, the flow pattern transition from wavy to annular flow for G = 500 kg/m²·s is somehow different from those of G = 200 kg/m²·s. The interface of air-water become so violent at G = 500 kg/m²·s that the rolling wave may touch the upper tube wall to become the annular flow. The liquid film thickness around the periphery is thicker and the gas superficial velocity is so high that the shear force can easily shed the liquid into the annular core, which make the flow pattern to become annular dispersed mist flow.

Figure 2 shows the observed flow pattern in this study. For comparison purpose, the boundaries of the flow patterns observed by Barnea et al. (1983, D = 4 mm) and Damianides (1988, 3 mm) are plotted in the figure. Basically, the observed flow pattern in this study is analogous to those by Damianides (1988, D = 3 mm) except the boundaries of stratified flow region.

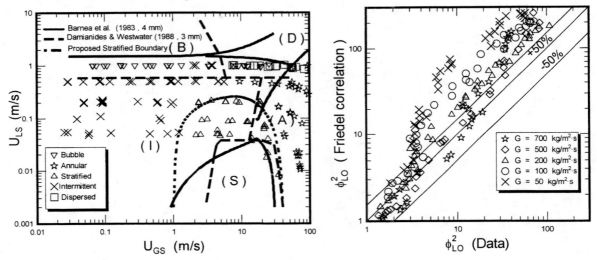

FIGURE 2. Flow regime map for the axes of gas and superficial velocities.

FIGURE 3. Comparison of the frictional mutiplier data and the Fridel's prediction.

Figure 3 shows the comparisons between the test results and those predicted by Fridel correlation (1979). Apparently, the predictions by Friedel considerably overpredict the test results at low gas quality and small mass flux. For larger mass flux like G = 500 and 700 kg/m²·s, predictive ability is comparative better than those of small mass fluxes. This is because that the Friedel correlation (1979) is more appropriate for larger diameter and larger mass velocities.

Figure 4 shows the measured data plotted in the form of ϕ_L vs. X and values predicted by Chisholm equation for C = 5 and 20. For comparison purpose, the factor C = 13.6 proposed by Mishima (1996) is also plotted. The bold line appeared in Figure 4 is generated from the best-fitted correlation for this data set. Apparently, neither C = 5 or C = 20 can describe the present results. The predictions by Mishima only fairly agree with the data in the range of X = 1 - 10. For X < 0.2, it corresponds to the higher vapor gas region of annular and stratified flows, the measured ϕ_L is considerably lower than the predictions. The multiplier data for G = 50 kg/m²·s are much lower than the predictions because the corresponding flow patterns are stratified flow. For better predictions of ϕ_L, the stratified flow model of Taitel & Dukler (1976) is suggested for the stratified multiplier predictions (Chen, 1991). For X >100, the measured ϕ_L is much higher than the predictions, a homogeneous flow model (Beattie, 1982) is also recommended for the intermittent flow prediction.

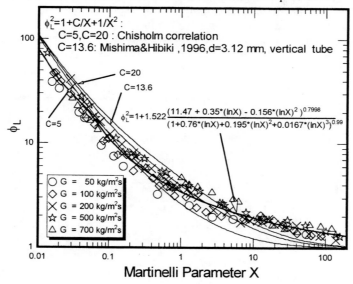

FIGURE 4. Frictional multiplier vs. Martinelli parameter.

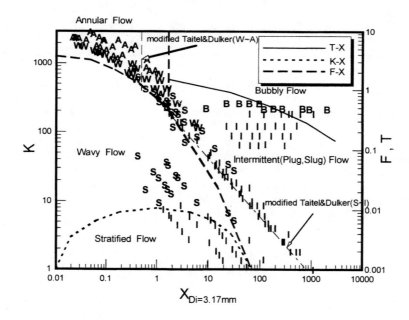

FIGURE 5. Taitel & Dukler flow regime map and the present data.

Because the empirical Chisholm and Fridel correlations can not fairly predict the present pressure drop data, it is necessary to identify the flow pattern by employing the flow regime dependent equations developed from analytical or empirical results. Taitel & Dukler (1976) presented a physical model with different force balance ratios for flow pattern transitions. These flow transitions were used to constructed the boundaries of different flow regimes, as shown in Figure 5 in which the observed flow pattern data are also included. The stratified flow data don't give good agreement with the transition, especially for the data at very low values of Froude number Fr $\{= Gx/(\rho_f-\rho_g)\rho_g gd\}^{0.5}$, where Fr is directly related to the total mass flux G and the gas quality. When the gas inertia force is very small at very low quality, the buoyancy force or surface tension force may dominate the flow. Ungar (1998) indicated that the criterion to satisfy this balance is Bond number, Bo $(= g(\rho_f-\rho_g)(D_i/2)^2/\sigma) = 1$, which is the ratio between the buoyancy force and the surface tension. For air-water flow in a 3 mm diameter tube, Bo $= 0.3$. Therefore, the effect of surface tension force should be considered in this study. The transition line between the separated flows and the intermittent flow is at X $= 1.6$. Most of the intermittent data are located in the Taitel and Dukler intermittent regime with X > 1.6. The annular and wavy (or semi-annular) data are slightly above the transition line between annular and wavy flow regimes, which can also be expressed as (Sardesair, 1981): Fr $= (0.7 X^2 + 2 X + 0.85)^{-1}$.

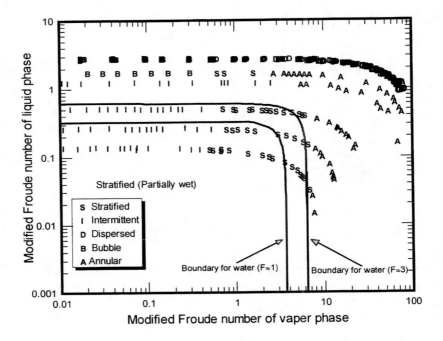

FIGURE 6. Comparison of the Klimenko criterion and the flow pattern data.

Since Taitel & Dukler flow regime map can not match the stratified flow pattern data in small tubes because the surface tension effect is not included in their map, the flow regime criteria developed by Klimenko (1990) is also used to compare the flow pattern data as shown in Figure 6. Notice that Klimenko (1990) developed a mechanistic criterion for determining the transition from stratified to un-stratified flow. The un-stratified flow, i.e., for the flow regimes with continuous wetting of the entire tube circumference was defined, and the following semi-empirical expression was proposed as:

$$F = 0.074\left(\frac{D_i}{b}\right)^{0.67} Fr_G + 8\left[1 - \left(\frac{\rho_G}{\rho_L}\right)^{0.1}\right]^2 Fr_L , \qquad (6)$$

$$b = \left[\frac{\sigma_L}{g(\rho_L - \rho_G)}\right]^{0.5} , \qquad (7)$$

where ρ_G and ρ_L are the gas density and liquid density, respectively, σ_L is the liquid surface tension, g is the earth gravity and D_i is the tube inside diameter. For $F > 1$, un-stratified flow exists, and for $F < 1$, stratified flow exists. They defined the Froude number for liquid and vapor phase using the superficial velocity,

$$Fr_L = \frac{\rho_L U_{LS}^2}{[(\rho_L - \rho_G)gD_i]}, \quad Fr_G = \frac{\rho_G U_{GS}^2}{[(\rho_L - \rho_G)gD_i]}. \tag{8}$$

A close examination of their database indicates that $F = 3$ is more appropriate for the water vapor system. Hence $F = 1$ and $F = 3$ were plotted in Fig. 6 for comparison. As seen, the $F = 3$ criterion is more appropriate to classify the present stratified and un-stratified pattern data.

CONCLUSIONS

For two-phase flow inside tubes, the frictional pressure drop is strongly affected by the flow pattern. If flow pattern in any two-phase flow condition could be predicted, then the flow pattern related two-phase multipliers should be able to give accurate prediction in the two-phase pressure drop. In summary, procedures for using the suitable flow regime map to identify flow pattern and calculate the pressure drop are described as follows:

1. The annular flow pattern can be fairly identified by the Klimenko criterion (1990) with $F > 3$, or by Taitel & Dukler criterion for the transition between annular and wavy-stratified flows (1976).
2. If it is not annular flow, and its $X > 5$, then this flow pattern is intermittent flow, otherwise it is stratified flow for $X < 5$. A criterion with the effects of surface tension and buoyancy force is need for the stratified flow regime prediction.
3. For $1 < X < 10$, the Chisholm equation (3) with factor $C = 13.6$ gives a fair prediction for ϕ_L.
4. Since the reported data is only came from air-water flow in a 3.17 mm tube at 20 °C., It is suggested to correlate a reliable multiplier equation for small horizontal tubes from a large database with various working fluids, and to develop the flow pattern related models for the prediction.

ACKNOWLEDGMENTS

The authors would like to acknowledge the support provided by National Science Committee (Research grant No. NSC88AFA2600074) and the Energy Commission of the Ministry of Economic Affairs, Taiwan, R.O.C.

REFERENCES

Barnea, D, Luninski, and Taitel, Y., "Flow Pattern in Horizontal and Vertical Two Phase Flow in Small Diameter Pipes." *Canadian J. of Chemical Engineering* **61**, 1983, pp. 617-620.

Beatie, D. R. H. and Whalley, P. B., *Int. J. Multiphase Flow* **8(1)**, 1982, pp. 83-87.

Chen, I. Y., Unger, E. K., Lee, D. Y. and Beckstrom, P. S. "Prototype Test Article Verification of the Space Station Freedom Active Thermal Control System," AIAA paper 93-2885, 1993.

Chisholm, D., *Int. J. Heat and Mass Transfer* **10**, 1967, pp. 1767-1778.

Daminanides, C. A. and Westwater J. W., "Two-phase Flow Patterns in Small Tubes," in Proceedings of the 2nd UK National heat transfer conference, paper C128/88, 2, 1988, pp. 1257-1268.

Friedel, L. "Improved Friction Pressure Drop Correlations for Horizontal and Vertical Two-phase Pipe Flow," Presented at the European Two-phase Group Meeting, Ispra, Italy, Paper E2, 1979.

Hsu, S. M. and Yang, C. Y., "Experimental Analysis of Boiling Heat Transfer for Referigerant HFC-134a in Small Tubes," Proceedings of 15th CSME Conference, Taiwan, 1998, pp. 213-219.

Klimenko, V. V., and Fyodorov, M.. "Prediction of Heat Transfer for Two-Phase Forced Flow in Channels of Different Orientation," Proceedings 9th Int. heat Transfer Conf., 6, 1990, pp. 65-70.

Mishima, K. and Hibiki, T., "Some Characteristics of Air-water Two-Phase Flow in Small Diameter Vertical Tubes," *Int. J. Multiphase Flow* **22**, 1996, pp.703-712.

Ungar, E. K., Chen, I. Y. and Chan, S.H., "Selection of a Gravity Insensitive Ground Test Fluid and Test Configuration to Allow Simulation of Two-Phase Flow in Microgravity," ASME Proceedings of the 7th AIAA/ASME Joint Thermophysics and Heat Transfer Conference, HTD 357(3), 1998, pp. 71-77.

Taitel, Y., and Dukler, A. E., "A Model for Prediction of Flow Regime Transitions in Horizontal and Near Horizontal Gas-Liquid Flow," *AIChE J* **22**, 1976, pp. 47-55.

Wambsganss, M. W., Jendrzejczyk, J. A., and France, D. M., "Two-Phase Flow Pattern and Transition in a Small, Horizontal Rectangular Channel," *Int. J. Multiphase Flow* **17**(3), 1991, pp. 327-342.

Wambsganss, M. W., Jendrzejcyk, J. A., France, D. M., and Obot, N. T., "Friction Pressure Gradients in Two-Phase Flow in a Small Horizontal Rectangular Channel," *Exp. Thermal and Fluid Sci.,* **5**, 1992, pp. 40-56.

Yang, C.Y., and Webb, R.L., "Friction Pressure Drop of R-12 in Small Hydraulic Diameter Extruded Aluminum Tubes With and Without Microfins," *Int. J. Heat Mass Transfer* **39**, 1996, pp. 801-809).

NOMENCLATURE

C	constant in Chisholm correlation
dP_f/dZ	measured two-phase frictional pressure gradient, N/m^2
$dP_{f,l}/dZ$	frictional pressure gradient for liquid flowing alone, N/m^2
$dP_{f,v}/dZ$	frictional pressure gradient for gas flowing alone, N/m^2
ΔP_a	pressure drop due to acceleration, N/m^2
ΔP_f	frictional pressure drop, N/m^2
D_i	inside diameter of the tube, m
F	constant in Klimenko and Fyodorov criterion
F	parameter used in Taitel and Dukler Map ($= \left(\dfrac{\rho_G}{(\rho_L - \rho_G)} \right)^{0.5} \dfrac{j_G}{(D_i g)^{0.5}}$)
Fr	Froude number
G	mass flux, kg/m^2·s
g	gravitational constant of acceleration
j	volumetric flux, m/s
K	parameter used in Taitel and Dukler Map ($= \left((\rho_G j_G^2 j_L)/(\rho_L - \rho_G)g v_L \right)^{0.5}$)
L	effective heating length, m
Re	Reynolds number for single phase fluid
Re_{LO}	Reynolds number for liquid flow only, dimensionless
T	parameter used in Taitel and Dukler Map ($= \left[(dP/dz)_f /(\rho_L - \rho_G)g \right]$)
U_{GS}	superficial velocity, vapor phase, m/s
U_{LS}	superficial velocity, liquid phase, m/s
v	kinematic viscosity, m^2/s
We	Weber number
x	vapor quality
X	Martinelli parameter

Greek Symbols

ρ	density of refrigerant, kg/m^3
ϕ_L^2	two phase friction multiplier for liquid flowing alone
μ	dynamic viscosity of refrigerant, N·s/m^2
σ_L	surface tension of water, N/m

Subscript

LO	evaluated as the total mass as the liquid phase
GO	evaluated as the total mass as the gas phase
L	liquid phase only
G	gas phase only

Critical Boiling Phenomena Observed in Microgravity

Yves Garrabos[*,1,2], Régis Wunenburger[1], John Hegseth[3], Carole Lecoutre-Chabot[1,2], and Daniel Beysens[4]

[1]ICMCB - CNRS - Université de Bordeaux I, Avenue du Dr. Schweitzer, F-33608 Pessac Cedex, France
[2]3AR - AEROSPATIALE – CNRS, BP11, F-33165 Saint Médard en Jalles, France
[3]Department of Physics, University of New Orleans, New Orleans, Louisiana 70148
[4]Service des Basses Températures, DRFMC, CEA Grenoble,
17 Rue des Martyrs, 38054 Grenoble Cedex 09, France

33- 5 56 84 63 37; garrabos@icmcb.u-bordeaux.fr

Abstract. Singular thermodynamic properties of pure fluids near their critical point (diverging specific heat and isothermal compressibility, vanishing thermal conductivity,...) lead to poor thermal diffusion, a large convective sensitivity, and a special heat transfer process through adiabatic compression, the so-called piston effect. The discovery and extensive study of the piston effect were performed in conditions of weightlessness to avoid convection. Although its mechanism in the supercritical range is now well understood, its coupling with an inhomogeneous density distribution and mass transport in the two-phase regime has been relatively less studied and remains puzzling. Recent experiments performed in the French Alice 2 facility built by CNES onboard the Mir station showed undoubtedly that when a liquid-vapor system of SF_6 near its critical temperature is heated in microgravity, the apparent contact angle becomes very large (up to 110°). In this slightly out-of-equilibrium configuration the gas appears to "wet" the solid surface. In addition, the temperature of the vapor becomes higher than that of the hot wall, whereas the temperature of the liquid evolves qualitatively as in the one-phase regime. Although the difference between the compressibility of liquid and vapor explains a higher vapor temperature compared to the liquid temperature, this paradoxical observation has not yet been modeled. Moreover, the phase distribution plays an important role in the efficiency of the heat transfer near the critical point. A three domains model valid at short time scale is presented. It is similar to the model of supercritical piston effect from Onuki, Hao and Ferrel (Onuki, 1990a), and takes into account the presence of a liquid wetting layer. Indeed, the thermal boundary layer only develops inside the liquid phase and compresses the vapor phase, in contrast to the situation in the one-phase regime. The leading characteristic time of the piston effect in the two-phase regime depends on the compressibility ratio between the two phases. It is larger than for a single liquid phase. Good agreement is obtained between the experimental and theoretical temperature evolution curves in both phases.

INTRODUCTION

The behavior of critical fluids is deeply influenced by hydrodynamics. Indeed, the behavior of such fluids on Earth is quite often dominated by gravity. This is because near the critical point (Stanley, 1971) key parameters like the isothermal compressibility, the capillary length, and the thermal diffusivity, exhibit extreme values (very large or very small) when compared to usual fluids. These severe anomalies, i) prevent the sample from being homogeneous in density since the fluid is compressed under its own weight and ii) prevent dynamical processes from being studied because of buoyancy-induced convection. Microgravity appears as a unique tool to allow valid experiments to be performed in fluids close to their liquid-gas critical point. The ALICE program is precisely a research program devoted to such studies in Space (Beysens, 1996 and Marcout, 1995). The goals are :
- to apply the results of experiments on specific fluids to a variety of substances that belong to the same class of universality. Such behavior, once appropriately scaled, can be extended to mixtures of simple liquids or to more complex liquids like polymers, micro-emulsions, liquid metals, etc. (Stanley, 1971),
- to establish the foundations of a mechanics of dense hyper-compressible fluids. This involves finding a bridge between hydrodynamics - the study of incompressible liquids - and aerodynamics - the study of low density compressible gases.

This paper is concerned with the boiling process (two-phase to one-phase transition), a complementary process to phase separation (one-phase to two-phase). In general, boiling is such an important process in industry and engineering that many heat transfer text books contain whole chapters on this phase transformation. Although there is much applied research in this area, a survey of the literature shows that there does not appear to be many fundamental studies on this topic which is a complicated task mixing fluid dynamics, heat transfer, and interfacial phenomena (Tong, 1997). Here the experiments were performed in orbit to suppress buoyancy driven flows and gravitational constraints on the liquid-gas interface. Moreover, the approach of the critical point is appealing to separate the different contributions of some important properties such as surface tension, latent heat, and density difference between gas and liquid, due to their divergence or convergence.

CP504, *Space Technology and Applications International Forum–2000*, edited by M. S. El-Genk
© 2000 American Institute of Physics 1-56396-919-X/00/$17.00

EXPERIMENTAL

Boiling experiments were performed with the ALICE 2 facility whose detailed description is given in (Marcout, 1995). This instrument associates a high performance system for managing diagnostics and stimuli with a regulation system that controls the temperature conditions on the sample cell unit (SCU) to within a few microKelvin. The SCU contains one or two optical sample cells. Optical cells are parallelepipedic in shape, and each cell body is made of a CuCoBe alloy. The internal fluid volume is a thin cylinder (12 mm internal diameter, 1.5 - 7 mm thickness) sandwiched between two parallel sapphire windows (12 mm external diameter) which are epoxied, using their cylindrical surface, to the cell wall. The fluid in the cell is studied here using light transmission, light scattering, field observation, microscopy, and grid shadow technique. Its schematic representation is given on Fig. 1. A thin layer of the pure fluid (SF_6 or CO_2) at a controlled slightly off-critical density (Morteau, 1997) is visualised through light normal to the windows. The initial two phase configuration in microgravity consists of a constrained gas bubble surrounded by the liquid that wets the walls of the cell. The corresponding liquid-gas interface observed in microgravity condition by light transmission is shown on Fig. 2a. The initial off-center position of the bubble with part of the bubble touching the CuCoBe ring is probably due to a small background acceleration from the rotation of the Mir Station about its center of mass and/or a slight tilting between the internal surfaces of each sapphire window.

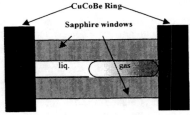

FIGURE 1. Transmission sample cell filled with SF_6 fluid. Schematic cross section of the constrained gas bubble inside the thin layer of liquid in weightlessness.

Our critical boiling experiments consist of a series of rapid increases in temperature (positive quenches) or a continuous increase in temperature (ramp) until the supercritical fluid region is reached. Note that the similar ground based experiments were done before, yielding forced Rayleigh-Taylor instabilities (Delville, 1997).

CRITICAL BOILING RESULTS

Fig. 2 shows several images obtained when the cells are heated linearly in time. After the temperature ramp was started, a dynamical change in the configuration of the liquid and gas regions appears almost instantaneously. As a first observable result still far from the critical temperature (Fig. 2b), the convex bubble shape changes (the liquid layers between the gas bubble and the sapphire windows also change). A second result is seen when approaching the critical temperature (Fig.2c): the gas phase appears to "*wet*" a large portion of the cell surface! Moreover, a fast spreading of the gas is seen as the bubble opens up to produce a double-tilde shape (Fig. 2d).

Even faster shape evolutions are observed when the cells are heated by temperature steps of +100 mK every 72 s. Fig. 3 shows the images obtained for a step that crosses the critical temperature. Here, the deformations of the gas - liquid interface are observed with the aid of the grid shadow technique (Gurfein, 1991). The relaxation process after each temperature quench explains the bubble shape in the stepping mode. Near the critical temperature the vapor bubble looses its convexity and rapidly evolves, as shown on Fig. 3b. The gas phase also "*wets*" very quickly a large portion of the cell wall. After having crossed T_c (Fig. 3c), the shape of the interface is quite different to that previously obtained in ramping. It looks more like a hat than a double-tilde.

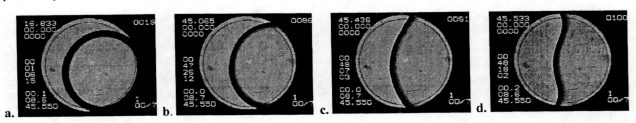

FIGURE 2. Bubble shape at various temperatures observed during continuous heating of a SF_6 cell. The rapid bubble shape changes that occur close to the critical temperature are clearly seen.

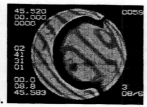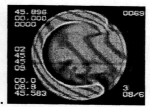

FIGURE 3. Bubble shape distortion during the +100 mK quench crossing the SF$_6$ temperature transition. Rapid changes during heating (b) and slow evolutions during step relaxations (c) are seen. Image (d) shows the remaining density gradient above T$_c$, evidenced by the grid shadow technique.

If the two phase fluid were at equilibrium, we recall that the contact angle at the triple contact point (liquid-gas-solid). would be zero (Ikier, 1996; Kayser, 1985; De Gennes, 1985), and the liquid phase would completely wet the solid. In our initial equilibrium configuration of the two-phase cell where the external bubble shape is constrained, see Fig. 2a, this liquid wetting condition corresponds to a wall-tangent elliptic profile. Fig. 2c shows that, as the critical point is approached, the contact angle rapidly increases up to 110°. A quantitatively similar result is obtained in step heating, see Fig. 3b. The apparent contact angle increase is undoubtedly caused by non-equilibrium effects.

DISCUSSION

The apparent shape change far below T$_c$, where the surface tension is still large, may be analyzed in terms of a temperature change along the liquid-gas interface since important thermal nonhomogeneities are present within the fluid cell. Such a temperature gradient will modify the bubble interface curvature. According to a local formulation of the Laplace law, it is possible to estimate the bubble deformation which corresponds to a decreasing curvature, as observed in the above experiments. However, this temperature gradient will create a surface tension gradient that could drive a thermal-capillary (Marangoni) flow in the bulk of both fluids (Hegseth, 1996; Davis, 1987; Palmer, 1976). The images obtained in our experiment are able to visualize convective flows by the shadowgraph effect. We have not seen any evidence of the steady convection that is required to create and maintain the observed bubble shape. We will thus neglect any fluid motion. We conclude from this analysis that should a surface tension change exists, it would diverge at the critical point with only a weak exponent.

Another possible source of bubble deforming stress does not require a temperature gradient along the interface. It can be shown (Garrabos, 1999) that the bubble may be deformed by the normal stress exerted on the interface by the recoil force from departing vapor (Straub, 1995 a; Nikolayev, 1999). In order to find the resulting shape deformation of the interface, it is necessary to solve the entire heat transfer problem. But this problem is complicated by the isentropic heat transfer process (Garrabos, 1998) where the isentropic compression by the boundary layer heats the gas more than the liquid, leading to quite large temperature differences as shown below. Here it is sufficient to note that, as T approaches T$_c$, the vapor mass growth follows the growth of its density (the vapor volume remains constant). The diverging vapor production near the critical point drives a strong diverging recoil force that should then modify drastically the resulting shape of the interface. This deformation is certainly magnified near the triple contact line and the large apparent contact angle can be seen as a sharp curvature increase from the solid wall contact. In fact, the largest mass transfer across the interface takes place near the triple contact line because the temperature varies sharply in the boundary layer adjacent to the cell wall (several more space experiments to be performed in the near future should further clarify this process). We note that this non equilibrium wetting of the vapor phase is similar to the situation of the « boiling crisis » in heat exchangers (Nikolayev, 1999). Although the boiling crisis usually occurs in liquids far from their critical point (Tong, 1997; Straub, 1995 a), the heat flux is also much higher so that the ratio of the recoil force to the surface tension scales similarly to the present problem, where the surface tension is very small.

For T ≥ T$_c$, the bubble's relaxation is quasi static (Fig. 3d). The evolution of the supercritical interface is then only defined by local diffusive effects. This reminiscent interface (in fact a sharp density gradient), maintained at 1K above T$_c$, disappears one hour after T$_c$ crossing. This period agrees with the well-known process where the density equilibration of the fluid cell at long times is uncoupled to the fast isentropic heating by the piston effect (Garrabos, 1998) and is only governed by the slow thermal relaxation at constant pressure.

TEMPERATURE EVOLUTION IN THE TWO PHASES

In order to understand the heat and mass transfer and the interface deformations observed during a temperature ramp or a positive quench, similar boiling experiments were also performed in cells equipped with in-situ temperature sensors. The temperature was measured at three points of the sample by means of thermistors (diameter 0.25 mm; time constant 10 ms). They were positioned so as to measure simultaneously the temperature in the liquid phase and in the gas phase. In Fig. 4 are shown two snapshots of the sample cell at two different temperatures below T_C, where the three thermistors are visible and labeled. In the absence of gravity the phase distribution depends on the wetting conditions and on the value of the interfacial tension. That is why the position of the thermistors with regard to each phase actually vary with the temperature and the history of the sample (g-gitter, residual gravity). At 10 K from T_C (Fig. 4a), the thermistor Th3 is surrounded by a thin liquid layer, whereas at 1K from T_C (Fig. 4b) it is located inside the gas bubble.

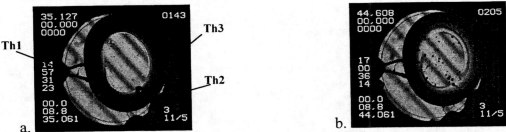

a.   b.

FIGURE 4. SF_6 cell observed by light transmission. Shown are the positions of the three thermistors at two temperatures: **a.** 10K below T_C and **b.** 1K below T_C. In both cases Th1 is in the liquid, Th3 is in the gas, and Th2 is at the liquid-gas interface.

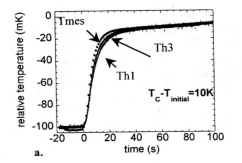 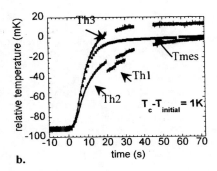

a.   b.

FIGURE 5. Experimental response of the thermistors when the cell is submitted to a +100mK quench : **a.** 10K and **b.** 1K below the critical temperature. We clearly see, at 1K below T_C, that before the cell reaches its equilibrium, the gas is hotter than the walls and the liquid that surrounds it.

The time evolution of the temperature of the cell wall and of the temperature measured by each thermistor during a positive quench of amplitude +100 mK, from two different initial temperatures, is shown in Fig. 5. During the first 100 seconds following the +100 mK quench from $T_C - 10$ K (where the effectiveness of the piston effect is weaker), the temperature measured by the thermistors located in the liquid phase near the interface (Th2 and Th3) remained below but very close to the temperature of the cell wall (Fig. 5a). During the first 100 seconds following the +100 mK quench from $T_C - 1$ K, the temperature increase measured by the thermistor located in the gas phase (Th3) is always much higher than the temperature increase of the thermostated cell wall, whereas the temperatures measured by the thermistor located in the liquid phase near the interface (Th2) and in the bulk of the liquid phase (Th1) remain well below the temperature of the cell wall (Fig. 5b). This apparent violation of the second law of thermodynamics is a consequence of the thermo-compressible nature of the energy transfer near the critical point as discussed below.

The prediction of the different phase behavior of the two-phase system was first reported by Onuki et al. (Onuki, 1990b). A 1-D numerical calculation of the heat transfer in near-critical fluids in the two-phase range was performed by Straub and Eicher (Straub, 1995b, c; Eicher,1996), and Zhong et al. (Zhong, 1996). Eicher observed numerically that the temperature evolution in both phases depends strongly on the phase distribution at the boundaries. In particular, the overheating in the gas phase is less pronounced when the gas phase is in contact with the thermostated cell wall than when liquid isolates the gas from them. Since the source of the piston effect is the expansion of the fluid heated by the cell wall, the efficiency of the piston effect depends crucially whether the thermal boundary layer develops in liquid or in gas, whose thermo-mechanical properties are very different.

Near the critical point, the liquid wets the wall, thus forming a thin layer. Moreover, in well prepared cells, in the absence of gravity and any other perturbation, the gas bubble can be completely surrounded by liquid, allowing the thermal boundary layer to expand in liquid only. But the present experimental conditions under which the in-situ temperature measurements were performed are not ideal since the bubble is constrained by the windows and slightly pushed toward the CuCoBe wall by a mechanical piston. Therefore only a thin wetting layer separates the gas from the cell wall at two quasi-circular surfaces on the windows and at a thin stripe on the CuCoBe wall. However, an important portion of the internal surface of the cell justifies the assumption that, with respect to the heat transfer, the gas is not in contact with the wall. In the spirit of Onuki et al. (Onuki, 1990a, b), we propose a symmetrical 1-D model with 3 domains, the liquid thermal boundary layer, the liquid phase as a part of the bulk, the gas phase as a bulk, to describe the main features of this system. The sketch of this phase distribution is drawn in Fig. 6 and during a positive temperature quench, the wall temperature varies as the ideal step function represented on Fig. 7. Practically, the two liquid boundary layer (LBL) are assumed to be the only fluid parts to receive heat from the thermostat.

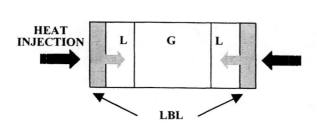

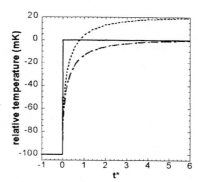

FIGURE 6. Symmetrical sketch of phase distribution for the three-domain 1-D model. The two boundary layers (LBL) are of the same liquid nature, expanding symmetrically and compressing isentropically the two remaining liquid bulks (L), and the entire gas phase (G).

FIGURE 7. Calculated liquid (dotted-dashed line) and gas (dotted line) temperature responses to an ideal temperature step of 100mK (full line) applied to the SF_6 sample cell initially at 1K below T_c, as a function of the reduced time $t^* = t/tc$ (tc is the typical time scale, see Eq.13). All the initial temperatures were shifted to zero.

The first important point to note is that when the cell wall is heated, a thermal boundary layer expands in the liquid, compressing isentropically the gas phase and the bulk of the liquid phase, that have different values for the isentropic thermodynamic derivative $(\partial T/\partial P)_S$. Neglecting the slow heat and mass transfer between the phases occurring at the interface and limiting the model to short times (times much shorter than the time that heat takes to diffuse on the length scale of the cell), it is straightforward to conclude that, under the same pressure increase, the gas is heated more than the liquid (Onuki, 1990b). The strong overheating of the gas can be explained by a second important point. In the phase distribution of Fig. 6, the expansion stops when the temperature of the bulk liquid reaches the temperature of the wall. The piston effect is homogeneous for the liquid, because its bulk part is heated due to the expansion of a liquid boundary layer. In this simple approach, the boundary layer is not in contact with the gas, thus the gas temperature does not influence the heat transfer. Its heating is just a consequence of the homogeneous piston effect taking place inside the liquid. This is why the heating of the gas also stops only when the bulk liquid temperature equals the temperature of the wall, allowing the gas to be strongly overheated above the cell temperature.

In the following the volume (V) and temperature (T) of the LBL are indexed by 1, and those corresponding to the bulk of the liquid phase (respectively gas phase) are indexed by 2 (respectively 3). Thermodynamic quantities concerning the liquid phase (gas phase) are labeled by L (G). The pressure (P) is supposed to be homogeneous. The temperature of the liquid bulk and gas bulk increase only because of the pressure increase, whereas the LBL temperature increases also because of the heating by the wall. Hence,

$$\frac{\partial V_1}{\partial t} = \frac{\partial v^L}{\partial P}\bigg|_S \frac{V_1}{v^L} \frac{\partial P}{\partial t} + \frac{\partial v^L}{\partial T}\bigg|_P \frac{1}{C_P^L} Q, \tag{1}$$

$$\frac{\partial V_2}{\partial t} = \frac{\partial v^L}{\partial P}\bigg|_S \frac{V_2}{v^L} \frac{\partial P}{\partial t}, \tag{2}$$

$$\frac{\partial V_3}{\partial t} = \frac{\partial v^G}{\partial P}\bigg|_S \frac{V_3}{v^G} \frac{\partial P}{\partial t}, \tag{3}$$

Where C_P is the specific heat at constant pressure, v is the volume per unit of mass, V_i is the volume of each subvolume i, and Q is the time derivative of the heat injected into the LBL. The conservation of the total volume V of the two-phase system can be expressed as

$$\frac{\partial V_1}{\partial t} + \frac{\partial V_2}{\partial t} + \frac{\partial V_3}{\partial t} = 0. \tag{4}$$

The adiabatic heating of the bulk part of the liquid and of the gas is written as

$$\frac{\partial T_2}{\partial t} = \left.\frac{\partial T}{\partial P}\right|_S^L \frac{\partial P}{\partial t}, \tag{5}$$

$$\frac{\partial T_3}{\partial t} = \left.\frac{\partial T}{\partial P}\right|_S^G \frac{\partial P}{\partial t}. \tag{6}$$

The same type of thermodynamic transformations as performed by Onuki et al. (Onuki, 1990b) leads to the following equation of evolution of the temperature in the bulk part of the liquid :

$$\frac{\partial T^L}{\partial t} = \left(\gamma^L - 1\right)\frac{\dot{Q}}{\rho^L C_V^L V} H^{-1}, \tag{7}$$

where γ is the ratio of the specific heat at constant pressure (C_P) to the specific heat at constant volume (C_V). The quantity H measures the gas character of the two-phase system and is defined as

$$H = 1 + \Phi^G \left(\frac{\kappa_S^G}{\kappa_S^L} - 1\right), \tag{8}$$

where Φ^G is the gas volume fraction, and β_S, κ_S are the isentropic thermo-elastic coefficients defined as

$$\beta_S = \left.\frac{1}{P}\frac{\partial P}{\partial T}\right|_S, \kappa_S = -\left.\frac{1}{v}\frac{\partial v}{\partial P}\right|_S. \tag{9}$$

Due to isentropic nature of the thermodynamic transformation associated with the piston effect, one can consider that at short times both the gas and the liquid evolve out of phase equilibrium, departing of the coexistence curve, thus behaving as if they were alone inside the cell. Hence the thermodynamic relation,

$$\left.\frac{\partial T}{\partial P}\right|_S \left.\frac{\partial P}{\partial v}\right|_S \left.\frac{\partial v}{\partial T}\right|_S = -1,$$

was assumed to be valid, as in the monophasic range. The equation of evolution of the temperature in the gas is :

$$\frac{\partial T^G}{\partial t} = \frac{\beta_S^L}{\beta_S^G}\frac{\partial T^L}{\partial t}, \tag{10}$$

in agreement with the prediction of Onuki et al. (Onuki, 1990b). Considering an ideal temperature step at the wall and assuming a pure diffusive temperature profile in the LBL (an approximation which does not take into account the temperature increase due to the small pressure increase inside the LBL), the time derivative of the total heat injected into the LBL is expressed as :

$$\dot{Q} = \lambda^L A \frac{T^W - T_2}{\sqrt{\pi D^L t}}. \tag{11}$$

where A is the surface of the heating wall, λ^L the thermal conductivity of the liquid, T^W the wall temperature, and

$$D^L = \frac{\lambda^L}{\rho^L C_P^L} \tag{12}$$

is the thermal diffusivity coefficient of the liquid. The typical time scale associated to the two-phase piston effect is equal to the classical one (Onuki, 1990a) applied to the liquid phase alone and corrected by the square of H :

$$t_C = \frac{\pi}{D^L \left(\gamma^L - 1\right)^2} \left(\frac{V}{A}\right)^2 H^2. \tag{13}$$

Note that H>1, whatever is the gas volume fraction, implying that the characteristic time of the piston effect is larger in the two-phase regime than in the monophasic liquid system at the same temperature. This is due to the larger expansion coefficient in the gas than in the liquid.

Exact solutions of the set of Eqs. 5 and 6 with heating conditions given by Eq. 11 are :

$$T_2 = T^W \left(1 - \exp\left(2\sqrt{\frac{t}{t_C}}\right)\right), \quad T_3 = \frac{\beta_S^L}{\beta_S^G} T_2, \tag{14}$$

with all the initial temperatures shifted to zero. These solutions are drawn in Fig. 7 as functions of the reduced time t/t_C, with initial temperatures shifted to −100 mK. The compressibility ratio was taken equal to 1.2 to fit to the overheating of the gas observed in Fig. 5b. Taking into account the amount of heat injected into the system, the departure of the real wall temperature profile from an ideal step, and the fact that the gas bubble is surrounded by only a thin wetting layer liquid in a large part of the cell, the agreement between the experimental and the theoretical temperature responses in the phases is satisfactory. A precise quantitative calculation with the help of an equation of state, realistic model of the heat injection, and numerical simulations assuming the particular phase distribution shown in Fig. 4 are now necessary to check the validity of this simple model.

CONCLUSION

Because of the high efficiency of the piston effect in a near-critical fluid, changing the fluid from a two-phase state to a one-phase state by the boiling process results in high temperature - low density *nonequilibrium gas* and low temperature - high density *nonequilibrium liquid* . The shape of the interface between gas and liquid quickly evolves in order to reduce these temperature non-homogeneities.

ACKNOWLEDGMENTS

Authors of this paper thank C. Brugne-Morteau, M. Salzmann, V. Auriac, A.C. Michels, and V. Nikolayev for their contributions. This work was supported by CNES and partly by NASA. We thank all the ALICE 2 teams, and especially J. F. Zwilling, for their technical support. We thank all individuals involved in the Cassiopee, Pegase French-Russian missions and GMSF US-French-Russian one on the Mir space station. We are particularly indebted to the crews of the missions, with special thanks to the cosmonauts C. André-Deshays, L. Eyharts, and S. Avdeiev.

REFERENCES

Beysens, D. and Garrabos, Y., "Critical Fluids Under Weightlessness," *1st Symposium on the Utilisation of the Int. Space Station*. ESA SP-385, pp. 647-665 (1996).

Davis, S. H., "Thermocapillary Instabilities,"Ann. Rev. Fluid Mech. 19, 403-435 (1987); and references therein.

De Gennes, P.G., "Wetting: Statics and Dynamics," Rev. Mod. Phys. 57, 827-862 (1985).

Delville, J. P., Salzmann, M., Garrabos, Y. and Beysens, D., "Gravitationnal Instabilities and Pattern Formation for 'Boiling'" in supercritical fluids", *Proc. of the 2nd European Symposium on Fluids in Space*, Ed Viviani, A. (J. Gilder Cong. srl, Roma) pp. 312-327 (1997).

Eicher, L., *"Equilibration in the Vicinity of the Critical Point in Pure Fluids,"* Thesis, Munchen (1996)

Garrabos, Y., Bonetti, M., Beysens, D., Perrot, F., Frohlich, T., Carles, P. and Zappoli, B., "Relaxation of A Supercritical Fluid After a Heat Pulse in Absence of Gravity Effects: Theory and Experiments," Phys. Rev. E 57, 5665-5681 (1998); and references therein.

Garrabos, Y., Chabot, C., Wunenburger, R., Delville, J.-P., Beysens, D., "Critical Boiling Phenomena Observed in Microgravity," J. Chim. Phys. 96, 1066-1073 (1999)

Gurfein, V., Beysens, D., Garrabos, Y., and Le Neindre, B., "Simple grid technique to measure refractive index gradients", Optics Communications, 85, 147-152 (1991).

Hegseth, J., Rashidnia, N., and Chai, A., "Natural Convection in Droplet Evaporation," Phys. Rev. E 54, 1640-1644, (1996).

Ikier, C., Klein, H., Woermann, D., "Optical Observation of the Gas-Liquid Phase Transition In Near-Critical SF_6 Under Reduced Gravity", J. Colloid Interface Sci. 178, 368-370 (1996).

Kayser, R. F., Schmidt, J. W., Moldover, M. R., "Wetting Layers and Dispersion Forces for a Fluid in Contact with a Vertical Wall," Phys. Rev. Lett. 54, 707-710 (1985).

Marcout, R., Zwilling, J. F., Laherrere, J. M., Garrabos, Y., Beysens,.D., "ALICE 2, an Advanced Facility for the Analysis of Fluids Close to Their Critical Point in Microgravity", IAF-94-J2-1100 (1995)

Morteau, C., Salzmann, M., Garrabos, Y., Beysens, D., "Density Characterization of Slightly Off-Critical Fluid Cells using the Alice2 facility," *Proc. of the 2nd European Symposium on Fluids in Space*, Ed. Viviani A. (J. Gilder Cong. srl, Roma), 327-333 (1997).

Nikolayev, V. S. and Beysens, D., "Boiling Crisis and Nonequilibrium Wetting Transition," *Europhysics Lett.* 47, 345-351 (1999)

Onuki, A., Hao, H., and Ferrel, R. A., "Fast Adiabatic Equilibration in a Single-Component Fluid Near the Liquid-Vapor Critical Point," Phys. Rev. A 41, 2256-2259 (1990a)

Onuki, A. and Ferrel, R. A., "Adiabatic Heating Effect Near the Gas-Liquid Critical Point," Physica A 164, 245-264 (1990b)

Palmer, H. J., "The Hydrodynamic Stability of Rapidly Evaporating Liquids at Reduced Pressure," J. Fluid. Mech. 75, 3, 487–511 (1976).

Stanley, H. E. *Introduction to Phase Transitions and Critical Phenomena*. Oxford, Clarendon Press (1971).

Straub, J., *Proc. of the IXth European Symposium on Gravity-Dependent Phenomena in Physical Sciences,* Ed. Rathe, L., Walter, H., Feuerbacher, B. (Springer, Berlin) pp. 351-359 (1995a).

Straub, J. and Eicher, L., "Density and Temperature Relaxation in the Two-Phase Region Near the Critical Point of a Pure Fluid," Phys. Rev. Lett. 75, 1554-1557 (1995b)

Straub, J., Eicher, L. and Haupt, A., "Dynamic Temperature Propagation in a Pure Fluid Near its Critical Point Observed in Microgravity during the German Spacelab Mission D-2," Phys. Rev. E 51, 5556-5563 (1995c)

Tong, L. S., *Boiling Heat Transfer and Two-Phase Flow*. Taylor & Francis, NY (1997).

Zhong, F., Meyer, H., "Density Equilibration Near the Liquid-Vapor Critical Point of a Pure Fluid. II. Coexisting Phases for $T<T_c$", Phys. Rev. E 53, 5935-5948 (1996)

Overview of Pool Boiling Heat Transfer Studies In Variable Gravity

Patricia Arlabosse[1], Christelle Reynard[2], and Lounes Tadrist[2]

[1] *École des Mines d'Albi Carmaux, Route de Teillet - 81013 Albi CT Cedex 09 (France)*

[2] *I.U.S.T.I., UMR 6595, 5 Rue Enrico Fermi - 13453 Marseille Cedex 13 (France)*

arlaboss@enstimac.fr ; ltadrist@iusti.univ-mrs.fr

Abstract. Boiling studies under microgravity conditions require a perfect control of all the involved parameters: gravity level, g-jitter, the time of experimentation, the heater size and geometry, the size of the boiling vessel, the fluid properties... In this paper, an attempt is made to expose the main works achieved over the past fifty years on boiling in variable gravity level. More particularly, the experimental conditions during these different experiments as well as a comparison of the different results and conclusions proposed by the authors are presented. According to this analysis, all the experimental conditions differ notably and no clear conclusion can be drawn about boiling phenomena under microgravity conditions. Nevertheless, it seems that the gravity level has little influence on the boiling heat transfer. Many questions are still open and further experiments are expected to yield improvements in the understanding of the basic boiling physics.

INTRODUCTION

Nucleate pool boiling is a very efficient mode of heat transfer in that relatively small temperature differences between the heated bodies and the liquids can provide large rates of heat transfer. Therefore, boiling is widely employed for thermal energy conversions, transport systems and component heating or cooling in on-earth technology. Since the first boiling curve (Nukiyama, 1934), many experiments have been performed to better understand the boiling process. Nevertheless, due to the complexity of the phenomena and the number of parameters involved, the physics of the boiling process is still poorly understood and the literature has been flooded with correlations, which provide quick input to design performance and safety issues for engineers. Dhir (1990) emphasized in his keynote presentation at the 9th International Heat Transfer Conference that *the usefulness of the correlations diminishes very rapidly as parameters of interest start to fall outside the range of physical parameters, for which the correlations are developed.* These correlations as well as the classical theories describing boiling on earth make an extensive use of parameter "gravitational acceleration" (Rohsenow, 1962), which could explain the high heat transfer coefficient owing to the bubble's detachment from the heater.

For quite a long time, no evidence had been obtained to experimentally check whether the influence of gravity was modeled correctly. Until the middle 1950's, all the experiments were devoted to analyze the effect of increased gravity on the boiling heat transfer curve by using centrifuges which provide high gravity conditions over the range a/g=1 and 100 (Merte, 1961; Costello 1961). The design of space devices sparked the interest in low gravity conditions and led to a rapid expansion of low gravity experimentation. During years 50-60, most of the low gravity experiments were initiated in the United States using ground-based facilities like drop towers (Siegel, 1968). With the change in the American space policy, the interest into microgravity experiments decreased in the middle 1970's. Little is known on the experiments performed in the ex-USSR to experimentally verified the influence of reduced gravity conditions, some attempts were conducted to use magnetic fields to compensate earth gravity (Verkin, 1976). This technique provides a large range of gravity field and can operate for long duration but some difficulties

CP504, *Space Technology and Applications International Forum–2000*, edited by M. S. El-Genk
© 2000 American Institute of Physics 1-56396-919-X/00/$17.00

are encountered in obtaining a perfectly unidirectional magnetic field. Since the end of the 1980's, Europe, specially Germany, and Japan took over the experiments using short duration microgravity facilities like drop towers (Oka, 1996), parabolic flights (Straub, 1990; Oka, 1992) and sounding rockets (Ohta, 1998). Nowadays, the shuttle is used to perform the experiments (Lee, 1997; Steinbichler 1998).

All the experiments were performed under special operating conditions, mostly related to the equipment necessary to vary the gravitational acceleration. According to these experimental constraints, several questions remain about the role of gravity, about the influence of the short time of experimentation, the residual convection and the heater size on boiling heat transfer and about the possibility to observe nucleate boiling and the critical heat flux in low gravity environment. In this paper, an overview of the boiling heat transfer studies performed in variable gravity level is proposed. In comparison with the previous bibliographical analysis (Siegel, 1968; Merte, 1990 and Di Marco, 1997), which presented the experimental results (and in particular the role of the gravity level on the boiling heat transfer), we focus our attention on the experimental conditions and on theirs influences on the experimental results.

EXPERIMENTAL CONDITIONS

A great number of experimental works aim at studying the influence of the gravity level on the boiling heat flux. Changing the gravity require the use of heavy facilities (drop towers, sounding rockets, shuttle, centrifuge...) which bring all the authors to reduce the size of the equipments and which leads to short time of experimentation. In the following, we reviewed the different parameters and their influence on the experimental results.

The Time of Experimentation

Drop towers, parabolic flights and sounding rockets enable short duration low gravity conditions ranging between a few seconds and a few minutes. With up to 10 seconds of low gravity conditions, drop towers provide the shortest experimental times. At the end of 50's, Siegel and Usiskin (Siegel, 1959) used a 2.7 meter high drop tower, this corresponds to a free fall time of 0.7 second with a residual acceleration of 0.014g. Since, other drop facilities were constructed like in Hokkaïdo: the drop tower, built in a 710 meter deep disused mine shaft, provides some 10 seconds at 10^{-5}g for experimentation (Oka, 1996). Aircraft flying parabolic trajectories (Straub, 1990; Abe, 1993) include periods of 20 seconds of reduced gravity (up to 10^{-2}g) and periods of 20 seconds at about 2g, which permits to study boiling phenomena at variable gravity levels. During a campaign, three flights with thirty parabolas each are performed. Sounding rockets give the highest quality microgravity conditions. Typically, the free fall time ranges between 360 and 420 seconds with residual acceleration of less than 10^{-4}g (Ohta, 1998; Straub, 1992).

As most of the facilities used to compensate earth gravity lead to short duration tests, except for the experiments performed with centrifuges or with the shuttle (where no problem exists to observe long term boiling), most of the results have not been considered as representative of a stationary boiling state. It is necessary to compare the time of experimentation with the characteristic boiling times: the time for the thermal boundary layer to establish, the time of the bubble to growth, to detach... For instance, the thermal boundary layers are of greater thickness in microgravity than in earth gravity and consequently a longer time is required for them to develop after a change in thermal condition. Siegel (1958) analyzed the transient free convection from a negligible heat capacity vertical plate in laminar flow. For a surface at constant temperature, the times for boundary layers to establish vary as $g^{-1/2}$ while, for uniform wall heat flux, they change with $g^{-2/5}$. For instance, for a plate that has been raised suddenly from 293 K to 393 K, the time for boundary layer to establish in earth gravity is about 2 second while it is 200 seconds if the gravity field is reduced to 10^{-4}g.

The Gravity Level

The gravity level encountered with the available facilities range between 10^{-5}g and 1.4 10^{-2}g. For the highest levels, convective motion can be generated in the liquid near the heater and thus doubts remain on the contribution of this residual convection to the heat transfer. In 1992, Oka carried out two parabolic flights to investigate pool boiling of water and CFC-113 under reduced gravity condition. He noted that some effect of residual convection on the behavior of the vapor bubbles was recognizable when no sign of mechanical effect of residual convection was observed in the experiments conducted with sounding rockets (Oka, 1996). In the same way, Ohta (1996) observed bubble's detachment from a plate heater immersed in water, which could be due to the residual gravity level, during a parabolic flight. This was confirmed later (Ohta, 1998) with experiments performed in sounding rockets where no bubble detached from the heater. Moreover, the g-jitter during the experiments, which induces large fluctuations of the capillary length $L_b = \sqrt{\gamma/(\rho_l - \rho_v)g}$, can cause bubble's detachment from the heater and thus enhance the heat transfer. During a parabolic flight, fluctuations around the average $10^{-2}g$ value can reach up to $3.5 \ 10^{-2}g$ (Abe, 1993) or worse $7.5 \ 10^{-2}g$ (Zell, 1991).

The Heater

Usually, two heater designs are found in the literature: flat plate heater or thin wires with diameters ranging between 25 (Sitter, 1998) and 500 µm (Siegel, 1965). Due to their negligible heat capacities, the temperature response after a change in operating conditions is very fast, what allows Straub (1996) to consider that steady state is reached after 0.7 second of microgravity for a 50 µm diameter wire.

Bakhru and Lienhard (1972) investigated pool boiling curves for very small diameter horizontal wires in a variety of fluids under earth gravity conditions. The classical boiling curves were found to be significantly altered when the characteristic length scale of the heater D_{wire} was small compared to the capillary length L_b: the classical nucleate boiling and transition boiling zones as well as the maximum heat flux were absent. Based on their experiments, they concluded that the local minimum and maximum on the boiling curve vanish for $D_{wire}/L_b < 0.02$. For $D_{wire}/L_b > 0.3$, the classical pool boiling curve is observed. The range $0.02 < D_{wire}/L_b < 0.3$ corresponds to a transition regime where hydrodynamic mechanisms responsible for the minimum and maximum heat fluxes establish themselves.

The size of commonly used earth equipment is so large that it is unlikely to encounter such characteristic boiling curve even though this case can be easily observed in microgravity as the capillary length increases. With thin wires, surface tension effects related to the heater curvature may overshadow other effects and all extrapolation or interpolation of the behavior from known points must be done cautiously. For example, using R113, a 200 µm diameter heater and a $10^{-4}g$ gravity level, the ratio D_{wire}/L_b is closed to 210^{-3} and according to the results of Bakhru and Lienhard, it doesn't make sense to speak about critical heat flux and the characteristic curve represents a borderline case. Such heater sizes, but also smaller one (Sitter, 1998; Straub, 1996), are commonly used in microgravity experiments to investigate the influence of gravity on the boiling curve. For instance, Steinbichler (1998) studied saturated boiling in Freon R134a around a 20 µm diameter platinum wire in a $10^{-4}g$ gravity field. The heat flux was increased up to 375 kW/m² and no critical heat flux was reached even when Lienhard hydrodynamic theory (1973) predicts that, for $10^{-4}g$, the critical flux is 140 kW/m².

The heater orientation is also an important parameter. Nishikawa (1983) plotted the boiling curves for different orientation of a plate heater at atmospheric pressure and under normal gravity condition. He brought to the fore that above a certain threshold, corresponding approximately to the transition between the nucleate isolated bubbles regime and vapor columns one, all the curves were identical. Below this threshold, the heater orientation systematically modifies the boiling curve. For a given heat flux, the heat transfer is higher for a downward-facing heater than for a upward-facing one. These results emphasize that the heat transfer can not be extrapolated to all the orientations at low heat fluxes. In microgravity, due to the contribution of the residual convection, the heater orientation has an influence on the boiling curve as found by Siegel (1965) and Merte (1990).

Fluid Properties

Keshock and Siegel (1964) measured the average bubble diameters at departure from the heater under reduced gravity in distilled water and a 60% by weight sucrose solution. For 60% sucrose solution, the departure diameter has hardly any variation with gravity whereas for water, the departure diameter increases when the gravity is reduced. To explain these different gravity dependencies, they investigated the forces acting on bubbles for saturated conditions. When the bubble's growth quickly on the heater, the inertial forces overcomes the surface tension force very rapidly and hence initiates the bubble detachment. When the growth rate is lower, the departure is dependent on an equilibrium of buoyancy and surface tension forces. Hence, as the growth rate of bubbles was very large in 60% sucrose solution, the departure was initiated by inertial force and the departure diameter does not exhibit a gravity dependence as well as the boiling curve.

According to the fluid physical properties, departure from the heater can be observed or not. Oka (1996) emphasized a significant difference in the gravity dependence of boiling heat transfer between oily fluids (CFC-113 and n-pentane) and water during drop shaft experiments at the Japan Microgravity Center. In water, detachment from the plate surface was observed almost immediately: the bubbles staid in the vicinity of the heater and continue to growth by pumping the primary bubbles. For CFC-113, most of the bubbles generated on the heater remain on the surface and coalesce into a single slug covering the heater in nearly saturated boiling conditions.

INFLUENCE OF THE GRAVITY LEVELON THE BOILING HEAT FLUX

Among the high number of experimentations performed in low gravity conditions, it is not easy to synthesize the results: all the operating conditions are different (the fluid used, the heater and the gravity level). We already discuss the problem related to the confinement and the heater curvature, which do not allow generalizing the results obtained with such special geometry. Finally, the time of experimentation, all except the shuttle, are short: the observed phenomena do not perhaps correspond to steady state boiling and it is difficult to extrapolate the behaviors observed in the short tests. The results obtained during microgravity experiments as well as the operating conditions are summarized in Table 1.

TABLE 1. List of Microgravity Experiments.

Name	Facility	Heater	Fluid	Flux (kW/m²)	Influence on Heat Transfer
Siegel (1965)	drop tower	horizontal 500μm diam. wire	water	$95 < q < 285$	Enhancement of heat transfer which decreases when q increases (0.014g)
	drop tower	horizontal 500μm diam. wire	alcohol	$90 < q < 197$	Enhancement of heat transfer which decreases when q increases (0.014g)
	drop tower	horizontal 500μm diam. wire	60% sucrose solution	$94 < q < 290$	No influence of gravity decrease (0.014g)
	drop tower	vertical 500μm diam. wire	water	$92 < q < 218$	Decrease of about 18% (0.014g)
	drop tower	vertical 500μm diam. wire	alcohol	$89 < q < 150$	Decrease of about 4% (0.014g)
	drop tower	vertical 500μm diam. wire	60% sucrose solution	$97 < q < 295$	Decrease of heat transfer which increases with q (0.014g)
Merte (1990)	drop tower	horizontal upward facing disk	Liquid nitrogen	$q < 110$	Slight enhancement (0.008g)
	drop tower	horizontal down-ward facing disk	Liquid nitrogen	$q < 110$	Slight decrease (0.008g)
	drop tower	vertical disk	Liquid nitrogen	$q < 110$	No influence (0.008g)
Oka (1992)	parabolic flight	40*20 mm² plate	pentane	$5 < q < 40$ $7 < \Delta T_{sub} < 10K$	Decrease of heat transfer which is lower when q increases (0.005g)
	parabolic flight	40*20 mm² plate	pentane	$20 < q < 50$ $20 < \Delta T_{sub} < 30$ K	Decrease of heat transfer which is lower when q increases (0.005g)
Straub (1992)	parabolic flight	200μm diam. wire	R12	$0 < q < 500$	Enhancement for q<275 and decrease for higher flux (0.02g)
	parabolic flight	40*20 mm² plate	R12	$0 < q < 100$	Decrease (0.02g)
	sounding rocket	200μm diam. wire	R113	$50 < q < 350$	No influence for q<200 and then slight decrease (10^{-4}g)

Abe (1993)	parabolic flight	50*50 mm^2 plate	n-pentane	$q < 120$ $0 < \Delta T_{sub} < 21K$	Slight degradation. CHF reached at 40% of earth CHF (10^{-2}g)
	parabolic flight	50*50 mm^2 plate	CFC-113	$q < 120$ $2 < \Delta T_{sub} < 15K$	Poor gravity conditions (1.5 10^{-2}g). No effect of subcooling
	parabolic flight	50*50 mm^2 plate	water	$q < 120$ $2 < \Delta T_{sub} < 16K$	Considerable heat transfer degradation (10^{-2}g)
	sounding rocket	30*30 mm^2 plate	n-pentane	$q < 50$	Slight heat transfer degradation. Near saturation, rapid dry-out
Straub (1996)	shuttle	200μm diam. wire	R134	$50 < q < 300$	Enhancement (10^{-4}g)
	shuttle	50μm diam. wire	R134	$50 < q < 300$	Significant decrease (10^{-4}g)
Oka (1996)	drop tower	40*40 mm^2 plate	water	$q < 400$	Heat transfer decrease. Nucleate boiling, no CHF reached (10^{-5}g)
	drop tower	40*40 mm^2 plate	CFC-113	$q < 45$	Heat transfer decrease. Nucleate boiling, CHF reached (10^{-5}g)
Ohta (1996)	parabolic flight	50 mm^2 diam. disk	water	$30 < q < 290$	Enhancement of heat transfer (even compared to 2g values) (10^{-2}g)
	parabolic flight	50 mm^2 diam. disk	ethanol	$10 < q < 100$ $3 < \Delta T_{sub} < 11$ K	Enhancement or degradation (10^{-2}g)
Lee (1997)	shuttle	19*38mm^2 plate	R113	$20 < q < 80$ $2 < \Delta T_{sub} < 11$ K	Enhancement, long term steady state nucleate boiling (10^{-5}g)
Ohta (1998)	sounding rocket	50 mm^2 diam. disk	ethanol	$3.5 < q < 80$ $0 < \Delta T_{sub} < 95$ K	Enhancement at low subcooling, steady state possible (10^{-4}g)
Steinbichler (1998)	shuttle	1.4mm diam. hemispherical heater	R123	$20 < q < 325$	High enhancement which decreases when the flux increases (10^{-4}g)
	shuttle	20μm diam. wire	R134a	$50 < q < 380$	Same tendency, no CHF reached.
Sitter (1998)	Drop tower	25μm diam. wire	FC72	$q < 40$	Decrease of heat transfer

According to these results, no clear quantitative conclusion can be drawn about the role of gravity in boiling heat transfer. It seems that the characteristic curve remains largely valid for nucleate pool boiling in low gravity fields. In enhanced gravity field, the boiling curve is only modified at low heat fluxes, where the contribution of natural convection to the heat transfer is still important. At low heat flux, the heat transfer is increased when the gravity level increases. For higher fluxes, the boiling curve is almost independent of the gravity level (Merte, 1961; Merte, 1990). These investigations were done with water and liquid nitrogen. The behavior was similar for both fluids, even if the heat transfer coefficient increase at low heat flux is smaller with liquid nitrogen.

CONCLUSIONS

A review of the main pool boiling experiments in variable gravity has been carried out and the main effects of experimental conditions, fluid properties and gravity level have been stressed. All the results found in the literature show that the role of the gravity in the boiling phenomena is still misunderstood and many questions are still open. Due to the facilities used to vary the gravity level, most of the results are contradictory and no quantitative conclusion can be drawn on the influence of a gravity change on the boiling heat transfer. Nevertheless, a general tendency gets clear of these works: it seems that the gravity level has little influence on the boiling heat transfer, which contradicts the extrapolation of the classical correlations used to describe boiling on earth outside their range of application. To confirm this tendency and to remove the uncertainties linked to the boiling characteristics in variable gravity, it is necessary to define standard simple experiments (one test fluid and one heater geometry, for instance) and to use different facilities to investigate the influence of the time of experimentation and the influence of the gravity level on the boiling phenomena.

ACKNOWLEDGMENTS

The authors acknowledge the members of the Heat and Mass Transfers team, who contribute to the realization of this study.

REFERENCES

Abe Y., "Pool Boiling Under Microgravity," *Microgravity Sci. Technol.* **6**, n°4, 229-238 (1993).

Bakhru N. and Lienhard J.H., "Boiling from Small Cylinders," *Int. J. Heat and mass transfer* **15**, 2011-2025 (1972).

Costello C.P. and Tuthill W.E., "Effects of Acceleration on Nucleate Pool Boiling," *Chem. Eng. Prog. Sym.* **57**, 189-196 (1961).

Dhir V.K., "Nucleate and Transition Boiling Heat Transfer under Pool Boiling and External Flow Conditions," in *9th Int. Heat Transfer Conference*, edited by G. Hestroni, Hemisphere Publisher, New York, 1990.

Di Marco P. and Grassi W., "Overview and Prospects of Boiling Heat Transfer Studies in Microgravity," Invited lecture – Int. Symposium in SPACE97, Tokyo Japan, pp. 14-41 (1997).

Keshock E.G. and Siegel R., "Forces Acting on Bubbles in Nucleate Boiling under Normal and Reduced Gravity Conditions," NASA Technical note, NASA TN D-2299 (1964).

Lee H.S., Merte H. Jr. And Chiaramonte F., "Pool Boiling Curve in Microgravity", *Journal of Thermophysics and Heat Transfer* **11**, n°2, 216-222 (1997).

Lienhard J.H. and Dhir V.K., "Hydrodynamic Predictions of Peak Pool Boiling Heat Fluxes from Finite Bodies," *Journal of Heat Transfer*, Transaction of ASME, Series C **95**, n°2, pp. 152-158 (1973).

Merte H. Jr and Clark J.A., "Pool Boiling in an Accelerating System," *Journal of Heat Transfer* **83**, 233-242 (1961).

Merte H. Jr., "Nucleate Pool Boiling in Variable Gravity," in *Low gravity fluid dynamics and transport phenomena* **130**, Progress in Astronautics and Aeronautics, edited by J.N. Koster and R.L. Sani, Washington DC, published by AIAA, 1990, pp. 15-72.

Nishikawa K., Fujita Y., Uchida S. and Ohta H., "Effects of Heating Surface Orientation on Nucleate Boiling Heat Transfer," in *Proc. ASME-JSME Thermal Eng. Joint. Conf.*, ASME, New York, **1**, 1983, pp. 129-136.

Nukiyama S., "The Maximum and Minimum Values of the Heat transmitted from Metal to bBoiling Water under Atmospheric Pressure," *Int J. Heat and Mass Transfer* **27**, n°7, 959-970 (1934).

Ohta H., Kawasaki K., Azuma H., Kakehi K. and Morita T.S., "Nucleate Pool Boiling Heat Transfer under Reduced Gravity Condition," *Microgravity Q* **6**, n°2-3, pp. 114-120 (1996).

Ohta H., Kawaji M., Azuma H., Inoue K., Kawasaki K., Okada S., Yoda S. and Nakamura T., "Heat Transfer in Nucleate Pool Boiling under Microgravity Condition," in *Heat Transfer 1998*, edited by G. Hestroni, Proceedings of 11th IHTC, Hemisphere Publisher, New York, Vol. 2, 1998, 401-406.

Oka T., Abe Y., Tanak K., Mori Y.H., Nagashima A., "Observational Study of Pool Boiling under Microgravity," *JSME Int. J.,* Ser. II **35**, n°2, pp. 280 (1992).

Oka T., Abe Y., Mori Y.H. and Nagashima A., "Pool Boiling Heat Transfer in Microgravity," *JSME Int. J.*, Series B **39**, n°4, 798-807 (1996).

Rohsenow W.M., "A Method of Correlating Heat Transfer Data for Surface Boiling of Liquids," *Trans ASME* **84**, pp. 969 (1962).

Siegel R., "Transient Free Convection from a Vertical Flat Plate," Trans. ASME **80**, p. 347 (1958).

Siegel R. and Usiskin C., "Photographic Study of Boiling in Absence of Gravity," *Journal of Heat Transfer* **81**, pp.3 (1959).

Siegel R. and Keshock E.G., "Nucleate and Film Boiling in Reduced Gravity from Horizontal and Vertical Wires," Nasa Technical Report, NASA TR R-216 (1965).

Siegel R., "Effects of Reduced Gravity on Heat Transfer," *Advances in Heat Transfer* **4**, 143–227 (1968).

Sitter J.S., Snyder T.J., Chung J.N. and Marston P.L., "Acoustic Field Interaction with a Boiling System under Terrestrial Gravity and Microgravity," *J. Acoust. Soc. Am.* **104**, n°5, pp. 2561-2569.

Steinbichler M., Micko S., Straub J., "Nucleate Boiling Heat Transfer on a Small Hemispherical Heater and a Wire under Microgravity," in *Heat Transfer 1998*, edited by G. Hestroni, Proceedings of 11th IHTC, Hemisphere Publisher, New York, Vol. 2, 1998, 539-544.

Straub J., Zell M. and Vogel B., "Pool Boiling in a Reduced Gravity Field," in *Heat Transfer 1990*, edited by G. Hestroni, Proceedings of 9th Int. Heat Transfer Conference, Hemisphere Publisher, New York, Vol. 1, 1990, pp. 91-112.

Straub J., Zell M. and Vogel B, "Boiling under Microgravity Conditions", in *Proceedings 1st European Symposium Fluids in Space*, ESA SP-353, 1992, p. 269.

Straub J. and Micko S., "Boiling on a Wire under Microgravity Conditions – First Results from a Space Experiment performed in May 1996," Eurotherm seminar n°48, 1996, pp. 275-283.

Verkin B.I. and Kirichenko Y.A., "Heat Transfer under Reduced Gravity Conditions," *Acta Astronautica* **3**, 471-480 (1976).

Zell M., "Untersuchung des Siedevorgangs unter Reduzierter Schwerkraft," Ph. D. Thesis, University of Munich, Germany, 1991.

Dynamics of Bubble Growth on a Heated Surface under Low Gravity Conditions

D.M. Qiu, S. Singh and V.K. Dhir

University of California Los Angeles, Department of Mechanical and Aerospace Engineering,
Los Angeles, CA 90095, U.S.A , e-mail: vdhir@seas.ucla.edu

Abstract. Experimental studies and numerical simulations of the single bubble growth and departure mechanisms under low gravity have been conducted. An artificial cavity of 10 μm in diameter was made on the polished Silicon wafer. The back surface of the wafer was electrically heated in order to control the surface nucleation superheat. The experiments were performed during the parabola flights of the KC-135. The test liquid was degassed water in the pressure range 14.7 –18.0 Psia. The data of bubble size and shape from nucleation to departure as well as bubble growth time were obtained for wall superheats between 2.5 and 6.5 °C under saturation and small subcooling conditions. Analytical/numerical models were developed to describe the heat transfer through the micro-macro layer underneath and around a bubble formed at a nucleation site. In the micro layer model the capillary and disjoining pressures were included. Evolution of the bubble-liquid interface along with induced liquid motion was modeled. The experimental data and the computational prediction showed a good agreement. From the comparison of the low gravity results and those at earth normal gravity it is found that at the same wall superheat and liquid subcooling the bubble departure diameter can be approximately related to the gravity level through the relation $D_d \propto 1/\sqrt{g}$. Liquid subcooling has an appreciable effect on the bubble growth period. A relatively small subcooling can significantly prolong the time a bubble stays on the heater surface.

INTRODUCTION

Boiling is known as a highly efficient mode of heat transfer. It is employed in component cooling and in various energy conversion systems. For space applications boiling is the heat transfer mode of choice since the size of the components can be significantly reduced for a given power rating. For any space mission, the size, and in turn the weight of the components plays an important role in the economics of the mission. Applications of boiling heat transfer in space can be found in such areas as thermal management, fluid handling and control, and power systems. The investigations of boiling heat transfer for space applications impose unique constraints in terms of the number of experiments that can be conducted under microgravity conditions, the duration of the experiments, the expense and the difficulties involved in performing the experiments. Thus, for space applications, it is even more important that a mechanistic understanding of the boiling process be developed and the performance features of the experiments be explored in advance. The low gravity environment in the aircraft KC-135 provides a less expensive means to accomplish these tasks.

Under microgravity conditions the early data of Keshock and Siegel (1964) and Siegel and Keshock (1964) on bubble dynamics and heat transfer show that the effect of reduced gravity is to reduce the buoyancy and inertia force acting on a bubble. As a result, under reduced gravity bubbles grow larger and stay longer on the heater surface. This leads to merger of bubbles on the heater surface and existence of conditions similar to fully developed nucleate boiling. Thus, under microgravity conditions partial nucleate boiling region may be very short or non-existent.

Merte et al., (1994) and Lee and Merte (1997) have reported results of pool boiling experiments conducted in the space shuttle on a gold film sputtered on a quartz plate of an area 19×38 mm². Subcooled boiling of R113 during long periods of microgravity was found to be unstable. The surface was found to dryout and rewet at higher surface heat fluxes. At 11.5 °C subcooling it was observed that after a period of growth of small bubbles and their coalescence a larger bubble was formed and stayed above the surface. The larger bubble acted as a reservoir sucking and removing the smaller bubbles from the surface. Average heat transfer coefficients during the dryout and

CP504, *Space Technology and Applications International Forum–2000*, edited by M. S. El-Genk
© 2000 American Institute of Physics 1-56396-919-X/00/$17.00

rewetting periods were found to be about the same. The nucleate boiling heat fluxes were higher than those obtained on a similar surface at earth normal gravity, g_e, conditions. It was concluded that subcooling has negligible influence on the steady state microgravity heat transfer coefficient. The effect of Maragoni convection (thermocapillary flow) on the heat transfer was considered to be reduced in microgravity conditions.

Straub, Zell and Vogel (1992, 1994) have conducted a series of nucleate boiling experiments using thin platinum wires and gold-coated plate as heaters at low gravity in the flights of ballistic rocket and KC-135. In the experiments R12 and R113 were used as test liquids at saturation and subcooled conditions. The results from the thin wires showed that at low gravity and low heat fluxes the heat transfer coefficients were slightly higher or were the same as those at normal gravity conditions for both saturated and subcooled liquids. Quasi-steady state was reached in these experiments. The translating movement of vapor along the wire was described. More nucleation sites and relatively stronger Marangoni convection were believed to be the mechanism of heat transfer at low gravity. For the flat plate heater with R12 as the test liquid, boiling curves similar to those at g_e were obtained when the liquid was saturated. Using R113, rapid bubble growth and large bubbles covering the heater surface were observed. However, neither temperature nor heat flux reached a steady state. For subcooled R113 and R12, a reduction in heat transfer coefficient of up to 50% in comparison to that at g_e was obtained. It was observed that larger bubbles occupied the surface and at their edges many smaller bubbles formed, coalesced and fed the larger ones.

Oka and Abe et al (1995) have studied pool boiling of n-Pentane, R113 and water under parabolic flight conditions with about 5 seconds of reduced gravity level of 0.02 g_e. For the first two liquids, bubble merger at the heater surface occurred by sliding of bubble along the surface. For water, the suction of smaller, newer bubbles into larger bubbles and the bubble coalescence were observed. The different bubble merger processes for the different liquid were postulated to relate to the vapor-liquid-solid contact behaviors. However, no quantitative parameters to explain this behavior were identified. In the low gravity period, no bubble detachment from the heater surface was observed.

From above description it appears that results from studies conducted till now are non-conclusive. Questions remain on the conditions of nucleation initiation, the stability of nucleate boiling, the roles of liquid microlayer underneath the stationary and translating bubble, the reasons for equivalence of magnitudes of heat transfer coefficients at normal gravity and low gravity conditions, the forces leading to bubble liftoff or sliding motion along the surface and on the physics that underlies the boiling phenomena at low gravity. As such, there is neither systematic experiments which can be used for scaling of the gravity effect, nor mechanistic models that describe the observed physical behavior and the dependence of nucleate boiling heat flux on gravity level and wall superheat.

When the dependence of cavity site density on wall superheat is known (true for designed surface with artificial cavities) the prediction of heat flux requires a knowledge of interfacial area per cavity, interfacial heat flux, and heat transfer on the unpopulated area of the heater. Size of bubbles at breakoff, bubble release frequency (growth period + waiting time) and the number of bubble release sites influence the time and area averaged heat transfer and also determine the vapor generation rate. As a first step in developing and validating a mechanistic model for prediction of nuclear boiling heat transfer, in this work complete process of nucleation inception to bubble departure for a single bubble formed at a well defined and controllable nucleation site is studied at the low gravity condition of the KC-135 aircraft. The numerical simulation model to study bubble dynamics (Son, Dhir and Ramanujapu (1999), Singh and Dhir (1999)) is exercised and the results are compared with the experimental data from KC-135 flights.

DESCRIPTION OF THE EXPERIMENTS

A schematic of the experimental apparatus for the KC-135 flights is found in (Qiu and Dhir, et al 1999). The apparatus is the same as used by Merte et al (1994) in their flight experiments. It consists of a test chamber (D=15 cm, H=10 cm), a bellow and a nitrogen (N_2) chamber. Three glass windows are installed on the walls of the test chamber for the optical measurements. For the control of the system pressure two pressure transducers are installed in the test chamber and the N_2 chamber respectively. The heater for studying nucleate boiling is installed at the bottom of the test chamber. In the vicinity of the heater surface a rake of six thermistors is installed in the liquid pool to measure the temperature in the thermal boundary layer, while another thermistor rake is placed in the upper portion of the chamber to measure the bulk liquid temperature. Degassed and filtered water was used as the test liquid. Two video cameras were installed to record the boiling processes. One of them operates at 250 frames/second and provides digitalized images with increased magnification. The second one is used for an overall view of the boiling process. The liquid temperature and pressure in the chamber were controlled according to the set-points established by the operator on board. A three-component accelerometer was installed on the frame on which the set-up is mounted.

Polished Silicon wafer 4 inch in diameter and 1000 μm in thickness was used as the test surface for the nucleate boiling experiments. From the manufacturer's specification the surface roughness is less than 5 Å. For single bubble studies, a 10 μm wide and 100 μm deep cavity was etched in the wafer center via the reactive ion etching technique. The wafer is cast with RTV, -Silicon rubber, on a Phenolic Garolite (G-10) base. The base in turn is mounted on the test chamber. At the back of the Silicon wafer the foil-like strain gages were bonded as heating elements. Each element was separately wired and the wires were taken out from the hole in the G-10 base. The heating elements are grouped in different regions. In each group a thermocouple is directly attached to the wafer. The heater surface temperatures in different regions are then separately controlled through a multi-channel feedback control system. As such, the surface superheat can be maintained constant during an experimental run and be automatically changed to the desired set-point.

Before each flight, the boiling experiments were conducted in the laboratory to ensure the existence of nucleate boiling on the cavity formed in Silicon heater and to examine the operation readiness of the system at the earth normal gravity. Before the plane took off, nucleation at the cavity was activated by energizing the heating elements underneath the cavity. During the parabolic flights of KC-135 the gravity level in the direction, z, normal to the heater surface basically varied within a range of $g_z \approx \pm 0.04\ g_e$, with the accidental increase up to 0.065 g_e. The period of this low gravity condition was 20 seconds. Before entering the low gravity condition, the high gravity condition was experienced ($g_z \approx 1.8\ g_e$). In between the parabolic flights, the bulk water was maintained at a uniform temperature and the system pressure was set to 17 Psia. Before entering the low gravity condition the stirrer was switched off and the pressure was decreased to the value corresponding to the desired subcooling or saturation temperature so that the liquid in the chamber calmed down and was in the state as specified in the test matrix. Once the gravity g_z normal to the heater started to decrease below the level ~ 1.8 g_e, the surface temperature of the Silicon wafer was set to the desired superheat. The control system simultaneously turned on the video camera and recorded pressure, temperature and acceleration data. Due to the high conductivity of the Silicon wafer and the fast response of the controller, it took less than 4 seconds to reach the desired superheat on the heater surface. In the low gravity condition, the surface superheat, water subcooling and the system pressure were maintained nearly constant.

ANALYTICAL/NUMERICAL SIMULATION MODEL

In the model a single bubble formed at a nucleation site is considered and the assumption of axis-symmetry is invoked. The domain of interest is divided into micro and macro-regions. The micro-region mainly encompasses the microlayer underneath the bubble. Heat from the solid surface is conducted through the microlayer and is utilized for evaporation at the interface. Forces acting on the liquid in the microlayer are those due to viscous drag, interfacial tension, long range molecular interactions and vapor recoil. Lubrication theory is used to analyze the microlayer and Radial thickness of the microlayer during evolution of the bubbles is calculated. For the macro-region complete conservation equations for both vapor and liquid are used. The shape of the interface obtained from the solution of the micro and macro is matched at the outer edge of the microlayer where the tangent to the interface yields macroscopic contact angle. For water at one atmosphere pressure boiling on Silicon surface the static contact angle was measured to be 54 °, see Ramanujapu and Dhir (1999). In the model, the effect of surface wettability is included through the specification of the value of the Hamakar constant. In the simulation of the bubble dynamics at low gravity conditions the temperature of the heater was assumed to increase to the desired value (superheat) two seconds prior to inception of the nucleation. As a result, the thickness of the thermal boundary layer adjacent to the heater was calculated by assuming transient diffusion of heat into the liquid.

EXPERIMENTAL RESULTS

Experimental data of single bubble growth and departure in low gravity conditions were obtained at different superheats of the heater surface (2.5-6.5 °C) and low liquid subcooling ($\Delta T_{sub} < 1.0$ °C). The results are described below:

Bubble Shape

Figure 1 shows the selected pictures of a complete cycle of bubble growth with the nucleation inception occurring at the designed cavity. The superheat, T_w-T_s, was 4.2 °C and the liquid subcooling, ΔT_{sub}, was 0.3 °C. The bubble is seen to acquire the shape of a truncated sphere during the growth period before the departure. The size of the bubble base first increased, reached a maximum value (Figure 1-b) and then decreased. At the moment of bubble departure the video image shows a point-like contact with the bubble to the surface. Bubble base diameter and bubble height are plotted in Figure 2 for a case in which liquid was saturated at wall superheat was 5.5 °C. It is noted that initially as the bubble height increases bubble base diameter also increases. However, after reaching a maximum value the bubble base diameter shrinks as bubble starts to detach from the heater surface and bubble height increases rapidly. Maximum bubble base diameter is about 40% of the bubble height. Since a microlayer forms underneath the bubble, an increase in bubble base diameter reflects a corresponding increase in the contribution of evaporation from the microlayer.

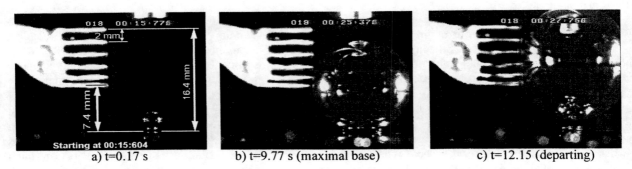

| a) t=0.17 s | b) t=9.77 s (maximal base) | c) t=12.15 (departing) |

FIGURE 1. Selected pictures of the single bubble during a growth-departure cycle Inception of nucleation at t=0 s), ΔT_{sub}=0.3 °C, T_w-T_s=4.2°C, $g_z\approx0.02$ g/g_e.

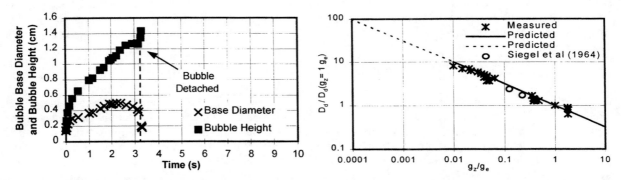

FIGURE 2. Bubble height and base diameter at ΔT_{sub}=0.0 °C, T_w-T_s=5.5 °C and $g_z\approx0.040$ g_e.

FIGURE 3. Bubble diameter at departure as a function of the gravity level.

Effect of Gravity Level

The bubble diameter was calculated by equating the volume of the actual bubble to that of a perfect sphere. Although during the parabola, the level of acceleration was reduced in all directions, the gravity level in the direction normal to the test surface was the largest. As such, bubble departure from the surface was determined primarily by the acceleration normal to the heater surface. The ratio of the observed bubble diameters at departure at any gravity level to that at g_z/g_e=1 is plotted in Figure 3 as a function of gravity level. The ratio of the bubble diameters at partial gravity ($g_z/g_e\approx0.35$) and high gravity ($g_z/g_e\approx1.8$) obtained during the KC-135 flights are also included. It is seen that for $10^{-2}\leq g_z/g_e\leq1.8$ the observed bubble diameters at departure can be approximately described as:

$$\frac{D_d}{D_{d(g_z/g_e=1)}} = \left(\frac{g_z}{g_e}\right)^{-1/2} . \tag{1}$$

Contact angle and temperature profile in the thermal layer at the test surface were also measured. During bubble growth, the contact angle (receding) was found to vary between 35 to 45°, where as that during the detachment phase (advancing angle) was about 45 to 60°. These values of contact angle are about those measured at earth normal gravity (Ramanujapu and Dhir (1999)).

Effect of Wall Superheat and Liquid Subcooling

Figure 4 shows bubble diameter as a function of time for wall superheats of 6.5, 5.5 and 3.7 °C. It is noted that with increase in wall superheat the bubble growth rate increases and bubble growth period decreases under low gravity environment of the KC-135 aircraft. This behavior is similar to that at earth normal gravity and is reflective of the increase in heat transfer rate through the microlayer and the vapor bubble boundary as wall superheat is increased. The bubble diameter at departure (≈ 1.4 cm) is found to be about the same for the three superheats when the gravity level at departure varied between 0.035 and 0.04 g_e.

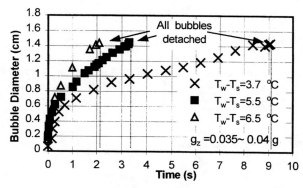

FIGURE 4. Bubble growth at different wall superheats in saturated water, g_z=0.035~0.04 g_e.

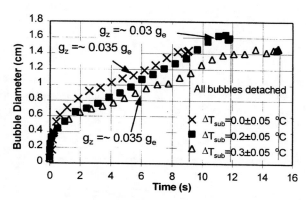

FIGURE 5. Bubble growth at slightly different subcoolings for T_w-T_s=3.5 °C.

Effect of liquid subcooling on bubble growth is shown in Figure 5. The data were obtained when g_z/g_e varied between 0.03 and 0.04 and for a wall superheat of 2.5 °C and the liquid subcoolings of 0, 0.2 and 0.3 °C. With increased liquid subcooling the bubble growth rate decreases and bubble growth period increases. However, the effect on bubble diameter at departure is small if any. With a subcooling of just 0.3 °C, the bubble growth period is increased from 9 to 15 seconds. Thus, in the experiments it is very important to know the liquid pool temperature precisely as slight subcooling can lead to large variation in bubble growth period without affecting the bubble diameter at departure.

NUMERICAL RESULTS

Using the numerical/analytical model described above, dynamics and heat transfer during single bubble nucleate boiling at low gravity were simulated. The results are described in the following:

Gravity Level

Figure 6 shows the predicted growth history of bubbles at gravity levels of g/g_e=1, 10^{-2} and 10^{-4} for saturated water at a wall superheat of 8 °C. The last point in the growth histories represents the bubble departure diameter and the growth period. It is seen that at earth normal gravity the bubble grows to about 3.2 mm in diameter before departure and the growth period is about 57 ms. The corresponding values for $g_z =10^{-2} g_e$ and $g_z =10^{-4}g_e$ are 2.9 cm and 5.5 seconds and 28 cm and 480 seconds respectively. Thus it is seen that with reduction of gravity both the bubble diameter at departure and the growth period increase. The bubble diameter at departure and the growth period are found to scale with gravity as

$$D_d \sim g_z^{-1/2}$$ (2)

and

$$t_d \sim g_z^{-0.98}.$$ (3)

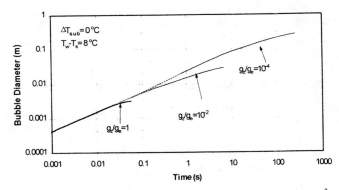

FIGURE 6. Numerical prediction of growth history of bubbles at gravity levels of $g_z/g_e=1$, 10^{-2} and 10^{-4} for saturated water at a wall superheat of 8 °C.

Wall Superheat and Liquid Subcooling

The effect of wall superheat on bubble growth history as obtained from the numerical model is shown in Figure 7 for a gravity level of $g_z=0.02\ g_e$. It is seen that with increase in wall superheat, the bubble growth rate increases, while the growth period of the bubbles decreases. The bubble diameter at departure remains almost constant with wall superheat.

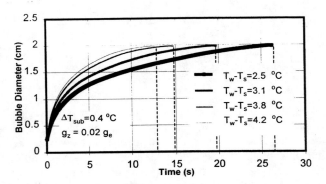

FIGURE 7. Bubble growth history for four superheats at $g_z=0.02\ g_e$.

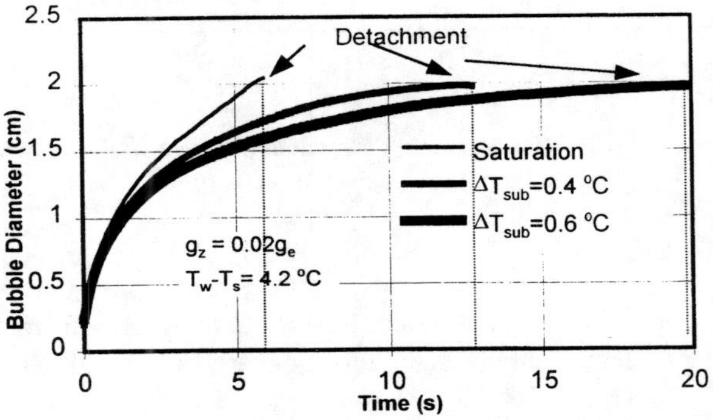

FIGURE 8. Bubble growth for different liquid subcooling at $g_z=0.02\ g_e$.

When the liquid is subcoooled, heat transfer to the liquid from the expanding interface of the bubble plays an important role in the bubble growth rate. Figure 8 compares the bubble growth rate at low gravity and at a wall superheat of 4.2 °C predicted for saturated liquid and for liquid subcoolings of 0.4 °C and 0.6 °C. It is noted that the effect of liquid subcooling is to reduce the bubble growth rate and to increase the bubble growth period. The bubble diameter at departure also reduces with liquid subcooling but the effect is small. Reduced bubble growth occurs because of loss of vapor as a result of condensation at the portion of the vapor-liquid interface. It is noted that in the later stages of the bubble growth as the interfacial area protruding into the subcooled liquid becomes large, the heat loss due to condensation is so large that the total heat transfer rate is smaller than the contribution from the microlayer as shown in Figure 9.

Figure 9 shows the variation of the heat flow into the bubble (total Q) for a subcooling of 0.4 °C and for wall superheats of 3.1 and 4.2 °C. During the early period of growth as the bubble base expands, the heat transfer rate increases, reaches its maximum value and gradually decreases. As the bubble starts to detach from the surface, the bubble base area shrinks causing the heat transfer rate to rapidly decrease. The negative value during the departure indicates that due to the condensation the bubble losses more heat to the surrounding liquid than that it gains from the evaporation and from the microlayer. The contribution of the microlayer to the total heat transfer rate is also shown in Figure 9. It is found that the microlayer contributes about 20 – 25% to the total heat removal rate from the wall and the magnitude of the contribution increases with wall superheat. Similar to that for the total heat transfer

756

rate, heat transfer rate in the microlayer decreases rapidly as bubble starts to detach, while the bubble base also rapidly shrinks.

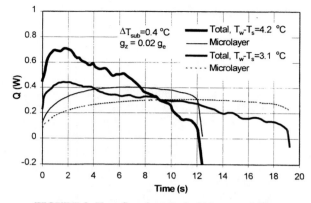

FIGURE 9. Heat flow into the bubble at $g_z=0.02\ g_e$.

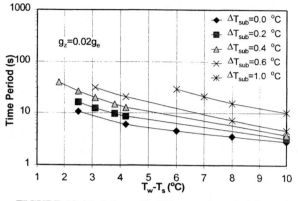

FIGURE 10. Variation of bubble growth period for various superheat and subcooling at $g_z=0.02\ g_e$.

Since bubble growth period is most sensitive to liquid subcooling, the effect of liquid subcooling on growth period was determined parametrically for the low gravity environment of KC-135. Figure 10 shows the bubble growth periods as a function of wall superheat for several liquid subcoolings when $g_z = 0.02\ g_e$. It is noted that for a wall superheat of 8 °C, just 1 °C subcooling of liquid can increase the growth period from about 3 to 16 seconds. With the duration for which low gravity is present in KC-135 flights being only about 20 seconds, small amount of subcooling can cause bubbles to depart only after the aircraft comes out of the low gravity parabola. This leads to observed bubble diameter at departure to be smaller than those expected under low gravity conditions.

COMPARISON BETWEEN NUMERICAL PREDICTION AND MEASURED DATA

As the results from both the numerical prediction and the measurements show, the scaling of the effect of gravity on bubble departure diameter can be described using Eq.(1) or (2). This is actually reflective of the fact that under slow bubble growth the bubble diameter at departure is given by the balance between surface tension and buoyancy forces, as is discussed in Singh and Dhir (1999). Equation (1) is also plotted in Figure 3 which shows a general agreement with the measured data. As an example at $g_z/g_e=10^{-2}$, bubble diameter at departure is ten times as large as that at normal gravity or bubble grows to about 28 mm before departure. Similarly, in the microgravity environment of the space shuttle ($g_z/g_e=10^{-4}$) single bubbles can be expected to grow to about 28 cm in diameter before departure (dotted line).

A comparison of the measured diameter of the bubble as a function of time in the low gravity environment of KC-135 with the prediction from the numerical simulations is made for cases when the liquid was saturated and had a subcooling of 0.4 °C (see Qiu and Dhir, et al, 1999). In both cases the agreement between the data and model prediction is generally good. The model not only correctly predicts the bubble diameter at departure but also the bubble growth period. In both cases the bubble diameter at departure is about the same but with slight subcooling and lower superheat the growth period is increased by a factor of 3.

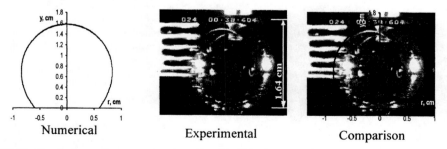

Numerical   Experimental   Comparison

FIGURE 11. Comparison of measured and predicted bubble shapes at $T_w-T_s=3.8$ °C, $\Delta T_{sub}=0.4$ °C and $g_z=0.02\ g_e$.

In Figure 11 the bubble size and the bubble shape at the moment when bubble base diameter reaches the maximum value are compared with those of the bubble measured in the KC-135 flight. The bubble shapes match quite well. The bubble base diameter is slightly over-predicted. The reason is believed in the difference of the contact angle of the bubble during experiment and that used in the numerical prediction. In the numerical prediction a constant angle was assumed both for the advancing and receding interfaces.

CONCLUDING REMARKS

By making artificial cavities on the polished Silicon wafer, well defined nucleation sites have been obtained. This approach has allowed experimental study of dynamics of a single bubble during nucleate boiling in the low gravity conditions of KC-135. Analytical/numerical model was developed to describe the bubble growth and departure process without using any empirical coefficients. Larger bubble departure diameters and longer bubble growth periods than those at earth normal gravity were predicted and measured. Bubble departure diameters as large as 20 mm were observed in the KC-135 experiments. Truncated sphere with a varying base area on the heater surface was found to be the typical shape of the single bubbles. The decrease of bubble base area is found to be associated with the bubble departure process. Bubble growth rate, growth period and bubble diameter from nucleation inception to departure were measured under subcooled and saturation conditions. Small subcooling in the liquid caused significantly prolonged bubble growth periods and reduced bubble growth rates.

The measured bubble departure diameters agree with the numerical predictions and confirm the fact that buoyancy and surface tension are the major forces during the bubble departure at low gravity. The results of both numerical prediction and measurements showed that the bubble departure diameter can be approximately related to the gravity level through the relation $D_d \propto 1/\sqrt{g}$ for single bubbles. A single bubble which grows to nearly 3 mm in diameter in about 50 ms before departure at earth normal gravity will have a departure diameter of about 30 cm and will take nearly 450 seconds to grow to this size at $g_e/g_e \approx 10^{-4}$. When experiments of bubble dynamics are conducted over short duration of low gravity, small subcooling in the liquid may cause bubbles not to grow to their predicted size before gravity level is raised.

REFERENCE

Keshock, E.G. and Siegel, R., "Focus Acting on Bubble in Nucleate Boiling under Normal and Reduced Gravity Conditions," *NASA TN-D-2999*, 1964.

Lee, H.S. and Merte, H., "Pool Boiling Curve in Microgravity," *Journal of Thermophysics and Heat Transfer*, Vol. 11, No.2, 1997, pp.216-222.

Merte, H., "Pool and Flow Boiling in Variable and Microgravity," *2nd Microgravity Fluid Physics Conference*, Paper No.33, Cleveland, OH, June 21-23, 1994.

Oka, T., Abe, Y., Mori, Y.H. and Nagashima, A., "Pool Boiling of n-Pentane, CFC-113, and Water Under Reduced Gravity: Parabolic Flight Experiments with a Transparent Heater," *J. of Heat Transfer*, H7:498-417, 1995.

Qiu, D.M., Dhir, V.K., Hasan, M.M. and Chao, D., et al., "Single Bubble Dynamics during Nucleate Boiling under Low Gravity Conditions," *Engineering Foundation Conference -Microgravity Fluid Physics and Heat Transfer, Sept. 19-24, 1999, Oahu, Hawaii.*

Ramanujapu, N. and Dhir, V.K., "Dynamics of Contact Angle during Growth and Detachment of a Vapor Bubble at a Single Nucleation Site", *Proceedings of the 5th ASME/JSME Joint Thermal Engineering Conference, March 14-19, 1999, San Diego, California*, Chapter G-01/627, 1999.

Siegel, R. and Keshock, E.G., "Effect of Reduced Gravity on Nucleate Bubble Dynamics in Water," *AIChE J.*, Vol. 10. 4, 1964, pp.509-516.

Singh, S. and Dhir, V.K., "Effect of Gravity, Wall Superheat and Liquid Subcooling on Bubble Dynamics during Nucleate Boiling," *Engineering Foundation Conference -Microgravity Fluid Physics and Heat Transfer, Sept. 19-24, 1999, Oahu, Hawaii.*

Son, G., Dhir, V.K. and Ramanujapu, N., "Numerical simulation of a single Bubble During Partial Nucleate Boiling on a Horizontal Surface," *J. Heat Mass Transfer*, 1999, in press.

Straub, J., "The Role of Surface Tension for Two-Phase Heat and Mass Transfer in the Absence of Gravity," *Experimental Thermal and Fluid Science*, Vol. 9, 1994, pp.253-273.

Straub, J., Zell, M. and Vogel, B., "Boiling under Microgravity Conditions," *1st European Symposium on FLUIDS IN SPACE, Ajaccio, France, Nov. 18-22, 1992.*

Design of Experiments for Thermal Protection System Process Optimization

Hans R. Longani

United Space Alliance, Kennedy Space Center, FL 32815

(407) 867-9751   Fax (407) 867-9852

Abstract. Solid Rocket Booster (SRB) structures were protected from heating due to aeroshear, radiation and plume impingement by a Thermal Protection System (TPS) known as Marshall Sprayable Ablative (MSA-2). MSA-2 contains Volatile Organic Compounds (VOCs) which due to strict environmental legislation was eliminated. MSA-2 was also classified as hazardous waste, which makes the disposal very costly. Marshall Convergent Coating (MCC-1) replaced MSA-2, and eliminated the use of solvents by delivering the dry filler materials and the fluid resin system to a patented spray gun which utilizes Convergent Spray Technologies spray process. The selection of TPS material was based on risk assessment, performance comparisons, processing, application and cost. Design of Experiments technique was used to optimize the spraying parameters.

INTRODUCTION

Components provided by USA for the Space Shuttle Solid Rocket Boosters (SRB) are protected from excessive heat during flight and SRB separation by a Thermal Protection System (TPS). MCC-1, a sprayable ablator coating, jointly qualified by USBI and NASA (MSFC) as an environmentally friendly TPS to replace the current solvent based ablator MSA-2.

Design of Experiments: Cork and glass particles are mixed together with the resins/catalyst in front of the spray gun and the mixture is sprayed on the structure by robot. Phase 1 sensitivity study consisted of changing "one variable at a time" while other factors were maintained at baseline to determine the effects of physical characteristics of MCC-1. This was a screening process for identifying factors for the Phase 2 study. The results indicate that there were eight statistically significant factors out of thirty four factors that affect the density of MCC-1. The objective of the Phase 2 was to find the main factors and suspected interactions by utilizing "Design of Experiments" techniques.

PROCEDURE

After examining the advantages and disadvantages of various Designs of Experiments, the MCC-1 team decided to utilize a Taguchi L-32 matrix for several reasons, including efficiency and a lower resource requirement. The eight factors were resin flow rate, catalyst flow rate, glass flow rate, cork flow rate, eductor air pressure, atomization air pressure, stand off distance, and spray cell temperature, each factor at two levels. Density was used as a response factor in the experiment.

CP504, *Space Technology and Applications International Forum–2000*, edited by M. S. El-Genk

TABLE 1. Taguchi Matrix of Factor Combinations.

Col No.	1	2	4	8	16	27	29	30
Factor Comb.	Resin Rate (g/min)	Catal. Rate (g/min)	Glass Rate (g/hr)	Cork Rate (g/hr)	Educt. Air Press. (psig)	Atom. Air Press. (psig)	SOD (Inches)	Spray Cell Temp. (F⁰)
1	228	228	2242	2769	29	34	6	74
2	228	228	2242	2769	35	48	8.7	92
3	228	228	2242	3281	29	48	8.7	92
4	228	228	2242	3281	35	34	6	74
5	228	228	2958	2769	29	34	8.7	92
6	228	228	2958	2769	35	48	6	74
7	228	228	2958	3281	29	48	6	74
8	228	228	2958	3281	35	34	8.7	92
9	228	244	2242	2769	29	48	6	92
10	228	244	2242	2769	35	34	8.7	74
11	228	244	2242	3281	29	34	8.7	74
12	228	244	2242	3281	35	48	6	92
13	228	244	2958	2769	29	48	8.7	74
14	228	244	2958	2769	35	34	6	92
15	228	244	2958	3281	29	34	6	92
16	228	244	2958	3281	35	48	8.7	74
17	244	244	2242	2769	29	48	8.7	74
18	244	244	2242	2769	35	34	6	92
19	244	244	2242	3281	29	34	6	92
20	244	244	2242	3281	35	48	8.7	74
21	244	244	2958	2769	29	48	6	92
22	244	244	2958	2769	35	34	8.7	74
23	244	244	2958	3281	29	34	8.7	74
24	244	244	2958	3281	35	48	6	92
25	244	244	2242	2769	29	34	8.7	92
26	244	244	2242	2769	35	48	6	74
27	244	244	2242	3281	29	48	6	74
28	244	244	2242	3281	35	34	8.7	92
29	244	244	2958	2769	29	34	6	74
30	244	244	2958	2769	35	48	8.7	92
31	244	244	2958	3281	29	48	8.7	92
32	244	244	2958	3281	35	34	6	74

Thirty-two sets of different spray combinations were used as per Table 1. The thickness, flatwise tensile strength and density measurements were taken per company standard procedure.

Analysis of Variance (ANOVA)

ANOVA was conducted on the data to investigate the main effects of resin flow rate, catalyst flow rate, glass flow rate, cork flow rate, eductor air pressure, atomization air pressure, stand-off distance, and spray cell temperature and their interactions on density (response factor). The ANOVA, as shown in Table 2, includes F ratio, sum of squares and percent contribution by each factor.

The analysis was performed at 95% confidence level with α = .05 level for testing the significance of main factors and interaction . As indicated in Table 2, the significant effects are: 1) cork flow rate, 2) spray cell temperature and 3) resin/catalyst (2216) flow rate.

It was determined that there were no interactions among factors from the analysis and engineering knowledge.

TABLE 2. ANOVA Table for MCC-1 Process Experiment.

Source of Variation	Degree of Freedom (df)	Sum of Squares	Mean Square	F-Ratio	Pure Sum of Squares	Percent Contribution
Resin Flow Rate	1	19.56	19.56	7.43	16.93	4.45
Catalyst Flow Rate	1	21.19	21.19	8.05	18	4.88
Glass Flow Rate	1	2.19	2.19	0.83	0.44	0.12
Cork Flow Rate	1	233.49	233.49	88.65	230.86	60.77
Eductor Pressure	1	7.33	7.33	2.78	4.7	1.23
Atomizing Air Pressure	1	0.21	0.21	0.08	2.41	0.63
Stand Off Distance	1	10.28	10.28	3.9	7.49	2.01
Spray Cell Temperature	1	24.99	24.99	9.49	22.35	5.88
All Other/Error	23	60.57	2.63		81.64	20.03

Analysis of Means (ANOM)

1) The y bars (average of 10 readings) were calculated for each run.
2) s (standard deviation of the response in any run) and the ln s were calculated.
3) The average of the averages $\overline{\overline{y}}$ ($\overline{\overline{y}}$), the average of s ($\overline{s}$) and the average of the ln s ($\overline{\ln s}$) were calculated and the results are as follows:

$$\overline{\overline{y}} = 35.06 \text{ pounds per cu. ft.}(561.66 \text{ kg/m}^3)$$
$$\overline{s} = 0.64 \text{ pounds per cu. ft.}(10.25 \text{ kg/m}^3)$$
$$\overline{\ln s} = -0.51 \text{ pounds per cu. ft.}(2.32 \text{ kg/m}^3)$$

The plots of y averages for the eight main factors are shown in Figure 1. The ANOM is given in Table 3.

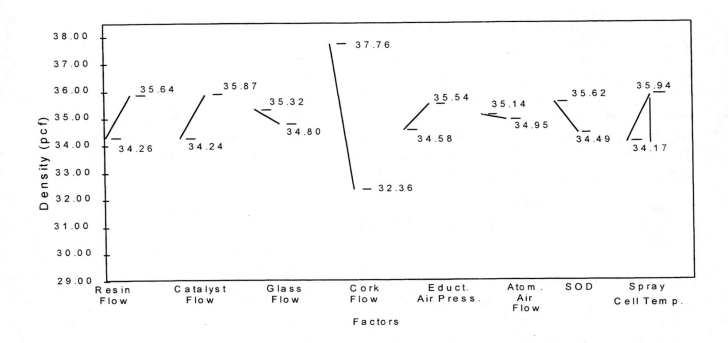

FIGURE 1. Plot of Averages (ANOM).

TABLE 3. Analysis of Means (ANOM).

Factors		Resin	Catalyst	Glass	Cork	Educ.	Atom.	SOD	Spr.temp
y bar	Ave.-	34.28	34.24	35.32	37.76	34.58	35.14	35.62	34.17
	Ave.	35.84	35.87	34.80	32.36	35.54	34.98	34.49	35.94
	Delta	1.56	1.63	-0.52	-5.40	0.96	-0.16	-1.13	1.77
	Delta/2	0.78	0.81	-0.26	-2.70	0.48	-0.08	-0.56	0.88
s	Ave.-	0.60	0.62	0.63	0.77	0.64	0.60	0.63	0.65
	Ave.	0.68	0.66	0.64	0.51	0.64	0.68	0.64	0.63
	Delta	0.08	0.04	0.01	-0.26	0.00	0.07	0.01	-0.02
	Delta/2	0.04	0.02	0.01	-0.13	0.00	0.04	0.00	-0.01
ln s	Ave.-	-0.57	-0.54	-0.51	-0.31	-0.53	-0.57	-0.50	-0.48
	Ave.	-0.45	-0.48	-0.51	-0.73	-0.50	-0.45	-0.52	-0.54
	Delta	0.12	0.05	0.00	-0.42	0.03	0.12	-0.01	-0.07
	Delta/2	0.06	0.03	0.00	-0.21	0.02	0.06	-0.01	-0.03

Prediction Equation

The prediction equation is as follows:

$\widehat{y}$ (Predic. Density=12.1207 (const.) +0.0977 x Resin Flow +0.1017 x Catalyst Flow - 0.0007 x Glass Flow - 0.010552 x Cork Flow +0.1595 x Educt. Press. -0.1178 x Atom. Air -0.4199 x Stand-off distance +0.09819 x Spray Cell Temp.

The forward skirt validation was sprayed in June 1995. The lessons learned from sensitivity studies were applied. The test results were as follows:

Cure Thickness = 0.163 - 0.174 in. (range), porta-pulls = 319 psi. (average), flatwise tensile strength = 408 psi (average), density = 35.68 lb_m/ft^3(average)

The averages of spray parameters acquired at every three seconds are as follows:

Glass flow rate = 2595.15 gms/hr.; cork flow rate = 3029.61 gms/hr.; resin flow rate = 235.96 gms/min.; catalyst flow rate = 235.99 gms/min.; atomizing air pressure = 45.00 psi; spray cell temperature = 90.35 degree F; stand-off distance = 8.0 in.

Confirmation Test

Substituting the average values of spray process parameters in the above prediction equation, $\widehat{y}$ predicted value comes out to be 35.41 pcf which is very close to actual measured average density value of 35.68 pcf. The delta is 0.7%! This approximately confirms the validity of our prediction equation.

PARETO CHART

The Avg.- is subtracted from corresponding Avg. + for each factor to obtain delta in density as given in Table 3. This delta is plotted against each factor in descending order of numerical value. It is evident that cork is a major contributor to the change in density followed in order by spray cell temperature and Resin/Catalyst (2216). The pareto chart is shown in Figure 2.

Main Effects

Figure 1 reveals the density changes (delta) with the corresponding change from low (Avg-) to high (Avg+) for each factor. The factors plotted were resin flow rate, catalyst flow rate, glass flow rate, cork flow rate, eductor air pressure, atomizing air pressure, stand-off distance and spray cell temperature.

The predicted values of density, high and low, for the three main factors are tabulated in Table 4 and were obtained by substituting appropriate values in a prediction (regression) equation. The model was constructed as shown in Figure 3.

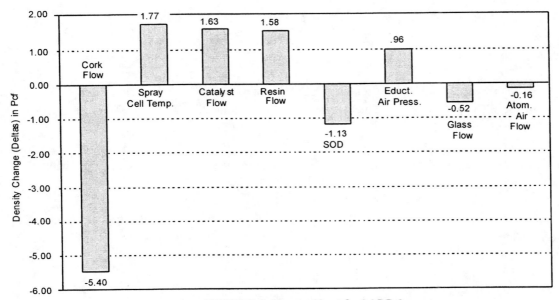

FIGURE 2. Pareto Chart for MCC-1.

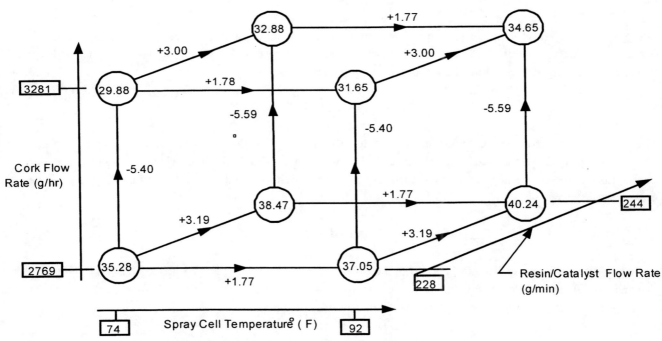

FIGURE 3. Model of Main Effects.

TABLE 4. Predicted Density Values Vs Main Factors (High and Low).

Cork Flow Rate (g/h)	Resin/Catalyst (g/min)	Spray Cell Temperature (^{0} F)	Predicted Density (lbs/ft^3)	Remarks
2769	228	74	35.28	
3281	228	74	29.88	Lowest density
2769	244	74	38.47	
3281	244	74	32.88	
3281	244	92	34.65	
2769	244	92	40.24	Highest density
2769	228	92	37.05	
3281	228	92	31.65	

CONCLUSION

The results indicate that there are three statistically significant factors that contribute 76% in controlling the density of the MCC-1 process, i.e., 1) cork flow rate, 2) spray cell temperature and 3) resin/catalyst (2216) flow rate. Also, the results did not reveal any statistically significant interaction among the factors. Although stand-off distance was not a very statistically significant factor contributing to density, it was, however, responsible for producing a waviness/devots pattern in MCC-1 at 6 inches, which was not acceptable. Therefore, it was recommended that the lower specification limit of 6 inches stand-off distance be increased to 7 inches.

The results obtained from the Phase 1 and Phase 2 Sensitivity Test will be used to establish Statistical Process Control (SPC) parameters to monitor and control the quality of MCC-1 on SRB Hardware. Besides the knowledge gained, this study will enable us to reduce MCC-1 development cycle for future programs, such as Titan and Atlas rockets and commercial applications.

Program Innovation/Lessons Learned: The Design of Experiments technique allows us to identify three main factors out of thirty four factors which contribute to the characteristics of the coatings. The results of the study concluded that we built the model which can predict and optimize the process. This technique has enabled us to understand the process better.

ACKNOWLEDGMENTS

I thank our members of the MCC-1 team, specifically Alan Kulchak, Samir Patel and Anil Patel for their invaluable assistance and Dr. F. Safie (NASA/MSFC) who provided his input in the analysis. My sincere thanks go out to Anthony Ulibarri for his encouragement and guidance.

REFERENCES

Longani, Hans R. 1995, "Environmentally Friendlier Thermal Protection System for Solid Rocket Booster," in *Proceedings of the Integrated Product and Process Design*, ASI press, pp 671-683.

Wu, Yuin ; Moor, Wille "*Quality Engineering*," American Supplier Institute, Inc., Dearborn, MI.

Feasibility Investigation on Potential Working Fluids for Electrohydrodynamically-Assisted CPLs

B. Mo, M.M. Ohadi, S.V. Dessiatoun, and M. Molki

Heat Transfer Enhancement Laboratory
Department of Mechanical Engineering
University of Maryland
College Park, MD 20742
Tel: (301) 405-5412; Email: bmo@eng.umd.edu

Abstract. The capillary pumped loop (CPL) is a state-of-the-art device for efficient cooling of electronic components used in spacecraft and telecommunications. CPL functions on a two-phase heat transport process in which a working fluid is driven by the pumping effect of the capillary action of a wick material imbedded in the evaporator. The wick structure is imbedded only in the CPL evaporator, and the rest of the loop is a simple wickless smooth tube. Therefore, compared to the widely used heat pipe, CPL provides a substantially higher heat transport capacity, more flexibility of installation, and a much greater distance of heat transport. The major disadvantages of the CPL, however, are the long and complicated start-up procedure, and the possibility of depriming at the high heat input and load variation. The main focus of this paper is on selection of potential working fluid candidates for use in Electrohydrodynamic (EHD)-enhanced CPLs. In this connection, a series of feasibility studies on various CPL working fluids have been performed and will be discussed. Among the working fluid candidates, propane and propylene are found to be the promising candidates. For appropriate electrode configuration, experimental results showed that the EHD pumping head of propane and propylene was high. Therefore, the two natural refrigerants, propane and propylene, are expected to improve the performance of EHD-enhanced CPLs.

INTRODUCTION

The capillary pumped loop, a wickless condenser heat pipe, is a passively pumped two-phase heat transport device that has demonstrated performance capabilities substantially greater than that of the conventional state-of-the-art heat pipes. Like other heat transfer devices, the CPL is limited by a maximum amount of heat that can be transported from one end to the other. The capillary limit is again a major concern similar to conventional heat pipes, expressed as,

$$(\Delta P_c)_{max} \geq \Delta P_v + \Delta P_l + \Delta P_w + \Delta P_g,$$ (1)

$$\Delta P_c = \frac{2\sigma}{r},$$ (2)

where ΔP_c is the capillary pressure drop; ΔP_v and ΔP_l are pressure drops in vapor and liquid lines, respectively; ΔP_w is the pressure drop in the wick; ΔP_g is the hydrostatic pressure drop (= $\rho_l g L_{eff} \sin\phi$, ϕ inclination angle). σ is the surface tension of a working fluid, and r is the porosity radius of the wick structure.

The other more specific limits to the operation of the CPL are the wick limitation and the liquid line boiling limitation. The wick limitation is related to a minimum pressure drop requirement between the evaporating and absorbing surface of the wick. This pressure drop is required for displacing the liquid from the vapor line and filling the liquid line and the compensation cavity. Maidanik et al. (1993) referred to this pressure drop as a small radial

CP504, *Space Technology and Applications International Forum–2000*, edited by M. S. El-Genk
© 2000 American Institute of Physics 1-56396-919-X/00/$17.00

temperature drop of the wick ΔT_v:

$$\frac{dP}{dT}\Big|_T (\Delta T_v) \approx \Delta P_c - \Delta P_w, \tag{3}$$

where the mean temperature T is taken as the average temperature between the liquid channel and vapor channel at the evaporator. Quantity $(\Delta P_c - \Delta P_w)$ is the pressure drop from the evaporating to the absorbing surface of the wick. dP/dT can be calculated from the Clausius-Clapeyron equation. The liquid line boiling limitation is a result of the liquid subcooling after condensation to avoid boiling in the liquid line. It is also related to the liquid pressure drop. Acquisition of heat at the evaporator provides the motivating force in any CPL system. Therefore, the capillary pump is the primary element for heat acquisition. For a given heat input to the CPL, the total pressure drop in the loop must not exceed the capillary pumping head if the CPL is to work properly. The wick meniscus will also adjust itself so that the capillary pressure rise equals the loop pressure drop. The screening of working fluids to generate pressure head is one of main concerns for the CPLs design.

As a result of the capillary pump limit, ensuring proper function of the loop or development of a vapor-tolerant CPL system becomes a challenge. Ohadi et al. (1997) and Mo et al. (1998) investigated the feasibility and performance improvements of the EHD-assisted CPL. This paper experimentally investigates the EHD pumping on potential working fluids for CPLs. EHD pumping is a complex phenomenon. It branches from electrodynamics, fluid mechanics, and heat transfer. It is based on the effect of electrical field on free charges, dipoles, and particles. The EHD pump has shown promising applications for industrial, terrestrial, and outer space applications.

EHD PUMPING PHENOMENON

The fact that dielectric liquids can be pumped by the injection of ions in an electric field has been known for some time (Chattock, 1899, and Harney, 1957). Since the early 1960s, theoretical and experimental investigations of the electrohydrodynamic (EHD) pump have been widely pursued. The study of ion-drag (injection type) EHD pump by Stuetzer (1960) and Pickard (1963), and induction EHD pump by Melcher (1966) prompted further studies of ion-drag EHD pump by many investigators such as Seyed-Yagoobi et al. (1994), and Ohadi et al.(1997). For the induction pump, the fluid temperature gradient model is an interesting topic and many researchers, such as Kelvin et al. (1981), Seyed-Yagoobi et al. (1994), Bart (1990), and Choi et al. (1995), conducted experiments based on Melcher's theory.

The EHD pump has many advantages. It has a small, simple, and stable structure without any rotating or moving parts and needs no maintenance. Low construction cost and low electric power consumption is additional advantages. Having no moving parts and no need for external pressure, EHD pumps are well suited for many applications such as circulating oil through heat exchangers. EHD pumping could also be used to increase the maximum heat transport capacity of heat pipes by exceeding the capillary pressure head and improving liquid-vapor separation. More importantly, the light weight, low maintenance, and vibration-free EHD pump makes them well suited for future outer space applications. It is well known that most pumping principles are based on the motion of mechanical parts that directly drive the fluids. The following mainly discusses the EHD pumping mechanisms, which is heavily relevant to our experimental studies.

The major driving force in ion drag pumps are ions. In such pumps, the liquid convection is induced by ions normally present in the liquid or injected by high voltage and a high gradient electrical field. The term "injection" needs to be clarified. The term "injection of charges" is widely used in the literature. Unfortunately, this term does not explain how charges are created, and what are the particular charges. The physical model of ion generation inside a neutral liquid can be described as follows: 1) since the liquid and vapor are not at the absolute zero temperature, a number of electrons exists inside the matter. Ionization of molecules by thermal energy and recombination of ions are in progress all the time. Also, the thermal energy of some electrons on the surface of metal is enough to free them from the metal surface. The electrical field accelerates this process substantially. 2)

767

When an electrical field is applied, electrons from liquid thermal ionization and electrode surfaces go into the liquid or vapor. Some of them are not lost in recombination. These electrons are the first charge generation in the fluid. 3) Electrons inside liquid or vapor are accelerated by the electrical field and break up the fluid molecules to create ions. Electrons must have a certain amount of energy to create ions. They have to be accelerated by a high electrical field or over a large distance before they collide with other molecules. Therefore, two conditions should be satisfied, the electrical field strength should be strong enough to accelerate the electrons and the distance between molecules should be large enough for the electrons to gain enough energy to ionize the molecules. Inside the vapor, this phenomenon occurs in a much lower electrical field, since the distance between the molecules is much larger.

When the electrical field is strong enough, the electron that has been liberated from the ionized molecule can be accelerated to ionize other molecules. This process can produce a growing "shower effect", which has been termed corona discharge. The Coulomb force that is produced by an external electrical field affects all the charges in the fluid. The resulting net force that acts on the electron and ion is the same, but the mass of the electron is negligible compared to the mass of the ion. Therefore, the major impact on the fluid movement is produced by ions (Stuetzer, 1960, and Crowley et al., 1990). The electrons, accelerated to high speed, act as ion producers.

The third EHD pump, based on the polarization of the media by an electrical field, is the Kelvin pump. The EHD Kelvin force affects the dipoles inside a liquid, moving the liquid in the direction of increasing electrical field. By applying a high electric field through the parallel electrodes over the two-phase medium, it causes the fluid to flow upward (Jones,1974). The dipole mobility that affects the electrical permittivity of liquids depends on temperature. Therefore, the pumping effect is increased in the single-phase fluid if a temperature difference exists. Also, this effect has been used in many other applications such as enhancement of heat transfer in single-phase liquid flow.

SCREENING OF WORKING FLUIDS

The requirements of a working fluid in heat transport systems are somehow different from those for refrigeration systems, due to no external power supply in heat transport systems. For a CPL system, the driving force comes from the pressure head generated in the wick. From the discussed equation 2, the surface tension of working fluid is a key factor for pressure generation. On the other hand, the good thermal characteristics of working fluid must be taken into consideration from the viewpoint of heat transfer for a CPL system. To improve the performance of an EHD-enhanced CPL, the electrical properties of working fluids becomes the third factor for the screening of potential working fluids.

Ammonia (NH_3) is commonly used as the working fluid for CPLs in space and aerospace industries. Ammonia has high surface tension, excellent thermal properties, and is very inexpensive. However, ammonia operates at high pressure and is toxic. Furthermore, the special safety shipment requirement of ammonia confines its promising application. Our preliminary experiments revealed that the high dielectric conductivity of ammonia made it inaccessible to implement performance improvements with EHD technique. Identifying other working fluid candidates for CPLs becomes a research topic for space and aerospace industries. For the time being, propylene is a potential fluid candidate due to its good thermal properties, especially its low freezing temperature for space applications.

McLinden (1988) made comprehensive studies on various working fluids for two-phase thermal systems. Ten of working fluids were screened out for possible applications in phase change thermal devices. They are ammonia, propane, methyamine, R-32, isobutane, R-152a, R-22, Halon1301, R-12, and R-11. We conducted a critical literature review on working fluids for EHD-friendly and two-phase thermal systems. From the three main factors to select working fluids that mentioned at the beginning of this section, Table 1 gives screening results of potential working fluids for EHD-assisted CPLs.

TABLE 1. Properties of the Potential Working Fluids at 25°C.

Refrigerant	Sat. pressure (MPa)	σ (mN/m)	h_{fg} (kJ/kg)	Viscosity of liquid (µPa·s)	Dielectric constant (κ)
HFC-134a	0.6457	8.35	178.88	215.4	9.51
HCFC-152a	0.5793	9.88	279.89	165.6	13.89
Ammonia	0.9726	21.53	1169.38	135.8	23
Propane	0.9997	6.86	333.3	110	1.89
Propylene	1.2118	6.68	331.2	-	2.16

As seen in the table, the natural refrigerants of propane (R-290) and propylene (R-1270) are included. Natural refrigerants are referred as substances comprised of molecules that occur in nature. The implication of the word "natural" is that the use of these refrigerants should have minimum adverse effects on the environment. The world atmospheric research community has focused its attention on quantifying the adverse effects of human produced chlorine, bromine, and their various components on depletion of the stratospheric ozone layer. From the viewpoint of ozone layer depletion potential (ODP) and global warning potential (GWP), natural refrigerants are attracting promising applications (Ohadi et al., 1998).

EXPERIMENTAL INVESTIGATIONS

After the screening of potential working fluids, we conducted experiments of the EHD pumping on selected working fluids for an EHD-enhanced CPL. The setup was designed and fabricated for the feasibility experiments of EHD pumping. For high-pressure flammable fluids, special safety protection needs to be considered for the setup design and fabrication. Different effective electrode configurations will be designed, tested and developed.

For the EHD pumping setup used in the current experiments, the high-pressure durable glass and pressure relieve valve were used for safety concern. The Teflon tube inside the setup was used for testing electrode chamber. The condenser at the top of the setup can control the operating conditions for EHD pumping experiments. A chiller with temperature controller for the condenser determines the operating temperature for the tested fluids. Pressure gauge at the top of the setup recorded the operating pressure accordingly.

The key component of the setup was electrode test chamber. Figure 1 sketches the details of two different electrode configurations, i.e., a needle electrode, and a parallel-plate electrode.

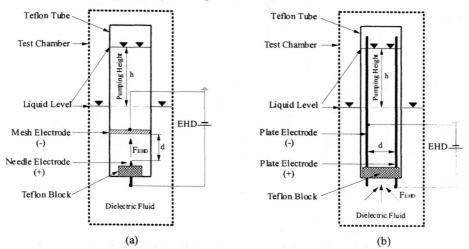

FIGURE 1. Details of electrode configurations, (a) Needle electrode, (b) Parallel-plate electrode.

The EHD pumping head (ΔP) can be calculated from the fluid rise,

$$\Delta P = \rho g h, \qquad \rho = \rho(T), \tag{4}$$

where $\rho(T)$ is the liquid density corresponding to the operating temperature (T), g is gravity, and h is the height of the pumping liquid. During experiments, the fluid interface inside the test chamber rose when an electric field was applied. By measuring the fluid height, the EHD pumping head is determined.

Figures 2(a) and 2(b) show the EHD pumping head with needle electrode for the tested fluids of R-134a, R-152a, propane, and propylene. As seen in Fig. 1(a), for the needle electrode, the needle is used as positive electrode, and the mesh is the ground electrode. The electrode gap is 5.5 mm. The operating temperature is 15°C. For the tested parameters, the highest pressure heads achieved for R-152a and R-134a were 230 Pa and 150 Pa (Fig. 2a), respectively. On the other hand, EHD pumping of R-152a showed the instability for long runs (5 hours) because the molecules of R-152a would change when they are subjected to an electric field for a long period of time. However, the highest pumping head achieved for R-290 and R-1270 was only about 10 Pa. As discussed in the EHD pumping section, the needle electrode belongs to the ion-drag pumping. The pumping effect is largely dependent on fluid molecules and ion generation when applying electric field. R-290 and R-1270 generated a very small amount of ions in the present of an electric field. Furthermore, R-1270 would generate fewer ions than R-290 (Fig. 2b). The results implied that the ion-drag EHD pump is not suitable for R-290 and R-1270.

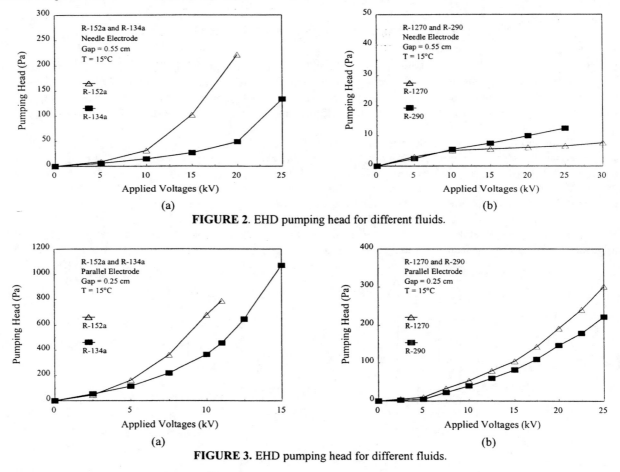

FIGURE 2. EHD pumping head for different fluids.

FIGURE 3. EHD pumping head for different fluids.

Figures 3(a) and 3(b) show the EHD pumping head with parallel electrode for tested fluids of R-134a, R-152a, propane, and propylene. As seen in Fig. 1(b), for the parallel-plate electrode, one parallel plate is used as positive electrode, and the other as the ground electrode. The electrode gap is 2.5 mm. The operating temperature is 15°C.

For the test parameters, Figure 5 shows that the highest pressure heads achieved for R-152a and R-134a were, respectively, 800 Pa and 1100 Pa. Figure 7 show that the highest pumping head of R-290 and R-1270 are 300 Pa and 200 Pa. From Figs. 5 and 6, under the same applied voltages, the results revealed that R-152a produced the highest pumping head, and R-290 gave lowest pumping head. This is because R-152a has the highest dielectric constant, and R-290 has the lowest dielectric constant (see Table 1). The function of parallel-plate electrode is based on the polarization pumping. The corresponding pumping effect is mainly due to the dipole moment and dielectric constant of fluids. Compared to Fig. 2(b), Figure 3(b) showed that propane and propylene could generate proper EHD pumping head with the parallel electrode. The results suggest that, with a proper electrode design, propane and propylene can provide a significant EHD pumping head.

CONCLUSIONS

Like other heat transfer devices, the CPL is limited by a maximum amount of heat that can be transported from one end to the other. The capillary limit is a major concern, as it is the case in conventional heat pipes. Properties of a working fluid play a great role on the CPL performance. EHD pumping effects would improve the performance of CPLs. An experimental investigation along with analysis of the controlling mechanism was employed to evaluate the EHD pumping on potential working fluids for EHD-assisted CPLs. Various pumping head data were addressed for the different fluids, and different electrode configurations. In general, the results of the present experiments suggest that with proper electrode design, propane and propylene can give proper EHD pumping effects.

ACKNOWLEDGMENTS

Financial support of this work by a consortium of sponsoring members is greatly acknowledged.

REFERENCES

Bart, S.F., "Microfabricated Electrohydrodynamic Pumps," *Sensors and Actuators*, A21-A23, 1990, pp.193-197.

Chattock, A.P., "On the Velocity and Mass of Ions in the Electric Wind in Air," *Philosophical Magazine Journal of Science*, Vol.48, 1899, pp.401-420.

Choi, J.W., and Kim, Y.K., "Micro EHD Pump Driven by Traveling Electric Fields," *IEEE Transactions*, 1995, pp.1480-1483.

Crowley J.M., Wright, G.S., and Chato, J.C., "Selecting a Working Fluid to Increase the Efficiency and Flow Rate of an EHD Pump," *IEEE Transactions*, Vol.26, No.1, 1990, pp.42-49.

Harney, D.J., "An Aerodynamic Study of the Corona Wind," *Ph.D. Thesis*, 1957, *CalTech*, Astia document AD 134400.

Jones, T.B., "An Electrohydrodynamic Heat Pipe," *Mechanical Engineering*, January, 1974, pp.27-32.

Kelvin, D., Crowley, J.M., and Chato, J.C., "Parametric Studies of a Large Thermal EHD Induction Pump," *IEEE Transactions*, 1981, pp.1015-1019.

Maidanik, Y., Solodovnik., N., and Fershtater, Y., "Experimental and Theoretical Investigation of Startup Regimes of Two-Phase Capillary Pumped Loops," *23rd International Conference on Environmental Systems, SAE Paper No. 932305*, 1993.

McLinden, M.O., "Working Fluid Selection for Space-based Two-phase Heat Transport Systems," *NASA Report, NBSIR 88-3812*, 1988.

Melcher, J.R., "Traveling-wave Induced Electroconvection," *Physics of Fluids*, Vol. 9, No.8, 1966, pp.1548-1555.

Mo, B., Ohadi, M.M., and Dessiatoun, S.V., "Liquid-vapor Separation and Thermal Management on EHD-enhanced Capillary Pumped Loop," *Proceeding of 11th International Heat Transfer Conference*, Korea, August, 1998, pp.63-68.

Ohadi, M.M., Mo, B., and Molki, M., "Natural Refrigerants -- Historical Development and an Updated Heat Transfer Research Progress," *Proceedings of International Conference on Energy Research and Development*, Kuwait, Nov., 1998, pp.i-xiv.

Ohadi, M.M., Dessiatoun, S.V., Mo, B., Kim, J., Cheung, K., and Didion, J., "An Experimental Feasibility Study on EHD-assisted Capillary Pumped Loop," *AIP Conference Proceedings 387*, 1997, pp.567-572.

Pickard, W.F., "Ion-drag Pumping I: Theory," *J. Appl. Phys.*, Vol.34, 1963, pp.246-250.

Seyed-Yagoobi, J., and Margo, B.D., "Effect of Frequency on Heat Transfer Enhancement in Temperature-induced EHD Pump," *IEEE Transactions on Dielectrics and Electrical Insulation*, Vol.1, No.3, 1994, pp.468-473.

Stuetzer, O.M., "Ion-drag Pumps," *J. Appl. Phys.*, Vol.31, 1960, pp.136-146.

Thermal Analysis of International Space Station Freeze-Tolerant Early External Active Thermal Control System Radiators

[1]Kambiz K. Andish, [2]Ing Youn Chen

[1]*Lockheed Martin Space Operations, Houston, TX 77258-8561*
[2]*Department of Mechanical Engineering, National Yunlin University of Science and Technology, Taiwan*
E-mail: kambiz.andish@lmco.com, cheniy@flame.yuntech.edu.tw

Abstract. In the early stages of International Space Station assembly, the photovoltaic (PV) radiators are utilized to dissipate the station's waste heat. To prevent radiator flow blockage that may occur by freezing of the radiator working fluid during certain cold orbits and low heat load conditions, a freeze tolerant radiator design was developed by NASA. The design not only alleviates the chance of radiator freezing, it can also tolerate high internal pressures that develop during local thawing of frozen flow tubes. To predict thermal performance of the PV radiator at freezing conditions SINDA/FLUINT program was used to model the freeze tolerant radiator design at one ORU level. The results of the analysis indicate that the radiator tubes will not freeze at the worst-case cold conditions specified as having a heat load of 4 kW and subjected to two consecutive cold sink orbits followed by two cold dips–one to -90°C for 0.25 hour and another to -98.3°C for 0.15 hour.

INTRODUCTION

In the early stages of the International Space Station (ISS) build-up process and before deployment of the External Active Thermal control System (EATCS) radiators, photovoltaic (PV) radiators are used to reject the ISS waste heat to deep Space. For this reason, the PV radiators at this stage are also called the Early External Active Thermal Control System (EEATCS) radiators. A schematic of the ISS at the assembly stage-7 5A with the orbiter attached is shown in figure 1 (NASA JSC, 1997).

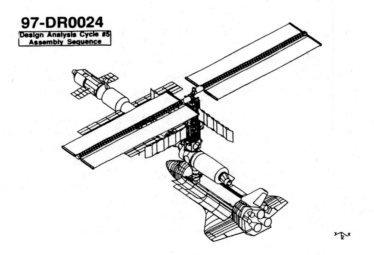

FIGURE 1. International Space Station configuration at stage-7 5A

CP504, *Space Technology and Applications International Forum–2000*, edited by M. S. El-Genk
© 2000 American Institute of Physics 1-56396-919-X/00/$17.00

The PV radiators are assembled in three Orbital Replacement Units (ORU) of seven radiator panels each. Each panel contains 24 parallel flow tubes with alternating connection to one of the two fluid loops. In the LVLH station attitude (flying in the positive X direction), the PV radiator ORUs located at the trailing edge (negative X direction) and starboard side (positive Y direction) are used as the EEATCS radiators. Only the third radiator ORU at the ISS leading edge (positive X direction) is used to cool the photovoltaic electric power system. In the LVLH flight attitude, the trailing edge ORU is maintained edge-to Earth and the starboard ORU has a very small view of Earth. The EEATCS radiators at stage 5A carry a very low heat load since they service only US LAB module. The combination of the low heat load, edge-to-Earth configuration, and open view of deep Space results in extreme cold environments for the EEATCS radiators, which may result in freezing of the working fluid (ammonia) in radiator flow tubes. After extensive studies, NASA concluded that the most reliable and cost effective alternative to address the radiator freezing issues was to design a freeze-tolerant radiator. Recently, a thermal analysis of the ISS EEATCS freeze-tolerant radiator original design was performed (Andish, 1997). The results of that analysis indicated that some flow tubes in each radiator panel could freeze when subjected to a constant sink temperature of –87°C (-125°F) and a heat load of 4kW, but would not interrupt the performance of the radiator. However, there were concerns about the radiator panel flow blockage that could occur by freezing of manifold tubes and/or the flex hoses connecting the radiator panels during certain critical orbital conditions.

Studies were performed to evaluate the effectiveness of the freeze-tolerant radiator design (Rocketdyne/Boeing, 1998). As a result of those studies, the original freeze-tolerant radiator design was improved to address issues such as manifold tube freezing, radiator panel blockage, and ORU blockage. To reduce the chance of manifold tube freezing, a design change was implemented to reduce the heat transfer between the radiator face sheet and the manifold covers. The 7th radiator panel in each ORU not only faces a colder environment due to its better view of deep Space, it also receives less flow from other panels because it is the last panel inline. Therefore, in a freezing environment the 7th radiator panel may freeze completely and disable the whole ORU. To prevent this scenario, the coating material of the last quarter of the face sheet in the 7th radiator panel in each ORU is changed from Silver-Teflon to Aluminized Kapton, which reduces radiative heat loss and prevents freezing of the flow tubes in that portion of the 7th panel.

DESCRIPTION OF PV RADIATOR

The flow tubes of PV radiators are constructed from Inconel and have an ID/OD of 1.7 mm/3.2mm (0.067"/0.125"). Inconel has a yield stress of 175 ksi (1.2×10^9 N/m²) and can withstand high internal tube pressures produced by local thawing of frozen ammonia. The flow tube is embedded in a two-piece extrusion constructed from 6061-T6 aluminum, as shown in figure 2.

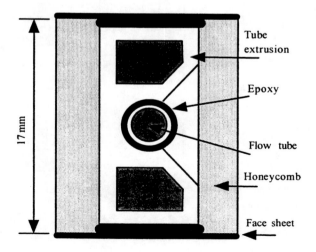

FIGURE 2. Radiator Tube Extrusion

A silver-filled epoxy EA1777, 0.09 mm (0.0035") thick is used to bond the tube to the extrusion. The extrusions are sandwiched between two 6061-T6 aluminum face sheets, 0.25 mm (0.01") thick. The face sheets have a silver-Teflon coating with an emissivity of 0.815. A film adhesive EA9689 with an average thickness of 0.04 mm (0.0015") is used to bond the face sheets to each extrusion. The void between the face sheets and the adjacent extrusions is filled with 5052-007 aluminum honeycomb for added panel strength. The edge distance at both edges of the panel is 0.0886 m (3.49"). The manifold covers consist of two pieces manufactured from 7075-T73511 aluminum and connected to each other by screws. The face sheets were originally bonded directly to the manifold covers by epoxy adhesive EA17777. In the redesigned radiator the epoxy bond was replaced with screws and stainless steel wire-mesh washers to reduce conductance between the face sheets and the manifold covers. A schematic of a PV radiator panel is shown in figure 3. The face sheets are 1.77 m wide, 3.15 m long, and 2.5×10^{-4} m thick (69.75"x124"x0.01"). The panel overall dimensions are 3.347 x 1.771 m (131.76" x 69.75") and 17.0 mm (0.67") thick. The PV panel contains 24 flow tubes, which are equally spaced 0.0673 m (2.65") apart, except near the corners and at the edges of the panel.

The ATCS loop containing the two EEATCS PV radiator ORUs is shown in Figure 4. Unlike the ATCS radiator that has rotating capability via the thermal radiator rotary joint (TRRJ), the EEATCS radiator array is fixed to the truss. There are seven radiator panels in each ORU arranged in a zigzag fashion by the deployment mechanism. Two independent flow loops A and B flow through alternating flow tubes of each radiator panel. Each loop has a constant flow rate of 0.214 kg/s (1700 lbm/hr). The setpoint temperature at the radiator return should lie between 1.67°C and 5.0°C (35°F and 41°F). To maintain the desire setpoint temperature, a radiator flow control assembly (RFCA) bypasses a portion of the warm flow and mixes it with radiator return flow before going to the pump. The RFCA is designed to maintain the flow temperature at the pump inlet at 2.22±1.67°C (38.0±3°F). The EEATCS radiators are required to reject a maximum of 14.0 kW at low beta angle orbits. At high beta angels the maximum rejection capacity is reduced to 9.5 kW.

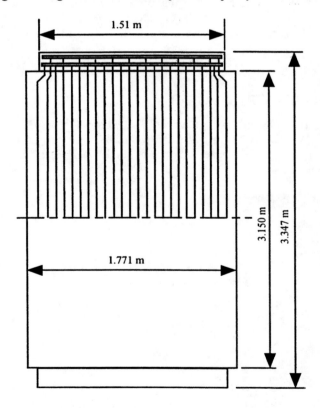

FIGURE 3. PV Radiator Panel.

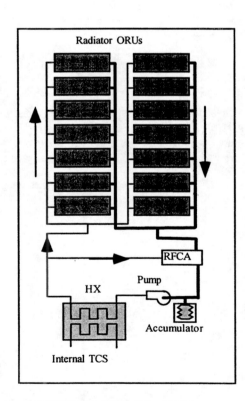

FIGURE 4. PV Radiator TCS Loop.

MODEL DESCRIPTION

On the basis of the PV radiator panel description presented above, SINDA/FLUINT (Cullimore and Ring, 1997) was used to develop a model of the EEATCS, including a detailed representation of the PV radiator. The seven radiator panels in one ORU are modeled for an accurate prediction of the ORU level heat rejection. To reduce the computer run time, the length of tubes, as well as the face sheet in the tube axial (X) direction are divided into 5 segments. In addition, the face sheet is divided into 24 strips in the Y direction, each strip corresponding to one flow tube. The effective sink temperature for each panel in each ORU differs from that of the adjacent panel due to their relative zigzag arrangement. To evaluate the ammonia freezing and thawing in tubes, a subroutine was developed that includes detailed derivations and associated assumptions for predicting freezing and thawing processes in tubes (Andish, 1996). The single PV radiator model is then expanded to 7 panels, and subsequently integrated to the ORU level model. The sink temperature profiles for each of the seven radiator panels vary from front to back of each panel due to shading and blocking effects. To simplify the ORU level model, the panel sink temperature is calculated by taking a 4th order average of the front and back temperature. The stage-7 5A sink temperature profiles for two consecutive cold orbits are used in the transient runs.

ANALYSIS AND RESULTS

A trial run was performed for the original PV radiator model to predict radiator freezing at cold environments (Andish, 1997). A single panel model was used to analyze a transient case using a simple sink temperature profile. A constant sink temperature of -79°C was used to initialize the solution at steady-state conditions. The initial temperature is cold enough to prepare the radiator for freezing but it does not cause any freezing at steady-state conditions. The total heat load on the loop was held constant at 5.0 kW. At time-0.1 hour, the sink temperature was dropped to -87°C and held constant for 1.8 hours to observe radiator freezing. Figure 5 shows the flow rates and the freezing sequence of tubes 1 to 6. The results indicate that during the 2-hour transient analysis, tubes 1 to 5 at the left edge, and tubes 20 to 24 at the right edge will freeze. The temperatures at the end of tubes 1 to 6 are shown in figure 6, which indicate that tubes 1to 5 and 20 to 24 the will freeze and the central tubes will not.

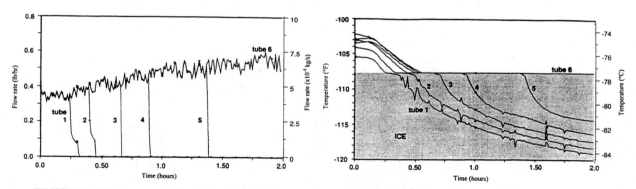

FIGURE 5. Flow rates of tubes 1- 6 of in the trial run. **FIGURE 6.** Temperature in tubes 1-6 in the trial run.

To predict the performance of the redesigned PV radiator, transient runs were performed using two consecutive cold sink orbits, followed by a cold dip down to –90°C for 0.15 hour and a second dip to –98.3°C for 0.25 hour. At the end of the cold dip period (at time=3.09 hr), the sink temperature was increased to –62.2°C in 0.92 hour to thaw any the frozen tube. Figures 7 and 8 depict the average sink temperature profiles for the 7th radiator panel in the starboard ORU and AFT ORU, respectively. The sink temperature at the 7th panel is colder than the other panels in each ORU. Overall, the AFT sink temperatures are slightly colder than the starboard sink temperatures. Using the ORU-level model for the redesigned PV radiators, four transient runs were performed at the maximum and minimum ATCS heat loads of 10kW and 4 kW for the starboard and AFT sink temperature profiles (Chen, 1998). The results indicate that all the radiator tubes in either starboard or AFT ORU will not freeze even at the minimum 4 kW EEATCS heat load. Only the results for the colder case with 4 kW heat load are discussed here.

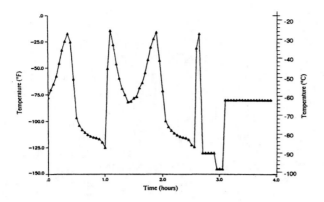 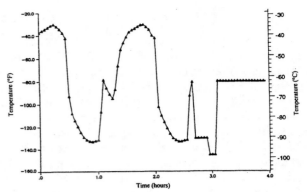

FIGURE 7. Sink temp. of radiator 7 in starboard ORU.   **FIGURE 8.** Sink temp. of radiator 7 in AFT ORU.

Figures 9 and 10 depict the liquid temperature profiles at the exit of tube 1 in panels 1 through 7, in starboard and AFT ORUs. The coldest temperature in tube 1 (at the tube exit) for 7 panels in starboard ORU drops to −74.4°C, while the coldest temperature at tube 1 exit in AFT ORU drops to −76.1°C. This indicates that tubes will not freeze since ammonia freezes at -77.8°C (-108°F).

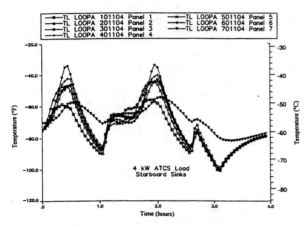 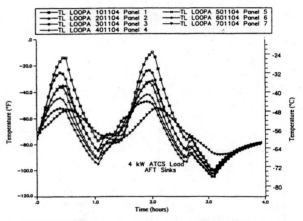

FIGURE 9. Temp. of tube 1 in starboard ORU panels.   **FIGURE 10.** Temp. of tube 1 in AFT ORU panels.

As shown in figures 9 and 10, the temperature for tube 1 of the 7th panel is warmer than the other panels because of the special surface coating of the last quarter face sheet of the 7th radiator pane. Figure 11 shows the fluid temperature profiles at the exit of tubes 1, 3, 12 and 24 in panel 1 of starboard ORU. The temperature of tube 12, located in the central position, is higher than the other tubes. Tubes 1 and 24 have the coldest temperatures because they are located at the panel edges. Figure 12 compares the rate of heat rejection from the radiator working fluid (calculated by the product of the radiator flow rate and the enthalpy difference between the radiator inlet and outlet) and the rate of heat rejection from the radiator to deep Space by radiation heat transfer. The results indicate that the rate of liquid sensible heat loss is maintained at 4 kW while the radiation heat loss varies around the 4 kW line to respond to changes in the sink temperature.

SUMMARY AND CONCLUSIONS

In the early stages of assembly of the ISS, and before the deployment of the ATCS radiators, two of the three PV radiator ORUs are utilized to dissipate the station's waste heat to deep Space. Under certain cold case conditions when radiator effective sink temperatures drop below the freezing temperature of the radiator working fluid (ammonia), radiators may freeze and result in reduced heat rejection capacity and flow blockage of the radiators. A freeze-tolerant radiator design has been developed by NASA to alleviate the impacts of

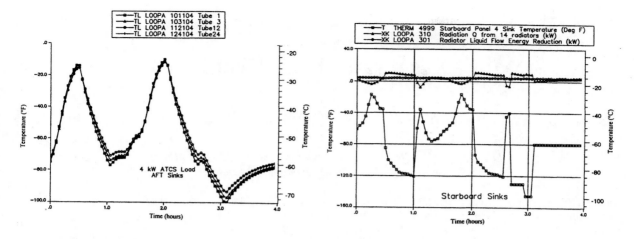

FIGURE 11. Tube exit temp. of panel 1 in starboard ORU.   **FIGURE 12.** Transient heat rejection at 4 kW.

radiator freezing on the ATCS performance. To predict the performance of the PV radiator at freezing environments, SINDA/FLUINT models were developed for the radiator system at the panel-level and ORU-level. The panel level model was analyzed under a simple two-step cold case to determine the flow tube freezing sequence and flow distribution. The results indicate that freezing initiates in the tubes at the panel outer edges and advances toward the panel center. Freezing of the panel outer tubes results in increased flow rate to the unfrozen tubes which in turn reduces their chance of freezing. The PV radiator design was later improved to address concerns about manifold tube freezing and ORU flow blockage. The radiator model was then modified to reflect the latest changes to the PV radiators, and expand the model to an ORU-level. The model was analyzed at the worst-case cold conditions. Four transient runs were performed at the maximum and minimum ATCS heat loads of 10kW and 4 kW for the starboard and AFT ORUs. The results indicate that the radiator tubes will not freeze even at the minimum 4 kW EATCS heat load. The results of the analysis confirm the effectiveness of the radiator redesign efforts to alleviate radiator freezing and minimizing the impact of frozen tubes on the performance of the thermal control system.

REFERENCES

Andish, Kambiz, " Freeze and Thaw Predictions of International Space Station Active Thermal Control System Radiators," Lockheed Martin, LMES-32153, 1996.

Andish, Kambiz, "Development of a SINDA/FLUINT Model for Freeze/Thaw Analysis of International Space Station EEATCS Radiators," Lockheed Martin, LMSMSS-32598, 1997.

Chen, Ing Youn, "ORU Level Freeze/Thaw Model of EEATCS Radiator," Lockheed Martin, LMSMSS-32948, 1998.

Cullimore and Ring Technologies, Inc., System Improved Numerical Differencing Analyzer and Fluid Integrator, Version 4.0, 1997.

NASA Johnson Space Center, "International Space Station on-orbit Assembly, Modeling, and Mass Properties Data Book,"JSC-26557, Rev. G, Vol. 1, LESC-31166, Rev. G, Vol. 1, 1997.

Rocketdyne/Boeing, "PV Radiator Redesign for the International Space Station Electric Power System," Presentation Charts (1/22/98) to NASA, 1998.

Multi-Evaporator Loop Heat Pipe

Konstantin A. Goncharov, Oleg A. Golovin, and Vladimir A. Kolesnikov

TAIS Ltd., Russia, Engelsa 21, Khimky, Moscow Region, 141400, Russia
telephone/fax:(095)573-63-74; tais.ltd@g23.relcom.ru

Abstract. This paper describes the last investigations of Loop Heat Pipes (LHPs) with branched configuration, i.e. LHPs with two, three and more Evaporators. These investigations were begun by development of "Push-Pull" LHP - LHP with two Evaporators. Successful performance of "Push-Pull" LHP proved the theoretical proposition that LHP with several Evaporators and arbitrary number of Condensers can be developed (Goncharov, 1998). Theoretical ground of the development of LHP with several Evaporators is presented here below. Each Evaporator is supplied with Compensation Chamber and can be placed at arbitrary distance from the other Evaporator in gravity conditions, heat power of arbitrary value can be applied to LHP Evaporators and each LHP Evaporator can even function as Condenser.

MAIN PECULIARITIES OF LHP AS COMPARED WITH CPL

Main peculiarities of LHP as compared with CPL can be formulated as following principles:
1. Thermal Connection between the Evaporator and the Working Fluid inside LHP Compensation Chamber (Reservoir).
2. Hydraulic (capillary) Connection between the Internal Surface of Evaporator Wick and the Liquid inside Compensation Chamber
3. Certain Ratio of LHP Components' Volume.

Certain ratio of LHP components' volume on the one hand and the Compensation Chamber volume and mass of charged working fluid on the other hand can be considered as, perhaps, main peculiarity of LHP as compared with other types of Two-Phase Loops with Capillary Pump. LHP design shall provide the minimum values of all LHP components' volumes and working fluid amount charged into LHP shall be enough to provide the following working fluid distribution: after working fluid is charged into Compensation Chamber and fills all other LHP components, at least a small part of working fluid shall remain in Compensation Chamber. If this requirement is met, then due to two LHP peculiarities mentioned above, it is impossible «to dry out» such two-phase loop and to define the situation when such device will be unable «to start-up». Evaporator wick is always saturated with liquid and LHP is always ready to start-up.

The most unfavorable situation of LHP operation can occur when LHP start-up at minimum temperature ("cold" critical state). In this case, working fluid volume is of minimum value. If working fluid amount in LHP is enough to fill all LHP components and a part of Compensation Chamber, LHP is ready to start-up and there will be no problems. Some amount of liquid in LHP is necessary to compensate LHP calculation, manufacturing and charging tolerance.

It ought to consider the other "hot" critical state when LHP is operating at limiting heat power and maximum temperature that are set for real LHP design on base of LHP performance conditions. In this case, Compensation Chamber volume shall be enough to contain:
- liquid from vapor and liquid lines;
- increased volume of liquid due to heating from minimum temperature $T_{min (start-up)}$ to T_{max};
- liquid from Evaporator wick pushed out due to vapor/liquid interface displacement;
- non-condensable gas;

CP504, *Space Technology and Applications International Forum–2000*, edited by M. S. El-Genk
© 2000 American Institute of Physics 1-56396-919-X/00/$17.00

- and to compensate LHP calculation, manufacturing and charging tolerance.

All LHP components' volume is usually determined on base of calculations performed for limiting transferred heat power and LHP thermal resistance. At the same time, LHP developers aspire to minimize the LHP components' volume (especially Vapor Line and Condenser volumes). Only the values of Compensation Chamber volume (V_{cc}) and mass of working fluid charged into LHP (M_f) are not defined by this calculation. Formula for Compensation Chamber volume (V_{cc}) calculation can be written as following:

$$V_{cc} \geq (V_{ev} + V_l) \cdot \frac{(\rho_{L_1} - \rho_{L_2})}{\rho_{L_2}} + (V_r + V_c) \cdot \frac{(\rho_{L_1} - \rho_{V_2})}{\rho_{L_2}} \tag{1}$$

where V_{ev} - capillary wick volume; V_l - liquid line volume; V_v - vapor line volume; V_c – condenser volume; ρ_{L_1} - liquid density at T=T$_{min}$; ρ_{L_2} - liquid density at T=T$_{max}$; ρ_{V_2} - vapor density at T=T$_{max}$;

Authors manufactured several tens of different LHPs developed on base of this relation. Authors never met any difficulties in definition of the working fluid mass charged into LHP. All LHPs display reliable operation.

DESIGN OF LHP WITH SEVERAL EVAPORATORS

Number of LHP Evaporators is limited just due to the last LHP peculiarity - certain ratio of LHP components' volume. Design schematic of LHP with n Evaporators is presented on Figure 1 and Figure 2. Each Evaporator of Multi-Evaporator LHP is supplied with its own Compensation Chamber.

It ought to note that Evaporators can be of different design and arbitrary position in relation to other Evaporators and g vector, different values of heat power are applied to Evaporators (even some Evaporators can function as Condensers). But all Compensation Chambers are of the same volume.

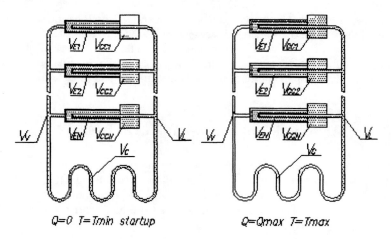

FIGURE 1. LHP "cold" critical state.   **FIGURE 2.** LHP "hot" critical state.

Critical cases of LHP operation are in main points analogous to situations described here above for LHP with a single Evaporator. For «cold» critical state (Figure 1), working fluid amount shall be enough to fill all LHP components except Compensation Chamber at Q = 0 and T = T$_{min\ (start-up)}$. For «hot» critical state (Figure 2), volume of each Compensation Chamber shall be enough to contain:
- liquid pushed out from Vapor Line and Condenser;
- liquid volume increase due to heating;
- additional liquid pushed out due to vapor/liquid interface displacement inside the wick at Q = Q$_{max}$;

- non-condensable gas;
- volume deviation due to calculation, manufacturing and charging accuracy.

It is necessary to analyze what the limiting number of Evaporators, n, depends on. For this purpose, equation for V_{cc_i} calculation shall be written taking into account the assumptions analogous to the assumptions set for equation (1) drawn for LHP with a single Evaporator. It is also assumed that: $V_{cc_1} = V_{cc_2} = ... = V_{cc_n}$

$$\Rightarrow V_{cc_i} = \frac{(V_v + V_c) \cdot (\rho_{L_1} - \rho_{V_2}) + (\sum_{i=1}^{n} V_{ev_i} + V_l) \cdot (\rho_{L_1} - \rho_{L_2})}{n \cdot \rho_{L_2} - (n-1) \cdot \rho_{L_1}} \tag{2}$$

To prove the theory of the development of LHP with several Evaporators several samples were manufactured and tested. Design description and test results and methods are presented here below for two samples of Multi-Evaporator LHP: LHP «Push Pull» with reverse direction of heat transfer for North-South radiators (radiators located at the opposite sides of Spacecraft) and LHP with three Evaporators «Zmey-Gorynych».

"PUSH-PULL" LHP

It is well known that LHPs inherently function as thermal diodes. This research was undertaken to demonstrate that LHPs can be designed to successful heat transfer in both directions. Many real world applications require bi-directional heat transfer to maintain very close temperatures at two or more points. One practical example is equalizing the temperatures between the north and south radiators on telecommunications satellites. The two radiators, which are located on opposite sides of the satellite some 2-3 meters apart, shall maintain nearly equal temperatures even when their radiated power may differ by thousands of watts. These systems shall be fully functional during ground testing of the spacecraft.

Conventional heat pipes provide excellent isothermalization, but their use is severely limited under conditions that demand high power, long distance thermal transport against significant gravitational heads. Thanks to the theoretical work begun in 1988, and extensive experimental development of Multi-Evaporator LHPs begun in 1992, bi-directional (Push-Pull) LHP can be designed to solve very complicated temperature tasks. The reversible (Push-Pull) LHP with two Evaporators, either of which may function as a Condenser, has been developed to provide temperature equalization under demanding conditions where conventional heat pipes are ineffective.

Description of "Push-Pull" LHP Design

The "Push-Pull" LHP consists of two identical Evaporators which may function as either Evaporators or Condensers depending upon their relative temperatures. The basic configuration of the device is shown in Figure 3. The evaporators are made of stainless steel tubing with diameter of 17.4 mm and a total length of 175 mm. Each Evaporator incorporates a capillary wick, which is fabricated of sintered nickel powder. The sintered wick has a porosity of 60 % and a pore radius of 0.6 μm (micrometer). The Compensation Chamber, which is attached directly to the body of the Evaporator, is made of stainless steel tubing with a diameter of 36 mm and a length of 50 mm. All internal surfaces of the Compensation Chamber are covered with a mesh wick, which is a part of the secondary wick structure. The secondary wick structure will assure that the main wick will be supplied with liquid regardless of adverse gravitational orientations or under microgravity. An aluminum saddle is soldered to the Evaporator surface. The dimensions of the aluminum saddle are 150 mm x 45 mm, and it has two flat contact surfaces for heat input or output. Vapor lines (internal diameter of 3 mm) and liquid lines (internal diameter of 2 mm) are made of thin-walled stainless steel tubing with an overall length of 4000 mm. The two Evaporators are connected in such a way that the vapor line carries only vapor and the liquid line carries liquid regardless of which Evaporator is heated or cooled. Ammonia is used as the LHP working fluid.

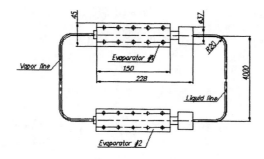

FIGURE 3. "Push-Pull" LHP Configuration.

Thermal Tests and Thermal-Vacuum Tests of "Push-Pull" LHP

Heat exchangers, which could accommodate both electric heating and liquid nitrogen cooling, were attached to the flat surfaces of the Evaporator No. 1 and Evaporator No. 2. Tests were performed both in atmosphere and in a vacuum chamber with 20 layers of Multi-Layer thermal Insulation (MLI). The precision of the temperature measurements was of ±0,1°C. Tests were performed in accordance with the following Test Program.

Test No. 1. Test was performed when LHP is oriented horizontally in vacuum chamber, without insulation. Input power of 360 watts was applied to the heater attached to the Evaporator No. 2 for one hour while no power was applied to Evaporator No. 1. Then the power was applied to the Evaporator No. 1 for one hour while no power was applied to Evaporator No. 2. This cycling continued for four hours.

The test results are presented in Figure 4.

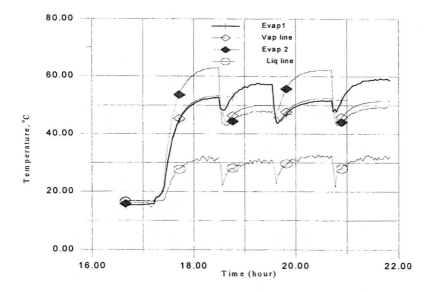

FIGURE 4. LHP test results in vacuum, horizon orientation.

Test No. 2. Test was performed when LHP is oriented horizontally in atmosphere without insulation. Input power of 200 watts was cycled between the two Evaporators at intervals of approximately 1.5 hours for 7.5 hours.

Test No. 3. Test was performed when Evaporator No. 1 was located 840 mm below Evaporator No. 2. The test was performed in atmosphere without insulation. Input power of 200 watts was cycled between the two Evaporators at intervals of 2.5 hours for a total of 10 hours. LHP successfully passed all tests.

"ZMEY GORYNYCH" LHP

LHP with three parallel Evaporators and two parallel Condensers, transport lines and vapor isolator was developed. Evaporators with outer diameter of 17.4 mm and length of 155 mm are fabricated of stainless steel tubes. Cylindrical capillary porous insert with outer diameter of 16 mm made of sintered nickel powder is pressed into the Evaporator envelope. Contact saddle with dimensions 150 x 50 mm made of aluminum alloy is soldered to the Evaporator envelope for heat input imitator mounting. Each Evaporator is supplied with Compensation Chamber and secondary wick. LHP Condensers are made of aluminum profiles with inner diameter of 4 mm. Vapor and Liquid Lines are made of stainless steel tubing with diameter of 4 and 3 mm, correspondingly. "Zmey Gorynych" LHP configuration and thermoelectric thermometers mounting schematic are presented in Figure 5.

LHP tests were performed in atmosphere at environmental temperature of 19...22 °C and relative humidity of 75%. LHP was mounted on special stand that provides LHP tilt in the range of ±75. Evaporator elevation above the Condenser was provided by the tilt of stand rotary platform. LHP was tested at three positions in gravity conditions:
- horizontal position (non-gravity conditions imitation) when all Evaporators and Condensers were in XY horizontal plane.
- position when all Evaporators were placed above the Condensers at the height of 1.5 m. Test results are presented in Figure 6.
- LHP Condensers and Evaporators were positioned at different heights (X axis up). Test results are presented in Figure 7.

Temperature measuring was performed at steady state mode of LHP operation.

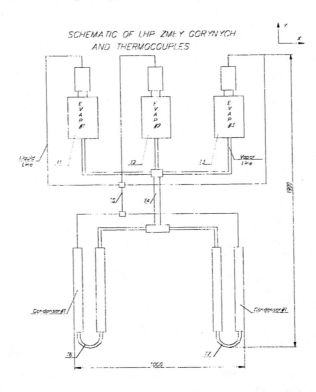

FIGURE 5. "Zmey Gorynych" LHP configuration and thermoelectric thermometers mounting schematic.

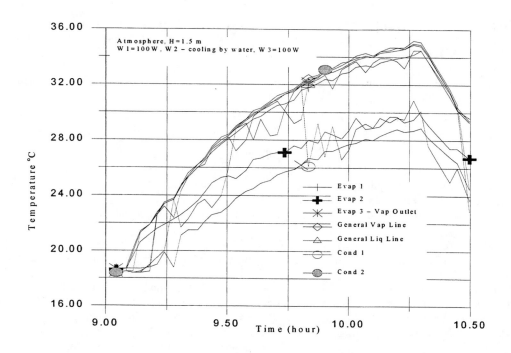

FIGURE 6. Test results when all Evaporators are placed above the Condensers at the height of 1.5 m.

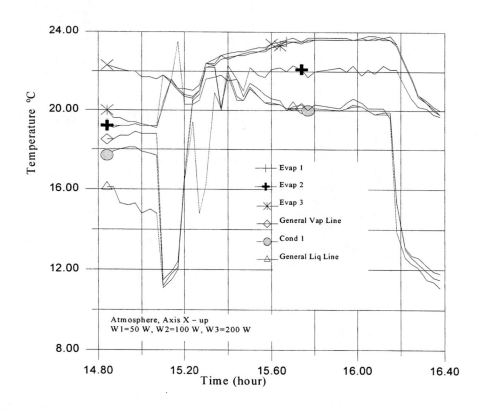

FIGURE 7. Test results when LHP Condensers and Evaporators are positioned at different heights (X axis up).

LHP successfully passed all tests. No deviation of LHP operation characteristics from forecasted values were observed when LHP testing. It allows to assert that theory developed for Multi-Evaporator LHP designing and calculation provides the reliable data.

Authors invite the partners interested in completion of Multi-Evaporator LHP investigation program. It is proposed to develop and test the LHP with five Evaporators.

CONCLUSIONS

1. LHP is a Two-Phase Loop with Capillary Pump and with following peculiarities:
    -   thermal connection between Evaporator and Compensation Chamber;
    -   hydraulic (capillary) connection between the internal surface of wick and the liquid inside Compensation Chamber;
    -   certain ratio between LHP components' volume on one hand and the Compensation Chamber volume and charged working fluid mass on the other hand. This ratio provides the LHP reliable start-up and stable operation.

LHP peculiarities mentioned above shall be taken into account when LHP designing, modeling and application.

2. Ratio between the LHP components' volume and the charged working fluid mass was explained and defined using simple formula. This formula allows to calculate the values of Compensation Chamber volume (Vcc) and mass of working fluid (Mf) for LHP with single and several evaporators with high accuracy.

3. It was shown that LHP with several Evaporators can be developed. Number of Evaporators, n, is limited by certain ratio between LHP components' volume and real temperature range, $T_{max_{Q_{max}}} >> T_{min_{startup}}$.

4. LHPs with two and three Evaporators were developed and tested. No deviation of LHP operation characteristics from forecasted values were observed when LHP testing. It allows to assert that theory developed for Multi-Evaporator LHP designing and calculation provides the reliable data.

REFERENCES

Goncharov K., and Kolesnikov V., "Components Volume Ratio in LHP with Single and Several Evaporators", TAIS Ltd. '98 International Workshop, Los-Angeles, March 6-9, 1998.

Goncharov K.A., Nikitkin M.N., Golovin O.A., Fershtater Yu.G., Maidanik Yu.F., and Piukov S.A., "Loop Heat Pipes in Thermal Control Systems for «OBZOR» Spacecraft", 25 ICES, San Diego, July 10-13, 1995.

Maidanik Yu.F. USA patent No. 4.515.209 from May 7, 1985.

Maidanik Yu.,. Fershtater Yu., Pastukhov V., Vershinin S.,. and Goncharov K., "Some Results of Loop Heat Pipes Development, Tests and Application in Engineering", 5IHPS, Melbourne, Australia, November 17-20, 1996.

Orlov A.A., Goncharov K.A.,. Kotliarov E.Yu.,. Tyklina T.A., Ustinov S.N., and Maidanik Yu.F., "The Loop Heat Pipe Experiment on board the «GRANAT» Spacecraft", CPL-97 Workshop, Noordwijk, The Netherlands, May 20-22, 1997.

Rosenfeld J.H., Anderson W.G., and North M.T., "Improved High Heat Flux Loop Heat Pipes Using Bidisperse Evaporator Wicks", DTX/Thermacore, Inc., 10th IHPC, Stuttgart, 1997.

Cryogenic Thermal Diodes

Brandon R. Paulsen, J. C. Batty, and John Agren

Space Dynamics Laboratory, 1695 North Research Park Way, North Logan, Utah, 84341

(435) 797-4600; Brandon.Paulsen@sdl.usu.edu

Abstract. Space based cryogenic thermal management systems for advanced infrared sensor platforms are a critical failure mode to the spacecraft missions they are supporting. Recent advances in cryocooler technologies have increased the achievable cooling capacities and decreased the operating temperatures of these systems, but there is still a fundamental need for redundancy in these systems. Cryogenic thermal diodes act as thermal switches, allowing heat to flow through them when in a conduction mode and restricting the flow of heat when in an isolation mode. These diodes will allow multiple cryocoolers to cool a single infrared focal plane array. The Space Dynamics Laboratory has undertaken an internal research and development effort to develop this innovative technology. This paper briefly describes the design parameters of several prototype thermal diodes that were developed and tested.

INTRODUCTION

As cryocooler technologies have improved significantly over the past few years, cryogenic sensor systems are no longer dependent on cryogen-based cooling systems. Cryocoolers offer extended mission lifetimes and reduced mass to these systems, but still have significant limitations. Most critical spacecraft components must be flown with redundancy. Cryocoolers, when used to cool primary sensors aboard multi-million dollar spacecraft, become the most critical failure mode of the spacecraft. This paper will discuss a recently developed technology that will provide the means to integrate multiple cryocoolers into redundant cooling systems.

Space qualified cryocoolers are typically expensive, on the order of $1.5 to $4 million, and are often sized to barely accommodate the heat lift requirements of the flight sensor systems. These coolers typically have features that limit the transmittance of vibration from the moving parts of the cooler to the sensitive sensors they are cooling. They are also capable of operating over long periods of time.

Tactical cryocoolers have also matured in the past decade. They are conventionally used on ground systems, but have recently been flown on advanced spacecraft programs such as Clementine. These tactical coolers are considerably less expensive than their space qualified counterparts, costing on the order of $10,000 to $50,000. However, they do not have the elaborate, vibration isolation features, and they do not offer the long operational lifetimes of the more expensive coolers.

A new technology developed by the Space Dynamics Laboratory, the cryogenic thermal diode, will allow the use of redundant tactical cryocoolers to essentially replace and out-perform the conventional high-performance cooler solution for a fraction of the cost. This technology could also be used to fly redundant long-life coolers, adding redundancy to the long list of features associated with these cryocooler systems.

The cryogenic thermal diode consists of a small, lightweight device; which is placed in the thermal path between a cryocooler and a sensor system. The device works analogous to a light switch. When the diode is activated, a low-impedance heat path connects a heat source (sensor system) to a heat sink (cryocooler). When the diode is deactivated, a high impedance heat path minimizes the amount of heat that is transferred from the heat sink (inoperative cryocooler) back into the heat source (sensor system). The thermal diode is quasi-autonomous because its actuation requires no active control to initiate the thermal contact; however, analysis shows that a small amount of energy may be needed to deactivate it.

CP504, *Space Technology and Applications International Forum–2000*, edited by M. S. El-Genk
2000 American Institute of Physics 1-56396-919-X

USES FOR CRYOGENIC THERMAL DIODES

There are several perceived uses for cryogenic thermal diodes. This section describes several of these uses; however, this technology may be a candidate for virtually any situation where a temperature difference of more than 60 Kelvin exists between the heat source and sink.

Cryogenic Sensor Applications

The cryogenic thermal diode is intended for application in cryogenically cooled spacecraft sensor systems. As mentioned previously, cryocooler-based sensor cooling systems are critical single-point failure modes for these spacecraft. Cryocooler designers attempt to minimize the parasitic heat load through cryocoolers in order to maximize the net cooling capacity available. Typical values for these parasitic loads are on the order of 0.3 to 0.7 W, while net cooling loads are typically 0.75 to 2.0 W.

When selecting cryocoolers, sensor systems are designed to minimize the required cryocooler cooling capacity. Additional cooling capacity comes with an increase in cryocooler cost and mass. In many cases, the entire parasitic heat load of a sensor system may be less than 0.7 W. Therefore, the redundant cryocooler solution may not be a feasible solution for many sensor systems. A means to lower the parasitic heat load of the redundant cooler is needed.

Figure 1 illustrates the placement of cryogenic thermal diodes within a redundant cryocooler system. This configuration prevents the parasitic heat load(s) from redundant cryocooler(s) from overloading the operational cryocooler(s). These devices are designed to provide a thermal impedance of 2500 K/W, which translates into less than 0.1 W for a thermal gradient between 293 K and 55 K.

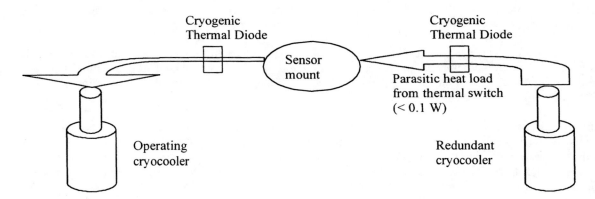

FIGURE 1. Redundant cryocoolers with cryogenic thermal diode.

Autonomous Radiator Switching

Another spacecraft thermal control problem is "radiator switching." Spacecraft attitude control systems tend to concentrate more on instrument pointing requirements than on optimizing the performance of thermal radiators. The autonomous thermal switch offers a new solution to performance problems that may arise.

Distinct thermal radiators may be placed on different sides of the spacecraft. These radiators are then coupled to the components that are dissipating heat via autonomous thermal switches. Cold radiators would cause the thermal switches to close, allowing heat rejection. Warm radiators would cause the thermal switched to open, preventing heat from being transferred from the warm radiators into the spacecraft.

PROTOTYPE DESIGN

The Space Dynamics Laboratory has funded an internal research and development task to design and fabricate a prototype thermal disconnect device. The prototype was designed to test the functionality of the switching mechanism. Mass and parasitic heat loads will be further reduced in future designs. Table 1 lists the design parameters of the prototype cryogenic thermal diode. A photo of the prototype thermal diode is seen in Figure 2.

TABLE 1. Prototype design parameters.

Design Parameter	Value
Mass (grams)	230
Envelope, cm (inches)	Diameter: 6.48 (2.55)
	Height: 14.0 (5.53)
Thermal impedance when actuated (K/W)	2
Thermal impedance when off (K/W)	2500
Prototype completion date	May 1999

FIGURE 2. Cryogenic thermal diode prototype.

FUNCTIONALITY

The thermal diode is actuated by mechanical displacements generated by a temperature reduction. Figure 3 illustrates the design of the linkage used in the first prototype. It is designed to convert a change in temperature into a useable displacement. The central bar of the linkage is made from a material with a very high coefficient of thermal expansion (CTE). The four outer linkages are fabricated from a material with extremely low CTE properties.

As the linkage assembly is cooled, the central bar of the linkage begins to contract. As it contracts, the distance between the pivots of the external linkage begins to decrease and the external linkages are forced out. Thermal contact surfaces mounted to the external linkages provide the thermal connection. When the linkage assembly is allowed to warm, the opposite occurs. The central bar of the linkage expands and the external linkages retract, breaking the thermal path through the switch.

FIGURE 3. Linkage assembly.

PROTOTYPE TESTING

Various tests have been carried out on the prototype thermal diode. The results of these tests are encouraging and indicate that the design of the linkage assembly is mechanically sound. The following sections describe the testing that has been completed, as well as testing to be completed in the near future.

Linkage Displacement Test

The linkage assembly was tested at liquid nitrogen temperatures to verify that there was sufficient displacement to actuate the thermal contact surfaces. The test began by measuring the displacement across the external linkage arms while at room temperature. The linkage was then submerged in liquid nitrogen and allowed to cool to 77 K. The displacement across the external linkage was quickly measured again. This test indicated that approximately 2.92 mm (0.115 in) of displacement occurred between 298 K and 77 K.

The results of this test were used in designing the rest of the thermal disconnect prototype. The gap between the external linkage and thermal contact surfaces is critical. If the gap is too large, contact will not occur; if the gap is too small, large contact pressures could cause the linkage arms to yield or fail. This gap also determines the switching temperature of the device. A smaller gap will produce a higher switching temperature.

Thermal Switching Test

Once the prototype was completed, a series of tests was conducted in order to verify the functionality of the device. Figure 4 depicts the test setup for this series of tests. The thermal disconnect device has a folded G-10CR thermal isolation base to isolate the cold linkage assembly from the mounting base. The base of the prototype is mounted to a warm vacuum flange (290 K) of a miniature nitrogen/helium dewar. The testing is conducted in a vacuum vessel in order to eliminate convection heat transfer, as convection would be a dominating heat transfer mechanism in the small gaps between the thermal contact surfaces. The cold side of the disconnect device is connected to a cryogen tank within the dewar via a flexible thermal linkage.

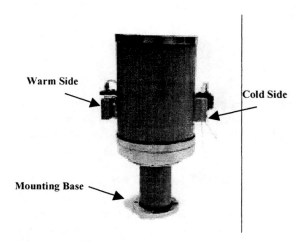

FIGURE 4. Thermal diode verification test setup.

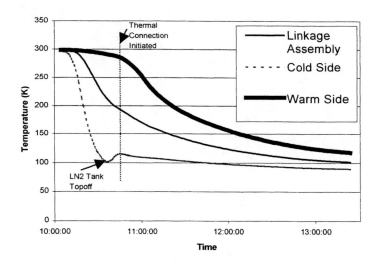

FIGURE 5. Thermal diode test results.

The results of this testing has validated the functionality of the overall prototype design. Figure 5 demonstrates the thermal performance of the thermal disconnect device as it switches. The switching temperature of the prototype device appears to be approximately 190 K. The prototype device took several hours to reach steady state because the test setup did not provide means for radiation shielding around the cold surfaces emerging from the device. Therefore, under worst case conditions, the maximum switching time could be up to two hours. However, with proper radiation shielding the switching time will be reduced considerably.

FUTURE DEVELOPMENT

Design optimization is underway to minimize the required mass and to minimize the parasitic heat loss through the device. Several flight programs are considering using the devices in the near future, which may expedite the development effort. The mass of the second prototype will be on the order of 65 grams. Fabrication of this prototype should begin in early 2000.

CONCLUSIONS

The prototype cryogenic thermal diode fabricated and tested by the Space Dynamics Laboratory has demonstrated that it is indeed feasible to produce such a device. While the operational parameters of the prototype demonstrate that the technology is still in its infancy, the basic design concepts have proven to be sound.

REFERENCES

Paulsen, B., Batty, J.C., and Agren, J., "Autonomous Thermal Disconnect Devices," in *Advances in Cryogenic Engineering*, edited by Mike DiPirro, 1999 Cryogenic Engineering Conference Proceedings, New York, Plenum Press, 1999.

Portable Infrared Reflectometer for Evaluating Emittance

Donald A. Jaworske[1] and Timothy J. Skowronski[2]

[1]NASA Glenn Research Center, 21000 Brookpark Rd., Cleveland, OH 44135, Phone:(216) 433-2312
[2]Cleveland State University, Cleveland, OH 44121

Abstract. Optical methods are frequently used to evaluate the emittance of candidate spacecraft thermal control materials. One new optical method utilizes a portable infrared reflectometer capable of obtaining spectral reflectance of an opaque surface in the range of 2 to 25 microns using a Michelson-Type FTIR interferometer. This miniature interferometer collects many infrared spectra over a short period of time. It also allows the size of the instrument to be small such that spectra can be collected in the laboratory or in the field. Infrared spectra are averaged and integrated with respect to the room temperature black body spectrum to yield emittance at 300 K. Integrating with respect to other black body spectra yields emittance values at other temperatures. Absorption bands in the spectra may also be used for chemical species identification. The emittance of several samples was evaluated using this portable infrared reflectometer, an old infrared reflectometer equipped with dual rotating black body cavities, and a bench top thermal vacuum chamber. Samples for evaluation were purposely selected such that a range of emittance values and thermal control material types would be represented, including polished aluminum, Kapton®, silvered Teflon®, and the inorganic paint Z-93-P. Results indicate an excellent linear relationship between the room temperature emittance calculated from infrared spectral data and the emittance obtained from the dual rotating black body cavities and thermal vacuum chamber. The prospect of using the infrared spectral data for chemical species identification will also be discussed.

INTRODUCTION

The optical properties of materials play a key role in spacecraft thermal control. In space, radiant heat transfer is the only mode of heat transfer for rejecting heat from a spacecraft. One of the key properties for defining radiant heat transfer is infrared emittance, a measure of how efficiently a surface can reject heat compared to a perfect black body emitter. Heat rejection occurs in the infrared region of the spectrum, nominally in the range of 2 to 25 microns. The emittance of a surface can be evaluated by either optical (Henninger, 1984) or calorimetric methods (Jaworske, 1993). Although steady state and transient calorimetric methods evaluate the emittance of a surface functionally at the desired temperature, the measurements are difficult to make, often require knowledge of the temperature dependent specific heat of the material, and are time consuming. Optical methods are preferred (Jaworske, 1996). The caveat is that the emittance is obtained with the surface at or near room temperature; and extending the emittance values to outlying temperatures by calculation is done so only at the risk of not accounting for a change in the physical properties of the surface at those outlying temperatures. For most materials, such a risk is small.

This paper describes a new portable infrared reflectometer designed to evaluate the emittance of surfaces and coatings in the laboratory or in the field. In addition, this paper reports the infrared emittance of a number of thermal control coatings, as measured by this new instrument, and compares their emittance values to values obtained on the same samples by different optical and calorimetric techniques. Emittance as a function of temperature was found from one infrared spectrum. The prospect of using the infrared spectral data for chemical species identification will also be discussed.

CP504, *Space Technology and Applications International Forum–2000*, edited by M. S. El-Genk
© 2000 American Institute of Physics 1-56396-919-X/00/$17.00

MATERIALS AND METHODS

Evaluating infrared reflectance over a given spectral range is common to all optical methods for measuring emittance. Spectral absorptance is calculated from spectral reflectance, and emittance is calculated using Kirchhoff's Law and the Stefan-Boltzmann equation. The differences in optical methods are the means of illuminating the sample with infrared energy and the means of detecting the energy after departing the surface. In the laboratory, it is anticipated that infrared reflectance measurements will be made on coupons of varying size. In the field, it is anticipated that infrared reflectance measurements will be made on large items, such as radiators and other spacecraft components that may already be installed on a spacecraft or in a test facility. A portable hand-held infrared reflectometer is needed that can make measurements on samples of virtually any size and in any orientation.

The new infrared reflectometer reported here, the SOC-400t, was developed by Surface Optics Corporation under a contract with NASA Glenn Research Center for the purpose of replacing our aging Gier-Dunkle DB-100 infrared reflectometer. The specifications for the new instrument were identified and include: a wavelength range of 2.5 to 25 microns, reflectance repeatability of ± 1%, self calibrating near normal spectral reflectance measurement type, full scan measurement time of 3.5 minutes, sample size of 1.27 cm (0.5 inches), spectral resolution selectable from 4, 8, 16, or 32 reciprocal centimeters, and optical property characterization utilizing an automatic integration to calculate total emittance in a selectable temperature range. The computer specified to drive the software is a laptop with a menu driven operating system for setup and operation, a full database manager, and a full data analysis capability through MIDAC Grams/32 software. Spectral scanning is achieved through the use of a FTIR Michelson-Type inteferometer. Samples are specularly illuminated from a near normal direction and the reflected signal is detected hemispherically.

The size and weight features were selected to enable portable operation. Although most of the planned uses for the instrument are expected to be in the laboratory, some field operations are anticipated. The only requirement for field operation is a source of power, at 115 volts AC.

The Gier-Dunkle DB-100 has been described and used extensively in the past (Henninger, 1984). Samples are illuminated by dual rotating black body cavities and the reflected signal is detected by a single detector at a near normal orientation. The instrument provides a total reflectance measurement in the vicinity of 9.7 microns and emittance is obtained by difference. No spectral data are available from the Gier-Dunkle DB-100. Hence, only room temperature emittance can be obtained from this instrument.

The bench top thermal vacuum chamber has also been described and used extensively (Jaworske, 1996). Samples are suspended in a vacuum chamber by thin thermocouple wires and are allowed to cool through radiant heat transfer to a surrounding cold wall. The low temperature calorimetric vacuum emissometer (LCVE) technique utilizes temperature-time data in an energy balance equation to calculate emittance. Cooling typically occurs in the temperature range of 400 to 150 K.

Materials selected for evaluation by the two infrared reflectometers and the calorimetric vacuum emissometer were common spacecraft thermal control materials and include: Polished aluminum, Kapton®, silvered Teflon®, anodized aluminum, and various thermal control paints and coatings such as Z-93-P. All samples were 2.54 cm in diameter. Results were tabulated for comparison.

RESULTS AND DISCUSSION

Samples for emittance evaluation were collected from a variety of sources over the years and kept in a NASA Glenn Research Center sample archive. Most samples had been placed previously in the LCVE to be evaluated by the calorimetric technique (Jaworske, 1996). The results from this study are presented in Table 1. Table 1 summarizes the results obtained from both infrared reflectometers and the calorimetric emissometer. A wide variety of samples were evaluated, including polished surfaces, aluminized surfaces, goldized surfaces, anodized surfaces, Teflon®,

Kapton®, aluminized beta cloth, various thermal control paints, coatings, polymers, and ceramics, yielding a wide variety in observed emittance values.

TABLE 1. Emittance of samples evaluated by the SOC-400t and DB-100 infrared reflectometers and low temperature calorimetric vacuum emissometer.

Coating Type	DB-100 Emittance	SOC-400t Emittance	LCVE Emittance
Aluminized Kapton®	0.020	0.017	0.057
Goldized Kapton®	0.021	0.018	0.064
Polished aluminum	0.021	0.021	0.090
0.025 mm (1 mil) Kapton® first surface mirror	0.022	0.000	0.106
Aluminized Kapton®	0.165	0.202	0.196
Aluminized Beta Cloth	0.178	0.115	0.305
0.013 mm (0.5 mil) Kapton® second surface mirror	0.517	0.605	0.527
0.025 mm (1mil) Kapton® on aluminum	0.616	0.696	0.616
Grit-blasted oxidized stainless steel	0.730	0.738	0.430
Aluminized Teflon®	0.784	0.832	0.833
H_2SO_4 anodized (.345 mil) aluminum	0.793	0.805	0.785
Al_2O_3-Y_2O_3-ZrO_2	0.795	0.793	0.741
Black film on 0.025 mm (1mil) Kapton®	0.836	0.822	0.836
H2SO4 anodized (.345 mil) aluminum	0.839	0.839	0.846
Black anodized aluminum	0.840	0.823	0.835
Silvered Teflon®	0.848	0.862	
Eccocoat® on Macor®	0.850	0.841	0.818
Eccocoat® on Bakelite®	0.856	0.849	0.785
0.025 mm (1 mil) carbon filled Kapton®	0.860	0.857	0.865
0.254 mm (10 mil) Teflon® FEP/silver/inconel	0.861	0.877	0.870
Clear anodized (hot water sealed) aluminum	0.871	0.861	0.837
MH55ICP	0.878	0.861	0.821
Pure Kapton®	0.880	0.871	0.821
Pure Macor®	0.885	0.864	
Macor®	0.890	0.879	0.833
Bakelite®	0.897	0.886	0.810
Z-306	0.903	0.890	0.912
Noryl®	0.905	0.896	0.862
Zelux®	0.910	0.893	0.887
Z-93	0.915	0.897	0.877

The data from the two infrared reflectometers may be plotted to show correlation. This is done in Figure 1. The line of perfect agreement is shown for reference. It is interesting to note that most of the SOC-400t emittance values agree with the DB-100 emittance values, particularly at the low and high ends of the range. However, there are some outlying points midrange. The agreement is noteworthy, given the differences in the design and operation of the two instruments. In the case of the SOC-400t, energy from an infrared source is directed onto the surface of the sample from a position that is approximately 20° from normal. Energy leaving the surface of the sample is collected hemispherically. In the case of the DB-100, energy from the dual rotating black body cavities is directed onto the surface of the sample hemispherically and energy leaving the surface of the sample is detected directionally, at an angle approximately 5° from normal. Under ideal circumstances, one would like to illuminate the sample hemispherically and detect the reflected signal hemispherically. However, such an arrangement is difficult to obtain in a portable instrument.

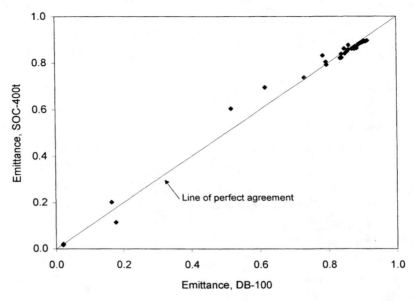

FIGURE 1. Emittance of a variety of samples as measured by both the SOC-400t and the DB-100.

Emittance data from previous calorimetric measurements are also available for comparison. Figure 2 shows the emittance as measured by the SOC-400t plotted against the emittance as measured by the LCVE.

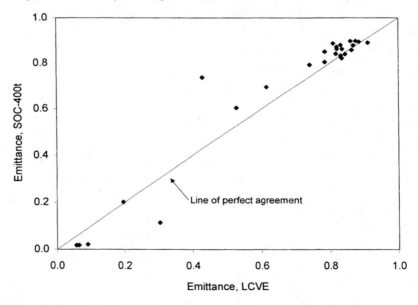

FIGURE 2. Emittance of a variety of samples as measured by both the SOC-400t and the LCVE.

Again, the line of perfect agreement is shown for reference. For the most part, good agreement is observed between the emittance values obtained from the SOC-400t and the emittance values obtained from the LCVE. Again, there are some outlying points midrange. Never the less, it is interesting that the emittance values are so similar, given that the values are calculated from fundamentally different principles, optically via Kirchhoff's Law and the Stefan-Boltzmann equation versus calorimetrically via a complicated set of energy balance equations. Figure 3 is included to complete the comparison. Again, the line of perfect agreement is provided for reference. Like Figure 2, there is good agreement among the data, with some outlying points midrange. And again, it is noteworthy that the values are so similar, given that they are calculated from fundamentally different principles.

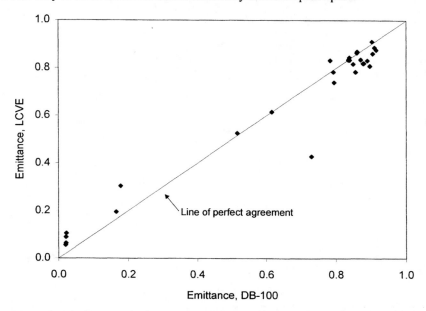

FIGURE 3. Emittance of a variety of samples as measured by both the LCVE and the DB-100.

It should be noted that the emittance of one sample (grit-blasted oxidized stainless steel), as measured optically, is substantially different from that measured calorimetrically, suggesting the need for occasional calorimetric measurements, particularly for those materials having a complex surface structure.

In addition to obtaining emittance values from the SOC-400t, one also obtains spectral data. Figure 4 summarizes spectral data from one of the samples, black anodized aluminum, and is very similar to the black anodized

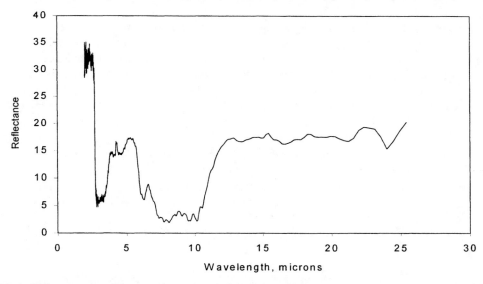

FIGURE 4. Reflectance as a function of wavelength for black anodized aluminum, as measured by the SOC-400t.

aluminum spectra reported previously (Jaworske, 1996). Infrared spectra of this type may be useful in identifying chemical species, not only the chemical species of a given substrate but also of chemical species of contaminants that may be found on the surface of the substrate. In such a case, it would be important to select a substrate that has a known and uniform infrared reflectance spectrum. Polished metal substrates or first surface metal vapor deposited mirrors would be ideal for this application. Polished gold surfaces are preferred. Species identification would be based on characteristic stretching bands found in the infrared spectrum and are best suited for contaminants of an organic nature.

Finally, reflectance as a function of wavelength can be used to calculate emittance as a function of temperature by convoluting the reflectance spectrum with respect to several black body spectra of the desired temperatures. This was done for the black anodized spectrum shown above and the results are summarized in Figure 5.

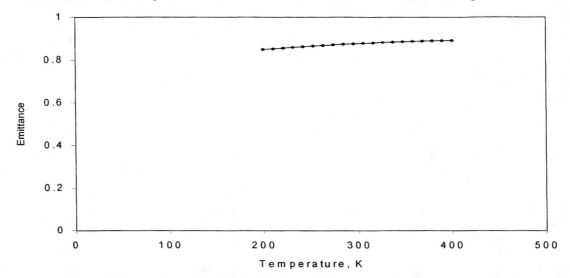

FIGURE 5. Emittance as a function of temperature, as calculated by convoluting the reflectance spectrum of black anodized aluminum with black body spectra in the range of 200 to 400 K.

CONCLUSIONS

A new infrared reflectometer was developed by Surface Optics Corporation under contract with NASA Glenn Research Center for the purpose of replacing our aging Gier-Dunkle DB-100 infrared reflectometer. Emittance values obtained from the new infrared reflectometer compare favorably with emittance values obtained from a DB-100 infrared reflectometer and calorimetric vacuum emissometer. Added features of the SOC-400t include the ability to collect reflectance data as a function of wavelength and the ability to convolute the reflectance data with respect to many black body spectra for the purpose of calculating emittance over a wide temperature range.

ACKNOWLEDGMENTS

The authors gratefully acknowledge Michael T. Beecroft, Surface Optics Corporation, for his undaunted efforts at manufacturing the instrument, writing the software, delivering it to NASA Glenn Research Center, and conducting the training needed to operate the instrument.

REFERENCES

Henninger, J. J., "Solar Absorptance and Thermal Emittance of Some Common Spacecraft Thermal Control Coatings," NASA Reference Publication 1121 (1984).
Jaworske, D. A., *Thin Solid Films*, **236**, 146-152 (1993).
Jaworske, D. A., *Thin Solid Films*, **290-291**, 278-282 (1996).

Electrochromic Emittance Modulation Devices For Spacecraft Thermal Control

C.L. Trimble[1], E. Franke[1], J.S. Hale[2], and J.A. Woollam[1]

[1] *Center for Microelectronic and Optical Materials Research, and Department of Electrical Engineering, University of Nebraska, Lincoln, NE 68588-0511. Phone: (402) 472-1975 email: ctrimble@engrs.unl.edu*
[2] *J.A. Woollam Co. Inc., Lincoln, NE 68588 Phone: (402) 477-7501 email:jhale@jawoollam.com*

Abstract. Electrochromic devices operating in the infrared have potential for control of spacecraft surface emissivity. These light-weight, thin film devices change emissivity due to optical property changes in Tungsten Oxide upon reversible intercalation or de-intercalation by Li^+ ions. Other contributors to thermal control performance are the properties of the underlying thin film layers and the transport of Li^+ ions between the layers. In this paper we discuss the potential for ultimate device performance and the present status of prototypical devices.

BACKGROUND

Electrochromic devices are stacks of thin films, mainly transition metal oxides, which can exhibit variable optical properties upon the application of a small voltage pulse. Much work has been done in this area in order to bring electrochromic devices to market in the form of "Smart Windows"(Monk, 1995 and Goldner, 1988). This application requires a device designed to work in the visible region of the spectrum. Another potential application of these thin film devices is as variable emittance modulators for thermal control surfaces on spacecraft. Their operation is analogous to charging a solid state battery. The application of an external potential difference causes small ions to move from one electrode to another. When the ions are in the electrochromic electrode the device will be in its 'high' emissive state. Figure 1 shows the mechanics of how the device works when it is in its high emittance state. The application of a voltage pulse moves the ions from the electrochromic film and into the charge storage electrode. In this state the device is in its 'low' emittance state. For a device to be useful as a thermal control surface it needs to be designed to work in the thermal IR (infrared) region of the spectrum. An electrochromic device has three critical layers. The top electrochromic layer contributes the majority of the optical switching. The ionic conductor needs to be able to conduct the ions between the storage and electrochromic layers while maintaining electrical isolation in order to prevent charge balancing electrons from traveling out of the electrochromic films. This allows the device to have a memory and eliminates the need to apply a constant voltage across the device. In this paper we use measured optical properties of the constituent layers to predict near-optimum performance of devices. We also summarize the state of the art in actual device performance.

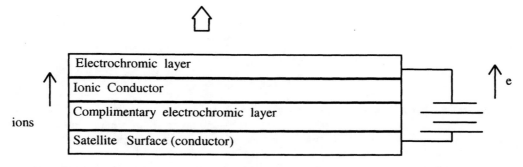

FIGURE 1. Important layers for an electrochromic device. The device is shown in its high emittance state, with the top electrochromic layer intercalated. The external voltage source is used to switch the device.

CP504, *Space Technology and Applications International Forum–2000*, edited by M. S. El-Genk
© 2000 American Institute of Physics 1-56396-919-X/00/$17.00

PREPARATION OF SPUTTERED THIN FILMS AND LITHIUM INTERCALATION

The thin films that are investigated for this work have all been reactively magnetron sputter deposited from metal targets. The tungsten oxide films were made in a 15/85 Oxygen to Argon environment. The nickel oxide films are used as charge storage layers and are made in an environment that had a 20/80 oxygen to argon ratio. Tantalum Oxide is studied as an ionic conductor for these devices. The environment for these films was also 15% oxygen and 85% argon.

In order to study the optical properties of the intercalated electrochromic films an electrochemical cell was used to insert lithium ions into the films. The films were deposited on Indium Oxide doped with Tin Oxide (ITO) so that they could be intercalated and bleached.

Spectroscopic Ellipsometry is a tool used to study these thin films. It is a technique that measures the change of polarization that a probe beam, with a known polarization, undergoes after being reflected from the sample surface. The measured parameters are the ellipsometric values ψ, and Δ, which are measured with respect to the frequency of the light, and the angle of incidence on the sample. Using the appropriate optical model allows the extraction of the optical constants, and micro-structural properties of the sample. More details of the technique can be found in (Woollam, 1999 and Johs, 1999).

Emissivity Calculations and the Blackbody Spectrum

The calculation for a 25°C blackbody was used to gauge the performance of a device. The curve for a blackbody is given by Planck's blackbody equation:

$$M(\lambda) = \frac{2\pi hc^2}{\lambda^5} \times \frac{1}{e^{hc/\lambda kT} - 1}, \tag{1}$$

where h is Planck's constant, c is the speed of light in vacuum, k is Boltzman's constant, T is the temperature in Kelvins, and λ is the wavelength. In order to determine the solar emittance for the devices the calculated reflectivity curves were integrated. This integration was weighted by the blackbody curve as follows:

$$\frac{\int_3^{30\mu m} (1 - R(\lambda)) \cdot M(\lambda) d\lambda}{\int_3^{30} M(\lambda) d\lambda}, \tag{2}$$

where $R(\lambda)$ is the calculated reflectivity of the devices, and M(λ) is the blackbody value.

RESULTS

In order to optimize layer thickness and evaluate the performance of proposed devices the optical constants of the constituent layers were measured. These optical constants extend the range of previous measurements (Hale, 1998). By using these optical constants in a device model the performance can be estimated.

Polycrystalline tungsten oxide is the layer that has the greatest effect on the modulation of the device. Amorphous tungsten oxide exhibits a broad polaron absorption in the near IR (Cogan, 1995); however, this absorption is not present in the polycrystalline oxide, this is the reason this material was selected for an IR modulating device. The crystallinity of the films was confirmed by x-ray diffraction measurements.

CP504, *Space Technology and Applications International Forum–2000*, edited by M. S. El-Genk
© 2000 American Institute of Physics 1-56396-919-X/00/$17.00

Nickel oxide is used as a charge storage layer in these devices. The intercalation ions are stored here when the device is in its low emittance state. The nickel oxide films show no significant change in IR reflectivity upon intercalation.

The optical constants of the ion-conducting layer also needed to be determined in order to complete the simulations for the devices. This layer is important for the device because it allows the mobile ions to pass from the electrochromic film to the charge storage film while maintaining electrical isolation of the two films. Electrical isolation is critical for the device to stay in the desired operating state without the maintenance of an external voltage.

Predicted Emittance Modulation

The devices that have been simulated are three layer structures. The simulation assumes that the substrate material is very reflective in the IR. To meet this end the optical constants of a gold substrate were used to represent the satellite surface, this substrate would also be used as an electrode for the device. The device requires separate electrodes to be in intimate contact with the charge storage layer (nickel oxide), and the electrochromic layer (tungsten oxide). These electrodes supply the electrical contact so that the voltage pulse used to switch the device can be applied. The other electrode was a simulated gold conducting grid. This grid should cover less than 10%(Cogan, 1995) of the surface of the electrochromic film to minimize its effect on the performance of the device. Figures 2a and Figure 2b show the structure of a simulated device, and its IR emissivity modulation. The emissivity modulation over the 25°C blackbody curve was calculated to be 40%. One of the reasons that this modulation is not greater is due to the material properties of both the tantalum oxide and tungsten oxide (Burdis, 1994) which both have lattice absorptions in the 12-17μm range.

Gold Conducting Grid 10% coverage	50 nm
Polycrystalline Tungsten oxide	150 nm
Tantalum Oxide Ion conductor	250 nm
Nickel Oxide	220 nm
Gold Substrate	1 mm

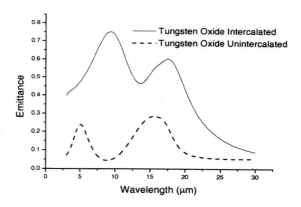

FIGURE 2a and 2b. The structure of the simulated device is shown on the left. The graph on the right gives the results of calculating the emissivity modulation based on the optical constants of the constituent layers in the device stack.

One potential method of improving the above device is to deposit an anti-reflective coating on top of it. The benefit of an antireflection coating comes from a minimization of reflection from the tungsten oxide/ambient interface. This is an effective way to maximize the results of the high emittance state of the device. Figure 3a and Figure 3b show the structure of the above device with a DLC antireflection coating. The AR coating broadens the emittance curve in the 6-11 μm region of the spectrum. The emissivity modulation of this structure was 51% over the region of the 25°C blackbody.

Measured Reflectance Modulation

Several Device structures have been made in order to gauge the validity of the modeling. In order to make a device that would work over several cycles a polymer electrolyte layer was used. This replaced the tantalum oxide layer in the simulated devices. The device construction and the optical constants of the electrolyte and ITO are described in previous work (Trimble, 1999). Figure 4a shows the structure of the device that had been fabricated, while Figure 4b shows the comparison between the measured device performance and the simulated device performance.

DLC Antireflection coating	900 nm
Gold Conducting Grid 10% Coverage	50 nm
Polycrystalline Tungsten Oxide	150 nm
Tantalum Oxide Ion Conductor	250 nm
Nickel Oxide Charge Storage Layer	220 nm
Gold Substrate	1 mm

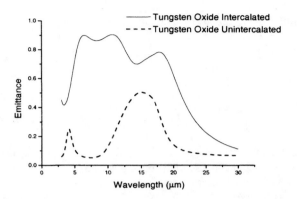

FIGURE 3a and 3b. The table on the left shows the device structure for the simulated device with the AR coating. The curves on the right show the performance of this device.

DLC AR coating	2.1μm
Low Dopant Concentration Si Wafer	125μm
Polycrystalline Tungsten Oxide	230nm
AMPS/DMA Polymer Electrolyte	25μm
Nickel Oxide Storage Layer	250nm
ITO Electrical Conductor	130nm
Glass Substrate	2mm

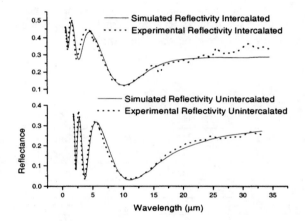

FIGUIRE 4. Comparison of a simulated device with experimentally measured device performance. The top curve shows the device with the ions in the charge storage layer. The bottom curve is the reflectance with the tungsten oxide intercalated.

The predicted emissivity modulation from this device was 9.6% while the actual measured modulation was 10%. The greatest contributor to the relatively low modulation in both the predicted and measured results is the use of a Si Substrate. The incident infrared light must penetrate through the low dopant concentration silicon substrate before it reaches the electrochromic device stack. Si has several phonon absorption modes in the region of the blackbody peak (Spitzer, 1967). These absorptions don't allow the probe beam to reach the reflective ITO layer of the device and then return to the IR detector. This is the main reason that the predicted results of Figure 4 do not show as much modulation as the theoretical devices shown in Figures 2 and 3.

There are several additional reasons for the low modulation performance in the measured device. The thick absorbing polymer layer prevents most of the radiation that makes it through the silicon substrate from reflecting off of the ITO layer. Also the lithium ions that cause the optical change do not reversibly intercalate and deintercalate from the films. In fact in order to fit the experimental data for the bleached state it was necessary to assume that the tungsten oxide layer was not fully deintercalated. Secondary neutron mass spectroscopy(SNMS) studies have shown that some Li^+ stays in the tungsten oxide layer, and some has also been found in the Si substrate. In order to fit for the intercalated state of the device, it was necessary to account for additional proton intercalation into the film. This is feasible because the polymer electrolyte is a mixed Li^+, and H^+ conductor. It is an acid based polymer so additional protons originating from this layer of the device can conceivably intercalate the tungsten oxide film.

DISCUSSION

This paper shows that using the measured optical constants of the constituent layers for an electrochromic device it is possible to build an optical model to predict performance. This can be beneficial in determining the feasibility of applying the device to a given application. This type of model is also a good way to optimize layer thickness so that device performance can meet necessary specifications.

The modeling will also help to find an AR coating with an index of refraction that matches, better than DLC, with the polycrystalline tungsten oxide is a means of improving the performance of the devices. As shown in Figure 3 an AR coating can tune the device so that it has an 80% emittance change in the spectral range of 5 to 10μm. This may lend itself to applying the device as an IR switch in a variety of applications.

The device simulation in this work shows that a maximum of 51% modulation is possible for the device structure and constituent materials that have been chosen. This may be adequate for its application to the surface of a spacecraft. The devices that we have fabricated to date show only a 8-10% modulation in emissivity, however we are optimistic in regards to improvements in these results.

There are several ways to improve current devices so that they are more similar to the simulated structure. The first is to use a conducting grid structure. The devices that have been made so far use a Si wafer that is only semi-transparent in the infrared region of interest. Another area for improvement is the ion conduction layer. This layer gets pinholes that allow the conduction of electrons by "short circuiting" the device. A vacuum intercalation process may help to solve this problem by eliminating the need for a damaging wet processes for the intercalation of the films.

One area of flexibility in the fabrication of the devices is the shape (geometries) of the thin films within the stack. Some devices have been made with the conducting electrode consisting of thick metal bands that are spaced far apart so that the majority of the electrochromic layer is exposed. These devices work fairly well but the operation lifetime is short due to the higher voltages used to switch the device. Taking advantage of this flexibility will allow optimization of device geometry so that a 50% ,or better, modulation may be realized.

ACKNOWLEDGMENTS

Research Supported by The Ballistic Missile Defense Organization #DASG60-98-C-0054 and The NASA Glenn Research Center, Grant Number NAG3-2219.

REFERENCES

Burdis, M.S., Siddle, J.R. and Taylor, S., "Fourier Transform Infrared Study of Tungsten Oxide Thin Films and Their Coloration by Lithium," in *Proceedings of Optical Materials Technology for Energy Efficiency and Solar Energy Conversion XIII,* edited by Wittwer, V., Granqvist, C., Lampert, C.M., Bellingham, Washington, The International Society for Optical Engineering 2255, 1994, pp. 371-383.

Cogan, S.F., Rauh, R.D., Klein, J.D.,Plante, T.D., "Electrochromic Devices For Optical Modulation in the Infrared," in *Proceedings of the Symposium on Electrochromic Materials II,* edited by Kuo-Chuan Ho, and Donald A. MacArthur, New Jersey, The Electrochemical Society, Inc. Proceedings Volume 94-2, 1995, pp. 269-277.

Goldner, R.B., "Electrochromic Smart Window Glass," in *Proceedings of the International Seminar on Solid State Ionic Devices,* edited by B.V.R. Chowdari and S. Radhakrishna, Singapore, World Scientific Publishing Co. Pte. Ltd, 1988, pp. 379-387.

Hale, J.S., DeVries, M., Dworak, B.,Woollam, J.A., "Visible and Infrared Optical Constants of Electrochromic Materials for Emissivity Modulation Applications," *Thin Solid Films* **313-314**, 205-209, (1998).

Johs, B., Woollam, J.A., Herzinger, C.M.,Hilfiker, J., Synowicki, R., and Bungay, C.L., "Overview of Variable Angle Spectroscopic Ellipsometry (VASE), Part II: Advanced Applications," in *Proceedings of Optical Metrology,* edited by G.A. Al-Jumaily, Bellingham, Washington, The Society of Photo-Optical Instrumentation Engineers CR72, 1999, pp 29-56.

Monk, P.M.S., Mortimer, R.J., and Rosseinsky, D.R., *Electrochromism: Fundamentals and Applications*, Weinheim, FRG, VCH, 1995, pp 1-8.

Spitzer, W.G., "Multiphonon Lattice Absorption," in *Semiconductors and Semimetals vol. 3,* edited by R.K. Willardson, and A.C.Beer, New York, Academic Press, 1967, pp. 17-33.

Trimble, C., DeVries, M., Hale, J.S., Thompson, D.W., Tiwald, T.E., Woollam, J.A., "Infrared Emittance Modulation Devices Using Electrochromic Crystalline Tungsten Oxide, Polymer Conductor, and Nickel Oxide," *Thin Solid Films* **347**, 1-9, (1999).

Woollam, J.A., Johs, B., Herzinger, C.M., Hilfiker, J., Synowicki, R., Bungay, C.L., "Overview of Variable Angle Spectroscopic Ellipsometry (VASE), Part I: Basic Theory and Typical Applications," in *Proceedings of Optical Metrology,* edited by G.A. Al-Jumaily, Bellingham, Washington, The Society of Photo-Optical Instrumentation Engineers CR72, 1999, pp 3-28.

Variable Emissivity Through MEMS Technology

Ann Garrison Darrin, Robert Osiander and John Champion[1],
Ted Swanson and Donya Douglas[2]

[1]*The Johns Hopkins University Applied Physics Laboratory, 11100 Johns Hopkins Road Laurel, MD 20723*
[2]*NASA Goddard Space Flight Center, Thermal Branch, Greenbelt, MD, 20771*

Ann.Darrin@jhuapl.edu, (443) 778 4952

Abstract. All spacecraft rely on radiative surfaces to dissipate waste heat. These radiators have special coatings that are intended to optimize performance under the expected heat load and thermal sink environment. Typically, such radiators will have a low absorptivity and a high infrared-red emissivity. Given the dynamics of the heat loads and thermal environment it is often a challenge to properly size the radiator. In addition, for the same reasons, it is often necessary to have some means of regulating the heat rejection rate of the radiators in order to achieve proper thermal balance. The concept of using a specialized thermal control coating which can passively or actively adjust its emissivity in response to such load/environmental sink variations is a very attractive solution to these design concerns. Such a system would allow intelligent control of the rate of heat loss from a radiator. Variable emissivity coatings offer an exciting alternative that is uniquely suitable for micro and nano spacecraft applications. This permits adaptive or "smart" thermal control of spacecraft by varying effective emissivity of surfaces in response to either a passive actuator (e.g., a bi-metallic device) or through active control from a small bias voltage signal. In essence the variable emittance surface would be an "electronic louver." It appears possible to develop such "electronic louvers" through at least three different types of technologies: Micro Electro-Mechanical Systems (MEMS) technology, Electrochromic technology and Electrophoretic technology. This paper will concentrate on the first approach using both MEMS and Micromachining technology to demonstrate variable emissivity.

BACKGROUND

All spacecraft and the instruments they support require an effective thermal control mechanism in order to operate as designed and achieve their expected lifetimes. In an increasing number of satellites, optical alignment and calibration require a strict temperature control. Traditionally, the thermal design is part of the spacecraft layout determined by all subsystems and instruments. Heat load levels and their location on the spacecraft, equipment temperature tolerances, available power for heaters, view to space, and other such factors are critical to the design process. Smaller spacecraft with much shorter design cycles and fewer resources such as heater power, volume, and surface, require a new, more active approach.

A number of active methods that vary the heat rejection rate in some controlled fashion are commonly used to maintain a reasonable thermal equilibrium. One such method is to cold bias the spacecraft and use simple electrical resistance make-up heaters to control the temperature. However, this can require considerable electrical power, which the spacecraft may not have available at all times. Another approach is to employ a radiator connected with variable conductive heat pipes, capillary pumped loops and/or loop heat pipes. Typically, this adds significant weight, cost and complexity to the systems and can, at least for heat pipes, introduce new issues concerning ground testing. Another approach is to use mechanical louvers that can be opened to expose a radiative surface. While functional, traditional mechanical louvers are bulky, expensive, subject to damage, and require significant thermal analysis to evaluate the effect of different sun angles.

The concept of using a specialized thermal control coating or surface which can passively or actively adjust its emissivity in response to variations in load and environmental conditions is a very attractive solution to these design

CP504, *Space Technology and Applications International Forum–2000*, edited by M. S. El-Genk
© 2000 American Institute of Physics 1-56396-919-X/00/$17.00

concerns. Variable emittance thermal control coatings have been under development at NASA-Goddard Space Flight Center (GSFC) since the mid-1990's. These coatings change the effective infrared-red emissivity of a thermal control surface to allow the radiative heat transfer rate to be modulated upon command. Two technologies have been under consideration, electrochromic and electrophoretic devices. The emittance modulation in electrochromic devices is achieved using crystalline electrochromic materials whose reflectance can be tuned over a broad wavelength (2 to 40 microns) in the infrared. Electrophoretic devices involve the movement of suspended particles through a fluid under the application of a small electrical field. When an electric field is applied the flakes, which are made of a material with high reflectivity, they align themselves and form an essentially flat reflective surface.

Another technique to vary the emissivity of a surface involves the use of mechanical louvers, where a mechanical vane or window is opened and closed to allow an alterable radiative view to space. (Gilmore, 1994). Current micro-machining techniques allow the designer to generate arrays of such structures with feature sizes on the order of micrometers (Helvajian, 1997). The three variable emissivity technologies have been chosen as a demonstration technology on NASA's New Millenium ST5 " Nanosat Constellation Trailblazer" mission. Each technology will control a 20 cm x 6 cm radiator area on one of the three nanosats.

MEMS LOUVERS

Micro-electro-mechanical (MEMS) louvers are similar to miniature venetian blinds that can be opened or closed to expose an underlying high emissivity radiator. The "effective emissivity" of the surface can be modulated in a controlled fashion by varying either the angle of the micro-louvers or the total number of micro-louvers that are opened. Figure 1 shows three different designs for the concept.

Louver Design

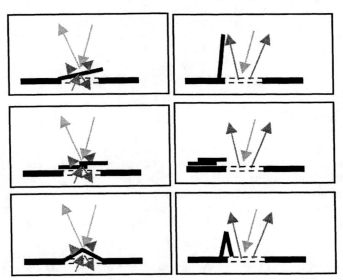

The simplest design is a single louver, which can be opened through 90 degrees. While a smaller opening angle would allow for more variation, it also requires different control due to the influence of the solar position. In the open position, the louvers will expose the high emissivity substrate material to space. In a second design, multiple levels of sliders move across each other. In this case, the total area which can be exposed depends on the number of layers available in the fabrication process. Advantages of this approach include the two-dimensional design, the linear variability of exposed area, and the sturdiness of the design. A third design looks more like a folding door. This design is more complicated than the other two since it uses more hinges, but we expect it to be more sturdy than the single louvers while providing the same active area. Preliminary experiments have shown that these devices are less likely to break during release and operation.

FIGURE 1. Different concepts for thermal control louvers. Top: Louver; Middle: Slider; Bottom: Folder.

Our early efforts focused on the design of the louvers for fabrication using the MCNC Multi-User MEMS Process (MUMPs). To date, two generations of prototype chips of MEMS louvers have been developed. Both sets were designed at the Applied Physics Laboratory (APL), fabricated at the MCNC Technology Applications Center under the MUMPs program, and subsequently released and tested at APL. The base material for the current devices is polysilicon and the exposed, top surface is coated with gold, which has a very low emissivity of 0.2.

FIGURE 2a. Images from video of MEMS louver opening.

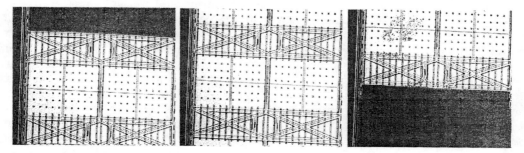

FIGURE 2b. Images from video of MEMS slider opening.

FIGURE 2c. Images from video of MEMS folder opening.

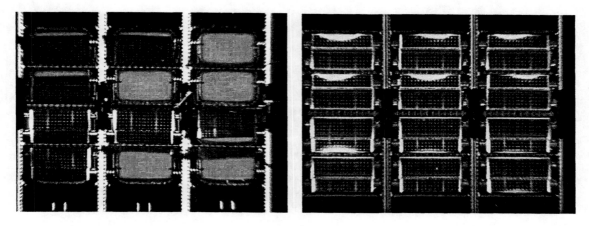

FIGURE 3. Optical Image of a louver array, with open louvers (left) and all louvers closed (right). The open louvers expose the background through the etched openings.

The coated areas are intended to cover an underlying, high emissivity surface. Images taken from videos of devices in operation are shown in Fig. 2. In the most recent generation, multiple sets of louvers have been grouped together to allow for measurements of emissivity variation using an infrared imager. Further, the silicon substrate under the louvers has been removed using reactive ion etching (RIE) to expose the high-emissivity substrate. This is desirable since silicon, while generally transparent in the IR, still retains a high reflectivity. Optical images of a group of louvers is shown in Fig. 3, with some louvers in the open position (left) exposing the bright background.

Infrared Emissivity Measurements

Although the louvers are not mounted on a radiator, infrared images taken at room temperature in the 8-12 μm wavelength range allow for a reasonable estimate of their performance to be made. Infrared images were taken using a Mikron Scanner, which scans the image with two mirrors onto a HgCdTe detector with a spatial resolution of 320x240 pixels. Using a close-focus attachment, the pixel resolution was on the order of 20 μm per pixel. Calibration was performed on an emissivity standard and the room temperature background was subtracted. An emissivity image of the MEMS louvers is shown in Fig. 4. The structures include multiple sets of louvers on the left, gold-coated "sliders" on the upper right, and the bare SiN layer lower right. All devices are on a Si substrate.

The measured emissivity varies from 0.3 for the SiN to 0 for gold (0.02 is the literature value (Wolfe, 1985)). The same measurements were performed for the louver arrays in the closed and open position. The emissivity images in the area denoted by the dashed line are for an array of louvers that are (a) closed, (b) partially open and (c) fully open. The average emissivity, ε, for the louver area is 0.18, 0.26, and 0.30, respectively. Future experimental setup improvements will allow measurements to be taken at increased temperatures with reduced background radiation

Based on the measurements, it was found that the variation in effective emissivity of the prototype devices was forty percent. Note that a sizeable fraction of the area over which this measurement was made (within the dashed lines of Fig. 4) is devoted to mechanical structures supporting the louver operations. Through design modifications, we believe that the ratio of louver area to support structure area can be increased by up to a factor of two. This would increase the variation in effective emissivity to eighty percent.

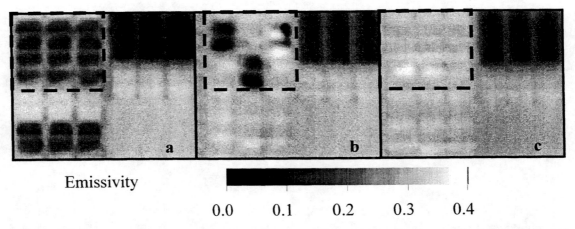

FIGURE 4. Emissivity image of MEMS louvers in various positions. The average emissivity in the region within the dashed lines is 0.18 (a), 0.26 (b), and 0.31 (c). Note that the bottom SiN layer has an emissivity of 0.30.

Louver Actuation

For a successful application of the louver concept for spacecraft thermal control, an actuation mechanism has to be identified which allows the highest individual louver control possible with a minimum of space necessary. Note, that all the space covered by the actuation is not active and presents an emissivity bias. Highly individual louver control provides the best accuracy in setting the emissivity and further allows increased control of the spatial emissivity variations and operational redundancy. Further, low power consumption and zero power in a static condition are required for small spacecraft applications. Several actuation mechanisms have been considered and in part, implemented. One implementation was the use for an electrostatic comb drive. While this is a low power, reliable and straight-forward designed MEMS actuation mechanism, disadvantages arise due to the large area requirement and, from a space-craft perspective, the relatively high driving voltages (10s V) required. Build up of static charges on the radiaator surface for space environmental effects may also be an issue. Another mechanism used in some prototype louvers is a "heatuator"(Butler, 1999), which does not have the high voltage requirement and takes up relatively less area on the louver chip. It is possible to locate both types of actuators outside of the active area above the radiator.

Another actuation mechanism under consideration involves coating the actuation structures with a metal other than gold to create a bi-morph that can be heated electrically to generate motion due to different thermal expansion coefficients. Such an actuation mechanism could be used to build a "smart" device where the surface temperature directly controls the louver actuation. Similarly, shape memory alloy coatings such as Nitanol^{TM} could also be used for the actuators (Seguin, 1999).

Reliability Aspects

There are many reliability issues surrounding the extended use of MEMS devices for spacecraft applications (Stark, 1999). The louvers must survive through the launch and operate in the harsh environment of space. In addition, the effects of pre-launch storage must also be taken into consideration. A non-exhaustive list of the of MEMS reliability concerns includes: stiction, wear, fatigue, contamination, and radiation effects. An extended evaluation of these issues is currently under study and only a brief overview follows.

Although stiction has not been observed in the prototype devices, the MEMS louvers are probably susceptible to this failure mechanism as a result of electrostatic interactions, capillary forces, or even localized cold welding (Patton, 1999). These concerns can be addressed in several ways. For example, proper ground design should minimize the potential mechanical seizure due to electrostatic clamping. Excessive condensation of moisture, especially during pre-launch storage, can be mitigated through the use of hydrophobic coatings and outgassing techniques. Furthermore, appropriate packaging could be employed to prevent the accumulation of water and other contaminants on critical surfaces of the devices.

While relative humidity (RH) levels in excess of 70% have been associated with been degraded mechanical performance attributed to increased stiction, elevated frictional wear between contacting parts has been observed in extremely low RH environments (Tanner, 1999). Due to the low RH of the intended operational environment, the possible degradation of the hinge joints over the device lifetime is an important issue. Minimum lifetimes will be on the order of 10,000 to 50,000 cycles. Various coatings and design modifications to minimize friction are being considered. Similarly, the effects of fatigue on hinges and actuators are also being examined.

Finally, in a space environment, the MEMS louvers will be subjected to high energy irradiation. As a result, charge buildup in dielectric layers could occur which may lead to inconsistent or degraded operation of either the louvers or electrostatic actuators. While this obstacle is not insurmountable, it further builds the case for passive louver actuation using "smart" materials.

SUMMARY AND FUTURE DIRECTIONS

To date, two generations of prototype MEMS louvers have been developed which clearly demonstrates the feasibility of using arrays of devices for miniaturized satellite thermal control. Successful actuation of the initial devices and the results of preliminary emissivity testing indicated the validity of the hinged louver concept for thermal control applications. After verification of space qualification of the louvers, the next step will be to fly one or more very small prototypes in a standard calorimeter as experiments on an upcoming spacecraft.

Numerous future NASA missions, such as the ST5 Nanosat Constellation Trailblazer, will undergo significant changes in its thermal environment and will require means of modulation in the spacecraft's heat rejection rate. Specifically, the Trailblazer spacecraft will undergo an approximately 2-hour or longer eclipse during which time the instruments must survive and possibly operate. Given their small capacity for power storage by batteries and low thermal capacitance, the best strategy will be to "close off" their radiator area and radically reduce their heat loss rate. One recent study by Aerospace Corporation[2] predicted heater power savings of 50 to 90% and a nearly 4:1 reduction in component temperature variations. In addition to the obvious weight and power savings, the technology of MEMS louvers for thermal control would greatly simplify spacecraft design and qualification testing and also allow adaptive response to changing power levels or unexpected thermal environments once on-orbit.

ACKNOWLEDGMENTS

The authors would like to express their gratitude to following individuals, who have supplied constant support for enabling technologies for Nanosat applications: Ms. Abby Harper, Deputy Program Manager, Solar Terrestrial Probes, Mr. Jan Gervin, Project Formulation Manager, Geospace Electrodynamics Connections, and Mr. Peter V. Panetta, STAAC, NASA Goddard Space Flight Center along with Dr. Richard Benson and Mr. Jay Dettmer of the Johns Hopkins University Applied Physics Laboratory.

REFERENCES

Butler, J.T., Bright, V.M., and Cowan, W. D., "Average power control and positioning of polysilicon thermal actuators," *Sensors and Actuators* **72**, 1999, pp 88-97.

Gilmore, D. G., *Satellite Thermal Control Handbook*, The Aerospace Corporation Press, El Segundo, CA, 1994.

Helvajian, H., Janson , S.W. and. Robinson, E. Y , "Big benefits from tiny technologies: Micro-nanotechnology applications in future space systems," *Critical Reviews of Optical Science and Technology*, Taylor, E.W., Ed., SPIE Vol. CR66, pp 22-23, 1997.

Patton, S.T., Cowan, W.D., and Zabinski, J.S., "Performance of a New MEMS Electrostatic Lateral Output Motor," *IEEE IRPS Proc.*, pp 179-188, 1999.

Seguin, J.L., Bendahan, M., Isalgue, A., Esteve-Cano, V., Carchano,m H., and Torra, V., "Low temperature crystallized Ti-rich NiTi shape memory alloy films for microactuators" *Sensors and Actuators* **74**, 1999, pp 65-69.

Stark, B., Ed., *MEMS Reliability Assurance Guidelines For Space Applications, JPL Publication 99-1, 1999.*

Tanner, D.M., Walraven, J.A., Irwin, L.W., Dugger, M.T., Smith, N.F., Eaton, W.P., Miller W.M., and Miller, S.L., "The Effect of Humidity on the Reliability of a Surface Micromachined Microengine," *IEEE IRPS Proc.*, 1999, pp 189-197.

Wolfe, W.L., and Zissis, G.J., Eds., *The IR Handbook,* Office of Naval Research, Dept. of the Navy, Arlington, VA, 1985.

Parametric Study of Variable Emissivity Radiator Surfaces

Lisa M. Grob[1] and Theodore D. Swanson[2]

[1]Swales *Aerospace, Beltsville MD 20705   lgrob@swales.com*
[2]*NASA Goddard Space Flight Center, Code 545, Greenbelt MD 20771*
Theodore.D.Swanson.1@gsfc.nasa.gov

Abstract. The goal of spacecraft thermal design is to accommodate a high function satellite in a low weight and real estate package. The extreme environments that the satellite is exposed during its orbit are handled using passive and active control techniques. Heritage passive heat rejection designs are sized for the hot conditions and augmented for the cold end with heaters. The active heat rejection designs to date are heavy, expensive and/or complex. Incorporating an active radiator into the design that is lighter, cheaper and more simplistic will allow designers to meet the previously stated goal of thermal spacecraft design. Varying the radiator's surface properties without changing the radiating area (as with VCHP), or changing the radiators' views (traditional louvers) is the objective of the variable emissivity (vary-e) radiator technologies. A parametric evaluation of the thermal performance of three such technologies is documented in this paper. Comparisons of the Micro-Electromechanical Systems (MEMS), Electrochromics, and Electrophoretics radiators to conventional radiators, both passive and active are quantified herein. With some noted limitations, the vary-e radiator surfaces provide significant advantages over traditional radiators and a promising alternative design technique for future spacecraft thermal systems.

INTRODUCTION

A parametric study of vary-e radiator surfaces is documented within. The scope of the study ranges from steady state heat balance equations to modified Earth Observing System (EOS-AM) Spacecraft Thermal Math Models (TMMs). Thermal performance, cost and weight of the MEMS, Electrochromics and Electrophoretics vary-e surfaces were compared with conventional satellite radiator surfaces. The significant savings in heater power, cost and weight illustrate the advantages vary-e radiator surfaces have over traditional surfaces. The savings these new technologies offer when a reduction in heat loss is required to maintain satellite temperatures can provide smaller, less complex and cheaper satellites. Detailed descriptions of the vary-e radiator surfaces and EOS-AM TMMs are beyond the scope of this paper.

DISCUSSION OF ANALYSIS

An incremental study approach was used to progressively assess the thermal performance of these vary-e radiators. The orbits and environments used in all phases of the study were selected to reflect the EOS-AM spacecraft design criteria since the final analysis included a comparison with the EOS battery radiators. EOS-AM mission orbit is a sun-synchronous 10:30 descending node, 98.2° inclination polar orbit at an altitude of 705 km.

Initially, steady state calculations, using orbital average fluxes were performed on a number of radiator surfaces. The calculated heat rejection capabilities of traditional radiators were compared against the three vary-e radiator surfaces. A "baselined" Silver Teflon (AgTeflon) radiator, sized at 350 in^2 (0.226m^2) was used to maintain a 30°C panel supporting a 75W "box". This design was evaluated using a Ag Teflon constant conductance heat-pipe panel

CP504, *Space Technology and Applications International Forum–2000*, edited by M. S. El-Genk
© 2000 American Institute of Physics 1-56396-919-X/00/$17.00

with Ag Teflon, a MEMS CCHP radiator, an Electrochromics CCHP radiator and an Electrophoretic CCHP radiator. All technologies were assessed using realistic designs based on the hot case parameters defined in Table 1. Radiator areas for traditional louvers and Electrochromics used in the comparison reflected area inefficiencies and non-optimized thermal optical properties. The actual radiator area of the traditional louvered radiator, was 425 in^2 (0.274m^2) due to area inefficiencies. Electrochromics radiator was evaluated at 400 in^2 (0.258m^2) because of the higher solar absorptance. Once the hot case was defined for all the surfaces, cold case calculations that included beginning of life (BOL) properties and no box dissipations, determined the heater power required to maintain the radiators at -10°C.

Of particular interest are the spacecraft "cold side" (+Y) radiators where most of the housekeeping equipment (including the nickel hydrogen batteries) resides and the instrument deck on the nadir side (+Z) for the typical LEO science mission. Both the nadir and cold side radiators experience mainly albedo and IR fluxes. The albedo is significant, however in the low beta orbits. During the year, the EOS-AM spacecraft has a beta range of 13° to 32°. The zenith (-Z) radiator was also studied since a significant portion of the EOS-AM housekeeping equipment is accommodated there. The zenith radiator experiences direct solar fluxes, which emphasizes the significance and/or limitations of the current vary-e technologies' absorptivities.

The analysis parameters and assumptions for the steady state calculations used to evaluate the different radiator technologies are found in Table 1. Figure 1 is an illustration of the radiator comparisons.

TABLE 1. Steady State Analysis Assumptions and Parameters.

Typical sun-synchronous LEO orbit / beta range 13° and 32°	Material	Radiator Properties		
		α		ε
		BOL	EOL	
Hot Case Models EOL environment and properties on the spacecraft cold side radiator	AgTeflon	0.08	0.18	0.79
75 watt box mounted on the 30°C panel	OSRs	0.08	0.14	0.80
Cold Case Models BOL environment and properties on the spacecraft cold side radiator	White Paint	0.15	0.40	0.85
	Electrochromics	0.30	0.30	0.29[1] 0.70[2]
Panel heaters maintaining a -10°C panel	MEMS	0.20	0.20	0.20[1] 0.80[2]
Effective emissivity of closed louver is 0.09	Electrophoretics	0.20	0.20	0.29[1] 0.80[2]

[1] light/closed/activated state
[2] dark/open/base state

In addition to the hot and cold case scenarios discussed above, heat rejection capability was calculated as a function of radiator area using the parameters outlined in Table 1. The resulting plots for the hot and cold cases, see Figures 2 through 9, reveal the heat rejection capabilities from the zenith, nadir and cold side radiators for areas up to 10 m^2. It is evident from Figures 2 and 4 that the Electrochromics and MEMS radiators will not run as cold as -10°C in these environments. Calculations were repeated for the zenith and nadir surfaces at achievable, cold end radiator temperatures, specifically 10°C and 1°C for the conditions specified.

The dynamic effects of the vary-e surfaces were included in this study, as well. TRASYS and SINDA TMMs of a simplified cube provided heater power requirements for the zenith, nadir and cold side radiators surfaced with the traditional and vary-e coatings. The parameters used in the transient analysis are listed in Table 2. Temperature set points of the proportional heater modeled in SINDA, were set below the vary-e light to dark set point in an attempt to minimize the number of cycles the variable emissivity surfaces would experience. This practice was especially evident when the EOS-AM battery charge/discharge dissipation profile, Figure 10 was applied to the cold side (+Y) of the cube model.

As a final exercise, the EOS-AM TMMs were modified to study the effects of a vary-e surface on the battery radiators. Incorporating the model changes required separating the TRASYS generated UV and IR incident environments on the radiators. The temperature specific absorbed fluxes were then defined in SINDA. Modifying

TABLE 2. Transient Analysis Assumptions and Parameters.

Typical sun-synchronous LEO orbit / beta range 13° and 32°	Material	Radiator Properties			
		α			ε
Hot Case Models EOL environment and properties on all cube surfaces		**BOL**	**EOL**		
	AgTeflon	0.08	0.18	0.79	
Panel temperatures maintained @ 30°C	OSRs	0.08	0.14	0.80	
Cold Case Models BOL environment and properties on all cube surfaces	White Paint	0.15	0.40	0.85	
	Electrochromics	0.30	0.30	0.29[1]	0.70[2]
Panel temperatures maintained @ -10°C	MEMS	0.20	0.20	0.20[1]	0.80[2]
Set Points -8°C to -10°C for proportional heater	Electrophoretics	0.20	0.20	0.29[1]	0.80[2]
-5°C for vary-e surfaces					

[1] light/closed/activated state
[2] dark/open/base state

the radiation couplings for the vary-e surfaces, also a function of temperature were straight forward for the space-viewing radiator surfaces.

RESULTS

The results of the steady state radiator comparison are listed in Table 3. The heater powers required to maintain the radiators at -10°C are listed for each configuration. With the exception of the 350 in^2 Electrochromics panel, all hot case radiator temperatures were calculated at 30°C. The hot case, 30°C Electrochromics radiator required an area of 400in^2. The peak auxiliary power for the VCHP configuration was defined from five, 5-watt reservoirs. Activating powers for the vary-e surfaces were considered negligible.

TABLE 3. Radiator Trade-offs Quantified.

Typical Radiation Configurations (Eff. Area .226 m^2/350 in^2)	Heater Power (W)	Peak Auxiliary Power (W)	Estimated Weight Difference (kg)	Cost (K$)
Ag Teflon CCHP	39	-	B/L	3
Ag Teflon VCHP	8	25	1.72	100
Louvered CCHP	6	-	1.36	80
Electrophoretics CCHP	14	-	.05	2.5
MEMs CCHP	9	-	.05	2.5
Electrochromics CCHP Eff. Area .258m^2 /400 in^2 Tradiator hot case = 30°C	15	-	.14	3
Electrochromics CCHP Eff. Area .226m^2/350 in^2 Tradiator hot case = 39°C	13	-	.05	2.5

The heat rejection capabilities for the zenith, nadir and cold side radiators are illustrated in Figures 2 through 9. It is evident from Figures 7 through 9 that an area penalty exists for the higher absorptivity surfaces; larger radiator areas are required to reject the heat due to the absorbed environments. The α/ε ratio drives the surfaces' heat rejection on the nadir and zenith faces because of the significant UV environments. The key parameter for the lower UV environment surface, cold side radiator is the coatings' emissivity. BOL heat rejection charts, Figures 2 through 6 show the reduced heater power requirements for the vary-e surfaces.

The transient results from the simple cube model are listed in Tables 4 and 5. Once again, the effects of the UV environment and radiator absorptivity are noted.

TABLE 4. Transient Cube Model Results.

Orbital Average Heater Power (W)			**Zenith Radiator**			
Orbit	Ag Teflon	OSRs	White Paint	Electrochromics	Electrophoretics	MEMs
13° EOL	32	37	28	9	10	6
Tmax/Tave	-8/-8	-8/-8	44/5*	34/3*	0/-6	0/-6
32° BOL	43	44	41	10	10	7
Tmax/Tave	-8/-8	-8/-8	-8/-8	19/-2*	-5/-7	-5/-7

* Elevated temperatures due to high α and UV environment.

			Nadir Radiator			
Orbit	Ag Teflon	OSRs	White Paint	Electrochromics	Electrophoretics	MEMs
13° EOL	10	11	6	2	2	1
Tmax/Tave	-8/-8	-8/-8	4/-5	1/-5	-4/-6	-4/-6
32° BOL	14	14	13	2	2	1
Tmax/Tave	-8/-8	-8/-8	-8/-8	0/-6	-4/-7	-4/-8

			Cold Side Radiator			
Orbit	Ag Teflon	OSRs	White Paint	Electrochromics	Electrophoretics	MEMs
13° EOL	39	40	41	13	14	9
32° BOL	40	40	43	14	14	10

Temperatures never go above heater set point temperatures.

TABLE 5. Battery Cube Model Comparison.

Radiator	Orbital Ave Heater Power (W)	Max Radiator Temperature (C°)
Ag Teflon	16	6
Electrochromics	3	12*
Electrophoretics	3	7
MEMs	2	7
*Elevated temperature due to higher α and non-optimized area		

The orbital average heater power savings realized for the EOS-AM batteries modeled with the MEMs radiating surface is 54 watts, essentially all the power needed for the batteries to maintain their minimum allowable temperature in the original configuration.

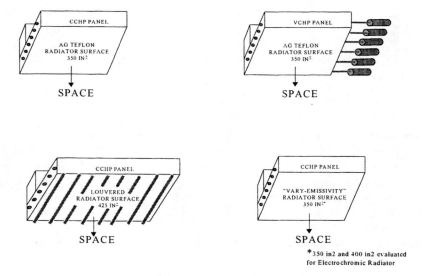

FIGURE 1. Radiator Trade-off Illustrations.

CONCLUSIONS

As illustrated throughout this study, vary-e technology offers significant advantages over current approaches for radiators in low UV environments. The heater power, mass, and cost savings that can be realized with these systems are potentially significant for many spacecraft design applications. For example, eliminating the EOS-AM battery heater power during on-orbit hot and cold cases by using a MEMs radiator demonstrated quite clearly the impact this technology may have on spacecraft design.

In order to make the benefits useful to all satellite radiating surfaces; a reduction in the vary-e's solar absorptivities is needed. Attention must also be paid to the vary-e maximum activation cycles, which may be the limiting factor for battery radiators and other dynamic dissipating systems. A reduction of cycles can be achieved with the proper set point temperature selection and the additional variable switch (i.e. battery pressure, time).

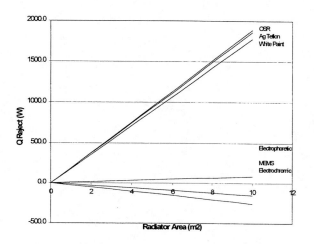

FIGURE 2. Heat Reject from Zenith at -10°C BOL / 32°β .

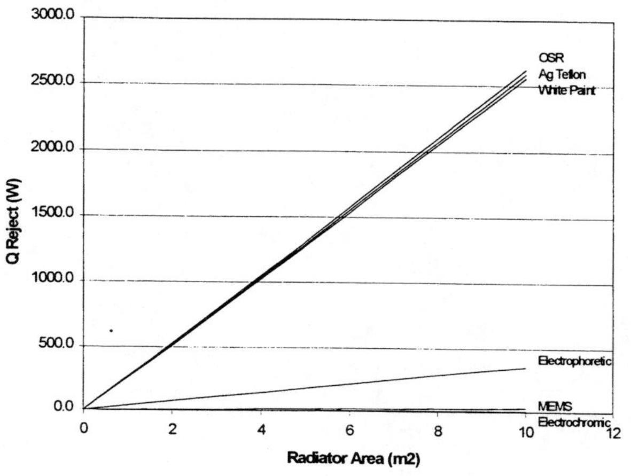

FIGURE 3. Heat Reject from Zenith at 10°C BOL / 32°β.

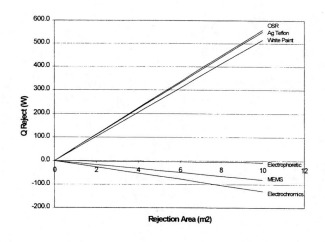

FIGURE 4. Heat Reject from Nadir at -10°C BOL / 32°β.

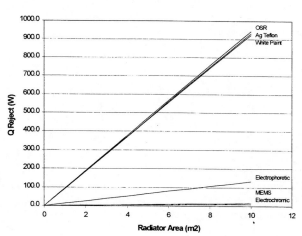

FIGURE 5. Heat Reject from Nadir at 1°C BOL / 32°β.

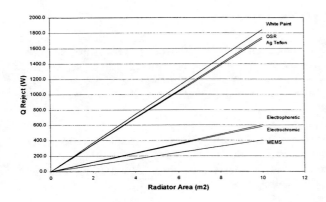

FIGURE 6. Heat Reject from Cold at -10°C BOL / 32°β.

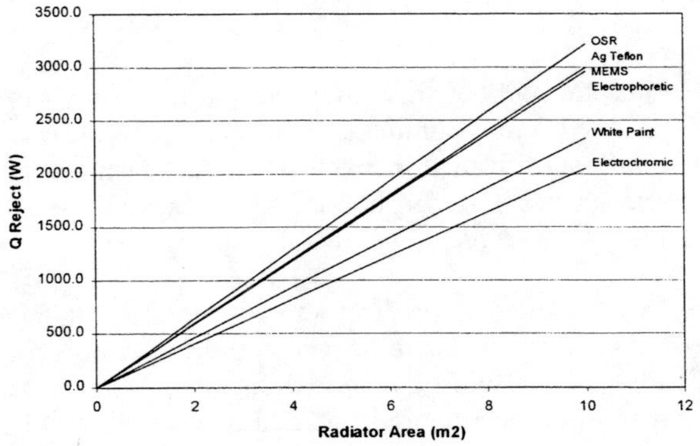

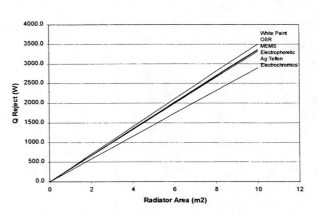

FIGURE 7. Heat Reject from Zenith at 30°C EOL / 13°β.

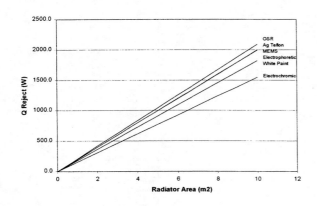

FIGURE 8. Heat Reject from Nadir at 30°C EOL / 13°β.

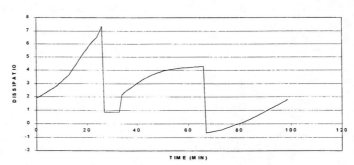

FIGURE 9. Heat Reject from Cold at 30°C EOL / 13°β.

FIGURE 10. EOS-AM Battery Dissipation Profile.

REFERENCES

J. L. Champion, R. Osiander, M. A. Garrison Darrin, T. D. Swanson, "*MEMS Louvers for Thermal Control*", 1999.

P. Chandrasekhar, "*Flexible, Flat Panel Visible/Infrared Electrochromics Based on Conduction Polymers for Space Applications*", Ashwin-Ushas Corp., Inc., 1998.

Thermal Management Approaches for
Large Planar Phased Array Space Antennas

Fred M. Jonas

Nichols Research, 2201 Buena Vista SE, Suite 203, Albuquerque, NM 87106
Air Force Research Laboratory, Phillips Research Site, Space Vehicles Directorate, Active Sensors Group,
AFRL/VSSS
(505) 853-6145, jonasf@nichols.com

Abstract. Large planar phased array antennas proposed for future space system applications in LEO offer unique challenges relative to the efficient management and control of thermal energy. Because of their proposed size, 100's of square meters, these antennas end up being excellent thermal radiators that must be kept warm enough to operate when needed in eclipse, but will most likely need to dump energy in full sunlight. Also, these antenna designs tend to be ultra-lightweight (low mass areal densities) which means that any proposed method of thermal control will not have any inherent system mass to work with in the basic antenna design. And, any proposed thermal management and control approach cannot add significant mass or power to the already tight electro-mechanical design power and mass goals for these future antennas. These opposing constraints mean that more traditional thermal management and control approaches such as heaters, heat pipes, and MLI may not be directly applicable to these antenna designs because of size, weight and/or power restrictions. This paper presents the results of a 1-D thermal model for examining the effects of proposed new thermal management approaches, such as variable emissivity coatings, on the expected thermal performance of proposed antenna concepts.

BACKGROUND

Current space based radar (SBR) efforts have led to the development of many proposed concepts. These concepts range from low earth orbit (LEO) constellations to medium earth orbit (MEO) and geosynchronous earth orbit (GEO) combinations, and concepts and proposed modes of operation from monostatic to bistatic to multi-static. Common to many of these SBR concepts are large planar phased array antennas, tens' of meters in size. As previously identified (Jonas, 1999; King, 1998), the development of these large aperture antennas is one of the pacing technical challenges to fielding an affordable SBR. One of the main challenges to this latter effort is making the antennas extremely lightweight while retaining the rigidity (flatness) necessary to perform precise antenna functions (radar, communications). The total antenna areal density goal for SBR and other space applications is 4-6 kg/m^2. Current technology for space-based phased array antennas is represented by Iridium at ~24 kg/m^2. Great strides have been made in meeting these areal density goals (Adler, 1998; Jonas/Staggs, 1998 and 1999).

It has been previously shown (Jonas, 1999) that these large flat phased array antennas make wonderful thermal radiators. This, combined with the distributed nature of phased array antennas (distributed power/energy sources) and relatively low power as compared to terrestrial/airborne applications, leads to the primary issue of the antenna being too cold to operate when in eclipse. The primary design driver of reduced weight (reduced mass) exacerbates this issue through removing the thermal mass one might use to retain and store energy for use in eclipse. Heat pipes, mechanical louvres and other traditional means of thermal management and control add significant mass and power requirements to already aggressive goals in these areas, and may not even prove to be feasible (or desired) for application. The challenge then is to develop an integral design that addresses all issues, including thermal, early in the design process to ensure synergisms occur minimizing weight and power impacts. This paper addresses the potential top level thermal challenges associated with these large antenna concepts in order to facilitate incorporation of thermal management concepts early in the design process. Specifically, the potential payoff of variable emissivity is presented along with the continuing need to maintain antenna temperatures in an appropriate range for operation of the electronics during the cold parts of an orbit (eclipse).

CP504, *Space Technology and Applications International Forum–2000*, edited by M. S. El-Genk

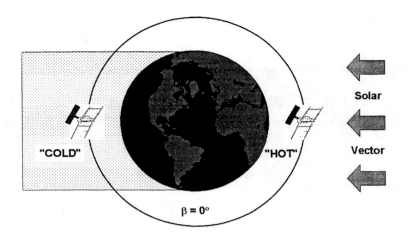

FIGURE 1. Hot and cold dwell for thermal analysis, full sun and in eclipse.

INTRODUCTION

Maintaining a reasonable thermal environment on board a spacecraft/satellite is an integral part of the space vehicle design. Mechanical and electronic subsystems and materials require certain temperature ranges be maintained over the life of the space vehicle in order to operate. Maintaining that environment, as well as controlling the temperature extremes, gradients and/or the effects of differential heating are the job of the thermal management system. As already discussed, to effect synergisms in the design process, the thermal issues should be considered early in the concept development to avoid major impacts after the design has solidified. This allows the system level designer to take advantage of incorporating thermal management concepts smartly in the design, and/or identifying challenges to be addressed, early in the design process. This paper examines those issues for proposed SBR LEO concepts employing large planar phased array antennas by analyzing two thermal extremes expected in the space environment. These extremes, steady state dwell, bound the problem (detailed transient and orbital analyses should of course be performed as the design progresses) and lead one to consider applicable thermal management approaches. Those extremes examined here are with the antenna exposed fully to the sun with full earth reflection (hot case) and in full eclipse (shadow, cold case). The two extremes are depicted in Figure 1. The goal is to maintain the temperature in a comfortable range for operation of the electronics on the antenna (typically -25 °C to +40 °C with the extreme range being -40 °C to +60 °C) by reducing both the overall temperatures and temperature extremes for these two cases.

THERMAL MODEL

A simplified thermal model employing the energy equation for both radiative and conductive (through the antenna) heat transfer was used to analytically determine the antenna surface temperature extremes. Sources and sinks are as shown in Figure 2. As shown in Figure 2, the energy equation is given by the following:

$$Q_{SUN} + Q_{ER} + Q_{INT} = Q_E + Q_R, \tag{1}$$

or, on a per unit area (A) basis, where $q \equiv Q/A$:

$$q_{SUN} + q_{ER} + q_{INT} = q_E + q_R. \tag{2}$$

SPACECRAFT ENERGY BALANCE

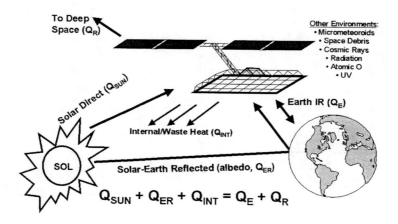

FIGURE 2. Spacecraft energy balance.

The following expressions were employed to analytically evaluate Equation (2) in order to determine the representative antenna surface temperatures (f is front, earth or nadir facing, and b is back of antenna):

$$q_{SUN} \approx [\alpha_f \, F_{S\text{-}f} \cos\theta_f + \alpha_b \, F_{S\text{-}b} \cos\theta_b] \, I_{SUN}, \tag{3}$$

$$q_{ER} \approx a \, \alpha_f \, F_{S\text{-}SE\text{-}a} \cos\theta_{ant} \, I_{SUN}, \tag{4}$$

$$q_{INT} \approx RF_{out} \, (1-e)/e + q_{heater}, \tag{5}$$

$$q_E \approx \sigma \, F_{S\text{-}E} \, (\varepsilon_f \, T_f^4 - \varepsilon_E \, T_E^4), \tag{6}$$

$$q_R \approx \sigma \, F_{S\text{-}Sp} \, (\varepsilon_f \, T_f^4 - \varepsilon_{sp} \, T_{sp}^4) + \sigma \, F_{S\text{-}Sp} \, (\varepsilon_b \, T_b^4 - \varepsilon_{sp} \, T_{sp}^4). \tag{7}$$

Note that the internal heat energy transfer, q_{INT}, is composed of two terms. The first is the "waste" energy from the conversion of energy into RF that needs to be disposed while the second term is a placeholder for any energy that might need to be added, by currently undefined means, in order to keep the antenna warm. Both terms combined also lead to the conductive term for the transfer of energy through the array (and enables the analytic solution),

$$q_{INT} \approx k \, (\Delta T) \, / \, t, \tag{8}$$

where (ΔT) is the temperature difference between the front and back of the antenna and t is the thickness. Note that for these future phased array antennas, thicknesses of these multi-layer antenna panels (Jonas/Staggs, 1998 and 1999) will be on the order of 1-2 cm, with the majority of the material being foam-like, thus very little thermal mass.

For the purposes of this analysis it was assumed that $a \approx 33\%$ (nominal value), $\varepsilon_E \approx 1$, $T_E \approx 290$ K, $T_{sp} \approx 0$ K, and that energy and properties are evenly distributed over the antenna surface. For example, this latter condition assumes that the energy dissipated from the electronic devices producing RF on the antenna, and/or energy added via "heaters" has been distributed evenly at the panel level. Analytic results are shown in Figure 3 for median antenna surface temperatures.

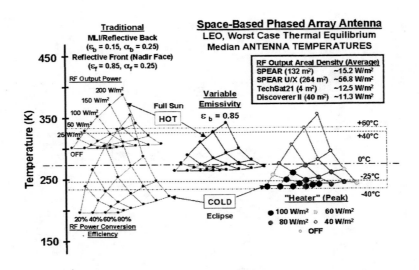

FIGURE 3. 1-D steady-state analytic results for hot and cold dwell, median antenna surface temperatures.

DISCUSSION

Shown in Figure 3 are representative median antenna surface temperatures versus RF power output (power areal density, W/m^2) and the associated input energy to RF power conversion efficiency at the antenna (20% being representative of today's technology for X-band, 40-50% projected for the near future). Four steady state dwell cases are shown: two in full sunlight and full earth reflection (Hot) and two in eclipse/shadow (Cold). The "traditional" hot and cold cases are representative of using some type of highly reflective insulator on the back (e.g., MLI with reflective surface) and a highly reflective and emitting surface on the front (nadir) face -- passive thermal control only, consistent with today's technology. The additional hot case depicts what happens if the back surface emissivity increased (variable emissivity) while the additional cold case depicts additional energy requirements to warm the antenna, both into the comfort zone for electronics. For reference purposes, the extreme and normal specification for the operation of electronic components is shown as –40 to + 60 °C and -25 to +40 °C respectively.

The main observation to be made concerning Figure 3 (Jonas, 1999) again is that these flat planar antennas are wonderful thermal radiators (the bigger the better) and have little difficulty radiating away the excess energy generated on the antenna up to extremely high RF output power areal densities. For comparison sake, the nominal emitted RF power areal densities projected for various SBR concepts are shown and are in the range of ~50 W/m^2 or less. Increasing the efficiency of RF power conversion simply means that there is less excess energy to get rid of (helps in the "hot" case but hurts in the "cold" case). The real payoff of increased efficiency is of course the reduction in prime power requirements (or more RF emitted for the same prime power).

Hot Case

Relative to thermal management, operation in full sunlight can apparently be maintained for the projected range of RF power output conditions with little effort. However, as shown in Figure 3, operation is at the hot end of the spectrum for the electronics. Even though this is within the specifications, cooler is better ("heat kills" according to one of my electrical engineering colleagues). I also anticipate that future RF power levels will increase since radar performance is dictated primarily by power and aperture. In order to dump more energy to cool off the antenna the backside emissivity was increased, from 0.15 to 0.85. As seen in the figure, this is more than sufficient and moves the antenna operation into a comfortable temperature range for nearly all operating conditions. Achieving variable emissivity is the issue. Today's technology is mechanical louvres that open or close to reject or retain heat energy. For proposed SBR antenna concepts this would add significant mass and is simply not a viable approach.

An emerging technology is variable emissivity coatings/films. Although not ready for application today in space, projections are that they should be ready for by the 2010 timeframe. Projected performance for these materials are emissivity ranges from 0.1 to 0.9 with low solar absorptivity, $\alpha < 0.4$. Proposed approaches include electrochromic thin films and conducting thin film, solid-state, polymers. All approaches involve negligible weight and power, on the order of 90% weight reduction over mechanical louvres while operating at power levels on the order of ~50 mW/m^2. This appears to be ideal for proposed SBR antenna concepts and other future space system applications.

Cold Case

The major problem will be keeping the antennas warm enough in eclipse, especially for power off conditions. This appears to be the main design challenge at the system level. And, the challenge is to keep the antenna electronics warm enough without adding significant mass beyond the overall areal density goal for the entire antenna structure, nor without adding significant power demands. Both of these constraints are in direct opposition to the need for keeping the antenna warm -- energy is either retained by thermal mass and/or added by "heaters" through providing additional power. Note that for the latter, simply keeping the antenna on does not solve the problem -- additional energy still needs to be added in order to keep the electronics warm enough for operation. The amount of energy that needs to be added (for this steady state dwell) is shown in Figure 3 as the "Heater" (Peak) case. "Heater" is in quotes to indicate that heat energy is being added by some undefined means. "Peak" is indicated in that in some sense this energy requirement at the coldest part of the orbit represents a peak power demand. The real problem is of course transient. For a satellite that comes out of sun into eclipse, there will be a transition of "heater" power demands as the satellite moves through the eclipse.

As can be seen in Figure 3, "heater" values range from 100 W/m^2 to power off depending on RF output power and conversion efficiency. Note that "heater" values chosen simply represent the energy requirements necessary to move antenna temperatures into the operating range for the electronics -- no attempt was made to optimize these values. The upper value of 100 W/m^2 represents a significant energy input, 13.2 kW total power required for a SPEAR concept (calculations by others for SPEAR-like concepts in LEO have indicated that average power required in eclipse is on the order of 2.5 to 5 kW depending on the materials). This is in addition to nominal antenna operation which is only 2 kW radiated for a 10 kW input, i. e., it takes as much peak energy to keep the antenna warm as it takes to nominally operate the radar (actual "average" power available to the radar in the SPEAR concept was proposed to be 25 kW during 20% of the orbit for radar operations with allowable 50 kW peak demands -- power numbers for conventional means continue to evolve but 13.2 kW is still a significant demand on top of the normal radar operations). Thus, even though the values presented in Figure 3 represent a cold soak and should be worst case, keeping the antenna warm during eclipse represents a significant challenge.

Current approaches for "heaters" are basically resistive heating which is typically not an energy efficient process. Another possibility may be represented by phase change materials (PCMs). PCMs are not new to thermal control but need to be developed and qualified for long-lived space applications. Liquid-solid PCMs present containment problems while solid PCMs are generally corrosive and adsorb/release water during phase change. There are potential dry materials (e.g., solid-solid organic phase change materials and micro-encapsulated solid-liquid phase change composites) that overcome these difficulties that need further investigation before application to space systems. Regardless, in this application, conceptually the PCM would store thermal energy from the "hot" part of the orbit for use during the cold parts of the orbit. The PCM would of course have to be tailored to each specific application. This is certainly an area for further research and development. While PCMs may appear attractive, they do add mass dependent on the overall energy requirements -- it is not clear what savings may be had at this time relative to resistive heaters, if any. The application of PCMs needs to be further investigated.

SUMMARY

The purpose of this paper was to examine the expected thermal environment extremes to be encountered by planar phased array antennas proposed for SBR concepts; and, examine design approaches to the thermal management approach that will take maximum advantage of available design synergisms now, early in the process. Specifically, "hot" and "cold" cases were examined along with potential approaches to mitigate the temperature extremes. The goal of course, is to bring the operating temperature into regions where the electronics can survive and operate. Considered for application were variable emissivity coatings/films, resistive heating and phase change materials. With little effort it appears that keeping the antenna cooler in the hot part of the orbit can be easily accomplished. Variable emissivity on the backside works, moving the operating temperature into the comfortable range for nearly all proposed operating conditions. Operation in the cold part of the orbit requires energy input to keep the antenna warm. This is in addition to simply keeping the antenna on. Input energy can be added by resistive heaters, and perhaps phase change materials that store energy during the hot part of the orbit. Both add either mass and power to the system, and the benefits of either are not clear at this point. This is an area for further investigation -- to find an elegant efficient means of keeping these antenna concepts warm enough for operation in eclipse.

NOMENCLATURE

$a \equiv$ albedo
$e \equiv$ efficiency of RF signal power conversion
$F \equiv$ view factor, sun/earth/space to front and/or back of antenna
$I_{SUN} \approx$ solar constant, 1400 W/m^2
$k \equiv$ thermal conductivity, $(W\text{-}m)/(m^2\text{-}K)$
$RF_{out} \equiv$ radiated RF energy from antenna per unit area (W/m^2)
$q_E \equiv$ antenna – earth energy exchange (W/m^2)
$q_{ER} \equiv$ sun-earth reflected energy input (W/m^2)
$q_{INT} \equiv$ waste energy from RF signal power conversion and added energy (W/m^2)
$q_R \equiv$ antenna – deep space energy exchange (W/m^2)
$q_{SUN} \equiv$ direct energy input from sun (W/m^2)
$T \equiv$ temperature (K)
$t \equiv$ thickness (m)
$\alpha \equiv$ absorptivity
$\varepsilon \equiv$ emissivity
$\sigma \equiv$ Stefan-Boltzmann constant, 5.67 (10^{-8}) $W/(m^2 – K^4)$
$\theta \equiv$ angle sun vector makes with antenna surface normal
$\theta_{ant} \equiv$ antenna orientation relative to earth normal

The author would like to acknowledgement the support of the Air Force Research Laboratory Active Sensors Group, AFRL/VSSS, 1Lt Eric Johnson and Dr. Ron Blackledge; and, the Space Based Radar team to include Mr. Jim Staggs, Mr. John Garnham and Mr. Denis Keiley.

REFERENCES

Adler, Mikulas, Hedgepeth, Stallard and Garnham, "Novel Phased Array Antenna Structure Design" in the IEEE Aerospace Conference Proceedings, 1998, CD-ROM Edition (Session 2.302, fp213.pdf), Snowmass at Aspen, CO, 21-28 Mar, 1998.

Jonas, "A Quick Look at the Expected Thermal Environment Extremes for SBR LEO Concepts," in the Space Technology & Applications International Forum Proceedings (STAIF-99), Albuquerque, NM, 31 Jan – 4 Feb, 1999, pp 753-758.

Jonas, Staggs and Johnson, "The Transmit/Receive Antenna Module II (TRAM II)", to be presented at the 1999 AIAA Space Technology Conference & Exposition, Paper No. AIAA-99-4562, Albuquerque, NM, 28-30 Sep, 1999.

Jonas, Staggs, King and Southall, "The Transmit/Receive Antenna Module (TRAM)", in the Proceedings of the Infrared Information Symposia, 3rd NATO/IRIS Joint Symposium, Vol. 43, No. 2, Quebec Canada, 19-23 Oct, 1998, pp 113-123.

King, Sherrod, Garnham and Blackledge, "Space-Based Radar for the Next Century", in the Proceedings of the Infrared Information Symposia, 3rd NATO/IRIS Joint Symposium, Vol. 43, No. 2, Quebec Canada, 19-23 Oct, 1998, pp 177-195.

Oscillatory Accelerations on Gas-Liquid Systems

Régis Wunenburger[1], Daniel Beysens[2], Carole Lecoutre-Chabot[1], Yves Garrabos[1], Pierre Evesque[3] and Stephan Fauve[4]

[1]*Institut de Chimie de la Matière Condensée de Bordeaux, UPR 9048 Centre National de la Recherche Scientifique, Avenue Dr. A. Schweitzer, 33608 Pessac Cedex, France*
[2] *Service des Basses Températures, DRFMC, CEA-Grenoble, 17 rue des Martyrs, 38504 Grenoble Cedex 09, France*
[3]*Laboratoire de Mécanique des Sols Structures Matériaux, URA CNRS D08500, Ecole Centrale de Paris, Grande Voie des Vignes, 92295 Chatenay Malabry Cedex, France*
[4]*Laboratoire de Physique Statistique, Ecole Normale Supérieure, 24 rue Lhomond, 75231 Paris Cedex 05, France*

+33-5.56.84.84.62; wunenbu@icmcb.u-bordeaux.fr

Abstract. We study the stability of liquid-gas interfaces with small density difference, subjected to high frequency vibrations (~60 Hz). We used CO_2 slightly below its critical point in order to vary in a scaled way the density difference and interfacial tension of the vapor and liquid via temperature. When the direction of vibrations is parallel to the interface, above a given velocity threshold an interface instability is observed, with the interface modulated as a "frozen" wave pattern. It is found that the wavelength and the amplitude of the stationary wave-like profile are both increasing functions of the frequency and amplitude of the vibration, and that they are proportional to the capillary length. Our measurements are consistent with a model of inviscid and incompressible flow averaging the effect of the vibration over a period and leading to a Kelvin-Helmholtz-like instability mechanism due to the two fluids relative motion. Under zero gravity (experiment onboard MiniTEXUS 5 ESA sounding rocket in Feb. 98), we observe that gas and liquid phases become structured into alternate layers perpendicular to the vibration. Preliminary results on the formation of alternate layers, and the influence of the vibrational parameters on the characteristic features of the multilayered structure are reported. The periodic acceleration seems to act as an "artificial" gravity.

INTRODUCTION

In presence of gravity the equilibrium shape of a fluid interface minimizes the sum of the fluid potential energy and interfacial energy. The characteristic length scale of the interface shape is the capillary length l_C, defined as

$$l_C = \sqrt{\frac{\sigma}{g(\rho_2 - \rho_1)}}, \qquad (1)$$

where σ is the interfacial tension, g is the gravity, ρ_2 (respectively ρ_1) is the density of the more dense fluid (respectively less dense fluid). When the typical length L of the fluid interface (e.g. the diameter of a gas bubble immersed in a liquid) is much smaller than l_C, the interface is nearly spherical. When L is much larger than l_C, the interface is flat, perpendicular to the gravity, except near the container walls where it is deformed on a length scale of the order of l_C. In absence of gravity, any fluid interface is spherical.

When oscillatory accelerations (vibrations) are applied to inhomogeneous fluid systems, their effects depend on the main characteristic features of the vibrations (frequency, amplitude) but also on their direction with respect to gravity. One of the most studied effects of vertical vibrations (i.e. parallel to gravity) is the parametric amplification of the interface oscillations of frequency equal to half the excitation frequency (Faraday instability). Under strong high frequency vertical vibration, the phase distribution of the less dense fluid below the more dense fluid is stabilized. This corresponds to the suppression of the Rayleigh-Taylor instability (Wolf, 1969).

Less studied are the effects of high frequency horizontal vibrations on fluid interfaces. Such vibrations applied to a vessel containing two immiscible fluids lead to a sine-like shape deformation of the interface, motionless in the

CP504, *Space Technology and Applications International Forum–2000*, edited by M. S. El-Genk
© 2000 American Institute of Physics 1-56396-919-X/00/$17.00

reference frame of the vibrated vessel (as opposed to the standing wave excited by the Faraday instability), as first reported by Wolf (Wolf, 1969). Lyubimov and Cherepanov (Lyubimov, 1987) studied the linear and weakly non linear hydrodynamic model of this instability and predicted its threshold. This instability mechanism is a Kelvin-Helmholtz type, because of the relative motion of the two fluids induced by vibrations. Ground experimental investigation on this instability was performed using a CO_2 gas-liquid mixture near its critical point, where the interfacial tension, the density difference between the two phases and the capillary length vanish according to the universal scaling laws of critical phenomena. Our experimental results presented in this paper are coherent with the model of Lyubimov and Cherepanov (Wunenburger, 1999). In the case of the Faraday instability (Fauve, 1992), as well as in the case of this Kelvin-Helmholtz-like instability, the most unstable mode of the deformation which grows at the instability onset scales with the capillary length, which diverges when gravity tends to zero. A question that naturally arises is the behavior of an inhomogeneous fluid system submitted to vibrations in absence of gravity. Indeed the capillary length is no more the pertinent length scale for the interface deformations. To answer this question, we have performed this vibration experiment in microgravity during a sounding rocket parabolic flight.

In part II of this paper we present the experimental setup. In part III we summarize the results of the experimental investigation of the effect of high frequency vibrations applied on a gas-liquid interface on the ground. The observations concerning the experiment performed in microgravity are analyzed and discussed in part IV.

EXPERIMENTAL SETUP

The experimental setup (TEM-FER module) was constructed by DASA (Bremen, Germany), Ferrari (Modena, Italy) and Techno System (Naples, Italy) as a space module to be flown onboard the MiniTEXUS 5 sounding rocket (from European Space Agency sounding rockets program). It provided 3 minutes duration of microgravity (10^{-5} g). The apparatus is composed of a thermostat containing a cell with three sample fluid volumes, vibrated by a shaker. The thermostat has a temperature accuracy of 1mK close to the critical temperature T_c with an operating range of [T_c, T_c - 150mK]. The shaker can apply vibrations of amplitude a and frequency f with the following values : a = 0.1mm, 0.3mm, 1mm, 2.5mm and f = 1Hz, 3Hz, 10Hz, 30Hz, 57.5Hz (a and f can be varied independently). It is noteworthy that the shaker produces vibrational energy in the vibration direction and with a smaller amount, approximately 10 times less, in the plane perpendicular to the vibration direction. The cylindrical fluid volumes (of diameter equal to 10.8 mm and of thickness equal to 10.8 mm) are machined in a CuCoBe alloy cell and closed by two parallel sapphire windows (of thickness equal to 9 mm). A photodiode continuously illuminates the samples by light transmission through the cell for vibration frequencies smaller than 30 Hz. For higher vibration frequencies the cell is illuminated by a stroboscope with a flash duration of 1ms. A stroboscopic frequency of 14 Hz (resp. 28 Hz) at a vibration frequency of 30 Hz (resp. 57.5 Hz) is chosen so that different phases of the vibration could be investigated at these frequencies, and also to check whether the observed patterns are stationary in the reference frame of the cell.

The critical point of CO_2 is defined by the critical temperature T_c=304.13K, the critical density ρ_C =468 kg.m^{-3} and the critical pressure P_C=7.37 MPa. The experimental study on the ground was performed near the critical point as characterized by the normalized density of each fluid sample and the normalized temperature distance:

$$0 \leq \delta\rho^* = \frac{\langle\rho\rangle - \rho_c}{\rho_c} \leq 0.05 , \tag{2}$$

$$4.10^{-6} \leq \tau = \frac{T_C - T}{T_C} \leq 5.10^{-4} , \tag{3}$$

where $\langle\rho\rangle$ is the mean density of the fluid sample. Three experimental fluid samples filled at three different mean densities $\langle\rho\rangle$ were used (see Table 1).

TABLE 1. Normalized density differences, T_{COEX} - T_C of the three fluid samples, and flight experiment parameters.

position on pictures of the fluid samples and labels	left 1	middle 2	right 3
$\delta\rho^*=(\langle\rho\rangle-\rho_c)/\rho_c$	0.05	0.028	0 ($\pm 10^{-3}$)
T_C-T_{COEX} (mK)	7.1	1.3	0
T_C-T (mK) (flight experiment)	15	2.3	7
volume fraction (flight experiment)	0.11	0.095	0.5

GROUND EXPERIMENT

A stationary sine-like shape profile in the reference frame of the cell was observed only for the following pairs of parameters (a,f) = (2.5mm, 30Hz), (1mm, 30Hz), (0.3mm, 57.5Hz), (1mm, 57.5Hz) and (2.5mm, 57.5Hz). On Fig. 1 are shown snapshots of a typical profile (Fig. 1a) and a typical top view (Fig. 1b) of the interfaces observed in the three fluid samples for a vibration at (a,f) = (1mm, 57.5Hz), at two fixed temperatures common to the three samples.

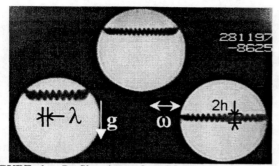

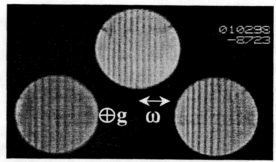

FIGURE 1a. Profile view of a CO_2 gas-liquid interface submitted to a vibration perpendicular to gravity. The left cell is 34 mK below the critical temperature, the middle cell at 13 mK, the right cell at 18 mK. The interface is steady in the reference frame of the cell.

FIGURE 1b. Top view of the gas-liquid interface. The left cell is 78 mK below the critical temperature , the middle cell at 57 mK, the right cell at 62 mK.

The variations of the wavelength λ (as defined on Fig. 1a) measured at different pairs of parameters (a,f) as a function of the distance to the critical point T_C-T are displayed in Fig. 2a. These variations are due to the evolution of the surface tension and of the density difference between both phases when T varies. It is noteworthy that λ is an increasing function of both f and a.

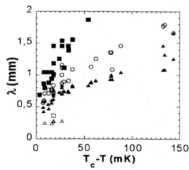

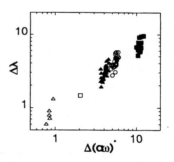

FIGURE 2a. Variations of the wavelength λ (mm) of the stationary profile as a function of the distance to the critical temperature T_C-T (mK) for various pairs of vibration parameters (f, a): f = 57.5 Hz, a = 2.5 mm (■); f = 30 Hz, a = 2.5 mm (O); f = 57.5 Hz, a = 1 mm (▲); f = 57.5 Hz, a = 0.3 mm (Δ); f = 30 Hz, a = 1 mm (□).

FIGURE 2b. Variations in logarithmic scales of $\Delta\lambda^*=(\lambda-\lambda_0)/\lambda_0$ as a function of $\Delta(a\omega)^*$. λ is the wavelength of the stationary profile observed at various temperatures for various pairs of vibration parameters (f, a): f = 57.5 Hz, a = 2.5 mm (■); f = 30 Hz, a = 2.5 mm (O); f = 57.5 Hz, a = 1 mm (▲); f = 57.5 Hz, a = 0.3 mm (Δ); f = 30 Hz, a = 1 mm (□). λ_0 is the predicted wavelength at the instability onset. $\Delta(a\omega)^*$ is the normalized distance to the instability onset, and $(a\omega)_0$ is the velocity threshold (see text and Eq. 5).

The Lyubimov and Cherepanov model assumptions are that the flow is irrotational, incompressible and vibrations are at high enough frequency to neglect viscous effects on the time scale of the vibration period. It predicts that a perturbation of wavelength λ becomes unstable when

$$(a\omega)^2 \geq \frac{1}{2}(a\omega)_0{}^2(\frac{\lambda}{\lambda_0}+\frac{\lambda_0}{\lambda})\tanh(\frac{\pi H}{\lambda}),\qquad(4)$$

with $\lambda_0 = 2\pi l_C$, H the height of the vessel, $\omega = 2\pi f$, and

$$(a\omega)_0^{\ 2} = \frac{(\rho_2 + \rho_1)^3}{\rho_1\rho_2(\rho_2 - \rho_1)}\sqrt{\frac{\sigma g}{\rho_2 - \rho_1}}\ , \tag{5}$$

is the threshold velocity where the instability onset is reached. As shown by Eq. 4, the first mode that is destabilized at the onset is of wavelength λ_0. The corresponding shape of the interface is sine-like and is steady in the reference frame of the vibrated vessel ("frozen wave"). For $\frac{\rho_2}{\rho_1} < 3.53$ the bifurcation is supercritical (Lyubimov, 1987), and the amplitude of the interface deformation is also proportional to the capillary length and should scale with the normalized distance to the onset defined as :

$$\Delta(a\omega)^* = \frac{\sqrt{(a\omega)^2 - (a\omega)_0^{\ 2}}}{(a\omega)_0}. \tag{6}$$

We note that contrary to the Faraday instability, the order parameter of the bifurcation is the vibration velocity and not the vibration acceleration.

Both λ_0 and $(a\omega)_0$ depend on the thermodynamical properties of the two-phase system. Considering the small distances to the critical point at which the experiment is performed, we assume that $\Delta\rho^*$ and σ both depend on the temperature according to the asymptotic scaling laws predicted from the critical phenomena theory (Stanley, 1971):

$$\Delta\rho^* = \frac{\rho_L - \rho_V}{2\rho_C} \approx B\tau^\beta\ , \tag{7}$$

$$\sigma \approx \sigma_0\ \tau^{2\nu}\ , \tag{8}$$

where $\Delta\rho^*$ is the normalized density difference between the liquid phase and the vapor phase, $\beta = 0.325$ and $\nu = 0.63$ are the universal critical exponents; $B = 1.6$ and $\sigma_0 = 7\ 10^{-2}$ N.m^{-1} are the respective values of the leading amplitudes for CO_2 (Herpin, 1974, Hocken, 1976, Sengers, 1978, Moldover, 1985). In the vicinity of the critical point, λ_0 and $(a\omega)_0$ take the exact form of a power law of τ :

$$\lambda_0 \sim \tau^{0.4675}, \tag{9}$$

$$(a\omega)_0 \sim \tau^{0.071}\ . \tag{10}$$

In our experimental temperature range λ_0 remains of the order of 0.1 mm, $(a\omega)_0$ remains of the order of 100 mm.s^{-1}, and the sound speed remains large enough to keep the acoustic wavelength corresponding to our vibration frequencies much larger than the vessel size. Due to the frequency range used in our experiment, the other assumptions of the model are also verified. Due to the discrete values of the vibrational parameters allowable by our apparatus, the instability onset could not be precisely detected, but no interface deformations were seen below the predicted onset. In order to take into account the variations of surface tension and density with temperature, we normalize the wavelength of the stationary profile to the capillary length. The reduced wavelength $\Delta\lambda^* = \frac{\lambda - \lambda_0}{\lambda_0}$ of the steady shape is plotted as a function of the normalized distance to the velocity threshold $\Delta(a\omega)^*$ in Fig. 2b. $\Delta\lambda^*$ seems to be proportional to $\Delta(a\omega)^*$. At the present time we have no explanation for this linear dependence.

MICROGRAVITY EXPERIMENT

During the flight experiment, the sample temperature was kept constant. Since the absolute value of the critical temperature is very sensitive to even minute impurities, during the ground tests the three samples were found to be at different distance from their own "critical" temperature while being thermostated at the same absolute temperature by the thermostat. Due to their different mean densities, the gas volume fraction was also different in the three sample cells. The flight experiment parameters are displayed in Table 1. We first present the main features of the initial state of the samples before the vibrations. After the rocket lift-off and its despinning at the end of the propulsive phase, the initially continuous fluid interface was broken into many small gas bubbles in samples 1 and 3, as indicated by the weakly transmitted light and the moving patterns of coalescing bubbles visible in Fig. 3 .

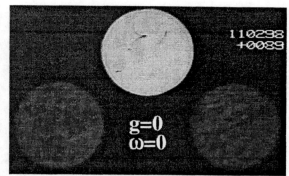

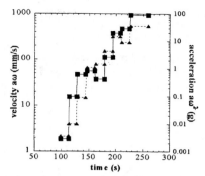

FIGURE 3. Cells 1, 2, 3, from left to right, after despinning of the rocket and before the beginning of vibrations under zero gravity.

FIGURE 4. Time evolution of the nominal vibration velocity $a\omega$ (■) and the vibration acceleration $a\omega^2$ (▲) during the parabolic flight. a is the amplitude and ω is the pulsation of the sinusoidal vibration imposed to the three sample cells.

The interface in sample 2 was deformed but not divided in many bubbles. This is because sample 2 exhibits a very weak density difference between the two phases making it less sensitive to the violent centrifugal acceleration during lift-off and despinning.

After 10 s at rest in microgravity, the samples were vibrated in a large amplitude and frequency range, corresponding to a large velocity and acceleration range, as showed in Fig. 4. Note that both the vibration velocity and acceleration increased monotonously during the flight. The parameters of the vibrations were changed every 15 s. Regardless of the initial state of the sample a periodic pattern of dark stripes perpendicular to the direction of the vibration appeared in the three samples at weak vibration intensity (a=2.5mm, f=3Hz). The pattern wavelength changed slightly with the vibration parameters during the flight. A typical array of stripes is visible in Fig. 5a in the three cells. In the samples 1 and 3, the dark stripes immediately reached their maximum length, i.e. the transverse dimension of the vessel. Simultaneously the turbidity decreased between the dark stripes, indicating that fewer gas bubbles were present between them. One can conclude that gas bubbles existing in the bulk of the sample concentrated in the dark stripes, where their coalescence was accelerated. In sample 2, for the vibration step (a = 0.3mm, f = 57.5Hz), the dark stripes appeared near the upper wall (see Fig. 5a) and their length was small. Considering that no initial (unstable) emulsion of gas bubbles was present in this sample, and that the gas volume was the smallest, one can deduce that in this sample the gas was located near the alloy walls of the vessel, separated from it by a liquid wetting layer. The length of the dark stripes in the sample 2 always increased with the intensity of the vibration up to their maximum length. The situation when they reached their maximum extension is visible in Fig. 5b. It is worthy to note that the evolution of the three samples was qualitatively the same, indicating that the effect of high frequency vibrations is very robust and does not depend on the initial state of the two-phase system.

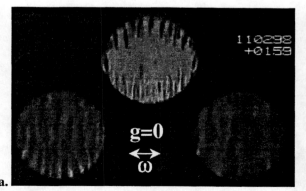
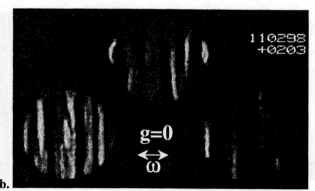

a.   b.

FIGURE 5. Alternate stratification of the gas and liquid phases when submitted to a periodic excitation (**a**): 57.5 Hz and 0.3 mm, (**b**): 57.5 Hz and 1 mm, directed left-right on the figure.

When vibration amplitude and frequency were changed, the gas inclusions moved inside the vessel and the period of the array of stripes (defined as the mean distance between two stripes) changed very slowly in time (despite the intensity of the vibrations). This long transitory regime can be due to the long time needed to transfer interstitial liquid through the thin wetting layers on the walls, when the distance between two inclusions of maximal length had to change. At the end of many vibration steps some gas inclusions were still moving slowly, indicating that the step duration may be too short to determine whether the steady state of the two-phase fluid system had been reached. The wavelength λ of the array of gas inclusions thus may not be the steady value. We measured λ by making a Fourier spectrum of a longitudinal cut of the image of the array, and by retaining the spatial period of the maximum of the spectrum. When plotted as a function of the vibrational velocity $a\omega$ (Fig. 6) λ is seen to increase with $a\omega$.

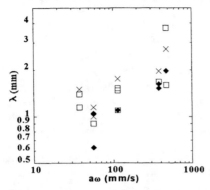

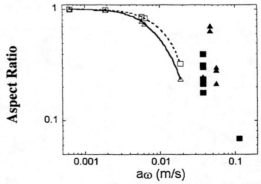

FIGURE 6. Wavelength of the periodic pattern observed in the three sample cells in microgravity as a function of $a\omega$ ($\square$ corresponds to cell 1, $\blacklozenge$ to cell 2 and $\times$ to cell 3).

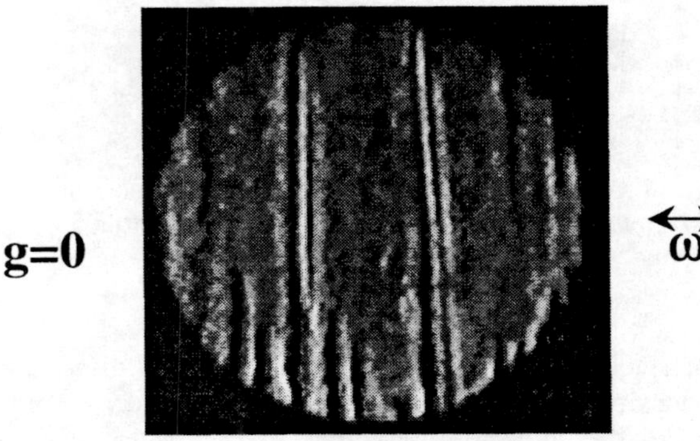

FIGURE 7. Snapshot of gas inclusions of ellipsoidal shape in sample 1 vibrated at $f = 57.5$ Hz and $a = 0.3$ mm.

When the contrast was good enough, some stripes also appeared to be transparent, and were found to be quasi-planar gas inclusions of ellipsoidal shape. The deformation of the gas inclusions can be seen in sample 1 at $f=57.5$ Hz and $a=0.3$ mm (Fig. 7). The gas inclusions become flatter when the vibration velocity increases. We define the aspect ratio of the gas inclusions as the ratio between the width (length in the direction of vibrations) and the height (length in the direction perpendicular to vibrations) of the gas inclusions observed as in Fig. 7. Lyubimov et al. (Lyubimov, 1997) predicted that the mean shape of a single bubble vibrated in microgravity is ellipsoidal, and that the aspect ratio of the bubble is a decreasing function of the vibrational parameter Q defined as

$$Q = a^2\omega^2 \frac{(\rho_L + \rho_V)R}{\sigma},$$
(11)

where R is the bubble radius. In Fig. 8 the aspect ratio of gas inclusions observed in sample cells 1 and 2 is plotted as a function of the vibration velocity $a\omega$, together with the predicted aspect ratio of a single bubble (of radius corresponding to a bubble volume equal to the total volume of gas existing in the sample cell at $T_C-T = 10$ mK) for smaller values of the vibration velocity (Roux, 1998).

FIGURE 8. Aspect ratio of gas inclusions observed in sample cells 1 ($\blacksquare$) and 2 ($\blacktriangle$) versus the vibration velocity $a\omega$. The predicted aspect ratio for a single bubble of volume equal to the total gas volume at $T_C-T = 10$ mK existing in cell 1 ($\square$) and 2 ($\triangle$) is also plotted.

We note that the experimental points are slightly above the extrapolation of the theoretical curves. This can be explained by noting that for a given value of the vibration velocity $a\omega$, and for a given temperature, the Q value of a bubble of small radius (like the gas inclusions observed in Fig 7) is smaller than the Q value of a single large bubble, which is the case considered by Roux et al. (Roux, 1998). Hence the aspect ratio for a bubble of small radius is larger than for a single large bubble. A quantitative comparison between the experimental values of the aspect ratio and the model of Lyubimov et al. was not possible for three reasons. First, Q depends on the radius of the bubble, which could not be determined precisely because we could not measure the volume of the gas inclusions. Second, the exact variation of the aspect ratio versus Q depends strongly on the ratio between the radius of the bubble and the size of the cell, as shown by the two different theoretical curves in Fig. 8. Third, the interaction between the gas inclusions may change their aspect ratio. Nevertheless Fig. 8 shows a good qualitative agreement between measurements and the model of Lyubimov et al. (Lyubimov, 1997).

CONCLUSIONS

Surprisingly, despite its rich phenomenology discovered by Wolf (Wolf, 1969), the effect of high frequency oscillatory accelerations on fluid interfaces has not been much studied yet, and the experimental results presented in this paper have to be considered as a first step in the comprehension of the observed phenomena. We studied the effect of high frequency horizontal vibrations on the interface in a near-critical CO_2 gas-liquid equilibrium, whose density difference and interfacial tension could be changed in a scaled way by tuning the temperature. The ground experiments confirmed the importance of the capillary length as the relevant length scale for the mean deformations of the interface, and of the vibrational velocity as the driving parameter of the interface instability. In particular, a velocity threshold for the appearance of a steady sine-like deformation at the interface was identified.

The microgravity experiment showed that vibrations have many effects on a gas-liquid system. First, they deform the mean shape of gas bubbles from a spherical to an ellipsoidal one. Second, they accelerate the coalescence of an emulsion by concentrating the bubbles. Third, they structure the system by inducing a periodic distribution of gas inclusions oriented perpendicular to the direction of vibrations, surrounded by a continuous domain of the liquid wetting phase. The periodicity of the bubbles distribution suggests that the stretched bubbles interact through long-range velocity and pressure perturbations induced in the liquid matrix by their rapid relative motion and their deformation. The influence of interfacial tension, density difference, viscosity and volume fraction of the gas phase have now to be investigated in order to identify the interactions between the bubbles. A long term objective of these studies is to achieve the control of the position of fluid interfaces in absence of stabilizing gravity by means of vibrations. Vibrations thus seem to act as an "artificial gravity" to induce phase ordering.

ACKNOWLEDGMENTS

The authors thank B. Roux, D. V. Lyubimov and T. Lyubimova for useful discussions. This work has been supported by the European Space Agency and CNES. We thank all the MiniTEXUS 5 teams, especially for the technical support of DASA (Bremen, Germany), Ferrari (Modena, Italy), Techno System (Naples, Italy) and Swedish Space Corporation (Esrange, Sweden).

REFERENCES

Fauve, S., Kumar, K., Laroche, C., Beysens, and D., Garrabos, Y., *Phys. Rev. Lett.* **68**, 3160-3163 (1992).
Herpin, J., and Meunier, J. C., *J. Phys. (Paris)* **35**, 847 (1974).
Hocken, R., and Moldover, M. R., *Phys. Rev. Lett.* **37**, 29-32 (1976).
Lyubimov, D.V., Cherepanov, A., *Izvestiya A. Akademii Nauk SSSR, Mekhanika Zhidkosti i Gaza* **6**, 8-13 (1986), translated in *Fluid Dynamics* **86**, 849-854 (1987).
Lyubimov, D.V., Cherepanov, A.A., Lyubimova, T., Beysens, D., Meradji, S., and Roux, B. "Equilibrium and Stability of a Drop in a Vibrational Field", *in Proc. of the Xth European and VIth Russian Symposium of Physics and Science in Microgravity*, edited by Avduyevsky, V.S., and Polezhaev, V.I., Institute for Problems in Mechanics, RAS, Moscow, 1997, pp. 66-73.
Moldover, M. R., *Phys. Rev. A* **31**, 1022-1033 (1985).
Roux, B., Lyubimov, D.V., and Lyubimova, T., Final Report of CNES Contract N°96/CNES/0279 (1998).
Sengers, J. V., and Moldover, M. R., *Phys. Lett.* **66 A**, 44-46 (1978).
Stanley, H. E. *Introduction to Phase Transitions and Critical Phenomena.* Oxford, Clarendon Press (1971).
Wolf, G. H., *Z. Physik* **227**, 291- (1969).
Wunenburger, R., Evesque, P., Chabot, C., Garrabos, Y., Fauve, S., and Beysens, D., *Phys. Rev. E* **59**, 5440-5445 (1999).

Development of Liquid Flow Metering Assemblies for Space

A.A.M. Delil[1], P. van Put[2], M. Dubois[3], and W. Supper[4]

[1]National Aerospace Laboratory NLR, Emmeloord, Netherlands, Fax +31 527 248210, E-mail adelil@nlr.nl;
[2]Bradford Engineering, Heerle, Netherlands; [3]SABCA, Brussels, Belgium; [4] ESA-ESTEC, Noordwijk, Netherlands

Abstract. As it is impossible to directly use commercial liquid flow meters in space, a study was done for the European Space Agency to adapt commercial flow meter assemblies for spacecraft applications. The activities led to the selection of two commercial units, which were re-designed and adapted for use in spacecraft single-phase (water) and two-phase (ammonia) thermal control loops. These flow meter assemblies were tested according to an agreed test programme, that included performance and calibration tests in a test bench (developed during the study), vibration and EMC/EMI testing. The results are discussed to assess to what extent the study objectives were met. Recommendations for future work are given.

PROJECT APPROACH

Fluid loops aboard spacecraft often need Flow Meters (FMs) to monitor or control the performance of the loop or its components. Typical examples are FMs in: 1) single-phase thermal control loops, to control pump speed and valve settings to have the right mass flow, hence the right amount of heat collection, transport and rejection; 2) two-phase thermal control systems, to control pump speed and valve settings to optimise system response and stability, in conjunction with vapour quality sensors; 3) propulsion systems to monitor (by bookkeeping) the total of propellant used, providing input for end-of-life determination, optimising mission strategy, or decision on satellite replacement.

FM technology has dramatically evolved in the last decades, producing a wide range of methods and systems, each of them being specifically related to specific fields of application. However, it is difficult to directly use one of these FMs for space applications, as the sensor, the measurement chain or both are not suitable for such conditions. Goal of the study (by prime contractor NLR, Bradford/SPPS, SABCA) was adaptation of commercial Flow Metering Assemblies (FMAs) for space applications. The project logic (Table 1) followed the Statement of Work (SOW).

TABLE 1. Project Logic.

Literature and Product Search & Selection of Concept	**Phase 1**
Market Survey of Commercial Items	
Trade-off Analysis & Pre-Selection of Five Commercial Flow Meters & Development of Test Bench	
Trade-off and Selection of Water and Ammonia Flow Meters	
Detailed Design & Adaptation for Space Use & Definition of Test Programme	**Phase 2**
Manufacturing and Assembly & Testing	**Phase 3**
Assembly & Calibration & Functional Testing	
Evaluation of Test Results & Conclusions	
Final Reporting	

The study started with an overview of various flow meter concepts, being:

- Weighing/Volumetric
- Ultrasonic Doppler
- Thermal Calorimetric
- Variable Area
- Turbine
- Differential Pressure
- Vortex Shedding
- Rotary Displacement
- Laser Doppler
- Coriolis
- Laser Time of Flight
- Thermal Time of Flight
- Ultrasonic Time of Flight
- Electromagnetic

An inventory of available FMs was made. T72 commercial units found were screened and evaluated with respect to performance, accuracy and space qualification prospects. A FM trade-off matrix form (including ranking figures) was established. Based on this form and on the evaluation criteria and ratings trade-offs were done, yielding rating lists for ammonia, water and propellants as working fluids. Five FMs were pre-selected out of these lists. It was agreed with ESA to focus on FMs for water and ammonia, as the primary study goal was applications for single- and two-phase thermal control systems. The FMs went through a preliminary test programme, that included functional performance, accuracy, calibration, and temperature tests in a test rig specially developed within this study, and also vibration tests of the meters, while in operation. Based on test results, the two best ones were re-designed for actual spacecraft applications foreseen: an 8-80 g/s water single-phase and a 2-15 g/s ammonia two-phase thermal control system. A critical analysis of these two FMs identified modifications of mechanical parts and electronics, needed to meet the specified requirements. The re-designed/modified FMAs were tested according to a test programme (defined in this study), that included calibration and accuracy, leak and proof pressure, and EMC tests. From the test results it was concluded to what extent the study objectives were met. Recommendations for further work are given.

EVALUATION OF COMMERCIAL UNITS

Based on the SOW and the Technical Requirements Document (TRD), the important parameters were identified and listed on the form to be used for the trade-offs. This form contains weight factors and evaluation criteria, which will yield, when applied to the commercial units, a list of FMs sorted such that the best one is on top. This not necessarily means that the top ones meet all requirements. The target is to identify the most promising FMs to be adapted for applications in space.

An inventory of commercial FMs was made next. Information on 72 commercial FMs, gathered from brochures and direct contacts with suppliers, was stored on FM Data Evaluation Forms, one form per fluid per FM.

The relative importance of parameters was expressed by assigning weight factors. The weight factor values indicate that "Flexibility on changes for space applicationson" is considered to be the most important (weighed 12). This weight factor accounts for efforts related to structure, electronics, vibration, configuration, accuracy, temperature and lifetime. Other important trade-off parameters are "Lifetime, maintainability, calibration sensitivity" and "Accuracy" (both weighed 10), "Flow range" and "Leakage" (both weighed 8), "Temperature range of fluid", "Pressure drop and response time", "Structural strength", "Gravity & orientation" (all weighed 6) and "EMI & EMC" (weighed 5). Lower values are assigned to parameters, which are most likely to be improved by a spatialisation: "Mass", "Vacuum " (compatibility), and "Power" (all weighed 3), and also "Volume", "Mechanical environment", "Environmental temperature" (all weighed 2). "Purchase costs" are considered to be minor also (2), certainly when compared to "spatialisation" efforts, hence costs. The criteria to rate the FMs range from 0 (unacceptable) to 10 (excellent).

More than 200 evaluation forms were filled out. The trade-offs were done according to the above guidelines. The final score for each FM was calculated by addition of all products of Rating (R) and Weight Factor (W), normalised (dividing the resulting number by the maximum possible value), and then multiplied by 10. Final scores range from 0 (unacceptable) to 10 (excellent). Based on the scores five FMs (Table 2) were pre-selected, while keeping in mind that 1) at this stage the study was restricted to ammonia and water only; 2) the pre-selected five had to represent different concepts; 3) availability, co-operation of supplier, delivery time and price were important also.

TABLE 2. Pre-Selected Flow Meters.

Type	Manufacturer/Distributor	Principle	Ammonia	Water	Notation
DS012S	Rosemount / Brooks	Coriolis	x		DS012S
FTO	EG&G	Turbine	x		EG&G-FTO
PT868/6068	Panametrics	Ultrasonic		x	PT868
7283 series	ITT Barton	Turbine		x	Barton
DS025S	Rosemount / Brooks	Coriolis		x	DS025S

TEST BENCH

In parallel to the pre-selection activities, a test bench was to be built to be used both for the preliminary testing of pre-selected units and for the final acceptance testing of the spatialised FMAs at a later stage. This chapter describes the test bench (Fig. 1), the design and the characterisation of performance.

Liquid flow, produced by moving a piston inside the cylinder via the actuator, flows via valves (2V4, 2V5, 2V6, 2V9 and 2V11) and through the FM test section to the exchangeable tank. The test section can contain two FMs in series, together with a reference FM. Pressure differences across the FM's can be measured by differential pressure sensors, one for each FM. At the end of each test, when the actuated cylinder is empty, it is automatically refilled by applying the maximum system pressure (20 bar) to the exchangeable tank. The liquid then flows back through the valves 2V11 and 2V10 (valve 2V9 closed), in this way moving the piston back downwards in the cylinder. The pressure on the exchangeable tank is controlled by inert gas and a floating piston.

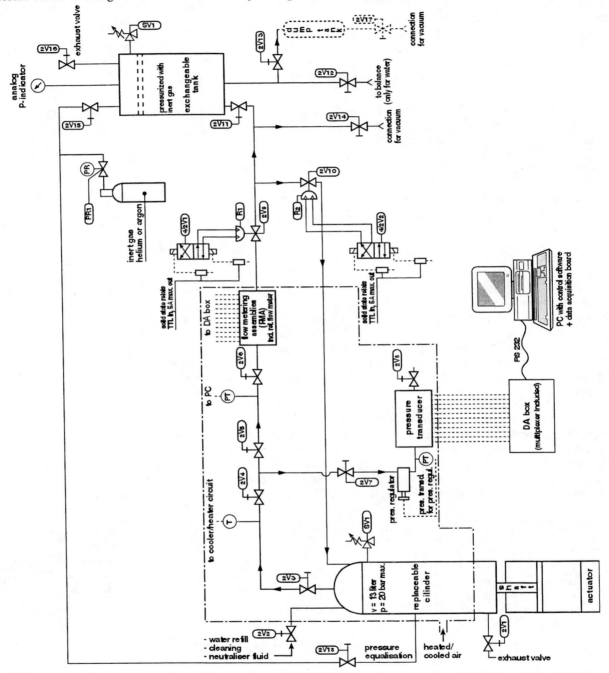

FIGURE 1. Test Bench.

The fluid temperature can be controlled by cooling or heating the test set-up, which is placed in a climat chamber (adjustable from -20 to +60 °C). The temperature of the FM section can be controlled by using an extra insulating module around this section. A vacuum connector, filters and a dump tank are integrated in the bench for tests with special fluids (e.g. propellants). The bench is partially automated to perform unattended tests, if required by safety.

The cylinder, designed for 20 bar (wall thickness 14 mm), consists of (AISI 304) steel to secure compatibility with the test fluids. The inner volume is 13 litres. The seals on the piston/shaft (minimum 3 in a row) are of Virgin PTFE on top of the cylinder. A pressure safety valve and an extra connection, for water refill and cleaning purposes, are also integrated in/on the cylinder. To vent the system a venting valve is installed on the cylinder and the cylinder is placed upside down.The exchangeable tank, also designed for 20 bar, is larger (20 litres) than the cylinder. The extra volume is used to provide a gas cushion on top of the floating (not sealed) piston, to give a good pressure distribution on the fluid. The tank (also AISI 304), has a spherical, seal-welded flange on the bottom where provisions are made for the entering liquid, a dump tank (for special fluids) and an exit to a balance (for weighing the fluid). The top flange has a PTFE seal and is bolted onto the cylindrical part of the tank. A venting valve, a pressure safety valve, a pressure indicator (safety) and a connection to a gas bottle, are integrated in the top flange. The bench structure, made of welded stainless steel profiles, is designed to handle all the expected forces and vibrations. All tubing is made of stainless steel, chosen to withstand at least 7 MPa. All couplings, elbows and valves (with PTFE seals) are standard Swagelok© items (stainless steel) and able to withstand the required maximum pressure. The isolation chamber is made of ISOCAP© (140 mm) panels compatible with the required temperatures. The extra insulation module around the FM section, is also made of these panels. The actuator is a stationary head with integrated drive, a travelling cross-head, two guiding columns and backlash free re-circulating ball screw bearings. The motor inside the actuator, equipped with a gearbox (ratio $1:10^6$), gives a pulsed output signal (encoder) to create an accurately feedback controlled movement.

Tests and computations were done to qualify the test bench. Qualification testing included: 1) Leak tests, proving that the bench is leak-tight up to 10^7 std cc/s Helium, consisted of a test at ambient temperature (Helium in the bench, sniffing outside) and a reverse test at -20, +20, and +60°C (Helium outside, vacuum inside); 2) Proof pressure tests up to 3.5 MPa, using nitrogen as pressurising gas, showing no physical damage, and the same performance before and after the testing: the bench meets the required operating pressure range; 3) Response time tests showing a test bench response time (time to establish a given flow, from zero flow, with accuracy better than ± 0.1 %) of less than 1 s (mostly an order of magnitude better) over the whole range; 4) Accuracy tests, showing that the test bench accurately handled flows between 10 and 256 g/s for water, between 0 and 15 g/s for ammonia, and yielding a flow accuracy of 0.025 % of reading. Therefore it was decided that the reference flow is not the flow measured by a reference FM (as originally planned), but will be the flow following from the displacement speed of the test bench piston, as the latter flow proved to be at least one order of magnitude more accurate than the flow derived from the best commercial FM.

PRE- & FINAL SELECTION AND RE-DESIGN OF SELECTED UNITS

All five pre-selected meters were compatible with water and ammonia, but they could not cover the mass flow rates required for both liquids. Therefore they were divided in meters for water and meters for ammonia, as shown in table 2. The five meters were subjected to a preliminary test programme: a check of the fluid compatibility, inspection of the physical dimensions, Helium leak test, proof pressure test and functional tests, which included measurement accuracy, response time, pressure drop, and power consumption. The latter tests were completed by vibration tests on meters in operation. For safety reasons the meters were tested with water only. All meters sustained the high level random vibrations. But they did not function properly during vibration. After vibration testing they again measured properly. But in some cases a rest time of more than 30 minutes or a re-set action became neccesary.

Trade-offs were done based on the test results for ammonia and for water. They showed that the EG&G-FTO turned out to be the most promising FM for ammonia, because of its flexibility with respect to "spatialisation", and the fact that, based on vibration test results of the Barton FM (being a more or less similar design), it can be assumed that the performance before and after vibration is the same. Also the repeatability and hysteresis were reported by the supplier to be small (0.1%), meaning that poor accuracy can be considerably improved by calibration.
Unfortunately after a first run with ammonia the meter broke down. The interior showed that rotating elements had disappeared due to corrosion, although the supplier had explicitly stated that the FM was ammonia compatible. Because of loss of credibility and the fact that the supplier could not deliver ammonia compatible items, it was decided to replace this meter by a similar ITT-Barton type, with stainless steel, ammonia compatible, internal parts. The results also showed that the Barton meter was most promising for water, because of its flexibility with respect to adaptation, extremely low pressure drop, acceptable accuracy, its identical performance before and after vibration,

and its relatively low response times. Identified actions for re-design were for both flow meters the investigation of lifetime and maintainability, extension of the flow range, and an electronics development for electrical interfacing and for fluid properties related correlations. In general it can be said that only minor hydraulical and mechanical changes were required (machining and surface treatment). The major effort concerned the electronics issues.

The final selection started with re-assessment of near-term FMA applications in space: in a single-phase water loop for COF or ISPR Rack (requiring flow rates ranging from 8 to 80 g/s, with a maximum pressure drop of 10 kPa), and two-phase ammonia loops with a heat transport range of 2.2 to 15 kW (ammonia flow rates from 2 to 15 g/s). The requirements were adapted as shown later in the verification matrices. It was concluded that: ITT Barton 7182 FM was the most appropriate for water (covering the 8-80 g/s range), ITT Barton 7506 FM the most appropriate for ammonia applications (2-15 g/s). Both FMs were to be adapted, provided that only minor development effort was expected for hydraulic parts of the water meter, and both ITT Barton meters could use the same electronics. Major effort was to be spent on the water meter, as near-term applications (COF and ISPR Rack) are foreseen.

ITT Barton 7506 (Fig. 2) is an electro-mechanical volumetric FM, using a Pelton wheel. Pulses are generated by the rotor, passing a pick-up coil, positioned above the wheel. The frequency of the pulses generated is proportional to the flow rate of the liquid. The FM consists of four parts: Housing, Pelton Wheel, Pick-up Coil and Conditioning Electronics. Internal and external parts are mounted on the housing. The main internal part is the rotor assembly, a rotor mounted on a sleeve bearing to allow free rotation. The bearing is kept in place by two spacer bushes, slid on a screw and locked up by the holding nut. The rotor is of AISI 430 to provide the magnetic properties required for the pick-up coil to function. The pick-up coil is not in direct contact with the fluid.

In **ITT Barton 7182** (Fig. 3), liquid (flowing first through a flow straightening section) is accelerated and forces a multi-blade, balanced, turbine rotor to rotate with a speed proportional to the flow. A pick-up coil senses the passage of each blade tip and generates a sine wave output, whose frequency is directly proportional to turbine speed, hence the flow rate. The meter consists of a housing body to which internal and external parts are mounted. The internal parts are the tube assembly and the bushing assembly, being the rotor shrunk on the bearing, mounted on a shaft centre, supported by the tube straightener assemblies and locked in place by the retainer ring. The straightener assemblies consist of three tube straighteners, held together by a welding cap end and a flow diffuser. Two assemblies are mounted to the FM, one at the inlet, one at the outlet. These assemblies were post-machined to meet the required accuracy. The only external part is the pick-up coil with a connector to the electronics. The rotor is of AISI 430, with magnetic properties necessary for the coil (which is not in direct contact with the fluid) to function.

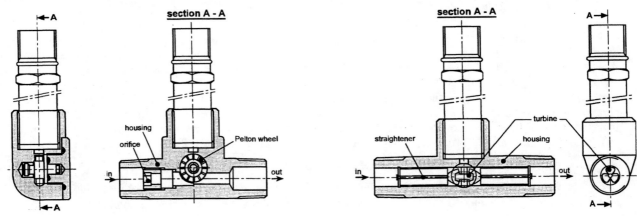

FIGURE 2. Pelton Wheel FM: ITT Barton 7506.   **FIGURE 3.** Turbine FM : ITT Barton 7182.

The **Electronic Control Unit** (Fig. 4) is a motherboard with three PCBs on it. This makes the configuration modular. The motherboard handles all data, power and command signal transport for the PCBs. It represents the internal bus. Connectors interface with the sensor control equipment for temperature and flow input/output signals and power input. The modular setup offers compact, lightweight, versatile electronics with minimum system modifications. The function of the electronics is to translate the FM (sensor) output pulse rate into an analogue output voltage representing the mass flow. Sensor non-linearities and temperature effects are compensated for. The FMA electronics are mounted on the analogue, digital and micro-controller PCBs. Figure 5 shows the setup.

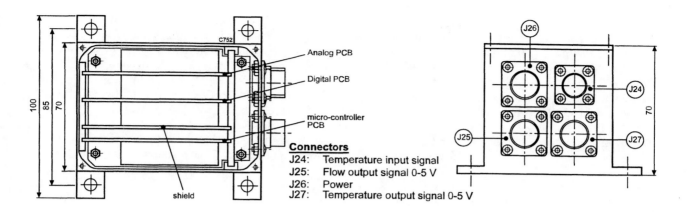

Connectors
J24: Temperature input signal
J25: Flow output signal 0-5 V
J26: Power
J27: Temperature output signal 0-5 V

FIGURE 4. Electronics box mechanical design.

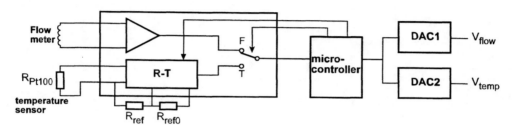

FIGURE 5. Basic set-up of the electronic configuration.

The flow signal is represented by the frequency of a sinusoidal signal, generated by a 7000 ITT Barton series turbine FM. The frequency ranges from 45-1500 Hz depending on the flow and slightly depending on temperature (second-order effect). To compensate this, a Pt100 temperature sensor measures the temperature of the fluid. The Pt100 is read according to the three-wire method, to ensure inherent good reliability and a good long-term stability. The measuring sequence is completed by measuring the reference resistors R_{ref} and R_{refo} for auto-correlation purposes. The measurements are executed by a micro-controller. With the R-T (resistance to temperature) converter, each of the three resistive signals is converted to a period length of a square-wave output signal for the micro-controller. Also the measured FM frequency is converted into a period length, so four period lengths/frequencies are measured. The micro-controller determines the frequency or period length and calculates the value of R_{Pt} and the temperature. From the calibration data, that characterise the flow meter behaviour (stored in an EPROM look-up table, containing 6 x 7 frequency data plus headers), the measured frequency can be translated to actual flow by linear interpolation between neighbouring flow frequency signals at measured temperatures.

TESTS AND TEST RESULTS

The results of the acceptance test programme, shown in the proper test sequence (Table 3), are given in the figures 6 to 9 and in the verification matrices (Tables 4, 5).

TABLE 3. FMA acceptance test programme and sequence.

No	Test	No	Test	No	Test
1	Mass/dimensions Check	10	Collapse Pressure Test	19	Time Drift Test
2	Flow Range Test	11	Functional Check	20	Response Time Test
3	Calibration Test	12	Pressure Surge Test	21	Radiated Emission
4	Repeatability Test	13	Mass/dimensions Check	22	Radiated Susceptibility
5	Time Drift Test	14	Functional Check	23	Power Leads
6	Response Time Test	15	External Leakage Test	24	Linear Acceleration/Sinusoidal
7	Proof Pressure Test	16	Reverse Leakage Test	25	Random Vibrations
8	Mass/dimensions Check	17	Calibration Test	26	Insulation
9	Functional Check orientations	18	Repeatability Test	27	Grounding & bonding

Mass and dimensions checks were done before and at the end of the test sequences. The values are 100.0 mm x 89.8 mm x 69.9 mm and, 565.5 g for the Electronics Box, 73.1 mm x 22.0 mm x 71.8 mm, 182.9 g for FMA7182, and 62.0 mm x 23.8 mm x 78.0 mm, 362.2 g for FMA7506. They did not change during the tests. EMC tests showed that the FMAs were only compliant with Conducted Emission CE01 requirements, and did not comply with the requirements for Radiated Emission RE01, Conducted Emission CE03, and Radiated Susceptibility RS03. This was expected because the commercial electronics components were non MIL-SPEC. Replacement by MIL-SPEC components (in a follow-up study) will most probably lead to compliance with all EMC requirements.

FMA 7182 was tested with the target fluid water only. A calibration test was done to fill the Look-Up Table (LUT). This table corrects the temperature dependence of the FMA. With this updated LUT the remaining tests were done.

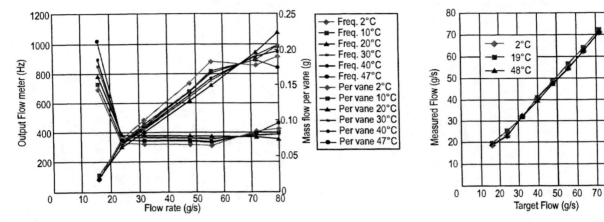

FIGURE 6. Flow range test FMA 7182.   **FIGURE 7.** FMA 7182 after 3 modifications.

Flow range test results (Fig. 6) were used to update the LUT. Figure 6 also shows the average mass per vane, deduced from the measured frequency. Mass transport per vane curves show considerable deviations from a linear relation between frequency and flow rate (due to slip) below 24 g/s. Update of the LUT had to correct this. With the above updated values implemented in the electronics, the FMA was calibrated in the following calibration test. The accuracy at 20 °C was encouraging: better than 3 % for flow rates bigger than 16 g/s.

Final results of the calibration tests of FMA 7182 with water, after three look-up table modifications, are shown in figure 7. From these results it could be derived that, for all temperatures, the accuracy is better than 2.5 % for a mass flow of 24 to 70 g/s. At 19 °C, the accuracy for mass flows bigger than 16 g/s is even better than 0.7 %.

FMA 7506 was tested for water and ammonia, but for ammonia not the full test programme, but only a reduced number of tests (as it had been agreed to put major effort on FMA 7182) was carried out.

First, a flow range test with water was done to compare with the ammonia calibration data. This first test was done with commercial electronics, this to obtain a first LUT. The obtained results are given in figure 8. The left ordinate gives the output frequency of the flow meter, the right ordinate gives the corresponding mass flow per vane. The figure shows that the frequency is almost linear with the mass flow within the entire mass flow range. Deviations occur only at small mass flows, as it is illustrated by the mass transport per vane that shows increased mass transport per vane with decreasing mass flow (due to slip). For this effect corrections were to be made via a LUT update.

Repeatability tests were done with water at 20 °C, to check whether FMA 7506 is able to reproduce. The results showed a repeatability better than 1 % in the whole flow range, even better than 0.3 % for large flow rates (>8 g/s).

The results of the calibration test of FMA 7506 for ammonia are given in figure 9. They show acceptable accuracy for flow rates above 7 g/s, for smaller flow rates the accuracy is unsatisfactory. Therefore the results of this test shall be used to create a new LUTto improve the accuracy of this FM to the values of FMA 7182 (2.5 % of reading). The second LUT update and the introduction of sophisticated interpolation algorithms is expected to realise the required accuracy. The latter activities were not carried out. They were considered to be part of a follow-up study, like the replacement of commercial components by MILSPEC components, following the agreement to focus on FMA 7182.

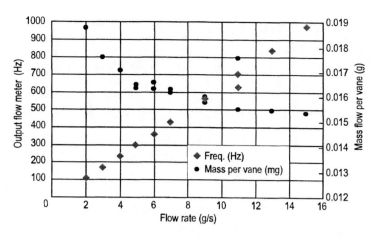

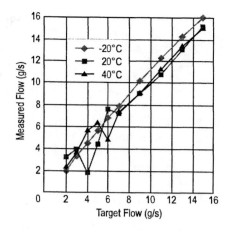

FIGURE 8. Flow range test FMA 7506 (water).

FIGURE 9. Calibration Test FMA 7506 (ammonia).

CONCLUSIONS

The conclusions, derived from the results given in the verification matrices (Tables. 4, 5), can be summarised by:
- The developed test bench can calibrate LFM accuracy better than 0.025% and response time smaller than 1 s. The bench is perfectly suitable to calibrate of LFM in the relevant mass flow ranges, for different liquids (water, ammonia, refrigerants, and propellants). Though this activity was originally not foreseen, the bench development became absolutely essential, as no commercial test bench and reference flow meters were available (especially taking into account the variety of fluids under investigation). The quality of the bench is such that it is investigated whether the bench can become a standard for FM calibrations, for different liquids and various flow rate ranges.
- For water meter FMA 7182, the majority of objectives are met. But it has, for a flow range from 24 to 72 g/s, an accuracy of 2.5% of reading in the whole temperature range (to be improved in a follow-up study). The time drift is less than 0.5%, which is very good. However, vibration and EMC tests indicate that there is more work to be done before FMA 7182 can be called fully applicable in space. FMA 7506, having a flow range up to 15 g/s for

TABLE. 4. Verification Matrix for Water FMA 7182.

Issue	Requirements from SOW & TRD	Updated Requirements	Delivered FMA 7182 Final Version	Remarks
Fluids Compatibility	Variety of fluids	Water	Water	
Operating Temperature	1.5 to 50°C		2 °C to 50 °C	
Operating Pressure	0.2 to 1.0 Mpa		0.2 to 1.0 Mpa	
Flow Rate Range	15 to 200 g/s		15 to 80 g/s	
Allowable Pressure Drop	700 Pa at 15 g/s, 10000 Pa at 200 g/s		27000 at 140 g/s	Specs. supplier
Accuracy (% of reading)	1.0 %		2.5 % for 24-70 g/s	To be improved
Response Time	< 1 s		> 15 s	To be improved
Maximum Over- Pressure	1.0 Mpa		Compliance met	
Proof Pressure	1.5 Mpa		Compliance met	
Burst Pressure	2.0 Mpa		Compliance met	
Orientation	In any orientation		Compliance met	
Leakage	External $<1 \cdot 10^{-9}$ mbar.l/s Reverse $<1 \cdot 10^{-8}$ mbar.l/s		$<1 \cdot 10^{-8}$ mbar.l/s	Detection limit of equipment
Humidity	20 to 100 %			No problems
External Pressure	115 kPa > P > 10^{-4} Pa		Compliance met	
External Temperature	2 to 50 °C		Compliance met	
Acceleration Requirements	Linear +/- 13 g/s		Compliance met	Model 7283 tested
Mass	< 3kg		182.9 g	Electronics 566 g
Volume	< 1400 cc		115.5 cc	Electronics 628 cc
Power	< 5 W		< 1 W	Specs. supplier
EMC-Conditions	MILSTD-462	MILSTD-461C	Not satisfactory	To be improved
Reliability	Operation (space) 5 years		MTTF 15.3 years	Reliability analysis

water and ammonia, shows an accuracy that is not satisfactory, especially at lower flow rates. Improvements can be realised by implementation of a new look up tables, insertion of more sophisticated interpolation algorithms, and replacement of commercial electronic components by MILSPEC components. This FMA meets most objectives, but the results of vibration and EMC tests lead to the same conclusions as for the FMA 7182: there is more work to be done in this area, before FMA 7506 can be called adequate for space applications.
- Both meters perform properly after vibration tests. Though proper performance during (e.g. launch) vibrations was not a requirement in this study, it can be expected that special mounting measures (dampers, isolators, etc.) will yield FMAs correctly measuring even in a vibration environment.
- The response time of both FMs. It is to be improved, but the required value < 1 s is to be questioned, as this comes from propellant systems requirements, and might be non-realistic for thermal or life science systems.
- EMC test results suggest that requirements will be met by replacing commercial by MIL-SPEC components.

TABLE 5. Verification Matrix for the Ammonia Flow Meter FMA 7506.

Issue	Requirements from SOW & TRD	Updated Requirements	Delivered FMA 7506 Final Version	Remarks
Fluids Compatibility	Variety of fluids	NH_3 and Water	NH_3 and Water	
Operating Temperature	NH_3 –20 to 40 °C Water 1.5 to 50°C	NH_3 -20°C to 40°C	NH_3 -20 °C to 40 °C	
Operating Pressure	NH_3 0.2 to 1.5 Mpa Water 0.2 to 1.0 Mpa	NH_3 0.2 to 1.5 Mpa	0.2 to 1.7 Mpa	
Flow Rate Range	NH_3 0.3 to 3.0 g/s Water 15 to 200 g/s	NH_3 0.3 to 3.0 g/s	2.0 to 15 g/s	
Allowable Pressure Drop	NH_3 50 Pa at 0.3 g/s, 250 Pa at 3 g/s. Water 700 Pa at 15 g/s, 10^4 Pa at 200 g/s	NH_3 50 Pa at 0.3 g/s, 250 Pa at 3 g/s	1400 Pa at 2.0 g/s 73000 Pa at 15 g/s	
Accuracy (% of reading)	NH_3 0.5 % Water 1.0 %	NH_3 0.5 %	2.5 % at large flows > 14 % small flows	To be improved
Response Time	< 1 s		> 15 s	To be improved
Maximum Over-Pressure	NH_3 2.0 Mpa Water 1.0 Mpa		Compliance met	
Proof Pressure	NH_3 3.0 Mpa Water 1.5 Mpa		Compliance met	
Burst Pressure	NH_3 4.0 Mpa Water 2.0 Mpa		3.0 MPa tested	No problem expected
Orientation	In any orientation		Compliance met	
Leakage	External <$1 \cdot 10^{-9}$ mbar.l/s Reverse: <$1 \cdot 10^{-8}$ mbarl/s		<$1 \cdot 10^{-8}$ mbar.l/s	Detection limit of equipment
Humidity	20 to 100 %			No problems
External Pressure	115 kPa > P > 10^{-4} Pa		Compliance met	
External Temperature	2 to 50 °C		Compliance met	
Acceleration Requirements	Linear acceleration +/- 13 g/s		Compliance met	Model 7283 tested
Mass	< 3kg		362.2 g	Electronics 566 g
Volume	< 1400 cc		115.1 cc	Electronics 628 cc
Power	< 5 W		< 1 W	Specs. Supplier
EMC-conditions	MILSTD-462	MILSTD-461C	Not satisfactory	To be improved
Reliability	5 years (space) operation		MTTF 11.8 years	Reliability analysis

Development and Testing of Advanced Cryogenic Thermal Switch Concepts

B. Marland, D. Bugby, C. Stouffer

Swales Aerospace, 5050 Powder Mill Road, Beltsville, MD 20705
[*Phone:(301) 595-5500; E-mail: bmarland@swales.com*]

Abstract. This paper describes the development and testing of two advanced cryogenic thermal switch (CTSW) options for use in long-life cryogenic space systems. The principal application for these two CTSW options in such systems is in implementing cryocooler redundancy with a minimum parasitic heating penalty. The two CTSW configurations covered in the paper are a hydrogen gas-gap (H_2-GG) design, flown on STS-95 in October 1998 as part of the CRYOTSU Hitchhiker flight experiment, and a differential thermal expansion (DTE) design. Both options are constructed primarily of beryllium for CTE compatibility with beryllium cryogenic components. The H_2-GG design utilizes a flat 2-mil gap between two cylindrical beryllium halves that are supported by a thin-walled titanium tube. A highly convoluted stainless steel bellows seals the unit. The H_2-GG CTSW is nominally "off" (evacuated) until actuated "on" by heating a metal hydride getter, which evolves hydrogen and provides thermal conductance across the gap. The H_2-GG design has demonstrated an "on" conductance of 1.0 W/K, an "off" resistance of 1000-1500 K/W and a range of operation from 15K-300K. The DTE design, which has just three parts, is very similar to the H_2-GG design except that a stainless steel tube replaces the titanium tube and the bellows and getter are no longer needed. The DTE CSTW is actuated "on" (both sides cold) by the higher CTE of stainless steel compared to beryllium and actuated "off" by temporarily applying power to a small heater on the stainless steel tube to expand the tube enough to open the gap. After the smaller of the two beryllium parts warms sufficiently, the heater is no longer needed and the DTE CTSW remains "off" (one side cold, one side warm). The DTE design has demonstrated the potential for an "on" conductance greater than 1.0 W/K, an "off" resistance of 1400 K/W and a range of operation from less than 4K to 300K. This paper describes the design of each CTSW option and the testing that was carried out to verify their performance.

INTRODUCTION

The reliability of today's cryocoolers remains a limiting design constraint for many long-life, low risk cryogenic space applications. As a result, near term cryocooler space applications will probably require cryocooler redundancy, which is accompanied by an additional parasitic heat load from the non-operating cryocooler. For typical space cryocoolers operating without a thermal switching device, the parasitic load due to the non-operating cryocooler is approximately 0.5 W at 60 K (Bugby, 99).

In order to minimize cooling and input power requirements, a reliable cryogenic thermal switch is desirable. A properly designed CTSW increases the thermal isolation between the instrument and expander body of the non-operating cryocooler and reduces the parasitic load from the non-operating cryocooler by 65-80% (Bugby, 99). As a result, cryocooler cooling and input power requirements are substantially reduced.

The benefit of a CTSW becomes increasingly pronounced as the instrument operating temperature and cryocooler efficiency are decreased. From 10-30K, the cooling and input power requirements needed to overcome an additional parasitic heat load of 0.5 W are prohibitive. In this operating regime, a CTSW becomes essential. Figure 1 illustrates a redundantly cooled dual CTSW system.

CP504, *Space Technology and Applications International Forum–2000*, edited by M. S. El-Genk

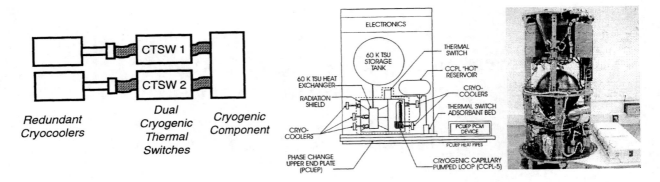

FIGURE 1. Dual CTSW System.　　　　FIGURE 2. Layout of the CRYOTSU Flight Experiment.

CRYOGENIC THERMAL SWITCH PERFORMANCE REQUIREMENTS

The development of the cryogenic thermal switches presented in this paper was funded by an AFRL-sponsored initiative to incorporate new and enabling cryogenic technologies into space systems. This initiative, dubbed the Integrated Cryogenic Bus (ICB), endeavors to combine a range of cryogenic integration solutions to meet the needs of future space applications. The design of the H_2-GG CTSW and Differential Thermal Expansion CTSW is driven by the ICB established CTSW performance requirements, outlined in Table 1.

TABLE 1. CTSW Performance Requirements.

Operating Temperature	Requirement Specification
Hot Side Nominal	300 K
Cold Side Nominal	35 K
Cold Side Range	35-60 K
Thermal Resistance	
Off Requirement	>1000 K/W
On Requirement	<2 K/W
Structural	
Side Load	>1 N

HYDROGEN GAS-GAP CRYOGENIC THERMAL SWITCH OVERVIEW

The H_2-GG CTSW was developed under the ICB initiative as one of four advanced cryogenic integration devices flown on the STS-95 CRYOTSU Hitchhiker flight experiment in October 1998. Figure 2 illustrates the layout of the CRYOTSU flight experiment. No degradation in switch performance due to launch loads was observed upon comparison of pre-flight and post-flight laboratory test results.

The concept for the hydrogen gas-gap cryogenic thermal switch utilizing a getter/hydride pump is based on a JPL design (Johnson, 1996) which used a zirconium-nickel hydride pump. A hydrogen gas-gap CTSW and hydride pump/getter system is shown in Figure 3. In principle, the H_2-GG CTSW is nominally "off" (evacuated) until actuated "on" by heating a metal hydride getter, which evolves hydrogen and provides thermal conductance across the gap. The design of the prototype and flight units is provided in Figures 4 and 5. For space applications, the H_2-GG CTSW is designed to mount the larger of the two beryllium parts to the instrument.

An alternative solution to the H_2-GG CTSW and getter/hydride pump system is presented in Figure 6. This system utilizes the H_2-GG CTSW design presented in Figure 5. However, the getter/hydride pump is replaced by two solenoid valves and a hydrogen/helium supply tank. In order to turn the switch "off" (evacuated), the supply tank solenoid valve is closed and the vent solenoid valve is opened. To actuate the switch "on", the vent solenoid valve is closed and the supply tank solenoid valve is briefly opened.

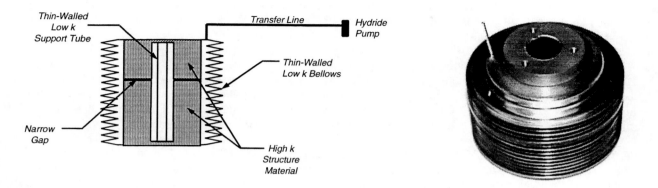

FIGURE 3. Hydrogen Gas-Gap CTSW and Hydride Pump System.

FIGURE 4. Prototype Hydrogen Gas-Gap CTSW.

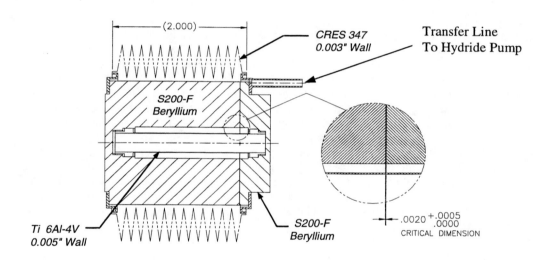

FIGURE 5. Hydrogen Gas-Gap CTSW Flight Unit Design.

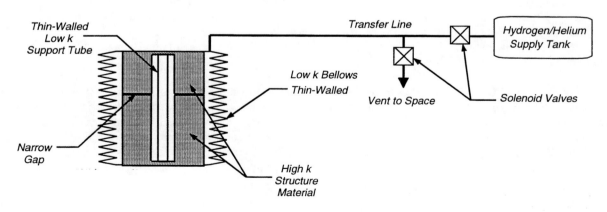

FIGURE 6. Hydrogen Gas-Gap CTSW and Solenoid Valve/Supply Tank System.

Hydrogen Gas-Gap CTSW Design

The critical design features of the flight unit are (1) the extremely narrow (0.05 mm) flat gap that separates the beryllium cylindrical parts, (2) the thin-walled titanium tube which aligns the beryllium parts and provides structural support, (3) the large size difference between the two beryllium components, (4) the two stainless steel weldment rings which are hermetically brazed to the beryllium and (5) the thin-walled, highly convoluted bellows which is welded to the rings to seal the unit.

Due to its high thermal conductivity, low density and thermal expansion compatibility with many cryogenic instruments, beryllium is the material of choice for the cylindrical main body of the H_2-GG CTSW. The diameter of the beryllium parts is sized in order to maximize the ratio of the "off"/"on" thermal resistance while satisfying the performance objectives outlined in Table 1. By altering the system sizing, mission peculiar requirements may be optimally satisfied. The mass of the flight unit, shown in Figure 5, is approximately 220 grams.

In the flight design, a narrow flat gap separates two beryllium parts, which are polished to lower the radiative coupling between the gap surfaces. The flat gap design allows gap width verification prior to final assembly. Compared to cylindrical or conical gaps, which cannot be easily verified, the flat gap design significantly reduces the risk of a thermal short when the switch is "off". Due to the nearly equal lengths of the titanium tube and larger beryllium part and the similarity in thermal expansion between these parts, gap stability is maintained during all modes of operation. In the "on" condition, all components of the H_2-GG CTSW are nearly isothermal and the titanium tube shrinks slightly more than the larger beryllium component, decreasing the gap width by approximately 0.02 mm. In the "off" condition, a large temperature gradient along the length of the tube and thermal contraction of the larger beryllium part increase the gap width from the room temperature condition of 0.05 mm.

While the switch is "off", the tube serves as a secondary heat path with a conductive resistance of approximately 3000 K/W for the operating range described in Table 1. Aside from providing critical alignment of the two beryllium parts, the titanium tube represents the primary support structure in the H_2-GG CTSW. For a given cross sectional area, the thin-walled design (0.13 mm as shown in Figure 5) maximizes the bending stiffness of the tube and provides significantly more structural support than a solid rod with equal cross section. The large size difference between the two beryllium components is structurally critical for space applications. By mounting the larger beryllium part to the instrument, the titanium tube only supports the mass of the smaller beryllium piece and a portion of the flexible cryocooler interface. Minimizing the mass of the smaller beryllium part was essential to the successful qualification of the H_2-GG CTSW for flight on STS-95. The tube diameter is sized to provide enough bending stiffness to avoid mechanical contact of the gap faces when the smaller beryllium part is subjected to a 10 g lateral load, approximately 4 N. A robust fundamental lateral frequency of 128 Hz and positive stress margins with a 40 g lateral load result from this sizing.

Providing relatively no structural support, the thin-walled, highly convoluted bellows and weldment rings are designed simply to seal the H_2-GG CTSW. During bellows welding operations, great care must be taken to avoid overheating the ring braze joints. Thermally, conduction through the bellows is unfortunately the dominant heat transfer mechanism while the switch is "off". To increase the thermal resistance of the bellows, the flight switch utilizes a 0.07mm foil instead of the nominal 0.13mm foil typically used to make welded bellows of this size. In addition, the number of convolutions in the bellows is maximized, thereby increasing the effective thermal length to about 350mm. Analytical models of the bellows predict a conductive resistance of approximately 2000 K/W for the operating range described in Table 1.

Although the H_2-GG CTSW design described above has functioned well in laboratory testing with a turbopump and a supply bottle of hydrogen, efforts to integrate the H_2-GG CTSW with a working getter/hydride pump are on-going. Before successful laboratory test methods and results are reviewed, some of the issues and lessons learned from these efforts are discussed below. In conjunction with this discussion, the feasibility of the H_2-GG CTSW and solenoid valve/supply tank system, shown in Figure 6, is briefly addressed.

Getter/Hydride Pump Design and Integration with the H$_2$-GG CTSW

After successful structural and thermal qualification of the H$_2$-GG CTSW design, the flight unit was filled and integrated into the STS-95 CRYOTSU Hitchhiker flight experiment in October 1998. Due to severe schedule restraints, development and thermal testing of a getter/hydride pump before the flight was not possible. As a result, the hydride pump used in the JPL thermal switch program was integrated with the H$_2$-GG CTSW. During flight, the hydride pump provided excellent "on" conductance, approximately 1.4 W/K, but was unable to evacuate the H$_2$-GG CTSW well enough to yield a significant "off"/"on" performance ratio.

Post-flight testing revealed the H$_2$-GG CTSW functioning nominally. However, "off" testing revealed that a low flow impedance path between the H$_2$-GG CTSW and evacuation source is needed (the impedance of 1 meter of 5 mm ID tubing is too great) for the switch to perform nominally. Based on a free molecular flow analysis, the switch requires a vacuum of less than 1×10^{-2} Pa for gas conduction across the gap to considered negligible. A high impedance evacuation path makes this standard extremely difficult, if not impossible, to achieve.

Testing on the flight and spare hydride pump revealed degraded pumping performance. Based on the marginal theoretical pumping ability of the flight zirconium-nickel hydride pump, approximately 1×10^{-2} Pa at room temperature, a new getter/hydride pump was located. The new zirconium-vanadium-iron getter was procured from SAES Getters. Operating below the hydrogen concentration level necessary to form hydrides, the SAES getter is not expected to experience the performance degradation that many hydride beds exhibit due to hydride expansion and contraction. In theory, the SAES Getter will provide a source evacuation pressure less than 1×10^{-4} Pa. One drawback to the SAES getter is that temperatures of at least 600°C will probably be required to provide full conduction in the "on" state (Giorgi, 1999). On-going efforts to produce an acceptable gas-gap cryogenic thermal switch system are focused on providing both a low flow impedance path and thermal isolation between the switch and getter.

Feasibility of H$_2$-GG CTSW with Solenoid Valve/Supply Tank System

With getter integration studies still underway, the feasibility of a system utilizing solenoid valves and a gas supply tank needs to be addressed (see Figure 6). Based on laboratory testing with hydrogen at 30 K, the H$_2$-GG CTSW achieves full conduction in the "on" condition at approximately 700 Pa. The measured volume of the switch is 75 cm^3. Therefore, a 5 standard cm^3 charge of hydrogen is sufficient to achieve full conduction. As a result, a properly regulated 100 cm^3 supply tank filled with 6×10^5 Pa of hydrogen would permit the H$_2$-GG CTSW to be actuated "on" nearly 100 times. By replacing hydrogen with helium as the working fluid, the H$_2$-GG CTSW operating range is extended to below 5 K. Latching solenoid valves are being used as part of the HST Nicmos Cooling System (Nellis, 1998). The ability of the space vacuum environment to evacuate the H$_2$-GG CTSW through such valves requires further study.

Hydrogen Gas-Gap CTSW Test Methodology and Results

To separate the performance of the hydride pump from the intrinsic thermal performance of the H$_2$-GG CTSW, laboratory "on" testing was carried out by filling the switch with hydrogen gas and "off" testing was carried out by evacuating the switch with a turbo pump. Various tests with the JPL hydride pump were also conducted, none of which evacuated the switch properly. The performance goals of this testing are outlined in Table 1. Figure 7 illustrates the H$_2$-GG CTSW test set-up.

The "on" test is conducted by simply filling the H$_2$-GG CTSW with hydrogen from either a hydrogen tank or a hydride pump. The cryocooler is turned on and the cold head is allowed to cool to 35 K. When steady state is reached the initial temperature difference (ΔT_o) across the switch is measured ($\Delta T = T_H - T_C$, where T_H is the temperature of the warm end and T_C is the temperature of the cold end). This step provides the datum from which the silicon diode (relative) measurement errors and the effects of external parasitics can be eliminated. Then, a known heater power Q_{on} is applied to the warm end and the new ΔT is measured. The "on" conductance is the heater power divided by the change in temperature difference. Equation (1) provides the analytical relationship.

The "off" test is conducted by simply evacuating the switch. The cryocooler is turned on and the cold head is allowed to cool to 35 K. The required heater power needed to equilibrate the warm end and vacuum chamber temperatures, Q_{Zero-P}, is then applied to the warm end. This heater power "zeroes out" the parasitics. The off resistance is the resulting ΔT divided by the heater power. Equation (2) illustrates the analytical relationship. Figures 8 and 9 illustrate results from two post-flight thermal tests. Note the "off"/"on" thermal resistance ratio of approximately 1450 indicated in the figures.

$$G_{on} = Q_{on} / (\Delta T(Q_{on}) - \Delta T_o).$$ (1)

$$R_{off} = \Delta T(Q_{Zero-P}) / Q_{Zero-P}.$$ (2)

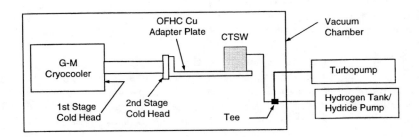

FIGURE 7. Hydrogen Gas-Gap CTSW Laboratory Test Set-Up.

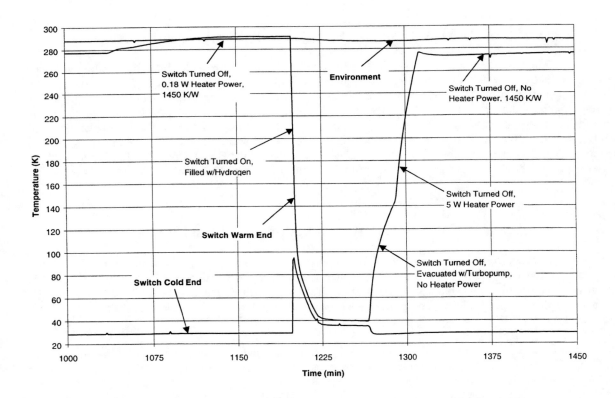

FIGURE 8. Hydrogen Gas-Gap CTSW "Off"/"On" Cycle Testing with Turbopump and Hydrogen Supply Tank.

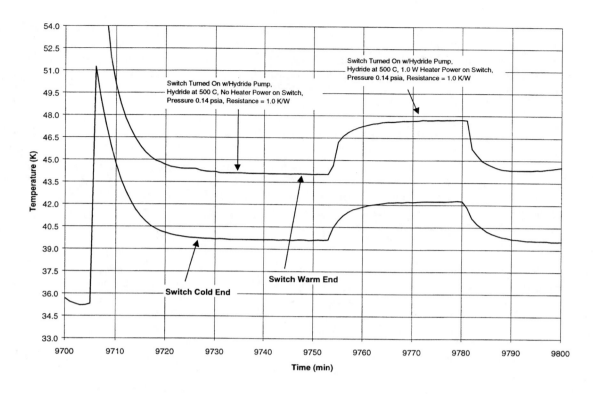

FIGURE 9. Hydrogen Gas-Gap CTSW "On" Testing with Hydride Pump.

DIFFERENTIAL THERMAL EXPANSION (DTE) CTSW OVERVIEW

In an effort to simplify the design of the H_2-GG CTSW while alleviating concerns about the reliability of hydride pumps/getters and hermetically sealed thermal switches, a novel and promising CTSW concept was invented. The DTE design, which has just three parts, is similar to the H_2-GG design except that a stainless steel tube replaces the titanium tube and the bellows and getter are no longer needed. Figures 10 and 11 illustrate the DTE CTSW design. Swales Aerospace is in the process of patenting this design.

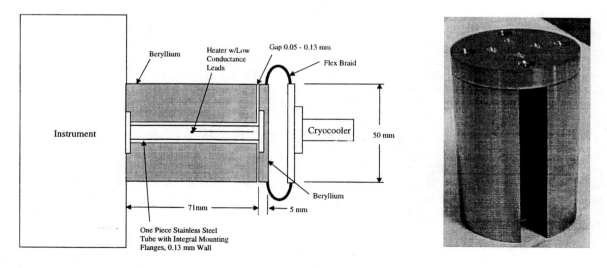

FIGURE 10. DTE Cryogenic Thermal Switch.

FIGURE 11. DTE Cryogenic Thermal Switch.

The DTE CSTW is actuated "on" when the smaller beryllium part is cooled sufficiently so that the higher CTE of the stainless steel tube compared with the beryllium causes the gap in the switch to close. By temporarily applying power to a small heater on the stainless steel tube, the stainless steel tube expands enough to open the gap and the DTE CTSW is actuated "off". After the smaller beryllium part warms sufficiently, the heater is no longer needed and the DTE CTSW remains "off" (one side cold, one side warm).

Differential Thermal Expansion CTSW Design

Although similar to the H_2-GG CTSW in design, the DTE CTSW has several distinguishing design features, which should be discussed. The critical design features of the DTE CTSW are (1) the extremely narrow (0.07 mm nominally), highly precise flat gap that separates the beryllium cylindrical parts, (2) the thin-walled stainless steel tube which aligns the beryllium parts and provides structural support, (3) the large size difference between the two beryllium components and (4) the proprietary gold plating process applied to each beryllium part to improve the contact conductance.

As with the H_2-GG CTSW design, the flat gap design permits gap width verification and reduces the risk of mechanical shorting. A major advantage of the DTE design over the H_2-GG design is that the gap width can be verified at *any* time, not just prior to final assembly. In addition, the nominal gap width of the DTE design is larger than in the H_2-GG design. The requirements for gap parallelism on the DTE design, however, are more stringent than for the H_2-GG design, due to the unforgiving nature of metal-metal contact.

The overall length of the DTE CTSW is approximately 75 mm, compared with approximately 50 mm for the H_2-GG CTSW, to provide an "off" thermal resistance greater than or equal to the H_2-GG design. The increased modulus of the tube, increased gap width and decreased mass of the smaller beryllium part more than compensate for the loss of structural stiffness due to the increased tube length. As a result, the structural performance of the DTE is better than that of the H_2-GG CTSW.

As with the H_2-GG CTSW design, the large size difference between the two beryllium components plays an important operational role. In the "on" condition, all components of the DTE CTSW are nearly isothermal and the stainless steel tube shrinks more than the larger beryllium component, decreasing the gap width enough to close the gap and generate approximately 450 N on the gap. While in the "off" condition, a large temperature gradient along the length of the tube and thermal contraction of the larger beryllium part maintain a gap in the system.

Baseline conductance tests have demonstrated that the two design features needed to minimize the contact resistance of the gap during "on" operation are (1) an extremely precise parallel gap and (2) a compliant coating on both beryllium parts. For a room temperature conductance test, a proprietary gold plating applied to the DTE CTSW (see Figure 12) was found to increase the contact conductance of the beryllium components by nearly an order of magnitude to greater than 2 W/K. The configuration for this test, a bolted together configuration with a 450 N preload, is shown in Figure 13. After the initial positive results, the components were cycled in this configuration 20 times without any noticeable performance degradation. The gap parallelism of the DTE CTSW is currently being improved upon in order to approach this level of performance with the stainless steel tube.

Differential Thermal Expansion CTSW Test Results

The test methodology applied to the DTE CTSW is identical to that of the H_2-GG CTSW. Figure 14 demonstrates an " on" test of the DTE CTSW. This test was conducted without gold plating on the switch and the smaller beryllium part was mounted to the coldhead. Figure 15 illustrates a cycle test for the DTE CTSW. The larger beryllium part was mounted to the coldhead in this test. A temperature controlled shroud was used to cycle the switch "on" and "off". The "off" resistance of the DTE CTSW, with a 300 K warm end and a 35 K cold end, has been experimentally quantified to be 1400 K/W, without the stainless steel tube heater. With the tube heater, the "off" resistance is expected to be at least 1200 K/W. An "on" conductance greater than 1 W/K is expected when DTE CTSW gap parallelism is properly honed in the near future.

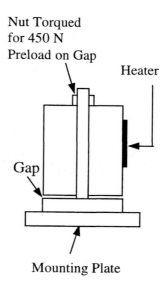

Nut Torqued
for 450 N
Preload on Gap

Heater

Gap

Mounting Plate

FIGURE 12. DTE CTSW with Gold Plating.   **FIGURE 13.** DTE CTSW Conductance Test.

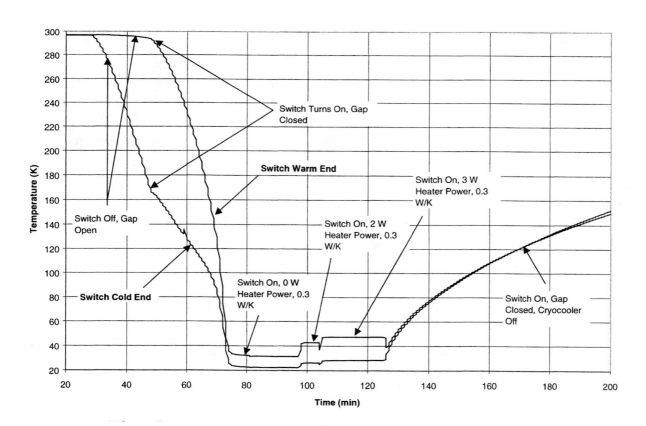

FIGURE 14. Differential Thermal Expansion CTSW "On" Testing without Gold Plating.

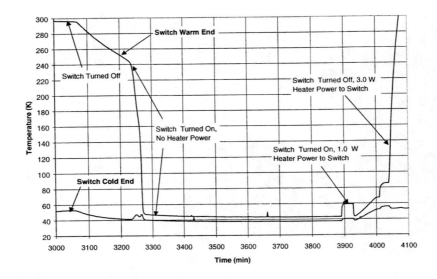

FIGURE 15. Differential Thermal Expansion CTSW "Off"/"On" Cycle Testing.

SUMMARY

This paper has described the design, operation, and test results of two advanced cryogenic thermal switches (CTSW), (1) the hydrogen gas-gap (H_2-GG) CTSW and (2) differential thermal expansion (DTE) CTSW. The hydrogen gas-gap (H_2-GG) CTSW was flown on STS-95 in October 1998 as part of the CRYOTSU Hitchhiker flight experiment. The H_2-GG design has successfully demonstrated an "on" conductance of 1 W/K an "off" resistance of 1450 K/W when evacuated with a turbopump. Efforts to achieve this level of performance with a getter are on-going. The DTE CTSW has demonstrated an "off" resistance of 1400 K/W and baseline conductance test results give promising indications that an "on" conductance greater than 1 W/K may be achieved in the very near future. The potential reliability, performance and simple three piece construction of the DTE CTSW, which enables gap verification at *any* time during testing and mission integration, presents a particularly intriguing promise for the near future implementation of cryogenic thermal switches in space applications. Cryogenic thermal switches are critical cooling and input power saving devices, which represent an important part of the Integrated Cryogenic Bus (ICB) initiative to incorporate new and enabling cryogenic technologies into space systems. This paper is intended to provide the cryogenic space applications community with a status report on these two promising designs.

ACKNOWLEDGMENTS

The authors would like to acknowledge the Air Force Research Laboratory for providing the funding for this work. We would also like to acknowledge the efforts of Thom Davis and B.J. Tomlinson of AFRL.

REFERENCES

Bugby, D., Stouffer, C., Hagood, B., et. al, "Development and Testing of the CRYOTSU Flight Experiment," Space Technology and Applications International Forum (STAIF-99), M. El-Genk editor, AIP Conference Proceedings No. 458, Albuquerque, NM, 1999, pp. 2-3.

Johnson, D. and Wu, J., "Feasibility Demonstration of a Thermal Switch for Dual Temperature IR Focal Plane Cooling," *Cryocoolers 9*, Plenum Press, New York (1996).

Giorgi, T. and Ferrario, B., "Non Evaporable Getters Activatable at Low Temperatures", SAES St 707™ Getter Informational Catalog, SAES Getters/U.S.A., Inc., Colorado Springs, CO (1999).

Nellis, G., Dolan, F., Swift, W., and Sixsmith, H., "Reverse Brayton Cooler for NICMOS," *Cryocoolers 10*, Kluwer Academic/Plenum Publishers, New York (1998).

Surface Reorientation Upon Step Reduction in Gravity

Jens Gerstmann, Michael E. Dreyer, Hans J. Rath

Center of Applied Space Technology and Microgravity (ZARM), University of Bremen, Am Fallturm, D-28359 Bremen, Germany, Tel. +49 421 218 3638, Fax +49 421 218 2521, E-mail: gerstm@zarm.uni-bremen.de

Abstract. The surface settling of a liquid meniscus upon step reduction in gravity in a right circular cylinder is calculated numerically. In the numerical simulations a slip boundary condition and a modified model for the dynamic contact angle are used. The influence of the boundary conditions on the interface oscillation is determined. The results of the numerical simulations reveal a significant influence of the dynamic contact angle boundary condition imposed on the moving contact line on the numerically obtained surface oscillations, i.e. on resonant frequencies and settling times.

INTRODUCTION

In this study we consider the effect of a sudden change in the acceleration level on the free surface in a right circular cylinder (Fig. 1). With vanishing hydrostatic pressure, interface oscillations are started depending on the damping of the system and the fluid moves to the zero gravity equilibrium shape (Fig. 2). The final curvature of the surface is constant and depends on the static contact angle γ_{stat}. During the reorientation we observe a moving contact line. The dynamic contact angle γ_{dyn} of the moving contact line is not equal to the static contact angle (Dussan, 1979; Ting, 1995). The chosen model for the dynamic contact angle in the numerical model has a significant influence on the behavior of the reorientation.

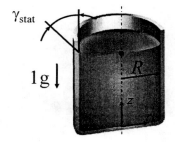

FIGURE 1. Normal gravity configuration.

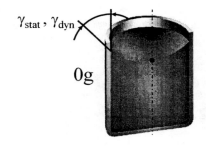

FIGURE 2. Low gravity configuration.

DIMENSIONAL ANALYSIS AND NUMERICAL MODEL

In our simulations we use the finite element code FIDAP. FIDAP solves the fully Navier Stokes equations. Axisymmetric reorientation of the fluid is assumed. The fluid is assumed to be incompressible and Newtonian with a passive overlaying gas phase. The calculations begin with a liquid/gas interface shape determined for the normal gravitational field instantaneously subjected to a zero gravity environment at a time t=0. From a scale analysis of the momentum equation we get the characteristic parameters for this problem. The velocity scale $u_c = (\sigma/\rho R)^{1/2}$ is chosen from the balance of the inertial force and the surface tension force. The time scale tc = $(\rho R^3/\sigma)^{1/2}$ results from comparing the pressure term and the unsteady term in the momentum equation. The pressure is scaled with the capillary pressure σ/R. The length are nondimensionalized by the radius R and the acceleration by the normal gravity. The dimensionless governing equations can then be written as:

CP504, *Space Technology and Applications International Forum–2000*, edited by M. S. El-Genk
© 2000 American Institute of Physics 1-56396-919-X/00/$17.00

$$\frac{1}{r}\frac{\partial}{\partial r}(ru_r)+\frac{\partial u_z}{\partial z}=0 \tag{1}$$

$$\frac{\partial u_r}{\partial t}+u_r\frac{\partial u_r}{\partial r}+u_z\frac{\partial u_r}{\partial z}=-\frac{\partial p}{\partial z}+\text{Oh}\left(\frac{\partial^2 u_r}{\partial r^2}+\frac{1}{r}\frac{\partial u_r}{\partial r}-\frac{u_r}{r^2}+\frac{\partial^2 u_r}{\partial z^2}\right) \tag{2}$$

$$\frac{\partial u_z}{\partial t}+u_r\frac{\partial u_z}{\partial r}+u_z\frac{\partial u_z}{\partial z}=-\frac{\partial p}{\partial z}+\text{Oh}\left(\frac{\partial^2 u_z}{\partial r^2}+\frac{1}{r}\frac{\partial u_z}{\partial r}+\frac{\partial^2 u_z}{\partial z^2}\right)+\text{Bog} \tag{3}$$

The nondimensional momentum equation written in cylindrical coordinates for the axial component z depends only on the Ohnesorge number Oh and the Bond number Bo,

$$\text{Oh}=\frac{\mu}{\sqrt{\rho\sigma R}}, Bo=\frac{\rho g R^2}{\sigma}. \tag{4}$$

The Ohnesorge number represents the ratio of the viscous to the surface tension force, and the Bond number Bo represents the ratio of the gravitational force to the surface tension force. The Bond number is a parameter for the start configuration of the surface. During the reorientation the term with the Bond number vanishes. Through the required boundary condition at the contact line we get further parameters. The static contact angle γ_{stat}, the dynamic contact angle γ_{dyn} and the slip length l_s. The slip length l_s is a defined region of slip at the moving contact line using the Navier slip condition to remove the stress singularity. The dynamic contact angle γ_{dyn} is calculated with the condition from Satterlee and Reynolds (Satterlee,1964). The dynamic contact angle depends on the z-coordinate of the contact line and a free parameter ξ. We modified this condition for static contact angles other than 90°.

$$\left.\frac{\partial z}{\partial r}\right|_{r=R}=-\xi\left.(z-z_\infty)\right|_{r=R}+\cot\left(\gamma_{stat}\right) \tag{5}$$

In this condition z_∞ represents the final coordinate of the contact line. The behavior of the fluid during the reorientation depends on the value of ξ for the contact line boundary condition. When $\xi = 0$, the contact angle is fixed and the contact line is free to move, whereas when $\xi \rightarrow \infty$, the contact line is fixed and the contact angle is allowed to vary. The following Figures 3 and 4 clarify the described influence of the parameter ξ.

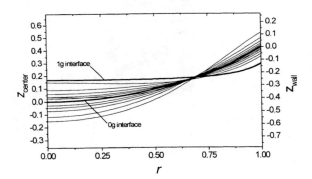 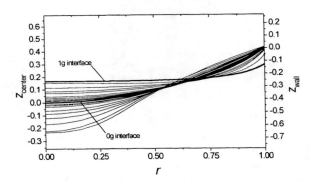

FIGURE 3. Free surface shape for $\gamma_{stat} = 44°$ with Oh $= 2.22\ 10^{-3}$ for several time steps between $t = 0$ and $t \rightarrow \infty$; $\xi = 0$.

FIGURE 4. Free surface shape for $\gamma_{stat} = 44°$ with Oh $= 2.22\ 10^{-3}$ for several time steps between $t = 0$ and $t \rightarrow \infty$; $\xi = 47,63$.

Figures 3 and 4 show the evolution of the shape of the free surface during the reorientation for different values of the parameter ξ. z_{center} and z_{wall} represent the coordinates of the free surface at the cylinder center and the cylinder wall. At the final position of the free surface z_{center} and z_{wall} are defined as equal to zero. Setting $\xi = 0$ (Fig. 3) the contact angle is fixed and the contact line is allowed to move without reservation. The result is an overshoot at the wall and a large-amplitude oscillation about the final value. During the whole oscillation the interface formed a single nodal point at $r = 0.67$. If $\xi = 47.63$ (Fig. 4), ξ restrains the free movement of the contact line at the wall and oscillations of this point are weak. The increasing dynamic contact angle in touch with the advancing contact line circumvents a remarkable overshoot at the wall. The interface is stronger deformed in comparison with the simulation with $\xi = 0$. The single nodal point of the interface moved out of the plotted range. For $\xi \to \infty$ the single point is the final position of the contact line.

NUMERICAL RESULTS

The principal objective was to determine the quantitative nature of the oscillations as a function of the parameter ξ for systems of various Ohnesorge numbers and static contact angles. The characteristics of the transient response are obtained from the surface oscillation at the centerline, i.e. the natural resonant frequency ω and the settling time t_s. The settling time is defined as the time required for the envelope curves of the transient response $z(t)$ to reach 2 % of the final amplitude.

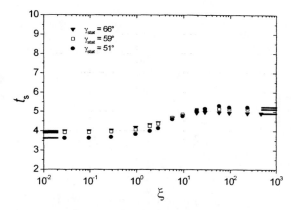

FIGURE 5. Effect of ξ on the settling time for varying γ_{stat}; Oh = 11.0 10^{-3}.

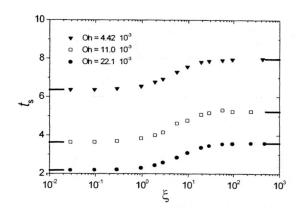

FIGURE 6. Effect of ξ on settling time for varying Oh; $\gamma_{stat} = 51°$.

Figure 5 (Oh = const) and 6 (γ_{stat} = const) show the dependence of the settling time t_s on ξ for different contact angles and Ohnesorge numbers, respectively. The horizontal lines on the abscissa on the left side represents the settling times for $\xi = 0$. On the right abscissa the horizontal lines indicates the behavior of the settling time for $\xi \to \infty$. In general we see, that the settling time increase with increasing ξ. For every fluid the settling time is lowest for $\xi = 0$ and highest for $\xi \to \infty$. Pronounced changes in settling time are only noticeable in the range of $\xi = 0$ to $\xi = 100$. The influence of varying the static contact angle in the range 51° to 66° on the settling time is weak. The dependence of the settling time on the Ohnesorge number is more distinct. Holding the contact angle constant in the range $51° \leq \gamma_{stat} \leq 66°$, the settling time increases with decreasing Ohnesorge number.

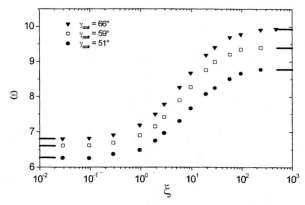

FIGURE 7. Effect of ξ on the resonant frequency for varying γ_{stat}; Oh = 11.0 10^{-3}.

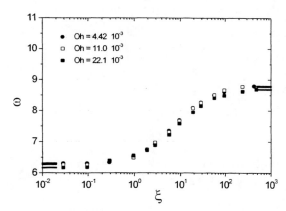

FIGURE 8. Effect of ξ on the resonant frequency for varying Oh; $\gamma_{stat} = 51°$.

Figures 7 and 8 show the dependence of the resonant frequencies for the different contact line conditions. In figure 7 again we hold the Ohnesorge number constant and vary the static contact angle. In Figure 8 we vary the Ohnesorge number and set the static contact angle as constant. The horizontal lines on the abscissa indicates also the limits of the value of the free parameter ξ. The line on the left side indicates the behavior of ω for $\xi = 0$ and on the right side for $\xi \to \infty$. For every fluid the frequency is minimized at $\xi = 0$ and maximized for $\xi \to \infty$. The resonant frequency increases with increasing static contact angle (Figure 7). The influence in the investigated range of the Ohnesorge number on the frequency (Figure 8) is weak.

The results of Figures 5 – 8 show that $\xi = 0$ and $\xi \to \infty$ depict the limits for the settling time and the resonant frequency. The following figures show the dependence of the settling time and the resonant frequency on the Ohnesorge number Oh and static contact angle γ_{stat} for the free-contact-line condition ($\xi = 0$) and for a approximately fixed-contact-line condition ($\xi = 95.25$).

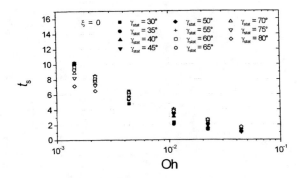

FIGURE 9. Effect of the Ohnesorge number on the settling time for varying γ_{stat}; $\xi = 0$.

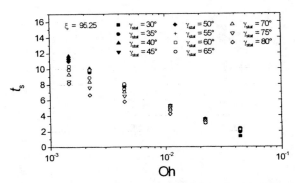

FIGURE 10. Effect of the Ohnesorge number on the settling time for γ_{stat}; $\xi = 95.25$.

Figures 9 and 10 show the dependence of the settling time on the Ohnesorge number and varying static contact angles for $\xi = 0$ and $\xi = 95.25$. In both figures the settling time decreases with increasing Ohnesorge number. For every Oh - γ_{stat} combination the settling time is lowest for $\xi = 0$. The decrement of the settling time with increasing Ohnesorge number decreases with increasing static contact angle.

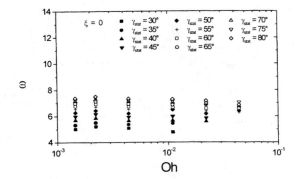

FIGURE 11. Effect of the Ohnesorge number on ω for varying γ_{stat}; $\xi = 0$.

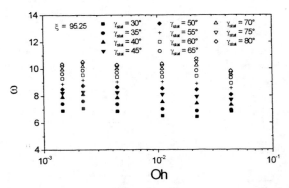

FIGURE 12. Effect of the Ohnesorge number on ω for varying γ_{stat}; $\xi = 92.52$.

In the Figures 11 and 12 the dependence of the settling time on the Ohnesorge number and varying static contact angles for $\xi = 0$ and $\xi = 95.25$ are illustrated. The influence of the Ohnesorge number on the resonant frequency is weak. The effect of the parameter ξ and the static contact angle are more distinct. The change of the resonant frequencies from $\xi = 0$ to $\xi = 92.52$ is in the range of 40 %. It is seen clearly that for a fixed value of the Ohnesorge number the resonant frequency increases with increasing static contact angle.

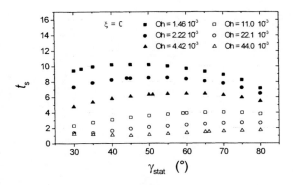

FIGURE 13. Effect of γ_{stat} on the settling time t_s for varying Oh; $\xi = 0$.

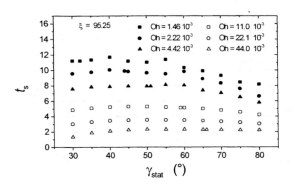

FIGURE 14. Effect of γ_{stat} on the settling time t_s for varying Oh; $\xi = 95.25$.

Figures 13 and 14 show the dependence of the settling time on the static contact angle and varying Ohnesorge numbers for $\xi = 0$ and $\xi = 95.25$. The influence of the static contact angle is much weaker as the effect of the Ohnesorge number. With the free-contact-line condition the settling time increases slightly until γ_{stat} reach 50° for low Ohnesorge numbers. For higher static contact angles t_s redescends. In contrast to this behavior the settling time increases continuously with increasing static contact angle for higher Ohnesorge numbers. The fixed-contact-line condition leads in the investigated range of Ohnesorge numbers at first to an increasing settling time with increasing γ_{stat}. Afterwards t_s decreases with increasing γ_{stat}. The turning point scrolls from 40° for lower Oh to 65° for higher Ohnesorge numbers.

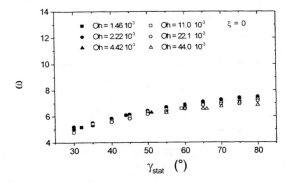

FIGURE 15. Effect of γ_{stat} on ω for varying Ohnesorge number; $\xi = 0$.

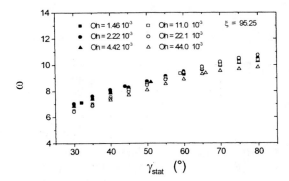

FIGURE 16. Effect of γ_{stat} on ω for varying Ohnesorge number; $\xi = 92.25$.

In the Figures 15 and 16 the dependence of the resonant frequency on the static contact angle and varying Ohnesorge numbers for $\xi = 0$ and $\xi = 95.25$ are illustrated. The resonant frequency increases with increasing static contact angle. The influence of the Ohnesorge number for a fixed static contact angle on the resonant frequency is weak.

We compare the numerical results with the experimental data from Weislogel & Ross (Weislogel, 1990). Weislogel & Ross investigated surface settling in a right circular cylinder in a drop tower experiment. Since ξ is a free parameter in the numerical study we have the possibility to tune the numerical data for the best fit with the experimental data. The fluid properties used in these calculations, the found value of ξ for the best fit and the numerical and experimental data for ω and t_s are listed are listed in Table1.

TABLE 1. Fluid properties; range of ξ, ω and t_s for the reorientation of a free surface compared with experimental results

Fluid properties					Numerical results			Experimental results (Weislogel, 1990)	
ρ (gcm⁻³)	$\mu\ 10^{-3}$ (gcm⁻¹s⁻¹)	σ (gs⁻²)	Oh 10^{-3}	γ_{stat} (°)	ξ	t_s	ω	$t_{s,exp}$	ω_{exp}
0.760	4.94	15.9	1.46	32	333	11,35	7,20	-	7,43
0.816	8.16	17.4	2.22	44	333	10,09	8,04	-	8,03
0.872	17.44	18.7	4.42	51	57	7,86	8,27	8,15	8,26
0.913	45.65	19.7	11.0	59	57	5,13	9,20	5,70	9,28
0.935	93.5	20.1	22.1	60	48	3,55	9,03	3,23	9,06
0.949	189.8	20.6	44.0	66	24	2,26	8,86	2,09	8,89

The experimental data of the settling time for the first two fluids could not be observed in the experiment. The settling time here is larger than the experiment time of 2.2 s in the used drop tower.

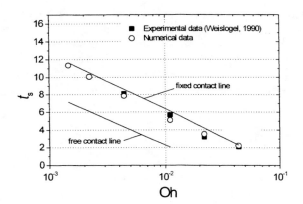

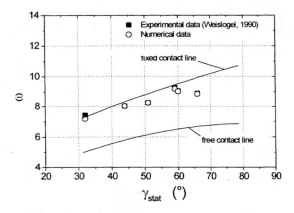

FIGURE 17. Comparison of the numerical and experimental data for the settling time t_s depend on the Oh.

FIGURE 18. Comparison of the numerical and experimental data for the resonant frequency depend on γ_{stat}.

Fig. 17 and 18 show the experimental results (Weislogel, 1990) for the settling time t_s and the resonant frequency versus the Ohnesorge number Oh and the static contact angle in comparison with the numerical results. The solid lines in Figure 17 and 18 represent the numerical limits of t_s and ω for the free-contact-line condition ($\xi = 0$) and the fixed-contact-line-condition (t → ∞). The experimental data from Weislogel & Ross (Weislogel, 1990) fall between the limits for the settling time and resonant frequency with the free-contact-line condition and fixed-contact-line condition. The experimental data agree well with the numerical data for the determined value of ξ (see Table 1). For low Ohnesorge numbers we found high values of the free parameter ξ. High values of ξ cause a fixed or pinned contact line in the numerical simulation. This behavior was observed by Weislogel & Ross in his experiments for low Ohnesorge numbers too.

CONCLUSIONS

Numerical investigations on the damped oscillations of a liquid/gas interface after step reduction in gravity were presented using the finite element code FIDAP. The results reveal the significance of the boundary condition at the moving contact line for numerically computed surface oscillations. The boundary condition imposed on the calculations incorporates a modified dynamic contact angle condition from which the parameter ξ arise as an adjustable parameter. The parameter ξ is correlated with the resonant frequency ω and the settling time t_s for a variety of Ohnesorge numbers Oh and static contact angles γ_{stat}. In each case, ω and t_s are minimized for $\xi = 0$ and maximized as $\xi \to \infty$. It is shown that the settling time primary depends on the Ohnesorge number. The settling time decreases with increasing Ohnesorge number. The influence of the static contact angle on the settling time is weak. In contrast the resonant frequency depends mainly on the static contact angle. The resonant frequency increases with increasing static contact angle. The resonant frequency is influenced only slightly by the Ohnesorge number. The experimental data from Weislogel & Ross (Weislogel, 1990) fall between the limits for the settling time and resonant frequency with the free-contact-line condition ($\xi = 0$) and fixed-contact-line condition ($\xi \to \infty$). For the right choice of the free parameter ξ in the numerical simulation we achieved good agreement with the experimental data.

ACKNOWLEDGMENTS

This work is supported by the German Science Foundation DFG through the grant number RA 352 / 11-2.

REFERENCES

Dussan V., E.B., "On the Spreading of Liquids on Solid Surfaces: Static and Dynamic Contact Lines," *Ann. Rev. Fluid Mech.* **11**, 371-400 (1979).

Satterlee, H.M., and Reynolds , W.C., "The Dynamics of the Free Liquid Surface in Cylindrical Containers under Strong Capillary and Weak Gravity Conditions," Dept. of Mechanical Engineering , Stanford Univ., TR LG-2, Stanford, CA (1964).

Ting, C.L., and Perlin, M., "Boundary Conditions in the Vicinity of the Contact Line at a Vertically Oscillating Upright Plate: An Experimental Investigation," J. Fluid Mech. **295**, 263-300 (1995).

Weislogel, M.M., and Ross, H.D., "Surface reorientation and settling in cylinders upon step reduction in gravity," *Microgravity sci. technol.* **3**, 24-32 (1990).

The Thermodynamics of Meniscus Formation on Wilhelmy Plate Immersion at the Air/water Interface and the Mechanics of Initial Film Deposition

Nicholas J. Tumavitch[1], D. Allan Cadenhead[1] and Brian J. Spencer[1]

[1]*Departments of Chemistry and Mathematics, University at Buffalo, Buffalo, NY. 14260.*
(716) 645-6800 X2123, checaden@acsu.buffalo.edu

Abstract. The meniscus which forms when a solid is immersed in a liquid is supported by the upward   acting surface tension of the liquid. Reducing the surface tension reduces the magnitude of the meniscus while reducing the gravitational pull will increase it. Edge effects can reduce the overall meniscus height, and the maximum height will only be achieved when the surface exceeds a minimum length. The shape of the meniscus at any point is given by an equation derived by minimizing the free energy of the system. From our observations of static menisci and our numerical study of the equations for the meniscus shape, we found that solids with identical sides will have identical menisci on each face, while solids where the width and the thickness differ will give a meniscus which will involve a contact angle change at and along the edge which can not be described by the classical theory. This observation has important implications for solid/liquid contact at corners in both normal and reduced gravity environments. Deposition of an insoluble film on the liquid surface results in a lowering of the surface tension and the size of the meniscus. If the solid surface is coated by a liquid film then the contact angle will not change during this process. Direct contact between the deposited film and the solid will give rise to a contact angle change.

INTRODUCTION

This project has focused on two areas: the wetting of edged solids by liquids and the deposition of lipid or oil films on differing solids. The two areas overlap in studies of film deposition on edged solids. In the first area we have developed a numerical scheme for solving the classical equations for an energy-minimizing surface and have compared the results with those observed experimentally. The results reveal a fundamental breakdown of the classical phenomenon in geometries with edges or corners. We are proposing an explanation of these effects based on "constrained energy minimization" and outline procedures to deal with such surfaces based on a modified theoretical treatment and a computer code for improved calculations. Understanding the local behavior near an edge or corner should allow us to apply the results to a wide range of solid geometries. Since a large number of solid shapes involve corners, the results have wide implications for describing the physics of wetting of solids by liquids.

The second area relates to the mechanics of film deposition. Deposition of films can take place by Langmuir-Blodgett or horizontal deposition, or by film compression techniques. Film compression results in a reduced meniscus size and deposition on a vapor covered surface. This may take place with or without contact angle change, depending on the strength of the interaction between the film and the solid, as well as on the rapidity of evaporation of the adsorbed vapor. Decompression was shown to reverse the process, increasing the meniscus and causing multiple film deposition accompanied by a contact angle change. While we understand the process of single-phase, single-component deposition rather well, we have yet to carry out a detailed study of multiple-phase and multiple-component film deposition.

The generality of the approach taken should allow us to investigate how reduced gravitational forces will affect meniscus height and shape. In general, the meniscus shape and size is set by both the length scale of the solid and the capillary length of the liquid. Since the capillary length is determined by the ratio of surface tension to gravitational acceleration, the capillary length can change by orders of magnitude in reduced gravity environments. Thus the meniscus on a fixed size object can be significantly different in reduced gravity relative to normal gravity.

CP504, *Space Technology and Applications International Forum–2000*, edited by M. S. El-Genk
© 2000 American Institute of Physics 1-56396-919-X/00/$17.00

RESULTS AND DISCUSSION

The process of meniscus formation on a hydrophilic plate can be envisaged as taking place in two distinct steps: firstly on making contact, a film of water will cover the hydrophilic surface; secondly, in order to minimize the free energy of the system a meniscus arises reducing the surface area. In reality the two processes take place simultaneously. Since the plate is hydrophilic it is completely wetted by the water and thus created from two hypothetically perpendicular water surfaces, the precise nature of the plate surface is immaterial and the meniscus shape is independent of that surface. Thus the surface tension measured, and the meniscus observed, on glass or platinum or even on filter paper is always the same. The classical equations for describing the meniscus shape in terms of its height $z = h(x,y)$ are obtained by minimizing the energy of the liquid/solid system at fixed liquid volume. The minimum energy state is then described by Laplace's equation for the surface of the liquid, and Young's equation for the contact angle of the liquid with the solid surface. Laplace's equation is

$$\gamma_L . \kappa(h_) + \rho g h = 0,$$

where γ_L is the surface tension of the liquid, $\kappa(h)$ is the curvature of the liquid surface, ρ is the density of the liquid, and g is the gravitational constant. Young's equation can be written:

$$\cos(\theta) = \Gamma,$$

where Γ is determined by the solid/liquid surface energies and the contact angle is $\theta = 0$. If the plate is much wider than its thickness then the meniscus can be approximated by the meniscus of an infinitely long plate. The equation for the meniscus shape in this case is (Batchelor, 1990):

$$x/l = \cosh^{-1}(2l/h) - \cosh^{-1}(2l/h_o) + [4 - (h_o/l^2)]^{1/2} - [4 - (h^2/l^2)]^{1/2},$$

where x is the distance from the plate, h_o is the length of the contact line, and $l = [\gamma_L/\rho g]^{1/2}$ is the capillary length. We have applied this equation to plates of different lengths and compared the results to observations (Tumavitch, 1999). The end effects cause a decrease in the height of the meniscus along the plate. We have demonstrated that the full meniscus height h_o will only be achieved with plates of width >2.5 cm and then only in the center of the plate. Edge effects will produce an overall reduced meniscus in all plates <2.5 cm and at each edge in plates >2.5 cm (Figure 1).

While the assumption of an infinitely long, infinitely thin plate makes the calculation of the surface tension fairly simple, it introduces a possible error due to the change of the meniscus near the end of the plate. To determine the magnitude of the error we solved Laplace's equation with Young's contact angle for different geometry plates using a finite difference method. We found that the predictions of classical theory can differ markedly from the actual shapes observed. If an object with a round cross section (such as a round rod) is immersed into a liquid, then the meniscus shape is predicted by classical theory. A distinguishing feature of the meniscus shape is an unbroken, smooth contact line between the meniscus and the rod, which extends around the perimeter of the object. However, if one dips an object which has a thin cross-section with corners (such as a thin plate) the meniscus shape is not predicted by classical theory. The observed shape has a contact line which curves as the edge is approached (Figure 1) and the contact line does not go smoothly around a corner. Instead, it runs vertically along the corner (Figure 2). More importantly, the observed shape clearly displays a non-equilibrium contact angle on the thin side of the plate. If Young's equation held, the contact angle would be zero, but we observe the contact angle to be close to 45°. Thus, the meniscus shape around a thin plate does not correspond to the energy minimizing solution predicted by classical theory.

The behavior of a meniscus near a corner has previously been studied by Orr *et al.* (1977). They studied the meniscus shape around a square pin dipped into a liquid by solving Laplace's equation and Young's equation numerically. They found that the contact line bent downward near the corners of the pin, but the contact line was continuous from one face of the pin to the next. Our calculations are in agreement with these observations. However, when we consider the case of a rod with a rectangular cross section, we found that the solutions developed singularities associated with a contact line that extended vertically along the corner of the plate (see Figure 3).

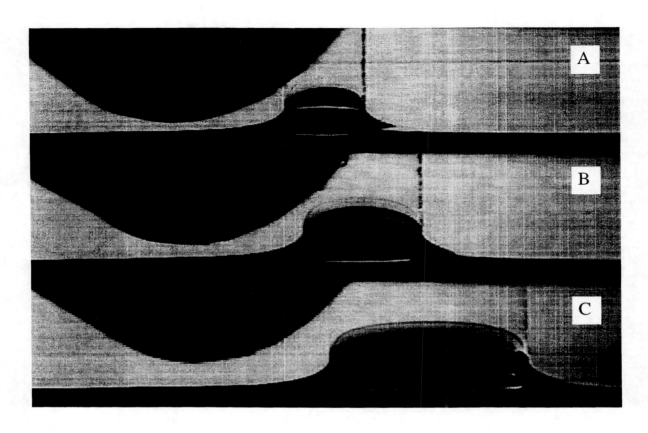

FIGURE 1. Meniscus edge contour as a function of plate length at 20.0°C; (A) 4.00mm (B) 6.00mm (C) 9.65mm.

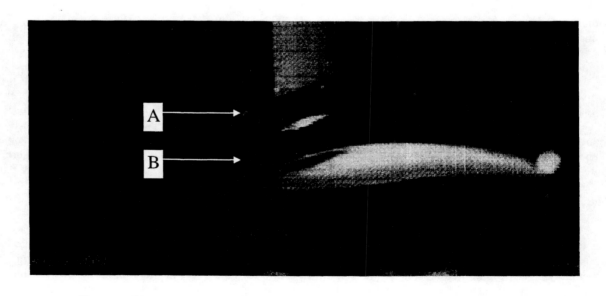

FIGURE 2. An image of the pocket between the planes of the edge and the face. The point of contact does not coincide with the intersection of the three-phase line and the plate edge.

Such a feature is not allowed in the classical model and indicates that the classical treatment breaks down at the corner. However, a close inspection of the actual contact line indicates that it indeed has a discontinuous character as it goes round a corner. To achieve the different meniscus heights on each of the adjacent faces to the corner, the contact line actually follows the corner vertically to join the two discontinuous meniscus heights. Such a feature cannot be treated with the standard classical equations for an energy-minimizing surface.

Another feature displayed by the meniscus for a thin plate is that Young's equation for the contact angle is not obeyed on the thin ends of the plate. If Young's equation holds, there should be a zero contact angle all the way round the thin plate in our experiment. Yet microscopic examination of the contact angle indicates that while Young's equation is obeyed along the faces of the plate, it is clearly not satisfied on the thin edge of the plate. Where the contact angle is observed to be about 45°.

Thus the meniscus around a thin plate displays two features that can not be explained by classical theories based on an energy-minimizing surface: (a) a vertical line at the corner of the plate, and (b) a non-equilibrium contact line at the plate edge. Such deviations from the classical theory do not violate any physical laws; the deviations are just a reflection of the fact that due to the geometry of the system, it is physically impossible for the meniscus to attain the theoretical shape which minimizes its total energy. Such deviations are commonplace when the system is constrained by geometrical considerations.

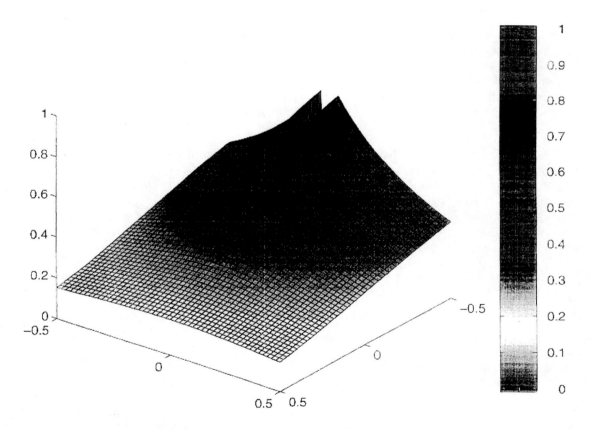

FIGURE 3. Calculation of the meniscus shape at the edge of a thin plate. The contact angle here is taken as 30°. The vertical and planar scales are reduced scales and can be adapted to specific plate dimensions.

A simple analogy would be the shape of a rope suspended between two poles. Gravity pulls the rope down, and the U-shape can be accurately predicted by minimizing the total energy of the system. However, such a U-shape is only relevant if the rope does not touch the ground. If the poles are too short, then the observed configuration has the rope running along a portion of the ground. The U-shape is still the theoretical shape that would minimize the total energy, however the system is constrained by geometry since the rope is not physically able to go underground to attain the U-shape. The configuration that the rope attains is that which does the best job at minimizing the energy subject to the constraint that the rope does not go underground. We believe that a similar **constrained energy minimization** is also responsible for the meniscus shapes in the case of a thin plate. We know from our numerical study of classical energy minimization theory that it predicts a meniscus shape which can not be physically attained. The meniscus shape actually observed is that which does the best job at minimizing the energy in the presence of the physical and geometrical constraints.

We can explain the deviations from classical theory by a mathematical model which relaxes the classical assumptions and then performs a constrained energy minimization. Of the two assumptions from the classical theory the Laplace equation is derived from minimizing the total energy of a surface independent of the geometry of the boundary, and thus will still apply even if the system can not physically attain its "global" energy minimum due to geometric constraints. Young's equation, on the other hand, is derived from minimizing the energy due to the contact of the meniscus with the boundary. This is precisely the condition that is not satisfied by the thin plate contact line. We thus suggest a mathematical model for the meniscus shape which relaxes the need for a constant contact angle imposed by Young's equation. Such a modification is not a novel idea in contact line problems; it is well known for liquid drops that the contact angle is not necessarily that predicted by Young's equation, and in fact there can be a range of contact angles.

By enforcing Laplace's equation for the meniscus but letting the contact angle vary, there will be many possible solutions because of the degree of freedom in the contact angle along the contact line. To determine which solution corresponds to the physical solution, we will determine which of the many solutions has the lowest energy. From a straightforward calculation of the total energy of each possible solution, we will use numerical optimization to find the solution which minimizes the total energy of the system, subject to contact angle freedom (Nedler and Meade, 1965, and Kirkpatrick, 1984). Our conjecture is that the system minimizes its energy by attaining the actual configuration observed in the thin plate (Figure 2), with the contact angle given by Young's equation along the long edge of the plate only. The current computer code has to be modified in order to incorporate these features. These are as follows:

- An adaptive gridding of the meniscus shape to cluster points where the meniscus height is changing rapidly (many node points near the plate and fewer node points away from the plate). This will be necessary in order to achieve an acceptable resolution of the meniscus shape within reasonable computing times.
- The capability of describing a zero contact angle
- The capability of describing a contact line which runs vertically along the corner of the plate.
- An efficient optimization method to find the distribution of contact angles along the plate that results in a minimum energy for the system.

An associated issue is that we can also investigate is the behavior of the contact angle near a "nearly-sharp" corner, by modifying the shape of the plate in our numerics, and also by the actual modification of actual plate edges.

It will also be of considerable interest to examine the behavior of different materials, especially those that exhibit contact angles other than zero. One such material would be SiO_2, which is of considerable interest in the semiconductor field and has been the subject of wetting, and film deposition studies [Spratte and Reigler (1994)]. The modification of such surfaces is a matter of great commercial interest and film deposition is one way in which this can be achieved. In addition to the standard deposition techniques of Langmuir-Blodgett and horizontal deposition we have used film compression-decompression cycles to achieve monolayer and mutilayer deposition. Since surface tension acts to support the meniscus and acts upwards, any reduction of the surface tension will reduce the size of the meniscus. Thus the compression of a film of dipalmitoylphosphatidylcholine (DPPC) reduces the meniscus to almost zero (Figure 4). A DPPC film is initially deposited on a thin liquid film adsorbed on the plate, but much of this water subsequently evaporates creating a more direct DPPC-film/plate contact. On decompression the surface tension again increases, as does the meniscus, but now the surface is hydrophobic due to the deposited film and a non-zero contact angle will now be evident.

Film deposition becomes increasingly more complex when the film is already biphasic or multiphasic prior to deposition. Figure 5 shows one such example for a myristic acid (MA) system being compressed and deposited on a wetted glass surface.

In summary, we have found a possibly important deviation from the classical equations for the meniscus shape that occurs at the corners of solid surfaces. We propose an explanation for this deviation based on "constrained energy minimization" and outline a method for verifying our conjecture. In addition to the importance of corner effects, we have also made headway in understanding the mechanics of surfactant film deposition on solid surfaces. We suggest that the corner-induced meniscus shape may be particularly important in reduced gravity environments because of the large change in the capillary shape, while an understanding of meniscus dynamics during film deposition processes might lead to improved techniques for coating materials.

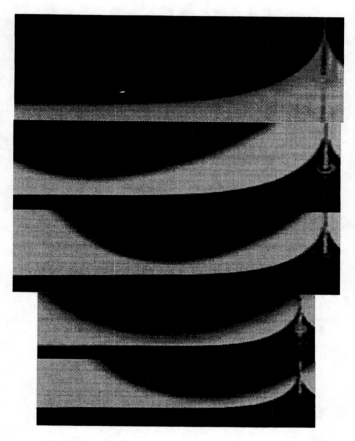

FIGURE 4. Meniscus contours of DPPC on water (top to bottom) 72.4, 54.8, 37.2, 22.9, 15.0 dynes/cm. T=22.2°C. The actual size of the meniscus in the two lowest images are actually 17% smaller than shown.

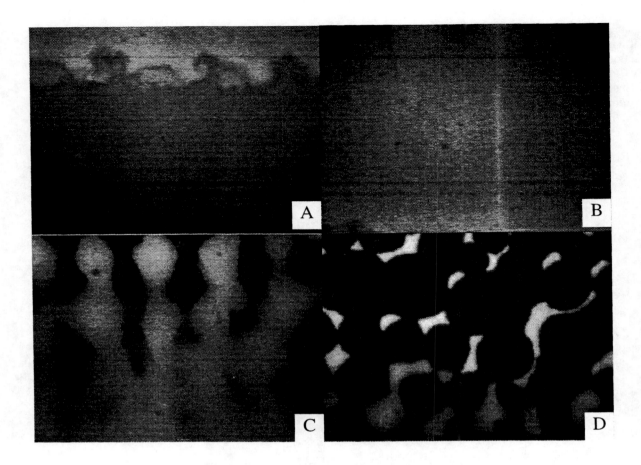

FIGURE 5. Monolayer deposition above the three-phase line for myristic acid. The region shown is near h_o, the original three phase line location. The swirls represent the G/LE transition region (top left). The LE region is homogeneous (top right), however, a monolayer film is deposited at increasing surface pressures (lower left). In the LE/LC transition region, domains span across the three-phase line (lower right). The working three-phase line is 0.710 mm from h_o.

REFERENCES

Batchelor, G.K. (1990) *Introduction to Fluid Dynamics* Cambridge University Press.

Kirkpatrick, S. (1984) *J. Statistical Physics,* 34, 975-986.

Nedler, J.A. and Meade, R. (1965) *Computer J.* 7, 308-313.

Orr, F.M. Scriven, L.E. and Chu, T.Y. (1977) *J. Colloid Interface Sci.,* 60, 402.

Spratte, K. and Reigler, H. (1994) *Langmuir,* 10. 3161-3173.

Tumavitch, N.J. "A New Approach to the Thermodynamics of Pure and Monolayer Covered Aqueous Interfaces", Ph.D. Thesis, University at Buffalo, Buffalo Ny, 14260.

Microgravity and Earth Thermal Diffusion in Liquids Holographic Visualization of Convection

Jean Colombani and Jacques Bert

Université Claude Bernard - Lyon I
Département de Physique des Matériaux (C.N.R.S. - UMR 5586)
43 Bld 11 Novembre 1918, F-69622 Lyon-Villeurbanne
Tel. +33 4 72 44 81 55; Jean.Colombani@dpm.univ-lyon1.fr; Jacques.Bert@dpm.univ-lyon1.fr

Abstract. We give a critical analysis of the last results obtained during two thermal diffusion experiments on molten salts performed during the flights D1 and D2 of the Space shuttle. Because of the limitation coming from the method of measurement (simple recording of the evolution of the thermoelectric power) without any visualization of the liquid stability, we used on earth an holographic technique in order to make new and improved experiments permitting both the measurement of the thermal diffusion phenomenon and the visualization of eventual convective disturbances in the cell. Consequently, convective oscillatory instabilities can then be evidenced on earth, as an intermediate regime between pure conduction and stationary convection. This up-to-date technique will be adapted to the microgravity environment of the International Space Station in the near future.

INTRODUCTION

A new scientific interest for the studies dealing with thermal diffusion has recently appeared (see for example the Proceedings of the 3rd International Meeting on Thermodiffusion, Mons, Belgium, September 1998). This undeniable emergence both lies on microscopic and macroscopic considerations which we have already analyzed in detail (Colombani, 1998). In addition, we must stress on the industrial impact of this effect which is connected to problems encountered in high purity crystal processing, in the solidification of alloys, for the obtainment of metals from the electrolysis of molten salts and about the behavior of hydrocarbons submitted to the geothermal gradient inside oil reservoirs. The difficult study of such cross effects necessarily involves the cooperation of several kinds of scientists from the physicist or chemist, to the fluid mechanics specialist because the major experimental problem for thermal diffusion measurements comes from the convective instabilities which spontaneously appear on earth if some conditions are overwhelmed and even in a microgravity environment. In this paper, we give the results obtained after a new data analysis on two thermodiffusion experiments performed in Space on molten salts systems. Due to insufficiencies coming from these blind experiments, we performed new experiments on Earth in liquids for which viscous forces could prevent convective motions, in order to evidence the presence or the absence of convective disturbances and to validate a visualization method (holographic interferometry) which could be applied to a microgravity environment.

MICROGRAVITY THERMODIFFUSION ON MOLTEN SALTS

The two Space long duration experiments performed by the group in the Space shuttle during the D1 and D2 flights have permitted for the first time, the microgravity measurement of the interdiffusion and thermal diffusion coefficients in a solution of molten salts. The choice of the materials for such experiments can be justified by their thermodynamic parameters; indeed, due to the value of their Rayleigh number Ra, thermal diffusion measurements

CP504, *Space Technology and Applications International Forum–2000*, edited by M. S. El-Genk

on molten salts are only possible in microgravity. This dimensionless number, is defined as: $Ra = \alpha \, g \, \Delta T \, h^3 / D_T \, \upsilon$, where α is the thermal expansion coefficient, g, the acceleration due to gravity, ΔT, the temperature difference between the two ends of the cell, h, the cell length (in a direction parallel to the temperature gradient), D_T, the thermal diffusivity and υ, the fluid kinematic viscosity. Its value is then about 50 times larger than that of a liquid metal in equivalent conditions, inducing spontaneous convection on Earth as soon as the tiniest disorientation between the gravity field and the temperature gradient exists. These unwanted disturbances explain the non-reproducibility of the experimental results obtained on earth by the various experimenters. This class of liquids also has well known characteristics, permitting a reliable predictive determination of the convective behavior when a cell with a given aspect ratio is submitted to a temperature gradient in various conditions of gravity. Hydrodynamic simulations then lead to choose the best cell geometry in order to measure the least-disturbed thermal diffusion effect in the microgravity environment. Furthermore, molten salts are ionic conductors, thus permitting the recording of the thermal diffusion effect by means of the induced thermoelectric power variation with time.

Among the various possibilities offered by molten salts, the choice of a silver iodide-potassium iodide mixture ($AgI_{0.72}/KI_{0.28}$ – eutectic mixture) was made because of the interesting phase diagram of this mixture, showing a deep eutectic point (Bert, 1997), together with the possibility of using convenient silver electrodes for the measurement of the thermoelectric power, image of the diffusion phenomenon.

A complete theoretical analysis of the mechanical phenomena was made prior to the experiments with the help of fluid mechanics specialists (Henry, 1987). We have considered the case of a vertical geometry with the temperature gradient parallel to gravity. As the cell orientation with respect to the residual gravity including the g-jitters cannot be controlled in spatial experiments, it has been necessary to deal with situations with non-zero inclination, for which convection cannot be avoided, in order to see how it influences the separation phenomenon. So, instead of searching for a stability threshold, we investigated the way in which the mass fraction profiles created by the thermal diffusion is distorted. The calculations have been performed with the molten salt numerical values (Prandtl number = 0.6 and Schmidt number = 60) and a 100°C/cm theoretical temperature gradient. This corresponds to Rayleigh number values of 4.34 and 0.434 for respective gravity values of 10^{-3} g_o and 10^{-4} g_o, g_o being the earth gravity. This preliminary study leads us to a cylindrical cell (diameter 3 or 5 mm, length 3 cm) with the two possible aspect ratios (length/diameter) 6 or 10 which were supposed to involve different hydrodynamic behaviors of the mixture in the microgravity environment, so permitting a validation of the calculation.

Although a few technical solutions might be improved for future flights, these first experiments can be considered as a good experimental success. The two main scientific results were the following (Bert, 1997):

• The direction of the thermodiffusion effect in the mixture was measured reliably for the first time: the Soret coefficient S_T (ratio of the thermal diffusion coefficient D' to the interdiffusion coefficient D_{Ag+}) is positive, i.e. conventionally, the silver (heavy) ions migrate towards the cold end of the cell; this direction can be determined either by the recording of the thermoelectric power measurement with time (Fig. 1), or by the chemical analysis of the sample which has been made after the flight;
• The quantitative value of the Soret coefficient determined from the thermoelectric power values is $S_T = (+0.9 \pm 0.2) \times 10^{-3}$ K^{-1}.

But beside these successes, we encountered the main following limitations:

• The interdiffusion coefficient (5.9×10^{-9} $m^2 s^{-1}$), calculated from the characteristic time of the Soret experiment, is five times smaller than expected from earth experiments (30×10^{-9} $m^2 s^{-1}$);
• The Soret coefficient calculated from the thermoelectric power variations with time, measured with silver electrodes in the cell, is not in agreement with the value coming from the chemical analysis of the sample after the flight: $S_T = (+1.4 \pm 0.1) \times 10^{-3}$ K^{-1};
• During intermediate times, the thermoelectric power variation has deviated from the expected exponential behavior;
• The variation of the convective behavior with the cells' aspect ratio has not been evidenced through the obtained diffusion results.

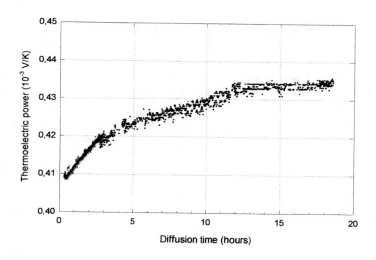

FIGURE 1. Variations of the thermoelectric power with time, image of the concentration evolution during D1 (5 hours) and D2 (20 hours) experiments (values obtained after ultimate corrections on the experimental data, particularly for the stray potentials existing inside the experimental cell and recently determined)

These four limitations might have different causes but what precisely handicaps the interpretation is that these measurements were practiced in the blind. The spatial environment should normally provide convection-free diffusion processes, but we had no possibility to check the actual absence of convection and the encountered troubles could originate in convective flows due to the residual gravity, or eventually in a slight Marangoni convection. If we take account for the care which has been taken by the experimenters in order to avoid this kind of disturbance (complete cell filling, absorption of the salt volume expansion when melting by means of a porous medium), this last type of perturbation should not however theoretically occur.

Thus has arisen the necessity of performing a new experiment enabling to visualize the mass fluxes during the diffusion. In this way, the spatial relevance of this kind of experiment could properly be demonstrated.

EARTH HOLOGRAPHIC MEASUREMENT OF THE THERMAL DIFFUSION

Among the possible visualization techniques, we chose holographic interferometry which enables the simultaneous access to physical information (thermal and solutal gradients) and hydrodynamic diagnostics (flows). So, we made a preliminary phase of experimentation on earth, prior to future microgravity experiments in the International Space Station in order to state the method feasibility. As specified above, thermal diffusion on earth is most of the time disturbed by gravity induced non-diffusive fluxes. So, in order to compensate the buoyancy effects and make measurements possible, we had to use a trick: taking advantage from our long experience on vitrifying liquids, we chose a ($LiCl$, H_2O) mixture also pertaining to the class of ionic liquids, but exhibiting a high viscosity increase at low temperature which delays the convection thresholds.

The main experimental doubts that had to be cleared up were the following (Colombani, 1999-1):

• Temperature and concentration variations with time are both diagnosed with the holointerferometry method, by the same fringes implying comparable refraction index changes, so the problem of decoupling arises. A physical solution to this problem is given by the fact that heat and mass transport exhibit characteristic times that have considerably different orders of magnitude. This is quantified by the Lewis number $Le = D/D_T$, D being the interdiffusion coefficient and D_T, the thermal diffusivity. In liquids, Le is of the order of 10^{-2}, so, according to this time scale difference, we can consider that the mixture remains homogeneous during the time corresponding to the establishment of the temperature gradient.

• Another point to deal with was the unusual simultaneity of the diffusion measurement and the hydrodynamic diagnostic. A pure diffusive Soret experiment is evidenced by a good parallelism of the interference fringes (perpendicular to the cell axis): this indicates the existence of a mass flux in the direction of the long axis of the diffusion cell. The development of an instability due to exceeding the temperature gradient threshold of thermal convection creates a distortion of the fringes, which can be easily detected on the fringe pattern.

Significant scientific results were inferred from these earth experiments both from the mechanical and from the physicochemical standpoints (Colombani, 1998, 1999-2):

• First, the reliable determination of the evolution of the Soret coefficient of (LiCl, H_2O) in a variety of thermodynamic conditions has been carried out. The observation of the continuous increase of S_T with temperature has strengthened this trend, already noticed in the literature in all other studied aqueous alkali halides. The study of the estimated temperature of inversion of the effect ($S_T=0$) has permitted to test the limits of Lin's volumetric model of thermal diffusion in aqueous solutions (Lin, 1991). The existence of a steep minimum of the evolution of S_T for this mixture with concentration had also been observed in aqueous sodium and potassium chlorides. The comparison of the behavior of these three salt solutions has led to modify Chanu-Gaeta's entropic theory of thermal diffusion (Chanu, 1967, Gaeta, 1982).

• Second, in order to test the ability of the procedure to highlight convective disturbances, we have studied the hydrodynamic behavior of the solution. In the portion of the stability diagram corresponding to a heating from below and salt migrating downwards, three convection regimes exist, as in a more extended geometry: stationary convection, oscillatory convection and steady overturning convection (Platten, 1984). The most amazing pattern is the second one where the restricted geometry impedes travelling waves to develop and where the interplay between the destabilizing thermal gradient and the stabilizing solutal gradient leads to a convection cell changing direction periodically; Fig. 2 gives an example of such patterns in the case of the mixture (LiCl, 9.7 H_2O)

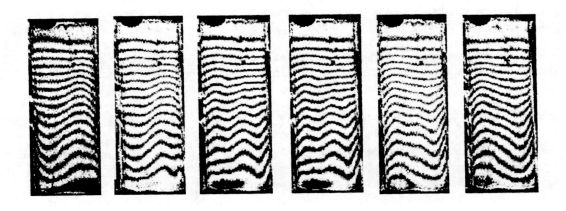

FIGURE 2. Evolution of isotherms during one half period in (LiCl, 9.7 H_2O), for T_{cold} =-31.9°C, Δc=0.54 wt% and ΔT=29.1°C. The period is 348 seconds and the various holograms are equally time separated.

These measurements have then yielded the confirmation that Soret experiments are extremely difficult to perform in liquids with standard viscosity on Earth, then confirming the necessity of the microgravity environment. In addition, we have shown that even in such conditions, the optical holography diagnostic is a necessity if one wants to know the reality of the disturbance-free thermal diffusion measurements.

CONCLUSION

The thermal diffusion experiments performed during two different flights of the shuttle have permitted us to obtain the first reliable values of the Soret coefficient on molten salts. We have shown that this coefficient is positive, which means that the heavier component migrates toward the cold end of the cell. In fact, this physical situation should permit a measurement of this effect on earth, if the experimenter takes care of setting a perfectly vertical cell with a heating from the top. However, as this perfect situation cannot experimentally be realized, the tiniest disorientation of the cell versus the gravity axis initiates hydrodynamic disturbances leading to an imperfect Soret measurement in the best case, or to a total obstacle to any possible measurement in the worst case. In addition, as radial gradients necessarily exist inside the cell, even if it was perfectly vertical, these stray gradients would also initiate high enough disturbances, acting as a total experimental obstacle. So, the obtained results should completely justify the spatial relevance of this type of experiment.

But problems remain, which are based on the main limitations encountered during the past spatial experiments: small measured value of the interdiffusion coefficient, disagreement between the Soret coefficient coming from the thermoelectric power and from the sample chemical analysis, and fluctuations of the thermoelectric power measurements during the flights. In order to determine the real validity of this type of experiments, we have proposed completely new thermal diffusion experiments to be performed in the International Space Station. These experiments are based on an optical approach of the diffusion with the possibilities offered by holographic interferometry, which permits the visualization of the diffusion phenomenon as well as the eventual existence of hydrodynamic disturbances. The earth experiments we performed on a viscous aqueous system have demonstrated the real feasibility of such measurements and the multiple possibilities of this kind of set-up, which can easily be adapted inside the Fluid Science Laboratory of the International Space Station.

AKNOWLEDGMENTS

The work on space experiments and on earth holographic interferometry was supported by the French space agency (CNES). We are also indebted to Daniel Henry and Hamda BenHadid for numerous and helpful discussions about the hydrodynamic behavior of the thermal diffusion cell.

REFERENCES

Bert, J., and Dupuy-Philon, J., "Microgravity Measurement of the Soret Effect in Molten Salts Mixtures," *J. Physics: Condensed Matter*, **9**, 50, pp. 11045-11060 (1997).

Chanu, J., "Thermal Diffusion in Halides in Aqueous Solutions," *Adv. Chem. Phys.*, **13**, p. 349 (1967).

Colombani, J., Dez, H., Bert, and J., Dupuy-Philon J., "Hydrodynamic Instabilities and Soret effect in an Aqueous Electrolyte," *Phys. Rev. E*, **58**, pp. 3202-3208 (1998).

Colombani, J., and Bert, J., "Holographic Convection Visualization during Thermotransport Studies – Application to Microgravity Experiments," *Meas. Sci. Tech.*, **10**, pp. 886-892 (1999).

Colombani, J., Bert, J., and Dupuy-Philon J., "Thermal Diffusion in (LiCl,RH$_2$O)," *J. Chem. Phys.*, **110**, pp. 8622-8627 (1999).

Gaeta, F.S., Perna, G., Scala, G., and Bellucci, F., "Nonisothermal Matter Transport in Sodium Chloride and Potassium Chloride Aqueous Solutions. 1. Homogenous System (Thermal diffusion)," *J. Phys. Chem.*, **86**, p. 2967 (1982).

Henry, D., and Roux B., "Three-dimensional Numerical Study of Convection in a Cylindrical Thermal Diffusion Cell: Inclination Effect," Phys. Fluids 30, pp. 1656-1666 (1987).

Lin, J.L., Taylor, W.L., Rutherford, W.M., and Millat, J., "Secondary Coefficients," in *Measurement of the transport properties of fluids*, edited by W.A. Wakeman, A. Nagashima and J.V. Sengers, Oxford, pp. 321-387 (1991).

Platten, J.K., and Legros, J.C., *Convection in liquids*, Berlin, Springer-Verlag, p. 566 (1984).

Periodic Oscillations of Low Prandtl-Number Fluids in Rectangular Enclosures

Daniel W. Crunkleton, Ranga Narayanan, and Timothy J. Anderson

Department of Chemical Engineering, University of Florida, Gainesville, FL 32611
(352) 392-0882; dcrunkle@che.ufl.edu

Abstract. Periodic oscillatory flows in low Prandtl-number fluids are numerically characterized in this study. A mathematical model is developed to solve the three-dimensional mass and momentum conservation equations using a control-volume algorithm. For various aspect ratios, the second critical Rayleigh number is determined. For an aspect ratio (length/height) of 0.25, macroscopic oscillations are unmistakably observed; while as the aspect ratio is increased, oscillations tend to dampen.

INTRODUCTION

Measurement of buoyancy-driven flow patterns in liquid metals and semiconductors is extremely difficult because of their opacity, low Prandtl-number, and high temperature. Moreover, these convective patterns markedly effect the quality of materials grown. Therefore, researchers must always be cognizant of the presence or absence of convective currents in low Prandtl-number fluids. From an experimental point of view, several techniques are currently used to sense convection in liquid metals. Moreno (Moreno, 1980) uses banks of thermocouples to detect convection patterns in silicone oil. However, temperature measurements have limited applicability to low Prandtl-number fluids because any heat flux within the fluid disperses quickly, thereby making any temperature resolution difficult. The use of tracer particles is another popular method for detecting convection currents (Ueda, 1982; Hsieh 1997); yet these techniques introduce contaminants into the growth system which is unacceptable for many crystal growth techniques. Therefore, a non-intrusive method that detects convective flows in low Prandtl-number *in situ* is needed. One such method is the subject of this research. This method uses electrochemical titration to introduce a specified amount of oxygen into a liquid metal system. This packet of oxygen is carried with the flow and is detected by various sensors located along the periphery of the enclosure. This technique is advantageous because of its fast response time and is readily adaptable to microgravity applications. A schematic of the prototype is shown in Figure 1.

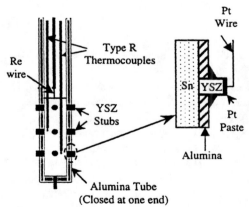

FIGURE 1. Schematic of electrochemical titration method.

The convection under investigation in this work is the classical Rayleigh-Bénard convection – an enclosure of fluid heated from below. As a low Prandtl-number fluid passes its first critical Rayleigh number, Ra_{c1}, steady convective

CP504, *Space Technology and Applications International Forum–2000*, edited by M. S. El-Genk
© 2000 American Institute of Physics 1-56396-919-X/00/$17.00

cells begin; while as the second critical Rayleigh number (Ra_{c2}) is surpassed, the cellular convection patterns oscillate. This work attempts to determine the point at which these periodic oscillations begin.

PREVIOUS WORK

Several authors have studied convection such as the type investigated herein. Computational studies are usually performed using either linear stability analysis or a direct numerical calculation. Linear stability analysis was initiated by the pioneering book of Chandrasekhar (Chandrasekhar, 1969); other analyses were performed by Catton (Catton, 1972) and Rosenblat (Rosenbalt, 1982). In these analyses, the study of small disturbances to the quiescent base states is considered where quadratic and higher order interactions are excluded. The only information that one can obtain is the onset convection condition (Ra_{c1}) and preferred structure – Ra_{c2} cannot be determined by linear stability analysis.

Ozoe and Hara (Ozoe, 1995) present calculation of 2D oscillations in a fluid with Pr=0.01 and where γ=4, and find a Ra_{c2} of 6000. They also make the important conclusion that the difference between Ra_{c1} and Ra_{c2} decreases as Pr decreases. Most of the three-dimensional numerical studies of natural convection in cavities heated from below have been restricted to moderate to high Prandtl number fluids (e.g. air, water, and silicone oil) in right circular cylinders (Neumann, 1990) and rectangular enclosures (Oertel, 1980; Pallarès, 1996; Gomiciaga, 1992; Tang, 1997). However, these solutions to the Rayleigh-Bénard problem for moderate- to high-Prandtl number fluids cannot be applied to low-Prandtl number regime for two reasons. First, for low-Prandtl-number fluids, the unsteady and the non-linear terms in the momentum equation become appreciable and this leads to time-dependent instabilities. Second, the inertial force dominates the flow, and the viscous effects are mainly confined to thin boundary layers near the walls.

This work seeks to fill a dearth in the literature concerning the oscillatory flow of fluids of a very low Prandtl-number in a variety of aspect ratios. In this work, the effect of the aspect ratio on the convection of liquid tin in rectangular enclosures is presented. Conclusions are made by observing the time-evolution of these convective cells in a macroscopic enclosure of fluid.

NUMERICAL MODEL

The mass, momentum, and energy conservation equations are (Bird, 1960)

$$\nabla \bullet v' = 0$$
$$\frac{\partial v'}{\partial t'} + v' \bullet \nabla v' = \nabla^2 v' - \nabla p' - \frac{Ra}{Pr}\left(T' - T_o'\right)$$
$$\frac{\partial T'}{\partial t'} + (v' \bullet \nabla)T' = \frac{1}{Pr}\nabla^2 T', \tag{1}$$

where v is the velocity, t is time, p is pressure, T is the temperature, and T_o is a reference temperature discussed below. In this equation, the prime " ' " indicated that the physical quantity has been made dimensionless. The dimensionless quantities are defined as

$$v' = \frac{v}{(v/H)}, \quad t' = \frac{t}{(H^2/v)}, \quad p' = \frac{p}{(\rho_o v^2/H^2)}, \quad \nabla' = H \cdot \nabla, \text{ and } \quad T' = \frac{T - T_C}{T_H - T_C}, \tag{2}$$

where H is the height, T_C is the cooler temperature, T_H is the hotter temperature, and ρ_o is a quantity to be defined later. In equation (1), the Boussinesq approximation has been made; that is, all physical quantities are assumed to be constant except for the body force term in the momentum conservation equation. These constant quantities are designated by a subscript "$_o$". Therefore, ρ_o is the (assumed constant) density, which is taken at the temperature T_o'. For this study, T_o' in equation (1) is assumed to be a constant value of 0.5.

The modeled geometry is a 3D rectangular box with the bottom plane and vertical walls to be no-slip surfaces, and the top (free) surface to be stress-free. The vertical walls are adiabatic and the top and bottom walls are isothermal. This is shown graphically in Figure 2.

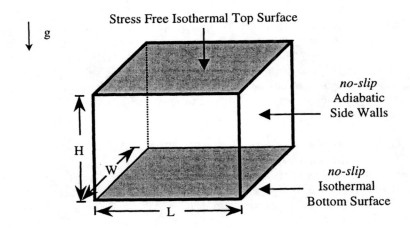

FIGURE 2. Modeled geometry and boundary conditions.

To carry out these calculations, a control volume algorithm was constructed using the SIMPLE algorithm as described by Patankar (Patankar, 1980). In this method, the space is discretized according to the following equations:

$$X_i = X_{i-1} + \Omega \sin\left(\frac{i \times \pi}{N+1}\right),$$

where X is the co-ordinate of the grid point being calculated, N is the number of grid points in a particular direction, i is a counting integer such that i=0,1,...,N, and π=3.14159. The function Ω is defined by

$$\Omega = \frac{L}{\sum\limits_{0}^{N} \sin\left((i \times \pi)\Big/(N+1)\right)},$$

where L is the length of the side of the box. This grid placement procedure provides more grid points near the edges of the domain and less towards the center.

To determine the exact point at which Ra_{c2} is surpassed, the amplitude of the flow oscillation, A, is calculated using fast Fourier transform and $Ra \Rightarrow Ra_{c2}$ as $A \Rightarrow 0$.

PERIODIC OSCILLATIONS

The control volume algorithm was verified in a variety of ways. Initially, all walls are taken to be stress-free and compared to the results of Rosenblat (Rosenblat, 1982). For aspect ratios of γ_x=0.4 and γ_y=0.25, and Ra=6000, Rosenblat's solutions show a (1,0) mode of convection (i.e., one roll in the x-direction). The algorithm used in this study duplicates this result. The algorithm was also verified by comparing it to the numerical results of Pallarès *et al.* (Pallarès, 1996). These calculations were performed for Ra=4000 and 9000 in a cubical container with all no-slip walls. The algorithm used in this study agreed very well with the results of Pallarès.

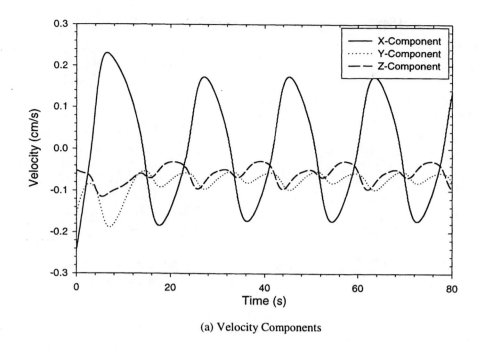

(a) Velocity Components

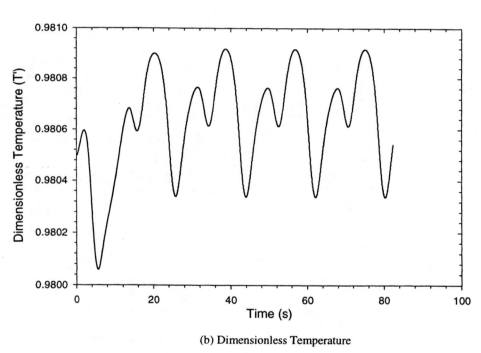

(b) Dimensionless Temperature

FIGURE 3. Time evolution of periodic oscillations for $\gamma=0.25$ and Ra=400,000.

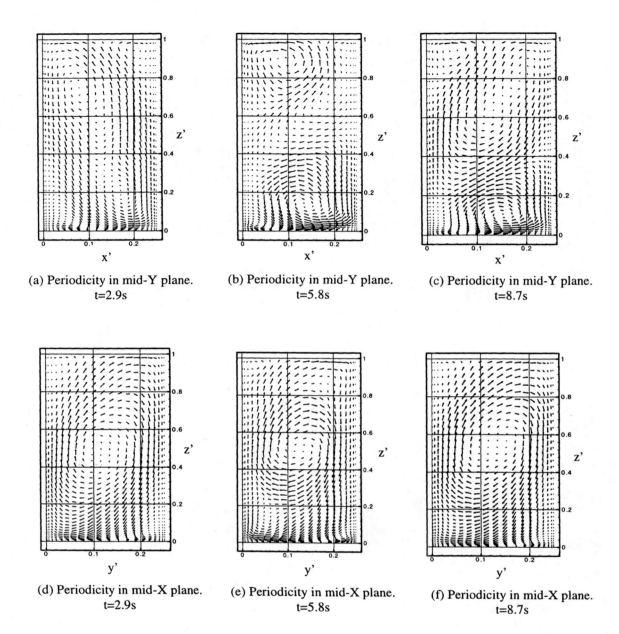

(a) Periodicity in mid-Y plane.
t=2.9s

(b) Periodicity in mid-Y plane.
t=5.8s

(c) Periodicity in mid-Y plane.
t=8.7s

(d) Periodicity in mid-X plane.
t=2.9s

(e) Periodicity in mid-X plane.
t=5.8s

(f) Periodicity in mid-X plane.
t=8.7s

FIGURE 4. Time evolution of periodic oscillations for $\gamma=0.25$ and Ra=400,000.

$$\gamma = 0.25$$

All calculations in the periodic oscillatory regime were calculated using the converged velocity profile for the Ra_{c1} calculation. A 20×30×30 mesh was used for these calculations. The Rayleigh number was slowly increased until oscillations were observed in the flow field at a point near the bottom of the computational domain (x'=0.125, y'=0.125, z'=0.00225). This point was chosen because it is close to the top of the growing crystal. Any oscillations that are present near the melt/solid interface will affect the morphology and physical properties of the proximity most and, therefore, were chosen to track.

The first critical Rayleigh number for $\gamma=0.25$ was calculated to be 250,000 and this was the starting point for the post-Ra_{c1} calculations. The Rayleigh number was slowly increased, using a Ra step of 50,000, until oscillations were detected. For this aspect ratio, Ra_{c2} was found to be 300,000. A trace of the velocity components and the dimensionless temperature for Ra=400,000 are shown in Figure 3a and 3b. To visualize the actual macroscopic structure of the flows, the Rayleigh number was increased to 400,000 so as to amplify the oscillations, and Figures 4a through 4f show the tracking of these oscillations. In the mid-Y oscillations (Figures 4a-4c), there is larger, oval-shaped convective cell that begins to split into two somewhat circular cells. Eventually, the two circular cells transform into another large, oval-shaped cell in the center of the domain. To get a better idea of the periodicity of this convection, a movie of the convection showing several individual stills will be shown during the presentation associated with this work.

Other Aspect Ratios

Calculations for $\gamma=1.0$ and $\gamma=2.0$ have been performed and oscillations have been detected; Ra_{c2} for $\gamma=1$ is calculated to be 83,500 and Ra_{c2} for $\gamma=0.25$ is found to be 43,500. However, when macroscopic time evolutions of the flow were constructed, cellular oscillations were not observed. It is possible that the larger aspect ratio (i.e., the flatter the box) may dampen the visualization of the oscillations. These two aspect ratios are, therefore, still under investigation.

REFERENCES

Bird, R., Stewart, W., Lightfoot, E., *Transport Phenomena*, New York, Wiley, 1960.

Catton, I., "The effect of insulating vertical walls on the onset of motion in a fluid heated from below." *Int. J. Heat Mass Transfer* **15**, 665-672 (1972).

Chandrasekhar, S.. *Hydrodynamic and Hydromagnetic Stability*. London, Oxford UP, 1969.

Gomiciaga, R., Ramos, E., Rojas, J., and Yianneskis, M., "Natural convection flow regimes in a cavity." *ICHEME Symposium Series*, **129**, 491-495 (1992)

Hsieh, S.-S., and Yang, S.-S., "Flow structure and temperature measurements in a 3-D vertical free convective enclosure at high Rayeigh numbers," *Int. J. Heat Mass Transfer*, **40**, 1467-1480, (1997).

Moreno, J., Jimenez, J., Córdoba, A., Rojas, E., and Zamora, M., New experimental apparatus for the study of the Bénard-Rayleigh problem," *Rev. Sci. Instrum.*, **51**, 82-85, (1980).

Neumann, G.. "Three-dimensional numerical simulation of buoyancy-driven convection in vertical cylinders heated fron below." *J Fluid Mech* **214**, 559-578 (1990).

Oertel, H., "Three-dimensional convection within rectangular boxes," *19th National Heat Transfer Conference, ASME*, **8**, 11-19 (1980).

Ozoe, H., and Hara, T., "Numerical analysis for oscillatory natural convection of low Prandtl number fluid heated from below," *Numerical Heat Transfer A.*, **27**, 307-317, (1995).

Pallarès, J., Cuseta, I., Grau, F.X., and Giralt, F., "Natural convection in a cubical cavity heated from below at low Rayleigh numbers." *Int J. Heat Mass Transfer* **39**, 3233-3247 (1996).

Patankar, S., *Numerical Heat Transfer and Fluid Flow*, New York, Hemisphere, 1980.

Rosenbalt, S., "Thermal convection in a vertical circular cylinder," *J. Fluid Mech.* **122**, 395-410 (1982).

Tang, L., Tate, T., and Tsang, T., "Temporal, spatial and thermal features of 3-D Rayleigh-Bénard convection by a least-squares finite element method." *Comput. Meth. Appl. Mech. Engrg.* **140**, 201-219 (1997).

Ueda, M., Kagawa, K., Yamada, K., Yamaguchi, C., Harada, Y., "Flow visualization of Bénard convection using holographic interferometry," *Appl. Optics*, **21**, 3269-3272, (1982).

Ground Based Experiments about Stability of Deformable Liquid Bridges

Shevtsova V.M., Mojahed M. and Legros J.C.

MRC, ULB, CP-165, 50 av. F.D.Roosevelt, 1050 Brussels, Belgium

e-mail: vshev@ulb.ac.be

Abstract. In this study the primary goal is to investigate the loss of stability of the steady convection in deformable liquid bridges under different surrounding thermal conditions (Pr=105). For the first time, it is experimentally obtained that the two branches of the function critical Marangoni numbers versus variation of the volume, correspond to different azimuthal wave numbers. The branch on which ΔT_{cr} grows with increasing volume belongs to m = 1, and the descending branch belongs to the azimuthal wave number m= 2. The influence of the mean temperature inside liquid bridge (temperature of the cold rod) on the onset of instability is experimentally considered. These experiments can display the role of temperature dependent viscosity and of the thermal conditions around the liquid bridge.

EXPERIMENTAL SET-UP

To study the thermal convection in silicone oil 10cSt (Pr = 105) two different set-up's were developed, the scheme of the last one is shown in Fig.1. The upper rod is fixed in such a way that it can be moved for adjustment in three directions. A heating element (Resistor Minco R $\approx$ 100 Ohms) is mounted around the upper rod. The lower rod is kept at constant temperature using a thermoregulated water cooling system. The rods to hold the liquid zone are made from Aluminum alloy (λ = 164W/mK) and have the same diameter 2R = 6mm. The present results correspond to liquid zone of the same length L = 3.6mm, the aspect ratio is thus Γ = L/R = 1.2. To establish a floating zone a dedicated electronic syringe pusher injects the liquid into a gap between the rods.

The temperature oscillations due to time-dependent convection have been measured by inserting five shielded thermocouples inside the liquid at the same radial and axial positions but different azimuthal angles. These thermocouples have been inbedded into the liquid through the upper rod to prevent the disturbance of the free surface. Unlike to general opinion, we did not observe strong influence of the amount of thermocouples on the critical ΔT. It should be mentioned that the free surface, which is responsible for the driving force, has not been disturbed. The choice of 5 thermocouples allows to determine without ambiguity the critical wave number and the type of hydrothermal wave; one should have an even number of thermocouples, but 3 were not enough to determine clearly m = 2.

All temperature signals given by thermocouples were amplified and band-pass filtered before being recorded by computer. To protect the liquid bridge from the ambient air fluctuations a special glass box of large volume has surrounded the entire set-up.

CP504, *Space Technology and Applications International Forum–2000*, edited by M. S. El-Genk
© 2000 American Institute of Physics 1-56396-919-X/00/$17.00

For some experiments the liquid bridge was also placed in a cylindrical volume with an internal diameter 2R = 12mm, the temperature of which can be easily controlled.

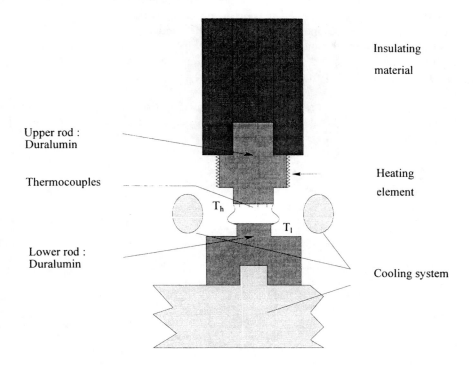

FIGURE 1. Sketch of the exprimental set-up.

STABILITY DIAGRAM

One aim of our experimental activity is to investigate the dependence of the critical temperature difference upon the liquid bridge volume and to give some physical explanations.

The stability diagram $(\Delta T_{cr}, V)$ in fig.2-a shows the transition from the axisymmetric steady flow to the oscillatory one. Two different oscillatory instability branches are found with respect to the variation of the volume. It is experimentally observed that those branches correspond to different azimuthal wave numbers. The branch, on which ΔT_{cr} grows belongs to m = 1, and the descending branch belongs to the azimuthal wave number m = 2.

Between these two oscillatory branches, there is a small range of volumes for which the steady flow is stable for very high values of the critical parameters. Our explanation follows the arguments proposed by Shevtsova and Legros, (1998) regarding the flow instability in deformable cylinders with wave number m = 0 and which could be partially extended for arbitrary m. The main reason for the instability is the existence of a cold finger near the hot disk. It appears that the stability of the system is very sensitive to the value of the upper contact angle. Surface deformation changes the distribution of the temperature near the hot wall. In the case of a straight cylinder the upper contact angle α = 90 and the maximum of the temperature gradient, at least for the basic flow is located on the hot wall. In the case of a deformed free surface the isotherms slide from the sharp corner into the interior (fig.2-b, left side). The local temperature gradient, driving the main flow, has smaller amplitude and is moving away from the hot wall when the upper contact angle α is decreasing. For instance, for a volume V = 0.8, the maximum of temperature gradient is located at 5-7% of the distance from the hot wall.

The positions of this maximum and of the steep branch of the temperature gradient, with a positive slope, coincide with the area of influence of the cold finger on the free surface. In this case, a small disturbance of the surface temperature has a very strong impact on the accelerations or decelerations of the main flow. On the other hand, the deformation of the liquid bridges produces a narrow neck in the shape near the hot wall. This is particularly true under gravitational conditions, when the contact angle is below 90°, even for liquid bridges with $V \geq 1$. As a result, the distance between the cold finger and the free surface is shorter and the sensitivity of the free surface is increased. With increasing volume, at a fixed aspect ratio, the neck widens and it is necessary to apply a higher ΔT to initiate oscillations.

From the above explanations we can draw a conclusion that the position of the gap between branches should depends upon aspect ratio. This is in agreement with what was observed by different experimental groups.

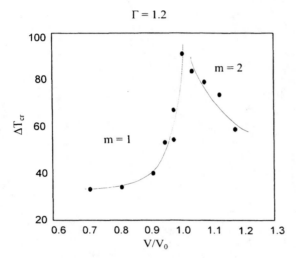

FIGURE 2-a. Stability diagram.

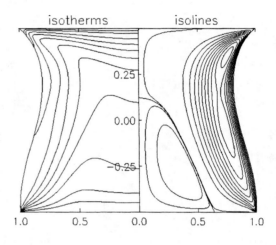

FIGURE 2-b. Typical flow pattern before the onset of instability.

ONSET OF INSTABILITY. TRAVELLING AND STANDING WAVES.

The time dependence of the temperature has been registered by 5 thermocouples equidistant in the azimuthal direction. The detailed analysis of the power spectrum and of the phase shift between thermocouples reveals that the instability begins as a mixed mode $m = 0$ and $m = 1$, followed by nearly standing wave with a wave number $m = 1$ which changes to a $m = 1$ travelling wave. To demonstrate this idea the dependence of the frequency upon the temperature difference in the supercritical area ($\Delta T > \Delta T_{cr}$) is shown in fig.3. The experiments have been carried out in the air at room temperature. Two independent fundamental frequencies appear just above the onset of time-dependent convection $f_0 = 0.449$Hz ($m = 0$) and $f_1 = 0.41$Hz ($m = 1$). The first one exists only in the narrow region $0. <\varepsilon = (\Delta T - \Delta T_{cr}) / \Delta T_{cr} < 0.05$, and the second one, by moving further into the supercritical area ($\Delta T > \Delta T_{cr}$) increases slowly. Above the secondary instability ($\varepsilon > 0.7$) an additional frequency appears, a $m = 2$ mode, which coexists with mode $m = 1$ within some range of ε ($0.7 < \varepsilon < 1.0$) and then becomes dominant.

The analysis of the time series and the of the corresponding phase shift for $\Delta T/\Delta T_{cr} = 1.2$, displays a standing wave with azimuthal wave number $m = 1$, which becomes a travelling mode for further increase of ΔT. The first super harmonic $f = 2f_1$ with a distinguishable tiny amplitude appears in the spectrum practically just after mode $m = 0$ vanished. The higher harmonics are exited with increasing of ΔT.

The critical ΔT which has been recorded is $33.2 \pm 0.1°C$. The Marangoni number is defined as $Ma=\gamma_*\Delta Td/\rho\nu k$, and gives $Ma_{cr} \approx 10200$, if we use the value of viscosity at temperature $25°C$. But if we take the viscosity at the mean temperature, following the dependence $\nu = \nu_{25}[1-8.135 \ 10^{-3}(T-25)]m^2s$, the critical Marangoni number will be $Ma_{cr}\approx11800$.

Surprisingly, the same frequency $f_1 = 0.41Hz$, corresponding to the $m = 1$ and $Ma_c \approx 13000$, was already recorded at the threshold of instability by Muehlner et al. (1997), although they use different silicone oil (Pr ≈ 40) and the aspect ratio was $\Gamma = 1$. The only common point between the present and their experiments is a total volume of high Prandtl liquid: $V = 81.4mm^3$ versus $V = 85mm^3$ (Muehlner et al., 1997),

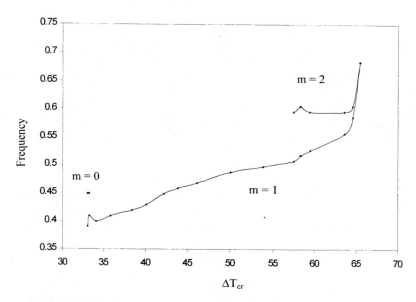

FIGURE 3. The dependence of the frequency upon the temperature difference in the supercritical area ($\Delta T > \Delta T_{cr}$).

Above the secondary instability with the mode $m = 2$ an additional fundamental frequency arises $f_2 = 0.605Hz$, when $Ma \approx 17700$ ($\nu = \nu_{25}$) or $Ma \approx 23200$ ($\nu = \nu_{av}$), to be compared with $f_2 = 0.65Hz$, $Ma = 19800$, (Muehlner et al., 1997),

Standing, mixed and travelling waves are studied experimentally and numerically by different groups of scientists in liquid bridges with straight free surfaces, but a coherent picture of the development of instability is not yet emerged. To shed the light on this question some results of the study of the amplitude equations near the threshold of instability in thin layer, heated from below, can be applied for the liquid bridges. Theoretically it is obtained by Colinet, (1994) that near the threshold of instability for a given set of parameters only one solution is stable, either standing or travelling wave. This is true under the following hypotheses: it should be supercritical type of bifurcation (small amplitude at onset of instability) and it should be Hopf bifurcation of the pure mode (no competition between modes). The oscillatory convection in liquid bridge always fulfils the first hypothesis and the second one is valid for some set of parameters. It is known that in some cases instability begins as a mixed mode.

Indeed, it can happens that due to some initial conditions we can observe travelling wave (TW), which is switched later to a standing one (or visa versa). It means that for this set of parameters only this type of hydrothermal wave (e.g. SW) is stable. If another initial conditions are chosen or the system is disturbed, upon some time we should have again a standing wave, which will remain standing until the surrounding conditions will be changed. The waiting time can be different and it depends how far the system is located from the final stable position. A special set of experiments was done to validate this theoretical prediction. For example, for the V = 0.8 and the temperature of the cold rod T ≈ 24°C the instability begins as a standing wave with mode m = 2, and if the experimental conditions remained constant, the SW was easily observed during 20 or 60 min or even much longer time, in reality, as long as we have been patient.

CONTROL OF CRITICAL MODE

Due to the non-linearity of the system a very complicated oscillatory structures can be observed if the driving force in the system is increased gradually. We have found experimentally one more possibility to change the critical mode m = 1 for m = 2. For this reason the experimental set-up has been surrounded by cylindrical volume with a diameter 2R = 12mm, further it is called shielding. The temperature of this cylindrical volume was regulated by the cooling system and was equal to the temperature of the bottom rod T ≈ 22.0°C. With this configuration the instability begins as a pure standing wave with a critical mode m = 2 for $\Delta T = 34.9$°C in comparison with m = 1 and $\Delta T = 33.2$°C for the experiment without shielding.

This experiment was repeated a few times and always the time-dependent temperature signals from 5 thermocouples and the phase shift between them, confirm m = 2 standing wave. The power spectrum of this mode, contains only the fundamental frequency f = 0.459Hz, the level of the noise is very low and there are no harmonic with distinguishable amplitude until $\Delta T = 45$°C.

To understand the influence of the thermal conditions around the liquid volume we have made an experimental study in which the temperature of the lower rod is varied in a large range 10°C < T_{cold} < 32°C. We have found that not only the critical Marangoni number strongly depends upon the mean temperature in the system, moreover the critical wave number is switched from m = 1 for the 10°C < T_{cold} < 20.7°C to m = 2, when T_{cold} > 20.7°C.

CONCLUSIONS

Time-dependent thermal convection has been investigated experimentally in deformed liquid bridges of silicone oil with Prandtl number Pr = 105. The signals from the five equidistant thermocouples have been taken for Fourier analysis and determination of the phase shift. The experimentally obtained stability diagram (ΔT_{cr}, V) shows that the transition from axisymmetric steady flow to the oscillatory one is very sensitive with respect to volume variations. It consists of two different oscillatory instability branches, belonging to different azimuthal wave numbers, and between them there is a small range of volumes for which the steady flow is stable for very high values of the critical parameters. It was found, that the position of gap is linked with the upper contact angle α.

The detailed analysis of power spectrum and phase shift between thermocouples reveals that instability begins as a mixed mode m = 0 and m = 1 within a narrow interval of ΔT, followed by nearly standing wave with a wave number m = 1 which changes to m = 1 travelling waves, when $\Delta T > 1.2\Delta T_{cr}$.

It has been proved experimentally that if standing (or travelling) wave is established in the liquid bridge it can exist infinitely long until experimental conditions are maintained constant. By surrounding the liquid bridge with another cylindrical volume with larger internal diameter, kept at constant temperature, the critical mode m = 1 can be switched to a m = 2. It is shown that the critical Marangoni number and critical wave number are very sensitive to the average temperature in the liquid bridge, for example, to the temperature of the cold rod.

ACKNOWLEDGMENTS

This text presents research results of the Belgian program on Interuniversity Pole of Attraction initiated by the Belgian state, Prime Minister's Office, Science Policy Programming. The authors assume the scientific responsibility.

REFERENCES

Colinet P. and Legros J.C. "On the Hopf Bifurcation Occuring in the Two-Layer Rayleigh-Benard Convective Instability", *Phys. Fluids* **6**, p.2631 (1994).

Muehlner K.A., Schatz M., Petrov V., McCormic W.D., Swift J.B., and Swinney H.L, "Observation of Helical Traveling-Wave Convection in a Liquid Bridge", *Phys. Fluids* **9**, p.1850-1852 (1997).

Shevtsova V.M. and Legros J.C., "Oscillatory Convective Motion in Deformed Liquid Bridge", *Phys. Fluids* **10**, 1621-1634 (1998)

The Experimental Study of The Periodic Instability of Thermocapillary Convection Around an Air Bubble

Christelle Reynard[1], Robert Santini[1], Lounès Tadrist[1], and Patricia Arlabosse[2]

[1] I.U.S.T.I, UMR 6595, Technopôle de Château Gombert, 5 Rue Enrico Fermi 13453 Marseille Cedex 13 (France)
[2] Ecole des Mines d'Albi Carmaux, route de Teillet-81013 Albi CT Cedex 09 (France)

reynard@iusti.univ-mrs.fr

Abstract. This paper concerns an experimental study on thermocapillary convection at the interface between an air bubble and a silicone oil with a low Prandtl number. The silicone oil layer is heated from above and cooled from below in order to obtain a vertical temperature gradient. When the temperature gradient is increased and a critical value is exceeded, oscillations appear in the liquid. This oscillatory thermocapillary convection is analyzed versus the temperature gradient and the bubble aspect ratio. A non intrusive optical measurement technique based on shadow technique is developed. Three oscillations modes are observed in a vertical section of experiment cell. The systematic study shows the influence of the temperature gradient and of the bubble size on their associated characteristic frequencies. The values obtained are close to 1 Hz.

INTRODUCTION

The boiling of a liquid allows to increase the heat transfer between a wall and a fluid. This technique is used a lot in many industrial processes and can be envisaged for applications in microgravity. Boiling phenomenon is still not enough understood. Recent boiling experiments performed in microgravity have shown that the influence of gravity on heat transfer mechanisms is not well explained. It appears that gravity has little influence on the nucleate boiling curve, and the reason for the equivalence of magnitude of the heat transfer coefficients at normal gravity and low gravity conditions has not been completely resolved (Straub, 1996), (Merte, 1990). In fact boiling brings into play several complex physical phenomena (natural convection, phase change, thermocapillary convection, microconvection, ...), consequently in order to have a better understanding, it is necessary to study separately each phenomenon. For instance to explain the heat transfer in microgravity, many authors have hypothesized that under microgravity conditions Marangoni convection is the dominant effect (Mc Grew, 1966), (Straub 1994).

This led in our laboratory to a basic study around an air bubble injected under the upper horizontal wall in a layer of silicone oil. This work consisted in studying experimentally and numerically the influence of the convection induced by Marangoni effect on the heat transfers between a plane walls and the liquid (Arlabosse, 1999). These first works concerned mainly the liquids with a high Prandtl number. Generally the liquids for which the boiling processus occurs have low Prandtl number. As a result other experiments have been carried out with silicone oil at low Prandtl number. In this case we have brought to the fore the existence of two regimes of thermocapillary convection. The steady flow of thermocapillary convection is followed by an oscillatory flow when a critical value of the temperature gradient is exceeded. The vortex of the stable thermocapillary convection oscillates (Arlabosse, 1998).

This periodic instability has been investigated by few authors in this configuration (Chun, 1991; Betz, 1997) and the frequency has not been studied systematically. Then we have developed an optical method in order to observe the phenomenon and measure the associated frequencies. In this paper the measurement technique and the different observed modes are exposed. We show the influence of the bubble size and the temperature gradient on the motion of the silicone oil. For this we determine the structure of the liquid flow and the typical frequencies of the oscillations.

CP504, *Space Technology and Applications International Forum–2000*, edited by M. S. El-Genk
© 2000 American Institute of Physics 1-56396-919-X/00/$17.00

MATERIALS AND METHOD

The experiment chamber used, in this study, is a cavity with horizontal brass walls with an inner square section (40mm*40mm) and an adjustable height H fixed at 5 mm in this case (Fig. 1).

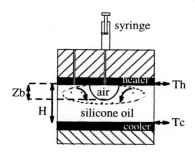

FIGURE 1. The experiment cell.

To obtain a vertical temperature gradient within the liquid layer, the upper horizontal wall is heated at the temperature T_h by an electrical heater and the lower one cooled at the temperature T_c with a Peltier element. In this configuration the surface tension and the buoyancy forces oppose one another. On the center of the upper wall, a hole with tube connection is made in order to inject the bubble by means of a syringe and another one is made for fluid volume compensation during heating. The vertical walls are made of polycarbonate for visualization of the phenomenon. The used test liquid is a silicone oil in order to decrease the pollution effects on the surface tension. The kinematic viscosity is equal to 1.10^{-6} m^2/s.

Measurement Technique

The measurement technique used is a non-intrusive optical method which yields an image of the bubble-silicone oil system and the frequency measurement of the oscillatory convection (Fig.2). A beam of light crosses the cell and then is divided with a separative slide. The reflected part of the beam crosses a split in order to focus on the oscillatory flow and is sent to a photodiode. The electrical signal delivered by the photodiode is analyzed to determine the frequency of the oscillations with a spectra analyzer. The other part of the signal is sent to a video-camera which allows the display of the phenomenon in order to measure or to check the bubble size, and to identify the oscillatory mode. An average of 4 power spectra during an acquisition time of 128 s is made. Then the frequency resolution is 3.10^{-2} Hz. This leads to the best results because the noise is enough decreased and the widening of the frequency is avoided.

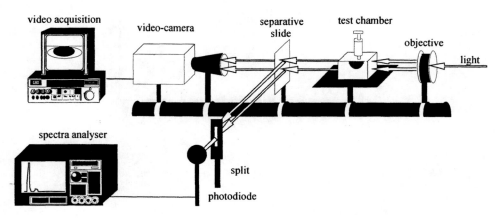

FIGURE 2. Schematic setup of the optical measurement technique.

RESULTS

The thermocapillary convection stays steady for low Marangoni numbers, denoted Ma (1) and for a high Prandtl number denoted Pr (2), (Arlabosse, 1999). On the other hand for low Prandtl number the steady flow is followed by an oscillatory flow.

$$Ma = -\left(\partial\gamma/\partial T\right)\left(T_h - T_c\right) Zb^2 \big/ \mu\, a\, H \qquad (1)$$

where γ is the surface tension, μ is the dynamic viscosity, a is the thermal diffusivity, H is the liquid layer thickness, and Zb is the bubble height.

$$Pr = \frac{\nu}{a} \qquad (2)$$

where ν is the kinematic viscosity. We have performed a systematic study of the influence of the bubble size and the temperature gradient on the oscillatory mode and its frequency for a silicone oil with a low Prandtl number (Pr = 13). We have studied the oscillatory flow for four bubble sizes. For each we have increased the temperature gradient of the liquid layer and a transition from a steady flow to an oscillatory mode has appeared. The results were correlated using the Marangoni number in order to take into account the bubble size and the temperature gradient.

Oscillations Threshold

The threshold for the appearance of the oscillatory modes depends on the bubble aspect ratio Rb/Zb (where Rb is the bubble radius). For each bubble size there is a particular critical value of the Marangoni number. This value corresponds to the beginning of the oscillatory flow. Using the same definition of the Marangoni number (equation (1)) for our results and ones of Chun and al. (Chun, 1991), the two curves have the same pattern. The critical Marangoni number (Ma_c) increases with the aspect ratio (Fig. 3). But our values of the critical Marangoni number at any bubble aspect ratio are bigger. This gap between their results and ours is may be due to viscous effects and enclosed effects. In fact their Prandtl number is equal to 6 and ours is equal to 13 and their silicone oil layer is ten times thicker than ours.

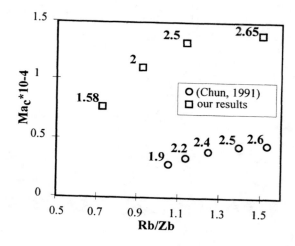

FIGURE 3. The critical Marangoni number versus the aspect ratio of the bubble (the number close to the square and circle symbols corresponds to the bubble height Zb in mm).

Oscillatory Modes

The flow is observed in a vertical section and by following the distortion of a line around the air bubble versus time,

880

the appearance of an oscillatory mode can be determined. This line is due to the variation of the refractive index of the silicone oil and represents the boundary of the thermocapillary vortex. During the experiments, the thermocapillary flow oscillates around the air bubble according to three modes depending on the aspect ratio of the bubble and/or the temperature gradient. We have observed and schematically represented (Fig. 4), a symmetric mode with a predominant horizontal move (a), an other symmetric one with horizontal and vertical move (b) and an asymmetric mode (c).

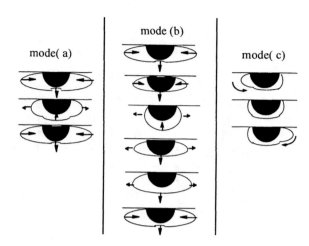

FIGURE 4. Schematic pictures of the three observed oscillatory mode.

For the mode (a), the line due to the variation of the refractive index has two extreme horizontal positions (one far from and one close to the bubble interface). The mode (b) is a more complex mode, the corresponding line oscillates simultaneously in the horizontal and vertical direction, the line passes by three characteristic horizontal positions. For the mode (c), in the same time the line is close to a bubble side and far for the other one and after the opposite. In this case the bubble oscillates horizontally on the upper wall.

The apparition of a particular mode depends on the temperature gradient and the bubble aspect ratio. In fact when we change the temperature gradient at a constant bubble size, different modes follow one after the other. We present an experimental acquisition of video images (Fig. 5). It shows the evolution of the modes during time for three temperature gradient values and a same fixed bubble size $Rb/Zb = 0.73$.

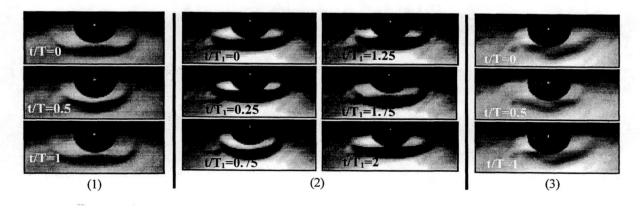

FIGURE 5. Representative video images of each oscillatory mode at three temperature gradients for $Rb/ZB = 0.73$. **(1)** mode (a) at 4.8 K/mm, **(2)** mode (b) at 4K/mm, **(3)** mode (c) at 3.6K/mm. (t is the time, T is the only period, T_2 is the lowest period).

For each aspect ratio a particular succession of modes exists before the flow becomes non periodic (Fig. 6). For the smallest aspect ratio the oscillatory modes are not steady. In fact a particular mode disappears and appears again versus the Marangoni number. In the case of the biggest aspect ratio, only a mode appears and the flow becomes fast non periodic. On the other hand for the two intermediate aspect ratios there is the same oscillatory behavior.

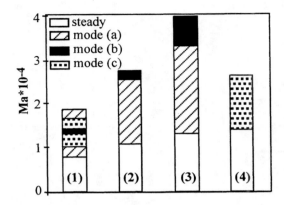

FIGURE 6. The succession of modes for four bubble aspect ratios (**(1)** Rb/Zb = 0.73 ; **(2)** Rb/Zb = 0.93 ; **(3)** Rb/Zb = 1.14 ; **(4)** Rb/Zb = 1.51).

Frequency Results

The oscillation frequency is less than 1.3 Hz. It depends on both the bubble aspect ratio and the Marangoni number. At any Marangoni number, the frequency increases with decreasing aspect ratio. The four curves (Fig. 7) show that there is only one or two frequencies for a fixed Marangoni number. The zones, where there is only one frequency correspond to the mode (a) or (c) and the ones where there are two frequencies, denoted f1 and f2, correspond to the mode (b). In fact for this mode the line oscillates vertically with the frequency f2 and horizontally with the two frequencies f1, f2.

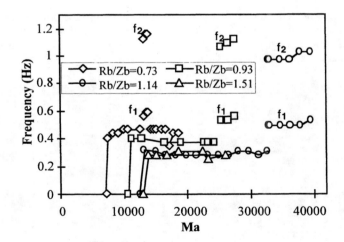

FIGURE 7. The variation of the frequency versus the Marangoni number for four bubble aspect ratios.

When increasing the Marangoni number we observe a frequency shift. This frequency shift corresponds to a change of oscillatory mode. When the mode (c) follows the mode (a) and vice versa, the frequency variation is light. In the other hand when the mode (b) appears, the frequency increasing is bigger and two frequencies coexist (Fig. 8).

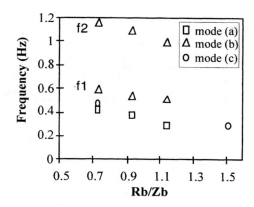

FIGURE 8. The average frequency of each mode versus the bubble aspect ratio.

CONCLUSIONS AND OUTLOOKS

Oscillatory thermocapillary convection around an air bubble immersed in a 5 mm thick silicone oil layer heated from above and cooled from below has been studied experimentally by a non intrusive optical method. Periodic instabilities occur after the steady thermocapillary convection and at a critical Marangoni number which increases with the bubble aspect ratio. Different oscillatory modes have been observed in a vertical section of the test chamber. The appearance of such modes depends on both the Marangoni number and the bubble aspect ratio. The frequency of the oscillations is a decreasing function of the aspect ratio and depends on the kind of the mode.

We will modify the apparatus in order to observe in vertical and horizontal cross sections. This will lead us to have a better understanding of this periodic instability. This will probably help us to explain the physical process of this " wave " . At last an experimental study in microgravity could determine if the periodic instability is due only to thermocapillary convection.

ACKNOWLEDGMENTS

The authors acknoledge the members of the Heat and Mass transfers team who contributed to the realization of this experimental study.

REFERENCES

Arlabosse, P., Tadrist, L., Tadrist, H., and Pantaloni, J., Instabilités thermoconvectives dans un liquide induites autour d'une bulle par effet Marangoni, *in congrès SFT98 Thermique et environnement,* Elsevier, 1998, pp 359-364.

Arlabosse, P., Tadrist, L., Tadrist, H., and Pantaloni, J., Experimental Analysis of The Heat Transfer Induced by Thermocapillary Convection Around a Bubble, Journal of Heat Transfer (to be published).

Arlabosse, P., Lock, N., Medale, M., and Jaeger, M., Numerical Investigation of Thermocapillary Flow Around a Bubble, Physics of Fluids, **11**,18-29, (1999).

Betz, J., Strömung und Wärmeübergang bei thermokapillarer Konvektion an Gasblasen, Thesis, Munich (Deutschland), (1997).

Chun, Ch.-H., Oscillating convection modes in the surroundings of an air bubble under a horizontal heated wall, Experiments in Fluids **11**, 359-367, (1991).

Mc Grew, J.L., Bamford, F.L., Rehm, T.R., Marangoni flow : an additional mechanism in boiling heat transfer, Science 153, n° 3740, 1106-1107, (1966).

Merte, H. Jr., Nucleate pool boiling in variable gravity, *in low gravity dynamics and transport phenomena* 130, Progress in Aeronautics and Astronautics, edited by J.N. Koster and R.L. Sani; Washington DC, published by AIAA, 1990, pp. 15-72.

Straub, J., The role of surface tension for two phase heat and mass transfer in absence of gravity, Experimental Thermal and Fluid Science **9**, 253-273, 1994.

Straub, J., Miko, S., Boiling on a wire under microgravity conditions - First results from a space experiment performed in May 1996, in *Proceeding of Eurotherm Seminar n°48,* Pool boiling, 2, Edition ETS (Pisa), 1996, pp. 275 - 283.

Theoretical Analysis of 3D, Transient Convection and Segregation in Microgravity Bridgman Crystal Growth

Andrew Yeckel, Valmor F. de Almeida, and Jeffrey J. Derby

Department of Chemical Engineering and Materials Science,
Army HPC Research Center, and Minnesota Supercomputer Institute,
University of Minnesota, Minneapolis, MN 55455-0132, USA

Fax: 1-612-626-7246; e-mail: derby@tc.umn.edu.

Abstract. We present results from simulations of transient acceleration (g-jitter) in both axial and transverse directions in a simplified prototype of a vertical Bridgman crystal growth system. We also present results on the effects of applying a steady magnetic field in axial or transverse directions to damp the flow. In most cases application of a magnetic field suppresses flow oscillations, but for transverse jitter at intermediate frequencies, flow oscillations grow larger.

INTRODUCTION

Growth of large single-crystals of exacting quality, needed in the fabrication of advanced optical, magnetic, and electronic devices, presents some of the most difficult challenges today in materials processing. Key to the advancement of crystal growth processing is a better understanding of the dominant influence of buoyancy-driven convection on segregation and morphological stability during crystal growth on earth. Since these effects in turn play a critical role in establishing the structure and properties of grown materials, considerable interest has developed for their study by use of the microgravity environment provided by space flight. However, space flight experiments remain subject to accelerations that vary both in magnitude and direction. These accelerations, referred to as g-jitter, complicate experiments by the introduction of significant three-dimensional effects.

Many past efforts have studied the general features of flow and its influence on segregation in microgravity systems using one-dimensional and two-dimensional analyses (Polezahev & Fedyoshkin, 1980; Polezahev et al., 1984; Kamotani et al., 1981; Heiss et al., 1987; Schneider & Straub, 1989; McFadden & Coriell, 1988; Griffin & Motakef, 1989a; Griffin & Motakef, 1989b; Alexander et al., 1989; Alexander et al., 1991; Garandet et al., 1996; Monti, 1990; Biringen & Danabasoglu, 1990; Shyy & Rao, 1994; Naumann, 1994; Naumann, 1996). Three-dimensional effects have only been more recently considered. Li (1996) and Baumgartl and Müller (1996) have both investigated use of a magnetic field to suppress g-jitter induced fluctuations in three-dimensional thermal convection. These studies have been performed on relatively coarse discretizations, however, and under only a few sets of conditions.

Here, we present results of a simplified model for three-dimensional, transient fluid flow for crystal growth by the vertical Bridgman method. The model, which accounts for g-jitter conditions in the presence of an applied magnetic field, is based on finite element methods that utilize state-of-the-art, parallel computational techniques for solution of three-dimensional phenomena (de Almeida & Derby, 1997; de Almeida & Derby, 1999a; de Almeida et al., 1999). Results from this study are discussed in the next section. Although our model represents a simplification of an actual crystal growth system, this study considers a wider range of conditions than previously investigated by Li (1996) and Baumgartl and Müller (1996). Also, we have performed tests to ensure that our results are converged with respect to discretization refinement.

CP504, *Space Technology and Applications International Forum–2000*, edited by M. S. El-Genk

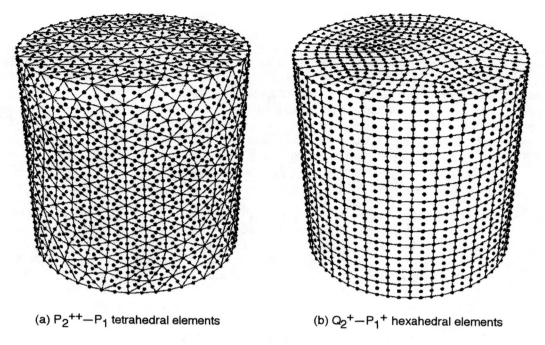

(a) $P_2^{++}-P_1$ tetrahedral elements   (b) $Q_2^+-P_1^+$ hexahedral elements

FIGURE 1. Sample discretizations used in this study.

NUMERICAL MODEL

The results presented here were obtained using a model based on the conditions set forth by Baumgartl and Müller (1996). The system consists of a cylindrical section 50 mm in diameter and 50 mm tall with fixed temperatures imposed at all boundaries: 1685 K at the cylinder bottom, 1710 K at the top, and parabolic along the sidewall (equal to 1725 K at cylinder mid-height). No-slip boundary conditions on flow and no-flux conditions on the electric potential are imposed at all boundaries. Fluid physical properties are those of silicon at its melting point. The flow is subjected to the following body force:

$$\mathbf{g} = (g_o + g_j sin(2\pi ft))g_e \mathbf{e}_g \tag{1}$$

where residual gravity $g_o = 10^{-6}$, and jitter amplitude $g_j = 10^{-4}$, are measured in units of earth gravity g_e (values typical of operating conditions on the US Space Shuttle). Results were computed for $\mathbf{e}_g$, the unit vector in the direction of gravity, in both axial and transverse orientations. Cases with and without applied magnetic fields (of strength $B = 53$ mT) were computed. Jitter frequencies, f, in the range 5.7×10^{-5} to 5.7 Hz were considered.

Two tests were made to ensure that the discretizations used in our calculations were adequate to resolve thermal and momentum boundary layers. In the first test, we compared solutions obtained with our three-dimensional code to solutions obtained with a two-dimensional code for an axisymmetric problem, the case of axial jitter without a magnetic field. The two-dimensional solutions were obtained using a series of progressively finer discretizations, until the velocity and temperature values at several monitoring points converged to 0.01%. These values were then compared with values at the same monitoring points in the three-dimensional solutions. In all cases the errors were less than 0.8%. The second test was performed by comparing various three-dimensional solutions to the transverse jitter case, with and without a magnetic field, obtained using three different types of finite elements. Figure 1 shows some of the discretizations used in this study. The element types tested include $P_2^{++}-P_1$ tetrahedral, $Q_2^+-P_1^+$ hexahedral, and Q_2-P_1 hexahedral, all second-order accurate elements (de Almeida & Derby, 1999b). Discrepancies of values at the monitoring points were within 1.2% for the various meshes used. Total number of degrees of freedom used in this study ranged from 48,429 to 57,809.

RESULTS

Figure 2 shows average velocity versus time for cases in which there is no magnetic field. Cases are shown for both axial jitter and transverse jitter at two different frequencies. Also shown for reference is the variation of gravity magnitude with time. Transverse jitter induces significantly stronger flow than axial jitter at both frequencies. Flow is considerably weaker at the higher frequency shown for both axial and transverse jitter. At $f = 0.057$ Hz, a primary maximum occurs after gravity reaches a maximum in the direction of residual gravity. A secondary maximum occurs after gravity reaches a maximum in the direction opposite to residual gravity. At a higher frequency of $f = 0.57$ Hz, there is no secondary maximum. In both cases, the maximum in flow strength lags the maximum in gravity force by a phase shift approximately equal to one-fourth $(\pi/2)$ of a jitter cycle. The phase shift is a function of frequency: at jitter frequencies lower than shown in the figure (< 0.01 Hz), the phase shift is near zero, whereas the shift increases for frequencies greater than those shown in the figure.

Figure 3 shows peak average velocity versus frequency, for axial jitter with and without a transverse magnetic field. Also included in the figure are curves indicating the limiting cases of very low and very high frequency without a magnetic field. These limiting cases were obtained by setting $\mathbf{g} = g_o g_e \mathbf{e}_g$, and $\mathbf{g} = g_j g_e \mathbf{e}_g$, respectively. The frequency range can be divided into three regimes. At lower frequencies, the peak velocity is nearly that of the zero frequency limit, whereas at higher frequencies, the peak velocity approaches the high frequency limit. Over an intermediate range of frequencies the peak velocity decreases rapidly as the frequency increases. One effect of the transverse magnetic field is to reduce flow strength at all jitter frequencies. Another effect is to extend the low frequency regime cutoff to a higher frequency than observed in the absence of a magnetic field. As a consequence, the relative degree of flow reduction is greatest in the lower frequency range.

Figure 4 shows peak average velocity versus frequency for transverse jitter. Both axial and transverse magnetic fields are considered. Comparison of Figure 4 to Figure 3 shows that transverse jitter induces stronger flow than axial jitter at all frequencies whether or not a magnetic field is imposed. As in the case of axial jitter, the magnetic field reduces flow strength at all jitter frequencies, and extends the low frequency regime cutoff to a higher frequency. In the case of transverse jitter, however, there is a frequency range (ca. 0.1–10 Hz) in which application of a magnetic field results in a stronger flow than without a magnetic field, regardless of field orientation.

DISCUSSION

The decrease in flow strength at higher frequencies, the appearance of a phase shift, and absence of a secondary maximum can be explained in terms of relative time scales. In the low frequency regime, gravity varies on a time scale much longer than momentum transport. A quasi-steady state is attained: there is no phase shift, and the flow reaches a fully-developed state when gravity reaches a maximum in either direction. In the high frequency regime, gravity varies on a time scale much shorter than that of momentum transport. The flow attains only a small fraction of its potential strength before gravity reverses direction, and a time lag develops to the point where maximum flow and maximum gravity are completely out of phase. A flow maximum occurs only when gravity is oriented in the direction of residual gravity. When gravity reaches a maximum value in the opposite direction, the effect of residual gravity overwhelms the transient effect of jitter. Consequently, flow reaches a minimum rather than a maximum. It is for this reason that the secondary maximum disappears at higher frequencies. The $f = 0.057$ Hz case in Figure 2 represents an intermediate condition in which gravity varies on a time scale comparable to momentum transport. The secondary peak is much smaller than the primary peak, because the effect of residual gravity is comparable to the transient effect of jitter.

The most surprising result of this study is that application of a magnetic field can result in a stronger flow at some jitter frequencies, as shown in Figure 4. At first glance this result is counter-intuitive, since a steady magnetic field should always have the effect of braking a steady flow. Indeed, results in Figures 3 and 4 are consistent with this observation in the low frequency (i.e. quasi-steady) limit. In the case of a time-dependent flow, however, the time scale for momentum transport also plays a critical role in determining flow dynamics. Figures 3 and 4 show that application of a magnetic field extends the low frequency regime to higher frequencies, which implies that the time scale for momentum transport decreases when a magnetic field is applied, no doubt due to changes in flow structure. Consequently, even though peak velocity in the low frequency limit is reduced when a magnetic field is applied, in the

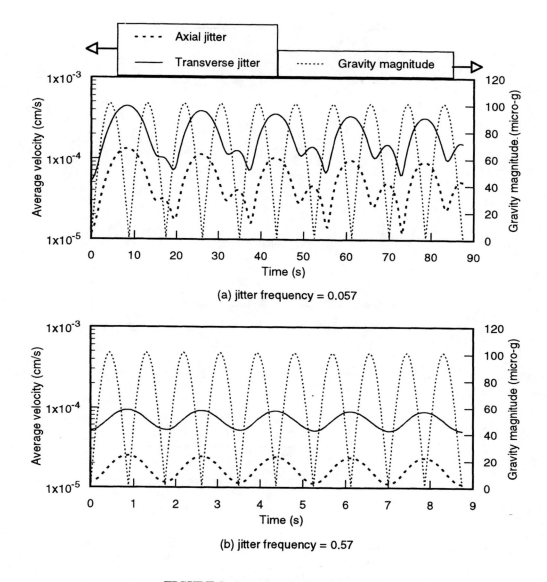

(a) jitter frequency = 0.057

(b) jitter frequency = 0.57

FIGURE 2. Average velocity versus time.

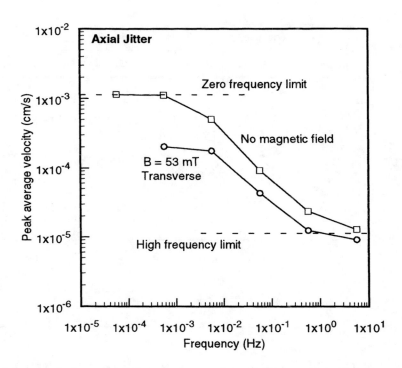

FIGURE 3. Peak average velocity versus jitter frequency, for axial jitter with and without an applied transverse magnetic field at 53 mT.

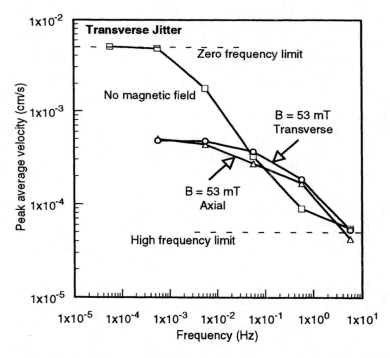

FIGURE 4. Peak average velocity versus jitter frequency, for transverse jitter with and without an applied transverse magnetic field at 53 mT.

intermediate frequency range of 0.1–10 Hz, the flow attains a peak velocity that is much closer to the low frequency limit than in the case of no magnetic field. The result is a higher peak velocity in the presence of a magnetic field at intermediate frequencies.

FINAL REMARKS

In this work we have presented preliminary results of an ongoing effort to better understand gravitational effects in melt crystal growth. These results have elucidated the critical influence that the time scale for momentum transport has on flow dynamics when a buoyancy driven flow is subjected to g-jitter in a microgravity environment.

Our long term goal is to develop computational tools that will enable the unambiguous interpretation of melt crystal growth flight experiments and will, perhaps even more importantly, provide a tool to design the conditions needed in future flight experiments to best study the effects of microgravity on segregation theory. This effort will dramatically advance the modeling capabilities and understanding of three-dimensional phenomena in melt crystal growth systems.

ACKNOWLEDGMENTS

This work was supported in part by the Microgravity Sciences Program of the National Aeronautics and Space Administration (NASA/NAG8-1474), the Minnesota Supercomputer Institute, and the U.S. Army, Army Research Laboratory, Army HPC Research Center. No official endorsement should be inferred.

REFERENCES

Alexander, J. I .D., Ouazzani, J., & Rosenberger, F., *J. Crystal Growth*, **97**, 285 (1989).

Alexander, J. I .D., Amiroudine, S., Ouazzani, J., & Rosenberger, F., *J. Crystal Growth*, **113**, 21 (1991).

Baumgartl, J., & Müller, G., *J. Crystal Growth*, **169**, 582 (1996).

Biringen, S., & Danabasoglu, G., *J. Thermophysics*, **4**, 357, (1990).

de Almeida, V.F., & Derby, J.J., *Page p. 800 of:* Atluri, S.N., & Yagawa, G. (eds), *Advances in Computational Engineering Science*. Tech Science Press, Forsyth, Georgia, (1997).

de Almeida, V.F., & Derby, J.J., *SIAM J. Scientific Computing*, accepted for publication (1999a).

de Almeida, V.F., & Derby, J.J. 1999b. *Int. J. Numer. Methods Fluids*, submitted.

de Almeida, V.F., Chapman, A.M., & Derby, J.J., *Numerical Methods for Partial Differential Equations*, submitted (1999).

Garandet, J. P., Corre, S., Gavoille, S, Favier, J.J., & Alexander, J. I. D., *J. Crystal Growth*, **165**, 471 (1996).

Griffin, P. R., & Motakef, S., *Appl. Microgravity Tech.*, **2**, 121 (1989a).

Griffin, P. R., & Motakef, S., *Appl. Microgravity Tech.*, **2**, 128 (1989b).

Heiss, T., Schneider, S., & Straub, J., *Page 517 of: Proc. 6th European Symp. on Materials Science under Microgravity Conditions*. Bordeaux, ESA SP-256 (1987).

Kamotani, Y., Prasad, A., & Ostrach, S., *AIAA J.*, **19**, 511 (1981).

Li, B. Q., *Int. J. Numer. Methods Fluids*, **39**, 2853 (1996).

McFadden, G. B., & Coriell, S. R., *Page 1572 of: Proc. AIAA/ASME/SIAM/APS 1st Natl. Fluid Dynamics Congr.* (1988).

Monti, R., *Progress in Astronautics and Aeronautics*, **130**, 275 (1990).

Naumann, R. J., *J. Crystal Growth*, **142**, 142 (1994).

Naumann, R. J., *J. Crystal Growth*, **165**, 129 (1996).

Polezahev, V. I., & Fedyoshkin, A. I., *Izv. Akad. Nauk SSR. Mekh. Zhidkosti i Gaza*, **3**, 11 (1980).

Polezahev, V. I., Lebedev, A. P., & Nikitin, S. A., *Page 237 of: Proc. 5th European Symp. on Materials Science under Microgravity*. Schloss Elmau, ESA SP-222 (1984).

Schneider, S., & Straub, J., *J. Crystal Growth*, **97**, 235 (1989).

Shyy, W., & Rao, M. M., *Microgravity Sci. Technol.*, **7**, 41 (1994).

Effect of Gravitational Potential
Energy on the Rate of Evaporation

P. Rahimi and C. A. Ward

Department of Mechanical and Industrial Engineering, University of Toronto, Toronto, Canada M5S 3G8

416-978-4807; ward@mie.utoronto.ca

Abstract. The statistical rate theory expression for the rate of evaporation indicates that this rate depends on the chemical potential and temperature discontinuities at the liquid-vapor interface. In an approximately isothermal system in which the vapor phase is in local equilibrium, the chemical potential discontinuity is predicted to increase as the gravitational potential energy at the interface increases, and as a result, to increase the evaporation rate. An experiment has been performed to investigate this prediction. The results are found to support the predicted increase in the evaporation rate with increases in the gravitational potential energy.

BACKGROUND

Phase change processes are fundamental to many of the experiments planned for the International Space Station. However, the conditions existing at the interface of the phases during such processes are not well understood. For example, the conventional assumption has been that the temperature is a continuous function across the liquid-vapor interface during evaporation. However, analyses based on classical kinetic theory of gases have predicted a temperature discontinuity to exist at the liquid-vapor interface during evaporation (Pao, 1971). This prediction was confirmed by a number of analytical studies. (Aoki, 1983, Cercignani, 1985, Cipolla, 1974). Each used more elaborate analytical procedures, but each was based on classical kinetic theory, and all of these studies predicted that at the interface of an evaporating liquid, the temperature was higher in the liquid than in the vapor. Recent experimental studies of evaporation have not supported this predicted direction of the temperature discontinuity. These experimental studies indicate that there is a temperature discontinuity at the liquid-vapor interface, that it is in the *opposite direction* to that predicted from the classical kinetic theory analyses, and that it is of a much larger magnitude (Fang, 1999 and 1999a).

A theory of kinetics based on quantum and statistical mechanics, called statistical rate theory (SRT), was recently used to analyze the evaporation measurements that contradicted the predictions of classical kinetic theory. The predictions from SRT were found to be in close agreement with the measurements (Ward, 1999). However, when SRT is applied to examine the effect of gravitational potential energy on the evaporation rate, an unexpected prediction is made. Statistical rate theory suggests that increasing the gravitational potential energy at an interface will increase the evaporation rate. If this prediction is correct, a number of questions are raised. For example, what will be the effect of g-jitter on phase change processes, such as crystal growth? We have conducted a series of experiments in a ground-based laboratory to evaluate the predicted effect of the gravitational potential energy on the rate of evaporation.

METHODOLOGY

To understand the theoretical basis for the predicted dependence of the evaporation rate on the gravitational potential energy, we first outline the SRT development of the expression for the evaporation flux (Ward, 1999). Consider a small, isolated volume that contains both fluid phases of a pure substance. The extent of each phase is sufficiently small so that its intensive properties are uniform. Suppose that thermodynamic equilibrium does not exist between the phases, but each phase may be approximated as being in a state of local equilibrium. Thus each phase may be viewed as a canonical ensemble system. If the average or internal energy of a phase were denoted as U^α, the temperature as T^a, the constant volume specific heat as C_v^a and the Boltzmann constant as k, the energy of the possible quantum states of each phase would lie within the energy range

CP504, *Space Technology and Applications International Forum–2000*, edited by M. S. El-Genk
© 2000 American Institute of Physics 1-56396-919-X/00/$17.00

$$U^\alpha \pm T^\alpha \sqrt{kC_V^\alpha}\,, \quad \alpha = L, V.$$ (1)

If the number of molecules at the time t is N^L, N^V, and this distribution is denoted as λ_j then the number of quantum states available to the system may be denoted as $\Omega(\lambda_j)$. If in the time interval δt one molecule evaporates, the molecular distribution would change to λ_k, or $N^L - 1$, $N^V + 1$. From the transition probability concept of quantum mechanics and the Boltzmann definition of entropy (i.e., $k \ln \Omega(\lambda)$), one may express the probability of this event occurring at any instant in the time interval δt as:

$$\tau(\lambda_j, \lambda_k) = \frac{2\pi |V_{v\varepsilon}| \varsigma}{\hbar} \exp\left(\frac{S(\lambda_j) - S(\lambda_k)}{k}\right),$$ (2)

where the matrix element for a change from a quantum state of λ_j to a quantum state of λ_k is denoted as $|V_{v_\varepsilon}|$ and the energy density of the states within the energy uncertainty as ς. We take the product of the latter two quantities to be constant for the states that lie within the energy uncertainty. For the isolated system the change in entropy may be expressed in terms of the chemical potential of each phase μ^L, μ^V, the temperatures and the enthalpy of the vapor phase h^V:

$$S(\lambda_k) - S(\lambda_j) = \left(\frac{\mu^L}{T^L} - \frac{\mu^V}{T^V}\right) + h^V\left(\frac{1}{T^V} - \frac{1}{T^L}\right).$$ (3)

We assume the transfer of a small number of molecules, δN from the liquid to the vapor phase would not change the intensive properties. Thus, the number of molecules that evaporate, δN during the period δt would be:

$$\delta N = \tau(\lambda_j, \lambda_k)\delta t.$$ (4)

Hence the unidirectional evaporation flux may be written

$$J_{LV} = \frac{2\pi |V_{v\varepsilon}| \varsigma}{\hbar} \exp\left[\left(\frac{\mu^L}{T^L} - \frac{\mu^V}{T^V}\right) + h^V\left(\frac{1}{T^V} - \frac{1}{T^L}\right)\right].$$ (5)

At the same instant that there is a probability for a transition from molecular distribution λ_j to molecular distribution λ_k there is also a probability of a transition to the molecular distribution corresponding to one molecule from the vapor condensing: $N^L + 1, N^V - 1$. Following the same procedure as that outlined to determine the unidirectional evaporation flux, one may obtain an expression for the unidirectional condensation flux:

$$J_{VL} = \frac{2\pi |V_{v\varepsilon}| \varsigma}{\hbar} \exp\left[-\left(\frac{\mu^L}{T^L} - \frac{\mu^V}{T^V}\right) - h^V\left(\frac{1}{T^V} - \frac{1}{T^L}\right)\right].$$ (6)

Since it has been assumed that $\varsigma |V_{v\varepsilon}|$ is constant for transitions between states that are within the energy uncertainty, we may take advantage of the fact that in equilibrium within the isolated system the two unidirectional rates would be equal, as would the chemical potentials and the temperatures. Under this condition, the two unidirectional fluxes would be equal to the equilibrium exchange rate, K_e. Thus:

$$\frac{2\pi |V_{v\varepsilon}| \varsigma}{\hbar} = \frac{2\pi |V_{v\varepsilon}| \varsigma}{\hbar} = K_e.$$ (7)

Under equilibrium conditions, we assume that each molecule from the vapor phase that collides with the interface condenses. Then under equilibrium conditions the unidirectional condensation flux would be equal to K_e and would be given by:

$$K_e = \frac{P_e}{\sqrt{2\pi mkT_e}}, \tag{8}$$

where P_e and T_e are the equilibrium pressure and temperature in the isolated system. They would in general be different than the instantaneous values of these parameters. The final expression for the net evaporation flux is the difference between the unidirectional fluxes:

$$J = \frac{P_e}{\sqrt{2\pi mkT_e}} (\exp[(\frac{\mu^L}{T^L} - \frac{\mu^V}{T^V}) + h^V(\frac{1}{T^V} - \frac{1}{T^L})] - \exp[-(\frac{\mu^L}{T^L} - \frac{\mu^V}{T^V}) - h^V(\frac{1}{T^V} - \frac{1}{T^L})]). \tag{9}$$

The expression for the evaporation flux given in Eq. (9) has been used to predict the steady state conditions at which three different liquids would each evaporate at a given rate. For all three liquids, (water, octane and methylcyclohexane) the predictions were found to be in close agreement with the measurements (Fang, 1999 and 1999a, and Ward, 1999). The interface was held at one position throughout each experiment, and thus the gravitational potential energy at each interface was constant.

Evaporation from a capillary

To examine the effect of the gravitational potential energy on the evaporation flux, consider the system shown schematically in Fig. 1. Three capillaries are present in a closed vessel that is surrounded by a temperature reservoir and is in a ground-based laboratory. The capillaries are closed at their lower end, but open at the other. Each is partially filled with water, and the liquid-vapor interfaces are at different heights and thus have different gravitational potential energies. We neglect any interaction between the capillaries, and assume the bulk liquid phase controls the properties in the vapor phase and that the reservoir imposes its temperature on the system, even during the evaporation process. In preparation for applying Eq. (9), we first establish the expression for the intensive properties in each phase at the interface within a capillary.

The interface within a capillary would be curved. We suppose the capillaries are sufficiently small so that the interface may be approximated as spherical. If W, g, z and D denote the molecular weight, the gravitational intensity, the vertical position above the bottom of the container and the diameter of a capillary, then since the vapor phase is in equilibrium (Ward, 1998):

$$\mu^V(T, P^V) + Wgz^V = \lambda, \tag{10}$$

where λ is a constant. To evaluate the constant, we apply Eq. (10) in the vapor phase at the bulk liquid (zero curvature) interface, z_B where the pressure is the saturation vapor pressure $P_{sat}(T)$, and at the interface within a capillary. If the properties at the latter interface are denoted with a subscript c, then:

$$\mu^V(T, P_c^V) = \mu(T, P_{sat}) + Wg(z_B - z_c). \tag{11}$$

To determine the pressure in the vapor at the interface within a capillary, we shall assume the vapor may be approximated as an ideal gas. Then from Eq. (11):

$$P_c^V = P_{sat} \exp[\frac{Wg}{kT}(z_B - z_c)]. \tag{12}$$

For the interface heights above the bulk liquid interface that we consider, one finds from Eq. (12) that P_c^v is negligibly different than the saturation pressure. If the liquid phase is approximated as incompressible with a specific volume v_{sat}^L, then for the liquid phase within a capillary:

$$\mu^L(T, P_c^L) = \mu(T, P_{sat}) + v_{sat}^L(P_c^L - P_{sat}). \tag{13}$$

To determine the expression for the difference in chemical potentials at the interface in the liquid and vapor phases within a capillary, we assume the Laplace equation is valid at the interface. If the surface tension, capillary diameter and contact angle are denoted as γ^{LV}, D, and θ, the Laplace equation may be written

$$P_c^V - P_c^L = \frac{4\gamma^{LV}\cos\theta}{D}.$$ (14)

From Eqs. (12), (13) and (14), one finds:

$$\mu^L(T, P_c^L) = \mu(T, P_{sat}) - v_{sat}^L(\frac{4\gamma^{LV}\cos\theta}{D}),$$ (15)

and after combing Eqs. (11) and (15):

$$\mu^L(T, P_c^L) - \mu^V(T, P_c^V) = \frac{-4\gamma^{LV}v_{sat}^L\cos\theta}{D} + Wg(z_c - z_B).$$ (16)

Equation (16) defines the difference between the intensive properties at the interface during evaporation from a capillary. If a small volume at the interface in a capillary that included both phases were considered, the volume would be viewed as open; thus *in general* Eq. (9) could not be used to predict the evaporation flux within such a volume, since Eq. (9) was obtained for an isolated system.

We propose to define an isolated system that has the same evaporation flux as that in a capillary at a time t. At this time, suppose that a small volume that includes the interface within one capillary is extracted and isolated. The intensive properties at this time in each phase of the open system are taken to be the same as those in the isolated system. We assume the rate of evaporation at this instant would be the same in the two systems. Afterwards, the evaporation flux in the open system would be very different from that in the isolated system. The latter would evolve to a final equilibrium state, and we may apply Eq. (9) to predict the evaporation rate.

We assume that when the element of volume is taken to form the isolated system, enough of the liquid phase is taken so that the final temperature in the isolated system would be near the temperature of the liquid at the instant t. Thus, T_e would be equal T. From Eqs. (8), (9) and (16), the SRT expression for the rate of evaporation is then given by:

$$J = \frac{P_{sat}(T^L)}{\sqrt{2mkT^L}}\{\exp[\frac{-4\gamma^{LV}v_{sat}^L\cos\theta}{kTD} + \frac{Wg}{kT}(z_c - z_B)] - \exp[\frac{4\gamma^{LV}v_{sat}^L\cos\theta}{kTD} - \frac{Wg}{kT}(z_c - z_B)]\},$$ (17)

An examination of Eq. (17) indicates that the larger the value of:

$$\frac{-4\gamma^{LV}v_{sat}^L\cos\theta}{kTD} + \frac{Wg}{kT}(z_c - z_B),$$ (18)

the larger the evaporation rate. Thus, for a given contact angle, the rate of evaporation should increase as the height of the interface above the bulk liquid interface is increased. Further, for water a difference in height of a few millimeters is predicted to produce a measurable difference in the rate of evaporation.

Experimental Procedure and Apparatus

The water to be used in the experiments was distilled, degassed and filtered. Its resistance was $18M\Omega$ per cm and its surface tension was within 2% of the documented value. When the capillaries were partially filled with water, each was filled to a different degree so that the interface within each had a different initial height. The length of the capillary tube above the interface of that capillary was initially the same for each capillary. This insured that the resistance to the flow of vapor out of the capillary tube was approximately the same for each capillary. Once the capillaries had been placed in the vessel, the vessel was sealed except for its connection to a vacuum pump. The water then was degassed. Afterwards the vessel was sealed and completely submerged in a water bath to prevent air leakage into the vessel. The water bath maintained the temperature within the vessel at $37.5 \pm 0.4°C$.

The pressure in the vessel was monitored with a pressure transducer intermittently during each experiment. No deviation from the saturation vapor pressure of the water was measured. In each experiment a fourth capillary that was open at both ends was present in the vessel (not shown in the schematic) and was used to monitor the surface tension.

RESULTS

The height of the interface within each capillary was monitored with a cathetometer from outside the vessel and the water bath. The height of an interface could be measured with an accuracy of ± 0.01mm. Each experiment was run for approximately 15 days. The amount of liquid evaporation from each capillary could be determined by simply measuring the height of the interface within each capillary. This allowed the initial average evaporation rate as a function of initial height to be determined. It was found that the initial rate of evaporation from the capillaries increased as the initial interface height within the capillary was increased. As may be seen in Fig. 2, an increase in interface height of approximately 10 mm was observed to increase the initial rate of evaporation by approximately 40%.

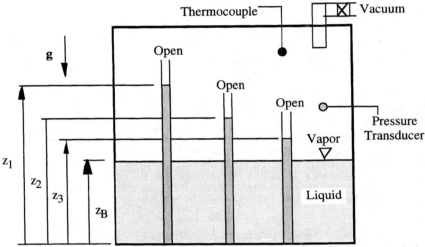

FIGURE 1. Schematic of the experimental apparatus. The vessel containing the capillaries was sealed and submerged in a water bath that maintained the system at constant temperature.

DISCUSSION

The observed dependence of the evaporation rate on the gravitational potential energy at an interface supports the dependence that is expected from SRT. The basic result from this theoretical approach is that the evaporation rate depends on the magnitudes of the chemical potential and temperature discontinuity at the interface. For the isothermal approximation that we consider, Eq. (11) indicates the chemical potential in the vapor at the interface in a capillary depends on the height of the interface above the bulk liquid interface. Equation (15) indicates that the chemical potential in the liquid phase at the interface of a capillary depends only on the contact angle. It has been assumed that the contact angle is independent of interface height. Thus, the dependence of the magnitude of the chemical potential discontinuity on interface position results from the dependence of the chemical potential in the vapor on position, Eq. (11). The experiments were not run for a sufficient time to allow equilibrium to be reached; however, the necessary condition for equilibrium can be determined from Eq. (16) by requiring J to be zero. One finds:

$$z_c - z_B = \frac{4\gamma^{LV} v_{sat}^L}{DgW}. \tag{19}$$

However, we emphasize that this is only the necessary condition for equilibrium. To date the stability of this equilibrium state has not been established. We note that a bubble immersed in a liquid phase has an unstable equilibrium state (Tucker, 1975). In many ways, a bubble immersed in a large volume of liquid is similar to the circumstance we have considered: the pressure in the vapor phase is greater than that in the liquid and the liquid phase properties are unchanging.

Graham (1991) measured the equilibrium height in a similar experimental apparatus but one with possibly important differences. The bulk liquid interface was curved in his case. He found the equilibrium height of an interface within a capillary that had one end closed was less than that within a capillary that had both ends open. If the same result were found in the apparatus we have used, then Eq. (19) would be invalid. Thus, at present the stable equilibrium configuration in an apparatus such as the one we have used has been established.

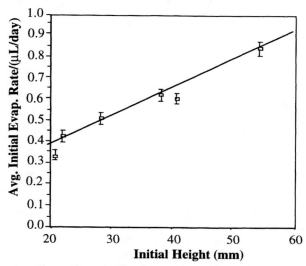

FIGURE 2. Summary of measured initial evaporation rates during three experiments with water.

ACKNOWLEDGMENTS

This work is supported by The Canadian Space Agency.

REFERENCES

Aoki, K. and Cercignani, C. "Evaporation and Condensation on Two Parallel Plates at Finite Reynolds Numbers," *Phys. Fluids* **26**, 1163 (1983).

Cercignani, C., Fiszdon, W. and Frezzontti, A. "The Paradox of the Inverted Temperature Profiles Between an Evaporating and a Condensing Surface," *Phys. Fluids* **8**, 3237 (1985).

Cipolla, J. W. Jr., Lang, H. and Loyalka, S. K. "Kinetic Theory of Condensation and Evaporation. II," *J. Chem. Phys* **61**, 69 (1974).

Fang, G. and Ward, C. A., "Temperature Measured Close to the Interface of an Evaporating Liquid," *Phys. Rev. E* **59**, 417-428 (1999).

Fang, G. and Ward, C. A., "Examination of the Statistical Rate Theory Expression for Liquid Evaporation Rates," *Phys. Rev. E* **59**, 441-453 (1999a).

Graham, M. "No Capillary Rise?" *Nature* **350**, 198 (1991).

Pao, Y.- P. "Application of Kinetic Theory to the Problem of Evaporation and Condensation," *Phys. Fluids* **14**, 306 (1971).

Tucker, A. and Ward, C. A. "Critical State of Bubbles in Liquid-Gas Solution," *J. A. Phys.* **46**, 4801 (1975).

Ward, C. A. and Fang, G., "Expression for Predicting Liquid Evaporation Flux: Statistical Rate Theory Approach," *Physical Review E,* **59**, 429-440 (1999).

Ward, C. A. and Sasges, M. R., "Effect of Gravity on Contact Angle: A Theoretical Investigation," *J. Chem. Physics,* **109**, 3651-3660 (1998).

Stress Singularities in Confined Thermocapillary Convection

Guillaume Kasperski[1], Eric Chénier[2], Claudine Delcarte[1],
Gérard Labrosse[1]

[1]*Université Paris-Sud XI, LIMSI-CNRS, UPR 3251, BP133, F-91403 Orsay cedex*
[2]*Université de Marne la Vallée, Cité Descartes, Bât. Lavoisier, F-77454 Marne la Vallée cedex 2*

[1]33/01.69.85.80.72, labrosse@limsi.fr; [2]33/01.60.95.73.12, chenier@univ-mlv.fr

Abstract. Axisymmetric thermocapillary convection is studied in a laterally heated liquid bridge. In this configuration, as in other wall-confined thermocapillary convection problems, a viscous singularity appears at the junction of the free and solid surfaces which any numerical approach of the problem must filter, either explicitly by smoothing the boundary conditions, or implicitly by using finite-precision discretization methods. Our approach is to filter the singularity explicitly, and to study the convergence properties of the solutions with the filter's characteristics, which cannot easily be done when using finite-precision methods. Results show both quantitative (on scales) and qualitative (on symmetry properties) effects of the filter. Based on observations of the problems encountered when treating the laterally heated case, we propose to compare the results supplied by different numerical approaches on a simple half-zone model. This is an important step to pass before running oscillatory and/or 3D computations.

INTRODUCTION

Thermocapillary convection is due, on a free surface, to thermally-induced inhomogeneities of surface tension. This phenomenon is of microscopic origin, but is generally treated macroscopically using the following equation, expressing the balance, at the free surface, between the thermocapillary and the corresponding viscous stress:

$$\vec{\nabla}_n(\vec{U}.\vec{\tau}) = \vec{\nabla}_\tau \sigma = \gamma \vec{\nabla}_\tau T, \tag{1}$$

where $\vec{U}$, T, σ, and $\gamma = \dfrac{\partial \sigma}{\partial T}$ respectively stand for the velocity, the temperature, the surface tension and the surface tension (negative) thermal coefficient. The unit vectors $\vec{\tau}$ and $\vec{n}$ are resp. tangent and normal to the free surface. This macroscopic writing generally leads, at the junction of free and solid surfaces, to a singular left hand side of equation (1), well known by physicists of the wetting (De Gennes, 1990). To numerically treat this problem, the singularity must be filtered, either by using a finite precision method, or by filtering it explicitly. Spectral methods, which must deal with regular problems, are generally considered as non-appropriate (Zebib, 1985). The use of finite precision methods alone nevertheless leads to mesh-dependant boundary conditions, which prevents easy tests of convergence with space discretization (Shen, 1998). In a simple configuration (the vicinity of the junction between a free surface and a cold wall), a scaling analysis was performed (Canright, 1994) in order to determine which small length scale, in the vicinity of the singularity, had to be resolved by numerical computations in order to get physically relevant solutions. The results show that this scale decreases when increasing the Marangoni number, characteristic of the thermal transport regime, and numerical calculations were led with meshes

CP504, *Space Technology and Applications International Forum–2000*, edited by M. S. El-Genk

in accordance to these theoretical predictions. Nevertheless, no a–posteriori study of the numerical results' convergence with the mesh refinement was proposed. Moreover, the asymptotic laws found in this simple configuration were not in agreement with those numerically found for a liquid bridge (Zebib, 1985): these results cannot be transposed directly to more complicated configurations.

When considering those last configurations, the convergence of the numerical results with mesh-refinement is seldom presented, the authors referring to oneanother, or to experimental measurements. Zebib et al. present convergence results for a half-zone configuration. They show a slow variation of the test variables, chosen there as stream function extrema. Which prevents from concluding about the convergence with the mesh refinement. The goal of this communication is to present a few results obtained for a laterally heated liquid bridge. The singularity was bypassed by the use of different filters with known characteristics. The convergence of the results with different filtering length scales has been analysed (Chénier, 1977; Kasperski, 1999; and Kasperski and Labrosse, 1999). High-Marangoni number flows converge very slowly and necessitate important space (spectral) discretizations. Due to the facts that, (1) the convergence properties of finite-precision methods are difficult to evaluate for singular problems, (2) it is difficult to make comparative studies of published results because the treated configurations generally differ (thermal boundary conditions on the free surface, deformable/undeformable free surfaces, ...), we propose a reference (simple) configuration, on which different spectral/finite-precision, regularized/non-regularized results could be compared.

THERMOCAPILLARY CONVECTION IN LATERALLY HEATED BRIDGE

Axisymmetric thermocapillary convection is studied in a (r,z) cylindrical liquid bridge of height $2H$ and radius R in a zero gravity environment, Figure 1.a. The liquid (density ρ, kinematic and dynamic viscosities υ and μ, thermal diffusivity and conductivity κ and k) is held between two planar isothermal solid surfaces, located at $z = \pm H$ and its free surface, at $r=R$, is submitted to a lateral heat flux $Q(z)=Q_0.q(z)$, scaled by Q_0. Heat is then driven from the free surface to the solid planes by convection and conduction, Figure 1.b,c.

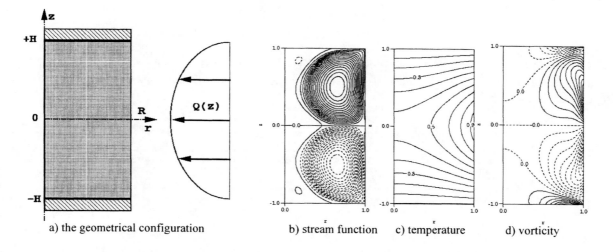

a) the geometrical configuration   b) stream function   c) temperature   d) vorticity

FIGURE 1. Problem statement, and flow fields isolevels for Ma=10, $\mathrm{Pr} = 10^{-2}$

The only considered source of motion is the thermocapillary stress. The classical Oberbeck-Boussinesq equations of motion and associated boundary conditions, once rendered non-dimensional with the scales presented in table I, are:

$$\nabla . \vec{U} = 0 \tag{2}$$

$$\frac{\partial u}{\partial t} + (\overset{\rho}{U}.\overset{\rho}{\nabla}).u = -\frac{\partial P}{\partial r} + \text{Pr}.(\Delta u - \frac{u}{r^2}) \tag{3}$$

$$\frac{\partial w}{\partial t} + (\overset{\rho}{U}.\overset{\rho}{\nabla}).w = -\frac{\partial P}{\partial z} + \text{Pr}.\Delta w \tag{4}$$

$$\frac{\partial T}{\partial t} + (\overset{\rho}{U}.\overset{\rho}{\nabla}).T = \Delta T \tag{5}$$

$$u_{z=\pm A} = w_{z=\pm A} = T_{z=\pm A} = u_{r=0} = \frac{\partial w}{\partial r}\bigg|_{r=0} = \frac{\partial T}{\partial r}\bigg|_{r=0} = u_{r=1} = 0, \tag{6}$$

$$\frac{\partial T}{\partial r}\bigg|_{r=1} - q(z) = 0 . \tag{7}$$

$$\frac{\partial w}{\partial r}\bigg|_{r=1} + \text{Ma}.\frac{\partial T}{\partial z}\bigg|_{r=1} = 0 \tag{8}$$

The problem depends on three non-dimensional parameters, the Prandtl number, $\text{Pr} = \nu / \kappa$, ratio of momentum to thermal diffusivities, the Marangoni number $\text{Ma} = -\frac{\gamma Q_0 R^2}{k\mu\kappa}$, characteristic of the thermal transport regime, and the aspect ratio of the cavity $A = H / R \equiv 1$. The heat flux is chosen as $q(z) = (1-(z/A)^2)^2$. The radial derivative of the axial velocity is singular at the junction of the free and solid surfaces (where $\partial T / \partial z$ generally does not cancel, but increases with Ma). The boundary condition (8) concerning this term is replaced by:

$$\frac{\partial w}{\partial r}\bigg|_{r=1} + \text{Ma}.\frac{\partial T}{\partial z}\bigg|_{r=1} f_n(z) = 0 . \tag{9}$$

$f_n(z) = (1-(z/A)^{2n})^2$ is a regularizing function, the shape of which depends on the integer parameter n. A filtering scale δ can be defined as the distance from the solid surfaces to the points where $f_n(z)=0.9$. This scale decreases as $1/n$.

TABLE 1. Variables and Scaling.

Variables	Length	Velocity	Temperature	Time	Pressure
Symbol	r,z	$\overset{\rho}{U} = (u,w)$	T	t	P
Scale	R	κ / R	$Q_0 R / k$	R^2 / κ	$\rho\kappa^2 / R^2$

Increasing n will lead to increase the averaged imposed vorticity on the free surface, and consequently the flow velocity. This in turn tightens the axial thermal gradient and increases back the vorticity on the free surface. This feed-back effect, already mentioned by Canright will be more efficient for high Ma regimes. The resolution of the equations is based on a projection-diffusion algorithm (Batoul,1994; Labrosse, 1997), with second order Adams-Bashforth Backward-Euler 2 time discretisation and a spectral Chebyshev collocation method using Gauss-Lobatto points for z and Gauss-Radau points for r. The results have been obtained either by a direct numerical simulation of the flows or by a continuation method coupled to a linear stability analysis.

Precision criteria

The precision criteria for the direct numerical simulations of the flows are the following. A solution is declared converged with the space resolution when the relative divergence of the velocity satisfies $\left\|\vec{\nabla}\vec{U}\right\|_{\infty}/\left\|\vec{U}\right\|_{\infty}<10^{-3}$.

The notation $\left\|\cdot\right\|_{\infty}$ denotes the maximum of the modulus of the considered field, over the collocation points. Indeed, the projection-diffusion method cancels the velocity divergence, but only asymptotically with the space resolution (Labrosse, 1997). This criterion supplies at least three exact digits of the vorticity maximum for a (Ma, Pr, A, n)

fixed configuration. A solution is declared stationary when: $\max_{\phi=u,w,T}\left(\dfrac{\left\|\phi^{n+1}-\phi^{n}\right\|_{\infty}}{\left\|\phi^{n}\right\|_{\infty}}\right)<10^{-6}\delta t$, with δt the

time step (about 10^{-5} for high Marangoni flows), the superscripts n and $n+1$ denoting two successive instants.

The n=1 case

The $n=1$ case was treated over large ranges of Ma values and for $\mathrm{Pr}\in\left[10^{-2},10^{2}\right]$ (Batoul, 1995; Chénier, 1998 and 1999; Kasperski, 1998 and 1999). All the steady solutions were found to be symmetrical about the mid-plane $z=0$ of the cavity. The main flow is composed of two contra-rotative cells, the upper one rotating counter-clockwise (Figure 1.b), the heat supplied on the free surface diffuses and is advected in the cavity. Positive vorticity, (Figure 1.d), in the upper part of the cavity, is imposed on the free surface, and diffuses along with the negative one generated by the flow at the solid boundaries. The momentum transport regime is very Pr-dependant, this transport being the more convective for lower Pr flows. Two kinds of destabilization of the steady flows have been observed, depending on Pr (Kasperski 1998 and 1999).

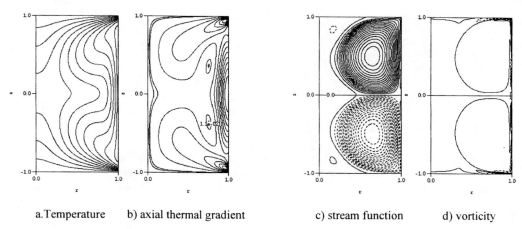

a. Temperature   b) axial thermal gradient   c) stream function   d) vorticity

FIGURE 2. The highest Ma steady flows obtained for n=1, (a.) and (b) for $Ma=1.2\,10^{5}$, $\mathrm{Pr}=10^{2}$, (c) and (d) for $Ma=10^{5}$, $\mathrm{Pr}=10^{-2}$.

High-Pr flows undergo a hydro-thermal instability, occurring with the appearance of a negative radial temperature derivative in the field (Figures 2.a,b). A criterion was found for the transition to unsteadiness of low-Pr flows, corresponding to a specific configuration of the vorticity field. The transition occurs when, for instance in the upper part of the cavity (Figure 2.c,d), the negative vorticity issued from the solid surface, advected by the flow, has turned around the positive vorticity main cell and comes back close to the solid surface, being then pinned against the positive vorticity imposed on the free surface. The flow being there nearly axial, locally fulfils the Fjørtøft criterion for instability, even if it is established only for an inviscid fluid. For liquids with $Pr=1$, no unsteady solution has yet been found at the highest attempted Ma values.

Results for n>1

Increasing the parameter of the regularizing function leads to quantitative and qualitative modifications of the flows. First, (Chénier, 1997; Kasperski, 1999; and Kasperski and Labrosse, 1999), as can be expected, the vorticity, velocity and axial thermal gradient scales of thermally convective flows are much dependant on the regularizing scale δ. The vorticity magnitude gets about ten times larger when going from $n = 1$ up to the convergence with δ. Consequently, decreasing δ lowers the transition thresholds to unsteadiness of low-Pr flows, according to the Fjørtøft criterion. The effect of δ is nevertheless not as important on the stream-function and temperatures scales. Those variables should therefore not be chosen to estimate the convergence of the solutions. Qualitative effects have also been observed (Chénier, 1997 and 1998). For $n>6$, $Pr = 10^{-2}$, $Ma>600$, non-symmetric stable solutions are allowed. The analysis of these new states, obtained up to now by a continuation method, is in progress. One should notice that to obtain those solutions by direct numerical simulation is very expensive, because of extremely small growth rates of the states associated with the pitchfork bifurcation. In the light of these preliminary results, the parameter n turns out to play an important role on the reliability of the solutions, axisymmetric and steady until now. What to say about oscillatory 3-D computed solutions? A particular numerical investment deserves to be proposed to clear up this interesting question.

PROPOSITION OF A TEST CONFIGURATION

As a first step, we propose to compare the results of various numerical approaches, applied on a simplified 2D half-zone configuration which has the advantage to avoid the symmetry breaking bifurcation observed in the previous configuration. The solid surfaces, distant by $2H$, and of radius R, are held at two different temperatures, T_{up} and T_{down}. The free surface is also straight and undeformable. The fluid motion is governed by the equations of motion (2-5), with the following boundary conditions:

$$u_{z=\pm A} = w_{z=\pm A} = T_{z=A} - T_{up} = T_{z=-A} - T_{down} = u_{r=0} = \frac{\partial w}{\partial r}\bigg|_{r=0} = \frac{\partial T}{\partial r}\bigg|_{r=0} = u_{r=1} = \frac{\partial T}{\partial r}\bigg|_{r=1} = 0, \qquad (10)$$

$$\frac{\partial w}{\partial r}\bigg|_{r=1} + Ma.\frac{\partial T}{\partial z}\bigg|_{r=1} f_n(z) = 0 \qquad (11)$$

with, now, $A = 1/2$ and $Ma = -\frac{\gamma |T_{up} - T_{down}| R}{\mu\kappa}$. The proposed regularization is not the only one that could be applied, and for instance it could be interesting to compare with the results of finite precision methods when they are applied replacing (11) by (8). Other approaches could be as well discussed and proposed. The proposed cases for the study are given in Table 2, a first set of cases is proposed for $Pr = 10^{-2}$, corresponding to liquid metals, and the second one to $Pr = 10^2$, corresponding to some silicon oils. As the latter remains steady, with a plane free surface (Shevtsova, 1998; Kasperski, 1999), higher Ma-values can be reached for stationary cases, making the convergence with space discretization and/or regularizing function more difficult. Spatial and temporal precision criteria used for the calculations are needed. We suggest that the convergence of the results with the used meshes and/or scale of the regularizing function be illustrated by the maxima of the stream function, temperature, velocity modulus, axial thermal gradient and vorticity. As a concrete result, the question of the existence (and evaluation) of a physical small scale, at the macroscopic level, required to get relevant results could be answered (and obtained).

TABLE 2. Proposition of test cases

$Pr=10^{-2}$	$Ma=1$ case 1	$Ma=10^2$ case 3	$Ma=10^3$ case 5	transition to unsteadiness case 7	-------
$Pr=10^2$	$Ma=1$ case 2	$Ma=10^2$ case 4	$Ma=10^3$ case 6	$Ma=10^4$ case 8	$Ma=10^5$ case 9

CONCLUSION

Axisymmetric thermocapillary convection is being studied for several years, in the full liquid bridge configuration, with straight and undeformable free surface. The treatment of stress singularity, localized where the free and solid surfaces meet, turns out to influence significantly the resulting numerical flows. It then seems important to conceptually master this aspect of the problem before attempting unsteady and/or 3D numerical simulations. For instance, should a small scale be explicitly taken into account in the models to get physically relevant results? What is its size as a function of Ma and Pr? A simple test configuration is therefore proposed to start with a comparison between various numerical approaches of the singularity and of their results.

AKNOWLEDGMENTS

The authors acknowledge support from the C.R.I. of the Paris-Sud University for C.P.U. allocation, and from E. Millour for reviewing the English.

REFERENCES

Batoul, A., "Simulation Numérique d'Ecoulements Thermocapillaires en Croissance Cristalline," *Thèse de Doctorat en Sciences, Université Paris XI, Orsay* (1995).

Batoul, A., Khallouf, H., Labrosse, G., "Une Méthode de Résolution Directe (Pseudo-Spectrale) du Problème de Stokes 2D/3D Instationnaire, Application à la Cavité Entrainée Carrée," *C. R. Acad. Sci. Paris,* **319**, Série I, 1455-1456 (1994)

Canright, D., "Thermocapillary Flow Near a Cold Wall," *Phys. Fluids,* **6**, 4, 1415-1424 (1994).

Chénier, E., "Etude de la Stabilité Linéaire des Ecoulements Thermocapillaires et Thermogravitationnels en Croissance Cristalline,"*Thèse de Doctorat en Sciences, Université Paris XI, Orsay* (1997).

Chénier, E., Delcarte, C., Labrosse ,G., "Solutions Multiples Thermocapillaires en Zone Flottante à Gravité Nulle," *Eur. Phys. J. AP,* **2**, 93-97 (1998).

Chénier, E., Delcarte, C., Labrosse, G., "Stability of the Axisymmetric Buoyant-Capillary Flows in Laterally Heated Liquid Bridge," *Phys. Fluids,* **11**, 3, 527-541 (1999).

De Gennes, P.G., Hua, X., Levinson, P., "Dynamics of Wetting: Local Contact Angles," *J. Fluid Mech.,* **212**, 55-63 (1990).

Kasperski, G., "Convection Thermocapillaire en Pont Liquide Chauffé Latéralement," *Thèse de Doctorat en Sciences, Université Paris XI, Orsay* (1999).

Kasperski, G., Batoul, A., Labrosse, G., "Up to the Unsteadiness of Axisymmetric Thermocapillary Flows in a Laterally Heated Liquid Bridge," *submitted to Phys.. Fluids,* Sept. 1998.

Kasperski, G., Labrosse, G., "On the Numerical Treatment of Viscous Singularities in Wall-Confined Thermocapillary Convection," *submitted to Phys. Fluids,* Jul. 1999.

Labrosse, G., Tric, E., Khallouf, H., Betrouni, M., "A Direct (Pseudo-Spectral) Solver of the 2D/3D Stokes Problem: Transition to Unsteadiness of Natural Convection Flow in a Differentially Heated Cubical Cavity," *Num. Heat Transfer, Part B,* **31**, 261-276 (1997).

Shen, J., "An Efficient Spectral-Projection Method for the Navier-Stokes Equations in Cylindrical Geometries I. Axisymmetric Cases," *J. Comp. Phys.* **139**,308-326 (1998).

Shevtsova, V. M., Legros, J. C., "Oscillatory Convection Motion in Deformed Liquid Bridges," *Phys. Fluids,* **10**, 7, 1621-1634, (1998).

Zebib, A., Homsy, G.M., Meiburg, E., "High Marangoni Number Convection in a Square Cavity," *Phys. Fluids,* **28**,12, 3467-3476 (1985).

Magnetic Compensation of Gravitational Forces in (-p) Hydrogen Near its Critical Point : Application to Weightlessness Conditions

Daniel Beysens [1], Régis Wunenburger [2], Denis Chatain [1], and Yves Garrabos [2]

[1] *Service des Basses Températures, DRFMC, CEA Grenoble,*
17, rue des Martyrs, 38504 Grenoble cedex 09, France
[2] *Institut de Chimie de la Matière Condensée de Bordeaux, UPR 9048 Centre National de la Recherche*
Scientifique, avenue Dr. A. Schweitzer, 33608 Pessac cedex, France

[+33.4.76.88.94.69; dbeysens@cea.fr]

Abstract. We report a study of the variation and compensation of gravitational forces in diphasic (p-) hydrogen. The sample is placed on the axis of a superconducting coil near one of the ends, in order to benefit from a nearly uniform magnetic field gradient. A variable reduced effective gravity g* can thus be applied to the fluid. It is shown that the exact compensation of gravitational forces ("weightlessness conditions") is limited by the inhomogeneity of the magnetic field gradient, which consists of a radial force field centered on the point of compensation and varying proportionally with the distance from this point. Such inhomogeneities can be quantified by an associated capillary length (l_C). The effect of this remaining field is negligible when l_C is smaller than the cell dimension. Near the critical point, the length l_C tends to zero and the gas-liquid interface is deformed in a paradoxical pattern, in which the liquid phase lies at the center of the container with the gas phase at the walls. We vary the effective gravity g* and determine in a wide range of reduced temperature $\tau = (T_C-T)/T_C = [10^{-4} - 0.02]$ the corresponding capillary length l_C from a fit of the interface profile. The data complement previous measurements in n-H_2 at g* = 1, performed further from T_C.

INTRODUCTION

Because of the weak diamagnetic susceptibility of many non-magnetic substances, important magnetic fields are needed for magnetic levitation. Such field intensities could be attained only recently (Beaugnon, 1991) and this technique was chiefly applied to the containerless handling of liquids, solids or living bodies (Beaugnon, 1993, Weilert, 1997) and to the simulation of low gravity or weightlessness conditions (Israelson, 1995, Valles, 1997). Handling applications require only a global compensation of the total weight of the body and a stable levitation equilibrium to be met, whereas gravity compensation needs more conditions to be fulfilled. Indeed, when a diamagnetic sample is placed in a magnetic field **B**, each of its molecules is subjected to a force which is proportional to its magnetic susceptibility χ and to the gradient of $\mathbf{B}^2$. The resulting magnetic acceleration $\mathbf{a}_m$ induced by the magnetic field **B** is

$$\mathbf{a}_m = \frac{\chi}{\mu_0 \rho} \vec{\nabla}(\frac{1}{2}\mathbf{B}^2) \,, \tag{1}$$

where ρ is the sample density and $\mu_0 = 4\pi\, 10^{-7}$ H/m. The total acceleration applying on the sample on the Earth is :

$$\mathbf{g} = \mathbf{g}_0 + \mathbf{a}_m \,, \tag{2}$$

where $\mathbf{g}_0$ is the gravitational acceleration vector. Let g* be the reduced acceleration (effective gravity level) :

$$g^* = \frac{\|\mathbf{g}\|}{\|\mathbf{g}_0\|} \,. \tag{3}$$

The acceleration field **g** is homogeneous on the length scale of the sample provided that the $\mathbf{B}^2$ gradient and the chemical composition is homogeneous on the sample length scale. Under these conditions, the Earth's gravitational acceleration can be increased, but also partially or exactly compensated in the same manner at each point of the sample. This leads to the accomplishment of a tunable artificial gravity, ranging from large to low values of g*, and

CP504, *Space Technology and Applications International Forum–2000*, edited by M. S. El-Genk

ultimately reaching weightlessness (g*=0). Since practically homogeneous acceleration field can only be achieved within a given precision, it is useful to evaluate the maximum amplitude of deviation Δg^* from its mean value within the sample :

$$\Delta g^* = \text{Max}_{\text{sample}}\left(g^* - \langle g^* \rangle_{\text{sample}} \right). \tag{4}$$

When global compensation of weight is achieved (levitation), Δg^* is called the "residual gravity". It is a quantitative evaluation of the "quality" of the weightlessness.

During the last decade, experimental investigation of the physics of pure fluids near their critical point benefited from weightlessness conditions, which prevented from convection and stratification (Beysens, 1996). Pure fluids exhibit universal behavior near their critical point, which means that some of their thermophysical properties diverge (e.g. the specific heat) or vanish (e.g. the thermal diffusion coefficient) at the critical point as universal power laws of the reduced distance to the critical point $\tau = |T_C - T| / T_C$ (T is the temperature of the fluid, and T_C is its critical temperature). This universal behavior allows a large variety of phenomena to be studied in any pure fluid just by changing its temperature. It allowed us to develop a magnetic levitation apparatus dedicated to para-hydrogen (p-H_2), chosen because of its high magnetic susceptibility ($\chi/\rho = -2.51\ 10^{-8}\ m^3.kg^{-1}$) and its low critical temperature ($T_C = 33$ K), easily compatible with the working temperature of a superconductive coil (2 K). The magnetic apparatus should be regarded more generally as a variable effective gravity facility which allows the effect of gravity to be studied in fluid physics, since it presently permits to apply a variable acceleration to a hydrogen sample in a range running from levitation to g*=2.2. In particular, the Moon and Mars gravity (g*$\approx$1/6 and 1/3) can be reproduced.

In previous works using magnetic weightlessness, precise measurements of the magnetic field (Israelson, 1995) as well as detailed analytical and numerical calculations of the force field applying on the sample were obtained (Beaugnon, 1993) but no measurement of the residual gravity was performed. We present here a simple method to check the homogeneity of the total acceleration field. It is based on the analysis of the shape of the gas-liquid interface and on the vanishing behavior of the capillary length near a critical point. A two-phase gas-liquid equilibrium is characterized by its interfacial tension σ. When only interfacial forces act, the bubble has a spherical shape, which minimizes its interfacial energy. When submitted to an acceleration field $\mathbf{g}$, it is deformed so as to minimize the sum of its potential and interfacial energies. One defines the capillary length ℓ_C as :

$$\ell_C = \sqrt{\frac{\sigma}{g(\rho_L - \rho_V)}} = \frac{1}{\sqrt{g^*}}\sqrt{\frac{\sigma}{g_0(\rho_L - \rho_V)}}, \tag{5}$$

where ρ_L (ρ_V) is the density of the liquid (gas) phase. The capillary length is the lengthscale of the deformations of the bubble. The shape of the interface then gives an information about the intensity of the acceleration field when the capillary length is kept smaller than the size of the bubble. The critical point is the endpoint at the top of the liquid-gas coexistence curve of a pure fluid in the P-T state diagram. At the critical point there is no more difference between the two phases and the interface disappears. The phase transition is of second order at the critical point, and along the coexistence curve the difference between the densities of the two phases varies as:

$$\frac{\rho_L - \rho_V}{\rho_C} = 2B\tau^{0.325}, \tag{6}$$

where ρ_C is the critical density. The amplitude B is system-dependent. B=1.61 for p-H_2 (Sengers, 1978) and 0.325 is a universal critical exponent. The surface tension σ between the two phases also vanish at the critical point as :

$$\sigma = \sigma_0\tau^{1.26}. \tag{7}$$

The amplitude σ_0 depends on the fluid and 1.26 is a universal critical exponent. According to Eq. 5, the capillary length ℓ_C vanishes at the critical point as :

$$\ell_C = \frac{1}{\sqrt{g^*}}\ell_{C0}\tau^{0.4675}, \qquad \ell_{C0} = \sqrt{\frac{\sigma_0}{2Bg_0\rho_C}}. \tag{8}$$

For n-hydrogen, $\ell_{C0} = 2.39$ mm (Moldover, 1985). When g* goes to zero, the capillary length increases, an increase which can be compensated by going closer to the critical temperature. Qualitatively speaking, the low intensity of the acceleration field is compensated by the fact that the near-critical interface is very deformable, thus very sensitive to the remaining accelerations.

The paper is organized as follows. In section II the experimental apparatus is presented in details. In section III is described a method to check the homogeneity of the acceleration field, and to deduce the capillary length and the acceleration field from the shape of the gas-liquid interface. In section IV the computation of the residual acceleration field in the levitation state is proposed.

EXPERIMENTAL

The apparatus consists of two parts, the cryostat containing the superconducting coil and the anticryostat containing the sample and the optics. It is schematically shown in Fig. 1. The cryostat is composed of two stages. A dewar containing liquid N_2 surrounds another dewar containing liquid He maintained at 2.17 K by a heat exchanger. A superconductive coil of height 200 mm is immersed inside the He bath. It supports a current up to 65 A and produces a magnetic field up to 10 T on its axis, which is parallel to Earth's gravity. The anticryostat is a cylindrical container partially immersed inside the Helium bath, standing along the axis of the coil, in which vacuum (10^{-4} Pa) is maintained for thermal isolation purpose.

The sample cell containing the H_2 sample is positioned along the axis of the coil. The altitude of the sample cell center along the coil axis corresponds to the maximum of intensity of the upward magnetic force, i.e. at the upper end of the coil, at 85 mm from its center. On the coil axis, g is vertical and Eq. 2 reduces to $g=g_0-a_m$ at the upper end ($g=g_0+a_m$ on the lower end). Levitation is attained when the electric current in the coil is equal to 63.1 A, corresponding to a magnetic field of magnitude 6.65 T in the cell. According to the linear relation between the electric current I in the coil and the magnetic field, g* is computed using the relation:
$$g^*(I)=g_0(1-(I/63.1)^2) \, . \tag{9}$$

The cell is made of electrolytic copper. The cavity containing the H_2 is cylindrical (3 mm in diameter) with its axis horizontal, i.e. perpendicular to the coil axis. It is closed by two parallel sapphire windows distant of 2 mm and sealed by indium rings.

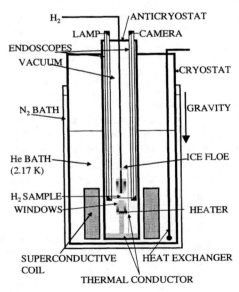

FIGURE 1. Experimental set-up.

The sample cell is in thermal contact with the He bath by means of a thermal conductor. Its temperature is measured with a Platinum thermometer and controlled via an electrical resistance supplied by a close loop temperature control device. The temperature control accuracy is ± 1 mK at 33 K and the working range is 15 – 40 K. The sample cell is filled with purified, pressurized H_2 at room temperature through a capillary. This capillary is closed by an H_2 ice floe, whose formation is locally allowed by a thermal conductor in contact with the helium bath (solidification temperature of H_2 : 14 K). The gas-liquid system is observed in light transmission. Parallel light is directed by means of mirrors from outside the anticryostat to the sample. The beam is parallel to the axis of the cylindrical cell, and is directed by mirrors to the camera outside the anticryostat. The cell is filled at nearly critical density as checked by the position of the meniscus in the middle of the vessel at the critical temperature. When H_2 is cooled in the cell, the n-H_2 - p-H_2 equilibrium is shifted and the percentage of p-H_2 increases from 25% at room temperature to 96% at 30 K. The critical coordinates of n-H_2 (p-H_2) are T_C=33.24 K, (32.976 K), P_C=1.298 MPa (1.2928 MPa), ρ_C=30.09 kg.m^{-3} (31.426 kg.m^{-3}). The slow evolution of the percentage of p-H_2 is followed in time. The equilibrium is supposed to be reached when the critical temperature of the H_2 sample is stationary.

ANALYSIS OF THE SHAPE OF THE MENISCUS

Measuring the Capillary Length from the Shape of the Meniscus

In Fig. 2 two typical pictures of the interface at various temperatures and effective gravity levels g* are shown. The dark stripe corresponds to the ascension of the liquid on the sapphire windows, which refracts the parallel light. Note that the low position of the meniscus in Fig. 2b is due to the increase of the wetting layer thickness on the cell walls when the magnetic field is applied. This phenomenon is fully reversible when the magnetic field is varied. Under gravity the capillary length can be extracted from the shape of the meniscus provided that it is much smaller than the horizontal extension of the meniscus. The lowest limit is given by the spatial resolution of the detection of the meniscus shape and the accuracy of the temperature control. Warren and Webb (Warren, 1969) used this method to deduce the interfacial tension from the capillary length measurements near the consolute point of near-critical cyclohexane-methanol mixtures under Earth's gravity. We inverted their method in order to deduce the effective acceleration from the capillary length measurements.

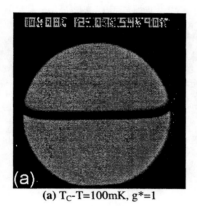 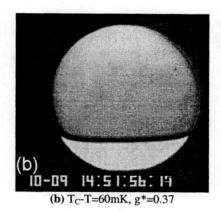

(a) T_C-T=100mK, g*=1   (b) T_C-T=60mK, g*=0.37

FIGURE 2. Video pictures of the interface.

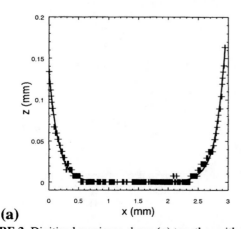 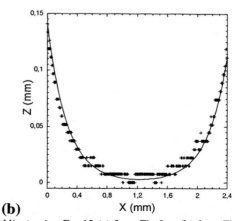

(a)   (b)

FIGURE 3. Digitized meniscus shape (+) together with the best fit (solid line) using Eq. 10 (a) from Fig 3a ; (b) from Fig. 3b.

When the meniscus is observed in transmission, the shape is the intersection of the interface with the plane perpendicular to the direction of observation and situated at half-width of the sample cell. As the length scale of the cell is large compared to ℓ_C, the analytic expression of the shape of the meniscus at half-width of the cell is

$$z(x) = z_{max}\left(\exp\left(-\frac{x}{l_c}\right) + \exp\left(-\frac{x-L}{l_c}\right)\right) \tag{10}$$

where z_{max} is itself a function of ℓ_C. This solution is in fact the sum of two independent menisci of opposite walls in semi-infinite vessels. The length ℓ_C is deduced from the best fit of the meniscus shape using Eq. 10 (Fig. 3).

Measuring the Effective Gravity from the Capillary Length

Due to the restricted domain of validity of the model of meniscus shape, the limited accuracy of the temperature regulation and the phenomenon of increased wetting under weak gravity, the measurements were performed in the temperature range $T_C-T = [12 \text{ mK} - 500 \text{ mK}]$, corresponding to the range $\tau = [5.10^{-4} - 2.10^{-2}]$ and in the gravity range $g^* = [0.25 - 1]$. To evaluate the homogeneity of the effective acceleration field, we have to check that : (i) the meniscus shape fits its theoretical shape in a homogeneous acceleration field (Eq. 10) and (ii) the effective gravity deduced from the measured capillary length is compatible with the effective gravity deduced from the value of the electric current.

(i) In Figure 5 are shown two typical meniscus shapes together with their best fit. Note that the agreement between the meniscus shape and the best fit is excellent all along the interface, indicating that the acceleration field is homogeneous.

(ii) We aim to compare the capillary length measurements performed at $g^* \neq 1$ to the measurements at $g^*=1$ (in absence of applied magnetic forces). According to the definition of the capillary length (Eq. 5), the quantity $\ell_C g^{*1/2}$ is independent of g^* and should behave as the capillary length at $g^*=1$ according to the critical scaling law (Eq. 8). In Fig. 4a, $\ell_C (g^*)^{1/2}$ is plotted as a function of τ together with the power law of Eq. 8 with $\ell_{C0} = 2.39$ mm (Moldover, 1985) and the exponent imposed to its theoretical value 0.4675. We note that the dispersion of the experimental data around the critical power law of Eq. 8 is the same at high g^* (where homogeneous Earth's gravity is dominant) than at low g^* (where magnetic forces are comparable to Earth's attraction). It follows that this dispersion is not due to inhomogeneities of the effective acceleration field.

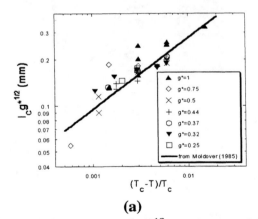
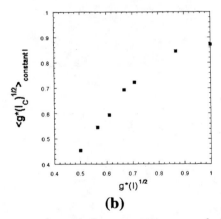

(a)   **(b)**

FIGURE 4. (a) the quantity $\ell_C g^{*1/2}$ as a function of the reduced distance to the critical temperature measured at various effective gravity levels g^*. The solid line is the critical power law for the capillary length defined in Eq. 8. (b) mean value of the data set $g^*(\ell_C)^{1/2}$ measured at a given electric current I as a function of $g^*(I)^{1/2}$.

The data set is well represented by the power law with $\ell_{C0} = 2.39$ mm, computed from previous measurements by Blagoi and Pashkov (Blagoi, 1966) in n-H_2 which shows critical parameters slightly different from p-H_2 data (to our knowledge there are no surface tension measurements reported near the critical point of p-H_2).

We also aim to compare the nominal values of $g^*(I)$ computed using Eq. 9 to the values of g^* deduced from the capillary length measurements

$$g*(\ell_C) = \left(\frac{\ell_{C0} \tau^{0.4675}}{\ell_C} \right)^2, \tag{11}$$

using $\ell_{C0} = 2.39$ mm. Since the capillary length is sensitive to the square root of g, we plotted in Fig. 4b the mean value of each data set $g*(\ell_C)^{1/2}$ measured at a given electric current I (or equivalently at a given value of the effective acceleration g*(I)) as a function of $g*(I)^{1/2}$. The departure of the data from the ideal line of slope 1 is more pronounced at high g* than at low g*, and then is not due to inhomogeneities of the effective acceleration field. The dispersion of the data has to be ascribed to the low frequency oscillations of the cell temperature induced by the closed loop regulation device and to the fact that in some cases the conditions of smallness of the capillary length needed to use the meniscus shape model were not accurately enough verified.

As a partial conclusion, our analysis of the meniscus shape shows that this magnetic levitation apparatus produces satisfactory homogeneous acceleration field on the length scale of the sample cell.

RESIDUAL ACCELERATION FIELD IN LEVITATION

At exact compensation of Earth's gravity, the residual acceleration field can be determined by simple considerations on the magnetic field. In the following, natural cylindrical coordinates (r,z) centered along the coil axis are used. The origin is the center of the sample cell, which is the point where the magnetic force exactly compensates the Earth's gravity. Noting $B_z(0,0) = b$ the magnetic field at the center, the characteristic length scale of the magnetic field gradient is

$$L = \frac{|\chi| b^2}{\rho \mu_0 g_0}. \tag{12}$$

With b = 6.65 T at levitation, $g_0 = 9.8$ m.s^{-2}, L = 8.9 cm for our apparatus. The expansion of the magnetic field around the center of the cell to the first non-zero order in r/L and z/L leads to the following expressions for the components of the effective acceleration field $\mathbf{g}$(r,z) :

$$g_r(r,z) = -g_0 \frac{3r}{4L} + O(x^3, x = r, z), \tag{13}$$

$$g_z(r,z) = g_0 \left(\frac{r^2}{4L^2} - \frac{z^2}{4L^2} \right) + O(x^3, x = r, z). \tag{14}$$

The radial acceleration is centripetal, which means that the levitation equilibrium is stable to radial perturbations of small amplitude. Note that the variations of the radial component of the effective acceleration around the cell center are much stronger than those of the axial component, since g_r is linear in r. In the levitation state, the residual gravity is roughly equal to the value of g_r when r equals the radius of the cylindrical cell, that is $g_r \approx 0.13$m.s$^{-2} \approx 10^{-2} g_0$ at r=1.5mm. It is of the same order of magnitude as the residual gravity achieved during parabolic flights in planes. When earth's gravity is only partially compensated (g*≠0) the acceleration inhomogeneity, which is proportional to the magnetic acceleration a_m, has a smaller amplitude than at g*=0. Consequently for g*<1 the maximum amplitude of the acceleration inhomogeneity occuring in this apparatus is less than $10^{-2} g_0$.

CONCLUSION

Suppressing the earth gravity by magnetic means is often cited as an alternative way to space microgravity. In order to qualify a magnetic "low gravity" apparatus dedicated to the study of near critical fluids, we used the vanishing behavior of the capillary length of a two-phase equilibrium of H_2 as the critical point is approached. This is a precise check of the homogeneity of the effective acceleration field. The shape of the meniscus and the derived capillary length were compared to the theoretical shape of the meniscus under homogeneous gravity and the predicted value of the capillary length. It follows that no perceptible acceleration inhomogeneities were detected within the accuracy of the temperature control. At exact compensation of the weight, the main residual acceleration field is centripetal and does not exceed $10^{-2} g_0$.

REFERENCES

Blagoi, Y. P., and Pashkov, V. V., "Surface Tension of Hydrogen near the Critical Point", *Sov. Phys. JETP* **22**, 999-1001 (1966).

Beaugnon, E., and Tournier, R., "Levitation of Water and Organic Substances in High Static Magnetic Field", *J. Phys. III France* **1**, 1423-1428 (1991).

Beaugnon, E., Bourgault, D., Braithwaite, D., de Rango, P., Perrier de la Bathie, R., Sulpice, A., and Tournier, R., "Material Processing in High Static Field. A Review of an experimental Study on Levitation, Phase Separation, Convection and Texturation", *J. Phys. I France* **3**, 399-421 (1993).

Beysens, D., and Garrabos, Y., "Critical Fluids under Weightlessness" in *Proc. of the 1st symposium on the Utilization of the International Space Station*, ESA SP-385, edited by ESA Publication Division, ESTEC, Noordwijk, 1996, pp.647-655.

Israelson, U.E., and Larson, M., "Transport Measurements near the Lambda Point of Liquid Helium in a Reduced Effective Gravity Environment on the Ground" in *Proc. of the 33rd Aerospace Sciences Meeting and Exhibit*, AIAA J. 95-0272, edited by the AIAA, Reno, 1995, pp. 1-6.

Moldover, M. R., "Interfacial Tension of Fluids near the Critical Points and Two-Scale-Factor Universality", *Phys. Rev. A* **31**, 1022-1033 (1985).

Sengers, J.V., and Levelt-Sengers, J.M.H., "Critical Phenomena in Classical Fluids" in *Progress in Liquid Physics*, edited by Croxton, C.A., New York, Wiley, 1978, pp. 103-174.

Valles, J. M. Jr., Lin, K., Denegre, J. M., and Mowry, K. L., "Stable Magnetic Field Gradient Levitation of *Xenopus Laevis* : Toward Low-Gravity Simulation", *Biophys. J.* **73**, 1130-1133 (1997).

Warren, C., Webb, W. W., "Interfacial Tension of Near-Critical Cyclohexane-Methanol Mixtures", *J. Chem. Phys.* **50**, 3694-3700 (1969).

Weilert, M. A., Whithaker, D. L., Maris, H. J., and Seidel, G. M., "Magnetic Levitation of Liquid Helium", *J. Low. Temp. Phys.* **106**, 101-131 (1997).

CONFERENCE ON ENABLING TECHNOLOGY AND REQUIRED SCIENTIFIC DEVELOPMENTS FOR INTERSTELLAR MISSIONS

NASA's Interstellar Probe Mission

P. C. Liewer[1], R. A. Mewaldt[2], J. A. Ayon[1], and R. A. Wallace[1]

[1]*Jet Propulsion Laboratory, Pasadena, CA 91109*
[2]*California Institute of Technology, Pasadena, CA 91125*
818-354-6538; **paulett.liewer@jpl.nasa.gov**

Abstract. NASA's Interstellar Probe will be the first spacecraft designed to explore the nearby interstellar medium and its interaction with our solar system. As envisioned by NASA's Interstellar Probe Science and Technology Definition Team, the spacecraft will be propelled by a solar sail to reach >200 AU in 15 years. Interstellar Probe will investigate how the Sun interacts with its environment and will directly measure the properties and composition of the dust, neutrals and plasma of the local interstellar material which surrounds the solar system. In the mission concept developed in the spring of 1999, a 400-m diameter solar sail accelerates the spacecraft to ~15 AU/year, roughly 5 times the speed of Voyager 1&2. The sail is used to first bring the spacecraft in to ~0.25 AU to increase the radiation pressure before heading out in the interstellar upwind direction. After jettisoning the sail at ~5 AU, the spacecraft coasts to 200-400 AU, exploring the Kuiper Belt, the boundaries of the heliosphere, and the nearby interstellar medium.

INTRODUCTION

Robotic exploration of our nearby galactic neighborhood will be one of the first great endeavors of the new millennium. NASA's Interstellar Probe mission, which will travel to 200-400 AU, will be the first spacecraft designed to travel beyond the solar system and sample the nearby interstellar medium. On its way, Interstellar Probe will explore the boundaries of the "heliosphere" -- the bubble in the interstellar medium that surrounds the Sun. The heliosphere is created by the solar wind plasma, which emanates from the corona and expands supersonically throughout and beyond the solar system. The interaction between the solar wind, flowing radially outward at 400-800 km/sec, and the local interstellar material, flowing at ~25 km/sec, creates a complex structure extending perhaps 200-300 AU in the upstream (towards the local interstellar flow) direction and thousands of AU tailward (Fig. 1). Voyager 1&2, now at approximately 76 and 60 AU respectively, should soon reach the first boundary in this complex structure, the solar wind termination shock, where the solar wind makes a transition from supersonic to subsonic flow. Beyond the termination shock lies the heliopause, which is the boundary between solar wind and interstellar plasma. The heliosphere shields the solar system from the plasma, energetic particles, small dust, and fields of the interstellar medium. To observe these directly it is necessary to go beyond the heliopause. Several recent estimates place the distance to the termination shock at ~80 to 100 AU, with the heliopause at ~120 to 150 AU. The Interstellar Probe Mission would be designed to cross the solar wind termination shock and heliopause and make a significant penetration into nearby interstellar space, with a minimum goal of reaching 200 AU, but with sufficient consumables (power, fuel) to last to 400 AU. The 150-kg spacecraft includes a 25-kg instrument payload.

The mission concept presented here was formulated by NASA's Interstellar Probe Science and Technology Definition Team (ISPSTDT) during the spring and summer of 1999 under sponsorship of the NASA Office of Space Science. The primary goal of this team was to develop a mission concept for the Sun-Earth-Connection Roadmap (http://www.lmsal.com/sec), as part of NASA's strategic planning activities. The resulting concept builds on a number of previous studies. In a 1990 study (Holzer et al., 1990), a 1000 kg spacecraft was to acquire data out to ~200 AU, exiting the solar system at ~10 AU/year using chemical propulsion coupled with impulsive maneuvers near the Sun. In a 1995 study of a smaller interstellar probe (Mewaldt et al., 1995), a ~200 kg spacecraft was to reach exit velocities of ~6 to 14 AU/year, depending on launch vehicle and trajectory, using chemical propulsion with planetary gravity assists or impulsive maneuvers near the Sun. Recent technological advances, notably lighter reflective sail materials (Garner et al., 1999) and lighter spacecraft designs, now make it feasible to accomplish essentially the same mission using a solar sail to accelerate a 150 kg spacecraft to ~15 AU/year, allowing the mission to reach ~200 AU in ~15 years and ~400 AU in ~30 years by following the trajectory in Fig. 2.

SCIENCE OBJECTIVES AND SCIENTIFIC PAYLOAD

Interstellar Probe's unique voyage from Earth to beyond 200 AU will enable the first comprehensive measurements of plasma, neutrals, dust, magnetic fields, energetic particles, cosmic rays, and infrared emission from the outer solar system, through the boundaries of the heliosphere, and on into the ISM. This will allow the mission to address key

CP504, *Space Technology and Applications International Forum–2000*, edited by M. S. El-Genk
© 2000 American Institute of Physics 1-56396-919-X/00/$17.00

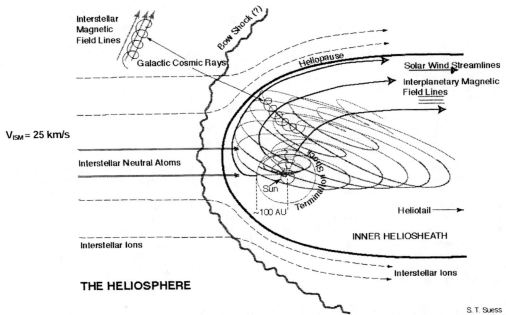

FIGURE 1. Schematic of the global heliosphere created by the supersonic solar wind diverting the interstellar flow around the Sun. The interstellar ions and neutrals flow at 25 km/s relative to the Sun. The solar wind, flowing outward at 400-800 km/s, makes a transition to subsonic flow at the termination shock. Beyond this, the solar wind is turned toward the heliotail, carrying with it the spiraling interplanetary magnetic field. The heliopause separates solar material and magnetic fields from interstellar material and fields. There may or may not be a bow shock in the interstellar medium in front of the heliosphere, depending on the strength of the unknown interstellar magnetic field.

questions about the distribution of matter in the outer solar system, the processes by which the Sun interacts with the galaxy, and the nature and properties of the nearby galactic medium.

The principal scientific objectives of the Interstellar Probe mission would be to

- Explore the nature of the interstellar medium and its implications for the origin and evolution of matter in our Galaxy and the Universe;
- Explore the influence of the interstellar medium on the solar system, its dynamics, and its evolution;
- Explore the impact of the solar system on the interstellar medium as an example of the interaction of a stellar system with its environment;
- Explore the outer solar system in search of clues to its origin, and to the nature of other planetary systems.

To achieve these broad, interdisciplinary objectives, the strawman scientific payload (Table 1) includes an advanced set of miniaturized, low-power instruments specifically designed to make comprehensive, in situ studies of the plasma, energetic particles, fields, and dust in the outer heliosphere and nearby ISM. These instruments will have capabilities that are generally far superior to those of the Voyagers. The wide variety of thermal and flow regimes to be encountered by Interstellar Probe will be explored by a comprehensive suite of neutral and charged particle instruments, including a solar wind and interstellar ion and electron detector, a spectrometer to measure the elemental and isotopic composition of pickup and interstellar ions, an interstellar neutral atom spectrometer, and a detector for suprathermal ions and electrons. Two cosmic ray instruments are included, one for H, He, electrons and positrons, and one to measure the energy spectra and composition of heavier anomalous and galactic cosmic rays. The magnetometer will make the first direct measurements of the magnetic fields in the ISM, and the plasma and radio wave detector will measure fluctuations in the electric and magnetic fields created by plasma processes and by interactions and instabilities in the heliospheric boundaries and beyond. As the spacecraft transits the inner solar system to the ISM, the energetic neutral atom (ENA) imager will map the 3D structure of the termination shock and the UV photometer will probe the structure of the hydrogen wall, a localized region of increased neutral hydrogen density just beyond the heliopause. Dust will be studied with in situ measurements of the dust distribution and composition and by a remote sensing infrared photometer that will map the dust distribution via its infrared emission. The infrared detector will also detect galactic and cosmic infrared emission. A partial list of additional candidate instruments is also included in Table 1, including a small telescope to survey kilometer-size Kuiper belt objects and additional particle instruments. The possibility of developing instrumentation to identify organic material in the outer solar system and the interstellar medium is also under study.

TABLE 1. Strawman Instrument Payload

Instruments	Additional Candidates
Magnetometer	Kuiper Belt Imager
Plasma and Radio Waves	New Concept Molecular Analyzer
Solar Wind/Interstellar Plasma/Electrons	Suprathermal Ion Charge States
Pickup and Interstellar Ion Composition	Cosmic Ray Antiprotons
Interstellar Neutral Atoms	
Suprathermal Ions/Electrons	
Cosmic Ray H, He, Electrons, Positrons	**Resource Requirements**
Anomalous & Galactic Cosmic Ray Composition	• Mass: 25 kg
Dust Composition	• Bit Rate: 25 bps
Infrared Instrument	• Power 20 W
Energetic Neutral Atom (ENA) Imaging	
UV Photometer	

The Interstellar Medium

Our present knowledge of the interstellar medium surrounding our heliosphere comes either from astronomical observations, measurements of sunlight resonantly scattered back towards us by interstellar H and He, or in situ measurements of the dust and neutral gas that penetrate the heliosphere. The Sun is thought to be located near the edge of a local interstellar cloud (LIC) of low density (~0.3 /cc) material blowing from the direction of star-forming regions in the constellations Scorpius and Centaurus. In situ observations of this local cloud by Interstellar Probe will provide a unique opportunity to derive the physical properties of a sample of interstellar material, free from uncertainties that plague the interpretation of data acquired over astronomical lines-of-sight, and from uncertainties arising from the exclusion of plasma, small dust particles and low energy cosmic rays from the heliosphere. Direct measurements would be made of the composition of interstellar dust, and of the elemental and isotopic composition of the ionized and neutral components of the interstellar gas and of low-energy particle components, including key isotopes such as ^{2}H, ^{3}He, ^{13}C, and heavier species. The local cloud is thought to be composed of younger material than that of the presolar nebula and is expected to be richer in heavier elements and neutron-rich isotopes as a result of continuing nucleosynthesis. A complete sample of the elemental and isotopic abundances of the LIC will provide a standard reference for the composition of plasma, neutrals and dust in diffuse interstellar material. Comparisons of this benchmark with the solar system abundances (representative of the presolar nebula) and with abundances from more distant galactic regions will provide important constraints on theories of galactic chemical evolution.

Measurement of the spectrum of cosmic ray nuclei and electrons, free from the influence of the heliosphere, will investigate astrophysical processes that include acceleration by supernova shock waves, interstellar radio and γ-ray emission, recent nucleosynthesis, and the heating and dynamics of the interstellar medium. Little is known about the properties of magnetic field in the local cloud or in the region beyond the termination shock. Interstellar Probe will make the first in situ measurements of interstellar magnetic fields and of the density, temperature, and ionization state of the interstellar gas, including studies of their variations over a variety of spatial scales. The possibility of identifying organic matter in the outer solar system and ISM is also under investigation.

Interaction between the Interstellar Medium and the Solar Wind

The solar wind, a continual low-density flux of charged particles, streams outward from the corona and expands supersonically throughout and beyond the solar system. The solar wind and the interstellar medium interact to create the global heliosphere, shown schematically in Fig. 1. It is primarily the balance between the solar wind ram pressure and the interstellar pressure which determines the size of the heliosphere. The ram pressure of the solar wind pushes the interstellar plasma away from the Sun, diverting the flow around it, creating an elongated "bubble" in the colder and denser ISM. The solar wind pressure decreases as the solar wind expands and, at some point (~100 AU), the solar wind pressure becomes comparable to the interstellar pressure and the solar wind makes a transition to subsonic flow at the "termination shock."

At present, there are no direct measurements of the size and structure of the heliosphere and our present understanding is based on theory and modeling, constrained by a few key measurements. The Voyager spacecraft have detected radio emissions, which are thought to be caused by interplanetary shock waves hitting the denser interstellar plasma. Voyager 1 should soon reach the termination shock, providing a first direct test of our current understanding of the size of the global heliosphere, although some of the Voyager instruments were not designed to explore the boundaries of the heliosphere and interstellar medium. Interstellar Probe's enhanced capabilities and lifetime will greatly extend Voyager's exploration of the structure and dynamics of the heliosphere. The Interstellar Probe Mission will answer questions relating to how the ISM influences the solar system and how the solar system influences the ISM.

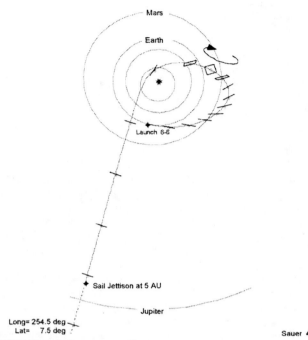

FIGURE 2. Interstellar Probe trajectory using a solar sail to reach a final velocity of 15 AU/year. The trajectory is towards the nose of the heliosphere, the shortest route to the interstellar medium. The orientation of the sail to achieve the proper thrust vector is also shown. The sail is jettisoned at 5 AU when further acceleration is negligible to avoid possible interference with some of the instruments.

Past the termination shock, in the region called the heliosheath, the solar wind flow is turned to match the flow of the diverted interstellar plasma, as illustrated Fig. 1. The spiraling solar magnetic field, frozen into the solar wind, is swept back with this flow. The heliopause is the boundary between the heated solar wind in the heliosheath and the interstellar plasma. Depending on the unknown interstellar magnetic field strength, there may or may not be a bow shock created in the interstellar medium ahead of the nose of the heliosphere. The interstellar neutrals which penetrate the heliosphere can charge exchange with the supersonic solar wind ions, creating energetic "interstellar pickup ions" which heat the solar wind. Interstellar Probe will pass through these boundary regions and make in situ measurements of the dust, plasma, fields and flows to answer questions regarding the size, structure and dynamics of the heliosphere and the processes occurring at the boundaries.

The termination shock is known to accelerate particles from keV to GeV energies, and in situ studies of shock structure, plasma heating, and acceleration processes at the termination shock will serve as a model for other astrophysical shocks. Energetic ions created by charge exchange in the heliosheath can be imaged to provide information on the 3D structure of the heliosphere. Charge-exchange collisions lead to a weak coupling between the neutral and ionized hydrogen in the interstellar medium causing a pile-up of neutral hydrogen at the heliosphere nose, referred to as the "hydrogen wall." Interstellar Probe will explore the structure of this wall with in-situ and remote-sensing observations, and relate its properties to observations of similar structures and winds observed in neighboring star systems. In general, the study of the structure and dynamics of our heliosphere will serve as an example of how a star interacts with its environment.

The Outer Solar System

Our Solar system is thought to be the end product of a common astrophysical process of stellar system formation from protoplanetary disk nebulae. Collisions play a central role in the formation and evolution of planetary systems, either increasing or eroding the mass of the bodies. The present interplanetary dust population is a result of collisional processes occurring in the solar system. Interstellar Probe will provide the opportunity for in situ and remote sensing of both interplanetary and interstellar dust in the heliosphere and the ISM. It will determine the composition and the mass and orbital distributions of dust in the outer solar system, study its creation and destruction mechanisms, and also search for dust structures associated with planets, asteroids, comets, and the Kuiper Belt. These studies will constrain theories of the collisional dynamics of the solar system and help us understand the origin and nature of our solar system and other planetary systems as well. Interstellar Probe can uniquely address the fundamental question of the radial extent of the primordial solar nebula, or, more precisely, the extent of the primordial planetesimal disk. This can be accomplished most directly by measuring the variation with heliocentric radius of the population of small bodies in the Kuiper Belt, or, less directly, by measuring the density distribution of dust grains derived from Kuiper Belt objects. Moreover, the Kuiper Belt is an analog for circumstellar disks around other stars and improved understanding of its properties will aid the interpretation of astronomical observations of planet-forming or planet-harboring disks in other stellar systems.

Organic material is found in both our solar system (in asteroids, comets, meteorites and dust) as well as the interstellar medium. It is not known if these non-terrestrial organic materials have a similar origin. Amino acids have been found in meteorites, but it is not known if they exist in the ISM. Organic material from both small bodies and the interstellar medium are known to reach Earth, but their possible role in the emergence of life on our planet is uncertain. A suitable instrument on the Interstellar Probe would search for organic material in the outer solar system, as well as the nearby ISM, in order to address questions about the nature and chemical evolution of this material.

The cosmic infrared background (CIRB) is the integrated light from all stars and galaxies that cannot be resolved into individual objects. Observations of the CIRB can determine how much energy was converted into photons during the evolution of galaxies, back to their formation. As a result, fundamental measurements about galaxy formation can be made even though individual protogalaxies cannot be seen. The CIRB spectrum provides information on how the first stars formed and how early the elements were formed by nucleosynthesis. The Cosmic Background Explorer (COBE) satellite detected the CIRB at wavelengths longer than 140 microns and established limits on the energy released by all stars since the beginning of time. COBE results at shorter wavelengths were not possible because of the very bright foreground emission from zodiacal light. The zodiacal dust is known to decrease in density with radius. ISP will map the radial distribution of zodiacal emission and beyond ~10 AU, it will be able to detect or limit the CIRB at wavelengths below 140 microns as the zodiacal background decreases.

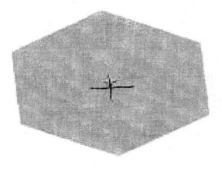 

FIGURE 3. Left: The hexagonal ~400 m diameter solar sail with the spin up booms still attached. Right: The spacecraft, whose 2.7 m dish antenna serves and the main structure, is supported by three struts in an 11-m hole in the center of the solar sail. Sail control is achieved by moving the spacecraft with respect to the center of mass of the sail. The instruments are attached near the rim of the antenna. The sail is spin-stabilized during sailing.

MISSION CONCEPT

Interstellar Probe mission requirements were defined by the ISPSTDT. To accomplish its science objectives, the probe should acquire data out to a distance of at least 200 AU, with a goal of ~400 AU. The trajectory should aim for the nose of the heliosphere, the shortest route to the interstellar medium. The average science data rate at 200 AU would be 25 bps; a lower data rate is acceptable at 400 AU. The instrument payload requires ~25 kg and ~20 watts of power. The spacecraft should spin to enable the in situ instruments to scan the particle, plasma, and magnetic field distributions and to permit the remote-sensing instruments to scan the sky.

JPL's mission design team developed mission and spacecraft concepts which met all requirements. The resulting spacecraft design is shown in Fig. 3 (right) in sailing configuration. The spacecraft is suspended inside an 11-m hole in the hexagonal sail. The instruments are placed around the rim of a 2.7-m dish antenna, which also functions as the main support structure. The spacecraft is designed for a mission to 200 AU with consumables to last to 400 AU (~30 year mission). Science and engineering data are gathered at an average rate of 30 bps. The telecommunications system uses Ka band to communicate with the Deep Space Network; data is stored and dumped using approximately 1 pass/week. The antenna is limited to 2.7 m to fit in the shroud of the Delta II launch vehicle. A downlink data rate of 350 kbps at 200 AU is achieved using a transmitter requiring 220 W. Power is provided by three next-generation advanced radioisotope power source (ARPS) units.

The total spacecraft mass (excluding sail) is ~150 kg including the instruments (Table 1). To achieve the 15 AU/year exit velocity, a solar sail with 1 gm/m^2 areal density (sail material plus support structure) and a radius of ~200 m is needed. The total accelerated mass (spacecraft plus sail system) is ~246 kg. The spacecraft initially goes in to 0.25 AU to obtain increased radiation pressure before heading out towards the nose of the heliosphere. The sail is jettisoned at ~5 AU when the further acceleration from radiation pressure becomes negligible, thereby avoiding potential interference with the instruments. Fig. 2 shows the orientation of the sail relative to the Sun to obtain the proper thrust vector for the trajectory shown. The total ΔV achieved is 70 km/s. In the sailing configuration, shown in Fig. 3 (right), the spacecraft is supported within a hole in the center of the sail by 3 struts. Sail control is achieved by offsetting the spacecraft with respect to the center-of-mass of the sail. The sail is deployed and stabilized by rotation; a number of mechanisms used to provide the initial spin up and deployment of the sail are jettisoned after sail deployment. Figure 3 (left) shows the sail after deployment, but with the spin-up booms still attached.

CONCLUSIONS

Although most of the instruments required for this mission have considerable flight heritage and could be built today, all of them would benefit from new technology in order to optimize the scientific return within the very restrictive weight and power resources. In addition, there are several exciting instrument concepts such as the molecular analyzer and the Kuiper Belt Imager that will require considerable development. The mission concept presented here also assumes a number of developments in spacecraft systems, including low-power avionics, advanced power systems, and phased-array Ka-band telecommunications. Many of these developments are also being counted on for other future NASA missions.

The most critical technology needed to carry out the mission described here is, of course, solar sail propulsion. Although solar sails have been studied extensively (Wright, 1992), they have never flown in space (although a large ~20 m sail was deployed on MIR). Indeed, spacecraft velocities of the kind envisioned here will require rather advanced sails, necessitating new, light-weight reflective material and developments in sail packaging, deployment and control. These developments will have to be tested in one or more flight demonstrations before a 400-m sail with an areal density of ~1 g/m^2 will be ready for flight, requiring an aggressive solar-sail development program (see, e.g., Wallace, 1999). Fortunately, there are also a large number of other missions that could benefit from solar-sail propulsion. If this program is successful, launch could be as early as 2010, and Interstellar Probe can serve as the first step in a more ambitious program to explore the outer solar system and nearby galactic neighborhood.

ACKNOWLEDGMENTS

A portion of this work was conducted at the Jet Propulsion Laboratory, California Institute of Technology under contract with the National Aeronautics and Space Administration. We wish to thank the other members of the Interstellar Probe Science and Technology Definition Team: E. Bakes, NASA Ames; P. Frisch, U. of Chicago; H. Funsten, LANL; M. Gruntman, USC; L. Johnson, MSFC; R. Jokipii, U. of Arizona; W. Kurth, U. of Iowa; J. Linsky, U. of Colorado; R. Malhotra, LPI; I. Mann, Caltech; R. McNutt, APL; E. Moebius, UNH; W. Reach, Caltech; S. Suess, MSFC; A. Szabo, GSFC; J. Trainor, GSFC/retired; G. Zank, U. of Delaware; T. Zurbuchen, U. of Michigan. Program Manager: S. Gavit, JPL. Program Scientist: V. Jones, NASA HQ. Deputy Program Scientist: J. Ling, NASA HQ. Program Executive: G. Mucklow, NASA HQ. NASA Transportation: D. Stone, NASA HQ. Interagency Representatives: D. Goodwin, DOE and E. Loh, NSF. Foreign Guest Participants: W. Druge, U. of Kiel, Germany, B. Heber, Max Planck, Germany; C. Maccone, Torino, Italy. JPL Support: C. Budney, S. Dagostino, E. De Jong, K. Evans, W. Fang, R. Frisbee, C. Gardner, H. Garrett, S. Leifer, R. Miyake, N. Murphy, B. Nesmith, F. Pinto, G. Sprague, P. Willis, and K. Wilson.

REFERENCES

Garner, C. E., Diedrich, B., and Leipold, M. "A Summary of Solar Sail Technology Developments and Proposed Demonstration Missions," AIAA-99-2697, presented at 35th AIAA Joint Propulsion Conference, Los Angeles, CA, June, 1999.

Holzer, T. E., Mewaldt, R. A., and Neugebauer, M., *The Interstellar Probe: Scientific Objectives and Requirements for a Frontier Mission to the Heliospheric Boundary and Interstellar Space, Report of the Interstellar Probe Workshop*, Ballston, VA, 1990.

Mewaldt, R. A., Kangas, J., Kerridge, S. J., and Neugebauer, M., "A Small Interstellar Probe to the Heliospheric Boundary and Interstellar Space", *Acta Astronautica*, **35** Suppl., 267-276 (1995).

Wright, J. L., *Space Sailing*, Gordon and Breach, Amsterdam, 1992.

Wallace, R. A., "Precursor Missions to Interstellar Exploration," Proc. IEEE Aerospace Conf., Aspen, CO, (1999) Paper 114.

A Realistic Interstellar Explorer

R. L. McNutt, Jr., G. B. Andrews, J. McAdams, R. E. Gold, A. Santo, D. Oursler, K. Heeres, M. Fraeman, and B. Williams

The Johns Hopkins University Applied Physics laboratory, Laurel, MD 20723

240-228-5435; ralph_mcnutt@jhuapl.edu

Abstract. For more than 20 years, an "Interstellar Precursor Mission" to ~1000 AU within the working lifetime of the initiators (<50 years) has been discussed as a high priority for multiple scientific objectives. During the last two years there has been renewed interest in actually sending a probe to another star system - a "grand challenge" for NASA - and the idea of a precursor mission has been renewed as a beginning step to achieve this goal. We revisit an old idea for implementing such a mission. The probe is launched initially to Jupiter and then falls to the Sun where a large propulsive ΔV maneuver propels it on a high-energy ballistic escape trajectory from the solar system. The implementation requires a low-mass, highly-integrated spacecraft to make use of moderate (Delta-class) expendable launch vehicles. We provide a first-order cut at many of the engineering realities associated with such a probe. We identify a mission concept that can link the required science, desired instruments, spacecraft engineering and the realities of the fiscal and technological milieu in which NASA must operate if such missions are to reach fruition and lay the groundwork for eventually realizing true starships.

INTRODUCTION

Travel to the stars is the stuff that dreams and science fiction novels are made of. However, there is also a very scientifically compelling and serious side of the concept as well. For more than 20 years, such an "Interstellar Precursor Mission" has been discussed as a high priority for multiple scientific objectives. These include (1) measuring the properties of the interstellar medium and understanding its implications for the origin and evolution of matter in the Galaxy, (2) determining the structure of the heliosphere and its interaction with the interstellar environment, and (3) studying fundamental astrophysical processes that can be sampled in situ. The chief difficulty with actually carrying out such a mission is the need for reaching significant penetration into the interstellar medium (~1000 Astronomical Units (AU)) within the working lifetime of the initiators (<50 years). During the last two years there has been renewed interest in actually sending a probe to another star system - a "grand challenge" for NASA - and the idea of a precursor mission has been renewed as a beginning step in a roadmap to achieve this goal.

We revisit an old idea for implementing such a mission. The probe is launched initially to Jupiter, using that planet's gravity to remove the probe angular momentum. The probe is then allowed to fall to the Sun - in this case to 4 solar radii - where a large propulsive ΔV maneuver is required over a very short time - taken here to be ~15 minutes to minimize gravity losses. At this time all acceleration is over and the probe continues on a high-energy ballistic escape trajectory from the solar system (the idea of using a high-I_{sp}, high-thrust maneuver close to the Sun was first identified in 1929 by rocket pioneer Hermann Oberth (*Ehricke*, 1972)).

In this scenario, the use of a carbon-carbon thermal shield and an exotic propulsion system capable of delivering high specific impulse (I_{sp}) at high thrust is enabling. For other ongoing Interstellar Probe studies at the Jet Propulsion Laboratory, the manufacture, deployment, and management of long-term, low-thrust maneuvering with a solar sail is the enabling propulsive technology. Both implementations require low-mass, highly- integrated spacecraft in order to make use of moderate (Delta-class) expendable launch vehicles (ELVs).

SCIENCE RATIONALE

A mission past the boundary of the heliosphere to only a fraction of a light year can yield a rich scientific harvest (*Jaffe and Ivie*, 1979; *Holzer et al.*, 1990; *Mewaldt et al.*, 1995; *McNutt et al.*, 1997; *McNutt*, 1998). The science goals include: (1) Explore the nature of the interstellar medium and its implications for the origin and evolution of matter in

CP504, *Space Technology and Applications International Forum–2000*, edited by M. S. El-Genk
© 2000 American Institute of Physics 1-56396-919-X/00/$17.00

TABLE 1. Example Interstellar Probe Science Payload.

Instrument	Identifier	Mass (kg)	Power (W)
Plasma waves/dust detection	PWD	1.5	2.5
Plasma/particles/cosmic rays composition and spectra	PPC	1.0	1.5
Magnetometer (w/boom)	MAG	3.0	0.5
Lyman-α imager	LYA	1.0	2.0
Infrared imager	IRI	1.5	1.5
Neutral atoms composition, density, speed, temperature	NAC	2.0	2.0
Totals	—	10.0	10.0

the Galaxy, (2) Explore the structure of the heliosphere and its interaction with the interstellar medium, (3) Explore fundamental astrophysical processes occurring in the heliosphere and the interstellar medium, and (4) Determine fundamental properties of the universe. A candidate payload to address these goals is listed in Table 1. Resource allocations are 10 kg and 10 W.

MISSION OVERVIEW

As a baseline we have chosen $\Delta V = 15$ km s^{-1} for the propulsion "target" to enable an asymptotic solar system escape speed of roughly 20 AU/yr. A 50-yr flight time will reach 1000 AU, a distance well outside of the projected influence of the heliosphere/interstellar medium interaction region. A probe lifetime of 500 years will reach 10,000 AU, a significant penetration into local interstellar space. We require that the probe should be launched with a Delta III-class launch vehicle and approach within 4 R_s of the Sun. With no other constraints, energy and momentum conservation set the required injection energy (C_3) to effect the near-Sun passage. The need to line up the outgoing asymptote of the trajectory toward the target direction while having Jupiter in the correct position to effect the gravity assist means that the required planetary alignment occurs roughly once every 13 years.

Perihelion Maneuver Requirements

As a specific example, a 2011 launch to Jupiter with a launch energy $C_3 = 117.1$ km^2/s^2 and, two years later, a 15.4 km/s perihelion burn near the Sun sends the spacecraft at 20.2 AU/year toward the Sun-similar star Epsilon Eridani (HD22049), a K2V dwarf main sequence star 10.7 light years from Earth. Launching toward such a star enables comparison with locally measured properties of the interstellar medium with integrated properties determined by detailed measurements of the target-star spectrum (*Frisch*, 1993; *Linsky and Wood*, 1996). A low-ecliptic latitude target minimizes the required perihelion burn for a given asymptotic escape speed. The details of Jupiter's orbit cause the launch energy requirements to change each launch window and return roughly to an optimum roughly once every 80 years for a launch toward ε Eridani. Additional planetary flybys will not help to ease the mass constraints of a given launch vehicle; they over-constrain the trajectory design problem. The trajectory is shown in Figure 1.

Chemical production of such high speed changes in deep space is not possible; a 19.7-metric ton IUS/PAM-D combination in Earth orbit was required to boost the 360-kg Ulysses spacecraft by 15.4 km s^{-1}. For the reference mission launched toward ε Eridani, the required ΔV is 15.393 km s^{-1}.

We have considered two concepts in some detail to identify a means of applying the ΔV required at solar perihelion: (1) an Orion-type nuclear drive and (2) solar thermal propulsion. The requirement is to provide a high thrust for a brief period of time similar to conventional rockets and upper stages, but, significantly, at a specific impulse about 3.5 times larger than with chemical systems, i.e. at $I_{sp} \sim 1000$s.

Injection Mass to Jupiter

The indirect launch mode (*Farquhar and Dunham*, 1999) - allowing an additional stage for the launch vehicle - is required to launch a reasonably-sized probe first to Jupiter and then to the Sun for the perihelion maneuver. Table 2 shows injection mass limits for various launch vehicle configurations. As an example, a Star 37 FM ($I_{sp} = 289.8$s, fuel fraction = 0.929, fuel loading = 1065.96 kg) is integrated with the probe assembly. By providing 1.08 km/s in Earth's gravity field at an altitude of ~300 km this stage provides extra lift capability (*Meissinger et al.*, 1997). Allowing ~17 kg for hardware associated with ejecting the empty motor casing and 81.5 kg for the empty casing itself, we obtain a net

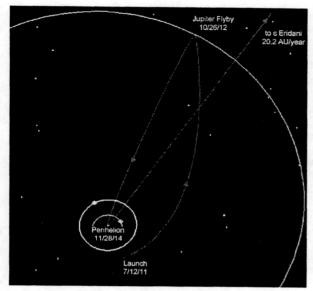

FIGURE 1. Ecliptic plane projection of an instellar probe trajectory toward ε Eridani. The probe reaches a heliocentric distance of 1000 AU on May 31, 2064.

TABLE 2. Injected Mass Limits

Launch Vehicle	Upper stage(s)	Injected mass (kg) at $C_3 = 117.1$ km^2 s^{-2} (ε Eridani)
Atlas IIAS	Star 48	342
Delta III	Star 48	256
Delta III	Star 48/Star 37FM	~475

mass of 475 kg injected into a trans-Jupiter orbit. A 30% mass reserve still leaves a design mass of 365 kg, larger than the injection mass provided by an Atlas IIAS/Star 48 with no reserves.

With a target probe final design mass of 50 kg we must add about 40 kg for the perihelion-maneuver Sun shield (six plys of carbon-carbon configured as a circular shield ~4.1 m in diameter). We add an additional 50 kg for the structure to hold the shield in place, including the means to jettison it and the perihelion propulsion module once the probe is outbound from the Sun to obtain a net mass of 140 kg. If we assume the perihelion propulsion module hardware has a mass of 60 kg, the total injected dry mass is 200 kg; including a 30% mass reserve brings the probe mass to 260 kg. The mass ratio for the perihelion propulsion system is then 1.83, and to achieve a ΔV of 15.4 km s^{-1}, the required I_{sp} is 2598 s. For a total dry mass of 475 kg at perihelion (including the probe, thermal shield, propulsion system, and all hardware) and the mass ratio of 1.83, we would require $475(1-1/1.83) = 215$ kg of fuel. The corresponding total energy required is $0.5 \times 215 \times 25486^2 = 6.98 \times 10^{10}$ joules. Allowing for a 15m = 900s burn time, the perihelion engine must develop a power of 77.6 MW while it is operating (for our assumed 60 kg propulsion system, this is a specific mass of 1.29 MW/kg). These numbers would be lower by a factor ~5 for an optimal mass ratio of 4.92.

Nuclear Pulse Propulsion

Fission may well provide the key element for the perihelion propulsion, but only in a pulsed mode with extremely-low fission yields per pulse. For the case mentioned previously, with ~7 x 10^{10} joules required, we need the fission energy of ~1.3 g of uranium - a total of about 13 tons of TNT equivalent (*Serber*, 1992). The problem is the coupling of the momentum into the ship over short time scales ~10^{-8}s. Transferring significant amounts of impulse to a structure over such short times typically causes stress to exceed the yield strengths of all known materials (*Dyson*, 1965; 1968). The Orion concept (*Boyer and Balcomb*, 1971) and its derivatives (*Solem*, 1994) require large masses for dealing with the release of ~1 to 10 kT explosions. The alternative is to go to much smaller explosive yields and pulse the system. Ideas adapted from inertial confinement fusion with yields ~0.01 kT have been adapted to a variety of systems studies; however, the spacecraft masses again tend to be large due to the power plant overhead, here the laser system, required to initiate the microexplosions (*Hyde et al.*, 1972). Ideally, a pulsed-mode autocatalytic reaction (*Serber*, 1992; *Winterberg*, 1981), similar to the operation of a pulse jet, e.g., the German V-1, is preferred. This type of rocket would represent the next step past a gas core nuclear engine, and, if doable, could perhaps bypass the engineering problems of the latter. This is a goal worth pursuing and would clearly be an enabling technology for this interstellar precursor mission as well as higher-speed missions to the stars.

Solar Thermal Propulsion

Solar thermal propulsion and a near-Sun maneuver seem to be "made for each other"; solar thermal propulsion uses thermal energy from the Sun to heat a low-molecular-weight working fluid to a high temperature (~2400K) and use the thermal energy to expel the propellant mass from the system, driving the probe forward. An "obvious" choice for a working fluid is liquid hydrogen (LH_2) due to its low molecular mass. However, realistic cryostats carry a substantial mass penalty for long-term LH_2 storage for use as a working fuel at perihelion.

The use of ammonia enables standard pressurized titanium tank technology (*Deininger and Vondra*, 1991). In this case dissociation of the ammonia requires higher temperatures and so the specific impulse is lower. Although the goal of reaching I_{sp}~1000s for an LH_2 system and ~600s for an ammonia system may be achievable (Lester, private communication) more work is required to examine the overall system. The bottleneck, in addition to mass, is providing sufficiently rapid heat transfer to the fuel in the near-Sun environment in order to provide efficient thrusting. While escape from the solar system at speeds as high as 20 AU/year with this technolnogy is unlikely, speeds of 10-15 AU/year may be achievable. More study of the systems aspect is required with emphasis on the mass of the overall propulsion system hardware plus fuel.

GENERAL MISSION ASPECTS

The extremely long duration mission that results from sending the Interstellar Probe to such enormous distances imposes a stringent set of environmental requirements on the Probe design. Hence a number of standard techniques and subsystems normally used on shorter missions within the solar system are not appropriate for this mission. This reality forced a thorough top-down redesign of the architecture of the spacecraft and its subsystems. For example, it is impractical to fly any components that use moving parts, such as momentum wheels, scan platforms or thrusters that incorporate valves. Additionally, any subsystem that required consumables was carefully evaluated. The only consumables permitted on the probe are blocks of solid Teflon used by the thrusters. Thrusters that require liquid or gaseous propellant were excluded.

The Interstellar Probe must be robust, reliable, and adaptable, since it must perform it's mission with minimal external help. The mission scenario which we have developed involves maintaining the probe in a dormant, cocooned mode during the early phase of the mission while the propulsion system carries the probe through the near-Sun swingby and perihelion burn maneuver. This phase, occurring well inside the solar system, includes the regions of highest radiation, acceleration, and temperature. The probe is protected by the substantial structure and thermal shield of the propulsion system. Once the propulsion system has completed the Sun swingby it is left behind and the probe continues on its journey. At this point the initial checkout of the probe commences. The duration of the Interstellar Probe shakedown period is dependent on how long two-way communications can be maintained. Two-way communications allow problems to be diagnosed and remedial action taken (such as a software upload).

This concept offers potential advantages - as well as an alternative - to low-thrust probes. Such systems require a great deal of control for their primary propulsion system. The approach here has only a few, short-duration critical operational periods. From a systems point of view this approach offers distinct advantages.

Prime Science Period

Once two-way communication is lost the Interstellar Probe enters an autonomous mode where only a communication downlink exists and, due to power limitations, transmissions occur fairly infrequently. The short transmission period is dedicated to downlinking acquired science data, as well as for health information. This is the prime science portion of the mission with a time line that runs from perihelion burn plus roughly 6 months to 50 years or greater. During this period:

(1) The probe maintains a slow spin with the spin axis pointing back at the Sun. The spinning spacecraft allows the science instruments to see the entire sky.

(2) The instruments collect and process their own data.

(3) At regular intervals the probe points accurately toward a Hubble-class receiving station which is orbiting the Earth and then transmits the science data. Unlike a typical "store and dump" spacecraft system which collects all the data from the instruments onto a data recorder and then dumps the data at a high rate to the ground at regular (daily) intervals, the Interstellar Probe is severely bandwidth limited so only highly reduced data can be transferred for short time periods at more infrequent intervals.

(4) Onboard processors continually monitor the health of the probe and take corrective action(s) as required.

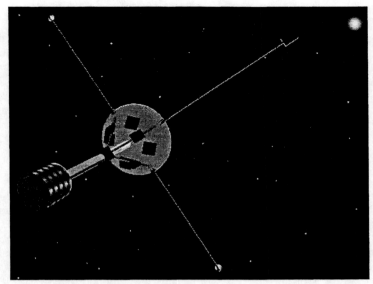

FIGURE 2. The Interstellar Probe in its post-perihelion flight configuration. The view is from the RPS toward the optical communications dish and toward out Sun. Flight electronics can be seen mounted on the back side of the optical reflector assembly.

Probe Architecture

Following the perihelion burn, the probe mechanical design consists of three main mechanical elements – a Radioisotope Power Source (RPS) assembly, a central support mast, and an optical dish. The RPS is placed at one end of the mast in order to minimize the radiation dose to other spacecraft components. At the other end of the mast is the large optical dish which faces away from the direction of travel and back toward the solar system.

Instruments and spacecraft electronics boxes are placed on the back of the optical dish and along the mast. Four booms, used for field measurements, are mounted orthogonal to one another at the perimeter of the optical dish. Since the spacecraft is a spinner the ratio of lateral to longitudinal moments of inertia of the probe is set to be roughly 1.25 to 1 for best stability. The spacecraft secondary battery and power control system are placed inside of the mast roughly at its midsection. The communication system laser is also placed inside the mast and points out the end of the mast, through a small hole in the one-meter optical dish, toward the hyperboloid reflector. The mass allocated for the structure is 15 kg with 5 kg and 5 W allocated for the spacecraft electronics (exclusive of the science payload but including the trickle charging of the secondary power system used for downlinking science and status data). With the allocations for the science instruments, the entire probe is allocated 50 kg and 15 W. The post-perihelion flight configuration is shown in Figure 2.

Power

The "obvious" choice for a power supply is a Radioisotope Power System (RPS). Assuming an RPS can supply ~4 electrical watts per kilogram and a continual power consumption of about 15 W, an RPS weighing approximately 4 kg is needed. Due to the radiation output, the RPS is mounted as far away from the other spacecraft components as possible: at the far end of the probe's central support mast, at a distance of 3 meters.

A secondary power system is baselined for to supply power during transient peak loads for both the communication and attitude control systems. A high-energy-density chemical battery is trickle charged by the primary system during the large fraction of time when the probe is neither communicating nor doing precision pointing. Power is drawn from the system as needed during the regular (weekly) telemetry communication periods.

Guidance and Attitude Control

The attitude determination and control system need only point the antenna boresight at a receiver on Earth ~once a week for ~15m to within the beamwidth of the laser signal ~0.15 arcsecond. This corresponds to an absolute pointing and control requirement of ±0.075 arcsecond (3-σ) and must include considertion of: (1) trajectory knowledge, (2) transmitter boresight pointing knowledge, and (3) boresight control.

Command and Data Handling

The probe does not have a subsystem on board that can be strictly identified as a "command and data handling system." Although it has a number of powerful processor modules which do perform command and data handling functions, these modules are not aligned or connected to any particular subsystem or task. All computing power and the processing that is done on them is distributed and all tasks are only loosely coupled with specific hardware. In the probe architecture all the identical processor modules are extremely powerful (each exceeding 200 million instructions per second (MIPS) in performance) so that any processor is capable of handling all the tasks on-board single-handedly (e.g., attitude control, instrument processing, etc.) if that were to become necessary. However, in general, a processor module handles the single task it has been assigned, but this situation is very fluid and flexible.

Each module contains a central CPU, memory, and a wireless communication module similar to processors being developed for today's cellular phone market. Current versions have 50 MIPS capability and only consume 18 mW. All spacecraft subsystems, including instruments, do not have any processors. Instead, they have a high rate wireless communication module which is used to broadcast data at their assigned frequency. All the subsystems are broadcasting their data so that any one or all of the processor modules can listen. Based on how the processor modules have decided among themselves to configure the tasks, the processor modules tune their receiver systems to the correct subsystem frequencies and collect data from one or more subsystems. In general, one subsystem (e.g., power management) is assigned to a single processor module. There are more processor modules than subsystems, so some of the modules are turned off for portions of the mission. If a processor fails, its task will be shifted to an unused processor. If no spare processors are available, then some of the remaining processors will have to handle more than one task. In the worst case scenario, when only one processor is still working, it alone can handle all tasks.

Communications

At a range of 100 AU electromagnetic waves will take 13.9 hours to travel from the spacecraft to Earth. Interactive control of the spacecraft becomes nearly impossible at these distances necessitating a highly autonomous craft and one-way communications to Earth. The main characteristics of the communications system are, therefore, that it needs to be low mass, low power and capable of one-way transmission at sufficiently high transmission rates. For a baseline design we have adopted a data rate of 500 bits per second (when transmitting) at an encoding-dependent bit error rate of 10^{-9}. Microwave communication techniques have a long and successful history and a large existing infrastructure such as ground-based arrays but are limited in bit-rate capability when compared to optical systems of comparable size. Hence, we have baselined an optical system for the primary science downlink.

While the spacecraft will have the ability to transmit and receive optical communications, the reception of signals from Earth is intended for use at close range, for a relatively brief time after the perihelion burn. This ability will allow controllers to conduct "wellness" tests on the probe after launch. Once the spacecraft has left the inner solar system communications will predominately concentrate on the one-way, Earth-bound, link. One possible system that meets the requirements would use a quantum cascade laser (~890 nm) with a 1-m optical aperture (1.13 μrad) on the probe broadcasting to a 4-m optical terminal near Earth. Earth broadcasts use binary amplitude operation in an incoherent mode while probe broadcasts use external binary phase shift modulation detected in a coherent mode at the Earth terminal.

Thermal System

During the perihelion passage a thermal shield must protect the probe and its perihelion propulsion stage from the intense solar illumination. The probe is not spinning and actively points the Sun shield toward the Sun during the perihelion passage. The umbra (shadow) of the shield does drive the mechanical configuration of the probe during the perihelion passage segment of the mission. A carbon-carbon spherical-segment shield spaced away from the spacecraft allows the backside of the shield to radiate to space. A secondary shield further reduces the thermal soak back. A standard MLI blanket is used to protect the spacecraft from the backside of the secondary shield. Although the shield temperature approaches 2700°C the spacecraft is maintained at normal operating temperatures.

Once the probe separates from the cocoon formed by the surrounding perihelion propulsion system, it quickly moves away from the Sun at 20 AU/yr, spending the majority of its mission in the cold of deep space. The Interstellar Probe contains a 1-m, highly polished, mirror with a graphite epoxy substrate providing thermal and structural stability to the mirror. Electronics boxes and instruments are mounted to the back of the mirror support structure. The design enables the mirror and its attached hardware is to operate between 75 and 125 K by utilizing ~85 W of heat rejected at the RPS locally (assumed to run at a surface temperature of 200°C).

Extension to Long Flight Duration

The Interstellar Explorer mission provides a significant science return, performing initial investigations of the totally unexplored realm of Interstellar Space, as well as the interesting regions at the boundary of our solar system. Additionally, a number of engineering milestones will be achieved. Perhaps the most important milestone is enhanced mission duration, since it is clear that some future missions, such as those that journey to the stars, will probably be more than an order of magnitude longer in duration. Indeed, this mission is the first small step toward realistically achieving a true interstellar capability.

There are two key aspects associated with performing extremely long-duration missions. Wars and other extremely disruptive events often occur, making it difficult to maintain any infrastructure for periods longer than several hundred years. However, there are a few buildings that have been continually maintained for over a thousand years, so maintaining contact with a probe travelling to the nearest stars may be possible for that length of time. In order to maximize the probability of continued contact, it is important to distribute the data-gathering knowledge among as many diverse peoples as possible to prevent a single catastrophe from wiping out all knowledge of the mission.

The second key aspect to a long duration mission is the selection of Interstellar Probe materials. Even though clever architectures can make systems robust and fault tolerant, long-term aging processes can reduce virtually any system to dust over time. One example is in the area of microcircuits, which are subject to electromigration, eventually causing the devices to fail. Recent developments, such as ultra-low power circuits (which have very low current densities), and the transition to copper and tungsten interconnects on sapphire substrates (instead of aluminum on silicon), may significantly improve the ultimate lifetimes of modern electronics.

SUMMARY AND CONCLUSIONS

We have provided a first-order cut at many of the engineering realities associated with sending a small interstellar precursor mission out of the solar system at a high speed. The primary engineering concern remains propulsion, but we have also identified other constraints and interactions that need to be approached in a systems manner:

To maximize the asymptotic escape speed from the Sun, a target direction near the plane of the ecliptic must be chosen; although the constraints may not be as great, this is probably the case for low-thrust schemes as well, e.g., nuclear-electric propulsion and solar-sail propulsion.

If a Jupiter flyby is required to reach the Sun to do a perihelion maneuver to boost the escape speed, then additional planetary flybys will not help to ease the mass constraints of a given launch vehicle; they over-constrain the trajectory design problem.

Some form of the indirect launch mode - allowing an additional stage for the launch vehicle - is doable and probably required to launch a reasonably sized probe first to Jupiter and then to the Sun for the perihelion maneuver.

The nuclear-pulse concept, *per se*, simply will not work with the limited mass available on a probe such as this. Fission may well provide the key element for the perihelion propulsion, but only in a pulsed mode with extremely low fission yields per pulse.

Solar thermal propulsion and a near-Sun maneuver seem to be "made for each other"; problems that require more study are heat transfer to the working fuel and how high a specific impulse can be obtained. Liquid hydrogen (and its dissociation at very high temperatures) appears to offer the best solution, but suffers from the mass associated with storing a cryogen for over three years in deep space. Ammonia offers an excellent storage solution, but appears to limit the specific impulse. More study of the systems aspect is required with emphasis on the mass of the overall propulsion system hardware plus fuel.

Thermal shielding of the probe near the Sun is not an issue as long as the probe is not spinning and can actively point a Sun-shield toward the Sun during the perihelion passage. The umbra (shadow) of the shield does drive the mechanical configuration of the probe during the perihelion passage segment of the mission.

Data downlink, attitude control and knowledge, and communications means and power are all intimately linked; the data downlink requirements tend to drive the entire probe architecture for the interstellar flight configuration. Implementation is simplified by using "fire and forget" operations - the probe requires an autonomous and self-healing character so that uplinks are no longer necessary following final departure outbound from the inner solar system.

Low-power operations will help to ensure longevity of the probe while minimizing the required mass for a radioisotope power system. Isotopes longer lived than plutonium have lower power densities and offer engineering challenges in providing efficient production of electricity; the subject requires further study.

Continued miniaturization of scientific instrument electronics and detectors is required to implement a mission with a reasonable science return - the reason for the mission in the first place.

The concept we are pursuing for a realistic Interstellar Explorer has applicability to any robotic interstellar precursor mission. In addition, this concept offers potential advantages - as well as an alternative - to low thrust probes. Such systems require a great deal of control for their primary propulsion system. The approach here has only a few, short-duration critical operational periods. From a systems point of view this approach offers distinct advantages.

ACKNOWLEDGMENTS

Many individuals in addition to those on the author list contributed to the effort briefly summarized here. Primary contributors include David R. Haley and Robert S. Bokulic. Useful discussions occurred with Paul E. Panneton, Judi I. Von Mehlem, Larry E. Mosher, Edward L. Reynolds, Robert W. Farquhar, David W. Sussman, David W. Dunham, Edmond C. Roelof, and Robert E. Jenkins of JHU/APL and with Roger Westgate of the JHU School of Engineering, Dean Lester of Thiokol Corporation, Dean Read at Lockheed-Martin and Daniel Doughty at Sandia National Laboratories. As always, Barbara A. Northrop of JHU/APL is thanked for many excellent editorial suggestions.

REFERENCES

Boyer, K. and J. D. Balcomb, "System Studies of Fusion Powered Pulsed Propulsion Systems," in *AIAA/SAE 7th Propulsion Joint Specialist Conference*, Salt Lake City, Utah, June 14-18, 1971, *AIAA Paper No. 71-636*, 1971.

Deininger, W. D. and R. J. Vondra, "Spacecraft and Mission Design for the SP-100 Flight Experiment," *J. Brit. Int. Soc.* **44**, 217-228 (1991).

Dyson, F. J., "Death of a Project," *Science* **149**, 141-144 (1965).

Dyson, F. J., "Interstellar Transport," *Phys. Today,* 41-45 (October, 1968).

Ehricke, K. A., "Saturn-Jupiter Rebound," *J. Brit. Int. Soc.* **25**, 561-571 (1972).

Farquhar, R. W. and D. W. Dunham, "Indirect Launch Mode: A New Launch Technique for Interplanetary Missions," IAA paper L98-0901, April 1998, in press *Acta Astronautica*, 1999.

Frisch, P. C., "G-Star Astropauses: A Test for Interstellar Pressure," *Astrophys. J.* **407**, 198-206 (1993).

Holzer, T. E., R. A. Mewaldt, and M. Neugebauer, editors, "The Interstellar Probe: Scientific Objectives for a Frontier Mission to the Heliospheric Boundary and Interstellar Space," Report of a Workshop held March 20-12, 1990 in Ballston, Virginia.

Hyde, R., L. Wood, and J. Nuckolls, "Prospects for Rocket Propulsion with Laser-Induced Fusion Microexplosions," in *AIAA/SAE 8th Joint Propulsion Specialist Conference*, New Orleans, Louisana, November 29-December 1, 1972, *AIAA Paper No. 71-1063*, 1972.

Jaffe, L. D. and C. V. Ivie, "Science Aspects of a Mission Beyond the Planets," *Icarus* **39**, 486-494 (1979).

Linsky, J. L. and B. E. Wood, "The α Centauri Line of Sight: D/H Ratio, Physical Properties of Local Interstellar Gas, and Measurement of Heated Hydrogen (the 'Hydrogen Wall') Near the Heliopause," *Astrophys. J.* **463**, 254-270, (1996).

McNutt, R. L., Jr., "A Realistic Interstellar Explorer," Submitted to *Proceedings of the Workshop Robotic Interstellar Exploration in the Next Century*, California Institute of Technology, Pasadena, CA, July 28-31, 1998.

McNutt, R. L., Jr., R. E. Gold, E. C. Roelof, L. J. Zanetti, E. L. Reynolds, F. W. Farquhar, D. A. Gurnett, and W. S. Kurth, "A Sole/Ad Astra: From the Sun to the Stars," *J. Brit. Int. Soc.* **50**, 463-474 (1997).

Meissinger, H. F., S. Dawson, and J. R. Wertz, "A Low-Cost Launch Mode for High-C_3 Interplanetary Missions," in *AAS/AIAA Astrodynamics Specialist Conference*, Sun Valley, Idaho, August 4-7, 1997, *Paper AAS 97-711*, 1997.

Mewaldt, R. A., J. Kangas, S. J. Kerridge, and M. Neugebauer, "A Small Interstellar Probe to the Heliospheric Boundary and Interstellar Space," *Acta Astron.* **35, Suppl.**, 267-276 (1995).

Serber R., *The Los Alamos Primer*, University of California Press, Berkeley, 1992.

Solem, J. C., "Nuclear Explosive Propulsion for Interplanetary Travel: Extension of the MEDUSA Concept for Higher Specific Impulse," *J. Brit. Int. Soc.* **47**, 229-238 (1994).

Winterberg, F., *The Physical Principles of Thermonuclear Explosive Devices*, Fusion Energy Foundation, New York, 1981.

The Science Case for In-Situ Sampling of Kuiper Belt Objects

Emma Bakes

MS 245-3, NASA Ames Research Center, Moffett Field, CA 94035-1000
604- 604-0787;bakes@shivakali.arc.nasa.gov

Abstract. This mission may be thought of as merely a first step in an overall initiative to push humanity's exploration beyond the Solar System and into interstellar space. By initiating a mission to the Kuiper Belt, we are poised to extend human space exploration beyond the Sun's family of planets. However, the ultimate goal of the mission is to better understand the formation and transformation of matter originating in the primordial interstellar medium (ISM) and how it relates in chemical composition and evolution to both the solar nebula and the cometary material which bombarded early Earth with prebiotic volatiles. We are planning an initial single in situ sampling mission of a range of classes of organic molecules and dust particles on two Kuiper Belt Objects (KBO) at around 40 AU, a flyby of an appropriate Centaur object (thought to be the evolutionary step between a KBO and a comet), plus we will perform accompanying cruise science to sample free flying interplanetary dust and impinging interstellar dust on the way to the Kuiper Belt. The cruise science will elucidate the broad physical properties of the dust grain population such as the variation of elemental and isotopic composition, size distribution and the variation in their energetic processing with radius from the Sun. In addition, we aim to quantitatively analyse the color variation of the KBO population. The mission will help establish a virtual presence throughout the Solar System and probe deeper into the mysteries of life on Earth and beyond. It also addresses the necessity to develop and utilize revolutionary enabling technologies for missions impossible in prior decades.

INTRODUCTION

Our Solar System was formed from the primordial interstellar medium when it collapsed to form a protoplanetary disk of gas and dust grains. Everything that comprises the solid material in our planetary system originates from primordial interstellar dust (Figure 1).

The recent discovery and study of rare interstellar grains preserved in meteorites (Anders and Zinner, 1993) has shown that interstellar grains do preserve excellent records about the nature of their parent stars, including details of the complex nuclear reaction processes that occur within them. Interstellar dust likely provided a source of prebiotic molecules to the primitive planetesimals that formed the early Earth billions of years ago and the cometary matter which bombarded its surface (Chyba et al.,1990; Oberbeck et al.,1991). Chemistry in the ISM is capable of producing organic molecules from which prebiotic molecules can form. This has been demonstrated by Snyder et al (1993), who found compelling evidence for the possible existence of the amino acid glycine in the interstellar region Sagittarius B2.

The Kuiper Belt is a region between 30-50 AU from the Sun. The interiors of Kuiper Belt Objects (KBOs) are composed of pristine interstellar material unchanged since the dawn of the Solar System. Within the structure of a KBO may lie the key to the connection between life on Earth and the interstellar medium. This key could unlock the secret of whether the Kuiper Belt provided a source of prebiotic organics for the seeding of terrestrial life via short period comets which bombarded the surface of early Earth. A recent article by Bernstein, Sandford and Allamandola (1999) details the nature of life's far flung raw materials, which appear to exist in such hostile environments as the interstellar medium and may be preserved intact in the solid material which forms planetary

CP504, *Space Technology and Applications International Forum–2000*, edited by M. S. El-Genk
© 2000 American Institute of Physics 1-56396-919-X/00/$17.00

systems like our own. This means that the stuff of life may be formed between the stars before a conducive planetary environment exists for it to initiate a flourishing biosphere.

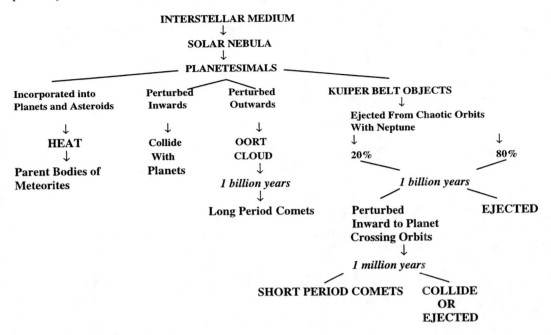

FIGURE 1. The connection between the interstellar medium and the solid bodies formed in the early Solar System (based on an article by Cruikshank published by the NRC (1998)).

In summary, the small volatile and organic rich bodies of the outermost solar system represent the frozen leftovers from planet formation in the solar nebula. As such, they hold clues to the origin of the outer planets and to the origins of Earth's inventory of volatiles. The incorporation of organic materials from the primordial ISM into the solid matter of our Solar System is of prime importance in the evaluation of the materials delivered to early Earth by comets, meteorites and asteroids, because these building blocks may have played an important role in the emergence of life (Pendleton and Chiar 1997). Since KBOs are relatively unmodified since their formation, then studies of their chemical composition will provide pathways of volatile and organic molecular materials from their interstellar origins to their deposition in Earth's hydrosphere, atmosphere and biosphere. Such knowledge may open the window to our understanding of the deepest and most compelling issues of the origins of life and its presence elsewhere (Cruikshank, 1998).

SCIENTIFIC JUSTIFICATION

The Kuiper Belt

The Kuiper belt can be divided into 3 overlapping components, namely the near zone, the far zone and the scattered disk. The near zone covers a range of 30-50 AU and its size distribution and compositional variations reflect the complex orbital dynamics and collisional evolution over 4 billion years. The far zone lies beyond 50 AU, and beyond the reach of the giant planets, so this part directly reflects the character of the outer solar nebula. Finally, there is the scattered disk, which is a small fraction of the Kuiper disk formed close to Neptune's orbit - i.e. transneptunian objects. This represents the transitional population between the classical KBOs and the Oort cloud. Ground based observations have currently identified 130 bright, slow moving KBOs (Malhotra, private communication). Very little is known about the physical properties of the Kuiper Belt population. The variations in their brightness is attributed to size but since little is known about their albedo, we cannot be certain. Our mission's primary focus will be on the chemistry and composition of KBOs. Following missions may include a secondary focus on their size distribution and cumulative spatial density.

Why Study Kuiper Belt Objects?

There are three main reasons to study the Kuiper Belt Objects.

- The Kuiper Belt is a remnant of the primordial planetesimal disk that gave birth to the planets and the Solar System. Because it is a large distance from the Sun, the Kuiper Belt preserves a relatively pristine mix of the original material from which the solar system formed.

- The Kuiper Belt is an analog of circumstellar disks around other stars. Improved knowledge of the properties of the Kuiper belt will serve in the interpretation of astronomical observations of planet-forming and planet-harboring regions around other stars.

- Knowledge of three fundamental characteristics, namely composition, orbital distribution and size distribution of the Kuiper Belt will significantly advance our understanding of the origin of the Solar System and the nature of the materials which bombarded early Earth with volatiles.

Why In Situ Sampling and Not a Fly By?

Infrared spectra of interstellar organic molecules cannot pinpoint individual molecules because organics tend to coexist as families or collections of similar molecular structures. The unique spectral signature of each large organic molecule thus blends in with that of other family members, making it very difficult to discern individual species within the population. A fly by mission would entail the same infuriating uncertainties incumbent in interstellar observations, namely a blended spectrum of a collection of unique emission signatures from a population of organic molecules. The ground truth of in situ sampling and analysis can chemically separate and identify individual molecules. This kind of chemical separation and analysis necessary for identifying individual molecules with certainty cannot be done from the ground. Nothing can replace ``being there'' for the first definitive identification of individual organic molecules in the remote and well preserved primordial interstellar material incorporated into the Kuiper Belt.

Centaur Objects

Centaur objects are believed to be the intermediate evolutionary step between KBOs and short period comets. Centaur objects appear to have originated from the Kuiper belt. Their volume is around 50 000 times that of a comet and they are broadly classified as minor planets, with the most evolved Centaur, Chiron, having the added classification of a cometary type body. Their individual surface chemical properties vary widely depending on their orbits and their periodic energetic processing during their proximity to radiation from the Sun. There are many outstanding issues connected with the Centaurs. For instance, how many Centaurs are there, what are their orbits and how did they get there, how did their orbits evolve from the Kuiper Belt and if so, what are the dynamical processes involved? The main focus for this mission will be on Centaur surface chemistry and composition and how it may relate to KBOs and the cometary material which bombarded early Earth. We need to predetermine promising Centaur objects via their degree of redness and low albedo, which is believed to be an indicator of complex organic refractory residues. We can easily do this via ground based observations, the current routine mechanism by which these objects are discovered and analysed.

Free Flying Dust Particles

Precisely what is dust? This seemingly simple question belies the truly complex nature of this material. The dust population is a highly efficient absorber of electromagnetic radiation and emits primarily in the infrared. The dust dominates the thermodynamics of an ambient environment. Additionally, the dust population acts as an efficient charge exchange agent for electrons and positively charged ions and this has a substantial effect on both the chemistry and the thermodynamics of its surrounding environment. Dust also provides a substantial surface area, acting as a powerful chemical catalyst to form molecules such as hydrogen (Duley 1996). Interstellar dust is actually a size distribution of particles, ranging from small organic (i.e. carbon containing) molecules such as polycyclic aromatic hydrocarbons (PAHs) to mineral aggregates of radius around 0.1-1 microns which may have refractory or icy surfaces.

Fomenkova et al. (1994) have analysed the composition of comet Halley and have identified many organic molecules embedded in a kerogen type matrix. The kerogen matrix is composed of organic refractory polymers of biogenic elements such as carbon C, hydrogen H , oxygen O and nitrogen atoms N (``CHON''s) and they conclude that the diversity of compounds is consistent with their origin in primordial interstellar dust. The substantially less abundant but highly significant organics include carbon grains, hydrocarbons, polymers of cyanopolyynes and multicarbon monoxides. Cronin and Chang (1993) reviewed the wealth of organic molecules in the Murchison meteorite. These include homologous series of alcohols, ketones, aldehydes, acids, dicarboxylic and hydroxy acids, polycyclic aromatic hydrocarbons, amines and amino acids (which are *not* terrestrial contaminants).

Because of the presumed close relationship between and meteoritic and cometary matter (albeit accounting for some degree of chemical processing between the two types of matter), one might expect that sampling the source of cometary material from the Kuiper Belt would parallel sampling meteoritic matter. There has always been heated debate about whether life arose from amino acids of a purely terrestrial origin or whether some amino acids came presynthesised, delivered to primitive Earth by cometary and meteoritic emissaries. An in situ sampling mission which reveals the presence of organic molecules like amino acids in the outer Solar System would have profound and far reaching implications for the origins of life, placing it outside of a merely terrestrial arena and into a much wider context. If amino acids can form successfully in the ISM and/or the outer Solar System and somehow remain intact, this would mean the formation of prebiological molecules would be much easier and more widespread than we ever imagined. We may be able to eventually conclude that life should not be a rarity, but in fact, a common phenomenon. Even if we do not find complex organics embedded in the selection of dust grains in our Solar System (both those embedded in KBOs and those in free flying dust particles), the usefulness and wide applicability of an inventory of elements and isotopes in these grains will allow us to better understand the fractionation of the solid materials in the protoplanetary disk and help us investigate how the terrestrial and gaseous planets ended up forming where they did and why their composition varies as it does. In short, much valuable information about the origin of our Solar System from the interstellar medium and its dynamical and chemical evolution can be garnered from the proposed cruise science involving interplanetary and interstellar dust grains.

SCIENCE GOALS

- To investigate the KBO surface and subsurface chemical composition. We will analyse the surface chemical composition and evolution, subsurface organics and mineralogy.

- To determine how the range of colors which Centaurs exhibit from ground based observations relates to their actual surface composition and dynamical evolution.

- To analyse the macroscopic variation in the radial distribution of free flying dust within the Solar System and quantify its chemical and isotopic composition, abundance, size distribution, degree of energetic processing and identify possible sources for predicted classes of dust grains.

- To identify and intercept a possible new comet and sample its pristine chemical composition (well before it reaches perihelion) as a representative sample of the primordial interstellar medium or solar nebula.

SCIENCE QUESTIONS ADDRESSED BY THE MISSION

- We intend that our science goals answer the following questions. The way in which this will be done is addressed in the following sections below.

- How do stars and planetary systems coevolve from the interstellar medium? Can we trace this via a chemical link in the processing and evolution of the material from which they are composed?

- What is the nature and history of our Solar System and how did Earth come to be similar to and different from its planetary neighbours?

- How did life originate and evolve on Earth; how might cometary impacts have affected the chemistry of the formation of life on Earth?

- Could life originate and persist beyond Earth?

- Could the Kuiper Belt be a valuable chemical resource for future use by humanity as it explores the outer Solar System? This exploratory mission to the Kuiper Belt will help lay the groundwork for further missions aiding human expansion into the outer Solar System and the local interstellar medium.

ONBOARD SCIENTIFIC ANALYSIS

- Onboard equipment for the mission's scientific analysis includes:

- A dust flux monitor package containing instrumentation that would record in-flight impacts by dust grains during the mission's cruise phase. This would require a variety of impact monitors and some sort of mass spectrometer. Given the wide range of dust relative velocities and trajectories that would likely be present, compositional information will probably be limited largely to elemental composition.

- A capillary electrophoresis system with fluorescence detection, capable of in situ chiral separation and analysis of femtomolar levels of amino acids, is currently under development and may be included in future missions to Mars. Sample extraction and processing are key concerns still being addressed, since the addition of water and/or heat may be necessary and yet may affect the chemistry of the sample.

- An instrument capable of UV Raman Spectroscopy, a nondestructive technique producing significantly sharper spectral bands than IR spectroscopy and other reflectance or emission technologies. The development of miniature spectrometers for in situ Raman spectroscopy on planetary surfaces has progressed to the point where an instrument of necessary size and mass (less than 100 grams) and sensitivity has been demonstrated. Excitation in the deep UV at 224 nm and 248 nm produces a resonance enhancement phenomena capable of detecting aromatic amino acids, and nucleobases. A key advantage of the UV Raman technique for the in situ analysis of primitive solar system bodies is that measurements can be carried out either in liquid (aqueous) solutions or on solid (i.e., mineral and/or ice) surfaces, and therefore it permits analyses of samples without requiring heating or chemical extraction which may alter the native chemistry of an object.

- A gas chromatograph/mass spectrometer (GC/MS) system equipped with a pyrolysis chamber would allow the characterization of compounds ranging from small volatiles to components of large polymers. Although GC/MS is less sensitive than capillary electrophoresis with fluorescence detection, it is a more inclusive technique ideal for detecting a wide spectrum of organic compounds.

- 5. Isotope analyses of D, $\delta^{13}C$, $\delta^{18}O$ and $\delta^{15}N$ are crucial to evaluating the fractionation of these elements relative to inner solar system objects and the interstellar medium and the nature of the parent stars which formed the dust grains. One of the anomalies carried in interstellar grains is a high D/H fractionation, purportedly due to cold ion-molecule reactions. This work could lead to links with the host of interstellar molecules identified in cold clouds by radio astronomical techniques. Thus, we will also try to determine D/H ratios to yield valuable information concerning the link between Solar System materials and their interstellar heritage.

MISSION DESCRIPTION

Implementation of a KBO-class mission is extremely challenging. Science exploration objectives drive mission and system performance into the advanced implementation (system technology) realm. The mission objectives and baseline mission characteristics can be summarized as follows:

Mission Objectives

1. In-situ exploration of a large (200 - 500 km) and a small (1 - 10 km) Kuiper Belt Object
2. Flyby of a Centaur (Nessus, Pholus, or Chiron)
3. Begin initial data return within 5 years of launch

Mission Characteristics

Launch Date: 2010
Launch Vehicle: Delta IV Heavy
Instrument Payload: 2 Landers, Fields and Particles, Imaging
Flight System: Nuclear Electric Propulsion
Data Acquisition: Cruise Science + 3 Encounters, Remote
Sensing of a Centaur, 1 Lander to a Big KBO, 1 Lander to a Small KBO and Cruise Science

The reference mission executes the flyby of a Centaur (Chiron) on the outbound trajectory, along with the post flyby, in-situ exploration of a large and a small KBO. At each KBO target, a lander is emplaced and a 40 day surface phase is planned. In addition, the carrier spacecraft is outfitted with instrumentation to conduct observations consistent with science goals 2, 3 and 4. Figure 2 illustrates a low-thrust trajectory used for the KBO mission concept and Figure 3 summarizes the reference mission scenario.

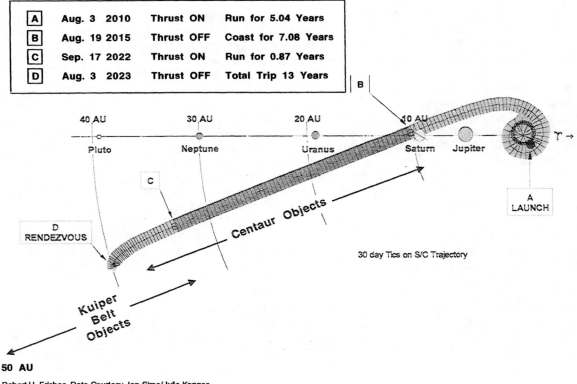

A	Aug. 3 2010	Thrust ON	Run for 5.04 Years
B	Aug. 19 2015	Thrust OFF	Coast for 7.08 Years
C	Sep. 17 2022	Thrust ON	Run for 0.87 Years
D	Aug. 3 2023	Thrust OFF	Total Trip 13 Years

Robert H. Frisbee, Data Courtesy Jon Sims/Julie Kangas

FIGURE 2. Representative NEP low-thrust trajectory for the KBO mission concept.

PROPULSION OPTIONS

Several propulsion options were selected for preliminary evaluation. The options considered include chemical propulsion with a ballistic trajectory, nuclear electric propulsion with a low-thrust trajectory, and solar electric propulsion options with a low-thrust trajectory. Solar sail implementations were eliminated from consideration due to the multiple rendezvous requirements. Fusion and antimatter implementations were eliminated because their technology readiness was considered to be too far beyond the mission time frame. Of those systems considered, the nuclear electric implementation seemed the most feasible from a performance viewpoint.

Various ballistic (chemical propulsion) options were examined for performance capability. Direct and multiple gravity assist trajectories were considered. As can be seen from Figure 4, none of the chemical options satisfy the flight time requirement to the first KBO target. In particular, a "direct" trajectory delivers no positive mass regardless of flight time, and trajectories delivering the required mass take in excess of 30 years flight time.

The nuclear electric system examined (100 kWe class), integrated a SNAP-derived nuclear reactor with a steam-Rankine power conversion system and krypton-fueled electric thrusters. The flight system was designed to deliver the payloads, rendezvous with required targets, and relay/return scientific data.

Two solar electric implementations, solar thermal/dynamic and solar photovoltaic, were considered. Both require a solar collector/concentrator to collect the required energy and focus it onto the implementation specific conversion system. In the solar thermal/dynamic implementation, collected light is focused onto the receiving end of a

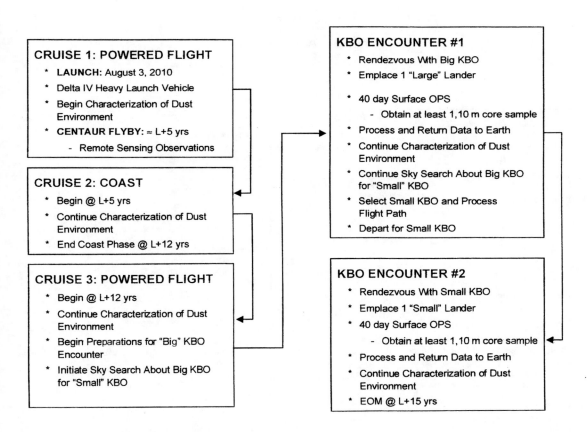

FIGURE 3. Reference misssion scenario for the KBO mission concept

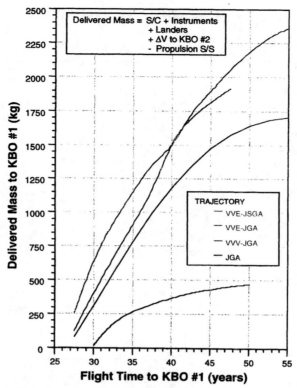

FIGURE 4. Comparison of chemical system performance for direct and multiple flyby trajectories (assumes a delivered mass of more than 1000 kg, [Figure-of-Merit]).

CONCEPTS	BENEFITS	ISSUES
Chemical with Gravity Assists	Least Technology Development Required	• Very long (>30 year) trip times required for modest payload • Requires launch approval for ARPS (high Curies at launch)
Electric Propulsion	• Electric Propulsion Tech Base (Already flying DS-1) • Faster Trip Times	
Solar	• Non-nuclear Option • Don't have shadow shield geometry constraints of NEP	• Concentrator/reflector technology development – High CR implies need for good shape control – serious mass impact compared to solar sail. Very large scale (km scale)
Solar Photovoltaic Cells	• Extensive technology base • Good specific mass and efficiency of cells	• Cost of 100 kWe of solar arrays • Technology of small, high-temp., cosmic-ray rad-hard, long-life cells
Thermal-to-Electric Power Conversion	• Technology base for Growth Space Station power • Dynamic system scales well at 100 kWe (use NEP power conversion system?) • Inherent radiation insensitivity of power system	• Mass, specific mass, efficiency, lifetime (reliability) competitive with solar cells? Many of the same issues associated with NEP: rotating machinery, radiators, etc.
Nuclear	• Most flexibility on payload delivery • Cold reactor at launch (low curies) • Technology can be leveraged for more ambitious missions	• Political sensitivity to space nuclear reactors

S. Leifer/R. Frisbee

TABLE 1: Summary comparison of the benefits and issues for the propulsion options considered for the KBO mission concept.

dynamic conversion system. In the solar photovoltaic implementation, light is focused upon a system of photovoltaic cells. Preliminary analysis implies that the solar thermal/dynamic system is probably the most feasible of the two options (assuming that the specific mass of the dynamic conversion system is equivalent to that of the nuclear option, further detailed analysis and design of both options is required). Unfortunately, either of these options strongly depend upon the ability to achieve a variable 1-km class optical structure (i.e. the collector/concentrator) with very low areal density (< 2 gm/m^2). This level of technology is beyond the technology readiness date of 2010 baselined for this mission concept. In addition, the photovoltaic option would require an extremely expensive solar array (100 kWe of Current Space-Qualified Cells = $300M at $3,000/Watt). Table 1 compares the propulsion options considered as part of the preliminary analysis of this mission concept.

The propulsion mechanism turns out to be the most challenging aspect of this mission. The mission is classified as a ``delta 4 heavy'' type. This, in tandem with the fact the spacecraft needs to decelerate when dropping a lander on a KBO, means nuclear electric propulsion is presently considered the most feasible mechanism for an adequate power supply. Alternatives include less feasible mechanisms such as a ballistic (chemical) propulsion system or Solar electric propulsion. Other propulsion candidates e.g. fusion and antimatter require technology breakthroughs.

CONCLUSIONS

An in situ sampling mission to the Kuiper Belt will provide us with unique and invaluable information on the primordial ISM from which our Solar System formed. The resulting investigation of chemical evolution may provide a vital link between the ISM, the Solar System and the formation of prebiotic molecules. It may help elucidate the origins of life on Earth and the possibilities of life on other worlds.

ACKNOWLEDGMENTS

The technical material in this paper was developed in support of an Advanced Technology Study at the Jet Propulsion Laboratory in cooperation with NASA Ames Research Center, NASA Glenn Research Center, and the Sandia National Laboratory. In particular, the author would like to thank Mr. Ron Lipinski and Mr. Roger Lenard of Sandia National Laboratory, the Advanced Projects Design Team (Team X) of JPL, Dr. Robert Frisbee, and the other members of the JPL design teams who supported the study but have not been identified here.

REFERENCES

Advanced Projects Design Team, "Kuiper Belt Object Rendezvous Mission 5-99," Jet Propulsion Laboratory, May 1999

National Research Council Space Studies Board,``*Exploring the Trans-Neptunian Solar System,*" Washington DC , National Academy Press, 1998, p12

Allamandola, L.J., Sandford, S.A. and Tielens, A.G.G.M., "*Infrared Spectroscopy of Dense Clouds in The C-H Stretch Region: Methanol and Diamonds,*" ApJ **399**, 134-144,(1992)

Anders E. and Zinner, E. , "*Interstellar grains in primitive meteorites: Diamond, silicon carbide, and graphite,*" Meteoritics **28**, 490-501, (1993)

Bernstein, M., Sandford, S.A. and Allamandola, L.J., "*Hydrogenated Polycyclic Aromatic Hydrocarbons and the 2940 and 2850 Wavenumber (3.40 and 3.51 Micron) Infrared Emission Features,*" ApJ **472**, L127, (1996)

Bernstein, M.P. and Sandford, S. and Allamandola, "*Life's Far Flung Raw Materials,*'' Scientific American, **281**, 42-55, (1999)

Chyba, C.F. , Thomas, P.J. and Brookshaw, L, "*Cometary Delivery of Organic Molecules to the Early Earth,*" Science, **249**, 366-370, (1990)

Cruikshank, D. P., Roush, T.L.,Bartholomew, M.J., Moroz, L.V, Geballe, T.R., White, S.,. Bell III, J.F., Pendleton, Y.J.,. Davies, J.K., Owen, T.C.,. deBergh, C.,. Bernstein, M.P., Brown, R.H. and Tryka, K.A., "*The Composition of Planetesimal 5145 Pholus,*" Icarus, **135**, 389-407, (1998).

van Dishoeck, E., Blake, G.A., Draine, B.T. and Lunine, J., "*Protostars and Planets III,*" Ed. J. Levy and J. Lunine, Arizona,UAP, 1993, pp 163-241.

Duley, W.W. "*The formation of H$_2$ by H-atom reaction with grain surfaces,*" MNRAS **279**, 591-598, (1996)

Fomenkova, M. N., Chang, S. and Mukhin, L. M. "*Carbonaceous components in the comet Halley dust,*" Geochimica et cosmochimica acta. **58**, 4503-4521, (1994)

Murayama, Takashi, Taniguchi, Yoshiaki and Arimoto, Nobuo, *"New Near-Infrared spectroscopy of the high-redshift quasar B1422+231 at z=3.62,"* Astronomical Journal, **117**, 1645-1650, (1999)

Oberbeck, Verne R. and Aggarwal, Hans, *"Comet Impacts and Chemical Evolution on the Bombarded Earth,"* OLEB **21**, 317-321, (1991).

Pendelton, Y.P.,and Chiar, J., "The Nature of Interstellar Organics,'' in *From Stardust to Planetesimals* , edited by. Y.P. Pendleton and A.G.G.M. Tielens, ASP Conference Series 122, San Francisco, 1997, pp179-200

Sandford, S.A., Allamandola, L.J., Tielens, A.G.G.M., Sellgren, K., Tapia, M and Pendleton, Y. , *"The Interstellar C-H Stretching Band Near 3.4 Microns: Constraints on the Composition of Organic Material in the Diffuse Interstellar Medium,"* ApJ **371**, 607-621, (1991)

Snyder, Lewis, Kuan, Y, Miao, Y and Lovas, Frank J. "Possible Detection of Glycine in Sgr B2," in *Habitable Zones*, edited by Seth Shostak, ASP Conference Series 74, San Francisco,1992, pp107-120.

Yen, C. L., KBO Chemical Mission Performance, PersonalCommunication, 28 May 1999 and 3 June 1999.

ronin, John R. and Chang, Sherwood, *"The Chemistry of Life's Origins,"* Dordrecht, Kluwer Academic Publishers , ed. J.M. Greenberg *et al.*,.1993, pp 112-138.

The Quest for Interstellar Exploration

Richard A. Wallace and Juan A. Ayon

Jet Propulsion Laboratory/California Institute of Technology, Pasadena, California 91109
*818-354-2797, **Richard.A.Wallace@jpl.nasa.gov;** 818-354-8643, **Juan.A.Ayon@jpl.nasa.gov***

Abstract. NASA strategic planning includes the objective for a set of missions that would begin exploration beyond our solar system. These missions would determine the nature of the interface between our solar system and the local interstellar medium, as well as directly sample the properties of the interstellar medium as an initial step in exploring the nearby Galaxy. Preliminary planning envisions four missions making up this initial quest for exploration. Each of the mission concepts is described in terms of science objectives, mission characteristics, and technology needs. The four missions are: 1) Interstellar Probe (reaching 200 AU in < 15 years via solar sail to explore the heliospheric boundaries and the interstellar medium); 2) Heliosphere Imager & Galactic Gas Sampler (delivery to a low inclination, 1 by 4 AU ecliptic orbit to establish the 3-D structure of the heliosphere and boundaries and sample Galactic material injected into the heliosphere); 3) Outer Heliosphere Radio Imager (16 subsatellites with mother spacecraft delivered to 20 to 30 AU via solar sail to radio-image the boundaries of the heliosphere); and 4) Interstellar Trailblazer (reach 2000 AU in ~ 30 years via advanced solar sail to explore the local interstellar cloud). Technology needs for these missions include solar sail propulsion, advanced instrumentation, autonomous spacecraft operations, advanced power, advanced telecommunication, and long life systems.

INTRODUCTION

NASA's Office of Space Science has recently completed a set of comprehensive strategic planning roadmaps for its four themes (Sun-Earth Connection, Exploration of the Solar System, Astronomical Search for Origins, and Structure and Evolution of the Universe). In the Sun-Earth Connection (SEC) Roadmap (NASA, 1999) a set of four missions is called out to answer the question of how the Sun and Galaxy interact. The four missions seek to take significant steps in three scientific endeavors:
1. Explore the structure and dynamics of the heliospheric boundaries
2. Determine the properties and composition of the interstellar medium
3. Explore the local galactic neighborhood
Each mission will address the above in both an evolutionary manner as well as with different investigative modes.

Figure 1 illustrates the scale of the missions in the sense of distances reached and the focus for their investigations. Although the entire region shown in Figure 1 is of interest, the focus for the four missions is on the interface between our solar system and the Galaxy/local interstellar medium. Two of the four missions penetrate the interface and carry out explorations into interstellar space. Interstellar Probe (ISP) will explore the heliospheric boundaries and the nearby interstellar medium in addition to preliminary investigations at the edge of our solar system. Interstellar Trailblazer (ITB) is expected to follow with an order of magnitude further exploration into the Galaxy. Heliosphere Imager & Galactic Gas Sampler (HIGGS), operating from well within the heliosphere, will sample material traveling to the inner solar system from interstellar space. Outer Heliosphere Radio Imager (OHRI) will travel to the region of the far outer planets to image the structure of the solar system-Galaxy interface. The overall strategy of these four missions is to significantly address how the Sun and our Galaxy interact via exploration of the heliospheric boundary and nearby galactic environment.

Three of the missions are enabled by application of solar sail propulsion. With recent technology advances in materials, solar sail technology has changed the concept of delivering, in reasonable flight times, scientifically significant payloads to the solar system-Galaxy interface and beyond, from conjecture into one of possibility (Garner, 1999) and (Sauer, 1999). The trajectory optimization process for the three solar sail-enabled missions was carried out by Carl Sauer in a similar manner to that in (Sauer, 1999).

CP504, *Space Technology and Applications International Forum–2000*, edited by M. S. El-Genk
© 2000 American Institute of Physics 1-56396-919-X/00/$17.00

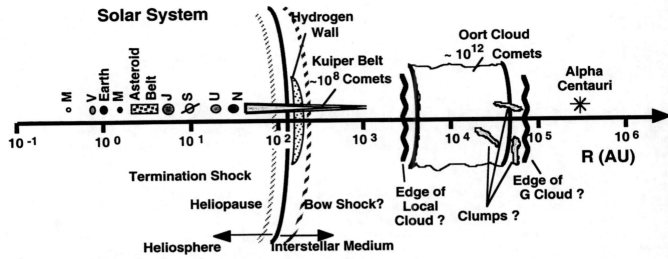

FIGURE 1. Scale of the Interstellar Medium (Mewaldt, 1998).

During the summer of 1999 the authors assembled a small team of JPL experts in space science mission and system design. After working intensively with this team during a two to three week period, a number of deep space mission concepts evolved and were defined that met science requirements in the then-developing new SEC strategic planning Roadmap. This paper describes a subset of the missions addressed, the four mission concepts defined for exploring the heliospheric boundary and the nearby galactic environment in the Roadmap's Quest III. Each of the four mission concepts is described in terms of science objectives, mission characteristics, and technology needs. The mission designs described, in most cases, are considered starting points for subsequent performance and technology trade studies. Where possible, preliminary trade study results are included.

INTERSTELLAR PROBE

Interstellar Probe will be the first mission to cross the heliosphere and begin exploring the interstellar medium. The mission concept has a long history of interest as a precursor to eventual travel to the stars (Wallace, 1999). The conceptual basis for the current design was laid out in 1990 (Holzer, 1990) when preliminary science objectives and a strawman science payload were described. Of the four mission concepts described in this paper the Interstellar Probe design is the most mature due to having science definition, mission/system design, and technology development teams dedicated to it.

Science objectives and mission concept description with key technology requirements for the Interstellar Probe are provided below. In addition the results of mission performance trades with sail technology capability are discussed.

Science Objectives & Mission Requirements

The Interstellar Probe science objectives listed below are taken from the SEC Roadmap (NASA, 1999):
- Explore interstellar medium and determine properties of plasma, neutral atoms, dust, magnetic fields, and cosmic rays;
- Determine structure and dynamics of heliosphere as an example of the interaction of a star with its environment;
- Study, in-situ, structure of solar wind termination shock and acceleration of pickup ions and other species;
- Investigate the origin and distribution of solar system matter beyond the orbit of Neptune.
- Measure in-situ, properties and composition of interstellar plasma, neutrals, dust, and low-energy cosmic rays;
- Determine heliospheric structure and dynamics by in situ measurements and global imaging;
- Map zodiacal dust cloud IR emissions; measure distribution of interplanetary dust & small Kuiper Belt Objects

The mission goal developed for the science objectives is relatively straightforward and evolved from analyses performed with inputs from the Interstellar Probe-dedicated science definition, mission/system design, and technology development teams: deliver a scientific payload of 25 kg to 200 AU in < 15 years.

Mission and Sail Technology Performance Trade

A baseline mission concept for the Interstellar Probe mission was developed for the application of solar sail technology that would be available for integration into a mission system launching in the 2010 time period (Garner, 1999). Other propulsion system candidates continue to be considered as options to this baseline.

A trade study was performed based on defining sail technology in terms of three parameters: sail size, sail areal density (the ratio of sail material and structure mass to sail area), and closest approach to the Sun. These parameters relate to a number of key solar sail technology developments and describe the capability of a particular sail quite well for early concept design studies:
- sail size: deployment, control, fabrication, packaging, structure
- areal density: materials, deployment, fabrication, packaging, structure
- solar closest approach: thermal management, materials, structure

Figure 2 illustrates a trade in solar closest approach, driving selection of materials and system configuration to withstand thermal input. To achieve low flight times it is necessary for the spacecraft to first travel inward towards the Sun (Sauer, 1999); the closer the pass, the faster the flight time (see Figure 2).

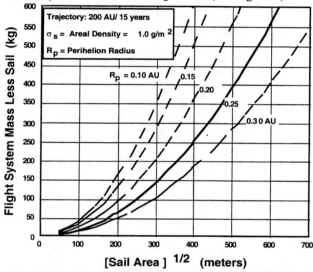

FIGURE 2. Sail Performance Trades: Radius of Perihelion Tradespace with Areal Density.

Trades similar to Figure 2 are possible varying areal density. Using such trades and estimating the level of sail technology development over the next 10 years, we have arrived at a summary chart for the performance of a sail that would allow trades in flight time and delivery mass. Figure 3 illustrates this trade for a circular sail of 200-m radius.

Preliminary system analyses indicates that spacecraft delivery mass will need to be in the range of 150 to 250 kg to support the required 25-kg science payload at 200 AU. This range of delivered spacecraft mass drives the sail size to circular areas with radii of from about 100 to 300 m (see Figure 2).

Mission Concept Selection & Key Technology Requirements

Sail definition parameters were varied with system design, and a 200-m radius sail with areal density of 1 g/m^2 was selected which could withstand a solar closest approach of 0.25 AU. The resulting spacecraft delivered to 200 AU is about 200 kg carrying a scientific payload of 25 kg and returning 25 bits per second (bps) of data from 200 AU.

A solar sail of 200-m radius, areal density of 1 g/m^2, and able to withstand a close solar approach of 0.25 AU is the key technology of the Interstellar Probe mission concept. An advanced Ka-band phased array system allowing 25 bps data return and spacecraft power system that can provide 275 W from 200 AU are the other two key technology developments needed for the baseline Interstellar Probe concept. Technology development in low mass/power instruments is a further need.

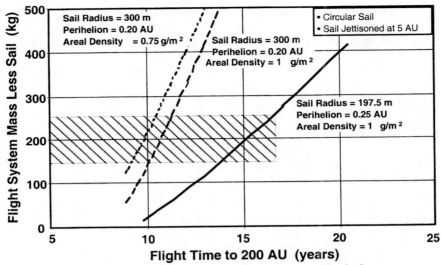

FIGURE 3. Mission Performance Trades: Flight Time to 200 AU Mission Trade Space.

HELIOSPHERE IMAGER & GALACTIC GAS SAMPLER

Heliosphere Imager & Galactic Gas Sampler (HIGGS) measures the elemental and isotopic composition of interstellar neutral atoms leaking deep into the heliosphere, allowing a determination of the size and shape of the heliosphere. These chemical messengers from interstellar space penetrate to within 3 to 4 AU of the Sun and provide us a direct sample of present-day galactic matter. This mission will take measurements from a relatively nearby orbit, making payload delivery/placement simplest of the four missions. The HIGGS mission concept is based on preliminary design; trade studies are a next step.

Science Objectives & Mission Requirements

The HIGGS science objectives listed below are taken from the SEC Roadmap (NASA, 1999):
- Establish 3-D structure of the interaction region between heliosphere and local galactic environment
- Determine elemental and isotopic composition of neutral atoms in a present-day galactic sample and explore implications for Big Bang cosmology, galactic evolution, stellar nucleosynthesis, and birthplace of the Sun
- Determine the shape of the heliosphere
- Measure precisely the cosmologically important abundances of ^{2}H and ^{3}He in local interstellar material
- Map the location and establish the characteristics of the extended inner source of neutral atoms in the heliosphere

An elliptical heliocentric orbit of 1 AU by 4 AU was selected for scientific observations. This orbit can be reached via chemical propulsion with Venus and Earth-return gravity assists with the required scientific payload mass after a flight time of one year. Delivery mass is driven by the specification of a science payload of 8 advanced space physics instruments with a total mass of 65 kg.

System Concept and Key Technologies

The system design includes solar power, chemical propulsion, one-day of data return per month at X-band at 100 bps, and 5-years of orbit observations. The total spacecraft mass delivered to the measurement orbit is 300 kg (including instruments and dry propulsion system); launch is via Delta II launch vehicle.

Advanced instrument developments will be required for this mission. Lower mass sensor structure and better sensitivity than current state of the art will be required. In particular, for two key instruments:
- EUV Spectrometer/Imager: low noise to < 1 micro-Rayleigh
- Pickup Ion Spectrometer: 100 times increased sensitivity

OUTER HELIOSPHER RADIO IMAGER

Outer Heliosphere Radio Imager (OHRI) measures radio emissions from the interaction of strong interplanetary shocks with the outer regions of the heliosphere. This mission will provide a way to sample the large-scale structure of the heliospheric boundary and measure its response to solar wind variations and disturbances.

Science Objectives & Mission Requirements

The OHRI science objectives listed below are taken from the SEC Roadmap (NASA, 1999):
- Determine the large scale structure of the heliospheric boundary
- Map 2-D shape of heliospheric boundary, including dynamic response to solar disturbances and to the solar cycle

A mother spacecraft with 16 subspacecraft are delivered to 20 AU via a trajectory similar to that of Interstellar Probe (initial perihelion of 0.25 AU). After about a two-year flight to 20 AU, the 16 subspacecraft are dispersed into an interferometric array. Each subspacecraft carries baseline orientation measurement equipment, as well as radio wave instruments, totaling about 7 kg. The total system mass delivered, including mother spacecraft and 16 subspacecraft, is about 650 kg. A further requirement is that the interferometric system of subspacecraft and its mother continue at about the same speed, 10 AU per year towards the nose of the heliosphere. A measurement operations objective is 5 years of observations after deployment of the subspacecraft, with a goal of as much as 11 years.

System Concept and Key Technologies

Our OHRI baseline design depends on solar sail technology application. The sail design must deliver a relatively large payload (650 kg) to orbit in a short period of time (20 AU in 2 years) and continue through regions of interest at this high rate of speed. The trajectory design is similar to that of Interstellar Probe with a close solar approach of 0.25 AU and high solar system departure speed. A sail design that meets the mission requirements has a sail radius of 300 m and areal density of 0.25 g/m^2. In addition to advanced solar sail technology, two other key developments are in space power (~ 25 W for each of the 16 subspacecraft) and in capability to place and keep the subspacecraft in positions for continued interferometric measurements (an interferometer multi-baseline acquisition and measurement system).

INTERSTELLAR TRAILBLAZER

The science rationale for the Interstellar Trailblazer mission that follows is taken from the SEC Roadmap (NASA, 1999). The solar system traverses a wide range of environments as it moves through the Galaxy. The Sun is presently located at the border of a great void in nearby interstellar matter known as the "Local Bubble", where we are embedded in a low-density cloud (our Local Interstellar Cloud, or LIC). We do not know the scale of density variations within our LIC, but it appears that the edge of the LIC in the direction of the Sun's travel is < 6000 AU away, possibly much less. Upon exiting the LIC the Sun may enter the hot, low-density Local Bubble, or it could enter a neighboring cloud. MHD simulations show that if the local interstellar density increased to that of a typical diffuse cloud (~10 cm^{-3}) the dimensions of the heliosphere would shrink by nearly an order of magnitude, which would undoubtedly have significant effects on the interplanetary environment at 1 AU. By targeting the Interstellar Trailblazer in the direction upstream of the Sun's motion, it will effectively blaze a trail through the future environment of the solar system over the coming centuries, scouting out the scale of density and other variations. Conversely, a downwind probe could explore our past.

Mission/System Concept and Key Technologies

The mission goal developed for the science objectives is relatively straightforward: deliver an advanced scientific payload of about 50 kg to 2000 AU in ~ 30 years, and communicate back to the Earth from 2000 AU at 25 bps. Figure 4 illustrates a trade in spacecraft mass as a function of sail size. An advanced sail of areal density equal to 0.1 g/m^2 and able to withstand a solar closest approach of 0.1 AU was selected for the baseline design. A net spacecraft system mass of about 225 kg (including 50 kg of instruments) resulted from our preliminary analysis. A circular sail of radius 600 m will be required to deliver this mass to 2000 AU in 30 years (see Figure 4).

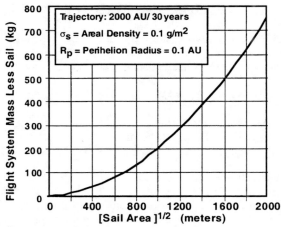

FIGURE 4. Interstellar Trailblazer Mission Performance Trade.

The key technologies and performance requirements for the above preliminary design are listed below:
- Solar Sail: Effective Sail Area of $\sim 10^6\,m^2$ (circular sail radius of 600 m); Areal Density 0.1 g/m^2
- Telecommunications/Power: 25 bps at 2000 AU/ $\sim$ 1 kW DC at 2000 AU
- Thermal Management: Survive Solar Closest Approach of 0.1 AU

SUMMARY

The quest for interstellar exploration has begun with a step by step outline of the objectives for a suite of four mission-based set of science investigations (NASA, 1999). Baseline mission designs have been defined in this paper for each of the four mission concepts, with mission requirements, mission/system designs, and technology needs.

ACKNOWLEDGMENTS

During the summer of 1999 the authors assembled a small team of JPL experts in space science mission and system design to assist us in defining a number of mission concepts suggested by the NASA SEC Roadmap Development Team. Those on the JPL team significantly aided in defining the four missions described in this paper: C. Budney, R. Oberto, K. Roust, G. Sprague, and T. Sweetser, all from Jet Propulsion Laboratory/California Institute of Technology. Science requirements integration into mission requirements were greatly assisted by R.A. Mewaldt, SEC Roadmap Team Quest III Leader, (California Institute of Technology) and P.C. Liewer from Jet Propulsion Laboratory/California Institute of Technology), JPL SEC Program Scientist.

The research described in this paper was carried out by the Jet Propulsion Laboratory, California Institute of Technology under contract with the National Aeronautics and Space Administration.

REFERENCES

Garner, C.E., Diedrich, B., and Leipold, M., "A Summary of Solar Sail Technology Developments and Proposed Demonstration Missions," AIAA-99-2697, presented at 35th AIAA Joint Propulsion Conference, Los Angeles, CA, June 1999.

Holzer, T.E., Mewaldt, R.A., and Neugebauer, M.M., "The Interstellar Probe: Scientific Objectives and Requirements for a Frontier Mission to the Heliospheric Boundary and Interstellar Space," Interstellar Workshop Report, Ballston, VA, March 1990.

Mewaldt, R.A, "Interstellar Mission Science Objectives," *NASA OSAT Ninth Advanced Space Propulsion Workshop Proceedings*, JPL (1998).

NASA, "Sun-Earth Connection Roadmap, Strategic Planning for 2000-2025," November 1999.

Sauer, C.G., "Solar Sail Trajectories for Solar Polar and Interstellar Probe Missions," AAS/AIAA Astrodynamics Specialist Conference, Girdwood, AK, Paper AAS 99-336, August16-19, 1999.

Wallace, R.A., "Precursor Missions to Interstellar Exploration", Proceedings of the 1999 IEEE Aerospace Conference, Aspen, CO, Track 2/Paper 114, March 10, 1999.

The Ultimate Performance of Gridded Ion Thrusters for Interstellar Missions

David G Fearn

Space Department, Defence Evaluation and Research Agency, Farnborough, Hants, GU14 0LX, UK
(44)-1252-392963
dgfearn@scs.dera.gov.uk

For a successful interstellar mission, the time taken to acquire useful data must be sufficiently short to maintain interest within the scientific community and associated funding agencies. Thus the mission must probably be productive within a few decades, implying a very high terminal velocity. Assuming that no new physical principles are employed, this can be accomplished only with propulsion systems which operate at ultra-high specific impulse (SI). In the near-term, advanced gridded ion thrusters offer a level of performance that may be of interest. This paper explores how far this technology can be taken to satisfy the demands of early interstellar missions, mainly by increasing the exhaust velocity through raising the ion accelerating potential and reducing the atomic mass of the propellant. It is concluded that an SI of perhaps 150 000 sec and a thrust in the 1 N range might be achieved, for a 10 cm ion beam diameter.

INTRODUCTION

Now that a considerable amount has been achieved in the initial exploration of the solar system, interest is increasing in precursor interstellar missions, which are becoming technically feasible. However, in such ventures, the time taken to acquire useful data must be sufficiently short to maintain interest within the associated scientific community and funding agencies. Therefore such missions must be productive within a few decades at most, implying a very high terminal velocity. If no new physical principles are employed, this requires propulsion systems which operate at ultra-high specific impulse (SI), defined as the ratio of thrust to the total rate of use of propellant in units of sea-level weight, and thus very large exhaust velocities.

Although photon rockets will eventually provide the maximum exhaust velocity that can be achieved, the velocity of light, very considerable research and development will be necessary before they can operate at thrust and efficiency levels of interest. In the meantime, advanced gridded ion thrusters offer a level of performance that may be of interest. This paper explores how far such technology can be taken to satisfy the demands of early interstellar missions. Present ion thrusters are designed for Earth orbital and interplanetary applications, where modest values of SI are needed. Thus devices intended for communications satellite station-keeping (Fearn, 1998) provide values of about 3500 sec, and laboratory tests have exceeded 5000 sec at thrusts of more than 300 mN (Martin, 1988).

The paper examines how the exhaust velocity of these devices can be increased, by raising the ion accelerating potential and reducing the atomic mass of the propellant. The latter process cannot go to the extreme of using hydrogen, because of storage problems over many years. Indeed, an easily-storable liquid propellant which can be vaporised readily is desirable. Viable possibilities are investigated, and it is concluded that an atomic mass improvement by about a factor of 30 can be realised. The accelerating potential is dependant on the design of the ion extraction grid system, including the necessary insulators, and current ideas suggests that at least a factor of 10 improvement should be achievable, perhaps as much as 30. This, together with the low atomic mass propellant, might allow an exhaust velocity of 1700 km/sec to be realised, giving an SI of perhaps 150 000 sec.

The thrust which can be achieved at such values of SI is determined by the design of the grid system and the

CP504, *Space Technology and Applications International Forum–2000*, edited by M. S. El-Genk
2000 American Institute of Physics 1-56396-919-X

plasma density within the discharge chamber. The high voltages mentioned above will improve the effective perveance of the grids, so that present current density limits can be exceeded. However, the reduction in ion mass, will reduce the achieved thrust simultaneously. The paper considers these various factors and concludes that a viable 10 cm diameter thruster can be developed which could produce a thrust in the 1 N range.

THE GRIDDED ION THRUSTER CONCEPT

In a gridded ion thruster, the propellant gas is ionised by an electrical discharge, which can be direct current (DC) or radiofrequency (RF). The resulting ions are extracted and accelerated by an electric field imposed between a set of perforated grids forming the downstream end of the discharge chamber. This concept permits the almost total separation of the plasma production process from the acceleration of the ions, aiding the design process considerably. A DC variant shown in figure 1a (Fearn, 1998), in which the various plasma regions are labelled, is of the Kaufman configuration (Kaufman, 1961). It uses xenon as the propellant, hollow cathodes to produce both the primary electrons and the electrons required to neutralise the space charge of the ion beam, and a magnetic field to maximise the ionisation efficiency. An RF device (Killinger, 1999) is similar in overall configuration, but with an insulating discharge chamber surrounded by a coil into which a powerful RF signal is fed.

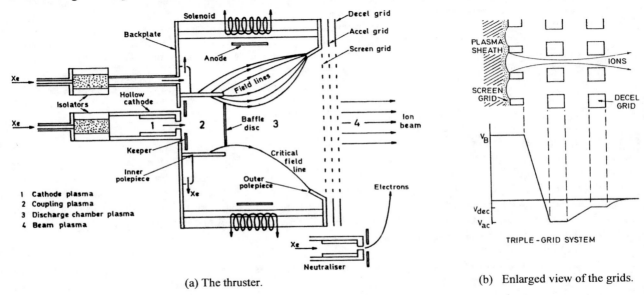

(a) The thruster.   (b) Enlarged view of the grids.

FIGURE 1. Schematic diagram of a gridded ion thruster.

Performance Limitations

Naturally, the gridded ion thruster has certain performance limitations which concern primarily the exhaust velocity and thrust density achievable, together with the lifetime implications of features designed to increase these parameters. Also of interest are thruster-spacecraft interactions, which include the ion beam divergence and the rate at which sputtered materials derived from the thruster are deposited upon sensitive surfaces.

Exhaust Velocity and Specific Impulse

From the point of view of interstellar missions, the most important parameter is the exhaust velocity, v_e, which determines the value of SI achieved. With reference to figure 1a, v_e is determined by the net ion accelerating potential, V_B, applied to the grids and by the ion mass, m_i. Here, V_B is the potential of the body of the thruster and of the inner, screen grid, which is typically 1 to 2.5 kV in present designs. The actual accelerating potential is larger than this because the second (accel) grid is at a negative potential, V_{ac}, of perhaps 200 to 500 V. The outer (decel) grid is at $V_{dec} \sim -50$ V and decelerates the ions. Further deceleration occurs as they enter the space plasma. If the charge on an electron is e, the ion velocity achieved is given by:

$$ve = \sqrt{\frac{2eVB}{mi}} \ .$$

(1)

In reality, this velocity is reduced to a value v_{eff} determined by the propellant utilisation efficiency, η_m, such that $veff = \eta mve$. Equating force to rate of change of momentum, the thrust, T, is given by $T = \dot{m}veff$, where $\dot{m}$ is the total propellant flow rate. If g_o is the acceleration due to gravity at sea level, the SI, denoted by I_{sp}, becomes:

$$I_{sp} = \frac{T}{\dot{m}g_o} = \frac{v_{eff}}{g_o} = \frac{\eta m v_e}{g_o} \ .$$

(2)

With grid separations of the order of 1 mm or less, operation at the few kV level is satisfactory, providing values of v_e for Xe of up to about 65 km/s, and giving an SI of between 3000 and 6000 sec (Martin, 1988). With slightly greater separations and appropriate mounting insulator designs, much higher values of V_B can be used, perhaps of the order of tens of kV. Thus, from equation 1, a factor of 3 to 4 improvement in SI is feasible by this means. However, a problem which must be overcome concerns the penetration of the inter-grid electric field into the discharge chamber plasma. Some penetration as depicted in figure 1b is desirable, but this will become severe at high voltages, influencing the ion trajectories adversely and perhaps causing unacceptable direct impingement on the accel and decel grids. As will be described below, this problem can be overcome by employing a 4-grid system.

Equation 1 shows that v_e is dependent on the propellant employed. Xe was originally chosen, as was mercury in earlier times, for its high atomic mass; this is beneficial in present applications because it provides an excellent momentum transfer and thrust for a given beam current. However, for high SI missions, a low m_i is required. The best that can be achieved is to use hydrogen, but this presents storage difficulties, since very high pressures or extreme cryogenic temperatures are required. However, suitable compounds of hydrogen with carbon and nitrogen exist with average atomic masses when completely dissociated of 4 to 5 atomic mass units (AMUs). Although dissociation must occur within the discharge chamber, the power required is not excessive in terms of the beam energy, and none of the chemical elements involved are harmful to thruster components. Using these compounds and assuming complete dissociation, the SI can be increased by a factor of about 5 compared to Xe.

Thrust and Thrust Density

The thrust and thrust density achievable are also of considerable interest, because spacecraft designers will not be able to make unlimited volume and surface area available for the propulsion system. As mentioned above, the thrust equals the rate of change of momentum so, noting that the beam current $IB = \eta m \dot{m}e / mi$:

$$T = \dot{m}veff = IB\sqrt{\frac{2VBmi}{e}} \ .$$

(3)

Thus, the thrust will increase with V_B, but decrease if the ion mass is reduced. Assuming the factors given above, of 3 to 4 for applied potential and 5 for the propellant, there will be a net decrease of thrust. Compensation for this can be obtained by extracting a larger beam current, which requires a higher plasma density in the discharge chamber and thus a greater discharge power. The limit here is the ability of the grids to pass this increased current. This is determined by the perveance, which can be defined as (Harbour, 1973):

$$\frac{IB}{V_T^{3/2}} = \frac{4\varepsilon_o}{9}\left(\frac{2e}{mi}\right)^{1/2}\frac{ATg}{d^2} \ ,$$

(4)

943

where $V_T = \left(V_B + |V_{ac}|\right)$, ε_o is the dielectric constant of free space, d is the ion acceleration distance, A is the area of the grids, and T_g is the effective transparency of the screen grid. This leads to the definition of a perveance parameter, P_g, which is a characteristic indicating the capabilities of the grid system. Thus

$$P_g = \frac{I_B m_i^{1/2} d^2}{A T_g} = \frac{4 \varepsilon_o \sqrt{2e}}{9} V_T^{3/2} \ .$$

(5)

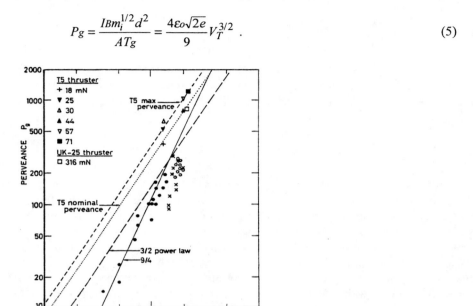

FIGURE 2. Values of P_g as a function of applied potential.

As can be seen in figure 2, a logarithmic plot of P_g against V_T is linear; here I_B is in amps, m_i is in AMU, d is in mm, and A is in m^2. Points from the T5 (Fearn, 1998) and UK-25 (Martin, 1988) Kaufman-type thrusters are shown which are consistent with the expected 3/2 power law. Earlier data (Martin, 1988) also included indicate a lower performance and a trend towards a 9/4 power relationship. From equation 4, I_B can be expected to fall if grids of a given design have to be separated by a greater distance to accommodate higher working potentials. However, increased values are readily accessible at the enhanced accelerating potentials discussed here, within limits set by the penetration of the field into the plasma, thereby minimising any loss of thrust density.

Lifetime and Thruster-Spacecraft Interactions

The requirement to operate at a higher plasma density will cause discharge chamber erosion to increase, leading to a reduced lifetime. This can be compensated for by protecting internal components, such as the baffle disc (figure 1a) with low sputtering rate materials, such as tantalum or carbon, or by manufacturing from those materials. Of course, only the protective route can be used where the magnetic circuit is concerned. As regards the cathodes, their lifetime can be made consistent with mission objectives by selecting the correct size and operating conditions.

The other life-limiting factor is grid erosion. This is due to direct ion beam impingement and to bombardment by slow charge-exchange ions (Fearn, 1993), which cause sputtering. The former problem should be alleviated by the higher axial accelerating electric fields to be used, which give a much lower beam divergence. It may also be possible to reduce the flux of charge-exchange ions by improving η_m. This can be done by operating at a relatively high discharge voltage, a change made possible by protecting internal discharge chamber components from sputtering damage. As regards interactions with the spacecraft, the beam divergence will be less than at present, and the deposition of sputtered material can be reduced if the utilisation efficiency can be improved.

PROPOSED THRUSTER DESIGN

The starting point for the proposed design is the grid system. As the triple-grid concept shown in figure 1a is probably not suitable for very high accelerating potentials, it should be replaced by the 4-grid arrangement depicted in figure 3. This originated in the development of ion accelerators for controlled thermonuclear reaction (CTR) experiments, normally with hydrogen as the working gas (Kim, 1979; Okumura, 1980). The underlying concept is that the ion extraction field between grids A and B is separated spatially from the acceleration zone B-C. Thus field penetration into the plasma is no greater than in lower SI thrusters.

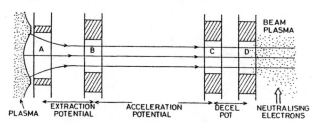

FIGURE 3. The 4-electrode grid system.

As indicated in table 1, some examples of this arrangement have operated at very high accelerating voltages, although these data refer to a pulsed mode dictated by power supply considerations. However, pulses in the case of the 10 cm diameter device were of 10 sec duration. Clearly, power levels in space in the range indicated could be provided only by nuclear reactors. Nevertheless, these results suggest that very high SI thrusters for interstellar missions should be feasible, with very low beam divergence. Grid open area ratios were small, causing operation to be at low efficiency, with severe grid heating in all cases; correcting this deficiency should not be difficult.

TABLE 1. Characteristics of Various CTR Hydrogen Ion Sources.

Authors	Diameter (cm)	Grid form	Open area ratio (%)	Beam energy (keV)	Beam current (A)	Beam power (kW)	Beam divergence (deg)	Current density (mA/cm^2)
Okumura (1980)	10	Flat	31	70	4-7	280-490	1.4	170-190
Ohara (1978)*	12	Flat	40	75	15	1125	0.6	133
Menon (1980)	18	Dished	51	35-65	7-20	245-1300	2	27-79

* Design study

There are many propellant candidates which are liquid at spacecraft temperature, but which can be vaporised easily; it would not be of any concern if dissociation occurred during the vaporisation process. Some are listed in table 2, from which it can be seen that it is reasonable to assume a mean atomic mass (AM) of 4.5 AMU. It would also be possible to employ a pressurised hydrocarbon gas fed by the same type of control equipment as for Xe.

TABLE 2. Possible Propellant Compounds.

Compound	Formula	Melting point (C)	Boiling point (C)	Mean AM
Cyclohexane	C_6H_{12}	6.5	80.7	4.7
Diethylamine	$C_4H_{11}N$	-50	55.5	4.6
Heptane	C_7H_{16}	-90.6	98.4	4.3
Hexane	C_6H_{14}	-95.3	68.7	4.3
Octane	C_8H_{18}	-56.8	125.7	4.4

The current densities given in table 1 suggest that 150 mA/cm^2 should be achievable, with a beam energy of 70 keV. For a 10 cm beam diameter, I_B becomes 11.8 A and the beam power 825 kW. Extrapolation of the curves in figure 2 suggests that this performance is on the very limit of what can be achieved with the 4-grid design, since a 1 mm separation is required between grids A and B in figure 3, together with a potential difference of 5 kV, to reach the value of P_g calculated from equation 5, which is 4.55×10^3.

This beam current indicates the need for an ion flux diffusing towards the grids of nearly 17 A if the open area ratio is 70%. If the design process for the T6 thruster remains valid (Wallace, 1998), this requires an anode current of about $6.5I_B$, which is 109 A. This should be achievable with a single large hollow cathode operating at a keeper potential of about 4 V. Assuming an anode potential of 30 V, the discharge power becomes 3.28 kW, and the electrical efficiency of the thruster, which is defined as beam power divided by total input power, becomes 99.6%. If the mean propellant atomic mass is 4.5, equation 1 gives $v_e = 1.73 \times 10^6$ m/s and, if $\eta_m = 0.85$, the SI is a very attractive 150 000 sec from equation 2. From equation 3, the thrust becomes 0.95 N and the propellant flowrate 0.65 mg/s. If a life of 10 000 hours can be achieved, the quantity of propellant used is 24 kg, and the total impulse provided becomes 3.4×10^7 Ns.

Thruster lifetime cannot be assessed without further work; if necessary, it can be improved by de-rating the performance. The performance values quoted above are such that this would seem to be an acceptable solution to any problems that might arise, such as the temperature of the thruster. Assuming that the complete discharge power must be radiated from the thruster surface of emissivity 0.5, and that this is of the same area as in the T5 design (Fearn, 1998), the external temperature must be about 840 C. While this may be acceptable for most of the constructional materials, provision of adequate insulation for the solenoid windings (figure 1a) will not be easy.

CONCLUSIONS

It has been shown in this paper that the design of gridded ion thrusters can be extended to the range of interest for precursor interstellar missions by using very high ion extraction potentials and hydrocarbon propellants. Simple calculations, based on the design process used for the T5, T6 and UK-25 thrusters, has shown that, in the limit, a 10 cm diameter thruster should be capable of reaching a thrust of nearly 1 N, with a specific impulse of 150 000 sec, an input power of almost 830 kW, and an electrical efficiency exceeding 99.5%. Assuming that a lifetime of 10 0000 hours can be achieved, the propellant consumption is under 24 kg for a total impulse of 3.4×10^7 Ns.

ACKNOWLEDGMENTS

© British Crown Copyright 2000. Published with the permission of DERA on behalf of the Controller of HMSO.

REFERENCES

Fearn, D. G., "Ion Thruster Lifetime Limitations Imposed by Sputtering Processes," International Electric Propulsion Conference Paper IEPC-93-177, (1993).

Fearn, D. G. and Smith, P., "A Review of UK Ion Propulsion - a Maturing Technology," International Astronautical Federation Paper IAF-98-S.4.01, (1998).

Harbour, P. J., et al., "Physical Processes Affecting the Design and Performance of Ion Thrusters with Particular Reference to the 10 cm, RAE/Culham T4 Thruster," American Institute of Aeronautics and Astronautics Paper AIAA-73-1112, (1973).

Kaufman, H. R., "An Ion Rocket with an Electron Bombardment Ion Source," NASA TN-585, (1961).

Killinger, R., Bassner, H. and Muller, J., "Development of an High Performance RF-Ion Thruster," American Institute of Aeronautics and Astronautics Paper AIAA-99-2445, (1999).

Kim, J., Gardner, W. L. and Menon, M. M., "Experimental Study of Ion Beam Optics in a Two-Stage Accelerator," Rev. Sci. Instrum., 50, 201-206, (1979).

Martin, A. R. et al., "A UK Large Diameter Ion Thruster for Primary Propulsion," J. Brit. Interplan. Soc., 41, 167-173, (1988).

Menon, M. M., et al., "Power Transmission Characteristics of a Two-Stage Multiaperture Neutral Beam Source," Rev. Sci. Instrum., 51, 1163-1167, (1980).

Okumura, Y., et al., "Quasi-dc Extraction of 70 keV, 5 A Ion Beam," Rev. Sci. Instrum., 51, 728-734, (1980).

Ohara, Y., "Numerical Simulation for Design of a Two-Stage Acceleration System in a Megawatt Power Ion Source," J. Appl. Phys., 49, 4711-4717, (1978).

Wallace, N. C., Fearn, D. G. and Copleston, R. E., "The Design and Performance of the T6 Ion Thruster," American Institute of Aeronautics and Astronautics Paper AIAA-98-3342, (1998).

A Solar Sail Design For A Mission To The Near-Interstellar Medium

Charles E. Garner,[1] William Layman,[1] Sarah A. Gavit,[1] and Timothy Knowles[2]

[1]*Jet Propulsion Laboratory, California Institute of Technology, Pasadena, Ca 91109*
[2]*Energy Science Laboratories Inc., 6888 Nancy Ridge Dr, San Diego, Ca 92121*

Abstract. Mission concepts to several hundred AU are under study at NASA Marshall Space Flight Center (MSFC) and NASA Jet Propulsion Laboratory (JPL). In order to send a scientific probe beyond the heliopause in a reasonable length of time—no more than 15 yr and preferably 10 yr—the ΔV requirements are approximately 70 km/s. The preliminary results of these mission studies indicate that a solar sail can provide a cumulative ΔV of over 70 km/s to send a probe to a distance of 200 AU from the Sun in under 15 years. This is done by using photon pressure on the sail to shape the trajectory in the inner solar system so that a perihelion of 0.25 AU is achieved. This paper presents the results of a design study for a solar sail to achieve the performance requirements identified in an interstellar probe (ISP) mission study to the near-interstellar medium. The baseline solar sail design for this ISP mission assumes an areal density of 1g/m^2 (including film and structure), and a diameter of ~410 m with an 11-m-wide central opening. The sail will be used from 0.25 to 5 AU, where it will be jettisoned. The total spacecraft module mass propelled by the sail is ~191 kg. The gores of the sail are folded together and wrapped around a small cylinder. Centripetal force is used for sail deployment. The spacecraft is moved off-center with booms for sail attitude control and thrust vector pointing.

INTRODUCTION

Solar sails have been studied in the literature for decades as a novel propulsion system for planetary and interstellar missions. Solar sail propulsion could enable missions never considered possible (McInnes, 1999, Leipold, 1996), such as non-Keplerian orbits around the earth or sun, or exciting commercial applications, such as polar communication satellites. NASA's drive to reduce mission costs and accept the risk of incorporating innovative, high payoff technologies into its missions while simultaneously undertaking ever more difficult missions has sparked a greatly renewed interest in solar sails. Solar sails are now included in National Oceanic and Atmospheric Administration (NOAA), National Aeronautics and Space Administration (NASA), Department of Defense (DOD), Deutsche Forschungsanstalt fur Luft-und Raumfahrt (DLR), and European Space Agency (ESA) technology development programs and technology roadmaps (Garner, 1999).

A solar sail is a large, flat, lightweight reflective surface deployed in space—essentially a large space mirror—that can propel spacecraft without the use of propellant. Propulsion results from momentum transfer of solar photons reflected off of the sail (photons have no rest mass, but they do have momentum). The concept of solar sailing is not new. Tsiolkovsky (Wright, 1992) proposed in 1924 that large spacecraft could be propelled through space using photon pressure, and in the same year Fridrikh Tsander (Wright, 1992) proposed the lightweight solar sail design that is discussed today-a metallized plastic film.

The technical challenge in solar sails is to fabricate sails using ultra-thin films, deploy these structures in space, and control the sail/spacecraft. For reasonable trip times the sail must be very lightweight-from 20 g/m^2 for missions that could be launched in the near-term to 0.1 g/m^2 for far-term interstellar missions. Modern sail designs make use of thin films of Mylar or Kapton® coated with about 500 angstroms of aluminum with trusses and booms for support structure. The thinnest commercially-available Kapton® films are 7.6 μm in thickness and have an areal density (defined as the total material mass divided by the material area) of 11 g/m^2. A propulsion trade study (Gershman, 1998) identified the benefits and sail performance required to provide significant advantages over other propulsion technologies. The study concluded that sails with areal densities (defined as the total sail mass divided by the sail area) of about 10 g/m^2 are appropriate for some "mid-term missions" such as a Mercury Orbiter (Leipold, 1996) or small spacecraft positioned between the sun and the earth. More "far-term" missions such as an Asteroid

CP504, *Space Technology and Applications International Forum–2000*, edited by M. S. El-Genk
2000 American Institute of Physics 1-56396-919-X

Rendezvous/Sample Return require sails with an areal density of 5-6 g/ m^2 and films with a thickness of approximately 1-2 μm. More advanced missions require sails with areal densities of under 3 g/m^2 for positioning spacecraft in non-Keplerian orbits or 1 g/m^2 for fast trip times to 200 AU.

Practical experience with solar sails is very limited. In the 1980's the World Space Foundation fabricated and deployed on the ground a 20-m (400 m^2) sail, and fabricated a 30-m (900 m^2) sail with an areal density of approximately 65 g/m^2 that was stowed in a deployment structure (Garner, 1999). In 1993 Russia deployed a 20-m-dia spinning disk solar reflector based on their Columbus 500 solar sail design with an areal density estimated to be 22 g/m^2 from a Progress resupply vehicle to provide sunlight to arctic regions in Russia. The sail, called Znamya 2, consisted of eight pie-shaped panels fabricated from 5μm-thick aluminized PETF film (a Russian version of Mylar) with no supporting structure. Deployment took three minutes; the sail remained attached to the Progress vehicle, which provided attitude control. In February 1999 the 25-m-dia Znamya 2.5 space reflector experiment failed due to a mission operations error and is discussed in more detail in this paper.

The Comet Halley Rendezvous mission studied by JPL in 1977 required a solar sail that had a total surface area of approximately 624,000 m^2 (790 m on a side) and weighed over 2,000 kg (Friedman, 1978); this enormous structure came to symbolize solar sail propulsion in the 1970's and 1980's. Despite advantages that could be obtained from using sails, deployment and control of sails of this magnitude in size and mass present a significant technical challenge and inhibit their application to NASA missions.

Recently NASA has encouraged programs to reduce the size and mass of spacecraft used for robotic exploration of the solar system (JPL, 1995). Spacecraft with masses below 100 kg are being studied for performing challenging missions, and microspacecraft technology is being developed that may result in robotic spacecraft with masses of 10 kg or less (Jones, 1991). Solar sail propulsion is synergistic with the new NASA approach to accomplish missions cheaper because the use of solar sails allows the use of smaller, cheaper launch vehicles. Solar sails have been studied in the literature for decades as a novel propulsion system for planetary and interstellar missions. Solar sail propulsion could enable missions never considered possible (McInnes, 1994; McInnes, 1994; McInnes, 1998; McInnes, 1998), such as non-Keplerian orbits around the earth or sun, or exciting commercial applications, such as polar communication satellites.

The basic idea behind solar sailing is simple, but there are difficult engineering problems to solve. The technical challenges in solar sailing are to fabricate sails using ultra-thin films and low-mass booms; package sails in a small volume; deploy these light-weight structures in space; and, understand the dynamics and have the ability to control of the sail/spacecraft. The solutions to these challenges must be demonstrated in space before solar sail propulsion is considered viable for any mission.

The feasibility of solar sail propulsion had been greatly enhanced by two recent developments: the successful deployment of an inflatable antenna from the space shuttle (Freeland, 1996), and reduction in spacecraft mass and sail size. For example, studies indicate that a main belt asteroid rendezvous and sample return mission (Leipold, 1995) can be accomplished within a seven-year trip time using a solar sail with an area of 90,000 m^2 (300 m on a side). Alternatively, a Geomagnetic Storm Warning mission (West, 1996) that would maintain a spacecraft at 0.98 astronomical units can be performed using a solar sail with an area of only 4,490 m^2 (67 m on a side).

NASA's Office of Space Science has developed four major themes for space exploration and a portrait of missions that are representative of the key technological challenges and scientific objectives that must be addressed. Two of these themes, Exploration of the Solar System and the Sun-Earth Connection, have identified solar sail propulsion as a technology that will enable or enhance portrait missions.

Progress in developing ultra-thin materials and lightweight carbon-fiber structures has made solar sails a feasible technology for high delta-velocity missions to Mercury, the outer planets and the local interstellar medium. Programs whose goals are to make solar sails a reality are now in place or planned. NASA programs include activities at Langley Research Center (LaRC), Jet Propulsion Laboratory (JPL), Marshall Space Flight Center (MSFC), Goddard Space Flight Center (GSFC), and the NASA Institute for Advanced Concepts (NIAC). There are National Oceanic and Atmospheric Administration (NOAA) and Department of Defense (DOD) activities as well.

The technical challenge in solar sails is to fabricate sails using ultra-thin films, deploy these structures in space, and control the sail/spacecraft. For reasonable trip times the sail must be very lightweight-from 20 g/m^2 for missions that could be launched in the near-term to 0.1 g/m^2 for far-term interstellar missions. Modern sail designs make use of thin films of Mylar or Kapton® coated with about 500 angstroms of aluminum with trusses and booms for support structure. The thinnest commercially-available Kapton® films are 7.6 μm in thickness and have an areal density (defined as the total mass divided by the area) of 11 g/m^2. A Propulsion Trade Study (5) performed in 1998 identified the benefits and sail performance required to provide significant advantages over other propulsion technologies. The Study concluded that sails with areal densities (defined as the total sail mass divided by the sail area) of about 10 g/m^2 are appropriate for some "mid-term missions" such as a Mercury Orbiter or small spacecraft positioned between the sun and the earth. More "far-term" missions such as an Asteroid Rendezvous/Return require sails with an areal density of 5-6 g/m^2 and films with a thickness of approximately 1-2 μm. More advanced missions require sails with areal densities of under 3 g/ m^2 for positioning spacecraft in non-Keplerian orbits or 1 g/m^2 for fast trip times to 200 AU (McInnes, 1999). Ultimately interstellar flyby and rendezvous missions require sails with areal approaching 0.1 g/m^2 (Gavit, 1999).

Among the challenging missions potentially enabled by solar sails are missions to the near-interstellar medium and interstellar precursor missions. Concepts for missions to several hundred AU are under study at NASA Marshall Space Flight Center (MSFC) and NASA Jet Propulsion Laboratory (JPL). The ΔV requirements are approximately 70 km/s to send a scientific probe beyond the heliopause in under 15 yr. The preliminary results of these mission studies indicate that a solar sail can provide a cumulative ΔV of approximately 70 km/s to send a probe to a distance of 200 AU from the Sun in under 15 years. This is done by using photon pressure on the sail to shape the trajectory in the inner solar system so that a perihelion of 0.25 AU is achieved (Gavit, 1999). Progress in developing ultra-thin materials and lightweight carbon-fiber structures has made solar sails a feasible technology for high delta-velocity missions to Mercury, the outer planets and the local interstellar medium. This paper presents the results of a design study for a solar sail to achieve the performance requirements identified in an ISP mission study (Gavit, 1999) to the near-interstellar medium.

MISSION DESCRIPTION

In 1999 the Jet Propulsion Laboratory (JPL) conducted a study in support of NASA Strategic Planning activities whose purpose was to develop a baseline architecture and technology list for the proposed "Interstellar Probe (ISP)" mission to the heliosphere and local interstellar medium with a launch date by 2010. This study was led by the JPL Interstellar and Solar Sail Technology Program with support from other NASA and government centers, private industry, and academia as needed. A full mission description is provided in the proceedings of the STAIF 2000 conference (Liewer, 2000). A brief summary of this mission is included to provide background information for sail requirements.

In order to send a scientific probe to the heliopause and beyond in a reasonable length of time—no more than 15 years—the ΔV requirements are such that advanced propulsion technology is required. To meet that challenge, the study baselined a solar sail propulsion system which uses the pressure of sunlight on a very large, light weight, shaped fabric to "sail" the spacecraft outside of our solar system. The mission baselined includes launching the spacecraft to a C3 = 0 using a Delta II rocket (7425). During launch, the sail is stowed inside of a canister. After the launch event, the 410 m diameter spinning sail is deployed and the deployment device is jettisoned. Solar photon pressure on the sail is then used to decrease the spacecraft velocity, such that the spacecraft swings into the inner solar system and around the sun with a perihelion of 0.25 AU. After accelerating the spacecraft away from the sun, the sail is then jettisoned at approximately 5 AU. This proposed design provides a cumulative ΔV of >70 km/s, propelling the spacecraft to a distance of 200 AU from the sun in less than 15 years.

The spacecraft, exclusive of the sail and its deployment hardware, can best be described as a "flying antenna". The 2.7 m rigid antenna functions as the main structure of the spacecraft with at least 12 instruments arrayed along its rim and the Ka-band telecommunications subsystem, Reaction Control System (RCS), and Alkali Metal Thermal to Electric Converters (AMTECs) located at the base of the antenna. At launch, this bus is attached to a simple sail control system, which is attached to a large deployment canister and the mechanisms for unfurling the solar sail. Once the solar sail is deployed, the majority of deployment mechanisms are jettisoned to minimize the mass that must be accelerated. While sailing, the movement of the sail is controlled by moving the spacecraft on a rail, which changes the center of mass of the sailcraft with respect to it's center of pressure. In its final configuration, the spacecraft is in a very slow spin with long deployed instrument booms. The spacecraft mass of 190.8 kg includes approximately 30.5 kg for sailcraft control hardware and a solar sail container adapter structure. The estimated accelerated mass (after sail deployment hardware is ejected) is estimated to be approximately 313 kg, which includes approximately 122.6 kg for the sail. A summary of key mission parameters is given in Table 1.

TABLE 1. Key Mission Parameters.

Launch vehicle	Delta IV or smaller
Technology cutoff date	2007
Launch date	2010
Arrival at 200 AU	< 15 years
Spacecraft mass, kg	190.8
Solar sail mass, kg	122.6
Total accelerated mass, kg	~ 313

SAIL DESIGN

The solar sail design is divided into the following categories:

- Requirements Summary,
- Trajectory Summary,
- Sail Structural Design,
- Gore Assemblies,
- Sail Deployment,
- Mass Summary,
- Sail Control

Requirements Summary

Sail design details to meet the ISP mission requirements are provided in this section. In order to establish the baseline mission design, an assessment of the key system parameters (distance of closest solar approach, sail radius and areal density, and time of flight) was performed. The distance of closest solar approach drives the spacecraft thermal and mechanical design and the time of flight. The sail size and areal density affects the following: time of flight, packaging and deployment, control and navigation, structural design, and materials and fabrication.

Trade studies (Gavit, 1999) indicate that the sail loading (defined as the total sail and spacecraft mass to be accelerated by the sail divided by the sail area) required to meet the mission requirements listed above is approximately 2.6 g/m^2. This requirement drives the design of the sail, including sail mass, sail areal density and sail diameter. The baseline solar sail design for this ISP mission assumes an areal density of 1g/m^2 (including film and structure), and a diameter of ~410 m with an 11-m-wide central opening. The sail will be used from 0.25 to 5 AU, where it will be jettisoned. The gores of the sail are folded together and wrapped around a small cylinder. Centripetal force is used for sail deployment, and a spin rate of 0.09 rad/s (0.9 rpm) is required for tension in the sail film. The spacecraft is moved off-center on a rail for sail attitude control and thrust vector pointing. A summary of sail requirements is provided in Table 2.

TABLE 2. Key Sail Design Parameters.

Sail diameter, m	410
Areal density, g/m^2	1
Sail loading, g/m^2	2.6
Perihelion radius, AU	0.25
Solar sail mass, kg	122.6
Spacecraft mass, kg	190.8
Total accelerated mass, kg	313
a_c, mm/s^2	3.0
Spin rate, rpm	0.9
Spin axis precession rate, deg/day	10

Trajectory Summary

A solar sail is used to increase the energy of the heliocentric orbit from a negative value (an elliptic orbit) to a positive value (a hyperbolic escape orbit). For any given delta-V, the energy change is proportional to the velocity at which the delta-V is applied and is maximized when the delta-V is parallel to the velocity vector. Furthermore, the acceleration of a solar sail increases when it gets closer to the Sun. Thus in order to maximize the energy gain for the ISP mission, the sail is first used to reshape the heliocentric orbit to lower perihelion, even at the expense of an initial energy loss, in order to increase both the delta-V which the sail supplies and the velocity at which that delta-V is applied. This increases the effectiveness of the sail so much that the heliocentric energy of the spacecraft can be increased from negative to positive (with an adequate departure velocity) in just one perihelion passage.

This baseline trajectory was optimized using a model of the sail as described below. With this assumption, the flight time is a function of only the characteristic acceleration, a_c, and the perihelion radius. A higher a_c or lower radius would shorten the flight time. The study assumed that the sail effectively reflects most of the light incident on it and that the rest of the light is transmitted through the sail without providing any thrust The effective thrust available as a function of the sun incidence angle (angle of the sun relative to the sail normal) was calculated with the following results: 91% at 0°, 90 % at 10°, 89% at 20°, 84% at 30°, 78% at 40°, and 60% at 50°. In this model the maximum incidence angle for which thrust can be achieved is 55°. The spin axis of the sail must be precessed up to 10 degrees per day for this trajectory. Once the solar sail is jettisoned, there is a separation maneuver performed to prevent the sail from obscuring the spacecraft's line of sight to the Earth. This maneuver has an insignificant effect on the trajectory. A plot of the trajectory as a function of time is given in Figure 1, and key angles of the spacecraft relative to the earth and sun are given in Figure 2.

Sail Structural Design

The solar sail structure includes the solar sail deployment module, the sail including the sail cylinder, and the launch vehicle interface structure. For launch, the spacecraft is mounted atop the control assembly and 1.5 m long x 1.5 m-diameter sail cylinder, which in turn is mounted on top of the sail deployment module housing the sail deployment hardware and booms. The structural load path passes from the 1.2-m launch-vehicle adapter through the sail deployment module, sail cylinder walls, and control assembly, picking up the ring at the base of the 2.7-m antenna (Figure 3).

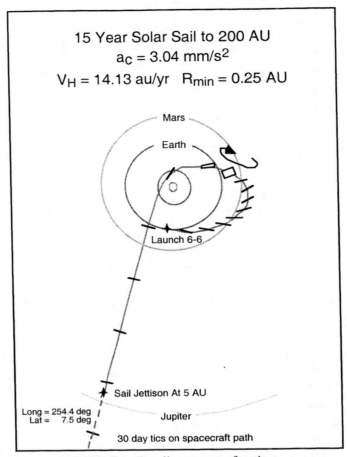

15 Year Solar Sail to 200 AU

$a_C = 3.04 \text{ mm/s}^2$

$V_H = 14.13 \text{ au/yr} \quad R_{min} = 0.25 \text{ AU}$

Mars

Earth

Launch 6-6

Sail Jettison At 5 AU

Long = 254.4 deg
Lat = 7.5 deg

Jupiter

30 day tics on spacecraft path

FIGURE 1. Baseline spacecraft trajectory.

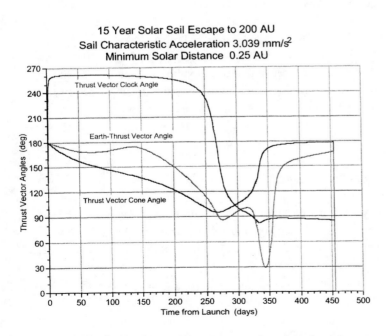

15 Year Solar Sail Escape to 200 AU
Sail Characteristic Acceleration 3.039 mm/s^2
Minimum Solar Distance 0.25 AU

Thrust Vector Clock Angle

Earth-Thrust Vector Angle

Thrust Vector Cone Angle

Thrust Vector Angles (deg) vs *Time from Launch (days)*

FIGURE 2. Thrust vector angles vs. time from launch.

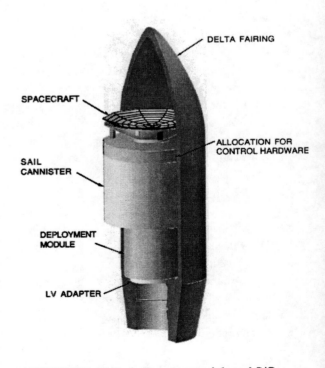

DELTA FAIRING

SPACECRAFT

ALLOCATION FOR
CONTROL HARDWARE

SAIL
CANNISTER

DEPLOYMENT
MODULE

LV ADAPTER

FIGURE 3. Sail, deployment module and S/C in Delta IV payload fairing.

Deployment hardware is located in the 1.5m x 1.5m deployment module to minimize the mass that must be accelerated by the deployed sail. Three each 10-m-long orthogonal booms are deployed from the deployment module, and thrusters at the ends of the booms are fired, spinning up the spacecraft and sail cylinder in a controlled fashion. The sail is comprised of 6 pie-shaped triangles which are folded into gores. At launch, the gores are wrapped around a 1.5 m long x 1.5 m diameter sail cylinder with an outer end diameter of approximately 2.2 m to allow for sail packaging. The sail cylinder is surrounded by a bellyband held in place with a marmon clamp. As the spacecraft spins, the bellyband is jettisoned and the sail is released by sequentially releasing restraining devices which allow the gores to slowly unfurl in a controlled manner to prevent snagging and collisions. Tethers holding the center of each gore to the sail cylinder are then played out to form a "wheel rim" which is 410 m in diameter. Each gore segment is then unfurled into a triangle by pulling tethers which connects each sail segment tip to the sail cylinder. After the deployment sequence is completed, the plane of the sail runs along the mid-section of the sail cylinder. After sail deployment, the sail deployment module, including the booms, spin-up assembly , and the launch vehicle interface structure are jettisoned from the spacecraft. Cartoons demonstrating the sail deployment sequence are shown in Figure 4, and the spacecraft with fully-deployed sail is shown in Figure 5.

Sail Film And Gore Design

Sail design for the ISP mission is complicated by the need for an extremely low areal density and high sail thermal loads due to sail operations within 0.25 AU of the sun. The conventional concept of a solar sail consists of thin sheets of Kapton® or other polyimide, metallized on one side to provide a reflective surface and the other an emissive surface, and seamed together to form the large mirror-like sail . Kapton® provides the mechanical strength needed to carry loads through the sail, including handling, deployment, and photon pressure, and serves as a base to support the thin reflective and emissive metal layers. The reflector, typically aluminum, provides photon pressure which accelerates the sail, and the emissive layer allows for temperature control.

In this study and areal density of 0.5 gm/m² was allocated for the sail fabric and 0.5 gm/m² for the rest of the sail structure to achieve a total sail areal density of 1 gm/m². Polyimide densities are typically about 1.4 gm/cm³, therefore to achieve a film areal density of 0.5 gm/m² requires a polyimide layer no more than 0.35 μm thick. If allowance is made for seaming, ripstops, and other structures then the polyimide thickness drops to approximately 0.2 μm. In addition, sail operations near 0.25 AU of the sun may result in sail temperatures at wrinkles and folds that exceed the glass transition temperature (~873 K) of polyimides (Wright, 1992). The requirements for the ISP sail mission stimulated investigations into alternatives to conventional sail material designs.

Recently a new material was proposed for use on solar sails (Knowles, 1997). Called a microtruss fabric, it is a porous, thick fabric with discontinuous carbon fibers joined to one another along nodes. Porosity allows light weight, and thickness allows intrinsic stiffness and strength. These fabrics can potentially be used at temperatures exceeding 1800 K or more (Knowles, 1997).

Carbon was selected because it has the following properties:

- High temperature tolerance,
- High emissivity,
- High tensile strength,
- Good radiation tolerance,
- Low Coefficient of Thermal Expansion (CTE),
- Minimal tear initiation capability, and
- Low areal density.

An overview of these parameters for carbon is given in Table 3.

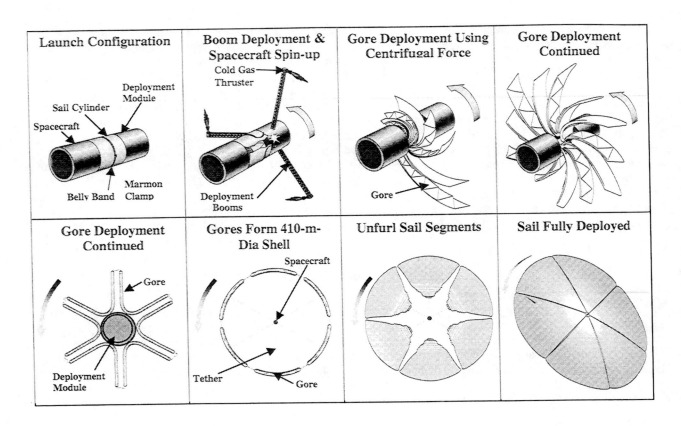

FIGURE 4. Sail deployment sequence.

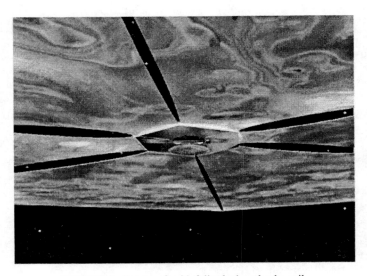

FIGURE 5. Spacecraft with fully-deployed solar sail.

TABLE 3. Carbon Material Properties.

Property	Value
Temperature range (degrees K)	70-2000
Hot spot temperature (degrees K)	753
Tensile strength at 300 K (MPa)	2205
Tensile strength at 500 K (MPa)	2205
Gamma radiation tolerance (Rad)	Excellent
UV radiation tolerance (Rad)	Good
CTE (per degrees C)	3×10^{-6}
Tear initiation at 550 K (Nt/μm)	3
Emissivity	0.4-0.9

The carbon microtruss fabric consists of 1 μm fibers and 10 μm fibers and is designed to support the aluminum reflective surface and the nanotube emissive surface (Figure 6). Nanotubes made from carbon provide for a high emissivity surface. A tension load of 6925 Pa ($\approx$1 PSI) in the aluminum film is assumed to remove wrinkles. The maximum microtruss span that can be bridged with aluminum film can be calculated using Roark's formula. For a 250 Å thick aluminum film subjected to a load of 6925 Pa, a 563 μm maximum span is achievable. Aluminum was selected as the reflective material because aluminum meets the temperature requirements, there is a substantial technology infrastructure for aluminum deposition and adhesion, aluminum is cost effective, and aluminum is non-toxic. For the purpose of this study, 27 μm gaps formed by the 1 μm fibers were assumed to provide margin, resulting in 74,075 gaps per square meter to support the aluminum film and nanotubes. This margin allows for ground handling of microtruss sail fabric. Samples of carbon mictrotruss fabrics have been fabricated at Energy Science Laboratories Inc. in San Diego, Ca (JPL 1998). A photograph of a microtruss fabric is shown in Figure 7 and a photograph of a microtruss fabric covered with a reflective aluminum layer is shown in Figure 8.

Sail Segments

Continuous rolls of 1-m-wide microtruss fabric consisting of 1-10 micron fibers, nanotube fibers, and the aluminum reflective film are assembled into sail segments supported by a net of 100 μm carbon ropes (Figure 9). The sail segment width is limited by the manufacturability and handleability of the microtruss fabric. Carbon fiber spun into 100 μm ropes of continuous length (no breaks) carry tension loads within the sail gores. Tension in the sail film is necessary to provide structural rigidity for handling, deployment, and flight, as well as to remove wrinkles which create hot spots and non-uniformity in the reflective surface. Each 100 μm rope is capable of sustaining a tension load exceeding 23.6 Nt. A continuous net of 100 μm carbon ropes spaced to approximately 14 per square meter are allocated to support the sail. The ropes have small carbon hooks on the side facing the microtruss fabric to affix the fabric to the ropes. A summary of elements which contribute to the thickness and areal density of each sail segment are given in Table 4.

Gore Assemblies

This study has baselined a 410 m diameter circular sail with an 11 m diameter opening at the center. It includes six pie-shaped gores which are $\approx$ 205 m long and are separated at a maximum distance of 7 m midway along the spokes. The separation distance between gores is driven by the desired sail fill fraction of ~ 90%, which is defined as the total actual physical area of the sail gores divided by the area subtended by the sail. The gores consist of approximately 199 ea, 1-m wide continuous rolls of microtruss fabric sail segments atop a net of 100 μm carbon ropes, and connected via hooks in the ropes. However, where folds are required, segment to segment connections are made using fibrous tape (Knowles, 1999). This planar array of tape consists of a ladder-like net of carbon fibers that connect to the 100 μm ropes or the microtruss fabric via a network of many interlocking, curled carbon fibers akin to Velcro®. The tape will be low density, extremely lightweight, and deformable in shear. This design has the advantage that it has built in ripstops, and can be folded without damage since folds are made along the carbon lines or the planar tape. The microtruss fabric connects to the 100 μm carbon net and fibrous tape as shown in Figure 10.

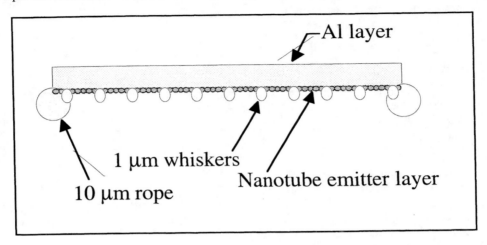

FIGURE 6. Schematic diagram of carbon fabric for the ISP sail.

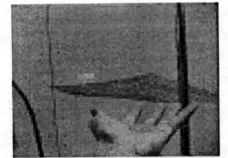

FIGURE 7. ESLI microtruss fabric, 1 gm/m².

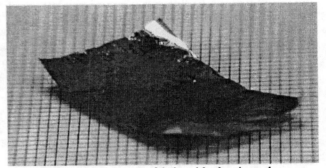

FIGURE 8. Microtruss fabric with aluminum layer.

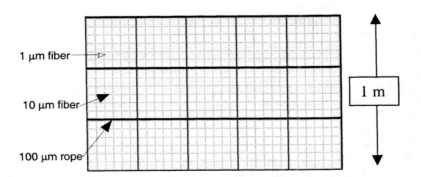

FIGURE 9. Top view of a continuous segment of sail material.

TABLE 4. Sail Segment Thickness and Areal Density.

Element	Thickness	Areal Density (g/m²)
Carbon Nano-Tube Layer	100 Å	0.003
Aluminum Layer	250 Å	.07
Carbon Whisker Net	1 μm	0.1
10 μm carbon rope	10 μm	0.1
100 μm carbon rope	100 μm	0.1
Contingency	NA	0.187
TOTAL	20 μm	0.5

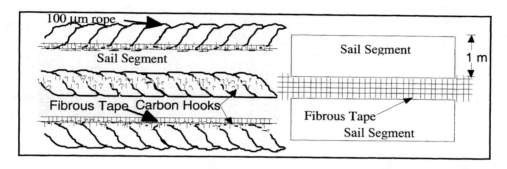

FIGURE 10. Gore seaming design. The left side shows how sail segments attach to the 100 μm ropes with velcro-like fibers; the right side shows folds between adjoining sail segments are made on the fibrous tape.

For packaging, deployment and control it is desirable to have a sail with the minimum sail diameter. The fill fraction can be increased by reducing the gap between gore edges, introducing rounded gore edges on the outer periphery (rounded gore edges will subtend more area than flat gore edges) and by adding mass to the perimeter of the gores to introduce radial tension into gores with rounded edges. The disadvantage of this design is the need for additional mass for these weights. A design study to optimize the trade between the amount of perimeter mass vs. fill fraction was not performed. For purposes of this study a fill fraction of 90% was assumed.

To reduce the gap between gores the radial tension must be increased (to "flatten" the edge from a rounder, Znamya-like gore) with masses located at the tip of each gore, or by increasing the spin rate. An optimization between the tip mass and spin rate was not performed, however a maximum spin rate under 1 round per minute (RPM) was assumed, for control reasons. This then drives the amount of tip mass required to flatten the edges of the gore. Each gore has 2 tip masses, located at each outer tip, for a total of 12 tip masses for the sail.

To obtain the maximum reflection capabilities the radial and tangential stresses in the sail should be equal and constant along the radius of the sail. Tip masses coupled with perimeter weights distributed along the outer perimeter of the sail gores are required to assure a minimum film stress of 6925 Pa (1 PSI). Perimeter weights, called "rim masses", are located along the perimeter of the curved outer edges of the gores. Since the tensioning characteristics of the microtruss fabric are unknown, we assume here that they are identical to an equivalent 0.33 μm Kapton® film which meets the film areal density requirement. It is assumed, based on the structure of the carbon microtruss fabric, that the actual stress required in this fabric will be less compared to Kapton® because the carbon fabric has stiffness intrinsically built into the fabric. The gores are connected to the sail cylinder using three ea, 1000 μm carbon tethers which run between the gores and the sail cylinder at the other end. Because there are sections of fabric that require additional strength (e.g.to provide reinforcement near the 11-m hole and yoke plates),

6 kg of additional mass has been allocated to the sail for additional material support such as ropes and plates. Diagrams of a section of a gore segment at the periphery and near the center are shown in Figure 11 and Figure 12 respectively.

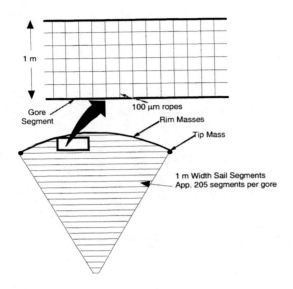

FIGURE 11. Blow-up of a sail segment.

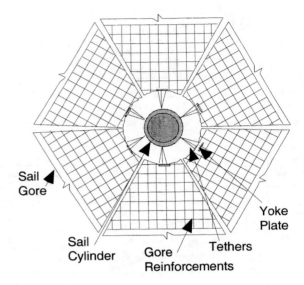

FIGURE 12. Blow-up of center section of the sail.

Sail Deployment

Prior to deployment, the sail is wrapped around the sail cylinder. The sail cylinder includes the following elements: central structure (1.5 m diameter X 1.5 m long, reinforced Al honeycomb); miscellaneous sail cylinder structures such as gore locks, rollers and tethers; undeployed sail with yoke plates, restraining plates, bolts and pyro devices; bellyband and marmon clamp to contain the sail prior to deployment.

To minimize the mass required to be accelerated by the sail, most of the sail deployment hardware is packaged in the Deployment Module which is ejected after deployment is complete. The Deployment Module includes the following elements: reaction control system (RCS) thrusters; three deployable booms (to provide the RCS thrusters a moment arm); pyrotechnic devices (for releasing gore restraining straps, cutting deployment tethers, and separating the deployment module from the sail canister); pulleys and motors (to deploy the gores); batteries (for electrical power to deploy the sail); and on-board computer (to control the sail deployment).

In their launch configuration, the gores are connected to each other at each tip, along their edges at various locations, and to the sail cylinder with 1000 µm tethers. The gores are Z-folded into pleats 1 m in width, which are then sandwiched between the yoke plates and restraining plates (Figure 13). The Z-folded gores are then folded in half (Figure 14) and wrapped around the sail cylinder. Each gore is then held to the cylinder with straps crossing every 10 meters or so, and the entire sail assembly is held down using a bellyband (Figure 15).

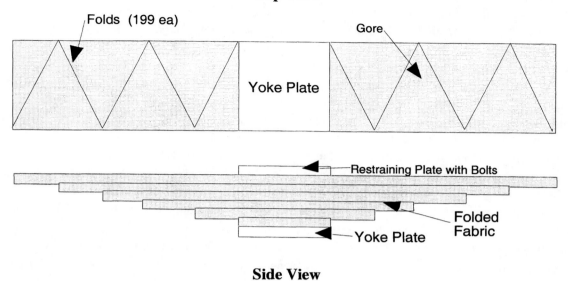

Top View

Folds (199 ea)

Gore

Yoke Plate

Restraining Plate with Bolts

Folded Fabric

Yoke Plate

Side View

FIGURE 13. Diagram of folded gores.

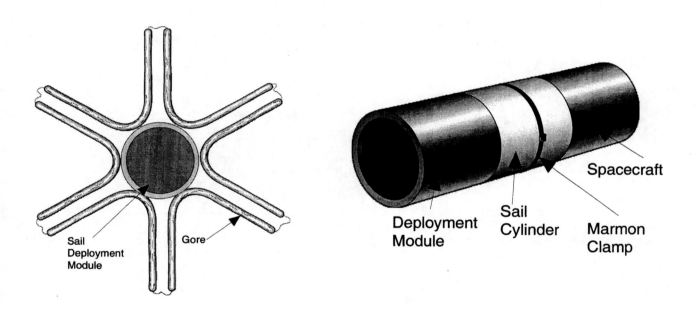

Sail Deployment Module

Gore

Deployment Module

Sail Cylinder

Marmon Clamp

Spacecraft

FIGURE 14. Each gore is folded in half before being wrapped around the sail cylinder.

FIGURE 15. Sail cylinder, deployment module, and spacecraft

The sail deployment sequence is shown in Figure 4. A discussion of this sequence is discussed below.

- The deployment module spins up the sail cylinder to an appropriate speed.
- The marmon clamp is cut and the bellyband securing the sail film is jettisoned.
- Pyros in the deployment structure are fired and motors begin retracting the gore restraining straps. The straps run over the sail film and wind up on rollers inside the deployment structure. Each gore unfurls in a "V" shape, restrained at the centers by tethers connected to yoke plates at the center of each gore. Sandwiching plates on the other side of the gore fabric form a "gore sandwich" with the fabric between the yoke plates and sandwiching plates. The tips of the gores are connected to adjoining gores with short tethers.
- Tether locks in the sail cylinder are opened along with actuation of motors in the deployment structure that release the straps securing the gores to the sail cylinder.
- Each gore is played out such that the 6 gores form a circle 1 m wide and 410 m in diameter. In this way the gores are deployed out of the sail cylinder in a controlled manner with less risk of gores shredding or entangling adjoining gores. The deployment module adjusts the spin rate to approximately 0.45 RPM.
- Pyros on the restraining plates fire, cutting bolts that release the restraining plates. Motors in the deployment module reverse to pull on the tethers between the yoke plates and the sail cylinder, pulling the gores inward.
- After unfolding the gores, tether locks in the sail cylinder lock, and pyros in the deployment structure fire, cutting the tethers between the deployment structure and sail cylinder. At the conclusion of this sequence, the spin rate is 0.9 RPM.
- Pyros fire which separate the deployment module from the sail cylinder. The sail is fully deployed.

Mass Summary

The mass of the sail (including all sail hardware carried by the sailcraft after the sail is deployed except for control) is ~ 122.6 kg for a total areal density of 1 g/m^2. A mass breakdown of the sail components is given in Table 5. Control hardware is included in the spacecraft mass allocation.

TABLE 5. Sail System Mass Summary.

System	Element	Mass (kg)
Sail	Sail fabric mass	61.4
	Cylinder, brackets, braces, rollers	15
	Seaming tapes, gore reinforcements	12
	Rim and tip masses	13
	Yoke plates, tether locks	9
	Gore deployment tethers	0.2
	Miscellaneous structures	12
	TOTAL	122.6
Deployment Structure	LV structure	30
	10-m booms, 3 ea	30
	Discarded sail structure	109
	Launch release mechanisms	15
	RCS to spin up sail	36
	Deployment Structure SubTotal	220
	Contingency (30%)	66
	Deployment Structure Total	286
Sail System Total		**408.6**

Sail Control

The spacecraft module and sun shade are centered in the 11-m-diameter central aperture of the sail. Attitude control and thrust vector pointing of the sail is provided by moving the spacecraft mass relative to the system center-of-pressure. This is accomplished by moving the spacecraft on a 2.7 m rail to precess the sailcrafts's spin rate by 10 degrees per day. The spacecraft can move up and down the rail, and the rail, on bearings that decouple the sail cylinder spin and spacecraft spin rates, can rotate to any position required for full sailcraft control.

ACKNOWLEDGMENTS

The authors thank the members of the NASA Interstellar Probe Science and Technology Definition Team whose efforts resulted in the material for this paper. The research described in this paper was carried out by the Jet Propulsion Laboratory, California Institute of Technology, under a contract with the National Aeronautics and Space Administration.

REFERENCES

Freeland, Robert et.al., "Inflatable Antenna Technology With Preliminary Shuttle Experiment Results And Potential Applications," 18th Annual Meeting and Symposium, Antenna Measurement Techniques Association, Seattle, Wa, Sept. 30-Oct. 3, 1996.

Friedman, L. et. al., "Solar Sailing-The Concept Made Realistic," AIAA-78-82, 16th AIAA Aerospace Sciences Meeting, Huntsville, Al, January 1978.

Garner, C.E. et. al., "A Summary of Solar Sail Technology Developments and Proposed demonstration Missions," AIAA-99-2697, 35th AIAA Joint Propulsion Conference, June 20-24, 1999.

Gavit, S.A., "Interstellar Probe Mission Architecture and Technology Report," Internal JPL Document, 1999.

Gershman, R. and Seybold, C., "Propulsion Trades for Space Science Missions, IAA-L.98-1001", 3rd IAA International Conference on Low-Cost Planetary Missions, Pasadena, April 1998.

Jones, R.M. et al., International Astronomical Federation Paper IAF-91-051, 1991.

JPL Document, "Dual Solar Sail Main Belt Asteroid Sample Return," Manfred E. Leipold, December 19, 1995.

JPL Purchase Order 1202226, December 16, 1998.

JPL Strategic Plan, JPL Internal Document, JPL 400-459, April 1995.

Knowles, Timothy, personal communication to Charles Garner, October 1997.

Knowles, Timothy, personal communication to Charles Garner, May 1999.

Leipold, M. et. al., "Mercury Sun-Sunchronous Polar Orbiter With A Solar Sail," Acta Astronautica V. 39, No. 1-4, pp. 143-151, 1996.

Liewer, P.C., Mewaldt, R.A., Ayon, J.A., and Wallace, R.A., "NASA's Interstellar Probe Mission," Space Technology and Applications International Forum (STAIF-2000), Jan. 30-Feb. 3, 2000, Albuquerque, NM.

McInnes, C.R.: 'Artificial Lagrange Points for a Non-Perfect Solar Sail'," Journal of Guidance, Control and Dynamics, Vol. 22, No. 1, pp.185-187, 1999.

McInnes, C.R.: 'Mission Applications for High Performance Solar Sails', IAF-ST-W.1.05, 3rd IAA Conference on Low Cost Planetary Missions, California Institute of Technology, Pasadena, 27th April - 1st May 1998.

McInnes, C.R.: "Artificial Lagrange Points for a Non-Perfect Solar Sail," Journal of Guidance, Control and Dynamics, Vol. 22, No. 1, pp.185-187, 1999.

McInnes, C.R., McDonald, A.J.C., Simmons, J.F.L. and MacDonald, E.W.:"Solar Sail Parking in Restricted Three-Body Systems," Journal of Guidance, Dynamics and Control, Vol. 17, No. 2, pp. 399-406, 1994.

McInnes, C.R.: "Dynamics, Stability and Control of Displaced Non-Keplerian Orbits," Journal of Guidance, Control and Dynamics, Vol. 21, No. 5, pp. 799-805, 1998.

McInnes, C.R.: "Mission Applications for High Performance Solar Sails," IAF-ST-W.1.05, 3rd IAA Conference on Low Cost Planetary Missions, California Institute of Technology, Pasadena, 27th April - 1st May 1998.

McInnes, C.R., McDonald, A.J.C., Simmons, J.F.L. and MacDonald, E.W.: "Solar Sail Parking in Restricted Three-Body Systems," Journal of Guidance, Dynamics and Control, Vol. 17, No. 2, pp. 399-406, 1994.

West, John, NOAA/DOD/NASA Geostorm Warning Mission, JPL Int. Document, JPL D-13986, October 18, 1996.

Wright, J. L., Space Sailing, Amsterdam, Gordon and Breach Science Publishers, 1992, pp. xi.

Mini-Magnetospheric Plasma Propulsion (M2P2): High Speed Propulsion Sailing the Solar Wind

Robert Winglee[1], John Slough[2], Tim Ziemba[2], and Anthony Goodson[3]

[1] Geophysics Program, University of Washington, Seattle WA 98195-1650,
ph: 1-206-685-8160, e-mail: winglee@geophys.wasington.edu
[2] Department of Aeronautics and Astronautics, University of Washington, Seattle WA 98195-2250
[3] The Boeing Corporation, P. O. Box 3999, MS 84-33, Seattle WA 98124

Abstract. Mini-Magnetospheric Plasma Propulsion (M2P2) seeks the creation of a magnetic wall or bubble (i.e. a magnetosphere) that will intercept the supersonic solar wind which is moving at 300 - 800 km/s. In so doing, a force of about 1 N will be exerted on the spacecraft by the spacecraft while only requiring a few mN of force to sustain the mini-magnetosphere. Equivalently, the incident solar wind power is about 1 MW while about 1 kW electrical power is required to sustain the system, with about 0.25 – 0. 5 kg being expended per day. This nominal configuration utilizing only solar electric cells for power, the M2P2 will produce a magnetic barrier approximately 15 – 20 km in radius, which would accelerate a 70 – 140 kg payload to speeds of about 50 – 80 km/s. At this speed, missions to the heliopause and beyond can be achieved in under 10 yrs. Design characteristics for a prototype are also described.

INTRODUCTION

Missions to travel to the outer rim of the solar system and beyond is no easy matter. The heliopause, which represents the boundary between the solar wind and the interstellar wind (Suess, 1990) is thought to be at about 140 +/- 20 AU, with 1 AU = 1.5 x 10^8 km being the distance from the Sun to the Earth (Baranov and Malama, 1993; Steinolfson et al., 1994; Linde et al., 1998). The solar wind itself moves out from the Sun with an average speed of 300 to 800 km/s (and on occasions even to 1000 km/s) and experiences essentially no deceleration until the termination shock, which is at approximately 80 +/- 10 AU. It is at this region where the flow changes from supersonic to subsonic. This deceleration is produced by interactions from reflected solar wind particles and transmitted interstellar particles from the interstellar medium as well as from interactions from neutrals from the interstellar medium.

For optimum scientific return, a spacecraft mission to these regions is presently required to be performed within a 10-yr period after launch. However, the distances involved are huge. For example Voyager 1, which was launched in 1977 with a present speed of about 17 km/s (twice the speed of the space shuttle) is only at about 70 AU at this time and is expected to encounter the termination shock in the next few years. The high speed of Voyager 1 has in part been facilitated by gravitational boosts from its planetary encounters but it is still not expected to cross the heliopause for another 20 yrs. A 10-yr mission to these regions requires a minimum speed of about 50 km/s or equivalently 10 AU/yr.

A critical factor in determining the speed of a spacecraft is the exit velocity of the propellant. For chemical thrusters the maximum exit velocity is only about 3-5 km/s. Because the fuel-to-payload ratio goes as the exponential of change in spacecraft velocity to the propellant velocity, a spacecraft cannot economically attain a speed of 50 km/s by this means. In order to increase speeds beyond existing chemical systems, the propellant speed must be increased substantially and the corresponding high temperatures means that the propellant must be in the form of a plasma. However, the energy requirements to produce sufficient plasma to propel a spacecraft to 50 km/s are enormous and cannot be sustained by solar electric cells alone. Nuclear power could be used but would greatly increase the mass

CP504, *Space Technology and Applications International Forum–2000*, edited by M. S. El-Genk
© 2000 American Institute of Physics 1-56396-919-X/00/$17.00

and cost of the spacecraft and at the same time pose a potential environmental hazard. The only other alternative is to utilize ambient energy through some form of sail technology. Solar sails have been proposed where the spacecraft picks up the momentum from the reflection of light. However, such sails face substantial technological difficulties in attaining sufficiently light material and being able to deploy the material over large distances in space.

An alternative that is described here is to tap the energy of the solar wind and as such would provide a specific impulse of $3 - 8 \times 10^4$ s. Initial proposals to harness this were through magnetic sails or magsail (Zubrin, 1993 and references therein). While different variants have been discussed, the basic concept was to deploy a superconducting magnet with a radius of 100 - 200 km to attain accelerations of the order of 0.01 m/s^2. The use of superconducting material would essentially eliminate all power requirements, except for initially setting up the currents and maintaining any cryogenic systems. However, the drawback of such a system was that it was fairly massive (of the order of a few metric tons) and the physical construction and cost of such a system presently limit its usage from technical and/or economical viewpoints.

The Mini-Magnetospheric Plasma Propulsion (M2P2) described here is analogous to the solar and magnetic sails in that it seeks to harness ambient energy in the solar wind to provide thrust to the spacecraft. However, it represents major advances in several key areas. First, the M2P2 will utilize electromagnetic processes (as opposed to mechanical structures) to produce the obstacle or sail. Thus, the technical and material problems that have beset existing sail proposals are removed from the problem. Second, because the deployment is electromagnetic large scale cross-sections (15 - 20 km for a prototype version) for solar wind interactions can be achieved with low weight (< 50 kg for the device) and low power (< 3 kW) requirements. Third, the M2P2 system acts similar to a balloon in that it will expand as the solar wind dynamic pressure decreases with distance from the sun. As such it will provide a constant force surface (as opposed to a mechanical structure that provides a constant area surface) and thereby provide almost constant acceleration to the spacecraft as it moves out into the solar system.

The physical principles and system requirements are discussed in Section 2 for a spacecraft that could move out of the solar system with speeds of 50 – 80 km/s after an acceleration period of about three months. Detailed numerical results from fluid simulations of the M2P2 system are given in Winglee et al. (1999). Section 3 describes developments in the making of a prototype of the M2P2 and the expected performance of such a system. A summary of results is given in Section 4.

CONCEPT DESCRIPTION

The basic objective of the M2P2 system is to deflect the solar wind particles by a large magnetic bubble or mini-magnetosphere whose field lines are attached to the spacecraft. As the charged particles of the solar wind are reflected by the magnetic field the force that they exert is transmitted along the fields lines to the spacecraft to produce its acceleration. In the simplest case in Figure 1a, the magnetic field is produced solely by a solenoid coil. In this case, the intensity of the magnetic field decreases as R^{-3} so that within about 10 coil radii the interaction the magnetic field is essentially zero and the interaction region with the solar wind is very restricted, and hence inefficient. However, with the injection of plasma, the plasma can drag the magnetic field outwards once it moves into a region where the thermal and/or dynamic plasma pressure exceeds the magnetic pressure. As a result the magnetic field can fall off very much more slowly than for a simple dipole, thereby facilitating much more efficient coupling to the solar wind as illustrated in Figure 1b.

This expansion of magnetic field by injection of plasma is seen in a variety of space plasmas. Examples include the formation of the heliopause itself, where the solar magnetic field is stretch out to 140 AU and are sufficiently strong to modulation the cosmic ray flux as a function of solar cycle. Closer to the Earth, magnetic clouds which associated with coronal mass injections are seen to erupt from the solar surface and seen to increase in size well beyond that of the Sun. Similarly the size of Jovian magnetosphere appears inflated relative to the terrestrial magnetosphere due to plasma from Io and its acceleration though the fast rotation of Jupiter.

The importance of the plasma injection can be also be seen from the following arguments. A dipole magnetic field decreases as R^{-3}. If the plasma energy density where made sufficiently high that it was able to freely expand and

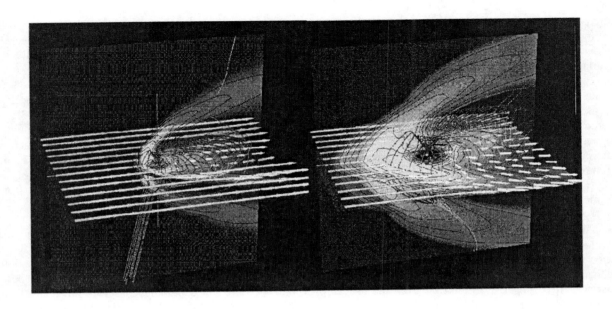

FIGURE 1. (a) The deflection of the solar wind by a simple dipole. (b) Enhanced interaction with the solar wind produced by the inflation of the magnetic field as plasma is injected into the magnetic field. The plasma flow is shown by the arrows with the size of the arrows being proportional to plasma speed. Contours of the plasma pressure are shown in the equatorial plane and in the noon-midnight meridian. The formation of a bow shock as the solar wind interacts with the magnetosphere is seen as a jump in the plasma pressure (yellow regions) around the magnetosphere. The magenta lines indicated some of the field lines attached to the magnetosphere.

carry the frozen-in magnetic field out with it then the magnetic field strength would decrease as R^{-2}. However due to the interaction with the solar wind this plasma is contained or compressed so that current sheets are formed within the plasma and the magnetic field decreases as R^{-1}. The numerical simulation results of Winglee et al. (1999) verify this result.

The size for an effective the mini-magnetosphere can be estimated as follows. If is assumed that the dragging of the field occurs very close to the coil (i.e., where speed of the injected plasma (V_{plas}) is close to the Alfven speed (V_A) near the magnet, then assuming a R^{-1} fall off in the magnetic field then the size of the mini-magnetopause (R_{MM}) would be given by

$$R_{MM} \sim R_{coil} \, (B_{coil}/ B_{MP}) \; , \tag{1}$$

where B_{MP} is the field strength so that the magnetic pressure balances the solar wind pressure. For a solar wind density of 6 cm^{-3} and a speed of 450 km/s, the solar wind dynamic pressure is equivalent to 2 nPa and a $B_{MP} = 50$ nT is sufficient. For a mini-magnetosphere extending to 15-20 km, the force exerted by the solar wind would be about $2.5 - 5$ N (and about a MW of incident power) . To produce this size magnetosphere for a coil radius of 10 cm, the minimum field strength that is required is about 100 G.

If $V_{plas} < V_A$, the coil magnetic field strength has to be increased by approximately the ratio V_A/ V_{plas}. Because the Alfven speed depends on the plasma density, the plasma that is supporting the mini-magnetosphere does not have to be energetic if it is sufficiently dense. As a consequence, the system attains a momentum and power leverage in effect by exchanging or reflecting high energy solar wind plasma with the lower energy plasma supporting the mini-magnetosphere.

A PROTOTYPE

In order to produce coupling with the solar wind over a region of 15-20 km, the M2P2 system has basically three components : (1) a relatively strong (~ 700 G) magnetic field on the spacecraft with the coil radius being about 10 cm, (2) plasma source to inflate the magnetic field with the plasma having an energy of a few eV and a density of the order of $5x10^{13}$ cm^{-3} for β ~ 1- 4 % (or equivalently V/ V_{plas} ~ 5 - 10) and (3) a power source sufficient to generate the magnetic field and plasma. For a 2.5 cm plasma source, the helicon can output several tends of mN of thrust for only about a kW of power.

There are only a few plasma sources that work in the presence of a strong magnetic field, and even fewer capable of producing the high density required. Plasmas generated using electrodes cannot tolerate the high heat load at the high energy densities. There are however inductively produced plasma sources (e.g. helicons) that not only produce high-density discharges but do it in the presence of kilogauss magnetic fields (Miljak and Chen, 1998; Gilland et al., 1998, Conway et al., 1998). It is for this reason that the prototype being developed at the University of Washington uses a helicon. The other advantage of the helicon is that it forms a closed system in the sense that the plasma is created on a field-line that is contained within the vacuum chamber and does not come into contact with the walls of the magnet or the plasma source.

FIGURE 2. The M2P2 prototype in a small (1m) vacuum chamber at the University of Washington. The magnets are encased in a stainless steel jacket so that will not be shorted by the plasma generated within the helicon (for a close-up see Figure 3). Gas and electrical feeds are passed through the copper pipes on the lower right of the figure.

A picture of the prototype is shown in Figure 2. Gas is fed into the vacuum chamber (1 m diameter) through the copper pipes in the lower right part of the figure, and enters into the center of the quartz tubes (5 cm diameter) seen in the center to the figure. Within the quartz tubes a Nagoya type III antenna is embedded to produce the radio frequency heating of the gas to create the plasma needed to support the mini-magnetosphere. The jacketing of the antenna on both sides by quartz tubes ensures that it is electrically isolated from the plasma. The magnetic field coils (20 cm diameter) are encased in a stainless-steel jacket to avoid potential shorting by generated plasma. The plasma source is placed off-axis to ensure that the plasma is created on closed field-lines. The electrical feeds for the helicon and magnets are run through the copper pipes.

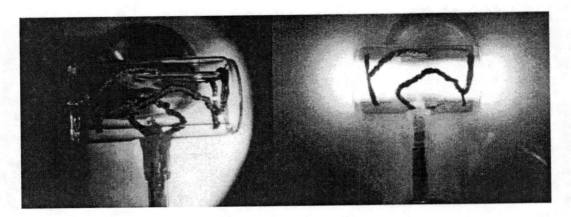

FIGURE 3. (a) Close up of the helicon in the vacuum chamber without the magnets attached. (b) Inductively coupled, high-density helium plasma discharge in the helicon. Approximately, 1 kW of RF power at 19 MHz was coupled to the plasma using a Nagoya type III antenna (copper braid in picture). The antenna is enclosed in an annular quartz envelope isolating it from the vacuum and plasma. The tube attached to the antenna enables the propellant and RF power to be delivered remotely.

FIGURE 4. The helicon being operated with magnets operating at 300 G with the chamber back-filled with argon at 5 mTorr. (a) The chamber illuminated by external lighting and (b) image of the plasma emissions alone. It is seen that the plasma trajectory tracks around the field lines to produce a loop-like structure faintly visible on the right hand side.

The performance and efficiency of the helicon source in creating plasma and moving magnetic field within the chamber are being evaluated at this time. Initial images are shown in Figure 3 of the plasma production by the helicon but without the magnets attached. The light emissions show the production of a dense column of plasma, which then expands as it moves out of the quartz tube. Figure 4 shows the difference when the solenoid coils are present, and producing 300 G fields. In this case, there is still strong plasma production but as it exits the helicon, the plasma is turned by the magnetic field to produce a loop-like structure to the right and which extends close to the chamber walls. This type of structure is one of the primary goals of the M2P2 system.

SUMMARY

In order to explore the outer solar system and nearby interstellar space spacecraft will have to travel at speeds in excess of 50 km/s. The power requirements needed to obtain such high speeds are much higher than can be presently supported by solar electric propulsion. Thus some other means of powering spacecraft must be developed. In this paper we have investigated one such system where energy from the solar wind is used to propel the spacecraft. This system which we call Mini-Magnetospheric Plasma Propulsion (M2P2) seeks to inflate a large magnetic bubble around the spacecraft to deflect and thereby pickup the momentum from the solar wind particles which are traveling at speeds of 350 to 800 km/s. A prototype for the device, consisting of a helicon plasma source inserted asymmetrically in a moderately strong magnetic field has been constructed in a laboratory vacuum chamber. Because the energy of the plasma used to produce the inflation is very much less than that of the solar wind, it is expected that only a few mN of force and a few kW of power is needed to sustain the magnetosphere, while intercepting nearly a N of force and a MW of power from the solar wind for a specific impulse of $3 - 8 \times 10^4$ s.

ACKNOWLEDGMENTS

This work was supported by a grant from NASA's Institute for Advance Concepts 07600-010, NSF Grant ATM-9731951 and by NASA grants NAG5-6244, NAG5-8089 to the Univ. of Washington. The simulations were supported by the Cray T-90 at the San Diego Supercomputing Center which is supported by NSF.

REFERENCES

Baranov, V. B. and Malama, Y. G., "Model of the solar wind interaction with the local interstellar medium: Numerical solution of the self-consistent problem, " J. Geophys. Res., 1993, 98, 15157.

Conway, G. D., Perry, A. J. and Boswell, R. W., "Evolution of ion and electron energy distributions in pulsed helicon plasma discharges, " Plasma Sources, Sci. and Tech., 1998, 7, 337.

Linde, T. J., Gombosi, T. I., Roe, P. L., Powell K. G. and DeZeeuw, D. L., "Heliosphere in the magnetized local interstellar medium: results of a three-dimensional MHD simulation, " J. Geophys. Res., 1998, 103, 1889.

Leipold, M., Borg, E., Lingner, S. , Pabsch, A. , Sachs, R. and Seboldt, W. , "Mercury orbiter with a solar sail spacecraft, " Acta Astronautica, 1995, 35, 635.

Miljak, D. G.,and Chen, F. F. "Density limit in helicon discharges, " Plasma Sources, Sci. and Tech., 1998, 7, 537.

Steinolfson, R. S., Pizzo, V. J. and Holzer, T., "Gasdynamic models of the solar wind/interstellar medium interaction, " Geophys. Res. Lett., 1994, 71, 245.

Suess S.T., "The heliopause, " Rev. Geophys., 1990, 28, 97-115.

Weaver, H A, "Comets, " Astronomical Society of the Pacific Conference Series., 1998, 143, pp. 213-26.

Weissman, P. R., "The Kuiper belt, " Annual Rev. Astron. Astrophys., 1995, 33, 327.

Winglee, R. M., et al., "Modeling of upstream energetic particle events observed by WIND, " Geophys. Res. Lett., 1996, 23, 12,276,.

Winglee, R. M., Slough, J., Ziemba, T., and Goodson, A., "Mini-magnetospheric plasma propulsion: Tapping the energy of the solar wind for spacecraft propulsion, " J. Geophys. Res., submitted, 1999.

Zubrin, R. M., "The use of magnetic sails to escape from low Earth orbit," J. British Interplanetary Society, 1993, 46, 3.

Sailcraft-Based Mission To The Solar Gravitational Lens

Giovanni Vulpetti

Advanced Space Mission Studies, Telespazio SpA, Via Tiburtina 965, 00156 Rome - ITALY

+39.06.4079.3896; giovanni_vulpetti@telespazio.it

Abstract. This paper shows that current or near-term technology together with the *H*-reversal mode for solar sailing would allow designing a 22-year flight-time mission to the minimum Solar Gravitational Lens (550 AU from the Sun). Sailcraft mass can be as low as 345 kg with a net payload of 100 kg. Al-Cr sail is 0.287 km^2 large and 0.14 μm thick. Sailcraft mass breakdown and trajectory profiles are detailed.

INTRODUCTION

In this decade, three new large classes of solar-sail trajectories have been found out (Vulpetti, 1996-99). Their excellent properties are based on a new solar-sailing concept: the sailcraft motion reversal or the *H*-reversal mode. The Aurora Collaboration (AC), that has been studying high-speed low-mass sailcraft since 1994, adopted such a potential strategy to propose new missions to the Italian Space Agency for scientific investigation beyond the planetary range. Wider aim is to try to make the solar-sail propulsion *feasible* for scientific, utilitarian and, subsequently, commercial use in space. In addition to spiral trajectories and halo-orbits (in the Earth-Moon system or inner/outer Solar System), the trajectory classes related to the *H*-reversal mode would dramatically allow expanding low-cost high-frequency low-mass scientific missions to very distant targets. It is based on the following properties, strictly proved from an astrodynamic viewpoint:

1. Sailcraft can escape the Earth-Moon system without the need of any other propulsion system or any lunar fly-by

2. If the sail loading is 2 grams/m^2 (or lower), then the sailcraft can decelerate down to reverse its orbital angular momentum; this allows the vehicle to point close to the Sun in acceleration mode for weeks and so getting the absolute (not a relative or strong) maximum of energy for a technology-given sailcraft

3. Perihelion distance and time can be tuned by design to meet different constraints, including temperature

4. The cruise phase is characterized by a speed practically independent of the distance from the Sun

5. High heliocentric speed is achieved by no planetary fly-by; sail is open at a parking orbit about the Earth, instead of at solar perihelion, thus avoiding strong jumps of thrust or impulses that may either damage the sail or induce unrecoverable attitude errors.

The first basic concept for performing smart sailcraft control was to turn the lightness-number concept as one fixed parameter into the concept of time-varying lightness vector. The second basic concept consisted of investigating what happens to a decelerating heliocentric sailcraft with a sail loading sufficiently low to allow a progressive decrease of the orbital angular momentum down to its reversal. Perihelion ranges from 0.30 to 0.15 AU whereas cruise speed spans in 11-17 AU/yr by 2 grams/m^2 (or 1.30 times the critical density) for a mission to the heliopause.

In the AC context, a further significant improvement of the *H*-reversal concept is in progress. On keeping the same range of admissible perihelion, an all-metal sail is under investigation with a sail loading as low as 1.20 grams/m^2 (or 0.78 times the critical density). A strong example mission such as the flight to the *minimum* Solar Gravitational Lens (SGL), or 550 AU, is the subject of this paper.

BASIC CONCEPTS OF SOLAR SAILING

We remind the reader a minimal set of solar-sailing concepts for a better understanding of the SGL mission. For a full treatment of dynamics and technology of solar sailing, one can refer to (McInnes, 1999) in general, and to (Vulpetti, 1996-99a) for high-speed trajectories, in particular.

There are two basic points: (*a*) in general, the force field actually acting on sailcraft is not conservative, (*b*) the sailcraft motion can be controlled by steering the normal-to-sail axis of symmetry. For our purposes, we need two frames of reference to describe the sailcraft motion: the heliocentric inertial frame (HIF) and the extended

CP504, *Space Technology and Applications International Forum–2000*, edited by M. S. El-Genk

heliocentric orbital frame (EHOF). HIF may be identified with the Dynamical-Equinox and Ecliptic at J2000. EHOF is the generalization of the usual orbital frame by including trajectory branches separated by a finite number of points where the sailcraft's orbital angular momentum per unit mass (**H**) vanishes. The EHOF axes are given by the columns of the matrix (**r** **h**×**r** **h**), where **r** denotes the direction of the sailcraft position vector, say, **R** in HIF and **h** is either the **H** direction for direct trajectory arc or the -**H** direction for retrograde trajectory arc or their (common) limit when **H**=0. (From here the phrase *H*-reversal mode). Motion reversal can take place even if **H** does not vanish. In such cases, the magnitude of **H** passes through a minimum (strictly, time of sailcraft motion reversal comes before this minimum of **H**, usually a few hours). The strict definition and properties of EHOF can be found in (Vulpetti, 1999a). One can write general sailcraft motion equations by introducing the time-dependent vector function named *the lightness vector* (Vulpetti, 1996-97), denoted by **L**, as follows:

$$\mathbf{L} \equiv \begin{pmatrix} \lambda_r & \lambda_t & \lambda_n \end{pmatrix}^T \qquad \lambda \equiv |\mathbf{L}| \qquad (1)$$

L is defined in EHOF. Its components (also called the *radial*, the *transversal* and the *normal* lightness numbers, respectively) represent the solar-pressure-induced vector acceleration in units of the *local* solar gravitational acceleration or μ/R^2, where μ denotes the solar gravitational constant. Thus, the classical-dynamics equations of sailcraft motion can be written as

$$\frac{d}{dt}\mathbf{R} = \mathbf{V}$$
$$\frac{d}{dt}\mathbf{V} = \frac{\mu}{R^2}\left[-(1-\lambda_r)\mathbf{r} + \lambda_t \mathbf{h}\times\mathbf{r} + \lambda_n \mathbf{h}\right] \qquad (2)$$

(Here, for simplicity, Equations (2) do not contain the mass rate equation; actually, if a sailcraft is controlled by small attitude rockets then $dM/dt < 0$, where M is the vehicle mass). One then realizes that sailcraft trajectory can be analyzed in terms of the **L** vector only, even though the actual control shall operate on the sail orientation. In order to highlight the different roles of the **L** components, we report the main equations for orbital sailcraft energy *E*, angular momentum and their time rates:

$$E = \tfrac{1}{2}V^2 - (1-\lambda_r)\frac{\mu}{R} , \qquad \frac{d}{dt}E = \frac{H}{R^2}\frac{d}{dt}H ,$$

$$\mathbf{H}\times\frac{d}{dt}\mathbf{H} = H\lambda_n \frac{\mu}{R}\mathbf{r} , \qquad \frac{d}{dt}\mathbf{H} = \frac{\mu}{R}\left(\lambda_t \mathbf{h} - \lambda_n \mathbf{h}\times\mathbf{r}\right) , \qquad (3)$$

$$\frac{d}{dt}H = \lambda_t \frac{\mu}{R} , \qquad \mathbf{H} = H\mathbf{h} .$$

An extended discussion of Equations (3) and their consequences for sailcraft dynamics can be found in (Vulpetti, 1996-97). The quantity *H* in Equations (3) is an invariant; it is the projection of **H** (defined in HIF) onto the Z-axis of EHOF. It can be of any sign and its derivative depends on the *transversal* lightness number. *dH /dt* drives the *E* change and determines the history of **H**. Note that the *normal* lightness number governs the bending of **H**. How to control a heliocentric sailcraft trajectory by *H* is explained in (Vulpetti, 1999a). In general, **L** is a complicated function of the sailcraft mass (M) on sail area (*S*) ratio (or the spacecraft *sail loading*, usually denoted by σ), the thermo-optical properties of the sail materials, the sail axis control angles, the spacecraft velocity and the source-of-light characteristics (McInnes, 1999) and (Vulpetti, 1999b). As usually conceived, a practical sail consists of a multi-layer film: the *reflective* layer, the *emissive* layer and the *substrate* the other two layers are deposited on. The reflective (or front-side) layer is always facing the Sun in a heliocentric trajectory, whereas the emissive (or backside) layer allows keeping the sail temperature sufficiently low. Although specular reflection is the dominant effect for photon-sail momentum exchange, other non-negligible effects have to be taken into account. If the substrate (which is the heaviest component of a sailcraft) is removed, then one has an all-metal sail capable to achieve very high speed. We shall consider this configuration for a fast mission to SGL.

Neglecting the sail irradiance reduction due to both the Sun finite-size and limb darkening - some models can be found in (McInnes, 1999) - and additionally retaining the linear terms in the sailcraft velocity results in the following simplified link between the direct control variables & parameters and the lightness numbers that enter the motion equations, or *the connection equations*:

$$\mathbf{L} = \lambda_0 \cos\alpha \; \cos\delta \left\{ \left[\begin{matrix} (2r\cos\alpha\;\cos\delta + \chi_f\,d + \kappa a)(1 - 2\beta_x) - \\ 2r\sin\alpha\;\cos\delta\,\beta_y \end{matrix} \right] \mathbf{n} + (a+d) \begin{bmatrix} 1 - 2\beta_x \\ -\beta_y \\ 0 \end{bmatrix} \right\}$$

where

$$\mathbf{n} \equiv \begin{bmatrix} \cos\alpha\;\cos\delta \\ \sin\alpha\;\cos\delta \\ \sin\delta \end{bmatrix}, \qquad \beta \equiv \frac{V}{C} \begin{bmatrix} \cos\varphi \\ \sin\varphi \\ 0 \end{bmatrix}, \qquad \varphi \equiv angle(\mathbf{R},\mathbf{V}), \qquad V \equiv |\mathbf{V}|,$$

(4)

$$\lambda_0 \equiv \tfrac{1}{2}\sigma_c \frac{S}{M}, \quad \sigma_c \equiv 2\frac{W_{1AU}}{g_{1AU}\,C} = 0.001539 \; kg\; m^{-2}, \quad W_{1AU} = 1368 \; W\,m^{-2}, \quad g_{1AU} = 0.00593 \; m\,s^{-2}.$$

In Equations (4), α and δ denote the azimuth and elevation of the sail axis $\mathbf{n}$ (oriented backward with respect to the reflective sail side) in EHOF. The set (r, d, a) denote the specularly reflected, the diffused reflected and the absorbed fractions of the solar incident flux by the sail materials, respectively. Their exact meaning, numeric handling and relationship to orbit determination can be found extensively in (Vulpetti, 1999b). χ_f denotes the surface coefficient of the front side, whereas κ is a known function of the sail temperature. The vector β (resolved in EHOF) accounts for the aberration effect, which is not negligible for a high-speed flight. The quantity σ_c represents the so-called *critical* density. One has the following important relationship

$$\sigma = \sigma_c\,\tau/\lambda,$$

(5)

where τ is the thrust efficiency. Note that g_{1AU} is the solar gravitational acceleration at 1 AU, whereas W_{1AU} denotes the solar constant; the above value is compliant with some other reference (Wright, 1993). W_{1AU} has to be measured accurately for space sailing (probably as part of the orbit determination process); this could be performed in experimental sailcraft missions.

SAILCRAFT FOR SGL MISSION

If the sailcraft sail loading is sufficiently low, then by appropriately orienting the sail axis it is always possible to meet the following conditions: $\tfrac{1}{2} \le \lambda < 1$, $\lambda_t \le 0$ for the first branch of heliocentric flight, even though, in principle, the sailcraft is able to accelerate considerably already at 1 AU. These necessary conditions plus a sufficiently long deceleration time are sufficient to force H to zero and then reverse it. This is the only way to approach the Sun closely in the continuous accelerating mode. Roughly, this may be thought as a very long burn around the perihelion. Solar sail, however, consume no propellant and can be controlled such a way the along-track component of the total vector acceleration is positive past the reversal point. The sailcraft energy no longer decreases and can achieve its *absolute* maximum for a given sailcraft technology. Such an overall non-linear effect is mathematically detailed in (Vulpetti, 1996-99a). The technology to achieve the H-reversal condition consists of in-orbit removing the plastic support from the multi-layer sail made, for instance, of Al-Kapton-Cr. Some successful methods are in progress in laboratory (Scaglione, 1999). Given a very light bare sail, the other systems, including the deployment & keeping sub-systems, of sailcraft should be comparable in mass while exhibiting mission-appropriate features. For the SGL mission, AC has selected and sized the following main systems:

A. Hydrostatic Beam-based Deployment System with load-supporting web (Genta, 1999)

B. Attitude Control System (ACS), based on the Field Emission Electric Propulsion developed by the European Space Agency. The current ACS has been sized by using a jet speed of 7 km/s for keeping power low

C. Main Power System (MPS), based on the Pu^{238} Radioisotope Thermo-Photo Voltaic Generator (Shock, 1997)

D. Small CO_2 Laser Power Transmission System (SLPTS) delivering part of the electric energy from MPS (located close to the sailcraft center-of-mass) to four two-engine packages the ACS consists of. They are positioned at the rim of the sail system. SLPTS can produce a regular input to ACS independently of the Sun-sailcraft distance

E. Communication System based on Nd-YAG. It is sized to yield a bit rate (with coding) of 200 baudes at 750 AU

F. Scientific Payload.

Mass values of the subsystems and systems composing the systems A-F plus other structures and contingency are reported in Fig. 1 for the SGL mission envisaged here. Note that the above system choice allows designing a sailcraft with a net payload of 29 percent.

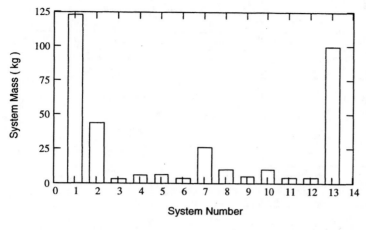

1. Bare Sail
2. Hydrostatic Beam Tube
3. Hydrostatic Beam Gas
4. Sail Load Paths
5. Power System
6. Power Conditioning
7. Attitude Control System
8. Telecommunication System
9. Computer
10. Thermal Control System
11. Other Structures
12. Contingency
13. Net Payload Mass

FIGURE 1. Sailcraft mass breakdown for the SGL mission.

H-REVERSAL FLIGHT TO SGL

One wants to reach the so-called galactic anti-center direction (Maccone, 1997), that has (approximate) Ecliptic Longitude = 86.83°, Latitude = 5.537°, with a minimum operational distance of 550 AU. These three numbers represent the current SGL-target. The flight design we are going to discuss is the solution to the following problem:

" *Given* the above sailcraft, find the three-dimensional L's history that minimizes the flight time to the SGL-target by using Sun fly-by *only* with the sail peak temperature not exceeding 60 percent of the aluminum melting point ".

The temperature constraint preserves the mechanical properties of the aluminum film and determines a lower limit on the reachable perihelion. We have used non-linear programming, in particular two versions of the Levenberg-Marquardt method, for minimizing the (Euclidean) norm of a penalty vector function. We shall show *nominal* time behaviors of meaningful quantities zoomed on the intervals where they vary appreciably. Outside, quantities are either constant or asymptotically flat. Figure 2 shows the time behavior of sail azimuth and elevation in EHOF. The duration of the thrusting arcs (t_1-t_0, t_2-t_1, t_3-t_2, t_4-t_3) is optimized according to the theory (Vulpetti, 1999a). Table 1 reports the complete optimal control and other meaningful flight information. How the above control affects the vector **L** is shown in Fig. 3. Note that at t_3 an attitude maneuver is performed such that $\lambda_r > 1$. That, induced by a

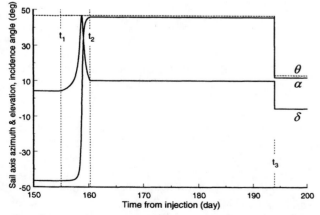

FIGURE 2. History of the direct control variables in the sailcraft orbital frame; also, the incidence angle (θ) behavior and trajectory arc time tags are shown.

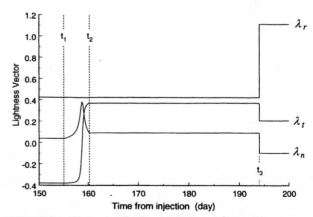

FIGURE 3. Profile of the lightness numbers, induced by the direct sail axis control, which minimize the flight time to the galactic anti-center direction in SGL.

sub-critical sail loading, causes a continuous increase of vehicle speed, thus avoiding a local maximum characterizing the *H*-reversal classes previously analyzed.

TABLE 1. Optimal control driving the initial state to the current SGL target by *H*-reversal mode (1 yr=365.25 d). Note that a control direction specified as constant in either frame is variable in the other one, thus producing the optimal history of Fig. 2. The initial sailcraft longitude has been optimized.

Arc	Reference frame	Duration	Sail azimuth	Sail elevation	Mode	Model of Sun
T1	EHOF	155.00 d	-46.50°	4.2299°	constant	point-like
T2	HIF (J2000)	5.2515 d	236.6055°	-7.2596°	constant, patched	point-like
T3	EHOF	33.7010 d	45.8728°	10.0913°	constant, patched	finite-size & limb-darkening.
T4	EHOF	21.300 yr	11.8516°	-5.5919°	constant	point-like

Mass [kg]	X [AU]	Y [AU]	Z [AU]	VX [AU/yr]	VY [AU/yr]	VZ [AU/yr]
344.4	-0.8704494	-0.4922579	0	3.1152469	-5.5086268	0
330.1	30.248	546.60	53.055	1.425	25.659	2.501

time to min-speed	time to reversal,	time to perihelion	perihelion distance&speed	perihelion long. & latitude	cruise speed	flight time
114.06 d	158.676 d	191.950 d	0.151 AU 22.04 AU/yr	194.61° -49.26°	25.82 AU/yr	21.831 yr

The above control generates the trajectory graphed in Fig. 4. Note that the sailcraft-to-Earth vector is always sufficiently distant from the Sun to allow safe communication. (The minimum of the Earth-sailcraft line-of-sight distance from the solar photosphere is 14.4 solar radii, 50.7 days past the sailcraft perihelion). Along such a trajectory, the vector **H** evolves a way shown in Fig. 5. According to the theory, **H** does not vanish at the reversal

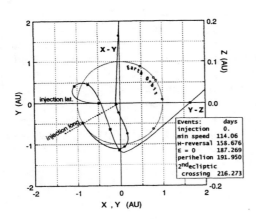

FIGURE 4. Zoomed projections of the sailcraft's 3D trajectory to the SGL target. Special events and the corresponding Earth positions are marked.

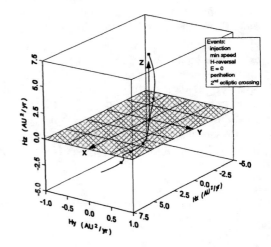

FIGURE 5. Hodograph of the orbital angular momentum in HIF. Flight begins with positive Hz. Special events are indicated.

point. Its magnitude is 313 times lower than the value at injection. Strong decrease is necessary to achieve the absolute maximum of energy. The stronger this reduction is the higher the cruise speed is achievable. Histories of the invariant and energy are shown in Fig. 6. Note the significant energy jump as a result of the optimal attitude maneuver (in intensity and time allocation). The overall effect of the *H*-reversal mode, a sub-critical sail loading and final attitude maneuver is displayed quantitatively in Fig. 7, in terms of Sun-sailcraft distance and speed. The shown information box is self-explaining; it adds to data reported in the last row of Table 1. This high cruise speed would allow the payload mission to extend from 550 AU over 200 AU (at least) in 7.75 years. Note that, strongly enhancing previous results, speed takes on a *square-root-like* profile; in other words, (not only energy but also) speed gets increased asymptotically. Such behavior can be considered the current dynamical output of the non-linear

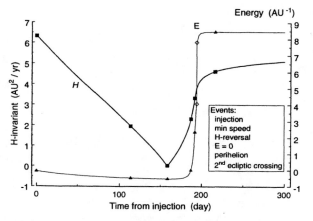

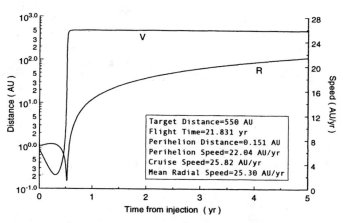

FIGURE 6. Time profiles of the sailcraft energy E and *H*-invariant with special events marked. Empty diamonds denote the E values before and after the optimal attitude maneuver.

FIGURE 7. Histories of Sun-sailcraft distance and speed in HIF. Meaningful flight values have been reported in the information box.

range of solar sailing. It may foster studying other sail materials and sailcraft systems capable to lessen the sail loading still further, once experiments on sailcraft begin in space. What ultimately matters for a mission to a very far target is the mean radial speed of the spacecraft. In our case, it is very close to the cruise speed (Fig. 7); this means that the time to achieve the acceleration point is rather short (114 days from Tab. 1, instead of 4-5 years by using planetary fly-by techniques). With regard to launch window, preliminary results indicate that the present SGL-target has a launch window at least of 16 days (from 11th to 27th April) <u>every</u> year. This baseline flight profile can be widely "deformed" to balance the off-optimal Earth longitude. Flight-time charge varies from few weeks to 1.2 yr.

CONCLUSION

This paper shows that *current* or near-term technology <u>and</u> *advanced* solar-sailing astrodynamics should be able to accomplish a highly scientific mission to the solar gravitational lens with a flight time less than 22 years with <u>no</u> planetary/lunar launch windows. Its perihelion distance (0.15 AU) is affordable, whereas the cruise speed (near 26 AU/yr) is sufficiently high to allow a scientific payload to get data for 8 years or, equivalently, up to 750 AU. Mission window spans 16 days every year.

REFERENCES

Genta, G., and Brusa, E., "The Aurora Project: a New Sail Layout", *Acta Astronautica*, **44**, No. 2-4, 141-146 (1999)

Maccone, C., *The Sun as a Gravitational Lens: Proposed Space Missions*, IPI Press, Colorado Springs, Colorado 1997, ISBN 1-880930-10-2

McInnes, C. R., *Solar Sailing: Technology, Dynamics, and Mission Applications*, Springer-Praxis Series in Space Science and Technology, Godalming (UK), May 1999, ISBN: 185233102X

Scaglione, S., and Vulpetti, G., "The Aurora Project: Removal of Plastic Substrate to Obtain an All-Metal Solar Sail", *Acta Astronautica*, **44**, No. 2-4, 147-150 (1999)

Shock, C. Or, and Kumar, V., "Design and Integration of Small RTPV with New Millennium Spacecraft for Outer Solar System", *Acta Astronautica*, **41**, No. 12, 1997

Vulpetti, G., "3D High-Speed Escape Heliocentric Trajectories by All-Metallic-Sail Low-Mass Sailcraft", *Acta Astronautica*, **39**, No. 1-4, 161-170 (1996)

Vulpetti, G., "Sailcraft at High Speed by Orbital Angular Momentum Reversal", *Acta Astronautica*, **40**, No. 10, 733-758 (1997)

Vulpetti, G., "General 3D H-Reversal Trajectories for High-Speed Sailcraft", *Acta Astronautica*, **44**, No. 1, 67-73 (1999a)

Vulpetti, G., and Scaglione, S., "The Aurora Project: Estimation of the Optical Sail Parameters", *Acta Astronautica*, **44**, No. 2-4, 123-132 (1999b)

Wright, J. L., *Space Sailing*, Gordon and Breach Science Publishers, Amsterdam, 1993

Fission-Based Electric Propulsion for Interstellar Precursor Missions

Ronald J. Lipinski[1], Roger X. Lenard[1], Steven A. Wright[1], Michael G. Houts[2], Bruce Patton[2], and David Poston[3]

[1]*Sandia National Laboratories, MS-1146, P.O. Box 5800, Albuquerque, NM 87185, rjlipin@sandia.gov*
[2]*Marshall Space Flight Center, Huntsville, AL, 35812, michael.houts@msfc.nasa.gov*
[3]*Los Alamos National Laboratory, Los Alamos, NM 87545, poston@lanl.gov*

Abstract. This paper reviews the technology options for a fission-based electric propulsion system for interstellar precursor missions. To achieve a total ΔV of more than 100 km/s in less than a decade of thrusting with an electric propulsion system of 10,000 s Isp requires a specific mass for the power system of less than 35 kg/kWe. Three possible configurations are described: (1) a UZrH-fueled, NaK-cooled reactor with a steam Rankine conversion system, (2) a UN-fueled gas-cooled reactor with a recuperated Brayton conversion system, and (3) a UN-fueled heatpipe-cooled reactor with a recuperated Brayton conversion system. All three of these systems have the potential to meet the specific mass requirements for interstellar precursor missions in the near term. Advanced versions of a fission-based electric propulsion system might travel as much as several light years in 200 years.

INTRODUCTION

Interstellar precursor missions are those which stretch our technical capabilities in the directions needed for later interstellar travel. Nominally they involve missions to beyond Pluto with trip time of less than 20 years and velocities of over 50 km/s. Two potential near-term interstellar precursor missions are a Kuiper Belt Object rendezvous mission and a Heliopause Probe mission. Kuiper Belt Objects (KBOs) are a recently-discovered set of solar system bodies which lie at about the orbit of Pluto (40 AU) out to about 100 astronomical units (AU). There are estimated to be as many as 100,000 KBOs with a diameter greater than 100 km (Malhotra, 1999, Jewitt, 1999). KBOs are postulated to be composed of the pristine material which formed our solar system and may even have organic materials in them. A rendezvous mission including a lander would be needed to perform chemical analysis of the surface and sub-surface composition of KBOs. Although the distance to the KBOs is not exceedingly large, the need to accelerate, coast, and then decelerate for rendezvous results in a total mission delta-V of about 50 km/s. This places it in the precursor category. The heliopause occurs between 100-200 AU, so a mission to there in under 20 years also would require a velocity of around 50 km/s or more. More challenging missions would require a ΔV's of well over 100 km/s.

To achieve a velocity of 100 km/s without a severe mass penalty for the propellant requires a specific impulse (Isp) of about 10,000 s. Such an Isp can be achieved with electric propulsion. To achieve this velocity in about a decade with this Isp requires a specific mass for the total spacecraft of less than about 45 kg/kWe (at 80% thruster efficiency). Subtracting off 10 kg/kWe for the thrusters, tankage, and remainder of the spacecraft leaves about 35 kg/kWe for the power system. Such a specific power can be provided by near-term fission-based power systems.

Several reactor and electrical conversion systems can be envisioned. Each has different advantages depending on the available launch mass, flight time, and development time and budget. One possible power system is a SNAP-10A derivative (using UZrH fuel) with a steam Rankine cycle conversion system. Another is a gas-cooled reactor (using SP-100 developed UN fuel) with a Brayton conversion system. A third is a heat-pipe cooled system (with

CP504, *Space Technology and Applications International Forum–2000*, edited by M. S. El-Genk
2000 American Institute of Physics 1-56396-919-X

UN or UO2- fuel) and a Brayton conversion system. None of these options require any new fuels or materials development. But they all require good engineering, testing, and integration. The various options and growth potential for the power system will be described and compared for the interstellar precursor missions.

SYSTEM ANALYSIS

To achieve the velocities needed for interstellar precursor missions (i.e., >100 km/s) in times acceptably short for such missions (i.e., < 20 years) requires both a high specific impulse and an adequately high acceleration (or thrust-to-mass ratio). The rocket equation dictates that the total mass of fuel needed increases exponentially with the ratio of the desired velocity increase (ΔV) over the specific impulse of the fuel (I_{sp}) times the gravitational acceleration on earth (g):

$$m_{fuel} = m_{nonfuel}\left(\exp(\Delta V /(gI_{sp}) - 1\right), \tag{1}$$

where m_{fuel} is the mass of the propellant, $m_{nonfuel}$ is the mass of everything that is not propellant (scientific payload, communications, guidance, rocket engines, power supply, etc.). One of the best current chemical propellants is liquid oxygen/liquid hydrogen which has a specific impulse of 460 seconds. To achieve a total ΔV of 100 km/s with a payload, power system, engines, and other inert masses ("non-fuel mass") of 5000 kg would require 21 billion tonnes of fuel. Clearly, chemical propellants cannot do the job. But NASA has recently developed very reliable electric propulsion units with a demonstrated specific impulse of 3300 s (Polk, 1998). An even higher Isp can be achieved either by using a larger voltage on the grids or by using a lighter gas. NASA is also developing a VASIMR electric thruster which is projected to achieve up to 20,000 s Isp (Chang Diaz, 1998). To reach a total ΔV of 100 km/s with a specific impulse of 10,000 seconds and 5000 kg of payload and structure requires only 8.9 tonnes of propellant. This is much more reasonable.

Electric propulsion requires very large amounts of electrical energy. Nuclear fission, in the form of a small research-sized reactor, is a near-term means for obtaining this energy. A fission reactor and power conversion system for space (excluding the radiator) would be about the size of an automobile and similar in power to research reactors found at many universities. The specific mass of the power system and other non-fuel portions of the spacecraft dictates how long it takes to accelerate to 100 km/s. This determines the time required to reach a given distance. For example, if the total non-fuel mass of the spacecraft is 5000 kg, and it delivers 250 kW of electrical power, the dry spacecraft has a total specific mass of 20 kg/kWe. If the electric thrusters are 75% efficient with an Isp of 10,000 s, the spacecraft would have a maximum acceleration of

$$a = \frac{2\eta P}{m_d gI_{sp}} = 7.6 \times 10^{-4} \text{ m/s}^2 = 78 \text{ microgees}, \tag{2}$$

where P is the electrical power, η is the thruster efficiency, m_d is the dry mass, g is the gravitational acceleration on earth, and I_{sp} is the specific impulse. With this acceleration (or thrust-to-weight ratio), it would take 4.2 years to accelerate up to 100 km/s (disregarding the retarding force of the sun and the initial earth-orbital velocity of 29 km/s around the sun). This suggests that the power system needs to have a specific mass of around 20 kg/kWe or less to obtain acceptable acceleration times.

There are two general approaches to fission-based electric propulsion precursor missions. The first approach is to launch the spacecraft into earth orbit and use the electric propulsion system to propel the spacecraft out of earth orbit and then on to interstellar space. The second approach is to launch the spacecraft into an earth-escape trajectory (called a C3=0 trajectory) and then use the electric propulsion system. The advantage of an earth-orbit launch is that it does not require as large a launcher for a given spacecraft weight because the highly-efficient electric propulsion system will be used to get the spacecraft out of earth's gravity well. That will reduce launch costs. A disadvantage is that it requires a longer total burn time for the reactor and electric thruster. It also requires the spacecraft to survive a relatively slow passage through the Van Allen radiation belt. Overcoming these difficulties is one reason to use a C3=0 launch. In addition, the safety analysis might be easier since reactor operation would not begin until an earth-escape trajectory is achieved. But the penalty for a C3=0 launch is that typically the mass that a given launcher can deliver to earth escape is about three times less mass than it can deliver to low-earth orbit.

A simple orbital mechanics code was developed and used to analyze the requirements for a KBO rendezvous mission, and the results are reported elsewhere in this symposium (Lipinski, 2000). That analysis showed that a KBO rendezvous could be achieved in 13.0 years with a 1000 kg science payload if the dry mass of the rest of the spacecraft (power, thrusters, navigation, communication, tankage, etc.) was 4000 kg and the electrical power was 100 kWe with a thruster total efficiency of 75%. The Isp for this system is 10,000 sec and the launch mass into a C3=0 trajectory is 7984 kg. If essentially the same spacecraft is launched into a 700-km low-earth orbit and an additional 634 kg of electric-thruster propellant is added, the spacecraft can spiral out of earth orbit and rendezvous in a total of 14.3 years. With a variable-Isp thruster (3,000 to 10,000 sec) such as VASIMR, the trip from LEO to KBO rendezvous can be made in 13.5 years with a LEO launch mass of 10,300 kg. The total ΔV for the trip are 45.9, 53.4, and 53.4 km/s, respectively, for the three options.

TUG DESIGN

As described previously, the specific mass of the electric power system for interstellar precursor missions should be about 35 kg/kW or less. This is higher in power and lower in specific mass (kg/kWe) than any space nuclear power system that has been fielded in the past, but quite reasonable for estimated masses of near-term space-reactor systems. The choice of technologies for the reactor system is fairly wide, although there is no off-the-shelf space-reactor system presently available. The SNAP program flew one space reactor in 1965 and built six other working reactors during the program. All of these used UZrH fuel with a NaK coolant (Anderson, 1983). The subsequent SP-100 program designed a reactor which used UN fuel with a Li coolant (Mondt, 1994; Mondt, 1995; El-Genk, 1994). The Russian Rorsat reactors (about 30 flown in space) used UMo fuel (Angelo, 1985). The Russian Topaz II reactor used UO_2 fuel and UZrH moderator. There are numerous other proposed designs in the literature (Angelo, 1985; El-Genk, 1994).

Given the need for a small reactor and shield mass, and a high conversion efficiency, we propose consideration of the following three options for the KBO mission: (1) a UZrH-fueled, NaK-cooled reactor with a steam Rankine conversion system, (2) a UN-fueled gas-cooled reactor with a recuperated Brayton conversion system, and (3) a UN-fueled heatpipe-cooled reactor with a recuperated Brayton conversion system. Table 1 summarizes the key features for 100-kWe systems. All three systems use the same assumed low-mass deployable thermal radiator to maximize the conversion efficiency. Subsequent sections describe the options in more detail, and examples for the KBO mission are given in a companion paper in this conference (Lipinski, 2000). There has not been sufficient detailing of the designs to perform an accurate weight estimate for each system, but rough estimates indicate that all of these systems have the potential to have specific masses less than 35 kg/kWe.

TABLE 1. Comparison of three electric power systems for the KBO mission.

Component	UZrH/NaK-cooled/Rankine	UN/gas-cooled/Brayton	UN/Heatpipe-cooled/Brayton
Electric power (kW)	100	100	100
Thermal power (kW)	345	220	333
Thermal efficiency (%)	29	46	30
Nuclear fuel	UZrH	UN	UN
Primary coolant	NaK	He/Xe	Na
Fuel clad material	Hastelloy	Nb1%Zr/Re	Nb1%Zr/Re
Vessel material	316 SS	Super Alloy	Mo
Conversion cycle	Rankine	Brayton	Brayton
Energy conversion working fluid	water	He/Xe	He/Xe
Thermal radiator type	heatpipe/fin	heatpipe/fin	heatpipe/fin
Radiator working fluid	ammonia	ammonia	ammonia
Reactor coolant exit temp (K)	723	1200	1200

ZrH-Fueled NaK-Cooled Reactor with Rankine Conversion System

The SNAP series of reactors used UZrH fuel and a "thermal" neutron spectrum. That is, the neutrons released by fission were slowed down by collisions with the hydrogen in the fuel so that they could interact with the uranium more easily. This results in a minimum amount of fuel needed to achieve a self-sustaining reaction, which allows the reactor, and also the radiation shield, to be near minimum mass. The SNAP program produced 6 complete operating reactor systems at various power levels in the 1960s (Angelo, 1985). One system, SNAP-10A, was flown in space (see Figure 1).

The fuel type used in the SNAP series (UZrH) is the same as is used in numerous research reactors throughout the U.S. It was specifically designed for this class of research reactor because of its inherently safe response to temperature changes, automatically reducing the number of fissions if the reactor temperature increases. This feature also allows the system to adjust for load fluctuations without having to move any control elements.

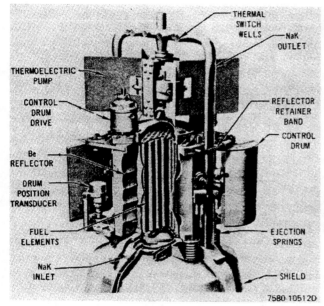

FIGURE 1. SNAP-10A Reactor

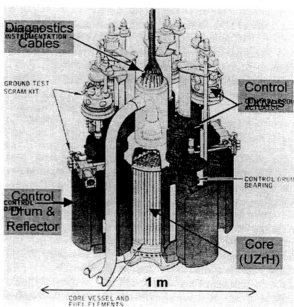

FIGURE 2. SNAP-8DR Reactor.

There were two SNAP reactors which produced 600 kW of thermal power (kWth): SNAP-8ER and SNAP-8DR. The presently envisioned design would be similar to SNAP-8DR (see Figure 2). We would operate the reactor with a peak fuel temperature of only 800 K to extend the lifetime of the fuel. An enrichment of 93% is preferred, but a lower enrichment is also possible. The SNAP-10A spaceflight reactor and shield weighed about 268 and 98 kg respectively, summing to nearly 366 kg. To allow for a larger power and total burnup capability, we estimate the reactor mass would be 500 kg. To allow for shielding a large radiator, we estimate the shield would be 400 kg.

UZrH fuel cannot be operated at as high a temperature as UN fuel. This necessitates the use of a low-temperature conversion system such as a steam Rankine conversion. Steam Rankine systems are a highly mature technology on earth with an extremely large industrial and extensive experience with reactor systems. However, they have never been tested or used in space, and this represents a major technical risk for this option. Figure 3 shows an overview of the conversion cycle. Heat is extracted from the reactor coolant (NaK) and converted to steam. The steam drives a turbine as it expands, and then condenses at a heat exchanger connected to the thermal radiator. A pump recirculates the condensed water back to the boiler. The turbine, pump, and alternator are all on a single shaft floating on a liquid bearing. The system has one moving element. The radiator consists of many parallel and separate capillary pumped loops with ammonia as the coolant. The temperature increase in the steam generator

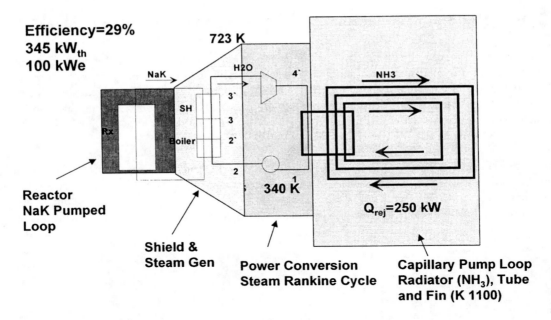

FIGURE 3. Overview of Rankine conversion system.

and superheater is about 383 K, but the temperature increase in the reactor coolant can be considerably less by designing the steam generator appropriately.

The Department of Energy and NASA'a Jet Propulsion Laboratory developed operational hardware for a 25-kW toluene Rankine system driven by solar thermal energy (Nesmith 1985). The program also generated several conceptual designs for 100 kWe systems. We estimate the weight of a TAP with a multi-stage turbine for the KBO mission would be about 90 kg. The additional components for conversion system components (excluding the radiator) are estimated to be about 160 kg, summing to a total of 250 kg.

A major key to success for this system is development of a large lightweight (deployable) radiator. This allows heat rejection to space at a low temperature, which can result in a high conversion efficiency overall. NASA and commercial firms have been working on such radiators (some of the designs and specific diagrams are proprietary). The most advanced designs use an array of parallel capillary pumped loops with thermal fins between them. Ammonia is typically used as the coolant. The use of many separate loops minimizes the need for shielding from micrometeor impact. The total mass of these radiators is about 2 kg/m^2 (which would be 1 kg/m^2 of radiating surface if used in a flat configuration with two-sided radiation). Capillary pumped loops, loop heat pipes, and heat pipes have all been demonstrated in space, but these new lightweight radiator configurations have not.

The total radiating area needed for the baseline design is about 360 m^2 at 330 K. Traditionally space reactor radiators have been designed as conical with only the outside surface radiating. This was done to maximize the radiator area behind the radiation shield. However, to allow easier deployment of the radiator, to minimize the radiator mass, and to help reduce the shield mass, we presently envision the radiator as being flat with both sides radiating. Thus the physical radiator would be 180 m^2 with a total radiating surface of 360 m^2. The radiator would deploy from a manifold extending along the boom. With this potential design and the mass numbers quoted for small advanced systems as background, the radiator system is estimated to weigh 360 kg.

The sum of all these estimated weights (reactor, shield, conversion, radiator) is 1510 kg. Structure, additional controls, other components, and contingency will add to this total estimate, but staying below the 2350-kg limit seems achievable.

UN-Fueled Gas-Cooled Reactor with Brayton Conversion System

The most recent U.S. space reactor power program, SP-100, developed detailed designs, advanced reactor fuel, a "zero-power" reactor critical assembly, radiation-hardened control drives, and various other hardware components in the 1980s. The baseline SP-100 was designed to produce 2400 kW thermal and 100 kW electric with a lifetime of 7 to 10 years (Mondt, 1994, Mondt 1995, El-Genk 1994). The projected specific power at program termination was about 42 kg/kWe. It used a high-temperature advanced fuel (UN) which was developed and proven with nuclear burn-up tests during the program. The fuel was not designed to slow down the fission neutrons, so the neutron spectrum was "fast" and the resulting core size and U-235 mass for the reactor to achieve criticality was thus larger than for the SNAP series.

The proposed design consists of a gas-cooled fission reactor with a closed Brayton cycle for power conversion at 100 kWe. The main difference is replacement of the 4.2% efficient thermoelectric conversion system with a closed Brayton cycle and generator to obtain about 46% conversion efficiency. Such a high efficiency is achieved by using a large thermal radiator, which allows a lower thermal sink temperature. There is a very extensive industrial data base and fabrication experience for open-cycle Brayton units: they form the basis for commercial and military jet engines as well as helicopter engines. Brayton conversion systems have one moving part: a single shaft connected to the turbine, the electrical generator rotor, and the compressor. In closed systems, this single shaft floats on a gas bearing bled off from the main gas flow and returned to it. A 52,000-hr ground test of a 10.7-kWe closed Brayton unit was conducted at NASA/LeRC in 1965.

The reactor is gas cooled (30/70 mole-% He/Xe) to couple better with the Brayton system. The fuel is uranium nitride (UN), which is the same fuel extensively tested for longevity in the SP-100 program. The active core is 0.40 m in diameter and 0.5 m long. The radial reflector is 0.15-m thick beryllium, and the axial reflector is 0.10-m thick BeO. The fuel rods are held in a lattice of BeO, which provides a small amount of moderation. There is a strong negative thermal feedback which allows the reactor power to naturally follow variations in load without needing adjustment of the control elements.

Figure 4 shows a schematic of the Brayton cycle and the associated state points. The reference design produces 100 kWe with 46% total thermal efficiency and has a specific mass of about 26 kg/kWe. A key feature is the heat

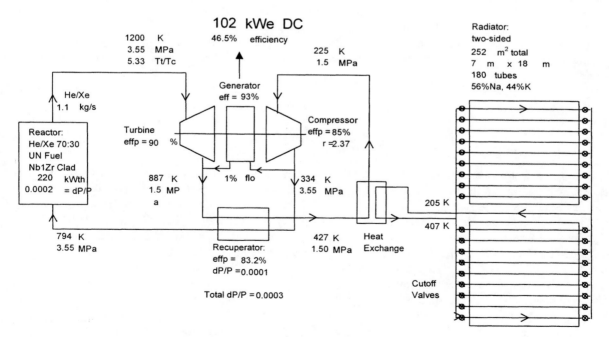

FIGURE 4. Brayton cycle schematic and associated state points.

exchanger which recuperates some heat from the turbine exit and uses it to preheat the gas returning from the heat sink before it re-enters the reactor. This recuperation step gives the cycle a greater conversion efficiency than a "non-recuperated" Brayton cycle. The radiator is opened out flat to radiate from two sides and is 126 m^2 per side. This allows a low heat-sink temperature and enables the high electrical conversion efficiency. Detailed weight estimates are not available for this configuration, but the larger reactor size will likely make it heavier than the UZrH Rankine system.

UN-Fueled Heatpipe-Cooled Reactor with Brayton Conversion System

Houts, Poston, and Emrich have reported on various designs for heatpipe cooled reactors (Houts, 1998; Poston, 1996). In the heatpipe power system (HPS) heatpipes are inserted into the reactor core at regular locations to remove heat by boiling and wicking of the coolant. This heat is then transferred via a heat exchanger to the working fluid of a conversion system. For our application, we would use UN fuel in the reactor, sodium in the heat pipe, and He/Xe in the Brayton conversion system. Thus the system state points would be very similar to the gas-cooled Brayton system. Figure 5 shows a fuel/heatpipe module and a cross section of the reactor.

There are numerous advantages and features of the HPS that make it an attractive near-term system:

1. **Safety**. The HPS is designed to be subcritical for all credible launch accidents.
2. **Reliability**. The HPS has no single-point failures.
3. **Long life**. The design lifetime is in excess of 10 years.
4. **Modularity**. The HPS consists of independent fuel/heatpipe modules which can be tested individually.
5. **Testability**. The HPS system launch hardware can be tested at full power using electrical heaters in place of fuel rods. Unirradiated fuel rods inserted before launch. Full-power nuclear tests might not be required.
6. **Versatility**. The HPS can use a variety of fuel forms, structural materials, coolants, and conversion systems.
7. **Scalability**. The HPS design scales well to beyond 1000 kW thermal power.
8. **Simplicity**. There are few system integration issues since there are no in-core shutdown rods, no hermetically sealed refractory metal vessel or flowing loops, no electromagnetic pumps, no coolant thaw systems, no gas separators, and no auxiliary coolant loop for decay heat removal.
9. **Fabricability**. Most of the fabricated parts are small modules with similar metals; there is no pressure vessel.
10. **Near Term**. The system needs no development of advanced materials or components. It can be developed quickly and inexpensively with few nuclear tests.
11. **Low Mass**. The HPS system has a high fuel fraction in the core since it uses no in-core shutdown rods. The potential for in-space fueling (because of no pressure vessel) allows a more compact form while still meeting launch safety requirements.

LAUNCH APPROVAL

There is a precedent for operating small research-sized reactors in space. There are presently over 30 shut-down nuclear reactors orbiting earth at about 600 km altitude. All but one of these are Russian reactors from Rorsat high-power radar satellites. The one U.S. reactor is SNAP-10A, launched in 1965. Every U.S. launch of a payload involving nuclear material must be reviewed by an Interagency Nuclear Safety Review Panel (INSRP) (Sholtis, 1994). The INSRP reviews the sponsors assessment of the risk and reports to the Office of Science and Technology Policy under the Executive Branch. The President or his designee (usually the Science Advisor) then decides whether to grant launch approval. This process has been followed for 25 launches of nuclear
materials over the past 40 years and approval has always been granted. All but one of these launches have involved radioisotopic power sources, but a space reactor would follow the same process.

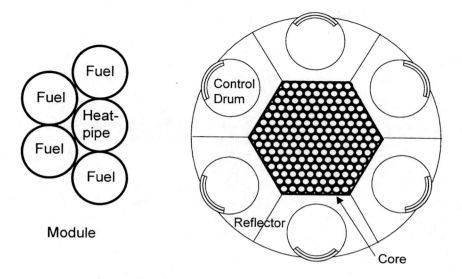

FIGURE 5. Module and reactor for heatpipe power system.

One key safety feature of a fission reactor is that it is barely radioactive before it is used. The radiological inventory in the fresh fuel is about the same as a truckload of uranium ore, with which uranium miners work safely in nations around the world. The space reactor can be tested prior to launch at essentially zero power to prevent any inventory buildup, so the primary concern for launch safety will be to assure that the reactor will not turn on for any conceivable launch accident scenario. This requires solid engineering, but is not difficult. In addition, once the reactor begins operation in space, it should be assured that it cannot reenter the earth's atmosphere. With high circular orbits and low-thrust ion-propulsion, this should be fairly straightforward. Once it is on its way away from earth, the safety issues vanish, and waste disposal is automatic.

POTENTIAL FOR INTERSTELLAR FLIGHT

To have ready access to nearby stars with travel times less than a few years will require speeds faster than the speed of light and "breakthrough" physics. We may all hope for that future day, but in the meantime, we should consider what is feasible using known laws of physics and extrapolations of present-day technologies.

For journeys far outside the solar system, the effect of the sun's gravity and the orbital velocity of the earth may be neglected. Then, for a spacecraft with electric thrusters running at a steady constant power, derivation of the velocity and distance traveled in a given time of thrusting is simply a matter of integrating equation (2) with m_d replaced with the total mass (including propellant), which is time dependent. The final velocity after a total thrusting time (or "burn" time) of t_b and the distance the spacecraft has traveled after thrusting for a duration of t_b and then coasting for an additional duration t_c are:

$$v_b = gI_{sp} \ln\left(\frac{t_b}{\tau} + 1\right) , \tag{3}$$

$$d = gI_{sp}\left(t_b - \tau \ln\left(\frac{t_b}{\tau} + 1\right)\right) + v_b t_c , \tag{4}$$

where τ is the time that it takes for the thrusters to consume an amount of propellant equal to the total dry mass of the spacecraft:

$$\tau = \frac{m_d \left(gI_{sp}\right)^2}{2\eta P} \ . \tag{5}$$

Note that equation (3) is the classical rocket equation in a different form. There is an optimum I_{sp} which results in a maximum d for a given set of t_b, t_c, η, P, and m_d.

We may use these relations to estimate what performance fission-based systems might be able to deliver. Since travel to the nearby stars without new physics is extremely challenging, we will allow a leisurely 200 years for the total mission. Figure 6 shows the total distance traveled after 200 years for various burn durations up to 50 years; the final velocity after 50 years of burn is also shown. The various curves represent different total specific masses for the spacecraft without propellant (m_d/P). Fission power systems at the 1-MW level using present technology should be able to reach 10 kg/kWe, and higher-powered systems with more advanced reactors and conversion systems should be able to reach 1 or 0.1 kg/kWe. A very large system with advanced technology might reach 0.01 kg/kWe. Each curve has a fixed specific impulse which was chosen to maximize the distance traveled for most of the span of the curve.

The figure shows that this technology can bring us a long distance toward interstellar flight. The number of AUs indicated on the right side of the figure is very large and shows just how far this system can penetrate very deep space. That we can even envision travel to a light year without tapping a large fraction of earth's resources and without postulating nation-sized propulsion systems is also impressive. The burn time is not unreasonable; terrestrial reactors are expected to last about 50 years. Refueling the reactor can be accomodated by inclusion of replacement nuclear fuel in the determination of the dry mass. Or, it may be considered as part of the overall "propellant", thrown overboard at zero velocity, and averaged in with the electric-propulsion propellant velocity to yield an effective specific impulse. The same can be done for the electric thrusters which would need replacement. (Most of the curves resulted in about five dry masses worth of propellent being consumed.) The specific impulses are challenging but not unreasonable for large systems.

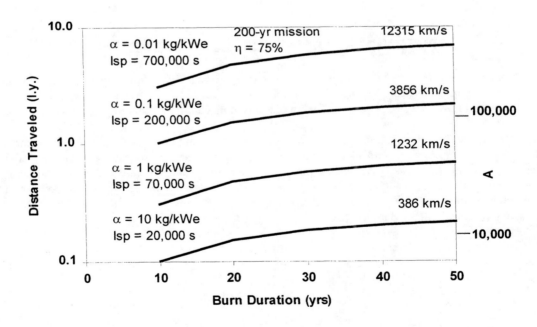

FIGURE 6. Distance traveled in 200 years by a fission-powered electric-propulsion spacecraft.

SUMMARY

Three possible power sources for a fission-based electric propulsion system are described: (1) a UZrH-fueled, NaK-cooled reactor with a steam Rankine conversion system, (2) a UN-fueled gas-cooled reactor with a recuperated Brayton conversion system, and (3) a UN-fueled heatpipe-cooled reactor with a recuperated Brayton conversion system. All three of these systems have the potential to meet the specific mass requirements for interstellar precursor missions capable of 100 km/s in the near term. Advanced versions of a fission-based electric propulsion system might travel as much as several light years in 200 years.

ACKNOWLEDGMENTS

This activity supported by Sandia National Laboratories. Sandia is a multiprogram laboratory operated by Sandia Corporation, a Lockheed Martin Company, for the United States Department of Energy under Contract DE-AC04-94AL85000.

REFERENCES

Anderson, R. V., et al, *Space-Reactor Electric Systems: Subsystem Technology Assessment*, ESG-DOE-13398, Rockwell International, Canoga Park, CA, chapter IV, 1983.

Angelo, Jr., J. A. and D. Buden, *Space Nuclear Power*, Orbit Book Co., Inc., Malabar, Florida, pp. 159-176, 1985.

Chang Diaz, F. R., "Recent Progress in the VASIMR", *1998 Div. of Plasma Physics Meeting, Nov. 16-20, 1998, New Orleans, LA*, 1998.

El-Genk, M., *A Critical Review of Space Nuclear Power and Propulsion, 1984-1993*, M. S. El-Genk, Ed., U. of New Mexico Intitute for Space Nuclear Power Studies, Albuquerque, NM, AIP Press, New York, NY, 1994.

Houts, M. G., D. I. Poston, and W. J. Emrich, Jr., "*Heatpipe Power System and Heatpipe Bimodal System Development Status*", *Space Technology and Applications Internations Forum 1998 (STAIF-98), January, 1998, Albuquerque, NM*, pp. 1189-1195, 1998.

Jewitt, D. C., "Kuiper Belt Objects," *Annual Review of Earth and Planetary Sciences*, **27**, 287-312. (1999).

Lipinski, R. J, R. X. Lenard, S. A. Wright, M. Houts, B. Patton, and B. Poston, "NEP for a Kuiper Belt Object Rendezvous Mission," *Space Technology and Applications Internations Forum 2000 (STAIF-2000), January, 2000, Albuquerque, NM*, Session E3, 2000.

Malhotra, Renu, "Migrating Planets," *Scientific American*, **281**, No. 3, 56-63 (1999).

Mondt, J. F., Truscello, V. C., and Marriott, A. T., "SP-100 Power Program," in *Eleventh Symposium on Space Nuclear Power and Propulsion, Albuquerque, NM 1995*, M. S. El-Genk, Ed., U. of New Mexico Intitute for Space Nuclear Power Studies, Albuquerque, NM, AIP Press, New York, CONF 940101, 1995 , pp. 143-155.

Mondt, J. F., V. C. Truscello, A. T. Marriott, "SP-100 Power Program," *11th Symp. on Space Nuclear Power and Propulsion, Albuquerque, Jan 1994*, CONF-940101, pp. 143-155, 1994.

Nesmith, B., B*earing Development Program for a 25-kWe Solar-Powered Organic Rankine-Cycle Engine*, DOE/JPL-1060-92, Jet Propulsion Laboratory, Pasadena, CA, 1985.

Polk, J., Anderson, J.R., Brophy, J.R, Rawlin, V. K, Patterson, M.J., and Sovey, J.S., "The Effect of Engine Wear on Performance in the NSTAR 8000 Hour Ion Endurance Test," *Joint Propulsion Conference, Cleveland, OH, July 14, 1998*, AIAA 97-3387, 1998.

Poston, D. I. and M. G. Houts, "*Nuclear and Thermal Analysis of the Heatpipe Power and Bimodal Systems*", *Space Technology and Applications Internations Forum 1998 (STAIF-96), January, 1996, Albuquerque, NM*, pp. 1083-1093, 1996.

Sholtis, Jr., J. A., Connell, L. W., Brown, N. W., Mims, J. E., and Potter, A., "U. S. Space Nuclear Safety: Past, Present, and Future," in *A Critical Review of Space Nuclear Power and Propulsion 1984-1993*, M. S. El-Genk, Ed., U. of New Mexico Intitute for Space Nuclear Power Studies, Albuquerque, NM, AIP Press, New York, NY, 1994, pp. 269-304.

MICF: A Fusion Propulsion System for Interstellar Missions

Terry Kammash[1] and Brice N. Cassenti[2]

[1]*Nuclear Engineering and Radiological Sciences Dept., University of Michigan, Ann Arbor, MI 48109*
[2]*United Technologies Research Center, 411 Silver Lane, MS 129-73, East Hartford, CT 06108*

[1]*(734) 764-0205 (voice), (734) 763-4540 (fax), tkammash@umich.edu*
[2]*(860) 610-7460 (voice), (860) 610-7536 (fax), cassenbn@utrc.utc.com*

Abstract. A very promising propulsion device that could open up the solar system and beyond to human exploration is the Magnetically Insulated Inertial Confinement Fusion (MICF) system. This scheme combines the favorable aspects of inertial and magnetic fusion into one where physical containment of the hot plasma is provided by a metal shell while its thermal energy is insulated from this wall by a strong, self-generated magnetic field. The fusion nuclear reactions in this device can be triggered by a beam of antiprotons that enters the target through a hole and annihilates on the deuterium-tritium (DT) coated inner wall giving rise to the hot fusion plasma. In addition to thermally insulating the plasma, the magnetic field helps to contain the charged annihilation products and allows them to deposit their energy in the plasma to heat it to thermonuclear temperatures. Preliminary analysis given in this paper shows that an MICF propulsion system is capable of producing specific impulses on the order of 10^6 seconds. Such capability makes not only the most distant planet in the solar system, but also the nearest star reachable in a human's lifetime. It also shows that a robotic mission to 10,000 AU can readily be achieved in less than 50 years.

INTRODUCTION

The Magnetically Insulated Inertial Confinement Fusion (MICF) concept (Hasegawa, 1988 and Kammash, 1989) illustrated in Fig. 1 represents a novel approach to fusion in that it combines the desirable confinement features of both magnetic and inertial fusions into one. In this scheme a core of plasma is created at the center of a hollowshell by bombarding the fuel-coated inner surface with a laser or particle beam that enters the shell through a hole. The ablated electrons form current loops (Sawanakamp, 1986) which in turn give rise to a strong magnetic field that serves to thermally insulate the plasma from the solid walls. Although different theories have been advanced to explain the mechanism for the generation of such a field a plausible explanation revolves around the formation of density and temperature gradients in the ablation region. According to the generalized Ohm's law these gradients, when at an angle with respect to one another , give rise to an electric field which in turn gives rise to a time-varying magnetic field.

The major advantage of MICF over conventional "implosion" type inertial fusion is the long lifetime of the plasma which allows it to burn longer and generate more fusion energy as represented by very large gain factors. This lifetime is dictated by the time it takes a shock wave to traverse the metal shell that surrounds the plasma once it is formed when the laser or particle beam strikes the inner wall. The sound speed in the shell is much slower than that in the plasma (which dictates the lifetime in the implosion type system) due to the larger atomic mass of the metal shell and its lower temperature that results from the thermal insulation provided by the self-generated magnetic field. Moreover the beam plasma coupling is much more efficient since the energy is put directly into the plasma rather than in an imploding pusher. The Rayleigh-Taylor instability which is known to plague the implosion type inertial fusion is totally eliminated in MICF due to the fact that the lighter fluid (the plasma) is supported by the heavier fluid (the shell) in the presence of a gravitational field. These unique properties allow for a significantly higher

CP504, *Space Technology and Applications International Forum–2000*, edited by M. S. El-Genk
© 2000 American Institute of Physics 1-56396-919-X/00/$17.00

energy multiplication factors making MICF a particularly attractive fusion power. As a propulsion system (Kammash, 1987) it is envisaged that MICF pellets injected into a burn chamber will be triggered by an incident laser or particle beam at the center, and the reaction products along with the ionized shell material exhausted through a magnetic nozzle which will be an integral part of an externally-applied magnetic configuration that also serves to cushion the chamber walls from the micro-explosion shock.

In this paper we will examine the propulsion capability of an anti-proton driven MICF and apply the results to an interstellar mission. Specifically we will address a robotic exploration mission to 10,000 AU and show that it can be done in less than 50 years.

PROPULSION ANALYSIS OF MICF

We consider a target such as that shown in Fig. 1 with a radius of 1 cm. and a core of 0.25 cm. whose wall is made up of 50% - 50% mixture of deuterium (D) and tritium (T). Through a hole with a diameter of 10 μm a beam of anti-protons is made to impinge on the wall ablating it and creating a hot dense plasma in the core. An appropriately chosen number of anti-protons and a pulse length are selected to create a plasma density of about 10^{23} cm^{-3} with a temperature of 20 keV in the core to initiate the burn in the target (Cassenti, 1997). With a metal shell such as tungsten surrounding the fusion fuel, and a magnetic field of several megagaus formed to provide the needed thermal insulation, the confinement time is found to be long enough for the fusion reactions to proceed and generate a large energy content in the pellet. The large energy production is attributed to the alpha particles generated by the DT reactions whose energy is deposited in the plasma to sustain the reactions. For a plasma with an ion density n_i, the number of alpha particles per unit volume generated by the fusion reactions is given by:

$$n_\alpha = \frac{n_i^2}{4} \langle \sigma v \rangle (t),$$

(1)

where $\langle \sigma v \rangle$ is the velocity averaged product of the fusion cross section σ and relative velocity v of the interacting particles. If we assume that the time t is equal to the reaction time $\tau_R = \left[n_i \langle \sigma v \rangle \right]^{-1}$, and maintain that the alpha energy E_α of 3.5 MeV does not get degraded in this

time period then:

$$n_\alpha = \frac{n_i}{4} = \frac{n_D}{2} = \frac{n_T}{2},$$

(2)

with n_D and n_T representing the deuterium and tritium densities respectively; then the energy content of the pellet can be primarily attributed to the alphas or:

$$E_T = n_\alpha E_\alpha V,$$

(3)

where V is the volume of the pellet. At 20 keV, $\langle \sigma v \rangle = 4.2 \times 10^{-16}$ cm^3/sec. and for $n_i = 10^{23}$, τ_R is found to be about 2.5×10^{-8} seconds which is about 0.250 times shorter than the plasma confinement time in MICF (~10^{-7} sec.). For a pellet of 3.5 gm. mass, the velocity with which is ejected is given by:

$$v_e = \sqrt{\frac{2 E_T}{m_T}},$$

(4)

which for the above scenario gives $v_e \cong 10^7$ m/sec. and a specific impulse $I_{sp} \sim 10^6$ seconds.

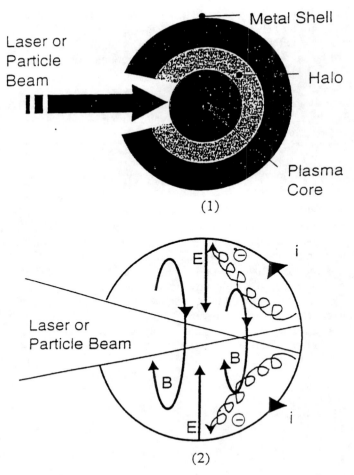

FIGURE 1. Schematic of (1) Plasma Formation and (2) Magnetic Field Formation in MICF.

The above result however may be somewhat optimistic since, as we shall see shortly, the alpha particle slows down in the plasma in a time which is much shorter than the reaction time. The slowing down process takes place as a result of interaction of the alpha particle with the electrons and ions of the plasma as represented by (Kammash, 1975).

$$\frac{dE}{dt} = C_1 E + C_2 / \sqrt{E} , \tag{5}$$

where C_1 and C_2 are given by:

$$C_1 = 2 \times 10^{-12} \ n_i \ (cm^{-3}) / T_e^{3/2} \ (keV) \tag{6}$$

$$C_2 = 9 \times 10^{-10} \ n_i \ (cm^{-3})/M_1 \ (amu) \tag{7}$$

When the plasma confinement time in the pellet is much longer than the reaction or slowing down times then the alpha particle density can be expressed by (Kammash, 1975)

$$N_\alpha = \int n_\alpha(E)dE,$$ (8)

where the alpha density per unit energy $n_\alpha(E)$ is given by:

$$n_\alpha(E) = \frac{\frac{n_i^2}{4}\langle \sigma v \rangle}{dE/dt},$$ (9)

with the denominator given by Eq. (5). Integration of Eq. (5) between the thermal energy E_{th} of 20 keV and the alpha birth energy of E_0 = 3.5 meV yields the thermalization time $\tau_{Th} \cong 2.32 \times 10^{-9}$ sec., which we observe to be much shorter than τ_R. This simply indicates that the alphas thermalize almost instantly after birth and their energy goes into heating the plasma. A quick estimate of this heating can be obtained by assuming alpha slowing down to be on the electrons only (i.e. ignoring the last term of Eq. 5). using Eqs. (9) and (5) into (8) we find that N_α= 2.44 x 10^{21} cm^{-3}. The change in the electron temperature T_e due to alpha heating can be written as (Kammash, 1992):

$$\frac{3}{2}\Delta T_e = \frac{RE_0}{1+R},$$ (10)

where $R = N_\alpha/n_i$. With the above result of 85 keV added to the initial electron temperature of 20 keV, the final electron temperature of 105 keV will also be the plasma temperature on the assumption that thermalization between the electrons and the ions of the plasma is also shorter then the confinement time. Using these results we find that E_T in this case is 5.3 x 10^9 joules, and that in turn gives a pellet exhaust velocity of v_e = 1.74 x 10^6 m/sec. resulting in an I_{sp} = 1.74 x 10^5 seconds.

MISSION ANAYLSIS

We consider a robotic fly-by exploration mission to the Oort cloud which is about 10,000 AU away using the MICF propulsion system examined above. If we denote by S_f the distance to the destination, and by t_f the time it takes to reach it, then from the standard (non-relativistic) rocket equation we can write (Cassenti, 1999)

$$t_f = \frac{M_i - M_f}{F}v_e,$$ (11)

$$S_f = \frac{M_i v_e^2}{F}\left[1 - \frac{M_f}{M_i} + \frac{M_f}{M_i}\ell n\left(\frac{M_f}{M_i}\right)\right],$$ (12)

$$v_f = v_e \ell n\left[\frac{1}{1 - Ft_f/M_i v_e}\right],$$ (13)

where M_I is the initial mass of the vehicle, M_f its dry mass, F the thrust, and v_f the final velocity of the vehicle when it reaches its destination assuming it started from rest. We assume that ten pellets are ejected per second; a rep rate of $\omega=10$ that is considered technologically feasible. We also take the dry mass to be 220 mT which is consistent with a laser driven inertial fusion propulsion system (Hyde, 1983) but without the laser driver on the assumption that the mass of an antiproton (or anti-hydrogen) source would be comparatively small to be neglected. With $S_f = 10,000$ AU, $F = \omega\, m_T v_e$ we calculate M_i from Eq. (12) and substitute it in Eq. (11) to calculate t_f. When these results are put in Eq. (13) we calculate v_f. These results for the mission in question are found to be:

$$F = 6.09 \times 10^4\, N,$$

$$t_f = 28.53\, Y,$$

$$v_f = 7.36 \times 10^6\, m/s \cong 0.025c,$$

where c is the speed of light.

CONCLUSION

We have shown that a novel fusion system that combines the favorable aspects of both magnetic and inertial fusions can produce very impressive propulsive parameters when used as a propulsion device. We have demonstrated that an MICF propulsion system driven by antiprotons is capable of making a fly-by mission to a 10,000 AU destination in about 28 years, well below the 50 year travel time often cited as desirable for such a mission.

ACKNOWLEDGMENTS

The work of one of us (TK) was supported by Universities Space Research Association.

REFERENCES

Cassenti, B. N., Kammash, T., and Galbraith, D. L., "Antiproton Catalyzed Fusion Propulsion For Interplanetary Missions," Journal of Propulsion and Power, **13**, pp. 428-434, (1997).

Cassenti, B. N., and Kammash, T., "Antiproton Triggered Fusion Propulsion For Interstellar Missions," Space Technology and Applications International Forum (STAIF-99) CONF-990103, held 31 January – 4 February 1999, AIP Conference Proceedings No. 458, Vol. 1: 1-760; Vol. II: 761-1666; CD ROM.

Hasegawa, A., Daido, H., Fujita, M., Mima, K., Murakami, M., Nakai, S., Nishihara, K., Terai, K., and Yamanaka, C., "Magnetically Insulated Inertially Confined Fusion – MICF," Nuclear Fusion, **28**, pp. 369-387, (1988).

Hyde, R. A., "A Laser Fusion Rocket For Interplanetary Propulsion," Lawrence Livermore National Laboratory Report, UCRL-88857, (1983).

Hyde, R. A., "A Laser Fusion Rocket For Interplanetary Propulsion," Lawrence Livermore National Laboratory Report, UCRL-88857, (1983).

Kammash, T., and Galbraith, D. L., "A High Gain Fusion Reactor Based On The Magnetically Insulated Inertial Confinement Fusion (MICF) Reactor," Nuclear Fusion, **29**, pp. 1079-1099, (1989).

Kammash, T., and Galbraith, D. L., "A Fusion Reactor For Space Applications," Fusion Technology, **12**, pp. 11-21, (1987).

Kammash, T., "Fusion Reactor Physics, Principles and Technology," Ann Arbor Science Publishers, Ch. 3, (1975).

Kammash, T., and Galbraith, D. L., "Antimatter-Driven Fusion Propulsion System For Solar System Exploration," Journal of Propulsion and Power, **8**, pp. 644-649, (1992).

Sawanakamp, S., and Kammash, T., "Estimate Of The Magnetic Field Generated In The Magnetically Insulated Fusion Reactor," Bulletin American Physical Society, **31**, p. 1411, (1986).

Dielectric Films for Solar and Laser-pushed Lightsails

Geoffrey A. Landis

Ohio Aerospace Institute, NASA Glenn Research Center mailstop 302-1, Cleveland, OH 44135

Abstract. This project analyzed the potential use of dielectric thin films for solar and laser sails. Such light-pushed sails allow the possibility of fuel-free propulsion in space. This makes possible missions of extremely high delta-V, potentially as high as 30,000 km/sec (0.1c), which is required for a fly-by mission to a nearby star.

INTRODUCTION

Beam-pushed propulsion systems, such as solar- or laser-pushed sails, allow the possibility of fuel-free propulsion in space. This makes possible missions of extremely high delta-V, potentially as high as 30,000 km/sec (0.1c), which is required for a fly-by mission to a nearby star. A sail pushed by an extremely high power laser was proposed by Robert Forward for interstellar flight (Forward, 1984), and is one of the only proposed technologies that can be used for this mission without either unrealistic travel time, or the requirement for new physical principles to be discovered. A workshop on "Robotic Interstellar Exploration in the Next Century" held at the California Institute of Technology in 1998 concluded that such sail technologies could be adapted for a number of missions, including Kuiper belt and outer planet missions as well as a prototype interstellar explorer (Landis, TBP). This workshop also looked at dust damage to the sail (Early and London, 1999).

LASER-PUSHED LIGHTSAIL

A laser-pushed sail is fundamentally different from a solar sail. Solar sail concepts typically use a thin reflective (metal) layer on a plastic film. The Forward laser sail concepts leaves the plastic film behind. The plastic is the major mass component, limits operating temperature, and is unnecessary; since the metal film itself can be made self supporting. Therefore, only the reflective layer is used. For the bare metal film proposed by Forward, the film thickness can be 20 nm or so (200 atoms thick). Film deposition on removable substrate has been demonstrated. The laser sail will need a supporting secondary structure, typically a fractal mesh. It cannot be rolled up and deployed, which means that it must be assembled in space.

The difficulty is that the sail and focusing lens required for an interstellar fly-by mission with an aluminum film sail, as proposed by Forward, is enormous, and the power required is huge. A lens is required to keep the beam spread due to diffraction at the aperture low. The fundamental diffraction-limit to beam spread is

$$\_y \_ 2.44 \lambda s /a, \tag{1}$$

where λ is the laser wavelength, a the effective laser aperture and s the distance. (The laser spot actually has an exponential tail outside this distance, but 84% of the light falls within the limit listed). To minimize the beam spread, a large lens is used. The effective aperture is then equal to the lens radius rather than the physical size of the laser. Forward proposed that extremely large lenses (thousands of kilometers) can be made using the "paralens" concept; alternating rings of thin material with refractive index n alternating with empty space to form a very large fresnel zone plate. Such a laser-pushed sail, as proposed, is in no way a "micro" mission.

CP504, *Space Technology and Applications International Forum–2000*, edited by M. S. El-Genk

The solution is that higher acceleration is required, to allow the sail to reach cruise velocity closer to the laser. By completing acceleration closer to the laser there is less beam divergence, and thus smaller lens and sail sizes are possible. In fact, the minimum diameter of lens or sail is inversely proportional to acceleration, and the minimum required laser power also reduces directly proportional to acceleration.

However, the acceleration is thermally limited-- the sail can't accelerate more with an Al sail, since the aluminum cannot take the high temperatures required. High acceleration sails require a better sail material. Metal films have low emissivity; and thus get hot in beam. A material with higher melting temperature is needed. Most preferably, we need a high- emissivity refractory material.

An earlier analysis (Landis, 1997) proposed a beryllium sail. Unfortunately, further analysis indicates that the infrared emissivity of beryllium is extremely poor, and thus that a beryllium sail will heat up in the beam more than indicated by the initial calculation.

DIELECTRIC SAIL FILMS

This project analyzed the potential use of dielectric thin films for solar and laser sails (Landis 1989). By use of the interference property of the sails, a high reflectivity (in principle over 50% for high refractive-index films) can be achieved. Reflectivity is maximum when the thickness of the film is one quarter the wavelength of the light measured inside the film, when the reflected light from the front and rear of the film interfere constructively:

$$t = \lambda/(4n), \tag{2}$$

where n is the index of refraction. The higher the refractive index, the thinner the film can be to provide maximum reflectivity. The reflectivity of a quarter-wave single-layer thin film of a dielectric in vacuum is:

$$R = [(n^2 -1)/(n^2 +1)]^2. \tag{3}$$

Table 1 shows some example reflectivities of quarter-wave films.

Advantages of dielectric films include refractory operating temperatures, very high emissivity, and low absorptivity. The absorptivity α should be <<1%. The emissivity ε will depend on thickness, but for all the films studied, the α/ε ratio should be <0.1 (compared to α/ε of 13%/6% for Al film) . Since thermally limited acceleration scales with α/ε ratio and with the fourth power of the temperature, these films are extremely attractive for high-acceleration sails.

The amount of power that can be radiated by the sail is proportional to the maximum temperature (Tm) raised to the fourth power. Assuming that the absorption and the emissivity are fixed, this sets the thermal limit on the amount of laser power per unit area that can be focused on the sail, and hence sets the maximum force per unit sail area that can be achieved. The maximum acceleration which can be achieved is equal to the maximum force per unit area divided by the sail mass per unit area, which is equal to the mass density (ρ) times the thickness. The acceleration will also be proportional to the reflectivity.

Putting all of these terms together, if we compare sails of equal thickness, the figure of merit for acceleration of the sail, Z, will be proportional to the product of the fourth power of the maximum temperature divided by the density:

$$Z = \alpha/\varepsilon \, RT^4/\rho. \tag{4}$$

Thus, along with the α/ε ratio, the maximum temperature T and the density ρ are thus the critical parameters to selecting the sail material. (Note that for a more detailed calculation, the reflectivity, emissivity, and absorptivity are also critical). Physical properties of some dielectric films are shown in table 2.

The advantages are extremely light weight and good high temperature properties, which are necessary for both for solar-sail missions inward toward the sun, for solar sail missions outward from the sun that use a close perihelion pass to build speed, and for high velocity laser-pushed missions for the outer solar system and for interstellar probes. Because of the higher temperature capability, the sails can operate under higher laser illumination levels, and hence achieve higher acceleration. This allows large decreases in the minimum size of the sail required, and makes the power requirement for the interstellar mission an amount that can be achieved in the reasonable future.

TABLE 1. Reflectivity of Representative Refractory Dielectric Materials, assuming a quarter wavelength film (maximum reflectivity)

Material	Reflectivity
Oxides	
Alumina (Al_2O_3)	26%
Tantalum Pentoxide	52%
Zirconium dioxide	42%
Fluorides	
Lithium fluoride	13%
Semiconductors	
Diamond	50%
Silicon	75%
Silicon Carbide	56%
Zinc sulfide	48%

TABLE 2. Physical Properties of Some Refractory Dielectric Materials.

Material	Max Temp (°C)	Density (g/cm³)	
Zirconium dioxide	2715	5.5	
Aluminum trioxide	2072	3.96	
Silicon Carbide	2000.*	3.17	(*sublimes)
Tantalum Pentoxide	1870	8.75	
Diamond	1800.†	3.5	(†graphite conversion)
Silicon dioxide	1600	2.7	
Lithium fluoride	820	2.6	

SUMMARY

The potential use of thin, self-supporting dielectric thin films as sail materials offers the possibility of improved performance from laser-pushed lightsails for high-performance missions, such as an interstellar flyby mission. The advantages of the dielectric films include high temperature capability, low absorption, and high emissivity. These properties allow higher laser intensities on the sail. The higher laser intensities allows the design of a high-acceleration system which can achieve the required acceleration closer to the laser, and thus permits a smaller and lower cost spacecraft.

ACKNOWLEDGMENTS

This study was funded by a grant from the NASA Institute for Advanced Concepts (NIAC). For more information about NIAC-funded research, see http://www.niac.usra.edu/

REFERENCES

Early, J.T., and London, R.A., "Dust Grain Damage to Interstellar Laser-Pushed Lightsail," Lawrence Livermore National Laboratory preprint number UCRL-JC-133687, March 25 1999. Submitted to *J. Spacecraft and Rockets.*

Forward, Robert L., "Roundtrip Interstellar Travel Using Laser-Pushed Lightsails," *J. Spacecraft and Rockets,* Vol. 21, Mar-Apr. 1984, pp. 187-195.

Landis, G.A., "Optics and Materials Considerations for a Laser-propelled Lightsail," Paper IAA-89-664, 40th IAF Congress, Torremolinos Spain, Oct. 7-13, 1989.

Landis, G.A., "Small Laser-propelled Interstellar Probe," *Journal of the British Interplanetary Society, Vol. 50,* No. 4, 149-154 (1997); Paper IAA-95-4.1.1.02, presented at the *46th International Astronautics Federation Congress,* Oslo Norway, 2-6 Oct. 1995.

Landis, G.A., "Beamed Energy Propulsion for Practical Interstellar Flight," to be published, *Journal of the British Interplanetary Society.*

Laser-Light Sailing and Non-Stationary Power Stations Applied to Robotic Star Probes

Gregory L. Matloff [1,2,3]

[1] *New York University, General Studies Program 50 West 4th Street, New York, NY, 10012-1165*
[2] *Dept. of Physical and Biological Sciences, New York City Technical College,, CUNY, 300 Jay Street, Brooklyn,NY,11201*
[3] *Summer 1999 Faculty Fellow, ASEE/UAH Program at NASA/MSFC, Alabama*

(718) 638-7586; gm21 @is3.nyu.edu

Abstract. The light sail has emerged as a leading contender to propel extrasolar expeditions. Because solar-sail performance is limited by the inverse-square law, one-way expeditions to other stars requiring voyage durations of a few centuries or less may be propelled by radiation pressure from a laser beam originating from a location closer to the Sun than the space probe. Maintaining a stationary laser power station in position between Sun and spacecraft for yearrs or decades presents many technical challenges. This paper presents a variation on the laser power station that may be simpler to implement, in which the Sun-pumped laser power station follows the spacecraft on a parabolic or slightly hyperbolic trajectory.

INTRODUCTION

At the conclusion of the British Interplanetary Society's Project Daedalus, project directors concluded that only nuclear-pulse and light-sail propulsion had the capability to eventually transfer human populations to nearly stars on voyages approximating a millennium (Bond, 1984). Perhaps because nuclear-pulse propulsion presents political and perceived ecological problems, the light-sail has emerged as a leading contender to propel early extrasolar expeditions (Johnson, 1999).

Researchers have long recognized interstellar solar-sail opportunities and limitations (Matloff, 1981 and Mallove, 1989). In an interstellar solar-sailing expedition, a highly reflective, strong, thermally resistant thin-film sail is unfurled as close to the Sun as possible. Travel times to nearby stars are governed by the inverse-square law of solar flux and many centuries of travel are required for both robotic probes and human-occupied "world ships."

A method of overcoming the limitations of the solar sail is to project a highly collimated laser beam from a Sun-pumped laser closer to the Sun than the spacecraft. The spacecraft is propelled by the radiation pressure of photons in the impinging laser beam. Variations of this approach have been suggested to propel both thin-film robotic "starwisp" probes and human-occupied interstellar expeditions of exploration / colonization (Forward, 1984 and 1985). But the technical challenges of maintaining beam-collimation over a distance measured in thousands of Astronomical Units (AU) for years or decades is enormous. It may not prove possible to maintain a power station in a stationary position between Sun and spacecraft for long time period. Even the slightest laser-beam jitter may result in loss of the mission.

We present here an alternative approach to power-station positioning (Matloff, 1996). Instead of attempting to maintain the laser power station stationary relative to the interstellar spacecraft and Sun, the power station follows the interstellar spacecraft on a parabolic or slightly hyperbolic trajectory. Theory of Non-stationary Laser Power Stations.

CP504, *Space Technology and Applications International Forum–2000*, edited by M. S. El-Genk

If we assume constant laser power (P_L) delivered to the interstellar spacecraft during the acceleration process,

$$P_L = \varepsilon \frac{1400\, A_{col}}{R^2_{max.\,au}}\,,$$

(1)

where ε = laser efficiency, A_{col} = power-station solar-collector area, and $R_{max,au}$ = maximum separation between Sun and laser during acceleration, in AU. The acceleration of the Interstellar spacecraft is expressed:

$$dV_{sc} = \frac{(1+k)P_L}{M_{sc}c}\,,$$

(2)

where k = sail reflectivity, M_{sc} = spacecraft mass, and c = the speed of light. Figure 1 presents the positions of the Sun, power station, and interstellar spacecraft.

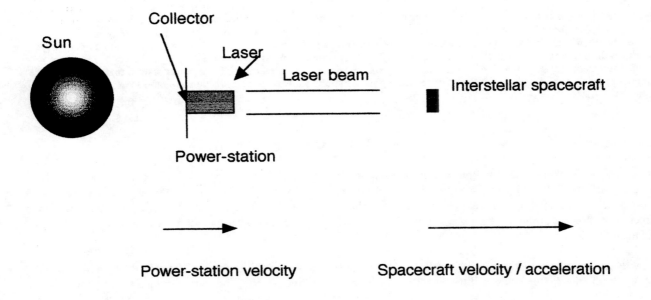

FIGURE 1.Representation of non-stationary laser power station and interstellar light -sail spacecraft.

If the power-station is in a near-parabolic orbit and commences operation when the interstellar spacecraft is released near the Sun, the time in years for the power station to reach a distance $R_{p,au}$ from the Sun, in Astronomical, Units can be written (Matloff, 1988).

$$T_{yr} = 0.077\, R^{1.5}_{p,au}$$

(3)

From elementary kinematics, the velocity of the spacecraft relative to the sun after T_{yr} acceleration can be written (assuming initial low spacecraft velocity relative to the Sun).

$$V_F = 3.15 \times 10^7 \, T_{yr} \frac{dV_{sc}}{dt} \, . \qquad (4)$$

During acceleration the spacecraft's distance from the Sun increases by

$$R_{sc} = 3.15 \times 10^7 T_{yr} (\frac{V_F}{2}) \, . \qquad (5)$$

The maximum separation R_{sep} between power station and interstellar spacecraft at the conclusion of the acceleration interval will be $R_{sep} = R_{sc} - 1.5 \times 10^{11} R_{max,au}$. Rayleigh's Criterion can be applied to relate R_{sep} to the laser wavelength in microns, λ_μ, laser-transmitting optics radius in meters, RAD_L, R_{sep} and the sail radius in meters, RAD_S (Matloff, 1996):

$$\frac{6.1 \times 10^{-7} \lambda_\mu}{RAD_L} = \frac{RAD_S}{R_{sc} - 1.5 \times 10^{11} R_{max,au}} \, . \qquad (6)$$

SOME MISSION SCENARIOS

Consider some missions based upon a 35-km diameter laser-power-station collector (A_{col} =9.6 X 10^8 m^2), similar to that discussed in a recent NASA "Interstellar Roadmap", (NASA, 1999). If the acceleration duration is 1 year, Eq.(3) reveals that the separation between power station and Sun after 1 year is about 5.5 AU.

Next assume a laser effficiency of 20%. Equation (1) can be used to calculate laser beam power as about 9 X 10^9 watt for maximum sun-laser separation of 5.5 AU. For a sail reflectivity of 0.9, substitution in Eq.(2) yields

$$\frac{dV_{sc}}{dt} = \frac{57}{M_{sc}} \, . \qquad (7)$$

Now consider two Interstellar spacecraft with masses of 50 and 100 Kg. The accelerations of these two spacecraft are respectively 1.14 and 0.57 m/sec^2. Applying Eq.(4), the final velocities of the 50 and 100 kg interstellar spacecraft after a 1 year acceleration period are respectively 3.59 X 10^7 m/sec (0.12c) and 1.8 X 10^7 m/sec (0.06c).

Applying Eq.(5), the 50-Kg spacecraft's distance from the Sun at the end of the acceleration interval is 5.65 X 10^{14} meters (3770 AU). At the end of the 1-year acceleration interval, the 100-kg spacecraft is at distance of 2.84 X 10^{14} meters (1890 AU) from the Sun.

If we next assume a 0.411μ laser and a 500-km radius-transmitter radius, substitution of there values and the values for R_{SC} and $R_{max,au}$ defined above into Eq. (6) allows us to calculate sail radius RAD_S for the 50-kg and 100-kg missions. For the 50-kg interstellar spacecraft, RAD_S = 0.28 km; RAD_S = 0.14 km for the 100-kg spacecraft. Assuming a 16-nm thick aluminum sail (NASA,1999), the sail masses for the 50-kg and 100-kg interstellar spacecraft are respectively 11 and 3 Kg. Both missions seem possible from the point of view of sail and payload mass budgets.

CONCLUSIONS

Although far from presenting an optimized mission design, this paper demonstrates that star probes requiring a human lifetime or less to fly through neighboring stellar systems are not impossible, if launched by beam pressure from non-stationary Sun pumped laser power stations. Many problems of beam collimation, orbital perturbations, etc, remain to be solved before the proposed approach can be considered feasible.

Improvements in space-payload miniaturization smaller laser wavelength, higher laser efficiency, etc. may simplify the probe-acceleration process. Use of a small thruster aboard the power station enables its reuse after the conclusion of probe acceleration.

ACKNOWLEDGMENTS

The author greatly appreciates the hospitality of Les Johnson and other NASA/MSFC personnel and the assistance of Gerald Karr and Martha Hammond of University of Alabama, Huntsville, during his Summer'99 tenure at Marshall Space Flight Center. He is also grateful for the technical assistance of Russell Lee and Alkesh Mehta, accompanying students from New York City Technical College City University of New York.

NOMENCLATURE

c = speed of light, 3×10^8 m/sec

k = interstellar spacecraft sail fractional reflectivity

A_{col} = laser-power-station solar-collector area, m^2

dV_{sc}/dt = interstellar spacecraft acceleration, m/sec^2

M_{sc} = interstellar spacecraft mass, kg

P_L = laser-beam power, watts

$R_{max,au}$ = maximum sun-laser separation, Astronomical units (AU)

$R_{p,au}$ = Sun-laser separation at time T_{yr}, AU

R_{sc} = Sun-spacecraft separation, m

R_{sep} = laser-spacecraft separation, m

RAD_L = laser transmitting-optics radius, m

RAD_s = interstellar spacecraft sail radius, m

T_{yr} = time when Sun-laser separation is $R_{p,au}$, years

V_F = velocity of interstellar spacecraft relative to Sun at time T_{yr}, m/sec

ε = laser fractional efficiency

λ_μ = laser wavelength, microns

REFERENCES

Bond, A and Martin, A.R., "World Ships—An Assessment of the Engineering Feasibility, "*JBIS, 37*, 254-266 (1984).

Forward, RL., "Roundtrip Interstellar Travel Using Laser-Pushed Lightsails," *J. Spacecraft and Rockets, 21*, 187- 195 (1984).

Forward, RL., "Starwisp: An Ultralight Interstellar Probe," *J. Spacecraft and Rockets, 22*, 187 - 195 (1985).

Jobnson, L., and Leifer, S., "Interstellar Exploration: Propulsion Options for Precursors and Beyond," IAA-99-IAA.4.1.05, 1999.

Mallove, E.F. and Matloff, G.L., *The Starflight Handbook*, Wiley, NY (1989)

Matloff, G.L., and Mallove, E.F.., "Solar Sail Starships, the Clipper Ships of the Galaxy," *JBIS, 34*, 371 -380 (1981).

Matloff, G.L., and Parks.K. ,"Interstellar Gravity Assist Propulsion: A Correction and New Application," *JBIS, 41*, 519-526 (1988).

Matloff, G.L., and Potter, S, "Near-Term Possibilities for the Laser-Light Sail," in *Missions to the Outer Solar System and Beyond: Proceedings of 1st IAA Symposium on Realistic Near-Term Advanced Scientific Space Missions*, ed. G. Genta, Levrotto & Bella, Turin, Italy, 1996.

Advanced Concepts Office, "Interstellar Science, Technology, Mission Roadmap, " NASA, 1999.

Experimental Investigation of 5-D Divergent Currents as a Gravity-Electromagnetism Coupling Concept

M. Tajmar

Vienna University of Technology, Institut für Allgemeine Elektronik und Quantenelektronik, Gußhausstrasse 27-29, 1040 Vienna, Austria, Email: tajmar@bigfoot.com

Abstract. The generation of thrust using corona discharges inside a sealed Faraday-cage box is investigated. A high-voltage power supply is connected to a box via strings which allow the box to move like a pendulum. The thrust is derived from its angular position using a laser displacement meter. A 5-dimensional relativistic Maxwell theory, which couples the electromagnetic and gravitational field and predicts the generation of thrust in such a configuration, is reviewed and discussed. Only oscillations during the corona discharge but no linear thrust as predicted was observed within the detectability of the instrumentation. A brief history of similar research which resulted in the claim of a force (Biefeld-Brown effect) is evaluated and discussed. The obtained results suggest that corona wind effects were misinterpreted as a connection between gravity and electromagnetism. If such an effect exists in the presented configuration, it must be below the detection threshold of the used instrumentation.

INTRODUCTION

Because the propellant onboard a spacecraft contributes to a large extend to the overall mass, propellentless propulsion with thrust levels at least comparable with existing electric propulsion thrusters could reduce current costs for space exploration dramatically. Conventional concepts along this goal use either electromagnetic tethers (utilizing the Earth's magnetic field) or photons (solar sails or laser propulsion). More exotic ideas speculate about a connection of gravitation and electromagnetism which could be used for propulsion purposes. However, such a connection which would allow to modify the gravitational field does not appear in nowadays physics, at least not for standard terrestrial laboratory conditions.

Reviewing unconventional approaches to this idea (Cravens, 1990) one finds the so called Biefeld-Brown effect, discovered by Dr. Paul Biefeld and Thomas Townsend Brown in the early 1920's. They found that if an asymmetrical electrode configuration (Figure 1) was charged up to high DC potentials, a thrust was created moving the electrode setup in the direction from the low-flux to the high-flux region (in this case from the plate/disc to the wire) with respect to the ambient dielectric medium (usually air). Brown claimed (Brown, 1965), that this effect remained even if the ambient medium was vacuum (up to 10^{-6} Torr). Hence, the thrust must be independent of electric wind effects usually created by corona discharges which are proportional to the air pressure (Loeb, 1965). Brown thought that this effect may show a possible connection between gravitation and electromagnetism. Unfortunately, nearly all his work is only summarized in patents (e.g. Brown, 1928, 1965) and not in scientific publications. Hence, these studies lack from detailed information about the behavior of this effect with respect to the voltage or current.

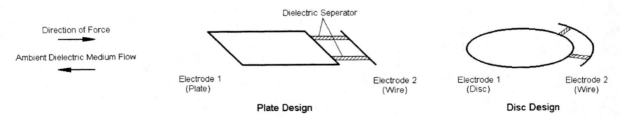

FIGURE 1. Brown's asymmetrical plate/disc-wire electrode configuration

CP504, *Space Technology and Applications International Forum–2000*, edited by M. S. El-Genk
© 2000 American Institute of Physics 1-56396-919-X/00/$17.00

Christensen and Møller (Christensen et al., 1967) built a similar electrode setup and published measurements of the obtained thrust in ambient air which were compared with theoretical predictions due to the electric wind. The agreement was very good tending to explain the Biefeld-Brown effect as a purely electric wind phenomena. Similar theoretical studies have been performed by Cheng (Cheng, 1962). However, Brown claimed that the effect remained in vacuum and therefore is not due to ionization of the ambient air. This was left unconfirmed.

Two decades later, Talley studied Brown's electrode configurations in vacuum chambers up to 10^{-6} Torr in great detail (Talley, 1988, 1991). He found no thrust in the case of a static DC potential applied to the electrodes. However he noticed an anomalous force during electrical breakdowns when a current was flowing.

This force due to currents in divergent electric fields (due to the asymmetrical electrode configuration) finds further support in 5-dimensional theories coupling the gravitational and electromagnetic field. Williams integrated a mass dependend 5th dimension into the relativistic Maxwell theory predicting a coupling between both fields (Cravens, 1990, Williams, 1983). In this theory, a divergent current flow results in an induced mass flow if the coupling constant is non-zero.

The following study concentrates on the investigation of corona discharges with a similar electrode configuration as used in the experiments of Brown and Talley which produces such divergent current flows. The experiment was carried out inside a Faraday-cage box to eliminate the effects of the electric wind under atmospheric pressure to a very large extent. The movement of the whole box was monitored using a laser displacement meter. This method allows to measure if a thrust exists additionally to conventional electric wind effects without using expensive vacuum facilities.

5-D THEORETICAL CONCEPTS

This section gives an overview of the concepts of 5-dimensional relativistic Maxwell theories. More details are given in the literature (Cravens, 1990, Williams, 1983). In 5-D space, the coordinates are given by x, y, z, t, and γ which indicates the 5th dimension. This added coordinate is assumed to be mass dependent.

The relativistic Maxwell equations originates from the antisymmetric electromagnetic field tensor $F_{\mu\nu}$ and the current density j_μ. Therefore, both have to be expanded to five dimensions. We start with the electromagnetic field tensor $F_{\mu\nu}$ which is calculated from the five dimensional vector potential A_μ which we express in five dimensions by

$$A_\mu = \left(\varphi, A_1, A_2, A_3, A_4 \right), \tag{1}$$

where φ is the scalar potential. The electromagnetic field tensor is now calculated from

$$F_{\mu\nu} = \partial_\mu A_\nu - \partial_\nu A_\mu, \tag{2}$$

and has the result

$$F_{\mu\nu} = \begin{pmatrix} 0 & E_1 & E_2 & E_3 & V_4 \\ -E_1 & 0 & B_3 & -B_2 & V_1 \\ -E_2 & -B_3 & 0 & B_1 & V_2 \\ -E_3 & B_2 & -B_1 & 0 & V_3 \\ -V_4 & -V_1 & -V_2 & -V_3 & 0 \end{pmatrix}, \tag{3}$$

where E is the electric field vector, B is the magnetic field vector (note that both depend only on 3 coordinates), and V is a new field originating from the 5th dimension which is related to the gravitational field g. The Maxwell equations are obtained from the Bianchi and current density equations

$$\partial_\alpha F_{\beta\gamma} + \partial_\gamma F_{\alpha\beta} + \partial_\gamma F_{\beta\alpha} = 0, \quad \partial^\nu F_{\mu\nu} = \frac{4\pi}{c} j_\mu, \tag{4}$$

where the five dimensional current density j_μ is expressed by:

$$j_\mu = (c\rho, j_1, j_2, j_3, j_4),$$ (5)

where ρ is the charge density, j_{1-3} is the three-dimensional current density and j_4 which can be interpreted as a mass-density flow (in relation to j_{1-3} which is the density of a flow of charged particles and we have assumed that the 5th dimension relates to mass). The new Maxwell equations contain additional (five dimensional related) terms. Of special interest for our problem is the continuity equation which we get from

$$\partial^\mu j_\mu = 0,$$ (6)

and results in

$$div\,\overset{\varpi}{j} + \frac{\partial \rho}{\partial t} + a_0 \frac{\partial j_4}{\partial \gamma} = 0,$$ (7)

where a_0 indicates the coupling constant between the electromagnetic and inertial field. For the case where a_0 is zero, Equation (7) is restricted to the classical continuity equation. But in five dimensions, the divergence can be nonzero and proportional to the 5D change ($\partial \gamma$) of the mass-density flow j_4.

In this case, a divergent current may induce a mass flow to conserve the charge.

If we consider an asymmetrical electrode configuration similar to Figure (1), and if we apply a sufficient high electric potential to initiate a corona discharge, a divergent current flows between both the wire and the plate/disc electrode. The 5D coupling in Equation (7) would then manifest in an additional mass flow, which would additionally accelerate the ions in the discharge proportional to the divergence of the current. This results in a force that would accelerate the whole configuration with respect to its surroundings being a possible explanation for the claimed Biefeld-Brown effect.

If a corona discharge is ignited inside a sealed Faraday cage box, the known side effects of a discharge like the corona wind would only contribute to oscillations of the box. However, a successful 5D coupling would result in a movement of the whole box with respect to its surroundings. This measurement can clarify, if a Biefeld-Brown type of effect exists under electric breakdown conditions as indicated by Talley's report without using expensive vacuum facilities.

EXPERIMENT

The design of the used electrode configuration is shown in Figure 2 and the box configuration in Figure 3 respectively. Contrary to the Biefeld-Brown plate/disc and wire shaped design, a cylinder and a ring shaped electrode have been used similar to Christensen and Møller (Christensen et al., 1967) to limit possible ion propulsion effects (electrons leaving the end of the wire electrode) and only concentrate on the corona discharge. Both cylinder and ring electrodes are made out of aluminium and are separated by four dielectric rods. The separation distance of 6 cm in air corresponds to the maximum applied potential of 40 kV to prevent sparks which would disturb the corona discharge.

The box is made out of wood and has the dimensions 50cm x 50cm x 50cm with a thickness of 5 mm at each wall. The walls are covered outside with an aluminium foil which is grounded and therefore acts as a Faraday cage. The electrode configuration is fixed to a bar made out of wood which is located in the middle of the box. The cables to connect the electrodes to the terminals outside are high voltage insulated. Both the cables as well as the box (through strings) are connected to two rings which are fixed through steel rods to a plate. This steel rods are finally connected to the terminals of a high voltage power supply (HEINZINGER HNCs 40 000-3ump). Therefore, the whole box including the cables can swing around the fixed rings.

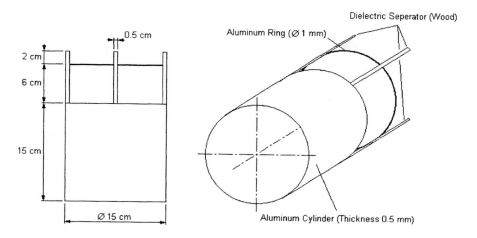

FIGURE 2. Electrode Configuration.

This movement is measured using a laser displacement meter (KEYENCE LC-2400 W) as shown in Figure 4. The laser was operated 105 cm away from the middle of the box to prevent possible electrostatic influence. The positioning data can be used to calculate, if the corona discharge inside the box produces a mass flow which causes the whole box to move with respect to its surroundings. The force of the box can be derived from

$$F = m.g.\sin \alpha = m.g.\frac{\Delta x}{l}, \tag{8}$$

where m is the mass of the box including the experiment and cables, g the acceleration in the Earth's gravitational field, l the length from the rings to the bottom of the box from where Δx, the difference from the box's position from it's zero position, is measured. The parameters during the measurement were **m=7.499 kg**, **l=70 cm** and **g=9.81 m/s²**. With a sensitivity of the laser unit giving positions of ± 0.1 μm, the achieved accuracy of the force measurement was 1.05×10^{-5} N. This is the lower limit of ultra-low thrust electric propulsion devices and is therefore representative to investigate, if an observed Biefeld-Brown effect could be utilized for space propulsion purposes. The potential on the HV power supply was manually increased from zero to approximately 38 kV and the positioning data as well as the potential and current information was transferred to a computer via a IEEE interface. The results are shown in Figure 5.

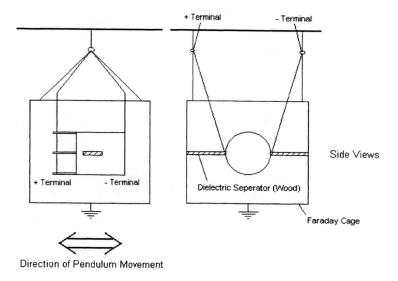

FIGURE 3. Box Configuration.

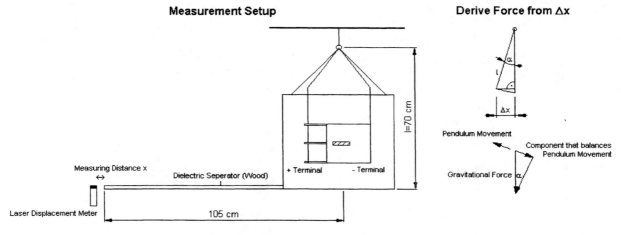

Measurement Setup

Derive Force from Δx

FIGURE 4. Geometry of Force Measurement.

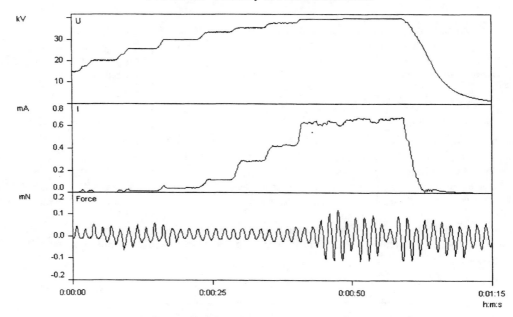

FIGURE 5. Measurement of Force, Current, and Potential.

DISCUSSION

The biggest difficulty during the experiment was to keep the oscillations from the box initially as small as possible. At the beginning of the measurement in Figure 5, the oscillations where limited to approx. 0.05 mN which corresponds to 0.95 μm. When the potential was increased from 0 to 20 kV, the corona discharge ignited and a small current of approximately 0.01 mA started to flow. However, the oscillations remained constant.

During the next increase of the potential to its maximum at 38 kV, the discharge current was increased accordingly resulting in a maximum current of 0.6 mA. At this maximum, the box oscillations were increasing from 0.05 mN to 0.1 mN. The corona discharge created a corona wind which resulted in an air flow (Loeb, 1965) circulating inside the box. This seems to explain the increase of the oscillations. After reducing the potential back to zero, the discharge current felt down very quickly and the oscillations were slowly reduced due to the conservation of energy. The most important result is, that the box always oscillated around the same mean position during the corona discharge with a maximum amplitude of 0.1 mN. This means, that no noticeable linear thrust was observed within the detectability of the used instrumentation. If a linear thrust exists in the presented configuration, the data suggests that it must be below 0.01 mN. Considering the maximum power used (38 kV x 0.6 mA = 22.8 W) we can express the specific power to thrust ratio as

$$\frac{P\,[\text{W}]}{F\,[\text{N}]} \geq 2.28 \text{x} 10^6 . \qquad (9)$$

Comparing this value to other highly efficient electric propulsion devices, i.e. Field-Emission-Electric-Propulsion (FEEP) or Xenon ion thrusters (Tajmar, 1997) with P/F=2x10^4, we note that this ratio is at least 2 orders of magnitude above current existing technologies. Therefore, even considering that such a linear thrust exists in the presented configuration and taking the advantage of a propellantless propulsion system into account, existing electric propulsion devices would be far superior. Calculating the maximum thrust due to corona winds, we use the measured expression by Christensen et al,

$$F = P \cdot \frac{l}{U} \cdot \frac{1}{b \cdot (1 + \phi)}, \qquad (10)$$

where F is the thrust, l the electrode separation distance, U the applied potential difference, b the ion mobility (b_{air}=2.15x10^{-4} m^{2}V^{-1}s^{-1}) and ϕ the fluid performance parameter (ϕ_{air}=2x10^{-2}). Using again the maximum input power during the measurement, the maximum thrust due to corona winds in dry air is 163 mN. Comparing this value to our obtained upper limit for an additional effect of 0.01 mN, a possible gravitation-electromagnetism interaction must be at least 5 orders of magnitude below the electric wind effects. Hence, the Biefeld-Brown effect in the claimed order of magnitude (movement of similar electrode configurations in vacuum and air) can not be confirmed. The results suggest that corona wind effects were misinterpreted as a connection between gravitation and electromagnetism.

CONCLUSION

No linear thrust was observed within the detectability of the used instrumentation. A possible connection between gravitation and electromagnetism which leads to a force due to the divergent currents used must be at least 5 orders of magnitude below the corona wind forces. A propulsion device based on this 5-dimensional concept would be at least 2 orders of magnitude less efficient than existing electric propulsion thrusters. Therefore, even if such a 5-dimensional connection would exist, the obtained forces would be too small to be utilized for space propulsion. The Biefeld-Brown effect could not be verified in the claimed order of magnitude. The results suggest, that corona wind effects were misinterpreted as a possible connection between gravitation and electromagnetism.

ACKNOWLEDGMENT

I would like to thank Prof. Becker from the International Space University for his support and the equipment to built a first setup. Many thanks go to Dr. Fehringer from the Austrian Research Center Seibersdorf were all measurements have been carried out. Many stimulating discussions with C. Belleval and C. de Matos were greatly appreciated.

REFERENCES

Brown, T.T., "A Method of and an Apparatus or Machine for Producing Force or Motion," British Patent Nr. 300.311, 1928

Brown, T.T., "Electrokinetic Apparatus," US Patent Nr. 3.187.206, 1965

Cheng, S.I., "Glow Discharge as an Advanced Propulsion Device," *ARS Journal*, Vol.12, 1962, pp.1910-1916

Christensen, E.A., Møller, P.S., "Ion-Neutral Propulsion in Atmospheric Media," *AIAA Journal*, Vol.5, No.10, 1967, pp.1768-1773

Cravens, D., "Electric Propulsion Study," *Final Report for Air Force Astronautics Laboratory*, AD-A227-121, 1990

Klein, O., "Quantentheorie und fünfdimensionale Relativitätstheorie," *Zeitschrift für Physik*, 1926, pp. 895-906

Loeb, L.B., "Electric Coronas, Their Basic Physical Mechanisms," University of California Press, 1965

Tajmar, M., "Backflow Contamination of a Cesium Field-Emission-Electric-Propulsion (FEEP) Thruster," M.Sc. Thesis, Department of Physics, Vienna University of Technology, May 1997

Talley, R.L., "21st Century Propulsion Concept," *Final Report for Air Force Astronautics Laboratory*, AFAL-TR-88-031, 1988

Talley, R.L., "Twenty First Century Propulsion Concept," *Final Report for Air Force Propulsion Directorate*, PL-TR-91-3009, 1991

Williams, P.E., "The Possible Unifying Effect of the Dynamic Theory," *Los Alamos Scientific Laboratory Report*, LA-9623-MS, May 1983

Simple Electrostatic Aether Drift Sensors (SEADS): new dimensions in Space Weather and their possible Consequences on passive field propulsion systems

Alexandre D. Szames[1], Patrick Cornille, PhD.[2],
Jean-Louis Naudin[3] & Christian Bizouard, PhD.[4]

[1]Head, Apex divison. Éditions A. Szames — 47-51 rue d'Aguesseau, 92100 Boulogne, France. aszames@yahoo.com [2]Senior Physicist. Advanced Electromagnetic Systems S.A. — 4, rue de la Pommeraie, 77130 Saint-Rémy-lés-Chevreuse, France. patrick.cornille@libertysurf.fr [3] Director of Research, Advanced Electromagnetic Systems S.A. — 4, rue de la Pommeraie, 77130 Saint-Rémy-lés-Chevreuse, France. Jnaudin509@aol.com [4]Astronomer. 6, rue Lacaze, 75014 Paris. France bizouard@danof.iap.fr

Abstract. The Trouton and Noble experiment, which was initially performed (and fiascoed) in 1903, has only been questioned recently. When correctly performed, this very simple electrostatic ætherdrift experiment gives unambiguous positive results: a suspended, parallel-plate capacitor charged at high voltage by means of lateral feeding wires exhibits a stimulated torque and tends to line up its plates in the East-West direction. Other tests by means of vertical feeding wires exhibited continuous rotations. This new class of elementary EM phenomena is described in the present paper. As far as it relates to the state of motion of the vacuum, it is our understanding that: (i) it serves as the physical basis for another class of electrostatic phenomena involved in the generation of linear thrust and technically referred to as "Biefeld-Brown effect" drives; (ii) it might be of tremendous importance for implementing "future flight" propulsion systems; (iii) it might add a new dimension in space weather. The present paper aims at clarifying these concepts. We shall only present the exploratory side of a wider, proprietary research and development effort, so as to encourage the replication of Trouton-Noble's experiment by academia and other members of the scientific and engineering community.

INTRODUCTION

In his kick-off essay, *The Challenge to Create the Space Drive*, Millis described several classes of speculative devices capable of generating propulsive thrust in empty space through the exchange of momentum with an external, *stationary* medium (Millis, 1997). This logical engineering requirement to eliminate on-board propellant mass on a space vehicle, may seem to reinvigorate the old concept of æther, which was rejected in 1905 when Einstein formulated his Special Theory of Relativity (STR). To the best of our knowledge, the concept of æther was fortunate enough to never completely vanish from the field of physics. Einstein himself would reintroduce it in the early 1920s as a spectral, mathematical, *gedanken* entity carving out the shape of the Universe: the "spacetime metric". Quite unfortunately, this newborn entity could not be subject to any direct means of detection. It is our belief that even ill-defined or ill-formalized, any conceptual "object" associated to a "reality-like" character of our material reality must be endowed with physical qualities, which can be subject to proper detection, quantitative measurements, manipulations, and ultimately *engineering*. The vacuum itself shall not be an exception to that rule. There are historic, pragmatic, theoretical and experimental reasons to question the apparent emptiness and inertness of vacuum. This analysis bears direct significance to: (i) the experimental foundations of the theory of relativity; (ii) the concept of "space weather"; (iii) the design of future means of space travel. This point paper is structured in three parts. In part one, we briefly review our understanding of the history of the concept of "æther" to enlight the reader on the nature and scope of our work. In part two, we concentrate on the Trouton-Noble (TN) ætherdrift experiment: its history, our experimental results and generic calculations of the "Target" the velocity vector the Earth aims at. The relevance of these results to the field of "breakthrough propulsion physics" is examined in part three. For a better clarity and understanding of this paper, we are compelled to work with CGS units rather than MKSA (SI) units.

CP504, *Space Technology and Applications International Forum–2000*, edited by M. S. El-Genk
© 2000 American Institute of Physics 1-56396-919-X/00/$17.00

WHERE WE STAND

Vacuum was initially devised as the logical siege of our universal reality. Rapid progress in fluid mechanics and field theories would force these thoughts into a testable, theoretical framework (section 1). The experimental quest for such an æther climaxed in the 19th century, along with the discovery of electromagnetism (and the formulation of incomplete "redressed" theories thereof). Experimental series of "null-results" called for new interpretations. Einstein's STR would finaly extrude the "æther" from physics (section 2). A renewed concept of "æther" would still be empowered within the framework of quantum physics, and progressively be expanded in physics through a growing (now *firm* and *massive*) body of experimental evidence STR cannot handle decently (section 3).

Vacuum as a traceable "æthereality"

By no means should the word "vacuum" be treated like a synonym of the biblical idea of Nothingness. The origin of this concept can be traced back to the observation and discussion of the motion of solid bodies in the air. Briefly stated, the Greek civilization fathered out two opposed "invisible colleges" and their associated party lines of reasoning. In his brilliant treatise *On the Heavens*, Aristotle favoured the idea that "nature abhors a vacuum": space was to be filled with "some substance" different in quality from that of the "elements". The property of mass itself was regarded as indestructible since it "could be accounted for at different points of any fluid, whether at rest or in steady motion" (Tokaty, 1971). Thus Aristotle introduced natural philosophy to the continuous, defined as "that which is divisible into parts which are themselves divisible to infinity, as a body which is divisible in all ways." Fluid mechanics are still based upon this principle. Aristotle also insisted that no motion, whether *natural* or *enforced,* would persist "unless sustained by the continued action of a motive power, applied directly to the moving body" (Nemenyi, 1962) This idea was proven "wrong" on a logical basis by Johannes of Alexandria, better known as Philoponus. On the other end of the spectrum, Leucippus and Democritus put forth vacuum as a "logical requirement" (B. Russell) to describe not only the empty space between atoms, but the very place deprived of matter where motion (of atoms) could happen without suffering from resistance. They had thus inaugurated kinematics, i.e. the "science of motions considered apart from causes — that is apart from *forces*" (Phipps, 1986b).

Aristotelian mechanics were still supported by the clergy when Galileo Galilei (1564-1642) fought them both. Galileo furthered the "atomistic school" ideas by asserting that: (i) objects undergoing rectilinear motions behaved as if they were at rest; (ii) changes of velocities, i.e. accelerations, only had to be accounted for in the "motional universe". Accelerations were supposed to break up with this newly inaugurated "principle of equivalence". Galileo had thus introduced the principle of relativity of motion in kinematics (and its underlying mathematical consistency, known as "Galilean covariance"). Rising off on the shoulders of his predecessors, Sir Isaac Newton established the basic laws of mechanics, and applied them to describe the motion of planets orbiting around the Sun. Behind the well-known law of inverse squares describing an electric-type gravitational action-at-a-distance — by means of a yet unknown mechanism — Newton also enforced what must be regarded today as his greatest achievement, the third law. Which was originally stated as follow: "To every action there is always opposed an equal reaction: or the mutual actions of the two bodies upon each other are always equal and directed to contrary parts."

Newton's third law would soon be merged with *hydrodynamics*, a newborn discipline technically implemented in the 18th century by the Bernouillis, Euler, d'Alembert, de Borda, du Buat and many others in their pioneering trails. Euler in particular, in his *Mechanica* (1736), played a key role for he turned *hydrodynamics* into the first field theory. "A field in mathematical physics is generally taken to be a region of space in which each point (with possibly isolated exceptions) is characterised by some quantities which are functions of the space coordinates and of time, the nature of the quantities depending on the physical theory in which they occur. The properties of the field are described by partial differential equations in which these quantities are dependent variables, and the space and time coordinates are independent variables" (Hesse, 1970). A variety of mathematical tools would spring up to account for both: (i) the behaviour of a variety of fluid media, and; (ii) the motion of projectiles in these media. While viscosity and elasticity had already been ascribed on Newton's mechanistic agenda (*Principia*, Book II) to "explain" the motion of bodies though "celestial space" devoid of any "corporeal fluid" — that very æthereal medium praised for by Descartes in France — Newton's demonstration was not convincing. On the other side of the spectrum, the father of the Method uttered that matter (*res extensa*) was embodied in the principle of continuity. Madelung's pioneering hydrodynamic interpretation of the quantum theory, de Broglie's "matter waves" would embrace this idea in the early 1920s.

Is the vacuum cleaner when it "exi(s)ts" ?

The discovery by Fresnel and Young that light had wave-like properties, that it could also propagate in the vacuum, recent advances in electrostatics (18th century), in electrodynamics, electromagnetism and electrokinetics (19th century), finally led James Clerk-Maxwell to empower Michael Faraday's grand concept of "magleticlines of force independent of matter or magnet" and synthesise Faraday's "experimental researches" in a contact-action, field-theory of known electromagnetic phenomena. Maxwell's legacy consisted in set of twenty equations carved out in a mathematical language known as quaternionic algebra (1864-1873). His theory was based on: (i) the concept of electronic state, defined as the "fundamental quantity in the theory of electromagnetism"; (ii) the existence of a fundamental continuous medium (continuum) characterized by polarization and strain. Among others, his theory implied that: (i) the electromagnetic field was only a manifestation of the electrotonic state of the continuum; (ii) light was a radiation, electromagnetic in origin, capable of propagating at finite speed (Hendri, 1986). This hypothesis would be experimentally "verified" by Hertz in 1888 (although Hertz used spark gaps). Maxwell's equations were subsequently rewritten and interpreted in a more fashionable manner by the "Maxwellians" (FitzGerald, Heaviside, Lodge, Hertz). In their hands, quaternions would be replaced by vectors, the electrotonic state by a mathematical artefact, the "vector potential," deprived of any physical significance. In his landmark paper, Barrett insisted that the British Maxwellians were still concerned, though, by "the dynamic state of the medium or æther" in their "redressed" interpretation of Maxwell's equations. This dynamical aspect of the medium, its stresses and strains, would be irradicated by Hertz. "… The Hertz orientation finally prevailed, and modern "Maxwell's theory" is today a system of equations describing electrodynamics which has lost its dynamical basis" (Barrett, 1993). Maxwell's "redressed" set of equations would generate what is today known as "classical electromagnetism".

This series of theoretical "achievements" had opened up new avenues in the pursuit of Newton's "corporeal fluid" hypothesis. They had triggered the quest for an all-pervading medium endowed with exotic qualities (viscosity, elasticity, electric permittivity, magnetic permeability, conductivity). The æther would thus become the game of a breathtaking snark-like hunt, organized in the last decades of the 19th century. The æther was envisioned as the propagating "medium" of all sorts of EM radiations, and as the ultimate siege, the "inductive essence" of reality. It was the arena where all known physical phenomena would happen. If the æther were real, and if the Earth travelled in it at a means approximate speed of 30 km.s^{-1} with respect a "stationary" reference frame anchored to the center of the Sun, an æther wind *should* be detectable. Three classes of ætherdrift experiments were finally devised to settle this matter: (i) optical devices; (ii) mechanical devices; (iii) electrostatic devices.

Though practical for electrical engineers, Maxwell's "redressed" equations were found to be "non-invariant" in a "Galilean transform". This discovery implied that the laws of nature governing known classical EM equations would not be identical in two "inertial frames of reference" (concept questioned further below). In the last decade of the 19th century, physicists would try to find an alternative to Galilean transforms compatible with the presumed "invariance" (in fact, "covariance") of Maxwell's "redressed" equations. These transforms were written up by Lorentz in 1904, acknowledged by Poincaré in 1905 and also employed in Einstein's 1905 article. Einstein's theory was based on two postulates supplemented with a third, "hidden" statement: (i) the principle of constancy of the light celerity; (ii) the principle of relativity; (iii) the formal correctness of Maxwell's "redressed" electromagnetism. Einstein's theory consisted in acknowledging the "fact" that none of the experimentalist who had tried to demonstrate the 'existence' of the æther in the past, would ever be successful reaching this objective in a close or distant future. He thus put a dogmatic theoretical limit to future empirical results. In Einstein's original theory, the physical æther was not required. To the best of our knowledge, the concept of æther was fortunate enough to never completely vanish from the field of physics. Einstein himself would reintroduce it in the early 1920s within the framework of the General Theory of Relativity (Kostro, 1994). The newborn æther would become a spectral, mathematical, *gedanken* entity carving out the shape of the Universe: the "spacetime metric". Quite unfortunately though, this newborn entity could not be subject to any direct means of detection, as far as it would wrap up the sensible word in its very definition, and would itself become embroidered in the supporting pillars of the relativistic edifice: (i) apparent "null-result" ætherdrift experiments; (ii) the principle of covariance and; (iii) the principle of equivalence. So as a result there is a quite annoying paradox: this fundamental essence of reality would pass unnoticed, and forever remain undetectable. Such a pessimistic picture cannot be drawn from physics, but from a self-fulfilling prophecy.

But is it so, really ?

A number of conceptual, pragmatic, theoretical, "techn(olog)ical," and experimental difficulties arose in the past century. We shall list a few of them here. First of all, Barrett insisted that a number of newly-discovered physical effects suggested that Maxwell's "redressed" theory of electromagnetism was incomplete. Among others, the Aharonov-Bohm, Altshuler-Aharonov-Spivak, Josephson, quantum Hall, Sagnac, and De-Haas-Van Alphen effects, had not been taken into account (Barrett, 1993). A series of intriguing electromagnetic transient phenomena not accounted for by Maxwell's "redressed" theory of electromagnetism was also reviewed (Hartmuth, 1993). The possibility that low-energy magnetic monopoles had indeed been discovered was experimentally investigated and reviewed (Mikhaïlov, 1995). There are other, deeper problems. The Lorentz force was not part of Maxwell's original set of equations. Our sensor technology, i.e. the very technology which has enabled our perception (and thus measurement) of the sensible micro and macroscopic world, is based on two enigmatic electromagnetic objects, the electron and the photon, of which no satisfying explanatory models exist today. As to why the electric charge is quantified or conserved remain two of the greatest mysteries in contemporary physics. Other experimental problems associated to electromagnetism have surfaced within the framework of quantum electrodynamics (QED), quantum field theory (QFT), and a new discipline known as stochastic electrodynamics (SED). Briefly stated, field fluctuations, which are known to persist at "zero point" (ZP) temperature, i.e. in the absence of any field excitation, and which happen to correspond to an energy of half a quantum per mode, can show up on macroscopic scales as they give rise to mechanical forces (Darrigol, 1984; Aitchinson, 1985). Since 1969, the static Casimir force has been interpreted this way. The ZP energy (ZPE) electromagnetic fluctuations are said to be distributed in such a way that their spectral energy density is isotropic throughout unperturbed regions of the surrounding "Lorentz-invariant" ZP field (ZPF). However, accelerated mobiles do perturb this "isotropicity", enabling said mobiles to radiate, and thus be detected. And "Lorentz-invariance" is too often taken, if not for granted, as "an oxymoron for covariance" (Phipps).

These remarks lead us to explore another range of problems. Covariance and invariance, figured out as (almost) self-explanatory in textbooks, are highlighted as two cornerstones of modern physical theories. But they seem to be the worst-defined and most poorly understood concepts in all classical, quantum and relativistic physics. This point has been emphasised by many authors already (Post, 1962; Anderson, 1971; Earman, 1974; Phipps, 1986a; Johnson, 1996; Arunasalam, 1997; Cornille, 1999). The debate is closely related to the existence of "inertial frames of reference" (IFRs). Unfortunately, such mathematical entities simply do not exist in nature. Indeed, the property of inertia is intertwined with the property of "mass" — whatever "mass" is — and defined as the resistance of said "mass" to accelerated motion with respect to a predefined frame of reference (FR). In Einstein's STR, when "observers" embarked onboard an IFR undergoing a uniform rectilinear motion, perform "measurements" against another IFR undergoing a different uniform rectilinear motion, the measurement process implies the recoil motion of the "measured" IFR. Thus enabling the latter to loose its "special" inertial status (Brillouin, 1970). There are three ways only to escape from this paradox: (i) physicists can postulate that physics as a discipline must only deal with fantomatic test-bodies, and carefully avoid verifying their "findings" with real experimental results (through implementation of virtual, *gedanken* process); (ii) physicists can define an IFR as a reference frame whose mass is infinite (so as to avoid recoil motions) or; (iii) physicists can negate the applicability of Newton's third law in relativistic physics. Needless to say, we must reject *gedanken* experiments since these "wishful thinkings" don't allow anybody to question (get a deeper understanding of) mother Nature, and thus discover new physical effects (Barrow, 1998). We must also reject the concept of IFR for the sake of sanity. Of course, we *must* also oppose any of those who proudly claim that Newton's third principle is outdated. "The equality of action and reaction has almost no place in relativistic mechanics. It must essentially be a statement about the forces acting on two bodies, as a result of their mutual interaction at a given instant. And, because of the relativity of simultaneity, this phrase has no meaning." (French, 1968). One shall be puzzled to learn that Newton's third law is still used in standard relativistic models to compute interaction forces emerging from the elastic collision process of two relativistic particles. One of us extensively worked out this paradox (Cornille, 1999). One author specifically insisted that Einstein's theory had "some explaining to do. For a theory that does not recognize the equality of action and reaction cannot, without apology, invoke the conservation of momentum" in such cases (Beckmann, 1987).

This *corpus* of elements lead us to conclude that there might be a fundamental problem with the underlying principles at work not only in Einstein's STR, but in any theory derived from it. An ever growing number of authors now feel the time is ripe to question, revise, reject and/or "falsify" Einstein's STR whose original inconsistencies are well-known (Cullwick, 1981; MacCausland, 1990). We reviewed the three families of ætherdrift experiments which supposedly confirm Einstein's STR. None of the original experimental result confirm in any manner this theory. It just happens that the history of physics might have been written somewhat hastily. Optical ætherdrift experiments were shown to require long series of recordings to give statistically acceptable results. Michelson-Morley did observe an interferometric pattern corresponding to a perceived motion of the Earth of 8 km.s $^{-1}$, a magnitude far smaller than what they expected (30 km.s $^{-1}$). Though discarded at the time as falling in the range of experimental errors, this discrepancy was revived by Vigier to account for the "missing mass" of the extended photon (Vigier, 1997). The reader is strongly urged to read the original source articles (Michelson-Morley, 1887; Miller, 1925, 1926, 1933; Allais, 1997b) and *not* their interpreted, downgraded versions. When they are not impossible to perform, such as the *gedanken* bent lever (Lewis-Tolman, 1909), mechanical experiments would either tend to reject the existence of predictable relativistic effects, such as the Thomas precession and the "Hergloz-Dewan-Beran" stress (Phipps, 1986b), or contradict relativistic predictions (Terrel, 1959; Allais, 1997a). These results confirm Crooks *et al*'s well-known statements. We are reminded that they employed a Faraday homopolar generator and noticed that "any moving magnets generated an induced electromotive force due to the presence of its own field: this generation leads to an apparent paradox in the case of translational motion for it implies the possibility that an observer in an inertial frame could measure its absolute velocity" (Crooks *et al.,* 1978). But there is one single electrostatic experiment who could have definitely settled the debate on the existence / non-existence of the æther in 1903: the Trouton-Noble (TN) experiment. Our team has spent considerable time analysing and exploring this experiment over the past three years. This experiment simply appeared to be one of the most formidable fiascoes in the history of physics.

HYPOTHESES FINGO ?

A short epistemological note on the TN experiment (section 1) is followed by an overview of our experimental results (section 2) which tend to indicate that it is possible to observe directly, if not the "absolute" motion of the Earth through space, the "Target" where the velocity vector of the Earth aims at (section 3).

Brief historical review of the Trouton-Noble experiment

Retarded Lorentz forces are known to violate Newton's third principle (Cornille, 1995). Therefore, and as predicted by FitzGerald and Lorentz themselves within the framework of Maxwell's "redressed" theory of electromagnetism, a charged parallel-plate capacitor supposedly "at rest" in the Earth frame *must* exhibit an observable rotational motion *because* the Earth moves through the "luminiferous ether," i.e. a preferred frame of reference against which one could measure "absolute" velocities. The Trouton-Noble (TN) experiment was specifically aimed at this objective. However, most textbooks "downgrade" its physical attributes to a much simpler, static, and naïve experimental device: two charges Q_1 and Q_2 fixed onto the ends of a solid rod of length R, capable of rotating around its center of mass (which is presumably moving with respect to the preferred frame of reference with a constant velocity U). We shall call this analysis of the TN experiment the "rigid model," because: (i) it can only account for surface effects; (ii) it involves no electrical current, even though static charges in motion can mathematically be computed as if they were elements of current. In this rigid model, Lorentz forces between the charges are given by the relation:

$$\mathbf{F}_{12} = Q_1Q_2[\mathbf{R} + 1/c^2\{\mathbf{U}\times(\mathbf{U}\times\mathbf{R})\}] /R^3 = - \mathbf{F}_{12} \quad , \tag{1}$$

which can be rewritten as:

$$\mathbf{F}_{12} = Q_1Q_2\{(1 - U^2/c^2)\mathbf{R} + 1/c^2(\mathbf{U}\cdot \mathbf{R})\mathbf{U}\} /R^3 \quad . \tag{2}$$

This force is not lined up with the direction of the vector R, thus generating a mechanical torque, electromagnetic in origin. When the charges are opposite in signs, and equal in magnitude, this torque tends to lock the solid rod *perpendicular* to the direction of motion, as described by Lorentz in 1904. For fixed static charges, the magnetic force follows Newton's third principle, as shown in equation (1). Specifically, an unbalanced torque:

$$\Gamma = (Q^2 U^2/2c^2 D) \sin(2\theta)\sin (2\varphi) \quad , \tag{3}$$

is computed, where Q is the charge of the capacitor, D the distance between the two plates (i.e. thickness of the dielectric material), U the velocity of the capacitor carried along by the Earth in its motion around the Sun, θ the angle between the velocity vector and the normal to the plates and φ the angle between the velocity vector and the suspending fiber. In their original experiment, Trouton and Noble tried to observe an effect resulting from the orbital velocity of the Earth around the Sun. In their minds, U = U $_{Earth\ @\ Sun}$ = 30 km.s^{-1}. However, it is our understanding that U must necessarily be the *absolute* velocity of the Earth with respect to a preferred frame of reference. This implies that it could / should be possible to demonstrate that both the magnitude and the direction of the torque depend on the position of the Earth on its motion in a preferred frame of reference if we perform continuous measurements along an appropriate period of time (i.e. several months or even years). Trouton and Noble failed to observe the effect they were looking for. But their amazing conclusions to support their "null" results would raise a huge doubt. "From experience gained with the apparatus, the deflections observed would appear to be attributable to small sparks or discharges taking place inside or over the condenser, causing slight heating and consequent perturbation of the surrounding air. This is further suggested by the fact that *when the condenser employed had become damaged by falls and other vicissitudes, so that audible sparkings occurred, the perturbations became so great as to prevent all possibility of observation.* There is no doubt that the result is a negative one" [our italics added] (Trouton & Noble, 1904).

TN's electrostatic ætherdrift experiment is still barely mentioned in textbooks today, and often highlighted as a "paradox" in the available literature (Jefimenko, 1999). It was only replicated three times over a 90-year period before we took the initiative to reproduce it. TN's experiment was somewhat hastily reviewed in a dissertation (Janssen, 1995). In a previous set of publications (Cornille, 1999; Cornille, Naudin, Szames, 1998, 1999; Szames 1998a), we demonstrated that none of TN's known replications had been convincingly performed, either because the apparatus could not mechanically produce the alleged torque (Tomaschek, 1925, Tomascheck, 1926), because the voltage was too low (Chase, 1926) or because the apparatus had been designed in such a way that it had strictly nothing to do with the TN experiment (Hayden, 1992, 1994). We also demonstrated that no author could decently claim that Chase and Tomaschek's papers supported TN's original findings since Chase refuted Tomaschek's investigations on an experimental basis. It is therefore impossible that some authors *read* or even *understood* the contents of the papers they quoted would be able to stand firm on their positions (Janssen, 1995; Yuan Zhong Zhang, 1997). Needless to say, TN's original null result was claimed to substantiate Einstein's Special Theory of Relativity (STR), according to which no absolute motion can be subject to direct detection. One of us reinvigorated in the mid-nineties the study of the TN-experiment, as far as it could be related to another class of electromagnetic phenomena involved in the generation of linear thrust, and technically referred to as "Biefeld-Brown effect" drives in the advanced field propulsion systems community (Cornille, 1995).

The experiments described below were performed in a laboratory located in the suburbs of Paris, France. Specifically, the series of experiments were performed in the northern hemisphere of the terrestrial globe, above the tropical line. These experiments were repeated dozens of time with similar test devices and different types of power supplies over a three-year period. The same types of results invariably came out. We shall only present here the exploratory side of a wider, proprietary research and development effort, so as to encourage the replication of the TN experiment by academia and other members of the scientific and engineering community. In the course of our investigations, TN's experimental set up was found to be incorrectly designed. The deficiencies might explain why this experiment and any of its subsequent replications patterned after the same original design, could only produce "null" results. A suspended, parallel-plate capacitor charged at high voltage does exhibit a stimulated torque and tends to line up its plates in the East-West direction (Cornille, Naudin, Szames, 1998, 1999).

Experimental results

Our basic experimental set up was patterned after Trouton-Noble's original experimental *intentions* and built along the following lines: a parallel-plate capacitor (≈ 500 pF) was manufactured by attaching two conducting aluminum foils (19×15 cm; 3,3 × 10^{-3} cm thick) on either side of a transparent, non-shielded insulating plate of Plexiglas (25 × 21 cm; 0,2 cm thick; ε$_r$ = 4). This capacitor was suspended from the ceiling of the laboratory by a thin nylon thread (1 = 150 cm). A hole (diameter: 5 × 10^{-2} cm) lined up with the center of mass of the capacitor, perforated the plate of Plexiglas 1 cm away from the top edge. Said thread was attached to the capacitor by means of a node, and fixed to the ceiling so as to reduce frictional forces. The suspended capacitor could rotate freely in either direction about its vertical axis, without being significantly affected in its rotational motion by the mechanical counter torque originating in the suspending fiber. The wires feeding the capacitor with high voltage were coated. Two variants of this basic device were used, thus enabling us to obtain three different manifestations of what seems to be the same underlying effect.

Test article TA#1: Trouton-Noble experiment. The wires feeding the capacitor were connected to the center of the plates. Our first qualitative experimental results were obtained back in 1996 and 1997 with a Wimshurst generator. This type of generator is known to produce high voltage (7×10^4 V) and very low currents ($\approx 1 \times 10^{-6}$ A). Any time the experiment was repeated, our capacitors sought a position of stable equilibrium in the East-West direction, where they remained "locked" until discharged. The total torque was subsequently demonstrated to be negligible for a symmetric and homogeneous distribution of charges over the plates, provided that the voltage used in the experiment were low. The observed effect was also demonstrated to be extremely weak if the charges distributed over the plates were the only charges involved in the calculation of the effect. Indeed, it appears that the structure of space charges *inside* the plates and the polarization of the dielectric material can significantly affect the generation of the observable torque when the two following criteria are met: (i) the applied voltage must be higher than $2{,}5 \times 10^4$ V and; (ii) a small leakage current must exist *inside* the dielectric material.

Test article TA#2: Augmented Trouton-Noble experiment / the "turbine". We employed a Wimshurst generator. The phenomenon of electrical influence was chosen as the primary mechanism for feeding the capacitor. A coated electrical wire, top bare, neighboured a rotating distributor, these two elements being distant from each other. The distributor was a flat, disc-shaped device attached to the suspending fiber, enabling us to benefit from the inertia effect. The capacitor would gain momentum whence charged, and discharged when it reached its calculated, stable position of equilibrium. We thus observed a "tricked" continuous rotation of ≈ 10 r.p.m.

Test article TA#3: Trouton-Noble experiment + leaky currents. Trouton-Noble experiment + leaky currents. Quantitative experimental results were obtained with a shielded, grounded bipolar power supply (two Glassman HT8 HV electronic generators). With such generators, voltage and potential differences were controlled, monitored and reached a maximum value of 5×10^5 V. Currents could only be monitored with an accuracy of 1%. TA#3 consisted in a slightly modified TA#2, where the segmented distributor was replaced by a continuous distributor. During the cruise regime, i.e. before the rotation of the capacitor was curbed and inhibited by the mechanical countertorque originating from the torsion fiber, the capacitor was found to rotate at 6 r.p.m. The outcome of this experiment seems to be related to the so-called "Biefeld-Brown effect." But the underlying effects still remain unclear today. One of us (Szames) has argued the torque could originate from the distributor itself, since it is almost patterned after Franklin's enigmatic electrostatic "flying top" device (Szames, 1998). One of us (Naudin) is currently exploring the efficiency of a "Poynting Flow Thruster" (PFT) resulting from pulsed, asymmetrical charging of a parallel-plate capacitor (Naudin, 1999a). A motor (PFT-M) capable of rotating at 1600 r.p.m. has also been investigated (Naudin, 1999b). One correspondent even argued that minute asymmetries in the design of our apparatus could stimulate a forced-flow electroaerodynamic regime, not to be misinterpreted as ionic wind effects (Galuchot, 1999).

Seven possible sources of experimental errors were investigated: (i) plausible interaction of the capacitor with the magnetic field of the Earth; (ii) interaction of the capacitor with the power supply device; (iii) interaction of the capacitor with the environment (laboratory, walls, etc.); (iv) interaction of the capacitor with the surrounding medium (electroaerodynamic regime) and; (v) an asymmetry of the voltage applied on each plate; (vi) an asymmetry of the feeding current; (vii) other unknown or unidentified physical parameters, mechanisms or causes.

Can the magnitude of the expected torque be computed ?

It is our understanding that three different classes of effects can be computed. We shall summarize them here so as to avoid any confusion: (i) surface *vs* volume effects; (ii) static *vs* dynamic effects; (ii) homogeneous *vs* inhomogeneous field effects. The expression "surface effect" refers to the potential difference between the plates of the capacitor. The expression "volume effects" both refers to: (i) the influence of surface effects on the structure of space charges inside the plates and inside the dielectric material; (ii) the possible modes of polarization of the dielectric material (no space charge, negative space charge, homocharge, heterocharge). The interplay between surface and volume effects can be best described as follows: while using the Wimshurst generator, we observed that the total torque originated from the heterogeneous distribution of charges inside bulk matter while a transient, almost negligible counter torque, opposite in direction, originated from the surface charges of the plates. The expression "static effect" refers to the static distribution of space charges inside the capacitor (plates and/or dielectric material). The expression "dynamic effect" refers to any physical mechanism involving the migration of charges inside the capacitor (plates and/or dielectric material).

The violation of Newton's third principle by stimulated forces was partly explored in our previous series of paper. Such a violation could also be explained by forces resulting from the reciprocal interactions of two elements of current. Gabillard demonstrated that when charges Q_1 and Q_2 were replaced by two elements of currents di_1 and di_2, the direction of which would form the respective angles α_1 and α_2 with the line between them, an unbalanced torque should appear in the most general case. This torque would reach its maximum magnitude when the parallel plate capacitor is charged in a direction *normal* to the velocity vector **U**. The magnitude of the maximum torque satisfies the following formula (Gabillard, 1999):

$$\Gamma_{max} = (\mu_o/4\pi) . (di^2/2D) \quad . \tag{4}$$

Since $di = Q\,v$, and $Q = CV$, (4) can be rewritten as:

$$\Gamma_{max} = (\mu_o/4\pi) . (C^2V^2/2r)\,U^2 \tag{5}$$

where $\mu_o = 4\pi\,10^{-7}$, D is the thickness of the dielectric material, **U** the velocity of the capacitor with respect to an heliocentric frame of reference, C its capacitance, and V its voltage. When the following set of numerical values is adopted ($U = 3 \times 10^6$ cm, $D = 2 \times 10^{-1}$ cm, $C = 5 \times 10^{-10}$ f, $V = 7 \times 10^4$ V), Γ_{max} appears to be equal to $16{,}5 \times 10^{-6}$ N-m, i.e. $\Gamma_{max} = 165$ dyne-cm. In a previous publication, three of us computed a theoretical $\Gamma_{max} = 2{,}5 \times 10^{-8}$ dyne-cm (Cornille, Naudin, Szames, 1999). Gabillard insisted that the charged parallel-plate capacitor would have to be installed at least 4 meters *away* from a Wimshurst generator to observe the expected Γ_{max} effect. Below this critical distance, parasitic electrostatic dipolar motion may seriously interfere, and even mask, the Γ_{max} effect sought for by the experimentalists. However, it is our understanding that the equation from which we derived the Γ_{max} effect, only took into account "surface effects." This point needs, of course, further refinements.

It has been argued that our experimental results indicated that it would only be possible to observe directly by means of a simple electromagnetic device the local rotational motion of the Earth around its polar axis. This statement tends to reduce the TN experiment to the mere status of the electromagnetic analogue of Foucault's mechanical pendulum. But the TN experiment was initially devised, and is still considered by the scientific community, as the electrostatic analogue of Michelson-Morley's optical device. When operated over long periods of time, such as in Miller's interferometric device, the TN experiment should enable us to observe *how* the plates line up in the East-West direction, i.e. enable us to find out if there are daily and/or seasonal variations of magnitude, orientation, and possibly direction of rotation, of the charged, suspended capacitor. Once available, these data will have to be compared to a mathematical model of the distant "target area" (the principles of which are given below) so as to validate or reject our claims.

Computing the coordinates of the distant "target" system (crude model)

In its orbital motion around the Sun, an Earth-based observer aims at an ever changing "target area" of the sky, hereinafter called the "Target," whose equatorial coordinates are (α_{Tgt}, δ_{Tgt}). It is our understanding that a correctly performed TN experiment should enable us to observe if not the "Target," a systematic deviation of the capacitor's orientation with respect to this calculated "Target" over long periods of time.

The problem we had to solve was twofold (Bizouard, 1999). As illustrated by Fig. 1, our primary goal consisted in formulating the "Target" coordinates as a function of the right ascension (α_{Sun}) and declination (δ_{Sun}) of the Sun (which can be found in any classical astronomical ephemerid logbooks, their values being given at 00:00 UT). The local coordinates of the Earth's orbital speed are easily computed provided we make the following assumptions: (i) the orbital motion of the center of mass of the Earth around the Sun is considered uniform on a circular trajectory so that the orbital velocity ($U_{Earth\,@\,Sun} = 30$ km.s^{-1}) remains normal to the direction Earth-Sun; (ii) for the sake of simplicity, the Earth's rotational irregularities (precession, nutation, polar motion, variations in length of day) are not taken into account; (iii) the values of the equatorial coordinates of the Sun (α_{Sun}, δ_{Sun}) remain unchanged during the day. λ_{Sun} is the ecliptic longitude of the Sun and $\varepsilon_{plane} = 23°26'$ the obliquity of the ecliptic plane. The above considerations allowed us to find out the following *equatorial coordinates* for the "Target":

$$\alpha_{Tgt} = \text{Arc tan}\{\sin(\varepsilon_{plane}).\cos(\varepsilon_{plane}).[\cos(\alpha_{Sun}).\cos(\delta_{Sun})/\sin(\delta_{Sun})]\} + [1 + \text{sign}(\sin\delta_{Sun})]\,\pi/2 , \tag{6}$$

$$\delta_{Tgt} = \text{Arc sin}[-\sin(\varepsilon_{plane}).\cos(\alpha_{Sun}).\cos(\delta_{Sun})] \quad . \tag{7}$$

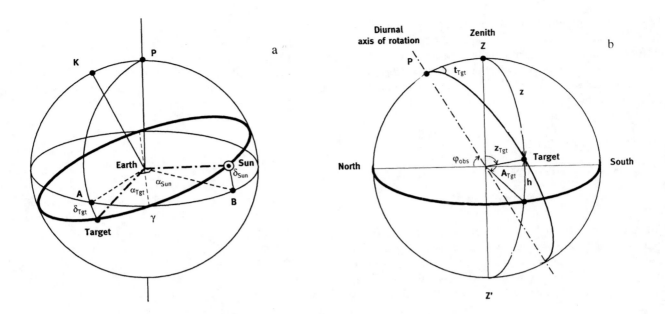

FIGURE 1. Parameters affecting the calculation of the distant "Target" coordinate system.

The second part of the problem, illustrated by Fig. 2, aimed at formulating the *horizontal coordinates* of the "Target" as a function of Universal Time (UT). For the sake of simplicity, the reader is invited to admit the following series of transforms between horizontal coordinates of the "Target" and its equatorial coordinates:

$$\cos(z_{Tgt}) = \sin(\varphi_{obs}).\sin(\delta_{Tgt}) + \cos(\varphi_{obs}).\cos(\delta_{Tgt}).\cos(t_{Tgt}) \quad, \tag{8}$$

$$\sin(z_{Tgt}).\sin(A_{Tgt}) = \cos(\delta_{Tgt}).\sin(t_{Tgt}) \quad, \tag{9}$$

$$\sin(z_{Tgt}).\cos(A_{Tgt}) = -\cos(\varphi_{obs}).\sin(\delta_{Tgt}) + \sin(\varphi_{obs}).\cos(\delta_{Tgt}).\cos(t_{Tgt}) \quad, \tag{10}$$

where A_{Tgt} represents the azimuth of the "Target," z_{Tgt} its zenithal distance, φ_{obs} is the latitude of the observer's laboratory and t_{Tgt} the "Target" horary angle, i.e. the angle between the of the observer meridian and the "Target" meridian. For a better understanding of the situation, Figs. 1 and 2a illustrate how these parameters interplay with each others. Since $0 \leq z_{Tgt} \leq \pi$, equation (8) can be rewritten as:

$$z_{Tgt} = \text{Arc cos } [\sin(\varphi_{obs}).\sin(\delta_{Tgt}) + \cos(\varphi_{obs}).\cos(\delta_{Tgt}).\cos(t_{Tgt})] \quad. \tag{11}$$

Thus enabling us to calculate the azimuth A_{Tgt}:

$$\sin(A_{Tgt}) = \cos(\delta_{Tgt}).\sin(t_{Tgt}) / \sin(z_{Tgt}) \quad, \tag{12}$$

$$\cos(A_{Tgt}) = -[\cos(\varphi_{obs}).\sin(\delta_{Tgt}) + \sin(\varphi_{obs}).\cos(\delta_{Tgt}).\cos(t_{Tgt})] / \sin(z_{Tgt}) \quad. \tag{13}$$

Our last unknown, illustrated by Fig. 2a, is the horary angle t_{Tgt}:

$$t_{Tgt} = -\text{ GST} + \lambda_{obs} - \alpha_{Tgt} \quad, \tag{14}$$

where GST stands for "Greenwhich Sideral Time," and λ_{obs} is the longitude of the observer. We shall first remind the reader that $GST_{00:00 \text{ UT}}$ is figured out in all ephemerid tables. We also insist that GST is described by this approximate function of Universal Time (UT): $GST = GST_{00:00 \text{ UT}} + 1{,}0027379 \text{ } TU_{obs}$. As a result,

$$t_{Tgt} = -\text{ GST}_{00:00 \text{ UT}} + 1{,}0027379 \text{ } TU_{obs} + \lambda_{obs} - \alpha_{Tgt} \quad. \tag{15}$$

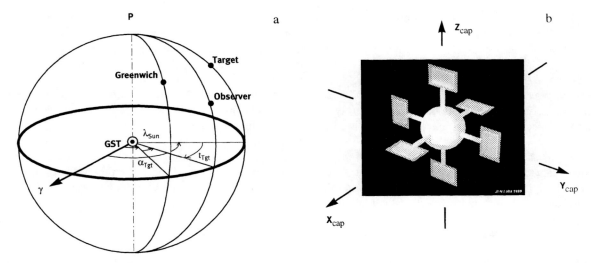

FIGURE 2. Parameters affecting the calculation of the distant "Target" horary coordinate (a) and conceptual SEADS craft (b)

These considerations enable us to determine the equatorial coordinates (α_{Tgt}, δ_{Tgt}), horizontal coordinates (A_{Tgt}, z_{Tgt}) and horary angle t_{Tgt} of the distant "Target," the velocity vector of the Earth aims at. If these parameters are known, and if the torque effect of the charged, suspended capacitor can be measured in a satisfying manner, it should be possible to compare the experimental results with the theoretical position of the Target and determine if the motion of the Earth influences the behaviour of said capacitor. However, in this crude model where $U_{absolute}$ is "reduced" to $U_{Earth @ Sun}$ as a first approximation, the "Target" can be demonstrated to remain in the ecliptic plane. This fact raised serious concerns in our team.

CONSEQUENCES

Our experimental results and theoretical analysis suggest that the "absolute" motion of the Earth can indeed be detected and observed directly. The debate as to what kind of variable would really be observed is now open (section 1). A dedicated orbital electrostatic sensor is proposed (section 2). Its applications to a better understanding of space weather and engineering of advanced, passive-type field propulsion systems are outlined (section 3).

Is the "absolute velocity" a composed set of "relative velocities" ?

As already noted in a previous paper (Cornille, Naudin, Szames, 1999), one of us (Naudin) discovered that the direction of rotation of the suspended, parallel-plate capacitor had changed between April-June and early September 1998. To our extreme puzzlement, a new change in the direction of rotation was observed "live" on September 23, 1998, i.e. the day of the autumnal equinox. A possible interpretation of this phenomenon has been given elsewhere (Naudin, 1998b). Contrary to what was incorrectly reported earlier to christen this enigmatic phenomenon, the "ecliptic crossover" is physically impossible to perform (since the Earth never crosses the plan its defines with the Sun). However, we did observe that the direction of the rotational motion of the suspended parallel-plate capacitor changed when the Earth crossed its equinoctial positions. However surprizing, this experimental result might be related, and therefore must be compared, to Allais' statistical analysis of Miller's chronological series. Among others, Allais demonstrated that Miller's optical ætherdrift observations would reach their climax when the Earth crosses its equinoctial positions (Allais, 1997c).

It is our understanding that Allais' analysis, together with our observed phenomena, can be explained if **U**, the velocity of the Earth being used in the calculation of the effects sought for, is *not* the velocity of the Earth around the Sun **U** $_{Earth @ Sun}$ = 30 km.s^{-1}, but the *composed* absolute velocity **U**$_{absolute}$ of the Earth with respect to a yet-to-be-defined, preferred frame of reference. It is our understanding that **U**$_{absolute}$ can be expanded in a series of constitutive, *relative* velocities added up according to the Galilean law of addition of speeds. This process does not imply that we are working within the framework of classical mechanics. In Cornille's crude model, on which this study is based, $U_{absolute}$ is demonstrated to be scalarly multiplied by another vector, namely the "relative" speed $U_{relative}$ (Szames, 1998b). One of these "relative" velocities would describe the diurnal motion of the Earth around its polar axis, another would represent the annual motion of the Earth around the Sun $U_{Earth\ @\ Sun}$, another the motion of the Sun with respect to the center of Milky Way $U_{sun\ /\ Galactic\ center}$, yet another may even result from the "relative" motion of the Milky Way with respect to other distant Galaxies, and so on. Logic forces us to admit that each of these aforementioned scalar products behaves differently as a function of time since each of the "relative" velocities is related to a specific rotational phenomenon having its own astronomical periodicity.

Interestingly enough, $U_{absolute}$ could be processed in such a way that its "harmonic" components (the "relative" velocities) are filtered out. FFT could be adopted as a first approximation, but it might be preferable to use more powerful analytic tools such as the KLT transforms (Maccone, 1994) if complex models of composed velocity of the Earth through "ætheral" space, are made available to researchers.

Towards a Simple Electrostatic Æther Drift Sensor (SEADS) ?

If our observations prove to be correct, this intricate set of relative motions might perturb the behaviour of a charged parallel-plate capacitor embarked on the Earth, either in its rotational motion, or in the translational motion of its center of mass. These considerations invited us to consider the necessity to manufacture a three-axis Trouton-Noble-like experimental device, which might enable us to determine how $U_{absolute}$ is composed over long periods of time. We are thus forced to introduce the *orbital* testability of the hypothesis of a velocity-dependent "space weather analogue" through: (i) dynamic characterization and computation of the observed phenomena and; (ii) three axis mapping of the associated, anticipated observed torque. This yet-speculative device, which might enable us to map this intricate composed velocity field, has been termed a "SEADS".

A conceptually Simple Electrostatic Aether Drift Sensor (SEADS) freeflying platform is depicted in Fig. 2b. This platform could be "launched" from the International Space Station (ISS) as part of a breakthrough propulsion physics effort. This platform would consist in a microsat-type, three-axis stabilized platform, outfitted with the following payload elements: (i) optical attitude sensor subsystem (star tracker type); (ii) conventional electric attitude control system; (ii) optical communication subsystem; (iii) radiation-hardened onboard systems and subsystems; (iv) shielded HV generators. In our preliminary conceptual design, six identical parallel-plate capacitors would be attached onto the ends of six non-conducting supporting arms, equal in length, fixed onto this platform along the platform's x, y, z axis. We shall identify any of these capacitors by their indexed position in space $\{(x_{cap.+}\ ;\ x_{cap.-});\ (y_{cap.+}\ ;\ y_{cap.-});\ (z_{cap.+}\ ;\ z_{cap.-})\}$. If both our theoretical analysis and experimental work is correct, couples of "paddles" identically charged at high voltage would tend to line up and lock themselves in a preferred direction in space (thus explaing the requirement for an embarqued optical star tracking system). Orientations before and after charging would be recorded and resulting data processed on Earth.

However, such a speculative design raises a number of serious engineering concerns. We still haven't solved the primary energy source issue. The possibility to design an electrodynamic tether-like system, or use beamed energy concepts is still left open to speculation. Among the many difficulties one might encounter to design a working SEADS platform, the following problems will have to be solved: (i) static electricity spacecraft charging (Maehlum & Troïm, 1990); (ii) availability of an electrofluidynamic model not only applicable to low-density plasmas, but to the entire superaerodynamic spectrum; (iii) a required vacuum of 10^{-7} Torr to avoid the so-called "Paschen effect" (which might generate an unwanted torque on the platform); (iv) studies of the interaction of a charged SEADS platform with the magnetic field of the Earth. Additionnaly, while vacuum is known to posess excellent dielectric properties, it will not protect HV charged SEADS-platforms from spurious, highly damaging breakdown phenomena. Some of these difficulties might prove to be unsurmountable with state-of the art technology.

Relevance to "space weather" and passive field propulsion systems

From the results gained with our crude model, apparatus, and a series of ongoing experimental research efforts, we inferred that, when a parallel-plate capacitor was charged at high voltage, both the direction and magnitude of the observed effects could be a function of an "absolute" velocity $U_{absolute}$ of which the "relative" velocity-dependent components could be identified. More specifically, these velocity-dependent terms could be traced back to periodic phenomena predictible by the proper use of orbital mechanics. These terms are yet to be defined, i.e. theoretically identified, experimentally measured, and analytically correlated to test the hypothesis of a space weather analogue. This area of research could be of potential immense interest to space weather. This discipline currently includes studies of the sun, the solar wind (and its evolution in the interplanetary "medium"), the magnetosphere, the ionosphere, and the upper atmosphere. Understanding how these regimes are coupled is of primary concern to space weather forecasters. We believe that while an "interdisciplinary approach" is required to unify this discipline, space weather still relies on classical electromagnetism and conventionnal plasma physics. It is our understanding that: (i) classical Maxwellian electromagnetism needs a complete refurbishement; (ii) the dynamics of the underlying interplanetary medium, which might just be another euphemism for "æther," might have to be studied to enforce or accelerate the pace of space weather regimes coupling. Should the above considerations prove to be correct, it is our understanding that SEADS, or SEADS derived sensors, could be helpful to test breakthrough-class propulsion systems. This novel class of sensor could possibly "map" the evolutions, pertubations, modifications of velocity vector fields around future field-dependent propulsion systems.

Innovative advanced propulsion systems have been designed making profitable use of space weather principles (Drell, 1965; Duchon, 1995). In his kick-off essay, *The Challenge to Create the Space Drive,* Millis described several classes of speculative devices capable of generating propulsive thrust in empty space through active exchange of momentum with an external, *stationary* medium. These exotic "warp drives" are essentially active devices (Mead, 1989): they *induce* local asymmetries enabling the exchange of momentum with the external, *stationary* medium. It is our understanding that Millis' concepts can be furthered out to other classes of speculative, passive devices capable of generating propulsive thrust in empty space by making profitable use of the *dynamical* properties of the aforementioned external medium, whose velocity could be measured, computed, modeled using a properly designed network of SEADS. We sincerely hope this study has established the possibility to investigate if not the logical feasibility of such a concept, its legitimacy.

CONCLUSIONS

However suprising, our work suggests that the behaviour of two electric charges in motion has not been completely understood, nor worked out in a satisfying manner, both from a theoretical perspective and from the standpoint of experimental practice. Correctly-performed Trouton-Noble-like experiments seem to be central to engineering future types of breakthrough-class space drives. A charged, suspended, parallel-plate capacitor has been demonstrated to exhibit, in certain experimental conditions, a stimulated torque whose magnitude and direction of rotation would not only support the existence of a directly observable electromagnetic "æther," but also help detect the proper "absolute" motion of the Earth. However, "absolute" may be overoptimistic as an adjective in the present context, since "absolute" velocities and motions *could* be interpreted as the end-products of an intricate, composed set of *dynamic* "relative" velocities and motions, endowed with proper qualities, features and periodicities.

"I play for seasons, not Eternities" — says Nature. George Meredith (1828-1909)

Should these æthereal motions (and periodicities) be detected by other, competing, investigative teams, "space weather" would have to be filled up with a new series of underlying, dynamic "dimensions". Should an adequate, "electromagnetic field disengagement system" be properly engineered, some of the foreseeable future speculative types of space drives could passively "ride" along these æthreal motions and "sail" to the farthest stars.

ACKNOWLEDGMENTS

This paper is not the result of any government-sponsored activity or research. It represents the views of the authors. We are especially grateful to our families and friends Barbara, Bernadette and Henri Szames, S. Blocher, J.-L. Delauche, J. Moulin (†), D. Pourquery, A. Gaubert, M. Touitou, as well as the Ast family for their exceptional support, to M.G. Millis, C. Yost and Pr. R. Gabillard for their generous and precious help. And of course, to Camille Saint-Saëns, Philipp Glass, John Adams, Steve Reich, Laurie Anderson, David Shea, Eric Satie, Jean-Luc Godard, Giovanni Baptista Piranesi for their unconditionnal support throughout the years, and especially while writting this paper.

NOMENCLATURE

Calculation of the torque

c	speed of light
C	capacity of the capacitor
D	distance between the electrodes
F_{ij}	stimulated force generated by two moving charges i and j $(i \neq j)$
i	current intensity
l	length of the suspending fibber
Q	charge of the capacitor
Q_i	point charges fixed onto the ends of a solid rod (i = 1,2)
R	length of the solid rod
U	velocity (general, see below)
V	potential difference between two electrodes
ε_r	dielectric constant (real part)
μ_o	magnetic permeability (vacuum)
θ	angle between velocity vector and normal to the plates
φ	angle between velocity vector and the suspending fibber
Γ	torque of the capacitor

Calculation of the distant "target" system

A_{Tgt}	azimuth of the "Target"
t_{Tgt}	horary angle of the "Target"
z_{Tgt}	zenithal distance
α_{Tgt}	right ascension of the "Target"
α_{Sun}	right ascension of the "Sun"
δ_{Tgt}	declination of the "Target"
δ_{Sun}	declination of the "Sun"
ε_{plane}	obliquity of the ecliptic
φ_{obs}	latitude of the observer's laboratory
λ_{obs}	longitude of the observer's laboratory
$U_{absolute}$	velocity of the capacitor with respect to a prefered inertial frame of reference
$U_{Earth @ Sun}$	relative velocity of the Earth around the Sun in a heliocentric frame of reference
$U_{relative}$	relative velocity (general)
$U_{Sun / Galactic center}$	relative velocity of the barycenter of the Solar System in a galactocentric frame of reference

REFERENCES

Aitchinson, I.J.R. "Nothing's Plenty: the Vacuum in Modern Quantum Field Theory," *Cont. Phys.*, **26**(4), 333-391 (1985).

Allais, M., *L'Anisotropie de l'Espace*, Clément Juglar Ed., Paris, France (1997a)

Allais, M., *Ibid.*, pp.331-372 (1997b).

Allais, M., *Ibid.*, pp.392-411, see in particular p.409 (n2) and pp.455-458 (1997c).

Anderson, J.L., "Covariance, Invariance and Equivalence: a Viewpoint," *General Relativity & Gravitation*, **2**(2), 161-172 (1971).

Arunasalam, V., "Einstein and Minkowski Versus Dirac and Wigner: Covariance Versus Invariance," *Physics Essays*, **10**(3), 528-532 (1997).

Barrett, T.W., "Electromagnetic Phenomena not Explained by Maxwell's Equations," in: A. Lakhtakia (Ed.), *Essays on the Formal Aspects of Electromagnetic Theory*, World Scientific, Singapore, pp.6-86 (1993)

Barrow, J.D., *Impossibility: The Limits of Science and the Science of Limits*, Oxford University Press, Oxford, USA, 1998

Beckmann, P., *Einstein Plus Two*, The Golem Press, Box 1342, Boulder, Colorado 80306, Boulder, Co. USA, p.77 (1987)

Bizouard, C., *Petite note concernant le calcul des coordonnées locales (ou horizontales) de la vitesse orbitale de la Terre en fonction de la date et de l'heure*, unpublished manuscript, Paris, France, 6 p. (1999)

Brillouin, L. *Relativity Reexamined*, Academic Press, New York, USA (1970)

Chase, C.T., "A repetition of the Trouton-Noble experiment," *Phys. Rev.*, **28**, 378-383 (1926)

Cornille, P., "The Lorentz force and Newton's third principle," *Can. Journ. Phys.*, **73**, 619-625 (1995)

Cornille, P., "Review of the Application of Newton's third law in physics," *Progr. En. Comb. Sci.*, **25**(2), 161 (1999)

Cornille, P., Naudin, J.-L. & Szames, A.D., " Way back to the future: Why did the Trouton-Noble experiment fail and how to make it succeed," in: *6th Conference on the Physical Interpretations of the Relativity Theory (PIRT-VI) 11-14/09/1998*, proceedings edited by the British Society for the Philosophy of Science (BSPS), Great Britain (1998)

Cornille, P., Naudin, J.-L. & Szames, A.D., "Stimulated Forces Demonstrated: Why did the Trouton-Noble Experiment Fail and how to Make it Succeed," *in:* M. EL'GENK (Ed.), *Proceedings of the Space Technology Applications International Forum (STAIF) 1999*, Albuquerque, NM, (1999)

Crooks, M.J., Litvin, D.B. Mathews, P.W. Macaulay J.R. & Shaw, J. "One piece Faraday generator: a paradoxical experiment from 1851," *Am. Journ. Phys.*, **46**(7), 729-ff (1978)

Cullwick, E.G. "Einstein and Special Relativity: Some Inconsistencies in his Electrodynamics," *Brit. Journ. Phil. Science*, **32**(2), 06/1981, 167-176 (1981)

Darrigol, O., "La Genèse du Concept de Champ Quantique," *Ann. Phys. (Paris)*, **9**(6), 433-501 (1984)

Drell, S.D. *et al.*, "Drag and Propulsion of Large Satellites in the Ionosphere: an Alfven Wave Propulsion Engine," *J. Geoph. Res.*, **70**, 3131-3145 (1965)

Duchon, P., *Satellite Artificiel Muni de Générateur de Moments Magnétiques et Aérodynamiques et Procédé de Commande d'un tel Satellite,* Patent No. WO 95 / 26905 filed 30/03/1994 (France), issued 12/10/1995 (world)

Earman, J., "Covariance, Invariance, and the Equivalence of Frames," *Found. Phys.,* **4**(2), 267-289 (1974)

French, A.P., *Special Relativity,* W. W. Norton & Co., New York USA (1968)

Gabillard, R., *Private communication,* Lille, France (1999)

Galuchot, B., *Private communication,* Courbevoie, France (1999)

Hartmuth, H.F., "Electromagnetic Transients not Explained by Maxwell's Equations," in A. Lakhtakia (Ed.), *Essays on the Formal Aspects of Electromagnetic Theory,* World Scientific, Singapore, pp.87-126 (1993)

Hayden, H.C. "High-Sensitivity Trouton-Noble Experiment," *Rev. Sci. Instr.,* **65**, 788-792 (1992)

Hayden, H.C. "Analysis of Trouton-Noble Experiment," *Gal. Electrod.,* **5**, 83-85 (1994)

Hendri, J., *James ClerkMaxwell and the theory of the Electromagnetic Field,* Adam Hilger Ltd., Bristol & Boston (1986)

Hesse, M.B., *Forces and Fieds (2nd ed.),* Greenwood Press Pub., Westport, USA, p.192 (1970)

Janssen, M.H.P., *A Comparion (sic) between Lorentz's ether theory and special relativity in the light of the experiments of Trouton and Noble,* Dissertation, University of Pittsburgh, Pittsburgh, USA, 1995

Jefimenko, O.D., "The Trouton-Noble paradox," *J. Phys. A: Math. Gen.,* **32**, 3755-3762 (1999)

Johnson, J.L., "Coordinate Transform Invariance," Foundations of Physics, **26**(11), 1529-1557 (1996)

Kostro, L., "The Physical Meaning of Albert EINSTEIN's Relativistic Ether Concept," in: M. Barone & F. Selleri (Eds.), *Frontiers of Fundamental Physics,* Plenum Press, New York, USA, pp.193-201 (1994)

Lewis, G.N., Tolman, R.C., "The Principle of Relativity and Non-newtonian Mechanics," *Phil. Mag.,* **18**, 510-523 (1909)

Maccone, C., *Telecommunications, KLT and Relativity,* IPI Press, Colorado Springs, USA, 1994 (not consulted)

MacCausland, I, "An Inconsistensy in Special Relativity," *Physics Essays,* **3**(2), 06/1990, 176-177(1990)

Maehlum, B.N. & Troïm, J., "Vehicle Charging in Low Density Plasmas," in: *Ionospheric Modification and its Potential to Enhance or Degrade the Performance of Military Systems,* NATO/AGARD-CP-485, Paper 24 (1990)

Mead Jr., F.B., "Exotic Concepts for Future Propulsion and Space Travel," *Advanced Propulsion Concepts,* 1989 JPM Specialist Session, (JANNAF), Chemical Propulsion Information Agency (CPIA) Publication 528, pp.93-99 (1989)

Michelson, A.A., Morley, E.W., "On the relative motion of the earth and the luminiferous ether," *Am. Journ. Sci.,* **24**, 333-345 (1887)

Mikhailov, V.F., "Six Experiments with Magnetic Charges," in T. Barrett & C. Grimes (Eds.), *Advanced Electromagnetism: Foundations, Theory and Experiments,* World Scientific, Singapore, pp.593-619 (1995)

Millis, M.G., "Challenge to Create the Space Drive," *Journ. Prop.Pow.,* **13**, 577-582 (1997)

Miller, D.C., "The Ether Drift experiment at Mount Wilson," *Proc. Nat Ac. Sci.,* **2**, 306-314 (1925)

Miller, D.C., "Significance of the Ether Drift experiments of 1925 at Mount Wilson," *Science,* **63**, 433-443 (1926)

Miller, D.C., "The Ether Drift Experiments and the Determination of the Absolute Motion of the Earth," *Rev. Mod. Phys.,* **5**(3), 203-242 (1933)

Naudin, J.-L., http://members@aol.com/jnaudin509/index.htm (1997)

Naudin, J.-L., http://ourworld.compuserve.com/homepage/jlnaudin/html/zpfmotor.htm (1998a)

Naudin, J.-L., "Discovery / Découverte : 15 - 24 Sept 1998," in Szames (1998), pp.343-368 (1998b)

Naudin, J.-L., http://www.fortunecity.com/tattooine/delany/256/html/pftm2.htm (1999a)

Naudin, J.-L., http://www.fortunecity.com/tattooine/delany/256/html/pftm2.htm (1999b)

Nemenyi, P.F., "The Main Concepts and Ideas of Fluid Dynamics in their Historical Development," *Arch. Hist. Exact Sci.* **2**, 52 (1962).

Phipps, Jr., T.E., *Heretical Verities: Recurring Themes in Mathematical Physics,* Classical Non-Fiction, Urbana, Illinois, USA, 1986a

Phipps, Jr., T.E., *Ibid,* p.19 (1986b)

Post, E.-J., *Formal Aspects of Electromagnetic Theory : General Covariance and electromagnetism,* North Holland (1962)

Szames, A.D., *L'Effet Biefeld-Brown : Histoire secrète de la plus grande découverte scientifique du XXe siècle* (2nd Print.), Éditions Alexandre Szames, 47-51 rue d'Aguesseau, F-92100 Boulogne, France, ISBN 2-913377-00-9 (1998a)

Szames, A.D., *Ibid.,* p.383 (1998b)

Terrel, J., "Invisibility of the Lorentz Contraction," *Physical Review,* **116**, 1041-1045 (1959)

Tokaty, G.A., *A History and Philosophy of Fluid Mechanics,* Dover, New-York, USA, p.18 (1971)

Tomascheck, R., "2. Über Versuche Zur Auffinding Elektrodynamischer Wirkungen der Erdbewegung in groben — Höhen I," *Ann. Phys,* **78**, 743-756 (1925).

Tomascheck, R., "3. Über Versuche Zur Auffinding Elektrodynamischer Wirkungen der Erdbewegung in groben — Höhen II," **80**, 509-514 (1926).

Trouton, F.T. & Noble, H.R., "The mechanical forces acting on a charged condenser moving through space," Philosophical Transactions of the Royal Society of London, *Philos. Trans. Roy. Soc. Lond.,* **202A**, 165-181 (1904)

Vigier, J.-P., "Relativistic Interpretation (with Non-Zero Photon Mass) of the Small Ether Drift Velocity Detected by Michelson, Morley and Miller," *Apeiron,* **4**(2-3), 04-07/1997, 71-76 (1997)

Zhong Zhang, Y., *Special Relativity and its Experimental Foundations,* World Scientific, Singapore, p.174 (1997)

Mass Fluctuations, Stationary Forces, and Propellantless Propulsion

James F. Woodward

Departments of History and Physics, California State University, Fullerton, California 92634;
e-mail: jwoodward@fullerton.edu

Abstract: The gravitational origin of inertial reaction forces is discussed and transient mass fluctuations that occur in accelerating objects are explained. It is pointed out that such mass fluctuations can be employed to produce stationary forces in special circumstances, and that those forces can, in principle, be used to produce propellantless propulsion. In laboratory tests the magnitude of the effects involved in this method are quite small, but since they are based exclusively on the known laws of physics the possibility exists that the method can be brought to practical fruition in the foreseeable future.

THE GRAVITATIONAL ORIGIN OF INERTIA

It has been known for two decades and more, at least to a small group of specialists in general relativity theory (GRT), that the cause of inertial reaction forces in our isotropic universe is gravity (Woodward and Mahood, 1999). Contention over the role of "Mach's principle", that is, the relativity of inertia, in GRT (Barbour and Pfister, 1995) has obscured this fact. Indeed, as long as a half century ago Dennis Sciama (1953) showed this to be the case using only the vector formalism of electrodynamics for gravity, which approximates GRT in many situations. For example, consider the translational acceleration of a test mass, located far from local concentrations of matter where the matter density of the universe can be considered uniform, by an external force. All of the other matter in the universe will exert a gravitational force on the test mass, and the gravito-electric force E per unit mass it exerts on the test mass will be given by:

$$E = - \nabla\phi - (1/c)\partial A/\partial t. \tag{1}$$

where ϕ is the gravitational scalar potential, c the vacuum speed of light, and A the gravitational three-vector potential. In analogy with electrodynamics Sciama too the scalar and three-vector potentials to be:

$$\phi = - \int(G\rho/r)dV, \tag{2}$$

$$A = - (1/c)\int(G\rho v/r)dV, \tag{3}$$

where G is the Newtonian constant of gravitation, ρ the matter density, ρv the matter current density in the integration volume element dV, r the distance from the test body to the volume element being considered, and the integration extends over all space (strictly speaking, space in the causally connected part of the universe). These potentials are attended by subtleties of choice of gauge that we ignore for the moment. The gravitational force produced by the matter in the universe on our test body is now easily computed. First we note that if our test body moves translationally with respect to the rest of the universe, the curl of A vanishes by symmetry. Direct gravito-magnetic forces are therefore absent in these circumstances. Since the density of matter is uniform and the test body, by assumption, doesn't contribute to the gravitational field, $\nabla\phi$ vanishes and $E = - (1/c)\partial A/\partial t$. We compute A with Equation (3) and take its time derivative. To do this Sciama notes that from the point of view of the test body every part of the universe appears to be moving rigidly with velocity $-v$ at each instant, so v can be removed from the integral in Equation (3). The remaining expression, save for the factor $1/c$, is the integral for the scalar potential ϕ, so in this case $A = \phi v/c$. With this expression for A Equation (1) becomes:

CP504, *Space Technology and Applications International Forum–2000*, edited by M. S. El-Genk

$$E = -(\phi/c^2)\partial v/\partial t. \tag{4}$$

It follows immediately from Equation (4) that if our test body moves with constant v with respect to the universe, it experiences no gravito-electric force. If it is accelerated, however, E isn't zero. And if $\phi/c^2 = 1$, then the gravito-electric force *is* the inertial reaction force per unit mass of the test body. Suffice it to say that the *locally measured* value of ϕ in GRT turns out to have the required value for inertial reaction forces to be gravitational.

Two points should be noted before turning to transient mass fluctuations. First, E in Equation (4) depends on $\partial A/\partial t$, not $\nabla\phi$, so E has the $1/r$ dependence that is characteristic of radiative interactions. Moreover, notwithstanding the trick of removing v from the integral for A by appealing to the appearance of the universe in the instantaneous rest frame of reference of the accelerating particle, the gauge used in this calculation is the Lorentz gauge. In this gauge all changes in the fields and all components of the vector potential propagate at vacuum light speed c. Since inertial reaction forces are felt at the instant of application of external forces, the only way they can arise from radiative coupling (via signals that propagate at light speed) to chiefly the most distant matter in the universe is through the mediation of both retarded (traveling forward in time) and advanced (traveling backward in time) waves. (This, by the way, is the only way instantaneous, "non-local" effects can be squared with relativity theory.) This behavior is characteristic of "radiation reaction". The second point is that inertial reaction forces are enormously larger than the gravitational forces of everyday experience. Indeed, they are comparable to electromagnetic forces. But one finds no force of this type in electrodynamics; electromagnetic radiation reaction forces are minuscule by comparison with inertial reaction forces. This is a consequence of the fact that $\phi \approx c^2$ for gravity, but for electrodynamics the corresponding potential is essentially zero because the net electric charge density at the cosmic scale is zero. These considerations suggest that if radiation reaction effects involving inertial forces can be found, they may be quite large and open novel propulsion possibilities.

TRANSIENT MASS FLUCTUATIONS

Space limitations do not permit presentation of the full, formal elaboration of transient mass fluctuations. That derivation can be found in Woodward (1995). Instead, I give a heuristic development that may be more helpful than the formal derivation. The simplest possible four-dimensional, thus relativistic generalization of Newton's law of gravity is:

$$\nabla \cdot f = \nabla^2\phi = 4\pi G\rho \quad \rightarrow \quad \nabla^2\phi - (1/\rho c^2)(\partial^2 E/\partial t^2) = 4\pi G\rho, \tag{5}$$

where f is the three-vector gravitational field strength and E the energy density created by the action of the field on its sources. The additional term in the relativistic version of this equation is a simple consequence of the fact that the time-like components of relativistically invariant four-forces are powers, that is, $\partial E/\partial t$, the rate at which the force [in this case exerted by the inertial/gravitational field on its sources] does work (Rindler, 1991). In order to write the expression in Equation (5) in the form of a relativistically invariant wave equation for ϕ we note that E is the gravitational energy density which is $\rho\phi$. Since $\phi \approx c^2$, this is not a negligible quantity. (Also note that since $\phi \approx c^2$, taking $E = \rho\phi$ amounts to nothing more than assuming that energy is equal to mc^2.) Substituting $\rho\phi$ for E in Equation (5) yields:

$$\Box\phi = \nabla^2\phi - (1/c^2)(\partial^2\phi/\partial t^2) \approx 4\pi G\rho + (\phi/\rho c^2)(\partial^2\rho/\partial t^2). \tag{6}$$

Note that a term involving a product of the time derivatives of ϕ and ρ has been suppressed on the RHS of this equation since $\partial\phi/\partial t$ is usually negligibly small when compared with $\partial^2\rho/\partial t^2$. [In the full derivation, given the condition $\phi \approx c^2$, this term is canceled by another term not recovered in this calculation.] The full formal derivation yields some additional transient source terms (Woodward, 1995). They, however, are unimportant in all but the most extreme and contrived circumstances (Woodward, 1997).

The transient term on the RHS of Equation (6) is a transient mass density:

$$\delta\rho(t) \approx (\beta\phi/4\pi G\rho c^4)(\partial^2 E/\partial t^2), \tag{7}$$

where $\rho = E/c^2$ has been used to express the mass density as an energy density. A dimensionless coefficient β has been introduced to allow for the possibility that real circumstances differ from the ideal conditions of this calculation where β would be of order one. In experiments the objective is to measure the value of β. A value of zero would mean no effect is present. From Equation (7) it follows that the total transient mass fluctuation δm induced in the volume V of some material subjected to an acceleration by an external force then is:

$$\delta m = \int \delta \rho dV \approx (\beta \phi / 4\pi G \rho c^4) \int (\partial^2 E / \partial t^2) dV. \tag{8}$$

Note that this transient mass fluctuation only occurs in massive objects subjected to accelerations by external forces. Energy density fluctuations of other origins do not necessarily lead to mass fluctuations of this sort. These are the transient mass fluctuations that may make possible the realization of the dream of propellantless propulsion. It is worth noting here that acceleration-dependent, transient mass fluctuations are known in GRT (Nordtvedt, 1988 and Woodward, 1992).

One might reasonably ask why these transient effects occur. Why is it reasonable for such effects to be recovered from relativistic calculations? I seems that the answer to this question is that since the forces involved are radiation reaction type forces, small phase shifts leading to small time-delays occur, leading to a transient unbalancing of forces that produce these effects – much as time-delays are an inherent part of the process of electromagnetic radiation reaction. *The most important point to be made here, though, is that all of this is based firmly on well-known principles of physics.*

STATIONARY FORCES

Transient mass fluctuations make possible the extraction of stationary forces in appropriate systems. The basic procedure involved is simple: One pushes (and pulls) synchronously on some object in which a periodic transient mass fluctuation is driven. If one always pushes in one direction when the mass fluctuation is positive, and pulls in the other when the mass fluctuation is negative (or at least less positive), then the inertial reaction force exerted by the object with the fluctuating mass on the system producing the pushes (and pulls) will have a stationary component. This follows formally in the case of sinusoidal mass fluctuations acted on by a sinusoidal excursion of the same frequency, for example, as a consequence of the fact that the product of two sinusoids of the same frequency yields a phase-dependent DC term.

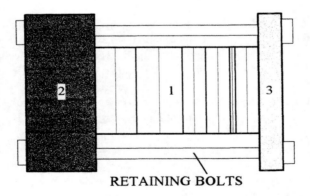

FIGURE 1. A schematic diagram (not to scale) of a device consisting of a stack of PZT disks excited by a two frequency waveform that should produce stationary forces.

These phenomena may be explored by driving mass fluctuations in, for instance, suitable capacitors mounted on (or between) piezoelectric transducers that shuttle them synchronously with the induced mass fluctuations (Woodward, 1996, 1998). An even simpler approach is to use a lead-zirconium-titanate (PZT) piezoelectric actuator to do both jobs at once. This can be achieved by driving a stack of PZT disks (see Figure 1) with a voltage signal that has two frequency components. The low frequency part of the signal drives the mass fluctuations. Since the mass fluctuations take place at the power frequency of the voltage signal, they occur at twice the frequency of the voltage signal. If the PZT stack is designed to mechanically resonate at the mass fluctuation frequency, then by adding a double frequency voltage signal to the low frequency signal (suitably phase-locked), the PZT stack can be driven into the excursions needed to generate a stationary force. A non-resonant reaction mass (2 in Figure 1) is used to make the acoustic waves in the PZT stack approximate a "fixed-free" configuration so that the largest accelerations occur in the "free" end of the stack. The interaction of the bulk mechanical excursion with the mass flcutuation in the "free" end of the stack yields the stationary force that acts on the reaction mass. The magnitude of the force exerted on the reaction mass can be adjusted by manipulating the amplitudes of the two signals and/or their relative phase.

How large are the forces produced in devices such as this expected to be? Well, a naïve calculation suggests that forces of thousands of dynes should be achievable with powers of hundreds of watts when these devices like are run in the 50 to 100 kHz range. Such forces have not in fact been seen, however, because no account of effects like lattice stresses and thermal dissipation have been included. Lattice stresses in particular can be expected to dramatically reduce any effect that might be present because they work against the accelerating effect of any external field in solid materials. So one should be prepared to look for a small effect. Should one be found, then one can explore ways to bring it up to practical levels for propellantless propulsion.

A few words on conservation principles seem in order at this point. After all, any device that permits one to exert a force on an object without expelling propellant, be it in the form of normal matter or photons, seems to violate at least the conservation of momentum, and perhaps the conservation of energy too. But the formalism of transient mass fluctuations and their propulsive applications outlined here is relativistically invariant. Accordingly, energy and momentum conservation are not violated. (This is guaranteed by "Noether's theorem", from which it follows that fields that are invariant under infinitesimal Lorentz transformations must have conserved sources.) Since the isolated system in which conservation laws apply in this case is the entire universe (because the gravitational/inertia interaction is with chiefly the most distant matter in the universe), it does not strain credulity that conservation laws are not compromised globally. But what about locally? How can a craft go motoring off in some direction just be wiggling some stuff, albeit in a very particular way? Evidently, the propagating distortions in the gravitational/inertial field created by the wiggling provide the energy-momentum flux needed to preserve local conservation principles.

EXPERIMENTAL CONFIGURATION

Over the past several years I have investigated the mass fluctuations that are critical to this method of generating stationary forces (Woodward, 1991 and 1996). And since the late spring of 1997, Thomas Mahood, a physics graduate student at CSUF, has worked closely with me on this project. Until the fall of 1998, the force effects sought in various experiments were detected with a commercial sensor based on a magneto-resistive Hall effect device wired as an active arm in a Wheatstone bridge. With the bridge voltage heavily amplified and signal averaging, minute forces could be detected with this device. However, during the fall of 1998 it became clear that this sensor could not be accurately calibrated in the conditions of its use: high frequency mechanical vibrations. In particular, in conjunction with another device specific quirk, it over estimated by orders of magnitude a small, seemingly genuine effect (Woodward, 1998). These insuperable problems led to the abandonment of electronic force sensors. They were replaced by very sensitive torsion pendula.

Use of torsion pendula for force detection obviates the problems associated with electronic force sensors. But torsion pendula are not problem free. The chief difficulty is delivering power on the order of 100 watts or more at ultrasound frequencies to the test devices mounted at the ends of a torsion beam without introducing spurious effects. Two approaches were pursued in this connection. One was based on making the meter long suspension fiber with multiple insulated conducting strands (of #34 AWG magnet wire) so that it could be used for the two leads of the power delivery circuit. In the other approach, Thomas Mahood developed a separate system using a single strand of copper magnet wire (#30 AWG) for the suspension as part of his thesis work [Mahood, 1999]. In his system the circuit was completed by a short length of wire hanging from the center of the beam that dipped into a mercury contact. Both of these systems were housed in Lucite vacuum vessels that could be pumped down to a few tenths of a Torr with a rotary vane vacuum pump. Motion detection in both cases was done by reflecting a laser beam off of a small mirror affixed to the axis of the suspension. Notwithstanding the very different construction of these two systems, fortuitously, they turned out to have nearly the same sensitivity. Figure 2 is a photo of the devices mounted on the torsion beam used in the multiple strand suspension system. The devices were attached to the beam with nylon screws with rubber washers to suppress the transmission of vibration from the devices to the suspension. As might be expected, trial and error testing was required to get these systems to work. For example, with the multiple strand suspension we found that no twist of the strands whatsoever could be tolerated. Thermal relaxation of any twist present during application of power to the devices would produce a torque like that sought.

The results reported below were obtained with a stack of 0.75" diameter PZT disks (1 in Figure 1) clamped with 4-40 stainless steel retaining bolts (to provide a preload on the stack) between a passive brass reaction mass (2) and a thin aluminum cap (3) (to keep the stack from tearing itself apart when activated). The PZT disks (Edo Ceramics material EC-64) were epoxied together with electrodes arranged so the ground planes correspond to the heavy lines, and the voltage waveform is applied to the planes represented by light lines in Figure 1. Thick disks (0.13") were located next to the reaction mass (2) because little effect is expected owing to the small mechanical excursion produced by the high frequency signal in this region. At the end near the aluminum cap (3) the disks are thin (0.06") so that the electric field strength will be high and large effects driven in this region where the mechanical excursion will be largest. Note also the pair of very thin disks included near the end of the stack. They were not electrically excited. Rather, they were used as an accelerometer to monitor the behavior of the stack when it was activated.

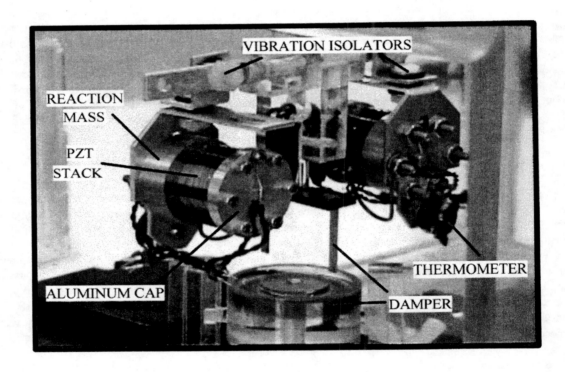

FIGURE 2. A photograph of two of the devices shown schematically in Figure 1 mounted on a torsion beam in a clear Lucite vacuum chamber.

To produce the effect sought, the devices must be driven with a voltage signal with an approximately sawtooth waveform. Sense resistors were placed in the high voltage part of the amplifier output circuit so that the waveforms of the instantaneous voltage and power being delivered to the devices could be monitored. The devices were typically run at a power level of about 50 watts each (100 watts total). Because at this power level heat evolved very quickly in the devices, they were fitted with spiral bimetalic thermometers so that high temperatures that might de-pole the PZTs could be avoided. Moreover, instead of simply applying power continuously and observing a deflection, the devices were run in pulsed mode, five second bursts of power being delivered at the resonant periods of the pendula (about 17 seconds). Only a few power bursts were needed to produce easily detected excursions of the pendula. The excursions were registered on a scale with the laser beam reflected off the mirror on the torsion beam. These were recorded on video tape with a camcorder so that the values of the endpoints of the oscillation of the beam could be easily read off (and checked later if desired).

RESULTS

When the devices shown in Figure 2 were driven with a sawtooth voltage waveform an unmistakable torque was exerted on the torsion beam, a torque that could be detected in the first five second power burst (of a sequence of such bursts). Knowing the period of the pendulum [about 17 seconds], its moment of inertia (calculable form the mass of the devices [about .12 kg each] and the length of the beam [0.06 m]), the angular excursion induced by each power pulse [about 0.003 m excursion of the laser beam with an optical lever arm of about 5 m], and the duration of the power pulse [5 seconds], it is a straight-forward matter (in the ballistic pendulum approximation) to calculate the force produced in each of the devices by the power pulses. At the 50 watt power level that force turned out to be roughly two hundreths of a dyne in each of the devices. Quoting a formal error in this result (perhaps 10% of the reported value) isn't especially interesting because of the several variables that can affect the measured value that in the present system are only rather crudely known. Ideally, one would also be able to compute a value for β; but this is not possible because the amplitude of the excursions of the active parts of the PZTs would have to be known. The apparatus needed to make such determinations *in situ* was not available during the course of these experiments. It was possible, however, to make interferometric measurements with the devices mounted in a jig. An excursion of the end of the PZT stack of several wavelengths of the red HeNe laser light was detected this way. A crude estimate of β thus can be made using the formula:

$$\beta \approx \pi G \rho c^2 F / \omega^3 P \delta l, \tag{9}$$

where ρ is taken to be about 5×10^3 kg/m^3, $F \approx 2 \times 10^{-7}$ Newtons, $\omega^3 \approx 1.4 \times 10^{16}$, $P \approx 50$ watts, and $\delta l \approx 10^{-7}$ to 10^{-6} m. With these assumptions β is found to lie in the range 10^{-7} to 10^{-8}. Evidently, the effect seen is very much smaller than a naïve calculation suggests might be present.

Much more interesting (and important) than the particular value of β is the question: Is the result reported here due to genuine mass fluctuations? Or is it the simple (or complicated) consequence of some pedestrian effect unrelated to the effect sought? Thomas Mahood and I have addressed this question in some detail, and a thorough account of Mahood's inquiries can be found in his masters thesis (Mahood, 1999). I report here a summary of the potential spurious effects we have investigated and what we have found out.

1. At ultrasound frequencies, vibrating devices in a fluid (like air) can produce a "sonic wind". This is a consequence of the non-linear response of the fluid molecules to the high velocity excursions of the device. An asymmetric sonic wind is to be expected from devices where some part undergoes larger excursions than the others, as is the case for the devices in question here. Sonic wind effects were easily detected when the apparatus was run in air at atmospheric pressure. Indeed, this effect could be seen all the way down to vacuum vessel pressures of about 10 Torr. Below 10 Torr, however, this effect quickly disappears. And at the normal operating pressure of a few tenths of a Torr this effect is absent.

2. When a vacuum vessel is exhausted down to a few Torr, the possibility of coronal discharge in the apparatus becomes a problem because of the moderately high voltages ($\pm$ roughly 200 volts) present. The presence of coronal discharge in the system is immediately obvious as a pronounced violet glow when the room is darkened. This effect can be suppressed by carefully insulating all of the parts of the system exposed to high voltages and by pumping the system to the limit of the rotary vane vacuum pump, a few tenths of a Torr in practice. No coronal discharge was present during the runs of the system where the forces mentioned above were seen.

3. In addition to the moderately high voltages just mentioned, alternating currents with an amplitude on the order of an ampere are present in the system during the power bursts. Given the high frequencies (38.0 and 76.0 kHz) of these voltages and currents, the likelihood of coupling to other parts of the apparatus in such a way that stationary forces would be produced is rather small. But to make sure that nothing of this sort was going on, tests were run. Moderately powerful ferrite magnets were brought into proximity of the apparatus as it was being run. And various

conducting and dielectric materials were introduced to see if their presence affected the behavior of the system. In no case was any effect found that might plausibly account for the forces seen.

4. Thermal relaxation effects proved more problematical than the effects considered so far. Thermal effects have their most pronounced consequences in two ways. On is relaxation of pre-existing stresses in the suspension fiber. These must be burned out by repeated cycling of the device (without de-poling the PZTs by overheating them) until they are no longer a serious problem. The other is changes in the mechanical behavior of the devices because of differential thermal expansion effects in the constituent parts made of different materials. Thermal effects of this sort could be detected by watching the degradation of the voltage waveform (since it depends on the electromechanical reaction of the PZTs as well as the power delivered by the power amplifier). Keeping the power pulses short, only delivering a few power bursts in a run, monitoring the temperature of the devices, and leaving plenty of cool-down time between runs kept thermal effects from seriously compromising the results. While thermal effects could be detected in the behavior of the system, it was ascertained that they were not responsible for the bulk of the force effect seen. This could be demonstrated, for example, by driving the units with a simple low frequency sine wave at a power level exceeding that used with the asymmetric waveforms. An excursion of the pendulum could be driven in this way. But it was much smaller than that recovered with the asymmetric waveform signals. (I should mention that this test is complicated by the fact that a simple sine wave applied voltage will almost certainly excite its second harmonic in the PZTs, especially as this is designed to be the lowest mechanical resonance frequency of the system. If mass fluctuations actually occur, then this inadvertently excited harmonic, in conjunction with the low frequency voltage signal driving mass fluctuations, will produce a stationary force despite attempts to suppress it. The only way to know if this sort of behavior is present is with interferometers and signal processing equipment unavailable to us at present.)

5. The last and most problematical source of potential spurious effects is mechanical vibrations that couple to the suspension fiber in some unspecified was so as to produce the torques seen. In principle, if everything is properly aligned, vibration in the beam should not produce a torque. But exact alignment is an ideal, so steps must be taken to insure that vibration is not the source of the effect seen. This was done by introducing vibration isolation in the form of nylon screws and rubber washers in two places at the joints where the devices were attached to the beam (visible in Figure 2). In addition, the devices were tested in the single filament suspension system with a different beam and suspension. Moreover, in this system vibration isolation was accomplished by strips of thin rubber [about 0.03 m long] attached with nylon screws at the top and bottom. The same effect was seen notwithstanding the differences between the systems.

CONCLUSION

Although it is always possible that the force effect seen in the devices described here may be due to some unidentified spurious source, the fact that the effect withstands the tests listed in section 5 suggests that it may indeed be due to transient mass fluctuations in the PZTs. This result accords with the results of earlier, published work that also found a seemingly real effect. The estimated value of β is consistent with expectation [allowing for the quirk in the Hall effect force sensor] based on earlier work. *And at the very least, this result shows that any experiment proposed to test the reality of transient mass fluctuations should be designed to accommodate a value of $\beta \approx 1 \times 10^{-7}$ or less (for the assumptions invoked above). Less sensitive experiments designed on the basis of simple applications of the above formalism will very likely fail.* Several system upgrades and optimization of the test devices should make possible the production of larger effects that will be easier to study. Should these advances confirm these preliminary torsion pendulum results, further optimization *may* bring this technology into the practical realm for propellantless propulsion.

ACKNOWLEDGMENT

A number of colleagues have made contributions to this work in the form of helpful comments and suggestions. Thomas Mahood's contributions have been extensive and in some cases quite important to the forwarding of this work. Equally important, he has been willing to cheerfully shoulder a reasonable share of the scut work that goes with any experimental project.

REFERENCES

Barbour, J.B. and Pfister, H., *Mach's Principle: From Newton's Bucket to Quantum Gravity* (vol. 6 of *Einstein Studies*, Birkhäuser, Boston, 1995).

Mahood, T.L., *A Torsion Pendulum Investigation of Transient Machian Effects*, CSUF physics masters thesis, in preparation (1999).

Nordtvedt, K., "Existence of the Gravitomagnetic Interaction," *Int. J. Theor. Phys.* **27**, 1395-1404 (1988), esp. p. 1401.

Rindler, W., *Introduction to Special Relativity* (2nd ed., Clarendon Press, Oxford, 1991) pp. 90-91.

Sciama, D.W., "On the Origin of Inertia," *Mon. Not. Roy. Astron. Soc.* **113**, 34-42 (1953).

Woodward, J.F., "Measurements of a Machian Transient Mass Fluctuation," *Found. Phys. Lett.* **4**, 407-423 (1991).

Woodward, J.F., "A Stationary Apparent Weight Shift from a Transient Machian Mass Fluctuation," *Found. Phys. Lett.* **5**, 425-442 (1992).

Woodward, J.F., "Making the Universe Safe for Historians: Time Travel and the Laws of Physics," *Found. Phys. Lett.* **8**, 1-39 (1995).

Woodward, J.F., "A Laboratory Test of Mach's Principle and Strong-Field Relativistic Gravity," *Found. Phys. Lett.* **9**, 247-293 (1996).

Woodward, J.F., "Twists of Fate: Can We Make Wormholes in Spacetime?," *Found. Phys. Lett.* **10**, 153-181 (1997).

Woodward, J.F. "Transient Mass Fluctuations and Spurious Signals," remarks posted to Peter Skeggs' gravity website (a response to remarks of David Cyganski posted there): <http://innetarena.com/~noetic/pls/gravity.html>, (1998).

Woodward, J.F. and Mahood, T., "What is the Cause of Inertia?," *Found. Phys.* **29**, 899-930 (1999).

Search for a Correlation Between Josephson Junctions and Gravity

Glen A. Robertson

Space Transportation Directorate
Propulsion Research Center
NASA-Marshall Space Flight Center
Huntsville, Alabama 35812

Tel. 1-256-544-7102, Fax 1-256-544-2590
Email: tony.robertson@msfc.nasa.gov

Abstract: Woodward's transient mass shift (TMS) formula has commonality with Modanese's anomalous coupling theory (ACT) and Woodward's capacitor experiment has commonality with Podkletnov's layered superconductor disk experiment. The TMS formula derives a mass fluctuation from a time-varying energy density. The ACT suggests that the essential ingredient for the gravity phenomenon is the presence of strong variations or fluctuations of the Cooper pair density (a time-varying energy density). Woodward's experiment used a small array of capacitors whose energy density was varied by an applied 11 kHz signal. Podkletnov's superconductor disk contained many Josephson junctions (small capacitive like interfaces), which were radiated with a 3-4 MHz signal. This paper formulates a TMS for superconductor Josephson junctions. The equation was compared to the 2% mass change claimed by Podkletnov in his gravity shielding experiments. The TMS is calculated to be 2% for a 2-kg superconductor with an induced total power to the multiple Josephson junctions of about 3.3-watts. A percent mass change equation is then formulated based on the Cavendish balance equation where the superconductor TMS is used for the delta change in mass. An experiment using a Cavendish balance is then discussed.

INTRODUCTION

In 1992 Eugene Podkletnov reported that a gravity shielding effect was seen above a YBCO superconductor being rotated by a magnetic field and irradiated by RF energy. The mechanism for this phenomenon is as debatable as the test itself. The claimed effect has not yet been credibly confirmed nor discounted although some investigations are underway (Li, 1997). Even if a genuine physical effect has been newly discovered, there is not yet enough evidence to determine if it is a "gravity shield" as speculated by Podkletnov, a reduction of local gravitational potentials or fields, some sort of directed-force effect, or something else. Also, prior to and following Podkletnov's 1992 paper, there have been a variety of theories published on related matters. To help resolve this uncertainty, recent theories were examined to find an empirical approach to further study these issues. From this examination, an interesting similarity between theories has been identified, and an experiment is now suggested to further explore these theories.

Independent of Podkletnov but about the same time, Li and Torr published several papers on the possible connection between gravity and superconductors. These papers (Li, 1990, 1992, and 1993) suggest that the gravity mechanism arise from the spin of the lattice ions aligned by an applied magnetic field. Contrary to the gravi-magnetic theory, Modanese has presented a theoretical model from Podkletnov's experimental results. The model (Modanese, 1997 and 1999), which is based on the "anomalous" coupling between Bose condensate and the gravitational field, suggests that the essential ingredient for the shielding is the presence of strong variations or fluctuations of the Cooper pair density in the disk. The behavior of the Bose condensate (Ross, 1983) or Cooper pairs within superconducting materials in an external gravitational field has been the subject of some study in the past. It was recently suggested (Casas, 1998) that Cooper pairs lead to Bose condensation at temperatures substantially greater than those of the BCS (Bardeen, Cooper, and Schrieffer) theory of superconductivity.

CP504, *Space Technology and Applications International Forum–2000*, edited by M. S. El-Genk
© 2000 American Institute of Physics 1-56396-919-X/00/$17.00

Like Modanese, this paper suggests that Cooper pairs fluctuating across Josephson junctions could produce the "anomalous" gravitational coupling seen by Podkletnov. It is also suggested that rf-modulation of magnetic flux at the pinning sites allow for the high internal current densities in the superconductor. Further, a commonality between a transient mass shift (TMS) (Woodward, 1996) and superconductor Josephson junctions provide for a equation that predicts a percent mass change (or shift) close to that seen by Podkletnov. This formula is then used to formulate a percent mass change equation based on the torsion (Cavendish) balance, which is routinely used to experimentally derive the gravitational constant G. A Cavendish balance experiment is then discussed, which could shed light on a transient mass shift near a Type II superconductor.

As a note, it is not the intent of this paper to justify any single theory, but only to show that these theories have commonalties. To this end, none of these theories are presented in full detail. The author suggests reading of the references for addition information.

JOSEPHSON JUNCTIONS

Of what is known of the type II, YBCO superconductor disk used in the Podkletnov experiments; it seems certain that a large number of superconductor-oxide insulator-normal conductors (S-I-N) Josephson junctions exist within the disk. In general, Josephson junctions (Tinkham, 1996) exhibit a unique property. They radiate RF energy when traversed by a current and generate a current when radiated with RF energy. This phenomenon indicates a fluctuation of energy across the junction. With this in mind, it must be noted that YBCO superconductors in general and the procedure for making the superconductor disk, including the pressing and sintering of varying size YBCO superconductor grains, produces a structure having many flux pinning sites around which exist the Josephson junctions. Flux pinning is a well-known phenomenon associated with Type II superconductors where magnetic flux is trapped inside a superconductor. Flux pinning results from any spatial inhomogeneity of the material, such as impurities, grain boundaries, voids, etc. To be most effective, these inhomogeneities (Tinkham, 1996) must be on the scale of the order of the penetration depth or the coherence length, i.e. $\sim 10^{-6}$ to 10^{-5} cm, rather than on the atomic scale where inhomogeneity causes electronic scattering which limits the mean free path. This so happens to be in the range of the holes that would form about the superconductor grains used to make Podkletnov's disks.

Further, the porosity of the sintered disk provides for the passing of magnetic flux throughout the disk. The magnetic flux remains trapped by inducing current loops around the holes created by the superconductor grains. The process used to create Podkletnov's disk (Podkletnov, 1997) would form Josephson junctions between the grains surrounding these holes. As previously noted, these currents would cause RF radiation to be emitted, but only in the junctions where current loops are persisting. It is then feasible to conceive that some junctions would not pass a current at the energy level provided by the magnetic field, say S-I-N junctions. Also as previously noted, RF radiation causes Josephson junctions to generate a current. It is then plausible that RF radiation and rotation of the superconductor acting through the pinned magnetic flux could modulate these Josephson junctions.

It can therefore be speculated that Podkletnov has produced a device that allows flux pinning to enhance the production of RF energy and RF energy to enhance the production of superconductor currents. These reinforcing phenomena should lead to high electron densities or super currents within the superconductor disk. These currents are focused between the superconductor grains, which form the Josephson junction sites.

TRANSIENT MASS SHIFT

Modanese's anomalous coupling theory attempts to connect Podkletnov's gravity experiments to the Bose condensate phenomenon (Modanese, 1999). Even the addition of the modulated flux effect previously discussed, the effects would be random in nature. This suggests that another phenomenon is at work in the Podkletnov experiment. In the case of Podkletnov's experiments where the maximum mass change was seen, the superconductor disk was composed of a dual layer, small grain YBCO system. The top layer was composed of YBCO superconductor grains (Typically -$YBa_2Cu_3O_7$) and the bottom was composed of YBCO grains in a normal conductor state (Typically $-Y_{1.5}Ba_{2.5}Cu_3O_{7-x}$). It is easy to see that the interface between the layers form a Josephson junction circuit as represented in figure 1.

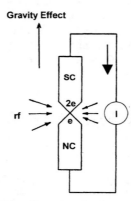

FIGURE 1. Representation of Josephson junction between normal and superconductor layers.

Tying this to proximity effects, which occur when Cooper pairs from a superconductor material in close proximity diffuse onto a normal material (Tinkham, 1996), the electrons must form pairs to cross the junction. That is, single electrons in the normal conductor would pair as they migrate into the superconductor and Cooper pairs would separate as they migrate into the normal conductor. This phenomenon would occur at the resonance frequency of the Josephson junction, which in turn would result in fluctuations of the Cooper pair density in the disk at the interface.

Then the Podkletnov gravity effect can be interpreted as being a function of the induced AC current across the Josephson junctions, which in turn is a function of the induced power. This is not however, to say that other conditions about the interface do not play a role. For example, electron resonance frequencies and densities in the normal conductor could be major contributors to this phenomenon.

During a review of current literature for clues on the connection between gravity and superconductors, it was found that Woodward, who has done some very interesting work with capacitors, both theoretically and experimentally (Woodward, 1991and 1996), has come up with an equation for a transient mass shift derived from Mach's Principle. Mach's Principle explains inertia – the tendency of an object to resist acceleration – by the sum of the gravitational attractions of all objects in the universe. Woodward presented the transient mass shift dm_0 as:

$$dm_0 = \frac{\beta w P_0}{2\pi G \rho_0 c^2},$$

(1)

where dm_0 is the transient mass shift; β is the ratio ϕ / c^2 (ϕ is the gravitational potential due to all the matter of the universe) and is approximately 1 and unitless; w is the frequency of the driving voltage into the capacitors in radians per second; P_0 is the power applied to the capacitors in Watts; G is the gravitational constant = 6.673 x 10^{-11} N m^2/kg^2; ρ_0 is the density of the capacitors; and c is the velocity of light = 2.9979 x 10^8 m/s.

It should be noted that Woodward's experimental results (Woodward 1991) have not yet been independently confirmed nor dismissed.

Woodward's transient mass shift (TMS) formula has commonality with Modanese's anomalous coupling theory (ACT) and Woodward's capacitor experiment has commonality with Podkletnov's layered superconductor disk experiment. The TMS formula derives a mass fluctuation from a time-varying energy density. The ACT suggests that the essential ingredient for the gravity phenomenon is the presence of strong variations or fluctuations of the Cooper pair density (a time-varying energy density). Woodward's experiment used a small array of capacitors whose energy density was varied by an applied 11 kHz signal. When these are vibrated up and down at the correct frequency so that they are going up when their mass is minimum and going down when their mass is maximum, then a small, constant, mass-force change is possible. Podkletnov's superconductor disk contained many Josephson junctions (small capacitive like interfaces), which were radiated with a 3-4 MHz signal. At the layered interface, the Cooper pairs are moving upward, while the electron pair separations are moving downward.

These commonalties allow for ease in rewriting Woodward's TMS (equation 2) in terms of a superconductor mass shift dM_{sc} as:

$$dM_{sc} = \frac{\beta f_{jj} P_{jj}}{G \rho_{sc} c^2},$$ (2)

where f_{jj} is the resonance frequency (in Hz) of the superconductor Josephson junctions, P_{jj} the effective combined power (in watts) of the junctions, and ρ_{sc} is the density of the superconductor.

In relationship to Podkletnov s experiment, this equation has but one unknown, P_{jj} the effective combined power of the junctions. However, by plotting equation (2) as a function of power, as represented in figure 2, it would seem that the maximized 2% mass change for a 2 kg superconductor and with a density of $\rho_{sc} = 48$ kg/m3 would occur at about 3.3 Watts for $f_{jj} = 3.5$ MHz.

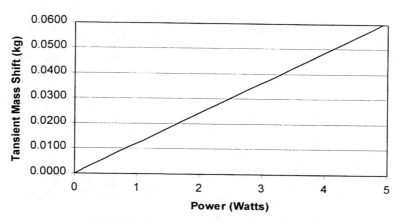

FIGURE 2. Power verse transient mass shift.

TESTING FOR A TRANSIENT MASS SHIFT

Further review of literature revealed an equation for the gravitational constant G that allows for the expression of G in terms of measurable quantities of a Cavendish balance (Tel-Atomic, 1999), which are:

$$G = \theta \frac{b^2 l}{8M} \left(\frac{2\pi}{T}\right)^2,$$ (3)

where M is the test mass, b is the distance between the center of a known mass m and the test mass M, l is the separation distance of the masses m on a torsion bar, T is the period of the damped harmonic motion of the torsion bar, and θ is the angular displacement of the torsion bar caused by the motion of the test masses. Note that the result is independent of the value of m and can be written in the following simplified form:

$$\theta = T_C GM;$$ (4)

where $T_C = (8/b^2 l)(T/2\pi)^2$ is a proportionality constant associated with the torsion balance characteristics.

If perceptible mass shift could be demonstrated with a Cavendish balance, the change in the measured angular displacement of the torsion fiber will be directly proportional to the test mass change, assuming the gravitational constant is unaltered. For example, the delta angular displacement $d\theta (=\theta - \theta_{new})$ associated with an effective delta in test mass dM due to gravitational modification is given by:

$$d\theta = T_C G dM.$$ (5)

Combining this equation with Woodward's transient mass shift (equation 2), the delta $d\theta$ in the angular displacement θ due to a mass change dM for a superconductor can be expressed in engineering terms as:

$$d\theta = \frac{T_C \beta f_{jj} P_{jj}}{\rho_{sc} c^2}.$$ (6)

By setting equation 4 equal to equation 5 through the constant $T_C G$, a relationship between the two angular displacements and the masses can be made. From this relationship, a percent change in mass can be expressed by:

$$M\% = \frac{dM}{M} \cdot 100 = \frac{d\theta}{\theta} \cdot 100.$$ (7)

Using $d\theta$ as defined by equation 6 and θ as defined by equation 4, where M_{sc} is the mass of the superconductor, the percent change in mass of the superconductor can be expressed as:

$$M\% = \frac{\beta f_{jj} P_{jj}}{G M_{sc} \rho_{sc} c^2} \cdot 100.$$ (8)

This equation allows for the experiment evaluation of the Podkletnov gravity experiment by using a sensitive Cavendish balance. A sketch of the proposed experiment is given in figure 3. As shown, U-shaped permanent magnets with poles of high field, rare-earth permanent magnets are arranged such that the magnetic field is through a high-T_c oxide superconductor. The RF coils are shown for representation. The actual configuration may vary dependent on the frequency required.

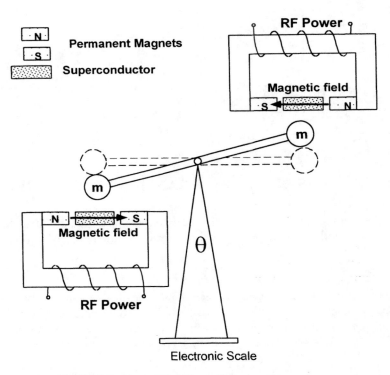

FIGURE 3. Cavendish Balance Experiment.

Making reasonable assumptions for the various balance parameters give an estimate of the sensitivity to a transient mass change. Based on the published characteristics of commercially available Cavendish balances, where $T = 120$ sec; $l = 30$ cm; $b = 3$ cm a numerical estimate for the sensitivity is given as:

$$\frac{\partial \theta}{\partial M} = 0.7 \text{ microradians / gram}.$$ (9)

Thus, the sensitivity of a Cavendish balance to small transient mass change effects is generally somewhat limited, and successful detection of these effects will require painstaking efforts in the design and fabrication of the instrument. It will be necessary to push every parameter toward its extreme values in order to achieve the highest possible sensitivity. For example, for a 500 gram test mass a 2% change in effective mass ($dM = 5$ grams) would yield a torsional displacement of $d\theta = 3.5$ microradians. This is within the established sensitivity of state-of-the-art torsional displacement transducers, which can easily resolve displacements of less than 1 microradian.

CONCLUSION

An equation that predicts a mass change in superconductors with multiple Josephson junctions was formulated from Woodward's transient mass shift. In relation to Podkletnov's gravity experiments, the equation has but one unknown – the additive power consumed by the junctions. The equation predicts that Podkletnov's 2% mass shift would occur at an internal combined power across the Josephson junctions of about 3.3 watts. This however, would be the mass change of the superconductor. How a percent mass shift of one mass can affect another with equal value is unclear. The torsion balance experiment proposed in this report could help resolve this issue. Furthermore, should a transient mass shift be seen by the described experiment, it would add supporting evidence that the Podkletnov effect is a genuine new physical effect and provide the necessary foundation for further work in this area.

ACKNOWLEDGMENTS

This research was supported in part by NASA's Advanced Space Transportation Research. The author wishes to thank R. Litchford for help in the sensitivity analysis of the Cavendish balance and R. Koczor, W. Brantley, and J. Cole for helpful discussions.

REFERENCES

Casas, M., Rigo, A., de Llano, M., Rojo, O., Solis, M. A., "Bose-Eistein condensation with a BCS model interaction," *Physics Letters A*, Vol. 245, pg. 55-61, 1998.

Li, N. and D. G. Torr, "Gravitational effects on the magnetic attenuation of superconductors," *Phys. Rev. B*, Vol. 46, 1992, p. 5489.

Li, N. and D. G. Torr, *Phys. Rev. D*, Vol. 43, 1990, p. 457.

Li, N. and D. G. Torr, "Gravito-Electric Coupling Via Superconductivity," *Found. Phys. Lett.,* Vol. 6, 1993, p. 371

Li, N., Noever, D., Robertson, T., Koczor, R., and Brantley, W. (1997) "Static Test for a Gravitational Force Coupled to Type II YBCO Superconductors," Physica C, 281:260-267.

Modanese, Giovanni, "On the theoretical interpretation of E. Podkletnov's experiment," LANL gr-qc/9612022, Presented for the World Congress of the International Astronautical Federation, 1997, nr. IAA-97-4.1.07.

Modanese, Giovanni, "Gravitational Anomalies by HTC superconductors: a 1999 Theoretical Status Report," Published on the web at http://www.gravity.org/nat.htm

D. K. Ross, J. Phys. A 16, 1331, 1983, *J. Anandan, Class. Q. Grav.* 11, A23, 1994 and references.

Podkletnov, E. and Nieminen, R. , "A Possibility of Gravitational Force Shielding by Bulk Yba2Cu3O7-x Superconductor," *Physica C*, C203, 441-444, 1992.

Podkletnov, E. E., "Weak gravitation shielding properties of composite bulk YBa2Cu3O{7-x} superconductor below 70 K under e.m. field," LANL cond-mat/9701074, 1997.

Tel-Atomic, Incorporated, "Computerized Cavendish Balance," TEL-RP2010," (1223 Greenwood Ave., P. O. Box 924, Jackson, Michigan 49204), 1999.

Tinkham, Michael and Gordan McKay, *Introduction to Superconductivity,* McGraw-Hill, Inc., pg. 196-286, 1996.

Woodward, James F., "Measurements of a Machian Transient Mass Fluctuation," *Foundation of Physics Letters,* Vol. 4, No. 5, (1991), pp. 407 – 423

Woodward, James F., "James F. Woodward: Mach's Principle Weight Reduction = Propellantless Propulsion," (A summary from *Foundation of Physics Letters,* Vol. 9, No. 3, (1996), pg. 247 – 293. Published on the web at http://www.inetarena.com/~noetic/pls/woodward.html

Candidate In-Orbit Experiment to Test the Electromagnetic Inertia Manipulation Concept

Hector Hugo Brito

Centro de Investigaciones Aplicadas, Instituto Universitario Aeronáutico
Ruta 20 Km. 6.5, 5022 Córdoba, Argentina
Phone: +54-351-4659017, Fax: +54-351-4333967; hbrito@make.com.ar

Abstract. Concerning the electromagnetic field-matter interactions in polarizable media, different formulations predict strongly different results, especially for quasi-stationary fields (Abraham-Minkowski controversy). It has recently been shown that under the assumption of Minkowski's Energy-Momentum tensor being the right one, by suitably manipulating those fields, the inertial properties of the generating device – the Electro-Magnetic Inertia Manipulation (EMIM) thruster - can be modified, which translates into a force-producing effect on the device as part of a closed system. An experimental setup was built up which consisted of mounting the device as a seismic mass atop a mechanical suspension, where the vibratory motion was detected via a piezoceramic strain transducer and the use of DSP techniques. Processed data pointed to a mechanical vibration induced by EM inertia manipulation, as predicted by Minkowski's formulation, after all other sources of vibration were taken into account, or removed when possible. However, no direct detection of the sought effect has been obtained up to now. A candidate experiment to be performed in orbit environment is subsequently proposed to assess, beyond all doubt, the validity of these ground testing results, which basically consists of an improved version of the thruster working in a "push and pull" mode onto an International Space Station based platform (Free-Flyer type), as part of the Advanced Propulsion Technology R&D foreseen activities. The existence and strength of EMIM propulsion effects can be assessed by laser interferometry tracking of the platform center of mass. As a bonus, the experiment rationale can be applied to any "propellantless" or breakthrough propulsion physics concept reducible to a force-producing device.

INTRODUCTION

One of the yet unsolved areas of Physics is that related to the tensor formalism of the electromagnetic field-matter interactions in polarizable media, where different formulations predict strongly different results, especially for quasi-stationary fields. The situations are seen as variants of the Abraham-Minkowski controversy (Brevik, 1979, 1982), while experiments to settle the question have been performed (James, 1968; Walker, 1975, 1977; Lahoz, 1979) without conclusive results. It has recently been shown that if Minkowski's Energy-Momentum Tensor is the right descriptor of the electromagnetic field-matter interaction in polarizable media, the inertial properties of the generating device can be modified, allowing to obtain mechanical impulses on the device as part of a closed system, not undergoing any exchange of mass-energy with the surrounding medium (Brito, 1999a). A propulsion concept, the Electromagnetic Inertia Manipulation (EMIM) thruster was subsequently drawn, which basically consists of suitably grouping the sources of electric and magnetic fields into a physically connecting device, as shown in Fig. 1. A stationary electromagnetic field momentum density ($D \times B$) can build up there thanks to the dielectric filled region; by controlling the intensities of these fields, the inertia properties of the system as a whole, when represented by its "matter" part – the device – are allowed to change so that EM field momentum is converted to mechanical momentum of the device.

To elucidate the issue, an experimental setup as shown in Fig. 2 was built up. By supplying a periodic voltage to the solenoid at a frequency close to the fundamental frequency of the seismic suspension, the expected mechanical effect would cause the fixture to resonate, adding up to the microseismic noise induced vibration and other unwanted sources of noise. The total vibratory motion was detected via a piezoceramic strain transducer; spectral analysis, system modeling & identification and adaptive filtering techniques were used, either independently or in combination to produce signal estimations from data gathered during the period 1993 - 1997.

CP504, *Space Technology and Applications International Forum–2000*, edited by M. S. El-Genk

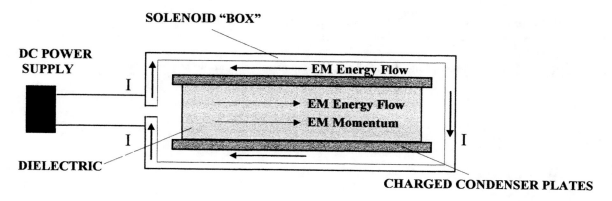

FIGURE 1. Electromagnetic Momentum Generator Schematics .

In practically all cases, the processed results pointed to a mechanical vibration induced by matter-electromagnetic field momentum exchange, as predicted by Minkowski's formulation,. Nevertheless, no direct detection of the sought effect has been obtained up to now; the experimental setup overall detectability needs further improvements, jeopardized by power supply induced EMI on the measurement channels, sharing the same spectral signature with the pursued effect.

The purpose of this work is to describe a candidate experiment to be performed in space environment to assess, beyond all doubt, the validity of these ground testing results; this can be achieved, in principle, by allowing an improved version of the thruster, to "push and pull" in an alternate mode, onto an International Space Station based platform (Free-Flyer type), being activated and monitored by Telemetry, Command and Control, as part of the Advanced Propulsion Technology R&D foreseen activities (Holt, 1999). If inertia manipulation is really achieved, it will show itself as a high frequency modulation of the platform center of mass (c.m.) acceleration, whose detectability should not be an issue for the expected high frequency EMIM thruster performances (tenths of N). Issues related to the hardware implementation and operation should be considered, instead.

THRUST BY ELECTROMAGNETIC INERTIA MANIPULATION

Studies about the inertia manipulation of the electromagnetic field for propulsion purposes are not new, a tentative explanation was already undertaken back in the sixties, based on the relativistic mechanics of extended bodies under

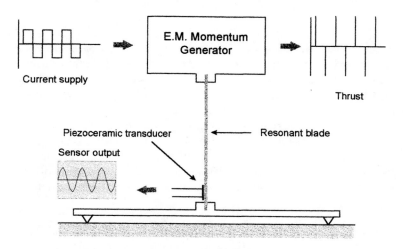

FIGURE 2. Micromotion Sensing Concept.

electrostatic pressures (Marchal, 1969). The Covariant Propulsion Principle (Brito, 1998) now allows for a general formulation of the problem, which consists of considering the whole system (Device + EM Field) as a single particle located at the "device" center of mass (or any "structural" point) so that a mass/inertia tensor is readily found as related to the whole system. Accordingly, the following geometrical relationship stands for the closed case ($\mathbf{f} = \mathbf{0}$)

$$d\left(\mathbf{M}\cdot\mathbf{v}\right) = \mathbf{0} \qquad \Rightarrow \qquad \mathbf{M}\cdot d\mathbf{v} = -d\mathbf{M}\cdot\mathbf{v}, \tag{1}$$

where

$$\mathbf{M} = \left(m_0 + m_{EM}^*\right)\mathbf{I} + \left(\mathbf{p}_{EM}\wedge\mathbf{v}\right)/c_0^2. \tag{2}$$

The 4-acceleration of the chosen "solidification point" can then be written as

$$m_0\,d\mathbf{v} = -\left(d\mathbf{p}_{EM}\wedge\mathbf{v}\right)\cdot\mathbf{v}/c_0^2. \tag{3}$$

Thus, the 4-thrust on the single particle, in any arbitrary frame, becomes

$$\mathbf{F} = -\frac{d\mathbf{p}_{EM}}{d\tau}. \tag{4}$$

The change of the mechanical (matter) momentum exactly balances the change of the EM field momentum; momentum is then being exchanged within the whole closed system. The device works as an EM momentum "accumulator" whereas the mechanical momentum that can be drawn from is, within the framework of present Physics paradigms, limited to the "accumulated" EM momentum amount.

GROUND EXPERIMENTS STATUS

An experimental thruster based on the schematics of Figure 1, was engineered up to the "proof of concept" level and an experiment was designed and implemented aimed to verify that a) The thruster is applying mechanical forces on the test stand without expenditure of mass, besides that equivalent to the radiant energy dissipated from the system (e.g., Joule heating), which cannot account for the observed effects. This implies b) Electromagnetic momentum in the matter rest frame of a closed system is being generated, i.e. a 4-mass tensor behaviour is being obtained, so that c) The Minkowski's Electromagnetic Energy-Momentum tensor does describe properly the electromagnetic field-matter interactions in polarizable media.

A supply of 6 A - AC (peak) to the coils and 4 kV - DC to the capacitors generates a total electromagnetic momentum (Minkowski's formulation) around 1.E-8 Ns (peak) by using BaTiO$_3$ ceramic dielectrics. A square wave activation of the device at a frequency close to the fundamental frequency of the seismic suspension, can alledgedly make the test fixture to resonate and a blade upper end displacement response in the range 10^{-8} - 10^{-7} m is obtained. Piezoceramic strain transducers are used to detect this range of displacements, taking into account technological as well as financial constraints. Since the full signal, as observed in preliminary testing, contained a significant amount of ground and environment induced noise, like the interaction between the coils and the Earth magnetic field, the magnetic interaction between the moving and the fixed parts of the AC and DC circuits (self magnetic interaction), air motion, electrostatic couplings, sound, radiometric effects, spherics, etc, amounting to an overall noise estimated effect   -60 dB < S/N < -40 dB at the transducer output, the need for further processing arose. To this aim, the analog transducer output signal is converted to digital through a 12 Bit data acquisition board.

Two series of tests were already conducted during the period 1993 - 1997. For the first series of tests, only one measurement channel was available and there was no vibration-free fixtures. The second series of tests included, besides the main transducer measurement channel, a dummy seismic fixture with an ancillary transducer and its measurement channel, a supply voltage measurement channel, and a vibration-free table.

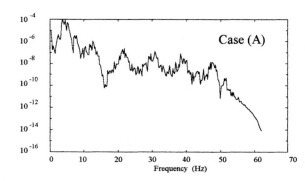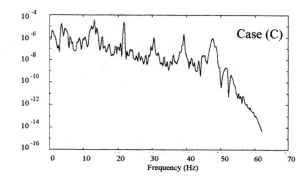

FIGURE 3. PSD of the Second Test Series Processed Results - [V^2/Hz].

The experimental data gathered during this series of tests, in 5000 samples sequences at a rate of 500 samples/sec, were firstly processed to carry on a system identification on the basis of the ground motion excitation only, an ARMA (Auto Regressive Moving Average) model structure being then identified; later, inverse filtering was performed for every output sequence in order to obtain the equivalent ground motion; then, filtering by the vibration isolation fixture led to the reconstruction of the sensing device base motion; finally, optimal filtering (Wiener filter) was performed on the resultant ouput, using the EMIM induced excitation as the "desired" signal.

A summary of the post-processed results is shown in Fig. 3 where Power Spectral Densities (PSD) were estimated over a 2048 length frequency interval, using Welch's averaged periodogram method. Spectrum (A) which relates to ground induced noise contains low level residuals induced by the Wiener filter - a sort of numerical artifact – while Spectrum (C) representing the case coils ON, capacitors ON & ground induced noise, shows a structure which strongly suggests an alternate impulsive excitation, as it turn out to be when a square wave EM field-matter momenta exchange is present. The figures are representative of around 16 sequences by case. Better detectability can be obtained by means of statistical analysis over the whole ensembles and by using adaptive noise cancellation procedures, either on the raw output data or on the inverse filtered output data. These activities are presently being carried on and no final results are available yet.

IN-ORBIT EXPERIMENT RATIONALE

The results of ground experiments are indeed promising. However, no direct detection of the sought effect by means of input reconstruction was obtained up to now and much work remains to be done to assess, beyond all doubt, their validity. One of the main concern is power supply induced EMI interference due to bad grounding and/or shielding, sharing the same spectral signature with the pursued effect, besides the unavoidable noise sources already mentioned, typical of lab environments. To get rid of most of the problems encountered when performing propulsion experiments on Earth, the best way is to do them in space, the natural propulsion arena. The International Space Station (ISS) offers this possibility as part of the Advanced Propulsion Technology R&D foreseen activities (Holt, 1999), being stated that it "can also be used to effectively conduct applications research on breakthrough propulsion physics effects and, later, to demonstrate associated performance capabilities".

The following approach is proposed: An improved and self-contained version of the EMIM thruster is allowed to "push and pull" in an alternate mode, onto an ISS based platform (Free-Flyer type), being activated and monitored by Telemetry, Command and Control. If inertia manipulation is really achieved, it will ideally show itself as a high frequency modulation of the platform center of mass (c.m.) acceleration. By applying Eq. 4 to a 30 kHz sine wave EM momentum 1. µN-s peak, an alternate thrust of 0.188 N is obtained. Primary detection by means of state-of-the art accelerometers should not be an issue for platform masses in the range 50-100 kg; pushing on the whole ISS would produce sub-µg accelerations instead. However, unless ideally rigid mounting of the thruster to the platform bus be obtained, bus accelerations will not necessarily represent force-producing effects coming from the thruster other than inner mutual interactions, as represented by $F_{T/B}$ and $F_{B/T}$. Accelerations of the thruster bed, sharing with the bus a flexible mechanical interface, will accordingly be needed in order to be able to estimate the c.m. acceleration of the closed system composed by the EMIM thruster and the platform

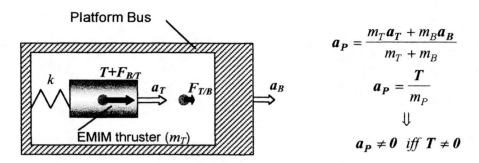

FIGURE 4. Experiment Schematics and Rationale.

bus, as illustrated in Fig. 4. Again, detection by means of state-of-the art accelerometers should not be an issue, since that related to the thruster bed is higher due to a lesser mass and can in addition be amplified by thrusting at the resonance frequency of the mechanical fixture. Independent confirmation of the accelerometric results will also be needed which can be achieved by means of laser-interferometry ranging of the two main components of the platform vibrating in resonance to take advantage of the mechanical Q. For the proposed thruster characteristics, displacements of the thruster bed around 0.05 nm (0.0001 He-Ne laser wavelength) should be expected, while bus displacements may still be an order of magnitude lower. It is to be noted that the two detection procedures have antagonist frequency requirements; the readings of the accelerometric one increase with the thrust which is frequency linear dependent, whereas those of the interferometric one decrease with $(1/\omega)$. A trade off shall be done to define nominal characteristics taking into account masses, detection thresholds of measurements procedures and growth potential of EM momentum/mass with present and foreseeable dielectric and/or ferromagnetic materials technologies, together with high intensity E and B fields power supplies.

Laser interferometry ranging requires mounting corner cubes on the tracked objects, i.e. the bus and the thruster bed. This will likely require station-keeping (in the range of mm's) and attitude control capabilities of the platform, with the experiments being performed during propulsion&control-free periods. By assuming extreme atmosphere mass density conditions (Smith, 1997) and the ISS working in µg mode (Del Basso, 1999), it is preliminarily found that the platform on-board propulsion system should be able to accommodate relative accelerations up to 1 µg and velocities up to 0.5 mm/s after an experimental period of 45 s. This set an upper limit of roughly 100 mN-s to the impulse control authority, while a 1% accuracy level is satisfied by a 500 µN-s impulse bit. These requirements can be achieved by using low power Ablative Pulsed Plasma Thrusters (APPT's) as the actuator function of a positioning control loop, as shown in Fig. 5, being operated out of the experiment timing.

Indeed, the same propulsion system could be used for maneuvering the platform from the station to a safe distance for performing the experiment and back. According to a preliminary mission analysis, on the basis of 100 sorties with safe

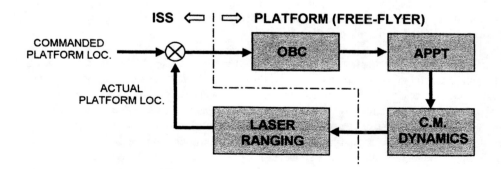

FIGURE 5. Platform Station-Keeping Control Loop.

maneuver relative velocities lower than 0.1 m/s and 100 station keeping cycles per sortie, total impulse requirements would amount to around 5000 N-s which approximately corresponds to the capabilities of APPT's already flight qualified units (Myers, 1995) or under development (Brito, 1999b). Automatic and/or human operation are alternatives that remain to be defined when addressing issues such as ISS safety, testing profile, contingencies management, quick-look data processing, a.s.o. It should be stressed that this rationale seems to be suitable for testing other advanced propulsion concepts, e.g. exploiting transient mass fluctuations (Mahood, 1999), using spinning gyros in free-fall environment (Hayasaka, 1997), using topological effects of circulating magnetic fluids (Hayasaka, 1999), or the alleged Biefeld-Brown effect (Talley, 1991), which at first sight look reducible to force-producing devices.

CONCLUSION

The demonstration that electromagnetic inertia manipulation can be achieved with present technology must be considered of paramount importance to the NASA Breakthrough Propulsion Physics Program and its challenge of discovering propulsion methods which eliminates or dramatically reduce the need for propellant (Millis, 1997). As already shown, electromagnetic inertia manipulation equals force-producing effect with no mass ejection from the propulsion system, although the ultimate achievable mechanical momentum in a given reference frame cannot be higher than the maximum EM field momentum in the same frame. However, a mechanism of momenta conversion would be available for specific uses (ultra-fine pointing/station keeping, "jerky" translations) until further breakthroughs in the understanding of the nature of inertia and motion make possible getting rid of that limitation. Ground testing of the concept has been performed with no conclusive results obtained as yet, because of unwanted noise inherent to the very nature of this kind of tests. A proposal is made to replicate the experiments in space environment to get rid of those ground/lab based sources of noise; an ISS based "free-flyer" type platform seems to be suitable for the task and functional as well as operational issues have preliminarily been addressed, like using ablative pulsed plasma thrusters for platform excursion and station keeping maneuvers. Additionally, the proposed in-orbit experiment could be applied to any "propellantless" propulsion concept reducible to a force-producing device.

ACKNOWLEDGMENTS

This work was carried out under grant FONCYT PICT-97-02205 of the National Science&Technology Office (Argentina) with the support and assistance of Instituto Universitario Aeronáutico and the patronizing of CONAE (Space Activities National Committee - Argentina).

REFERENCES

Brevik, I., "Definition of some Energy-Momentum Tensors," *Experiments in Phenomenological Electrodynamics and the Electromagnetic Energy-Momentum Tensor*, PHYSICS REPORT (Review Section of Physics Letters) **52**, No. 3, 1979, p. 139.

Brevik, I., "Comment on "Electromagnetic Momentum in Static Fields and the Abraham-Minkowski Controversy"," *Physics Letters* **88 A**, 335-338 (1982).

Brito, H. H., "A Propulsion-Mass Tensor Coupling in Relativistic Rocket Motion," in *Space Technology Applications International Forum – 1998*, edited by Mohamed S. El-Genk, AIP Conference Proceedings 420, American Institute of Physics, Albuquerque, NM, 1998, pp. 1509-1515.

Brito, H. H., "Propellantless Propulsion by Electromagnetic Inertia Manipulation: Theory and Experiment," in *Space Technology Applications International Forum – 1999*, edited by Mohamed S. El-Genk, AIP Conference Proceedings 458, American Institute of Physics, New York, 1999a, pp. 994-1004.

Brito, H. H., De Alessandro R. O., Dominguez, C. A., "Preliminary Development Status of the IUA's P4S-1 Ablative Pulsed Plasma Thruster," accepted for publication in the proceedings of *26th International Electric Propulsion Conference*, October 17-21, 1999b, Kokura-kita, Kitakyushu, Japan.

Del Basso, S., "International Space Station – Microgravity Environment Design & Verification," in *Space Technology Applications International Forum – 1999*, edited by Mohamed S. El-Genk, AIP Conference Proceedings 458, American Institute of Physics, New York, 1999, pp. 465-470.

Hayasaka, H., Tanaka, H., Hashida, T., Chubachi, T., Sugiyama, T., "Possibility for the existence of anti-gravity: evidence from a free-fall experiment using a spinning gyro," *Speculations in Science and Technology*, **20**, 173-181 (1997).

Hayasaka, H., Minami, Y., "Repulsive Force Generation due to Topological Effect of Circulating Magnetic Fluids," in *Space Technology Applications International Forum – 1999*, edited by Mohamed S. El-Genk, AIP Conference Proceedings 458, American Institute of Physics, New York, 1999, pp. 1040-1050.

Holt, A. C., "International Space Station Advanced Propulsion Technology R&D," in *Space Technology Applications International Forum – 1999*, edited by Mohamed S. El-Genk, AIP Conference Proceedings 458, American Institute of Physics, New York, 1999, pp. 1021-1026.

James, R. P., "Force on Permeable Matter in Time-Varying Fields," *Ph.D Thesis*, Dept. of Electrical Engineering, Stanford University, 1968.

Lahoz, D. G., Graham, G. M., "Observation of Electromagnetic Angular Momentum within Magnetite," *Physical Review Letters* **42**, 137-1140 (1979).

Mahood, T. L., "Propellantless Propulsion: Recent experimental results exploiting transient mass modification," in *Space Technology Applications International Forum – 1999*, edited by Mohamed S. El-Genk, AIP Conference Proceedings 458, American Institute of Physics, New York, 1999, pp. 1014-1020.

Marchal, R., "Sur l'inertie électromagnétique," *Comptes Rendus* **268 A**, 299-301 (1969).

Millis, M. G., "Challenge to Create the Space Drive," *Journal. of Propulsion and Power* **5**, 577-582 (1997).

Myers, R., et al, "Pulsed Plasma Thruster Technology for Small Satellite Missions," NASA-CR-198427, 1995

Smith, O. E., Adelfang, S. I., Smith, R. E., "Neutral Orbital Altitude Density Effects on the International Space Station," *Jnl. of Spacecrafts and Rockets* **34**, 817-823 (1997).

Talley, R. L., "*Twenty-First Century Propulsion Concept*," Phillips Laboratory (Propulsion Directorate), Air Force Systems Command, Final Report No. PL-TR-91-3009, Project 3058 (1991).

Walker, G. B., Lahoz, D. G., Walker, G., "Measurement of the Abraham Force in Barium Titanate Specimen," *Canadian Jnl. of Physics* **53**, 2577-2586 (1975).

Walker, G. B., Walker, G., "Mechanical forces in a dielectric due to electromagnetic fields," *Canadian Jnl. of Physic* **55**, 2121-2127 (1977).

NOMENCLATURE

c	= velocity of light in an arbitrary medium, m/s
a	= classical acceleration vector, m/s^2
B	= Magnetic induction vector, T
D	= Electric displacement vector, C/m^2
E	= Electric field strength, N/C
f	= external 4-force, N
I	= 4-space metric tensor
M	= 4-space mass tensor, kg
m	= classical mass, kg
m_{EM}	= mass of the electromagnetic field, kg
m_0	= spaceship rest mass, kg
p_{EM}	= 4-momentum of the electromagnetic field, kg · m/s
t	= time, s
T	= Thrust, N
v	= 4-velocity, m/s
τ	= proper time, s

Subscripts

B	= bus
P	= platform
T	= thruster
0	= at rest, in vacuum

Superscripts

*	= spaceship center of mass rest frame

Theoretical and Experimental Investigations of Gravity Modification by Specially Conditioned EM Radiation

H.D. Froning[1]   T.W. Barrett[2]

[1]Flight Unlimited, 5450 Country Club Drive, Flagstaff, AZ 86004
[2]BSEI, 1453 Beulah Road, Vienna, VA 22182

Abstract. Ordinary electromagnetic (em) fields do not couple significantly with those that underlie gravitation and give rise to inertia, and it is suggested that this is because they are of different field essence and form. It is also suggested that a much stronger electro-gravitic coupling might be accomplished for effective propulsion by conditioning ordinary em fields into configurations that are similar to those which underlie gravitation and give rise to inertia. This paper summarizes theoretical and experimental em field conditioning work.

INTRODUCTION

Although there is no scientific unanimity as to the origin of gravity and inertia, Puthoff (1989); Haisch, Rueda, Puthoff (1994); and Rueda, Haisch (1998) suggest that they are - at least in part - a consequence of an electromagnetic interaction between the charged substructure of matter and the zero-point fields. This would suggest that gravity and inertia might be modified (reduced) in a ship's vicinity by generation of appropriately conditioned electromagnetic (em) fields. And Alqubierre (1994) has shown one solution of General Relativity that would enable vehicle impulsion by generation of "exotic" fields. But, ordinary em fields may not be the exotic fields envisioned by Alqubierre because the fields that underlie ordinary electromagnetism and gravitation are of completely different essence and configuration. We therefore suggest that an electro-gravitic coupling/interaction for effective propulsion requires the conditioning of ordinary em fields into fields whose forms that are similar to those which underlie gravity and give rise to inertia.

In this respect, behavior of matter and radiation can be described in terms of "gauge" fields, whose sources are conserved quantities. Gauge fields that describe electricity, magnetism, and ordinary em radiation are abelian (obey commutation rules involving ableian algebra) while fields that describe weak and strong interactions within nuclei are nonabelian (obey commutation rules involving nonabelian algebra). The more intricate configurations and increased internal degrees of freedom of nonabelian fields result in higher group symmetries. Thus, abelian fields associated with ordinary em radiation are of relatively modest U(1) symmetry, while nonabelian fields associated with weak and strong interactions are of higher SU(2) and SU(3) symmetry respectively. Yang (1993) suggests that a nonabelian field of group symmetry, higher than U(1) underlies gravitation. If so, a significant interaction between nonabelian gravitational fields and abelian electromagnetic fields would not be expected because of their different essence and form. But one of us (Barrett 1993a, 1993b) has devised special methods to generate em fields that have nonabelian form and higher symmetry. This paper summarizes theoretical and experimental work that explores the gravity modification possibilities of such em fields.

THEORETICAL FIELD CONDITIONING WORK

We are employing two approaches to identify higher order em field modulations that: (a) show some promise of coupling favorably with fields which may be associated with gravity and inertia, and (b) could be generated by suitable refinement or modification of existing em devices. One describes higher order em modulations in terms of various

CP504, *Space Technology and Applications International Forum–2000*, edited by M. S. El-Genk
© 2000 American Institute of Physics 1-56396-919-X/00/$17.00

nonabelian symmetry groups. The other characterizes higher order em modulations by topological descriptions involving spatial mappings to a manifold. The results of these two approaches are being combined into gauge theory and topological descriptions of candidate em modulations that could conceivably possess the same symmetry group and topology as that of the gravitational field and which can be described in terms of the same gauge group algebra as well. Group symmetries, that are being considered in the context of both Lorentz and Ampere force laws include: SU(2); SU(2)/Z2; SU(2,2); SL(2,c); SU(3); SO(3); and GL(3,C). Topological analysis employs: differential topology (to simplify Maxwell equations); tangent spaces and covariant derivatives (for gravitational field description); affine connectivity (to relate neighboring points of spacetime); and parallel transport) (to describe gravitational differences throughout spatial regions) Objectives of this work are identification of candidate em modulations that could be embodied within existing em devices for gravity-modification testing and (if successful) within airborne or spaceborn microwave or laser transmission systems for operational use.

ACCOMPLISHED THEORETICAL FIELD CONDITIONING WORK

Specially conditioned em radiation that embodies SU(2), SU(2,2), or SU(2)/Z_2 group symmetry have been examined in some degree, and their topological configurations have been defined in Barrett (1993, 1999). And specific modulations of ordinary em radiation that would be associated with these 3 groups have been identified in Barrett (1993, 1999). Thus, if the non-abelian fields associated with gravity and inertia possess SU(2), SU(2,2) or SU(2)Z_2 group symmetry, a significant electro-gravitic might be achieved. Expanded Maxwell equations, which are applicable to SU(2), SU(2,2) and SU(2)/Z2 radiation have also been derived in Barrett (1993). Electromagnetic emanations of higher SU field symmetry are solutions to Maxwell equations of more expanded and symmetrical form, and as shown in Figure 1, these equations include existing Maxwell terms plus additional terms that involve the coupling of electric and magnetic fields through the action of the A vector potential. In ordinary situations, the A-vector and scalar potential are used only to compute the E and B fields. But for cases of SU field symmetry, they have a significance of its own. And because the dot and cross products within the terms that include the A vector scalar potential obey the commutation relations of non-abelian algebra, they are never equal to zero if SU em phenomena are present. Derivation of em modulations and expanded Maxwell equations for SL(2,C) radiation fields are beginning but such derivation for SU(3) and SO(3) radiation has not yet been attempted.

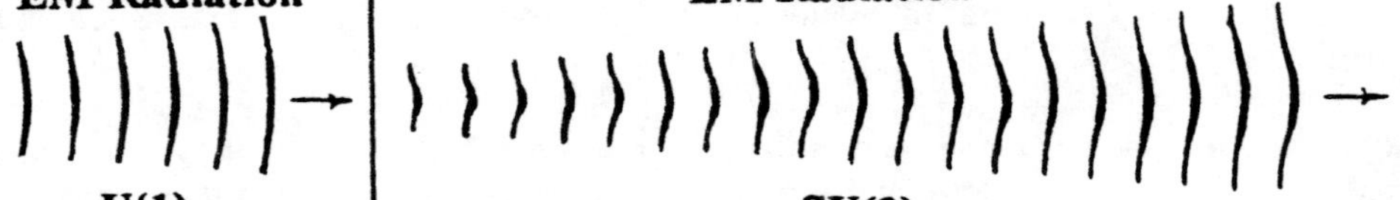

	Ordinary EM Radiation U(1)	Specially Conditioned EM Radiation SU(2)
Coulomb's Law	$\nabla \cdot E = J_0$	$\nabla \cdot E = J_0 - iq(A \cdot E - E \cdot A)$
Ampère's Law	$\dfrac{\partial E}{\partial t} - \nabla \times B + J = 0$	$\dfrac{\partial E}{\partial t} - \nabla \times B + J + iq(A_0 E - E A_0) - iq(A \times B - B \times A) = 0$
Gauss's Law	$\nabla \cdot B = 0$	$\nabla \cdot B + iq(A \cdot B - B \cdot A) = 0$
Faraday's Law	$\nabla \times E + \dfrac{\partial B}{\partial t} = 0$	$\nabla \times E + \dfrac{\partial B}{\partial t} + iq(A_0 B - B A_0) + iq(A \times E - E \times A) = 0$

$A =$ vector potential. $A_0 =$ scalar potential

FIGURE 1. Expanded Maxwell equations for SU(2) electromagnetic fields

Although even the most complex combinations of frequency, amplitude, or pulse width modulation do not transform ordinary abelian em fields into nonabelian em fields of higher symmetry, Barrett (1993a, 1993b, 1998) shows two ways of accomplishing such transformations. These ways are indicated in Figure 2. One way is modulating the polarization of em wave energy emitted from microwave or laser transmitters. The other is driving alternating current through toroidal coils - from which primarily A vector fields are emitted when resonant frequencies are reached. And, for either type of modulation, specially conditioned em radiation of SU(2), SU(2,2), or SU(2)/Z2 is obtained - depending upon the degree of polarization modulation or toroid transmitter sophistication.

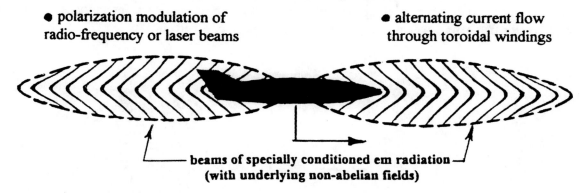

● polarization modulation of
radio-frequency or laser beams

● alternating current flow
through toroidal windings

— beams of specially conditioned em radiation —
(with underlying non-abelian fields)

FIGURE 2. Two examples of specially conditioned SU(2) radiation

Creation of specially conditioned radiation of SU(2) symmetry and non-abelian form by means of polarization modulation is shown in Figure 3. One fraction of input wave energy has its polarization plane rotated 90 degrees orthogonal to that of the other fraction of wave energy, while wavelengths of the two waveforms are made to continually differ by small amounts of phase modulation. Both fractions of wave energy are combined at a "mixer" and emission of specially conditioned em radiation whose electric and magnetic fields are rotated through all possible polarizations - with continually varying polarization with respect to time resulting. Upon a plane orthogonal to the axis of an em beam, only a single line or a single circle is traced out by the arrow of the electric field vector of linear or circularly polarized em radiation, Figure 3 also shows the more variegated path traced out upon the orthogonal plane by the "arrow" of the electric field vector of polarization modulated em radiation after the radiation has traveled 50 wavelengths from the emitter. A phase modulation frequency that is 0.1 the frequency of the input waveform is used in this example.

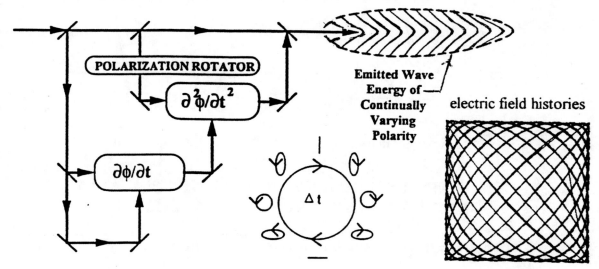

POLARIZATION ROTATOR

$\partial^2 \phi / \partial t^2$

$\partial \phi / \partial t$

Emitted Wave
Energy of
Continually
Varying
Polarity

Δt

electric field histories

FIGURE 3. Polarization modulation of ordinary electromagnetic radiation

Specially conditioned em fields of SU(2) symmetry and nonabelian form can also be created, as shown in Figure 4, by driving alternating current through a toroid with single windings. The resulting magnetic and electric fields do not extend significantly outside the toroid, but its geometry and the alternating current flow produces overlapping A vector potential patterns which extend outward from the toroid over significant distances and combine into "phase factor" waves which represent disturbances in A vector potential. The pattern of these disturbances is also shown in Figure 4 with maximum disturbance in A vector potential occurring as phase factor wave intensity peaks at the resonant frequencies where A vector potential patterns are exactly out-of-phase. Resonant frequencies are determined by the dimensions of the toroid's geometry and the propagating speed of the alternating electric current thru it. And if electro-gravitic interactions are possible with such A-vector fields, they would be the most intense at the resonant frequencies.

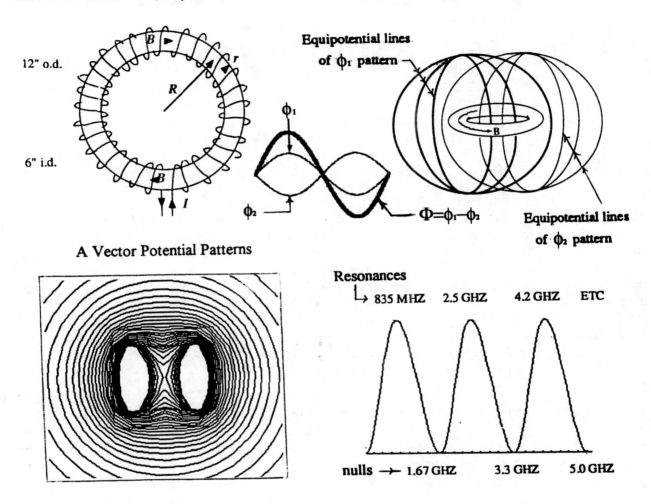

FIGURE 4. A-vector field modulation by a.c. current driven through toroidal windings

PRELIMINARY EXPERIMENTS INVOLVING SPECIALLY CONDITIONED ELECTROMAGNETIC FIELDS

Tests Involving Polarization Modulated Em Radiation

Mikhailov (1996) has provided a degree of confirmation of Barrett's ideas as to the possibility of em fields with different form and higher field symmetry than that of ordinary em fields, and as to the possibility of coupling with such em fields with specially conditioned em fields of comparable form and symmetry. Here, experiments by Mikhailov , which used apparatus similar to that shown in Figure 5, have shown that ferromagnetic aerosols in solution behave as if they possess a magnetic charge, reversing their motion with reversal in magnetic field direction (while falling under the influence of gravity) - as shown in Figure 5. Barrett (1994) noted that such behavior could be due to spherical boundary (cavity) conditions and global ordering of electron spins within the ferromagnetic particle ensemble causing it to react to magnetic influences just as many isolated magnetic monopoles would. He also determined that a nonabelian em field of SU(2) symmetry would be associated with such electromagnetics and that such an em field should therefore interact with that of polarization modulated em radiation of similar SU(2) form.

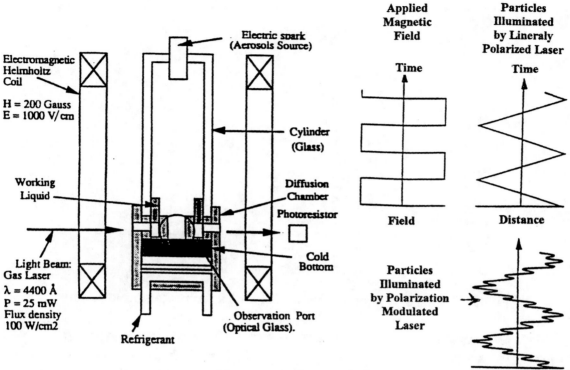

FIGURE 5. Ferromagnetic aerosol experiments

Taking this suggestion, Mikhailov modulated the polarization of the laser beam used to track particle trajectories. As mentioned, fixed polarization of the laser beam caused negligible perturbation of the linear (straight line) motion of the falling particles as they responded to the magnetic field, but the polarization modulated laser light caused the particles to oscillate with respect to their linear motion (Mikhailov, 1994) - as shown in Figure 5. Unfortunately Mikhailov (1994) contains insufficient information for ascertaining what degree of polarization modulation was actually achieved. But Mikhailov reported that particle oscillations maximized at a given polarization frequency, and oscillations appeared to be in synchrony with the polarization modulation frequency.

Tests Involving A Vector Field Radiation from Toroid Antennas

Initial experimental work that involved fabrication and testing of toroid antennas that emanate em radiation in the form of A-vector potential disturbances was accomplished at the laboratories of Hathaway Associates in Toronto Canada during 1998. This initial work: perfected techniques for fabricating toroidal coils; verified ability to achieve A vector potential resonances with such coils; determined (with only moderate precision) signal strength attenuation in air and; identified problems associated with driving alternating currents through toroidal coils over wide frequency ranges at modest power levels. Most of the goals of the initial work were achieved. Toroid antennas with conventional and caduceus windings were successfully fabricated, and A vector potential resonances were detected at frequencies that closely corresponded - as indicated in Figure 6 - to predicted values (Barrett 1998). Toroid signal strength variation with distance was as expected at small distances from the toroid center. But signal strength diminishment with distance was less than expected at longer distances. Lack of power amplification capability at very high and low alternating current frequency prevented the detection of A vector potential resonances in these regimes and lack of test time obviated tests to determine signal strength attenuation by metal structures.

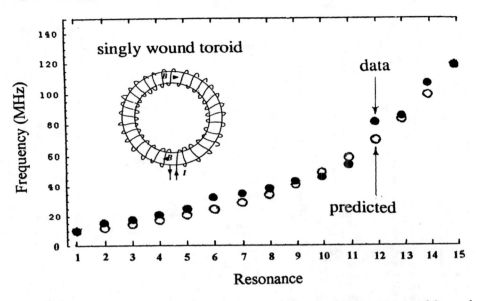

FIGURE 6. Correlation of theoretical and experimental toroid results

PLANNED EXPERIMENTS WITH SPECIALLY CONDITIONED (POLARIZATION MODULATED) ELECTROMAGNETIC RADIATION

We have defined an experiment involving generation of specially conditioned em radiation by polarization modulation of the output from a pair of existing carbon dioxide lasers operating at 30W in a CW mode at the Directorate for Applied Technology and Simulation at the U.S. Army White Sands Missile Range. And any gravity modification occurring within the single polarization modulated laser beam that is emitted from the laser pair will be detected by the beam's effect upon the transit time of ordinary em radiation emitted by another laser. Since the ordinary laser radiation will traverse some of the space affected by the specially conditioned laser radiation, any gravity modification within the affected space will change the time for the ordinary radiation to transit it, and a resulting phase shift will be detected by interfometric techniques.

The output of the two orthogonally polarized carbon dioxide (CO_2) lasers will be combined such that they emit specially conditioned em radiation in the form of a single polarization modulated laser beam that is aligned to overlap one of the helium-neon laser interferometer beams over a distance of approximately 10 meters of its length - as shown in Figure 7. And variation in the amount of polarization modulation will be accomplished by increasing or decreasing the wavelength of one of the CO_2 lasers by as much as 3 percent and of increasing or decreasing the wavelength of the other by as much as 15 percent.

Using an existing low power helium-neon (HeNe) laser and appropriate beam splitters and mirrors, a laser interferometer circuit will be set up - as shown in Figure 7 - to measure differences in the phases of clockwise and counterclockwise beams that impinge upon the location indicated in Figure 7. Phase differences between these beams will manifest as fringe shifts imaged upon a screen. Fringes will be photographed using a variety of cameras to achieve an accuracy in the 1 to 5 degree range (approximately .003 to .014 of the wavelength of the helium-neon laser light).

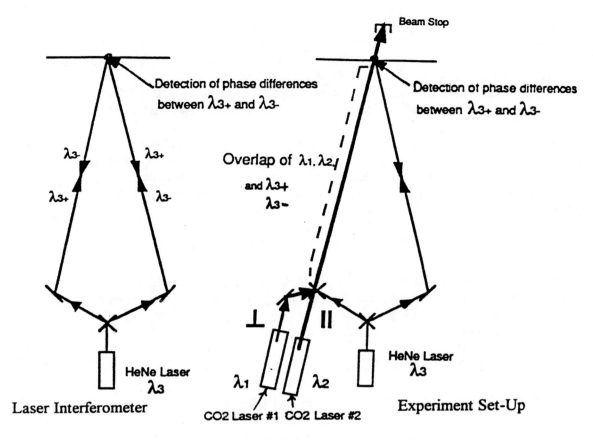

FIGURE 7. Planned test involving polarization modulated laser light

The electric and magnetic fields within the polarization modulated carbon dioxide laser beam will rotate through all possible polarizations during approximately each wavelength of their travel, affecting a cylindrical region of air that is approximately 5mm wide. And, the ordinary HeNe laser radiation, undulating with a wavelength of the order of 10^{-6} meter and a wave period of the order of 10^{-14} sec, will be traversing a segment of the cylindrical region of possible gravity modification - a segment that is approximately 10 meters long and 5mm wide. Since phase changes in the 1 to 5 degree range appear achievable, changes in the time or distance traveled by the radiation as small as one hundredth of a wave period or wavelength will be detected within the laser interferometer. And since the HeNe laser radiation travels approximately 10^7 wavelengths or wave periods over a 10 meter distance, changes in light speed (due to gravity modification) as small as the order of 10^{-7} percent should be detectable. Thus, if polarization modulated laser light which is of the order of 300 watts per cm^2 in intensity can cause a gravity modification that results in more than a 10^{-8} percent change in light speed per meter of interaction length, this should be detectible in our experiment.

PLANNED EXPERIMENTS WITH SPECIALLY CONDITIONED (TOROID ANTENNA) ELECTROMAGNETIC RADIATION

We plan to enhance power amplification capability at Hathaway Associates in order to detect A vector potential resonances and measure signal variation from very high to very low a.c. frequencies and we have determined that a frequency band width of 0.2khz-200Mhz can be achieved. Transparency of various metallic, composite and ceramic materials to specially conditioned radiation from toroid antennas will be determined to assess the possibility of inclosing toroid antennas within aerospace vehicle structures and of toroid radiation penetrating gravitometer enclosures and changing the weight of test masses within them. Heat generated by current flow within the tested toroids limited their steady state power carrying capability to about 100 watts in the initial experiments. Since much higher levels of power will surely be required to cause any significant gravitational interaction, we plan to incorporate oil cooling in future toroid designs to enable peak power levels as much as 10kw. We also plan to test "flattened" disc-shaped toroids with symmetrical and asymmetrical windings in order to focus A vector field radiation into narrower beams of higher energy density.

If A vector potential wave patterns are found to penetrate gravitometer enclosures, we will use the precision gravitometer available at Hathaway Associates to look for gravitational perturbations caused by specially conditioned em radiation from toroid antennas. Here, a successful electrical toroid/gravitometer interface with test equipment was accomplished during initial work. A ring laser gyro, as a "back-up" means of detecting gravitational perturbations, is also being considered.

SUMMARY AND CONCLUSIONS

Significant gravity modification for breakthrough propulsion may be achievable with specially conditioned em radiation if such radiation possesses the same field symmetry and topology as the fields which underlie gravity and give rise to inertia. Electromagnetic modulations that might enable such radiation have been identified, together with experiments using such specially conditioned em radiation for discriminating tests.

ACKNOWLEDGMENTS

The described work was performed in part under the sponsorship of Flight Unlimited with internal funds.

REFERENCES

Alqubierre, M.(1994) "The Warp Drive: Hyper-Fast Travel within General Relativity," Class 1 Quant Grav, Vol 11, IOP Publishing Ltd.

Barrett, T.W. & Grimes, D.M.(1993b) "Advanced Electromagnetism: Foundations, Theory Applications," pp. 278-313, World Scientific Publ Co.

Barrett, T.W.(1993a) "Electromagnetic Phenomenon not Explained by Maxwell's Equations," Essays on Formal Aspects of Electromagnetic Theory, p 6 World Scientific Publ Co.

Barrett, T.W.(1994) "The Ehrenhoft-Mikhailov Effect Described as Behavior of a Low Energy Density Magnetic Monopole-Instanton," Annales de la Foundation Louis de Broglie, 19, p 291

Barrett, T.W. (1998) "The Toroid Antenna as a Conditioner of Electromagnetic Fields into (Low Energy) Gauge Fields," Proceedings of the *Progress in Electromagnetics Research Symposium 1998, (PIERS' 98)*, 13-17 July, Nantes, France

Haisch, B., Rueda, A., Puthoff, H. (1994) "Inertia as a Zero-Point Field Lorentz Force," Phys. Rev. A, Vol. 49, p678

Mikhailov, V.F.(1996) "Experimental Detection of Dirac's magnetic change?" J. Phys. D: Apll. Phys.29, p801

Mikhailov, V.F.(1994) "Experimental Detection of Discriminating Magnetic Charge Response to Light of Various Polorization Modulations," Annales de la Foundation Louise de Broglie, 19 p 303

Puthoff, H.E (1989) "Gravity as a Zero-Point Fluctuation Force," Phys, Rev. A, Vol. 39, p 2333

Rueda, A., Haisch, B.(1998) "Contribution to inertial mass by reaction of the vacuum to accelerated motion," Phys. Rev. A, Vol 39, p 2333

Yang, C.N.(1993) "Gauge Theory," McGraw-Hill Encyclopedia of Physics, 2nd Edition, p 483

TOWARD AN INTERSTELLAR MISSION:
ZEROING IN ON THE ZERO-POINT-FIELD INERTIA RESONANCE

Bernhard Haisch[1] and Alfonso Rueda[2]

[1] *Solar & Astrophysics Laboratory, Lockheed Martin, L9-41, B252, 3251 Hanover St., Palo Alto, CA 94304*
[2] *Dept. of Electrical Eng. and Dept. of Physics & Astronomy, California State Univ., Long Beach, CA 90840*

[1] *haisch@starspot.com*   [2] *arueda@csulb.edu*

Abstract. While still an admittedly remote possibility, the concept of an interstellar mission has become a legitimate topic for scientific discussion as evidenced by several recent NASA activities and programs. One approach is to extrapolate present-day technologies by orders of magnitude; the other is to find new regimes in physics and to search for possible new laws of physics. Recent work on the zero-point field (ZPF), or electromagnetic quantum vacuum, is promising in regard to the latter, especially concerning the possibility that the inertia of matter may, at least in part, be attributed to interaction between the quarks and electrons in matter and the ZPF. A NASA-funded study (independent of the BPP program) of this concept has been underway since 1996 at the Lockheed Martin Advanced Technology Center in Palo Alto and the California State University at Long Beach. We report on a new development resulting from this effort: that for the specific case of the electron, a resonance for the inertia-generating process at the Compton frequency would simultaneously explain both the inertial mass of the electron and the de Broglie wavelength of a moving electron as first measured by Davisson and Germer in 1927. This line of investigation is leading to very suggestive connections between electrodynamics, inertia, gravitation and the wave nature of matter.

BACKGROUND

Although at this time there are no known or plausible technologies which would make interstellar travel possible, the concept of an interstellar mission has recently started to become a legitimate topic for scientific discussion. In July 1996 NASA formally established a Breakthrough Propulsion Physics (BPP) Program. The initiation of a program operating under the auspices of NASA provides a forum for ideas which are required to be both *visionary* and *credible*. This approach was designed by program manager Marc Millis to provide a valuable and necessary filter for ideas and their proponents. The first BPP activity was a workshop held in August 1997 to carry out an initial survey of concepts; this is now available as a NASA conference proceedings (NASA/CP—1999-208694). Following additional funding, the BPP program issued a NASA Research Announcement (NRA) in November 1998 soliciting proposals. Out of 60 submissions, six studies were selected for funding.

Three other events relevant to interstellar exploration but not directly connected with the BPP program have also taken place. In February 1998, the Marshall Space Flight Center (MSFC) hosted a four-day workshop on "Physics for the Third Millenium" directed by Ron Koczor of the MSFC Space Science Laboratory. In July 1998 a four-day workshop on "Robotic Interstellar Exploration in the Next Century" sponsored by the Advanced Concepts Office at JPL and the NASA Office of Space Science was held at Caltech. In September 1998 an Independent Review Panel was convened to assess the NASA Space Transportation Research Program (STRP). The STRP supports such areas of investigation as fission- and fusion-based advanced propulsion and even a serious attempt to replicate a claimed "Delta-g Gravity Modification Experiment." A report was issued in January 1999 entitled: "Ad Astra per Aspera: Reaching for the Stars."

CP504, *Space Technology and Applications International Forum–2000*, edited by M. S. El-Genk
© 2000 American Institute of Physics 1-56396-919-X/00/$17.00

The report had the following to say about the Delta-g Gravity Modification Experiment:

An experimental demonstration of gravity modification, regardless of how minute, would be of extraordinary significance. The Delta-g experiments now being carried out at MSFC are an attempt to duplicate a claimed anomalous weight loss of up to two percent in objects of various compositions suspended above a rotating 12-inch diameter Type II ceramic superconductor. This apparent phenomenon was discovered by accident during superconducting experiments by a Russian scientist, Dr. Eugene Podkletnov, then working at Tampere University in Finland. Initial results of the MSFC replication have been published in Physica C, 281, 260, 1997. As of November 1998 the group, led by David Noever and Ron Koczor, had made a 12-inch YBCO disk that survived pressing and heat treating in one piece. This is now being characterized and cut up to do mechanical testing. The next step is to make a 2-layer disk of the sort used by Podkletnov.

The review committee was impressed with the high quality of the researchers and the careful and methodical approach. We believe that this research is a prime candidate for continued STR support and urge that funding is adequate to permit a definitive replication to be carried to completion."

In all of these activities (in which the first author participated) it became clear that there are two approaches to conceptualizing an interstellar mission. The first is to extrapolate certain relevant present-day technologies by orders of magnitude and then see what possibilities might emerge. One example — discussed at the Caltech workshop — would be a craft propelled by a laser-pushed lightsail... but this would require a 1000 km-diameter sail and a 10 km-diameter lens having an open-loop pointing capability of 10^{-5} arcsec (given a feedback time of years): quite a formidable challenge!

Another "known technology pushed to the limit" example is based on production and storage of huge amounts of anti-matter. This makes a good example of overwhelming technical difficulties in pushing present-day technology. Based on energy arguments alone (i.e. ignoring the issue of specific impulse) one can achieve a speed of $0.1c$ by annihilating 0.5 percent of the mass of a starship; this is simply calculated by equating the final kinetic energy of the starship, $m_s v^2/2$, where $v = 0.1c$, to the rest energy of the propellant, $m_p c^2$, a good approximation in this regime. It would take an equal amount of energy to stop, and similarly for the return to Earth. Under perfectly ideal conditions then, one percent of the mass of the starship in antimatter and one percent in ordinary matter would, in principle, suffice as propellant. One hundred percent efficiency for conversion of rest mass energy into kinetic energy is out of the question. Let us take an optimistic 10 percent. This means that for a starship as modest as the space shuttle in size — about 100 tons, hardly adequate for a century-long out and back mission to Alpha Centauri — one would require 10 tons of antimatter.

The manufacture of antiprotons is extremely inefficient. Techniques for creating antiprotons at CERN require approximately two and one-half million protons each accelerated to an energy of 26 GeV to create a single antiproton. This amounts to an energy efficiency of 3×10^{-8}. This is further reduced by a factor of 20 or so for the efficiency of the proton accelerator, leaving a net efficiency of perhaps 1.5×10^{-9}, i.e. about one part in a billion! At a cost of 5 cents per kilowatt-hour of electricity the cost of 10 tons of antiprotons would be 1.4×10^{21} dollars. A good way to imagine this is to say that it represents the total current U. S. federal budget (app. \$1.2 trillion per year) spent every year for the next 1.2 billion years (cf. M. R. LaPointe, NASA SBIR Phase I Final Report for contract NAS8-98109). Other technological extrapolations for propulsion discussed in the various workshops and reviews suffer from similar tremendous order-of-magnitude problems. Building a starship based on extrapolation of known technology might be likened to insisting on building some kind of sailing ship capable of crossing the Atlantic in 6 hours, when one really has to discover flight to do that.

The second approach is to try to find new regimes in physics or perhaps even new laws of physics. When Alcubierre (1994) published his article, "The Warp Drive: Hyper-fast Travel within General Relativity," this aroused considerable enthusiasm since it demonstrated that, in principle, general relativity allowed for local metric modifications resulting in expansion of space faster than the speed-of-light. Indeed, as is well known in cosmology, there is no speed limit to the expansion of space itself; in conventional inflationary big bang theory there must be regions of the Universe beyond our event horizon whose Hubble speed (relative to us) is greater than c and this in no way conflicts with special relativity. Relativity merely forbids motion *through* space faster than c. Alcubierre demonstrated that the mathematics of general relativity allowed for the creation of what might be termed a "bubble" of ordinary flat (Minkowski) space — in which a starship could be situated — that could surf, so to speak, at arbitrarily large speeds on a spacetime distortion, a faster than light stretching of spacetime; this would indeed be a warp drive. The Alcubierre bubble was soon burst, though, by Pfenning and Ford (1997) who showed that it would require more energy than that of the entire Universe to create the extremely warped space necessary. However recently Van den Broeck (1999) has shown that this energy requirement can be reduced by 32 orders of magnitude via a slight change in the Alcubierre geometry. While this still leaves us a long way from a feasible interstellar technology, warping space or creating wormholes are physics possibilities meriting further theoretical exploration.

Another regime of "new physics" is in actuality almost a century old. The concept of an electromagnetic zero-point field was developed by, among others, Planck, Einstein and Nernst. If an energetic sea of electromagnetic fluctuations comprising the electromagnetic quantum vacuum fills the Universe — for reasons discussed below — then this suggests the possibility of generating propulsive forces or extracting energy anywhere in space. Even more intriguing possibilities are opened by the linked proposed concepts that gravitation and inertia may originate in the zero-point field. If both gravitation and inertia are manifestations of the vacuum medium and in particular of its electromagnetic component, the ZPF, they can be treated by the techniques of electrodynamics, and perhaps they can be manipulated. The concept of gravity manipulation has been a staple of science fiction, but in fact inertia manipulation would be even more far reaching. As exciting as it would be to reduce the (gravitational) weight of a launch vehicle to zero, this would merely set if free from the gravitational potential well of the Earth and of the Sun. The problem of adding kinetic energy to reach high interstellar velocities would remain... unless one can modify inertia. Modification of inertia would (a) reduce energy requirements to attain a given velocity and (b) possibly allow greatly enhanced accelerations. The latter would open many possibilities since it would be far more efficient to have a perhaps enormously large acceleration device that never has to leave the ground and needs to act over only a short distance to rapidly impart a huge impulse, slingshotting a starship on its way. (We assume that the inertia of everything inside the starship would be modified as well. For the time being we overlook the problem of deceleration at the end of the journey, it being prudent to tackle only one apparent impossibility at a time.)

The concept of inertia modification may forever remain a *practical impossibility*. However at the moment it has become a legitimate *theoretical possibility*. In the following sections we summarize a recently developed theoretical connection between the ZPF and inertia and report on the discovery that a specific resonance frequency is likely to be involved. It is shown that such a resonance would simultaneously offer an explanation for both the inertia of a particle and the de Broglie wavelength of that particle in motion as first measured for electrons by Davisson and Germer (1927).

THE ELECTROMAGNETIC ZERO-POINT FIELD

The necessary existence of an electromagnetic zero-point field can be shown from consideration of elementary quantum mechanics. The Heisenberg uncertainty relation tells us that a harmonic oscillator must have a minimum energy of $\hbar\omega/2$, where $\hbar$ is the Planck constant divided by 2π and ω is the oscillation frequency in units of radians per second. (Expressed in cycles per second, Hz, this minimum energy is $h\nu/2$.) This is the zero-point energy of a harmonic oscillator, the derivation of which is a standard example in many introductory quantum textbooks.

The electromagnetic field is subject to a similar quantization: this is done by "the association of a quantum-mechanical harmonic oscillator with each mode of the radiation field" (cf. Loudon 1983). The same $h\nu/2$ zero-point energy is found in each mode of the field, where a mode of the field can be thought of as a plane wave specified by its frequency (ν), directional propagation vector ($\hat{\mathbf{k}}$), and one of two polarization states (σ). Summing up over all plane waves one arrives at the zero-point energy density for the electromagnetic field,

$$\rho_{ZP}(\nu) = \int_0^{\nu_c} \frac{4\pi h\nu^3}{c^3} d\nu \ , \tag{1}$$

where ν_c is a presumed high-frequency cutoff, often taken to be the Planck frequency, $\nu_P = (c^5/G\hbar)^{1/2} = 1.9 \times 10^{43}$ Hz. (See the appendix for a brief discussion of the Planck frequency). With this assumed cutoff, the energy density becomes the same (within a factor of $2\pi^2$) as the maximum energy density that spacetime can sustain: with $\nu_c = \nu_P$, the ZPF energy density is $\rho_{ZP} = 2\pi^2 c^7/G^2\hbar$. This is on the order of 10^{116} ergs cm^{-3} s^{-1}. The term "ZPE" is often used to refer to this electromagnetic energy of the zero-point fluctuations of the quantum vacuum. Note that the strong and weak interactions also have associated zero-point energies. These should also contribute to inertia. Their exact contributions remain to be determined since we have yet to consider these in the present context. For now we restrict ourselves to the electromagnetic contribution.

Can one take seriously the concept that the entire Universe is filled with a background sea of electromagnetic zero-point energy that is nearly 100 orders of magnitude beyond the energy equivalent of ordinary matter density? The concept is inherently no more unreasonable in modern physics that that of the vast Dirac sea of negative energy anti-particles. Moreover the derivation of the zero-point energy from the Heisenberg uncertainty relation and the counting of modes is so elementary that it becomes convoluted to try to simply argue away the ZPF. The objection that most immediately arises is a cosmological one: that the enormous energy density of the ZPF should, by general relativity, curve the entire Universe into a ball orders of magnitude smaller than the nucleus of an atom. Our contention is that the ZPF plays a key role in giving rise to the inertia of matter. If that proves to be the case, the principle of equivalence will require that the ZPF be involved in giving rise to gravitation. This at least puts the spacetime-curvature objection in abeyance: in a self-consistent ZPF-based theory of inertia and gravitation one can no longer naively attribute a spacetime curving property to the energy density of the ZPF itself (Sakharov, 1968; Misner, Thorne and Wheeler, 1973; Puthoff, 1989; Haisch and Rueda, 1997; Puthoff, 1999).

One might try taking the position that the zero-point energy must be merely a mathematical artifact of theory. It is sometimes argued, for example, that the zero-point energy is merely equivalent to an arbitrary additive potential energy constant. Indeed, the potential energy at the surface of the earth can take on any arbitrary value, but the falling of an object clearly demonstrates the reality of a potential energy field, the gradient of which is equal to a force. No one would argue that there is no such thing as potential energy simply because it has no well-defined absolute value. Similarly, gradients of the zero-point energy manifest as measurable Casimir forces, which indicates the reality of this sea of energy as well. Unlike the potential energy, however, the zero-point energy is not a floating value with no intrinsically defined reference level. On the contrary, the summation of modes tells us precisely how much energy each mode must contribute to this field, and that energy density must be present unless something else in nature conspires to cancel it.

Another argument for the physical reality of zero-point fluctuations emerges from experiments in cavity quantum electrodynamics involving suppression of spontaneous emission. As Haroche and Raimond (1993) explain:

> These experiments indicate a counterintuitive phenomenon that might be called "no-photon interference." In short, the cavity prevents an atom from emitting a photon because that photon would have interfered destructively with itself had it ever existed. But this begs a philosophical question: How can the photon "know," even before being emitted, whether the cavity is the right or wrong size?

The answer is that spontaneous emission can be interpreted as stimulated emission by the ZPF, and that, as in the Casimir force experiments, ZPF modes can be suppressed, resulting in no vacuum-stimulated emission, and hence no "spontaneous" emission (McCrea, 1986).

The Casimir force attributable to the ZPF has now been well measured, the agreement between theory and experiment being approximately five percent over the measured range (Lamoreaux, 1997). Perhaps some variation on the Casimir cavity configuration of matter could one day be devised that will yield a different force on one side than on the other of some device, thus providing in effect a ZPF-sail for propulsion through interstellar space.

Independent of whether a propulsive force may be generated from the ZPF is the question of whether energy can be extracted from the ZPF. This was first considered — and found to be a possibility in theory — in a thought experiment published by Forward (1984). Subsequently Cole and Puthoff (1993) analyzed the thermodynamics of zero-point energy extraction in some detail and concluded that, in principle, no laws of thermodynamics are violated by this. There is the possibility that nature is already tapping zero-point energy in the form of very high energy cosmic rays and perhaps in the formation of the sheet and void structure of clusters of galaxies (Rueda, Haisch and Cole, 1995). Another very useful overview is the USAF

Report by Forward (1996) and also the discussion of force generation and energy extraction in Haisch and Rueda (1999).

THE INERTIA RESONANCE AND THE DE BROGLIE WAVELENGTH

Can the inertia of matter be modified? In 1994 we, together with H. Puthoff, published an analysis using the techniques of stochastic electrodynamics in which Newton's equation of motion, $\mathbf{F} = m\mathbf{a}$, was *derived* from the electrodynamics of the ZPF (Haisch, Rueda and Puthoff [HRP], 1994). In this view the inertia of matter is reinterpreted as an electromagnetic reaction force. A NASA-funded research effort at Lockheed Martin in Palo Alto and California State University in Long Beach recently led to a new analysis that succeeded in deriving both the Newtonian equation of motion, $\mathbf{F} = m\mathbf{a}$, and the relativistic form of the equation of motion, $\mathcal{F} = d\mathcal{P}/d\tau$, from Maxwell's equations as applied to the ZPF (Rueda & Haisch, 1998a; 1998b). This extension from classical to relativistic mechanics increases confidence in the validity of the hypothesis that the inertia of matter is indeed an effect originating in the ZPF of the quantum vacuum. Overviews of these concepts may be found in the previous STAIF proceedings and other conference proceedings (Haisch and Rueda, 1999; Haisch, Rueda and Puthoff, 1998; Haisch and Rueda, 1998; additional articles are posted or linked at http://www.jse.com/haisch/zpf.html).

In the HRP analysis of 1994 it appeared that the crucial interaction between the ZPF and the quarks and electrons constituting matter must be concentrated near the Planck frequency. As discussed in the previous section (and in the appendix), the Planck frequency is the highest possible frequency in nature and is the presumed cutoff of the ZPF spectrum: $\nu = (c^5/G\hbar)^{1/2} \sim 1.9 \times 10^{43}$ Hz. In contrast, the new approach involves the assumption that the crucial interaction between the quarks and electrons constituting matter and the ZPF takes place not at the ZPF cutoff, but at a resonance frequency. We have now found evidence that, for the electron, this resonance must be at its Compton frequency: $\nu = m_e c^2/h = 1.236 \times 10^{20}$ Hz. This is 23 orders of magnitude lower (hence possibly within reach of electromagnetic technology) than the Planck frequency.

In Rueda and Haisch (1998a, 1998b) we show that from the force associated with the non-zero ZPF momentum flux (obtained by calculating the Poynting vector) in transit through an accelerating object, the apparent inertial mass derives from the energy density of the ZPF as follows:

$$m_i = \frac{V_0}{c^2} \int \eta(\nu)\rho_{ZP}(\nu)d\nu \ . \tag{2}$$

where $\eta(\nu)$ is a scattering parameter (see below).

It was proposed by de Broglie that an elementary particle is associated with a localized wave whose frequency is the Compton frequency, yielding the Einstein-de Broglie equation:

$$h\nu_C = m_0 c^2. \tag{3}$$

As summarized by Hunter (1997): "...what we regard as the (inertial) mass of the particle is, according to de Broglie's proposal, simply the vibrational energy (divided by c^2) of a localized oscillating field (most likely the electromagnetic field). From this standpoint inertial mass is not an elementary property of a particle, but rather a property derived from the localized oscillation of the (electromagnetic) field. De Broglie described this equivalence between mass and the energy of oscillational motion...as *'une grande loi de la Nature'* (a great law of nature)." The rest mass m_0 is simply m_i in its rest frame. What de Broglie was proposing is that the left-hand side of eqn. (3) corresponds to physical reality; the right-hand side is in a sense bookkeeping, defining the concept of rest mass.

De Broglie assumed that his wave at the Compton frequency originates in the particle itself. An alternative interpretation is that a particle "is tuned to a wave originating in the high-frequency modes of the zero-point background field" (de la Peña and Cetto, 1996; Kracklauer, 1992). The de Broglie oscillation would thus be due to a resonant interaction with the ZPF, presumably the same resonance that is responsible for creating inertial mass as in eq. (2). In other words, the ZPF would be driving this ν_C oscillation.

We therefore suggest that an elementary charge driven to oscillate at the Compton frequency by the ZPF may be the physical basis of the $\eta(\nu)$ scattering parameter in eqn. (2). For the case of the electron, this would imply that $\eta(\nu)$ is a sharply-peaked resonance at the frequency, expressed in terms of energy, $h\nu = 512$ keV. The inertial mass of the electron would physically be the reaction force due to resonance scattering of the ZPF at that frequency.

This leads to a surprising corollary. It can be shown (Haisch and Rueda, 1999; de la Peña and Cetto, 1996; Kracklauer, 1992) that as viewed from a laboratory frame, the standing wave at the Compton frequency in the electron frame transforms into a traveling wave having the de Broglie wavelength, $\lambda_B{}' = h/p$, for a moving electron. The wave nature of the moving electron appears to be basically due to Doppler shifts associated with its Einstein-de Broglie resonance frequency.

The identification of the resonance frequency with the Compton frequency would solve a fundamental mystery of quantum mechanics: Why does a moving particle exhibit a de Broglie wavelength of $\lambda = h/p$? It can be shown that if the electron acquires its mass because it is driven to oscillate at its Compton frequency by the ZPF, then when viewed from a moving frame, a beat frequency arises whose wavelength is precisely the de Broglie wavelength (Haisch & Rueda, 1999; de la Peña & Cetto, 1996). Thus within the context of the zero-point field inertia hypothesis we can simultaneously and suggestively explain both the origin of mass and the wave nature of matter as ZPF phenomena. Furthermore, the relative accessibility of the Compton frequency of the electron encourages us that an experiment to demonstrate mass modification of the electron by techniques of cavity quantum electrodynamics may soon be feasible.

ACKNOWLEDGMENTS

We acknowledge support of NASA contract NASW-5050 for this work.

REFERENCES

Alcubierre, M., "The Warp Drive: Hyper-fast Travel Within General Relativity," *Class. Quantum Grav.*, **11**, L73 (1994).

Cole, D.C. and Puthoff, H.E., "Extracting Energy and Heat from the Vacuum," *Phys. Rev. E*, **48**, 1562 (1993).

Davisson, C.J. and Germer, L.H., "Diffraction of Electrons by a Crystal of Nickel," *Phys. Rev*, **30**, 705 (1927).

de la Peña, L. and Cetto, A.M., *The Quantum Dice: An Introduction to Stochastic Electrodynamics*, (Kluwer Acad. Publ.), (1996).

Forward, R., "Extracting electrical Energy from the Vacuum by Cohesion of Charged Foliated Conductors," *Phys. Rev. B*, **30**, 1700 (1984).

Forward, R., "Mass Modification Experiment Definition Study," *J. of Scientific Exploration*, **10**, 325 (1996).

Haisch, B. and Rueda, A., "Reply to Michel's 'Comment on Zero-Point Fluctuations and the Cosmological Constant'," *Astrophys. J.*, **488**, 563 (1997).

Haisch, B., and Rueda, A., "The Zero-Point Field and Inertia," in *Causality and Locality in Modern Physics*, (G. Hunter, S. Jeffers and J.-P. Viger, eds.), (Kluwer Acad. Publ.), 171, (1998). xxx.lanl.gov/abs/gr-qc/9908057

Haisch, B., and Rueda, A., "Progress in Establishing a Connection Between the Electromagnetic Zero-Point Field and Inertia." AIP Conference Proceedings No. 458, p. 988 (1999) xxx.lanl.gov/abs/gr-qc/9906069

Haisch B. and Rueda, A., "On the Relation Between Zero-point-field-induced Inertial Mass and the Einstein-de Broglie Formula." *Phys. Lett. A*, in press (2000) xxx.lanl.gov/abs/gr-qc/9906084

Haisch, B., Rueda, A. and Puthoff, H.E. (HRP), "Inertia as a Zero-point-field Lorentz Force," *Phys. Rev. A*, **49**, 678 (1994).

Haisch, B., Rueda, A. and Puthoff, H.E, "Advances in the Proposed Electromagnetic Zero-point-field Theory of Inertia." AIAA 98-3143 (1998) xxx.lanl.gov/abs/physics/9807023

Haroche, S. and Raimond, J.M., "Cavity Quantum Electrodynamics," *Scientific American*, **268**, No. 4, 54 (1993)

Hunter, G., "Electrons and photons as soliton waves," in *The Present Status of the Quantum Theory of Light*, (S. Jeffers et al. eds.), (Kluwer Acad. Publ.), pp. 37–44 (1997).

Kracklauer, A.F., "An Intuitive Paradigm for Quantum Mechanics." *Physics Essays*, **5**, 226 (1992).

Lamoreaux, S.K., "Demonstration of the Casimir Force in the 0.6 to 6 μm Range," *Phys. Rev. Letters*, **78**, 5 (1997)

Loudon, R., *The Quantum Theory of Light*, chap. 1, (Oxford: Clarendon Press) (1983).

McCrea, W., "Time, Vacuum and Cosmos," *Q. J. Royal Astr. Soc.*, **27**, 137 (1986).

Misner, C.W., Thorne, K.S. and Wheeler, J.A., *Gravitation*, (Freeman and Co.), pp. 426–428 (1973).

Pfenning, M.J. and Ford, L.H., "The Unphysical Nature of 'Warp Drive'," *Class. Quant. Grav.*, **14**, 1743-1751 (1997).

Puthoff, H.E., "Gravity as a Zero-point-fluctuation Force," *Phys. Rev. A*, **39**, 2333 (1989).

Puthoff, H.E., "Polarizable-Vacuum (PV) Representation of General Relativity," preprint, (1999). xxx.lanl.gov/abs/gr-qc/9909037

Rueda, A. and Haisch, B., "Inertia as Reaction of the Vacuum to Accelerated Motion," *Physics Lett. A*, **240**, 115 (1998a). xxx.lanl.gov/abs/physics/9802031

Rueda, A. and Haisch, B., "Contribution to Inertial Mass by Reaction of the Vacuum to Accelerated Motion," *Foundations of Physics*, **28**, 1057 (1998b). xxx.lanl.gov/abs/physics/9802030

Rueda, A., Haisch, B. and Cole, D. C., "Vacuum Zero-Point Field Pressure Instability in Astrophysical Plasmas and the Formation of Cosmic Voids," *Astrophys. J.*, **445**, 7 (1995).

Sakharov, A.D., "Vacuum Quantum Fluctuations in Curved Space and the Theory of Gravitation," *Sov. Phys.-Doklady* **12**, No. 11, 1040 (1968).

Van den Broeck, C. "A 'warp drive' with Reasonable Total Energy Requirements," preprint, (1999) xxx.lanl.gov/abs/gr-qc/9905084

APPENDIX: THE PLANCK FREQUENCY

The Planck frequency is assumed to be the highest frequency that spacetime itself can sustain. This can be understood from simple, semi-classical arguments by combining the constraints of relativity with those of quantum mechanics. In a circular orbit, the acceleration is v^2/r, which is obtained from a gravitational force per unit mass of Gm/r^2. Letting $v \to c$, one obtains a maximum acceleration of $c^2/r = Gm/r^2$, from which one derives the Schwarzschild radius for a given mass: $r_S = Gm/c^2$. The Heisenberg uncertainty relation specifies that $\Delta x \Delta p \geq \hbar$, and letting $\Delta p \to mc$ one arrives at the Compton radius: $r_C = \hbar/mc$, which specifies the minimum quantum size for an object of mass m. Equating the minimum quantum size for an object of mass m with the Schwarzschild radius for that object one arrives at a mass: $m_P = (c\hbar/G)^{1/2}$ which is the Planck mass, i.e. 2.2×10^{-5} g. The Compton radius of the Planck mass is called the Planck length: $l_p = (G\hbar/c^3)^{1/2}$, i.e. 1.6×10^{-33} cm. One can think of this as follows: Due to the uncertainty relation, a Planck mass cannot be compressed into a volume smaller than the cube of the Planck length. A Planck mass, m_P, in a Planck volume, l_P^3, is the maximum density of matter that can exist without being unstable to collapsing spacetime fluctuations: $\rho_{P,m} = c^5/G^2\hbar$ or as an energy density, $\rho_{P,e} = c^7/G^2\hbar$. The speed-of-light limit constrains the fastest oscillation that spacetime can sustain to be $\nu_P = c/l_P = (c^5/G\hbar)^{1/2}$, i.e. 1.9×10^{43} Hz.

Precision Measurements of the Material and Boundary Geometry Dependence of the Casimir Force

Anushree Roy, Chiung-Yuan Lin and U. Mohideen

Department of Physics, University of California, Riverside, CA 92521

(909)787-5390; umar.mohideen@ucr.edu

Abstract. The Casimir force results from the modification of the electromagnetic zero point energy which prevades all of space as predicted by quantum field theory. It depends on the conductivity, dielectric properties and shape of the metal boundaries. For example while the Casimir force is attractive between two flat metal plates it can be repulsive between two hemispheres. In order to measure such material and boundary dependences of the Casimir force, precision measurements of the force and the separation between the two surfaces is necessary. We will discuss experimental techniques to perform such precision measurements of the Casimir force. We will present some of our recent measurements to probe such material and shape dependences.

PRECISION MEASUREMENTS OF THE MATERIAL AND BOUNDARY DEPENDENCE OF THE CASIMIR FORCE

The Casimir force (Casimir,1948; Milonni, 1994; Mostepanenko, 1997; Plunien, 1986) results from the alteration by metal boundaries of the zero point electromagnetic energy $E = \sum_{n}^{\infty}(1/2)\eta\omega_n$, where $\eta\omega_n$ is the photon energy in each allowed photon mode n. Lifshitz (1956) generalized the Casimir force to any two infinite dielectric half-spaces as the force between fluctuating dipoles induced by the zero point electromagnetic fields and obtained the same result as Casimir for two perfectly reflecting (infinite conductivity) flat plates. The Casimir force between metallic surfaces has been demonstrated between two flat plates (Sparnaay, 1958, 1989) and a large sphere and a flat plate (van Blockland, 1978; Lamoreaux, 1997). Spring balances were used by Sparnaay (1958, 1989) and van Blockland (1978) to measure the Casimir force. A torsion pendulum was used by Lamoreaux (1997). We have measured the Casimir force using an Atomic Force Microscope (AFM) and shown its value to be in agreement with the theory to an average rms deviation of 1% (Mohideen, 1998; Klimchitskaya, 1999). Theoretical treatments of the Casimir force have shown that it is a strong function of the boundary geometry and spectrum (Boyer, 1968; Balian, 1978; Golestanian, 1997; Ford, 1998). Through experiments with periodically corrugated boundaries we have demonstrated the nontrivial boundary dependence of the Casimir force (Roy, 1999). The experiment with corrugated boundaries will not be discussed here. An important application of Casimir force measurements, is the ability to place strong limits on hypothetical long range forces and light elementary particles like those predicted by supersymmetric theories (Fienberg, 1979; Bordag, 1998).

The Casimir force for two perfectly conducting parallel plates of unit area separated by a distance z is: $F(z) = -\dfrac{\pi^2 \eta c}{240} \dfrac{1}{z^4}$. It is strong function of 'z' and is measurable only for $z < 1$ μm. Experimentally it is hard to configure two parallel plates uniformly separated by distances less than a micron. So the preference is to replace one of the plates by a metal sphere of radius R where $R \gg z$. For such a geometry the Casimir force is modified to

(Derjaguin, 1956; Blocki, 1977): $F_c^0(z) = \frac{-\pi^3}{360} R \frac{\eta c}{z^3}$. This definition of the Casimir force holds only for hypothetical metals of infinite conductivity, and therefore a correction due to the finite conductivity of Al has to be applied. Such a correction can be accomplished through use of the Lifshitz theory (Lifshitz, 1956; Schwinger, 1978). For a metal with a dielectric constant ε the force between a large sphere and flat plate is given by (Lifshitz, 1956):

$$F^0(z) = -\frac{R\eta}{\pi\, c^3} \int dz \int_0^\infty \int_1^\infty p^2 \xi^3 dp\, d\xi \left\{ \left[\frac{(s+p)^2}{(s-p)^2} e^{\frac{2p\xi z}{c}} - 1 \right]^{-1} + \left[\frac{(s+p\varepsilon)^2}{(s-p\varepsilon)^2} e^{\frac{2p\xi z}{c}} - 1 \right]^{-1} \right\}, \tag{1}$$

where 'z' is the surface separation, R is the sphere radius, $s = \sqrt{\varepsilon - 1 + p^2}$, $\varepsilon(i\xi) = 1 + \frac{2}{\pi} \int_0^\infty \frac{\omega \varepsilon''(\omega)}{\omega^2 + \xi^2} d\omega$ is the dielectric constant of Al and ε'' is its imaginary component (Lifshitz, 1956). ξ is the imaginary frequency given by $\omega = i\xi$. Here the complete ε'' extending from 0.01eV to 1000eV from Palik (1985), augmented with the free electron model below 0.01eV is used to calculate $\varepsilon(i\xi)$. Al metal was chosen because of its high reflectivity at short wavelengths (corresponding to close surface separations).

There are also corrections to the Casimir force resulting from the roughness of the surface due to the stochastic changes in the surface separation (Mohideen, 1998; Klimchitskaya, 1999, Schram, 1974; Maradudin, 1980; Mazur, 1981). The roughness of the metal surface is measured directly with the AFM. The metal surface is composed of separate crystals on a smooth background. The height of the highest distortions are 14nm and intermediate ones of 7nm both on a stochastic background of height 2nm with a fractional surface areas of 0.05, 0.11 and 0.84 respectively. The crystals are modeled as parallelepipeds. This leads to the complete Casimir force including roughness correction given by: $F^r(z) = F^0(z) \left[1 + 0.86 \left(\frac{A}{z} \right)^2 + 1.02 \left(\frac{A}{z} \right)^3 + 1.9 \left(\frac{A}{z} \right)^4 \right]$ (Klimchitskaya, 1999). Here,

$A=11.8$nm is the effective height defined by requiring that the mean of the function describing the total roughness is zero and the numerical coefficients are the probabilities of different distance values between the interacting surfaces. The roughness correction here is $\leq 1.3\%$ of the measured force. There are also corrections due to the finite temperature (Mehra, 1967; Brown, 1969) given by:

$$F_c(z) = F^r(z) \left(1 + \frac{720}{\pi^2} f(\eta) \right), \tag{2}$$

where $f(\eta) = (\eta^3 / 2\pi) \zeta(3) - (\eta^4 \pi^2 / 45)$, $\eta = 2\pi k_B T z / hc = 0.131 \times 10^{-3} z$ nm^{-1} for T= 300°K, $\zeta(3)=1.202\ldots$ is the Riemann zeta function and k_B is the Boltzmann constant. The temperature corrections are less than 1% of the Casimir force for the surface separations reported here.

A schematic diagram of the experiment is shown in fig. 1. The force is measured at a pressure below 50mTorr and at room temperature. As in the previous version (Mohideen, 1998) the experiments were done on a floating optical table. Additionally the vacuum system was mechanically damped and isolated to decrease the vibrations coupled to the AFM. Polystyrene spheres were mounted on the tip of 320μm long cantilevers with Ag epoxy. A 1cm diameter optically polished sapphire disk is used as the plate. The cantilever (with sphere) and plate were then coated with 250nm of Al in an evaporator. Both surfaces are then coated with 7.9±0.1nm layer of 60%Au/40%Pd (measured >90% transparency for λ<300nm and is thus a transparent spacer for surface separations being considered here). This coating was necessary to provide a non-reactive surface and to prevent any space charge effects due to patch oxidation of the Al coating. The sphere diameter was measured using the Scanning Electron Microscope (SEM) to be 201.7±0.5μm. The rms roughness amplitude of the Al surfaces was measured using an AFM to be 3nm. The decrease in roughness was achieved with controlled metal evaporation and reduced coating thickness. The roughness of the metal coating prevents *a priori* knowledge of the average separation on contact of the two surfaces. Here an independent and exact measurement of the average surface separation on contact of the two surfaces is done by electrostatic means.

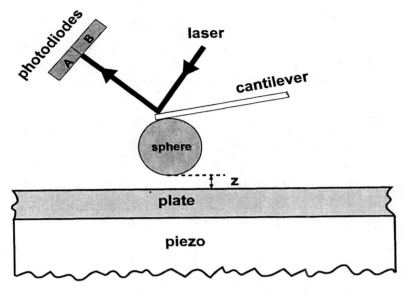

FIGURE 1. Schematic of the experimental setup.

In the schematic shown in figure 1, a force on the sphere causes the cantilever to tilt. This tilt is detected by the deflection of the laser beam leading to a difference signal between photodiodes A and B. This force and the corresponding cantilever deflection are related by Hooke's Law: $F = -k \Delta z$, where 'k' is the force constant and 'Δz' is the cantilever deflection. As reported in ref.7 the cantilever is calibrated and the residual potential difference between the grounded sphere and plate is measured using the electrostatic force between them. The force constant was measured as k =0.0169±0.0003N/m from the electrostatic force for surface separation > 2μm as reported in

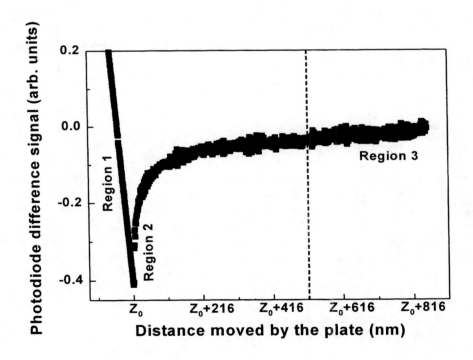

FIGURE 2. Raw data from one scan.

Mohideen (1998). Next the residual potential of the grounded sphere was measured as $V=7.9\pm0.8$mV by the AC measurement technique reported earlier (Mohideen 1998). This residual potential is a contact potential that arises from the different materials used to fabricate the sphere and the flat plate. The corrections due to the piezo hysteresis and cantilever deflection were applied as reported in Mohideen (1998) to the sphere-plate separations.

To measure the Casimir force between the sphere and flat plate they are both grounded together with the AFM. The plate is then moved towards the sphere and the corresponding photodiode difference signal was measured (approach curve). The raw data from one scan is shown in Figure 2. Region-3 is the flexing of the cantilever resulting from the continued extension of the piezo after contact of the two surfaces. In region-2 (z_0+516nm>surface separations>z_0+16nm) the Casimir force is the dominant characteristic far exceeding all systematic errors. The systematic effects are primarily from the residual electrostatic force (<1.5% of the force at closest separation) and a linear contribution from scattered light and drift (<1% of the largest forces). The linear contribution is observed and measured at large surface separations as in region-1. Next we use of the electrostatic force between the sphere and flat plate to arrive at an independent and consistent measurement of z_O, the average surface separation on contact of the two surfaces. This is done immediately following the Casimir force measurement without breaking the vacuum and no lateral movement of the surfaces. The flat plate is connected to a DC voltage supply while the sphere remains grounded. The applied voltage V is so chosen that the electrostatic force is >10 times the Casimir force. The experiment is repeated for voltages between 0.3-0.8 V leading to an average value of $z_O=49$nm. Given the 7.9nm Au/Pd coating on each surface this would correspond to a average surface separation 49+16= 65nm for the case of the Casimir force measurement.

The electrostatically determined value of z_O can now be used to apply the systematic error corrections to the force curve of figure 2. The force curve in region-3, is fit to a function: $F= F_c(z+65\text{nm}) +F_e(z+49\text{nm}) + Cz$. The first term is the Casimir force contribution to the total force in region-1. The second term represents the electrostatic force between the sphere and flat plate due to the residual potential difference 7.9mV. The third term C represents the linear coupling of scattered light from the moving plate and experimental drift and corresponds to a force <1pN (<1% of the forces at closest separation). The difference in z_O in the electrostatic term and the Casimir force is due to the 7.9nm Au/Pd coating on each surface. The value of C is determined by minimizing the χ^2. The value of C determined in region-1 and the electrostatic force corresponding to the potential difference $V=7.9$mV is used to subtract the systematic errors from the force curve in region-3 and 2 to obtain the measured Casimir force as: $F_{c-m}= F_m-F_e- Cz$ where F_m is the measured total force. Thus the measured Casimir force from region-2 has no adjustable parameters.

The experiment is repeated for 27 scans and the average Casimir force measured is shown as open squares in figure 3. The error bars represent the range of data from the 27 scans at representative data points. The theoretical curve given by eq.2 is shown as a solid line. The theory has no adjustable parameters.

The precision of the experiment can be evaluated using a variety of statistical measures. A key point to note is that the Casimir force is generated for the whole range of separations and is compared to the theory with *no adjustable parameters*. Here we restrict the measurement to surface separations corresponding to wavelengths where Al can be considered a highly reflective metal and the Au/Pd cap layer is largely transparent. Thus we check the accuracy of the theoretical curve over the complete region between 100-500nm with $N=443$ points (with an average of 27 measurements representing each point) with no adjustable parameters. Given that the experimental standard deviation over this range is 7pN from thermal noise, the experimental uncertainty is $\leq \dfrac{7}{\sqrt{27}} =1.3 pN$ leading to a precision which is better than 1% of the largest forces measured. If one wished to consider the rms deviation of the experiment ($F_{experiment}$) from the theory (F_{theory}) in eq.3, $\sigma = \sqrt{\dfrac{\left(F_{theory} - F_{experiment}\right)^2}{N}} =1.9$pN as a measure of the precision, it is also on the order of 1% of the forces measured at the closest separation. Thus by any of the above definitions, the statistical measure of the experimental precision is of order 1% of the forces at the closest separation.

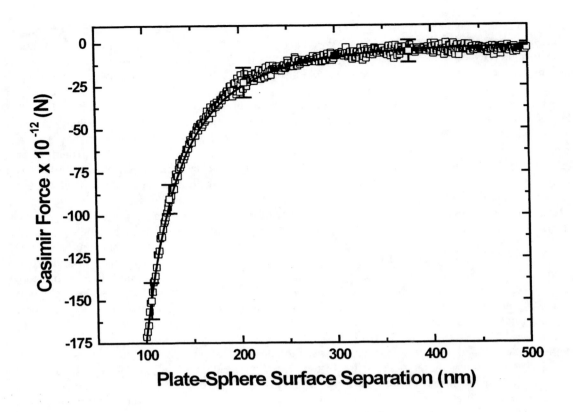

FIGURE 3. The average Casimir force between flat plate and large sphere. The solid line is the theory from eq.2 with no adjustable parameters.

In conclusion, we have used an Atomic Force Microscope to perform precision measurements of the Casimir force between a large Al coated sphere and flat plate. The average precision is around 1% of the forces measured at the closest separation. The same method has been used to measure the Casimir force between corrugated plates and plates and sphere coated with different materials. In particular with the case of corrugated plates (Roy, 1999) we have shown that the measured Casimir force deviates significantly from the theory even for small amplitude perturbations. Given the strong dependence of the Casimir force on the properties and shapes of the boundaries, experiments such as above would hold many surprises.

REFERENCES

Balian, R. and Duplantier, B., "Electromagnetic waves near perfect conductors II, Casimir effect," Ann. Phys. (N.Y) 112, p.165 (1978).

Blocki, J., Randrup, J., Swiatecki, W.J. and Tsang, C. F., "Proximity forces," Ann. Phy., 105, p. 427 (1977).

Bordag, M., Geyer, B., Klimchitskaya, G.L., and Mostepanenko, V.M., Phys. Rev. D, 56, p. 075003 1 (1998).

Boyer, T. H., "Quantum electrodynamic zero-point energy of a conducting spherical shell and the Casimir model for a charged particle," Phys. Rev., 174, p.1764 (1968).

Brown L.S., and Maclay, G.J., "Vacuum stress between conducting plates," Phys. Rev. 184, 1272 (1969).

Casimir H.B.G., " On the attraction between two perfectly conducting plates," Proc. Kon. Ned. Akad. Wet., 51, p. 793 (1948).

Derjaguin, B.V., Abrikosova, I.I., and Lifshitz, E.M., "Direct measurement of molecular attraction between solids separated by a gap," Q. Rev. Chem. Soc., 10, 295 (1956).

Golestanian, R. and Kardar, M., "Mechanical Response of Vacuum," Phys. Rev. Lett., **78**, p. 3421 (1997).

Ford, L.H., "Casimir force between a dielectric sphere and a wall: A model for amplification of vacuum fluctuations," Phys. Rev. A, **58**, p. 4279 (1998).

Feinberg, G., and Sucher, J., "Is there a strong van der Waals force between hadrons," Phys. Rev. D, **20**, p. 1717 (1979).

Israelachvili J.N. and Tabor, D., "The measurement of van der Waals dispersion forces in the range 1.5 to 130nm," Proc. Roy. Soc. A, **331**, p. 19 (1972).

Klimchitskaya, G.L., and Pavlov, Yu. V., " The corrections to the Casimir forces for configurations used in experiments: the spherical lens above the place and crossed cylinders," Int. J. Mod. Phys. A, **11**, 3723 (1996).

Klimchitskaya G.L., Roy A., Mohideen, U., and Mostepanenko, V.M., "Complete roughness and finite conductivity corrections for the recent Casimir force measurement," Phys. Rev. A, **60**, 3487 (1999)

Lifshitz, E.M., "The theory of molecular attractive forces between solids," Sov. Phys. JETP, **2**, p.73 (1956).

Lamoreaux, S.K., "Demonstration of the Casimir effect in the 0.6 to 6µm range," Phys. Rev. Lett., **78**, p. 5 (1997); Erratum, **81**, 5475 (1998); "Calculation of the Casimir force between imperfectly conducting plates," Phys. Rev. A, p. R3149 (1999).

Maradudin, A.A., Mazur, P., "Effects of surface roughness on the van der Waals force between macroscopic bodies," Phys. Rev. B, **22**, p. 1677 (1980);

Mazur, P., Maradudin, A.A., "Effects of surface roughness on the van der Waals force between macroscopic bodies II," Phys. Rev. B, **23**, p.695 (1981)

Mehra, J., "Temperature correction to the Casimir effect," Physica, **37**, p.145-152 (1967).

Milonni P.W., "*The Quantum Vacuum*", Academic Press, San Diego, CA (1994)

Mohideen, U. and Roy A., "Precision measurement of the Casimir force from 0.1 to 0.9µm," Phys. Rev. Lett., **81**, p. 4549 (1998).

Mostepanenko, V.M. and Trunov, N.N. , "*The Casimir effect and its Applications*," Clarendon Press (1997)

Plunien, G., Muller B., and Greiner, W., "The Casimir effect," Phy. Rep. **134**, p 87-193 (1986).

Palik, E.D., edited, *Handbook of Optical Constants of Solids*, Academic Press, New York (1985).

Roy, A., and Mohideen, U., "Demonstration of the nontrivial boundary dependence of the Casimir force," Phys. Rev. Lett., **82**, p.4380 (1999).

Sparnaay, M.J., "Measurement of the attractive force between plates," Physica, **24**, p.751 (1958); "Physics in the Making," edited by Sarlemijn, A. and Sparnaay, M.J., Published by North-Holland, Amsterdam, (1989).

Schwinger, J., DeRaad, Jr. L.L., and Milton, K.A., "The Casimir effect in dielectrics," Ann. Phys. **115**, 1 (1978)

Tabor, D., and Winterton, R.H.S., "Surface forces: Direct measurement of normal and retarded van der Waals forces," Proc. Roy. So. Lond., A **312**, 435 (1969).

van Bree J.L.M.M, Poulis, J.A. and Verhaar, B.J. and Schram, K., "The influence of surface irregularities upon the van der Waals force between macroscopic bodies," Physica **78**, p. 187 (1974).

van Blokland, P.H.G.M., and Overbeek, J.T.G., "van der Waals forces between objects covered with a Chromium layer," J. Chem. Soc. Faraday Trans., **74**, p. 2651 (1978).

A Design Manual for Micromachines using Casimir Forces: Preliminary Considerations

G. Jordan Maclay

Quantum Fields LLC
20876 Wildflower Lane, Richland Center, WI 53581
PH: 608-647-6769; FAX: 608-647-5494; e-mail: jordanmaclay@quantumfields.com

Abstract. General properties of the Casimir force are reviewed, with particular attention to the effects of geometry. Using the conservation of energy, the forces in several simple idealized structures are derived, including the lateral forces on partially overlapping parallel plates, the force in a Casimir "comb" drive with several interleaved surfaces, and the average force when a plate is inserted into a rectangular cavity. The properties of rectangular cavities with a moveable piston are discussed, and illustrated with numerical computations. An oscillating structure is proposed.

INTRODUCTION

Casimir predicted the existence of an attractive force between two infinite parallel uncharged perfectly conducting plates in vacuum almost 50 years ago (Casimir, 1948). This force arises because of the boundary conditions that the quantized source-free electromagnetic field must meet at the metal surfaces as discussed in an extensive review (Plunian et al, 1986). The prediction came very shortly after (Bethe,1948) and (Welton,1948) explained the Lamb shift in the hydrogen atom as due to interaction of the electron with the quantized vacuum electromagnetic field. The Casimir force was a startling and unexpected mesoscopic phenomena arising from the presence of surfaces in the quantized vacuum field. The force was predicted to vary as the inverse fourth power of the separation between the plates. At a separation of 100 nm the predicted force/area was equivalent to about 10^{-4} atm; at 10 nm it was about one atmosphere. The Casimir force has also been computed using the alternative language of source theory and radiative reaction, without explicit reference to vacuum fluctuations by (Schwinger, 1978) and (Milonni, 1994). Since Sparnaay's first attempts in 1959, various measurements have been made on dielectrics (Sparnaay, 1989) that have generally verified the theory of Casimir forces as developed for dielectrics by (Lifshitz, 1956), but not until quite recently was the existence of this attractive Casimir force between metallized surfaces verified, in two separate experiments. (Lamoroux, 1997) used a torsion pendulum with an electromechanical feedback system to measure the Casimir force between a metallized spherical lens and a flat plate to an accuracy of about 5-10% for separations of about 0.6 micrometer to 6 micrometer. (Mohideen, 1998) used an Atomic Force Microscope to measure the force between a metallized optical flat and a metallized ball mounted on the AFM cantilever, obtaining a precision of about 1% for separations of 0.1 to 0.9 micrometers. As measured by the AFM, the forces on an effective area of approximately 10 micrometers^2 were in the picoNewton range. It should be mentioned that if the distance between the surfaces is much less than the wavelength corresponding to the plasma frequency of the metal, then the force becomes the unretarded van der Waal's force.

With the advent of improved methods of making micron and submicron structures, such as microfabrication technology, it has become possible to explore forces arising from quantum fluctuations in greater detail. For example, the cantilever used by (Mohideen, 1998) is a silicon micromachined device, often called a MEMS device (Micro-electromechanical System). The small separation between neighboring surfaces in various MEMS devices means it is possible that the presence of Casimir forces may result in adjacent surfaces being attracted to or sticking to each other (Serry, 1998). Using micromachining methods, a variety of MEMS structures in which vacuum stresses are present can be fabricated. A harmonic oscillator with a Casimir interaction has been modeled but not yet built (Serry, 1995). Some structures, including cavities, can be built to investigate Casimir forces especially using AFM methods (Maclay, 1998).

CP504, *Space Technology and Applications International Forum–2000*, edited by M. S. El-Genk

TABLE 1. Casimir Forces at zero Kelvin for Different Perfectly Conductive Geometries (only cut-off independent, geometry dependent terms are included)

Parallel Plate (area = L x L, spacing=a)	Cube (side=a)	Sphere (radius=r)
$-0.0411\hbar cL^2/a^4$	$+0.0305\hbar c/a^2$	$+0.0462\hbar c/r^2$

In this paper, we give some preliminary considerations about possible structures and approaches that one might employ in designing MEMS devices that utilize Casimir forces, particularly with parallel plates and rectangular cavities. The analysis includes the role of the forces and the energy balance, assuming the ambient temperature is absolute zero and the ambient pressure is zero.

GENERAL COMMENTS ON VACUUM ENERGY AND CASIMIR FORCES FOR DIFFERENT GEOMETRICAL STRUCTURES

Casimir forces have been computed for only a few common geometries (See Table 1). The most common one is the infinite, parallel plate geometry that produces an attractive force between the plates. If we examine the infinite parallel plate geometry more fully, and imagine placing perfectly conductive, metal surfaces normal to the parallel plates in order to enclose the volume between the plates, then there would be outward forces on these additional four infinitely long, narrow, surfaces (Brown, 1969). For a conducting spherical shell (Milton, 1978) and (Boyer, 1968) have predicted outward or repulsive Casimir. (Lukosz, 1971) has predicted repulsive forces for a conducting hollow cube. The vacuum stress on two intersecting planes is attractive. For conductive rectangular cavities with square cross section ($1 \times 1 \times c$) the Casimir energy and the forces on the cavity walls have been computed by (Ambjorn, 1983), (Haycan, 1993), and (Mostepanenko, 1997) with the result that the forces can be inward or outward depending on the specific dimensions. (Ambjorn, 1983) has computed the constant energy contours for a $a1 \times a2 \times a3$ geometry for the region $a2, a3 > 1$. (Maclay, 1999) has computed the energy and force for rectangular cavities of arbitrary dimension.

For the parallel plate, sphere, cube, and rectangular cavity expressions for the stress-energy tensor $T^{\mu\nu}$ have been derived, giving separate expression for the stresses T^{ij} and energy density T^{00}. It is very important to note that for all geometries for which energy densities and forces have been computed, the expression for the force can also be obtained from the expression for the energy density by the principal of virtual work, which is based on the conservation of energy. By this principal, the force F_x in the x-direction equals $-\partial E/\partial x$, where E is the total energy in the volume. Thus in microdevices utilizing Casimir forces, we expect to conserve the total energy, which includes the mechanical energy and the field energy.

Geometries with curved surfaces or intersecting planes as discussed by (Deutsch, 1979) and (Balian, 1978) present special problems with respect to vacuum energy which we mention briefly. For gently curved conductors, (Deutsch, 1979) has shown that the stress-energy tensor is approximately proportional to the sum of the reciprocals of the two principal radii of curvature and varies inversely as the cube of the distance to the surface. It follows that the total vacuum energy in any compact region that actually contains part of the curved conducting surface is infinite.

Infinities of this type represent a breakdown of the perfect-conductor approximation. The ether cannot store an infinite amount of energy (whether positive or negative) in a compact region, nor can the conductor support the infinite stresses (DeWitt, 1989). The perfect-conductor boundary conditions are pathological, and lead to an infinite physically observable gravitational field (Deutsch, 1979). For a real metal the electrons are unable to follow an applied electromagnetic field at frequencies above the plasma frequency of the metal. Thus for frequencies above the plasma frequency, the zero-point electromagnetic field does not meet the boundary conditions for an ideal conductor. Also the corners of a real conductive surface are not infinitely sharp, but rounded, eliminating divergences.

SOME SIMPLE DYNAMICAL STRUCTURES

We consider the forces and energy balance in several simple structures: 1.with moving parallel plates; 2. with moving plates inserted in rectangular cavities, and 3. with pistons moving in rectangular cavities.

Structures using Parallel Plates

Consider two conducting, square, parallel plates, a distance L on each side, that are a distance a apart, with $a<<L$. If we allow the upper plate to approach the lower (fixed) plate quasistatically, then the attractive force $F_{pp}(a)= -KL^2/a^4$ does positive mechanical work during this reversible thermodynamic transformation. We are neglecting edge effects by assuming that the force is proportional to the area. Also during the transformation, the vacuum energy $E_p(a) = -KL^2/3a^3$ between the plates will be reduced, conserving the total energy in the system. If the separation decreases from a_i to a_f, then the energy balance is

$$E_p(a_f) - E_p(a_i) = - \int_{a_i}^{a_f} F_{pp}(a)da \ . \tag{1}$$

If we then separate the plates quasistatically, letting a increase from a_f to a_i, we do work on the system to restore it to its initial state. Over the entire cycle no net work is done, and there is no net change in the vacuum energy.

Consider an alternative cycle that numerous investigators have proposed in order to extract energy from the vacuum fluctuations. After the plates have reached the point of minimum separation, slide the upper plate laterally until it no longer is opposite the lower plate, then raise the upper plate to its original height, and slide it laterally over the lower fixed plate. Then we allow the plates to come together as before, extracting energy from the vacuum fluctuations and doing mechanical work. If no energy were expended in moving the plate laterally, then this cycle would indeed result in net positive work equal to the energy extracted from the vacuum. Although no one has yet computed in detail the lateral forces between offset parallel plates, it is probable that such forces are not zero, and that no net extraction of energy occurs for this cycle. We do know that the vacuum energy is not altered by a single infinite conducting plate (see DeWitt(1989)). If we neglect Casimir energy "fringing fields," and assume that the energy density differs from the free field density only in the region in which the square (LxL) plates overlap a distance x, where $0<x<L$ (see Fig. 1a), then we can compute the lateral force F_{L2} between the two plates using the conservation of energy (principal of virtual work):

$$F_{L2}(x) = -d[-KLx/a^3]/dx = KL/a^3 \ . \tag{2}$$

When one includes the work done by this constant attractive lateral force, which tends to increase x or pull the plates so they have the maximum amount of overlap, then there is not net change in total energy (mechanical plus field) as x goes from 0 to L. The normal Casimir force between these plates when they are directly opposite, with complete overlapping ($x = L$), is L/a times larger than the constant lateral force given by Eq. (2).

Consider the case of two fixed, square (LxL), parallel plates separated by a distance a, with a third moveable plate that slides in between the two parallel plates, separated by a distance a/2 from each plate (Fig. 1b). If we neglect vacuum energy "fringing fields", as before, that the vacuum energy is different from zero only in regions between directly opposing plates, and we can compute the lateral force on the moveable plate in the middle as minus the derivative of the vacuum energy. The energy, as a function of the overlap x of the fixed and moveable plate, is

$$E(x) = -2KLx/(a/2)^3 - KL(L-x)/a^3 \ , \tag{3}$$

which yields a lateral force $F_{L3} = -dE(x)/dx$ equal to

$$F_{L3} = -15 \ KL/a^3 \ . \tag{4}$$

This force is $15a/L$ times the normal Casimir force between the plates separated by a distance a. For a device with a = 0.1 micrometer, L = 1 mm, the lateral force would be an easily measurable 31 nanoNewtons. This structure is analogous to the electrostatic comb drive that is used extensively in MEMS (microelectromechanical systems) devices. One key operational difference between the Casimir and electrostatic drives is that the Casimir force drive always yields an inward or attractive force, whereas the voltage on the electrostatic comb drive can be reversed in polarity, reversing the direction of the force. Another difference is that the Casimir force comb drive requires no

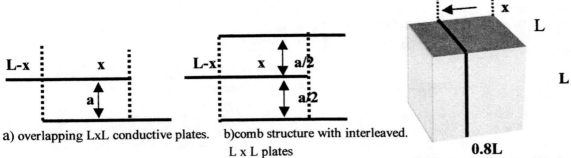

a) overlapping LxL conductive plates.　　b)comb structure with interleaved.
L x L plates

FIGURE 1 . Some structures with Casimir forces.

c) conductive moveable piston (solid black line) in a conductive rectangular box L x L x 0.8L.

external power source, whereas the electrostatic drive does. One could imagine a second set of parallel plates with a variable spacing a to generate a force in the opposite direction. Instead of applying a variable voltage as in the electrostatic case, one would need to alter the spacing between the plates to control the direction and magnitude of the net force. This device could be operated as a motor if at least one set of plates had an adjustable spacing. Because of the conservative nature of the electromagnetic field, one would not expect to extract a net energy from the vacuum.

Casimir force comb type structures with parallel plates have many of the same design issues as electrostatic type comb structures. For example, the structures are unstable with respect to movement of the plates in the direction normal to the plate movement. In other words, the normal forces increase if one of the normal separation distances decreases.

Structures using rectangular cavities and sliding plates

The mechanical behavior of the parallel plate configurations is determined by the negative energy density that arises if $a << L$. If we consider cavities that have dimensions in orthogonal directions that are within about a factor of about 3 of each other, then we can have regions with positive or negative energy density and can obtain both attractive and repulsive forces on sliding plates.

For example, consider a rectangular cavity $L \times L \times a$ formed from conductive plates. Imagine that the side of length a is constructed so that we can insert an additional plate (assumed to have zero thickness) in a direction normal to the a direction, dividing the cavity into identical rectangular regions with sides $L \times L \times a/2$. By the conservation of energy we can compute the average force required to insert this moveable plate. If we assume that the vacuum energy density is altered only in the region within the cavity as we insert the plate, then the change in vacuum energy is equal to minus the average force $<F>$ present during insertion of the movable plate times the distance L. Defining en(a1, a2, a3) as the vacuum energy of a rectangular cavity with sides (a1,a2,a3), we can express the average force as

$$<F(L,L,a)> = - [2en(L,L,a/2) - en(L,L,a)]/L . \qquad (5)$$

Depending on the ratio of L/a we can obtain positive, zero, or negative average forces.

As discussed by (Maclay, 1999), the energy for a rectangular cavity is a homogeneous function of the dimensions: $en[\lambda a1, \lambda a2, \lambda a3] = \lambda^{-1} en[a1,a2,a3]$. With this information we can evaluate the average force for several examples. Assume a/L=0.816, then by numerical computation we have the final state en(L,L,0.408 L)=0, and for the initial state en(L,L,0.818L) = 0.1ℏc/L. For this geometry, the mean force is therefore:

$$<F(L,L,0.816L)> = - [0 - 0.1ℏc/L]/L = 0.1ℏc/L^2 . \qquad (6)$$

This is a attractive force, pulling the moveable plate inwards with an average force approximately equal in magnitude to the force on a cube of side L. If we start with a cavity of zero energy, namely en(L, L, 3.4L) =0, then we have a final state with two cavities each with positive energy $en(L,L,1.7L) = 0.072\hbar c/L$ (Maclay, 1999). This configuration yields a positive or outward mean force during insertion of the plate equal to $<F(L,L,3.4L)> = -0.144\hbar c/L^2$.

Rectangular structures with a moveable piston

Consider a rectangular conductive cavity *(L x L x a)* with a moveable piston that moves along the *a*-direction, dividing the cavity into two regions, each with its contribution to the total vacuum energy. We assume the piston is infinitely thin and normal to the *a*-direction (Fig. 1c). We can then numerically compute the total vacuum energy *Ep (L,L,x)* of the structure as a function of the distance x of the piston from one end of the cavity. From our definition of *en(a1,a2,a3)*, and the definition $\xi = x/L$, we have

$$Ep(L,L,x) = [en(1,1, a/L - \xi) + en(1,1,\xi)]/L \quad . \tag{7}$$

If we differentiate this with respect to x, we obtain an expression for the force *F(x)* due to the vacuum stresses on the moveable plate. Consider an example in which *a = 0.8L*, so $0 \le \xi \le 0.8$. Figure 2 shows the dimensionless energy and force respectively *LEp(L,L,x)* and $L^2F(x)$ as functions of *x*. For values of x near the center (x ≈ 8.0), the force on the piston is approximately directly proportional to x, and the energy is approximately a negative parabola with negative curvature. A small deflection from x=8.0 leads to a force causing an increased deflection. Thus Figure 2 shows the state of the system with the piston near the center is unstable: the piston would be pushed to the closest end of the cavity.

However, if we include the restoring force that arises from the small deflection of a deformable membrane as given by Hooke's Law, then this configuration might become stable if the material force constant exceeds that for the Casimir force. These results suggest the intriguing possibility of making a structure that displays simple harmonic motion for small displacements by employing two adjacent cubical cavities, with a common face that can be deflected.

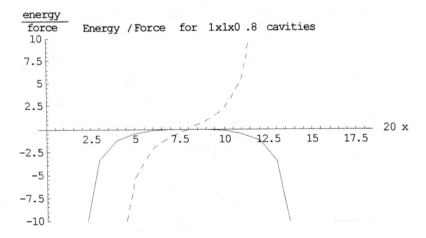

FIGURE 2. Plot of the vacuum energy (___solid line) and Casimir force (_ _ _ dashed line) for a 1 x 1 x .8 cavity that is divided into two rectangular cavities by a sliding piston that moves along the 0.8 direction. The maximum energy is at abscissa of 8 which corresponds to the center of the cavity, x = 0.4. For these calculations, we have set $\hbar c=1$ and set L equal to 1 unit. To obtain a numerical result, we use the MKS value of $\hbar c$. If we let L=0.5 micron, then the abscissa is in units of 0.025 micron, and the energy scale is in units of 6.3×10^{-20} joule and the force scale is in units of 1.26×10^{-13} joule. Forces of this magnitude are just measurable using AFM technology.

CONCLUSIONS

Some basic features of structures using parallel plates and rectangular cavities experiencing Casimir forces have been explored. Material properties need to be included in more detailed considerations, as well as corrections for finite temperatures. As the level of sophistication in MEMS fabrication increases, we may expect to see some of these structures realized. The utility of such devices is still unknown.

ACKNOWLEDGMENT

We would like to thank Carlos Villarreal, Robert L. Forward, Peter Milonni, Bryce DeWitt, Rod Clark, and Jay Hammer for helpful discussions, and NASA for funding the continuation of this effort.

REFERENCES

Ambjorn, J., and Wolfram, S., "Properties of the Vacuum, I. Mechanical and Thermodynamic," Annals of Physics **147**, 1-32 (1983).

Balian, R., and Duplantier, B., "Electromagnetic Waves Near Perfect Conductors. II. Casimir effect," Ann.Physics **112**, 165-208 (1978).

Bethe, H., "The Electromagnetic Shift of Energy Levels," Phys. Rev. **72**, 339-341(1948).

Boyer, T., Phys. Rev. **174**, 1764 (1968).

Brown, L., and Maclay, J. "Vacuum Stress Between Conducting Plates: an Image Solution," Phys. Rev. **184**, 1272-1279 (1969).

Casimir, H.B.G., Koninkl. Ned. Akad. Wetenschap. Proc. **51**, 793 (1948).

Deutsch, D., and Candelas, P., "Boundary Effects in Quantum Field Theory," Phy. Rev. D **20**, 3063-3080 (1979).

DeWitt, B., "The Casimir effect in Field Theory," from *Physics in the Making*, ed. By A Sarlemijn and M. Sparnaay, (Elsevier Science Publishers B.V., 1989).

Hacyan, S., Jauregui, R., Villarreal, C., "Spectrum of Quantum Electromagnetic Fluctuations in Rectangular Cavities," Phys.Rev A **47**, 4204-4211 (1993).

Lamoroux, S., " Measurement of the Casimir Force Between Conducting Plates," Physics Review Letters, **78**, 5-8 (1997).

Lifshitz, E.M., "The theory of Molecular Attractive Forces Between Solids," Soviet Physics JETP **2**, 73-83 (1956).

Lukosz, W., "Electromagnetic Zero-Point Energy and Radiation Pressure for a Rectangular Cavity," Physica **56**, 109-120 (1971).

Maclay, J., Serry, M., Ilic, R., Neuzil, P., Czaplewski, D., "Use of AFM to Measure Variations in Vacuum Energy Density and Vacuum Forces in Microfabricated Structures," pp. 247-256, Proceedings of the NASA Breakthrough Propulsion Workshop, NASA Lewis Research Center, Cleveland OH, August, 1998; Document No. NASA/CP-1999-208694.

Maclay, J., "Analysis of Zero-Point Electromagnetic Energy and Casimir Forces in Conducting Rectangular Cavities," submitted to Physical Review A, 1999.

Milonni, P., p. 239, *The Quantum Vacuum* (Academic Press, San Diego, CA, 1994).

Milton, K., DeRaad, L., and Schwinger, J., "Casimir Self-Stress on a Perfectly Conducting Spherical Shell," Annals of Physics (N.Y.) **115**, 388-403 (1978).

Mohideen, U., Anushree, Roy, "Precision Measurement of the Casimir Force from 0.1 to 0.9 micron", Physical Review Letters **81**, 4549 (1998).

Mostepanenko, V., Trunov, N., p. 40, *The Casimir Effect and Its Applications*, (Oxford Science Publications, Clarendon Press, Oxford, 1997).

Plunian, P., Muller, B., Greiner, W., "The Casimir Effect," Physics Reports (Review Section of Physics Letters) **34**, 2&3, 87-193 (1986).

Schwinger, J., "Casimir Effect in Dielectrics," Ann. Phys. **115**, 1-23 (1978).

Serry, M., Walliser, D., Maclay, J., "The Role of the Casimir Effect in the Static Deflection and Stiction of Membrane Strips in Microelectromechanical Systems (MEMS)," Journal of Applied Physics. **84**, 5, 2501-2506(1998).

Serry, M., Walliser, D., Maclay, J., "The Anharmonic Casimir Oscillator," IEEE-ASME Journal of Microelectromechanical Systems **4**, 193-205 (1995).

Sparnaay, M., "The Historical Background of the Casimir effect," from *Physics in the Making*, edited by A. Sarlemijn and M. Sparnaay (Elsevier Science Publishers B.V., 1989). See also D. Tabor and R.H.S. Winterton, "The Direct Measurement of Normal and Retarded van der Waals Forces", Proc. Royal Soc. **A312**, 435-450 (1969).

Welton, T., "Some Observable Effects of the Quantum Mechanical Fluctuations of the Electromagnetic Field," Phys. Rev. **74**, 1157-1167 (1948).

Relating Work, Change in Internal Energy, and Heat Radiated for Dispersion Force Situations

Daniel C. Cole

Dept. of Manufacturing Engineering, Boston University, Boston, MA 02215
Phone: (617) 353-0432, E-mail: dccole@bu.edu

This article describes how Casimir-like forces can be calculated for quasistatic situations of macroscopic bodies composed of different materials. The framework of stochastic electrodynamics (SED) is used for much of this discussion in an attempt to provide a very clear physical picture when considering quantities like forces, work done, changes in internal energy, and heat flow. By relating these quantities, one can readily understand why the different methods of calculating dispersion forces agree, such as when obtaining forces via changes in electromagnetic zero-point energy versus computing the average of the Maxwell stress tensor. In addition, a number of physical subtleties involving dispersion forces are discussed, that were certainly not recognized in early work on blackbody radiation, and that still may not be fully appreciated.

INTRODUCTION

The intent of the present article is to provide a physical perspective and explanation behind the often very complicated calculations encountered when evaluating van der Waals and Casimir forces. In the present section, motivation for this examination is provided. Briefly stated, for complex electromechanical situations such as in the atomic force microscope or sonoluminescence, it is easy to lose track of the relationships between work done, changes in stored field energy, and radiated energy; this article is aimed at helping to clarify these relationships.

The physical significance of electromagnetic zero-point (ZP) radiation has certainly been recognized for quite some time, not just in terms of its occurrence in quantum field theory, but in terms of verifiable experiments. During the late 1940s, the Lamb shift (energy shift in hydrogen between the $2s_{1/2}$ and $2p_{1/2}$ levels) and the Casimir force between neutral conducting plates, became recognized as key examples where electromagnetic ZP radiation has a definite physically measurable effect. Prior to that, most physicists seemed to feel that ZP energy had no real observable consequences and could simply be discarded via taking a different lowest energy level; specifically, the infinite energy term of $\sum_{\mathbf{k},\lambda} \frac{1}{2}\hbar\omega_{\mathbf{k},\lambda}$ for electromagnetic ZP radiation could simply be dropped. Actually, perhaps a better description of the prevailing attitude was that there was no obvious reason for retaining what was felt to be a rather ugly infinity occurring in the existing field theory, as physicists were not yet aware of physical effects caused by this term. Moreover, if the term was retained, there would be difficulty in reconciling it with appropriate sources for gravitation.

We now know that changes in position of macroscopic bodies and even atoms and molecules, can change the effective state of equilibrium that exists between matter and radiation. For cavities without matter inside, or with a few molecules introduced, as now studied extensively in cavity quantum electrodynamics (QED), the following procedure is often followed: namely, the change in normal modes of the standing waves of the radiation in the cavity is found and compared with the situation where the cavity is very large, and/or when no particles have been introduced. The apparent singularities attributed to vacuum energy fluctuations is now well recognized to be an important contributor to mass and charge renormalization in QED. Why these infinities cancel during renormalization is still not really understood at a fundamental level. However, following this renormalization procedure is well known to yield reliable and accurate predictions of experimental situations.

CP504, *Space Technology and Applications International Forum–2000*, edited by M. S. El-Genk
© 2000 American Institute of Physics 1-56396-919-X/00/$17.00

There is little question that the long-range Casimir and van der Waals forces are critically important in our everyday lives. Indeed, in a review article, Elizalde and Romeo wrote (Elizalde, 1991), "van der Waals forces play a very important role in biology and medical sciences. They are in general particularly significant in surface phenomena, such as adhesion, colloidal stability, and foam formation. One could dare say that they are the most fundamental physical forces controlling living beings and life processes." Indeed, these intermolecular forces are extremely important for gases and liquids. Properties such as surface tension, capillary action, solubility, viscosity, heat of evaporation, and the deviation from the ideal gas law, are largely explained by these forces.

There are three classes of van der Waals forces, namely: (1) an orientation force, involving molecules with permanent dipole moments, (2) an induction force, involving a dipole or quadrupole moment in one molecule, that induces a moment in a second molecule, and (3) a dispersion force, involving molecules/atoms without permanent moments, that induce moments in each other due to correlation effects in their fluctuating electromagnetic interactions. This last category is our main interest here. Generally, this force is quite weak, at least in comparison with the other interatomic interactions (*e.g.*, ionic and covalent binding). However, despite this weakness, in a sense, this interaction is fairly universal, in that two atoms will always induce correlated interactions with each other, at essentially all distances where the two systems can be considered separated. Why this universality is typically not recognized or considered, is that the interaction is usually completely masked by ionic and covalent interactions, so that van der Waals interactions often become quite secondary. Nevertheless, for neutral atomic and molecular systems, where ionic forces will not play a role, and where electrons in each system are all paired so that covalent bonds will not form, then the relatively weak van der Waals interaction can become the dominant interaction.

When combining systems of molecules and atoms into, for example, uncharged dielectric walls, the forces do not add vectorally, but combine in a much more complicated manner. Nevertheless, even though the forces are not vectorally additive, in general the more atomic systems, the larger will be the net force that the structure will exert on another structure separated in space from the first one. Forces between such macroscopic, geometry dependent structures are often referred to as Casimir forces, due to Casimir's prediction in 1948 of the attraction of two uncharged conducting parallel plates (Casimir, 1948). Thus, the forces can become reasonably large enough to be measurable. However, the measurements are far from trivial, as nicely described by Derjaguin (Derjaguin, 1960). Making surfaces sufficiently smooth, keeping them free from electrostatic charge and dust particles, and accurately monitoring distances in spite of vibrations, all make complications for experimentalists. For these reasons, and others, fairly definitive experimental confirmation between theoretical predictions and experiment for conducting plates didn't become available until Lamoreaux's work in 1997 (Lamoreaux, 1997). Sparnaay's work in 1958 certainly showed agreement, but was not sufficiently accurate to really pin down the specific power law dependence of the force upon the plate separation (Sparnaay, 1958).

Despite these complications, and despite the weakness of the force, there is still considerable interest in Casimir and van der Waals forces, perhaps for several reasons. Some of these reasons will be the motivation for this article. First, as mentioned, van der Waals forces are very important for biological systems, liquids, and gases, and some solids. Second, ingenious measurement techniques have been developed that rely heavily on van der Waals and Casimir forces, namely, the atomic force microscope, and its related family of measurement techniques. Also, new effects are being investigated in the area of cavity quantum electrodynamics (QED) that involve the interaction of particles with walls and radiation (Sukenik, 1993). Third, the fact that at temperature $T=0$, van der Waals and Casimir forces are naturally described in terms of interactions between atoms and molecules, mediated via the electromagnetic ZP radiation, continues to interest physicists. Careful attention needs to be paid to the singularities that arise in the calculations, to work out the small, but finite forces that result. How and why these infinities cancel out has interested physicists for some time, as has the related singularities that must be dealt with in the program of renormalization for mass and charge in QED. Fourth, many physicists have done extensive investigations into whether the relatively weak Casimir effect in atomic and molecular systems may become critically more important in fundamental particle physics. All quantum fields have vacuum fluctuations and zero-point energies. For quark and gluon fields, this Casimir-like effect has been thought to become much more significant, with the idea that it may play an important role in subatomic particle structure, such as with regard to the bag model of hadrons (Milton, 1983). Finally, scientists are interested in Casimir forces, again, because of the tight relation to ZP radiation, but also because ZP fields are becoming the subject of other areas of physics. In particular, recent observations in astronomy have strongly indicated that our universe is undergoing an accelerated rate of expansion, to the surprise of most scientists (Perlmutter, 1998a and 1998b). The vacuum may have a role in this phenomena (Rueda, 1999), although much work remains to be done to fully analyze this possibility. Other astrophysical phenomena may also owe their origin, at least in part, to effects from electromagnetic ZP radiation, such as cosmic rays and cosmic voids (Rueda, 1990a, 1990b, 1995; Cole, 1995).

Understanding the connection between Casimir forces and ZP radiation more deeply for atomic and molecular systems, has served, and will probably continue to serve, as a testing ground for the effects of ZP radiation in other physical areas. For that reason, many physicists have spent considerable effort reexamining the theoretical description of Casimir forces from many different angles, as nicely described in Milonni's text on the vacuum (Milonni, 1994). In particular, Milonni's book analyses the relation between vacuum and source fields, explaining how phenomena such as the Lamb shift, the Casimir and van der Waals forces, and even the fundamental linewidth of a laser, can be explained from both of these perspectives.

A rather amazing observation regarding Casimir forces in QED is that these forces are generally quite weak, but they originate from changes in infinite, or extremely large quantities. Indeed, in Ref. (Cole, 1999b), a discussion was given on whether there were restrictions on extracting energy from the ZP radiation. Two such constraining conditions were identified that largely applied to reversible thermodynamic operations (Cole, 1990b). For irreversible thermodynamic operations, restrictions for energy extraction are far less apparent. As described elsewhere in Refs. (Cole, 1993b, 1999a, and 1999b), energy contained within electromagnetic zero-point radiation has become of increasing interest to engineers, wondering whether this direction could lead to greater energy sources. Indeed, many of the normal irreversible thermodynamic energy extraction mechanisms that we are familiar with, such as chemical reactions involving batteries, lighting a match, chemical explosions, etc., may be attributed at least in part to electromagnetic ZP radiation. Undoubtedly this viewpoint seems surprising to most, as our normal description of such phenomena is not usually expressed in this manner. However, the vacuum field can be shown to be formally necessary for stability of atoms, as briefly discussed in (Cole, 1999a) and in much more detail in (Milonni, 1994). Without taking the fluctuating vacuum into account, radiation reaction will cause canonical commutators like $[x, p_x]$ to decay to zero. In this sense, the vacuum acts as, roughly stated, a "stabilizer" for physical systems. Changes in physical constraints, such as changes in mechanical, thermal, or chemical equilibrium conditions, temporarily changes stability conditions and enables irreversible processes to occur. The radiation of a metal surface with light (i.e., the photoelectric effect) or the chemical reactions that result from striking a match or closing a circuit switch containing a chemical battery, are examples of such changes in equilibrium conditions. Operations like these result in heat flow and work being done by or on the system.

Two key aspects that are important for making further progress in creating, manipulating, and experimenting with interesting and useful devices that utilize aspects of dispersion forces, are (1) the physical understanding of the phenomena, and (2) improved and versatile means of carrying out detailed calculations. Interestingly enough, both aspects are presently affecting each other considerably. For example, until the conducting sphere problem was fully carried out (Boyer, 1968), physicists did not know whether the net force from electromagnetic ZP radiation would be attractive or repulsive; most everyone undoubtedly assumed it would be attractive, as in the case of two parallel conducting plates. As the shapes of conducting shells are changed, such as from a rectangular parallel-piped, to a sphere, the signature and magnitude of the force can change completely. Our intuition, or what we feel is our understanding of physical situations, is often built up by doing repeated examples, such as for a sphere, an ellipsoid, a cylinder, a box, etc., and examining what happens as these shapes are manipulated. However, our intuition cannot be built up easily if the calculations and experiments are difficult to carry out, as they are for Casimir forces. Still, researchers continue to investigate further, since modification of shapes, materials, and placement, including dynamic variations (e.g., as in the fluctuating lever of the atomic force microscope), promise to yield very interesting insights and applications.

CASIMIR FORCE CALCULATIONS

An illuminating place to begin with regard to Casimir force calculations is simply the conventional uncharged, conducting parallel plate situation (Casimir, 1948). The calculations have been carried out in several different ways. Figure 1 illustrates the situation often treated [see, for example, p. 480 in (Boyer, 1970) or p. 167 in (de la Peña, 1996)]. The force existing between the two plates separated by a distance d is what is desired; the distance R is taken to be quite large, as is the area $L \times L$ of each plate. Let $E\left(L_x, L_y, L_z\right)$ be the average electromagnetic ZP energy in a box of size $L_x \times L_y \times L_z$, surrounded by perfectly conducting walls. The standard calculation often performed is then

$$U = E(R - d, L, L) + E(d, L, L) - E(R, L, L) \ , \tag{1}$$

so that the difference is found between the sum of the two energies in both regions, minus the energy in a single region where the center plate is not present. What is interesting here, is that each of the three terms above is infinite, but the combination of the three yield a finite, measurable result. Each term in Eq. (1), in SED, is equal to the

ensemble average of the electromagnetic ZP energy, $\left\langle \int_V d^3x \left(\mathbf{E}^2 + \mathbf{B}^2 \right) \right\rangle$, taken over the respective volume. This

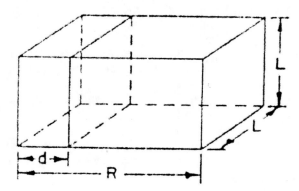

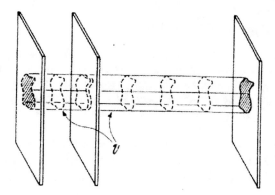

FIGURE 1. Three plates, separated by distances d and (R-d).

FIGURE 2. A volume V cutting through the three plates in Fig. 1.

quantity turns out to equal the sum over all electromagnetic modes, of the average energy existing per mode: $\sum_{k,\lambda} \frac{1}{2} \hbar \omega_{k,\lambda}$. Alternatively to Eq. (1), and perhaps more expediently, one can simply calculate:

$$\frac{\Delta U}{\Delta d} = \frac{\Delta}{\Delta d}\left[E(R - d, L, L) + E(d, L, L) \right] \ . \tag{2}$$

This quantity is taken to be equal to the negative of the average force that would be measured as existing between the plates. Each of the two terms on the right side of Eq. (2) is infinite, but, again, the singularities cancel, leaving a finite result.

Other related ways for finding the force are via calculating the radiation pressure on the plates (Milonni, 1988) and via the closely related way of making use of the Maxwell stress tensor (Lifshitz, 1956; Brown, 1969). Each of these methods encounter infinities, but the singularities cancel out during the calculations.

Here we will relate these methods via making use of the classical electrodynamic theory often called stochastic electrodynamics (SED) (de la Peña, 1996). Analyzing Casimir force situations via the use of SED has the advantage of providing a more intuitive and physical description for many engineers and scientists. SED describes nature via Maxwell's equation, with the relativistic version of Newton's second law of motion for the trajectory of particles. The key difference of this theory versus conventional classical electrodynamics is that the restriction is not made that radiation must vanish at a temperature T=0. The relaxation of this subtle restriction enables the deduction to be made that classical electromagnetic ZP radiation must be present to satisfy a number of thermodynamic conditions (Cole, 1990a and 1990b). In turn, the inclusion of this property enables a more accurate description of our physical world, while still retaining a classical view.

However, a cautionary comment should of course be made here: research *to date* has concluded that agreement between QED, which scientists clearly trust, and SED, only holds for the situation involving linear equations of motion for the modeled atomic systems and linear equations governing the electromagnetic fields for the modeled macroscopic systems (Boyer, 1975; Cole, 1993a; de la Peña, 1996). Thus, if one treats the macroscopic materials as being composed of a set of electric dipole oscillators, or as macroscopic bodies with linear dielectric properties, then we can have confidence in the agreement. Of course, this situation is quite limiting in many critically important ways, but what is interesting is that even in QED, handling the general situation of nonhomogeneous, dispersive, and absorbing material has, to my knowledge, not been successfully tackled yet. The problem is far from trivial, both from the physical description point of view (*i.e.*, appropriate quantization method for this complicated many-particle system, and relation to measurable physical quantities), as well as the mathematical complication of solving the resulting governing equations. Much of the research in this area still relies on solving the problem in QED by precisely the model where QED and SED agree, namely, where atomic systems are treated as being composed of simple harmonic electric dipole oscillators. Shortly we will discuss why this situation exists, but for now, with this reassurance that the agreement between the two theories of SED and QED lies precisely in the area that to date has been quite important, provides incentive for much of the following discussion.

Now, to be precise, we really need to follow a procedure more along the lines of (Cole, 1993a), where renormalization in the classical equations of motion in the charged particles are considered. The expressions, as presented there, become quite lengthy and fairly involved. For the present purposes, where we are trying to reveal more of the physical aspects of a procedure such as Eq. (2), let us proceed much more roughly, without worrying about mass renormalization of charges.

In the subsequent discussion, Maxwell's microscopic equations will be assumed, so that $\mathbf{E}$ and $\mathbf{B}$ represent the microscopic electric and magnetic fields, respectively, while $\rho(\mathbf{x},t)$ and $\mathbf{J}(\mathbf{x},t)$ are the microscopic charge and current densities. Following treatments such as in (Jackson, 1998), one obtains:

$$\int_V d^3x \left[\rho(\mathbf{x},t)\mathbf{E}(\mathbf{x},t) + \frac{1}{c}\mathbf{J}(\mathbf{x},t) \times \mathbf{B}(\mathbf{x},t) \right]_i = \oint_S d^2x \sum_{j=1}^{3} n_j T_{ij}(\mathbf{x},t) - \frac{1}{c^2}\frac{d}{dt}\int_V d^3x S_i(\mathbf{x},t) , \qquad (3)$$

where
$$T_{ij} = \frac{1}{4\pi}\left[E_i E_j + B_i B_j - \delta_{ij}\frac{1}{2}(\mathbf{E}\cdot\mathbf{E} + \mathbf{B}\cdot\mathbf{B}) \right] \qquad (4)$$

is the Maxwell stress tensor, and
$$\mathbf{S}(\mathbf{x},t) = \frac{c}{4\pi}(\mathbf{E}\times\mathbf{B}) \qquad (5)$$

is the Poynting vector. Now, in SED, each of the quantities in these equations are very rapidly varying in time. The average of them, either a time average, or an ensemble average, correspond to quantities most accessible to experimental measurement. For perfectly conducting plates, there would be rapidly fluctuating currents on the surface of the plates, that would vary in just such a way as to make the fields equal to zero within the plates. If V in Eq. (3) is taken to be the region illustrated in Fig. 2, then the ensemble average of Eq. (3) yields, for the left side, the average Lorentz force acting on the region of the center plate within the volume V.

As for the right side, if the plates are being held fixed, then
$$\frac{d}{dt}\int_V d^3x \left\langle S_i(\mathbf{x},t) \right\rangle = 0 , \qquad (6)$$

since $\left\langle S_i(\mathbf{x},t) \right\rangle$ must be independent of time. Thus, we obtain a direct relationship between the Lorentz force and the average of the surface integral over the Maxwell stress tensor in Eq (3).

To relate this result to Eq. (2), we can proceed in the following way. Again from Maxwell's equations, one can derive the microscopic formulation of Poynting's theorem:

$$\int_V d^3x \mathbf{J}(\mathbf{x},t)\cdot\mathbf{E}(\mathbf{x},t) = -\frac{\partial}{\partial t}U(\mathbf{x},t) - \int_S d^2x \hat{\mathbf{n}}\cdot\mathbf{S}(\mathbf{x},t) , \qquad (7)$$

where
$$U(\mathbf{x},t) = \frac{1}{8\pi}\int_V d^3x(\mathbf{E}\cdot\mathbf{E} + \mathbf{B}\cdot\mathbf{B}) , \qquad (8)$$

and S is the surface of V, and $\hat{\mathbf{n}}$ is the outward surface normal. The ensemble average of the quantity in Eq. (8) is tightly connected with the quantities in Eqs. (1) and (2). In particular, if we quasistatically displace the center plate in Fig. 2, so that at each instant of time, the system can be treated as being in thermodynamic equilibrium, and if we integrate the ensemble average of Eq. (7) in time over the period of displacement, then:

$$\left\langle U(\mathbf{x}, t_{\mathrm{II}})\right\rangle - \left\langle U(\mathbf{x}, t_{\mathrm{I}})\right\rangle = -\int_{t_{\mathrm{I}}}^{t_{\mathrm{II}}} dt \int_V d^3x \left\langle \mathbf{J}(\mathbf{x},t) \cdot \mathbf{E}(\mathbf{x},t)\right\rangle - \int_{t_{\mathrm{I}}}^{t_{\mathrm{II}}} dt \int_S d^2x \, \hat{\mathbf{n}} \cdot \left\langle \mathbf{S}(\mathbf{x},t)\right\rangle \quad . \tag{9}$$

In this way, the left side becomes equal to the quantity in brackets on the right side of Eq. (2). The first term on the right becomes equal to the work done by the electromagnetic fields of the system, acting on the charges in the center plate, as the plate is displaced. This quantity is approximately equal to the negative of the work done by external forces while quasistatically displacing the plate. The last term of $\int_{t_{\mathrm{I}}}^{t_{\mathrm{II}}} dt \int_S d^2x \, \hat{\mathbf{n}} \cdot \left\langle \mathbf{S}(\mathbf{x},t)\right\rangle$ equals the ensemble average of the electromagnetic energy that flows *out* of the volume V.

Thus, Eq (9) *roughly* becomes equal to the first law of thermodynamics,

$$\Delta U_{\mathrm{internal}} = W + Q \quad , \tag{10}$$

where $\Delta U_{\mathrm{internal}}$ is the change in the internal energy of this electromagnetic system enclosed within the volume V, W is the work done *on* the system while moving the plates, and Q is the heat flow *into* the volume V. The reason for the qualification of "roughly," is that Eq. (9) has not taken into account all of the internal energy of the system, nor accounted for mass renormalization in the case of point charges. These specific steps are taken into account in (Cole, 1993a), but, as mentioned, certainly complicate the description. Nevertheless, it is important to at least point out that a full description does require these aspects, and that if one is to compare these terms to experimentally measured quantities, then the correct understanding of these terms needs to be addressed. In particular, the internal energy of a system also needs to include the internal kinetic energy due to charge motion, as would take place in the plate to ensure the good conductivity of the plate. More specifically, our rough description above ignored the specific heat of the section of the conducting plate contained in Fig. 2. In the same vein, some of the work done on the plate could end up as work being done on internal vibratory motion, and would be contained in the first terms on the right of Eqs. (9) and (10). These aspects are certainly included in Refs. (Cole, 1993a, 1990a, and 1990b). Brown (1969) contains a related discussion on the first law of thermodynamics, as applied to the present problem.

Ignoring these complications, the left side of Eqs. (9) and (10) becomes equal to ΔU in Eq. (2), for a displacement Δd. The first term on the right of these equations becomes roughly equal to the average force acting on the plate, as obtained in the discussion involving Eq. (3) and the Maxwell stress tensor, times the displacement of the plates.

As will be noted, in order for this change in internal energy to equal the work done on the system, requires that the last term on the right side of Eqs. (9) and (10), namely, Q, must equal zero. If the temperature of the system is held fixed during this quasistatic operation, so that T neither increases nor decreases, then the condition that $Q=0$ for any displacement is only satisfied at a temperature of $T=0$. This was the condition that was imposed in Refs. (Cole, 1990a, 1992a, and 1992b) to provide derivations for the classical electromagnetic ZP radiation spectrum.

Now, what is also useful about Eq. (10), is that it actually applies for any geometry, rather than being applicable only for the parallel plate situation. This rough description enables us to understand different means for calculating forces existing on, for example, a spherical conducting shell, as discussed in (Boyer, 1970), or rectangular parallel-piped shapes, as discussed in (Ambjørn, 1983). One could carry this calculation out at $T=0$ using either a change in internal energy calculation, where the sphere or cube is made to grow or shrink slightly [left side of Eq. (9)], or via the average of the Maxwell stress tensor integrated over a volume enclosing the structure. Moreover, the equation can be applied to an unusual geometrical structure, such as a conducting shell with irregular and unsymmetrical shaped walls, as opposed to the highly symmetrical case of the parallel plates, or the spherical shell. Also, an odd shaped cavity in a block of material might be considered. A slight displacement of any small area of the wall, such as might be either imagined (thought experiment), or possibly experimentally carried out by applying external stress forces in that area [perhaps originating from a piezioelectrical apparatus, as might possibly be constructed using micro-electromechanical structures (MEMS)], enables Eq. (10) to be applied to determine the force required to make the quasistatic, infinitesimal displacement. The change in internal energy due to an infinitesimal quasistatic displacement of a section of a wall, must equal the work done during the displacement at T=0, via the definition of the T=0 state (Cole, 1990a).. In this way, the component of the force acting in the direction of the displacement can be obtained. This method should agree with the equally valid method using the evaluation of the surface integral of the average of the Maxwell stress tensor over the region of the wall displaced in the previous hypothetical situation.

Regarding difficulties of calculations, however, much of what was said above certainly drastically oversimplifies the required steps. The symmetrical cases of two parallel plates, a spherical conducting shell, or conducting rectangular parallel-pipeds, are all treatable by nearly fully analytical methods. However, even these are far from trivial, as evidenced by the time delay between 1953, when Casimir published the suggestion of computing the spherical conducting shell situation (Casimir, 1953), and 1968, when Boyer solved the proposed problem (Boyer, 1968). Treatment of unusual geometries would be even more difficult. However, certainly the same basic methods should

apply. In particular, it is interesting to note that the ensemble average of Eq. (8) will reduce to a sum, $\sum_i \frac{1}{2}\hbar\omega_i$, for

the electromagnetic energy in an unusual shaped conducting cavity, where the sum is over all the electromagnetic modes, which are in turn dependent on the basis functions used to describe the fields. Section 3.1.2 in Ref. (de la Peña, 1996) provides background here. This observation does not necessarily simplify the problem, as one still needs to find all the modes, but at least it places the problem into a more familiar framework. Indeed, once the modes are found, then perturbation methods can be applied to calculate changes in energy and applied forces. For example, in Ref. (Casimir, 1951), a perturbation technique is described for finding the changes in the modes associated with a resonant cavity containing electromagnetic energy, when one of the walls of the cavity is dented or a particle is introduced into the cavity.

This general discussion now brings us to the consideration of the more general problem of treating nonhomogeneous, dispersive, and absorptive dielectric materials. Lifshitz first treated the case of two semi-infinite dielectric walls, separated by a small distance, using a macroscopic electromagnetic treatment (Lifshitz, 1956). The results there have been generally found to be valid, as compared with more fundamental treatments. However, the extension of these methods to more general situations in QED of arbitrary geometry and nonhomogeneous conditions, has been quite problematic. Kupiszewska summarizes much of the issues here in Ref. (Kupiszewska, 1992). The issues largely revolve around the appropriate conditions to apply for quantization rules, when the medium may be nonlinear, nonhomogeneous, dispersive, and dissipative. Consequently, a number of articles have appeared in the QED literature that model the medium by treating it as being composed of electric dipole harmonic oscillators. These oscillators model the actual atoms existing in real materials. By placing them in varying density arrangements, and of course coupling all their interactions appropriately, one can treat the case of a nonhomogeneous material, as well as one that is dispersive and dissipative. Renne made the connection (Renne, 1971b) between this more microscopic approach of treating a medium to the more macroscopic electrodynamic treatment by Lifshitz .

Fortunately, Renne's basic approach (Renne, 1971a) involving full retarded van der Waals forces between oscillators, was shown to yield identical results to those of SED by Boyer (Boyer, 1973) for the case of two oscillators, and later for an arbitrary number of oscillators (Cole, 1986). Indeed, Kupiszewska's treatment of this problem in QED can be cast entirely within the description of SED, as described in Sec. 6.1 of Ref. (de la Peña, 1996). Finally, it should be mentioned that provided one composes a macroscopic body by an arbitrary number of electric dipole oscillators, then quasistatic displacements of any of them will yield the general ideas described here, namely, no net radiation will flow out of a large volume containing these oscillators during the quasistatic displacement operation, provided T=0 (Cole, 1990a).

CONCLUDING REMARKS

This article traced some of the physical issues involving Casimir forces between macroscopic materials. The physical description of SED was used in much of this discussion, because the language is familiar to most engineers and scientists. Moreover, the model that is presently used to treat nonhomogeneous, dispersive, dissipative materials is precisely the area that QED and SED have agreement, thereby providing additional incentive for this treatment. The discussion in this article was kept rather general, in order to emphasize some of the physical points; however, references were also provided that contain more detailed description.

A number of physical issues were not discussed here that will now be briefly mentioned. Reference (Cole, 1992a) analyzed thermodynamic issues associated with operations performed on blackbody radiation, which includes the case of radiation at a temperature T, within a cavity, with moveable walls. There, a generalized Wien's displacement law was deduced. An early analysis by Planck in 1913, as republished in (Planck, 1988), contains a very deep physical analysis of the thermodynamic aspects of radiation, that is still quite illuminating today. Indeed, with the emphasis now being on cavity QED, some of these arguments are well worth reexamining. For example, one generalization necessary in Planck's work, is that the electromagnetic energy density varies as a function of position within a cavity, and is highly dependent on the shape and size of the cavity. This is well appreciated now in Casimir related problems. A second key point in Planck's analysis was the use of hypothetical "stopcocks," "carbon particles," and rough wall surfaces to scatter radiation; the latter two were used in Planck's analysis to "thermalize" the radiation after thermodynamic operations. In cavity QED, where highly nonthermal states of radiation can be maintained for long periods of time, such as in cavities with very smooth, reflective walls, then deeper analyses needs to be considered for these nonequilibrium situations. Third, it is very interesting that one of the arguments made by Planck and others was that it should not matter what the piston or cylindrical walls were composed of in the Wien's displacement operation; they assumed the force acting on the piston should be the same [see Sec. 66 in Ref. (Planck, 1988)]. We now know that is not the case, as emphasized by Casimir forces being different for parallel plates composed of different materials. Of course, the force on one plate is equal and opposite to the force on the other plate, in order to keep them separated, but if one of the plates is replaced by a plate of another material, the force magnitude on each plate will change. This recognition was missed in the early blackbody radiation work. Of course, it is easy to understand why, since the cylinders being considered were taken to be quite large, so that this material dependent force contribution would then be negligible. For cavity QED situations, however, such factors can be essential.

ACKNOWLEDGMENTS

This work was supported in part by the California Institute of Physics and Astronomy (CIPA).

REFERENCES

Ambjørn, J. and Wolfram, S., "Properties of the Vacuum. I. Mechanical and Thermodynamic," Annals of Phys. **147**, 1-32 (1983).

Boyer, T. H., "Quantum Electromagnetic Zero-Point Energy of a Conducting Spherical Shell and the Casimir Model for a Charged Particle," Phys. Rev. **174**, 1764-1776 (1968).

Boyer, T. H., "Quantum Zero-Point Energy and Long-Range Forces," Annals of Phys. **56**, 474-503 (1970).

Boyer, T. H., "Retarded van der Waals Forces at All Distances Derived from Classical Electrodynamics with Classical Electromagnetic Zero-Point Radiation," Phys. Rev. A **7**, 1832-1840 (1973).

Boyer, T. H., "General Connection between Random Electrodynamics and Quantum Electrodynamics for Free Electromagnetic Fields and for Dipole Oscillator Systems," Phys. Rev. D **11**, 809-830 (1975).

Brown, L. S. and Maclay, G. J., "Vacuum Stress between Conducting Plates: An Image Solution," Phys. Rev. **184**, 1272-1279 (1969).

Casimir, H. B. G., "On the Attraction Between Two Perfectly Conducting Plates," Koninkl. Ned. Adak. Wetenschap. Proc. **51**, 793-795 (1948).

Casimir, H. B. G., "The Theory of Electromagnetic Waves in Resonant Cavities," Philips Res. Rep. **6**, 162-182 (1951).

Casimir, H. B. G., "Introductory Remarks on Quantum Electrodynamics," Physica **19**, 846-848 (1953).

Cole, D. C., "Correlation Functions for Homogeneous, Isotropic Random Classical Electromagnetic Radiation and the Electromagnetic Fields of a Fluctuating Classical Electric Dipole," Phys. Rev. D **33**, 2903-2915 (1986).

Cole, D. C., "Derivation of the Classical Electromagnetic Zero-Point Radiation Spectrum via a Classical Thermodynamic Operation Involving van der Waals Forces," Phys. Rev. A **42**, 1847-1862 (1990a).

Cole, D. C., "Entropy and Other Thermodynamic Properties of Classical Electromagnetic Thermal Radiation," Phys. Rev. A **42**, 7006-7024 (1990b).

Cole, D. C., "Reinvestigation of the Thermodynamics of Blackbody Radiation via Classical Physics," Phys. Rev. A **45**, 8471-8489 (1992a).

Cole, D. C., "Connection of the Classical Electromagnetic Zero-Point Radiation Spectrum to Quantum Mechanics for Dipole Harmonic Oscillators," Phys. Rev. A **45**, 8953-8956 (1992b).

Cole, D. C., "Reviewing and Extending Some Recent Work on Stochastic Electrodynamics," in *Essays on Formal Aspects of Electromagnetic Theory*, edited by A. Lakhtakia, Singapore, World Scientific, (1993a) pp. 501-532.

Cole, D. C. and Puthoff, "Extracting Energy and Heat from the Vacuum," Phys. Rev. E **48**, 1562-1565 (1993b).

Cole, D. C., "Possible Thermodynamic Law Violations and Astrophysical Issues for Secular Acceleration of Electrodynamic Particles in the Vacuum," Phys. Rev. E **51**, 1663-1674 (1995).

Cole, D. C., "Calculations on Electromagnetic Zero-Point Contributions to Mass and Perspectives," in Proceedings of NASA conference, *"Breakthrough Propulsion Physics Program,"* Cleveland, Ohio, Aug. 12-14, 1997, NASA/CP-1999-208694, January (1999a), pp. 72-82.

Cole, D. C., "Energy and Thermodynamic Considerations Involving Electromagnetic Zero-Point Radiation," in *Space Technology and Applications International Forum – 1999*, edited by M. S. El-Genk, AIP Conference Proceedings 458, American Institute of Physics, New York, (1999b) pp. 960-967.

de la Peña, L. and Cetto, A. M., *The Quantum Dice. An Introduction to Stochastic Electrodynamics*, Boston, Kluwer, 1996.

Derjaguin, B. V., "The Force Between Molecules," Scientific American **203** (203), 47-53 (1960).

Elizalde, E. and Romeo, A., "Essentials of the Casimir Effect and Its Computation," Am. J. Phys. **59** (8), 711-719 (1991).

Jackson, J. D., *Classical Electrodynamics*, 3rd ed., New York, Wiley, 1998.

Kupiszewska, D., "Casimir Effect in Absorbing Media," Phys. Rev. A **46**, 2286-2294 (1992).

Lamoreaux, S. K., "Demonstration of the Casimir Force in the 0.6 to 6.0 μm Range," Phys. Rev. Lett. **78** (1), 5-8 (1997).

Lifshitz, E. M., "The Theory of Molecular Attractive Forces between Solids," Soviet Phys. – JEPT **2**, 73-83 (1956), English translation of Zh. Eksperim. i. Teor. Fiz. **29**, 94 (1955).

Milton, K. A., "Fermionic Casimir Stress on a Spherical Bag," Annals of Phys. **150**, 432-438 (1983).

Milonni, P. W., "Radiation Pressure from the Vacuum: Physical Interpretation of the Casimir Force," Phys. Rev. A **38**, 1621-1623 (1988).

Milonni, P. W., *The Quantum Vacuum. An Introduction to Quantum Electrodynamics*, San Diego, Academic Press, 1994.

Planck, M., *The Theory of Heat Radiation*, Vol. 11 of the series, *The History of Modern Physics, 1800-1950*, New York, AIP, 1988.

Perlmutter, S. et al., "Measurements of Ω and λ from 42 High-Redshift Supernovae," LBNL, preprint 41801 (1998a). For all relevant detailed updated information, see: www-supernova.gov .

Perlmutter, S., Aldering, G. et al., "Discovery of a Supernova Explosion at Half the Age of the Universe," Nature **391**, 51-54 (1998b). Also see references therein.

Renne, M. J., "Retarded van der Waals Interaction in a System of Harmonic Oscillators," Physics **53**, 193-209 (1971a).

Renne, M. J., "Microscopic Theory of Retarded van der Waals Forces Between Macroscopic Dielectric Bodies," Physics **56**, 125-137 (1971b).

Rueda, A., "Survey and Examination of an Electromagnetic Vacuum Accelerating Effect and its Astrophysical Consequences," Space Science Reviews **53**, 223-345 (1990a), and references therein.

Rueda, A., "Electromagnetic Vacuum and Intercluster Voids: Zero-Point-Field-Induced Density Instability at Ultra-Low Densities," Phys. Lett. A **147**, 423-426 (1990b).

Rueda, A., Haisch, B., and Cole, D. C., "Vacuum Zero-Point Field Pressure Instability in Astrophysical Plasmas and the Formation of Cosmic Voids," Astrophys. J. **445**, 7-16 (1995).

Rueda, A., Sunahata, H., and Haisch, B., "Electromagnetic Zero Point Field as Active Energy Source in the Intergalactic Medium," in *35th AIAA/ASME/SAE/ASEE Joint Propulsion Conference and Exhibit 20-24 June 1999, Los Angeles,* paper 99-2145 (1999).

Sparnaay, M. J., "Measurement of Attractive Forces Between Flat Plates," Physica **24**, 751-764 (1958).

Sukenik, C. I., Boshier, M. G., Cho, D., Sandoghdar, V., and Hinds, E. A., "Measurement of the Casimir-Polder Force," Phys. Rev. Lett. **70** (5), 560-563 (1993).

Simplistic Propulsion Analysis of a Breakthrough Space Drive for Voyager

Malcolm D.K. Boston

Tennessee State University, College of Engineering & Technology, 3500 John A. Merritt Blvd. Nashville, TN 37209-1561

(615) 963-5401, Fax: (615) 963-5397, bostonm@harpo.tnstate.edu

Abstract. When considering exploration beyond our solar system, speed is a critical factor. With the speeds achievable with current propulsion technology, interstellar distances cannot be traversed within a human life span. For example, the Voyager spacecraft would take approximately 80,000 years to traverse 4.3 light-years - the distance to our nearest neighboring star. In 1996 NASA established the Breakthrough Propulsion Physics program to search for further advances in physics to circumvent these limitations. One of the goals of this program is to discover a new method of propulsion that eliminates the need for propellant. A simplistic analysis is offered in this paper to assess the trip-time benefit of this single goal, using the Voyager spacecraft as a basis. The existing propulsion performance of the Voyager spacecraft is compared to its performance assuming it was equipped with a breakthrough physics space drive that could convert energy directly into kinetic energy. Given that the physics does not yet exist for such a space drive, these comparisons are at the most rudimentary level, based on energy comparisons. Specifically, the velocity and distance covered by the Voyager spacecraft are compared to that achievable by a hypothetical space drive using the same energy available to the existing Voyager. The additional benefit of having the Voyager's Radioisotope Thermoelectric Generators (RTGs) supply propulsion power is also considered.

INTRODUCTION

This paper examines the potential benefits of a breakthrough in propulsion. The performance of a hypothetical new device (BPP Device) will be measured against current technology for a deep space probe.

Currently, a one way trip to the moon would take approximately three days; a one way trip to the heliopause would take approximately 32 years (JPL, 1999), and, at current speed, it would take approximately eighty *thousand* years for the voyager space probes to reach our nearest star, Proxima Centauri. These speed limitations restrict the bounds of space exploration. In order to make space voyages to other planets both within our solar system and beyond, speeds will have to be increased.

NASA established the Breakthrough Propulsion Physics program in 1996 to research new physics that could eliminate or significantly reduce the limitations on interstellar travel (Millis,1999). Of the three goals of the program, listed below, only the first is addressed in this paper.

(1) Vehicle propulsion without propellant mass
(2) Transit at maximum physical speeds possible
(3) New methods of energy production to power such devices.

This paper examines the benefit if goal one is achieved. Specifically it assumes that a breakthrough propulsion device (BPP device) exists that can convert stored energy directly into kinetic energy of the spacecraft. The feasibility of such a BPP device is not discussed in this paper. The possibilities and critical issues for such a device have been introduced and discussed by others (Millis,1997).

CP504, *Space Technology and Applications International Forum–2000*, edited by M. S. El-Genk
© 2000 American Institute of Physics 1-56396-919-X/00/$17.00

Since the physics does not exist for such a propulsion device, the performance comparisons offered here are quite rudimentary. The calculations are simply based on energy and how energy affects the final velocity of the spacecraft, with and without a BPP device. Specifically, the following calculations are made:

(1) The burnout velocity for the Centaur upper stage, carrying Voyager, is compared to the velocity achievable with a BPP device that assumes the same energy level and empty mass as the Centaur / Voyager.
(2) An additional velocity boost is calculated assuming that the energy from the Voyager's RTG's is applied to a BPP device on Voyager for 50 years.
(3) The reduced trip-time benefits from the higher velocities achievable from the cases above are calculated considering a mission to our nearest neighboring star, 4.3 light-years distant.
(4) In addition to the velocity and trip-time comparisons, the energy to propel a Voyager-sized spacecraft to our nearest neighboring star, with an arbitrary one-way mission time of 50 years, is calculated.

EXPLANATION OF ASSUMPTIONS, PARAMETERS, AND OTHER CONSTRAINTS

The Voyager probes are used in this comparison because they are a real and reasonable example of a deep space probe mission. The two probes were launched over twenty years ago and are only now approaching the edge of the solar system (JPL, 1999). The craft were subjected to multiple methods of acceleration to get them up to their current velocity of approximately 17 km/s (38,000 mph) including using the thrust from a Centaur upper stage and a gravity-assist maneuver around Jupiter. Although both probes used a gravity assist maneuver to attain their final cruising speed, the gravity assist could be equally available to a future BPP device and thus is not considered in these comparisons. Only the Centaur's performance is used as the baseline for comparison.

The calculations assume the following mission profile: A probe has been boosted to its starting point by whatever convenient apparatus exists. (i.e. in 1977, a Titan multi-stage rocket launched Voyager) Next, some form of upper stage is used to propel the vehicle toward its destination. Please note that the numbers generated by these calculations represent an isolated case, based on the Voyager mission. The results could vary greatly based on a larger payload mass, more available energy in the form of rocket fuel, etc. The following data are assumed for this mission:

TABLE 1. Data used in the Calculations

Parameter	Value
Mass of Voyager Spacecraft	722 kg
Empty Mass of Voyager plus Centaur Upper Stage (m_e)	3353 kg
Gross Mass of Voyager plus Centaur Upper Stage (m_g)	16960 kg
I_{sp} of Centaur Upper Stage	444 s
Acceleration due to gravity (Earth)	9.8 m/s^2
Thrust from Centaur Upper Stage (T_{vac})	131,134 N
Burn Time for Centaur Upper Stage (t_{burn})	454 s
RTG power available	420 W at t = 0 yrs, and 322 W at t = 22 yrs
Distance to nearest Star (Proxima Centauri)	4.3 ly = 4.1E+16 m

CENTAUR / VOYAGER PERFORMANCE

Equation 1 is used to determine the burnout velocity of the probe propelled by *a Centaur upper stage*. (Berman, 1961):

$$V_{bo} = I_{sp}*g*\ln(m_g/m_e). \tag{1}$$

For the Centaur, this results in a burnout velocity of 7.1 km/s. At this speed it would take 182,893 years to travel 4.3 light-years. Equation 2 is used to calculate the energy of a rocket which will then be used in a later calculation. (Berman, 1961)

$$E_r = 1/2*T_{vac}*I_{sp}*g*1/eff*t_{burn},$$ (2)

This results in a rocket energy (Centaur) of 1.3E+11 J.

UPPER STAGE BPP DEVICE PERFORMANCE

It is now assumed that a BPP Device exists that is capable of converting some form of stored energy directly into kinetic energy of motion with 100% efficiency. It is assumed that the energy of the Centaur, as calculated previously is converted, by our BPP Device, directly into kinetic energy. It is further assumed that this BPP device and the stored energy requires no propellant mass, and hence, is assumed to have the same mass as the empty mass of Voyager plus the Centaur Upper Stage. From this we can calculate a new final velocity for the BPP Device upper stage as follows. This is done by simply solving for velocity (eq 3) from the kinetic energy equation.

$$V = \sqrt{\frac{2KE}{m}},$$ (3)

This results in a final velocity of 8.8 km/s and a 4.3 light-year trip-time of 148,000 years.

RTG BOOST BPP DEVICE PERFORMANCE

In addition to the upper stage comparison, the possibility of using Voyager's RTG's to power a BPP Device is considered. For this analysis, it is assumed that a second BPP Device (BPP$_2$) exists on the Voyager that converts the RTG energy directly into kinetic energy with 100% efficiency. It is further assumed that the mass of this BPP Device is included in the total mass of Voyager. (BPP$_1$ is jettisoned just as a spent Centaur would be). The ultimate final velocity of the probe is determined after factoring in the burnout velocity plus the acceleration over a fifty-year period where the RTG's are supplying BPP$_2$ with power. BPP$_2$ will draw on power that might otherwise be wasted as electrical systems and instruments are shut down during the transit phase of the mission.

For this calculation, the power available from the RTG's is assumed to fall off exponentially in time, and the specific quantity of power is calculated for the two data points listed previously. Specifically, equation 4 is used to calculate the energy available, where the constant , k , was calculated from the two data points shown earlier using equation 5.

$$E = \int_{t_o}^{t_{final}} Pe^{-kt} dt,$$ (4)

$$k = -\ln(\frac{P(t = 22\, years)}{P(t = 0)})/22\, years.$$ (5)

Assuming now that this energy was available to a BPP$_2$ device operating just on the Voyage probe mass of 722 kg, the velocity boost would be 37 km/s. This calculation uses the kinetic energy equation (eq 3) as used previously and results in a 4.3 light-year trip-time of 28,118 years.

INTERSTELLAR VOYAGER ENERGY REQUIREMENTS

To reflect the difficulty of "practical" interstellar exploration, even given a space drive, the total energy necessary to travel 4.3 light-years in 50 years with constant acceleration is calculated. Equation 6, the kinematic equation for a particle in one dimensional motion with constant acceleration, is used to calculate the required acceleration. Then, Newton's Second Law (Eq 7), and the energy equation of work, (Eq 8) are used to calculate the required energy.

$$X = X_o + V_o t + 1/2at^2 \,, \tag{6}$$

$$F = ma, \tag{7}$$

$$W = F * dX. \tag{8}$$

From this, the energy to propel a 722 kg probe across 4.3 light-years in 50 years is found to be 9.61E+17 J.

CONCLUSIONS

Current technology is inadequate for practical interstellar missions. A BPP Device, if ever found to be possible, would reduce trip times significantly. For example, a BPP Device upper stage using the same amount of energy as a Centaur could improve the burnout velocity from 7 km/s to 9 km/s. Furthermore, if the RTG energy could be applied for 50 years, a total velocity of 37 km/s could be achieved. This would reduce the trip time from 183-thousand to 28-thousand years; a factor of 6.5 improvement. Even with such a Breakthrough Device, however, interstellar flight is still enormously difficult. With a 100% efficient propellant-less space drive, it would still take 10E+17 Joules to propel a voyager sized (722 kg) probe to our nearest star within a 50 year time span. It is reiterated that these findings represent only an isolated test case and the calculations could vary by changing the mass of the spacecraft, the amount of energy available to the BPP Device, etc. These findings are significant because, assuming such a perfect propulsion device exists, they set an upper limit for comparisons of actual inventions or findings to current technology.

ACKNOWLEDGMENTS

This paper was conceived and supported through the Breakthrough Propulsion Physics Program, NASA Glenn Research Center, Lewis Field.

REFERENCES

Berman, A. *Astronautics: Fundamentals of Dynamical Astronomy and Space Flight.* New York. John Wiley & Sons, Inc., 1961.

Millis, M. "Challenge to Create the Space Drive," *Journal of Propulsion and Power,* **13**, 577-582. (1997)

Millis, M. "NASA Breakthrough Propulsion Physics Program," *Acta Astronautica,* **44,** 175-182. (1999)

NASA Jet Propulsion Laboratory (JPL), ON-LINE: "http://vraptor.jpl.nasa.gov/voyager/voyager.html" (1999)

Engineering Warp Drives

Brice N. Cassenti[1] and Harry I. Ringermacher[2]

[1]*United Technologies Research Center, 411 Silver Lane, MS 129-73, East Hartford, CT 06108*
(860) 610-7460 (voice), (860) 610-7526 (fax), cassenbn@utrc.utc.com
[2]*General Electric Company, Corporate Research and Development, Bldg. KW, D254, PO Box 8, Schenectady, NY 12301*
(518) 387-5469 (voice), (518) 387-5752 (fax), ringerha@crd.ge.com

Abstract. Warp drives are widely perceived by the public to be the propulsion system of choice in the not too distant future. Engineers and scientists usually ignore the possibility because of the apparently insurmountable obstacles required to create a warp drive. The promise and the problems of warp drives are examined in three parts: 1) a review of some of the current literature, 2) a summary of the physics, and 3) the engineering developments required to develop warp drives.

INTRODUCTION

Warp drives are a popular device used in science fiction to circumnavigate the speed of light as the maximum possible speed. This restriction follows from the special theory of relativity. Relieving it allows unlimited access to the galaxy and the universe. But the physics of warp drives is not well understood and the engineering prospects appear to place their development well into the future.

The existing literature consists mostly of the description of the physics of spacetime associated with wormholes and is not suited to developing actual systems, but does establish the possibility. A summary of the physics of wormholes can be found in an excellent book by Visser (1996). Millis (1996) has described several conceptual approaches to develop warp drives. A general description of general relativity and space warps can be found in Forward (1989), and includes descriptions of rotating and charged black holes. A description of the effects of magnetic fields on spacetime can be founds in Maccone (1995), and Landis (1997). The most directly applicable work on warp drives can be found in Alcubierre (1994) and Van Den Broeck (1999).

This paper first summarizes the Theory of General Relativity, and then applies the equations of General Relativity to motion near the surface of the earth. The next section consists of a detailed discussion of the Alcubierre warp drives, and is followed by an extension of General Relativity that includes electromagnetic forces (Ringermacher, 1994 and Ringermacher, 1997). The last two sections consist of a brief description of the engineering aspects and conclusions.

SUMMARY OF GENERAL REALTIVITY

The Theory of Special Relativity (Einstein, 1905) was proposed to reconcile experimental and theoretical conflicts that existed in physics at the beginning of the twentieth century. Experimental evidence indicated that the speed of light in a vacuum was not a function of the speed of the source or the observer. For this to remain true in all constant velocity reference frames the interval, $d\tau$ below, must remain constant, where

$$c^2 d\tau^2 = c^2 dt^2 - dx^2 - dy^2 - dz^2. \tag{1}$$

In eq. (1) c is the speed of light, t is the time, and x, y, z are the Cartesian coordinates of a point. Equation (1) can be written in a more general form by setting

CP504, *Space Technology and Applications International Forum–2000*, edited by M. S. El-Genk
© 2000 American Institute of Physics 1-56396-919-X/00/$17.00

$$ds = cd\tau, \ dx^0 = cdt, dx^1 = dx, \ dx^2 = dy, \ dx^3 = dz \ , \tag{2}$$

equation (1) can now be written as

$$ds^2 = \sum_{\mu=0}^{3} g_{\mu\nu} dx^\mu dx^\nu = g_{\mu\nu} dx^\mu dx^\nu \tag{3}$$

where superscripts are for contravariant tensors and subscripts are for covariant, The tensor $g_{\mu\nu}$ is the metric, and the last equality in eq. (3) indicates repeated indices (one contravariant and one covariant) are summed unless otherwise noted. For eq. (2) the metric is given by

$$g_{00} = 1, \text{ and } g_{11} = g_{22} = g_{33} = -1 \tag{3'}$$

All other components of the metric vanish for the interval given by eq. (1). In relativity the metric is symmetric (i.e., $g_{\mu\nu} = g_{\nu\mu}$). Although Einstein's original papers on Special Relativity are easy to read, good physical insight can obtained from the text by Taylor and Wheeler (1966).

General Relativity is an extension of eqs. (3) and (3') to spacetimes that are not flat (Einstein, 1956, and Bergmann, 1976), where the metric cannot be transformed to (3') by any coordinate transformation. Such spacetimes are Riemannian and posses an intrinsic curvature, $R_{\mu\nu\sigma}{}^{\lambda}$ (the Riemannian tensor), given by:

$$R_{\mu\nu\sigma}{}^{\lambda} = \frac{\partial \Gamma_{\sigma\nu}^{\lambda}}{\partial x^\mu} - \frac{\partial \Gamma_{\sigma\mu}^{\lambda}}{\partial x^\nu} + \Gamma_{\tau\mu}^{\lambda}\Gamma_{\sigma\nu}^{\tau} - \Gamma_{\tau\nu}^{\lambda}\Gamma_{\sigma\mu}^{\tau}, \tag{4}$$

where the connection, $\Gamma_{\mu\nu}^{\sigma}$, is given by

$$\Gamma_{\mu\nu}^{\sigma} = \frac{1}{2} g^{\lambda\sigma} \left(\frac{\partial g_{\lambda\nu}}{\partial x^\mu} + \frac{\partial g_{\lambda\mu}}{\partial x^\nu} - \frac{\partial g_{\mu\nu}}{\partial x^\lambda} \right), \tag{5}$$

and,

$$g^{\lambda\sigma} g_{\sigma\mu} = \delta^{\lambda}{}_{\mu} = \begin{cases} 1 & \lambda = \mu \\ 0 & \lambda \neq \mu \end{cases}. \tag{6}$$

The connection is sometimes referred to as the Riemann-Christoffel symbol of the second kind, and vanishes only if the metric is constant.

The Theory of General Relativity does not use the rank four Riemann curvature tensor directly but instead uses the rank two Ricci tensor, $R_{\mu\nu}$, given by:

$$R_{\nu\sigma} = R_{\lambda\nu\sigma}{}^{\lambda} = \frac{\partial \Gamma_{\sigma\nu}^{\lambda}}{\partial x^\lambda} - \frac{\partial \Gamma_{\sigma\lambda}^{\lambda}}{\partial x^\nu} + \Gamma_{\tau\lambda}^{\lambda}\Gamma_{\sigma\nu}^{\tau} - \Gamma_{\tau\nu}^{\lambda}\Gamma_{\sigma\lambda}^{\tau} = R_{\sigma\nu}. \tag{7}$$

The equations of motion are found by 'minimizing' the interval, ds, between two events as

$$\frac{d^2 x^\sigma}{ds^2} + \Gamma_{\mu\nu}^{\sigma} \frac{dx^\mu}{ds} \frac{dx^\nu}{ds} = \frac{e}{mc^2} F^{\sigma\mu} \frac{dx_\mu}{ds}, \tag{8}$$

where $x_\mu = g_{\mu\nu} x^\nu$, e is the charge on the particle, m is the mass of the particle, and $F^{\sigma\mu}$ contains the electromagnetic forces. In Cartesian coordinates :

$$F^{\sigma\mu} = \begin{bmatrix} 0 & -E_x & -E_y & -E_z \\ E_x & 0 & -H_z & H_y \\ E_y & H_z & 0 & -H_x \\ E_z & -H_y & H_x & 0 \end{bmatrix} \tag{9}$$

where (E_x, E_y, E_z) is electric field intensity, and (H_x, H_y, H_z) is the magnetic field intensity.

The equations for the fields can now be written as

$$G_{\mu\nu} = \frac{8\pi G}{c^2} T_{\mu\nu} \tag{10}$$

where the Einstein tensor, $G_{\mu\nu}$, is given by:

$$G_{\mu\nu} = R_{\mu\nu} - \frac{1}{2} g_{\mu\nu} R \quad . \tag{11}$$

R is the Ricci scalar and is given by

$$R = g^{\mu\nu} R_{\mu\nu} \tag{12}$$

$T_{\mu\nu}$, is the energy density tensor, which can be found from the relation:

$$T_{\mu\nu} = (p + \rho c^2) \frac{dx_\mu}{ds} \frac{dx_\nu}{ds} - p g_{\mu\nu} + \frac{1}{4\pi^2} \left(\frac{1}{4} g_{\mu\nu} F_{\sigma\lambda} F^{\sigma\lambda} - F_{\mu\lambda} F_\nu^\lambda \right) \quad , \tag{13}$$

where p is the pressure of the matter, ρ is the density of matter, and G is the gravitational constant.

MOTION NEAR THE SURFACE OF THE EARTH

The motion of a particle near the surface of the earth can provide some insight into the conservation of momentum and energy in relativity theory. In special relativity the momentum and energy are the components of a four dimensional vector exactly analogous to the coordinates x, y, z, and t of an event. The length (i.e. the interval) of the vector is invariant with a coordinate system rotation or translation (including systems with a relative velocity) and is given by

$$m_0^2 c^4 = E^2 - p_x^2 c^2 - p_y^2 c^2 - p_z^2 c^2 \quad . \tag{14}$$

The quantity $m_0 c^2$ is the rest mass energy, where m_0 is the rest mass, and is an invariant. Equation (14) connects the energy, E, which contains the kinetic energy, and the momentum, $\vec{p} = (p_x, p_y, p_z)$. Hence, energy and momentum are components that can be transformed into one another by a coordinate system translation.

For a particle near the surface of the earth the metric is given by (e.g. see Cassenti and Ringermacher, 1996),

$$ds^2 = \left(1 - \frac{2g_0R^2}{c^2} + \frac{2g_0R^2z}{c^2R^2}\right)d(ct)^2 - \left(1 + \frac{2z}{R}\right)dx^2 - \left(1 + \frac{2z}{R}\right)dy^2 - \left(1 + \frac{2g_0R^2}{c^2} - \frac{2g_0R^2z}{c^2R^2}\right)dz^2 . (15)$$

where z is the altitude, R is the radius to the surface of the earth, and g_0 is the acceleration of gravity at the surface of the earth. Using the metric in eq. (15), the equations of motion become:

$$\frac{d^2t}{d\tau^2} + \frac{2g_0}{c^2}\frac{dt}{d\tau}\frac{dz}{d\tau} = 0, \frac{d^2x}{d\tau^2} = 0, \frac{d^2y}{d\tau^2} = 0, \text{ and } \frac{d^2z}{d\tau^2} + g_0\left(\frac{dt}{d\tau}\right)^2 - \frac{g_0}{c^2}\left(\frac{dz}{d\tau}\right)^2 = 0 \quad . \quad (16)$$

These equations can be integrated once to yield

$$\frac{dt}{d\tau} = 1 + \frac{g_0R^2}{c^2} + \frac{v^2}{2c^2} - \frac{g_0R^2z}{c^2R}, \frac{dx}{d\tau} = 0, \frac{dy}{d\tau} = 0, \text{ and } \frac{dz}{d\tau} = -g_0\frac{dt}{d\tau} \quad , \quad (17)$$

where $\left(\frac{dz}{cd\tau}\right)^2 \ll 1$, $\left(\frac{z}{R}\right)^2 \ll 1$ and the initial conditions were taken as $\frac{dt}{d\tau} = 1$ and $\frac{dx}{d\tau} = \frac{dy}{d\tau} = \frac{dz}{d\tau} = 0$.

Making use of eq. (15) and neglecting terms that are small relative to one, the first and last of eqs. (17) become:

$$\frac{1}{2}v^2 + g_0z = \frac{E}{m_0} = const. \text{ and } v + g_0t = \frac{P}{m_0} = const. \quad , \quad (18)$$

where $v=dz/dt$. Using the momentum $p=m_0v$ eqs.(18) can be written as

$$\frac{p^2}{2m_0} + m_0g_0z = E \text{ and } p + m_0g_0t = P \quad . \quad (19)$$

Equations (19) are the conservation of energy and momentum. The common explanation is to note that the center of mass of the earth and the particle continues to move at a constant velocity thus conserving the energy and momentum, but eqs.(19) indicate a different interpretation. Momentum and energy are transferred locally from the gravitational field to the particle and then changes in the field due to the particle are propagated to the source (i.e., the earth). With this interpretation the source still cannot create a change in spacetime that also changes the total momentum and its energy. But note that if only a local field is necessary to transfer energy and momentum then a warp drive is possible. The local field may then propagate the back to other sources such as the vacuum fluctuations or the universe as a whole.

ALCUBIERRE WARP DRIVE SOLUTION

Alcubierre (1994) has proposed a metric that appears to allow travel faster than the speed of light. The metric proposed by Alcubierre results in the interval

$$ds^2 = c^2dt^2 - (dx - v_s(t)f(r_s)dt)^2 - dy^2 - dz^2, \tag{20}$$

where $v_s(t) = \dfrac{dx_s}{dt}$ is the source speed, $r_s = \sqrt{[x - x_s(t)]^2 + y^2 + z^2}$, and

$$f(r_s) = \frac{\tanh[\sigma(r_s + R)] - \tanh[\sigma(r_s - R)]}{2\tanh(\sigma R)}. \tag{21}$$

The parameters σ and R are constants. Note that as $\sigma \to \infty$

$$f(r_s) \to \begin{cases} 1 & -R \le r_s \le R \\ 0 & otherwise \end{cases}.$$

Alcubierre shows that the space behind the source is expanding while the space is contracting ahead of the source. An observer outside the source region can see the source as moving faster than then speed of light. The metric when analyzed does not satisify the Weak Energy Condition and the Strong Energy Condition (see Visser, 1996). Van Den Broeck (1999a) modified the metric proposed by Alcubierre to

$$ds^2 = c^2dt^2 - B^2(r_s)(dx - v_s(t)f(r_s)dt)^2 - dy^2 - dz^2. \tag{22}$$

The proper choice for $B(r_s)$ greatly reduces the energy requirements and satisfies the Strong Energy Condition.

THE ELECTRODYNAMIC CONNECTION

The Alcubierre warp drive requires a method for distorting the space. Exotic matter is needed but electromagnetic fields could provide a more realistic method for distorting the spacetime. This requires a more direct connection between gravitational and electromagnetic forces and would make the development of the necessary fields clearer. Ringermacher has proposed a change to the connection (Ringermacher, 1994), and the associated field equations (Ringermacher, 1997) that creates a geometry that can include electromagnetic forces. The modified connection, $\widetilde{\Gamma}^{\sigma}_{\mu\nu}$, is the sum of the connection, $\Gamma^{\sigma}_{\mu\nu}$, in eq.(5) and terms containing the electromagnetic fields, $F^{\sigma\mu}$, in eq.(9), and reads

$$\widetilde{\Gamma}^{\sigma}_{\mu\nu} = \Gamma^{\sigma}_{\mu\nu} - \frac{e}{2mc^2} g^{\sigma\lambda}[v_\mu F_{\lambda\nu} + v_\nu F_{\lambda\mu} - v_\lambda F_{\mu\nu}]. \tag{23}$$

Note that $F_{\mu\nu} = g_{\mu\sigma}g_{\nu\lambda}F^{\sigma\lambda}$. Using the connection in eq.(23), the resulting field equations (with $T_{\mu\nu}=0$) can be written as (see Ringermacher, 1997)

$$G_{\mu\nu} = \frac{e}{mc^2}v^\sigma(F_{\nu\sigma;\mu} + F_{\mu\sigma;\nu} + g_{\nu\sigma}j_\mu + g_{\mu\sigma}j_\nu - 2g_{\mu\nu}j_\sigma), \tag{24}$$

where

$$F_{\mu\nu;\sigma} = \frac{\partial F_{\mu\nu}}{\partial x^\sigma} - \Gamma^{\lambda}_{\mu\sigma}F_{\lambda\nu} - \Gamma^{\lambda}_{\nu\sigma}F_{\lambda\mu}, \tag{25}$$

$$j_\mu = \begin{Bmatrix} c\rho_e \\ -j_x \\ -j_y \\ -j_z \end{Bmatrix} \quad . \tag{26}$$

ρ_e is the charge density and $\vec{j}$ is the current density in a flat spacetime. Since the theory includes currents which are higher order contributions than the energy densities, and, hence, experimental tests of the theory appear to be feasible (Ringermacher, 1998).

CONCLUSIONS

A warp drive would, not only, allow travel faster than the speed of light but warping spacetime could also reduce or cancel gravity, provide artificial gravity, and allow for propulsion without the use of a propellant. The first steps must include an examination of the physics in detail, and any extensions of physics that can unite electromagnetism with matter and gravity. If a source cannot pick up, or dump, momentum to the universe, or the vacuum, then it may not be possible to construct warp drives, and if electromagnetism cannot be readily used to warp space then the construction of a warp drive will require enormous energy densities over significant volumes. Hence, if the physics works out, the best path to a warp drive includes using the vacuum as source of momentum and applying electromagnetic fields to warp the spacetime.

REFERENCES

Alcubierre, M., "The Warp Drive: Hyper-Fast Travel Within General Relativity," Classical and Quantum Gravity, 11, pp. L73-L77,(1994).

Bergmann, P.G., Introduction to the General Theory of Relativity, New york, Dover Publications, 1976.

Cassenti, B.N., and H.I. Ringermacher, "The 'How To' of Antigravity," Paper No. AIAA-95-2788, presented at the 32nd AIAA/ASME/SAE/ASEE Joint Propulsion Conference, Lake Buena Vista, FL, July, 1-3, (1996).

Einstein, A.,"On the Electrodynamics of Moving Bodies" and "Does the Inertia of a Body Depend on its Energy Content?," Annalen Der Physick, 17, (1905). See The Principle of Relativity, New York, Dover Publications, Inc. , 1952.

Einstein, A., The Meaning of Relativity, Princeton, The Princeton University Press, 1956.

Forward, R.L., "Space Warps: A Review of On Form of Propulsionless Transport;" Journal of the British Interplanetary Society, 42, pp. 533-542, (1989).

Landis, G.A., "Magnetic Wormholes and the Levi-Civita Solution to the Einstein Equations," Journal of the British Interplanetary Society, 50, pp. 155-157, (1997).

Maccone, C., "Interstellar Travel Through Magnetic Wormholes," Journal of the Interplanetary Society, 48, pp. 453-458, (1995).

Millis, M.G., "The Challenge to Create the Space Drive," Journal of the British Interplanetary Society, (1996).

Ringermacher, H.I., "An Electrodynamic Connection," Classical and Quantum Gravity, 11, pp. 2383-2394, (1994).

Ringermacher, H.I. , et al " Search for the Effects of an Electrostatic Potential on Clocks in the Frame of Reference of a Charged Particle," presented at the Breakthrough Propulsion Physics Workshop, Cleveland, August 12-14, (1997).

Ringermacher, H.I., and B.N. Cassenti, "An Experiment to Verify a Relativistic Field Theory Based on an Electrodynamic Connection," Paper No. AIAA-98-3136, presented at the 34th AIAA/ASME/SAE/ASEE Joint Propulsion Conference, July 13-15, (1998).

Taylor, E.F., and A. Wheeler, Spacetime Physics, San Francisco, W.H. Freeman and Company, 1966.

Van Den Broeck, C., "A 'warp drive' with reasonable total energy requirements," to be published in General Relativity and Quantum Cosmology, (1999).

Visser, M., Lorentzian Wormholes - From Einstein to Hawking, Woodbury, New York, American Institute of Physics, 1996 .

Gravitational Radiation and its Application to Space Travel

Giorgio Fontana

University of Trento, Faculty of Science, 38050 Povo, TN, Italy
fontana@science.unitn.it

Abstract. Gravitational radiation is an elusive form of radiation predicted by general relativity, it is the subject of intense theoretical and experimental research at the limit of the sensitivity of today's instrumentation. In spite of the fact that no direct evidence of this radiation now exist, observed astrophysical phenomena have given convincing proofs of its existence. Theories predict that gravitational radiation may also be employed for propulsion, moreover the nonlinear behaviour of spacetime may permit the generation of spacetime singularities with colliding beams of gravitational radiation, this phenomenon could become a form of propellantless propulsion. Both applications would require gravitational wave generators with high power and appropriate optical properties. Among the proposed techniques that could be applicable to the production of gravitational waves, a promising one is the possible emission of gravitons by quantum systems. A hypothesis describing the production of gravitons in s-wave/d-wave superconductor junction is presented.

INTRODUCTION

Everybody can have a personal experience with mass, gravity, inertia, and the effects of a reaction force. From this point of view, modern space propulsion is high technology applied to very old concepts. Propulsion by reaction is a well established technology but it is not suitable for interstellar space travel because the total amount of propellant required would become unacceptable, moreover the related speed limitations would require missions lasting many decades if not centuries.

Many interesting techniques have been proposed in the literature, which could improve the various functional elements of a space propulsion system, but no radically new approach appeared until 1994. In that year Miguel Alcubierre described a theoretical approach to what appears to be a form of propellantless propulsion capable of reaching the highest speeds (Alcubierre, 1994). Alcubierre's analysis did neither addressed the problem of the energy required for his propulsion system, nor he explained the precise nature of the warp drive engine.

The estimations of the energy required for a warp drive have changed from an amount comparable to ten times the total energy content of the universe to an amount of few solar masses. The hardware of the warp drive engine is still a mystery. For a warp drive, negative energy densities are required and the associated exotic matter is forbidden classically. Negative energy densities may exist in quantum field theories, nevertheless it is not know if these favourable conditions can be created in a non transient form and, with more emphasis, it is not known how they could be created in the space surrounding the vehicle. Fortunately a simpler approach to propellantless propulsion exists and it can be derived from general relativity like Alcubierre's warp drive.

This paper describes how spacetime could be manipulated with gravitational radiation and how gravitational radiation could be generated. The important issues of the amount of energy required and the detailed structure of the propulsion system are still an open question, nevertheless these issues appear within our research abilities. Our approach is based on some aspects of general relativity, specifically the existence of gravitational radiation, the properties of colliding beams of gravitational radiation and the possible mechanisms for the generation of gravitational radiation.

CP504, *Space Technology and Applications International Forum–2000*, edited by M. S. El-Genk

ON THE EXISTENCE OF GRAVITATIONAL RADIATION

At the beginning of the twentieth century Albert Einstein developed the concepts of relativity and newtonian gravity into a more complete and credible theory of gravitation, general relativity. In general relativity there are 10 quantities that can create gravity: the energy density, three components of momentum density, and six components of stress. There are 10 unknowns, represented by the components of the metric tensor.

The field equations can be written in terms of a set of 10 fields that are components of a symmetric 4x4 matrix $h_{\alpha\beta}$ representing the deviation of the metric tensor from that of special relativity, the Minkowsky metric of flat spacetime.

$$\left[\nabla^2 - \frac{1}{c^2}\frac{\partial^2}{\partial t^2}\right]h_{\alpha\beta} = \frac{G}{c^4}(source) \tag{1}$$

The source represents the set of energy densities and stresses that can create the field. This expression substantially describes the gravitational field as distortion of the geometry of spacetime as a function of energy, momentum densities and stresses in a source. If velocities in the source are much smaller than c, and h is small compared to 1 (non linear terms in the source negligible) then the Einstein equations reduce to the Newton's equation in and near the source.

General relativity admits solutions of the field equations in the form of waves. Einstein himself calculated the emission of gravitational waves from various sources under a number of restrictive conditions. We have indeed numerous exact solutions obtained using the linearized equations. These solutions are employed for gravitational wave research from astrophysical sources in order to study the emission, propagation and the detection of gravitational waves. They have been also employed for the study of the emission of gravitational radiation from experimental devices.

If the above mentioned linearization is not applicable we are dealing with a problem that could be solved with an ad hoc approximation or by very complex numerical methods. For instance the full scale highly relativistic simulation of inspiralling and merging black holes might require 10 years of supercomputer operations. Usually a problem can be divided into a number of partial problems to which different techniques apply. For instance the emission of gravitational radiation by an astrophysical source may require ad hoc methods with a nonlinearity expansion, but wave propagation and detection can be studied with the linearized approximation (Thorne, 1980).

Gravitational wave are transverse wave like electromagnetic waves, they differ from e-m waves also because of their quadrupolar nature. With a set of free test particles, a passing gravitational wave will produce a small relative acceleration of the particles and of their local inertial frames. The relative acceleration is described by quadrupole-shaped lines of force. The two possible polarizations are "+" and "×" separated by a $\pi/4$ angle. If a set of test particles is distributed along a circle, with the plane containing the particles orthogonal to the direction of propagation of the wave, the passing wave will change the shape of the circle to that of an ellipse, then to a circle, then to an ellipse rotated by $\pi/2$ respect to the previous one, etc. The transverse plane gravitational wave are area preserving and the amplitude of the deformation of the circle of test particles is h.

In 1918 Albert Einstein derived the expression for the gravitational wave field as a function of the second time derivative of the quadrupole moment of the source.

$$h_{jk}^{TT}(t,x) = (2/r)(G/c^4)[\ddot{q}_{jk}(t-r/c)]^{TT} \tag{2}$$

Where x is the location of the observer in a coordinate system centered on the source, r is the distance between source and observer and $q_{jk}(t') = \int \rho(x',t')\cdot(x'_j x'_k - \frac{1}{3}\delta_{jk}r'^2)d^3x'$ is the source mass quadrupole moment, with ρ the mass density.

The dimensionless amplitude h of the gravitational waves of astrophysical origin that could be detected on earth and with a frequency of about 1 kHz is between 10^{-17} and 10^{-22}. Gravitational waves can be detected by measuring the effects of spacetime distortion on a beam of laser light (interferometric sensor) or the resulting stress on a mechanical resonator (Weber bar detector) (Thorne, 1980). The predicted amplitude for astrophysical sources is so small that researchers are not certain that a direct gravitational wave detection has ever been made. The detection of artificially generated gravitational waves is an even worse challenge, because only ultradense materials rotating at near relativistic regimes could efficiently emit gravitational radiation detectable at great distances. In spite of the difficulties a near field test has been succesfully made (Astone, 1991) using a 8.75 kg rotor with a quadrupole moment of $5.5 \cdot 10^{-3}$ kg m^2 rotating at 30000 rpm and located at a distance of 3.5m from the center of the Weber bar gravitational wave antenna Explorer at CERN.

As gravitational radiation transfers energy and momentum, its existence could be inferred by the back-reaction on the source. It has been determined that the back-reaction on a possible binary pulsar could change the orbital period of the system in a detectable and very peculiar way. The study of such a system could also provide a validity check for general relativity. Fortunately in 1973 Taylor and Hulse discovered the first binary pulsar, they discovered the signature of the emission of gravitational radiation in the emitted radio pulses, and verified the validity of general relativity. This work has been recognized with the Nobel prize in Physics in 1993 (Hulse, 1994), (Taylor, 1994).

ON THE PROPERTIES OF GRAVITATIONAL RADIATION

Two properties of gravitational radiation are of interest for us. The first property is that gravitational radiation can be directly employed for propulsion. The second property is a consequence of the non-linearity of Einstein equations. If the amplitude of the gravitational wave is sufficiently high, this non-linearity is the source of harmonics and coulomb-like components. Moreover, for colliding beams, spacetime may act as a mixer, again with the appearance of coulomb-like components. By comparison, the propagation of an electromagnetic wave in a non linear medium, for instance the ionosphere, may cause charge separation which is the origin of ionization and electric discharges, but it is not useful for propulsion, this is because electromagnetism admits positive and negative charges, but macroscopic objects are usually neutral.

We do not discuss the amplitude of the gravitational wave required for the generation of evident non-linearity effects, we simply observe that these phenomena are compatible with a highly relativistic system, for instance a binary system of neutron stars.

About the possibility of accelerating an object only by its internal motion a very interesting paper has recently been published (Bonnor, 1997). Here the motion of a rocket accelerated by an anisotropic emission of gravitational waves has been studied using approximation methods. These methods do not assume conservation of momentum or ad hoc formulae, the equation of motion are obtained by directly solving the field equations. The energy loss of the rocket by the emission of gravitational waves has been found to be in agreement with the quadrupole formula. The power loss is:

$$P = \frac{\dddot{Q}_{xx}^2 + \dddot{Q}_{yy}^2 + \dddot{Q}_{zz}^2}{45(c^5/G)}, \tag{3}$$

where Q_{ii} are the quadrupole moment of the source. For a rod with mass M and length L rotating ω times a second the power P emitted with gravitational radiation is:

$$P = \frac{2}{45} \frac{M^2 L^4 \omega^6}{(c^5/G)}, \tag{4}$$

With the rocket at rest for t=0, the velocity V acquired at t=t1 has been found to be:

$$V = \frac{mG a^5}{315 c^7} \int_0^{t1} \dot{p} \dot{q} \, dt, \tag{5}$$

where m is the initial mass of the rocket, a some convenient length associated with it and p, q are function of t, the quadrupole moment is here $Q(t) = ma^2 h(t)$, and the octupole moment is $O(t) = ma^3 k(t)$, moreover $p(x) = d^2 h(x)/dx^2$ and $q(x) = d^3 k(x)/dx^3$. Both expressions contain terms indicating that, with today's knowledge, only astrophysical objects could emit gravitational radiation capable of producing a detectable effect, but we must take into account that in general relativity the definition of the energy transported by a plane gravitational wave is an open question.

About the second property of gravitational radiation which could be of interest for space travel, the results obtained after 30 years of research may be briefly described as follows. The interaction of two impulsive plane waves with infinite wavefronts starts a self focusing process which is believed to end with the creation of a spacetime singularity regardless of the amplitude of the wave (Szekeres, 1972). The time required for the creation of the singularity is a function of the amplitude of the waves A and the relative polarization α of the two waves (Ferrari, 1988).

$$\Delta t = \frac{1}{A^2} \sqrt{1 + sin\alpha} \qquad (6)$$

These results can also be applied to the more realistic case of beam-like gravitational waves (Ferrari, Pendenza, Veneziano, 1988), and are confirmed by a work describing the interaction of two graviton beams (Veneziano, 1987). The collision of gravitational plane wave may produce a curvature singularity or a coordinate singularity, where the radiation is completely converted in a coulomb-like gravitational field. We may now make the conjecture that a single perfectly focused beam of gravitational radiation might produce a spacetime singularity at the focus regardless the amplitude of the wave, moreover in the eventuality of optical imperfections of the focused beam, they could be spontaneously reduced by the behaviour of the collision process, this property should improve with the amplitude and the high frequency content of the wave.

The solutions of the collision problem correspond to a class of black hole solutions, the similarity of the solutions does not imply that the two physical system are identical or identically stable when the external conditions are changed, but we might expect similar effects. Although the similarity could be simply due the precise mathematical description of the collision problem, the possibility of creating a mini black hole cannot be excluded with today knowledge, and this is the main safety concern if attempts are made for creating these conditions in a laboratory.

Again, theories give curious results, in general relativity non-linearity is associated to highly relativistic systems like extreme astrophysical objects, but the interaction of gravitational waves seems capable of reaching these extreme conditions regardless the amplitude of the wave.

We have seen that the mutual interaction of gravitational waves would cause the appearance of a rectified wave, accompanied by a coulomb-like gravitational field. If this field is created outside a spacecraft, the spacecraft would free-fall towards the distorsion. Using the classical picture for the spacetime distortion created by a mass, our spacecraft would follow a depression in spacetime. The moving depression would in turn emit energy as gravitational waves like a moving mass.

Using the famous Einstein equation we observe that energy density (matter) is not an efficient source of gravity because of $m=E/c^2$, and for two equal massive particles the internal rest energy is much higher than the integral gravitational energy, moreover moving this source of gravity to a different location requires the transportation of this energy density. Instead, gravitational waves can be directly and completely converted into a gravitational field which should follow the focus point of the beam(s). The source(s) of the beam(s) could be onboard the spacecraft.

LABORATORY GENERATION OF GRAVITATIONAL RADIATION

Historically, the emission of gravitational radiation has been studied in astrophysical systems, as reported in the second section of this paper, and a quite large literature exists on the subject. Instead, the laboratory generation of gravitational radiation is still in its theoretical stage of development except, perhaps, the rotor employed for the calibration of a GW antenna (Astone, 1991), which has indeed produced experimental results for a near field detection.

Because we expect that gravitational radiation should be generated starting from a known energy source, we observe that an important parameter certainly is conversion efficiency i.e. the ratio between the output power in gravitational radiation and the input power. The difference between the input and output power has to be dissipated and this could be a serious collateral problem.

Moreover, from eq. 4 we observe that among the many parameters, the amplitude is influenced by ω^6, leading to the idea that a higher frequency could improve the output of the generator. Combining the request of higher frequencies and low losses leads to the idea that a microscopic source of gravitational radiation could be the preferred source for our application. The possible structure, the computed output power and conversion efficiency of classical and quantum sources of gravitational radiation are shortly described in the following sub-sections.

Gravitational Radiation from Classical Systems

Three non-quantum mechanism can be identified for the possible laboratory generation of high frequency gravitational radiation (Pinto, 1988): the coherent EM-GW converter, the EM-pulsed source and the photon/phonon pumped generator. The possibility of arranging several elementary generators in an array will give a beamed emission.

In EM-GW converter the stress-energy tensor of the EM field is the source of the gravitational radiation. The EM-GW converter has been studied on the basis of a cylindrical EM resonator with a static axial bias magnetic field H_0.

Using the TE_{111} resonant mode the power dissipated by the walls and the power emitted by gravitational waves has been computed (Pinto,1988).

$$P_{walls} = \frac{\omega_{111} W_{em}}{Q_{111}} \tag{7}$$

$$P_{gw} = \frac{G(H_0 H_{111})^2 \sin\theta_c \Delta\theta}{(2\pi R / p'_{11})^2 \left(1 + 2.912(R/d)^2\right)^{-1}} \tag{8}$$

In which it has been assumed the gw radiation contained in the conical beam $\theta_c - \Delta\theta/2 < \theta < \theta_c + \Delta\theta/2$. Using $H_0 = H_{111} = 10^5$ Gauss, $\lambda = 1m$, $d/R = 10$, we obtain $P_{gw} \approx 10^{-17}$ W, with $P_{input} \approx P_{walls} \approx 10^9$ W.

The EM-pulsed source consists in a short solenoid or permanent magnet and a TEM transmission line which traverse the magnetic field generated by the solenoid. Pulses of EM energy are then sent along the transmission line. The amplitude $h \approx 10^{-34}$ is expected from this generator if employing ring transmission lines with a radius of 10 km and EM pulse generators with power of several MW.

The photon/phonon pumped GW generator consists of an array of piezoelectric plates excited by a UHF-SHF modulated laser beam. An expression of the GW output power of this generator has been obtained (Pinto, 1988):

$$P_{gw} \approx 4 \cdot 10^{-21} \left[\frac{v_s}{c}\right]^3 \left[\frac{1(cm)}{l}\right] \left[\frac{S_w}{10(cm^2)}\right] \left[\frac{\rho_0}{5(g/cm^3)}\right] \left[\frac{Q_{ac}}{100}\right] \left[\frac{P_{ac}}{10^4(W)}\right] W \tag{9}$$

Where Q_{ac} is the quality factor of the plates, P_{ac} the input power, S_w and l respectively area and length of the plates, v_s and c the speed of sound and the speed of light respectively. The computed conversion efficiency for typical v_s/c ratio is about four order of magnitude lower than that of a pure EM converter. If $v_s \approx c$ than the efficiency of this converter could become about 13 orders of magnitude higher than the pure EM converter. By comparison to the EM-GW converter we have, for coherent sources:

$$\frac{h_{mech}}{h_{em}} = \frac{\rho_{mech}}{\rho_{em}} \left[\frac{v_{sound}}{c}\right]^5 \tag{10}$$

For completeness, we have that typical $v_{sound}/c \approx 10^{-5..-6}$, and we see that EM fields of about 10^8 V/cm produce a mass density equivalent of that of water.

Gravitational Radiation from Quantum Systems

Like electromagnetic radiation, gravitational radiation could be emitted by quantum transitions. We introduce this concept with the simple analogy to the well known binary pulsar GW source. By this analogy the ideal laboratory source of gravitational radiation could be a couple of almost identical orbiting objects with nuclear matter density and with electric charge, which will give us the ability of controlling them with an electromagnetic field. Being the scale factor in principle not relevant for the efficiency of the EM-GW converter, and looking for a high frequency array of such objects, we could reduce the scale factor reaching, for instance, the size of Cooper's pairs, which certainly satisfy our initial requirements.

At atomic scale the emission of a quantum of gravitational radiation, the graviton, is accompanied by a $L=2$ transition in the quantized angular momentum of the emitting system. In (Halpern, 1964) the investigation of the interaction of the gravitational fields with microscopic systems has been extended to the nuclear and molecular phenomena, with the interesting result that the gravitational interactions have here a greater significance than at a macroscopic level, where gravitational radiation is extremely difficult to generate and detect. The multipole expansion of the gravitational radiation field resulting from periodically oscillating sources has been performed in full analogy with the method used for the electromagnetic fields, thus formally reproducing a successful and experimentally tested methodology.

According to (Halpern, 1964) and (Halpern, 1968) atomic transitions for which the orbital quantum number L changes by + or − 2, and for which the total quantum number J changes by 0 or + or − 2 are gravitational quadrupolar transitions and are permitted for the emission of gravitons, while the emission of photons is forbidden. It has been found that atomic transitions from orbitals 3d to 1s, 3d to 2s and 3d to 3s are possible candidates for transitions, which may be applicable for the generation of gravitational radiation by atoms of a suitable material. The material could be pumped by photons and let decay gravitationally. Unfortunately the gravitational transition probability is much lower than in the electromagnetic case; this ratio for matrix elements of equal structures is of the order of $\chi^2(mc)^2/e^2$, where χ^2 is defined in (Halpern, 1964), e is the charge and m the mass of the emitting particle. For a proton this ratio is $1.6 \cdot 10^{-36}$, but it could be about 10^4 times larger for molecular transitions. Halpern and Laurent first looked for a natural source of high frequency gravitational radiation, they computed that the energy of gravitons involved in some possible stellar processes was very high, 14.4 keV for ^{57}Fe in the sun and 16.1 MeV from supernovae (Halpern, 1964).

Discussing the possibility of a stimulated emission, they also suggested the physical structure of the gravitational counterpart of the laser, called a "gaser". The device structure was a cavity-less single pass device similar to X-ray lasers of today. Again, the extremely low probability of graviton absorption and emission indicates that gasers might not be possible within simple atomic systems.

We have seen that if we pump a given quantum system to a gravitationally excited state with electromagnetic radiation, the resulting state is also an electromagnetically excited state and the probability of electromagnetic emission is higher than the gravitational one by a factor of about 10^{36}, leading to the conclusion that a different approach is required.

We now imagine a binary quantum system in which we abruptly change the attraction force between the two equal particles composing the system, the resulting probability of states changes accordingly, and if the appropriate change has taken place, an induced emission of a graviton could result. Measurements of the quantized angular momentum of Cooper's pairs in different superconducting materials have been made, and they are compatible with above mentioned possibility. In fact, recently (Harlingen, 1995), (Kouznetsov, 1997), (Sigrist, 1995), (Ding, 1996), (Barret, 1991) it has been experimentally observed the existence of only two different symmetries of the order parameter in low T_c and high T_c superconductors, a symmetry with a s-wave component (LTSC and YBCO HTSC) and a symmetry with a d-wave component (HTSC) (Kouznetsov, 1997), therefore we know that Cooper-pairs are in s-orbitals and d-orbitals respectively. We can also predict that when Cooper-pairs move under non equilibrium conditions, i.e. under the effect of magnetic fields, from a superconductor where the symmetry of the order parameter is of type d to a superconductor where this symmetry is of type s, they are subject to transition and loose energy by the emission of a particle with a spin of 2.

A related phenomenon, which is important for the estimation of the binding energy, is the observed emission of wide band THz electromagnetic radiation in a YBCO/insulator/normal-metal junction. This emission is not related to the Josephson effect, and it has been found to be originated by the recombination of Cooper-pairs and quasiparticles at the interface with the non-superconducting material that gives a channel to photon and phonon recombination (Lee, 1998); although the power of the emission was very small, the experiment shows that the electron binding energy is actually released at the interface and a measure of this energy is given. A comprehensive discussion on the symmetry of pairing states in both conventional and high Tc superconductors can be found in (Harlingen, 1995) and therein references.

A junction between a s-wave and a d-wave superconductor (SDS junction) can be here defined with the purpose of inserting the junction in an electrical circuit and studying the maximum possible emission of radiation. We could estimate the maximum amplitude of the radiation emitted by a SDS junction employing an energy balance equation. Keeping the SDS junction at a temperature much below Tcs and Tcd, the binding energy that is released by a single transition could be a fraction of: $\left|T_{cd} - T_{cs}\right| k_B$, where T_{cd} is the critical temperature of the d-wave superconductor, T_{cs} is the critical temperature of the s-wave superconductor and k_B the Boltzmann constant, considering that the electron binding energy is proportional to the critical temperature.

If we make the hypothesis that this fraction is a factor of one, we may write the usual energy balance equation: $\left|T_{cd} - T_{cs}\right| k_B = h\nu$, obtaining frequencies of the order of hundreds of GHz, which are near the frequencies observed in (Lee, 1998). The maximum power emitted by the process could be found with the hypothesis of currents about the critical currents of most superconductors, and of about 10 kA/cm^2, this current density may also suppress the Josephson effect on the Bose condensate. Introducing the charge of the electron we obtain a power density of:

$$\left|T_{cd} - T_{cs}\right| k_B \left(10^4 / 3.2 \cdot 10^{-19}\right) \quad ; \tag{11}$$

which is of the order of ten W/cm^2.

A detailed analysis of SDS junctions has been not yet performed, mainly because no satisfactory theory exists to explain electron pairing in high T_c superconductors. Therefore whether these induced $L=2$ transitions are associated to gravitons remains to be investigated both theoretically and experimentally.

CONCLUSION

This paper has shortly described the most relevant elements of what the author believes could become a new propulsion technique. The theoretical background is that of general relativity and the main subject of the paper is gravitational radiation from man made sources. With reference to a wide collection of theoretical papers and few experimental one, it has been shown that gravitational radiation could be employed for space propulsion. Moreover it has been shown that gravitational radiation can be generated by artificial means and a new hypothesis on its possible emission from a quantum system has been proposed. Further developments are expected by a deeper analysis of the collision problem and the study of gravitational transitions in some promising materials.

ACKNOWLEDGMENTS

The author wishes to thank professor Antonio Zecca for his encouragement and help in doing this work.

REFERENCES

Alcubierre Miguel, "The Warp Drive: Hyper-Fast Travel Within General Relativity", *Class. Quantum Grav*, **11**, 1994, pp. L73-L77

Astone et…,"Evaluation and Preliminary Measurement of the Interaction of a Dynamical Gravitational Near Field with a Cryogenic G.W. Antenna", *Zeischrift fuer Physik C*, **50**, 1991, pp. 21-29

Barret S.E., Martindale J.A., Durand D.J., Pennington C.P., Slichter C.P., Friedmann T.A., Rice J.P., and Ginsberg D.M., "Anomalous Behavior of Nuclear Spin-Lattice Relaxation Rates in YBa$_2$Cu$_3$O$_7$ below Tc", *Phys. Rev. Lett.* 1991 , **66**, pp. 108-111.

Bonnor W. B. and Piper M. S., "The Gravitational Wave Rocket", *Class. Quantum Grav.* ,**14**, 1997, pp. 2895-2904.

Ding H. et al. , "Angle-resolved Photoemission Spectroscopy Study of the Superconducting Gap Anisotropy in Bi$_2$Sr$_2$CaCu$_2$O$_{8+x}$. ", *Physical Review B.*, **54**, N. 14, 1 Oct. 1996-II, pp. R9678-R9681

Ferrari V., Pendenza P., and Veneziano G., "Beam-like Gravitational Waves and Their Geodesics", *General Relativity and Gravitation*, **20**, No 11, 1988, pp. 1185-1191

Ferrari Valeria, "Focusing Process in the Collision of Gravitational Plane Waves", *Physical Review D*, **37**, No10, 15 May 1988, pp. 3061-3064

Halpern L., Jouvet B., "On the Stimulated Photon-Graviton Conversion by an Electromagnetic Field", *Annales H. Poincare*, **VIII**, NA1, 1968 pp. 25-42.

Halpern L., Laurent B., "On the Gravitational Radiation of Microscopic Systems", *IL Nuovo Cimento*, **XXXIII**,

N. 3, 1964, pp. 728-751

Harlingen D.J., "Phase-sensitive Tests of the Symmetry of the Pairing State in High-temperature Superconductors- Evidence for d Symmetry", *Reviews of Modern Physics*, **67** No. 2, 1995, pp. , 515-535.

Hulse Russel A., "The Discovery of the Binary Pulsar", *Reviews of Modern Physics*, **66**, No. 3, July 1994, pp. 699-710,.

Kouznetsov K.A. et al., "c-axis Josephson Tunneling between $YBa_2Cu_3O_{7-x}$ and Pb: Direct Evidence for Mixed Order Parameter Symmetry in a High Tc Superconductor", *Physical Review Letters*, **79**, 20 Oct. 1997, pp. 3050-3053,.

Lee Kiejin, Iguochi Ienari, Arie Hiroyuki and Kume Eiji, "Nonequilibrium Microwave Emission from DC-Biased High Tc $YBa_2Cu_3O_{7-y}$ Junctions", *Jpn. J. Appl Phys.* ,**37** Part 2, No. 3A, 1 March (1998) pp. L 278-280.

Pinto I.M. and .Rotoli G, "Laboratory Generation of Gravitational Waves ?" *Proceedings of the 8th Italian Conference on General Relativity and Gravitational Physics",* Cavalese (Trento) August 30 – September 3, 1988 World Scientific – Singapore p 560-573

Sigrist M. and Rice T.M., "Unusual Paramagnetic Phenomena in Granular High-temperature Superconductors- A Consequence of d-wave Pairing ? ", *Reviews of Modern Physics*, **67**, No. 2, 1995, pp. 503-513.

Szekeres P., "Colliding Plane Gravitational Waves", *J. Math. Phys,* **13**, No. 3, March 1972, pp. 286-294

Taylor Joseph H., Jr, "Binary Pulsar and Relativistic Gravity", *Reviews of Modern Physics*, **66**, No. 3, July 1994, pp. 711-719.

Thorne Kip S., "Gravitational-wave Research: Current Status and Future Prospects", *Reviews of Modern Physics,* **52**, No. 2, Part I, April 1980, pp. 285-297.

Thorne Kip S., "Multipole Expansion of Gravitational Radiation", *Reviews of Modern Physics*, **52**, No. 2, Part I, April 1980, pp. 299-339

Veneziano G., "Mutual Focusing of Graviton Beams", *Modern Physics Letters A,* **2**, No 11, 1987, pp. 899-903

Frequency- and Time-Domain Detection of Superluminal Group Velocities in One Dimensional Photonic Crystals (1DPC)

Mohammad Mojahedi[1], Kevin J. Malloy[1] and Raymond Chiao[2]

[1]Center for High Technology Materials (CHTM), 1313 Goddard SE, Albuquerque NM 87106
[2]Department of physics, University of California at Berkeley, Berkeley CA 94720

email: mojahed@chtm.unm.edu; malloy@chtm.unm.edu; chiao@physics.berkeley.edu

Abstract. In this work we discuss two recent experiments in the frequency- and time-domain for electromagnetic waves tunneling through an optical multilayer also known as 1DPC. These experiments are intended to demonstrate measurability and hence the physical reality of the superluminal (exceeding the speed of light in vacuum) group velocities for evanescent modes. Despite these anomalous velocities, Einstein causality is not violated since the front (Sommerfeld forerunner) remains luminal.

SUPERLUMINALITY AND BREAKTHROUGH PROULSION PHYSICS (BPP) PROGRAM GOALS

In recent years, the subject of superluminal propagation has received much attention. A review of this field is provided in reference (Chiao, 1997) and a short summary will be given in the next section. Unfortunately, as is the case with any new subject, there are misinterpretations and misrepresentations which deserve a closer examinations. For example it has been suggested that superluminal group velocity is indeed the information velocity (Heitmann, 1994) and concepts such as Sommerfeld forerunner are irrelevant, or that evanescent modes are not necessarily Einstein causal (Nimtz, 1999). Aside from these controversies, the BPP program has challenged researchers with the need to attain the ultimate transit speed to dramatically reduce travel times. In this regard, the possibility of superluminal velocities and the role of the Sommerfeld forerunner (or the front) deserve much attention. If the measured superluminal group velocities are indeed the information velocity, then one must conclude that Einstein causality has been violated. On the other hand, if as we believe the critical concept is the propagation of the Sommerfeld forerunner, which under all circumstances shall remain luminal, then there is no violation of special relativity. The issue of superluminal propagation and Sommerfeld forerunner and their connection with the well known requirements of special relativity is closely associated with the BPP "maximum transit speeds" goal and in particular they relate to critical "make-or-break issues".

INTRODUCTION TO SUPERLUMINAL PROPAGATION

Chiao and his co-workers (Steinberg, 1993) have demonstrated that single photons generated in the process of spontaneous parametric down-conversion can tunnel through an optical multilayer with group velocities 1.7 times faster than c. Additionally, in a series of experiments Ranfagni and his coworkers have used undersized waveguides, slightly misaligned horn antennas and two side-by side prisms (resulting in frustrated total internal reflection) to demonstrated superluminal velocities (Ranfagni, 1991a; Ranfagni, 1991b; Ranfagni, 1993; Mugnai, 1998). Nimtz and his coworkers have also studied the undersized waveguide and were able to improve on Ranfagni's results (Enders, 1992; Enders, 1993a; Enders, 1993b; Nimtz, 1994; Nimtz, 1997). They (Nimtz) also briefly considered tunneling through an optical multilayer inserted in an undersized waveguide (Nimtz, 1994). Unfortunately, this experiment in particular, and their interpretation in general, is incomplete. For example, to assign the superluminal tunneling time to the 1DPC is misleading since by their own admission the tunneling is due to both the undersized waveguide and the 1DPC. Moreover, in Refs. (Nimtz, 1994), and in fact in all of their frequency-domain analysis, they used the Fourier transform to extend the frequency-domain network analyzer (NA) results to the time-domain. In their use of the Fourier integral, they had to replace the $-\infty$ to $+\infty$ limits of the integral with the bounded limit of v_1 to v_2. This meant that their *assumed* incident wave packet had zero frequency components outside (v_1, v_2) frequency interval [8.7±0.5 GHz in their particular case (Nimtz, 1994)]. This simple point can lead to misinterpretation when presenting frequency-domain results as direct time-domain measurements, particularly in light of the fact that these neglected large frequency components are essential in understanding the Sommerfeld forerunner. Even more troubling, in their analysis they only used the transmission function *amplitude* to calculate the time-domain signal. This is equivalent to assuming a *constant phase* for the transmission function, which is strictly true only for an infinitely long, undersized waveguide or 1DPC, and in fact is erroneous in the case of the 1DPC with few periods. In this light, a correct and reliable measurement of the tunneling times and superluminal group velocities for 1DPCs in the microwave region is in order.

Before closing this section, let us briefly discuss the previous understanding of evanescent wave propagation and electromagnetic wave tunneling. Historically, evanescent propagation of the tunneling wave packet was thought to distort the transmitted pulse to the extent that the known theoretical superluminal or even negative group velocities were rendered unphysical or meaningless (Brillouin 1960) (Landau, 1984) (Born, 1970) (Brillouin, 1946) (Jackson, 1975). However, most recently this trend is beginning to change as evidenced by the latest edition of "Classical Electrodynamics" by John Jackson (Jackson, 1998) or the review paper by Chiao and Steinberg in E. Wolf "*Progress in optics*" (Chiao, 1997).

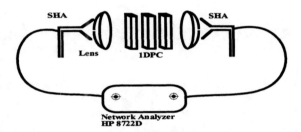

FIGURE 1. Frequency-domain experimental setup.

DESCRIPTION OF THE EXPERIMENTAL PROCEDURES

Figure 1 shows the free-space experimental setup. It consists of two K-band standard horn antennas (SHA), connected to ports 1 and 2 of an HP 8722D NA, configured to measure the transmission coefficient. The setup is enclosed in a anechoic chamber to reduce stray signals.

With recent advances in non-coaxial (free-space) calibration techniques for NA such as the "Thru-Line-Reflect" (TRL), it is possible to remove most of the systematic errors and the influence of the microwave components on the measured transmission coefficient. Performing the TRL calibration in free-space allows us to measure the transmission coefficient ($T = |T| e^{-i\varphi}$) solely due to the 1DPC and eliminates dispersion and dissipation losses associated with inserting the 1DPC inside a waveguide. After calibrating the system (without the 1DPC), a reference plane of unit magnitude for $|T|$ and zero phase for φ is established midway between the two SHAs. At this point, the 1DPC is inserted and the receiver horn is moved back exactly by a length equal to the thickness of the 1DPC (L_{pc}).

Within the stationary phase approximation, the concept of group delay, given by the angular-frequency derivative of the transmission phase ($-\partial\varphi/\partial\omega$), is a natural approach to understanding propagation through a finite, dispersive structure. This concept can be extended in order to obtain the group velocity as a function of frequency, according to;

$$\frac{v_g}{c} = \frac{L_{pc}}{c\tau_g} = \frac{-L_{pc}}{c\left(\partial\varphi/\partial\omega\right)}. \tag{1}$$

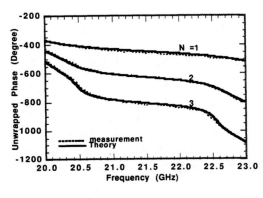

FIGURE 2. Theoretical and measured phase.

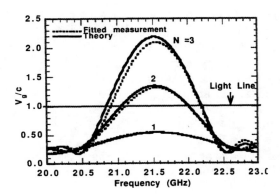

FIGURE 3. Normalized group velocity.

Figure 2 is the calculated (solid line) and measured (dotted line) unwrapped phase for a 1DPC with three, two and one dielectric slabs (the spacer is always air). The theoretical calculation is based on the diagonalization of one period matrix, and is presented in Ref. (Mojahedi, 1999), as space limitations prevent our repeating it here.

According to Eq. (1), the data presented in Fig. (2) must be differentiated with respect to frequency. However, applying differentiation to noisy data amplifies the noise and may lead to spurious effects. To avoid the arbitrariness associated with smoothing the data, we have chosen to obtain the best nonlinear least square fit of the experimental phase data to the equation for the transmission phase (φ) presented in Ref. (Mojahedi, 1999). The parameters used in this fit are the dielectric thickness (a_j), spacer thickness (d_i) and the real part of the index of refraction (n'_j). Figure 3 shows the result of the least square fit to the phase data of Fig. 2 together with applying Eq. (1), in order to determine the normalized group velocity in a 1DPC with one, two, and three dielectric slabs.

Along with the velocities derived from the fit (dotted curves), the theoretical group velocities calculated from measured values of the thicknesses and indices are also shown (solid curves). The fitting parameters and the measured values for these parameters are shown in Table 1. As Fig. 3 indicates, in the case of $N=3$, a maximum superluminal group velocity 2.1 times c is observed.

TABLE 1. Measured and fitted parameters for the 1DPC with Eccostock® slab and air spacer.

	Fitting, $N = 3$	Fitting, $N = 2$	Fitting, $N = 1$	Measured
d_i	1.794 cm	1.825 cm	——————	1.76 cm
a_j	1.399 cm	1.366 cm	1.396 cm	1.33 cm
n'_j	3.216	3.288	3.245	3.40

Superluminal tunneling can also be directly demonstrated in the time-domain. Figure 4 is the experimental setup used to compare the time-of-flight for a single microwave pulse tunneling through a 1DPC as compared with a companion wave packet propagating in free space. A backward-wave-oscillator (BWO) in conjunction with a conical horn antenna (CHA) is used to generate a narrow single-microwave pulse centered at 9.68 GHz (100 MHz bandwidth) of approximately 10 ns duration. The microwave pulse is sampled at two distinct points of the antenna's radiation intensity pattern, hereon referred to as "side" and "center". The delay between these two paths due to different cable lengths, internal differences of the scopes' display units (Tektronix SCD-5000), or any other mechanism was measured in the absence of the 1DPC. This delay was then removed electronically such that the peaks of the two "side" and "center" pulses

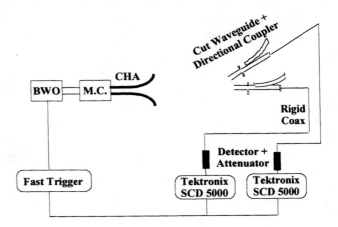

FIGURE 4. Time-domain Experimental setup.

arrive at the same time. Figure 5 shows the result. At this point, a 1DPC consisting of alternating layers of polycarbonate and air, designed to have minimal dispersion at 9.68 GHz, was inserted in the "center" path. The results are depicted in Fig. 6.

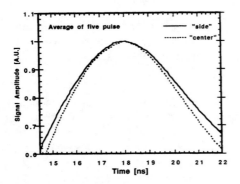

FIGURE 5. Reference Pulses.

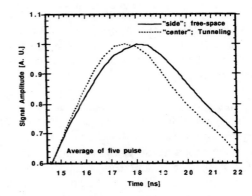

FIGURE 6. Shift to earlier times.

The peak of the pulse tunneling through the 1DPC (dotted line) is clearly shifted to an earlier time compared to the free space pulse (solid line). Since group velocity is the velocity by which the peak of a wave packet travels, it is evident that the tunneling pulse propagated superluminally. The measured advance in Fig. 6 is 440 ± 20 ps, corresponding to a velocity of $(2.38 \pm 0.15)\, c$.

The traditional view asserts that the tunneling wave packet is distorted to the extent that the comparison between the incident and transmitted wave packet is rendered meaningless (Brillouin, 1960, pp. 22) (Landau, 1984). Therefore, it is important to compare the transmitted tunneling pulse with the incident pulse propagating along the same path. Repeated measurements indicated that the FWHM of the "center" tunneling pulse depicted in Fig. 6 is only increased by 2.2% as compared to the "center" free-space pulse depicted in Fig. 5 (Mojahedi, 1999, pp. 41). In light of this, one must accept the fact that if group velocity is a good parameter describing the propagation of a free-space wave packet, it must also be a good variable describing the propagation of the tunneling wave packet.

WHY EINSTEIN CAUSALITY IS NOT VILOATED

At this point the reader may ask: How are the results presented in the last section are in agreement with Einstein causality? The answer to this question rests in the fact that under all circumstances the velocity by which the front or the Sommerfeld forerunner travels remains luminal. This compels us to associate the information velocity with the velocity of these points of non-analyticity (Chiao, 1997). This idea, although in complete agreement with our understanding of theory of special relativity and electromagnetism, is perhaps not a practical definition under all circumstances. The reason lies in the fact that the forerunner's field is usually of very high frequency and very small amplitude.

It must be emphasized that at the observation point x, no detection of the signal (the front) can be made prior to time x/c. This can be seen via contour integration, in the upper half plane, of the expressions such as Eq. (2) below, which describes the field at the position x and the time t for a medium with index or effective index n (Jackson, 1998, pp. 336)

$$u(x,t) = \int_{-\infty}^{+\infty} \frac{2}{1+n(\omega)} A(\omega) e^{ik(\omega)x - i\omega t} d\omega, \qquad (2)$$

$$A(\omega) = \frac{1}{2\pi} \int_{-\infty}^{+\infty} u(x=0, t) e^{i\omega t} d\omega. \qquad (3)$$

At the time $t = t_0 = x/c$ the earliest part of the signal (Sommerfeld forerunner) can be detected. The frequency of oscillation and the field amplitude for these forerunner's fields are discussed in Ref. (Mojahedi, 1999, pp. 60-69). To summarize those results, the frequency of oscillation is given by

$$\omega_s = \sqrt{G'(0)} \Big/ \sqrt{2\left(\frac{t}{t_0} - 1\right)}, \qquad (4)$$

where $G'(0)$ is the time derivative of the susceptibility kernel (Jackson 1998, pp. 332) evaluated at $t = 0$. Furthermore, for the incident wavepackets proportional to t^m (m is an integer) the Sommerfeld forerunner is described by a Bessel function of order m according to

$$u(x,t) \approx a \left(\frac{t - t_0}{\gamma}\right)^{m/2} J_m\left(2\sqrt{\gamma(t - t_0)}\right); \qquad \gamma = \frac{G'(0) t_0}{2}; \qquad \text{for } t > t_0. \quad (5)$$

In light of importance of the Sommerfeld forerunner and its connection to Einstein causality, we are currently in the process of devising a procedure which allows experimental detection of this and the Brillouin forerunner for a 1DPC.

CONCLUSIONS

In this manuscript we have discussed frequency- and time- domain experiments which demonstrate superluminal group velocities. Unfortunately, the response to such phenomenon has been from two opposite points of view. While traditionally it has been believed that superluminal group velocities are unphysical (see the references in the text) the experiments presented here along with many others cited in the text have demonstrated the measurability and hence the physical reality of such abnormal behavior. On the other hand some authors have argued that evanescent modes used in these experiments are not necessarily Einstein causal and that the measured superluminal group velocities are indeed information velocities and that the notion of the front (Sommerfeld forerunner), which must remain luminal under all circumstances, is not relevant. Clearly, such assertions (if true) can result in violation of special relativity, which requires the speed of light in vacuum to be the ultimate information velocity.

In this work in addition to presenting two experiment which demonstrate superluminal (but Einstein causal) group velocities, we have discussed the role, the frequency, and the functional form of the Sommerfeld forerunner. Currently, within the NASA-BPP program we are investigating the experimental conditions under which the theoretical predictions regarding Sommerfeld forerunner can be tested for optical barriers. The subject of superluminal velocity and whether or not this is a genuine information velocity (in violation of Einstein causality) is closely related to BPP challenges set forth as "Discovering methods for achieving the shortest possible travel times," and is "aimed to advance physics to address critical unknowns, make-or-break issues, or curious effects."

ACKNOWLEDGMENTS

This material is based on work supported by, or in part, by the US Army Research Office under grant number DAAH04-96-1-0439.

REFERENCES

Born, M. and Wolf, E. *Principles of optics; electromagnetic theory of propagation, interference and diffraction of light*, Oxford, Pergamon, 1970, pp. 23.

Brillouin, L., *Wave propagation in periodic structures; electric filters and crystal lattices*, New York, McGraw-Hill, 1946, pp.75.

Brillouin, L., *Wave propagation and group velocity*, New York, Academic Press, 1960, pp. 74-79.

Chiao, R. Y., and Steinberg, A.M., "Tunneling Times and Superluminality," *Prog. Optics* **37**, 345-405 (1997).

Enders, A., and Nimtz., G., "On Superluminal Barrier Traversal," *J. Phys. I France* **2**, 1693-1698 (1992).

Enders, A., and Nimtz, G., "Photonic-Tunneling Experiments," *Phys. Rev. B* **47**, 9605-9609 (1993).

Enders, A., and Nimtz, G., "Zero-Time Tunneling of Evanescent Mode Packets." *J. Phys. I France* **3**, 1089-1092 (1993).

Heitmann, W., and Nimtz, G., "On Causality Proofs of Superluminal Barrier Traversal of Frequency Band-Limited Wave-Packets," *Phys. Lett. A* **196**, 154-158 (1994).

Jackson, J. D., *Classical electrodynamics*, New York, Wiley, 1975, pp. 302.

Jackson, J. D., *Classical electrodynamics*, New York, Wiley, 1998, pp. 325-326.

Landau, L. D., Lifshittz, E. M., et al. *Electrodynamics of continuous media*, Oxford, Pergamon, pp. 285 (1984).

Mojahedi, M., *Superluminal group velocities and structural dispersion*, Albuquerque, University of New Mexico, 1999, pp. 16.

Mugnai, D., Ranfagni, A., et al. "The Question of Tunneling Time Duration : a New Experimental Test At Microwave Scale," *Phys. Lett. A* **247**, 281-286 (1998).

Nimtz, G., "Evanescent modes are not necessarily Einstein causal," *Eur. Phys. J. B* **7**, 523-525 (1999).

Nimtz, G., Enders, A., et al. "Photonic Tunneling Times," *J. Phys. I France* **4**, 565-570 (1994).

Nimtz, G., and Heitmann,W., "Superluminal Photonic Tunneling and Quantum-Electronics," *Prog. Quant. Electr.* **21**, 81-108 (1997).

Ranfagni, A., Fabeni, P., et al. "Anomalous Pulse Delay in Microwave Propagation : a Plausible Connection to the Tunneling Time," *Phys. Rev. E* **48**, 1453-1460 (1993).

Ranfagni, A., Mugnai, D., et al. "Delay-Time Measurements in Narrowed Wave-Guides As a Test of Tunneling," *Appl. Phys. Lett.* **58**, 774-776 (1991).

Ranfagni, A., Mugnai, D., et al. "Optical-Tunneling Time Measures : a Microwave Model," *Physica B* **175**, 283-286 (1991).

Steinberg, A. M., Kwiat, P. G., et al., "Measurement of the single-photon tunneling time," *Phys. Rev. Lett.* **71**, 708-711 (1993).

Computational Tools

for Breakthrough Propulsion Physics:

State of the Art and Future Prospects

Claudio Maccone

Alenia Spazio, Via Martorelli 43, I-10155 Torino (TO), Italy
Phone: +39 011 71 80 313, Fax: +39 011 71 80 012, E-mail: cmaccone@ to.alespazio.it

Abstract. To address problems in Breakthrough Propulsion Physics (BPP) one needs sheer computing capabilities. This is because General Relativity and Quantum Field Theory are so mathematically sophisticated that the amount of ***analytical*** calculations is prohibitive and one can hardly do all of them by hand. In this paper we make a comparative review of the main ***tensor calculus*** capabilities of the three most advanced and commercially available "symbolic manipulator" codes: Macsyma, Maple V and Mathematica. We also point out that currently one faces such a variety of ***different conventions in tensor calculus*** that it is difficult or impossible to compare results obtained by different scholars in General Relativity and Quantum Field Theory. Mathematical physicists, experimental physicists and engineeres have each their own way of customizing tensors, especially by using the different metric signatures, different metric determinant signs, different definitions of the basic Riemann and Ricci tensors, and by adopting different systems of physical units. This chaos greatly hampers progress toward the chief NASA BPP goal: the design of the NASA Warp Drive. It is thus concluded that NASA should put order by establishing ***international standards in symbolic tensor calculus*** and enforcing anyone working in BPP to adopt these NASA BPP Standards.

INTRODUCTION

NASA firstly inaugurated the use of symbolic manipulators (variously ascribed as "computer algebra" codes, or "symbolic mathematics" codes, and so on) in the early 1960s. NASA's main problem at that time was to check the validity of a number of analytical results in Celestial Mechanics that had been found ***by hand*** in the previous 300 years. NASA decided to face the problem by creating a new code from scratch: "Macsyma", a lisp-written symbolic manipulator developed by the Artificial Intelligence Laboratory of MIT from 1965 on to 1982, and later taken over by private corporations for further developments. So, in the 1980s Macsyma was endowed with a ***tensor calculus package***, the first software in history capable of handling long analytical expressions for the Riemann, Ricci and stress-energy tensors required to compute analytical solutions of both the Einstein and combined Einstein-Maxwell equations.

When C became the stardard programming language, two new research companies were established to re-write "the mathematical knowledge of humankind" in C rather than in lisp. The outcome of this new effort were two more codes:
1) "Mathematica", developed by Wolfram Research Inc.;
2) "Maple V", developed by a team of the University of Waterloo (Canada).

Initially, both these codes did ***not*** have a tensor package, but several tensor packages were created in the 1990s and they are, respectively:
1) "MathTensor", written by Leonard Parker and Steven M. Christensen, which runs on Mathematica;
2) "GRTensorII", written by Kayll Lake and co-workers Peter Musgrave and Denis Pollney of Queens University in Kingston, Ontario, which runs on Maple V, currently under development (a smaller version of it runs on Mathematica also);
3) A further, separate tensor package is "Ricci", developed by John M. Lee of Washington University for Mathematica 3.0 (more details are included in the Ricci User Manual listed among the References).

CP504, *Space Technology and Applications International Forum–2000*, edited by M. S. El-Genk
© 2000 American Institute of Physics 1-56396-919-X/00/$17.00

This paper tries to determine which one of the above-mentioned codes posesses mathematical capabilities fitting best to the needs of BPP. The comparison is based on the experience of explicit calculations in the field of General Relativity and Quantum Field Theory that the author has performed over the last ten years, as soon as the latest releases of the above codes became available. Though MathTensor currently appears to be the "superior" code to face BPP problems, by no means does this mean that other codes cannot solve the same problems. In particular, GRTensorII has now reached such a degree of sophistication that it may well catch up with MathTensor soon, and prove most capable to do mathematical research in General Relativity.

Aside from the mathematical depth of these codes, at least two more issues appear to be of great importance for their present and future use to address BPP problems:
1) the *cost*, that varies enormously from one code to another; and
2) the question of the *conventions* that the international scientific community, and the BPP community in particular, might like to adopt universally in order to avoid the mess of different hypotheses and conventions currently plaguing research in General Relativity and Quantum Field Theory.
Let us briefly review these two topics.

As for the cost, it is pretty evident that a commercial "war" is now going on between the producers of the codes: while MathTensor is sold at at least $536.00 in the US (and more abroad), Macsyma costs around $300, while GRTensorII and Ricci come for free! How will this situation change when it is realized that all codes could help finding important breakthroughs in BPP ?

Should NASA "adopt" its own symbolic manipulator with strong tensor capabilities, just as, long ago, it created "Nastran" (i.e. NAsa's STRuctural ANalyzer, now commercially available and universally adopted by all aerospace corporations all over the word) ? And, in case NASA one day took such a decision, how long would we have to wait for such a "universal" tensor code? Wound't it be cheaper for NASA to extend and adapt to the needs of BPP one of the codes mentioned above, presumably MathTensor or GRTensorII ?

The last question is intimately related to one of the worst problems plaguing current reasearch in General Relativity and Quantum Field Theory: which *conventions* are going to be adopted by NASA's BPP leaders to simplify BPP research as much as possible? To understand this point a little better a comparison should be made with the situation in Europe before the French revolutionaries adopted (and imposed) the "metric system": that is, each country had its own "system of units", and a considerable amount of time had to be wasted just to make the various results numerically comparable! Nowadays, a rather similar situation appears to plague both General Relativity and Quantum Field Theory: different *conventions* on the metric signature, on the definition of which indices are contracted in the Riemann tensor to yield the Ricci tensor, and, of course, different unit systems other than the Système International units system (like esu, emu, Gaussian, Heaviside-Lorentz, gravitational and Planck systems) simply complicate life to such an extent that a number of excellent results in General Relativity and Quantum Field Theory obtained by either pure mathematicians, or theoretical physicists, or applied technologists and engineers are hardly comparable to each other, thus hampering their applications to BPP.

These and other problems are reviewed in this paper with the goal of finally paving the way to the design of the NASA Space Drive in the next millenium.

MACSYMA: THE FIRST SYMBOLIC MANIPULATOR ENDOWED WITH A TENSOR PACKAGE

The first "elementary math" packages of Macsyma were written since 1965 at the Artificial Intelligence Laboratory of the Massachusetts Institute of Technology. They were designed to provide NASA and the then-ongoing Apollo Program with a software capable of checking the analytical results in Celestial Mechanics that had been piling up in the previous 300 years with no practical possibility of checking them by hand again. For instance, the French astronomer Charles E. Delaunay (1816-1872) had spent ten years of his life (1847-1857) to calculate new and more accurate analytical results for the motion of the Moon, then spent ten more years (1857-1867) to check them, and finally published them in 1867. His results were taken for granted for over a century, basically because nobody who was going to devote twenty years of one's life to check Delaunay's results! But, in 1970, the advent of Macsyma allowed this "impossible" task to be performed: in just 20 hours of computer time André Deprit, Jacques Henrard and Arnold Rom, at the Boeing Scientific Research Labs in Seattle, checked all Delaunay's results! They found only one analitycal mistake: on page 234 of the second volum of his book titled "Théorie de la Lune" he incorrectly wrote one fraction as 13/16, whereas the correct fraction is 33/16. Two further errors were just a consequence of this one. This error in Delaunay's calculations turned out not to be vital for the Apollo flights to the Moon. But this success made of Macsyma the preferred world-wide symbolic computational tool for at least 25 years: from 1965 to 1990 and beyond.

Though the Macsyma source files are written in lisp, the user needs not know lisp at all, but, rather, he must know a sort of self-describing mathematical language to let Macsyma perform the requested calculations. This "Macsyma language" is described in both paper manual form and on-line help form. However, this programming language may not be immediately helpful to the user because the names for the commands are invented a bit "randomly", i.e. without strict logical rules that the programmer's mind may memorize immediately (see the "Macsyma Mathematics Reference Manual" and "Macsyma System Reference Manual" listed among the References).

GRTENSOR II:
THE MOST UPDATED TENSOR SYMBOLIC MANIPULATOR
RUNNING ON MAPLEV

Prof. Kyall Lake of Queen's University in Kingston, Ontario, Canada, and co-workers Peter Musgrave and Denis Pollney, created a new tensor package specifically designed to tackle research problems in General Relativity. This package is called "GRTensor II" and runs on Maple V. A limited version it runs on Mathematica also.

It can be downloaded *for free* from the web site: http: //www.astro.queensu.ca/~grtensor/GRHome.html. Very recently (October 1999) the main GRTensorII site has been changed to :
http://130.15.26.63/~grtensor/NewGRTensorII/
GRTensorII is a wonderful code from which newcomers to General Relativity can learn a lot. The leading creator, Prof. Kyall Lake, long ago had been working on the Tensor Package of Macsyma, identified the latter's limitations, and then went on to create GRTensorII.

To operate GRTensorII one has to learn:
1) The language of Maple V;
2) A dedicated "General Relativity" language to handle tensors, both old and "new", i.e. tensors at reasearch level like the Bel-Robinson tensor, and the like.
3) Many more advanced features of GRTensorII are quite remarkable: just to mention a few, the automatic Petrov Classification of Einstein spaces, the possibility of handling immediately solutions of the full set of Einstein-Maxwell equations, a number of worked-out-in-detail exact solutions (Bondi metric, Stephani solution for perfect fluids, Tomimatsu-Sato solution, Cosmological metrics, etc.). Five-dimensional metrics of the Kaluza-Klein type are also starting showing up among the worked examples, and this paves the way to future generalizations to N=11 dimensions as typical of supergravity, string theories, etc.
4) Last but by no mean least, GRTensorII has a unique package called "GRJunction" that allows the symbolic computations of exact solutions made up by "cutting, pasting and joining together" already known, easier exact solutions. This is a unique feature of GRTensorII, that no other tensor package has, including "MathTensor" described in the next section.

GRTensorII does, however, have some drawbacks, to our opinion: there is no way for the user to change the definitions of
1) the signature of the metric (assumed to be space-like);
2) the sign of the determinant of the metric (assumed to be negative);
3) the definition of the Riemann tensor in terms of the Christoffel symbols and their derivatives;
4) the definition of the Ricci tensor from the Riemann tensor, i.e., which indexes of Riemann one is going to contract.

This is serious inasmuch as one may get an annoying −1 sign when comparing the same exact solutions of the Einstein or Einstein-Maxwell equations by using different tensor packages, like GRTensorII and Macsyma, or GRTensorII and MathTensor.

Finally, GRTensorII does not use the MKS (or SI = Système International) system of units, and, there is no way for the user to change the definition of tensors accordingly.

In conclusion, GRTensorII is an excellent code (see the "GRTensor II User Manual" as from the References) currently under construction, that, with a little more work and willingness from the authors, could perhaps be selected by NASA to become the basic symbolic computational tool for facing all the BPPP problems. It must be noted, however, that the authors showed no intertest in adapting their code to the NASA BPPP needs.

MATHTENSOR 1.2:
THE MOST UPDATED TENSOR SYMBOLIC MANIPULATOR RUNNING ON MATHEMATICA

Professors Leonard Parker of the University of Wisconsin at Milwaukee and Steven M. Christensen of the University of North Carolina at Chapel Hill are the creators of MathTensor 1.2, the most udated tensor symbolic manipulator running on Mathematica, and one of the best such codes if not the best ! They also published a great book (listed among the References) as the MathTensor user manual.

Unlike GRTensorII, MathTensor 1.2 does not come for free: it cost \$536.00 in the US at the cheapest dealers and much more abroad, even twice as much.

The main advantage of MathTensor over its rivals is that definitions of :
1)   the signature of the metric;
2)   the sign of the determinant of the metric;
3)   the definition of the Riemann tensor in terms of the Christoffel symbols and their derivatives;
4)   the definition of the Ricci tensor from the Riemann tensor, i.e., which indexes of Riemann one is going to contract;
can be choosen independently by the user by just assigning four parameters, respectively:
MetricSign = 1, DetgSign = -1, RmSign =1 and RcSign = 1.

From this example one immediately remembers that both Mathematica and MathTensor 1.2 are both case-sensitive, so that great care must be paid by the user. Also the convention adopted in Mathematica of using long, compound input words, where upper case letters denote the beginning of a new word within the compound word, is kept in MathTensor just the same. This appears to be both bad and good because:
1)   It forces the author to input very long, case-sensitive words (bad);
2)   The meaning of this very long words is self-explanatory (good, to the contrary of what happens in both Macsyma and Maple V).

One more unique advantage of MathTensor 1.2 is that the user can freely adopt the preferred unit system: one just has to set the variable "Units" equal to either of the following names in order to have all equations computed in the relevant units system, respectively: emuUnits, esuUnits, GaussianUnits, HeavisideLorentzUnits, RationalizedMKSUnits or SIUnits. This is a non-trivial feature of MathTensor, since the definition itself of some tensors, like the stress-energy tensor of the electromagnetic field, depends on conventions adopted about the units system. So the amount of time saved by the user changing results from one unit system to another one is remarkable. Unfortunately, MathTensor is not currently capable of handling Greek letters, and so symbols like ε_0 and μ_0 have to be written eps0 and mu0. Worse still, the user interface of MathTensor is very primitive: one has to edit the file Comventions.m in order to set up all the above conventions once and for all.

Still MathTensor is indeed a mathematical treasure that must have taken its two authors years of work. As a final example, there is one topic that neither Macsyma nor GRTensorII can face at the moment: deriving the set of tensor equations (like the Einstein equations) as the Euler-Lagrange equations from a suitable Variational Principle (like the Einstein-Hilbert Action). This can be accomplished by MathTensor by partial integration firstly, and computation of the Euler-Lagrange equations, secondly. The result is that, even for Lagrangians that are quadratic in the Riemann tensor, obtaining the relevant differential equations is virtually immediate.

TENSOR SYMBOLIC MANIPULATORS IN THE YEAR 2000, AND THE NEED FOR A NASA UNDERTAKING IN THIS FIELD

At the beginning of the new millenium, one can thus summarize the situations as follows:
1)   Two symbolic tensor manipulators are taking the lead: GRTensorII and MathTensor 1.2, run on Maple V Release 5 and Mathematica 3.0, respectively.
2)   In principle, all the knowledge piled up in a century of Relativity and Quantum Field Theory could be re-written by virtue of either of these codes. However,
3)   Only an extremely small amount of relativistic knowledge has been transcribed so far, and even that was done with different conventions about the unit system, metric, metric determinant, and definitions of the Riemann and Ricci tensors.

In order to end the confusion, and stimulate the development of the most effective computational tools, I believe NASA should consider imposing a *standard* in symbolic tensor calculations to achieve the supreme goal of *designing the space drive.*

Suggested conventions that NASA might adopt seem to fall into two categories: obvious and less obvious.

Obvious conventions are:
1) The adoption of the Rationalized MKS System of Units (Système International) as presently done by most engineers all over the world. This implies rejecting all "clever" systems used by physicists and mathematicians where $c = 1$ and/or $G = 1$ and/or $h = 1$ and the like, and instead, to use the physically observable units and values.
2) The adoption of a standard language and a standard mathematical notation for tensor calculus, where everybody can understand what is what.
3) The paper format and the fonts adopted for all tensor symbols;
4) The introduction of different colours on both computer screens and printouts might greatly help distinguishing between mathematical input code, mathematical output code, and comments.

In turn, all this leads to adopting the less obvious convention:
5) Should GRTensorII or MathTensor be adopted ? In other words,
6) Should Maple V or Mathematica be adopted ?

CONCLUSIONS

The last question – should Maple V or Mathematica be adopted ? - is admittedly a selection difficult to make, because there are large economic interests behind the selection.

NASA should probably start up preliminary contacts with all creators of the codes. It seems clear, however, that such an endevour as designing the Space Drive should rely on a large team of experts, rather than on a small one. Now, the overall amount of symbolic software already available in the year 2000 seems much larger for Mathematica than for Maple V. So, in the view of the present author, Mathematica and MathTensor should unavoidably become the standards to be adopted by NASA to design the Space Drive. However, NASA should enforce major improvements on MathTensor, especially in the User Interface, that should be much more friendly than the current interface which requires frequent modification of the Conventions.m file and the like.

The terrible task of putting order into a century of wonderfully messy results awaits the current and future designers of the NASA Space Drive.

DISCLAIMER

The views expressed in this paper are exclusively those of the author.

ACKNOWLEDGMENTS

The author is grateful to Dr. Richard Petti of Macsyma Inc. for a years-long cooperation in the development of this code, with special reference to handling basic tensors for General Relativity. Occasional exchange of viewpoints also occurred with Prof. Kyall Lake, main creator of GRTensor II, and Prof. Steve Christensen, one of the two co-authors of MathTensor. The key role that these symbolic codes may succesfully play in the development of BPP research was firstly recognized by the pioneer leader of NASA BPP, the late Dr. John Anderson of NASA HDQ. His successor, Dr. Marc Millis of NASA Glenn Research Center at Lewis Field, has then continued John Anderson's work by organizing BPP Sessions at several recent STAIF Conferences in Albuquerque, NM. The author of this would like to acknowledge Marc Millis for his interest to a question as vital as "how will BPP benefit from symbolic tensor manipulators ? ". Finally, Professor Emeritus Jordan Maclay was the wonderful co-chair of Session C7 that this author run for STAIF2000: his deep insight and great help was very much appreciated.

REFERENCES

John M. Lee, "Ricci – A Mathematica Package for Doing Tensor Calculations in Differential Geometry – Version 1.33," User Manual downloadable for free from the site http://www.math.washington.edu/~lee/Ricci/.

Kayll Lake, Peter Musgrave and Denis Pollney, "GRTensor II User Manual," downloadable for free from the site http://130.15.26.62/NewDemo/frame.html.

Leonard Parker and Steven M. Christensen, "MathTensor – A System for Doing Tensor Analysis by Computer," Addison Wesley Publishing Company, Reading, MA, 1994.

Various Authors, "Macsyma Mathematics Reference Manual" and "Macsyma System Reference Manual," Macsyma Inc., Arlington, MA, 1993.

Alcubierre's Warp Drive: Problems and Prospects

Chris Van Den Broeck

Starlab, Excelsiorlaan 40-42, 1930 Zaventem, Belgium
E-mail: vdbroeck@starlab.net

Abstract. Alcubierre's warp drive geometry seemingly represents the ultimate dream for interstellar travel: there is no speed limit, the passengers are weightless whatever the acceleration, and there is no time dilation. However, in its original form, the proposal suffers from several fatal flaws, such as unreasonably high energies, energy moving in a locally spacelike direction, and a violation of the energy conditions of classical Einstein gravity. I present a possible solution for one of these problems, and I suggest ways to at least soften the others.

INTRODUCTION

Alcubierre introduced an entirely novel concept for superluminal travel (Alcubierre, 1994). Put simply, the idea was to start from Minkowski spacetime, choose an almost arbitrary curve, and deform spacetime in the immediate vicinity in such a way that the curve became a timelike geodesic. This implied that anyone traveling on such a curve would be weightless, whatever the acceleration with respect to observers in the flat space outside. The metric is

$$ds^2 = -dt^2 + (dx - v(t)f(r(t,x,y,x))dt)^2 + dy^2 + dz^2, \tag{1}$$

where $r = \sqrt{(x - x_s(t))^2 + y^2 + z^2}$, with $x_s(t)$ the path of the central geodesic; $v(t) = \frac{dx_s(t)}{dt}$ is the apparent velocity of the warp drive. The function $f(r)$ is zero outside the warp bubble and 1 inside. The passengers of a warp drive would live in a bubble of flat spacetime, surrounded by a 'wall' in which space is distorted in such a way that it contracts in the front and expands in the back, thus at the same time 'pushing' and 'pulling' the bubble. The proper time of the central geodesic would be the same as the external coordinate time, so that there would be no time dilation. And the most interesting feature was, of course, the apparent absence of any speed limit.

This sounded promising for interstellar travel, but the warp drive has its share of troubles. I will try to provide a comprehensive outline of the problems that have been identified over the past five years, and look at possible solutions. I will keep the discussion on a descriptive level, and not overload it with formulas. The detailed calculations can be found in the references.

PROBLEMS

Alcubierre himself immediately pointed out that his geometry suffered from a problem that also plagues traversable wormholes: a violation of at least three of the energy conditions of general relativity, specifically the weak, strong, and dominant energy conditions (WEC, SEC, and DEC). These conditions essentially impose positivity on the energy density/pressures; for instance, the WEC demands that the energy density measured locally by any observer should be positive at any time. See (Hawking, 1987) for an overview. Energy that violates the energy conditions is usually called 'exotic matter'.

CP504, *Space Technology and Applications International Forum–2000*, edited by M. S. El-Genk
© 2000 American Institute of Physics 1-56396-919-X/00/$17.00

Although the energy conditions have been used as input for many important theorems of classical general relativity in the past ((Hawking, 1987) provides a host of examples), their physical validity is no longer credible. For example, the WEC, SEC and DEC are violated by the well–known Casimir effect (Casimir, 1948), although the violations are tiny. In its simplest realization, the Casimir effect entails a difference between the zero point energy of the electromagnetic vacuum energy in between and outside two parallel conducting plates. The boundary conditions imposed by the plates restrict the number of zero modes in between, resulting in an energy density that is lower than that of the vacuum in the absence of the plates. One consequence of the effect, an attractive force between the plates, was first shown experimentally more than forty years ago (Spaarnaay, 1958). Other quantum effects are known to violate energy conditions (Visser, 1995).

Even though the energy conditions are thought to be violated in Nature, there might be restrictions on the extent of the violations. Ford and Roman (Ford, 1996) have argued that WEC violations originating from vacuum fluctuations of quantum fields are subject to an uncertainty principle: if $\hat{\rho}$ is the energy density measured by an inertial observer, averaged over a 'sampling time' τ_0, where τ_0 is much smaller than the local curvature radius of spacetime divided by c, then

$$\hat{\rho}\tau_0^4 \geq -\frac{3}{32\pi^2}\frac{\hbar}{c^3}. \tag{2}$$

Considering that $\frac{\hbar}{c^3} \sim 3 \times 10^{-60}\frac{\text{J}}{\text{m}^3}\text{s}^4$, and taking into account that the fourth power of the sampling time appears, this is a severe restriction.

Using this so–called quantum inequality (QI), Ford and Pfenning (Ford, 1997) showed that the warp bubble wall should be no thicker than a hundred Planck lenghts for velocities in the order of c; otherwise, inertial observers crossing the wall would measure negative energy for too long. In addition, they calculated that the total energy of a macroscopic warp bubble is given by

$$E = -\frac{1}{12}v^2\frac{c^2}{G}\frac{R^2}{\Delta}, \tag{3}$$

where R is the bubble radius, Δ its wall thickness, and v the apparent velocity of the warp drive. For a bubble with a radius of 100 m, a wall thickness of 100 l_P and a velocity $v \sim c$, one gets roughly

$$E \sim -10^{63}\text{kg} \times c^2, \tag{4}$$

which, in absolute value, is ten orders of magnitude larger than the energy content of the entire visible Universe.

Some physicists might raise yet another objection, namely that the basic warp drive geometry can easily be used for creating a spacetime with closed timelike curves (CTCs) (Everett, 1996). Indeed, it is easy to see that any means of superluminal travel within an asymptotically flat spacetime will appear to be time travel for some observers in the quasi–Minkowskian region: when an interval is crossed which is considered spacelike by a quasi–Minkowskian observer, there will always be an observer for whom the arrival takes place before the departure. Hawking's Chronology Protection Conjecture (Hawking, 1991) forbids CTCs, but it has never been derived from first principles; rather, it is based on a number of examples suggesting that quantum fluctuations in the energy density of fields will diverge in the presence of CTCs. An explicit counterexample – at least for the case of non–gravitational fields – is provided in (Li–Xin Li, 1994). Of course, the absence of divergences in itself does not guarantee that CTCs are allowed by Nature. The debate continues.

However, the original Alcubierre geometry has a much more fundamental problem. This is the behaviour of negative energy density in the outer layers of the warp bubble. It is easy to see that, outside some surface in the warp bubble wall (the 'critical surface'), the energy will move with spacelike velocity, i.e. it will move *locally* faster than light (Krasnikov,1998; Coule, 1998). This is forbidden both by classical general relativity and quantum field theory. For reasons that will become clear later on, it has (rather euphemistically) been referred to as the 'control problem' (Everett, 1997; Krasnikov, 1998); here, I will call it the 'tachyonic motion problem' (TMP). It would seem that solving the TMP is the make–or–break issue for 'warp drive engineering'.

TENTATIVE SOLUTIONS

By now it should be clear that Alcubierre's proposal, when taken literally, is not realistic. However, Alcubierre himself considered his geometry to be no more than an ansatz. I will try to show that altering the spacetime in a judicious way can bring us a long way in solving the problems.

First, let us have a look at the total energy problem. In (Van Den Broeck, 1999a), it was shown that altering the Alcubierre geometry in a simple way could dramatically reduce the energy required. Basically, the idea was to keep the warp bubble subatomically small, so that its energy would be small, but 'blow up' its interior, so that a large spaceship would fit inside. The modified geometry is defined by

$$ds^2 = -dt^2 + B^2(r)[(dx - v(t)f(r(t, x, y, z))dt)^2 + dy^2 + dz^2], \tag{5}$$

where the function B determines the space blow–up. B is chosen such that it is identically 1 outside the warp bubble and in the wall, becoming very large towards the center, and then leveling off to allow for a flat interior. The general structure of such a modified warp drive is shown in Figure 1.

Calculating the energy density as measured by inertial observers and integrating over a constant time hypersurface, one finds that the blow–up requires an amount of positive energy of the order of a solar mass, and an amount of negative energy of the same magnitude. This puts the modified Alcubierre geometry in the energy bracket of a large traversable wormhole (Visser, 1995).

An important problem is that the energy densities (in absolute value) of both the bubble wall and the space blow–up region, attain values of no less than 10^{94} kg/m^3. In (Van Den Broeck, 1999a), great care was taken that the QI would not be violated. Even so, a mechanism for creating such large densities is sorely lacking; all the effects we know of that violate energy conditions are microscopic quantum effects in *density* as well as in size. For the blow–up, it might still be possible to lower the density by making a better choice for B than (??), but this is not the case for the bubble wall, since its energy density in e.g. a plane through

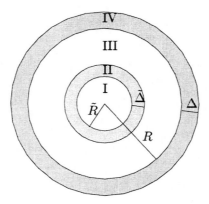

FIGURE 1. Region I is the 'pocket', which has a large inner metric diameter. II is the transition region from the blown–up part of space to the 'normal' part. It is the region where B varies. From region III outward we have the original Alcubierre metric. Region IV is the wall of the warp bubble; this is the region where f varies. Spacetime is flat, except in the shaded regions.

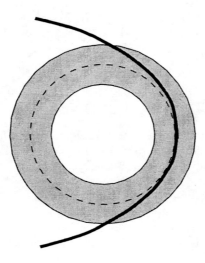

FIGURE 2. A schematic representation of a warp bubble. The function f varies in the shaded region. The dotted circle represents the critical surface, while the thick line is the horizon. The region where B varies is not drawn.

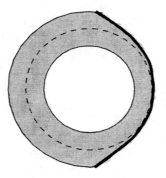

FIGURE 3. An independently moving warp bubble. The dotted line again represents the critical sphere; the thick line now indicates a singular surface.

the center and perpendicular to the direction of motion is

$$T_{00} = -\frac{v^2}{32\pi} \left(\frac{df(r)}{dr}\right)^2,$$ (6)

and because of the fact that the wall is constrained to be so thin, *df/dr must* become large somewhere.

In the previous section, I referred to the TMP as the most important problem of warp drive geometries. There exists a 'critical surface' such that all negative matter outside the sphere moves tachyonically. A solution was hinted at in (Van Den Broeck, 1999b). There, it was noted that superluminal motion can be simulated. A pulsar provides a simple example: the spot where the pulsar beam intersects a sphere with a sufficiently large radius will appear to move faster than light. In the warp drive, a signal could be sent to a point in the bubble wall, diminishing the exotic matter density, and another signal to a point further ahead, spacelike separated from the first, where the density should be increased. Now, some parts of the warp bubble wall are inaccessible for signals sent by the 'pilot' of the warp bubble, due to a horizon that cuts him off from the front of the bubble (see Figure 2).

One might draw the conclusion that the warp drive will need continuous support from outside, possibly by devices placed in advance along the path of the bubble. However, this is not necessary; if the pilot keeps controlling the parts of the bubble that are accessible to him, the Alcubierre geometry will certainly be maintained behind the horizon. The exotic matter in front of the horizon will, in a short time, be overtaken by the superluminally moving bubble. This will lead to a surface singularity; at the horizon, there will be a discontinuous transition from Minkowski spacetime (in front) to Alcubierre spacetime (behind). The result is shown schematically in Figure 3.

The surface singularity will not inconvenience the pilot; once the flight is over, he can 'engineer' the back of the warp bubble in such a way that he drops out of it, never encountering the singularity (which he wouldn't be able to reach, even if he wanted to). It should be emphasized that at present, there is no practical proposal for implementing the control process described above. This is work for the future.

CONCLUSIONS

Apparently, the theoretical problems with the warp drive do not seem to be insurmountable. Now, how soon will we be able to build one? I've discussed the TMP in some detail above, so here I'll focus on the other problems.

To avoid unphysically large energies, the warp bubble has to be kept subatomically small, so we would have to learn how to manipulate spacetime on such small scales. Let me add immediately that the energy problem is a consequence of the fact that the bubble wall needs to be thin, which is imposed by the QI. However, the validity of the QI in highly curved spacetimes is not beyond dispute; after all, it was not proven rigorously. Should the QI be false, then it is clear from expression (3) that the absolute value of the total energy can be made as small as one pleases, and the construction of (Van Den Broeck, 1999a) would be unnecessary. However, if they are valid, it seems we would need to have access to large amounts of positive as well as negative energies.

To generate the negative energy densities, we will need some quantum effect, and it will need to be a macroscopic one. It is not clear how e.g. the Casimir effect can be made large. The quantum vacuum has not yet yielded to our desires, but there is no reason why that couldn't happen in the foreseeable future.

What all this boils down to is: chances are slim that the warp drive will be built in our lifetime. But science has a whole new millennium ahead of it...

ACKNOWLEDGMENTS

I would like to thank D.H. Coule, E.W. Davis, F. Denef, P.–J. De Smet, and L.H. Ford for very helpful comments.

REFERENCES

Alcubierre, M., "The Warp Drive: Hyper-Fast Travel within General Relativity," Class. Quantum Gravity **11**, L73–L77 (1994).

Casimir, H.B.G., "On the Attraction between two Perfectly Conducting Plates," Proc. Kon. Nederl. Akad. Wet. **51**, 793-796 (1948).

Coule, D.H., "No Warp Drive," Class. Quantum Grav. **15**, 2523–2527 (1998).

Everett, A.E., "Warp Drive and Causality," Phys. Rev. D **53**, 7365–7368 (1996).

Everett, A.E., and Roman, T.A., "A Superluminal Subway: the Krasnikov Tube," Phys. Rev. D **56**, 2100–2108 (1997).

Ford, L.H., and Roman, T.A., "Quantum Field Theory Constrains Traversable Wormhole Geometries," Phys. Rev. D **53**, 5496–5507 (1996).

Ford, L.H., and Pfenning, M.J., "The Unphysical Nature of 'Warp Drive'," Class. Quantum Grav. **14**, 1743–1751 (1997).

Hawking, S.W., and Ellis, G.F.R., *The Large–Scale Structure of Space–Time*, Cambridge, England, Cambridge University Press, 1987.

Hawking, S.W., "The Chronology Protection Conjecture," Phys. Rev. D **46**, 603–611 (1991).

Krasnikov, S.V., "Hyperfast Interstellar Travel in General Relativity," Phys. Rev. D **57**, 4760–4766 (1998).

Li-Xin Li, "New Light on Time Machines: against the Chronology Protection Conjecture," Phys. Rev. D **50**, 6037–6040 (1994).

Olum, K.D., "Superluminal Travel requires Negative Energies," Phys. Rev. Lett. **81**, 3567–3570 (1998).

Spaarnaay, M.J., Physica **24**, 751-764 (1958).

Van Den Broeck, C., "A 'Warp Drive' with reasonable total energy requirements," to be published; http://xxx.lanl.gov /abs/gr-qc/9905084 (1999a).

Van Den Broeck, C., "On the (im)possibility of Warp Bubbles," to be published; http://xxx.lanl.gov/abs/ gr-qc/9906050 (1999b).

Visser, M., *Lorentzian Wormholes*, New York, American Institute of Physics, 1995.

Wormholes and Time Travel

P. C. Aichelburg

Institut für Theoretische Physik, Universität Wien
Boltzmanngasse 5, A-1090 Wien, Austria

pcaich@doppler.thp.univie.ac.at

EXTENDED ABSTRACT

In one of his books, the science fiction author Stanislaw Lem (1978) describes an astronaut travelling alone in his spaceship with a broken steering which he is unable to repair. The spacecraft is sucked into a strong gravitational field and suddenly there is a second person in the spaceship who turns out to be the same astronaut but from the next day. Then a discussion starts whether or not it makes sense to try to repair the spaceship together.

Already Einstein (1914) worried that his theory of relativity might allow for spacetimes with so-called closed timelike curves. Gödel (1949) constructed a cosmological model where this phenomena can happen, however at the cost of an enormous amount of energy for the journey. More recently renewed interest focussed on the possibility of constructing such time machines with the help of "wormholes". Wormholes are spacetimes with nontrivial topology in which a kind of tunnel exists connecting distant parts in the universe. These wormholes may not only serve as shortcuts in space but also for timetravel.

That the famous Schwarzschild black hole contains a wormhole was realized already by Flamm (1916) and later discussed by Einstein and Rosen (1935). This wormhole however is non-traversable. The problem is that before a light ray or an observer reaches the other side of the throat the wormhole shrinks under its own gravitational force to a singularity. In order to keep the throat open one has to introduce "exotic" matter which violates the classical energy conditions. Such wormholes were considered by Morris and Thorne (1988) (inspired by Sagan (1985)), Visser (1995) and others. There is another problem of a more technical nature: if the wormhole is to connect different parts in the same asymptotic region of the universe, gravitational attraction will pull the mouths together and spacetime will not be static, making the explicit construction of such a solution extremely difficult. In the paper "String-supported wormhole spacetimes containing closed timelike curves" (1996), we were able to overcome this difficulty by anchoring two strings to the wormhole mouths held at infinity to prevent them from falling towards each other. We thus obtained an axisymmetric static spacetime containing a traversable wormhole. However, this construction requires, like others, exotic matter. Recently Schein and the author (1996); (1998) constructed a wormhole with the following properties: spacetime is asymptotically flat, axisymmetric and static. It contains a one-way traversable wormhole connecting to the same asymptotic region. Moreover, it is not necessary to violate the energy conditions. On this spacetime closed timelike curves exist from any point in the asymptotic region.

Two important theorems about wormhole spacetimes are known: Hawking (1992) in his paper on "Chronology projection conjecture" showed, loosely speaking, that for the construction of a time machine one necessarily needs to violate the energy conditions. Friedman *et al.* (1993), on the other hand, proved a "topology protection theorem" by which it is impossible, under certain assumptions, to probe the nontrivial topology, i.e. travelling or sending light rays through the wormhole from the asymptotic region. Neither of these theorems applies to our construction: Hawking's theorem refers to spacetimes where closed causal curves exist from a certain time on (or up to a certain time), while our solution is an eternal time machine. Friedman's conclusion requires that spacetime is globally hyperbolic, a requirement which is not met by our construction. Whether or not this is physically acceptable is open.

CP504, *Space Technology and Applications International Forum–2000*, edited by M. S. El-Genk
© 2000 American Institute of Physics 1-56396-919-X/00/$17.00

REFERENCES

Aichelburg, P.C. and Schein, F. 1998, Acta Phys. Pol. **29B**, 1025

Einstein, A. 1914, Berliner Berichte **29**, 1030

Einstein. A. and Rosen, N. 1935, Phys. Rev. **48**, 73

Flamm, L. 1916, Physik. Zeitschr. XVII, 448

Friedman, J.L., Schleich, K. and Witt, D.M 1993, Phys. Rev. Lett. **71**, 1486

Gödel, K. 1949, Rev. Mod. Phys. **21**, 447

Hawking. S. W. 1992, Phys. Rev. **D46**, 603

Morris, M. and Thorne, K. 1988, Am. J. Phys. **56**, 395

Lem, S. 1978, *"Sterntagebücher"*, Suhrkamp

Sagan, C. 1985, *"Contact"*, Simon & Schuster

Schein, F., Aichelburg, P.C. and Israel, W. 1996, Phys. Rev. **D54**, 3800

Schein, F. and Aichelburg, P.C. 1996, Phys. Rev. Lett. **77**, 4130

Visser, M. 1995, *"Lorentzian Wormholes: From Einstein to Hawking"*, AIP Press

Toward a Traversible Wormhole

Serguei Krasnikov

The Central Astronomical Observatory at Pulkovo, St. Petersburg, 196140, Russia

Gennady.Krasnikov@pobox.spbu.ru

Abstract. In this talk I discuss pertinence of the wormholes to the problem of circumventing the light speed barrier and present a specific class of wormholes. The wormholes of this class are static and have arbitrarily wide throats, which makes them traversable. The matter necessary for these spacetimes to be solutions of the Einstein equations is shown to consist of two components, one of which satisfies the Weak energy condition and the other is produced by vacuum fluctuations of neutrino, electromagnetic (in dimensional regularization), and/or massless scalar (conformally coupled) fields.

WORMHOLES AND THEIR RELATION TO HYPER-FAST TRAVEL

Wormholes are geometrical structures connecting two more or less flat regions of a spacetime. This of course is not a rigorous definition, but, strange though it may seem, there is no commonly accepted rigorous definition of the wormhole yet. Normally, however, by a wormhole a spacetime is understood resembling that obtained by the following manipulation:
1) Two open balls are removed each from a piece of approximately flat 3-space (the vicinities of thus obtained holes we shall call *mouths* of the wormhole);
2) The boundaries (2-spheres) of the holes are glued together, and the junction is smoothed. In the process of smoothing a kind of tube arises interpolating the spheres. We shall call this tube the *tunnel* and its narrowest part the *throat*.

The resulting object (its two-dimensional version to be precise) is depicted in Fig.1. If in the course of evolution the spacetime surrounding such an object remains approximately flat (which may not be the case, since flatness of each 3-dimensional section does not guarantee that the 4-dimensional space formed by them is also flat) we shall call the object a wormhole.

Wormholes arise in a natural way in general relativity. Even one of the oldest and best-studied solutions of the Einstein equations — the Schwarzschild spacetime — contains a wormhole, which was found at least 80 years ago (Flamm, 1916). This wormhole (also known as the Einstein-Rosen bridge) connects two asymptotically flat regions ('two universes'), but being non-static is useless in getting from one of them to the other (see below).

Depending on whether the tunnel connects distant regions of 'the same' universe or these regions are otherwise not connected the wormholes fall into two categories (Visser, 1995). To which category a wormhole belongs depends on how the vicinities of the mouths are extended to the full spacetime:

It may happen that the mouths cannot be connected by any curve except those going through the tunnel (as it takes place in the Einstein-Rosen bridge). Such wormholes are called *inter-universe*. A simplest static spherically symmetric inter-universe wormhole can be described (Morris, 1988) by a manifold $R^2 \times S^2$ endowed with the metric

CP504, *Space Technology and Applications International Forum–2000*, edited by M. S. El-Genk
© 2000 American Institute of Physics 1-56396-919-X/00/$17.00

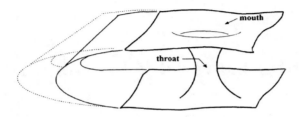

FIGURE 1. The sketch of a wormhole with the mouths in motion. One dimension (corresponding to the coordinate ϑ) is omitted. The ways in which the upper and the lower parts are glued at $t = 0$ and at $t = 1$ are depicted by thin solid lines and by dashed lines respectively. Though the geometry of the wormhole does not change, the distance (as measured in the outer, flat space) between mouths increases with time.

$$ds^2 = -e^{2\Phi}\, dt^2 + dr^2/(1 - b/r) + r^2(\mathrm{d}\theta^2 + \sin^2\theta\, d\phi^2), \tag{1}$$

where $r \in (-\infty, \infty)$ (note this possibility of negative r, it is the characteristic feature of the wormholes), $\Phi(r) \to 0$ and $b(r)/r \to 0$, when $r \to \pm\infty$.

Alternatively as shown in Fig.1 it may happen that there are curves from one mouth to another lying outside the wormhole. Such a wormhole connects distant parts of a 'single' universe and is called *intra-universe*. Though intra-universe wormholes are in a sense more interesting most papers deal with inter-universe ones, since they are simpler. It does not matter much, however. The distant regions of the 'universes' are taken to be approximately flat. And it is usually implied that given an inter-universe wormhole we can as well build an intra-universe one by simply gluing these distant regions in an appropriate way.

It is stable intra-universe wormholes that are often used for interstellar travel in science fiction (even though they are sometimes called 'black holes' there). Science fiction (especially Sagan's novel *Contact*) has apparently acted back on science and in 1988 Morris and Thorne pioneered investigations (Morris, 1988) of what they called *traversible wormholes* — wormholes that can be (at least in principle) traversed by a human being. It is essential in what follows that to be traversible a wormhole should satisfy at least the following conditions:

(C1). It should be sufficiently long-lived. For example the Einstein-Rosen bridge connects two asymptotically flat regions (and so it is a wormhole), but it is not traversible — the throat collapses so fast that nothing (at least nothing moving with $v \leq c$) can pass through it.

(C2). It should be *macroscopic*. Wormholes are often discussed [see (Hochberg, 1997), for example] with the radius of the throat of order of the Plank length. Such a wormhole might be observable (in particular, owing to its gravitational field), but it is not obvious (and it is a long way from being obvious, since the analysis would inevitably involve quantum gravity) that any signal at all can be transmitted through its tunnel. Anyway such a wormhole is impassable for a spaceship.

Should a traversible wormhole be found it could be utilized in interstellar travel in the most obvious way. Suppose a traveler (say, Ellie from the above-mentioned novel) wants to fly from the Earth to Vega. One could think that the trip (there and back) will take at least 52 years for her even if she moves at a nearly light speed. But if there is a wormhole connecting the vicinities of the Earth and Vega she can take a short-cut by flying through it and thus make the round trip to Vega in (almost) no time.

Note, however, that such a use of a wormhole would have had nothing to do with circumventing the light barrier. Indeed, suppose that Ellie's start to Vega is appointed on a moment $t = 0$. Our concern is with the time interval Δt_E in which she will return to the Earth. Suppose that we know (from astronomical observations, theoretical calculations, etc.) that if in $t = 0$ she (instead of flying herself) just emit a photon from the Earth, this photon after reaching Vega (and, say, reflecting from it) will return back at best in a time interval Δt_P. If we find a wormhole from the Earth to Vega, it would only mean that Δt_P actually is small, or in other words that Vega is actually far closer to the Earth than we think now. But what can be done if Δt_P is large (one would hardly expect that traversible wormholes can be found for *any* star we would like to fly to)? That is where the need in hyper-fast transport comes from. In other words, the problem of circumventing the light barrier (in connection with interstellar travel) lies in the question: how to reach a remote (i. e. with the large Δt_P) star and to return back sooner than a photon would have made it (i. e. in $\Delta t_E < \Delta t_P$)? It makes sense to call a spaceship faster-than-light (or *hyper-fast*) if it solves this prolem.

A possible way of creating hyperfast transport lies also in the use of traversible wormholes (Krasnikov, 1998). Suppose that a traveler finds (or builds) a traversible wormhole with both mouths located near the Earth and suppose that she can move the mouths (see Fig.1) at will without serious damage to the geometry of the tunnel (which we take to be negligibly short). Then she can fly to Vega taking one of the mouths with her. Moving (almost) at the speed of light she will reach Vega (almost) instantaneously by her clocks. In doing so she *rests* with respect to the Earth insofar as the distance is measured through the wormhole. Therefore her clocks remain synchronous with those on the Earth as far as this fact is checked by experiments confined to the wormhole. So, if she return through the wormhole she will arrive back to the Earth almost immediately after she will have left it (with $\Delta t_E << \Delta t_P$).

Remark 1. The above arguments are very close to those showing that a wormhole can be transformed into a time machine (Morris, 1988), which is quite natural since the described procedure is in fact the first stage of such transformation. For, suppose that we move the mouth back to the Earth reducing thus the distance between the mouths (in the ambient space) by 26 light years. Accordingly Δt_E would lessen by ≈ 26 yr and (being initially very small) would turn *negative*. The wormhole thus would enable a traveller to return before he have started. Fortunately, $\Delta t_E \approx 0$ would fit us and we need not consider the complications (possible quantum instability, paradoxes, etc.) connected with the emergence of thus appearing time machine.

Remark 2. Actually two *different* worlds were involved in our consideration. The geometry of the world where only a photon was emitted differs from that of the world where the wormhole mouth was moved. A photon emitted in $t = 0$ in the latter case would return in some $\Delta t_{P'} < \Delta t_E$. Thus what makes the wormhole-based transport hyper-fast is changing (in the causal way) the geometry of the world so that to make $\Delta t_{P'} < \Delta t_E << \Delta t_P$.

Thus we have seen that a traversible wormhole can possibly be used as a means of 'superluminal' communication. True, a number of serious problems must be solved before. First of all, where to get a wormhole? At the moment no good recipe is known how to make a *new* wormhole. So it is worthwhile to look for 'relic' wormholes born simultaneously with the Universe. Note that though we are not used to wormholes and we do not meet them in our everyday life this does not mean by itself that they are an exotic rarity in nature (and much less that they do not exist at all). At present there are no observational limits on their abundance [see (Anchordoqui, 1999) though] and so it well may be that there are 10 (or, say, 10^6) times as many wormholes as stars. However, so far we have not observed any. So, this issue remains open and all we can do for the present is to find out whether or not wormholes are allowed by known physics.

CAN TRAVERSIBLE WORMHOLES EXIST?

Evolution of the spacetime geometry (and in particular evolution of a wormhole) in general relativity is determined via the Einstein equations by properties of the matter filling the spacetime. This circumstance may turn out to be fatal for wormholes if the requirements imposed on the matter by conditions (C1,C2) are unrealistic or conflicting. That the problem is grave became clear from the very beginning: it was shown (Morris, 1988), see also (Friedman, 1993), that under very general assumptions the matter filling a wormhole must violate the Weak Energy Condition (WEC) The WEC is the requirement that the energy density of the matter be positive in any reference system. When the stress-energy tensor T_{ik} is diagonal the WEC may be written as

$$\text{WEC:} \qquad T_{00} \geq 0, \quad T_{00} + T_{ii} \geq 0, \qquad i = 1, 2, 3 \tag{2}$$

Classical matter always satisfies the WEC (hence the name '*exotic*' for matter violating it). So, a wormhole can be traversible only if it is stabilized by some quantum effects.

Candidate effects are known, indeed [quantum effects can violate any local energy condition (Epstein, 1965)]. Moreover, owing to the non-trivial topology a wormhole is just a place where one would expect WEC violations due to fluctuations of quantum fields (Khatsymovsky, 1997a). So, the idea appeared (Sushkov, 1992) to seek a wormhole with such a geometry that the stress-energy tensor produced by vacuum polarization is exactly the one necessary for maintaining the wormhole. An example of such a wormhole (it is a Morris-Thorne spacetime filled with the scalar non-minimally coupled field) was offered in (Hochberg, 1997). Unfortunately, the diameter of the wormhole's throat was found to be of the Plank scale, that is the wormhole is non-traversible.

The situation considered in (Hochberg, 1997) is of course very special (a specific type of wormholes, a specific field, etc.). However arguments were cited [based on the analysis of another energetic condition, the so called ANEC (Averaged Null Energy Condition)] suggesting that the same is true in the general case as well (Flanagan, 1996, see also the literature cited there). So an impression has been formed that conditions (C1) and (C2) are incompatible, and TWs are thus impossible.

YES, IT SEEMS THEY CAN

The question we are interested in is whether such macroscopic wormholes exist that they can be maintained by the exotic matter produced by the quantum effects. To put it more mathematically let us first separate out the contribution T^Q_{ik} of the 'zero-point energy' to the total stress-energy tensor T_{ik}:

$$T_{ik} = T^Q_{ik} + T^C_{ik}. \tag{3}$$

In semiclassical gravity it is deemed that for a field in a quantum state $|\Psi>$ (in particular, $|\Psi>$ may be a vacuum state) $T^Q_{ik} = <\Psi| T_{ik} |\Psi>$, where T_{ik} is the corresponding operator, and there are recipes for finding T^Q_{ik} for given field, metric, and quantum state [see, for example, (Birrel, 1982)]. So, in formula (3) T^Q_{ik} and T_{ik} are determined by the geometry of the wormhole and the question can be reformulated as follows: do such macroscopic wormholes exist that the term T^C_{ik} describes usual non-exotic matter, or in other words that T^C_{ik} satisfies the Weak Energy Condition, which now can be written as

$$G_{00} - 8\pi T^Q_{00} \geq 0, \quad (G_{00} + G_{ii}) - 8\pi(T^Q_{00} + T^Q_{ii}) \geq 0, \qquad i = 1, 2, 3. \tag{4}$$

(we used the formulas (2,3) here)?

One of the main problem in the search for the answer is that the relevant mathematics is complicated and unwieldy. A possible way to obviate this impediment is to calculate T^Q_{ik} numerically (Hochberg, 1997; Taylor 1997) using some approximation. However, the correctness of this approximation is in doubt (Khatsymovsky, 1997b), so we shall not follow this path. Instead we shall study a wormhole with such a metric that relevant expressions take the form simple enough to allow the analytical treatment.

The Morris-Thorne wormhole is not the unique static spherically symmetric wormhole (contrary to what can often be met in the literature). Consider a spacetime $R^2 \times S^2$ with the metric:

$$ds^2 = \Omega^2(\xi)\,[-d\tau^2 + d\xi^2 + K^2(\xi)(d\theta^2 + \sin^2\theta\,d\phi^2)], \qquad (5)$$

where Ω and K are smooth positive even functions, $K = K_0 \cos\xi/L$ at $\xi \in (-L, L)$, $K_0 \equiv K(0)$ and K is constant at large ξ. The spacetime is obviously spherically symmetric and static. To see that it has to do with wormholes consider the case

$$\Omega \sim \Omega_0 \exp\{B\xi\}, \qquad \text{at large } \xi. \qquad (6)$$

The coordinate transformation

$$r \equiv B^{-1}\Omega_0 \exp\{B\xi\}, \qquad t \equiv Br\tau, \qquad (7)$$

then brings the metric (5) in the region $t < r$ into the form:

$$ds^2 = -dt^2\,2\,t/r\,dt\,dr + [1 - (t/r)^2]dr^2 + (BK_0 r)^2(d\theta^2 + \sin^2\theta\,d\phi^2). \qquad (8)$$

It is obvious from (7) that as r grows the metric (5,8) becomes increasingly flat (the gravitational forces corresponding to it fall as $1/r$) in a layer $|t| < T$ (T is an arbitrary constant). This layer forms a neighborhood of the surface $\tau \equiv t \equiv 0$. But the spacetime is static (the metric does not depend on τ). So, the same is true for a vicinity of *any* surface $\tau \equiv const$. The spacetime can be foliated into such surfaces. So this property (increasing flatness) holds in the whole spacetime, which means that it is a wormhole, indeed. Its length (the distance between mouths as measured through the tunnel) is of order of $\Omega_0 L$ and the radius of its throat $R_0 = \min(\Omega K)$.

The advantage of the metric (5) is that for the electro-magnetic, neutrino, and massless conformally coupled scalar fields $T\mathcal{Q}_{ik}$ can be readily found (Page, 1982) in terms of Ω, K and their derivatives [actually the expression contains also one unknown term (the value of $T\mathcal{Q}_{ik}$ for $\Omega = 1$), but the more detailed analysis shows that for sufficiently large Ω this term can be neglected]. So, by using this expression, calculating the Einstein tensor G_{ik} for the metric (5) and substituting the results into the system (4) we can recast it [the relevant calculations are too laborious to be cited here (the use of the software package *GRtensorII* can lighten the work significantly though)] into the form:

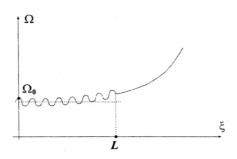

FIGURE 2. A conformal factor Ω satisfying requirements (i)–(iii).

$$E_i \geq 0 \qquad i = 0, 1, 2, 3, \tag{9}$$

where E_i are some (quite complex, e. g. E_0 contains 40 terms; fortunately they are not all equally important) expressions containing Ω, K, and their derivatives and depending on what field we consider.

Thus if we restrict ourselves to wormholes (5), then to answer the question formulated above all we need is to find out whether such Ω exist that it
i)   has appropriate asymptotic behavior [see (6)],
ii) satisfies (9) for some field,
iii) delivers sufficiently large R_0.

It turns out (Krasnikov, 1999) that for all three fields listed above and for arbitrarily large R_0 such Ω do exist (an example is sketched in Fig.2) and so the answer is positive.

ACKNOWLEDGMENTS

I am grateful to Prof. Grib for stimulating my studies in this field and to Dr. Zapatrin for useful discussion.

REFERENCES

Anchordoqui A., Romero, G. E., Torres, D. F., and Andruchow, I., *Mod. Phys. Lett.* **14**, 791 (1999)

Birrell, N. D., and Davies, P. C. W., *Quantum fields in curved spacetime*, Cambridge, Cambridge University Press, 1982.

Epstein, H., Glaser, V., and Jaffe, A., *Nuovo Cimento* **36**, 1016 (1965).

Flamm L., *Physikalische Zeitschrift* **17**, 448 (1916)

Flanagan, E. E., and Wald, R. M., *Phys. Rev. D* **54**, 6233 (1996).

Friedman, J. L., Schleich, K., and Witt, D. M., *Phys. Rev. Lett.* **71**, 1486 (1993).

Hochberg, D., Popov, A., and Sushkov, S., *Phys. Rev. Lett.* **78**, 2050 (1997).

Krasnikov, S., *Phys. Rev. D* **57**, 4760 (1998).

Krasnikov, S., Eprint *gr-qc* 9909016.

Khatsymovsky, V., *Phys. Lett. B* **399**, 215 (1997a).

Khatsymovsky, V., 'Towards possibility of self-maintained vacuum traversible wormhole' in *Proceedings of the II Int. Conference on QFT and Gravity*,TGPU Publishing, Tomsk, 1997b.

Morris, M. S. and Thorne, K. S., *Am. J. Phys.* **56**, 395 (1988).

Page, D. N., *Phys. Rev. D* **25**, 1499 (1982).

Sushkov, S. V., *Phys. Lett. A* **164**, 33 (1992).

Taylor, B. E., William A. Hiscock W. A., and Anderson P. R., *Phys.Rev. D* **55**, 6116 (1997)

Visser, M., *Lorentzian wormholes — from Einstein to Hawking*, New York, AIP Press, 1995.

A Viable Superluminal Hypothesis: Tachyon Emission from Orthopositronium

M. Skalsey[1], R.S. Conti[1], J.J. Engbrecht[1], D.W. Gidley[1], R.S. Vallery[1], P. W. Zitzewitz[2]

[1]*Randall Laboratory of Physics, University of Michigan, Ann Arbor, MI*
[2]*Department of Natural Sciences, University of Michigan: Dearborn, Dearborn, MI*

(734)763-3464; skalsey@umich.edu

Abstract. Tachyons are hypothetical particles that travel faster than the vacuum speed of light. Previous experiments have searched for, but have not found evidence of tachyons. Long-standing, anomalous measurements of the orthopositronium (o-Ps) decay rate are interpreted as evidence for two tachyons being occasionally emitted when o-Ps decays. Restricting the coupling of tachyon pairs to a single photon (no tachyon coupling to matter) yields a new theory where tachyons are only observed in o-Ps decay and not in the previous tachyon experiments. Combining the single photon coupling theory with all previous experiments predicts that these tachyons must deposit energy while traversing scintillator detectors. A new tachyon search experiment will use this energy loss prediction to attempt to find tachyons passing through the apparatus or set limits disproving the original o-Ps to tachyon hypothesis. Viewing an intense o-Ps source, a time-of-flight spectrometer uses the superluminal property of tachyons for identification. Several months of continuous data acquisition will be necessary to completely eliminate the o-Ps to tachyon hypothesis.

INTRODUCTION

The velocity of light in vacuum, c, is the ultimate speed for material objects. This limit follows from the theory of special relativity and is confirmed by observations. The possibility of accommodating faster-than-light particles into special relativity is a 40-year-old theoretical idea (Bilaniuk, 1962). The name tachyon (Feinberg, 1967), from Greek ταχνζ (swift or fast), is now applied to any proposed superluminal particle. Special relativity reveals nothing about the dynamical properties of tachyons, simply energy and momentum kinematics. Hence, for freely propagating tachyons, performing experimental searches are difficult and in recent years, have become unfashionable.

Previous tachyon experiments, primarily cosmic ray and accelerator searches, have set limits on tachyons and their interactions with ordinary matter (Hikasa, 1992). Most of these tachyon limits are several decades old. The current lack of interest stems from philosophical problems with freely propagating tachyons and causality. The principle of causality states, in the language of special relativity: For any given inertial observer, cause must chronologically precede its effect. Examples of how tachyons might violate causality have been published (Rolnick, 1969) and discussed extensively (Recami, 1986). Tachyon theories embracing causality [e.g. extended relativity (Recami,1986)] lead to no observable consequences for experimentation.

In spite of these philosophical barriers, a new tachyon search experiment is beginning based on the hypothesis that tachyons do not interact directly with ordinary matter, but just with light (photons). This new experiment has its origins in a long-standing anomaly in low energy atomic physics: the disagreement (Dobroliubov, 1993) between theory and experiment for the orthopositronium annihilation decay rate in vacuum (λ_T). While the decay rate disagreement is small, $\Delta\lambda_T^{exp-th}/\lambda_T = (0.14 \pm 0.02)\%$, it is statistically significant and also very important because the theoretical prediction comes from quantum electrodynamics (QED).

CP504, *Space Technology and Applications International Forum–2000*, edited by M. S. El-Genk

QED is the fundamental theory of the interaction of light and matter. The theoretical predictions of QED are enormously successful with numerous experimental verifications using many different systems and energy scales (Kinoshita, 1990). The lone, long-standing QED discrepancy is the λ_T problem. Orthopositronium (o-Ps) is the $\overset{\rho}{S} = 1$ (spin one) bound state of an electron and a positron (Rich,1981). QED predicts that, in vacuum, o-Ps annihilates into three photons with a lifetime of about 142 ns ($\lambda_T = {}^1/\tau$). High precision λ_T measurements are 0.14% above the current QED theory.

Since $\lambda_T^{exp} > \lambda_T^{th}$, a possible explanation is that there is another decay mode for o-Ps, besides three photon (3γ), not included in the QED calculation. This idea is at least 20 years old (Mikaelian, 1978), but has never been applied to tachyons until now. In general, adding another non-3γ decay channel to o-Ps increases the observed decay rate, λ^{obs}, thusly:

$$\lambda^{obs} = \lambda_T + \lambda_{new},$$

$$(1)$$

where λ_{new} is the decay rate of o-Ps into the new channel. The λ_T experiments measure the total disappearance rate of o-Ps, regardless of decay products. Examples of λ_{new} include normal processes like scattering, but also exotic particle or forbidden decays (Skalsey, 1997A). A branching ratio for decays into the new channel can be defined:

$$B.R. = \lambda_{new} / \lambda^{obs} .$$

$$(2)$$

Notice only 0.14% of the o-Ps decays need to proceed through the new channel to obtain agreement between QED theory and λ_T experiments. Specifically, an o-Ps decay to a pair of tachyons is hypothesized:

$$o\text{-}Ps \rightarrow T + \bar{T}$$

$$(3)$$

where T ($\bar{T}$) is a tachyon (anti-tachyon or a second tachyon). The choice of a two-tachyon final state for the hypothesized decay is arbitrary. Numerous other possible final states have been hypothesized previously and many of these have been investigated experimentally (Skalsey, 1997A).

Directly coupling the tachyons to ordinary subluminal matter (o-Ps) is inconsistent with previous experiments. Existing limits and measurements preclude tachyon-matter interactions at a level necessary to explain the o-Ps λ_T discrepancy. Therefore, along with the hypothesized decay in Eqn. (3), a new theory of tachyon interactions is developed. This new theory couples the pair of tachyons to a single photon, no direct tachyon-matter interactions. The tachyon connection to o-Ps is established through a process called virtual, single photon annihilation of o-Ps, a unique property of positron-electron annihilation from the triplet $\overset{\rho}{S} = 1$ state. This new theory of tachyons appears to be consistent with all available experimental information.

Conservation of momentum and energy applied to Eqn. (3) require back-to-back tachyon emission, each with an energy of 511keV, the electron rest mass energy. The branching ratio for the decay in Eqn. (3) is taken to be 0.14% to match the λ_T problem. A sufficiently precise experiment, capable of detecting tachyons at this level, can therefore disprove this hypothesis. But, how does one detect tachyons without knowing their properties?

One approach, taken in early tachyon accelerator experiment (Baltay, 1970), is to look for missing energy and momentum in collision events. If a tachyon is produced in collision and then leaves the apparatus without interacting, an imbalance will appear in the final state products with a characteristic tachyon signature. The tachyon limits obtained from this early experiment (Baltay, 1970) are not sensitive enough to exclude the o-Ps to tachyon decay in Eqn. (3) (by a factor of roughly 100 to 1000). But this general technique has merit for experimental tachyon searches with o-Ps.

Consider an experiment with an o-Ps source surrounded by many photon detectors in all directions. If the detectors are thick enough, every time o-Ps decays, some photon energy must be observed. This type of set-up is referred to as a hermetic photon detector. Fortunately, this very difficult and expensive experiment has already been performed at the Univ. of Tokyo (Mitsui, 1993). The Tokyo experiment used one ton of scintillator material for the hermetic photon detector. Their result is a 3×10^{-6} limit on the branching ratio of o-Ps to "invisible decays" (Mitsui, 1993), which are interpreted here as tachyon pair decays.

To retain the o-Ps to tachyons hypothesis with 1.4×10^{-3} branching ratio, the Univ. of Tokyo experiment must have somehow missed the tachyons that are assumed to leave their apparatus without interacting. If the tachyons had deposited some energy in their hermetic detector, the tachyons could have been overlooked in (Mitsui, 1993). The minimum deposited energy is about 300 keV total or 150 keV per tachyon. Combining the 150 keV with the detector size, a range of energy deposition rates for tachyons is deduced. To summarize the conclusions, if the hypothetical decay o-Ps$\rightarrow$T + $\overline{\text{T}}$ exists with B.R. = 0.14%, then the tachyons must deposit energy that can be observed in scintillator detectors.

Energy deposition by tachyons is the basis for the new tachyon search experiment. Tachyons are supplied by an intense o-Ps source. On one side of the source, a time-of-flight (TOF) spectrometer is being constructed to detect tachyons. The TOF of tachyons is shorter than any other correlated background traversing the apparatus, allowing unique identification. Limits on the tachyon branching ratio will be obtained in reasonable running times sufficient to disprove the o-Ps to tachyon explanation for the o-Ps λ_T problem.

STATUS OF THE o-Ps DECAY RATE PROBLEM

The decay rate, $\lambda = {}^1/\tau$, of o-Ps is expressed in QED as a series in α, the fine structure constant ($\alpha \cong {}^1/_{137}$):

$$\lambda = \frac{4}{9} \, (\pi^2 - 9) \, \frac{\alpha^6 m_e c^2}{h} \, [1 + A\frac{\alpha}{\pi} + B\alpha\ln\alpha + C(\frac{\alpha}{\pi})^2 + ...]. \tag{4}$$

At present, the A and B coefficients are well known (Adkins, 1996; Karshenboim, 1993) as also some very small, even higher order terms. Without C, the theoretical result is $\lambda^{th} = 7.03820\mu s^{-1}$. The C coefficient is a very challenging calculation. At this time, only parts of the calculation have been published (Milstein, 1994; Burichenko, 1993; Labelle, 1994; Adkins, 1995A; Adkins, 1995B; Faustov, 1995). There is intense interest in finishing the C calculation (Adkins, 1998) because of the experimental situation.

Two recent experiments at the University of Michigan are both in significant disagreement with λ^{th}. The first (Westbrook, 1989) used low-density gases to form o-Ps and found $\lambda^{gas} = 7.0514 \pm 0.0014\mu s^{-1}$. In subsequent investigations (Skalsey, 1998), we have found λ^{gas} needs a small revision downward, but not down to theory. The second λ experiment (Nico, 1990) used a slow e^+ beam (< 1 keV) to form o-Ps in an evacuated, MgO-lined cavity. This systematically different experiment yielded $\lambda^{cavity} = 7.0482 \pm 0.0016\mu s^{-1}$, more than 6$\sigma$ from theory. The difference between theory and experiment (Nico, 1990) is $(0.14 \pm 0.02)\%$. λ^{cavity} (Nico, 1990) has withstood rigorous systematic testing (Gidley, 1991) and continues to be the major focus in our positronium program.

In 1995, a group at the Univ. of Tokyo published a new experimental λ result (Asai, 1995), done in fine-grained SiO_2 powders, and in agreement with theory: $\lambda^{powd} = 7.0398 \pm 0.0029\mu s^{-1}$. Unfortunately, their analysis ignored the Stark effect on o-Ps. Large electric fields are known to be present in these insulating powders and the potential effect on λ is discussed in detail in (Ford, 1976). The effects estimated in (Ford, 1976) are large compared to the error bar of (Asai, 1995) and correcting will raise their decay rate value. This shift can be understood classically as the electric field pulling the e^+ and e^- farther apart, reducing the overlap and the annihilation rate. The quantum mechanical argument involving P states is given in (Skalsey, 1997A). We have discussed the Stark effect with the head of the Tokyo group, Prof. T. Hyodo and they now agree that the Stark effect must be accounted for (Hyodo, 1997). As an aside, the parapositronium [p-Ps: e^+e^- spin zero ground state] decay rate has just recently been calculated including all the α^2 correction terms (Czarnecki, 1999). Measurements of the p-Ps decay rate (Al-Ramadhan, 1994) done at the Univ. of Mich. are in excellent agreement with theory, and the fractional experimental uncertainty is the same as the o-Ps studies, 0.02%. This somewhat surprising result is consistent with the o-Ps tachyon decay hypothesis.

The o-Ps decay rate remains a serious problem for QED. Most of our positronium program is dedicated to checking and rechecking our decay rate results. The theorists are busy finishing the C calculations, but no really big, resolve-the-discrepancy terms have been found yet. Further the convergence of the QED series has recently been questioned (Adkins, 1998), leaving future resolution of this problem an open question. We, as experimentalists can only check our apparatus and then search for other, non-QED explanations, like tachyons.

PROPOSED o-Ps → T T̄

The proposed decay branch to be considered here is o-Ps →T + T̄ (T-tachyon, T̄ -antitachyon or a second tachyon), occurring at 1.4×10^{-3} of the dominant rate, o-Ps→3γ. Since o-Ps decays essentially at rest in the laboratory, the T T̄ products are emitted back-to-back, each with a total energy of 511 keV (=$m_e c^2$). To obtain these results, conservation of momentum and energy are assumed during the decay. Tachyons with real energies and momenta can be included in special relativity if tachyons possess an imaginary rest mass: M = i μ, μ - positive real and called proper mass [see Feinberg's Scientific American article (Feinberg, 1970)]. Kinematics results, like equal energy, back-to-back tachyon emission from o-Ps, can then be derived. Off hand it may seem likely that such a simple decay should have been seen in some previous experiment. It will be shown that this is not the case. The special properties of o-Ps give the proposed tachyon decay unique characteristics that set it apart from all previous tachyon searches and simultaneously maintain consistency with high precision tests of QED. A useful property of o-Ps is called single photon virtual annihilation, an effect that is known to occur in QED as a radiative correction term. The particular use for this situation is shown in the Feynman diagram in Fig.1c depicting o-Ps decaying into a single photon that decays into two tachyons. [This type of use of single photon virtual annihilation is not unprecedented. It was used in discussing a hypothesized charge conjugation odd boson participating in o-Ps decay (Skalsey, 1997B) and S. Glashow used it in discussing possible mirror universes (Glashow, 1986; Gninenko, 1994)].

Charge conjugation invariance requires e^+e^- annihilation from the singlet state (e.g. p-Ps) coupling to an even number of photons (usually two Fig.1a). Conversely, triplet states (e.g. o-Ps) couple to an odd number of photons, real decays require 3γ (Fig.1b) to conserve momentum and energy or virtual decays to 1γ (Fig.1c, Fig.1d will be discussed later.) This even/odd property makes other QED tests less sensitive to tachyons since they would appear only in higher order correction terms and, given their inherently weak coupling, render them undetectable. While it is not possible to prove the previous statement for all cases here (there are numerous, high precision QED tests and each must be addressed), the example of the electron g-factor will be roughly worked out. A crude scaling scheme labels each electron-photon vertex with amplitude e and squaring gives the rate α, the fine structure constant, roughly 10^{-2}. In Fig.1b, there are three such vertices giving 10^{-6}. If Fig. 1c is to have a rate of 10^{-3} compared to Fig.1b, then an overall rate of 10^{-9} is estimated for Fig. 1c. Since there is only one e-γ vertex in Fig. 1c, the tachyon vertex corresponds to a 10^{-7} rate. To insert this vertex into the electron g-factor, a closed tachyon loop with two 10^{-7} vertices is the lowest order contribution. This loop can be put in the g-factor diagram itself or the lowest order anomaly diagram. Either way the tachyon contribution to the electron g-factor is at least 100 times smaller than the present experimental sensitivity, making the tachyon component negligible.

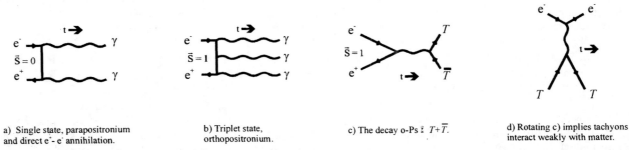

a) Single state, parapositronium and direct e⁻- e⁻ annihilation.

b) Triplet state, orthopositronium.

c) The decay o-Ps ≥ $T+\bar{T}$.

d) Rotating c) implies tachyons interact weakly with matter.

FIGURE 1. Feynman Diagrams for normal QED decays [a) and b)] and processes with tachyons [c) and d)].

The same conclusion pertains to cosmic-ray searches for tachyons. Positronium, binding energy 6.8eV, is exclusively a low energy phenomenon. In cosmic-ray showers, e^+ annihilation-in-flight (Kantele, 1973) is the pertinent process, which is predominately (99%) 2γ annihilation from the singlet state. The quote from (Kantele, 1973): "the three-quantum annihilation can always be neglected" overstates the situation but shows the 3γ component from the triplet state is small. High-energy accelerator experiments also find a very few e^+e^-→3γ events at 30 and 80 GeV and the rate agrees with QED theory when detector thresholds are properly accounted for (Fernandez, 1987; Martyn, 1990; L3 Collab, 1990; Opal Collab, 1991) . The complication of detector threshold sensitivity combined with the complexity of cosmic-ray showers precludes a quantitative analysis of triplet state e^+e^- annihilation rates in the previous tachyon cosmic-ray searches (Marini, 1982; Bhat, 1979; Bartlett, 1978; Smith,

1977; Prescott, 1976; Hazen, 1975; Clay, 1974). However, we have discussed this problem with Prof. Wayne Hazen (Hazen, 1975) who explained that these experiments were not analyzed for the fundamental processes (as we desire) but rather, to simply disprove the Clay and Crouch (Clay, 1974) positive tachyon result. Prof. Hazen's feeling was that the previous tachyon cosmic-ray searches lacked the sensitivity to observe tachyons from triplet state e^+e^- annihilations-in-flight. Therefore previous cosmic-ray tachyon searches do not seem to exclude our proposed tachyon experiments, which will also attempt to directly detect the tachyons.

Direct tachyon detection is a formidable task considering nothing is known about tachyons except superluminal speed. The alternate approach, pioneered by Feinberg (Baltray, 1970), produces the hypothesized tachyon in a collision between a high-energy proton and an atom in a bubble chamber. The collision products are analyzed event-by-event, searching for missing momentum and energy due to a tachyon departing the apparatus undetected. This scheme makes very mild assumptions about tachyon interactions in detectors, unfortunately, the limits are not very restrictive. An equivalent o-Ps minimal assumption experiment (Mitsui, 1993) has already been performed, searching for o-Ps → "nothing", i.e. undetected decay products, which are interpreted here as tachyons. This monstrous and expensive experiment used about one ton of CsI and NaI scintillator detectors surrounding a well-tagged o-Ps source. A limit of 2.8×10^{-6} at 90% confidence was thereby derived on the branching ratio for o-Ps decays to undetectables (Mitsui, 1993) disproving that, if tachyons are produced, then they leave the apparatus without interacting. This very important result implies that if the decay o-Ps → $T + \bar{T}$ exists, the tachyons deposit energy while leaving their apparatus. The deduced conclusion is that the tachyons must lose energy in traversing scintillator detectors, a conclusion that is actually not surprising. The original hypothesis, that o-Ps → $T + \bar{T}$ exists through a presumed coupling, predicts that tachyons will interact with detectors through the same coupling. Fig. 1d is a rotated version of Fig. 1c with the same couplings and displays a propagating tachyon interacting with an electron. The interpretation is that propagating tachyons infrequently strike electrons and ionize atoms, the principle of scintillation detection. Tachyons traversing detectors would lose at least some of their energy, probably as a long weak ionization trail. The rules for calculations from rotating Feynman diagrams that involve tachyons are unclear at this time. So one should not expect quantitative agreement between this tachyon theory and (Mitsui, 1993). The energy loss units measure essentially the number of electrons, no material dependence will be assumed. Normal charged particles traversing material roughly follow this simple approximation. The tachyon stopping power assumption could be criticized thusly: Suppose all or most of the tachyon energy is lost in a single interaction (photons frequently interact with matter in this manner.) Another tachyon decay model for o-Ps can be constructed with only one tachyon and one photon in the final state. In this alternative model, coupling between the initial state(o-Ps) and the final state (tachyon and photon) is through the same virtual single photon annihilation as with the two-tachyon final state theory. The one-tachyon theory predicts full energy deposition in a local region. Previous o-Ps experiments place restrictions on these tachyons (Gidley, 1991), leaving the stopping power assumption as the most plausible. Details of the analysis and limits for the single tachyon theory will be published elsewhere.

The o-Ps → "nothing" experiment (Mitsui, 1993) places a lower limit on the stopping power of tachyons $S_T > 1 \, keV - cm^2/g$, i.e. values smaller than this are excluded by (Mitsui, 1993). Further, an upper limit of $S_T < 500 \, keV - cm^2/g$ is also established from (Mitsui, 1993). If tachyons lose energy too fast, they do not escape the source region. The upper limit of $500 \, keV - cm^2/g$ is only a factor of two smaller than minimum ionizing particles (e.g. a low energy muon). These stopping power limits are the starting point for our proposed experiments, which will survey S_T values in the primary range $1-500 \, keV - cm^2/g$ and, if possible, higher values.

PROPOSED EXPERIMENTS

Figure 2 shows the experimental time-of-flight (TOF) set-up to investigate the lower range of S_T values. The major components are detectors and a thin ^{64}Cu e^+ source, with activity of up to 2Ci and a half-life of 12.6 hr. Activation of ^{64}Cu e^+ sources in the Ford Nuclear Reactor on the Univ. of Mich., campus uses a maximum in-core thermal neutron flux of 3×10^{13} n/cm²-sec. A thin, natural Cu foil is held in a polyethylene tube while in the reactor. The thin e^+ source thus produced is placed between two compressed pellets of fine-grained MgO powder, where about 20-25% of the e^+ stop and form positronium. Paramagnetic O_2, which rapidly quenches o-Ps (Kakimoto, 1990) is excluded from the source MgO region by using a pure N_2 atmosphere. With the N_2 pressure just above 1atm, thin plastic walls can contain the source foil and MgO pellets. Making the o-Ps source region as massless as possible is important for investigating the larger values of S_T: $100 \, keV - cm^2/g \, < S_T < 500 \, keV - cm^2/g$, so the tachyon can get out of the source and deposit energy in the detectors. The large S_T experiment is shown in Fig. 3. Note, the detectors are much thinner than those in Fig. 2, the thickness being only 3.5mm.

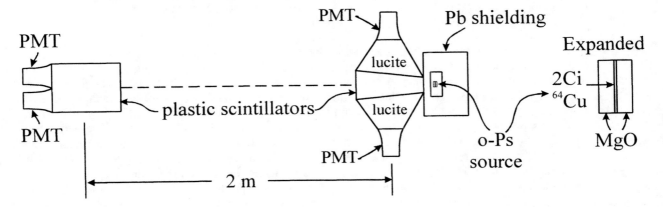

FIGURE 2. Tachyon TOF experiment using thick detectors and large activity sources.

From solid angle constraints in Fig. 2, the second, further-from-the-source, detector is much larger in volume than the close detector. The large size scintillators need several 4" photomultipliers on each to obtain good light collection efficiency. For large pieces of scintillator, the timing resolution is degraded simply by the large spatial extent of the detector and the finite speed of light. For smaller detector volumes, better time resolution can be obtained and faster 2" phototubes are appropriate.

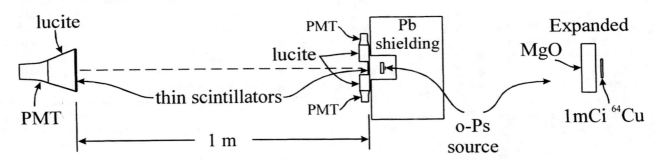

FIGURE 3. Another tachyon TOF experiment using thin detectors and low activity sources.

The important points in obtaining fast time resolution from plastic scintillator detectors are 1) good light collection efficiency, 2) good phototubes and, finally 3) signal processing with constant fraction timing discriminators. Constant fraction refers to timing derived in such a way as to be independent of the signal pulse heights, which span a wide range in plastic scintillator detectors. The timing signals from the discriminators are fed to a high-rate, time-to-digital converter (CAMAC-LeCroy 4204) interfaced to a PC computer, which records time spectra. The time resolution for the best detectors is expected to be 0.5 ns FWHM. The time resolution for the largest detectors will be a real concern if greater than 3 ns FWHM; sensitivity to tachyons could be severely reduced. To compensate for bad time resolution, the flight path between the detectors would need to be increased, reducing the solid angle roughly as the square of that increase.

One final correction must be considered, i.e. the scattering angle of tachyons when depositing energy. In normal multiple or plural scattering theory, an average or r.m.s. scattering angle is calculated after many collisions. It can be shown that larger tachyon masses will deflect less for a given energy deposition. Without more knowledge of tachyon dynamics, only a crude estimate can be made of how many tachyons will scatter enough in the first detector to miss the second detector, a conservative estimate is 90% will miss, i.e. a factor of ten loss in event rate. In Table I, tachyon event rates are calculated for the lowest and highest S_T configurations shown in Fig 2 and Fig 3, respectively. Two more configurations with sizes between the two shown here are also required to complete the experiment, but they are omitted from Table I. The detected tachyon event rate is expressed in terms of the scintillator detector efficiency ε. With hundreds of keV deposited energy, (also estimated in Table 1) ε will approach unity. At 40 keV deposited energy, however, ε will be quite diminished, to probably about 0.1. Tachyon event rates in the range 0.004-4s^{-1} are therefore predicted, taking $\varepsilon^2 = 0.01$. From these numbers, one concludes that much longer running times are required with weak sources. Extra small sources can be activated at the outside reactor surface and retrieved without turning the reactor off, greatly increasing the small source running time.

TABLE 1. Estimates of Signal and Noise Event Rates and Energy Loss.

Figure 2 Setup S_T between $1\text{-}4 \frac{keV-cm^2}{g}$		Specification		Figure 3 Setup S_T between $100\text{ - }500 \frac{keV-cm^2}{g}$
$7.4\times10^{10} \frac{decays}{s}$ (2 Ci)		source activity		$3.7\times10^{7} \frac{decays}{s}$ (1 mCi)
$1.41\times10^{10} \frac{e^+}{s}$	19%	positron fraction from source	19%	$7.0\times10^{6} \frac{e^+}{s}$
$8.4\times10^{9} \frac{e^+}{s}$	60%	positrons entering MgO	30%	$2.1\times10^{6} \frac{e^+}{s}$
$1.27\times10^{9} \frac{o\text{-}Ps}{s}$	15%	positrons forming o-Ps	15%	$3.2\times10^{6} \frac{o\text{-}Ps}{s}$
$2.0\times10^{6} \frac{T\overline{T}}{s}$	1.6×10^{-3}	o-Ps decaying into $T+\overline{T}$	1.6×10^{-3}	$500 \frac{T\overline{T}}{s}$
$400 \frac{incident\,T}{s}$	4×10^{-3}	Tacyons entering detectors	8×10^{-4}	$0.4 \frac{incident\,T}{s}$
$400\,\varepsilon^2 \frac{detected\,T}{s}$		Tacyons detected (ε – detector efficiency)		$0.4\,\varepsilon^2 \frac{detected\,T}{s}$
0.25 s^{-1} & 4 s^{-1}		uncorrelated and correlated noise rates		0.014 s^{-1} & 0.9 s^{-1}
negligible		Tachyon energy deposited in source		20 – 100 keV
60 – 240 keV		Tachyon energy deposited in Pb		0
40 – 160 keV		Tachyon energy deposited in detector 1		40 – 200 keV
40 – 111 keV		Tachyon energy deposited in detector 2		40 – 200 keV
0.2 – 0.8 keV		Tachyon energy deposited in air		10 – 50 keV

The final issue concerns predicting noise rates, correlated and uncorrelated, in the tachyon TOF spectrum. The most prevalent correlated noise source will be a single γ-ray Compton scattering in the first TOF detector and then the scattered γ-ray detected in the second detector. Of course, the transit time will correspond to $v = c$. All sources of correlated noise that have been considered result in $v \le c$, with the exception of $t < 0$ events e.g. cosmic ray muons going backwards through the apparatus. This muon noise can be measured with no source present. Any tachyon signal ($v > c$) should have a different transit time than any correlated noise source. The necessary time resolution (FWHM: 0.5 to 2.5 ns) of the TOF system has been discussed previously. Separating the tachyon signal from correlated noise will be a strong function of the tachyon velocity, and consequently the assumed mass. Tachyons with large proper masses will be the easiest to set limits on, the limits will worsen as $m_T \rightarrow im_e$. Uncorrelated noise in a TOF spectrum can arise when, say, two γ-rays from two, almost simultaneous decays fire the detectors. Since there is no time correlation in such an event, it leads to a relatively flat background over the entire time spectrum. Any tachyon peak, separated from the correlated noise peak, will still be situated on top of this uncorrelated background. Using estimated γ-ray rates in the detectors, the uncorrelated noise rate has been calculated for a peak-sized time window. The results are shown in Table I along with correlated noise rate estimates derived by considering Compton scattering only, the dominant correlated noise source. The choice of the ^{64}Cu positron source maximizes the signal to noise ratio in this experiment.

From the estimates of correlated noise rates, it is clear that the time resolution of the TOF system must be smaller (a factor of three is used here) than the photon flight time to separate the tachyon and photon time peaks. Hence, the observed time resolution in the largest detectors will determine the final, longest flight path distance, assumed now to be 2m. The correlated noise will limit the experimental sensitivity as the assumed tachyon mass becomes as small as im_e. For larger masses, the uncorrelated noise (flat background) in the time spectrum will limit the tachyon sensitivity. In the large mass limit, the sensitivity to small S_T tachyons (Table Ia) will enable branching ratio limits of better than 10^{-5}, assuming ten of the hottest (2Ci) irradiations, giving about 5 days of running with >1Ci. Again in the large tachyon mass limit, large S_T tachyons (Table Ib) will have branching ratio of limits of 10^{-4} from 100 days of running many 1mCi sources. To exclude tachyons from causing the o-Ps decay rate discrepancy, branching ratio limits of 10^{-4} are sufficient. Naturally, if a positive tachyon signal were observed, systematic tests of the result would be performed. Examples of these systematic tests are: 1) change the flight path between the detectors using as wide a variation in detector size as possible. 2) change the MgO powder to a SiO_2 powder with similar properties or other Ps formation media like gases. 3) change the processing and data acquisition electronics, etc. Modifications and additions to this apparatus could investigate tachyon properties, if observed.

At this point, it is important to restate the required assumptions for this investigation: 1) Relativistic energy and momentum are conserved; 2) The exotic decay branch o-Ps→T $\bar{T}$ occurs at the 1.4×10^{-3} level, from assumption 1, yielding back-to-back two tachyon emission, each starting with 511 keV energy. Notice that the conclusion that these tachyons deposit energy in scintillator is not an assumption, this is a deduced fact from the previous o-Ps→ "nothing" experiment. However, further assumptions are still required about the energy deposition process, i.e., it is uniform and does not greatly deflect the tachyons.

CONCLUSIONS

To conclude, a discrepancy between QED theory and an atomic physics experiment is interpreted as indirect evidence for tachyon emission. The hypothesized tachyon emission process is analyzed and found to be consistent with (i.e. not excluded by) previous experiments. An existing search for direct evidence of the tachyon emission process implies that a tachyon, if it exists, deposits measurable amounts of energy in scintillator detectors. Using this fact, an experiment has been designed to observe tachyons moving faster than the speed of light through an apparatus. A real, positive result would be a major breakthrough. More likely, a negative result, no tachyons or other anomalies observed, will be sufficiently precise to address the atomic physics discrepancy over a wide range of possible tachyon masses. In the large mass limit, all possible stopping powers will be excluded and the original hypothesis is discredited. In the unlikely event that tachyons were discovered, many questions would surface and more experimentation would follow. The relation of tachyons to spacecraft propulsion is still speculative. Perhaps, a useful long-range communications device could be devised using tachyons. More importantly, if tachyons were discovered, it would give real hope that someday faster-than-light travel might be possible

ACKNOWLEDGMENTS

We wish to acknowledge very useful discussions with G.W. Ford, W.E. Hazen, J.C. Lee, R.R. Lewis, and J.C. Zorn. The theoretical analysis in preparation for this experiment has been funded by the National Science Foundation Grant PHY-9731861 and the University of Michigan.

REFERENCES

Adkins, G.S., and Shiferaw, Y. "Two-Loop Corrections to the Orthopositronium and Parapositronium Decay Rates Due to Vacuum Polarization," Phys. Rev. A **52**, 2442 (1995)B.

Adkins, G.S., and Lymberopoulos, M., "Light-by-Light Scattering Contribution to the Decay Rate of Orthopositronium," Phys. Rev. A. **51**, R875 (1995)A, "Contribution of Light-by-Light Scattering to the Orders 0 ($m\alpha^8$) and 0($m\alpha^8 \ln\alpha$) Orthopositronium Decay Rate," Phys. Rev. A **51**, 2908 (1995)A.

Adkins, G.S., and Sapirstein, J. private communication, 10/1998.

Adkins, G.S. Analytic Evaluation of the Orthopositronium-to-Three-Photon Decay Amplitudes to One-Loop Order" Phys. Rev. Lett. **76**, 4903 (1996).

Al-Ramadhan, A.H., and Gidley, D.W. "New Precision Measurement of the Decay Rate of Singlet Positronium." Phys. Rev. Lett. **72**, 1632 (1994).

Asai, S., Orito, S., and Shinohara, N. "New Measurement of the Orthopositronium Decay Rate." Phys. Lett. B **357**, 475 (1995).

Baltay, C., Feinberg, G., Yeh, N., and Linsker, R. "Search for Uncharged Faster-than-Light Particles." Phys. Rev. D **1**, 759 (1970).

Bartlett, D. F., Soo, D., and White, M.G. "Search for Tachyon Monopoles in Cosmic Rays," Phys. Rev. D. **18**, 2253 (1978).

Bhat, P.N. et al., "Search for Tachyons in Extensive Air Showers." J. Phys. G: Nucl. Phys **5**, L 13 (1979).

Bilaniuk, O.M, Deshpande, V.K., and Sudarsham, E.C.G. ""Meta" Relativity," Am. J. Phys. **30**, 718 (1962).

Burichenko, A.P. "Large contribution to the correction ~α^2 to the width of orthopositronium," Yad. Fiz. **56**, 123 (1993) Phys. Atom. Nuclei. **56**, 640 (1993).

Clay, R.W., and Crouch, P.C. "Possible observation of tachyons associated with extensive air showers," Nature **248**, 28 (1974).

Czarnecki, A. , Melnikov, K. and Yelkhousky, A., "α^2 Correction to Parapositronium Decay," Phys. Rev. Lett. **83**, 1135 (1999).

Dobroliubov, M.I., Gninenko, S.N., Ignatiev, A.Yu., and Matveev, V.A. "Orthopositronium Lifetime Problem," Int. J. Mod. Phys. A **8**, 2859 (1993).

Faustov, R. N., Martynenko, A.P., and Saleev, V.A. "O(α^2) corrections to the orthopositronium decay rate," Phys. Rev. A **51**, 4520 (1995).

Feinberg, G. "Possibility of Faster-than-Light Particles," Phys. Rev. **159**, 1089 (1967).

Feinberg, G. "Particles that go faster than light," Scientific American **222** No. 2 (Feb), 68 (1970).

Fernandez, E. et al., "Tests of quantum electrodynamics with two-, three-, and four-photon final states from e^+e^- annihilations at $\sqrt{s}$ = 29 GeV," Phys. Rev. D **35**, 1 (1987).

Ford, G.W., Sander, L.M., and Witten, T.A. "Lifetime Effects of Positronium in Powders." Phys. Rev. Lett. **36**, 1269 (1976).

Gidley, D.W., Nico, J.S., and Skalsey, M. " Direct Search for Two-Photon Decay Modes of Orthopositronium." Phys. Rev. Lett. **66**, 1302(1991).

Glashow, S.L. "Positronium versus the Mirror Universe," Phys. Lett. B **167**, 35 (1986).

Gninenko, S.N. " Limit on "disappearance" of orthopositronium in vacuum," Phys. Lett. B. **326**, 317 (1994).

Hazen, W.E. et al., "A search for Precursors to Extensive Air Showers," Proc. 14th Inter'al Conf. on Cosmic Rays, Munich, (1975) p. 2485.

Hikasa, K, et al. "Review of Particle Properties," Particle Data Group, Phys. Rev. D **45**, S1 (1992).

Hyodo, T. priv. comm. (1997).

Kakimoto, M., Hyodo,T., and Chang, T.B. "Conversion of ortho-positronium in low-density oxygen gas," J. Phys. B: At. Mol. Opt. Phys. **23**, 589 (1990).

Kantele, J. , and Vakonen, M. " Corrections for Positon Annihilation In Flight In Nuclear Spectroscopy," Nucl. Instr. Meth. **112**, 501 (1973).

Karshenboim, S.G. "New logarithmic contributions in muonium and positronium," JETP **76**, 541 (1993).

Kinoshita, T. "Quantum Electrodynamic,s" (World Scientific, Singapore, 1990).

L3 Collab. "Test of QED in $e^+e^- \rightarrow \gamma\gamma$ at LEP." Phys. Lett. B **250**, 199 (1990).

Labelle, P., Lepage, G.P., and Magnea, U. "Order $m\alpha^8$ Contribution to the Decay Rate of Orthopositronium," Phys. Rev. Lett. **72**, 2006 (1994).

Marini, et al., "Experimental limits on quarks, tachyons, and massive particles in cosmic rays," Phys. Rev. D. **26**, 1777 (1982).

Martyn, H.U. "Test of QED by High Energy Electron-Positron Collisions," in (Kinoshita, 1990)

Mikaelian, K.O. "Orthopositronium Decay into Axions," Phys. Lett. B **77**, 214 (1978).

Milstein, A.I., and Khriplovich, I.B. "Large relativistic corrections to the positronium decay probability," JETP **79**, 379 (1994).

Mitsui, T. et al., "Search for Invisible Decay of Orthopositronium." Phys. Rev. Lett. **70**, 2265 (1993).

Nico, J.S., Gidley, D.W., Rich, A., and Zitzewitz, P.W. "Precision measurement of the Orthopositronium Decay Rate using the Vacuum Technique," Phys. Rev. Lett. **65**, 1344 (1990).

OPAL Collab. "Measurement of the cross sections of the $e^+e^- \rightarrow \gamma\gamma$ and $e^+e^- \rightarrow \gamma\gamma\gamma$ at LEP," Phys. Lett. B **257**, 531 (1991).

Prescott, J.R. "Tachyons revisited-comments on a search for Faster-than-Light Particles," J. Phys. G: Nucl. Phys. **2**, 261 (1976).

Recami, E. "Classical Tachyons and Possible Applications," Riv Nuovo Cim. **9**, 1 (1986).

Rich, A. "Recent Experimental Advances in Positronium Research," Rev. Mod. Phys. **53**, 127 (1981).

Rolnick, W.B. "Implications of causality for Faster-than-Light Matter," Phys. Rev. **183**, 1105 (1969).

Skalsey, M. "Exotic decays of Positronium and C-Odd bosons," Mater. Sci. Forum **255-7**, 209 (1997)A.

Skalsey, M., and Conti, R.S. "Search for very weakly interacting, short-lived, C-odd bosons and the orthopositronium decay rate problem," Phys. Rev. A **55**, 984 (1997)B.

Skalsey, M., Engbrecht, J.J., Bithell, R.K., Vallery, R.S., and Gidley, D.W. "Thermalization of Positronium in Gases," Phys. Rev. Lett. **80**, 3727 (1998).

Smith, G.R., and Standil, S. "Search for tachyons preceding cosmic ray extensive air showers of energy $\geq 10^{14}$ eV," Can. J. Phys. **55**, 1280 (1977).

Westbrook, C.I., Gidley, D.W., Conti, R.S., and Rich, A. "Precision measurement of the orthopositronium vacuum decay rate using the gas technique," Phys. Rev. A **40**, 5489 (1989).

Extended Relativity

Eduardo Valencia

Advanced Gravity, AC
Cerrada Pradera 9-1, La Pradera
Cuernavaca, Morelos 62170, Mexico
zane@mor1.telmex.net.mx

Abstract. An extended set of Lorentz transformations is proposed, allowing real valued superluminal transformed quantities. This set serves to formulate the extended version of the Special and General relativity.

INTRODUCTION

The laws of physics must be valid in all systems of coordinates; tensor equations expressing these laws guarantee this requirement, Lorentz invariance of tensor equations is the key principle of relativistic physics, as Galilean invariance is to Newtonian physics. The nature of Lorentz transformations, since they were formulated, exhibits the light speed limit for propagation for non-zero rest mass particles and fields. Since these transformations are the building blocks of special and general relativity, space-time defined by such theories inherits this limit as a fundamental feature.

The first postulate of relativity stated at the beginning of this section, in order to hold, regardless of the relative speed between a given pair of reference frames, requires an extended definition of Lorentz invariance which, in turn, serves to redefine special and general relativity theories. We have then a natural theoretical framework allowing superluminal geometries and dynamics in space-time.

Such redefinition must exhibit Einstein theories as a special case of the extended theory, where until now "forbidden" processes can occur under the light of new effects. The historical formulation of relativity will be performed here in terms of these extended transformations, arriving to a natural labeling of the action as a consequence of the definitions, instead of deriving the theory from an EIH action quantity and then labeling the action in terms of characteristic topologies of given geometries.

This extended formulation yields a spectrum of metric fields, each of which labels a sector of the configuration space of the theory, that is the space of solutions of the classical theory, defining the superluminal processes and limits for a given sector. Such labeling occurs in terms of a topological index, playing the role of a quantum number for field states, described by wave functionals in the corresponding quantum theory, which happens to be the canonical formulation of quantum gravity.

A first consequence can be deduced from the extension proposed. Black holes, quantum black holes, black string or black brane solutions, arise from the actual status of Lorentz invariance, but such solutions or states become *relative* within the extended framework; a black hole is not black, if superluminal light or particles can come through.
The whole dynamics are to be redefined in terms of this new invariance; leading to new effects, in the same manner when the transit from Newtonian to Relativistic dynamics was done.

How the whole set of physical quantities such as energy, momentum, wave functions, fermionic, bosonic and electromagnetic fields will behave at superluminal speeds in space-time, is determined by this extended set of transformations, and experimental verification can be conducted.

Superluminal processes have been formulated before in terms of imaginary masses and velocities or negative energies. This phenomenology is not to be ruled out; however, the reality condition for metric fields requires real valued quantities. Since the metric translates physical measured quantities between observers, it is natural to demand the same behavior at superluminal speeds.

The classical warp solutions satisfying the Einstein equations have already been analyzed, and the energy required for these metric geometries is negative, therefore classical relativistic dynamics seems to close the door, unless we resign to the positivity condition, find a wider principle for energy, or the existence of such type of matter is found in a given experiment. However, these types of solutions and matter configurations are in consequence a special case of the extended theory.

It is the aim of this work to show that the definition of the Extended Relativity allows superluminal classical geometries and processes, and the relationship of the metric spectrum with canonical quantum gravity.

EXTENDED LORENTZ INVARIANCE

Let there be two frames one of which moves in the x direction, then the following set of coordinate transformations:

$$x_r = [x + v(t/N)]\gamma_N \ , \tag{1}$$
$$y_r = y \ ,$$
$$z_r = z \ ,$$
$$t_r = [t + (vx/Nc^2)]\gamma_N \ ,$$

where,

$$\gamma_N = \frac{1}{\sqrt[2]{1-\beta_N^2}} \ , \tag{2}$$
$$\beta_N^2 = \frac{v^2}{N^2c^2} \ ,$$

keep the four dimensional distance invariant:

$$x_r^2 + y_r^2 + z_r^2 - c^2t_r^2 = x^2 + y^2 + z^2 - c^2t^2 \ . \tag{3}$$

The set is labeled by an integer, defining a collection of Lorentz transformations.

Proper time and length are then:

$$\Delta t = \Delta t' / \sqrt{1-\beta_N^2} \ , \tag{4}$$
$$\Delta x = \Delta x' \sqrt{1-\beta_N^2} \ . \tag{5}$$

This set reduces to the usual Lorentz transformations for N=1, setting the speed of light limit for non-zero rest mass moving objects. However, for higher values of the label, spatial like superluminal speeds are allowed. Therefore, light cone opens, and then Lorentz invariance is fully defined, once the state of the light cone defined by N is known.

The above transformations exhibit a periodicity in terms of the light speed limit given by Nc, the integer labeling the collection, which can be thought as a quantum number revealing the periodicity of occurring infinities or barriers for a transformed quantity, say A:

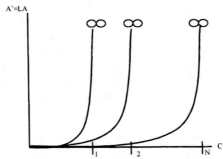

FIGURE 1. Light barrier periodicity in Minkowskian space-time.

Therefore the light barrier exists at speeds greater than C, by virtue of the label, and Lorentz invariance is kept in the usual way. Such periodicity exhibits then sectors in Minkowskian space-time exhibited by extended Lorentz topology.

At a given moment in time with N=1, there will be causally disconnected sectors of space which can be connected by means of a light cone structure transition, opening the light cone taking higher values of N.

The usual prescription for the horizon problem in standard cosmology, is to connect such regions opening the light cone by a redefinition of the speed of light as a function of a cosmological variable, breaking Lorentz invariance and therefore violating the postulate of special relativity that the speed of light is constant for all observers.

EXTENDED LORENTZ GROUP

The Lorentz boost group element Λ^{m}_{j} in an arbitrary direction becomes:

$$\Lambda^{m\ N}_{j} = \begin{pmatrix} \gamma_N & 0 & 0 & \beta_N\gamma_N \\ 0 & 1 & 0 & 0 \\ 0 & 0 & 1 & 0 \\ \beta_N\gamma_N & 0 & 0 & \gamma_N \end{pmatrix} \tag{6}$$

with m, j=0,1,2,3 and N=1,2,3...n-1.

The labeling of the Lorentz group allows the definition for Extended General Relativity equations as follows.

Space-time metric can be defined as:

$$g^{N}_{\mu\nu} = \Lambda^{m\ N}_{j} e^{j}_{\mu} \Lambda^{k\ N}_{i} e^{i}_{\nu} \eta_{mk} \quad , \tag{7}$$

then Einstein equations become:

$$G_{\mu\nu}^N = -kT_{\mu\nu}^N + \Theta_{\mu\nu}^N \ , \tag{8}$$

where

$$\Theta_{\mu\nu}^N = \lambda^N g_{\mu\nu}^N \ , \tag{9}$$

for N=1 these equations set the Einstein General Relativity as usual.

The action is therefore:

$$S_N = \int \sqrt{-g_N} \left(\frac{R^N - 2\lambda^N}{16\pi \, G_{Nw}} + L_{matter}^N \right) d^4x \ . \tag{10}$$

it is independent of any given spatial topology within the four-dimensional manifold, and reflects the periodicity of light barriers occurring in Minkowskian space-time given the labeling.

SUPERLUMINAL CONES

The Minkowski vacuum characterized by $F_{\mu\nu}^{vac} = 0$ is decomposed into topologically distinct vacua A classified by an integer index, transitions between two Minkowski vacua labeled by k and j are interpolated by an instanton with winding number $n = k - j$ where n is the Pontryagin index.

Here we are not using imaginary time (in order to properly use the term instanton) and given the fact that the topology is fixed (is SO (3)) the labeling of Minkowski space time vacua and metric, occurs in terms Lorentz labeled transformed quantities, belonging to causally connected sectors of maximal speed Nc for a given vacua or gauge potential A, corresponding to a solution for a given N value of Einstein equations.

The transition between two given lightcones can be viewed as the opening of the base light cone (maximal speed c) into the final cone (maximal speed Nc), which is equivalent to tunneling between to equal time light cones (the horizon problem).

The amplitude for the light cone opening is given by:

$$< N \,|\, e^{iHt} \,|\, j >= \int dg_{N=k-j} \exp\{-iS_N\} \ , \tag{11}$$

where H is the Hamiltonian of the theory. In Euclidean quantum gravity, vacua are labeled by two integer numbers because of the symmetry group used by the approach that is $SO(4) = SU(2) \times SU(2)$ each integer corresponding to the two factors.

Given the fact that the symmetry group in this approach is the $SO(3)$ group, only one integer is needed to label the gravitational vacua, the metrics, described by the labeled action S_N, therefore tunneling between different vacua will be caused by the first Pontryagin class yielding the number N:

$$S^P = \frac{\theta}{4} \int g^{\frac{1}{2}} \varepsilon_{\mu\nu\alpha\beta} R_{\mu\nu\rho}{}^{\sigma} R_{\alpha\beta\sigma}{}^{\rho} d^4x \ . \tag{12}$$

The curvature tensor $R_{\mu\nu\rho}{}^{\sigma}$ is the $R^N{}_{\mu\nu\rho}{}^{\sigma}$ of the extended theory with N=1.

This is a feature derived from the CP problem in quantum field theory, it is used to derive quantum theories of gravity labeled by an angle θ because of the Lorentz symmetry, given the fact that the topology of the connected Lorentz group is contained in its maximal subgroup $SO(3)$.

From this feature quantum gravity states can be built in terms of the action previously labeled, these states will identify the n sectors of the theory, characterizing each sector with a maximal speed. Therefore, spatial-like superluminal speeds are a consequence of the periodic nature of the wave functionals describing space-time, which is a new feature, derived from quantum gravity.

The periodicity of the wave functional defined by N vacua, corresponds to the periodicity of light barriers described by the action labeled by N, then factoring:

$$\Psi(S_N) \text{ as } \Psi_N \equiv \Psi(S_N) = \Psi(S_1)e^{-iS^P\theta} , \tag{13}$$

this wave functional describe states yielding superluminal speeds where S_1 is the standard EIH action corresponding to the setting N=1 of the extended formulation.

The energy conditions such as the Null Energy Condition and the Strong Energy Condition **hold** within the definition of Extended Relativity, this is worth to be mentioned because violations of such conditions stand against any form of faster than light dynamics.

The state of Lorentz invariance is the key to allow causal spatial like superluminal transformations of matter and fields, as well as extended relativistic motion in space-time. Actually the violation of Lorentz invariance or the energy conditions is used to explain the horizon problem in VSL cosmologies; this can be solved using extended invariance formulation.

CONCLUSIONS

The formulation of the General relativity in terms of extended Lorentz group allows superluminal dynamics. The extension of the relativistic principle and the nature of the topological factors defining the theory can lead to experiments where the physical possibility of superluminal dynamics is a fact. Astronomical observations of galaxy clusters motion could be used to verify the transformations proposed here, as well as radiation emission rates from black holes exceeding black hole theoretical predictions estimated by means of Hawking's radiation. Such confirmations of this Extended Theory are the first step towards the Superluminal Physics Era.

ACKNOWLEDGMENTS

I would like to thank G. Torres and J.Escalona for discussions, comments and references supplied and to everyone at AGAC.

REFERENCES

Alcubierre, M., The Warp Drive: Hyper-Fast Travel Within General Relativity,*Class. And Quantum Grav.* **11 L73** (1994)

Ashtekar, A., The CP Problem in Quantum Gravity, *Syracuse University preprint* **SU-4228-386**, June 1988

Balachandran, A.P., What becomes of the Vacuum Angle in Generally Covariant Gauge Theories, *Syracuse University preprint* **SU-4228-388**, Aug. 1988

Giulini, D., No-Boundary Theta-Sectors in Spatially Flat Quantum Cosmology ,Phys.Rev. D46 (1992) 4355-4364

Hawking, S., Euclidean Quantum Gravity, in *Recent Developments in Gravitation.* Plenum Press, 1979

Misner, C.W., Thorne, K.S. and Wheeler, J.A., *Gravitation.* Freeman, San Francisco, 1973

Valencia, E., Quantum Light Cones, *Advanced Gravity e-print*, www.angelfire.com/nj/FTLphysics

Visser M., Superluminal Censorship, Washington U., St. Louis, e-Print Archive: **gr-qc/9810026,** Oct. 1998

CONFERENCE ON COMMERCIAL/CIVIL
NEXT GENERATION
SPACE TRANSPORTATION

The Space Taxi™ Transportation System

Douglas Stanley

Orbital Sciences Corporation, 21700 Atlantic Ave., Dulles, VA 20166

Abstract. This paper summarizes the results of recent studies by Orbital to significantly reduce NASA's future launch costs and improve crew safety through the implementation of a low-risk, evolutionary space transportation architecture. These studies were performed as a part of NASA's Space Transportation Architecture Studies (STAS) and subsequent internally-funded efforts. A large number of vehicles and architecture approaches were examined and evaluated. Orbital's recommended architecture includes a small, multifunctional vehicle, referred to as a Space Taxi™, which would serve as: an emergency crew return vehicle for the International Space Station (ISS), a two-way human space transportation system, a small cargo delivery and return vehicle, and as a passenger module for a future Reusable Launch Vehicle (RLV). The Space Taxi™ would initially be launched on a heavy-lift Evolved Expendable Launch Vehicle (EELV), currently under development by U.S. industry and the U.S. Air Force. Together with a small cargo carrier located behind the Space Taxi™, this combination of vehicles would be used to meet future ISS servicing requirements. Later, a two-stage, commercially developed RLV would replace the EELV in launching the Space Taxi™ system at a significantly lower cost.

INTRODUCTION

Over the past several years NASA and industry have been actively developing new technologies and system approaches to dramatically reduce the cost of access to space through the introduction of Reusable Launch Vehicles (RLVs). Through the innovative X-34 flight vehicle and recently completed Space Transportation Architecture Studies (STAS), Orbital is leading the way in the development and application of RLV technologies. The STAS studies were commissioned by NASA Headquarters to advise NASA and the White House on the future direction of U.S space transportation. A large number of vehicles and architecture approaches were examined and evaluated. This paper summarizes Orbital's approach to significantly reducing NASA's future launch costs and improving crew safety through the implementation of a low-risk, evolutionary space transportation architecture. In particular, this paper will provide a detailed description of one key element of Orbital's recommended architecture: the Space Taxi™ Transportation System.

STUDY OBJECTIVES

The purpose of the STAS studies was to advise NASA and the U.S. government on a preferred architecture approach to augment and eventually replace the current Space Shuttle, while satisfying a number of objectives:

1) Meeting NASA's launch and human rating requirements
2) Providing significant cost savings
3) Providing resiliency against uncertain events
4) Minimizing government ownership
5) Maximizing industry's role
6) Encouraging competition
7) Benefiting U.S. military and economic security

The time period during which the architecture would be developed and operated was assumed to be from 2000 to 2020. Because of the goals of reducing costs by minimizing government ownership and maximizing industry's role, an extensive study of commercial launch requirements was also included, and the preferred architecture was designed to satisfy NASA, military, and commercial requirements.

CP504, *Space Technology and Applications International Forum–2000*, edited by M. S. El-Genk

STUDY APPROACH

In order to accomplish this multifaceted study effort, Orbital performed a wide variety of analyses, including:

1) Evaluating key government and industry mission databases and developing a reference mission model and excursions.
2) Examining a wide variety of old and new space transportation concepts and developing a very large number of candidate architectures using combinations of these elements.
3) Creating a detailed Master Investment Model for examining the economic payoff of these architecture approaches to the government and potential equity investors.
4) Evaluating the various architectures against criteria and sub-criteria of interest to NASA
5) Selecting and analyzing in detail a preferred architecture that best satisfies these criteria.
6) Developing implementation roadmaps for the preferred architecture, including technology requirements, milestones, transition opportunities, and off-ramps.
7) Examining a wide variety of regulatory and policy issues and potential government financial incentives to enable the implementation of the preferred architecture approach.

STUDY RESULTS

The STAS studies were quite extensive in their scope and depth. They included detailed mission and cost analyses, system definitions, development plans, technology roadmaps, and policy and regulatory recommendations. A summary of all of these results would be significantly beyond the scope of this paper. Instead, the results discussed below will focus on the definition of a new Space Taxi™ Transportation System and related findings and recommendations.

Space Taxi™ Transportation System

Orbital's proposed Space Taxi™ configuration, shown in **Figure 1**, provides a low-risk architecture solution to satisfying NASA's future human spaceflight needs with significant reductions in cost and order-of-magnitude improvements in crew safety and reliability.

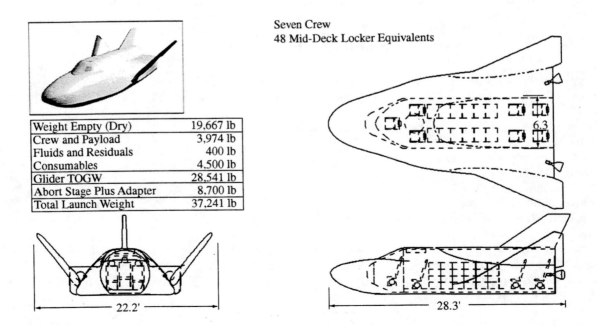

Seven Crew
48 Mid-Deck Locker Equivalents

Weight Empty (Dry)	19,667 lb
Crew and Payload	3,974 lb
Fluids and Residuals	400 lb
Consumables	4,500 lb
Glider TOGW	28,541 lb
Abort Stage Plus Adapter	8,700 lb
Total Launch Weight	37,241 lb

FIGURE 1. Space Taxi reference configuration.

Space Taxi™ Configuration

The reference Space Taxi™ vehicle configuration is about 28 ft long, with a body width of less than 15 ft. It is a lifting body shape with canted fins and a small vertical stabilizer. When the canted fins are folded, the Space Taxi™ can fit in the Space Shuttle payload bay. The empty weight of the vehicle is 20,000 lb, and the total gross launch weight is approximately 28,500 lb. A 76-inch diameter pressurized cabin provides room for seven crew members, 48 Shuttle mid-deck locker equivalents (MLEs), and two extravehicular activity (EVA) suits, simultaneously. The Space Taxi™ is designed to perform the International Space Station (ISS) crew rotation mission currently performed by the Space Shuttle. Normal access to the cabin is through a cabin roof hatch, sized to mate with the ISS Pressurized Mating Adapters (PMAs). A larger access hatch at the aft end of the cabin can be used to berth the Space Taxi™ to Common Berthing Mechanism (CBM) ports on the ISS, allowing the loading of International Standard Payload Racks (ISPR) for return to Earth. The aft hatch also provides a second emergency crew egress exit in the event of a water landing.

As shown in **Figure 2,** the reference Space Taxi™ would initially be launched on top of a new heavy-lift Delta IV Evolved Expendable Launch Vehicle (EELV). Later it would be launched on top of Orbital's two-stage Reusable Launch Vehicle (RLV), which is described in more detail below. A key feature of the Space Taxi™ launch configuration is the use of a Launch Escape System (LES) composed of solid rocket motors attached to the Space Taxi™ aft launch adapter. During a successful mission, these solid rocket motors would nominally be jettisoned from their location on the adapter cone after second stage ignition. If required for abort, the solid rockets have been sized to separate the Space Taxi™ from the launch vehicle with an acceleration of 8 g's to allow intact recovery of the Space Taxi™ vehicle from a catastrophic booster failure at any time during ascent. A set of optional sustainer motors can be used to provide a higher probability of reaching a runway for safe landing at the launch site or downrange. In the event that a runway can not be reached, the Space Taxi™ is designed for a safe water landing with provisions for Apollo-style parachutes and flotation devices (Ehrlich, 1991). This launch escape system was baselined to ensure that Orbital's Space Taxi™ would meet all of NASA's human-rating requirements and provide an order-of-magnitude improvement in crew safety.

The Space Taxi™ configuration and approach has a tremendous amount of heritage in a very detailed series of studies

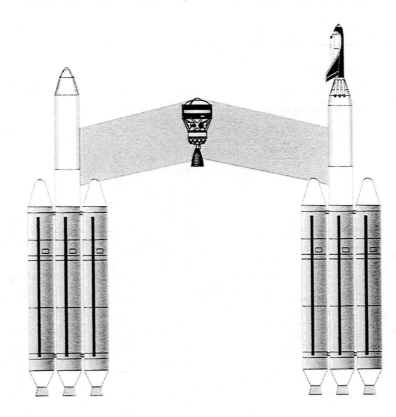

- Separate Crew and Most Cargo

- Launch on Delta IV Heavy EELV

- Delta III Upper Stage

- Abort Throughout Flight Regime

- Synergy with Commercial GEO

- Cargo Carrier Behind Space Taxi

FIGURE 2. Space Taxi Approach.

that was performed by NASA in the early-90's of a vehicle referred to as the HL-20 (Ehrlich, 1991)(Urie, 1992). In fact, an entire issue of the *AIAA Journal of Spacecraft and Rockets* was dedicated to this HL-20 vehicle (Stone, 1993). The aerodynamic shape of the HL-20 was based on a Russian-designed vehicle known as the BOR-4 (Clark, 1998). This vehicle flew numerous sub-orbital trajectories from Russia in the 80's and was recovered in the Indian Ocean. It was used to test thermal protection systems and control system approaches for the Russian Buran Orbiter vehicle. The reference Space Taxi™ version has a trimmed subsonic lift-to-drag ratio (L/D) of 3.8 and a trimmed hypersonic L/D of greater than 1.3, thus providing 1,100 nmi. of re-entry crossrange capability. The only major difference in external geometry between the Space Taxi™ and the original HL-20 aerodynamic configuration is the larger base area of the Space Taxi™, caused by a necessary extension of the internal crew cabin to provide extra volume to meet the ISS resupply requirements. The use of a modified HL-20 configuration allows Orbital to draw on tremendous databases of detailed analysis in the areas of: subsystem weights, aerodynamics, computational fluid dynamics (CFD), 6-degrees-of-freedom trajectories and flight control, thermal protection system sizing, structures and loads, and human factors.

The Space Taxi™ uses a metallic pressure vessel to carry crew and supplies. All other primary and secondary structures are made of graphite composites. Durable ceramic thermal protection system (TPS) tiles and blankets are bonded to the composite panels. The nose, canted-fin leading edges, and aerosurfaces are made of a high-temperature carbon silicon-carbide hot structure with insulation. The major subsystems are located outside the pressurized cabin and are accessible through panels on the upper body surface. The orbital propulsion system is a storable, non-toxic hydrogen peroxide/JP-4 system. The guidance, navigation, and control system of the Space Taxi™ was designed to operate in an autonomous mode to allow the safe return of the ISS crew in an emergency or to allow the Space Taxi™ to be operated uncrewed as a cargo-only delivery and return vehicle.

As shown in Figures 3 and 4, the Space Taxi™ vehicle is normally operated to and from the ISS as part of a system that acts like a cargo-carrying "trailer", attached to the rear, and actually towed by the Space Taxi™. This "trailer" can be configured in two ways: as a Pressurized Logistics Vehicle (PLV) or as an Unpressurized Logistics Vehicle (UPLV). The personnel and cargo carrying capacities of the Space Taxi™, PLV, and UPLV have been sized to make full use of the delivery capability of the heavy-lift Delta IV to the vicinity of the ISS. Therefore, the Space Taxi™ propulsion system only has to perform ISS approach, rendezvous, and later, de-orbit of the Space Taxi™ and PLV combination.

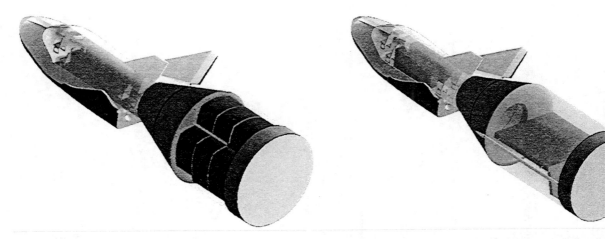

FIGURE 3. On-orbit Configuration of Space Taxi/PLV.   **FIGURE 4.** On-orbit Configuration of Space Taxi/UPLV.

The PLV is a pressurized cylinder capable of carrying launch loads and moments. Its interior is sized to accommodate 12 ISP Racks, which are loaded and removed through a forward Common Berthing Mechanism (CBM) hatch. The UPLV while still carrying launch loads and moments, provides a bay for carrying an unpressurized logistics carrier (UPLC). At least six flights of the Space Taxi™ and cargo carriers are required to satisfy the annual servicing requirements of the ISS. Four of these flights would be crewed, while the remaining flights would be uncrewed.

The Space Taxi™ could potentially be launched on multiple boosters. As shown in Figure 2, the reference Space Taxi™ would initially be launched on top of the new heavy-lift Delta IV EELV. Later it would be launched on top of Orbital's two-stage Reusable Launch Vehicle (RLV), which is described in more detail below. Once the two-stage RLV, described below, is in operation, the proven Delta IV would provide an alternate access capability for the Space Taxi™ if the RLV were not available for an extended period of time due to a catastrophic failure. Before the RLV becomes operational, the Space Taxi™ could be designed to be launched on an Atlas V or on an Ariane V if the Delta IV were not available for an extended period of time.

As a part of these STAS studies Orbital developed a two-stage RLV concept, which could provide unprecedented reductions in cost and improvements in reliability, safety, and performance. As shown in Figure 5, developing two identical vehicles to operate together in a two-stage-to-orbit (TSTO) "Bimese" configuration provides a large reduction in development cost and risk over a single-stage-to-orbit (SSTO) approach because of the much smaller size and weight of the vehicle that must be developed. This RLV is specifically designed to be multi-functional and to provide the widest possible range of mission flexibility. As shown in Figure 6, development of a single vehicle could provide the flexibility of launching almost any weight payload. In addition, since the payloads are mounted externally, a wide range of payload volumes can be accommodated. Two vehicles launched together would provide the same capability as a heavy-lift EELV; a single vehicle could provide a significant sub-orbital trajectory capability; a single vehicle, augmented by varying numbers of solid rockets or liquid stages, could be used to launch small and medium-class payloads; and a single vehicle could also be used as a booster for a heavy-lift core stage to provide the capability of launching very large payloads. This flexibility is very important to its commercial viability because the development cost can be amortized over a significant number of flights of payloads from various weight classes. Because of the potential cost savings that it would provide to a range of customers, this RLV would be commercially developed, owned and operated.

The nominal mission profile for launching the Space Taxi™ on this two-stage RLV is shown in Figure 7. The mission sequence for launching the Space Taxi™ on a heavy-lift EELV is similar. The Space Taxi™ is launched by two RLVs

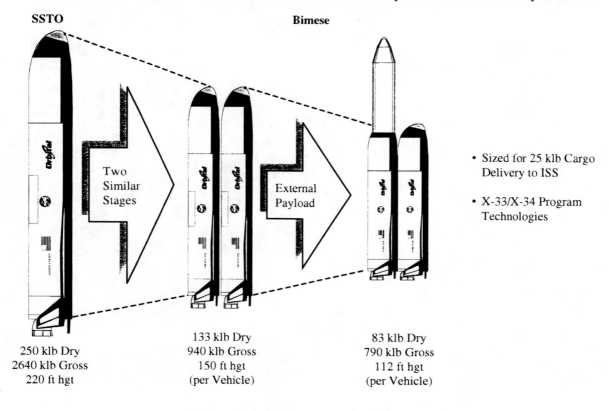

SSTO   **Bimese**

Two Similar Stages

External Payload

- Sized for 25 klb Cargo Delivery to ISS

- X-33/X-34 Program Technologies

250 klb Dry
2640 klb Gross
220 ft hgt

133 klb Dry
940 klb Gross
150 ft hgt
(per Vehicle)

83 klb Dry
790 klb Gross
112 ft hgt
(per Vehicle)

FIGURE 5. SSTO versus Bimese approach.

1139

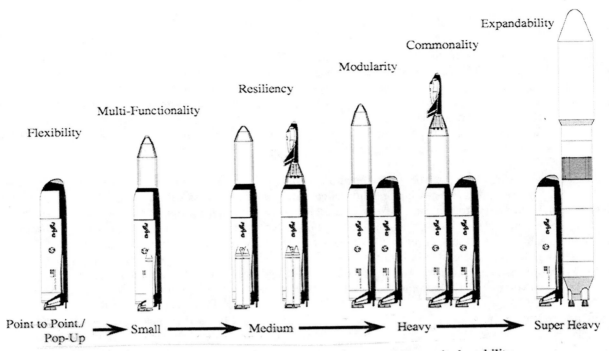

FIGURE 6. Orbital's RLV approach maximizes mission flexibility and adaptability.

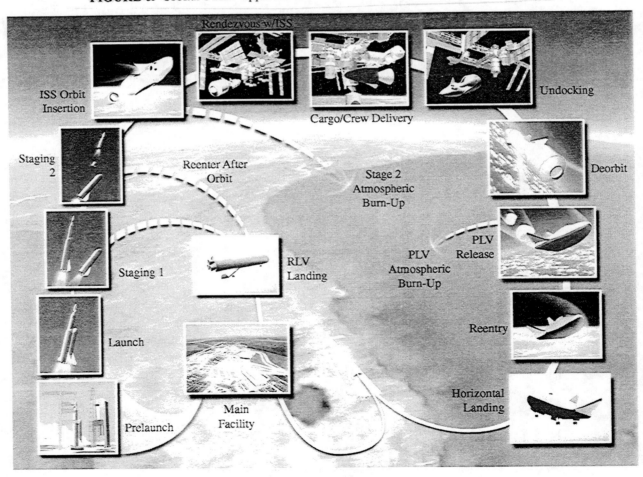

FIGURE 7. Space Taxi/RLV mission profile.

in a parallel burn configuration with cross-feeding of propellant. In this configuration, propellant from the booster is cross-fed to the orbiter, which carries the Space Taxi™ system on top. When the booster's propellant is exhausted (at Mach 3.4 and 100,000 ft.), it separates and glides back unpowered to a runway landing at the launch site. The orbiter then completes the ascent trajectory using its own propellant and inserts into a low-Earth orbit. The 5.1-meter Delta III second stage then separates and begins a burn to raise the Space Taxi™ system to an ISS orbit. Meanwhile, the orbiter RLV jettisons the mating collar between itself and the second stage, performs a de-orbit burn, re-enters, and glides to a horizontal landing at the launch site. The combined Space Taxi™, adapter, and PLV approach the ISS and dock, using the top hatch of the Space Taxi™ for the docking interface. The ISS manipulator arm is then used to detach the PLV and adapter from the Space Taxi™ and attach them to a pressurized port for unloading. At this point, the Space Taxi™ can begin serving as a Crew Return Vehicle (CRV) for the next three months.

ISS operations include egress of crew from the arriving Space Taxi™, unloading cargo from the arriving Space Taxi™ and the PLV, loading returnable cargo into the returning Space Taxi™, loading disposable trash into the PLV, and mating the PLV to the returning Space Taxi™. Return of the Space Taxi™ with the deconditioned crew consists of undocking and departing from the ISS, de-orbit, release of the PLV and adapter to burn up in the atmosphere, and re-entry, resulting in a controlled glide to the landing area. Actual touchdown proceeds much as the Shuttle lands today. When the vehicle is crewed, final guidance and landing can be performed under control of the pilot. For an uncrewed flight, primary guidance is autonomous, with provisions for emergency override if ground controllers detect a problem.

Space Taxi™ Advantages

The Space Taxi™ was specifically designed to be multi-functional and to provide a wide range of mission flexibility. The reference Space Taxi™ would serve as an ISS Crew Return Vehicle (CRV), a two-way human transportation system, a small cargo return vehicle and an RLV passenger module. Because the Space Taxi™ is designed to serve as a CRV, the true cost to NASA of the Space Taxi™ is only the marginal cost over what the Agency would have paid for CRV development and implementation. When a Space Taxi™ brings a fresh crew to the ISS every three months, the deconditioned crew members return in the Space Taxi™ that was left at the ISS three months earlier. Hence, similar to the Soyuz approach, there is always a Space Taxi™ docked to the ISS to serve as a CRV. Because of its small size and potential to be launched on different boosters, Orbital's Space Taxi™ approach provides a very high degree of operational flexibility. Future commercial variants of the Space Taxi™ might be used for satellite servicing and repair, satellite re-boost, space transfer, in-space construction, and even space salvage. In addition, variants of the reference Space Taxi™ could satisfy a number of military missions as a Space Maneuvering Vehicle (SMV), including: orbital and sub-orbital reconnaissance activities; global force projection and suppression; and satellite re-boost, orbit transfer, repair, and servicing.

Orbital's reference Space Taxi™ system would provide an order of magnitude improvement in crew safety because of its use of a Launch Escape System (LES) that would allow intact recovery of the Space Taxi™ vehicle from a catastrophic booster failure at any time during ascent. With its use of state-of-the-art technology and a new, highly reliable RS-68 propulsion system, the Delta IV system should prove to be the most reliable ELV ever constructed. Coupled with the inherent reliability of the Space Taxi™, the probability of crew survival on a single mission could be greater than .9995. The Space Taxi™ system provides an autonomous flight control capability that does not require crew members on cargo transfer flights. This separation of crew and cargo reduces the number of annual crewed flights to four, thereby further reducing the probability of crew loss during the life of the program. The Space Taxi™ configuration also provides a low-g entry for injured, sick, or deconditioned crew members, coupled with a smooth runway landing with direct pilot visibility.

The technologies employed by the Space Taxi™ system are highly consistent with the timing and content of NASA's on-going X-38/CRV technology program, currently being implemented by NASA's Office of Space Flight. The only major technology requirement of the Space Taxi™ that is not included in the existing X-38/CRV program is the advanced development of an integrated propulsion system that uses JP fuel and hydrogen peroxide. A similar system is currently under advanced development as a part of a joint NASA/DoD technology program (Lewis, 1998). Other elements, such as the adapter/LES and PLV, use predominately state-of-the-art technologies. Hence, there is very little risk to the architecture implementation caused by technology availability for the Space Taxi™-related elements.

STUDY FINDINGS

As noted above, the STAS studies were quite extensive in their scope and depth. They included detailed mission and cost analyses, system definitions, development plans, technology roadmaps, and policy and regulatory recommendations. Hence, a large number of findings and recommendations were generated. A summary of all of these findings would be beyond the scope of this paper. Instead, this section will summarize the key technical findings and recommendations related to the development and implementation of the proposed Space Taxi™ Transportation System.

1) Upon full implementation of the Space Taxi™ Transportation System and retirement of the Space Shuttle, NASA would save over $1 billion per year in launch costs under very conservative assumptions. These savings would represent a Net Present Value of $10-15 billion to NASA in current dollars if NASA were to invest in the development of this architecture approach.

2) NASA would recover its investment in the Space Taxi™ Transportation System within two to three years after full system implementation and retirement of the Space Shuttle, and the required annual investment would not exceed NASA's current budget wedge for space transportation development.

3) Implementation of the Space Taxi™ Transportation System would provide an order-of-magnitude improvement in crew safety because of its use of a Launch Escape System that would allow intact recovery of the Space Taxi™ vehicle in the event of a catastrophic booster failure at any time during ascent.

4) Because the Space Taxi™ is designed to serve as a CRV for the International Space Station, the true cost to NASA of the Space Taxi™ is only the marginal cost over what the Agency would have paid for CRV development and implementation. However, in order to provide such an emergency crew escape capability in a timely fashion, NASA must decide soon whether to continue with the current CRV approach or to develop a multifunctional Space Taxi™, which can greatly reduce NASA's human space transportation costs and improve human safety.

5) The Space Taxi™ Transportation System relies on existing or evolutionary technologies, rather than revolutionary technology advances. This greatly reduces schedule and cost risk.

6) Because of its small size and potential to be launched on different boosters, the Space Taxi™ provides a high degree of operational flexibility. Future variants of the Space Taxi™ might be used for human exploration, satellite servicing and repair, satellite re-boost, space transfer, in-space construction, and even space salvage. Variants of the reference Space Taxi™ could satisfy a number of military missions as a Space Maneuvering Vehicle, including: orbital and sub-orbital reconnaissance activities; global force projection and suppression; and satellite re-boost, orbit transfer, repair, and servicing.

REFERENCES

Clark, P., *The Soviet Manned Space Program*, New York, Salamander Books, 1998.

Ehrlich C. F. et al, "Personnel Launch System Study Final Report," Rockwell International Corporation, Contract NAS1-18975 (1991).

Lewis, T., "Interim Design Review for the Upper Stage Flight Experiment Program," Orbital Sciences Corporation, Contract NAS8-98-040 (1998).

Stone, H. et al, "Special Section: HL-20 Personnel Launch System," *AIAA Journal of Spacecraft and Rockets* **30**, 521-634 (1993).

Urie, D. et al, "PLS Feasibility Study Final Report," Lockheed Advanced Development Company, NASA Contract NAS1-18570 (1992).

ROCKOT – An Available Launch System for Affordable Access to Space

U. de Vries, M. Kinnersley, P. Freeborn

EUROCKOT Launch Services GmbH, P.O.Box 28 61 46, 28361 Bremen

Abstract. The *Rockot* launcher will perform its fifth launch, the first commercial launch, in Spring 2000 from the Plesetsk Cosmodrome in Northern Russia carrying two American satellites into a LEO orbit. In preparation for that a launch pad verification flight will be carried out in November this year to prove the functionality of the adapted facilities at the Plesetsk launch site and by placing a Russian satellite into a highly inclined orbit. The results of the launches will be described in detail in the paper as well as the installations at the launch site.

Eurockot, the German-Russian joint-venture company marketing and managing the *Rockot* launch vehicle is meanwhile an integral part of the space launch community. Eurockot was formed by DaimlerChrysler Aerospace and Khrunichev State Research and Production Space Center. A brief overview of its activities, the commercial program and the performance/services offered by Eurockot is presented.

Rockot can launch satellites weighing up to 1850 kg into polar or other low earth orbits (LEO). The *Rockot* launch vehicle is based on the former Russian SS-19 strategic missile. The first and second stages are inherited from the SS-19, the third stage *Breeze* which has already been developed has multiple ignition capability. The *Breeze* upper stage is under production at Khrunichev in Moscow.

The *Rockot* launch system is flight proven and is operated from the Plesetsk as well as from the Baikonur launch site.

GERMAN-RUSSIAN MARKETING COMPANY

In March 1995 DaimlerChrysler Aerospace AG (DASA) of Germany and the Russian Khrunichev State Research and Production Space Center have formed a jointly owned company named EUROCKOT Launch Services GmbH. Both companies are leaders in the German and Russian space industries respectively. They contribute their comprehensive expertise and their experience of international co-operation into the new EUROCKOT marketing company.

EUROCKOT is a single point of contact for the customers and ensures the following launch services whereas Khrunichev is responsible for all technical provisions, as is shown in FIGURE 1.

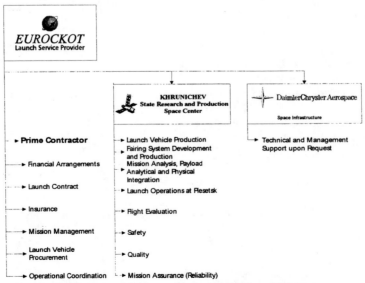

FIGURE 1: Share of Roles and Responsibilities

CP504, *Space Technology and Applications International Forum–2000*, edited by M. S. El-Genk

PREPARATION FOR ROCKOT'S COMMERCIAL DEBUT

Rockot, produced by Khrunichev, aims at the market of small and medium size satellites weighing up to 1850 kg, to be launched into circular and elliptical LEO orbits including polar orbits and sun-synchronous orbits (SSO). These comprise scientific, earth observation and polar meteorological satellites as well as the new generation of small communication satellites in LEO known as the "Constellations". The *Rockot* performance capabilities allow both single and multiple launches depending on payload dimensions.

As can be seen in the table below, the *Rockot* launch system has an extensive flight heritage in the past. This includes the 138 successful SS-19 (*Rockot* booster stage) flights as well the three successful *Rockot* missions from the Baikonur launch site.

TABLE 1: Past and Projected Flight History of the *Rockot* and *Rockot* Family of Launch Vehicles

Vehicle Variant	Launch Pad Type	Locations	Reliability of Last 25 Flights	Overall Flight Record
RS-18/ Rockot Booster	Silo	Various	100%	138/ 141
Rockot / Breeze-K Standard Fairing	Silo	Baikonur/ LC175/1	n.a.	3/ 3
Rockot / Breeze-K Standard Fairing	Pad/ Surface Type	Plesetsk/ 11P865PR	n.a.	Nov 1999
Rockot / Breeze-KM Commercial Fairing	Pad/ Surface Type	Plesetsk/ 11P865PR	n.a.	2Q 2000

In preparation for entry of the *Rockot* launch vehicle into the commercial launch services arena, several areas regarding the *Rockot* launch system have been addressed. This includes the building-up of a western standard launch center for the vehicle at the Plesetsk Cosmodrome in Northern Russia. Furthermore, the basic *Rockot* vehicle configuration has been 'commercialized' via the introduction of large commercial payload fairing and strengthened equipment bay. The configuration changes to 'commercialize' the vehicle are described below. They are purposely restricted to structural changes to the upper composite of the vehicle so as to retain the impressive flight heritage gained with this vehicle.

VERIFICATION OF THE ROCKOT LAUNCH SYSTEM FOR COMMERCIAL LAUNCH SERVICES

Prior to the flight debut of the commercial *Rockot* launch system slated for 2Q 2000, several major tasks need to be completed. These tasks cover the full series of verifications and tests for the vehicle as well as the completion of the supporting infrastructure. These tasks include the following:

- 4Q 1999: Completion of Plesetsk launch vehicle pad and associated launch vehicle infrastructure
- 4Q 1999: Pad Verification Flight (*Rockot / Breeze*-K)
- 4Q 1999: Completion of ground qualification tests for the 'commercialized' *Rockot* configuration, i.e. commercial payload fairing, strengthened equipment bay and payload dispenser systems (*Rockot/ Breeze*KM)
- 1Q 2000: Completion of Eurockot Payload Processing Facilities in Plesetsk
- 1Q 2000: Completion of Plesetsk launch site mission pathfinder tests
- 2Q 2000: First Commercial Flight

These tasks are described in brief below:

Completion of Plesetsk Launch Vehicle Pad and Associated Launch Vehicle Infrastructure

Goal: To have a completed launch vehicle infrastructure to support commercial launch services with the *Rockot* vehicle from the Plesetsk launch site.

Description/Status: The infrastructure to support the launch of the *Rockot* vehicle is near to completion in Plesetsk and will be verified through the so-called pad verification flight (PVF) planned for November 1999 and discussed in detail below. This infrastructure includes the *Rockot* vehicle control room located within the bunker, the launch pad, mobile service tower and booster and *Breeze* fuelling facilities. Photos of these facilities can be seen at the end of this paper.

Pad Verification Flight (Rockot / Breeze-K)

Goal: To verify through an actual *Rockot* flight article the readiness of the launch site infrastructure for start of commercial operations.

Description/Status: A 'pad verification flight' (PVF) will take place in November 1999 to conclude the verification of the above mentioned launch vehicle infrastructure in Plesetsk. This will complete the transfer of the *Rockot* launch vehicle operations from its original silo base in Baikonur to the surface facility in Plesetsk. This flight will also allow Eurockot to validate the accuracy of launch vehicle performance predictions as well as the flight environment modeling, in particular the acoustic noise environment. In addition the existing assets of the launch range infrastructure will be tested with a *Rockot* flight vehicle. The PVF will launch a Russian test payload of designation RVSN-40.

Success Criteria: Successful operation of launch vehicle infrastructure via analysis of post launch data and internal reviews.

Completion of Ground Qualification Tests for the 'commercialized' Rockot Configuration

Goal: The whole range of qualification tests for the commercial version of the *Rockot* launch vehicle shall be completed in time for the commercial service entry in 2Q 2000.

Description/Status: The qualification tests include both launch vehicle specific tests to account for the new *Breeze* and payload fairing configurations (see FIGURE 2 below) and also mission peculiar tests (separation system). A summary of the tests is given below: The tests are scheduled to be completed by 4Q 1999.

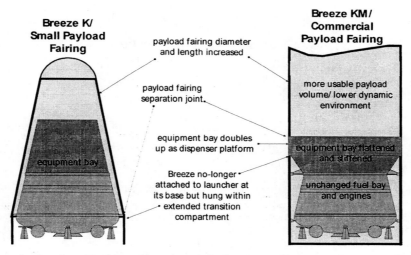

FIGURE 2: Differences in Configuration between the Commercialized *Rockot* Vehicle and the Standard *Rockot* Vehicle.

Qualification Test Summary:

- Model Tests,
  - Aerodynamics of new fairing
- System Level Tests:
  - Payload Fairing Acoustic Noise Tests
  - Dynamic and Static Loading Tests on *Breeze*-KM Structure and Fairing
  - Fairing Separation Tests
- Component and Subsystem Level Tests
  - Mechanisms and components, i.e. Fairing lock tests, separation system components
  - Spacecraft Separation System
  - Separation tests

Criteria to meet goal: Successful close-out of qualification with the Russian regulatory authority TSNiimash as well as internal and external (with first commercial customer) post qualification review boards. The photographs contained within the ANNEX show the *Breeze* static and dynamic model tests underway at the Khrunichev facilities.

Completion of Eurockot Payload Processing Facilities in Plesetsk

Goal: To have fully completed spacecraft payload preparation facilities (see technical description later in this paper) by 4Q 1999 in time for the spacecraft pathfinder tests prior to the first commercial launch.

Status: The *Rockot* dedicated spacecraft processing facilities are close to completion with final equipment installation into the class 100 000 clean rooms. The spacecraft processing facility building will also include the appropriate infrastructure that western customers are accustomed to including offices and high standard communication links within the site and to the outside world. A western style hotel complex (The Zarya hotel in the nearby town of Mirny) is also nearing completion and will be available for commercial customers.

Criteria to meet goal: The completion of the facilities will be closed out by successful completion of a facility acceptance review. The correct functioning of the facilities and correction of problems will be covered in a subsequent spacecraft launch site pathfinder test described in the next section.

Completion of Plesetsk SV Launch Site Pathfinder Tests

Goal: To run through all potential spacecraft operations and combined operations from transportation up to launch vehicle stacking using the newly completed payload preparation facilities and launch pad infrastructure prior to commercial flight use.

Status: This event is currently planned for 1Q 2000 following completion of the pad verification flight and the payload preparation facilities.

Criteria to meet this goal: The completion of the tests shall be closed out by successful conclusion of a post test review board between Eurockot/ Khrunichev, the Plesetsk range and the first commercial customer

Global Schedule for Rockot Program

Following successful completion of the above mentioned activities including the pad verification flight as well as the specific mission integration activities for this first commercial mission, the first commercial launch from Plesetsk is slated to take place in 2Q 2000.

Eurockot is also currently re-activating the Baikonur launch site for a firm customer (see later section on this) and is scheduled to be operational by 4Q 2001.

Eurockot expected launch capacity is estimated at 6 launches per year or more.

ROCKOT CONFIGURATION AND PAYLOAD ENVELOPE

Rockot is a three stage all liquid launch system. The launch vehicle configuration consists of a booster unit containing the first and second stages and an upper composite unit consisting of the *Breeze* upper stage, the interstage, the fairing, the payload adapter and the payload which can be seen in FIGURE 9 of the ANNEX. The overall launch vehicle length is about 29 meters and its external diameter is 2.5 meters. The launch mass is approximately 107 metric tons.

The booster unit is an adaptation of the highly reliable SS-19 ICBM tested in flight over 140 times (see earlier table). A multiple re-ignitable and highly manoeuvrable third stage *Breeze* is added to provide orbital capability. The complete *Rockot* system including *Breeze* has been proven in flight three times with a 100 % success rate. The last flight transported an amateur radio satellite from the Baikonur launch site into a 1881 km x 2163 km x 64.8° Orbit. In addition to this test flight series, all components have undergone an extensive ground qualification test program. Furthermore, many of the components such as the *Breeze* upper stage engine have a proven heritage from other former Soviet space program. Current Eurockot activities include additional verification and qualification tests for the commercial payload fairing and strengthened *Breeze* equipment bay (see earlier section).

LAUNCH CAMPAIGN AT THE PLESETSK RANGE

An existing launch pad which was formerly used for the COSMOS launch vehicle has been adapted exclusively for *Rockot* (see FIGURE 3). The integration facilities including clean rooms for payload preparation as well as ground measurement infrastructure are currently under construction at the Plesetsk range.

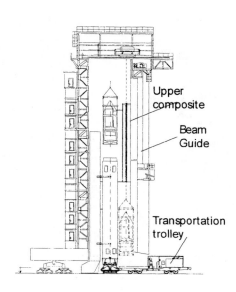

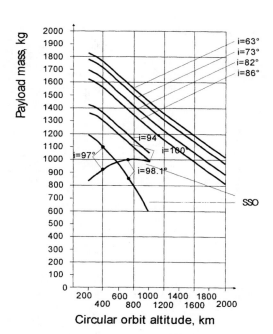

FIGURE 3: *Rockot* Launch Pad at Plesetsk

FIGURE 4: *Rockot* Performance Capabilities from Plesetsk

ROCKOT PERFORMANCE CAPABILITIES

Orbits with inclinations ranging from 63° to 94° can be achieved within the allowed launch azimuth range without plane change prior to the payload deployment. To achieve SSO orbits, a plane change from two optional parking orbits with an inclination of either 94° or 100° is to be performed. The full range of performance capabilities associated with the corresponding circular orbits is given in FIGURE 4.

ROCKOT LAUNCHES OUT OF BAIKONUR COSMODROME

In spring 1998 EUROCKOT together with their partner Khrunichev started preliminary activities in order to prepare for launches with *Rockot* out of Baikonur from 4Q 2001 onwards.

Baikonur is very well equipped due to ongoing launch campaigns by other well established launchers, eg Proton.

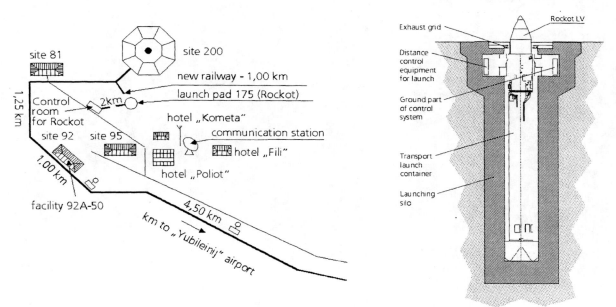

FIGURE 5: Sites of Baikonur Cosmodrome used for *Rockot* Launches

FIGURE 6 : *Rockot* Silo

The main infrastructure at Baikonur Cosmodrome , requiring only few modifications, is in place for *Rockot*.

Khrunichev integration facilities which are world standard and are in condition will be available for *Rockot* after having been used for current large scale programs, minor modifications in the infrastructure only are necessary. *Rockot* will be launched out of Baikonur from a silo, adjustments to the current silo allocated for *Rockot* are being initiated. The *Rockot* performance and accuracy from Baikonur are shown below:

TABLE 2. Performance from Baikonur Cosmodrome at 920 km altitude.

Inclination, deg	50°	51°	52°	53°
Performance, kg	1220	1270	1320	1370
Altitude accuracy = +/-18.4 km				
Inclination accuracy = +/-0.05°				
All 3 sigma, based on analysis and past flight performance				

SUMMARY

The advantages of using the *Rockot* launch system are as follows:

- *Rockot* is a flight proven reliable launch vehicle with highly manoeuvrable upper stage

- *Rockot* provides for a wide range of payload masses up to 1850 kg to be launched into LEO and MEO orbits with a wide inclination range

- *Rockot* offers a dual/multiple launch concept

- EUROCKOT secures low prices tailored to payload mass

- EUROCKOT provides all services from a single source.

ANNEX

FIGURE 7. Dynamic Test Model of *Rockot* Upper Composite consisting of *Breeze*-KM Stage and commercial Payload Fairing

FIGURE 8. *Rockot Stages* Stored at the Khrunichev Facility in Moscow.

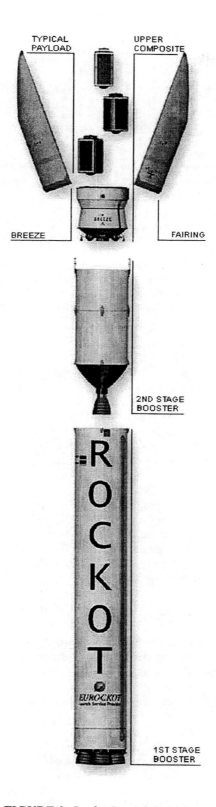

FIGURE 9: *Rockot* Launch Vehicle

Opening the Spaceways

Jess M. Sponable

Universal Space Lines, 1501 Quail St., Suite 103, Newport Beach, CA 92660
(949) 476-3594; sponable@spacelines.com

Abstract. Identifies spaceplane technical concepts proposed by both large and small companies. Highlights that the size and complexity of spaceplanes can be bounded by two well-defined concepts: the Lockheed Martin X-33/Venturestar concept and the Boeing Military Spaceplane vehicle. Also identifies a number of spaceplane concepts being proposed by small commercially financed companies. Reviews possible policy, regulatory, technology, financing and market catalysts that the government and Congress can establish to improve the business climate and encourage investment in spaceplane ventures. Argues that the government has a role and an obligation to help open the spaceways through a prudent mix of government investments and business incentives.

THE VISION

If one knows where to look, history often provides the answers to many of today's problems. Understanding the imperative for future commercial spaceplanes can be illuminated through that same lens of history. Imagine yourself one hundred years ago in 1899 discussing the future of flight. If the experts of that era had such a discussion, it would have focused on the future of balloons with the visionaries talking about dirigibles. Kittyhawk, man's first powered flight, won't happen for another four years. Commercial aviation will take decades to grow into a profitable industry. And as for the U.S. military, it will take half a century, two world wars, the wholesale export of American aviation technology overseas, and an Act of Congress creating the United States Air Force before airpower is accepted.

We're in the same situation today. Even defining what commercial spaceplanes will look like is difficult. The only thing we can say for sure is that if spaceplanes are to follow the growth path of the aviation industry, then they will eventually have to demonstrate the operating characteristics of aircraft. Truly viable commercial spaceplanes won't come into existence overnight. Indeed, our nation has yet to fly a spaceplane that could enable this exciting future. But again, history teaches us that just as someone eventually built a DC-3 that opened the airways, someone will build the first DC-3 for space and open the spaceways.

THE CONTENDERS

Two classes of spaceplanes are being developed today. NASA is championing the first. Their characteristics generally include heavy lift payloads; liquid oxygen and hydrogen powered rocket engines; single stage to orbit (SSTO) capability (and in some cases Two Stage To Orbit—TSTO); sophisticated new technologies; and a price tag to first flight of several billion dollars. Small entrepreneurial companies and, hopefully the larger companies as well, are developing the second class of vehicles. Their characteristics are generally small to medium lift payloads; liquid oxygen and kerosene propellants; TSTO flight; available off-the-shelf technologies; and a price tag to first flight of several hundred million dollars.

A key difference between the two classes of vehicles stems from their choice of new undeveloped versus available technologies. To mitigate risk and cost the entrepreneurial companies usually use available Russian liquid oxygen and kerosene (Jet Propellant, JP-8) rocket engines. These Russian rockets are easily two generations ahead of equivalent kerosene rockets made in the United States. They have better specific impulses and higher thrust to weight ratios. Moreover, there are a host of alternative engines to choose from, many still in production. Use of available technology also allows the entrepreneurs to focus on demonstrating reliable, safe flight and "aircraft-like" operations: essential ingredients for viable commercial spaceplanes. NASA generally has the same long-term goal

CP504, *Space Technology and Applications International Forum–2000*, edited by M. S. El-Genk
© 2000 American Institute of Physics 1-56396-919-X/00/$17.00

although their needs don't require the same degree of operability or reliability as the commercial sector. Moreover, the development of many new technologies on NASAs X-Vehicles necessarily limits their focus to getting the hardware out the door. Operability is a luxury often precluded by NASAs charter to develop new unproven technologies.

As with any generalization there are exceptions to the above categories. Some of the entrepreneurs are using higher risk approaches such as SSTO and Rocket Based Combined Cycle airbreathing RAM/SCRAMJET engines. And some of the NASA contractors including Orbital Sciences Corporation's (OSCs) X-34 and Boeing's X-37 are developing the technologies for TSTO smaller payload vehicles. Nonetheless, categorizing the two classes of vehicles is useful because it enables the comparison shown in Figure 1. The figure graphically illustrates the size difference between large SSTO vehicles using voluminous hydrogen and oxygen propellants with an average density of 23 lbs/ft^3 and the much smaller two stage vehicles using JP-8 and oxygen propellants with a bulk density nearly three times as great. The two anchor points in the figure are the Boeing Military Spaceplane using the high-density propellants and the Lockheed-Martin Venturestar using liquid oxygen and hydrogen. Over the past 5-10 years many tens of millions of dollars have been invested in these concepts, making them the best-defined spaceplane configurations in industry today.

The size difference is misleading as it is only partially caused by the propellant density, the other factor is different payload sizes. Most of the entrepreneurial vehicles such as Universal Space Lines Space*Clipper* shown in the figure

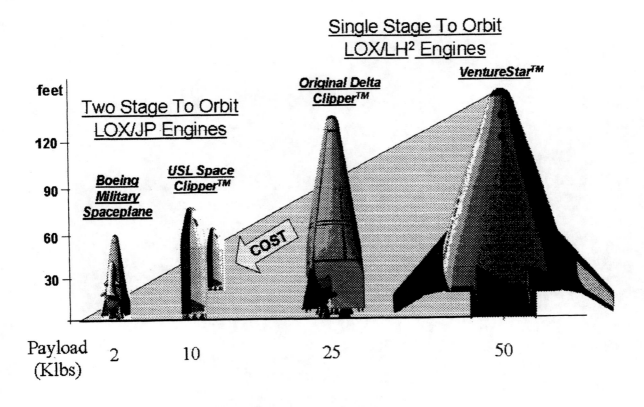

FIGURE 1. Size Comparison.

are designed to launch relatively small payloads. Approximately 70% of all payloads projected to fly during the next decade weigh less than 10,000 lbs. The affordability of large expendable boosters depends on launching multiple satellites on each flight. Conversely, a reliable spaceplane can capture the majority of satellite traffic more cost efficiently and without the complexity of stacking payloads. Figure 2 highlights that the Space*Clipper* is only one of many entrepreneurial spaceplanes proposed by industry; most having smaller payloads. The Starbooster, Astroliner and Rocketplane all require an expendable second stage, while the K-1 and SpaceClipper have a reusable second stage. The Roton is designed as a SSTO vehicle although it could deploy larger payloads by staging at high mach using an expendable rocket stage. This kind of "pop-up" staging maneuver is employed by several of the vehicles shown including the Astroliner, Rocketplane and the SpaceClipper.

THE REALITY

Judging by the variety of spaceplane concepts proposed; one gets the impression the industry is vibrant and healthy. Unfortunately, that's not true. NASA is investing significant funds in advanced spaceplane technologies, but only token amounts in the kind of "aircraft-like" operability features fundamental to safe, reliable and routine flight--the issues potential business investors care about. The private sector is in even worse shape. Many of the entrepreneurial companies have secured accredited "angel" investors willing to invest a few million dollars, but none have convinced institutional investors to put significant funds on the line.

In an era where billions of dollars are routinely invested in internet vaporware, it is ironic that the current business environment won't support investments in next generation transportation. There are many good reasons for the attitudes on Wall Street ranging from a speculative market to a healthy desire not to compete with the government. Rockets and space historically are places you spend money, not make money. And the current financial woes of Iridium's satellite-based communications further erode investor confidence.

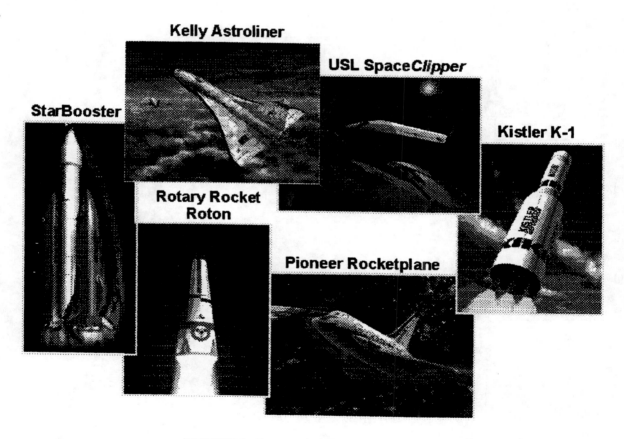

FIGURE 2. Representative Entrepreneurial Spaceplane Concepts.

But perhaps most fundamental is a long American tradition, indeed an unwritten expectation, that the government must take the lead developing next generation transportation infrastructure. The government played a key role developing every modern form of transportation: from transcontinental trains to highways and aircraft. Even Wall

Street's darling, the internet, was created by the Defense Department. That doesn't mean the government needs to pay all the costs for next generation spaceplanes, it does mean the government has a responsibility to create an environment where space transportation investments can compete for institutional Wall Street money. Unfortunately, today there are many policy, regulatory, technology, financing and market impediments to the creation of a spaceplane transportation industry.

THE GOVERNMENT ROLE

The U.S. government can approach the emerging spaceplane industry from two perspectives. It can ignore the emerging industry. In that case the industry will likely struggle forward similar to the barnstormers of the 1920's. Like the barnstormers, many of the company's will meet an untimely death, but emerging technology will eventually ensure that spaceplane transports are built. Unfortunately, in this scenario American companies probably won't build them, nor will America control next millennium space markets. As the recent China-Loral incident highlights, business reality will drive American companies to work with whatever supplier offers the best business deal, whether American or foreign.

In the second scenario the government becomes an advocate and establishes policy, regulatory, technology, financing and market catalysts that expand the industry. Between the 1930's and 1950's similar government catalysts such as the airmail service, NACA investments, and military purchases before, during and after WWII created an environment increasingly supportive of commercial aviation. Yet other analogies can be drawn from the introduction of transcontinental railways and interstate highways.

This section will review some of the catalysts the government can potentially introduce as incentives for creating a new space transportation industry. A few possibilities include: R&D tax credit legislation for commercial space transportation ventures; streamlining regulatory processes; providing commercial launch services like the airmail of the 1930's; "Zero Gee–Zero Tax" legislation to encourage new space markets; government loan guarantees; and balancing today's scarce R&D funds to fly both new technology vehicles such as the current fleet of X-Vehicles, and "operations demonstrators" aimed at enabling the first DC-3s of space.

Tax Credits: Prospace recently authored and proposed a legislative Bill that provides Research and Development (R&D) Tax Credits for investors in space transportation ventures. Similar tax credits helped OSC launch their Pegasus booster during the 1980's. American investors will go to almost any length to reduce their tax burden; a heritage whose origins may date to the Boston Tea Party and the inception of the American Revolution. For many American investors saving money in taxes is worth the risk of a bad investment. Tax credits are one sure way to increase investment in spaceplane ventures.

Streamlining Regulations: The spaceways will consist of commercial spaceports geographically dispersed around the United States and the world, as well as the regulatory processes and space traffic control systems needed to open a new frontier of commerce. Unfortunately, no single agency today is tasked with building the spaceways, or even planning for the spaceways. Indeed, there isn't even consensus on what the spaceways are. Instead, there are currently Federal Aviation Administration (FAA) regulations that govern the aviation industry. Many of these regulations are not directly applicable to spaceplanes, yet better regulations have not been developed.

The FAA could be tasked to work with state spaceport authorities to develop ten, twenty or thirty year plans for developing the spaceways. Burdensome regulations can be identified and streamlined. The FAA/AST office has taken a first step by commissioning an industry RLV working group to recommend regulations that will allow spaceplanes to safely fly from the United States. Such regulations are essential lest the industry go overseas to fly their vehicles. A case in point is Kistler Aerospace, which is hoping to fly initially from Australia, and Beal Aerospace that plans to fly its expendable launch vehicle from the Caribbean.

Commercial Launch Services

Both NASA and the Air Force purchase launches today from established providers. Although new company's with untried hardware can legally bid on launch service proposals, they have little credible chance of winning. Just as airmail created a government market for aviation services during the 1930's, the government could award launch service contracts to multiple competitors. One example might be logistics resupply of the International Space Station. The vast majority of station traffic will consist of small items such as water, food, supplies, and propellants that can be launched by many of the entrepreneurial spaceplanes. If safety is an issue, then "park" the delivery near the station until NASA has time to retrieve the shipment using the Shuttle or alternative transfer vehicles.

A portion of these contracts could also be awarded to higher risk companies that haven't yet flown vehicles. By establishing and tracking key milestone payments the government can protect its "investment" even as it encourages the birth of a new industry. Such contracts have occasionally been used in the past at places like the Defense Advanced Research Projects Agency (DARPA) and the Strategic Defense Initiative Organization (SDIO). Specifically, OSCs Pegasus and Taurus launch vehicles may never have flown without the help of visionary personnel at DARPA willing to lend a helping hand. And don't expect success all the time. The failure of the Conestoga and ORBEX launchers in the early 1990's illustrates both the technical and business risks inherent in today's launch market.

Kennedy Space Center has a Request For Proposal (KSC-RFP-10-99-0021) on the street that could provide the nucleus for both launch service contracts with established providers, as well as incentives for new start companies. Key to success is ensuring that launch service contracts are routinely competed every few years and widely awarded to both new and old launch providers. Launch service contracts executed as a consistent and measured policy of the U.S. Government could provide a powerful spur for opening the space frontier.

Zero Gee–Zero Tax Legislation

Congressman Dana Rohrabacher of California has on several occasions urged adoption of legislation that encourages investment in space enterprises by limiting the tax liability or establishing a "tax holiday". Again, American investors will likely respond favorably, and with their pocket books, to such legislation. And spaceplane companies will be one of the many beneficiaries. Low cost space transportation will be fundamental to the success of any large-scale space business enterprise. In some respects this legislation is akin to the transcontinental railway companies receipt of large tracts of worthless government land adjoining the railways in the last century. Ultimately, the growth of businesses around the railways made the land far more valuable than the rail infrastructure itself, and the government also benefited from the larger tax base. Zero-Gee Zero-Tax legislation can potentially expand new space enterprises in a similar fashion; filling the vacuum with revenue.

Loan Guarantees

Senator John Breaux of Louisiana has introduced the "Breaux Bill" seeking to provide loan guarantees for companies introducing new space transportation ventures. Such guarantees are common in many government sectors, but aren't normally used in the aerospace industry. As currently structured the bill caters to large companies willing to invest some of their funds in exchange for much larger government guarantees. In this capacity, the Bill will help large established companies, but it probably won't noticeably expand the fledgling spaceplane industry.

Today, the largest aerospace companies have little incentive to introduce truly low cost space transportation. Indeed, each has billions of dollars of annual revenue and hundreds of millions of dollars of profit at stake. Any Chief Executive Officer who suggests the revenues in hand should be replaced by speculative new spaceplane markets would be violating their fiduciary responsibility to stockholders. The danger of government loan guarantees is that in combination with market guarantees (i.e., guaranteeing existing revenue streams) a de facto government backed monopoly could be created that could squelch all competition for a generation.

In my view a better loan guarantee program is to issue guarantees in conjunction with local state spaceport authorities, and incorporate a provision where not more than 10-25% of the appropriated funds can be used to support any one company or venture. The key to building a space transportation infrastructure is establishing a robust competitive spaceplane industry, not simply investing in established companies. A cap will ensure multiple worthy ventures are funded, and won't prevent a future Congress from adjusting the cap or expanding the capital base by adding appropriated funds.

Operations Demonstrators

In 1936 the Douglas DC-3 entered commercial service with a payload of 6,600 lbs. Donald Douglas was quoted as saying he'd be happy if he sold ten aircraft. Over 14,000 DC-3's were eventually built serving as the primary air cargo carrier used in both WWII and in the emerging commercial air transport market. What made the DC-3 so successful was not the use of new technology, but rather a well thought out design that emphasized operational simplicity, redundancy and cost efficiency. The DC-3 offered a quantum jump in flight safety, comfort and reliability to the paying public.

Drawing an analogy to the 1930's, America's space launch industry needs more than newer launch technology. Instead, we must learn to apply today's technology to spaceplanes designed to fly with "aircraft-like" operational efficiencies. Figure 2 portrays several such vehicles proposed by industry. The challenge is not building these spaceplanes, but getting them financed. Wall Street is rightly concerned given the history of low cost to orbit promises. My experience in both the private and government sectors is that finding financing on Wall Street, or acceptance at the Pentagon, will not happen until operability demonstrators are routinely and reliably flying with fundamentally lower cost "aircraft-like" efficiencies and durability's.

Operability demonstrators need not fly to orbit, but they should prove the viability of orbit capable vehicles with durable and long lifetimes. More important, they must fly routinely (turnaround in hours), reliably (hundreds of times), and at very low cost (100X lower). NASA has the charter to contract with industry and build such vehicles, but must balance its investment in new technology X-Vehicles with investments in operability demonstrators. NASA can't build commercial spaceplanes anymore than the NACA could invest in the DC-3. But NASA certainly can build the precursor vehicles that demonstrate the kind of routine, reliable, safe and affordable flight attractive to commercial investors.

THE WAY AHEAD

Several spaceplane ventures proposed today are technically viable and could potentially be profitable. Unfortunately, the current business environment is not very supportive of such investments. Many policy, regulatory, technology, financing and market incentives could be established to catalyze the creation of a new spaceplane industry. This paper reviews some of those options and many others could be developed. But first the government must decide if it has a role helping to encourage the development of a new industry. My own belief is that the government has a role and an obligation to help open the spaceways. A prudent mix of several of the above catalysts could provide the investment incentives that open a new frontier of business commerce.

If we can successfully navigate the treacherous path to commercially viable reusable spaceships—the first DC-3's of space—then America will lead the world into a new millennium of global security and vast wealth on the space frontier. But we must start with government policy and legislative changes that create a business environment where investments are deemed both prudent and potentially profitable.

Deriving an Acceptable Level of Reusable Launch Vehicle Flightworthiness

René J. Rey

The Boeing Company, 5301 Bolsa Avenue, MC H014-C441, Huntington Beach, CA 92647

(714) 372-1740; rene.j.rey@boeing.com

Abstract. Second-generation reusable launch vehicles (RLVs) will become catalysts for the codification of RLV-specific flightworthiness standards—their potential to evolve into globe spanning space transportation systems operating on a daily basis is immeasurable. To enable these operationally prevalent RLVs to achieve their maximum potential, they will need to function within the confines of an international regulatory framework, and an established airworthiness code as provided by Annex 8 to the Convention on International Civil Aviation. Hence, certification and licensing standards should not be tailored to accommodate the uniqueness of each RLV design concept, as proposed by some of today's hopeful developers and operators. Rather, an RLV design that will achieve aircraft-like operations *without* compromising aircraft safety standards should be established as the benchmark upon which the legal regime for *follow-on systems* will be based.

BACKGROUND

Aircraft are designed to meet airworthiness and vehicle certification requirements in accordance with well established national and international standards that have evolved over the past 70 years (beginning with the Warsaw Convention, which was signed in 1929). These standards are designed to promote safety and reliability in aircraft systems; and apply to all aircraft designs, whether they are operational or developmental.

The emergence of commercial reusable launch vehicle systems, in contrast, will not have the benefit of specific, codified standards. Yet, developmental and operational RLVs are expected to operate like conventional aircraft from multiple sites around the Earth, carry cargo, and overfly population centers across a broad range of azimuths. And, like aircraft, RLVs will come in a variety of shapes and sizes that employ a wide range of operational concepts. These concepts include single-stage-to-orbit (SSTO) and multi-stage-to-orbit (MSTO) vehicles, each incorporating a unique combination of structure, propulsion, and flight profile to achieve orbital injection requirements (while carrying a meaningful payload mass).

However, it is doubtful that many second-generation RLV systems will achieve their performance objectives without taking some major short cuts that could compromise the same safety and reliability standards that commercial air transports are required to adhere to. In fact, maximum allowable dry-mass-fractions for RLVs using current technology propulsion systems, may preclude incorporation of the robust structures and subsystems necessary for *reliable*, reusable operations involving *the **safe** over-flight of population centers*.

RELEVANT LIABILITY DOCTRINES

The doctrines of *absolute liability* and *product liability* are the most relevant to reusable launch vehicle flightworthiness. Within the purview of these doctrines there are at least three entities which may be held liable: the launching state, the appropriate state party, and non-governmental entities (Gorove, 1991). Since this paper is being written from a commercial RLV enterprise perspective, only the liability for a non-governmental entity will be addressed.

CP504, *Space Technology and Applications International Forum–2000*, edited by M. S. El-Genk
© 2000 American Institute of Physics 1-56396-919-X/00/$17.00

The reusable launch vehicle enterprise may be held liable under the Liability Convention, the Outer Space Treaty, or domestic law for damage caused by an RLV. The basis for liability under domestic law is likely to be either negligence or strict liability—the standards for product liability. However, should a court regard the RLV operation as ultra-hazardous, absolute liability will be imposed (Gorove, 1991).

Absolute Liability

The basis for absolute liability is *ultra-hazardous activity*, which is defined as "an act or conduct, not of common usage, which necessarily involves a risk of serious harm to the person or property of others which cannot be eliminated by the exercise of utmost care (Haley, 1963)." Absolute liability is a product of the machine age, whose evolution produced numerous instances of serious property damage and personal injury. As a result, it became necessary to place these losses "on those who, though free from negligence or tortious intent, had control over the instrumentality causing the harm and who, in most cases, were better able to foresee the possibility of financial loss and protect through insurance techniques against it (Haley, 1963)."

The history for liability of damage to property caused by crashing aircraft may well be a harbinger for the course the law will take regarding reusable launch vehicle accidents. This precedent was established during the early days of aviation, when aircraft and balloon flights were held to be ultra-hazardous activities (Haley, 1963). In *The Law of Aviation*, a treatise on air law written in 1938, Mr. Hotchkiss stated:

> It has been generally recognized that where an aircraft descends on a person or property on the ground beneath, or where objects thrown from the aircraft cause damage, the owner or operator of the aircraft should be held to the strictest accountability (Haley, 1963).

During the time period immediately preceding this treatise, many states passed laws making an aircraft owner absolutely liable for any damage or injury caused by the crash of his aircraft. In fact, twenty-one states and territories in the United States adopted the Uniform Aeronautics Act in the period from 1920 to 1930 (Haley, 1963). Section 5 of this statute reads:

> The owner of every aircraft which is operated over the lands or waters of this state is absolutely liable for injuries to persons or property on the land or water beneath, caused by the ascent, descent or flight of the aircraft or the dropping or falling of any object therefrom, whether such owner was negligent or not, unless the injury is caused in whole or in part by the negligence of the person injured or the owner or bailee of the property injured (Haley, 1963).

This statute aptly fits the flight operations of second-generation reusable launch vehicles. The complex interaction of rocket propulsion, structures, avionics and software—and the myriad possibilities for malfunction—naturally place RLV systems at great risk.

Product Liability

U.S. Aviation Product Liability Law comprises three *foundations for claims* categories which are differentiated by the requirements (e.g., willful misconduct, rightful claimants, etc.), and the legal consequences (e.g., scope of compensation, etc) of each case (Fobe, 1994). These categories include claims derived from *warranty*, *negligence* and *strict liability* (Fobe, 1994). Variations in national law on product liability, however, differ from jurisdiction to jurisdiction. Aviation product liability law relates to reusable launch vehicle flightworthiness because RLVs will have to be operated and maintained like the state-of-the-art jet aircraft in service today. Similarity in operations implies that a similar product liability regime will be applied as well. The following discussion concerning RLV product liability, however, is limited to negligence and strict liability in tort.

Negligence

In U.S. tort law an action in negligence against a manufacturer has been possible since 1916, when Judge Cardozo wrote his landmark opinion in *MacPherson* v. *Buick Motor Co.* (Haanappel, 1979):

> We are dealing now with the liability of the manufacturer of the finished product, who puts it on the market to be used without inspection by its customers. If he is negligent, where danger is to be foreseen, a liability

will follow...There is no break in the chain of cause and effect. In such circumstances, the presence of a known danger, attendant upon a known use, makes vigilance a duty.

***

Precedents drawn from the days of travel by stagecoach do not fit the conditions of travel today. The principle that the danger must be imminent does not change, but the things subject to the principle do change. They are whatever the needs of life in a developing civilization require them to be (Bostwick, 1994).

After *MacPherson*, within the law of product liability the concept of strict liability was developed, first in contract for breach of warranty, express or implied, and later strict liability in tort for physical harm to persons and tangible things (Bostwick, 1994).

Liability for the negligence of a reusable launch vehicle enterprise can be invoked directly by a third party if the enterprise fails to take every reasonable measure to avoid any foreseeable risk in the manufacture and handling of products. The prerequisites of liability are (Fobe, 1994):

1. a duty of care;
2. a breach of this duty;
3. an adequate causal connection between the damage sustained and the negligently constructed or operated product; and
4. damage sustained by the plaintiff.

Once these elements are met, the liability for negligence can include liability for improper design and faulty manufacturing, the duty of product control, and inadequate warning, instructions for use, etc (Fobe, 1994). In addition, U.S. aviation product liability law places considerable importance on the various sections of the "Restatement, Torts, Chapter 14, 'Liability of Persons Supplying Chattels for the Use of Others,' Paragraph 388-408," and compliance with the certification and airworthiness standards of the Federal Aviation Regulations (FARs) (Fobe, 1994). Hence, second-generation RLV enterprises would do well to design and manufacture their flight vehicles with an eye toward eventually complying (at some level) with the appropriate FAR standards.

Strict Liability

A significant problem presented by the prospect of property damage and personal injury caused by a negligently constructed aircraft has been the difficulty of proof—causing a shift towards the theory of strict liability (Fobe, 1994). The decision that gave birth to this doctrine in aviation product liability law was *Greenman* v. *Yuba Power Products* in 1963, where Chief Justice Traynor provided the following opinion on the definition of strict liability:

A manufacturer is strictly liable in tort when an article he places on the market, knowing that it is to be used without inspection for defects, proves to have a defect that causes injury to a human being (Fobe, 1994).

Strict liability requires that the injured third party only show that the product itself is defective to ensure recovery. For this standard, only three elements need to be proven (Fobe, 1994):

1. the existence of a defect;
2. the defect existed at the time the product left the manufacturer's control; and
3. the defect caused the injury.

Two years following the historic *Greenman* v. *Yuba Power Products* decision, strict liability was adopted, in amended form, in Section 402A of the Restatement (Second) of Torts. This section provides that (Fobe, 1994):

1. One who sells any product in a defective condition unreasonably dangerous to the user or consumer or to his property is subject to liability for physical harm thereby caused to the ultimate user or consumer, or to his property, if
   (a) the seller is engaged in the business of selling such product, and
   (b) it is expected to and does reach the user or consumer without substantial change in the condition in which it is sold.
2. This rule applies although
   (a) the seller has exercised all possible care in the preparation and sale of his product, and
   (b) the user or consumer has not bought the product from or entered into any contractual relationship with the seller.

FACTORS CONTRIBUTING TO RLV LIABILITY

The focus of many present-day reusable launch vehicle concepts revolves solely around the technical performance, ease of manufacture and the minimal amount of testing and analysis required to "validate" a flight vehicle system. The primary impetus for this focus is the RLV operator's need to drive launch costs down to less than $1,000 per payload pound to orbit, which in turn forces many tradeoffs among a flight vehicle's subsystems, propulsion, structure and design for operability (Kaplan, 1997). Needless to say, design tradeoffs must not only take into account the technical performance issue of getting a specific payload mass into orbit; but also address the fact that "...the concept of low-cost immediately imposes the requirements of high usage rates and fast turnaround times with minimum maintenance (Kaplan, 1997)." Engineering reviews for most RLV development programs address system requirements, preliminary design and critical design milestones as the predominant criteria for determining whether their launch systems are "viable." However, these reviews fail to acknowledge the operational liabilities inherent in space launch vehicles designed to be operated and maintained like traditional high performance, heavy jet aircraft.

At the heart of a reusable launch vehicle's technical performance is its *dry mass fraction*, which is the ratio of an RLV's structural mass to its total mass (with a full propellant load and payload).

> For example, VentureStar is projected to have a gross lift-off weight of 2,186,000 lbs, with a propellant load of 1,929,000 lbs. Hence, the propellant mass fraction is 88.2%, which leaves 11.8% for dry mass and payload. Assuming a payload mass fraction of 2.7%, this leaves a structural mass fraction of 9.1% (Kaplan, 1997).

Due to the inherent limitations of state-of-the art propulsion systems, any increase in an RLV's dry mass fraction will directly reduce the payload mass it can deliver to orbit (which in turn has an affect on an RLV venture's revenue per flight). In order for an RLV to perform the same mission as an expendable launch vehicle, assure very high reliability and return to its spaceport for a quick turnaround, its dry mass must accommodate many additional requirements. These requirements include, but are not limited to: retro/maneuvering engines; additional propellants and tanks; return maneuvering structures and mechanisms; reentry thermal protection systems; reusability modifications to structures, engines, tanks and avionics; health monitoring systems; safe return-from-abort equipment; and landing gear/supports (Kaplan, 1997).

> Current dry mass fractions for expendable launch vehicles fall into the range of about 10% to 13%. Without the introduction of new technologies, the addition of improvements for reusability must add an additional several per cent to these fractions (Kaplan, 1997).

One way to offset the additional mass resulting from RLV-unique requirements is through the innovative use of composite materials, which in turn will add complexity to an RLV's fabrication processes and increase manufacturing costs. Another way is to develop new generation propulsion systems capable of producing the requisite thrust level and specific impulse parameters necessary to reduce an RLV's mass fraction to insignificance. However, reducing vehicle mass fractions is only one aspect of reducing an RLV's exposure to liability.

Imposing the requirements of *high usage rates* and *fast turnaround times with minimum maintenance* on reusable launch vehicles further exacerbates their dry mass growth propensities. This is because enhanced factors-of-safety, and higher levels of redundancy must now be designed into an RLV system to allow it to safely and reliably perform 100+ sorties between major overhauls (the operational goal of the Space Access Launch System—a fully reusable, three-stage to GTO system). Hence, it would appear that *high usage rates* and *fast turnaround times with minimum maintenance* are the predominant factors in assessing a reusable launch vehicle's exposure to liability. However, it is a combination of these attributes, an RLV's performance margin (as determined by the interaction of the vehicle mass fraction and propulsion performance) and its operational over-flight corridor(s) that will determine a reusable launch vehicle's over-all exposure.

The key *operational over-flight corridor* contributors to an RLV's over-all liability exposure include (but are not limited to):

- flight vehicle controllability, intact abort and emergency landings;
- departure/approach corridor deviations to designated flight paths;
- re-entry of customer payloads following flight vehicle separation; and
- the presence of damage-prevention mechanisms (passive and/or active).

Within the context of these contributors, the U.S. Federal Aviation Administration (FAA), Office of Commercial Space Transportation (AST), will currently license reusable launch vehicles based on whether they are (1) designed to be safe; (2) built to design; and (3) capable of safe operation. These criteria naturally imply that RLVs will be either flying from a remote location, or a government-licensed range. Because there are no common flightworthiness standards against which different RLVs can be assessed, it is up to the individual RLV operators to "prove" their vehicles are *safe for each flight*—using procedures and requirements which are open to the variations inherent in the RLV operator/FAA negotiation process.

Title 14 of the Code of Federal Regulations (CFR), Chapter III, contains the procedures and requirements that govern the authorization and supervision of all space activities conducted from United States territory, or by citizens of the United States. These procedures and requirements examine four areas of concern that directly impact the potential liability of a reusable launch vehicle enterprise: site location safety, operating procedures accuracy, personnel qualifications and equipment adequacy. In addition, RLV system safety and mission reviews are conducted. This licensing process establishes a level of environmental safety that is appropriate for commercial expendable launch vehicles and the research, design and development of second-generation reusable commercial launch vehicles. Although the licensing requirement of not jeopardizing public health and safety and the safety of property is appropriate for the initial development of RLV systems, it should also encourage industry growth on a global scale, and provide consistency. A certification path will allow the RLV industry to establish an initial set of repeatable standards and practices, and build around a logical progression as it expands into passenger-rated operations.

INDEMNIFICATION CONSIDERATIONS

Insuring reusable launch vehicle risks will present unique challenges that are materially different from those of aircraft. For example, aircraft insurers are able to assess their risks, and charge premiums (Bender, 1995):

- based on historical data showing few losses;
- placed on thousands of identical aircraft sold and operated around the world;
- relying on proven technology which is not subject to extraordinary forces;
- taking advantage of long accepted risk management techniques including limitations of liability in the Warsaw and related conventions; and
- acting on voluminous and meaningful readily accessible information.

Needless to say, none of these factors will apply to second-generation RLVs. In fact, second-generation RLV insurers will have nothing but the dismal reliability record of the expendable launch vehicle industry to rely upon. With failure rates averaging 12% for U.S. licensed launches between 1989–1999 (Historical Launch Activity, 1999), RLV risks will:

> ...present a series of...problems, such as critically low predictability, almost complete lack of risk spreading through homogeneous units, technological volatility, inability to exercise meaningful risk control and containment, and the nearly absolute asymmetry of information (Bender, 1995).

In order to mitigate second-generation RLV-unique risks, insurers will rely on the historical practice of providing indemnity for a multi-phase ELV operation. However, four separate phases of operation will need to be considered in the case of RLVs (Gimblett, 1997):

1. pre-launch, which will cover the RLV during integration, testing and other launch site operation phases prior to launch;
2. launch cover, which generally begins at launch ignition and lasts through the orbit attainment phase of the RLV system;
3. in-orbit cover, which will begin simultaneously with the termination of the launch cover, and is relevant to the on-orbit operation of an RLV system; and
4. re-entry cover, which will begin at retrograde ignition (de-orbit burn) and end with recovery of the RLV system on the ground.

Within the boundaries of these operational phases, assessing the risk for second-generation reusable launch vehicles will focus on the types of product (e.g., subsystems, components, etc.) comprising the RLV system and the inherent dangers associated with them (Fobe, 1994). In addition, insurers will take into account the RLV system manufacturing process; system ground and flight-testing; and flight vehicle insurable performance parameters (e.g., the flight vehicle's ability to deliver a payload at the correct state-vector) for each mission.

Financial responsibility and allocation of risk requirements for commercial RLV space launch activities authorized under an AST launch license are contained in 14 CFR Part 440. These requirements include:

- Sec. 440.7, Determination of maximum probable loss (MPL);
- Sec. 440.9, Insurance requirements for licensed launch activities;
- Sec. 440.17, Reciprocal waiver of claims requirements; and
- Sec. 440.19, United States payment of excess third-party liability claims.

In addition, the U.S. Government has traditionally encouraged industry to provide space launch capability for Government purposes through Public Law (PL) 85-804 indemnification. However, the commercial launch industry is now being exposed to excessive and potentially uninsurable risks by reasons of:

- NASA's view that the protection of PL 85-804 does not apply to launches it procures because there is no national defense nexus; and
- NASA's position that the Commercial Space Launch Act (and its indemnification protection) does not apply to launches it procures where there is "substantial Government involvement." Meanwhile, NASA accepts no part of the risk for required unusually hazardous activities.

The result of the application of the regulatory guidelines for FAA licensed launches vs. Government controlled launches and the interpretation of PL 85-804 is that the RLV industry may be denied indemnification under both the Commercial Space Launch Act (CSLA) and PL 85-804 for Government-procured commercial launch services.

With the foregoing in mind, it is doubtful that the U.S. Government will continue to cover excess third-party liability claims for commercial RLV operators once the industry matures (CSLA indemnification provisions for U.S. Commercial Launchers expire in December 1999, unless extended by pending legislation). Hence, the international regulation of RLV flightworthiness will favor the space indemnity market with significantly lower insurance rates by improving the commercial viability and safety of reusable launch vehicles, as manifested in three fundamental areas: reusability, reliability and quality control. Most importantly, it should be noted that aircraft liability insurance coverage is directly tied to an airworthiness certificate—the same requirement would eventually apply to RLVs in the absence of a Government guaranty.

RLV CERTIFICATION BASIS

The certification of reusable launch vehicles should be conceptually based upon the systems engineering approach evolved for the commercial aircraft industry. A key principle of this systems engineering approach is that an RLV design should be considered holistically, and not as the mere sum of its parts. Another principle is that the design criteria for an RLV and its subsystems should emanate from a logical set of performance requirements and operability attributes, and comply (at some level) with an appropriate set of standards for certification. These standards should then form the basis against which the system will be flight-tested.

TABLE 1. Federal Aviation Regulation (FAR) Guidelines Applied to Present-day Launch Vehicles.[1]

• **Accidents Not Caused By:**	• **Accidents Caused By:**
– Flight Rules » *airspace problems, mid-air collision, right-of-way*	– Demonstrable Flight Characteristics » *lack of envelope expansion flight test*
– Operators » *highly trained, current, and qualified* » *no operator-caused accidents[2]*	– Design and Construction, Equipment and Systems » *material flaws in structure or equipment, non-redundant*
– Operating Limits » *limits not intentionally violated*	– Structural Failure » *limit loads exceeded*

[1] Courtesy of Space Access, LLC.
[2] With the exception of Challenger.

Table 1 offers some insight into how flightworthiness standards will benefit reusable launch vehicles. This table graphically illustrates the predominant failure modes of present-day launch vehicles when Federal Aviation Regulation (FAR) certification guidelines are applied. This would imply that FAR-qualified RLVs will not fail routinely, and that an ample margin for recovery will exist should an anomaly occur. Naturally, the only way to certify the flightworthiness of reusable launch vehicles is through extensive flight testing, and the collection of time-age-cycle data on RLV subsystems, propulsion and structural components.

Evolving the Standards for RLV Certification

Regulations and minimum standards relating to the manufacture, operation and maintenance of aircraft are resident in Title 14 of the U.S. Code of Federal Regulations, Chapter I, Parts 1 through 199 (14 CFR, Chapter I). These regulations and standards have their legacy in the Air Commerce Act of May 20, 1926, as amended by the Federal Aviation Act of 1958, and Public Law 103-272 in 1994—and have evolved considerably since the introduction of jet airliners. In fact, the Air Commerce Act "was passed at the urging of the aviation industry, whose leaders believed the airplane could not reach its full potential without Federal action to improve and maintain safety standards (A Brief History of the FAA, 1999)." Likewise, it would be in the reusable launch vehicle industry's best interest for the FAA to begin formulating a reusable launch vehicle certification process. The legal authority for formulating this process exists within the guidelines of FAR Part 1, Section 1.1; and Part 21 for aircraft.

The Federal Aviation Regulations (FARs) have evolved since their inception to accommodate the introduction of new aviation technologies—and are inherently flexible enough to address RLV-unique attributes. This is evident in FAR Part 1, Sec. 1.1, which defines the certification of aircraft by *Category*, *Class* and *Type*. Specifically:

- "Category:" ...(2) As used with respect to the certification of aircraft, means a grouping of aircraft based upon intended use or operating limitations. Examples include: transport, normal, utility, acrobatic, limited, restricted, and provisional.
- "Class:" ...(2) As used with respect to the certification of aircraft, means a broad grouping of aircraft having similar characteristics of propulsion, flight, or landing. Examples include: airplane; rotorcraft; glider; balloon; landplane; and seaplane.
- "Type:" ...(2) As used with respect to the certification of aircraft, means those aircraft which are similar in design. Examples include: DC-7 and DC-7C; 1049G and 1049H; and F-27 and F-27F...(3) As used with respect to the certification of aircraft engines means those engines which are similar in design. For example, JY8D and JT8D-7 are engines of the same type, and JT9D-3A and JT9D-7 are engines of the same type.

Furthermore, Section 1.1 defines "Rocket" as "an aircraft propelled by ejected expanding gases generated in the engine from self-contained propellants and not dependent on the intake of outside substances. It includes any part which becomes separated during the operation." This definition, in combination with the definition for "Powered-lift," will readily allow reusable launch vehicles—differentiated by category, class, and type—to be included in the FAR certification process.

In the lexicon of the FARs, *airworthy* is defined as "an aircraft that meets its type design and is in a condition for safe operation." By similarity then, a reusable launch vehicle will be considered flightworthy if it also meets these requirements. Hence, by following a similar process to what aircraft use to acquire a *standard airworthiness certificate*, RLVs can qualify for the "equivalent" *standard flightworthiness certificate*. The legal basis for this can be found in FAR Part 21, Section 21.17(b):

> For special classes of aircraft...[e.g., other non-conventional aircraft], for which airworthiness standards have not been issued under this subchapter, the applicable requirements will be *the portions of those other airworthiness requirements* contained in Parts 23, 25, 27, 29, 31, 33, and 35 found by the Administrator to be appropriate for the aircraft and applicable to a specific type design, *or such airworthiness criteria as the Administrator may find provides an equivalent level of safety to those parts* [italics added for emphasis].

Likewise, an applicant is entitled to a type certificate for a reusable launch vehicle if, in accordance with FAR Part 21, Section 21.21(b):

> The applicant submits the type design, test reports, and computations necessary to show that the product to be certificated meets the applicable airworthiness, aircraft noise, fuel venting, and exhaust emission requirements of the Federal Aviation Regulations and any special conditions prescribed by the Administrator...

Like its aircraft counterpart, a reusable launch vehicle's flightworthiness certificate would be effective as long as the maintenance, preventive maintenance, and alterations were performed in accordance with the RLV-unique portions of, or counterparts to, FAR Parts 21, 43, and 91; and the RLV was registered in the United States.

Within the guidelines previously discussed, the FAA should establish a *Directorate for Space Flight*, patterned after the existing FAA Directorates (i.e., Transport Aircraft, Engine, Small Aircraft, and Helicopter). Staffed with dedicated "subject matter" experts, this new *Directorate* could work with the reusable launch vehicle industry to formulate flightworthiness standards that complement existing FARs only to the extent necessary to regulate RLVs. These FARs, 14 CFR 1 – 199, can be synthesized into a separate "FAR Part 4XX for Reusable Launch Vehicles" that addresses the major areas listed in Table 2.

TABLE 2. Major Areas Addressed by FAR Part 4XX for RLVs.

- **RLV Design and Maintenance Standards**
  - Demonstrable Flight Characteristics
  - Structural Capability
  - Manufacturing and Materials
  - Equipment and Systems
  - Operating Limits
  - Instructions for Continued Flightworthiness

- **Airspace Requirements**
  - Flight Rules
  - Air Traffic Control

- **Provisions for Recognizing New Technologies**

- **Operations Requirements and Training**
  - Flight Operations
  - Ground Operations
  - Maintenance Operations
  - Personnel Training, Currency, Medical

- **Facilities and Ground System Requirements**
  - Mission Control
  - Maintenance
  - Support Equipment

The standards and requirements of Table 2 should be defined with an eye toward how they will be employed in the design and verification of third generation, or follow-on, reusable launch vehicle systems. Hence, the commercial RLV development process should emphasize a methodology for integrating the systems engineering design and verification process with the certification process.

> The FAA, in cooperation with the Society of Automotive Engineers (SAE), has taken a major step in incorporating system engineering principles into the certification process by the publication of ARP 4754 [*Certification Considerations for Highly-Integrated or Complex Aircraft Systems*]…it represents a look into the future of certification and demonstrates the FAA's and SAE's commitment to the [systems engineering] process (Jackson, 1997).

The Role of Flight Testing

Flight-testing will be critical to the validation of "derived" RLV flightworthiness standards, and the certification process. In addition to validating the requisite performance and operational capabilities of commercial RLVs—including safety compliance—second-generation systems will establish the precedent for successfully operating in the present-day air and space legal regimes.

Flight-testing of second-generation reusable launch vehicles is readily accommodated in the appropriate subsections of FAR Part 21, Section 21.191, which state:

> Experimental certificates are issued for the following purposes:
> (a) Research and development. Testing new aircraft design concepts, new aircraft equipment, new aircraft installations, new aircraft operating techniques, or new uses for aircraft.
> (b) Showing compliance with regulations. Conducting flight tests and other operations to show compliance for issuance of type and supplemental type certificates, flights to substantiate major design changes, and flights to show compliance with the function and reliability requirements of the regulations…

Second-generation reusable launch vehicles will operate under experimental certificates until compliance with their type certificate requirements is proven. In addition, the collection of time-age-cycle data on RLV subsystems, propulsion and structural components will be instrumental in the validation of RLV-unique, flightworthiness standards.

CONCLUSIONS

The reusable launch vehicle industry, as a whole, has not acknowledged the underlying theme of liability that is inherent in space launch vehicles designed to be operated and maintained like traditional high performance, heavy jet aircraft. This paper has examined the factors contributing to an RLV's exposure to liability, which include performance shortfalls, high usage rates, fast turnaround times with minimum maintenance, and operational over-flight corridors. Liability doctrines and indemnification considerations relevant to reusable launch vehicle flightworthiness were also discussed. Finally, a method for formulating RLV flightworthiness standards was introduced that is conceptually patterned after the FAA standards used to certify aircraft airworthiness. As flight-testing of second-generation reusable launch vehicles continues, these systems will become catalysts for the eventual codification of RLV-specific flightworthiness standards.

The international regulation of reusable launch vehicle flightworthiness would ensure that RLV operators, contractors and subcontractors implemented the necessary measures for managing risk and preventing failures in reusable launch vehicles. However, mere adherence to these standards may not be enough, because they would represent the "minimum" requirements for mitigating liability. The key to reliable, reusable launch vehicles is the robust systems, subsystems and components capable of withstanding the rigors of reuse in the air and space environments, with minimal maintenance and downtime—this implies a duty of care requiring the strict implementation of quality control principles.

ACKNOWLEDGMENTS

The author would like to acknowledge the assistance of Messrs. Scott Jackson and Walt Smith of The Boeing Company, Airworthiness and Systems Engineering Group. An additional thanks is given to Mr. Robert Catania, Counsel for The Boeing Company.

REFERENCES

"A Brief History of the Federal Aviation Administration," *Federal Aviation Administration Homepage*, Online @ http://www.faa.gov/history.htm, 1999.

Bender, R., *Space Transport Liability – National and International Aspects*, The Hague: Kluwer Law International, 1995.

Bostwick, Phillip D., "Liability of Aerospace Manufacturers: 'MacPherson v. Buick' Sputters into the Space Age," *22 Journal of Space Law*, University, Mississippi: University of Mississippi Law Center, 1994, pp. 75–96.

Fobe, Jean-Michel, *Aviation Products Liability and Insurance in the EU*, Deventer: Kluwer Law and Taxation Publishers, 1994.

Gimblett, Richard, "Space Insurance into the Next Millenium," *Outlook on Space Law Over the Next 30 Years*, edited by G. Lafferranderie and D. Crowther, The Hague: Kluwer Law International, 1997, pp. 163–172.

Gorove, Stephen, *Developments in Space Law*, Dordrecht: Martinus Nijhoff Publishers, 1991.

Haanappel, P.P.C., "Product Liability in Space Law," *2 Houston Journal of International Law*, Houston, Texas: University of Houston Law Center, 1979, pp. 55–64.

Haley, Andrew G., *Space Law and Government*, New York: Appleton-Century-Crofts, 1963.

"Historical Launch Activity," *Associate Administrator for Commercial Space Transportation (AST) Homepage*, Online @ http://ast.faa.gov/launch_info/launch/history.cfm, 1999.

Jackson, Scott, *Systems Engineering for Commercial Aircraft*, Brookfield: Ashgate Publishing Company, 1997.

Kaplan, Marshall H., "The Reusable Launch Vehicle: Is the Stage Set?" *Launchspace Magazine*, March 1997, pp. 26–30.

Title 14 – Code of Federal Regulations, Chapter I – Federal Aviation Administration, Department of Transportation (Parts 1 through 199).

Title 14 – Code of Federal Regulations, Chapter III – Office of Commercial Space Transportation, Federal Aviation Administration, Department of Transportation (Parts 400 through 499).

The Virginia Space Flight Center Model for an Integrated Federal/Commercial Launch Range

Billie M. Reed

Virginia Commercial Space Flight Authority, Norfolk, VA 23508

Abstract. Until 1998, the federal government has been the predominant purchaser of space launches in the U.S. through the purchase of hardware and services. Historically, the government provided the necessary infrastructure for launches from the federal DoD and NASA launch ranges. In this historical model, the federal government had complete ownership, responsibility, liability, and expense for launch activities.

In 1998, commercial space launches accounted for 60% of U.S. launches. This growth in commercial launches has increased the demand for launch range services. However, the expense, complexity of activities, and issues over certification of flight safety have deterred the establishment of purely commercial launch sites, with purely commercial being defined as without benefit of capabilities provided by the federal government. Provisions of the Commercial Space Launch Act have enabled DoD and NASA to support commercial launches from government launch ranges on a cost-reimbursable, non-interference basis. The government provides services including use of facilities, tracking and data services, and range and flight safety.

In the 1990's, commercial space market projections indicated strong potential for large numbers of commercial satellites to be launched well into the first decade of the 21st century. In response to this significant opportunity for economic growth, several states established spaceports to provide the services necessary to meet these forecast commercial needs. In 1997, NASA agreed to the establishment of the Virginia Space Flight Center (VSFC), a commercial spaceport, at its Wallops Flight Facility. Under this arrangement, NASA agreed to allow the Virginia Commercial Space Flight Authority (VCSFA) to construct facilities on NASA property and agreed to provide launch range and other services in accordance with the Space Act and Commercial Space Launch Act in support of VSFC launch customers. A partnership relationship between NASA and VCSFA has emerged which pairs the strengths of the established NASA Test Range and the state-sponsored, commercial launch facility provider in an attempt to satisfy the needs for flexible, low-cost access to space.

The continued viability of the VSFC and other commercial spaceports depend upon access to a space launch and re-entry range safety system that assures the public safety and is accepted by the public and government as authoritative and reliable. DoD and NASA budget problems have resulted in deteriorating services and reliability at federal ranges and has caused fear with respect to their ability to service the growing commercial market. Numerous high level studies have been conducted or are in progress that illuminate the deficiencies. No federal agency has been provided the necessary funding or authority to address the nations diminishing space launch capability. It is questionable as to whether the U.S. can continue to compete in the global space launch market unless these domestic space access problems are rapidly corrected.

This paper discusses a potential solution to the lack of a coordinated response in the U.S. to the challenge presented by the global market for space launch facilities and services.

CP504, *Space Technology and Applications International Forum–2000*, edited by M. S. El-Genk

17th SYMPOSIUM ON
SPACE NUCLEAR POWER
AND PROPULSION

Future Planetary Missions Potentially Requiring Radioisotope Power Systems

Jack F. Mondt[1] and Bill J. Nesmith[2]

Jet Propulsion Laboratory, 4800 Oak Grove Drive, Pasadena, CA 91109-8099
[1]jack.mondt@jpl.nasa.gov; [2]bill.j.nesmith@jpl.nasa.gov

Abstract. This paper summarizes the potential Radioisotope Power System, (RPS), technology requirements for future missions being planned for NASA's Solar System Exploration (SSE) theme. Many missions to the outer planets (Jupiter and beyond) require completion of the work on advanced radioisotope power systems (ARPS) now underway in NASA's Deep Space Systems Technology Program. The power levels for the ARPS can be divided into four classes. Forty to one hundred milliwatt-class provides both thermal and electric power for small in situ science laboratories on the surface of bodies in the solar system. One to two watt class for surface and aerobot science laboratories. Ten to twenty-watt class for micro satellites in orbit, surface science stations and aerobots. One hundred to two hundred watt class for orbiter science spacecraft, for drilling core samples, for powering subsurface hydrobots and cryobots on accessible bodies and for data handling and communicating data from small orbiters, surface laboratories, aerobots and hydrobots back to Earth. Using the most optimistic solar-based power system instead of advanced RPSs pushes the launch masses of these missions beyond the capability of affordable launch vehicles. Advanced RPS is also favored over solar power for obtaining comet samples on extended-duration missions.

INTRODUCTION

A brief description of each mission concept and a summary of the corresponding potential RPS technology needs are presented and discussed. The NASA Solar System Exploration, (SSE), theme is divided as follows: 1) Exploring Organic Rich Environments, 2) To Build Planet... 3) Bring Mars to Earth... 4) Robotic Outposts, and 5) Exploration of the far Outer Solar System the Kuiper Belt and Beyond. Studies of mission concepts to satisfy the science objectives of the SSE theme are being carried out at JPL to assess feasibility and to investigate the potential benefits of advanced technology. Science objectives and priorities for the studies were obtained from the SSE Roadmap (Elachi 1996) and from the SSES Campaign Strategy Working Groups. Results of these studies are reflected in this paper. The desired power levels, efficiency, mass and voltage projected for missions that are enabled or strongly enhanced by an advanced RPS are described. These results should be considered preliminary since further studies including additional missions are currently under way.

MISSION CONCEPTS

The objectives of Exploring Organic-Rich Environment are; 1) Inventory the organic materials found in diverse planetary environments across the outer solar system, 2) Compare these materials with those on which Earthly life is based, 3) Explore these environments to search for insights into the nature and variation of prebiotic chemistry. The Present/Planned missions for exploring organic-rich environments are Galileo Europa, Cassini/Huygens, Europa Orbiter and Pluto/Kuiper Express. The Near-Future missions are Europa Lander, Titan Organics Explorer and Neptune Orbiter with Triton Flybys. Example Farther-Future missions are Triton Lander, Europa and Titan Sample Return and Kuiper Belt Explorer.

The objectives of To Build a Planet... are; 1) analyze the building blocks of which planets are made, 2) observe the dynamics processes involved in forming planets and planetary systems and 3) study the diverse outcomes of planet formation. The Present/Planned missions to begin to meet these objectives are NEAR, Lunar Prospector,

CP504, *Space Technology and Applications International Forum–2000*, edited by M. S. El-Genk
© 2000 American Institute of Physics 1-56396-919-X/00/$17.00

Stardust, Deep Impact and CONTOUR. The Near-Future missions are Comet Nucleus Sample Return, Saturn Ring Observer, Venus Sample Return. Example Farther-Future missions are Asteroid sample returns, Lunar giant basin sample return, Mars's seismic network, Multiplanet probes and Mercury in situ explorer.

The objectives of Bring Mars to Earth... are; 1) determine the biological history and potential of Mars, 2) understand the evolutionary history and resource inventory of Mars, 3) establish a continuous robotic presence and prepare for human exploration, and 4) engage the public in the challenge and excitement of Mars exploration. The Present/Planned missions are Mars Global Surveyor, Mars Climate Orbiter/Polar Lander, Mars Surveyor 2001, Mars Sample return, Mars mciro-misions and telecom satellites. The Near-Future missions are subsurface sampling, *In-Situ* analysis of sites of biological interest, coordinated exploration networks, global telecom/navigation networks. Examples of Far-Future missions are self-sustaining interactive networks, seamless Earth-Mars Internet and Human-assisted labs.

Robotic Outpost is a new paradigm of exploration. The objectives of the paradigm are exploration of surface and subsurface via long-life robots, surface dynamics and tides subsurface search for prebiotic/biotic signature and interactive studies among spaceborne, surface and subsurface sensors. Europa may one day become a target for a comprehensive "Mars-like exploration program.

Exploration of the outer solar system the Kuiper Belt and beyond is emerging as an important step toward a complete understanding of the evolution of our solar system and the development of life. Studying these type of missions will focus far-term mission and technology planning. The required technology capabilities for Interstellar Exploration missions will greatly enhance near-term solar system missions, i.e. solar sail, optical communications, autonomy, power systems, structures, propulsion, etc.

The Near-Future and Farther-Future missions that are potential users of radioisotope power sources are outlined below.

Solar System Exploration Mission Set for RPS Technology Planning

Near-Future Missions
- Solar Probe
- Mars Weather/Seismic Stations
- Europa Lander, including Europa subsurface concepts
- Titan Organic Explorer
- Neptune Orbiter with Triton Flyby
- Comet Nucleus Sample Return
- Saturn Ring Observer
- Mars Robotic Outpost and Future Sample Returns

A summary description of the Near-Future missions is given below the following Farther-Future missions.

Farther-Future Mission Examples
- Advanced Outposts
- Robotic Outpost
- Kuiper Belt Explorer
- Europa Ocean Science Station
- Sample all Accessible Bodies
- InterStellar Probe

Solar Probe

Solar Probe will deeply penetrate our nearest star's atmosphere to make local measurements of the birth of solar wind, and to remotely image features as small a 60 kilometers across on the sun's surface. The primary science objectives of the mission are: 1) to determine the mechanisms that accelerate the solar wind, 2) find the source and trace the flow of energy that heats the million-degree corona, 3) determine the three-dimensional structure of the inner corona above the polar region and the equatorial belt, 4) map the configuration and state of the magnetic field, and the pattern of the surface and subsurface flows from pole to pole, and 5) find the origin of the fast and slow solar wind near the surface of the sun.

Solar Probe launch is planned on a Delta III class launch vehicle with a Jupiter Gravity Assist. This trajectory uses a retrograde swingby of Jupiter causing it to almost "fall" directly toward the sun. The resulting highly elliptical orbit will have perihelion close to the Sun and aphelion at more than 5 AU. With a February 2007 launch, first perihelion for Solar Probe will occur near maximum solar activity. A second perihelion pass will be possible 4.5 years later, near solar minimum. Because the spacecraft goes out beyond 5 AU, then into 4 solar radii and back out beyond 5 AU a 150 to 200 watt, <20 kg, 18% efficient advanced RPS is potentially required. It requires multiple very different solar power systems to provide power at all times between 5 AU and 4 solar radii. The 75,000 to 1 insolation would require multiple solar arrays. The large solar array required at 5AU has to be either stowed behind a small spacecraft sun shield or discarded at 4 solar radii. Then the large solar array deployed or a third power system deployed for the 4.5-year flight time back out to 5 AU.

Mars Weather/Seismic Stations

Mars Weather/Seismic Stations mission will establish a global network of 24 stations on the surface of Mars that will record the planet's weather and seismic activities for ten Mars years (18.8 Earth years). This long-term continuous presence will define the nature of the variability in the Martian climate system including the behavior of the polar reservoir of carbon dioxide. It will also provide weather-monitoring infrastructure for future Mars missions and of great scientific benefit to any simultaneous observations from orbiters, landers or Earth. The internal structure of Mars can also be studied by seismological measurements. At multiple sites, a minimum of three, the seismological analysis will be able to determine precisely the location of the seismic source. Passive seismic sounding will be used to determine depth profile of seismic velocities.

A low power, 40milliwatt, long-life, 20 year, advanced radioisotope power system is enabling for this mission. The power system will be used to collect atmospheric pressure, temperature, opacity and seismic data and charge ultra-capacitors to store energy for communicating this data to a Mars orbiter for communications back to Earth.

Europa Lander

A Europa Lander will conduct chemical analyses of near-surface ice and organics and will study the interior structure of Europa. In the most ambitious concepts, a "cryobot" would melt or burrow through the ice to explore the postulated underlying ocean. The trajectory being considered would insert into Jupiter orbit and use a series of satellite flybys lasting approximately 1 year to remove energy from the orbit prior to a descent to the surface. During the one-year Jupiter/Satellite flybys a 150 to 200 watt ARPS enables the spacecraft to obtain science data and to operate the communications system in the high radiation environment. Regardless of the main propulsion system used to reach Jupiter, a significant portion of the launch mass would be allocated to transporting a chemical propulsion system to Jupiter for these operations.

Technology advances are needed on a broad front to enable a landed mission on Europa. The mission is very demanding energetically, calling for a combination of lightweight, radiation-tolerant systems and improvements in the performance and hardware mass of chemical propulsion. Many of the lander concepts examined would benefit from availability of small radioisotope based power systems (<10 watts). Navigation to the landing site is also a significant challenge, but perhaps the most critical area is for development of systems to perform the desired science. This includes, in most concepts, systems to acquire samples of ice from a meter or so below the surface, to concentrate the samples, and to perform a broad range of organic chemical analyses. In the long term, it also may include a radioisotope power system to power a "cryobot" through the ice with a small RPS to power "hydrobot" systems for ocean exploration.

Titan Organic Explorer

A Titan Explorer would primarily study the distribution and composition of organics on the Saturnian moon, as well as look at the dynamics of the global winds. Aerocapture at Titan, avoiding a Saturn orbit insertion, is currently the most attractive trajectory option. A 150 to 200 watt advanced RPS that is 18% efficient and has mass of 20 kg or less for the Orbiter will be enabling.

A variety of mission profiles have been proposed for Titan based on a variety of models of surface and atmospheric states. Cassini data will shed light on the validity of these models, but in the mean time, because of the importance of Titan as a potential host for prebiotic chemistry, it makes sense to take the early steps toward a quick follow-on to Cassini. This includes work on organic chemistry analysis systems (some overlap with work needed for Europa) and on delivery systems including aerocapture, balloon systems, and landers. For long life (>3 months) lander concepts, small radioisotope power sources, 1 watt to 10 watt, will also be enabling.

Neptune Orbiter with Triton Flybys

This mission would use a full complement of remote sensing instruments to characterize both the planet and its largest moon. To accomplish this with affordable launch vehicles and acceptable mission duration we need a very low mass spacecraft (as envisioned in the current work on "system-on-a-chip" technology), advanced solar powered propulsion systems (using SEP or solar sail in one or more close in orbits of the sun to accelerate the spacecraft for a quick trip to Neptune), and aerocapture into Neptune orbit.

Return of a high volume of science information from the distance of Neptune requires a 150 to 200 watt, 20 kg 18% efficient long life radioisotope power system. Low Mass, small antenna optical communication system is also required for the orbiter spacecraft. The study emphasized use of optical communication along with advanced techniques for selection, editing, and compression of the data. This mission would be enhanced with a 20 kWe, 500-kg nuclear electric propulsion system, (nuclear reactor powered propulsion would also be applicable but is not currently being considered for NASA science missions), for propulsion to Neptune, retro-propulsion into a Neptune/Triton orbit and for power communicating with Earth.

Comet Nucleus Sample Return

A Comet Nucleus Sample Return mission would obtain 200 gram of the comet nucleus taken from multiple sampling sites - using a subsurface sampling apparatus such as a drill or a tethered penetrator. Challenging science goals for the mission include deep drilling (10 to 100 m) and obtaining samples from multiple sites. A mother ship would return the samples to Earth. A wide range of mission profiles including variations of the relative intelligence of the mother ship and the surface elements have been suggested.

Because of the large propulsive energy (Delta V) requirements associated with first rendezvousing with a comet and then returning to Earth, advanced solar-powered propulsion technology is enabling for all comets of interest to the science community. (Nuclear reactor powered propulsion would also be applicable but is not currently being considered for NASA science missions.) The most likely form of this would involve advances relative to the current state-of-the-art of solar electric propulsion (SEP) with a specific weight goal of 30 kg/kWe (including the power system). Improvements on solar array performance can contribute to this goal.

Techniques for approach, landing, anchoring, sample collection, and sample preservation were also identified as enabling for this mission. Many of these are well along the technology development path because of the development activities that were started for the cancelled ST-4 (Champollion) mission. An advanced 50 to 100 watt, 20% efficient 10 kilogram advanced radioisotope power system for deep drilling to obtain subsurface samples would enhance this mission. In the far term, a solar sail and a very low-mass radioisotope power system would offer the capability of accomplishing the mission with a smaller launch vehicle and potentially a shorter flight time.

Saturn Ring Observer

The Saturn Ring Observer science objectives are; 1) to understand the ring process and evolution as a model for the origin of planetary systems, 2) to determine ring particle physical properties, dynamics and spatial distribution, and 3) to determine the composition of the ring particles. The mission would be accomplished by a 100 kg class spacecraft launched on a Delta III class launch vehicle. The mission would be an 8-10 year flight to Saturn with ballute or aeroshell aerocapture at Saturn to get into a low-inclination Saturn orbit with periodic plane changes to avoid ring-plane crossing.

Advanced solar electric propulsion or solar sail technology and a high efficiency 150 to 200 watt 18% efficient 20-kg advanced radioisotope power system are enabling for this mission. This mission would be strongly enhanced with a 20 kWe, 500-kg nuclear electric propulsion system for propulsion to Saturn, retro-propulsion into a Saturn orbit, maneuvering at Saturn and for power to obtain scientific data at Saturn and high-speed communication with Earth.

Mars Robotic Outposts and Future Sample Returns

Mars Robotic Outpost and Future Sample Returns will provide a continuous robotic presence on Mars and prepare for future human exploration. These outposts will determine resource inventory on Mars and help scientists and the public understand the evolutionary history of Mars. The continuous presence on Mars with frequent reports and new information will engage the public in the challenges and excitement of Mars's exploration.

Mars polar terrain science laboratory would be strongly enhanced with a 50 to 100 watt 20% efficient; 8-kg advanced RPS with a different radioisotope as the heat source for long life (>4 years) and continuous day and night surface and subsurface investigations. Central labs for analyses and coordination of all activities on Mars would be strongly enhanced with a long life 150 to 200 watt advanced RPS. Large distributed robots, wide area 3D exploration and sample acquisition and handling systems would be strongly enhanced with an advanced radioisotope system especially in the Polar regions of Mars.

RPS TECHNOLOGY REQUIREMENTS

NASA JPL has further studied and refined the potential missions to the Solar System as outlined in the Solar System Roadmap (Elachi, 1996). These further mission studies have more specifically defined all the technology requirements. The entries in Table 1 are quantitative; giving a value or range of values derived from the particular mission concept as shown.

The technology requirements for advanced radioisotope power source based on the mission studies outlined above are more specific as shown in Tables 1 than the general requirements (Mondt, 1997) based on the original roadmap.

The first priority requirement for all radioisotope power systems is that the RPS must meet all nuclear safety requirements.

The criticality level for needing an advanced radioisotope power system for each mission shown in the table is defined as follows:

    (1) Enabling - provides for achieving the science objectives of the mission with an affordable launch vehicle (Delta 3 class or smaller).

    (2) Strongly enhancing - provides substantial increase in payload or reduction in cost or risk.

    (3) Enhancing - provides increase in payload or reduction in cost or risk.

TABLE 1. Advanced RPS Top Level Requirements for Near-Future Missions

Power (We)	Mass (kg)	Lifetime (Yr.)	Efficiency (%)	Voltage (Vdc)	Potential Missions
0.040 to 0.100	0.25	20	4 - 5	5	(1) Mars Weather/ Seismic Stations
1 to 2	0.5	5-10	5 - 10	5	(1) Europa Lander Surface Laboratory
10 to 20	2	15-20	15	5	(2) Surface In-situ Laboratories Aerobots or Aero-rover
50 to 100	7 to 10	4 to 5	18 to 20	28	(2or3) Rover and Sample Retriever
100 to 200	8 to 20	10	18 to 21	28	(1) Europa Lander
100 to 200	8 to 20	15	18 to 21	28	(2) Titan Explorer
100 to 200	8 to 20	15	18 to 21	28	(1) Neptune Orbiter
100 to 200	8 to 20	15	18 to 21	28	(1) Saturn Ring Observer
100 to 200	8 to 20	15 to 30	18 to 21	28	(1) Interstellar Probe

CONCLUSIONS

Near-Future and Farther-Future missions Exploring Organic-Rich Environments, To Build a Planet, Bring Mars to Earth, Robotic Outposts, Sampling all Accessible Bodies and Interstellar Exploration require completion of NASA's Deep Space Systems Technology present work on advanced radioisotope power systems (ARPS). Accomplishing all these missions also requires new further advanced, innovative, dual-use, smaller radioisotope power systems.

Using the most optimistic solar-based power system causes the launch masses of these missions to grow beyond the capability of affordable launch vehicles. Advanced radioisotope power systems are also lower mass and easier to land and use than solar power systems for extended-duration comet sample acquisition activities. Innovative dual-use smaller radioisotope power systems and heaters are needed for the future surface, subsurface, orbiting and airborne science laboratories. Solar power and other power systems alternatives will continue to be considered and evaluated for these missions.

ACKNOWLEDGMENTS

The work described in this paper was performed at the Jet Propulsion Laboratory, California Institute of Technology, under contract with the National Aeronautics and Space Administration. The JPL Mission and System Architecture Section under the guidance of the JPL Solar System Exploration Office developed the mission concepts described in this paper. The authors acknowledges and thanks Robert Gershman and Doug Stetson for their work in developing and presenting these missions concepts and the preliminary requirements for the advanced radioisotope power systems technology.

REFERENCES

Elachi, C, et al, 1996, "Mission to the Solar System Exploration and Discovery: A Mission and Technology Roadmap", September 27, 1996, Version B.

Mondt, J., M. Underwood and B. Nesmith (1997) "Future Radioisotope Power Needs for the Missions to the Solar System", IECEC 97, July 27 – August 1, 1997, Aerospace Power System and Technologies, Honolulu, HI, Volume I: page 461-464.

Exploring the Ocean of Europa: Reactor or RHU?

David Poston and Andrei Belooussov

Los Alamos National Laboratory, MS-K551, Los Alamos, NM 87545
(505)667-4336; poston@lanl.gov

Abstract. This paper examines the heat transfer characteristics of a probe (cryobot) penetrating through the ice layer of Europa. Initially, simple 1D calculations are used to predict the ideal (no heat losses or temperature limitations) penetration rates for various size cryobots. Next, a detailed 2D model is used to more realistically model penetration rates. It is found that for small, low power density systems, conductive losses can cause the penetration rate to be significantly lower than the ideal rate. The results of these calculations are meant to establish rough limits on the size of cryobot that can be powered by an RHU (Radioisotope Heater Unit), and at what sizes a reactor becomes enabling. It is concluded that if an RHU system (that delivers almost all power to the bit) can be developed with an overall, fully-engineered power density of ~1 W/cm³, then an RHU system may be suitable for some mission scenarios, although slow penetration times (which increase mission risk) and/or high Pu-238 requirements (cost and availability) may still make a reactor a more optimal choice. If there is a requirement for a large payload and/or a rapid penetration time (~months), then a reactor will probably be required. The final portion of the paper examines potential reactor designs that could be used to power a cryobot. Two potential reactor designs are discussed – a near-term, low-cost heatpipe cooled system and a conductively-cooled metal-fueled reactor.

INTRODUCTION

Evidence continues to mount that a liquid ocean lies under the icy surface of Europa. One of the most recent estimates is that the ice may be between 10 to 30 km thick (Pappalardo, 1999) – although estimates range from 1 km to 100 km (which is the estimated extent of the entire water layer). One of the most talked about and exciting planetary missions is to send a probe (or cryobot as it is often referred to within NASA) to penetrate the ice and explore this ocean. The cryobot will have 3 important tasks: melt through the ice, maintain surface/space communications, and most importantly perform the science of the mission. One of the keys to making this mission a reality is the development of a compact reliable power system that can support these 3 tasks. The level of science of the mission will depend heavily on the performance of the power source. Depending on the ice thickness, a small power supply (~1kWt) may be able to deliver a small scientific payload to the ocean/ice interface (at a rather slow penetration rate), and provide 10s of Watts of electrical power. A large power supply (10 kWt to 100 kWt) could deliver a large power-rich payload, and potentially a "loaded" submarine (or hydrobot) for exploring the ocean and potentially the ocean floor to look for evidence of chemosynthesis.

This paper examines the heat transfer characteristics of the cryobot. The keys to successful penetration are high power density and the ability to deliver the power to the desired location (mostly the front face) while avoiding unwanted heat loss. Initially, simple calculations are used to predict the ideal (no heat losses or temperature limitations) penetration rates of various size cryobots. Next, a detailed 2D model is used to more realistically model penetration rates. The results of these calculations are meant to establish rough limits on the size of cryobot (if any) that can be powered by an RHU, and at what sizes a reactor becomes enabling. The final portion of the paper examines potential reactor designs that could be used to power a cryobot. The reactors considered range from a near-term low-cost heatpipe cooled system, to a conductively-cooled metal-fueled reactor.

SIMPLIFIED THERMAL ANALYSIS

The simplest method of estimating ice penetration capabilities is to assume that all system energy goes directly into melting the ice below the cryobot. For this analysis, the Europan ice is assumed to be pure H_2O. The temperature at the surface is assumed to be 100 K, and is assumed to increase linearly to 273 K at the ice/ocean interface. There are

CP504, *Space Technology and Applications International Forum–2000*, edited by M. S. El-Genk
© 2000 American Institute of Physics 1-56396-919-X/00/$17.00

no assumptions made regarding pressure, since there is no attempt to model what happens to the water after it is formed (i.e. no vapor production). The ideal penetration rate is determined by a very simple formula:

$$\text{Penetration Rate} = \frac{\text{Power into ice}}{A_{melt}\,\rho\,(\,h_{fus} + C_p\,(273 - T_{ice})\,)}\quad, \tag{1}$$

where A_{melt} is the area of ice melted, ρ is the ice density, h_{fus} is the heat of fusion, C_P is the specific heat, and T_{ice} is the initial ice temperature. The values used in this analysis are $\rho = 920$ kg/m^3, $h_{fus} = 335$ kJ/kg, and $Cp = 2000$ J/kg-K. These parameters cause the ideal penetration rate to be approximately two times slower through 100 K ice (near the surface) as through 273 K ice (at the ocean/ice interface). Figure 1 plots the ideal ice penetration rate vs power density for .25 m and .5 m diameter cryobots in 100 K and 273 K ice. The larger diameter cryobots have a faster rate because the length-to-diameter (L/D) of the power source is assumed to be 1, thus more power per unit area reaches the ice. In this case the curve for the .25 m, 273 K case almost exactly overlaps the .5m, 100 K case. Figure 2 plots the ideal penetration time vs power through 10 km of ice, assuming the average ice temperature is 200 K. The smaller cryobots penetrate at a much faster rate per unit power because they have a much higher power density (although at higher powers they are reaching very unrealistic power densities).

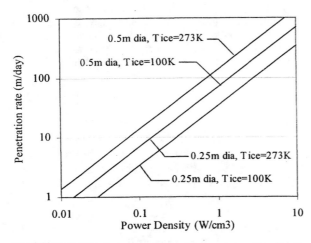

FIGURE 1. Ideal penetration rate vs power density for .25 m and .5 m cryobots in 100K and 273K ice.

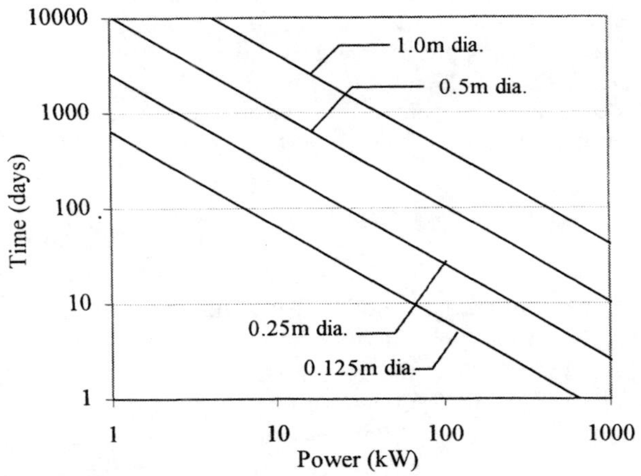

FIGURE 2. Ideal time to penetrate 10 km of ice vs power for various cryobot sizes (ave. ice T=200K).

TWO-DIMENSIONAL THERMAL ANALYSIS

There are several limitations in the simplified thermal analysis. The most important factor missing is engineering reality; the calculations do not consider heat transfer mechanisms and material temperature limitations. Another important factor missing is heat losses caused by conduction radially outward into the ice. Because of these limitations, a 2D FORTRAN finite-difference model was developed to explicitly model the system. The cryobot is modeled in a cylindrical geometry – r and z (azimuthal symmetry). A schematic of the model is shown in Figure 3. The model contains 4 regions: power source, payload, vessel (the bottom of which is referred to as the bit), and ice/water. Each region can be discretized to any desired level, and each region is assigned a density, thermal conductivity, specific heat, and power density. The gap-conductance between all regions can be specified, i.e. if the power source-to-vessel and power source-to-payload gap conductances are set to zero (perfect insulation), then all energy must flow to the bit. The most difficult part of modeling this system is deciding what to do with water that is generated under the bit. In the calculations presented in this report, all water that is generated below the bit is immediately transferred to the top of the vessel (in the future, an attempt will be made to model the water transfer in a more realistic fashion). After the water is "removed", the vessel then moves downward into the space vacated by the melted ice. The coordinate system stays with the cryobot; therefore, every time the cryobot penetrates downward the enthalpy of the ice/water is convected upward (to simulate the movement).

The model was benchmarked to the ideal results by only allowing conduction to the ice below the bit, and not allowing conduction radially outside of the vessel diameter. This ensured that all power went to the ice, and the results matched exactly with the ideal solution. To reduce the parameter space in this analysis, several assumptions were made. Some assumptions might favor a low power density design, while others might favor high power densities. The assumptions below were chosen in an attempt to make it a "fair" comparison between all ranges of powers.

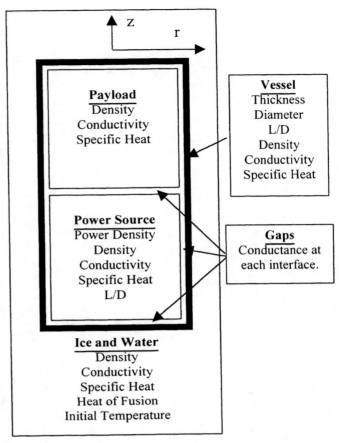

FIGURE 3. Input parameters for 2D heat transfer model.

- The conductivity of the power source is set to infinity. This allows the power to travel unimpeded to the regions of least resistance. This in effect assumes that the system has a perfect heat removal and transfer system.
- The gap-conductance between the power source and the bit (bottom of vessel) is set to infinity. The bit is assumed to be 1 cm of stainless-steel; this provides the only resistance between the power and the ice.
- The radial gap-conductance between the power source and the vessel wall is set such that the minimum side-wall temperature is 273 K. It is assumed that the wall temperature will need to be at least 273 K to keep the side of the vessel from freezing with the ice.
- The ice properties are constant with temperature: thermal conductivity = 0.2 W/m-K, density = 920 kg/m^3, specific heat = 2000 J/kg-K, heat of fusion = 335 kJ/g. The vessel is given the properties of stainless-steel at 300 K.
- All power is deposited within the power source; i.e., there is no energy leakage due to neutron and gamma escape (would probably require a few percent increase in total power for a reactor).
- No power is diverted to an electric power conversion system.

Several cases were run that varied the ice temperature, and the power and dimensions of the cryobot. Unless otherwise specified, each cryobot has a L/D of 1 for the power source, and L/D of 2 for the vessel. For comparison, the ideal penetration rate is plotted on these figures with a faint line. In all cases the penetration rate at high powers is very close to the ideal rate. At lower powers, losses conducted outward into the ice begin to effect the penetration rate. These losses are evident on Figure 4 below about 0.1 W/cm^3. The impact of these losses is highly dependent on the cryobot diameter. The losses are relatively small at the 0.5-m diameter; however, at the 0.125-m diameter these losses become so large that the penetration rate goes to zero just below 0.1 W/cm^3. Figure 5 plots the temperature of the power source for the same cases represented in Figure 4. This plot may be very important because some cases may exceed the temperature limitations of a fully-engineered cryobot. On this plot it is very evident at what power density penetration stops because the temperature falls below 273 K. Each curve flattens out as the temperature approaches 273 K and then drops more quickly once no ice is being melted. Figure 6 plots the same data as Figure 4 vs power instead of power density.

Figure 7 shows the effect of the ice temperature on penetration rate for the 0.25-m diameter cryobot. A lower ice temperature lowers the rate for 2 reasons. First, more energy is required to melt the ice because it is sub-cooled (as seen in the ideal analysis). Second, colder ice draws more of the heat away from the cryobot, thus increasing the fraction of wasted power. Therefore, the penetration in 200 K ice remains much closer to the ideal value than in 100

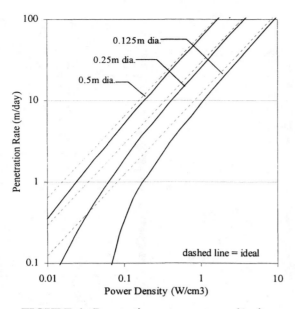

FIGURE 4. Penetration rate vs power density.

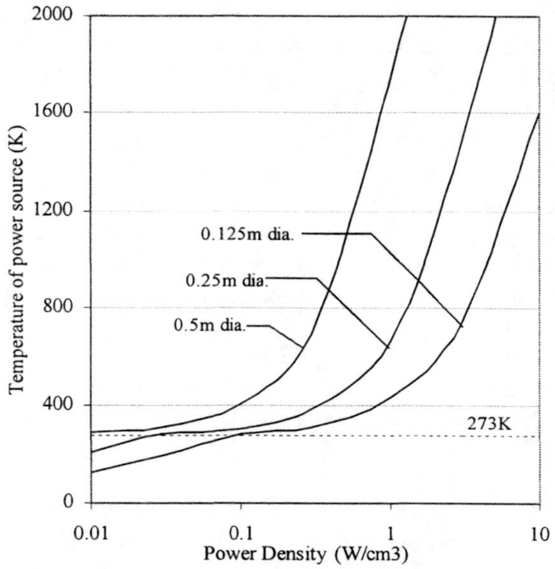

FIGURE 5. Power source temperature vs power den.

K ice (which is expected near the surface). This difference would be much more evident if the plot was not on a logarithmic scale. Figure 8 offers a more dramatic representation of how the ideal and 2D results vary. This plot shows the fraction of the ideal penetration rate (or fraction of power that goes into melting the ice). As was noted on Figure 4, the 0.125-m cryobot loses all of its power through conduction just below 0.1 W/cm^3; thus, it does not penetrate into the ice. Another parameter that can significantly effect penetration rate is the L/D of the vessel. Figure 9 plots fraction of ideal penetration rate vs L/D for a 0.25-m dia cryobot with a power density of 0.1 W/cm^3 (the baseline case has L/D = 2). As the L/D increases, there is more heat conducted outward to the ice. The difference between the actual and ideal penetration rate becomes even larger as L/D increases. This is because the extra losses from the side-wall lower the temperature of the power source, thus less heat is conducted downward into the ice.

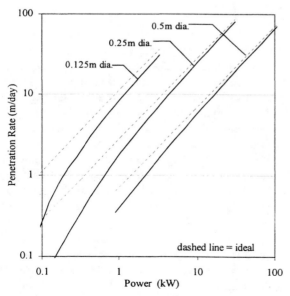

FIGURE 6. Penetration rate vs power for 3 cryobot diameters.

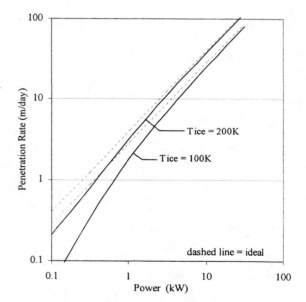

FIGURE 7. Penetration rate vs power for 2 ice temperatures.

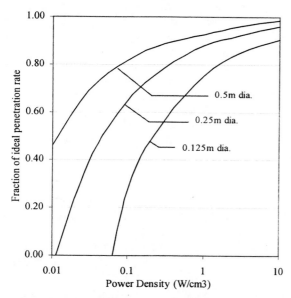
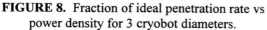
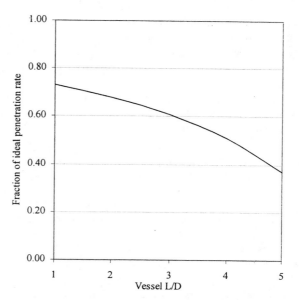

FIGURE 8. Fraction of ideal penetration rate vs power density for 3 cryobot diameters.

FIGURE 9. Fraction of ideal penetration rate vs L/D for 0.25-m diameter cryobot operating at 0.1 W/cm^3.

Finally, it is important to reiterate that these calculations are far from realistic. An actual system will not perfectly shunt heat away from the power source, and the water produced under the bit will not magically transport to a region above the vessel. Also, impurities or dust in the ice could severely hamper cryobot penetration, and the cryobot may not take a direct path straight down through the ice. Detailed phenomenological modeling, and a better understanding of the properties of the ice, will be required to get a more definitive answer.

OTHER ISSUES

One way to improve heat transfer to the ice is to use a conical (or other) shaped bit. A conical bit can increase surface area, thus increase penetration rates – provided that heat can be delivered effectively to all parts of the cone. This could improve the results of the lower power density cases, thus making them closer to the ideal case. A cone with L/R = 3 has twice the surface area of a flat disk. In the most optimistic scenario, this would cut the heat losses to the ice by a factor of 2, thus bringing the penetration rate ½ way toward the ideal penetration rate; although, distributing the heat uniformly throughout the long cone would be difficult. Performance could also possibly be enhanced by altering the dimensions of the cryobot so that the side-walls have a smaller diameter than the bit. In this scenario it may not be required to keep the side-walls at 273 K.

The issue of maintaining surface communications is one of the most difficult aspects of this mission. The technique used will depend on the condition of the hole after the cryobot passes through it. There are several possibilities of what will happen to the water generated by the cryobot. It may flow around or though the core (if permitted) leaving a pool above the reactor, it may vaporize and escape the hole, or it may vaporize and then condense and freeze along the bore hole – eventually clogging and pressurizing the hole, which could then lead to a pool above the vessel. In the cases evaluated here, the water was place above the vessel, and the resulting pool froze very quickly. This would eliminate the option of reeling a cable from the surface, and any attempt to heat the cable in order to maintain a liquid interface would probably have prohibitive power losses. Reeling a cable from the cryobot itself is an option, but this would only make sense for a very large, high-power system (due to size constraints). For most systems, this only leaves the option of leaving a trail of communication pods frozen in the ice, or some other innovative technique.

REACTOR VS RHU

Radioisotope power sources are a well proven technology and should be seriously considered for this mission. The power density of Pu-238 decay is 0.558 W/g. Assuming that the Pu is in oxide form, the resulting power density of the PuO$_2$ (assuming fully enriched in Pu-238) is ~5 W/cm^3. Based on cladding and heat removal requirements, plus decay en-route, it will be very difficult to produce a fully-engineered power source (that can transport all of the power to the bit) with a power density greater than 1 W/cm^3. If we assume a 1 W/cm^3 system, a 0.125-m diameter cryobot will have a penetration rate of ~9 m/day. This would require ~3 years to penetrate 10 km of ice. If the ice thickness is in fact 10 km or less, this may be an option, although the long penetration time places considerable risk on the mission. A 0.25-m cryobot would have a penetration rate of ~25 m/day. This lowers the penetration time to ~1 year, which may be acceptable; but is still rather long. The 0.5-m diameter cryobot would have an optimum penetration rate of ~59 m/day. This will make the mission time very reasonable; however, to produce the required 90 kW of power would require ~160 kg of Pu-238. If all the assumptions leading up to these conclusions are correct, then an RHU may be a viable option for this mission. However, as mentioned earlier, the penetration rates calculated in this report may be very conservative, plus the design of an RHU source for this application that can provide a fully-engineered system power density of 1 W/cm^3 may prove very difficult.

In addition to heat-transfer concerns, there are several other factors that may discriminate a reactor or RHU for this mission. Some factors that are in favor of an RHU are: several systems built and launched, launch approval recently obtained, and a proven containment mechanism. Some factors that weigh against an RHU system are: radioactivity at launch, lower penetration rates, high expense of Pu-238 (if available), higher mass per unit power, and the need for heat removal prior to deployment. This last issue is a major concern for this system. Prior to deployment, an RHU design will need a heat sink mechanism that will simulate the heat removed by the ice. This may prove to be the most difficult aspect of the design, and will be a significant safety concern – ensuring that this removal mechanism is reliable prior to and during launch.

Some factors in favor of reactor systems are: potential for very high power densities (higher penetration rates), virtually non-radioactive until deployed, established technology base, lower mass per unit power, and the power does not decrease with time. Some factors that weigh against reactor designs are: no recent systems flown or launch approval, the design will have to contain fission products, more shielding may be required (increasing L/D thus decreasing penetration rate), very small cryobots may approach criticality limits, neutronic design may be difficult – may need control worth to handle ice, water, or void in various locations. One other advantage of the reactor system is that is could evolve into a high temperature rock melting/drilling system. This could enable many lunar/planetary missions, and have many uses on Earth as well.

In conclusion, these heat transfer calculations do not provide a definitive answer to the reactor vs RHU question. The decision will be dictated by the level of scientific payload desired and the mission time allowed for the penetration. If there is a requirement for a large payload and/or a rapid penetration time (~months), then a reactor will probably be required.

POTENTIAL REACTOR DESIGNS

The reactor requirements for the Europa cryobot are significantly different that for most electric power producing space reactors. The list below contains some of the most significant differences.

Performance – As with any space reactor, low-mass is very important. For the cryobot, a compact size is also extremely important because the larger the footprint of the system, the more ice that has to be melted. In addition, the size of auxiliary components is also important. High power density is critical, which is why a reactor is generally more desirable than an RHU, and the reactor must have a mechanism to transport most of the heat to the bit surface.

Operating Environment – The environment on Europa is very harsh. Not only are temperatures very cold, but there is a significant radiation field from Jupiter. These will effect operation and particularly startup (although a reactor is better equipped to deal with this environment than most other power sources). Operating in ice/water will cause corrosion concerns, although the ice/water could possibly be used productively as a neutron reflector.

Shielding – Shielding requirements may be significantly different than for a typical space reactor. There will be no boom – thus no r-squared reductions; however, the components to shield may be relatively small and localized.

Reliability – The system will need to be highly reliable. It will have very little contact with the surface of the moon, let alone Earth. Fortunately, the system will have a short lifetime requirement due to the high rate of penetration.

<u>Safety</u> – If a reactor is chosen for this mission, then environmental impact will be probably become a major issue (as it will be for an RHU system as well). The clad and/or other containment mechanism will have to ensure fission product containment. Safety during launch and transport to Europa can easily be achieved, but will be scrutinized (as it will be for any space reactor).

Several fission reactor designs were initially considered to power the drill. Actively cooled systems were ruled out because of concerns about reliability, size of components, and system development cost; although, if very large systems with fast penetration rates are required, this type of system may be required. Two systems were found to be best suited for this application. A heatpipe cooled reactor may be best for large higher power drills (10 - 100 kWt), and a conductively cooled reactor is best suited for more compact lower power drills (1 – 10 kWt). The heatpipe reactor under consideration is the Heatpipe Power System (HPS) – a concept under development by LANL (Poston and Houts, 1999). The HPS is a safe, simple reactor that is designed to have a low development time and cost. The reactor design would be very similar to HPS designs for space applications, except that heatpipes would transport most of the energy to the bit, instead of a power conversion system. Perhaps the heatpipes would transfer their energy to a pool of liquid metal (or a solid conductive block) just above the bit. A nice feature of the HPS is that the system can be fully tested with electric heating, so it should be simple and safe to test the system on ice on Earth. In fact, a 30 kWt HPS core is currently being constructed, and it may be possible to use this core (with resistance heating) to power an experiment on a frozen lake. The conductively cooled concept is a compact uranium metal fueled system (95% enriched). The core is configured in a geometry that produces most of the power close to the drill bit; there is a cylindrical upper half and a hemispherical lower half (in which most of the power is generated). The diameter of the entire system is ~12 cm. Ice and/or water must surround the core in order to reach criticality. The core is Inconel clad, and lead (or lead-bismuth) fills the gap between the fuel and the bit to enhance the heat transfer. A helium-filled gap along the core edge is used both to control heat transfer to the side wall and to control the fission reaction by introducing boron poison. Preliminary heat transfer calculations show that the system can produce close to 10 kWt thermal before the fuel temperature reaches 1350 K. If this system can be successfully engineered, it could have a penetration rate of close to 100 m/day (based on the assumptions used in the calculations above). This concept may require more development time and cost than the HPS concept. In conclusion, these two designs look very promising for this application, but considerably more work is required to make a full evaluation.

SUMMARY

Several heat transfer calculations were performed for various cryobot dimensions and powers. Penetration rates were obtained for an ideal system (all energy melts ice) and for a more detailed 2D model. It was found that for small, low-power density systems, conductive losses can cause the penetration rate to be significantly lower than the ideal rate. Based on these calculations, it is concluded that if an RHU system can be developed with an overall, fully-engineered power density of ~1 W/cm^3, then an RHU system may be suitable for some mission scenarios, although slow penetration times (which increase mission risk) and high Pu-238 requirements (cost and availability) may still make a reactor a more optimal choice. If there is a requirement for a large payload and/or a rapid penetration time (~months), then a reactor will probably be required. The final portion of the paper examines potential reactor designs that could be used to power a cryobot. Two potential designs are discussed – a near-term, low-cost heatpipe cooled system and a conductively-cooled metal-fueled reactor.

ACKNOWLEDGMENTS

This work was supported by the Los Alamos National Laboratory Center for Space Science and Exploration (CSSE). The authors would like to thank Jim Blacic, Ted Mochler, and Bob Reid of LANL for their support and ideas.

REFERENCES

Pappalardo, R. T., Head, J. W., Greeley, R. , "The Hidden Ocean of Europa," *Scientific American,* October, 1999, pp. 34-43.

Poston, D. I., Houts, M. G., Emrich, W. J., "Development Status of the Heatpipe Power and Bimodal Systems*,*" *Space Technology and Applications International Forum-1999 (STAIF-99),* edited by Mohamed S. El-Genk, AIP Conference Proceedings 458, 1999, pp. 1197-1204.

Utilizing Fission Technology to Enable Rapid and Affordable Access to any Point in the Solar System

Mike Houts, Joe Bonometti, Jeff Morton, Ivana Hrbud, Leo Bitteker, Melissa Van Dyke, Tom Godfroy, Kevin Pedersen, Chris Dobson, Bruce Patton, James Martin, Suman Chakrabarti

NASA MSFC, TD40, Marshall Space Flight Center, Alabama, 35812
michael.houts@msfc.nasa.gov / (256)544-7143 / Fax: (256)544-5926

Abstract. Fission technology can enable rapid, affordable access to any point in the solar system. Potential fission-based transportation options include bimodal nuclear thermal rockets, high specific energy propulsion systems, and pulsed fission propulsion systems. In-space propellant re-supply enhances the effective performance of all systems, but requires significant infrastructure development. Safe, timely, affordable utilization of first-generation space fission propulsion systems will enable the development of more advanced systems. First generation space systems will build on over 45 years of US and international space fission system technology development to minimize cost.

INTRODUCTION

Fission technology can enable rapid, affordable access to any point in the solar system. Advanced concepts (i.e. the "Medusa" concept (Solem, 1993) and vapor or droplet core fission systems driving high-efficiency thrusters (Anghaie, 1999)) could reduce trip time to Mars, Jupiter, and beyond by an order of magnitude compared to today's systems. In the mid-term, bimodal nuclear thermal rockets with liquid oxygen afterburners (LANTR) could reduce earth-lunar transit time to 24 hours, enable affordable six-month transits to Mars, and explore much of the inner solar system utilizing in-situ propellant re-supply (Borowski, 1999). In-space propellant re-supply could greatly enhance the effective performance of all propulsion systems.

Compared to other advanced propulsion options, fission systems are conceptually quite simple. All that is required is for the right materials to be placed in the right geometry - no extreme temperatures or pressures required - and the system will operate. In addition, the fuel for fission systems (highly enriched uranium) is virtually non-radioactive, containing 0.064 curies/kg. This compares quite favorably to radioisotope systems (Pu-238 contains 17,000 curies/kg) and D-T fusion systems (tritium contains 10,000,000 curies/kg). At launch, a typical space fission propulsion system would contain an order of magnitude lower onboard radioactivity than Mars Pathfinder's Sojourner Rover. The primary safety issue with fission systems is avoiding accidental criticality – addressing this issue through proper system design is quite straightforward.

The potential of space fission systems is illustrated in Figure 1. As shown in the Figure, the energy density in fission systems is seven orders of magnitude greater than that of the best chemical systems. Put another way, completely fissioning a piece of uranium the size of a coke can would yield two orders of magnitude more energy than burning all of the chemical fuel contained in the space shuttle main tank. If properly harnessed, the energy density in fissile fuel far exceeds that required to enable rapid access to any point in the solar system. Additionally, the technology readiness level (TRL) of space fission systems is much higher than that of nuclear fusion, matter annihilation, and hot isomeric transition. Fission systems are the nearest-term option for high efficiency, high thrust in-space propulsion.

CP504, *Space Technology and Applications International Forum–2000*, edited by M. S. El-Genk
© 2000 American Institute of Physics 1-56396-919-X/00/$17.00

Potential Propellant Energy Sources

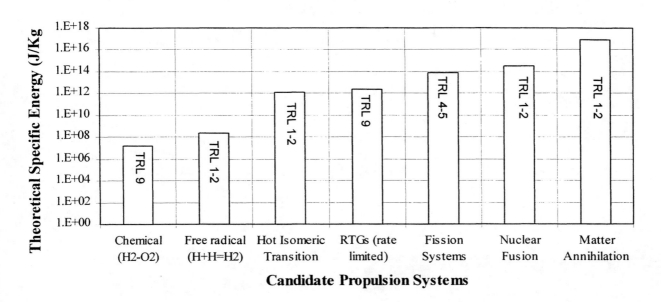

FIGURE 1. Energy density of candidate propellant energy sources.

FIRST GENERATION SPACE FISSION SYSTEMS

Despite the relative simplicity and tremendous potential of space fission systems, the development and utilization of these systems has proven elusive. The first use of fission technology in space occurred 3 April 1965 with the US launch of the SNAP-10A reactor. There have been no additional US uses of space fission systems. While space fission systems were used extensively by the former Soviet Union, their application was limited to earth-orbital missions. Early space fission systems must be safely and affordably utilized if we are to reap the benefits of advanced space fission systems.

Table 1 gives a partial list of major US space fission programs that have failed to result in flight of a system. There are a variety of reasons why these programs failed to result in a flight. The fact that so many programs have failed indicates that a significantly different approach must be taken if future programs are to succeed.

Terrestrial fission systems have been utilized by the government, universities, industry, and utilities for over 50 years. In addition, technology development directly related to space fission systems has been progressing for over 40 years. The next generation fission system should capitalize on this experience. Nuclear testing can be one of the most expensive and time consuming aspects of space fission system development. If a system can be designed to operate within established fuel burnup and component radiation damage limits, the requirement for nuclear testing can be minimized. Designing the system such that resistance heaters can be used to closely simulate heat from fission will also facilitate development and allow extensive testing of the actual flight unit.

Additional innovative approaches will have to be used to ensure that the next space fission system development program results in system utilization. Safety must be the primary focus of the program, but cost and schedule must also be significant drivers. System performance must be adequate, but the desire to make performance more than adequate should not be allowed to drive system cost and schedule. The next generation space fission system must be safe, simple, and as inexpensive to develop and utilize as possible.

One option for a first generation fission propulsion system is the First Generation Least Expensive Approach to Fission (FIGLEAF) system. A high temperature FIGLEAF module is currently on test, and full core FIGLEAF testing is scheduled to begin in FY 2000. Figure 2 is a photograph of the FIGLEAF module operating at 1750 K. Figure 3 shows the FIGLEAF module operating at 1750 K with heat pipe simultaneously operating at 1450 K. Additional details on the FIGLEAF are presented in (Van Dyke, 2000).

TABLE 1. Partial list of major US Space Fission Programs that Have Failed to Result in Flight of a System.

• Solid-Core Nuclear Rocket Program	• SNAP-50 / SPUR	• Advanced Liquid Metal Cooled Reactor
• Medium-Power Reactor Experiment (MPRE)	• High-Temperature Gas-Cooled Electric Power Reactor (710 Reactor)	• Advanced Space Nuclear Power Program (SPR)
• Thermionic Technology Program (1963-1973)	• SPAR / SP-100	• Multi-Megawatt Program
• Space Nuclear Thermal Rocket Program	• Flight Topaz	• Thermionic Fuel Element Verification Program
• SP-100	• DOE 40 kWe Thermionic Reactor Program	• Air Force Bimodal Study

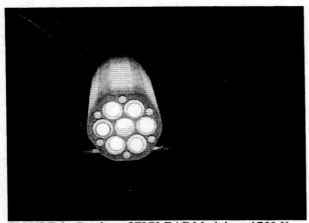

FIGURE 2. Portion of FIGLEAF Module at 1750 K.

FIGURE 3. FIGLEAF Module.

PULSED FISSION (OR FISSION/FUSION) PROPULSION SYSTEMS

Pulsed propulsion systems have been under consideration since the late 1940's. In a pulsed propulsion system, fission or fission/fusion pulses that release between 10^{13} and 10^{15} Joules of energy are used to propel a vehicle. The primary difficulty with pulsed propulsion systems is coupling the pulse to the vehicle without exceeding limits on acceleration. If adequate coupling schemes are devised, effective specific impulses exceeding 50,000 seconds with vehicle thrust-to-weight ratios exceeding 1.0 are feasible. Nearer-term systems would be more likely to have specific impulses on the order of 4000 s, still nearly an order of magnitude greater than the best chemical propulsion systems (Dyson, 1999).

In the energy range of interest, the equivalent specific impulse of pulsed propulsion systems increases with the magnitude of the pulse. The primary obstacle to the utilization of large pulses is devising a method for spreading the acceleration out over a period of several seconds. Large pulses could deliver total impulses on the order of 10^8 kg-m/s or more. Assuming a 200 MT vehicle and a maximum acceleration limit of 250 m/s^2 thus requires that the impulse be delivered to the spacecraft over a minimum of two seconds. Another concern is the cost of the pulse unit. Because cost is not strongly dependent on pulse unit size, the use of large pulses may also result in less expensive missions. A 10^{14} J pulse requires the fissioning of roughly 1.25 kg of uranium. Future research related to pulsed propulsion systems should focus on methods for utilizing large pulses, on the order of 10^{14} J or higher. Systems utilizing large pulses could enable rapid access to any point in the solar system.

Most previous work related to pulsed fission propulsion systems focused on earth-to-orbit systems. This focus drove vehicle designs to those capable of utilizing a rapid series of pulses, on the order of one pulse per second. A system requiring several minutes to reconfigure between pulses may be acceptable for in-space transfer applications. Dealing with variability in pulse sizes and the occasional failure of a pulse unit may also be simplified for in-space transfer applications.

HIGH SPECIFIC ENERGY FISSION PROPULSION SYSTEMS

The specific energy of fissile fuel is 8×10^{13} J/kg. For systems requiring a year of operation at full thrust without refueling, the minimum theoretical specific mass is thus 4×10^{-4} kg/kW. In an actual system, structure, heat removal, energy conversion, waste heat rejection, radiation shielding, and other subsystems will significantly increase specific mass. However, it may still be possible to devise high efficiency (Isp > 3000 s) fission propulsion systems with a specific mass in the 0.1 to 1.0 kg/kW$_{propellant}$ range. Such systems would enable rapid access to any point in the solar system.

Initial research on these systems could involve non-nuclear simulations of vapor or droplet core fission reactors. Advanced energy conversion subsystems including MHD energy conversion and high-temperature Brayton cycles could be investigated. Flowing UF_4 (or other fuel-form) loops could be constructed (using natural or depleted uranium) to validate thermal hydraulic predictions and investigate high temperature materials compatibility.

IN-SPACE PROPELLANT RE-SUPPLY

The performance of any space propulsion system can in theory be enhanced with in-space propellant re-supply. Propellant re-supply increases the effective specific impulse of a given propulsion system compared to the same propulsion system without propellant re-supply, in some scenarios by nearly an order of magnitude. Several challenges must be overcome before in-space propellant re-supply can be utilized.

1. A source of propellant must be available at a desired location outside of a large gravity well. In a few instances the propellant source may already be at the desired location, otherwise it must be moved there.
2. A method for collecting and processing the propellant must be devised.
3. A method for storing the propellant until needed must be devised.
4. A method for effectively utilizing the propellant must be devised.

Perhaps the best potentially available propellant source is water. Water can either be electrolyzed and used in chemical propulsion systems or used directly in nuclear steam rockets (Zuppero, 1999). Unfortunately, there are no known in-space reservoirs of relatively pure water in the inner solar system. Numerous individuals have proposed bringing icy bodies from the outer solar system into the inner solar system – some of those proposals are discussed in Fogg, 1995. The discovery of the Kuiper Belt in 1992 further strengthened arguments for utilizing icy bodies from the outer solar system. It is now believed that there are over 30 billion Kuiper Belt Objects, the majority of them with a radius less than 1 km (Weissman, 1999). Centaurs (believed to originate in the Kuiper Belt) are also a potential source of water in the outer solar system. One Centaur, Chiron, has a radius of 150 km and an orbit that crosses that of Saturn.

The difficulty associated with moving propellant into a desired orbit can be thought of in terms of the required propulsive delta-V. A very significant advantage of obtaining water from the outer solar system is that the required propulsive delta-V can be quite small, on the order of a few hundred meters per second (Fogg, 1995). The remainder of the required velocity change can be accomplished via gravity assist maneuvers. Imparting a velocity change of a few hundred meters per second to a large, icy object could be accomplished with a relatively near-term nuclear thermal rocket. For example, suppose a 500 m/s velocity change is to be imparted on a 1 billion kg icy object. If a nuclear steam rocket with an exhaust exit velocity of 1500 m/s is used (attainable with stainless steel or superalloy clad fission systems) then a total of 2.84×10^8 kg of steam would need to be exhausted. Assuming that the reactor provides all of the kinetic energy for the steam, and then adding 25% to the energy requirement to account for thermal losses, results in a requirement that 5 kg of uranium be fissioned. Because of the relatively low core temperatures required by the steam rocket, a stainless-steel clad UO_2 core should be viable. The propulsive delta-V could be imparted over a period of 50 days given a 100 MW core, and for reasonable fuel loadings, established fuel burnup limits would not be exceeded. There would still be significant challenges to developing the steam rocket, including propellant acquisition, impurity limits, and engine control.

One scenario for obtaining water ice in the inner solar system could thus be to heat a Kuiper Belt Object, Centaur, or other icy body (in-situ) and evaporate steam from the body. The steam would then be condensed inside a membrane, and the resulting ice balloon placed in the desired inner solar system orbit through a series of propulsive maneuvers and gravity assists. Although extremely challenging from an engineering standpoint, the physics required for obtaining water in this fashion is known.

Once water is in the desired orbit, it can either be used directly in steam rockets (Zuppero, 1999) or electrolyzed for use in chemical engines. If orbits are properly chosen and in-space propellant re-supply is used, fast planetary transfers can be accomplished using currently available rocket technology. For example, calculations (Kos, 1999) show that properly placed propellant re-supply stations can enable missions to Mars with one-way transit times of less than 70 days. These missions can be performed by any high thrust system, independent of the system's specific impulse and without the requirement for an aerobrake. Very robust, reusable vehicles could be utilized. Vehicles with propellant mass fractions of less than 50% would not be out of the question. Without in-space propellant re-supply, high thrust engines with a specific impulse greater than 5000 s could be required to perform an equivalent mission. With in-space propellant re-supply, engines with a specific impulse of 400 s (or less) could still accomplish the 70 day Earth-Mars transit.

While high specific impulse is not required for performing fast missions using propellant re-supply, the use of high specific impulse systems (such as nuclear thermal rockets) reduces the required number of propellant re-supply stations. Calculations also show that missions to the asteroid belt are feasible with 4 month transit times and missions to Jupiter are feasible with 12 month transit times, again using any high-thrust propulsion system (Kos, 1999).

A schematic for performing orbital changes via in-space propellant re-supply is shown in Figure 4. In the scenario shown, a Mars-bound spacecraft would be assembled in orbit, then fueled using propellant obtained from space. The spacecraft would then propel itself to earth escape, obtain propellant, propel itself into orbit 1, obtain propellant, propel itself into orbit 2, obtain propellant, then propel into a slightly hyperbolic trajectory towards Mars. Upon reaching Mars, a similar procedure would be used to slow down. All of the outbound propulsion would be performed near earth in as short of time period as possible (preferably a few hours). All propulsion required to rendezvous with Mars would likewise be performed in a short period of time.

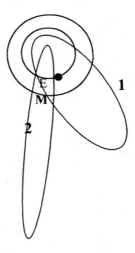

FIGURE 4. Schematic for orbit changing via propellant re-supply.

There are numerous challenges with the in-space propellant re-supply scenario. First, propellant re-supply stations must be in resonance orbits to allow periodic alignment of the re-supply stations. For example, in Figure 4 all of the re-supply stations must be near point "E" before the fast transit can be accomplished. Second, highly-automated spacecraft would be required to obtain the propellant, direct it into the correct orbit, and then maintain the correct orbit. Third, the window for performing the fast transit might last only a few days at each alignment, and a great number of re-supplies could be required by low performance propulsion systems. Fourth, rapid transfer of propellant from the re-supply station to the spacecraft could be difficult.

The primary advantage of using in-space propellant re-supply is that while logistically complex, all of the required physics is known. Once an in-space propellant re-supply infrastructure was in place, existing LOX/hydrogen engines would have the equivalent of an order of magnitude greater specific impulse for certain missions. Solid-core nuclear thermal rockets would have similarly improved effective performance.

RECOMMENDATIONS FOR FUTURE RESEARCH

Research should continue on a first generation fission propellant energy source. The focus of this research should be on demonstrating that fission propulsion systems can be developed and utilized in a safe, timely, and affordable fashion. Research on pulsed propulsion systems should focus on methods for utilizing large (1×10^{14} J or greater) pulses. Research on high specific energy systems can be focused in a variety of areas, but should lead to systems capable of providing on the order of a kilowatt of power into the propellant for every kilogram of system mass. Research related to in-situ propellant re-supply can also be focused in a variety of areas, but should lead to the capability to place propellant re-supply stations where they are most needed for a given mission. Research needs related to the LANTR system are detailed elsewhere (Borowski, 1999).

ACKNOWLEDGMENTS

Unless otherwise referenced, the research reported in this paper was funded by and performed at NASA's Marshall Space Flight Center.

REFERENCES

Anghaie, Samim, Personal Communication, Innovative Nuclear Space Power and Propulsion Institute (INSPI), 1999.

Borowski, S.K., Dudzinski, L.A., and McGuire, M.L. "Bimodal NTR and LANTR Propulsion for Human Missions to Mars/Phobos," in Space Nuclear Power and Propulsion, edited by Mohamed S. El-Genk, AIP Conf 458, American Institute of Physics, New York, 1999, pp. 1261-1268.

Dyson, F., Personal Communication, Princeton, NJ 1999.

Fogg, M.J. *Terraforming: Engineering Planetary Environments*, Society of Automotive Engineers, Inc. Warrendale, PA 15096-0001, 1995.

Kos, L., Personal Communication, Huntsville, AL 1999.

Solem, J.C. "Medusa: Nuclear Explosive Propulsion for Interplanetary Travel," Journal of the British Interplanetary Society, Vol. 46, pp. 21-26, 1993.

Vandyke, Melissa "Results of the Propellant Energy Source Module (PESM) Testing at the Marshall Space Flight Center: Non-Nuclear Testing of a Fission System" to be published in Space Nuclear Power and Propulsion, edited by Mohamed S. El-Genk, American Institute of Physics, New York, 2000, within these procedings.

Weissman, P.R., McFadden, L.A., and Johnson, T.V. *Encyclopedia of the Solar System*, Academic Press, San Diego, CA, 1999, pp. 557-583.

Zuppero, A., Larson, T.K., Schnitzler, B.G., Werner, J.E., Rice, J.W., Hill, T.J., Richins, W.D., and Parlier, L. "Origin of How Steam Rockets can Reduce Space Transport Cost by Orders of Magnitude" in Space Nuclear Power and Propulsion, edited by Mohamed S. El-Genk, AIP Conf 458, American Institute of Physics, New York, 1999, pp. 1211-1216.

The Ph-D Project: Manned Expedition to the Moons of Mars

S. Fred Singer

Science and Environmental Policy Project, 4084 University Dr., Fairfax, VA 22030
ssinger1@gmu.edu; (703) 934-6940

Abstract: The Ph-D (Phobos-Deimos) mission involves the transfer of six to eight men (and women), including two medical scientists, from Earth orbit to Deimos, the outer satellite of Mars. There follows a sequential program of unmanned exploration of the surface of Mars by means of some ten to twenty unmanned rover vehicles, each of which returns Mars samples to the Deimos laboratory. A two-man sortie descends to the surface of Mars to gain a direct geological perspective and develop priorities in selecting samples. At the same time, other astronauts conduct a coordinated program of exploration (including sample studies) of Phobos and Deimos. Bringing men close to Mars to control exploration is shown to have scientific and other advantages over either (i) (manned) control from the Earth, or (ii) manned operations from Mars surface. The mission is envisaged to take place after 2010, and to last about two years (including a three- to six-month stay at Deimos). Depending on then-available technology, take-off weight from Earth orbit is of the order of 300 tons. A preferred mission scheme may preposition propellants and equipment at Deimos by means of "slow freight," possibly using a "gravity boost" from Venus. It is then followed by a "manned express" that conveys the astronauts more rapidly to Deimos. Both chemical and electric propulsion are used in this mission, as appropriate. Electric power is derived from solar and nuclear sources. Assuming that certain development costs can be shared with space-station programs, the incremental cost of the project is estimated as less than $40 billion (in 1998 dollars), expended over a 15-year period. The potential scientific returns are both unique and important: (i) Establishing current or ancient existence of life-forms on Mars; (ii) Understanding the causes of climate change by comparing Earth and Mars; (iii) Martian planetary history; (iv) Nature and origin of the Martian moons. Beyond the Ph-D Project, many advanced programs beckon; discussed here are exploitation of Martian resources, Martian "agriculture", and the possibility of planetary engineering experiments that can benefit survival on the Earth.

INTRODUCTION

The last Moon walk took place a quarter-of-a century ago. Fewer than half of today's Americans saw the first Apollo landing in 1969. Skylab, a functioning space station, was allowed to die a fiery death as it entered the Earth's atmosphere in 1974. The dreams of visiting Mars died with Apollo and Skylab, at least as far as the government was concerned. Public interest, however, has never really waned. The "Mars Underground" came to life in 1981, together with the Planetary Society sparked by the charisma of Carl Sagan. Just in the past few years, the Mars Society, founded by Robert Zubrin, has gained popularity. But can this grassroots interest survive and be translated into action?

It may be taken for granted that Mars is the most interesting and worthwhile target for planetary exploration. However, it will take two concrete steps to bring men to Mars:

- Step 1: Convince decision-makers, and also planetary and life scientists, that unmanned missions are not giving us the results we want. Manned Martian missions can solve the fundamental scientific issues of planetary evolution, climate change, and especially the origin of life within a reasonable time frame. And only manned missions can further international collaboration, enhance national prestige, and create the public interest and excitement that will produce the funding for Mars projects.

- Step 2: Choose the first Manned Martian Mission with great care. If it is too ambitious, it will not be funded. A case in point is the Space Exploration Initiative (SEI) which NASA presented to then-

CP504, *Space Technology and Applications International Forum–2000*, edited by M. S. El-Genk

President Bush. A package costing around $450 billion, it featured manned habitation of the Moon as a "stepping stone" to Mars, when in fact it is likely to be a detour. And it has provided grist for the mills of opponents of manned spaceflight. On the other hand, one should not automatically choose the simplest and least costly of the possible missions. A manned flyby of the planet has returns that are likely to be so small, it may well turn out to be the first and last Manned Martian Mission. Yet, with a small increase in cost, a Flyby can become a Mars Orbiter, with an orbit that matches that of Deimos, the outer Martian moon.

THE PH-D MISSION

In its simplest terms, the Ph-D mission is nothing more than the transfer of a manned habitat from near-Earth orbit to a circular Mars orbit at a distance of 6.9 Martian radii. After tying up to Deimos, a tiny body with negligible gravity, the astronauts start their program of exploration, beginning with Deimos. There is no need to set up a base; the habitat and laboratory of the space ship is the base. There follows a manned sortie to Phobos, the other small moon, to obtain samples, investigate its structure, and eventually figure out if the two moons are related and puzzle out their origin. Are they of uniform composition or are they mixtures of minerals? Are they similar – chemically and petrologically? Were they both captured after Mars formed? Or did they originate with Mars? Are they the remaining fragments of a much larger body? Or are they samples of the original planetesimals of the early solar system, left over when the planets formed? So many fundamental questions that can only be answered by direct exploration.

The high point of the mission would be a manned sortie (of perhaps two astronauts) to the surface of Mars. Even if they only spend a couple of days on the surface, exploring the immediate vicinity of the landing site with a rover vehicle, selecting interesting samples and perhaps digging below the surface, it would be a great human adventure.

All the while, the remaining astronauts would launch robotic rovers to promising locations on Mars, as far as the ice caps, to take measurements and recover samples from the surface and subsurface. Deimos is close enough so that the vehicles can be operated in a telepresence mode in real time. Based on the data received from the initial rovers, others would follow up, leading to a system of "sequential experimentation", a most efficient way of getting important scientific results. A laboratory analysis of a sample sent back from the rover to Deimos could trigger a request to get more material from the same location, or to dig deeper.

All of this would be impossible to do if the rovers were to be controlled from Houston. The average time delay of about 40 minutes would make telepresence impractical. Autonomous rovers, if they existed, might help but could never fully take the place of human control. Sequential exploration that included sample examination would stretch over decades not hours. Moreover, the volume of data that would have to be transmitted to Earth is so large as to overload any reasonable communication system.

How to Carry out the Ph-D Project

There is one point to keep in mind: It can all be done with existing proven technology: chemical propulsion and solar electric power supplies. But it may be possible to reduce cost without undue delay by using improved rocket boosters to put the nominal 300-ton payload into low earth orbit (LEO); a nuclear reactor to supply electric power; and low-thrust but high-efficiency ion propulsion. Power reactors have already been tested in space and several schemes of ion propulsion have been successfully flown. Intriguing proposals abound for a cheaper "space truck" to provide the heavy lift. Such proposals include the use of the large Russian SS-18 missile booster and other surplus ICBMs, to schemes for a series of reusable rocket planes launched from large carrier aircraft flying at 40,000-ft altitude.

Such decisions can only be made in detailed engineering studies that try to optimize the mission. First and foremost, a decision must be made on whether to divide the mission into a "slow freight" (SF) that is prepositioned on Deimos, followed by a "manned express" (ME) that carries the astronauts in their habitat. By increasing chemical propellant weight, or with ion-propulsion during the coast phase, or with the use of both, the ME can cut the time for reaching Deimos, and reduce mission duration, weight of consumables (atmosphere, water, and food supply), and overall risk. The duration may become short enough to eliminate the need for an artificial-G environment.

Another decision concerns near-Mars operations. Should one use aerobraking in the Martian atmosphere and then boost the spacecraft into a Deimos orbit? Alternatively, is may be cheaper and/or safer to ease into the orbit directly by the use of propulsion? Much will depend on whether ion propulsion is available.

Electric power is a must for all spacecraft operations, for communications, and for ion propulsion. It may be advisable to consider both a solar photovoltaic and a nuclear supply. The nuclear unit could be carried with the SF, the solar with the ME. Nuclear power would overcome the inevitable shadow problem that turns solar into an intermittent supply source. The nuclear supply could be left behind on Deimos to continue resource recovery experiments and await the next mission, while the astronauts return to Earth with their solar supply.

Ph-D versus Surface Operations

It is necessary finally to compare the costs and benefits of a Ph-D project with a mission where the spacecraft lands on Mars, with perhaps a half-dozen astronauts who remain there for several months. What would they do; how effective would they be; and what kind of problems would they face?

First of all, operating on the planet, driving rover vehicles, involves risk and safety problems, especially if one is distant from the base. It is far safer and also more efficient to operate rovers by telepresence and radio control. The astronauts can then remain at the Mars base and conduct a virtual operation. However, with the difficult Martian topography, a satellite communication system would seem to be essential. Control from a base on Mars surface would thus have no advantage over human control of rovers from Deimos.

How would samples get from distant rovers, perhaps operating near the ice caps, back to the Mars base, near the equator? Rocket delivery would be the answer, allowing rapid transport, which would be especially important for volatile samples. Again, no special advantage over a Deimos base. But when if comes to scientific examination of the samples, Deimos has a real advantage: It furnishes a ready-made natural vacuum, while the Martian laboratory needs to provide a full-blown vacuum system to operate its mass spectrometers and scanning electron microscope.

And how to provide electric power? An obvious point: On Deimos, there are no dust storms that can cover up solar cells or even damage them. A solar supply on Mars would have a 24-hour day-night cycle, just like on Earth. It would require costly (in terms of weight) battery storage. A similar problem might occur on Deimos, whose orbit is nearly synchronous with Mars' rotation. However, one might circumvent the problem there by dividing the photovoltaic supply and disposing it around Deimos so as to always receive solar illumination (albeit reduced). (Using the same technique, it would be much easier to stay warm on Deimos.) By moving the habitat around Deimos, the astronauts can also protect against solar cosmic rays and against the impacts of meteor streams.

MARS RESOURCES AND PLANETARY ENGINEERING

The Ph-D project is conceived as the initial mission to Mars, to be followed by more ambitious undertakings. The exploitation of mineral resources on the planet has been widely discussed. The Martian moons offer an even more exciting potential for economic returns. Phobos and Deimos have as of now an unknown resource potential. However, they do contain minerals of some sort and can provide the cheapest source of materials for solar system operations, far cheaper than the Moon. It's not distance that counts but the micro-gravity of the Martian moons. It's cheaper in terms of propulsion to transport material to Earth from Deimos than from the Moon.

It is generally agreed that the preconditions for some kind of life are present on the surface of Mars. Even if no evidence is found for living material, it might still be possible to implant organisms which can exploit the various ecological niches available in the different regions of Mars. It may be possible to use existing organisms from the Earth, such as algae, bacteria, or combinations such as lichens. It may even be possible to develop or engineer specific organisms which could prosper under Martian conditions. Once we can simulate the Martian environment on the Earth, we may be able to proceed more rapidly with the design of such organisms. If one succeeds in populating the surface of Mars with organisms that multiply rapidly, and if one can develop higher organisms which can efficiently harvest the lower forms, then we may have the basis for a renewable resource which takes advantage of the ecological opportunities of Mars. It would be fascinating to study this problem further, even up to the point of devising a self-sustaining system of <u>Martian "agriculture"</u> which could support eventual human settlements on the surface of Mars.

A different investigation might be along the line of biological evolution. Certain kinds of experiments may not be appropriately done on the Earth because of the danger of interaction with existing life forms. They could be done more easily on Mars, which might serve as an appropriate laboratory for such work. The Martian environment with its higher radiation levels (or by increasing radiation artificially) may speed up the process of evolution and allow us to study the evolution process under different environmental conditions.

From many points of view, climate studies are most fascinating and perhaps most useful. We have already discussed earlier the possibility of unraveling the history of climate change on Mars and comparing it with that of the Earth, thereby elucidating the different causes for climatic change. Here we can speculate about the future possibility of being able to modify the Martian atmosphere artificially and carry on other climate control experiments on Mars. These could be done without any danger to the climate on the Earth, but would provide important lessons that could be applied to the Earth. As we know, climate changes constantly on the Earth; but we are not certain about the causes nor about the magnitude of the effects. It is particularly worrisome that human activities may be influencing climate in ways which we do not fully understand.

The Martian atmosphere provides us with a laboratory where many scientific ideas about climate can be tested. The thinness of the Martian atmosphere may be of some advantage in these studies. It should be possible to modify the atmosphere either by adding trace gases which have important radiative properties or by releasing gases from the polar cap. One experiment, often suggested for the Earth's Arctic Ocean, is to sprinkle carbon black on the ice so as to speed up melting and evaporation. Another might be to promote volcanic eruptions which would put large quantities of dust as well as gases into the atmosphere. For obvious reasons, one may not wish to go ahead with such experiments on the Earth for fear of causing irreversible changes. Such fears may not be as worrisome when we consider the planet Mars; yet the information gained can be of tremendous importance to us on the Earth in understanding climate change and especially the human role in climate change.

CONCLUSION

The Ph-D project has a clear scientific and operational advantage over manned surface operations and a huge cost and safety advantage. We need to consider then whether having six people on Mars for 200 days is so much better for public interest than two men on the surface for two days. I think not.

REFERENCES

The Case For Mars (P.J. Boston, ed.) American Astronautical Society, Vol. 57, San Diego, CA, 1984
Strategies For Mars: A Guide to Human Exploration (R.C. Stoker and C. Emmart, ed.) American Astronautical Society, Vol. 86, San Jose, 1998

NEP for a Kuiper Belt Object Rendezvous Mission

Ronald J. Lipinski[1], Roger X. Lenard[1], Steven A. Wright[1], Michael G. Houts[2],
Bruce Patton[2], and David I. Poston[3]

[1]Sandia National Laboratories, MS-1146, P.O. Box 5800, Albuquerque, NM 87185, rjlipin@sandia.gov
[2]Marshall Space Flight Center, Huntsville, AL, 35812, michael.houts@msfc.nasa.gov
[3]Los Alamos National Laboratory, Los Alamos, NM 87545, poston@lanl.gov

Abstract. Kuiper Belt Objects (KBOs) are a recently-discovered set of solar system bodies which lie at about the orbit of Pluto (40 AU) out to about 100 astronomical units (AU). There are estimated to be about 100,000 KBOs with a diameter greater than 100 km. KBOs are postulated to be composed of the pristine material which formed our solar system and may even have organic materials in them. A detailed study of KBO size, orbit distribution, structure, and surface composition could shed light on the origins of the solar system and perhaps even on the origin of life in our solar system. A rendezvous mission including a lander would be needed to perform chemical analysis of the surface and sub-surface composition of KBOs. These requirements set the size of the science probe at around a ton. Mission analyses show that a fission-powered system with an electric thruster could rendezvous at 40 AU in about 13.0 years with a total DV of 46 km/s. It would deliver a 1000-kg science payload while providing ample onboard power for relaying data back to earth. The launch mass of the entire system (power, thrusters, propellant, navigation, communication, structure, science payload, etc.) would be 7984 kg if it were placed into an earth-escape trajectory (C=0). Alternatively, the system could be placed into a 700-km earth orbit with more propellant, yielding a total mass in LEO of 8618 kg, and then spiral out of earth orbit to arrive at the KBO in 14.3 years. To achieve this performance, a fission power system with 100 kW of electrical power and a total mass (reactor, shield, conversion, and radiator) of about 2350 kg. Three possible configurations are proposed: (1) a UZrH-fueled, NaK-cooled reactor with a steam Rankine conversion system, (2) a UN-fueled gas-cooled reactor with a recuperated Brayton conversion system, and (3) a UN-fueled heatpipe-cooled reactor with a recuperated Brayton conversion system. (Boiling and condensation in the Rankine system is a technical risk at present.) All three of these systems have the potential to meet the weight requirement for the trip and to be built in the near term.

INTRODUCTION

Kuiper Belt Objects (KBOs) are a recently-discovered set of solar system bodies which lie at about the orbit of Pluto (40 AU) out to about 100 astronomical units (AU). The Harvard Minor Planet Center reports that as of September 6, 1999, 191 KBOs ("transneptunian objects") have already been discovered (http://cfa-www.harvard.edu/cfa/ps/lists/TNOs.html). These are typically 100 to 300 km in diameter, and there are estimated to be approximately 100,000 KBOs 100 to 1000 km in diameter (Malhotra, 1999, Jewitt, 1999), and possibly as many as 10,000,000,000 KBOs with a diameter greater than 1 km. Clearly, this is a significant fraction of the solar system which has not yet been explored. KBOs are postulated to be composed of the pristine material which formed our solar system and may even have organic materials in them. A detailed study of KBO size, orbit distribution, structure, and surface composition could shed light on the origins of the solar system (Malhotra, 1999) and perhaps even on the origin of life in our solar system. A rendezvous mission including a lander would be needed to perform chemical analysis of the surface composition of KBOs. Since cosmic rays and solar wind have altered the chemistry of the outer 10 m of the KBOs, the lander also would need to have a drill to obtain a deep core sample. These requirements set the size of the science probe at around a ton. To deliver a one-ton payload to rendezvous with a KBO in 10-15 years requires acceleration to and from high velocities (about 25 km/s). A fission

CP504, *Space Technology and Applications International Forum–2000*, edited by M. S. El-Genk
2000 American Institute of Physics 1-56396-919-X

reactor with electric propulsion can deliver such performance in the near future without the excessive weight penalty associated with chemical or solar propulsion.

SYSTEM ANALYSIS

In order to rendezvous with a KBO at a nominal distance of 40 AU in a minimum time, one must accelerate to a substantial velocity, perhaps coast for awhile, and then decelerate to rendezvous speed. The total velocity capability of the spacecraft (ΔV) is the sum of the acceleration, deceleration, and rendezvous adjustment velocities. Typically, the total ΔV for a rendezvous in 12 years is about 50 km/s.

The rocket equation dictates that the total mass of fuel needed increases exponentially with the ratio of the desired velocity increase (ΔV) over the specific impulse of the fuel (I_{sp}) times the gravitational acceleration on earth (g):

$$m_{fuel} = m_{nonfuel}\left(\exp(\Delta V /(gI_{sp})) - 1\right), \tag{1}$$

where m_{fuel} is the mass of the propellant, $m_{nonfuel}$ is the mass of everything that is not propellant (scientific payload, communications, guidance, rocket engines, power supply, etc.). One of the best current chemical propellants is liquid oxygen/liquid hydrogen which has a specific impulse of 460 seconds. To achieve a total ΔV of 50 km/s with a payload, power system, engines, and other inert masses ("non-fuel mass") of 5000 kg would required 324,000 tonnes of fuel. Clearly, chemical propellants cannot do the job. But NASA has recently developed very reliable electric propulsion units with a demonstrated specific impulse of 3300 s (Polk, 1998). An even higher Isp can be achieved either by using a larger voltage on the grids or by using a lighter gas. NASA is also developing a VASIMR electric thruster which is projected to achieve up to 20,000 s Isp (Chang Diaz, 1998). To reach a total ΔV of 50 km/s with a specific impulse of 10,000 seconds and 5000 kg of non-fuel mass requires only 3.3 tonnes of propellant. This is much more reasonable.

Electric propulsion requires very large amounts of electrical energy. Nuclear fission, in the form of a small reactor, is the only credible near-term means for obtaining this energy, especially in deep space. Solar energy is not practical because the spacecraft needs to decelerate by about 25 km/s at a distance of about 35 au where the solar intensity is 1000 times dimmer than at earth. A fission reactor and power conversion system for space (excluding the radiator) would be about the size of an automobile and similar in power to research reactors found at many universities.

A system analysis was performed based on the following assumptions: (1) the design point goal is a rendezvous with a KBO at 40 AU, (2) the reactor and propulsion system operates at steady power and accelerates at a fixed thrust for a specified duration (typically many years), (3) the spacecraft then coasts for awhile, (4) the reactor switches to low-power mode during the coast period for spacecraft maintenance, (5) the spacecraft then decelerates at full power and fixed thrust to achieve rendezvous velocity with the KBO, and (6) some propellant is reserved for maneuvering and rendezvousing with additional KBOs. In deference to historical rockets, the propulsion duration will be called a "burn," even though the propellant is not actually burned.

There are two general approaches to the mission scenario. The first approach is to launch the spacecraft into earth orbit and use the electric propulsion system to propel the spacecraft out of earth orbit and then on to the Kuiper belt. The second approach is to launch the spacecraft into an earth-escape trajectory (called a C3=0 trajectory) and then use the electric propulsion system. The advantage of an earth-orbit launch is that it does not require as large a launcher for a given spacecraft weight because the highly-efficient electric propulsion system will be used to get the spacecraft out of earth's gravity well. A disadvantage is that it requires a longer total burn time for the reactor and electric thruster. It also requires the spacecraft to survive a relatively slow passage through the Van Allen radiation belt. Overcoming these difficulties is one reason to use a C3=0 launch. In addition, the safety analysis for a C3=0 launch might be easier since reactor operation would not begin until an earth-escape trajectory is achieved. But the penalty for a C3=0 launch is that typically the mass that a given launcher can deliver to earth escape is about three times less mass than it can deliver to low-earth orbit.

There are many approaches that can be used in comparing options. One could assume fixed spacecraft mass, fixed spacecraft power, fixed launcher type or launcher cost, fixed travel time, etc. One could minimize travel time, launch mass, total mission cost, etc. by varying specific impulse, burn duration, coast duration, etc. We will not present a comprehensive span of possibilities in this paper. Rather, we will present some representative cases that show the potential feasibility of this mission.

A simple orbital mechanics code was developed using the fundamental laws of gravity and acceleration along with the algebraic and Runge-Kutta solution packets in the Mathcad (MathSoft, 1999) commercial software. Three cases were considered. For case 1, the spacecraft was assumed to be launched into a C3=0 trajectory; for case 2, the spacecraft was identical to that in case 1 except that it was launched into low earth orbit (LEO) and additional propellant and tankage to get the spacecraft to C3=0. The added tankage (small) was jettisoned after earth escape. This allowed a simple comparison of the C3=0 and LEO options. Case 3 is like Case 2 except it assumes a lower Isp during earth escape, such as might be available from a VASIMR thruster. This reduces the time needed for earth escape, but increases the mass required in LEO. For both LEO cases, the initial orbit is still high enough that the orbital decay time would be long enough to allow reactor fission products to decay back down to background levels before reentry (about 600 years).

Table 1 shows the key parameters. The "science payload" is assumed to be 1000 kg. Everything else (i.e. everything except the science payload and propellant) is called the "tug" and has a dry mass of 4000 kg. The tug consists of the reactor, the reactor shield, the power conversion system, the thermal radiator, the electric thrusters, the guidance and control system, the communication system, propellant tankage, structures, etc. The tug electric power is 100 kW. The total efficiency of the thrusters (propellant power over electrical power) is assumed to be 75%. Acceleration, coast time, and deceleration duration were adjusted so as to end up at 40 AU with the proper rendezvous velocity (4.7 km/s around the sun). It was determined that the trip time was not a strong function of coast time if the burn time could be adjusted accordingly. A coast time of 7 years yields an attractive configuration.

TABLE 1. Reference mission configurations.

Parameter	Case 1 (C3=0)	Case 2 (LEO)	Case 3 (LEO)
Power in the tug (kWe)	100	100	100
Mass of the tug (kg)	4000	4000	4000
Mass of the science payload (kg)	1000	1000	1000
Specific impulse (s)	10000	10000	10000
Efficiency of converting electrical to exhaust power (%)	75	75	75
Acceleration duration (yr)	3.79	3.79	3.79
Coast duration (yr)	7.00	7.00	7.00
Deceleration duration (yr)	2.24	2.24	2.24
Propellant mass used from C3=0 to KBO rendezvous (kg)	2984	2984	2984
Altitude of initial LEO (km)	N.A.	700	700
Isp for trip from LEO to C3=0 (s)	N.A.	10000	3000
Propellant mass from LEO to C3=0 (kg)	N.A.	634	2315
Duration from LEO to C3=0 (yr)	N.A.	1.29 yr	0.42 yr
Total spacecraft wet mass at launch (kg)	7984	8618	10299
Total mission ΔV to rendezvous (km/s)	45.9	53.4	53.4
Total burn time to arrival (yr)	6.03	7.32	6.45
Total flight time to KBO rendezvous (yr)	13.03	14.32	13.45

The total trip duration is about 13 years, which is not unreasonable for a rendezvous trip as far out as Pluto. The total amount of high-power burn time for both the reactor and thrusters is 6 years for case 1, which is a reasonable lifetime for these systems. There will need to be some low-power maintenance during the 7-year coast and the several-year science mission after arrival, so component longevity needs to be considered for this mode also.

The total mass for the C3=0 option is 7984 kg, which would require a launcher such as a Delta IV-M (under development). However, if the launch is only to LEO and the tug uses its own power to escape earth orbit, smaller launchers can be used. The mass launched to LEO for Cases 2 and 3 are 8618 kg and 10299 kg, respectively.

However, the trip time and the burn time increases by 1.3 yr and 0.4 yr, respectively, so the savings might not be worth it, especially since this increases the time in the Van Allen radiation belt.

Figure 1 shows the trajectory for Case 1. The trajectory is plotted on a polar plot with the radius on a logarithmic scale. The dots in the trajectory each represent about 7 days. The trajectory from about 5 au (Jupiter) to 40 au is essentially straight. The code does not explicitly calculate the final maneuver to match orbits with the KBO (which consists of turning a residual 4.6 km/s velocity vector 90 degrees), so instead we allow an additional 4.6 km/s of ΔV at the end for this maneuver. Thus the velocity in the figure ends up at zero, rather than 4.6 km/s.

Figure 2 shows the total velocity (relative to the sun) vs. time. The total velocity plot shows the three thrust regimes (accelerate, coast, and decelerate). After earth escape, the spacecraft has a total velocity of about 30 km/s, which is the velocity of the earth around the sun. Then, as the spacecraft slowly thrusts along its velocity vector, it puts itself into an elliptical orbit in which the sun's gravity slows it down as it travels outward, even though it is still thrusting. The spacecraft is gaining energy at this time, but it is gaining potential energy rather than kinetic energy (like climbing a mountain). Eventually the continuous thrust starts to overcome the sun's pull and the spacecraft begins to pick up speed. After 4 years, it still does not have as much total velocity as it started out with, but that velocity vector is now pointed almost directly away from the sun. At this point the spacecraft enters the coast phase for a period of 5 years. Finally, 9 years after launch, it begins deceleration so that it can match velocity with the KBO.

Figure 2 also shows the spacecraft distance from the sun vs. time and along with the orbital distances of the outer planets. (Pluto orbits between 30 and 50 au.) This spacecraft, if directed in the right direction, could fly be Jupiter in 3 years, Saturn in 4 years, Uranus in 7 years, Neptune in 9 years, or Pluto in 9 to 10 years after launch (but not all of them in one trip). This would be an impressive capability.

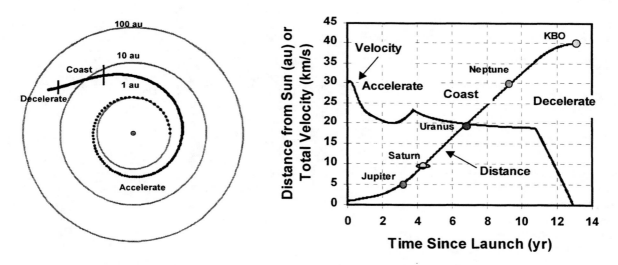

FIGURE 1. Trajector in log-radius coordinates.   **FIGURE 2.** Distance from sun and total velocity vs. time.

TUG DESIGNS

To achieve the performance described, the total tug dry mass must be about 4000 kg with an electric power of 100 kW. The thrusters and power conditioners are estimated to weight about 500 kg, and the tankage for 3000 kg of propellant is estimated to weigh about 150 kg. Communications, guidance, structures, etc. might be an additional 1000 kg. That leaves a total of about 2350 kg for the power source which consists of the reactor, radiation shield, conversion system, and thermal radiator. The specific mass of the power system is thus about 23 kg/kWe. This is higher in power and lower in specific mass (kg/kWe) than any space nuclear power system that has been fielded in the past, but reasonable for estimated masses of near-term space-reactor systems.

The choice of technologies for the reactor system is fairly wide, although there is no off-the-shelf space-reactor system presently available. The SNAP program flew one space reactor in 1965 and built six other working reactors during the program. All of these used UZrH fuel with a NaK coolant (Anderson, 1983). The subsequent SP-100 program designed a reactor which used UN fuel with a Li coolant (Mondt, 1994; Mondt, 1995; El-Genk, 1994). The Russian Rorsat reactors (about 30 flown in space) used UMo fuel (Angelo, 1985). The Russian Topaz II reactor used UO_2 fuel and UZrH moderator. There are numerous proposed designs in the literature (Angelo, 1985; El-Genk, 1994).

Systems for conversion to electricity (for the electric thrusters) include thermoelectric (TE), thermionic, Rankine, Stirling, and recuperated Brayton cycles. Thermoelectrics have been the most used in space and typically result in only 5% conversion efficiency. Thermionics are a bit more efficient but require a higher temperature. Brayton, Rankine, and Stirling achieve typically 20-30% efficiency or more but require heavier parts and moving parts. Higher efficiencies can be achieved if a large radiator is available to provide a low-temperature heat sink for the cycle. There has been extensive development of all of these conversion systems on earth, but no dynamic conversion system has been used in space to date.

Given the need for a small reactor and shield mass, and a high conversion efficiency, we propose consideration of the following three options for the KBO mission: (1) a UZrH-fueled, NaK-cooled reactor with a steam Rankine conversion system, (2) a UN-fueled gas-cooled reactor with a recuperated Brayton conversion system, and (3) a UN-fueled heatpipe-cooled reactor with a recuperated Brayton conversion system. All three systems use the same assumed low-mass deployable thermal radiator to maximize the conversion efficiency. Table 2 compares the three systems. Subsequent sections describe the options in more detail. There has not been sufficient detailing of the designs to perform an accurate weight estimate for each system, but rough estimates indicate that all of these systems have the potential to be below the weight limit of 2350 kg.

TABLE 2. Comparison of three electric power systems for the KBO mission.

Component	UZrH/NaK-cooled/Rankine	UN/gas-cooled/Brayton	UN/Heatpipe-cooled/Brayton
Electric power (kW)	100	100	100
Thermal power (kW)	345	220	333
Thermal efficiency (%)	29	46	30
Nuclear fuel	UZrH	UN	UN
Moderator	Be	BeO	N/A
Reflector	Be	Be	Be
Primary coolant	NaK	He/Xe	Na
Number of fuel pins	211	361	138
Enrichment (%)	93	93	93
Total mass of ^{235}U (kg)	8	70	75
Fuel clad material	Hastelloy	Nb1%Zr/Re	Nb1%Zr/Re
Vessel material	316 SS	Super Alloy	Mo
Core vessel outer diameter (m)	0.235	0.40	0.20
Core + reflector outer diam. (m)	0.40	0.70	0.42
Conversion cycle	Rankine	Brayton	Brayton
Working fluid	water	He/Xe	He/Xe
Thermal radiator type	heatpipe/fin	heatpipe/fin	heatpipe/fin
Working fluid	ammonia	ammonia	ammonia
Two-sided radiating area (m^2)	360	252	252
Reactor coolant exit temp (K)	723	1200	1200
Average radiator temp (K)	330	306	306

UZrH-Fueled NaK-Cooled Reactor with Rankine Conversion System

The SNAP series of reactors used UZrH fuel and a "thermal" neutron spectrum. That is, the neutrons released by fission were slowed down by collisions with the hydrogen in the fuel so that they could interact with the uranium more easily. This results in a minimum amount of fuel needed to achieve a self-sustaining reaction, which allows the reactor, and also the radiation shield, to be near minimum mass. The SNAP program produced 6 complete operating reactor systems at various power levels in the 1960s (Angelo, 1985). One system, SNAP-10A, was flown in space.

The SNAP-8 reactors were derivatives of SNAP-10A, and the candidate reactor design would be similar to SNAP-8 and SNAP-10A. We would operate the reactor with a peak fuel temperature of only 800 K to extend the lifetime of the fuel. An enrichment of 93% is preferred, but a lower enrichment is also possible. The SNAP-10A spaceflight reactor and shield weighed about 268 and 98 kg respectively, summing to nearly 366 kg. To allow for a larger power and total burnup capability, we estimate the reactor mass would be 500 kg. To allow for shielding a large radiator, we estimate the shield would be 400 kg.

UZrH fuel cannot be operated at as high a temperature as UN fuel because of hydrogen dissociation. This necessitates the use of a low-temperature conversion system such as a steam Rankine conversion. Steam Rankine systems are a highly mature technology on earth with an extremely large industrial and extensive experience with reactor systems. However, they have never been tested or used in space, and this represents a major technical risk for this option. Figure 3 shows an overview of the conversion cycle. Heat is extracted from the reactor coolant (NaK) and converted to steam. The steam drives a turbine as it expands, and then condenses at a heat exchanger connected to the thermal radiator. A pump recirculates the condensed water back to the boiler. The turbine, pump, and alternator are all on a single shaft floating on a liquid bearing. The system has one moving element. The radiator consists of many parallel and separate capillary pumped loops with ammonia as the coolant. The temperature increase in the steam generator and superheater is about 383 K, but the temperature increase in the reactor coolant can be considerably less by designing the steam generator appropriately.

The Department of Energy and Jet Propulsion Laboratory developed operational hardware for a 25-kW toluene Rankine system driven by solar thermal energy (Nesmith 1985). The program also generated several conceptual designs for 100 kWe systems. We estimate the weight of a TAP with a multi-stage turbine for the KBO mission would be about 90 kg. The additional components for conversion system components (excluding the radiator) are estimated to be about 160 kg, summing to a total of 250 kg.

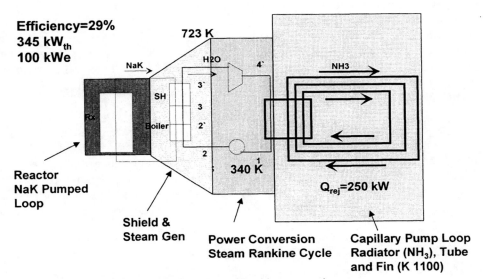

FIGURE 3. Overview of Rankine conversion system.

A major key to success for this system is development of a large lightweight (deployable) radiator. This allows heat rejection to space at a low temperature, which can result in a high conversion efficiency overall. NASA and commercial firms have been working on such radiators (some of the designs and specific diagrams are proprietary). The most advanced designs use an array of parallel capillary pumped loops with thermal fins between them. Ammonia is typically used as the coolant. The use of many separate loops minimizes the need for shielding from micrometeor impact. The total mass of these radiators is about 2 kg/m^2 (which would be 1 kg/m^2 of radiating surface if used in a flat configuration with two-sided radiation). Capillary pumped loops, loop heat pipes, and heat pipes have all been demonstrated in space, but these new lightweight radiator configurations have not. The heat load required for each individual heat pipe has been demonstrated on earth, but not at for the required lengths yet.

The total radiating area needed for the baseline design is about 360 m^2 at 330 K. Traditionally space reactor radiators have been designed as conical with only the outside surface radiating. This was done to maximize the radiator area behind the radiation shield. However, to allow easier deployment of the radiator, to minimize the radiator mass, and to help reduce the shield mass, we presently envision the radiator as being flat with both sides radiating. Thus the physical radiator would be 180 m^2 with a total radiating surface of 360 m^2. The radiator would deploy from a manifold extending along the boom. With this potential design and the mass numbers quoted for small advanced systems as background, the radiator system is estimated to weigh 360 kg.

The sum of all these estimated weights (reactor, shield, conversion, radiator) is 1510 kg. Structure, additional controls, other components, and contingency will add to this total estimate, but staying below the 2350-kg limit seems achievable.

UN-Fueled Gas-Cooled Reactor with Brayton Conversion System

The most recent U.S. space reactor power program, SP-100, developed detailed designs, advanced reactor fuel, a "zero-power" reactor critical assembly, radiation-hardened control drives, and various other hardware components in the 1980s. The baseline SP-100 was designed to produce 2400 kW thermal and 100 kW electric with a lifetime of 7 to 10 years (Mondt, 1994, Mondt 1995, El-Genk 1994). The specific power at program termination was about 42 kg/kWe. It used a high-temperature advanced fuel (UN) which was developed and proven with nuclear burn-up tests during the program. The fuel was not designed to slow down the fission neutrons, so the neutron spectrum was "fast" and the resulting core size and U-235 mass for the reactor to achieve criticality was thus larger than for the SNAP series.

The proposed design consists of a gas-cooled fission reactor using the SP-100 developed UN fuel and a closed Brayton cycle for power conversion at 100 kWe. There is a very extensive industrial data base and fabrication experience for open-cycle Brayton units: they form the basis for commercial and military jet engines as well as helicopter engines. Brayton conversion systems have one moving part: a single shaft connected to the turbine, the electrical generator rotor, and the compressor. In closed systems, this single shaft floats on a gas bearing bled off from the main gas flow and returned to it. Space Brayton units would be weightless and at constant power. They should easily last for many years without any maintenance. A 52,000-hr ground test of a 10.7-kWe closed Brayton unit at NASA/LeRC in 1965 supports these expectations.

The reactor is gas cooled (30/70 mole-% He/Xe) which directly drives the Brayton system. The active core is 0.40 m in diameter and 0.5 m long. The radial reflector is 0.15-m thick beryllium, and the axial reflector is 0.10-m thick BeO. The fuel rods are held in a lattice of BeO with a 2.0-mm thick flow channel around each rod for the gas coolant. The BeO provides a small amount of moderation. There is a strong negative thermal feedback which allows the reactor power to naturally follow variations in load without needing adjustment of the control elements. There is a thick radiation shield on top of the reactor to shield the payload from the reactor neutron and gamma radiation.

Figure 4 shows a schematic of the Brayton cycle and the associated state points. The reference design produces 100 kWe with 46% total thermal efficiency and has a specific mass of about 26 kg/kWe. A key feature is the heat exchanger which recuperates some heat from the turbine exit and uses it to preheat the gas returning from the heat sink before it re-enters the reactor. This recuperation step gives the cycle a greater conversion efficiency than a "non-recuperated" Brayton cycle. The radiator is opened out flat to radiate from two sides and is 126 m² per side. This allows a low heat-sink temperature and enables the high electrical conversion efficiency. Detailed weight estimates are not available for this configuration, but the larger reactor size will likely make it heavier than the UZrH Rankine system.

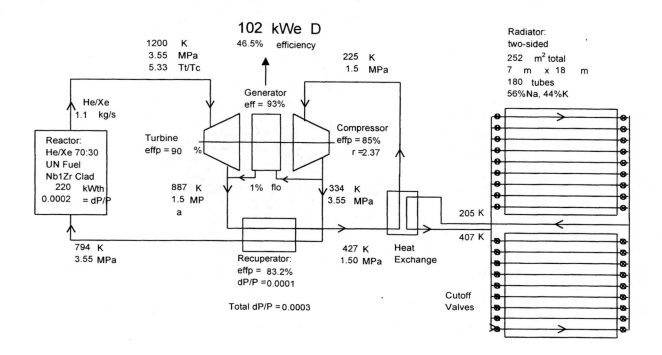

FIGURE 4. Brayton cycle schematic and associated state points.

UN-Fueled Heatpipe-Cooled Reactor with Brayton Conversion System

Houts, Poston, and Emrich have reported on various designs for heatpipe cooled reactors (Houts, 1998; Poston, 1996). In the heatpipe power system (HPS) heatpipes are inserted into the reactor core at regular locations to remove heat by boiling and wicking of the coolant. This heat is then transferred via a heat exchanger to the working fluid of a conversion system. For our application, we would use UN fuel in the reactor, sodium in the heat pipe, and He/Xe in the Brayton conversion system. Thus the system state points would be very similar to the gas-cooled Brayton system. Figure 5 shows a fuel/heatpipe module and a cross section of the reactor.

There are numerous advantages and features of the HPS that make it an attractive near-term system:

1. **Safety**. The HPS is designed to be subcritical for all credible launch accidents.
2. **Reliability**. The HPS has no single-point failures.
3. **Long life**. The design lifetime is in excess of 10 years.
4. **Modularity**. The HPS consists of independent fuel/heatpipe modules which can be tested individually.
5. **Testability**. The HPS system launch hardware can be tested at full power using electrical heaters in place of fuel rods. Unirradiated fuel rods inserted before launch. Full-power nuclear tests might not be required.
6. **Versatility**. The HPS can use a variety of fuel forms, structural materials, coolants, and conversion systems.
7. **Scalability**. The HPS design scales well to beyond 1000 kW thermal power.
8. **Simplicity**. There are few system integration issues since there are no in-core shutdown rods, no hermetically sealed refractory metal vessel or flowing loops, no electromagnetic pumps, no coolant thaw systems, no gas separators, and no auxiliary coolant loop for decay heat removal.

9. **Fabricability**. Most of the fabricated parts are small modules with similar metals; there is no pressure vessel.
10. **Near Term**. The system needs no development of advanced materials or components. It can be developed quickly and inexpensively with few nuclear tests.
11. **Low Mass**. The HPS system has a high fuel fraction in the core since it uses no in-core shutdown rods. The potential for in-space fueling (because of no pressure vessel) allows a more compact form while still meeting launch safety requirements.

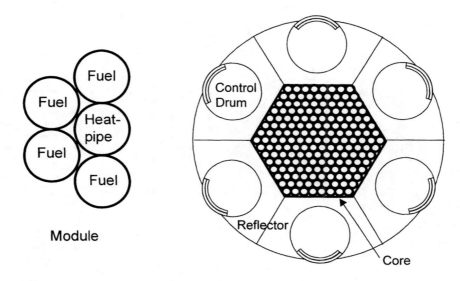

FIGURE 5. Module and reactor for heatpipe power system.

LAUNCH APPROVAL

There is a precedent for operating small research-sized reactors in space. There are presently over 30 shut-down nuclear reactors orbiting earth at about 600 km altitude. All but one of these are Russian reactors from Rorsat high-power radar satellites. The one U.S. reactor is SNAP-10A, launched in 1965. Every U.S. launch of a payload involving nuclear material must be reviewed by an Interagency Nuclear Safety Review Panel (INSRP) (Sholtis, 1994). The INSRP reviews the sponsors assessment of the risk and reports to the Office of Science and Technology Policy under the Executive Branch. The President or his designee (usually the Science Advisor) then decides whether to grant launch approval. This process has been followed for 25 launches of nuclear
materials over the past 40 years and approval has always been granted. All but one of these launches have involved radioisotopic power sources, but a space reactor would follow the same process.

One key safety feature of a fission reactor is that it is barely radioactive before it is used. The radiological inventory in the fresh fuel is about the same as a 100 cubic yards of Florida dirt or beach sand. The space reactor can be tested prior to launch at essentially zero power to prevent any inventory buildup, so the primary concern for launch safety will be to assure that the reactor will not turn on for any conceivable launch accident scenario. This requires solid engineering, but is not difficult. In addition, once the reactor begins operation in space, it should be assured that it cannot reenter the earth's atmosphere. With high circular orbits and low-thrust ion-propulsion, or with a launch to an earth-escape trajectory before the reactor is turned on, this should be fairly straightforward. Once it is on its way away from earth, the safety issues vanish, and waste disposal is automatic.

SUMMARY

A mission to rendezvous and land on one a Kuiper Belt Object is stressing and requires a large amount of total ΔV. A fission-powered system with an electric thruster could rendezvous at 40 AU in about 13.0 years with a total DV of 46 km/s. It would deliver a 1000-kg science payload while providing ample onboard power for relaying data back to earth. The launch mass of the entire system (power, thrusters, propellant, navigation, communication, structure, science payload, etc.) would be 7984 kg if it were placed into an earth-escape trajectory (C=0). Alternatively, the system could be placed into a 700-km earth orbit with more propellant, yielding a total mass in LEO of 8618 kg, and then spiral out of earth orbit to arrive at the KBO in 14.3 years. To achieve this performance, a fission power system with 100 kW of electrical power and a total mass (reactor, shield, conversion, and radiator) of about 2350 kg. Three possible configurations are proposed: (1) a UZrH-fueled, NaK-cooled reactor with a steam Rankine conversion system, (2) a UN-fueled gas-cooled reactor with a recuperated Brayton conversion system, and (3) a UN-fueled heatpipe-cooled reactor with a recuperated Brayton conversion system. All three of these systems have the potential to meet the weight requirement for the trip and be built in the near term.

ACKNOWLEDGMENTS

This activity supported by Sandia National Laboratories. Sandia is a multiprogram laboratory operated by Sandia Corporation, a Lockheed Martin Company, for the United States Department of Energy under Contract DE-AC04-94AL85000.

REFERENCES

Anderson, R. V., et al, *Space-Reactor Electric Systems: Subsystem Technology Assessment*, ESG-DOE-13398, Rockwell International, Canoga Park, CA, chapter IV, 1983.

Angelo, Jr., J. A. and D. Buden, *Space Nuclear Power*, Orbit Book Co., Inc., Malabar, Florida, pp. 159-176, 1985.

Chang Diaz, F. R., "Recent Progress in the VASIMR," *1998 Div. of Plasma Physics Meeting, Nov. 16-20, 1998, New Orleans, LA*, 1998.

El-Genk, M., *A Critical Review of Space Nuclear Power and Propulsion, 1984-1993*, M. S. El-Genk, Ed., U. of New Mexico Institute for Space Nuclear Power Studies, Albuquerque, NM, AIP Press, New York, NY, 1994.

Houts, M. G., D. I. Poston, and W. J. Emrich, Jr., "*Heatpipe Power System and Heatpipe Bimodal System Development Status,*" *Space Technology and Applications Internations Forum 1998 (STAIF-98), January, 1998, Albuquerque, NM*, pp. 1189-1195, 1998.

Jewitt, D. C., "Kuiper Belt Objects," *Annual Review of Earth and Planetary Sciences*, **27**, 287-312. (1999).

Malhotra, Renu, "Migrating Planets," *Scientific American*, **281**, No. 3, 56-63 (1999).

Marriott, A. T., and T. Fujito, "Evolution of SP-100 System Designs," *11th Symp. on Space Nuclear Power and Propulsion, Albuquerque, Jan 1994*, CONF-940101, pp. 157-169, 1994.

MathSoft, Inc., 101 Main St., Cambridge, MA 02142-1521.

Mondt, J. F., Truscello, V. C., and Marriott, A. T., "SP-100 Power Program," in *Eleventh Symposium on Space Nuclear Power and Propulsion, Albuquerque, NM 1995*, M. S. El-Genk, Ed., U. of New Mexico Institute for Space Nuclear Power Studies, Albuquerque, NM, AIP Press, New York, CONF 940101, 1995 , pp. 143-155.

Mondt, J. F., V. C. Truscello, A. T. Marriott, "SP-100 Power Program," *11th Symp. on Space Nuclear Power and Propulsion, Albuquerque, Jan 1994*, CONF-940101, pp. 143-155, 1994.

Nesmith, B., B*earing Development Program for a 25-kWe Solar-Powered Organic Rankine-Cycle Engine*, DOE/JPL-1060-92, Jet Propulsion Laboratory, Pasadena, CA, 1985.

Polk, J., Anderson, J.R., Brophy, J.R, Rawlin, V. K, Patterson, M.J., and Sovey, J.S., "The Effect of Engine Wear on Performance in the NSTAR 8000 Hour Ion Endurance Test," *Joint Propulsion Conference, Cleveland, OH, July 14, 1998*, AIAA 97-3387, 1998.

Poston, D. I. and M. G. Houts, "*Nuclear and Thermal Analysis of the Heatpipe Power and Bimodal Systems,*" *Space Technology and Applications Internations Forum 1998 (STAIF-96), January, 1996, Albuquerque, NM*, pp. 1083-1093, 1996.

Sholtis, Jr., J. A., Connell, L. W., Brown, N. W., Mims, J. E., and Potter, A., "U. S. Space Nuclear Safety: Past, Present, and Future," in *A Critical Review of Space Nuclear Power and Propulsion 1984-1993*, M. S. El-Genk, Ed., U. of New Mexico Institute for Space Nuclear Power Studies, Albuquerque, NM, AIP Press, New York, NY, 1994, pp. 269-304.

Space Rocket Engine on the Base of the Reactor-Pumped Laser for the Interplanetary Flights and Earth Orbital Applications

Andrey V.Gulevich, Peter P.Dyachenko, Oleg F.Kukharchuk, and Anatoly V.Zrodnikov

Federal Scientific Center, Institute for Physics & Power Engineering, 1, Bondarenko Sq., Obninsk 249020 Russia
Fax: +7 (095) 8833112, Phone: +7 (08439) 98351, E-mail: gulevich@ippe.rssi.ru

Abstract. In this report the concept of vehicle-based reactor-laser engine for long time interplanetary and interorbital (LEO to GEO) flights is proposed. Reactor-pumped lasers offer the perspective way to create on the base of modern nuclear and lasers technologies the low mass and high energy density, repetitively pulsed vehicle-based laser of average power 100 kW. Nowadays the efficiency of nuclear-to-optical energy conversion reached the value of 2-3%. The demo model of reactor-pumped laser facility is under construction in Institute for Physics and Power Engineering (Obninsk, Russia). It enable us to hope that using high power laser on board of the vehicle could make the effective space laser engine possible. Such engine may provide the high specific impulse ~ 1000 – 2000 s with the thrust up to 10 – 100 n. Some calculation results of the characteristics of vehicle-based reactor-laser thermal engine concept are also presented.

INTRODUCTION

The idea of using high power ground based lasers for the propulsion of space vehicles was proposed by Kantrowitz and coworkers at Arco-Everett Research Laboratory about 30 years ago (Douglas-Hamilton, 1978). The main attractive applications of this idea such as space propulsion, beaming power to orbital transfer vehicles, removing space debris, LEO to GEO orbit changes and etc. has been already discussed in the scientific literature (Caveny, 1984, Phipps, 1992, Campbell, 1998) .It should be noted that all of this applications concern the using of ground based power lasers of the multi megawatt level. It is supposed that the most acceptable laser for this goal is a free-electron laser.

The alternative type of high power lasers is reactor-pumped laser (RPL) which offer unique advantages for different space applications (Lipinski, 1995, Pashin, 1997, Petra, 1998, Gulevich, 1999). The main attractive properties of RPLs are huge power capacity, compactness, autonomy, possibility to pump almost unlimited volumes with active media due to the big penetrating capacity of neutrons in multiplicative media and etc. These properties allow us to hope to create on the base of nuclear reactor pumping powerful land-based and space-based laser systems for the perspective space applications.

In this paper we will consider the concept of the traditional space laser engine which is powered by laser beam from the reactor-pumped laser of moderate power ~ 100 kW and mass of 5 tons. In our concept the RPL system may be placed on the board of the space vehicle or on the wire rope with the length that provides the effective and safe functioning of the vehicle and RPL. In both of cases the source of laser energy moves together with the vehicle.

The advantages of this concept are obvious:
- no problem of transportation the laser beam from the Earth to the vehicle,
- no necessity to place the large-scale optics into space,
- no any multi-megawatt level ground-based laser is needed, and etc.

The disadvantages of the concept are: the necessity to place the source of energy on the board of vehicle and the absence of the real working reactor-pumped laser of 100 kW power, the necessity to remove the vast heat from the reactor laser into the space. It should be noted that all disadvantages could be solved or significantly reduced on the present level of laser and reactor technologies.

CP504, *Space Technology and Applications International Forum–2000*, edited by M. S. El-Genk

PRESENT STATUS OF NPL RESEARCH

The first demonstration of a nuclear pumped laser was in mid 1970's. Thanks to the activity of scientists from different laboratories, a laser generation of more than thirty types of gas media pumped by the nuclear reaction products has been demonstrated. Encouraging data about the possible use of uranium containing laser liquids for NPL systems had also been obtained. Substantial advancements have been made in understanding the physical processes occurring in nuclear-pumped lasers and technology of NPL. Two specialist conferences on Physics of Nuclear Induced Plasmas and Problems of Nuclear Pumped Lasers were held in Russia in 1992 (Obninsk) and 1994 (Arzamas-16).

The most effective nuclear-into-laser energy conversion (up to the 2 - 3%) had been demonstrated on the Ar-Xe, He-Ar-Xe gas media on the transitions with the wavelengths 1733, 2032 nm (Hebner, 1993, Lipinski, 1995).

One of the most perspective conceptual design of the high power reactor pumped laser system based on the optical quantum nuclear reactor-pumped amplifier (RPA) had been proposed in the Institute of Physics and Power Engineering several years ago (Dyachenko, 1991). The optical scheme of the RPL system is based on the "master oscillator - two round trip amplifier" principle (Dyachenko, 1993, 1995) and presented at Fig.1.

Scheme of RPA is similar to the scheme that was considered in the report (Dyachenko, 1993) and presented in Fig.1. RPA is a coupled reactor system consisting of pulse periodic fast reactor and deep-subcritical (in neutron sense) laser module (LM) with the multiplication factor ~ 0.8 – 0.9. Reactor and laser modules are principal components of RPA. Periodic pulsed fast reactor (Shabalin, 1976) is equipped with a liquid metal cooling system of the active core. If necessary a several pulsed reactors can be used.

LM acts as a multiplier of reactor neutrons. It is shaped as a cylindrical structure sized to provide the space for housing the active core and reactivity modulator of pulse reactor.

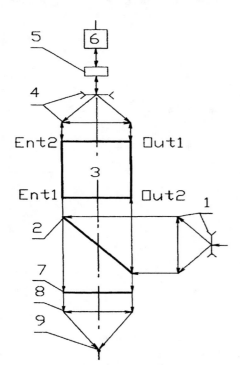

FIGURE 1. Optical scheme of the RPL system based on RPA: 1 - beam expander, 2 - polarizer, 3 - RPA, 4 - beam expander, 5 - Faradey cell, 6 - conjugation cell, 7 - harmonic generator, 8 - lens, 9 - target.

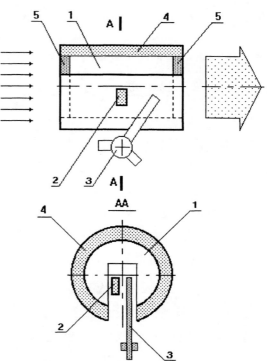

FIGURE 2. Scheme of the optical quantum RPA.:1 - laser module, 2 - pulse reactor, 3 - reactivity modulator, 4 – neutron reflector, of neutrons, 5 - optical windows.

The pumping section of the laser module is filled with a gaseous or liquid laser active medium containing a fissile material and includes the elements of a neutron moderator. The construction of laser-active element with gas medium is detailed described in the report (Gulevich, 1999).

This section is surrounded with reflector of neutrons, flanks are prepared from an optically transparent material to provide the input and output of the laser beam. The fission energy of uranium in the laser module can be utilized for pumping of laser. The liquid metal coolant system may be used to remove the surplus heat energy from the LM.

In the Table 1 the results of the mathematical modeling of the characteristics of the periodically acting RPLS with average laser power ~ 100 kW are presented. The different types of gas active media were considered (Lipinski, 1995).

The energy model of a pulse nuclear reactor pumped laser system on the base of the two core "Bars-6" fast burst reactor (Snopkov, 1992) is created at IPPE and presented at fig.3. It was started up in 1999. Nuclear pumped amplifier (NPA) is basis of the model.

LASER ENGINE ON THE BASE OF THE REACTOR-PUMPED LASER

The principal scheme of the space rocket engine on the base of the reactor-pumped laser is presented in fig. 4. In this concept a repetitively pulsed laser may be suited on the board of the vehicle or fasten on the wire rope to the vehicle as shown on the fig.4. Reactor-pumped laser is a source of energy for space laser engine.

The placement of RPL on the board of the vehicle enable to solve the numerous problems connected with a transportation of laser beam from the source of energy to laser engine and necessity to suit the large-scale optics in space. No land-based huge and expensive laser systems are needed to construct on the Earth. Having a high-graded laser energy on the vehicle the numerous additional possibilities for space applications become are available.

TABLE 1. Characteristics of RPL system for space laser engine.

Lasing atom	Xe	Xe	Ar
Buffer gas Wavelength (nm) Efficiency of nuclear-to-laser conversion (%)	He, Ne, Ar 2032 3.0	He, Ne, Ar 1733 3.0	He, Ne 1271 1.1
Efficiency of the surface source pumping (%)	30	30	30
Laser energy per pulse (kJ)	90	90	60
Laser pulse duration (ms)	0.5	0.5	1
Maximal laser power (MW)	180	180	60
Frequency of pulses (Hz)	1	1	1
Energy of reactor pulse (MJ)	5.0	5.0	10.0
Effective multiplication coefficient of LM	0.8	0.8	0.9
Energy of pumping in laser module (MJ)	10.0	10.0	20.0
Mass of uranium in LM (kg)	50	50	70
Length of laser module (m)	2.0	2.0	2.5
Total mass of RPL system with irradiator (ton)	5	5	7

FIGURE 3. General view of the energy model of RPL (IPPE).

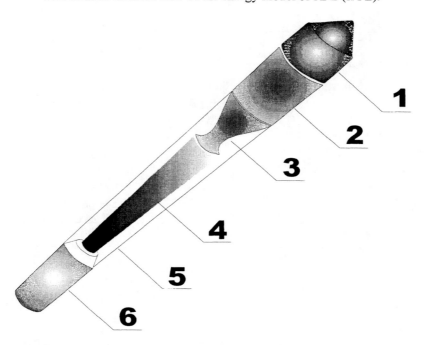

FIGURE 4. Principle scheme of the space rocket engine on the base of the RPL
1- payload; 2 – fuel; 3 – laser engine; 4- laser beam; 5 – rope; 6 - RPL

The disadvantages of this concept are low efficiency of nuclear-into-laser conversion, additional mass of RPL on the board of the vehicle and problem of cooling the RPL in the space. All problems mentioned above, however, may be solved on the base of present and future nuclear, laser and space technologies.

We will assume that the output laser power of RPL is equal N = 100 kW. Vehicle receives this energy on an ablation disk and propels itself by the ablation of the disk mass. In this case the material of the disk is a fuel for the spacecraft.

Note that we may use also the traditional thermal laser engine scheme. In this case the vehicle should be fueled (for example) by the liquid He. The stationary regime of functioning of RPL is also possible for this case.

In the case of laser propulsion the exhaust velocity as well as a specific impulse may be 3-4 times larger than that available from the chemical reaction.

The laser coupling coefficient C and the ratio of incident laser energy to mass of fuel removed from the disk Q characterizes the efficiency of the engine.

The typical value of C is within the range of 0.05 - 1 n/kW (Phipps, 1993). In our case one can expect that the maximal thrust for the engine with RPL is

$$F = C \cdot N = 100 \text{ n}.$$

Using the value Q=20 MJ/kg for ablated material we can estimate the maximal expected exhaust velocity and specific impulse

$$Ve = C \cdot Q = 20 \text{ km/s} \quad \text{and} \quad Isp = 2000 \text{ s}.$$

Thus the perspective laser engine with RPL will be of small thrust and high specific impulse. It may be used for the Earth orbital applications and interplanetary flights.

1) At first we will consider of the LEO-to-GEO transportation of the vehicle (10 ton of payload) with use of the laser engine powered from RPL (see the Table 1). Total mass of RPL with the irradiator is 5 ton.

To decide this problem of LEO-to-GEO transportation we have to provide the additional velocity ~ 5 km/s (Caveny, 1984) to the vehicle. The mathematical modeling of the LEO-to-GEO mission was carried out. The estimated characteristics of laser engine are presented in the Table 2.

TABLE 2. Characteristics of the LEO-to-GEO mission using the RPL.

Peak beam power (MW)	180
Beam diameter at fuel disk (cm)	5
Peak intensity at Target (MW/cm^2)	10
Laser Coupling coefficient, C (n/kW)	1
Specific Impulse, Isp (s)	2000
Heat of Ablation, Q (MJ/kg)	20
Ablation efficiency (%)	100
Initial mass (kg)	20000
Ablated mass (kg)	4300
Mass to GEO (Payload + RPL+laser engine), (kg)	15700
Time (days)	10
Total Laser Energy to GEO (GJ)	84

The burn-up of uranium in the laser module of RPL will be only 0.1 kg. Thus reactor-laser system may be used as a shuttle for LEO-to-GEO mission with the resource of ~ 50 - 100 flights.

2) The efficiency of application of space laser engine with RPL for the interplanetary flights will depend mainly on the total mass of fuel (mass of the ablation disk) on the board of the vehicle. For the case when the initial mass of fuel is 30 ton and parameters of the RPL system are the same as considered above (see the Table 1), we may provide the additional velocity of ~ 20 km/s to the spacecraft after a 2,5 months of functioning the RPL.

CONCLUSION

Thus, the concept of laser engine powered by the vehicle-based reactor-pumped laser for Earth orbital applications (LEO to GEO) and long-time interplanetary flights is proposed. RPL on the base of the nuclear reactor-pumped amplifier is a promising source of high-graded laser energy for such engine. It was displayed that after solving some scientific and technical problems the space laser engine with the high specific impulse ~ 1000 – 2000 s and thrust ~ 10 – 100 n may be created oh the present space, laser and nuclear technologies.

NOMENCLATURE

English

C:	laser coupling coefficient (n/kW)
Q:	heat of Ablation (MJ/kg)
N:	average power of laser beam (kW)
F:	thrust (n)
Isp:	specific impulse (s)
Ve:	exhaust velocity (km/s)

REFERENCES

Campbell J and Taylor C. "Cround-Based Laser Propulsion for Orbital Debris Removal," in Proc. of STAIF-98, 1998.

Caveny, L.H. et al. "Orbit-rasing and Maneuvring Propulsion: Researsh Status and Needs," Vol. 89, Progress in Astronautics and Aeronautics, Published by the AIAA, Inc. New York, 1984, chapter 1.

Douglas-Hamilton D, Kantrowitz A. and Reilly D. "Laser-Assisted Propulsion Research," Progress in Astronautics and Aeronautics, Published by the AIAA, Inc. New York, 1978, 61, p.271.

Dyachenko P.P. et al., Fusion Technology 20, 969-976 (1991).

Dyachenko P.P. et al., "Energy Model of a Pulse Nuclear Reactor Pumped Laser System," in Proc.of 7 th Intern. Conf. on Emerging Nuclear Energy Systems, World Scientific. Singapore-New Jersey-London-Hong Kong, 1993, 1, pp.372-376.

Dyachenko P.P., "Nuclear-Pumped Lasers and Problems of ICF," in 16th IEEE/NPSS Symposium on fusion engineering-95, edited by G.H.Miley et al., 1, 1995, pp.78-81.

Gulevich A.V. et al, "Concept of the Power Reactor Laser for Space Energetic Needs," in Proc. of STAIF-99, 1999.

Hebner G.A. and Hays G.N., J.Appl.Phys., 73, pp.3614 (1993).

Lipinski, R. and Mcarthur D., "Application for Reactor-Pumped Lasers," in Proc. of 2nd Int. Conference on Physics of Nuclear-Induced Plasmas and Problems of Nuclear-Pumped Lasers-94, Arzamas-16, v.1, 1995, pp.44-51.

Pashin, E.A. et al., Communications of Higher Schools. Nuclear Power Engineering 4, 31-35, Russia, (1997).

Phipps, C.R., "Space Propulsion Concept Using High-Energy, Pulsed Laser Ablation," in Proc. of Conf. on Physics of Nuclear-Induced Plasmas and Problems of Nuclear-Pumped Lasers, IPPE, Obninsk, Russia, 3, 1992, pp. 196-205.

Petra M. and Miley G. "Nuclear Pumped Lasers for Space Power Beaming," in Proc. of STAIF-98, 1998.

Shabalin E.P. Fast burst reactors, Moscow, Atomizdat, 1976.

Snopkov A.A. et al., in Proc. of Conf. on Physics of Nuclear-Induced Plasmas and Problems of Nuclear-Pumped Lasers. IPPE, Obninsk, Russia, 1, 1992, pp. 144-156.

Realistic Development and Testing of Fission Systems at a Non-Nuclear Testing Facility

Tom Godfroy, Melissa Van Dyke, Ricky Dickens, Kevin Pedersen, Roger Lenard, Mike Houts

Propulsion Research Center, Marshall Space Flight Center, National Aeronautics and Space Administration, TD 40, Huntsville, Alabama, 35812. (256) 544-5720, melissa.vandyke@msfc.nasa.gov

(256) 544-5720; melissa.vandyke@msfc.nasa.gov

Abstract. The use of resistance heaters to simulate heat from fission allows extensive development of fission systems to be performed in non-nuclear test facilities, saving time and money. Resistance heated tests on a module has been performed at the Marshall Space Flight Center in the Propellant Energy Source Testbed (PEST). This paper discusses the experimental facilities and equipment used for performing resistance heated tests. Recommendations are made for improving non-nuclear test facilities and equipment for simulated testing of nuclear systems.

INTRODUCTION AND BACKGROUND

The discovery of fission was reported in February 1939. On December 2, 1942 the world's first self-sustaining fission chain was realized at the University of Chicago. Fission reactors have since been used extensively by the Navy (powering submarines and surface ships) and the commercial power industry (20% of US electricity provided by fission reactors, much higher percentage in other countries). Operating experience in the US alone totals thousands of reactor-years.

The potential for using space fission systems to open the solar system to extensive exploration, development, and settlement has been recognized for decades. However, despite numerous US programs aimed at developing and utilizing fission systems, the only US flight of a fission system occurred over three decades ago on April 3, 1965 (SNAP10A). Although the Former Soviet Union (FSU) successfully utilized over 30 fission systems in space, all US programs since 1965 have failed to fly.

Previous space fission system development programs have failed primarily because of heavy reliance on nuclear testing for system development and because of pressure to develop systems that serve every potential customer or mission need. Heavy reliance on nuclear testing increases cost and makes it very difficult to achieve significant milestones early in a program. Recently, there has been great interest in first-generation flight demonstrator concepts whose main purpose is to demonstrate space fission systems which can be developed and utilized safely in an affordable and timely fashion. A successful first-generation flight demonstrator would address many of the programmatic and technical issues associated with the use of space fission systems. Lessons learned, technical issues resolved, and the data gathered from a first-generation flight demonstrator would be used in the design of the second and third generation systems. If fission systems are to be utilized for any flight program in the US, the next fission flight system must be safe, simple, robust and inexpensive to develop. Technology risks must be kept at a minimum, development and utilization should be inexpensive and timely, and the experience from existing nuclear databases should be utilized. This would mean that no new nuclear development would be required for full confidence in a flight demonstrator, thus, development challenges (if any) would be related to thermal-hydraulics, structures, and "balance-of-plant".

CP504, *Space Technology and Applications International Forum–2000*, edited by M. S. El-Genk
2000 American Institute of Physics 1-56396-919-X

Benefits of a Simulated Nuclear Ground Test Program

Inexpensive development of fission systems can be accomplished through a strong non-nuclear ground test program.
- All non-nuclear ground tests are directly applicable to nuclear system development
- Technology issues can be demonstrated/resolved faster and cheaper with ground based hardware testing rather than paper studies.
- Program success does not hinge on performing nuclear testing. Issues can be resolved and program paybacks can be discovered long before a nuclear test has to be performed.
- Robustness of the system can be demonstrated. This results in a high confidence of the probability of flight success.
- Realistic margins of safety can be established through failure testing with no "nuclear" issues involved.
- Significant milestones can be achieved within modest budgets and schedules.
- Significant technical progress can be made with minimal risk of being "squashed" politically.
- Extensive tests can be performed on an actual flight unit.

NON-NUCLEAR TEST FACILITIES AT MSFC: THE PROPELLANT ENERGY SOURCE TESTBED (PEST)

At the Marshall Space Flight Center in Huntsville, Alabama, simulated nuclear testing is being conducted with a refractory metal module in the PEST. The first test article to operate in the PEST is the First Generation Propellant Energy Source Module (FiGPESM) test article. This test article, a 2-inch diameter, 17.75" long pure tungsten "block", represents a module consisting of 6 fuel pins surrounding a central Molybdenum-Lithium heatpipe. The fuel pins are boron-nitride resistance heaters that are used to simulate the heat of a nuclear fission reaction. Figures 1 and 2 show an internal and external view of the PEST.

FIGURE 1. External View of The PEST.   **FIGURE 2.** Internal View of the PEST.

PEST consists of a 24" diameter, 6 foot long, cylindrical water jacket cooled stainless steel vacuum chamber capable of operating at better than 1.0×10^{-7} Torr. The vacuum chamber is able to rotate vertically about its center axis and is mounted on a mobile support frame that facilitates quick and easy movement of the entire chamber. There are 16 2-3/4" Conflat Flange ports, 9 6" Conflat, and 1 8" Conflat ports located symmetrically about the chamber. These ports are used as viewports for visual inspection of the article or for optical data collection as well as feedthrough ports for data, gas, and power. The vacuum system is connected to the largest port. Due to the high temperature / high power nature of most simulated nuclear testing, the chamber is equipped with a water jacket to remove heat from the chamber walls produced during testing. Flow rates to both the chamber and the pumps are measured with flowmeters and monitored by a control system.

PEST was designed from the initial stages to be mobile, fully computerized, and equipped with standard off the shelf components. LabView software and National Instruments hardware was selected as the data acquisition and control (DAC) software due to its high level of industry implementation and high level of versatility. LabView is highly modular and has been customized to perform most all the routine operations standard to PEST. Additionally, each experiment is customized with its own LabView module for specific data collection or control needs. The data acquisition and control hardware consists of a SCXI chassis outfitted PCB cards specific to experiment needs. Additionally, if the DAC hardware needs to be replaced, for future experiment needs, the customized software can be retained without needing to be re-coded, greatly reducing the initial costs and time losses. All electronic controls and data acquisition devices are located on a rack that is also mobile.

The vacuum system, capable of reaching better than 1×10^{-7} Torr, consists of a water-cooled Alcatel Turbo molecular pump staged with a TriScoll 600 dry scroll vacuum pump. Pressure is monitored using multiple vacuum TC gauges for pressures above 1×10^{-3} Torr and both a Cold Cathode and Baypert-Albert ion gauge for pressures below TC gauge capability. Real-time pressure data is gathered both by a stand-alone "Varian Vacuum multi-guage controller" and LabView. The fully automatic vacuum system valves and the turbomolecular pump are controlled by LabView with coded-in procedures for vacuum chamber pump down, vacuum chamber pressurization, and fail-safe checks and routines. All valves fail closed in the event of loss of power or pneumatic pressure. Additionally, switches are located on the instrumentation and control cabinet for manual override.

31,360 Watts are currently available to PEST via 480 three phase. In the PESM test article, this power was routed through a power feedthrough to the six heaters. The heaters were connected with 2 heaters in series to create a pair, and the 3 pairs connected in parallel to create the system. The temperature needs were met with approximately one third of the available power supplied to the heaters. Multi-meters whose output signal is read by LabView measured the current and voltage. Power is manually controlled.

Additional PEST Capability Modifications

By the end of 1999, the ability to introduce cold gas and extraction of hot gas in the PEST should be complete.

RECOMMENDATIONS FOR FUTURE

Any future fission program, whose goal is a flight system, should investigate the use of non-nuclear testing where appropriate to significantly decrease programmatic costs. The ability to test using resistance heaters to closely simulate heat from fission should be a primary design goal. Data gained from such tests may be more thorough (i.e. failure testing and margin testing) since a great deal of the safety issues associated specifically with nuclear testing will not have to be addressed.

Results of a First Generation Least Expensive Approach to Fission Module Tests: Non-Nuclear Testing of a Fission System

Melissa Van Dyke[1], Tom Godfroy[1], Mike Houts[1], Ricky Dickens[1], Chris Dobson[1], Kevin Pederson[1], Bob Reid[2], J. Tom Sena[2]

[1] *Propulsion Research Center, Marshall Space Flight Center, National Aeronautics and Space Administration*
TD 40, Huntsville, Alabama, 35812; (256) 544-5720, email: melissa.vandyke@msfc.nasa.gov
[2] *Los Alamos National Laboratory, University of California, US Department of Energy*
PO Box 1663, MS J576, Los Alamos, New Mexico 87545; (505) 667-2626, email: rsr@lanl.gov

Abstract The use of resistance heaters to simulate heat from fission allows extensive development of fission systems to be performed in non-nuclear test facilities, saving time and money. Resistance heated tests on the Module Unfueled Thermal-hydraulic Test (MUTT) article has been performed at the Marshall Space Flight Center. This paper discusses the results of these experiments to date, and describes the additional testing that will be performed. Recommendations related to the design of testable space fission power and propulsion systems are made.

INTRODUCTION

Successful development of space fission systems will require an extensive program of affordable and realistic testing. In addition to tests related to design/development of the fission system, realistic testing of the actual flight unit must also be performed. Testing can be divided into two categories, non-nuclear tests and nuclear tests.

Full power nuclear tests of space fission systems are expensive, time consuming, and of limited use, even in the best of programmatic environments. Factors to consider when performing nuclear tests include the following:
1. Time and cost associated with fabricating and handling the test article;
2. Non-flight-prototypic modifications to the test article required to enable ground testing;
3. Required modifications to existing nuclear facilities to enable testing;
4. Time and cost associated with testing the article at a nuclear facility;
5. Time and cost associated with radiological cool down and transfer/shipping to a hot cell;
6. Expense and slow pace of assessing failures in a hot cell environment; and
7. Limited ability to correctly identify failure mechanisms in a hot cell environment.

History provides examples related to the seven concerns listed above. During the highly successful Rover Nuclear Rocket Development Program, it still took nearly four years to move from the Pewee ground nuclear test (1968) to the follow-on nuclear test, the Nuclear Furnace 1 test in 1972 (Koenig, 1986). The first five full ground nuclear power tests of the program (Kiwi A, Kiwi A', Kiwi A3, Kiwi B1A, Kiwi B1B, total cost >$1B FY00 equivalent) all resulted in massive fuel damage due to thermal hydraulic problems and flow-induced vibrations. These problems were not resolved until non-nuclear cold-flow tests were performed. During the SP-100 program, tens of millions of dollars were spent attempting to modify the Hanford Site 309 Building to allow a full ground nuclear test of an SP-100 system (Carlson, 1993). In addition, the system to be tested (SP-100 Ground Engineering System) was significantly different from the SP-100 Generic Flight System (Fallas, 1991). The Hanford Site 309 Building was selected in 1985 to be the site of the Ground Engineering System test (Baxter, 1991). At the end of the SP-100 program (nearly 10 years later) significant modifications still remained before nuclear tests could be performed in the building. During the Thermionic Fuel Element Verification Program it frequently took more than a year for thermionic fuel elements (TFEs) and TFE components to be removed from the test reactor, shipped, and readied for post-irradiation examination (PIE). When PIE was performed, limited data was obtained due to the expense, time, and limited equipment availability associated with working in a hot cell (Ranken, 1994). Neither the Rover program, nor the SP-100 program, nor the TFEVP led to the flight of a space fission system.

CP504, *Space Technology and Applications International Forum–2000*, edited by M. S. El-Genk
2000 American Institute of Physics 1-56396-919-X

Non-nuclear tests are affordable and timely, and the cause of component and system failures can be quickly and accurately identified. The primary concern with non-nuclear tests is that nuclear effects are obviously not taken into account. To be most relevant, the system undergoing non-nuclear tests must thus be designed to operate well within established radiation damage and fuel burn up limits. In addition, the system must be designed such that minimal assembly is required to move from non-nuclear testing mode to a fueled system operating on heat from fission. If the system is designed to operate within established radiation damage and fuel burn up limits while simultaneously being designed to allow close simulation of heat from fission using resistance heaters, high confidence in fission system performance and lifetime can be attained through a series of non-nuclear tests. Any subsequent operation of the system using heat from fission instead of resistance heaters would then be viewed much more as a demonstration than a test - i.e. the probability of system failure would be very low.

All future space fission system development programs could benefit from optimizing the use of realistic non-nuclear tests. First-generation systems will benefit the most, as they are most likely to operate within established radiation damage and fuel burn up limits. Although advanced fission systems will require extensive nuclear testing, experience and support gained from the in-space utilization of earlier systems should facilitate their development. Testing of the MUTT at the Marshall Space Flight Center is a first step towards the testing of nuclear systems in a non-nuclear test facility. The MUTT is the first test in a series of tests for the First Generation Least Expensive Approach to Fission (FiGLEAF) program proposed by the Propulsion Research Center (PRC) at NASA/MSFC.

The MUTT test series has five top-level goals:
1. Demonstrate that realistic non-nuclear testing can be used to resolve thermal hydraulic and other issues associated with space fission system development.
2. Demonstrate that the eventual user of space fission systems (in this case NASA) can be heavily involved in all aspects of space fission system development.
3. Demonstrate the desirability of a modular core design that allows issues to be resolved on a module level prior to fabrication and test of a full core.
4. Demonstrate the superiority of hardware-based technology assessment over the never-ending cycle of paper studies often associated with advanced system development.
5. Experience gained from the MUTT test series will be directly applicable to full-core tests slated to begin later in FY00.

Specific technical goals of MUTT test series include the following:
1. Gain experience using resistance heaters to realistically simulate heat from fission. Test module to thermal design limits by demonstrating capability of module to operate at 1477°C. (1750 K).
2. Demonstrate energy transfer capability of the heat pipe (greater than 1 kW) Test heat pipe to thermal design limits by demonstrating a heat pipe operating temperature of 1027°C (1300 K).
3. Demonstrate heat pipe operation at extreme transients (fast start followed by instantaneous shutdown).
4. Demonstrate direct thermal propulsion by introduction of cold gas (ambient conditions) and extraction of hot gas (900°C) from the chamber.
5. Development of instrumentation techniques for flow, temperature, and other measurements in a simulated fission system.

EXPERIMENTAL APPARATUS

This MUTT is a 5.08-cm diameter, 45-cm long pure tungsten "block", which represents a module with 6 "fuel" pins surrounding a central molybdenum-lithium heat pipe. It is supported at each end by stainless steel end caps that are insulated with a molybdenum foil to prevent reaction with the block, see Figure 1. A support member, mounted to an extension elbow, holds the two end caps. Fingers from the elbow capture the internal diameter of two opposing viewports to hold the MUTT in place. The block is insulated with graphfoil insulation (not shown).

FIGURE 1. Module Unfueled Thermal-hydraulic Test Article in PEST.

The tungsten block is heated with 6 resistance heaters (simulating "fuel" pins) 50 to 53 cm long and 1.17-cm diameter to simulate the heat produced by nuclear fuel elements. The high-temperature boron nitride heaters, capable of reaching over 2000 K, were designed and produced by Advanced Ceramics Inc, of Lakewood, OH. They are connected in two heater pairs, which are connected in parallel to an electrical feed through in the chamber. Fourteen gauge copper is connected the heaters to the feed through. This provides MUTT with a maximum available power of 3 kW to each heater (operating temperature limit, not power available limitation). Digital output multimeters deliver total heater current and voltage information to the data acquisition system. Temperature readings are obtained with an optical pyrometer and thermocouples. Representative interstitial holes run parallel to the "fuel pins" for direct thermal heating of gases. Gaseous helium passing through module simulates direct heating.

A molybdenum-lithium heat pipe, developed at Los Alamos National Laboratory (Reid, 1999), is inserted in the center hole of the tungsten block and supported at the far end by a stainless steel support bar. The heat pipe is 145-cm long, 1.27-cm outer diameter, and has a crescent-annular wick structure consisting of 7 layers of 400 mesh sintered molybdenum screen. Before delivery to MSFC, the heat pipe was tested at Los Alamos where it demonstrated radiation coupled operation to the environment of 1 kW at 1450 K. The heat pipe is instrumented with 9 type C thermocouples tack welded to the heat pipe on a nickel foil interlayer. The distance between the first 8 thermocouples is approximately 10 cm and beginning 10 cm from the end of the block. The distance between the last two thermocouples is about 20 cm. One thermocouple was attached to the tungsten block between the block and one thermocouple was attached to the chamber wall of PEST. An optical pyrometer is used to verify the accuracy of the thermocouple data. The thermocouple temperature data was directed to the data acquisition system.

Helium is injected through a gas feed through to a manifold that distributes the gas into six feeds that connect to the inlet side end cap of the tungsten block. The gas is then heated by the block and vented into the chamber where it is pumped out. The exhaust end cap is outfitted with thermocouples positioned over the gas exhaust holes to record change in temperature. Inlet temperature of the gas is measured prior to injection into the chamber. Gas flow rate is monitored and controlled by an MKS flow control unit

Pressure in the chamber was monitored using multiple vacuum thermocouple gauges for pressures above 10^{-3} Torr. For pressures below the capability of the thermocouple gauge, a cold cathode and Baypert-Albert ion gauge were used. Real-time pressure data was gathered both by a stand-alone Varian vacuum multi-gauge controller and LabView.

LabView software and corresponding National Instruments hardware was selected as the data acquisition and control (DAC) software due to its high level of industry implementation and versatility. LabView is highly modular and has been customized to perform most all the routine operations standard to PEST. The data acquisition and control hardware consisted of a SCXI chassis outfitted with cards specific to MUTT needs. The chassis contained a thermocouple card, a control card for operation of valves and switches, and card to handle the pressure information. Interface with the SCXI chassis was by computer running LabView software. LabView collected and assembled the data as well as monitored most aspects of the experiment. All electronic controls and data acquisition devices were located on a rack next to the chamber.

RESULTS OF EXPERIMENT

The first test determined the ability of the heaters to heat the module (neither gas flow or heat pipe were included in this test). The heaters were set at a constant power level and the uninsulated module temperature was recorded using an optical pyrometer. The power level was kept at this constant level until it appeared that module temperature reached a steady state. The terminal voltage across each heater was then increased by 20 V and kept at the constant level until the module again reached steady state. This procedure continued until the maximum available current that could be delivered by the power supply was reached. This corresponded to a maximum power of approximately 7 kW delivered to the heaters and a module maximum temperature of 1663 K. Figure 2 shows the time-temperature profile for this test. Although the curve shape is similar for each power level, at higher power levels (temperatures), the module temperature had a larger slope and reached steady state fairly quickly. Figure 3 shows the module at approximately 7.2 kW at 4000 s.

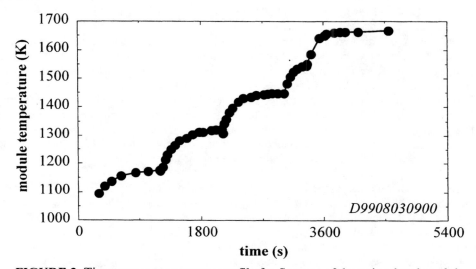

FIGURE 2. Time versus temperature profile for first test of the uninsulated module.

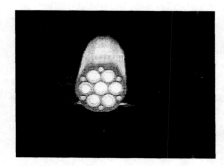

FIGURE 3. The uninsulated module at 7.2 kW and 4000 s.

Radiation calculations verified that the heat rejected from the module was approximately equal to that delivered to the module from the heaters. A second test, carried out with the identical settings and procedures as the first test, yielded the same results as the first test. These two tests verified that the heaters could be used to realistically simulate heat from fission. In an effort to increase the power available to the heaters, the power supply was rewired so an increase in current, resulting in an increase in available power, could be delivered to the heaters. The third test showed that at the same power levels, the time-temperature profiles were identical to the first two tests. The maximum power delivered by the heaters for the third test was approximately 9.2 kW corresponding to a maximum module temperature of 1754 K. This temperature is higher than that required for a potential first fission propulsion flight demonstrator.

The next series of tests were to verify the heat pipe's ability to operate under desired conditions. Type C thermocouples were installed on both the heat pipe and on the module to record temperatures. The thermocouple on the module served both to verify the optical pyrometer readings from earlier tests and to serve as a frame of reference for the heat pipe thermocouples.

The first heat pipe test was to verify heat-pipe operation, instrumentation hook-up, and test procedure. The first test ran for a total of 115 min and showed successful operation of the heat pipe. Since a slow start-up of the pipe was desired, the power supply was initially set to deliver 60 V (0.12 W), and increased at approximately 10 V increments every 10 min. This brought the heat pipe to a maximum operating temperature of 1220 K after 115 min. Figure 4 shows the thermocouples instrumented heat pipe. Figure 5 shows the thermocouple data over the period of the test.

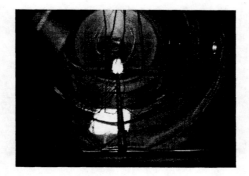

FIGURE 4. Instrumented molybdenum-lithium heat pipe.

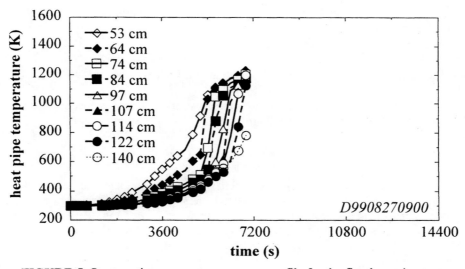

FIGURE 5. Start-up time versus temperature profile for the first heat pipe test.

At the end of the first test, air leaked into the chamber though a defective sight glass. The chamber was flooded with gaseous helium and kept at 1 Torr as the module and heat pipe cooled to ambient conditions. The module was hydrogen cleaned and a second heat pipe test was conducted again to determine the operational capability of the heat pipe and to verify that no damage had occurred. Since a slow start-up of the heat pipe was desired, the power supply was set to deliver 60 V (0.15 W), increasing approximately 15 V every 10 min. This brought the heat pipe to a maximum operating temperature of 1395 K after 245 min. Figure 6 shows the thermocouple data over the period of the test. The data showed successful heat pipe operation with the entire heat pipe at an operating temperature greater than that of the first test (>1220 K). At the end of the 245 min, the heat pipe was isothermal and the test terminated. This demonstrated that the heat pipe was able to operate successfully, even when exposed to worst case conditions. Both an optical pyrometer and a thermocouple were used for measuring the temperature of the thermocouple on the heat pipe that was closest to the module (TC1). The difference between both methods varied by only a maximum of 1.5 %, verifying the "goodness" of the data from the first three tests which used only the optical pyrometer.

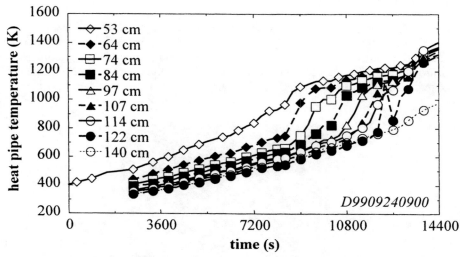

FIGURE 6. Start-up time versus temperature profile for the second heat pipe test.

Additional MUTT Tests Planned

Several more test series are scheduled for the MUTT and should be completed by early 2000. Specifically the following areas to be addressed are:

- Insulation of module to reduce radiation losses to chamber;
- testing to thermal design limits (i.e. capability for extremely high temperature (>2000 K))
- Testing of direct thermal propulsion, including introduction of cold gas and extraction of hot gas from the chamber;
- Testing of a fission system that allows simultaneous testing of thermal propulsion and heat pipe operation; and
- Investigation of the feasibility of using laser diagnostics to determine temperatures.

CONCLUSIONS

Full power nuclear tests of space fission systems are expensive, time consuming, and of limited use, even in the best of programmatic environments. Non-nuclear tests are affordable and timely, and the cause of component and system failures can be quickly and accurately identified. If the system is designed to operate within established radiation damage and fuel burnup limits while simultaneously being designed to allow close simulation of heat from fission using resistance heaters, high confidence in fission system performance and lifetime can be attained through a series of non-nuclear tests.

The MUTT was successful at demonstrating the use of resistance heaters to realistically simulate heat from fission. The MUTT demonstrated the ability to use several different instrumentation techniques for measuring temperature and pressure in a simulated fission (thermal hydraulic) environment. Finally, the MUTT was able to demonstrate the energy transfer capability and operation of a heat pipe under worst case operating conditions.

Additional testing will be completed shortly which should demonstrate the capability to test direct thermal propulsion and heat pipe operation simultaneously.

RECOMMENDATIONS

This test series demonstrated that some aspects of fission system operation can be simulated using non-nuclear test facilities. Any future fission program, whose goal is a flight system, should investigate the use of non-nuclear testing where appropriate to significantly decrease programmatic cost. Data gained from such tests may be more thorough (i.e. failure testing and margin testing) since a great deal of the safety issues associated specifically with nuclear testing, such as hot cells, will not have to be addressed.

REFERENCES

Baxter, W. F., Burchell, G. P., Fitzgibbon, D. G., and Swita, W. R. (1991) "SP-100 Ground Engineering System Test Site Description and Progress Update" in Space Nuclear Power and Propulsion, edited by Mohamed S. El-Genk and Mark D. Hoover, DOE Conf 910116, American Institute of Physics, New York, pp. 1329-1334.

Carlson, W. F., and Bitten, E. J. (1993) "A Facility for Testing 10 to 100-kW$_e$ Space Power Reactors" in Space Nuclear Power and Propulsion, edited by Mohamed S. El-Genk and Mark D. Hoover, AIP Conference Procedings 271, American Institute of Physics, New York, pp. 867-872.

Fallas, T. T., Gluck, R., Motwani, K., Clay, H., and O'Neill, G. (1991) "SP-100 Nuclear Assembly Test Assembly Functional Requirements and System Arrangement" in Space Nuclear Power and Propulsion, edited by Mohamed S. El-Genk and Mark D. Hoover, DOE Conf 910116, American Institute of Physics, New York, pp. 1323-1328.

Koenig, D. R. (1986) Experience Gained from the Space Nuclear Rocket Program (Rover), LA-10062-H, Los Alamos National Laboratory, Los Alamos, NM.

Ranken, W. A. (1993) Personal Communication, Los Alamos National Laboratory.

Reid, R. S., Sena, J. T., and Merrigan, M. A. (1999) "Transient Tests of a Molybdenum-Lithium Heat Pipe," 11th International Heat Pipe Conference, Tokyo, Japan, September 12-16.

Proposed Experiment on a Controlled Orbital Mass

Thomas E. Lett, III

2240 Kenwood Drive, Lexington, KY 40509

(606) 263-9410   tel@aol.com

Abstract. The experiment should determine whether a toroid orbital mass could be used in a propulsion system. The experiment would prove or disprove that a mass ring, or toroid, would exhibit displacement in a gravity field due to its velocity. If the results are positive, a vehicle could be produced for continuous use with relatively low energy consumption. The purpose for this proposed experiment is to open discussion and pursue a fundamental change in man's view of how objects behave in space.

PROBLEM STATEMENT

The basis of this experiment is a 'three body' problem. Body One determines the field of operation such as the earth or moon. Body Two is a toroid placed horizontally in the gravity field of Body One. Body Three is the containment or 'platform' for Body Two. Body Two and Body Three are the assembly components and have the same center of gravity. The experiment determines if displacement occurs between Body One and the combined masses of Body Two and Body Three when the amount of energy in Body Two exceeds tangential escape velocity. Higher efficiency is obtained when the mass of Body Three is proportionally less the mass of Body Two.

The experiment determines whether a force is obtained in the opposite direction of gravity. This event should occur when the ring reaches orbital velocity. At this point, a change in weight should occur. The total assembly is placed on scales and the weight should remain constant until just below orbital velocity. If the toroid and chamber are of equal mass, the total system should approach zero weight as the toroid reaches escape velocity. The effective acceleration in a gravity field of the system should be zero. This balance determines whether the displacement is positive or negative.

Experiment

Displacement opposite gravity could be demonstrated with a rotating toroid mass enclosed in an evacuated chamber. This toroid mass should be designed to withstand centrifugal forces when the angular velocity reaches 'escape velocity' of the earth's gravity field. The minimum diameter should be at least 156 meters and parallel to the surface. The toroid would consist of alternating magnets along its perimeter reinforced by a connecting frame (see figure 1). The toroid could be propelled and levitated by magnets and use electricity as an energy source. Levitation would maintain a separation of the toroid and containment chamber. The evacuated containment chamber and motors should be designed to be equal to or less than the mass of the rotating mass ring. The toroid is the only moving mechanism.

Toroid

The toroid "floats" on permanent magnets within the evacuated chamber (see figure 2). Steel and composites could be designed as a superstructure to hold the magnets in place. Current technology would allow angular accelerations of over two million m/s*s. The desired threshold is 2.4 million m/s*s for this experiment.

CP504, *Space Technology and Applications International Forum–2000*, edited by M. S. El-Genk

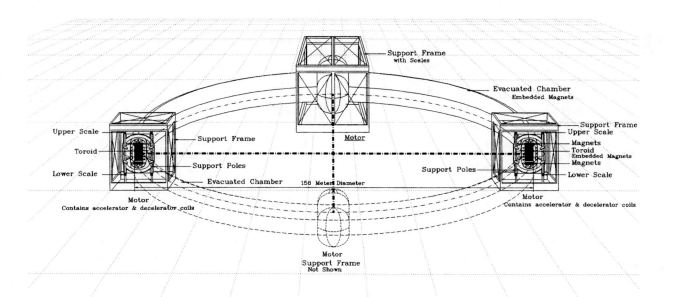

FIGURE 1. Controlled Rotating Mass Diagram.

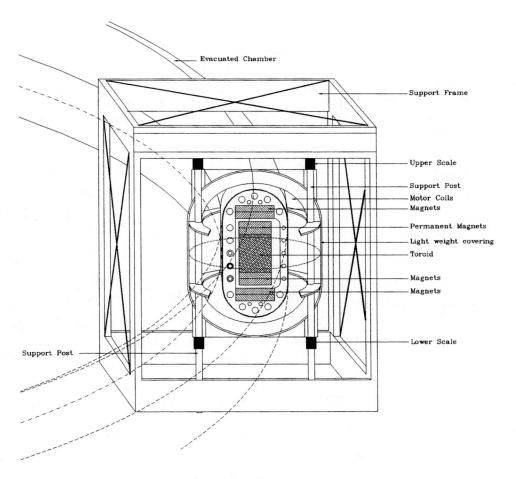

FIGURE 2. Motor.

Future designs might allow a smaller radius when materials are developed to withstand the generated angular forces. Considerations must also be made regarding expansion of materials, static charges and heat dissipation.

TABLE 1. The following are design criteria of the toroid.

Outside Diameter (m)	Inside Diameter (m)	Depth (m)	Matl. Volume (m³)	Total Weight (N)	Total Mass (Kg)
156.0	155.0	1.0	264	334,000	34,096

At orbital velocity, rate of rotation is around 16 rotations per second. The rotation rate is 23 rps at escape velocity and 30 rps to acquire an effective +1g.

Evacuated Chamber

The evacuated chamber is used as containment and control of the toroid. The frame should be designed as light as possible with all inside air removed. The chamber also has embedded magnets and is attached at the motors to make the platform.

TABLE 2. The following are design criteria of the Chamber.

Outside Diameter (m)	Inside Diameter (m)	Depth (m)	Matl. Volume (m³)	Total Weight (N)	Total Mass (Kg)
156.25	154.25	2.0	320	228,000	23,275

Motor

The motor is used to accelerate and decelerate the toroid. The motors are the points where the support system comes in contact with the platform. Vertical posts pass through the motor frame. Spring scales are attached above and below the motor assemblage around the posts. Any change in the scales will indicate whether a force and displacement occur at the orbital threshold. The current design requires eight motors with a minimum capacity of 20,000 watts each.

TABLE 3. The following are design criteria of the motor:

Outside Diameter (m)	Inside Diameter (m)	Depth (m)	Matl. Volume (m³)	Total Weight x 8 (N)	Total Mass x 8 (Kg)
3.0	3.0	3.0	12.0	106,000	10,821

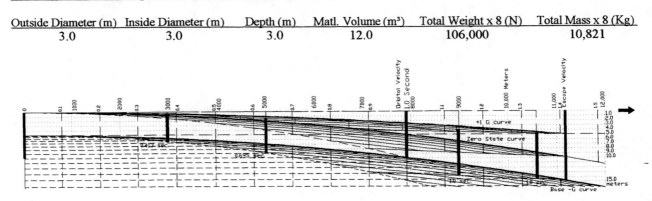

FIGUIRE 3. One second diagram section.

The total assemblage weighs 668,000 newtons with a mass of 68,192 kg.

RESULTS

The properties of the toroid should emulate an object in an elliptical orbit. The force derived from the gravity field is proportional to the tangential velocity of the toroid. If escape velocity is achieved, the total weight of chamber and ring should be zero. Velocities of the toroid above "escape" would induce displacement opposite the gravity field (see figures 3 and 4). The toroid should emulate the force on a mass at the perigee. At escape velocity, the resultant acceleration is positive 1g for the toroid. If the toroid were not in containment, it would 'fall up' 4.9 meters in one second. Since the platform is of equal mass, its weight offsets the vector force of the toroid.

If a force of 1.098 m/s*s is applied by the motors to the toroid, the energy to obtain orbital velocity starting from zero is about 1.1 trillion joules. This process should require about two hours. No change in displacement will occur until this level of energy is achieved. After another fifty minutes, escape velocity is reached by the toroid. The amount of energy required to change the velocity from orbital to escape is 1.1 trillion joules. At this point, the entire platform should weigh zero. The total energy applied to toroid is about 2.2 trillion joules before a positive displacement is performed. The platform should have a resultant of +1g by applying another 1.1 trillion joules to the toroid making its velocity 13,691 m/s. The +1g force will be maintained indefinitely if the velocity is maintained, and the platform is not allowed to displace. If the toroid is allowed to seek the higher orbit, it should decrease in velocity. As long as the toroid is held in place, it should maintain the angular velocity applied to it.

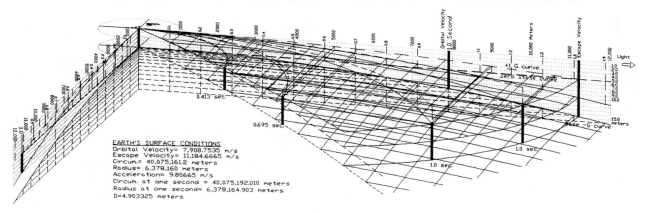

FIGURE 4. One second diagram.

The amount of force the toroid induces on the platform will decrease as the distance from Body One increases (see figure 5). The rate of change would be proportional to the acceleration of the toroid as it exceeds escape velocity. The more intense the gravity field is, the greater the energy required for the toroid to do work.

The earth might not be the best place to test this problem. The moon's escape velocity is 2,375 m/s. A smaller platform would be only 8 meters in diameter to determine the same effects if this experiment is performed on the moon. At that size the angular accelerations in the toroid would be about the same as the earth model. The velocities in the moon's vicinity are slower than those required in the earth model (see figure 5).

A moon experiment would pose unique problems such as energy supplies, lateral movement, site control and transport. This apparatus is more complex mechanically but does allow a freedom of movement not proposed for the earth experiment. To make the platform independent of a grounded site, a series of three stacked toroids could be assembled. The center toroid would be twice the mass of the toroids above and below and would rotate in the opposite direction of the other two. All three toroids would need to maintain the same velocities. Separate motors for each toroid at the quadrants would allow energy to be applied in equal and opposite directions to the toroids without imposing an angular force to the platform. The platform would remain stable while accelerating and decelerating the toroids. Since energy is more difficult to obtain on the moon, the toroids could be primed before transport. The rotations could be synchronized on earth with energy only required to maintain the set velocities.

The proposed experiment is based on a spherical gravity field. Classic Newtonian physics is applied because of relatively low velocities. The addition of lateral dimension is an extension of Newton's laws. The question remains

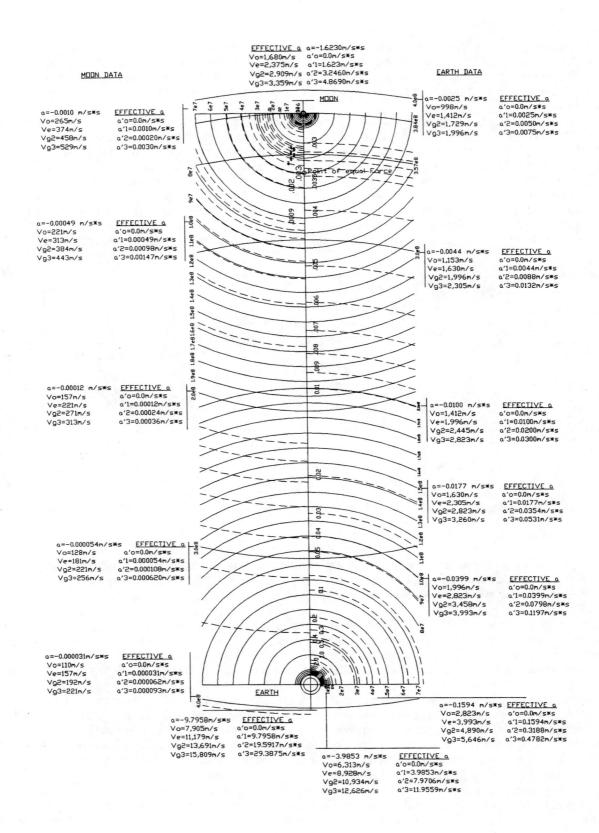

FIGUIRE 5. Earth and Moon fields.

as to the existence of a force opposite the force of gravity. This experiment proposes a different way of viewing a gravity field and would illustrate that any mass in an elliptical orbit will maintain radius distance in any lateral direction as long as it maintains its tangential velocity. Each particle in the toroid is independent and derives angular force from the toroid's shape. If the particle becomes unattached, it will observe a classic orbit. If that particle has a tangential velocity greater than orbital, it will increase in radius and decelerate. If the particle is below orbital velocity, it will decrease in radius and accelerate. This should remain true for particles in the toroid. If the toroid does not displace when the velocity exceeds orbital, the consequence is more than one velocity for the same radius in gravity field.

CONCLUSION

Unlike conventional systems, the toroid would contain the energy to perform work without starting from a zero state. No large fuel tanks with heavy fuel are needed since the energy to orbit the system is applied while at a zero state. The energy to move the system is the same in any direction.

If the experiment proves positive, a mechanism of multiple toroids rotating in opposite directions could be assembled for practical applications. Applications include the ability to create a stationary platform at any given point in a gravitational field, low acceleration and deceleration of a platform in atmospheres, extended acceleration and deceleration of a platform over large distances in space with use of electricity as an energy source.

ACKNOWLEDGMENTS

Man's search for knowledge in a curious and evolving environment.

REFERENCES

Considine, D.M. *Van Nostrands's Scientific Encyclopedia,* Litton Educational Publishing, Inc., 1976
Manual of Steel Construction, American Institute of Steel Construction, Inc., 1970

Recent Advances in Inertial-Electrostatic Confinement (IEC) Fusion for Space Power and Propulsion

J. Nadler[1a], G.H. Miley[1b], M. Coventry[1c] and H. Momota[2]

[1a-c]*NPL Associates, Champaign, Illinois 61821*
[2]*Fusion Studies Laboratory, University of Illinois, Urbana, IL 61801*
[1a]*nadrad@fgi.net;* [1b]*g-miley@uiuc.edu*

Abstract. Concept Studies have shown the IEC to be one of the most attractive approaches to fusion propulsion, provided the physics and technology involved can be scaled-up to high power levels. A key step involves development of a pulsed IEC that can obtain high ion currents along with good ion confinement. Results from initial pulsed IEC experiments are described here. The obtainment of a D-D fusion neutron yield of 8×10^8 n/s at a peak pulse current of 17 A demonstrates that the dynamic formation of the required accelerating fields with the IEC discharge is possible. A next step involves improvement in confinement.

INTRODUCION TO IEC-FUSION AND SPACE PROPULSION

Inertial-Electrostatic Confinement of a fusion plasma can occur when high energy ions converge in either a spherical, or cylindrical, geometry. The converging ions build up a positive space charge that attracts electrons; the resulting dual species plasma will oscillate, forming a series of "virtual" cathodes and anodes. The positive space charge attracts the electrons, and where the electrons slow and turn around the corresponding negative charge will attract ions. In this manner, a relatively high density of ions can be confined at non-Maxwellian energy distributions for the purpose of controlled nuclear fusion. The ions may be introduced via either the ionization of background neutral gas or the injection of ions from an external source. The electrons, conversely, can be supplied through ion collisions with the cathode, or (again) from the ionization of the background neutral gas. Usually, to initiate IEC operation, a highly transparent cathode-grid is placed inside and concentric to a larger, anode (e.g., a grounded vacuum vessel). This cathode will attract ions when a sufficiently high voltage is applied (~30 kV or higher).

The use of IEC fusion has been studied for space propulsion through a series of conceptual design studies by R. W. Bussard and L.W. Jameson (Bussard, 1993, 1994) and by G. Miley et al. (Miley 1995). These studies confirmed that an ultra-high specific impulse could be achieved because, unlike magnetic fusion devices, heavy magnets are not required in the IEC, and unlike inertial confinement fusion (ICF) devices, a heavy laser or accelerator system is avoided. Indeed, C. Williams and S. Borowski (Borowski, 1997) confirmed these advantages in a recent NASA study. The devices discussed in this paper are small-scale experimental devices that are employed to test the viability of the concept. Future devices, if employed for space propulsion and power, will be substantially larger, operating at much higher power levels.

Background of the Concept

The approach to fusion using IEC was originally conceived by P. Farnsworth (Farnsworth, 1962) (inventor of electronic television in the US), and later studied experimentally by Hirsch (Hirsch, 1967). However little was done to study this concept until R. W. Bussard (Bussard, 1991) and G. H. Miley (Miley, 1991) renewed studies in the early 1990s.

CP504, *Space Technology and Applications International Forum–2000*, edited by M. S. El-Genk
© 2000 American Institute of Physics 1-56396-919-X/00/$17.00

The basic IEC approach pursued here (spherical devices with grids operating in the star mode) was subsequently conceived by Miley, et al. (Miley, 1994, 1997) as part of an IEC-based fusion neutron source project at the U. of Illinois (UI). This configuration is now being commercialized by Daimler-Chrysler Aerospace Corp. (the first commercial application of a confined, fusing plasma!) (Sved 1997). This basic concept was significantly extended by the development of pulsed operation at high peak currents as described here.

DESIGN OF EXPERIMENTS

The IEC experimental reactor vessel uses a 61-cm spherical vacuum vessel with various 2.75-inch ports to accommodate power feed-through's and diagnostics. A high-voltage power feed-through for the cathode-grid is positioned on a larger 8-inch port since voltages on the order of 50-80 kV are employed. One, two, or sometimes three grids are used inside the vessel; the inner most grid is the main cathode; additional grids are used for controlling electron clouds for background gas ionization enhancement. A photograph of the IEC reactor in Star Mode operation with one grid is given in Fig. 1.

Prior operation of IEC experiments has concentrated on steady-state operation. To achieve the high currents required for future propulsion systems, pulsed operation appears desirable. Thus the experiments described here concentrated on the development and study of pulsed IEC discharges. A transmission line-type pulser was developed for these experiments, and is depicted in Fig. 2. It consists of a high-voltage DC power supply (0 – 5 kV) grounded at one end and connected in series with charging "choke" inductors (~8 H total). The chokes limit the rate at which the energy storage capacitors are charged. Two energy storage capacitors are then wired in parallel and placed in series with a 30-mH inductor and connected to a pulse transformer (1:10 step-up ratio). The opposite lead of the step-up transformer leads to a diode array, which prevents possible reflections from damaging either the pulser or the steady-state power supply. A hydrogen-thyratron switch is used to initiate the pulse and discharge the energy storage capacitors. Figure 3 shows the voltage and current output of a typical pulse, as recorded by an oscilloscope. The scaling is such that one division on the voltage trace is equal to 1-kV; one division on the current trace is 1 A. Note how the voltage drops from ground (which is at the top), while the current rises from the middle of the scope. The time scale is 50 µs per division, so the entire pulse lasted on the order of 100 µs.

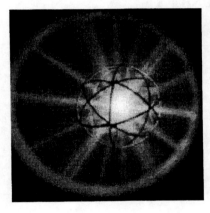

FIGURE 1. IEC Reactor Core in Star Mode Operation.

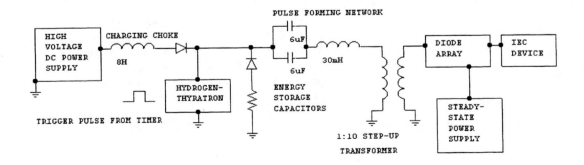

FIGURE 2. Electronic schematic of the pulser and IEC device.

In order to measure nuclear fusion reaction rates, a neutron diagnostic was developed to count the 2.54-MeV neutrons released following deuterium-deuterium (D-D) fusion. Use of conventional steady-state is prevented by their inherent "dead time. To overcome this problem, a silver-foil (Ag-foil) activation detector was designed and built. This detector employs several Geiger-Muller tubes, covered in thin silver foil and set at the center of a block of polyethylene. Fast neutrons first interact with the plastic, slowing down in energy so that they can be absorbed by the Ag-foil. The half-life of the activated Ag is on the

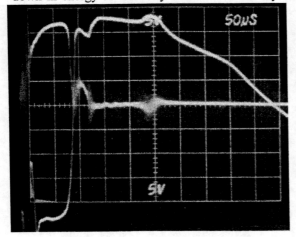

order of a half of a minute. Therefore, the IEC is operated in a continuous pulsed fashion at 10 Hz for several half-lives to "saturate" the activated Ag. Then the IEC is rapidly turned off and the gamma rays from decay of the activated Ag are counted. Absolute neutron yields (and hence, total fusion production rates) can be determined based on a calibration with a radioisotope (Pu-Be) neutron source.

FIGURE 3. Oscilloscope trace of typical IEC Reactor pulse.

RESULTS OF PULSING EXPERIMENTS

A key objective of these experiments was to demonstrate that dynamic evolution of accelerating fields to generate energetic ions required for fusion can be achieved in the pulse mode. Neutron production as a function of pulsed current for several different cathode-grid designs is presented in Fig. 4. Data was collected for pulsed operation for cathode voltages of 40 - 50 kV, and currents up to 17 A; a peak yield of 8 x 10^8 n/s was obtained at the maximum current.

The data in Fig. 4 was taken with two different grids. The first, a relatively small reference grid (called Grid A), was used for the first set of measurements; a second, larger grid was used for the remaining points in the figure (Grid B). Operation with the smaller Grid A produced a slight, but significantly lower fusion reaction rate compared to the larger Grid B. This can be explained by considering the background pressure during operation. The larger grid allows for a lower neutral background pressure in accordance with Paschen theory of breakdown. This lower pressure reduces charge exchange, and hence, allows for longer ion confinement times and higher reaction rates (see discussion below).

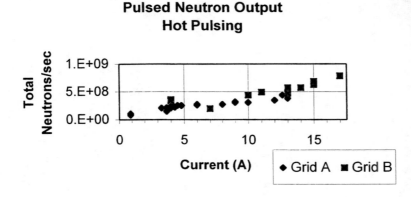

FIGURE 4. Pulsed Neutron Output at peak of pulse vs. Peak Cathode Current for IEC Reactor Operation.

Electron Emitter Source Design

Electron emitters were incorporated into the pulsed design to provide an enhanced ion source. They offer three key benefits: 1) ionization of ions around the vessel edge where the electric potential energy is the highest; 2) reduced background neutral gas pressure during IEC operation; and 3) removal of the need for operation of a pre-pulse steady-state discharge to create ions.

For the electron emitters to work most efficiently, an extraction grid with somewhat large holes (~1-cm^2 plus) is placed within a few centimeters of the emitter. The large holes allow electrons to circulate about the grid wires before actually coming in contact with the wires. Because the emitters are often negatively biased in order to give electrons higher initial kinetic energy, they must also be strategically placed in order to ionize the greatest amount of gas without disturbing the potential-well formation within the plasma core. In these experiments four were located symmetrically around the spherical chamber.

The scaling of the neutron yield with peak current is almost linear over the range covered from 7 to 17 A. The achievement of a four-fold increase in neutron yield over this range is viewed as a very positive result from this preliminary study of pulsed operation. The physics issue of concern in such operation is that electron currents generated during the voltage rise could cancel the local accelerating potential, preventing the generation of the high-energy ions required for fusion reactions. The detailed electron-ion dynamics involved n such pulses is complex. However, as illustrated from these results, approximate ion source generation via pre-pulse discharge generation or by electron-emitter assistance permits the desired evolution of accelerating fields in the plasma discharge.

The linear yield-current relationship observed is expected since the present high pressure operation results in a predominance of beam-background neutral fusion reactions (vs. beam-beam). Future experiments are planned to study methods to enhance ion confinement via virtual electrode production, moving reactions into a beam-beam regime.

THRUST GENERATION METHODS

A key issue for use of the IEC for propulsion (vs. electrical power) is the development of an efficient method for generating directed thrust. There are several different possible approaches, which require study before selecting an optimal approach. In one, used by Bussard (Bussard, 1993) in the QED conceptual design, employs conversion to electrical output, which drives electron beam emission from high voltage diodes. The resulting MeV-beams are used to vaporize liquid hydrogen, which is then expanded through a nozzle to provide the desired thrust. Miley et al. (Miley 1995) studied an alternate method where the electrical output was first conditioned and then used to drive an ion beam thruster. Yet another important category of techniques would employ a bending electrostatic or magnetic field to bunch the energetic fusion products and direct them to provide thrust.

The fusion product thruster results in an extremely high specific impulse. However, for some missions variable impulse-thrust is desired. For that purpose a plasma jet might be added to the IEC. Such a jet would add additional mass, hence thrust to the exhaust. Further mass could be added by mixing H_2 with the jet. Thus, the thrust could be varied over a wide range, using the fusion product exhaust as the "base" load, while varying the contribution of the plasma jet and varying the ratio of H_2 to fusion ions in it. Since the plasma jet is a basic component in this concept, experiments on jet formation have been performed as described next.

Under select conditions, high-density ion and electron beams will form in the IEC device, initiating the "star mode" of operation, as shown earlier in Fig. 1. In this operational mode, high-density space-charge-neutralized ion beams, termed "microchannels," are created that pass through the open spaces between the grid wires. These microchannels significantly reduce ion-grid bombardment and erosion and substantially increases the power efficiency. For conventional star mode operation the IEC grid is designed to be highly symmetric so that the microchannel beams are also symmetric, providing good convergence at the center of the grid. However, enlarging one of the grid openings distorts the potential surfaces, resulting in the creation of a very intense, tightly coupled space-charge-neutralized ion jet directed outward from the central core plasma region (see Fig. 5) (Gu, 1994). It is this mode that would be employed in combination with the directed fusion product jet for the proposed thruster.

A cylindrical "channel" grid with the same electrostatic potential as the central spherical grid could be added to create a passage through which ions can escape to the outside of the vessel. To ensure complete neutralization of the plasma jet, additional electron emitters are attached close to the jet discharge hole. In addition, there is the issue of redirecting the energetic fusion products such that they also flow parallel to the exhaust jet. For this purpose a series of magnetic coils or added grids would be employed to redirect the fusion products, e.g. the alphas and protons from D-^{3}He reactions.

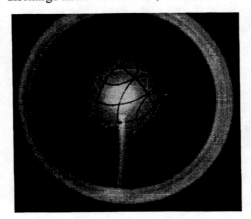

This would enable an extremely high specific impulse device, but for some missions a variable thrust-specific impulse could be added. Possibilities include directing the energetic ions into a travelling wave decelerator, directing them into a flowing plasma, or adding a gas such as H_2 into the exhaust plasma jet, there by increasing the momentum of the exhaust.

FIGURE 5. Jet operational mode in experimental IEC device.

FUTURE WORK

In order to achieve a higher Q in the IEC reactor, the ion confinement time must be further increased. To increase ion confinement, improved potential well trapping is required. This in turn requires reduction of fast ion charge exchange. The most significant reduction in charge exchange will occur with a several orders of magnitude reduction in background pressure. Several operating parameters and designs of the experiment can be adjusted to effect this change: 1) Improved vacuum in experimental chamber; 2) Differentially-pumped ion sources; and 3) Operation at a higher cathode potential.

If the background pressure is reduced a means of generating ions (other than the collisional ionization) must be introduced, e.g., the use of ion sources mounted on the edge of the reactor vessel. Analysis of past results with electron emitters and ion guns indicate that this type of source could be made successful if differential pumping ports were added to the "barrel" of the ion guns. This additional pumping would reduce the leakage of background gases from the ion sources. This would be a first step towards a higher power IEC experiment. Additional steps include addition of actively cooled grids, improved pulsing circuits, pulsed ion sources, thrust coupling and direct energy conversion.

SUMMARY

Various conceptual design and mission studies have confirmed the attractiveness of the IEC for future fusion propulsion. Due to the simplicity of the IEC a very high specific power is possible. The scale up of present laboratory type IEC experiments to a propulsion-level source will require high ion currents, best achieved by pulsed-operation. The preliminary study of such pulsed operation described here was designed to demonstrate suitable dynamic acceleration of ions to energy level required for fusion. Experimental results were quite positive with fusion rates corresponding to 8 x 10^8 n/s at 17 A being obtained in 10 Hz pulsing. A next step involves modification of the experimental design to improve ion confinement, hence the fusion gain.

ACKNOWLEDGMENTS

The authors wish to recognize the encouragement and technical discussions with Drs. George Schmidt, Francis Thio, and Ivana Hrbud of NASA's Marshall Space Flight Center, and of Mr. Mike Williams of the Fusion Studies Laboratory, University of Illinois, in the completion of this work. This work was supported under a Phase I Small Business Innovative Research Contract with NASA's Marshall Space Flight Center, Contract number NAS8-9904.

REFERENCES

Bussard, R. W., *Fusion Technology*, Vol. 19, No. 2, p. 273-293 (1991).

Bussard, R.W., "The QED Engine System: Direct-Electric Fusion-Powered Rocket Propulsion Systems," *Proceedings of the 10th Symposium on Space Nuclear Power and Propulsion*, American Institute of Physics, Albuquerque, NM, 1993, pp. 1601-1611.

Bussard, R.W. and L.W. Jameson, AIP Conf. Proc. 301: *11th Symp. On Space Nucl. Power and Propulsion*, eds. M.S. El-Genk and M.D. Hoover, Conf. 940101, AIP Press, 1994, pp. 1289-1296 (1994).

Gu, Y.B., Nadler, J.H, Miley, G.H., et al."Physics and Effects of Grid-Electric Field Perturbation on Spherical Electrostatic-Inertial Confinement Fusion," *21st EPS Conference on Controlled Fusion and Plasma Physics* eds. Joffrin, E., Platz, P. and Stott, P.E., Montpellier, France, 436, (1994).

P.T. Farnsworth, "Electric Discharge Device for Producing Interactions Between Nuclei," U.S. Patent No. 3,358,402, issued June 28, 1966, initially filed May 5, 1956, rev. Oct. 18, 1960, filed Jan. 11, 1962.

R. Hirsch "Inertial-Electrostatic Confinement of Ionized Fusion Gasses," *J. Appl. Phys.*, 38, 11, 4522 (1967).

Miley, G.H., "Dense Core Plasma in an Inertial-Electrostatic Confinement Device," *1991 U.S.-Japan Workshop on Nuclear Fusion in Dense Plasmas*, Austin, TX (1991).

Miley, G.H., Nadler, J., et al., "An Inertial Electrostatic Confinement Neutron/Proton Source," *Third International Conference on Dense Z Pinches*, eds. Haines, Malcolm, and Knight, Andrew, AIP Conference Proceedings 299, AIP Press, 675-689 (1994).

Miley, G. H., et al., "Innovative Technology for an Inertial Electrostatic Confinement Fusion Propulsion Unit," Fusion Energy in Space Propulsion, (T. Kammash, ed.) Vol. 167, *Progress in Astronautics and Aeronautics*, American Institute of Aeronautics and Astronautics Press, Washington DC, 1995, pp. 161-178.

Miley, G.H., Nadler, J, et al., "Discharge Characteristics of the Spherical Inertial Electrostatic Confinement (IEC) Device," *IEEE Transactions of Plasma Science* (Oct. 1997).

Sved, J., "The Commercial IEC Portable Neutron Source," *Trans. of the ANS*, 77, 504, (1997).

Williams, C.H. and S. K. Borowski, "An Assessment of Fusion Space Propulsion Concepts and Desired Operating Parameters for Fast Solar System Travel," *33rd AIAA/ASME/SAE/ASEE Joint Propulsion Conference & Exhibit*, Seattle, WA, 1997.

Development of High-Capacity Antimatter Storage

Steven D. Howe and Gerald A. Smith

Synergistic Technologies, Inc., 19 Karen Circle, Los Alamos, NM 87544
505-672-0676, steve@synergistictech.com

ABSTRACT Space is vast. Over the next few decades, humanity will strive to send probes farther and farther into space to establish long baselines for interferometry, to visit the Kuiper Belt, to identify the heliopause, or to map the Oort cloud. In order to solve many of the mysteries of the universe or to explore the solar system and beyond, one single technology must be developed -- high performance propulsion. In essence, future missions to deep space will require specific impulses between 50,000 and 200,000 seconds and energy densities greater than 10^{14} j/kg in order to accomplish the mission within the career lifetime of an individual, 40 years. Only two technologies available to mankind offer such performance -- fusion and antimatter. Currently envisioned fusion systems are too massive. Alternatively, because of the high energy density, antimatter powered systems may be relatively compact. The single key technology that is required to enable the revolutionary concept of antimatter propulsion is safe, reliable, high-density storage. Under a grant from the NASA Institute of Advanced Concepts, we have identified two potential mechanisms that may enable high capacity antimatter storage systems to be built. We will describe planned experiments to verify the concepts. Development of a system capable of storing megajoules per gram will allow highly instrumented platforms to make fast missions to great distances. Such a development will open the universe to humanity.

INTRODUCTION

Perhaps for the first time in history, humanity's exploration of the physical frontier is limited by our state of technology. Throughout history, human exploration has utilized whatever was available to push the frontiers of habitation Over the past 30 years, human exploration of space has for all practical purposes stopped at the edge of space -- low earth orbit. Consequently, humanity has continued the exploration of space using unmanned, instrumented platforms. These small craft have performed remarkably well and provided a rich storehouse of discoveries and scientific data to the world. Looking at the entire wealth of this data, however, one conclusion seems clear -- every time we send out a new probe into unknown territory, we discover far different conditions than we expected to find. In addition, detailed telescopic observations now indicate that the interstellar neighborhood around our solar system is far from empty and is, in fact, a complex time-varying region awash with dust storms, intense radiation, complex molecules, and primordial lumps of matter of all sizes. Consequently, the importance of accurately exploring the region of space outside the solar system has greatly increased in priority.

Studying the heliopause where the interstellar wind meets the solar wind, the Kuiper belt where many comets are born, and even the Oort cloud where the residue of the original solar system still resides, has become important to understanding the dynamics of our solar system -- even to understanding the fate of our race. The possibility exists that large-scale future events, such as an increase in local interstellar gas density caused by an approaching dust cloud, may impact human existence. The change in gas density may alter galactic cosmic radiation levels reaching the Earth or cause perturbations in the cometary clouds that might eventually lead to fragments diving down into the solar system, i.e. toward Earth. Such possibilities necessitate an active exploration of deep space to understand just what is out there. Unfortunately, we do not as yet have the ability to mount such an exploration.

To answer many of the major questions in astrophysics, instrumented probes will have to be sent deeper into space, carry more instruments, and make faster trips-- in essence they must achieve much higher velocities than currently deliverable with today's technology. Missions to the Kuiper Belt, the heliopause, or the Oort cloud will require a revolutionary advance in propulsion to achieve their goals within reasonable time frames. In essence, future missions to deep space will require specific impulses between 50,000 and 200,000 seconds and specific masses of .1 to 1 kg/kw to be considered.

CP504, *Space Technology and Applications International Forum–2000*, edited by M. S. El-Genk
© 2000 American Institute of Physics 1-56396-919-X/00/$17.00

Antimatter has the highest specific energy of any source known to man. At nearly 10^{17} J/kg, it is three orders of magnitude larger than nuclear fission and fusion and ten orders of magnitude larger than chemical reactions. Currently, antiprotons are produced and stored in small quantities (10^8) in Penning Traps. Penning Traps are a mature technology but are limited in the particle density, thus the energy density, which they can contain. Antimatter must be stored in much higher densities to be applicable for missions into the outer realms of our solar system.

We intend to investigate two concepts to increase the storage capacity of antimatter beyond that of Penning Traps. The first method entails the use of a electromagnetic trap that is coated with either solid hydrogen or solid ammonia to reduce annihilation at the walls of the trap. The other concept relies on storage of antihydrogen. Similarly, if the concept of quantum reflection proves viable for antimatter, a coating of liquid helium may reduce the annihilation rate of the antihydrogen on the walls of the trap. Both of these concepts can be demonstrated with simple experiments.

MISSION REQUIREMENTS

In order to reach destinations of a few hundred astronomical units (AU) in a few decades, velocities of 100s to 1000s of km/s must be achieved. The kinetic energy of every kilogram of mass that acquires that velocity at the end of the propulsion phase will be significant. Table I shows some the missions currently being discussed by NASA as part of the planning of future space science programs. The Table also depicts the "characteristic" velocity that must be given to the platform to achieve the mission. This velocity may be equivalent to the average velocity for a mission in which the propulsion phase is the entire mission. The kinetic energy of each kilogram of mass is then shown. In order to produce these levels of kinetic energy, the potential-energy density of the entire ship will have to be greater than these levels at the beginning of the mission. Therefore, the energy density of the "fuel" will have to be substantially greater than these levels to account for the payload, structural, and engine masses. Energy densities of known sources are shown in Table II.

In addition, current estimates of specific mass and specific impulse for representative systems utilizing these sources are also shown. Comparing Table II and Table I, nuclear electric systems might be able to accomplish the 250 AU mission but not much further. The only systems that use on-board energy sources that can go into deep space are fusion or antimatter.

TABLE I. Velocities required for deep space missions

Mission	Velocity (km/s)	Energy Density (J/kg)
250 AU in 10 yrs	60	1.8e09
10,000 AU in 40 yrs	1200	7.2e11
_Centauri in 40 yrs	30,000	4.5e14

TABLE II. Energy Densities of Known Sources

Reaction	specific energy (J/kg)	specific mass (kg/kw)	specific impulse (s)
Chemical	1.5e07	?	470
Fission (100%)	7.1e13	35	5,000-10,000
Fusion (100%)	7.5e14	1	40k-60k
Antimatter	9.0e16	.01-.1	40k-100k

Typically, fusion systems are envisioned to be large, massive and complicated. Thus, the overall specific energy will be below the levels necessary for interstellar precursor missions. Alternatively, antimatter is considered to be expensive and difficult to produce. A recent investigation [Schmidt,1999] at the NASA Marshall Space Flight Center (MSFC) indicates that sufficient levels of production could be present within the next few decades. The issues of production and conversion into thrust are also currently being pursued by Synergistic Technologies under SBIR and STTR grants. At this point, the other major issue that is required to make future deep space missions possible is the need for high-density storage if the required specific energy levels are to be achieved.

ANTIPROTON STORAGE -- STATUS

Currently, antiprotons are produced at several high energy accelerator facilities around the world. One such facility in Europe, CERN, captures and decelerates the antiprotons down to energies where they can be injected into long term storage devices called Penning Traps [Holzscheiter,1996]. Over the past tens years or so, great success has been achieved in collecting and holding antiprotons in Penning Traps. This technology now appears to be capable of storing antiprotons in densities up to 10^{11} per cm^3. With funding from the Jet Propulsion Laboratory, Dr. G. A. Smith and colleagues at Pennsylvania State University have built a portable Penning Trap that can contain 10^8 antiprotons. The 1/e storage time for the trap is about one week. A second generation trap [Smith,1998], the High Performance Antimatter Trap (HiPAT), is under development at the NASA MSFC that has a design goal of holding 10^{12} antiprotons with a 1/e time of a few weeks (Figure 1). By using these traps as a source of low energy antiprotons, we intend to experimentally investigate other concepts that may store antimatter at densities 1000 times higher. We foresee in the next ten years great progress toward the confinement of antiprotons at much higher densities. This goal will be reached by a carefully planned and executed series of experiments.

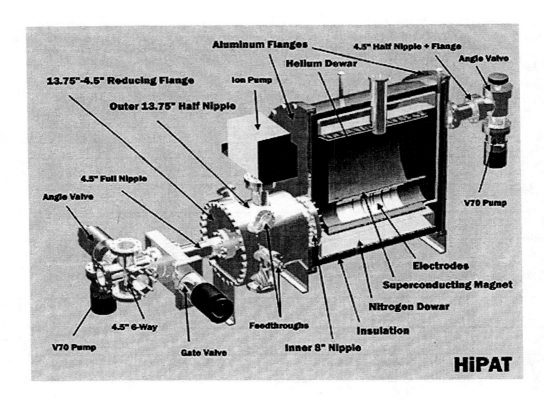

FIGURE 1. Design of a high capacity antimatter trap (courtesy of NASA MSFC [Smith,1998].

HIGH CAPACITY STORAGE OF ANTIPROTONS

The limiting factor in a non-neutral plasma such as that held in a Penning Trap is the Brillouin Limit. This is the density (about 100 billion antiprotons/cc for a 5T magnetic field) at which the stored space charge forces overcome the confining magnetic forces. Any truly significant steps beyond this limit would have to come from other physical phenomena. However, if a mechanism can be found that would reduce the losses of antiprotons as they hit the walls of the containers, then the capacity of traps could be increased substantially even though the Brillouin Limit is not exceeded.

Parelectricity

Recent publications predict the possibility of a phenomenon called Parelectricity [Chaio,1994]. Parelectricity predicts that illumination of a layer of solid ammonia by 9.56 micron microwaves will create a population inversion in the material. One result of this inversion is that an image charge will be induced by an approaching charged particle that is of the same sign. Thus, the approaching particle will be repelled by the image charge. A thin layer of ammonia on the surfaces of the trap may reflect any approaching charged particle with near unity probability. Therefore, the original idea of making the walls of the trap reflective is still valid — just the methodology has changed.

It appears rather straightforward to experimentally check the validity of these considerations. Cool antiprotons could be injected from the MSFC HiPAT presently under construction, or the Penn State/JPL Mark I trap, into a container which is laser cooled to a temperature below 0.77 K for the hydrogen reflection concept. Similarly, the cavity can be illuminated with a microwave source to confirm Parelectricity. Measurements of reflection coefficients on the walls, and overall lifetimes of antiprotons, could be done fairly simply once the containment vessel is built, "iced", and magnetically or laser excited.

Antihydrogen

We have examined another very promising advanced technology, i.e. synthesis and storage of atomic antihydrogen in Ioffe-Pritchard traps, which is capable of achieving very high density storage (10^{14}-10^{17}/cc). This technology is currently being developed at the Antiproton Decelerator facility at CERN, in Geneva, Switzerland, by the ATHENA experiment [ATHENA,1998], of which one of us (G.A. Smith) is a collaborating member.

The major advantage of this technique is that one is working with electrically neutral atoms, so all space charge problems are eliminated. Furthermore, the magnetic properties of the atoms make them controllable under the influence of external electromagnetic fields. In the ATHENA trap antiprotons and positrons (antielectrons) are made to cohabitate a common volume using nested double-well potential barriers.

The ATHENA experiment at CERN will attempt to synthesize atomic antihydrogen next year. Very dilute mixtures of antiprotons and positrons (10^7 each) will be synthesized into antihydrogen atoms at 4K and below by spontaneous radiative recombination at densities of roughly 10^7/cc. Since the rate for spontaneous radiative recombination scales as $1/T^{.5}$, every attempt will be made to achieve sub-K temperatures in this experiment. Nonetheless, at about 1K, the expected recombination rate is 9,000 per second. The atoms are to be confined in an Ioffe-Pritchard trap which in the case of hydrogen has confinement densities approaching 1 microgram/cc for minutes.

Once formed, the antiatoms are confined in an Ioffe-Pritchard trap, a technology which has been used successfully for many years to confine hydrogen atoms at high densities for fundamental physics measurements. A gradient magnetic field is provided by current-carrying quadrupole coils. A force due to the gradient magnetic field is imposed on the magnetic moment of the atom.

Since the moment results from the combined spins of the constituents, and these in turn follow the rules of quantum mechanics, there are four hyperfine spin states possible. The low field seeking state, with the positron and antiproton spins parallel and pointing up, feels an attractive force into the center of the trap, and is confined. The state with the positron and antiproton spins antiparallel (high field seeking) is expelled from the trap.

By spontaneous radiative recombination, the particles bind to form an antihydrogen atom, which has all the properties of the hydrogen atom apart from the opposite charges of the constituents. The recombination rate is estimated by ATHENA at 9,000 per second at a temperature of 1 K. Since the recombination rate scales as $(1/T)^{.5}$ the rate could be increased one thousand-fold by laser cooling the atoms, for example, to a temperature of 10^{-6} K. At this rate, one could form 10^{14} antiatoms in about 10^7 seconds, or 4 months.

As with all electromagnetically-confined systems, ultimately regions of instability are found. In the case of the Ioffe-Pritchard traps, when confined atomic densities increase, interatomic scattering becomes important, resulting in electron (positron) spin flips, taking the atom from a low field seeking state to a high field seeking state. The newly created high field seeker then jumps out of the trap, lost forever. At densities approaching 10^{14}/cc, lifetimes have been reduced to minutes, which is obviously unacceptable for long-term space propulsion applications.

Quantum Reflection

Substantial work has been performed in the past few years on creating ultra cold neutral atoms. Laser cooling of cesium atoms has been achieved down to temperatures well below a meV. The DeBroglie wavelengths at these temperatures is hundreds of angstroms. Thus, the neutral atom interaction with the surface atoms in a wall will be a complex many body interaction.

This interaction was first examined in 1936 [Lennard-Jones,1936]. They predicted at the time the possibility of "quantum reflection." Quantum reflection predicts that in the limit as the energy of a particle approaching the wall of a container gets near zero, the probability of "sticking" to the wall approaches zero. Classically, the sticking probability, S, is predicted to approach unity but quantum mechanics predicts just the opposite solution, i.e that S is proportional to the square root of the particle's energy. Measurements [Yu,1993] have been made for atomic hydrogen at millikelvin temperatures that support the quantum reflection theory [Carraro,1998]. Whether ultracold antihydrogen atoms will reflect from a wall is an open question but could dramatically decrease loss rates in a trap. In addition, we have pursued the idea of creating a two-dimensional Bose-Einstein Condensate [Fried,1998; Safanov,1998] on the surface of a liquid helium layer to enhance the quantum reflection mechanism. Whether quantum reflection will occur for an antihydrogen atom is not clear.

PROPOSED EXPERIMENTS

Synergistic Technologies has completed the design of a solid-state degrader system that will accept a high energy antiproton beam and output a low-energy beam. Computational simulation of a variety of configurations of the device over a range of incident energies shows that production efficiencies as high as 10^{-5} can be expected. Current accelerator technology could produce a low energy beam with near 100% efficiency but would require $10M and five years to build. The degrader system will allow a source of trappable antiprotons to be available within two years. The objectives of our current Phase II SBIR are to construct the degrader system matched to the beam conditions existing at the Fermi National Accelerator Laboratory and to confirm and verify performance and operations. Based on the operational conditions at FNAL, we expect to provide around 1.7×10^6 antiprotons per hour for research directly or for injection into portable Penning Traps for off-site research. By the end of the project, we intend to provide a potentially commercial source of low energy antiprotons in portable traps to the research community.

Using the degrader system and the HiPAT from NASA MSFC, we propose to perform proof-of-concept experiments of the Parelectricity and Quantum Reflection concepts. The HiPAT will be used to produce a low temperature source of antiprotons or antihydrogen atoms. The particles will be transferred into a test cavity which has a thin layer of either solid ammonia or liquid helium covering the walls. Lifetimes and annihilation location will easily reveal if the particles are being reflected or are annihilating with the wall material.

SUMMARY

We have identified two mechanisms that may enable high capacity storage of antimatter. The first indicates that if antiprotons are kept in a container lined with frozen ammonia that is excited with an imposed microwave field, the particles may be repelled by induced parelectric repulsion. Similarly, using a concept theorized decades ago, the possibility of quantum reflection may reduce the probability of annihilation of antihydrogen atoms on the surface of any container.

In both approaches, we have adopted a slightly altered strategy. Originally, our goal has been to find methods to increase the particle density in order to increase storage. This was the motivation in examining the storage of antihydrogen. However, the idea of developing a reflecting wall also has tremendous potential. By removing the possibility of loss of the antiparticles by wall collisions, we may be able to make the entire volume of the traps into active storage region. This allows a tremendous increase in storage capacity of the Penning Trap even though the particle density has not changed over current levels. Both of these concepts could enable systems with ultra-high energy density to be developed. Proof of concept experiments have been designed and may be completed within the next few years.

ACKNOWLEDGMENTS

The authors would like to acknowledge the NASA Institute of Advanced Concepts (NIAC), directed by Dr. Robert Cassanova, without whose support this project would not have been possible.

REFERENCES

ATHENA (AnTiHydrogEN Apparatus), CERN proposal SPSLC 96-48/P303

Carraro, C. and M. Cole, "Sticking Coefficient at Ultralow Energy: Quantum Reflection," Prog. Surface Sci. 57, 61 (1998).

Chiao, R.Y. and J. Boyce, "Superluminarity, Parelectricity and Earnshaw's Theorem in Media with Inverted Populations," Phys. Rev. Lett. 73, 3383 (1994).

Fried, D.G. et al., "Bose-Einstein Condensation of Atomic Hydrogen," Phys. Rev. Lett. 81, 3811 (1998).

Holzscheiter, M., et al., "Are Antiprotons Forever?", Physics Letters A214, 279, 1996.

Lennard-Jones, J.E. and A.F. Devonshire, Proc. R. Soc. London A156, 6 (1936).

Safanov, A.I. et al., "Observation of Quasicondensate in Two-dimensional Atomic Hydrogen," Phys. Rev. Lett. 81, 4545 (1998).

Schmidt, G.R., H.P Gerrish, J.J. Martin, G.A. Smith, and K.J. Meyer, "Antimatter Production for Near-Term Propulsion Applications," PSU LEPS 99/03, Jour. Propulsion and Power (to be published).

Smith, G.A. and Meyer, K.J., "Preliminary Design for the High Performance Antimatter Trap (HiPAT)," NASA Marshall Space Flight Center, 1998.

Yu, Ite A. et al., "Evidence for Universal Quantum Reflection of Hydrogen from Liquid He4," Phys. Rev. Lett. 71, 1589 (1993).

External Pulsed Plasma Propulsion
And its Potential for the Near Future

J. A. Bonometti, P. J. Morton and G. R. Schmidt

NASA MSFC, TD40, Marshall Space Flight Center, Alabama, 35812

Joe.Bonometti@msfc.nasa.gov/ (256) 544-4019 / Fax: (256) 544-5926
Phillip.Morton@msfc.nasa.gov/ (256) 544-4613 / Fax: (256) 544-5926
George.Schmidt@msfc.nasa.gov/ (256) 544-6055 / Fax: (256) 544-5926

Abstract. This paper examines External Pulsed Plasma Propulsion (EPPP), a propulsion concept that derives its thrust from plasma waves generated from a series of small, supercritical fission/fusion pulses behind an object in space. For spacecraft applications, a momentum transfer mechanism translates the intense plasma wave energy into a vehicle acceleration that is tolerable to the rest of the spacecraft and its crew. This propulsion concept offers extremely high performance in terms of both specific impulse (Isp) and thrust-to-weight ratio, something that other concepts based on available technology cannot do. The political concerns that suspended work on this type of system (i.e. termination of Project ORION) may now not be as insurmountable as they were in 1965. The appeal of EPPP stems from its relatively low cost and reusability, fast interplanetary transit times, safety and reliability, and independence from major technological breakthroughs. In fact, a first generation EPPP system based on modern-day technology (i.e., GABRIEL - an evolutionary framework of EPPP concepts) may very well be the only form of propulsion that could realistically be developed to perform ambitious human exploration beyond Mars in the 21st century. It could also provide the most effective approach for deterrence against collision between earth and small planetary objects - a growing concern over recent years.

INTRODUCTION

NASA is currently conducting research on advanced propulsion technologies capable of supporting ambitious human exploration of the solar system in the early part of the next century. Most research to date has been geared towards concepts that offer tremendous performance improvements over current systems. The only problem is that virtually all of these technologies, such as fusion, antimatter and beamed-energy sails, have fundamental scientific issues and practical weaknesses that must be resolved before they can be seriously considered for actual applications. For instance, fusion is limited by the fact that we are still far away from demonstrating a device having energy gains sufficient for commercial power, let alone space applications. Antimatter has much appeal because of its high energy density, but it is severely hampered by extremely low propulsion efficiencies and high costs of current production methods. Beamed energy offers great potential too, but requires materials far beyond current state-of-the-art and tremendous investment in ground/space-based power beaming infrastructure.

Although we are optimistic that some of these issues will eventually be overcome, there is no guarantee that any of these technologies will be available by the first half of the next century. This state-of-affairs points to the disappointing fact that none of the advanced, high-power density propulsion concepts being considered by NASA could, with any degree of certainty, meet the goals and timetables of NASA's own Strategic Plan. This is especially true in light of the conservative fiscal environment of the post-Apollo era, which could limit the sizable investment needed to resolve the fundamental issues associated with these concepts. Moreover, the cost for developing actual vehicles based on these technologies and their required infrastructure could realistically be on the order of hundreds of billions of dollars.

To obtain a quantum jump in propulsive capability by the early part of the next century, we must have safe, affordable systems with very high-power densities. Precedents suggest that any device engineered within the next 30 to 50 years should be based on the well-understood physics of today. The need for high power densities eliminates all but nuclear energy sources. The emphasis on known physics and affordability limits the scope still further to fission processes. Of the fission-based concepts that have been considered in the past (e.g., solid-core nuclear thermal, gas-core, internal and external nuclear pulse), only external nuclear pulse circumvents the Isp constraints imposed by containment of a heated gas, and provides the very high power densities needed for ambitious space transportation.

CP504, *Space Technology and Applications International Forum–2000*, edited by M. S. El-Genk
2000 American Institute of Physics 1-56396-919-X

In the past, both internal and external pulse-engine concepts have been considered. Comparisons between these two approaches pointed to external pulse as the best candidate mainly because of its higher temperature limits and lower inert mass (Martin and Bond, 1979, Nance 1965). In addition, several researchers have investigated various forms of external momentum coupling. The most prominent examples are the standard pusher plate (Reynolds, 1972), the large lightweight sail/spinnaker (Solem, 1993), the rotating cable pusher (Cotter, 1971), and the combined pusher plate/magnetic field (Martin and Bond, 1979).

The most familiar effort in the area of external pulse-engines was Project ORION, which took place between 1958 and 1965. The Air Force spent approximately 8 million dollars on the program over its first 6 years (Prater, 1963). ORION, which was classified throughout most of its brief lifetime, engaged an impressive group of physicists and engineers who carried out numerous studies and tests on most aspects of the vehicle. The basic ORION design is shown in Figure 1. The proposed ships were large (from 10 to 30 meters in diameter) since performance tended to increase with diameter of the ship's pusher plate. This was due to the higher specific yields (i.e., burn up fractions) of larger pulse units, and the wider propellant interception angles at the minimum standoff distances allowed by material strength considerations. NASA funded several additional studies until 1965 when the entire effort was terminated - primarily for political reasons. The extensive analyses and experiments performed for ORION and subsequent studies indicate that spacecraft with high thrusts (~1 to 10 g accelerations) and high Isp's (~10,000 sec) could be built, even with 1960's materials technology.

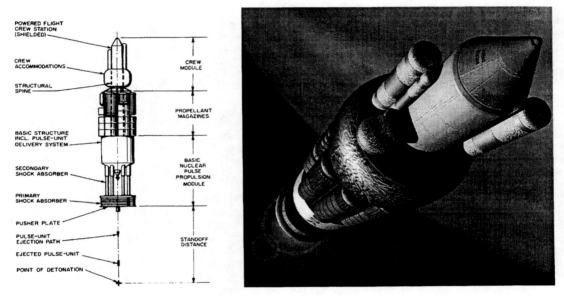

FIGURE 1. 1960 ORION concept.

CONCEPT OVERVIEW

At first glance, a nuclear pulse rocket appears to be quite radical, although it is conceptually very simple. Thrust is produced by ejecting and detonating small, fission-driven, pulse units at the aft end of the vehicle. This "external" engine operation, where the fission process is unconfined by material walls, is relatively independent of the reaction rate, temperature, pressure and other characteristics of the fuel. In practice, the system must be operated in a pulsed mode to allow the transfer of energy into a practical acceleration of the ship, which is limited by human and equipment tolerances. The physics behind creating a highly efficient fission burst is well understood, and in a vacuum, it produces a shell of ionized particles with an extremely high radial velocity. Thus, this concept of "riding on a plasma wave" is appropriately termed External Pulsed Plasma Propulsion or EPPP.

Key to EPPP's extraordinary performance are the facts that: (1) common materials can withstand an intense nuclear environment for very brief periods of time (i.e., nanoseconds), and (2) nuclear detonations are not only well understood, but also come much closer to achieving the maximum power density available from the fission process. Also, high thrust over a relatively short time imparts nearly optimum impulse to the vehicle for fast, efficient trajectories. In sharp contrast to the original ORION approach, recent analyses based on present-day considerations and technologies (e.g., dedicated in-space operation, low-energy pulse unit yields, low-ablation pusher plate materials) indicate that the performance advantages of EPPP could be applied to relatively small vehicles. If this is the case, then it is possible to develop small spacecraft that could carry human crews between Earth and Mars in just 1 to 3 months, as opposed to 6 to 12 months with chemical or nuclear thermal propulsion technology. In addition, EPPP would permit much more flexible return windows and eliminate the need for long stay times in the vicinity of Mars. Most importantly, EPPP provides a technology path leading to much higher Isp's (~100,000 sec) using larger vehicles and more energetic detonations (e.g., fission/fusion and fusion) which could ultimately be used to open up the entire solar system to human exploration.

The main objection to EPPP has been the concern over nuclear contamination. Since modern-day practices would assuredly limit this concept strictly to space, radioactive contamination may not be as serious of issue as with ORION. Furthermore, the harsh environment of space has far more background radiation (particularly in the form of harmful gamma rays) than that produced by very small pulse units. Within 24 hours, the pulse unit's ionized mass dissipates completely into the background of the nominal space plasma density. Depending on the pulse unit efficiency, the exhaust velocities of the radioactive particles could exceed solar escape velocity (certainly beyond that of earth escape). Thus, there is no residue or permanent contamination to the environment beyond the natural sun's radiation.

Application #1: Human Interplanetary Exploration

There are two reasons for seriously considering EPPP as an option for future development. The first is its potential for human exploration. Since the early years of the space program, most human exploration studies have concentrated on either the Moon or Mars. Although it is recognized in NASA's Strategic Vision that the ultimate goal is to extend human presence throughout the solar system and eventually the stars, only a negligible amount of effort has been devoted to these type of missions. EPPP provides a technology that would allow us to seriously consider missions to the outer planets. It would also enable dramatically shorter trip times to Mars and other nearer-term destinations.

The propulsion concepts that have been traditionally considered for Mars missions are chemical propulsion based on O2/H2 combustion and solid-core nuclear thermal propulsion. Although the Isp of nuclear thermal (~900 sec) is approximately twice that of chemical (~450 sec), both systems suffer from the same limitations with regards to trip time and mission planning. The main advantage of nuclear thermal is its potential to reduce vehicle mass in low-earth orbit, thus reducing the number of heavy-lift vehicle launches.

The performance that characterizes these two concepts favors Hohmann-type transfers into very slow heliocentric orbital trajectories. This narrows the available trajectories for return and necessitates long stays on the Mars surface while awaiting favorable return windows. This leaves the crew and equipment exposed to an extremely hostile environment for long periods of time - nominally 560 days surface stays with 170 to 200 day transit times (Kos, 1998). Cost is also significant, since earth launches are about half the mission budget in most conventional scenarios. Longer missions translate to larger payloads and more expendables, both of which increase launch requirements.

EPPP can solve this problem with its much higher Isp (5,000 to 10,000 seconds), while still providing the high-thrust needed for fast orbit transfers. The result is higher energy transfer orbits, which could greatly reduce not only transit time, but permits broader return windows. This provides much more flexibility in mission planning and would not constrain the crew to long stay times on the Martian surface. It would also reduce the crew's exposure to the highly radioactive space environment and long periods of weightlessness.

Application #2: Comet/Asteroid Deflection

The other and perhaps most compelling application for EPPP is its use in asteroid or comet defense. Collisions between the Earth and small planetary objects occur frequently, with the typical result being that the objects burn up in the atmosphere. However, there is a low, but not negligible, probability of a collision with objects of sufficient size to cause catastrophic damage or an extinction-scale event. Good risk management would dictate that some effort be placed on devising countermeasures, if possible. Past studies identified a number of possibilities, almost all of which entailed ground and space-based infrastructure more extensive than that envisioned for ballistic missile defense. Because of the limitations of current propulsion technology, these systems would require permanent deployment of interceptors in deep space in order to allow engagement at a sufficient distance from Earth. In addition, the low-impulse methods of altering the object's trajectory, such as sails or electric thrusters, would probably not provide enough time for adequate trajectory alteration between detection and impact - especially in the case of a comet.

EPPP could be applied to the development of a much less expensive, purely ground-based deterrence system. If a likely catastrophic collision were identified, an EPPP-propelled interceptor could be launched into space using a conventional chemical launcher. It would have the power density necessary to rapidly travel to the target in time to force the threatening object from its collision course. The object's course change might be performed using sails or electric thrusters. However, these schemes are very risky since their effectiveness depends on the body's size, shape, speed, trajectory and many other properties. There is little room for error once the target is engaged, and the propulsion systems must operate reliably for very long durations to effect the change.

Alternatively, the same EPPP system that propelled the interceptor could be used to move the target. Single or successive pulse detonations at a predetermined distance from the asteroid's surface could be used to easily "nudge" the planetesimal and alter its course. The first wave of X-rays from the pulse would illuminate the planetesimal's surface causing ablation and thrust parallel to the object's projected area. The second wave of pulse fission products would produce another impulse in the same direction.

This approach has important advantages. It does not require asteroid capture or attachment of a propulsion unit to a highly variable surface. Since the "thrust" is parallel to the object's projected area, this approach is independent of the object's relatively indeterminate mass distribution and angular momentum. Also, the amount of impulse delivered can be easily tailored to any asteroid by the number of pulses, detonation standoff distance, and type of pulse unit.

DESIGN CONCEPTS UNDER STUDY

The realistic maximum Isp obtainable with fission-based EPPP is ~100,000 seconds. However, this type of performance would only be possible with very large spacecraft. Such vehicles would be impractical until the cost of access to space dropped substantially or in-space manufacturing became available. Therefore, a more conservative approach has been taken by considering smaller vehicles with lower performance (Isp   10,000 seconds) using technology available in the near-term. This concept has been informally termed "GABRIEL." The GABRIEL series includes an evolutionary progression of vehicle concepts that build upon the nearest-term implementation of EPPP. This concept roadmap eventually culminates in larger systems that employ more sophisticated methods for pulse initiation and momentum transfer. GABRIEL is characterized by the following four levels:

1.   Mark I: Solid pusher plate and conventional shock absorbers (small size)
2.   Mark II: Electromagnetic coupling incorporated into the plate and shocks (medium size)
3.   Mark III: Pusher plate extensions such as canopy, segments, cables (large size)
4.   Mark IV: External pulse unit driver such as laser, antimatter, etc. (large size)

All of these levels, besides the GABRIEL Mark I, require technology that is not currently available, but may be attainable for a second-generation vehicle. The Mark I (Fig. 2) is also the smallest and least expensive version, but suffers from the poorest performance (nominally 5,000 seconds and 4 million newtons of thrust). Nonetheless, the Mark I has better Isp and thrust than any other known rocket system that could be reasonably developed within the next 20 years. Its heavy payload capacity and short trip times would significantly reduce the development challenges associated with manned spacecraft, as well as add extra safety margins through redundant systems, large reserve supplies and increased robustness. Interestingly, the same shielding used to protect the astronauts from solar flares could be used during engine operation (usually only a few hours at most), and the resulting radiation dose received would be much less than conventional multi-year missions. It is even conceivable that a vehicle with a performance as high as 4,000 seconds and 2 million newtons of thrust could be deployed and assembled in orbit using several Titan IV launch vehicles.

Several technical issues and trades must be addressed in order to define even a Mark I vehicle. These are the type of pulse unit, its degree of collimation, detonation position and fissile burn-up fraction. These issues dictate propulsion efficiency and drive design of the vehicle's mechanical elements. Another issue is the pusher plate-plasma interaction. The amount of ablation experienced during each pulse could be significant and would dramatically affect Isp and thrust levels. Other issues include shock absorber efficiency, timing and dynamic response. Reusability will be important, so component wear must be kept to a minimum. In-space assembly, earth-to-orbit launch packaging and pulse unit safety and loading also must be addressed. Most of these issues have been investigated in the past and, although engineering challenges still remain, there are no formidable technical problems to overcome.

The ultimate hurdle in developing EPPP would be political in nature. Although GABRIEL does not face any insurmountable technical or financial obstacles, it does face one of perception. Use of nuclear material is almost always met by vehement opposition. However, there have been some important changes in the political landscape that may afford EPPP a chance where ORION failed. The Cold War is over and the fears of a large-scale nuclear conflict have abated somewhat. The existing ban on nuclear weapons in space actually has provisions that may allow peaceful uses of EPPP-type techniques below certain energies.

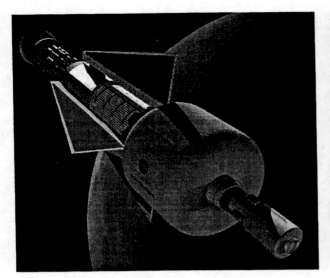

FIGURE 2. GABRIEL Mark I vehicle.

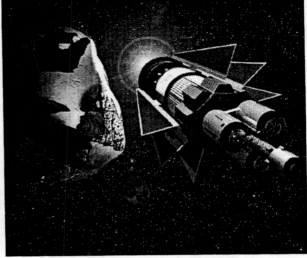

FIGURE 3. Asteroid deflection maneuver.

Even if EPPP is still viewed as too controversial for development in the near future, it would be worthwhile to begin reexamining it within the context of modern technologies and capabilities. Unlike physics, the sociopolitical environment does change, and a propulsion system with this tremendous capability may be needed - possibly on rather short notice (Fig. 3). The fact that many of the advanced propulsion concepts being researched now may never move beyond the "proof-of-principle" phase suggests that EPPP may be the only option we have for very ambitious human exploration of space in the foreseeable future.

SUMMARY

The case for reexamining nuclear pulsed propulsion and more modern embodiments of the EPPP concept has been made. The modern version of this propulsion concept, GABRIEL, is distinguished by its superior performance (i.e., both high Isp and high thrust-to-weight), its practicality (borrowing from only existing technologies), benign environmental impact (i.e., dedicated in space operation and reduced crew radiation exposure) and its economics (i.e., small size and reusability). More advanced systems with much better performance could be developed as technology in key areas mature. Improved performance can be achieved through advanced materials, magnetic fields (both on the pusher plate and along the shocks), novel momentum transfer schemes, and pulse unit drivers.

However, it is the rationale for considering EPPP that is most important. EPPP offers a highly effective method for deflecting comets or asteroids. Trips to and from Mars may be significantly shorter and safer than with conventional propulsion concepts. The flexibility of missions employing EPPP is enormous, allowing massive payloads, emergency return capability and routine transit from a reusable vehicle. Beyond Mars, missions to the asteroid belt, Jupiter and other planets are possible with the same basic system.

Timing for development of EPPP may also be better than during the days of ORION. In many ways, international cooperation is more prevalent, and could conceivably be extended to the peaceful application of unused nuclear material. Stockpiles of fissionable material can be permanently disposed of and environmental contamination is negligible if used outside the earth's magnetosphere. Finally, the human race is at the threshold of truly exploring, developing resources and permanently inhabiting space. GABRIEL may provide the best means of accomplishing this in the near future.

ACKNOWLEDGMENTS

The authors would like to thank George Dyson for providing much of the background documents on the ORION program, and Peggy for providing the many final proof readings and grammatical corrections.

REFERENCES

Cotter, T. P., "Rotating Cable Pusher for Pulsed-Propulsion Space Vehicle," Los Alamos Scientific Laboratory, LA-4666-MS UC-33, Propulsion Systems and Energy Conversion TID-4500, 1971.

Kos, L., "Human Mars Mission: Transportation Assessment," AIAA-98-5118, 1998.

Martin, A. R., Bond, A., "Nuclear Pulse Propulsion: A Historical Review of an Advanced Propulsion Concept," *Journal of the British Interplanetary Society*, Vol. 32, pp.283-310, 1979.

Nance, J. C., "Nuclear Pulse Propulsion", *IEEE Transactions on Nuclear Science*, February, 1965.

Prater, Lt., F. A. Gross, *Nuclear Impulse Propulsion, Project 3775, (ORION)*, USAF document, 1963.

Reynolds, T. W., "Effective Specific Impulse of External Nuclear Pulse Propulsion Systems", *NASA Technical Note*, NASA TN D-6984, 1972.

Serber, R., *The Los Alamos Primer*, Berkley CA, University of California Press, 1992.

Solem, J. C., "Medusa: Nuclear Explosive Propulsion for Interplanetary Travel," *Journal of the British Interplanetary Society*, Vol. 46, pp. 21-26,1993.

Self-Supporting Radioisotope Generators With STC-55W Stirling Converters

C. Or, V. Kumar, R. Carpenter, and A. Schock

Orbital Sciences Corporation (OSC), 20301 Century Boulevard, *Germantown, MD 20874*
Phone: (301) 428-1168; Fax (301) 353 8619; Email: or@oscsystems.com

Abstract. Previous Orbital Stirling generator designs rely on the spacecraft mounting structure to fasten the radiators, the converters, and the heat source assembly. This paper describes a self-supporting generator concept with a 1-piece honeycomb panel serving as a radiator and a rigid platform for the converters and the heat source assembly to be bolted on. This self-supporting generator allows for simpler mounting structure and more options on mounting location. Using this self-supporting generator concept, we derived four different design options to connect the converter pair rigidly to reduce vibration. This paper describes the four design options and their assembly procedure.

REVIEW

The performance characteristics of a 4-converter, 4-quadrant radiator, and 3-GPHS (General Purpose Heat Source) Stirling generator system shown in Figure 1 were published in January 1999 (Schock, Or, and Kumar, 1999a). This design allows for 1 converter or 1 heat pipe failure. Since the publication was submitted, the Jet Propulsion Laboratory (JPL) has increased the end-of-mission (EOM) power goal from 150 We to 210 We. To meet the higher power demand utilizing the same STC-55W converters (STC, 1998), two generator systems were investigated.

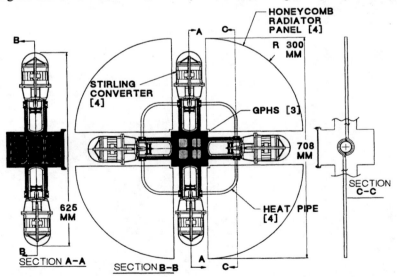

FIGURE 1. Stirling Generator Design with 4-Converters, 4-Radiator Quadrants, and 3-GPHS Blocks.

The first system, shown in Figure 2, uses the same layout as before with 4 STC-55W converters. The differences are 4 GPHS blocks instead of 3 are used, and the radiators are larger to accommodate the extra heat rejection. The second system uses 2 generators. Each generator, shown in Figure 3, has 2 converters, 2 semicircular radiators, and 2 GPHS blocks.

CP504, *Space Technology and Applications International Forum–2000*, edited by M. S. El-Genk
© 2000 American Institute of Physics 1-56396-919-X/00/$17.00

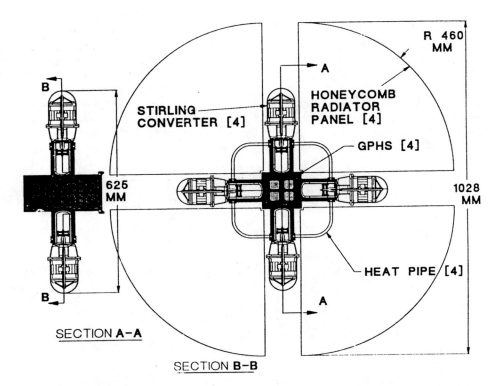

FIGURE 2. Stirling Generator Design with 4-Converters, 4-Radiator Quadrants, and 4-GPHS Blocks.

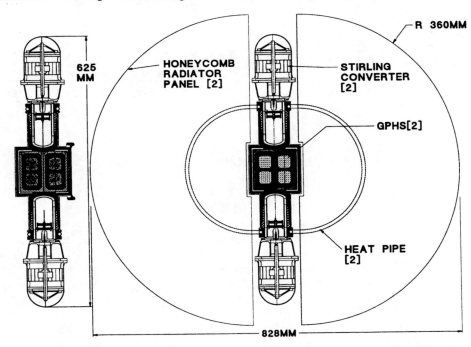

FIGURE 3. Stirling Generator Design with 2 Converters, 2 Semicircle Radiators, and 2-GPHS Blocks.

In both systems, all the converters and heat pipes have to be working to meet the power requirement. The performance characteristics with spacecraft blockage of the 1-generator system and the 2-generator system were reported in the Stirling Information Exchange Meeting at the Department of Energy (DOE) in Germantown, Maryland in December 1998 (Schock, 1998).

SELF-SUPPORTING RADIOISOTOPE STIRLING GENERATOR

Previous OSC Stirling generator designs rely on the spacecraft mounting structure to fasten the radiators, the converters, and the heat source assembly. A self-supporting generator would have more flexibility on the mounting structure and the mounting location. Figure 4 shows a self-supporting generator with a 1-piece honeycomb panel serving as a radiator and a rigid platform for the converters and the heat source assembly to be bolted on. The generator can be mounted to the spacecraft through the heat source-housing cap. The smaller mounting area to the spacecraft and the more rigid generator should result in a smaller mounting structure. Subsequent design options also employ this dual-purpose radiator and rigid mounting platform concept.

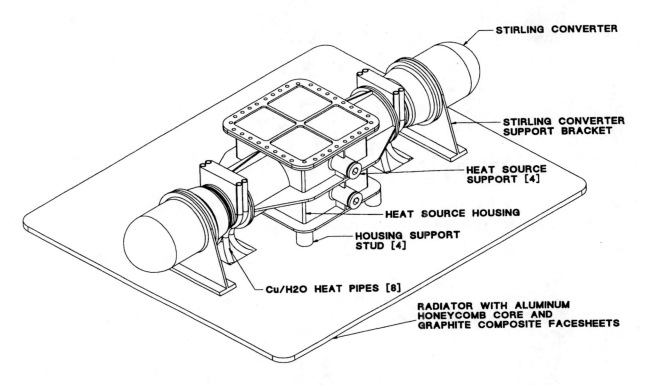

FIGURE 4. Self-Supporting Radioisotope Stirling Generator.

FACESHEET MATERIALS

Figure 5 shows the cross-section of the radiator facesheet material TC1050 made by Advanced Ceramics Corporation (ACC, 1996). The material is chosen because its thermal pyrolytic graphite (TPG) core has an in-plane thermal conductivity 8 times that of aluminum, and a density 20% lower than aluminum. However, TPG has a low planar shear strength. It needs to be encapsulated in a high strength material. The encapsulant can be metal, ceramics, or composite.

Carbon fiber composite (CFC) K13C2U/RS-3 is chosen as the encapsulant material in this study for its high strength and lightweight. Figure 6 shows the exploded view of TC1050-CFC. The fiber directions are parallel in each layer, and perpendicular between layers. ACC tested 10 TC1050-CFC panels. The TPG thickness ranged from 0.2 to 1.4 mm. All 10 panels have 4 layers of CFC on each side with a thickness of 0.2 mm. The average bending stiffness of the panels tested is 3.7 times that of aluminum. Thermal conductivity measurements before and after thermal cycling (-125 to +125 °C, 200 cycles) had no measurable change. This confirms that thermal cycling would not cause delamination within the TPG or between the TPG core and the CFC encapsulant.

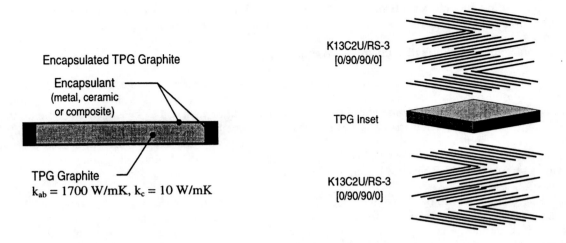

FIGURE 5. Cross-section of TC1050 Facesheet.

FIGURE 6. Exploded View of TC1050-CFC Facesheet.

ASSEMBLY PROCEDURE OF THE SELF-SUPPORTING GENERATOR

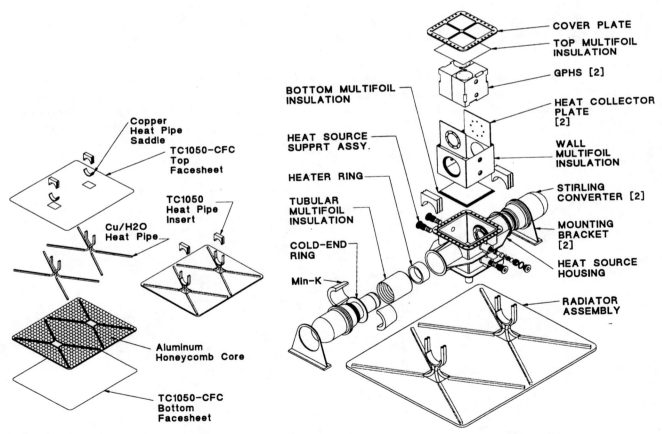

FIGURE 7. Exploded View of Radiator Subassembly.

FIGURE 8. Exploded View of Self Supporting Stirling Generator.

Figures 7 and 8 show the components of the self-supporting Stirling generator and its assembly procedure. The honeycomb panel and the 8 heat pipes are glued onto the bottom facesheet of the radiator. The top facesheet is glued onto the honeycomb panel. The heat pipe saddles are then soldered onto the heat pipes. The above description completes assembling the radiator subassembly.

In Figure 8, the bottom, wall, and tubular multifoil insulations are first inserted into the heat source housing. The heater ring is brazed onto each Stirling converter. Each of the Stirling converters with the 2 Min-k halves held against it is inserted into the heat source housing. The cold-end ring of each converter is then brazed onto the heat source housing. Each heat collector plate is slid into the heat source housing and bolted onto the heater ring. The mounting bracket is spot welded onto each converter. The assembled converters and heat source housing are lowered and soldered onto the heat pipe saddles. The heat source housing and the mounting brackets are bolted onto the radiator assembly. The heat pipe inserts are soldered onto the heat pipes and the converters.

When the generator is ready to be fueled, the 2 GPHS modules are lowered into the heat source housing. The 4 heat source support assemblies are then inserted against the indentations of the GPHS modules to hold them in place. The top multifoil insulation is put in place and the cover plate is then bolted onto the heat source housing. The above description completes assembling the self-supporting generator.

MECHANICAL COUPLING OPTIONS OF STIRLING GENERATOR

Vibration reduction is a primary design concern in the Stirling generator. In all the design options, the converters are arranged in opposing pair configurations. When the 2 converters are completely synchronized, the vibrations from one converter are balanced by the vibrations from its opposing unit. In order to balance vibrations, the converter pair must be connected rigidly. This report shows 4 different design options to connect the converter pair rigidly.

In the first option, shown in Figure 9, the mechanical coupling is carried out in the hot-end through the connection of the heater head shell, heater ring, and the I-shaped hot-frame. The heater ring is brazed onto the heater head shell, and the hot-frame is bolted onto the heater ring. The I-shaped hot-frame replaces the open-top-box hot-frame in the previously reported design. This would stiffen the coupling of the converters and allow for the GPHS blocks to separate in reentry.

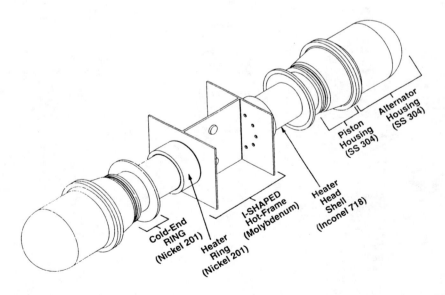

FIGURE 9. Mechanical Coupling through Hot Ends.

On the outside, the coupling through the housing is made elastic by adding bellows to the heater head housings as shown in Figure 10. This is needed to account for differential thermal expansion between the inside hot-end couplings and the outside housing. The 2-GPHS blocks are now supported by 4-spring loaded supports (Schock, Noravian, Or, and Kumar 1999b). In this design, the heater head shell needs to be thickened to accommodate the vibration load. A thicker shell leads to more conduction heat loss, which would lower the conversion efficiency.

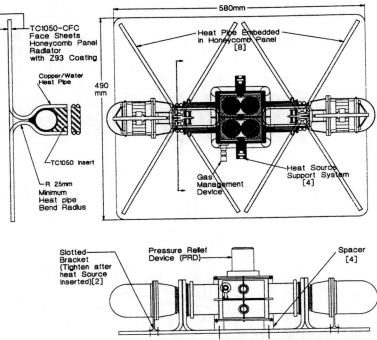

FIGURE 10. Stirling Generator Design with Mechanical Coupling through the Hot Ends.

In the second option, shown in Figure 11, the mechanical coupling is through the heat source housing. We thickened and extended the housing flange to enhance connections. We also extended and thickened the cold-end ring to join the heat source housing to the piston housing and the alternator housing, thus completing the connection between the 2 converters.

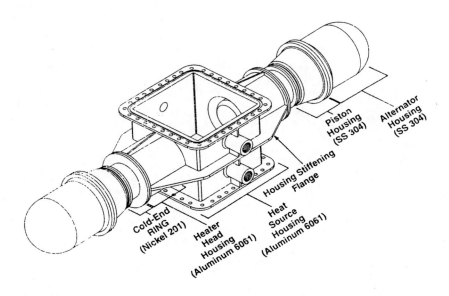

FIGURE 11. Mechanical Coupling through the Heat Source Housing.

Figure 12 shows that in this option the hot-end of the converter with the heat collection plate is free floating. There is no differential thermal expansion problem. Heat is transported from the heat source to the heat collection plate by radiation. The layout is similar to the first option shown in Figure 10.

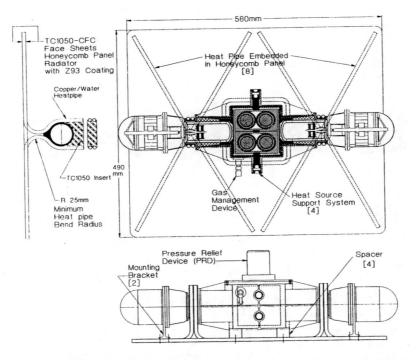

FIGURE 12. Stirling Generator Design with Mechanical Coupling through the Heat Source Housing.

In the third option shown in Figure 13, the heat source assembly is moved to the side. The mechanical coupling is through the cold-end bracket, converter housing and hot-plate housing.

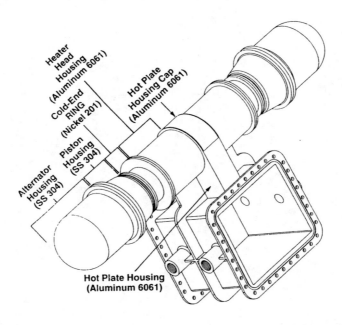

FIGURE 13. Mechanical Coupling through Hot-Plate Housing and Heater Head Housing.

As shown in Figure 14, heat is transported from the heat source through a hi-conductivity graphite plate between the GPHS blocks, through the gap by radiation onto the heat collection plate to the hot-ends of each converter. In this option, the mechanical coupling is better than that through the heat source housing in the second option. However, the off-center heat source assembly may require a larger mounting structure to the spacecraft. Having to go through an extra step through the hot-plate for heat transfer, the GPHS temperature is higher than other designs. Consequently, thermal loss would be higher. Preliminary calculations show that the clad temperature is close to the 1330°C limit. The clad temperature limit may end up a limiting factor of this design.

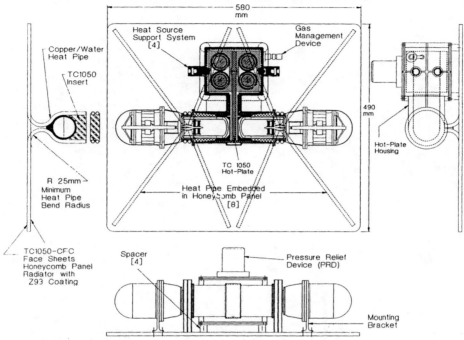

FIGURE 14. Stirling Generator Design with Mechanical Coupling through the Hot-Plate Housing.

In the fourth option shown in Figure 15, the converters are flipped with the two alternator housings facing each other. The alternator housings are mechanically coupled by a pipe.

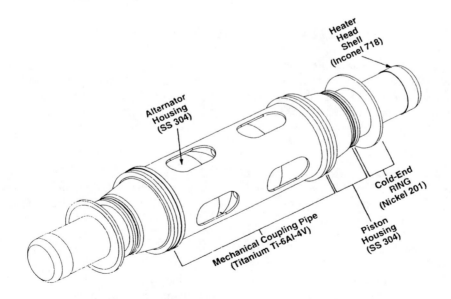

FIGURE 15. Mechanical Coupling through Alternators Connection Pipe.

As shown in Figure 16, each converter has its own heat source assembly on the hot-end. Heat is transported by radiation through the gap from the GPHS block to the heat collector plate. To prevent the heat source from rotating, square cross-section zirconia studs are used instead of the zirconia balls used in previous design options. In this option, the mechanical coupling is the most rigid of all. Another advantage is, when one converter fails, the failure would not propagate to the other converter. The problem is, separate heat source assemblies result in heavier mass and more thermal loss. Another problem is that the heavy heat source assemblies are on the two far ends. The increased bending inertia may require a larger mounting structure to the spacecraft.

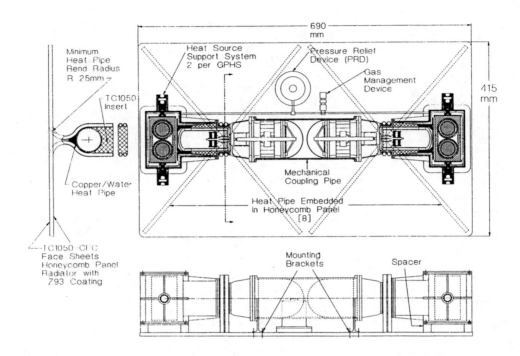

FIGURE 16. Stirling Generator Design with Mechanical Coupling through Alternators Connection Pipe.

MATERIAL COMPATIBILITY

The difference in coefficient of thermal expansion between Aluminum-6061 (23.4×10^{-6} /°C) and Nickel-201 (13.1×10^{-6} /°C) is large. It may pose a problem at the junction between the Al-6061 housing and Ni-201 cold-end bracket. Current designs use Aluminum-Silicon alloy as brazing material. The highest differential thermal expansion stresses occur when the junction is brazed at 580°C and cooled down to room temperature, although some of these stresses would be relieved at system operating temperatures. At the junction dimensions the highest differential thermal expansion is about 0.25 mm.

Another area of concern is the high brazing temperature of 580°C relative to the 616°C melting point of Al-6061. At this brazing temperature, the Al-6061-T6 heat treat condition will be annealed out, lowering the alloy strength near the braze area. More investigations will be performed to ensure the differential thermal expansion and the high brazing temperature can be tolerated.

An alternative to the Al-6061 housing material is Ti-6Al-4V titanium alloy. The titanium alloy's coefficient of thermal expansion of 8.6×10^{-6} /°C is much closer to Ni-201's 13.1×10^{-6} /°C. The 3.8 times higher ultimate tensile strength than Al-6061 enable a thinner housing which would more than compensate the effect of higher mass from the 1.6 times higher density. However, the titanium alloy's higher melting point of 1590°C could impose a problem. In inadvertent reentry in case of accident, the heat source housing is required to melt to enable the heat source modules to separate and reenter individually. Before the material can be adopted, analysis is needed to ensure that the reentry heat pulse would melt the titanium alloy housing.

CONCLUSIONS

Each design has its own merits. Analyses and trade-off studies are needed to compare the performance of each design. So far the generator design with mechanical coupling through the hot-ends shown in Figure 10 is least attractive. It requires thickening of the heater head shell, which results in a lowering of converter efficiency. The generator design with mechanical coupling through the heat source housing shown in Figure 12 is most attractive. It has good mechanical coupling, probably the highest thermal efficiency, the lowest generator mass, and requires the smallest mounting structure.

ACKNOWLEDGMENTS

This work is supported by DOE/Office of Space and Defense Power systems. The authors thank Maurice White and Songgang Qiu of STC for their valuable comments and feedback.

REFERENCES

Advanced Ceramics Corporation, " TC1050-CFC Carbon Fiber Encapsulated TPG Materials," work completed under NASA SBIR, Contract # 96-1 07 03 8681, 1996.
Stirling Technology Company, " Final Design Review Report for DOE 55-W Technology Demonstration Converter Program," June 9, 1998.
Schock, A., " Stirling Information Exchange Meeting," presented at DOE-Germantown, MD, December 1998.
Schock, A., C. Or, and V. Kumar, "Radioisotope Power System Based on Improved Derivative of Existing Stirling Engine and Alternator," in *Space Technology and Applications International Forum* 1999, edited by M.S. El-Genk, AIP Conference Proceedings 458, American Institute of Physics, New York, 1999a.
Schock, A., H. Noravian, C. Or, and V. Kumar, "Recommended OSC Design and Analysis of AMTEC Power System for Outer-Planet Mission," in *Space Technology and Applications International Forum* 1999, edited by M.S. El-Genk, AIP Conference Proceedings 458, American Institute of Physics, New York, 1999b, pp. 1534 - 1553.

Performance of the Preferred Self-Supporting Radioisotope Power System With STC 55-W Stirling Converters

C. Or, R. Carpenter, A. Schock, and V. Kumar

Orbital Sciences Corporation (OSC), 20301 Century Boulevard, Germantown, MD 20874
Phone: (301) 428-1168; Fax (301) 353 8619; Email: or@oscsystems.com

Abstract. Orbital has designed various self-supporting radioisotope power system options utilizing STC 55-W Stirling converters for possible application to NASA's Europa Orbiter mission. The preferred generator design with mechanical coupling though the heat source housing was analyzed. The various parameters studied include radiator facesheet thickness, radiator size, separation distance between spacecraft and radiators, Sun angle, and mission phase with different solar constants and radioisotope thermal power. Analytical results show that the Europa EOM power goal of 210 W_e can be met comfortably with the preferred power system of 2 generators, each with 2 GPHS modules and 2 STC 55-W Stirling converters.

INTRODUCTION

A companion paper presented here (Or, Kumar, Carpenter, and Schock, 2000) described a self-supporting generator with 4 mechanical coupling options. The paper concluded that mechanical coupling through the heat source housing is preferred. Figure 1 shows the preferred self-supporting generator with 2 Stirling Technology Company (STC) 55-W converters (STC, 1998) and 2 GPHS modules. To meet the EOM power goal of 210 W_e the power system requires 2 generators.

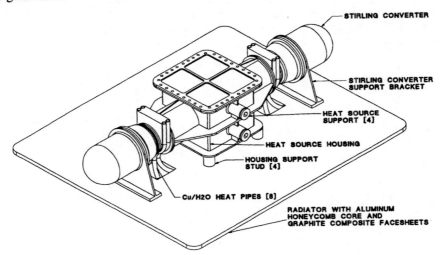

FIGURE 1. Self-Supporting Radioisotope Stirling Generator.

Figure 2 shows the 2-generator system integrated with an illustrative spacecraft. The paper describes the analytical performance of the power system. The varied parameters include radiator facesheet thickness, radiator size, separation distance between spacecraft and radiators, Sun angle, and mission phase with different solar constants and radioisotope thermal power.

CP504, *Space Technology and Applications International Forum–2000*, edited by M. S. El-Genk
2000 American Institute of Physics 1-56396-919-X

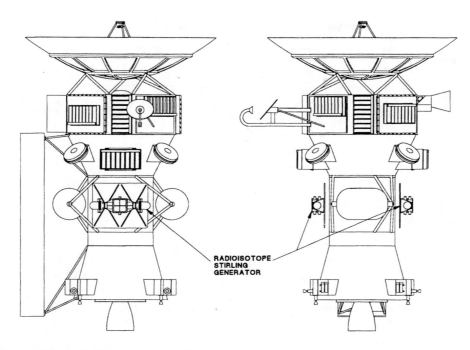

FIGURE 2. Radioisotope Stirling Power System Integrated with an Illustrative Spacecraft.

MODEL DESCRIPTION AND PHYSICAL PROPERTIES

Detailed finite difference models in SINDA (Gaski, 1987) were created to predict the Stirling power system performance. In the model, radiation exchange between surfaces were computed separately by a code called SSPTA (Little, 1986). The surface properties used are shown in Table 1. A typical model consists of 980 surfaces and about 2700 radiation couplings. In the SINDA models, the mass of each component used is shown in Table 2. The conductivity of all materials used is temperature dependent for more accurate predictions. A typical model consists of 1700 nodes and 7300 conductors.

TABLE 1. Surface Properties for Radiation Exchange Calculations.

Component	Absorptivity	Emissivity
Spacecraft Body (Kapton)	0.45	0.82
Heat Source Housing (Bare Al-6061)	0.44	0.11
Radiator (Z-93 coating)	0.17	0.92
Converter Housing (Grit Blasted Stainless Steel)	0.58	0.38

For both the SSPTA and SINDA models, 1-generator models were created. The 2-generator system results were then derived by adding the 1-generator results in their corresponding circumstances. This scheme greatly reduced computation time without sacrificing accuracy.

RESULTS AND DISCUSSIONS

As described in the companion paper and illustrated in Figure 1, the generator's waste heat is transported from the cold-end of the 2 Stirling converters to the radiator by 8 heat pipes. The waste heat is then rejected by the radiator to space by radiation. How effectively the waste heat is rejected depends on various parameters. Each of these parameters was investigated and will be discussed below.

TABLE 2. Mass Breakdown of Radioisotope Stirling Generator.

Generator Component	Mass (kg)
Heat Source – 2 GPHS	**2.890**
Housing, Insulations, and Supports	**2.015**
Housing and Cover Plate (Al-6061)	0.980
Multifoil Insulation (Mo/Zr O_2)	0.450
Min-K Insulation (2)	0.051
Heat Source Support (Graphite, Zr O_2, SS, Al-6061) (4)	0.130
Converter Mounting Bracket (Ti-6Al-4V) (2)	0.392
Housing Support Stud (Al-6061) (4)	0.012
Stirling Converter and Accessories	**8.436**
Stirling Converter (2)	6.000
Cold-End Ring (Ni-201) (2)	0.502
Heater Ring (Ni-201) (2)	0.537
Heat Collector Plate (Ni-201) (2)	0.377
Adaptive Balancer	0.820
Converter Controller (Inc. AC-to-DC Converters)	0.200
Heat Rejection System	**1.947**
TPG Core in TC1050-CFC Facesheet (t = 0.3 mm x 2)	0.390
CFC Encapsulant in TC1050-CFC Facesheet (t = 0.2 mm x 4)	0.415
Honeycomb Core (Al) and Adhesive	0.238
Heat Pipe (Cu) (8)	0.641
Heat Pipe Saddle (Cu) (2)	0.091
Heat Pipe Insert (TC1050) (2)	0.172
Generator Accessories	**0.790**
Pressure Relief Device	0.430
Gas Management valve	0.160
C-Seals, Fasteners, and Miscellaneous	0.200
Total	**16.078**

Effect of Radiator Facesheet Thickness

TC1050-CFC (ACC, 1996) was chosen as radiator facesheets (Or, Kumar, Carpenter, and Schock, 2000). The TC1050-CFC is made of high conductivity thermal pyrolytic graphite (TPG) core encapsulated by carbon fiber composite (CFC). The encapsulant thickness is fixed at 0.2 mm. The facesheet thickness is increased by increasing the TPG thickness.

Figure 3 shows that for a thicker radiator facesheet, the cold-end temperature is lower. It is because thicker radiator facesheet spread out heat more effectively in the radiator. The temperature difference between the hot and the cold extreme is smaller. It also implies that the temperature difference between the hottest and the average is smaller. For the same heat rejection, the average radiator temperature is roughly the same. Consequently, a thicker radiator facesheet results in a lower maximum radiator temperature. The radiator area in contact with the heat pipe is the hottest. Going upstream of the heat conduction path, lower maximum radiator temperature results in lower heat pipe temperature, which results in lower converter cold-end temperature.

For a fixed heat input to the converters and a fixed hot-end temperature, a lower cold-end temperature results in higher conversion efficiency and higher power output. A thicker radiator facesheet also leads to a higher generator system mass. Figure 3 presents a quantitative trade-off between system performance and system mass.

When the TPG thickness is small, a small increase would lower the cold-end temperature a lot and increase the power output rapidly. Beyond 0.7 mm TPG thickness, the curve has reached a plateau. Further increase in TPG thickness shows little improvement in performance.

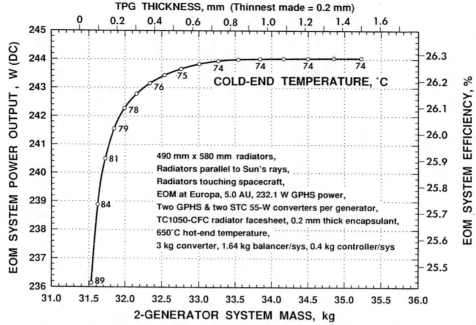

FIGURE 3. Effect of Radiator Facesheet Thickness on EOM Performance of OSC Self-Supporting Generators.

Effect of Radiator Size

Other than thickening the radiator facesheet, the cold-end temperature may also be lowered by increasing the radiator size. Figure 4 shows the system performance for 3 radiator sizes. As shown in the curves, increasing the radiator size decreases the cold-end temperature and increases the power output. Again, the price to pay is an increase in system mass. All 3 curves have similar shapes with rapid increases at first and leveling off soon after. The rate of increase in power output versus mass determines the specific power shown by the dotted diagonal lines.

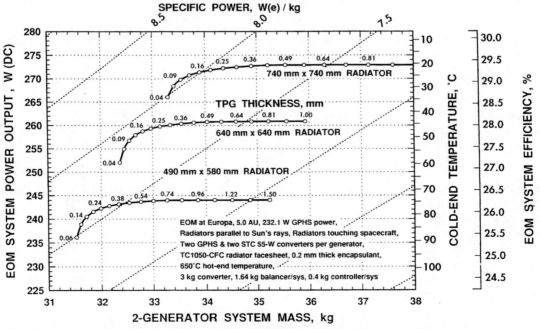

FIGURE 4. Effect of Radiator Size on EOM Performance of OSC Self-Supporting Generators.

Effect of Spacecraft Blockage

With the presence of the spacecraft, part of the radiator's view to space is blocked. Thus, heat rejection by radiation is more restricted. The closer the spacecraft is to the radiator, the more pronounced the blockage effect. Figure 5 shows the spacecraft blockage effect on the EOM performance. When the separation distance between the spacecraft and the radiators increases, the radiator temperatures and hence the cold-end temperatures decreases. Consequently, the power output and the efficiency increase.

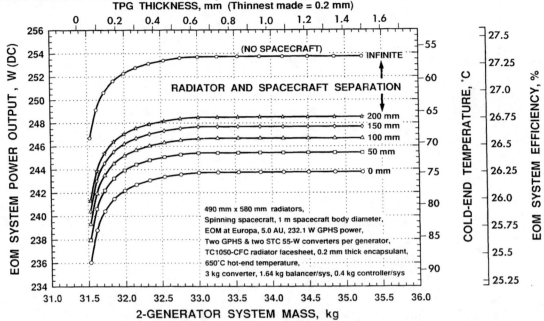

FIGURE 5. Effect of Spacecraft Blockage on EOM Performance of OSC Self-Supporting Generators.

Effect of Sun Angle

The amount of solar flux impinging on the radiators depends on the solar constant and the Sun angle relative to the radiator. On top of the waste heat from the Stirling converters, the radiators have to reject the extra solar power absorbed. More solar power absorbed results in higher radiator temperature, and hence higher cold-end temperature and lower power output.

Figure 6 shows the Sun angle effect on the Stirling system performance. Three mission phases were analyzed: at BOM when the spacecraft is in Earth orbit; one year after BOM at Venus flyby; and, at EOM when the spacecraft is in Europa orbit. Each mission phase analyzed includes 3 Sun angles: radiators parallel to the Sun's rays, radiators perpendicular to the Sun's rays, and, spinning spacecraft.

For each mission phase, when the radiators are parallel to the Sun's rays, even though there is no direct solar flux impinging onto the radiators, there is a relatively small amount of solar flux reflected from the spacecraft. When the radiators are perpendicular to the Sun's rays, even though one of the 2 radiators is blocked by the spacecraft, the total solar power absorbed is still the highest. When the spacecraft is spinning, the total power absorbed is in between the 2 extreme cases of radiators parallel to the Sun's rays and radiators perpendicular to the Sun's rays. As shown in the figure, when the radiators are parallel to the Sun's rays, the power output is the highest. When the radiators are perpendicular to the Sun's rays, the power output is the lowest. When the spacecraft is spinning, the power output is in between.

The solar constant is proportional to the inverse square of the distance to the Sun. At Venus flyby at 0.6 AU, the Sun's effect is high. Therefore, the power output shows the biggest difference at different Sun angles. At Europa at 5.0 AU, the solar flux is low. Therefore, the power output shows very little difference at different Sun angles.

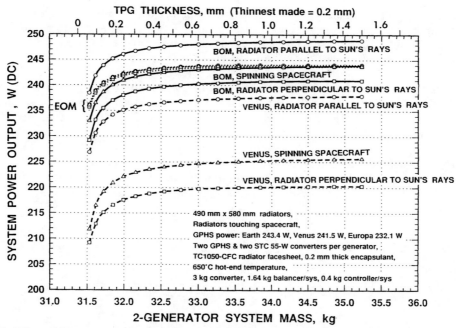

FIGURE 6. Effect of Sun Angle on BOM, Venus Flyby, and EOM Performance of OSC Self-Supporting Generators.

Worst-Case Performance

The preceding parameters showed that the power output is the lowest when the radiators are touching the spacecraft, and when the radiators are perpendicular to the Sun's rays. Figures 7, 8, and 9 show the system performance under this worst-case configuration at BOM, at Venus flyby, and at EOM respectively.

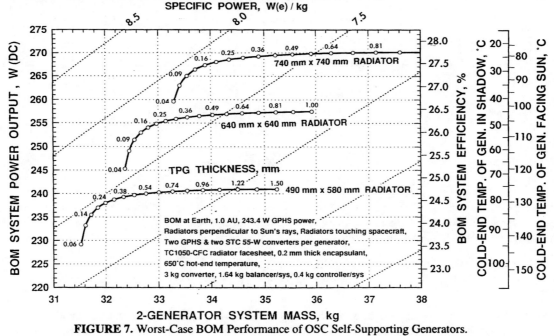

FIGURE 7. Worst-Case BOM Performance of OSC Self-Supporting Generators.

In each mission phase, the generator facing the Sun has a higher radiator temperature than the generator behind the spacecraft. Hence the Sun-facing generator has a higher cold-end temperature. Consequently, it has a lower power output for approximately the same thermal power through the converters and the same hot-end temperatures.

In Figure 7 at BOM, for any point on the 3 curves, *i.e.*, any radiator size and facesheet thickness, it corresponds to 2 cold-end temperatures indicated on the vertical axis on the right. The Sun facing generator has a higher cold-end temperature. Conversely, the spacecraft shadowed generator has a lower cold-end temperature. The system power output in the graph is the sum of the power output from the exposed and the shadowed generators.

In Figure 8 at Venus flyby, the solar constant is higher than in Earth orbit at BOM. The difference in cold-end temperatures is higher than that in Figure 7. For the smallest radiator studied, the cold-end temperature of the exposed generator is around 180°C. The corresponding magnet temperature should be checked to make sure that the magnet is operating within the safety limit.

In Figure 9 at Europa, the solar constant is very low. The difference in cold-end temperatures is small. As always, a larger radiator or thicker radiator facesheet produces more power output. The price to pay is higher system mass. As shown in Figure 9, any radiator size and radiator facesheet thickness produces significantly more power than the goal of 210 W_e.

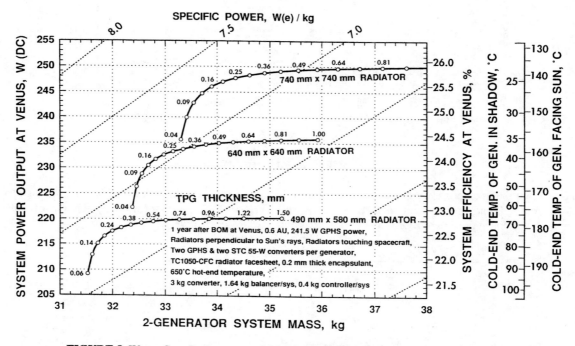

FIGURE 8. Worst-Case Performance of OSC Self-Supporting Generators at Venus Flyby.

CONCLUSIONS

This paper presents the performance of the self-supporting radioisotope Stirling power system. The effects of various parameters were investigated. Analytical results show that even in the worst-case scenario, the Europa EOM power goal of 210 W_e can be met with more than 15% margin.

ACKNOWLEDGMENTS

This work is supported by DOE/Office of Space and Defense Power systems. The authors thank Maurice White and Songgang Qiu of Stirling Technology Company for their Stirling converter performance information.

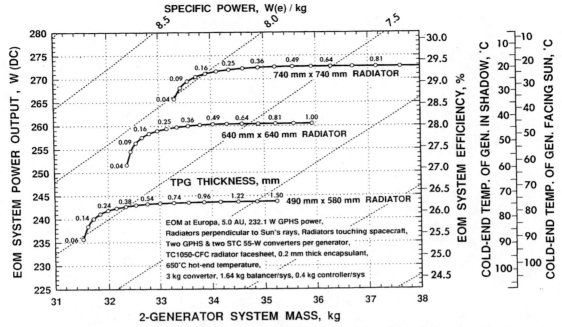

FIGURE 9. Worst-Case EOM Performance of OSC Self-Supporting Generators.

REFERENCES

Little, A.D. SSPTA (Simplified Space Payload Thermal Analyzer), version 3.0/VAX, for the NASA/Goddard, Arthur D. Little Inc., Cambridge, MA, 1986.

Gaski, J. SINDA (System Improved Numerical Differencing Analyzer), version 1.315 from Network Analysis Associate, Fountain Valley, 1987.

Advanced Ceramics Corporation, " TC1050-CFC Carbon Fiber Encapsulated TPG Materials," work completed under NASA SBIR, Contract # 96-1 07 03 8681, 1996.

Stirling Technology Company, "Final Design Review Report for DOE 55-W Technology Demonstration Converter Program", June 9, 1998.

Or, C., V. Kumar, R. Carpenter, A. Schock, "Self-Supporting Radioisotope Generators with STC-55 W Stirling Converters," to be presented at the Space Technology and Applications International Forum 2000.

Technology Development for a Stirling Radioisotope Power System

Lanny G. Thieme[1], Songgang Qiu[2], and Maurice A. White[3]

[1]NASA John H. Glenn Research Center at Lewis Field, MS 301-2, 21000 Brookpark Road, Cleveland, OH 44135
phone: 216-433-6119; fax: 216-433-8311; E-mail: Lanny.Thieme@grc.nasa.gov
[2]Stirling Technology Company, 4208 West Clearwater Avenue, Kennewick, WA 99336
phone: 509-735-4700; fax: 509-736-3660; E-mail: sqiu@stirlingtech.com
[3]Stirling Technology Company, 4208 West Clearwater Avenue, Kennewick, WA 99336
phone: 509-735-4700; fax: 509-736-3660; E-mail: mawhite@stirlingtech.com

Abstract. NASA Glenn Research Center and the Department of Energy are developing a Stirling convertor for an advanced radioisotope power system to provide spacecraft on-board electric power for NASA deep space missions. NASA Glenn is addressing key technology issues through the use of two NASA Phase II SBIRs with Stirling Technology Company (STC) of Kennewick, WA. Under the first SBIR, STC demonstrated a synchronous connection of two thermodynamically independent free-piston Stirling convertors and a 40 to 50 fold reduction in vibrations compared to an unbalanced convertor. The second SBIR is for the development of an Adaptive Vibration Reduction System (AVRS) that will essentially eliminate vibrations over the mission lifetime, even in the unlikely event of a failed convertor. This paper presents the status and results for these two SBIR projects and also discusses a new NASA Glenn in-house project to provide supporting technology for the overall Stirling radioisotope power system development. Tasks for this new effort include convertor performance verification, controls development, heater head structural life assessment, magnet characterization and thermal aging tests, FEA analysis for a lightweight alternator concept, and demonstration of convertor operation under launch and orbit transfer load conditions.

INTRODUCTION

NASA Glenn Research Center and the Department of Energy (DOE) are developing a Stirling convertor for an advanced radioisotope power system to provide spacecraft on-board electric power for NASA deep space missions. Stirling is being evaluated as an alternative to replace Radioisotope Thermoelectric Generators (RTGs) with a high-efficiency power source. The efficiency of the Stirling system, about 25% for this application, will reduce the necessary isotope inventory by a factor of 3 or more compared to RTGs. Stirling is the most developed convertor option of the advanced power concepts under consideration (Frazier, 1998 and Mondt, 1998).

FIGURE 1. Two opposed 55-We convertors on test.

DOE is developing the radioisotope Stirling convertor under contract with Stirling Technology Company (STC) of Kennewick, WA (White, 1998 and 1999). Two 55-We convertors are now being tested by STC in a dynamically-balanced opposed arrangement, as shown in figure 1. Both design convertor power and efficiency have been demonstrated. NASA Glenn is providing technical consulting for this effort under an Interagency Agreement with DOE.

The design of the 55-We Stirling convertor is based on previous successful STC development efforts, particularly those for the 10-We radioisotope terrestrial convertor, RG-10, and the 350-We RG-350 aimed at commercial cogeneration and remote power (Erbeznik, 1996). One RG-10 has now been on life test for over

CP504, *Space Technology and Applications International Forum–2000*, edited by M. S. El-Genk
© 2000 American Institute of Physics 1-56396-919-X/00/$17.00

50,000 hours (5.7 years) with no convertor maintenance and no degradation in performance. Another RG-10 convertor has recently been fueled with isotope and field testing is beginning. Multiple units of the RG-350 and their companion BeCOOL cryocoolers have accumulated over 80,000 total hours of operation, much of this at independent third-party test sites. STC has also completed extensive component life testing and, in particular, has over 1000 years of total test time (230 flexures) on the critical flexural bearings. These numerous test hours on various systems and on key components provide a high confidence that the 55-We convertor will meet its life and reliability goals.

NASA Glenn has been investigating Stirling radioisotope power systems for deep space missions since about 1990. This work grew out of earlier Stirling efforts conducted for DOE for a Stirling automotive engine and for the NASA Civil Space Technology Initiative (CSTI) to develop Stirling for a nuclear power system to provide electrical power for a lunar or Mars base (part of the SP-100 program). NASA Glenn also provided technical management for DOE for the Advanced Stirling Conversion System (ASCS) terrestrial dish Stirling project. Overall, NASA Glenn has been developing Stirling technologies since the mid-1970's.

As part of the overall Stirling radioisotope convertor development, NASA Glenn is addressing key technology issues through the use of two NASA Phase II Small Business Innovation Research (SBIR) contracts with STC. NASA Glenn has also just started an in-house project to provide additional supporting technology developments. This paper will discuss the status and results from these SBIR contracts and describe the tasks to be accomplished as part of the new NASA Glenn efforts.

Systems using Stirling convertors are being analyzed by NASA Glenn for other space applications in addition to Stirling radioisotope power for deep space missions. These include solar dynamic power systems for space-based radar (Mason, 1999) and as a deep space alternative to the radioisotope system, a combined electrical power and cooling system for a Venus lander, and lunar/Mars bases and rovers.

SYNCHRONOUS OPERATION OF OPPOSED STIRLING CONVERTORS

STC, as part of a NASA Phase II SBIR contract, has successfully demonstrated synchronous operation of two thermodynamically independent free-piston Stirling convertors with linear alternators connected electrically in parallel. Previous Stirling development had focused on single convertors and had not addressed how to connect multiple convertors in a system. However, in most potential space applications, multiple convertors are important for redundancy and modularity. Thermodynamically independent convertors allow one convertor to fail without affecting the performance of the other. Finally, the use of multiple convertors is important to controlling vibrations, a critical issue for a dynamic space power system. Synchronization of convertor pairs operating in an opposed configuration provides balanced operation with minimal vibration.

Two RG-350 convertors were used for this development. Initial efforts included computer simulations of multiple-convertor connection methods and single-convertor baseline testing of each of the RG-350 convertors. Each convertor was tested separately over a range of hot-end and cold-end temperatures and charge pressures. Approximately 5 g's vibration was measured at the nominal conditions for a single convertor.

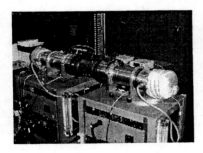

FIGURE 2. Stirling convertors synchronized for system operation with low vibrations.

During multiple-convertor testing, synchronization was achieved with the two convertors (see figure 2) operating over a wide range of conditions. The frequency of each convertor was identical, and the pistons operated nominally 180 degrees out-of-phase mechanically. The convertors were connected electrically in parallel and mechanically through external attachments on the cold-end pressure vessels. A mechanical coupler was developed that aligns the two convertors and can compensate for any inherent misalignments. Synchronization produced a 40 to 50 fold reduction in vibrations compared to an unbalanced convertor, a value that appears to be well below pixel smear limits for deep space sensing. Equal power generation between the two convertors was demonstrated under nominal conditions. This connection method is now being used to connect the DOE/STC 55-We convertors.

The synchronization was shown to be very robust by testing at conditions of simulated degradation and by transient testing. For one set of runs, the average hot-end temperature of one convertor was varied widely while holding the other convertor hot-end temperature nearly constant at about 600°C. Synchronization was maintained over this range, and figure 3 shows the effects on the phase difference between the power pistons and the maximum vibration measured on the two convertors. Zero-degree phase lag indicates the pistons are operating 180 degrees out-of-phase mechanically. The maximum vibration increased from 0.12 g to just 0.48 g over this wide range of operation; this is compared to about 5 g's vibration for an unbalanced convertor. The convertor power output at the lowest temperature was about 20% of the power output for the convertor operating at about 600°C.

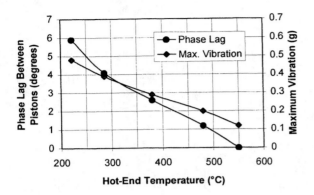

FIGURE 3. Response of the maximum vibration and phase lag between pistons when the hot-end temperature of one convertor of a fully coupled pair is varied.

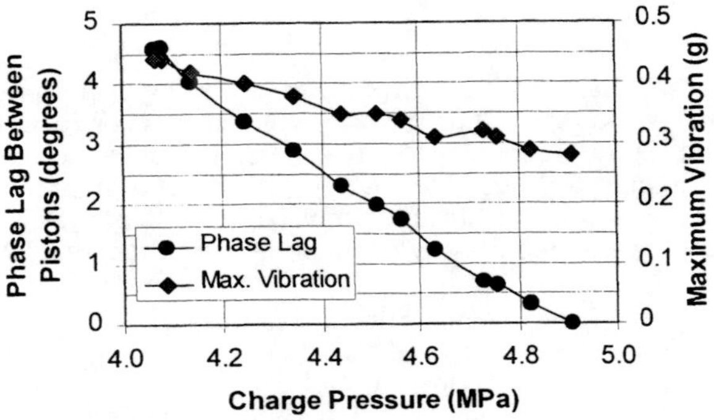

FIGURE 4. Response of the maximum vibration and phase lag between pistons when the charge pressure of one convertor of a fully coupled pair is varied.

Charge pressure variations of up to 20% for one convertor while maintaining constant charge pressure in the other and constant power input to each were also tested. Results for this case are shown in figure 4. Again, synchronization was maintained, and only small increases in vibration were measured. These tests were run with an earlier version of the mechanical coupler so the vibration level at the nominal design condition is somewhat higher than that shown in figure 3. Power output remained essentially the same for each convertor over this range; this was as expected from the single-convertor test results.

Transient data taken during various electrical connections and disconnections of the two convertors showed the ability to achieve synchronization reliably and rapidly. No significant transient overstrokes were seen or any other potentially damaging results. Figure 5 shows transient traces for the piston and displacer motions, voltage, current, and vibrations from one convertor when the other convertor is disconnected and shut down. The convertors are synchronized at nominal conditions at the beginning of the transient. The only noticeable effect is the increase in the vibrations of the now electrically uncoupled convertors.

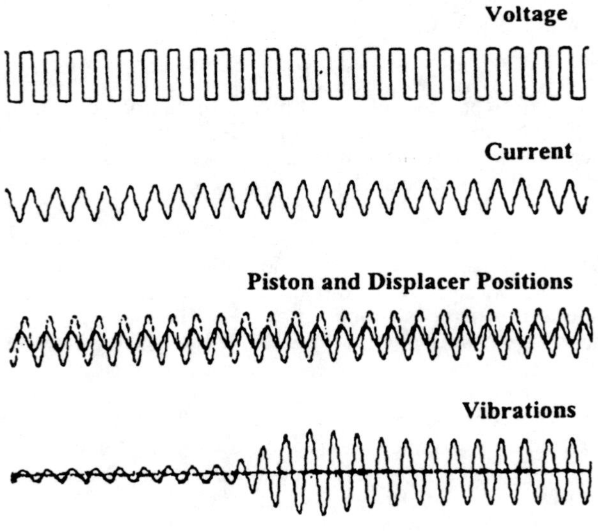

Some transient effects were found when the two convertors were electrically connected while operating at nominal conditions. Typically, an acceleration spike to about 6-10 g's was seen and may be due to the two convertors being briefly in-phase mechanically after the coupling. This transient settled out in about 15 cycles (1/4 sec.), after which the vibration level was stable at the very low values achieved with synchronization. If necessary, this transient could be mitigated by adding an extra load briefly on the alternator output while connecting the convertors to dampen the piston motions during the transient.

Voltage

Current

Piston and Displacer Positions

Vibrations

FIGURE 5. Transient response when the parallel electrical connection is broken and one convertor is shut down.

Successful system operation was demonstrated with the two synchronized convertors feeding a battery charger load, as would most likely be used in a radioisotope power system. Four standard automotive batteries were connected in series and tests run over a range of convertor hot-end temperatures and battery state-of-charge. Operation was found to be essentially the same as when dissipating power to the controller internal load resistors.

Tests were also run with the electrical coupling only and with the mechanical coupling only. With only the electrical coupling, the convertors synchronized as before; however, there was no reduction in vibrations as there was no mechanical connection between the convertors. With only the mechanical coupling, there was at most a weak synchronous connection, and the vibrations were similar to a single unbalanced convertor.

A further innovation during this SBIR project was the demonstration of an artificial neural network (ANN) that could potentially be used to monitor the health of a convertor using only non-invasive instrumentation that does not penetrate the convertor pressure vessel. The ANN successfully predicted piston and displacer amplitudes and phasing for a 10-We RG-10 convertor using voltage, current, and rejection temperature as the only inputs. Simulated pressure degradation for one of the fully coupled RG-350 convertors was also successfully tracked using current, current-voltage phasing, and output power for each convertor as inputs. It is felt that the ANN has a high probability of detecting any convertor degradation that may occur without needing any internal instrumentation that would decrease the convertor reliability. This could then allow the system controller to adjust operation to maximize system performance.

ADAPTIVE VIBRATION REDUCTION SYSTEM

Under a second NASA Phase II SBIR, STC is developing an Adaptive Vibration Reduction System (AVRS) that will further reduce vibration levels by a factor of 10 or more under normal operating conditions. It will achieve this with an active balance system with feedback from a vibration signal and will cancel the fundamental vibration and up to 10 harmonics. Even more importantly, the AVRS will be adaptive and will add the ability to adjust to any changing convertor conditions over the course of a mission. Thus, it should allow successful dynamic balancing over the mission lifetime and will be able to demonstrate its adaptive ability through up-front testing. The AVRS is now being developed on two RG-350 convertors and will also be demonstrated on the DOE/STC 55-We convertors.

The AVRS will use a balance mass driven by a separate linear motor; only one balance mass and motor are needed for two opposed Stirling convertors. A balance mass and motor being used in the first AVRS testing with the RG-350 convertors is shown in figure 6. The vibration signal will be measured with either a load cell or an accelerometer. A fast Fourier transform of this signal will then be used to construct a compensation signal that will be sent to the balance motor through a power amplifier. Both the amplitude and phase of each harmonic will be adjusted. The motion of the balance mass center-of-gravity will be opposed to and proportional to the motion of the center-of-gravity of the combined system of two pistons and two displacers. The AVRS will adjust to any change in convertor operating conditions, any convertor degradation that may occur over a mission, or even in the unlikely event of a convertor failure.

FIGURE 6. AVRS balance mass and motor for RG-350 testing.

Stirling cryocoolers are currently used to cool vibration-sensitive sensors in space applications. STC has demonstrated a cryocooler vibration level of only 0.007 g's using similar technology to the AVRS. This technique has been shown to be effective with reasonable power and mass budgets. One key difference for balancing power convertors is that the frequency is not fixed as it is in coolers. Thus, the frequency must be measured on a continual basis and factored into the control algorithm.

Initial tests of the AVRS on the RG-350 have shown a 500 fold reduction in unbalanced vibrations under normal operating conditions with two synchronized convertors in an opposed configuration. This compares to the 40 to 50 fold reduction with just the synchronized convertors and was accomplished with only 2 W of power dissipation. Testing was also done with a simulated failed convertor, and a 50-fold vibration reduction was obtained with only 7 W of power dissipation; this power dissipation scales to less than 2-3 W for a 55-We convertor. These initial tests only balanced the fundamental of the vibration signal (no harmonics) and, thus, vibration levels should be even further decreased as development of the full control algorithm is completed.

A further task of this contract will demonstrate a passive heat rejection system for both the RG-350 and the 55-We convertors. Initial system studies completed by Orbital Sciences Corporation (Schock, 1999) used a heat pipe to transport the convertor waste heat to the radiator. Design and fabrication are currently being completed for an aluminum hub clamped to the Stirling rejector with multiple water heat pipes inserted into the hub to demonstrate this cooling concept on the RG-350. Thermacore, Inc. is consulting on this effort and providing the heat pipes.

NASA GLENN SUPPORTING TECHNOLOGY DEVELOPMENT FOR THE STIRLING RADIOISOTOPE CONVERTOR

NASA Glenn has recently initiated an in-house technology development project in support of developing the Stirling radioisotope power system for deep space missions. The project tasks build on NASA Glenn expertise developed as part of previous Stirling research, especially for the Stirling space power development during the NASA CSTI project (Niedra, 1994; Abdul-Aziz, 1995; and Rauch, 1995). Tasks were identified in the appropriate areas where value-added development is provided as part of the overall Stirling radioisotope effort. Each task is described below.

Two RG-350 convertors (see figure 2) and two 55-We convertors (see figure 1) will be tested in synchronous opposed configurations. This will provide an independent performance verification for the 55-We convertors to be used for the radioisotope application. The primary emphasis will be on technology development for a reliable, lightweight integrated control system for multiple convertors and multiple system loads. Heat transfer characteristics of high porosity regenerators will also be determined using an existing heat transfer test rig.

Heater head life is a critical element for achieving the 100,000+ hour life of the convertor. NASA Glenn materials and structures personnel will perform an independent thermal and structural analysis; characterize creep properties for the Inconel 718 heater head material in the final condition that will be used in the convertor; adapt, verify, and characterize life prediction models; and then use these models to assess heater head life. Heater head tests to failure will also be conducted. Successful completion of these efforts should give high confidence in the ability of the heater head to meet the operational life requirements. A materials review and an evaluation of joining methods will also be included. A test rig conducting biaxial creep testing is shown in figure 7.

FIGURE 7. Biaxial creep testing.

FIGURE 8. Magnet characterization test rig.

FeNdB permanent magnets are currently being used in the 55-We convertors. Specific magnet choice is dependent on the final convertor cold-end temperature chosen based on system analyses (now expected to be between 80 and 120°C), magnet temperature calculations, and the safety margin necessary to prevent demagnetization. Characterization testing of appropriate FeNdB magnets will first be done using an existing test rig that was developed for testing SmCo magnets during the previous NASA CSTI research for SP-100 (see figure 8). Selected FeNdB and possibly SmCo magnets will then be put on aging tests at various temperature levels and in a demagnetizing field to quantify any potential magnet degradation with time and temperature. Such degradation could affect both the remanent magnetization and the demagnetization resistance. This task will also include a finite element analysis (FEA) of a lightweight linear alternator concept that has been previously tested by STC and had performance shortfalls. The FEA will be used to evaluate the alternator flux profiles and potential design modifications to improve performance.

The two 55-We convertors will be tested in the vibration test facility of the NASA Glenn Structural Dynamics Laboratory, shown in figure 9, to demonstrate their ability to operate under launch and orbit transfer load conditions. The Structural Dynamics Laboratory will provide analysis, mounting structure, and testing for this effort. The convertor load-carrying capability for these loads is provided by an aft bearing that has been added to the design for this purpose. NASA Glenn tribology personnel will provide consulting relative to understanding the characteristics of this bearing.

As part of the Interagency Agreement with DOE, NASA Glenn conducted an extensive conceptual screening of heat pipe and non-heat pipe (figure 10) radiator concepts for both two- and four-convertor radioisotope Stirling power systems. The starting point for the studies was based on the Orbital Sciences Corporation system layout using heat pipe radiators (Schock, 1999). Various material options were analyzed including standard state-of-the-art materials and more advanced high thermal conductivity materials. A further task in this new NASA in-house effort is expected to continue this radiator evaluation and include both view factor and FEA analyses.

FIGURE 9. Vibration Test Facility at NASA Glenn.

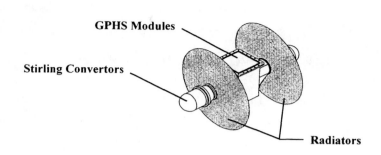

FIGURE 10. Non-heat pipe radiator concept.

CONCLUDING REMARKS

NASA Glenn Research Center and the Department of Energy are developing a Stirling convertor for an advanced radioisotope power system to provide spacecraft on-board electric power for NASA deep space missions. Stirling is being evaluated as an alternative to replace RTGs with a high-efficiency power source. STC, under contract to DOE, is making rapid progress in developing and demonstrating the 55-We convertor. Under two NASA Phase II SBIRs, STC has developed a synchronous system connection for multiple convertors and demonstrated very low vibration levels based on this synchronous connection and an adaptive vibration reduction system. NASA Glenn is now starting an in-house supporting technology development to provide further key inputs, including a heater head life assessment, permanent magnet aging characteristics, and demonstration of convertor operation under launch and orbit transfer loads.

REFERENCES

Abdul-Aziz, A., Bartolotta, P., Tong, M., and Allen, G., "An Experimental and Analytical Investigation of Stirling Space Power Convertor Heater Head", NASA TM-107013, 1995.

Erbeznik, R.M. and White, M.A., "Test Results and Commercialization Plans for Long Life Stirling Generators", Proceedings of the 31st Intersociety Energy Conversion Engineering Conference, edited by P.R.K. Chetty et al, Institute of Electrical and Electronics Engineers, 1996, Vol.2, pp. 1265-1270.

Frazier, T.A., "Advanced Conversion Technology Review Panel Report", in *Meeting Global Energy and Environmental Needs, Proceedings of the 33rd Intersociety Energy Conversion Engineering Conference*, American Nuclear Society, 1998, Paper IECEC-98-398.

Mason, L.S., "Technology Projections for Solar Dynamic Power", NASA TM-1999-208851, 1999.

Mondt, J.F. and Nesmith, B.J., "Advanced Convertor Technology Evaluation and Selection for ARPS", in *Space Technology and Applications International Forum*, edited by M.S. El-Genk, AIP Conference Proceedings 420, American Institute of Physics, 1998, Part Three, pp. 1098-1105.

Niedra, J.M., "Comparative M-H Characteristics of 1-5 and 2-17 Type Samarium-Cobalt Permanent Magnets to 300 C", NASA - CR-194440, 1994.

Rauch, J.S. and Kankam, M.D., "Transient and Steady-State Tests of the Space Power Research Engine with Resistive and Motor Loads", NASA TM-106832, 1995.

Schock, Alfred, Or, Chuen, and Kumar, Vasanth, "Radioisotope Power System Based on Improved Derivative of Existing Stirling Engine and Alternator", in *Space Technology and Applications International Forum 1999*, edited by M.S. El-Genk, AIP Conference Proceedings 458, American Institute of Physics, 1999.

White, M.A., Qui, S., Erbeznik, R.M., Olan, R.W., and Welty, S.C., "Status of an Advanced Radioisotope Space Power System Using Free-Piston Stirling Technology", in *Meeting Global Energy and Environmental Needs*, Proceedings of the 33rd Intersociety Energy Conversion Engineering Conference, American Nuclear Society, 1998, Paper IECEC-98-417.

White, M.A., Qui, S., Olan, R.W., and Erbeznik, R.M., "Technology Demonstration of a Free-Piston Stirling Advanced Radioisotope Space Power System", in *Space Technology and Applications International Forum 1999*, edited by M.S. El-Genk, AIP Conference Proceedings 458, American Institute of Physics, 1999.

Preliminary Test Results from a Free-Piston Stirling Engine Technology Demonstration Program to Support Advanced Radioisotope Space Power Applications

Maurice A. White, Songgang Qiu, Jack E. Augenblick

Stirling Technology Company
4208 West Clearwater Avenue
Kennewick WA 99336-2626 USA
509/735-4700 phone; 509/736-3660 fax
<mawhite@stirlingtech.com> <sqiu@stirlingtech.com> <jack@stirlingtech.com>

Abstract. Free-piston Stirling engines offer a relatively mature, proven, long-life technology that is well-suited for advanced, high-efficiency radioisotope space power systems. Contracts from DOE and NASA are being conducted by Stirling Technology Company (STC) for the purpose of demonstrating the Stirling technology in a configuration and power level that is representative of an eventual space power system. The long-term objective is to develop a power system with an efficiency exceeding 20% that can function with a high degree of reliability for up to 15 years on deep space missions. The current technology demonstration convertors (TDC's) are completing shakedown testing and have recently demonstrated performance levels that are virtually identical to projections made during the preliminary design phase. This paper describes preliminary test results for power output, efficiency, and vibration levels. These early results demonstrate the ability of the free-piston Stirling technology to exceed objectives by approximately quadrupling the efficiency of conventional radioisotope thermoelectric generators (RTG's).

INTRODUCTION

The baseline 210-W(e) class system incorporates two independent 105-W generators. Each generator integrates two general purpose heat source (GPHS) units with two 55-W(e) RG-55 free-piston Stirling convertors (integral engine/alternator that converts heat to electricity) deployed as a balanced pair. The reference design uses one balanced pair of convertors on each side of the spacecraft. Overall objectives and approach were described at STAIF 99 [White, 1999]. This paper documents the early test results from a single pair of technology demonstration convertors (TDC's) with demonstrated performance that closely correlates with projected performance in [White, 1999]. The TDC's were developed under DOE Contract No. DE-AC03-97SF21317. A NASA Phase II SBIR contract to develop an active adaptive vibration reduction system has resulted in successful balancing demonstrations on two 350-W(e) RG-350 convertors as described in [Thieme, 1999] and another STAIF 2000 Stirling session paper from NASA Glenn Research Center. Application of that technology to the DOE-funded TDC's will be demonstrated in early 2000. Some preliminary results from balancing two TDC convertors by electrical and mechanical coupling without the active balancer are highlighted here. Concepts for integrating direct derivatives of the TDC into a Europa Orbiter or Pluto Express class of spacecraft are described in [Or, 1999]. More updated system integration concepts are described in other STAIF 2000 Stirling session papers from Orbital Sciences Corporation (OSC) and Lockheed Martin Astronautics (LMA).

TECHNOLOGY LEGACY

The RG-55 is one of a family of similar generator sets that includes the 10-W(e) RG-10 radioisotope-fueled generator [Montgomery, 1996] and the 350-W(e) RG-350 remote power generator [Erbeznik, 1998]. These three generators, plus the 1-kW RG-1000 and the 3-kW RG-3000 generators currently under development, all use the same technology approaches. Technologies in common for all of these free-piston Stirling engine/linear alternator generator sets include

CP504, *Space Technology and Applications International Forum–2000*, edited by M. S. El-Genk
© 2000 American Institute of Physics 1-56396-919-X/00/$17.00

a monolithic Inconel 718 heater head, flexure bearings that support pistons with non-contacting clearance seals (no rubbing or wearing parts), and linear alternators that incorporate moving iron laminations and stationary permanent magnets. All the convertors are hermetically-sealed, helium-charged units. The only pressure vessel penetrations are electrical feedthroughs for delivering generated electrical power to the load. This common technology approach was described in detail in other publications [White, 1996 and White, 1999].

Demonstrated endurance testing that comprises a technology legacy for the space power engines includes an on-going test of a single convertor with more than 54,000 maintenance-free, degradation-free hours. Composite operating times include more than 150,000 hours on all related system test units. Much of this system testing was conducted at independent third party test sites, and there has been only one failure of an optional add-on component when operated within specifications. Additional critical component and subsystem endurance tests include more than 1,000 flexure-years of operation on the flexure bearings and more than 50,000 hours on each of two linear alternators.

TDC CONFIGURATION DESCRIPTION

The TDC final design cross-section layout is shown in Figure 1. This is a refined design that integrates the long-life, maintenance-free technology legacy described above with innovative implementation of weight reduction and ruggedization features. The basic configuration, consisting of a posted displacer with internal flexure bearings and a flexurally supported moving iron linear alternator, is the same as for the RG-10 and the RG-350. These units, from which the endurance test legacy is derived, bracket the TDC in power level. Refinements in the TDC relative to the earlier designs include weight reductions and an improved streamline flow path through the internal heat exchangers that improved efficiency. Further significant weight reduction is possible by material substitutions, removal of heavy sections used for invasive instrumentation such as pressure and position transducers that will not be used on flight systems, and removal of heavy flanges used for easy assembly and disassembly of demonstration hardware.

One unique feature in the TDC is the aft backup bearing, which was added to accommodate launch load vibrations. The backup bearing normally operates with a clearance and no rubbing contact just like the power piston. It is not, however, required to provide the seal function that is inherent with the power piston. The flexure bearing stiffness is sufficient to maintain operating clearances for the piston and backup bearing under normal operation, which includes any orientation in an earth gravity field. In the event of more severe side loads, the piston in all STC engines is designed to rub and provide additional bearing support to augment the flexure bearings. The piston utilizes contact surface treatments that are sometimes used in free-piston engines for wear couples that are capable of operating for thousands of hours. For relatively high vibration loads, such as will be experienced during spacecraft launch, there is the risk that side loads on the linear alternator mover (which is cantilevered out a significant distance from the piston) could force the mover over against the stator and damage the system. The clearance between mover and stator is much greater than that between the piston and cylinder, but magnetic side loads increase with eccentricity of the mover and could potentially force the mover over far enough to result in a magnetic lockup. The function of the aft backup bearing is to provide a support wheelbase, in concert with the piston, that spans the entire linear alternator region and minimizes lateral movement to prevent significant magnetic side loads from ever coming into play. The efficacy of this approach will be evaluated in a planned launch load characterization assessment using the vibration testing facility at the NASA Glenn Research Center (GRC) in December 1999.

SPACE POWER SYSTEM GOALS

The primary goal of the DOE contract is to demonstrate TDC operation with acceptable performance in a size appropriate for a 210-W spacecraft power system. This is consistent with the larger goal of developing an advanced radioisotope power system that can enable deep space missions with far less radioisotope inventory than the existing technology as implemented with RTG's. The specific objective of demonstrating an overall thermal to electric conversion efficiency of at least 20% was significantly exceeded as described below. The TDC's also demonstrated the ability to meet the goal of a weight comparable to or less than equivalent RTG's, as well as demonstrating vibration levels well below the target goals identified by the Jet Propulsion Laboratory (JPL). The ability of the power system to function with a high degree of reliability in a deep space environment for up to 15 years is critical.

The NASA SBIR goal is to support the Stirling space power activity by demonstrating an active vibration reduction system that can provide up to two orders of magnitude reduction in vibration, first on existing 350-W engines and then on the DOE 55-W TDC's. Early performance results described in this paper have demonstrated vibration levels below identified target levels. The lowest measured vibration levels provide a factor of 100 or more safety margin below the limit that can cause pixel smear, and even in the worst case scenarios with one convertor in a balanced pair not operating, the active balancer still provides significant margins. System integration design and analyses are also being conducted in collaboration with OSC, LMA, and GRC. Those results are being separately reported in companion papers from GRC, LMA, and OSC. In general, the LMA designs are relatively conservative with contingencies built in at various levels and approaches such as a solid aluminum radiator for heat rejection. The OSC designs use more advanced technology such as heat pipes with carbon composite radiators that are lighter in weight and provide lower heat rejection temperatures for better engine efficiency. As a consequence, the LMA designs result in small power margins while the OSC designs project substantial margins.

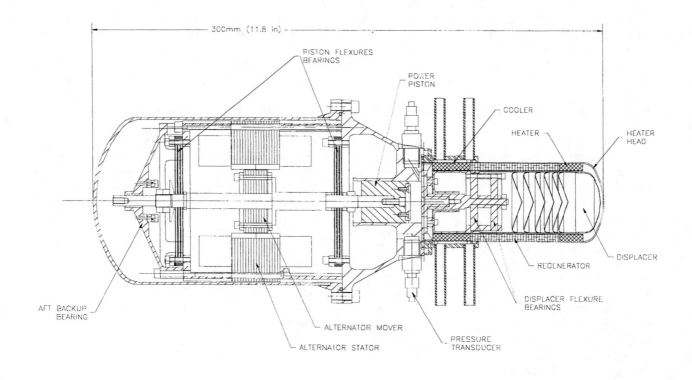

FIGURE 1. Cross-Section of RG-55 Technology Demonstration Convertor (TDC).

PERFORMANCE MEASUREMENTS

Most of the TDC testing to date has supported design iterations typically needed to approach predicted performance levels with a new design free-piston Stirling engine. Achievement of the target performance objectives has been realized in a relatively short time span, with measured power output and efficiency at essentially the levels that were predicted over 1 year earlier for the preliminary design review and published in [White, 1999]. The previously published performance trends were for an earlier 150-W class generator configuration that incorporated three GPHS heat sources and four convertors, with either three or four convertors operational for redundancy. The present baseline generator design uses two GPHS units with two convertors and no redundancy to provide the nominal 105-W output with some margin. Two generators would be used for a typical spacecraft mission with a nominal power requirement of 210 watts. A common reference point is the individual performance of a single convertor.

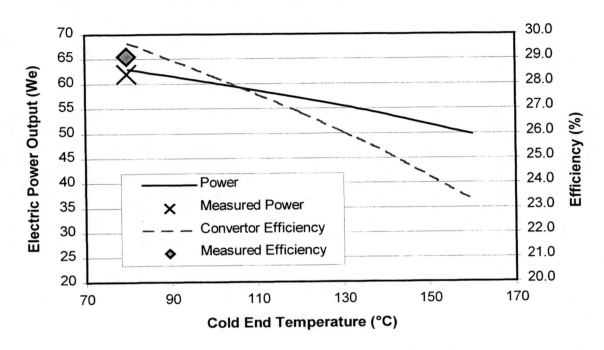

FIGURE 2. TDC Performance as a Function of Cold End Temperature for a Hot End Temperature of 650° C.

Figure 2 is adapted from projected performance plots in [White, 1999] to show the power output and efficiency of a single convertor. It illustrates how the output power and system efficiency vary with rejection temperature at the nominal heat acceptor or hot end convertor temperature of 650° C. These power and efficiency levels are derived from some preliminary reference conditions, which are subject to change as final reference conditions are developed. The present conditions include a reference thermal output from a GPHS heat source of 226 watts at the 9-year point in a mission, where the beginning of mission output is baselined at 243 watts. An assumed multifoil insulation loss of 6% results in 213 watts of thermal input provided to the convertor. Projected efficiency drops rapidly from over 29% at a rejection temperature of 80°C to about 23% at 160° C. It is about 26.5% at the nominal LMA rejection temperature of 120° C. The efficiency reduction is only moderately non-linear, but it shows an increasing rate of degradation as the rejection temperature increases.

Performance mapping of a TDC convertor over a wide range of operating conditions is planned for the near future, but the present data cover only a very limited range of conditions. Prior experience provides a high level of confidence that performance trends over a substantial range of conditions will follow analytical predictions. The important and successful result obtained to date is that the measured data fall essentially on the predicted curves.

TDC test data at a hot end temperature of 650 C and a cold end temperature of 80 C, with actual fiber insulation losses corrected to the 6% assumed multifoil insulation loss, produced 62 watts output at 29.1% efficiency. This compares with 63 watts and 29.6% efficiency for the analytical predictions made for the preliminary design review over 1 year earlier. This level of performance is outstanding, particularly at this early stage in the implementation of a new convertor configuration. It provides a high level of confidence that target performance goals can be met or exceeded for any reasonable set of operating conditions selected for specific mission needs.

VIBRATION BALANCING RESULTS

One factor of concern with Stirling machines in space is that they have reciprocating pistons that can generate undesirable oscillations in a spacecraft. A completed NASA SBIR contract successfully demonstrated a factor of 50 reduction in vibration oscillation levels by coupling two convertors in a balanced mode where they are coupled electrically in parallel and mechanically coaxially [Thieme, 1999]. This is substantially below the target vibration levels provided by JPL, which are in turn a factor of 5 to 10 below the level that can cause pixel smear. The current NASA SBIR contract has demonstrated an additional factor of 15 vibration reduction using an active adaptive vibration reduction system. The primary reason for incorporating an active vibration reduction system is to provide the necessary balancing in the mode where one engine of a balanced pair is not operating. In that condition, the active balancer will also reduce vibration levels far below the target objectives with minimal power consumption.

Vibration imparted to a spacecraft is determined by the convertor unbalanced residual force, the spacecraft moment of inertia, and the moment arm from the spacecraft center of gravity normal to the convertor unbalanced axis of vibration. A worst-case reference scenario has a 1-meter moment arm and 145 kg-m^2 for the lightest dry spacecraft moment of inertia. At the 80 Hz TDC operating frequency, an unbalanced force of 183 newtons would result in a 5-microradian oscillation of the spacecraft. This is below the level that can cause pixel smear, but a further safety factor of two to five is desired. Thus the target force should be in the range of 35 to 90 newtons. Measured vibration force levels for axially coupled TDC's by comparison include the following. When both are operating in the normal mode, the force level is 3.1 newtons. If one is stalled and the other keeps operating, the net unbalanced force is 58.3 newtons. Even with no active vibration balancing this is intermediate to the range of safety margin identified by JPL. Therefore the need for an active balancer is an open issue, especially if efforts are made to tilt the convertor axis more toward the spacecraft center of gravity.

If it is determined that greater safety margins are needed, even the preliminary active balancer results obtained to date provide a high level of confidence that the issue can be put to rest. Early results with a developmental active balancer applied to various implementations of a balanced pair of 350-W RG-350 convertors have demonstrated the basic viability of the approach. When tested with the normally operating balanced pair, vibration forces were reduced by an additional factor of 15. With one of the pair of convertors not operating, the active balancer reduced the vibration level to essentially the same as that for the normally balanced pair. The former case required about 2 watts of power to drive the active balancer, and the latter case required less than 7 watts. When the relevant parameters are scaled from the RG-350 case to the RG-55, balance motor power consumption is less than 1 watt for the normally balanced condition and about 2 watts for the case of one operating convertor. These results indicate that the option is available to reduce vibration to a level as low as may be desired for any conceivable operating condition. The tradeoff is maximum simplicity for worst-case vibration levels that are satisfactory versus some further complexity for worst-case vibration levels orders of magnitude below a potential compromise of data acquisition.

CONCLUSIONS

A hardware demonstration effort, aimed at assessing whether the long-life, maintenance-free Stirling technology developed by STC in 10-W and 350-W sizes could be adapted to meet advanced radioisotope space power system needs, has successfully achieved technology demonstration goals. Ongoing endurance testing of the 10-W terrestrial radioisotope convertor has demonstrated more than 6 years of operation with no maintenance or degradation, and no failure or degradation mechanisms that would prevent several times that period of reliable operation have been identified. Further evaluation is being conducted to determine how a Stirling power system would be integrated with a spacecraft and when and if it will be integrated into future mission programs. Results to date strongly support the conclusion that Stirling is a viable, high performance alternative technology for deep space power systems.

NOMENCLATURE

DOE	Department of Energy	OSC	Orbital Sciences Corporation
GPHS	general purpose heat source	RTG	radioisotope thermoelectric generator
JPL	Jet Propulsion Laboratory	SBIR	Small Business Innovative Research
LMA	Lockheed Martin Astronautics	STC	Stirling Technology Company
NASA	National Aeronautics & Space Administration	TDC	Technology Demonstration Convertor

ACKNOWLEDGMENTS

The authors express their appreciation for programmatic support provided by Richard Furlong, Department of Energy Germantown, and Lanny Thieme, NASA Glenn Research Center.

REFERENCES

Erbeznik, R.M. and Colenbrander, K., "Generators that Won't Wear Out for Biomass Applications," in *Proceedings of the 10th European Conference and Technical Exhibition,* Würsburg Germany, 1998.

Montgomery, W.L., Ross, B.A., and Penswick, L.B., "Third Generation Development of an 11-Watt Stirling Converter," in *Proceedings of the 31st Intersociety Energy Conversion Engineering Conference,* Institute of Electrical and Electronics Engineers, Washington DC, 1996.

Or, C., Carpenter, R., Schock, A., and Kumar V., "Performance of the Preferred Self-Supporting Radioisotope Power System with STC 55-W Stirling Converters," in these *Proceedings from STAIF-2000, Space Technology and Applications International Forum,* Albuquerque, New Mexico, 2000.

Or, C., Kumar, V., Carpenter, R., and Schock, A., "Self-Supporting Radioisotope Generators with STC 55-W Stirling Converters," in these *Proceedings from STAIF-2000, Space Technology and Applications International Forum,* Albuquerque, New Mexico, 2000.

Thieme, L.G., Qiu, S, and White, M.A., "Technology Development for a Stirling Radioisotope Power System," in *Proceedings of the 34th Intersociety Energy Conversion Engineering Conference,* Society of Automotive Engineers, Vancouver, British Columbia, Canada, 1999.

White, M. A., Colenbrander, K., Olan, R.W., and Penswick, L.B., "Generators That Won't Wear Out," *Mechanical Engineering,* pp. 92-96, 1996.

White, M.A., Qiu, S, Olan, R.W., and Erbeznik, R.M., "Technology Demonstration of a Free-Piston Stirling Advanced Radioisotope Space Power System," in *Proceedings from STAIF-99, Space Technology and Applications International Forum,* Albuquerque, New Mexico, 1999.

Solar Stirling for Deep Space Applications

Lee S. Mason

NASA Glenn Research Center, Cleveland, OH 44135
Lee.Mason@grc.nasa.gov, (216) 977-7106

Abstract. A study was performed to quantify the performance of solar thermal power systems for deep space planetary missions. The study incorporated projected advances in solar concentrator and energy conversion technologies. These technologies included inflatable structures, lightweight primary concentrators, high efficiency secondary concentrators, and high efficiency Stirling convertors. Analyses were performed to determine the mass and deployed area of multi-hundred watt solar thermal power systems for missions out to 40 astronomical units. Emphasis was given to system optimization, parametric sensitivity analyses, and concentrator configuration comparisons. The results indicated that solar thermal power systems are a competitive alternative to radioisotope systems out to 10 astronomical units without the cost or safety implications associated with nuclear sources.

INTRODUCTION

The traditional means to satisfy electrical power requirements for outer planetary space probes is through Radioisotope Thermoelectric Generators (RTGs). RTGs were most recently used on the Galileo, Ulysses and Cassini spacecraft (Kelly, 1997). A joint DOE/NASA program is in place to develop an improved radioisotope power system to replace RTGs. The higher efficiency, Advanced Radioisotope Power System (ARPS) will reduce the required plutonium inventory providing cost and safety benefits (Herrera, 1998). ARPS will utilize an Alkali-Metal Thermal-to-Electric Converter (AMTEC) combined with three General Purpose Heat Source (GPHS) modules to produce 92 watts beginning-of-mission (BOM) in a single power unit. The AMTEC system is projected to offer at least a 2x improvement in conversion efficiency as compared to conventional thermoelectric converters used in RTGs. This technology is planned for use on future deep space science missions such as Europa Orbiter ('03) and Pluto-Kuiper Express ('04). The projected specific power for the ARPS system is 6.2 W/kg and the efficiency is 12.6% at BOM (Lockheed Martin, 1998)

While reducing the amount of plutonium reduces the health risk associated with an accidental orbital reentry and provides substantial system cost savings, it would be desirable to have a non-nuclear option for deep space missions. However, typical planar photovoltaic (PV) arrays are not effective for space probes traveling beyond Mars (1.5 astronomical units, or AU) due to the decrease in insolation with the square of the distance from the sun.

FIGURE 1. Lightweight Inflatable Concentrator (courtesy of SRS Technologies)

CP504, *Space Technology and Applications International Forum–2000*, edited by M. S. El-Genk
2000 American Institute of Physics 1-56396-919-X

Solar thermal power systems offer a potential alternative. Progress in advanced lightweight concentrator technology provides a necessary first step toward making solar thermal power for deep space missions a viable option. Companies such as L'Garde, SRS Technologies, ILC Dover, United Applied Technologies, and Harris Corporation are developing concepts for large, lightweight solar concentrators. Figure 1 shows an example of a lightweight concentrator using thin-film, inflatable technology. This advanced concentrator technology offers a factor of five improvement in aerial density (kg/m^2) over conventional rigid panel concentrators (Mason, 1999). The other key elements to a mass competitive solar power system for far-sun missions are high efficiency secondary concentrators and high efficiency, free-piston Stirling convertors.

Secondary concentrators can provide an increase in the overall geometric concentration ratio as compared to primary concentrators alone. This reduces the diameter of the receiver aperture and the associated infrared cavity losses, thus improving overall efficiency. The use of a secondary concentrator also eases the pointing and surface accuracy requirements of the primary concentrator, making inflatable structures a more feasible option. Typical secondary concentrators are hollow, reflective parabolic cones. Recent studies at Glenn Research Center have investigated the use of a solid, crystalline refractive secondary concentrator for solar thermal propulsion which may provide considerable improvement in throughput efficiency by eliminating reflective losses (Wong, 1999). The refractive secondary concept, shown in Figure 2, also offers the benefit of directed flux tailoring within the receiver cavity via a unique "flux extractor." Such a device has the potential to improve the energy transfer to the Stirling heater head.

FIGURE 2. Refractive Secondary Concentrator

Stirling convertors have the potential to provide very high thermal-to-electric conversion efficiency. Stirling Technology Company (STC) in Kennewick, Washington has successfully designed, built, and operated free-piston convertors at 10 watts and 350 watts for terrestrial applications. The 350 watt STC convertor is pictured in Figure 3. STC is also developing a space-rated, 55 watt unit for radioisotope applications designed to provide system conversion efficiencies of greater than 24% (White, 1999). All of these engines share common technology characteristics including flexure bearings and linear alternators. The 10 watt engine has undergone endurance testing to over 50,000 hours in order to demonstrate long life and reliability.

FIGURE 3. Free-Piston Stirling Convertor (courtesy of STC)

STUDY GROUNDRULES

The overall study objectives were three-fold: 1) determine the feasibility of a solar Stirling power system for deep space missions using various advanced component technologies, 2) determine the key parameters which most influence system performance, and 3) compare system performance to other deep space power system options. The analysis evaluated system mass and deployed area for solar thermal power systems out to Pluto (about 40 AU).

Some of the key study assumptions are provided in Table 1. A reference electrical power level of 200 watts was chosen as typical of future deep space missions. The insolation and effective sink temperature were varied with distance from the Sun. Several of the component metrics were derived from design work performed by Orbital Sciences Corporation (OSC) in support of a Stirling concept for ARPS (Schock, 1999). The solar heat receiver was envisioned as a simple structure which supports the secondary concentrator and provides a thermal interface to the Stirling heater head, similarly to the GPHS container for the radioisotope Stirling concept. The waste heat radiator and the power management and distribution (PMAD) system were also derived from the OSC Stirling concept. A 10% system mass margin was included to account for interface structure and other unknowns.

TABLE 1. Key General Assumptions

Element	Assumptions
System	200 W electric power, 10% mass margin
Primary Concentrator	3.5% pointing loss, 10% wrinkle loss, 10% area margin
Secondary Concentrator	7:1 concentration ratio, reflective – 85% efficiency, refractive – 95% efficiency, passive cooling
Receiver	No thermal energy storage, 6 kg/kWt, 5% insulation loss
Stirling Convertor	STC design, 43 kg/kWe (includes active balancer), 2 convertors/system, temperature ratio <4.5, heater head temperature <1300 K, convertor efficiency = f(Trat)
Radiator	Heat pipe with C-C facesheets, 2 sided, 75% effective area, 6.4 kg/m², sink temperature = f(AU)
PMAD	28 Vdc bus, 150 W/kg, 95% efficiency

In order to determine the most promising component technologies, several different representative concentrator configurations were compared. These included a parabolic, thin-film inflatable system having a total reflection/transmission (R/T) efficiency of 63%, an areal density of 2 kg/m² (which includes the gas make-up system), and an Earth geometric concentration ratio (GCR) of 1600:1. The Earth GCR is defined as the ratio of the primary concentrator area to the receiver aperture area (or secondary entrance area) as required at 1 AU and provides a measure of the concentrator's overall surface accuracy. The theoretical maximum GCR for a solar concentrator at 1 AU having a focal distance-to-diameter (f/d) ratio of 1 is about 12000:1. This ratio varies with distance from the sun in relation to the subtended angle of the Sun relative to the concentrator. A second primary concentrator concept employing inflatable structure and a flat, fresnel reflector was assumed to have a combined R/T efficiency of 85%, an areal density of 0.5 kg/m², and an Earth GCR of 1000:1. Three different secondary concentrator options were considered: no secondary, a reflective secondary, and a refractive secondary. The masses of the secondary concentrators were scaled based on previous designs, the refractive crystal having a mass of over four times that of the reflective option for the same entrance diameter.

The Stirling temperature ratio (Trat, defined as Thot/Tcold) and heater head temperature (Thot) were concurrently optimized for minimum system mass. Generally, higher temperature ratios relate to higher conversion efficiency (smaller primary concentrators) at the expense of lower waste heat rejection temperatures (larger radiators). Consequently, a mass optimized temperature ratio results from the trade-off of concentrator mass and radiator mass. Varying the heater head temperature yields a minimum system mass based on a balance of infrared cavity loss and Stirling efficiency. Higher temperatures result in greater receiver losses, but allow the Stirling to operate at higher efficiency (higher Trat) without adversely effecting radiator size. Like the temperature ratio optimization, the heater head temperature optimization also results from a compromise between concentrator mass and radiator mass. For this study, Stirling temperature ratio was limited to 4.5 and heater head temperature was limited to 1300 K.

STUDY RESULTS

Figure 4 illustrates the Stirling optimization process showing system mass as a function of Stirling temperature ratio for a Jupiter mission (5.2 AU) using a Fresnel primary and a refractive secondary. Local minimum mass points for each of three different heater head temperatures are indicated by asterisks. Higher temperatures result in greater optimum temperature ratios. The global minimum mass design point was achieved at a heater head temperature of 1150K and a temperature ratio of 3.8 resulting in a system mass of 33.6 kg. The optimum heater head temperature and temperature ratio varied greatly with mission destination (i.e. solar distance). For Mars, minimum system mass was achieved at a heater head temperature of 1300K, while a Pluto system resulted in a minimum mass heater head temperature of 600K. The key factor in determining the optimum heater head temperature was the size of the receiver aperture and the associated infrared losses. At near-Earth distances, the receiver aperture was relatively small so a high temperature cavity did not produce excessive losses. However, as the primary concentrator increased for greater solar distances, a corresponding increase in receiver aperture size necessitated a lower cavity temperature to control receiver losses.

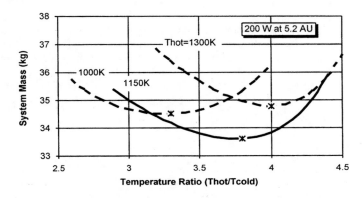

FIGURE 4. Stirling Mass Optimization for Jupiter Mission

A comparison of concentrator configurations for the 5.2 AU Jupiter mission is provided in Table 2. The Fresnel primary and refractive secondary combination resulted in the lowest system mass. The higher mass for the thin-film cases, was primarily a result of the 4x increase in areal density relative to the Fresnel. The Fresnel/refractive system also corresponded to the highest system efficiency, defined as the ratio of electric power produced to solar power collected by the primary. System efficiency was found to be a good indicator of system mass since the primary concentrator tended to be the dominant mass component. The primary concentrator was about 35% of the system mass at 5.2 AU, and beyond 10 AU, the mass fraction increased to greater than 50%.

TABLE 2. Concentrator Comparison at 5.2 AU

Primary	Secondary	Mass (kg)	Diam (m)	Thot (K)	Sys Eff (%)
Fresnel	None	40.9	6.1	750	13.5
	Reflective	34.4	5.6	1150	15.9
	Refractive	33.6	5.3	1150	18.2
Thin-film	None	90.2	6.0	900	13.9
	Reflective	83.4	6.0	1150	13.9
	Refractive	76.5	5.7	1150	15.7

The performance metrics assumed for the two primary concentrator options were chosen by projecting present day performance toward future systems. Uncertainty in those projections makes it appropriate to evaluate performance sensitivities. Figure 5 compares system mass versus primary diameter at 5.2 AU with parametric variations in Earth GCR, areal density, and R/T efficiency. The reference point represents the baseline assumptions for the Fresnel primary: 1000:1 Earth GCR, 0.5 kg/m^2, and 85% R/T efficiency. In general, the Earth GCR and the R/T efficiency parameters have a greater influence on the primary diameter size. Conversely, the concentrator areal density has a dramatic effect on system mass. These same trends were consistent over the entire range of solar distances considered in the study.

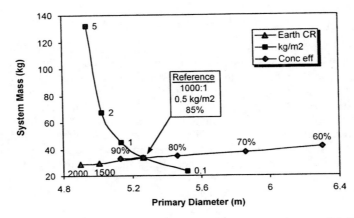

FIGURE 5. Primary Concentrator Sensitivity Analysis, 200 W at 5.2 AU

The variation in system mass and primary diameter with increasing solar distance is presented in Figure 6. Since areal density was determined to be a key system mass driver, values from 0.1 kg/m² to 5 kg/m² were considered. Based on the entrance diameter and the corresponding mass of the refractive crystal, it was desirable to use a reflective rather than a refractive secondary for missions beyond 10 AU. Below 10 AU, reasonable system mass was achievable with primary concentrators of less than 10 m and areal densities of less than 2 kg/m². Systems for missions beyond 10 AU required primary concentrators greater than 20 m and areal densities below 0.5 kg/m² in order to achieve reasonable system mass.

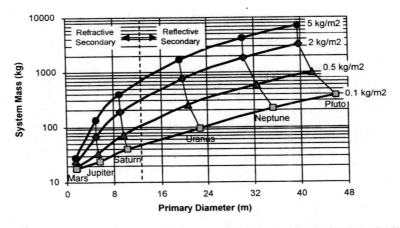

FIGURE 6. Performance Variations with Distance from Sun for 200 W Solar Stirling

SYSTEM COMPARISONS

Coincidentally, the 10 AU breakpoint serves as a reasonable upper limit for this technology as compared to radioisotope power systems. Table 3 compares system performance of 200 watt solar Stirling power systems at 1.5 AU (Mars), 5.2 AU (Jupiter), and 9.5 AU (Saturn) with two different radioisotope options: ARPS and small RTG. The solar power systems utilize the Fresnel/refractive concentrator configuration and vary in specific power from just under 3 W/kg for Saturn to almost 11 W/kg for Mars. The radioisotope systems require two units to approach the 200 watt end-of-mission (EOM) requirement resulting in specific power levels between 4 and 5 W/kg. A 200 watt solar Stirling for Jupiter has about the same mass as two ARPS units providing 150 watts EOM. The ARPS configuration would require 6 GPHS modules for the two units, while the small RTGs would require a total of 12 GPHS modules. In reference to the size of the solar collector, the 5.3 m primary concentrator diameter for the Jupiter system is similar to one Tracking and Data Relay Satellite System (TDRSS) antenna.

The natural decay of the radioisotope source causes a decrease in electrical power output with time. The EOM power and efficiency estimates for the radioisotope systems in Table 3 are based on the GPHS providing 232 watts per module after six years of operation (BOM thermal power from a GPHS module is 243 watts). Definition of the BOM power level for the solar systems requires further study. The large primary needed to collect power at the outer planets causes excessive power to be collected at Earth orbit. Some form of energy management would be required. Options include: 1) off-pointing of the primary, 2) adaptive focusing of the primary, 3) variable diameter shutter on the receiver aperture, or 4) a high temperature radiator for the receiver. An additional option might be to use a small solar panel for initial power and deploy the concentrator at further distances from the Sun.

TABLE 3. Comparison to Radioisotope Power Systems

	Solar Stirling	Solar Stirling	Solar Stirling	ARPS AMTEC	Small RTG
Distance (AU)	1.5	5.2	9.5	variable -	6 yr life
EOM power (W)	200	200	200	150	182
EOM system efficiency (%)	17.8	18.2	18.1	10.8	6.5
Total mass (kg)	18.6	33.6	74.4	29.8	45.2
Sp power (W/kg)	10.7	5.9	2.7	5.0	4.0
Prim dia (m)	1.6	5.3	9.7	-	-

CONCLUSION

The results of this study indicate that a solar Stirling power system is a feasible alternative for deep space applications. The key technology elements include a lightweight primary concentrator, a high efficiency secondary concentrator, and a high efficiency Stirling convertor.

System mass and deployed area were characterized out to 40 AU. Various concentrator configurations were considered included Fresnel and thin-film primaries, and reflective and refractive secondaries. Earth geometric concentration ratio and reflection/transmission efficiency of the primary were found to have a major effect on concentrator deployed area. Concentrator areal density was a key system driver, having a dominant influence on overall system mass. The flexibility of the Stirling convertor to operate at variable heater head temperatures, depending on the mission destination, helped to control infrared losses and maintain high overall system efficiency.

The solar Stirling system compared favorably with other deep space power options. Mass was competitive with radioisotope power systems out to 10 AU. A 200 watt solar power system for Jupiter offered 18.2% system efficiency at a specific power of 5.9 W/kg, a modest improvement over projected ARPS performance without the complications brought on by nuclear sources.

REFERENCES

Herrera, L., "The U.S. Department of Energy Advanced Radioisotope Power System Program," *Proceedings of the 33rd Intersociety Energy Conversion Engineering Conference*, 1998.

Kelly, C.E., and Klee, P.M., "RTG Performance on Galileo and Ulysses and Cassini Test Results," *Proceedings of the Space Technology and Applications International Forum (STAIF-97)*, 1997.

Lockheed Martin Astronautics, "ARPS Quarterly Program Review, DOE Headquarters," September 17, 1998.

Mason, L., "Technology Projections for Solar Dynamic Power," *Proceedings of the Space Technology and Applications International Forum (STAIF-99)*, NASA/TM-1999-208851, 1999.

Schock, A., Or, C., and Kumar, V., "Radioisotope Power System Based on Improved Derivative of the Existing Stirling Engine and Alternator," *Proceedings of the Space Technology and Applications International Forum (STAIF-99)*, 1999.

White, M.A., Qui, S., Olan, R.W., and Erbeznik, R.M., "Technology Demonstration of a Free-Piston Stirling Advanced Radioisotope Space Power System," *Proceedings of the Space Technology and Applications International Forum (STAIF-99)*, 1999.

Wong, W.A, and Macosko, R.P, "Refractive Secondary Concentrators for Solar Thermal Applications," *Proceedings of the 34th Intersociety Energy Conversion Engineering Conference*, NASA/TM-1999-209379, 1999.

Pulsed Laser Thermal Propulsion For Interstellar Precursor Missions

Jordin T. Kare

Kare Technical Consulting, 222 Canyon Lakes Place, San Ramon, CA 94583
(925) 735-8012; jtkare@ibm.net

Abstract. A laser-thermal propulsion system is proposed for launching large numbers of small interstellar precursor probes at velocities up to ~300 km/s (0.001 c). This system uses a stationary pulsed laser, based on Earth or in near-Earth space, to beam energy to probe vehicles during their initial acceleration. Each vehicle collects laser energy using a deployable reflector, and focuses the laser energy into a thruster. The focused laser pulses ablate and heat an inert propellant, which expands to produce thrust at a selectable specific impulse up to of order 20,000 seconds (exhaust velocity up to 200 km/s). This technology permits the vehicles to be simple and light, while allowing much higher acceleration than alternative propulsion systems. The laser system is ideal for launching large numbers of flyby probes, for example to examine many objects in the Oort cloud. A laser system with 30-meter-class transmitting optics and a 100-MW laser is capable of launching 100 kg payloads to 50 km/s, with payload mass fraction (probe payload / probe initial mass in Earth orbit) of 10-20%. The same system can launch much larger payloads to lower velocities for Solar System exploration. Scaling relationships are derived and scaling options discussed, along with possible near-term development and proof-of-concept tests.

INTRODUCTION

Interstellar precursor missions such as Kuiper object flybys, Solar Focus missions, and the TAU (Thousand Astronomical Unit) mission involve launching spacecraft to ranges of 50 – 1000 AU. To complete these missions in reasonable time (5 – 20 years) requires velocities of 10 – 50 AU per year, or 50 – 300 km/second mean mission velocity.

Current trends in miniaturization are shifting mission designs toward small spacecraft (100 – 1000 kg) and microspacecraft (<100 kg). For many missions, such as exploration of the Kuiper belt and Oort cloud, large numbers of such small vehicles may be inherently more productive than single large spacecraft, since no single "target" exists. A high ΔV propulsion system matched to small vehicles would thus be highly desirable, especially if it permitted launching many missions at low per-mission cost.

PROPULSION CONCEPT

Laser propulsion is a beamed-energy propulsion concept in which a stationary ground- or space-based laser transmits energy to the spacecraft. Because the expensive components of the system – the laser and transmitting optics -- remain stationary, they can be reused over many missions. Laser-electric propulsion, in which the laser energy is converted via photovoltaic (PV) cells to electricity to drive an electric thruster, has been evaluated for both near-Earth and interstellar missions (Forward 1985, Jackson 1978, Landis 1989), but has limited thrust-to-weight and high per-mission cost due to the large PV cell area needed.

In laser thermal propulsion, the laser energy is used directly to heat a propellant, which is exhausted to produce thrust. Laser thermal propulsion has been considered mainly for Earth launch and near-Earth maneuvering, using either continuous (CW) or pulsed lasers and various types of thruster (Glumb 1984, Kare 1990, Kare 1995, Phipps 1994). The exhaust velocities required for interstellar precursor missions are too high for practical operation of a

CP504, *Space Technology and Applications International Forum–2000*, edited by M. S. El-Genk
© 2000 American Institute of Physics 1-56396-919-X/00/$17.00

CW thruster with a material nozzle. CW thrusters with specific impulse (I_{sp}) >> 1000 s may be possible using magnetic nozzles, but this author is not aware of any studies of this option.)

Pulsed ablation thrusters use the laser energy to ablate a thin layer of a (usually solid) propellant. If the laser flux and fluence are sufficiently high, the ablated material will form an absorbing plasma which can be further heated by the beam to extremely high temperatures. The actual plasma temperature can be controlled by varying the laser flux and fluence, and to some extent by varying the laser pulse shape.

A pulsed thruster may or may not use a divergent nozzle to constrain the plasma expansion. For sufficiently short pulses, the plasma is created in a layer that is thin compared to the width of the thruster, and expansion is effectively one-dimensional over most of the thruster even without a nozzle. Such a nozzleless thruster is particularly simple, and has the advantage that the laser illumination can come from any angle, while the thrust is always normal to the propellant surface.

The primary components of a laser thermal propulsion system for in-space use are shown in Figure 1. The laser produces an average power P, and may be continuous (CW) or pulsed. The laser beam is expanded through various optics and finally focused by a transmitting aperture of diameter D_t, which may be a large diffractive element or concave mirror. We assume for this paper that the beam leaving the transmitting aperture is diffraction-limited. The beam is received at the vehicle through by a collector of diameter D_c, which may also be a diffractive lens or mirror, and can generally be of much poorer accuracy. The collector focuses the beam onto (or into) the thruster.

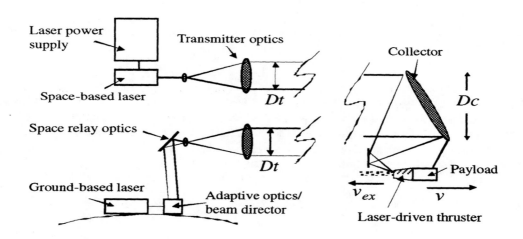

FIGURE 1. Laser Propulsion System Concepts for In-Space Propulsion.

Pulsed Ablation Thruster Characteristics

A laser pulse with energy E at the thruster ablates a mass δm of propellant. In vacuum, the resulting plasma expands, reacting against the remaining propellant and nozzle, if any, and eventually reaching a mean directed exhaust velocity v_{ex} relative to the vehicle. By conservation of momentum, this results in an impulse of $\delta m * v_{ex}$, and a specific impulse of v_{ex}/g, where g is 9.8 m/s^2. v_{ex} depends on the material ablated, the laser flux and fluence, and to some extent the laser wavelength and pulse shape.

The thruster efficiency η_{thr} is defined as:

$$\eta_{thr} = \delta m \, v_{ex}^2 / 2E, \tag{1}$$

and represents the fraction of laser energy that is coupled into useful exhaust kinetic energy (as opposed to radiation, lateral motion, or internal energy of the exhausted gas). Thruster efficiency is also dependent on the ablated material, pulse properties, etc. Simple ablation of common materials (aluminum, plastics, etc.) can produce exhaust velocities of 10^3 to 10^5 m/s with efficiencies of $10 - 50\%$, although with considerable scatter in the data (Phipps 1994). With optimized material and pulse properties, we assume $\eta_{thr} = 0.2$ will be consistently achievable up to $v_{ex} = 200$ km/s While this efficiency may seem low, it is comparable to the efficiency of a laser-electric propulsion system:

$$\eta(\text{PV cell}) \times \eta(\text{power conversion}) \times \eta(\text{electric thruster}) \approx 0.4 \times 0.95 \times 0.5 = 0.19. \tag{2}$$

For convenience, we define an exhaust power $P_{ex} = \eta_{thr}\, \eta_{opt} P$, where η_{opt} is the optical transmission efficiency of the laser beam, including diffraction, scattering and reflection losses, etc.

SYSTEM SCALING

The laser propulsion thruster exhaust velocity can be chosen by the system designer; for a given laser power, higher v_{ex} yields lower thrust (as $1/v_{ex}$, assuming fixed η_{thr}) and lower propellant consumption (as $1/v_{ex}^{2}$). Assuming both P and v_{ex} are held constant over the acceleration period, the mass ratio and relative time to accelerate a given final mass m_f to a final velocity v_f varies with v_{ex} as shown in Figure 2. Acceleration time (and therefore total laser energy used) are minimized at $v_f/v_{ex} \sim 1.63$, but varies slowly for $1 < v_f/v_{ex} < 2$.

Starting from rest, the range at "burnout" is given by:

$$R = \frac{m_f v_f^{3}}{2 P_{ex}} \left(\beta^{-3}\right)\left(e^{\beta} - \beta - 1\right), \tag{3}$$

where $\beta = v_f/v_{ex}$. The variation in range with v_f/v_{ex} is also shown in Figure 2; the minimum range occurs at slightly lower exhaust velocity and higher mass ratios than the minimum time. Slightly shorter times can be achieved by varying the exhaust velocity over time to approximate $v_{ex}(t) = v(t)$ (which gives maximum energy efficiency) at the expense of substantially higher mass ratios. Actual systems may choose to use higher-than-optimum-range exhaust velocities to minimize propellant mass launched from Earth, although the mass ratios involved are small by chemical rocket standards. We assume in scaling estimates below that $v_f/v_{ex} = 1$, so that the final mass is ~35% of the initial mass in Earth orbit.

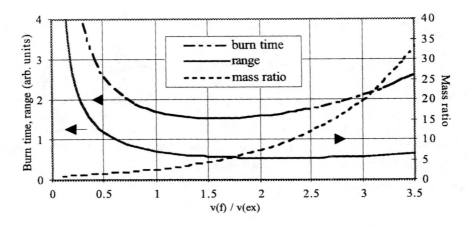

FIGURE 2: Variation of acceleration (burn) time, range, and mass ratio with v_f/v_{ex}.

The range at burnout is limited by diffraction of the transmitted beam. This limit is given by:

$$R_{max} = \frac{D_t D_c}{f_{opt} \lambda}. \tag{4}$$

Traditionally, f_{opt} is taken to be 2.44, which corresponds to the collector just spanning the first null of the far-field Airy pattern for an initially uniform beam and collecting 84% of the beam energy. In fact, a useful fraction of the transmitted beam power can be collected at substantially longer range, and η_{opt} should be calculated as a function of range. For simplicity, we take $f_{opt} = 2$ and fix η_{opt} at 0.5 in the system calculations below.

(The effective value of R_{max} can be increased – in the limit, doubled – by starting acceleration "uprange" of the transmitter aperture, rather than starting from the vicinity of the transmitter. This requires that the transmitter be able to track the vehicle over a large angle at a significant angular rate, and that the vehicle be able to accept a beam from in front as well as behind. This factor is _not_ included in the system performance calculations.)

Setting $R = R_{max}$ and rearranging, we get:

$$m_f = \left(\frac{D_t D_c}{f_{opt} \lambda} \right) \left(\frac{2 P \eta_{thr} \eta_{opt}}{v_f^3 f(\beta)} \right), \tag{5}$$

where $f(\beta) = \beta^{-3} \left(e^\beta - \beta - 1 \right) \approx 0.52$ (minimum) to 0.72 ($\beta = 1$). All of the terms on the right are independent variables except the collector diameter D_c. As the final mass increases, clearly the mass available for the collector also increases. Assuming a fixed fraction f_c of m_f is allocated for a collector (and its support structure) with a mean areal density σ:

$$f_c m_f = \frac{\pi}{4} \sigma D_c^2 \quad \text{or} \quad D_c = \sqrt{\frac{4 f_c m_f}{\pi \sigma}}. \tag{6}$$

Defining a "payload" mass (vehicle dry mass excluding collector) $m_{pay} = (1 - f_c) m_f$, we have:

$$m_{pay} = (1 - f_c) \left(\frac{4 f_c}{\pi \sigma} \right) \left(\frac{D_t}{f_{opt} \lambda} \right)^2 \left(\frac{2 P \eta_{thr} \eta_{opt}}{f(\beta)} \right)^2 v_f^{-6}, \tag{7}$$

which is obviously maximized at $f_c = 0.5$, i.e., roughly half the mass remaining at burnout should be collector.

A variation of v_f^{-6} is quite impressive, suggesting that, for example, a system capable of launching 1 kg to 300 km/s would at least in principle be able to launch 1000 tons to 30 km/s. In practice, collector mass is likely to vary somewhat faster than D_c^2, and eventually reach a practical limit for a given technology, but m_f will still vary rapidly with v_f.

D_t is limited by cost and available space optics technology. In general, telescope apertures vary in cost approximately as $D^{2.5}$. Assuming P is proportional to the laser cost, it can be shown that the optimum allocation of cost is 2/7 to the transmitter aperture and 5/7 to the laser, and in this simple model mf will vary approximately as the system capital cost to the 2.8 power, or, for fixed payload mass, the system cost will vary roughly as v_f^2.

PROJECTED SYSTEM CHARACTERISTICS

Existing high average power lasers operate primarily in the range between 1 and 10 μm. Future high power lasers may be expected to operate between 0.3 and 1 μm. Current costs for ground-based high-average-power lasers are well over $1000/watt (Phipps 1996). However, both diode-pumped solid state lasers and free-electron lasers have the potential to produce power levels of 10 – 100 MW at ~$100/watt or less. Reaching $10/watt, even using ground-based laser hardware, will require significant progress or breakthroughs such as low-cost phase-locking of large arrays of diode lasers, but seems plausible in 20 to 30 years.

Space optical systems for the near-IR planned for the next 10 years, such as the 8-meter Next Generation Space Telescope, have masses of ~15-20 kg/m^2 (Stockman 1997); advanced reflective optics technologies are under development with predicted masses ranging from 10 to 1 kg/m^2. Diffractive optics, which are well suited for use with monochromatic light, are proposed with aperture masses well below 1 kg/m^2 at 25-50 m aperture diameter (Hyde 1998). We can therefore reasonably expect a 10,000-kg-class, $1-B class optical aperture to evolve from ~10 to ~100+ m diameter over the next few decades.

Collector optics may be extrapolated to a small multiple of the mass of a reflective film, assuming either a very lightweight (e.g., gas-inflated) concave reflector or an embossed holographic concentrator. In the former case, the short operating life may allow lighter structure than would be required for, e.g., a long-lived solar concentrator. With current technology, inflatable solar concentrators have areal densities of <0.2 kg/m^2 (5 m diameter, 3.3 kg (SRS Technologies 1999)). An aggressive collector technology, using, e.g., embossed aluminum film 1 µm thick as a holographic reflector, could reduce this by an order of magnitude, with corresponding tenfold increase in payload. This is still several times the areal density proposed for aggressive solar or laser sail designs (Matloff 1984).

Figures 3a and 3b shows the laser power and transmitter aperture diameter for various mission velocities for a nominal mission with m_{pay} = 100 kg. Figures 3c and 3d give the system range and the time to accelerate one payload. Note that these are still "first approximation" values, assuming free-space acceleration, and do not account for (or take advantage of) orbital dynamics. Figure 3e shows a calculated "system cost" based on simple scaling of the laser and telescope costs only. This cost is obviously only a rough indicator of what a real system would cost, but does indicate how the cost scales with mission velocity and technological progress. The assumptions used to derive Figure 3 are given in Table 1.

TABLE 1. Nominal Parameters for Laser Propulsion System Scaling.

Parameter	Near term (2010)	Mid term (2020)	Far term (2030)
Laser cost ($/watt)	100	30	10
Telescope cost baseline (Dia. of $1B telescope, m)	10	30	100
Collector areal density σ (kg/m^2)	0.2	0.1	0.02
Wavelength (µm)	1	0.5	0.33

Common parameters: η_{thr} = 0.2, η_{opt} = 0.5, f_{opt} = 2, β = 1, payload mass 100 kg.

Several features stand out in Figure 3. First, laser power levels and transmitting apertures for interstellar precursor missions, while large by current standards, are well within reach of foreseeable engineering. Lasers with power levels of 100 MW to 1 GW have been proposed (and analyzed in some detail) for strategic defense. 100-meter-class optics are well beyond the current state of the art, but are plausible with thin-film mirrors or diffractive optics, especially for a narrow-band, narrow-field-of-view system. Second, a laser propulsion system for interstellar precursor missions benefits strongly and synergistically from progress in several technology areas, notably lasers, space telescopes, and thin-film collectors. System performance improves essentially as the product of the performance of these individual technologies, rather than being driven by a single "long pole". Thus, despite the extremely steep relationship between mission velocity and system parameters, the mission velocity achievable at a given level of system cost will almost certainly increase steadily over the next few decades, and in the long run, laser propulsion is likely to be superior to other propulsion approaches which are dependent on progress in a single technology.

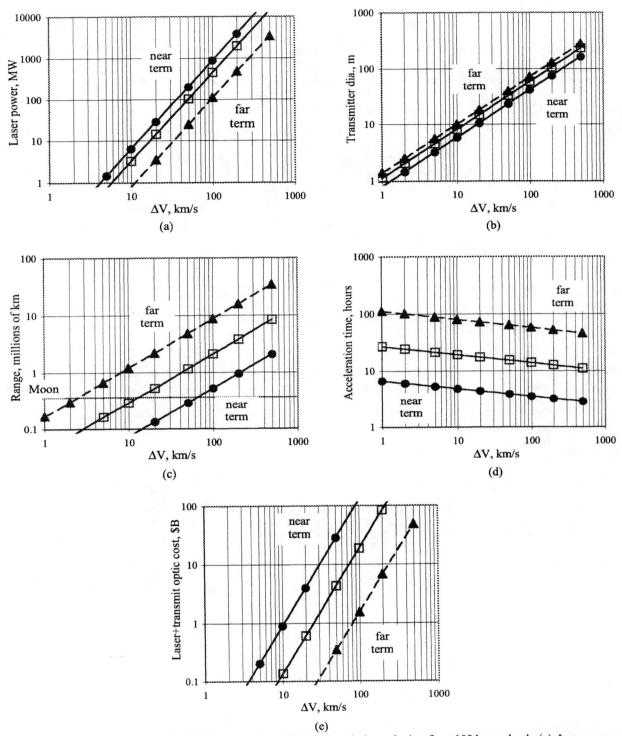

FIGURE 3. Variation in laser propulsion system characteristics vs. mission velocity, for a 100 kg payload: (a) Laser power, (b) Transmitter aperture, (c) Range at burnout, (d) Acceleration time, and (e) Estimated basic system (laser+transmitter optics) cost. Parameters for near-, mid-, and far-term system are given in Table 1.

OPTIONS FOR DEVELOPMENT AND DEMONSTRATION

Near term development of high-I_{sp} pulsed ablative propulsion requires mainly incremental extension of existing laser-ablation research, including laser lethality studies, to repetitive pulses, and to materials chosen for optimum propellant properties. Repetitive pulses are needed to measure the efficiency and exhaust velocity in steady state. Models of laser ablation processes and gas/plasma flow can be used to predict efficiency in operation and investigate, e.g., required beam uniformity and optimum laser pulse profile.

Demonstration of propulsion in space will require either a laser with high pulse energy and reasonable (kilowatt-class) average power in space, or such a laser on the ground with suitable adaptive and relay optics to propagate a high quality beam to at least low Earth orbit. A plausible "technology demonstration" system would have a 100 kW-class 1 µm laser, presumably ground-based, and 3- to 4-meter transmitting aperture. Such a system has sufficient optical range (~30,000 km, to an 18 meter, 100-kg collector) to provide high-I_{sp} propulsion anywhere in sub-geosynchronous space., At a thruster I_{sp} of 1000 s, it would take 14.6 hours of thrusting to accelerate a 100 kg payload by 1 km/s. A demonstration system could thus provide useful orbital maneuvering propulsion as well as serving as a testbed for a larger system for interstellar precursor missions.

REFERENCES

Forward, R. L., "Advanced Propulsion Concepts Study: Comparison of SEP and Laser Electric Propulsion", Final Report, JPL Contract 954085, 1985.

Glumb, R. J., and Krier, H., "Concepts and Status of Laser-Supported Rocket Propulsion," *J. Spacecraft and Rockets* **21**, 70-79 (1984).

Hyde, R. A., "Large Aperture Fresnel Telescope," UCRL 131320, Lawrence Livermore National Laboratory, 1998.

Jackson, A.A., and Whitmire, D.O, "Laser Powered Interstellar Rocket," *J. British Interplanetary Soc.* **31**, 335-337 (1978).

Kare, J. T., "Pulsed Laser Propulsion for Low-Cost High-Volume Launch to Orbit," *Space Power* **9**, 67-76 (1990).

Kare, J. T., "Laser Powered Heat Exchanger Rocket for Ground-to-Orbit Launch," *J. Propulsion and Power* **11**, 535-543 (1995).

Landis, G. A., Flood, D.J., and Bailey, S.J., "Advances in Thin-Film Solar Cells for Light Weight Space Photovoltaic Power," NASA TM-102017, *Space Power* **8**, 31-50 (1989).

Matloff, G. L., "Interstellar Solar Sailing: Consideration of Real and Projected Sail Materials," *J. British Interplanetary Soc* **37**, 135-141 (1984).

Phipps, C. R., "LISP: Laser impulse space propulsion," *Laser and Particle Beams* **12**, 23-54 (1994).

Phipps, C. R., "Comments on the Beamed Energy Technology Meeting," in *Proceedings of the Beamed Energy Transportation(BET) Workshop*, edited by G. W. Zeiders, NASA Marshall Space Flight Center, 1996, pp. 10-12.

SRS Technologies System Technology Group., Huntsville AL, http://www.stg.srs.com/larc-cp.htm, 1999.

Stockman, H. S., Ed., *Next Generation Space Telescope: Visiting a Time When Galaxies Were Young*, The Association of Universities for Research in Astronomy, Inc., Washington, D.C., 1997, pp. 64-75.

Advanced Plasma Propulsion for Human Missions to Jupiter

Benjamin B. Donahue[1], J Boise Pearson[2]

[1]Boeing Space Systems, 499 Boeing Blvd., Huntsville, AL 35824
[2]Propulsion Research Center/TD40, NASA Marshall Space Flight Center, AL 35812

Abstract. This paper will briefly identify a promising fusion plasma power source, which when coupled with a promising electric thruster technology would provide for an efficient interplanetary transfer craft suitable to a 4 year round trip mission to the Jovian system. An advanced, nearly radiation free Inertial Electrostatic Confinement scheme for containing fusion plasma was judged as offering potential for delivering the performance and operational benefits needed for such high energy human expedition missions, without requiring heavy superconducting magnets for containment of the fusion plasma. Once the Jovian transfer stage has matched the heliocentric velocity of Jupiter, the energy requirements for excursions to its outer satellites (Callisto, Ganymede and Europa) by smaller excursion craft are not prohibitive. The overall propulsion, power and thruster system is briefly described and a preliminary vehicle mass statement is presented.

FUSION PROPULSION SYSTEMS

Mission and system studies have shown that fusion reactors potentially may be the most attractive energy source for space power and propulsion systems. Multiple magnetic confinement concepts have been examined for mission applications. These systems offer very high thrust and specific impulse, but significant engineering implementation challenges exist that are largely inherent to magnetically confined plasma fusion systems whether terrestrial or space based. These include; further development of superconducting magnet technology, development of long-lived first surface material (That material next to the hot plasma), removal of heat flux from the first surface and the protective blanket for superconducting coils, development of a suitable blanket to shield superconducting magnets from the neutron flux, conversion of the plasma enthalpy to electrical power for onboard and propulsion system use, and radiation damage of wall materials by neutron flux.

Most of these challenges can only be met by adding significant amounts of weight to the propulsion system. An alternative fusion power concept exists that offers potential for high performance without the mass intensive engineering solutions to these challenges..

Inertial Electrostatic Confinement

A brief investigation was conducted of several fusion power and propulsion system concepts. An advanced fusion power source identified by its Inertial Electrostatic Confinement (IEC) scheme for containing the fusion plasma, was chosen for further consideration. It was judged as offering potential for delivering the performance and operational benefits needed for high energy human expedition missions to Jupiter. IEC reactor technology provides electrostatic heating and confinement of the fusion plasma, as well as direct energy conversion to electrical energy. It therefore does not require the heavy magnets, high-power drivers, thermodynamic electrical plant and other massive components required of thermal fusion propulsion devices typically analyzed in advanced propulsion studies (Miley, 1995).

These IEC fusion systems would enable a much lower weight for the propulsion system when compared to those fusion concepts requiring heavy magnetic systems for confinement of the plasma. Coupled with state-of-the-art magnetoplasmadynamic (MPD) electric thruster technology, this propulsive concept appears to offer an attractive solution for a fast transfer mission to Jupiter.

CP504, *Space Technology and Applications International Forum–2000*, edited by M. S. El-Genk
© 2000 American Institute of Physics 1-56396-919-X/00/$17.00

IEC System Description

The term IEC describes a family of fusion reactor concepts in which concentric electrostatic fields are used to accelerate ions radially into the center, where the charged particles converge and react, yielding energetic charged fusion products (Bussard, 1995, Miley, 1995). A very high fusion power energy density (of roughly mono-energetic ions) is achieved as these ions converge at the center of the vessel, forming a dense central core region where fusion occurs. Fuel combinations of D-T, D-D, D-^{3}He and p-^{11}B are being investigated. The D-^{3}He and p-^{11}B reactions are aneutronic, eliminating or greatly reducing the need for neutron shielding. Rather than utilizing fusion energy to directly heat a propellant, a Direct Energy Converter (DEC) can be used to obtain electric power from the radially moving, positively charged fusion products as they escape the active region. This DEC is built by surrounding the electric confinement volume with an electrically biased grid system that decelerates these ions, where the ions lose energy to the electric field. The ions can be made to arrive at the collecting grids with little or no residual energy (Bussard, 1995, Post, 1969, Roth, 1970). The general feasibility of this means of direct conversion has been proven by work at the Lawrence Livermore National Laboratory in earlier (1973-83) research studies (Bussard, 1995). This DEC can produce megavolt dc current at levels required for the MPD thrusters. Specifics on the DEC, cooling system, power conversion and electric thrusters have been developed in some detail (Miley, 1993). Since the IEC is used as an electrical source, the fusion fuel need not be exhausted for propulsion, hence fuel recycle is possible.

MPD THRUSTERS

The MPD thruster dates back to the 1960s and is a promising candidate for interplanetary missions because it has the potential to run at multi-megawatt power levels and produce 100's of Newtons of thrust. By operating at high current levels, the MPD thruster can accommodate high input power levels without requiring complex power conditioning to boost voltage (Cassady, 1988). The inherent simplicity of the device, coupled with a high power handling capability makes it attractive for energetic missions. In a simple MPD thruster, a plasma current is generated by the arc between the tip of an axial conducting bar and a coaxial cylindrical wall. The current along the conducting bar creates an azimuthal magnetic field that interacts with the plasma current. The resulting Lorentz force has two components; a radially inward force that constricts the flow, and a driving force along the axis that produces the directed thrust. Since ions and electrons are produced and accelerated in bulk, the accelerated stream as a whole is neutral and unpolarized, (Demetriades, 1962). This is in contrast to the Ion engine, wherein ions and electrons are produced, but only the ions are accelerated by electrostatic means. The neutral plasma exhaust of the MPD thruster eliminates the need for a separate neutralizer.

MPD Thruster Demonstrations

Both steady state and pulsed MPD systems have been developed. Megawatt pulsed MPD systems have demonstrated efficiencies above 50 percent with hydrogen propellant. Noble gas propellants can also be used, but with lower specific impulses. Princeton University recently presented results of lab tests with a Russian thruster. This thruster is a Lorentz Force Accelerator (a variant on MPD thrusters) which uses lithium propellant. Adding a little barium to the mix increases thruster lifetime to thousands of hours. The thruster has operated at 121 kW, 44.2% efficiency, and 3568 sec Isp. Their partners at Energia have run a similar unit a 500 kW input power with 12.5 N thrust for 500 hours (Choueiri, 1998). Other testing at Princeton University (Ziemer, 1998), the Jet Propulsion Laboratory (King, 1985) and at other private and government laboratories have characterized the performance of MPD thruster technology, including thrusters of the 70 Newton capability chosen for this study.

Propulsion System For Demanding Missions

The IEC / MPD thruster combination represents an attractive propulsion system for demanding, high delta velocity human exploration transfers to the outer planets. Though the IEC fusion technology would require a very significant technology development program, the interplanetary transfer capability enabled by it would be immensely more robust than the chemical, solar and nuclear fission concepts currently advocated for human Earth-Mars missions.

The 12 MW IEC fusion reactor chosen for this mission is taken from the literature (Miley, 1995). The MPD thruster chosen is within the range of demonstrated laboratory performance, and is used in a cluster of six thrusters.

Table 1 lists characteristics of the IEC reactor (Miley, 1995), MPD thrusters and combined propulsion system chosen for the mission in this study.

TABLE 1. IEC / MPD Propulsion System Characteristics.

IEC Fusion Reactor		Single MPD Thruster		Total Propulsion System	
Fusion power	12 MW	Thruster Isp	3,000 s	Thrust (6 thrusters)	408 N
Reactor mass	5 mt	Thruster efficiency	50 %	Isp	3,000 s
Reactor diameter	5.3 m	Thruster specific mass	0.1 kg / kW	Propellant flowrate	.0138 kg/s
Specific power	2.4 kW$_e$ / kg	Thrust per thruster	68 N	Thrust/Weight	.008
		Propellant feed pressure	1 atm	Specific jet power	1.2 kW/kg
		Propellant	Krypton		
		Propellant flowrate	0.0023kg/s		

MISSION DESCRIPTION

A four year round trip trajectory to Jupiter was chosen for analysis. This direct Earth-Jupiter-Earth mission is characterized by an outbound leg of just over 2 years, a one month stay time, and a 2 year inbound leg. Transfer vehicle operations are described as follows: about one month before the Trans-Jovian Injection window opens, a test crew will board the vehicle for final tests and pre-orbital-launch checkout. One week before the window opens the mission crew will board the vehicle in LEO; after a tie-in period the test crew will return to Earth on the shuttle that delivered the mission crew. The vehicle departs low Earth orbit (LEO) and conducts a slow spiral out typical of low thrust electric propulsion vehicles. (Vehicle thrust to weight at departure is on the order of 10^{-4}.) The spiral out continues until the vehicle nears Earth escape velocity, then commences on a heliocentric transfer trajectory to Jupiter, at 5.2 astronomical units (AU) from the Sun. After arrival and matching of Jovian velocity, a high altitude parking orbit is established, conducive to exploration of one of the outer satellites. Transfer and descent preparations begin, including high-resolution imagery and viewing of the planned landing sites. Several members of the crew enter the excursion craft and transfer to one of the Jovian satellites for a one month stay, while the remaining crew stays onboard the transfer vehicle directing the operations of unmanned probes sent to Jupiter and its other moons.

After the surface mission is accomplished, the crew returns to the orbiting transfer vehicle, jettisons the ascent stage and departs for Earth on a two year inbound trajectory. About 16 hours before Earth arrival, the crew enters the small Crew Return Vehicle (CRV). At entry minus 12 hours the CRV separates from the rest of the transfer stage. The transfer vehicle is to be expended, and thus is not on an Earth atmosphere intercept path, and the CRV makes a burn of about 10 m/sec delta velocity to place it on its entry path. The transfer stage is expended as it swings by Earth, continuing on a long lived heliocentric orbit. The CRV enters Earth's atmosphere, decelerates, and either rendezvous with the space station, or deploys parachutes, and makes a landing to complete the mission.

Jovian Destinations

Due to its immense mass and gravity, a manned descent into, and return from Jupiter itself was deemed inadvisable considering the modest efficiencies of current high thrust propulsion systems. The mass of Jupiter is 340 times greater than the Earth. Even the mass benefit associated with utilization of *in-situ* resources for ascent was considered inadequate for returns from descents deep into the planet's atmosphere. The outer Jovian satellites are however, accessible. Once the transfer stage has matched the heliocentric velocity of Jupiter, or entered an appropriate high altitude parking orbit around it, the energy requirements for excursions to these satellites are not prohibitive. The actual excursion / lander craft would use traditional high thrust-to-weight (t/w) chemical propulsion, or fission nuclear thermal propulsion for ascent/descent propulsion (Donahue, 1995). These could provide the modest delta velocities (2.1-2.8 km/s) needed for the near vicinity orbital transfers to one of Jupiter's outer satellites, such as Callisto, Ganymede or, perhaps most interesting, Europa. It is one of these to which future Jovian exploration missions are most likely to be targeted (Beebe, 1997).

Delta Velocity Requirements

Mission trip times and delta velocity increments are provided in Table 2. Low-thrust departures from within a sizable gravity field are relatively inefficient when compared to high thrust, impulsive departures. For comparison, the delta velocities for a high thrust transfer vehicle flying a comparable 4 year round trip trajectory are also shown in Table 2.

TABLE 2. Jovian Mission Duration and ΔV s.

Mission Leg	Months	Event	Low-Thrust (m/s)	High-Thrust (m/s)
LEO to High Earth Orbit spiral	3	LEO to near escape velocity	6118	2600
High Earth Orbit to Jupiter	24	Trans Jupiter injection	20280	8000
at Jupiter	1	Jovian injection	13520	5500
Jupiter to Earth flyby	24	Trans Earth departure	13520	5500
		Earth arrival burn (aerobrake)	10	10
Total	51	Totals	53448	21610

JOVIAN TRANSFER VEHICLE

The Jovian transfer vehicle (JTV) consists of the IEC fusion power system and MPD thrusters at one end and the crew habit and lander excursion craft at the other end, separated by a truss section with attached propellant tanks. Though hydrogen would produce the highest specific impulse (Isp), xenon or krypton would be used as propellant due to boiloff concerns with hydrogen. Krypton and xenon are inert, and neither will explode or corrode fuel lines or tanks. At typical storage pressures xenon is about 60 percent denser than water, thus the tankage fraction could be considerably less than it would be if hydrogen were utilized.

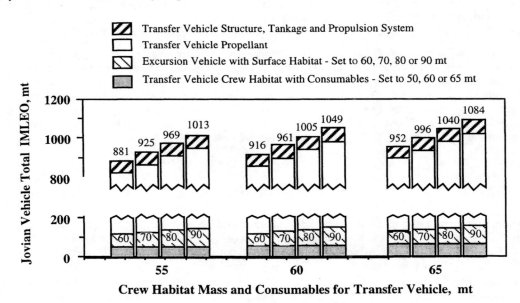

FIGURE 1. Human Jovian Mission Vehicle Mass Estimates

Jovian vehicle Initial Mass in Low Earth Orbit (IMLEO) values were calculated for several values of excursion vehicle masses (lander) and transfer vehicle crew habitat masses (Donahue, 1998). Excursion vehicle masses of 60, 70, 80, and 90 mt were utilized. This mass includes the ascent stage, the descent burn tankage, the crew surface habitat, the surface exploration equipment and rovers. JTV crew habitat masses (including consumables) of 55, 60, and 65 mt were utilized. The habitat mass will determine the number of crew the vehicle can accommodate. The crew has four separate habitable volumes that can serve as living, working and recreational space on the outbound journey; (1) the large transfer vehicle habitat, the lander's (2) surface habitat and (3) small ascent crew cab, and the

(4) CRV capsule. On the return leg only the large habitat and the CRV volumes would be available. IMLEO estimates are shown in Figure 1. The relative proportions of the propellant mass and inert vehicle mass (structure, propulsion and tankage) are given for the 12 combinations of habitat and lander masses. A propellant reserve allotment of 1 percent for each burn was allocated. As a weight saving measure, the Trans-Jupiter injection propellant tank is jettisoned after it is emptied.

CONCLUSIONS

The combination of an IEC nuclear fusion power source and a cluster of high power MPD thrusters represents a mass efficient propulsion system capable of providing the high delta velocities required for fast trip human exploration missions to Jupiter. The tasks needed to bring such a system up to the technology level necessary for in-space operation principally involve the development and testing of electrostatic confinement of fusion plasmas and direct electrical energy production characteristics of the selected IEC system. Larger scale follow on tests would be necessary to demonstrate the systems capability for multiyear power production at multimegawatt power levels. Jovian satellite excursion vehicles could utilize traditional chemical propulsion or nuclear thermal propulsion systems for their less energetic landing and ascent maneuvers.

ACKNOWLEDGMENTS

This paper is largely based on work done in support of the Affordable In-Space Transportation (AIST) concept definition study funded by the NASA Marshall Space flight Center and supported by Boeing (contract NAS8-5000 Schedule F, TOF-028).

REFERENCES

Beebe, Reta, *Jupiter, the Giant Planet*, Smithsonian Institution Press, 1997

Bussard, R. W. and Jameson, L. W., "Inertial-Electrostatic-Fusion Propulsion Spectrum: Air-Breathing to Interstellar Flight," *Journal of Propulsion and Power*, Vol. 11, No. 2, March-April 1995.

Cassady, R. J., "An MPD Thruster Driven Cargo Ferry for Support of the Manned Mars Mission," AIAA Paper 88-2896, July 1988

Choueiri, Edgar, Princeton University, Presentation to Ninth Annual NASA-JPL Advanced Propulsion Research Workshop, Jet Propulsion Lab, March 11-13, 1998.

Demetriades, S. T., *Plasma Propulsion* Technical Memorandum NSL 62-14, Northrop Space Laboratories, Jan 1962

Donahue, B. B., Editor, *Affordable In-Space Transportation Concepts, Special Studies: Advanced Plasma Propulsion Applications to Human Missions to Jupiter*, Contract NAS8-50000, Schedule F, TOF-028, Task 3 Final Report, Boeing Space Systems, Huntsville, Alabama, January 1998.

Donahue, B. B., "Mars Ascent-Stage Design Utilizing Nuclear Propulsion," *Journal of Spacecraft and Rockets*, Vol. 32, No. 3, May-June 1995.

Donahue, B. B., "Self Assembling Mars Transfer Vehicles: The Preferred Concept of the Space Transfer concepts and Analysis for Explorations Missions Study," AIAA paper 94-2761, July 1994

Donahue, B. B., Editor, Space Transfer Concepts and Analyses for Exploration Missions, Contract report, Nuclear Thermal Propulsion Implementation Plan and Element Description Document, NAS8-37857, Document No. D615-10069, Boeing Civil Space Product Development Group, Huntsville, Alabama, Sept. 1993.

King, D. Q., "Feasibility of Steady-State, Multi-Megawatt MPD Thruster," AIAA Paper 85-2004, 1985.

Miley, G. H., Burton, R. L., Javedani, J., Gu, Y., Satsangi, A., Heck, P., Nebel, R. A., and Schulze, N., "Inertial Electrostatic Confinement as a Power Source for Electric Propulsion," *Vision-21 Conference Proceedings*, NASA Conference Publication. 10129, 1993

Miley, G. H., Satsangi, A. J., DeMora, J., Javedani, J. B., Gu, Y., Burton, R. L. and Nakashima, H., "Innovative Technology for an Inertial Electrostatic Confinement Fusion Propulsion Unit,". *Fusion Energy in Space Propulsion*, Vol. 167, Progress in Astronautics and Aeronautics, 1995

Post, R. F., "Direct Conversion of Thermal Energy of High Temperature Plasma," *Bulletin of American Physics Society*, Vol. 14, No. 11.November 1969.

Roth, J. R., Rayle, W. D., and Reinmann, J. J., "Technological Problems Anticipated in the Application of Fusion Reactors to Space Propulsion and Power Generation," NASA TM X-2106, NASA Washington, D. C. October 1970.

Ziemer, John K., Private communication, Princeton University, March 1998

Dual-Mode, High Energy Utilization System Concept For Mars Missions

Mohamed S. El-Genk

Institute for Space and Nuclear Power Studies and Dept. of Chemical and Nuclear Engineering
University of New Mexico, Albuquerque, NM 87131
(505) 277-5442, mgenk@unm.edu

Abstract. This paper describes a dual-mode, high energy utilization system concept based on the Pellet Bed Reactor (PeBR) to support future manned missions to Mars. The system uses proven Closed Brayton Cycle (CBC) engines to partially convert the reactor thermal power to electricity. The electric power generated is kept the same during the propulsion and the power modes, but the reactor thermal power in the former could be several times higher, while maintaining the reactor temperatures almost constant. During the propulsion mode, the electric power of the system, minus ~ 1-5 kW_e for house keeping, is used to operate a Variable Specific Impulse Magnetoplasma Rocket (VASIMR). In addition, the reactor thermal power, plus more than 85% of the head load of the CBC engine radiators, are used to heat hydrogen. The hot hydrogen is mixed with the high temperature plasma in a VASIMR to provide both high thrust and $I_{sp} > 35,000$ N.s/kg, reducing the travel time to Mars to about 3 months. The electric power also supports surface exploration of Mars. The fuel temperature and the inlet temperatures of the He-Xe working fluid to the nuclear reactor core and the CBC turbine are maintained almost constant during both the propulsion and power modes to minimize thermal stresses. Also, the exit temperature of the He-Xe from the reactor core is kept at least 200 K below the maximum fuel design temperature. The present system has no single point failure and could be tested fully assembled in a ground facility using electric heaters in place of the nuclear reactor. Operation and design parameters of a 40-kW_e prototype are presented and discussed to illustrate the operation and design principles of the proposed system

INTRODUCTION

To minimize the astronauts' exposure to space ionizing radiation, future manned missions to Mars require high specific impulse propulsion ($I_{sp} > 30,000$ N.s/kg) powered by nuclear fission reactor systems. (1.0 N.s/kg = 9.806 $lb_f.s/lb_m$). Such high I_{sp}, however, is not achievable with solid core nuclear reactors and direct hydrogen thermal propulsion (~3000 K) (Ludewig et al., 1993; Harty and Brengle, 1993; and Morley and El-Genk, 1995). Although other options such as the gas core and nuclear fusion (Howe et al., 1998; and Kammash and Emrich, Jr., 1998) could potentially operate at significantly higher I_{sp} (up to 50,000 N.s/kg), they would require major investment and a long lead-time to develop and demonstrate the technology (> 15 years). The projected manned missions to Mars in 2012-2016 would need to rely on off-the-shelf technologies and those requiring short lead time (<5-10 years) to develop, and a great deal of engineering innovation in system design and operation. Current consensus among astronauts is to reduce the travel time to Mars to about three months, with access to electric power for house keeping while in route and on the surface of Mars for exploration and experimentation.

These challenging operation and performance requirements could be met using high performance plasma thrusters that could utilize both the electric power and a large fraction of the thermal power of a nuclear reactor system. A very promising plasma propulsion device for interplanetary exploration missions is the Variable Specific Impulse Magnetoplasma Rocket (VASIMR) (Chang-Diaz et al., 1995; and LaPointe and Sankovic, 2000). This open-ended, RF-heated, mirror-like plasma device could vary the thrust and the I_{sp} via mixing neutral gas, such as hot hydrogen, with the high temperature exhaust plasma, providing high thrust for orbit maneuvering, planetary escape, as well as high I_{sp} for shorter travel time. The VASIMR, currently being developed at NASA Johnson Space Flight Center, could cut travel time to Mars to 96

CP504, *Space Technology and Applications International Forum–2000*, edited by M. S. El-Genk
© 2000 American Institute of Physics 1-56396-919-X/00/$17.00

days. The projected performance of a 10 MWe VASIMR, having an exhaust diameter of 10 cm and efficiency of 60%, when using hydrogen varies from 4,500 N to 420 N at specific impulse of 50,000 and 300,000 N.s/kg, respectively (Change-Diaz,1999).

Recent analysis of manned missions to Mars using VASIMR, however, necessitates the use of nuclear reactor power systems with a very high specific electric power of 200-250 We/kg. To achieve such high specific power, the efficiency of converting the reactor's thermal power to electricity would need to be more than 60%, which far exceeds the capabilities of current energy conversion technology and any projected in the intermediate future. However, energy utilization, including both electric and thermal, of more than 70% is possible using a dual-mode (DM) electric power and propulsion system.

This paper describes a DM, high-energy utilization (HEU) system concept for meeting the aforementioned ambitious performance for manned missions to Mars. Also presented and discussed are the calculated performance parameters of a 40-kW$_e$ prototype that could be used for ground testing, in conjunction with a scaled down VASIMR, for concept verification. In such testing, the nuclear reactor could be replaced with electric heaters and the entire system would be fully assembled and instrumented. The data generated in these tests could be used in developing and verifying models for scaling the system design to the multi-megawatt electric level.

SYSTEM DESCRIPTION

The proposed DM-HEU system employs a gas cooled Pellet-Bed Reactor (PeBR) as the heat source (El-Genk et al., 1994). The reactor core could be divided into three or more equal, neutronically and thermally coupled sectors. Each sector is self contained, independently cooled, and hydrodynamically and thermally coupled to one Closed Brayton Cycle (CBC) engine or more for generating equal fraction of the nominal electric power of the system. This design approach eliminates single point failure, hence enhancing the system's reliability.

The PeBR core is packed with spherical fuel elements (mini-spheres ~2 mm in diameter (see Fig. 1)) comprised of ZrC coated (U, Zr)C fuel microspheres (~ 500 μm in diameter) dispersed in graphite at a volume fraction as high as 50 % (El-Genk et al., 1994). The packing fraction of the fuel elements in the reactor core is about 62.5 %. A gas mixture of 70-mole % Helium and 30-mole % Xenon (He-Xe) cools the reactor core. This gas is also the CBC engines' working fluid. The technology of CBC engines is mature and some units have undergone thousands of hours of testing without failure (Table 1) (Ashe, 1993). Separate bleed lines in the coolant loops control the amount of He-Xe circulating through the core sectors during the power and propulsion modes. During the propulsion mode, the fraction of the working fluid flowing through the reactor core increases commensurate with the increase in the reactor thermal power. The bleed fraction of the working fluid decreases to almost zero at the peak thermal power of the reactor during the propulsion mode, and could be less than 0.20 during the power mode. During this mode, the reactor thermal power is also much lower (< 20%) than the nominal design value during the propulsion mode. The coolant bleed lines concept help maintain the temperatures of the fuel and structure materials in the PeBR core almost the same during both the power and propulsion modes, hence avoiding any thermal or structure stresses, for excellent system reliability.

Safety and Design Features

The PeBR, DM-HEU system offers many safety and operations advantages:
  (a) Avoids design and startup complexities associated with liquid-metal cooled reactors due to the need to incorporate a subsystem to thaw the working fluid at startup.
  (b) Could be operated at high temperatures (> 2000 K) not achievable with liquid metal cooled reactors.
  (c) Could be launched fueled or empty and can be fueled, emptied, and refueled while in space, an advantage that could be exercised very readily with mini-spheres fuel elements (see Figure 1) with an easily sliding graphite surface (El-Genk et al., 1992).

(d) Dividing the reactor core into equal sectors that are hydrodynamically coupled to separate CBC engines eliminates single-point failure in the reactor coolant loop.

(e) The inert He-Xe working fluid poses no compatibility concerns with reactor core structure materials of the fuel elements.

(f) Maintaining the reactor temperature almost constant during both the electric power and propulsion modes, avoid causing thermal stress in the reactor core and structure materials.

(g) The PeBR is passively cooled after shutdown with the aid of liquid metal heat pipes inserted in the radial Be reflector (see Figure 1) (El-Genk et al., 1994).

(h) Despite the complexity of the hydrogen flow and heating loop, all the hydrogen flow control is while in the liquid phase, upstream of the heat rejection radiator.

(i) The heat rejection radiator is sized only for the electric power level of the system, which is a small fraction of the system's total power in the propulsion mode. As a result, the mass of the heat rejection radiator is a small fraction of the total system mass. The radiator size and mass, however, would increase as the electric power of the system increases.

To better illustrate the design and operation principle of the proposed DM-HEU system for manned missions to Mars, the following sections describe the design and present the performance parameters of a 40 kW$_e$ prototype based on the gas cooled PeBR. Detailed neutronics, thermal-hydraulics, and safety analyses of this system have been performed (El-Genk et al., 1994; and Liscum-Powell and El-Genk, 1995a,b). It is worth noting that the DM-HEU system could utilize any gas cooled solid core nuclear reactor with similar operation, safety, and redundancy features.

DESCRIPTION OF THE 40 kW$_e$ PROTOTYPE SYSTEM

The 40 kW$_e$ prototype system employs three 13.33 kW$_e$, He-Xe CBC engines of the Rotating type (BRU in Table 1) that have slightly different operation parameters than the one listed in Table 1. The inlet temperatures of the working fluid to the reactor and the CBC engines of 900 K and 1144 K, respectively, are dictated by the design of the latter (Ashe, 1993). The inlet and exit temperatures of the He-Xe gas mixture in the nuclear reactor core and the inlet temperature to the CBC engines are also kept constant during both the electric power and propulsion modes. A radial cross-section of the PeBR core for the 40-kWe prototype system and a schematic of the 13.33 kW$_e$ CBC units are shown in Figures 1 and 2, respectively.

TABLE 1. Operation parameters of CBC engine designs for space applications.

Operation parameter	Brayton Cycle Demonstrator (BCD)	Brayton Rotating Unit (BRU)	High Performance Engine	Dynamic Isotope Power System (DIPS)
Cycle Efficiency (%)	27	28.5	31	31
Power Output (kW$_e$)	3	10.5	30	1.4
Thermal Power Input (kW$_{th}$)	11.1	32.8	96.7	4.8
Shaft Speed (rpm)	64,000	36,000	52,000	52,000
Compressor Inlet Pressure (kPa/psia)	3.1/4.5	16.3/23.7	21.8/31.6	26.2/38
Compressor Inlet Temperature (R/K)	500/277	540/300	473/262	546/303
Compressor Pressure Ratio	2.3	1.9	1.61	1.89
Turbine Inlet Temperature (R/K)	2000/1111	2060/1144	1860/1033	1960/1089
Recuperator Efficiency (%)	90	95	96	84
Working Fluid Flow Rate (kg/s)	0.0771	0.38	0.0862	0.685
Working Fluid type (gm/mole)	Argon (39.9)	He-Xe (83.3)	He-Xe (83.8)	Argon (39.9)

A number of these CBC units, designed by Allied Signal, were tested successfully (see Figure 2) in the seventies at NASA Glenn Center for 36,000 hours without any performance degradation (Ashe, 1993). All technologies required to implement these CBC engines at turbine inlet and exit temperatures of 1144 K and 1070 K, respectively, are proven and mature, and no technology "*breakthrough*" is required; only engineering implementation (Ashe, 1993; and Juhasz, et al., 1993).

Table 1 lists the operation characteristics of other more efficient engines also designed and fabricated by the Allied Signal Aerospace Company (Ashe, 1993). The specific power of some of the BRU type engines are compared in Figure 3 at turbine inlet temperatures of 1144 K and 1400 K (Ashe, 1993).

As shown in Figure 3, a 100 kW$_e$ CBC engine operating at a turbine inlet temperature of 1144 K has 30% higher specific power (47 W$_e$/kg) than the 13.33 kW$_e$ engines (36 W$_e$/kg) used in the 40-kW$_e$ prototype. Increasing the inlet turbine temperature also increases the CBC engine specific power. Figure 3 also shows that for a 50 kW$_e$ engine, increasing the inlet turbine temperature from 1144 K to 1400 K increases the engine's specific power by ~ 35%, from 43 to 58 W$_e$/kg.

Power Mode Operation

During the power mode, a fraction (x) of the He-Xe returning from the compressor at 900 K is bled off before entering the PeBR core to the mixing chamber located at the exit of the reactor core. The remainder of the incoming working fluid (1-x) flows through the reactor core sectors (see Figure 4) (Liscum-Powell and El-Genk, 1995a) to remove the heat generated by nuclear fission. The He-Xe gas mixture flows radially through the reactor core and exits through three separate central channels, each connected to a core sector and a separate CBC engine. The temperature of the coolant exiting from the reactor core is kept at least ~ 100-200 K below the design maximum fuel temperature. This temperature could be as much as 2000-2400 K, for long operation lifetime (Morley and El-Genk, 1995), since the melting temperature of the (U-Zr)C fuel in the coated microspheres is more than 3500 K.

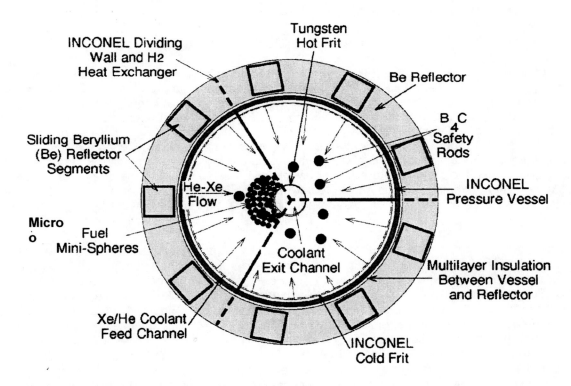

FIGURE 1. A radial cross-section of the PeBR core for the 40 kW$_e$ prototype DM system.

Upon exiting the reactor core, the hot He-Xe is mixed with the cooler bled fraction in the mixing chamber to lower its temperature to the turbine inlet temperature of 1144 K (see Figure 4). This temperature is maintained constant during both the power and propulsion modes of operation. In the propulsion mode, the temperature of the He-Xe gas mixture exiting the reactor core and its bled fraction are higher than during the power mode. Therefore, the hydrogen flowing through exchangers HX$_2$ is used to lower the temperature of the He-Xe exiting the mixing chamber, to the turbine inlet temperature of 1144 K.

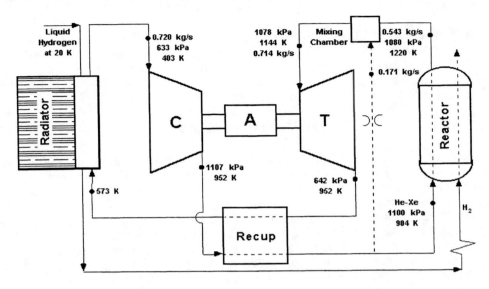

FIGURE 2. A schematic and operation parameters of a BRU-type CBC engine.

Propulsion Mode Operation

During the propulsion mode, liquid hydrogen from the storage tank at 20 K is boiled off and heated in exchangers (HX$_1$), which are integral parts of the heat rejection radiators of the CBC engines (see Figure 4). In these exchangers, the hydrogen captures > 70% of the radiators' heat load. In the 40-kW$_e$ prototype, since the CBC efficiency is 22.5 %, the fraction of the heat rejection that is recovered by the hydrogen propellant in exchangers HX$_1$ equals ~96 kW$_{th}$. The thermal effectiveness of exchangers HX$_1$ for heating the hydrogen could be as much as 95%, increasing this thermal energy recovery to 123 kW$_{th}$. This hydrogen preheating feature is a key to increasing the thermal energy utilization of the system from only 22.5% during the steady-state electric power mode to > 70% during the propulsion mode (El-Genk, 1995). Another advantage of this design feature is that hydrogen boiling occurs external to the nuclear reactor core, hence avoiding affecting the reactor operation and control by the hydrogen two-phase flow in exchangers HX$_1$.

The hot hydrogen gas exiting exchangers HX$_1$ enters exchangers HX$_2$ to be heated further by the hot He-Xe exiting the mixing chamber (see Figure 4). The temperature of the hydrogen exiting HX$_2$ could be as high as 1600-2000 K, depending on the maximum fuel design temperature in the core and the hydrogen mass flow rate. Upon exiting exchangers HX$_2$, hydrogen is heated further in the three heat exchangers incorporated into the dividers of the reactor core sectors (see Figure 1). Such in-core heating of hydrogen could increase its temperature by an additional 200-400 K, depending on the reactor thermal power and the hydrogen flow rate. For reliable long time operation, the temperature of the hydrogen exiting the PeBR core sectors is limited to about 200 K below the maximum fuel design temperature. As indicated earlier, such fuel temperature could be as high 2400 K. The calculations presented in the following section, however, were performed at conservative maximum fuel design temperatures of 1600 K and 2000 K (Liscum-Powell and El-Genk, 1995b).

The hot hydrogen exiting the reactor core heat exchangers could be allowed to expand through a high thrust nozzle in case of selecting a thermal propulsion only option. In this option, the electric power of the system, minus 1-5 kW$_e$ for house keeping, could be used to increase the hydrogen temperature before entering to thruster in order to achieve higher I$_{sp}$ (see Figure 4). Owing to the low I$_{sp}$ of the thermal propulsion option (< 750 N.s/kg), we propose mixing the hot hydrogen exiting the PeBR core with the high temperature plasma exhaust of a VASIMR, to achieve both high thrust and significantly higher I$_{sp}$. In this case, the system's electric power, minus ~ 1-5 kW$_e$ for house keeping, would be used to operate the VASIMR at I$_{sp}$ ≥ 3500 s, thus reducing the travel time to Mars to about three months.

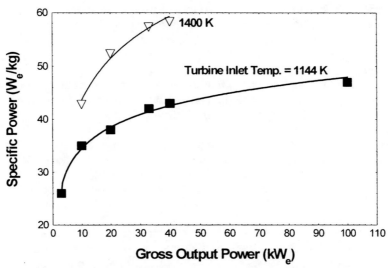

FIGURE 3. Specific electric power of a CBC-BRU type engines.

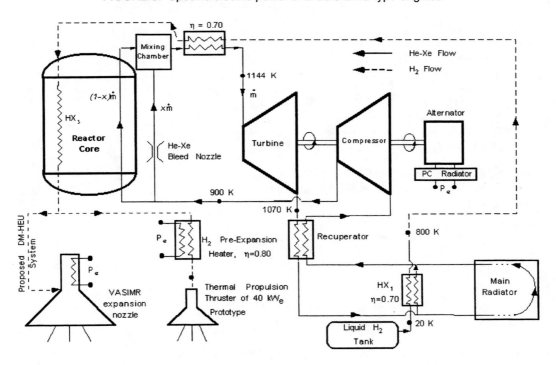

FIGURE 4. Line diagram of an energy conversion and thermal-hydraulic loop of the 40-kWe prototype system, with in-core heating of the hydrogen.

The following section presents and discusses the performance parameters of the 40-kWe prototype *for the thermal propulsion only option*, with and without pre-expansion electric heating of the hydrogen exiting the PeBR core sectors (see Figure 4). Although the results are not directly applicable to the high performance option using VASIMR, they clearly demonstrate that high energy utilization fractions can be achieved in the present PeBR DM system (see Figure 4).

PERFORMANCE PARAMETERS OF THE 40-kW$_e$ PROTOTYPE

An integrated kinetics, thermal-hydraulics, and performance model of the 40-kWe prototype, DM-HEU system has been developed and used to investigate the system's steady-state operation during both the power an the propulsion modes as well as its response during operation transients. These transients include nuclear reactor start up from a cold state and changing the reactor operation from the power to the propulsion mode (Liscum-Powell and El-Genk, 1995b). The system model consists of several interactive building blocks, namely:
   (a) Six-group reactor point kinetics model with linear temperature reactivity feedback.
   (b) Lumped transient thermal model of the nuclear reactor core and hydrogen heat exchangers.
   (c) Energy conversion and thermal propulsion performance model.

During both the power and propulsion modes of operation, the temperature of the He-Xe working fluid at the inlet of the CBC turbine is maintained the same at 1144 K. The total flow rate of the working fluid is also maintained constant during both modes at 2.444 kg/s (0.81 kg/s per CBC engine) (see Figure 4). Table 2 lists the operation and performance parameters of the 40 kW$_e$ prototype for DM electric power and thermal propulsion, with and without hydrogen pre-expansion electric heating (see Figure 4).

TABLE 2. Performance parameters of 40-kW$_e$ prototype for DM electric power and thermal propulsion with and without pre-expansion electric heating of the hydrogen.

Steady-State System Parameters	H$_2$ Flow Rate 5.0 g/s	H$_2$ Flow Rate 10.0 g/s
Reactor thermal power in power mode (kW$_{th}$)	178	178
Maximum fuel design temperature (K)	1610	1636
Electric output in power & propulsion modes (kW$_e$)	40	40
Reactor thermal power in Propulsion mode (kW$_{th}$)	251	355
In-core mass flow rate of He-Xe gas (%)	35.7	24.5
Calculated maximum fuel temperature (K)	1610	1636
Temperature of He-Xe exiting reactor core (K)	1590	1615
Temperature of H$_2$ entering reactor core HXs (K)	520	270
Temperature of H$_2$ exiting reactor core HXs (K)	1425	1381
Temperature of H$_2$ at inlet of expansion Nozzle		
With pre-expansion electric heating (K)	1425	1381
Without pre-expansion electric heating (K)	1777	1557
Specific Impulse (lb$_f$.s/lb$_m$) / (N.s/kg)		
With pre-expansion electric heating	658 / 6,452	647 / 6,344
Without pre-expansion electric heating	735 / 7,207	687 / 7,737
Thrust (N)		
With pre-expansion electric heating	32	63
Without pre-expansion electric heating	36	67

The calculations presented in this section and the performance parameters listed in Table 2 were based on the following conservative assumptions:
   (a) In-core hydrogen flow during the propulsion mode absorbs only 10% of the reactor thermal power;
   (b) Thermal effectiveness of exchangers HX$_1$ and HX$_2$ is only 70%. As indicated earlier, heat exchangers effectiveness as high as 95 % is readily attainable for gas/gas heat exchangers.
   (c) Maximum hydrogen exit temperature from the reactor core is limited to 200 K below the maximum design temperature of the fuel in the reactor core (< 1650 K). This temperature for the PeBR fuel, however, could be raised safely to 2000-2500 K, resulting in a much higher performance that reported in Table 2.

(d) Hydrogen flow rate is limited by the design exit temperature from the reactor core. Therefore, raising the maximum fuel temperature in the PeBR during the propulsion mode to 2500 K would significantly increase the hydrogen flow rate and hydrogen temperature, resulting in significantly higher thrust and I_{sp} as much as 800 s.

(e) During the propulsion mode, all electric power, minus 1.0 kWe for house keeping, is used for further heating of the hydrogen exiting the reactor core, at an efficiency of 80%. This efficiency of could be as much as 98%.

Performance Envelope

The performance envelopes of the 40-kWe prototype for maximum fuel design temperatures of 1600 K and 2000 K are compared in Figure 5. The reactor thermal power during the power mode, $P_o = 177$ kW$_{th}$, increases during the propulsion mode by 250-450%, depending on the hydrogen flow rate, without exceeding the maximum fuel design temperatures of 1600 and 2000 K, respectively. The left and right boundaries of the operation envelopes in Figure 5 represent the performance parameters during the thermal propulsion mode, without and with pre-expansion electric heating of hydrogen, respectively. The difference in performance (thrust and I_{sp}) between the left and right boundaries of the envelopes decreases as the reactor thermal power in the propulsion mode increases, or the electric power becomes a smaller fraction of the system's total power. Note that in-core heating of the hydrogen slightly increases both the delivered thrust and I_{sp}.

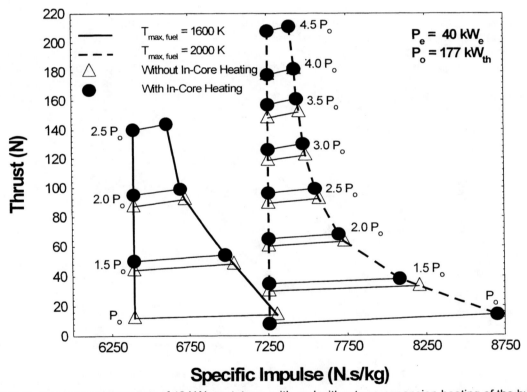

FIGURE 5. Performance parameters of 40 kW$_e$ prototype, with and without pre-expansion heating of the hydrogen.

For the cases when the reactor thermal power in the propulsion mode was identical to that during the electric power mode ($P_o = 177$ kW$_{th}$), only results with in-core heating of hydrogen are presented in Figure 5. The corresponding propulsion parameters are represented by the lower boundary of the operation envelopes. Similarly, the top boundary of the operation envelopes represents the propulsion parameters corresponding to the highest reactor thermal power achievable during the propulsion mode with in-core heating of hydrogen. Note that increasing the maximum fuel design temperature in the core from a modest 1600 K to 2000 K shifts the performance envelope of the system to the right (high I_{sp}) and raises the top boundary (high thrust).

Energy Utilization Fraction

Increasing the maximum fuel design temperature in the reactor core and implementing in-core heating of hydrogen, increases the fraction of the working fluid flow through the PeBR core (see Figure 6) and the utilized fraction of the reactor thermal power during the propulsion mode (see Figure 7). The results presented in Figure 6 show the effect of the maximum fuel design temperature on the mass flow rate fraction (1-x) of the He-Xe gas mixture in the reactor core during the propulsion mode. With in-core heating of hydrogen, the fraction of He-Xe flowing through the reactor core is lower than without in-core heating. In the latter, 10% of the reactor thermal power in the propulsion mode is assumed transported to the hydrogen flowing through the three in-core heat exchangers (see Figure 1). This percentage, however, could be higher, depending upon the design of the in-core heat exchangers of hydrogen.

Figure 6 shows that for a maximum fuel design temperature of 1600 K, the fraction of the mass flow rate of He-Xe in the reactor is 0.41 and 0.37, with and without in-core heating of hydrogen, respectively. In these cases, the reactor thermal power in the propulsion mode was the same as in the power mode ($P_o =$ 177 kW$_{th}$). At the highest reactor thermal power in the propulsion mode, the fractions of He-Xe flowing through the PeBR core, with and without in-core heating of hydrogen, increases to 0.92 and 0.82 respectively.

Figure 7 shows that for a maximum fuel design temperature of 1600 K, the system's energy utilization fraction during the propulsion mode, where the reactor thermal power was 442 kW$_{th}$, was about 0.54 and 0.51, with and without in-core heating of hydrogen, respectively. However, when the maximum fuel design temperature was raised to 2000 K, the energy utilization fraction of the reactor thermal power (800 kw$_{th}$) increased to 0.62 and 0.59 during the propulsion mode, with and without in-core heating of the hydrogen, respectively (see Figure 7). These fractions could exceed 0.85, as the effectiveness of exchangers HX$_1$ and HX$_2$ increases from 70% to 95%, which is technically possible.

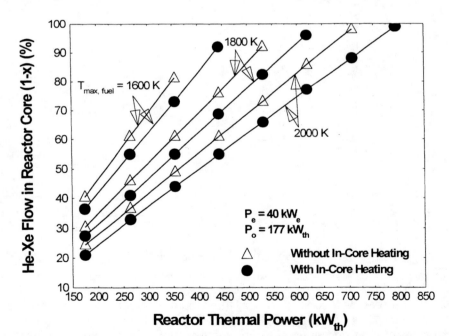

FIGURE 6. Change in the fraction of the working fluid flow rate in the PeBR with the reactor thermal power in the propulsion mode and the maximum fuel design temperature.

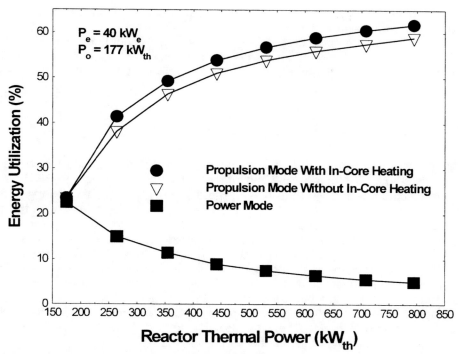

FIGURE 7. Estimates of the energy utilization in the 40 kW$_e$ prototype system.

Mass Estimates

Table 3 lists the mass estimates of the components of the 40 kW$_e$ prototype. The reactor mass of 336 kg had been determined based on detailed neutronics calculations (Liscum-Powell and El-Genk, 1995b). The radiation shadow shield consists of a thin tungsten front layer for gamma shielding followed by a thick layer of depleted LiH neutron shield. This shadow shield, with a half-cone angle < 10°, is contained in a stainless-steel canister. The estimated mass of the shadow shield is based of those reported for similar systems (El-Genk et al., 1989).

TABLE 3. Mass estimates of the 40 kW$_e$ prototype system.

System Component	Mass (kg)
Nuclear Reactor	336
Radiation Shadow Shield	350
CBC engines, power conditioning, radiator and heat exchangers	930
Additional structure and subsystems	84
TOTAL	1700.0

The estimated shield mass is 350 kg for a reactor thermal power of as much as 1000 kW$_{th}$, and separation distance of the payload of ~ 10 meters. The mass estimates in Table 3 exclude those of the propulsion subsystem and of the hydrogen storage tank and associated piping and liquid hydrogen injection pump. The total system mass of 1700 kg in Table 3 could be shown later to be on the conservative side.

For a maximum fuel design temperature of 2000 K and exchangers HX$_1$ and HX$_2$ effectiveness of only 0.7, the specific electric power of the 40 kW$_e$ system is estimated at ~23.5 We/kg. Also, the system's specific energy utilization is as much as 440 W$_{th}$/kg during the propulsion mode. Increasing the exchangers' effectiveness to 0.9 - 0.95, which is technically achievable at present for gas-gas heat exchangers, could increase the specific energy utilization of the system beyond 500 W$_{th}$/kg. Using such DM-HEU, PeBR system, given that mixing hot hydrogen with exhaust plasma in VASIMR or a similar plasma propulsion subsystem is possible, a reliable, redundant, and light-weight systems can be designed for future manned mission to Mars. The high thrust and very high I_{sp} (> 35,000 N.s/kg) of this DM-HEU system could cut down the projected travel time to Mars to about three months. This is only half the projected travel time for systems which use solid core nuclear reactors and hydrogen thermal propulsion (El-Genk et al., 1992; Morley and El-Genk, 1995; Ludewig et al., 1993; and Harty and Brengle, 1993).

SUMMARY AND CONCLUSION

A dual-mode, high-energy utilization nuclear reactor system concept is proposed for potentially reducing the travel time to Mars to ~ 3 month. The system employs a gas-cooled PeBR heat source and generates the same electric power during the power and the propulsion mode. During both modes, the maximum fuel temperature in the reactor core and the inlet temperature of 1144 K to the turbine of the CBC engine are kept constant. The working fluid of the CBC engines for converting the reactor thermal power to high voltage, AC electric power is a He-Xe gas mixture, which is also serves as the reactor coolant. This inert gas is chemically compatible with the reactor core and CBC structural materials. The DM-HEU system does not have a single point failure. The reactor core is divided into three self-contained, but neutronicaly and thermal-hydraulically coupled sectors. The dividers of the core sectors contain heat exchangers for heating hydrogen propellant, as an option. Each core sector has an independent flow loop for the working fluid, a separate set of CBC engines, and a separate heat rejection radiator and attached heat exchanger. The latter is used to boil off and convert the liquid hydrogen coming from the storage tank to hot gas during the propulsion mode.

The hydrogen gas exits the radiator's preheating heat exchangers and is then heated further by the hot He-Xe gas exiting the reactor core during the propulsion mode. In this mode, the mass flow rate fraction of the He-Xe in the reactor core increases from < 0.2, during the power mode, to close to 0.95, commensurate with the increase in the reactor thermal power, while maintaining the core temperature almost constant. The hydrogen exiting the high temperature He-Xe heat exchangers could be heated further in the reactor core by flowing through in-core heat exchangers placed in the dividers of the core sectors. The hydrogen exits the PeBR core between 1600-2300 K, depending on the reactor thermal power, hydrogen flow rate, and the maximum fuel design temperature. The exiting hot hydrogen could be mixed with the high temperature plasma exhaust of the VASIMR to achieve high thrust and $I_{sp} \geq 35,000$ N.s/kg.

The performance parameters of the 40 kW_e-prototype DM system, which uses *a thermal propulsion* only option, have shown that energy utilization fraction of 0.6-0.8 is possible when the maximum fuel design temperature is only 2000 K. At this temperature, the reactor thermal power could be raised ~ 450% during the thermal propulsion mode, resulting in a system specific energy utilization in excess of 400 W_{th}/kg. Increasing this value to more than 500 W_{th}/kg is possible by simply increasing the effectiveness of the hydrogen heat exchangers in the system from 70% to 85-95%, which is achievable with today's technology.

In conclusion, the DM-HEU PeBR system offers unique safety and operation features for future manned missions to Mars. When coupled to a VASIMR, or similar high performance plasma thruster that allows mixing neutral hot hydrogen with the high temperature plasma exhaust, high thrust and very high I_{sp} (> 35,000 N.s/kg) could be achieved. Such high I_{sp} would cut the travel time to Mars to 96 days, instead of the six months projected for solid-core nuclear reactors and hydrogen thermal propulsion. The DM-HEU/VASIMR system could be developed and tested fully assembled in a ground facility, by replacing the nuclear fuel in the PeBR core with electric heaters, within the next 5-10 years. The PeBR and CBC engines technologies are proven, mature, and no technology *breakthrough* is required. Most of the development needs, however, are related to the VASIMR technology, which is currently underway at NASA Johnson and Glenn Research Centers.

ACKNOWLEDGMENTS

This research is sponsored by the University of New Mexico's Institute for Space and Nuclear Power Studies (ISNPS). The author thanks Hamed Saber and Jeff King of ISNPS for helping with the final preparation of the figures and drawings.

REFERENCES

Ashe, T., Personal Communications, Allied-Signal Aerospace Company, Phoenix AZ, October 1993.

Chang Diaz, F., Braden, E., Johnson, I., Hsu, M., and Yang, T. F., "Rapid Mars Transits with Exhaust-Modulated Plasma Propulsion," *NASA Technical Paper* No. **3539**, NASA Johnson Space Center, 1995.

Chang Diaz, F., "Research Status of the Variable Specific Impulse Magnetoplasma Rocket," *Transactions of Fusion Technology*, **35**, 87 - 93 (1999).

El-Genk, M., Carre, F., and Tournier, J.-M., "A Feasibility Study of using Thermoelectric Converters for the LMFBR Derivative ERATO-20 kWe Space Nuclear Power System," *in Proceedings of 24th IECEC Meeting*, 2, Paper # 899311, Washington, DC, 6-11 August 1989, pp. 1281-1288.

El-Genk, M., Buden, D., and Mims J., *Nuclear Reactor Refuelable in Space*, United States Patent No. 5,106,574, April 1992.

El-Genk, M., Guo, Z., Morley, N. and Liscum-Powell, J., "Passive Cooling of Pellet Bed Reactor Concepts for Bimodal Applications," *AIAA/SAE/ASME/ASEE, 30th Joint Propulsion Conference and Exhibit*, Paper No. AIAA 94-3168, Indianapolis, IN, 27 - 29 June 1994.

El-Genk, M., *Apparatus and Method for Nuclear Power and Propulsion*, United States Patent No. 5,428,563, June 27, 1995.

Harty, R. and Brengle, R., "Wire Core Reactor for Nuclear Thermal Propulsion," *in Proceedings of 10th Symposium on Space Nuclear Power and Propulsion*, edited by M. El-Genk, AIP Conference Proceedings 271, 1, 1993, pp. 571 – 577.

Howe, S., DeVolder, B., Thode, L., and Zerkle, D., "Reducing the Risk to Mars: The Gas Core Nuclear Rocket," *in Proceedings of Space Technology and Applications International Forum*, edited by M. El-Genk, AIP Conference Proceedings 420, 3, 1998, pp. 1138 - 1143.

Kammash, T. and Emrich, Jr., W., "Interplanetary Missions with the GDM Propulsion System," *in Proceedings of Space Technology and Applications International Forum*, edited by M. El-Genk, AIP Conference Proceedings 420, 3, 1998, pp. 1145 – 1150.

Juhasz, A., El-Genk, M. S., and Harper, W., "Closed Brayton Cycle Power System with a High Temperature Pellet Bed Reactor Heat Source for NEP Applications," *in Proceedings of 10th Symposium on Space Nuclear Power and Propulsion*, edited by M. El-Genk, AIP ConferenceProceedings 271, 2, 1993, pp. 1055 - 1064.

LaPointe, M. R. and Sankovic, J. M., "High Power Electromagnetic Propulsion Research at the NASA Glenn Research Center," *in this Proceedings of the Space Technology and Applications International Forum (STAIF) 2000*, edited by M. El-Genk, 2000.

Liscum-Powell, J. and El-Genk, M., "Options for Enhanced Performance of Pellet Bed Reactor Bimodal Systems," *in Proceedings of 12th Symposium of Space Nuclear Power and Propulsion*, edited by M. El-Genk, AIP Conference Proceedings 324, 2, New York, 1995a, pp. 845 - 855.

Liscum-Powell, J. and El-Genk, M., "Transient Analysis of a Bimodal of Pellet Bed Reactor System," *in Proceedings of 12th Symposium of Space Nuclear Power and Propulsion*, edited by M. El-Genk, AIP Conference Proceedings 324, Vol. 2, New York, 1995b, pp. 857 - 869.

Ludewig, H., Powell, J., Lazareth, O., Todosow, M., "A Particle Bed Reactor Based NTP the 112,500 N Thrust Class," *in Proceedings of 10th Symposium on Space Nuclear Power and Propulsion*, edited by M. El-Genk, AIP Proceedings 271, 3, 1993, pp. 1737 - 1742.

Morley, N. and El-Genk, M., "Neutronics and Thermal-Hydraulics Analyses of the Pellet Bed Reactor for Nuclear Thermal Propulsion," *J. Nuclear Technology*, **109**, 87 - 107 (1995).

Development Of A Cylindrical Inverted Thermionic Converter For Solar Power Systems

Holger H. Streckert[1] , Lester L. Begg[1], Yuri V. Nikolaev[2], David L. Tsetskhladse[2], Stanislav A. Eriomin[2], Oleg L. Izhvanov[2], Nikolai V. Lapochkin[2]

[1]GENERAL ATOMICS 3550 GENERAL ATOMICS COURT, San Diego, CA 92121-1194
[2]State Research Institute of SIA LUTCH Podolsk, Moscow Region, 142100 Russian Federation
[1](858) 455-2911; [2] (095) 137-9876

Abstract. The design, development and fabrication of a single-cell cylindrical inverted converter are presented. The converter is a coaxial system of two electrodes, in which the outer clad heats the emitter directly and a collector is set inside the emitter. The materials and design elements of a testable prototype are described. Different design versions of structural system components are considered. The results of preliminary tests of the main system components are presented.

INTRODUCTION

The use of thermionic converters for solar powered satellites or space platforms has attracted considerable interest. The Air Force is pursuing a Solar Orbit Transfer Vehicle (SOTV), which requires up to 100 kW electric power produced by thermionic converters. Planar converters typically have small emitter surface areas and consequently have relatively small power output. For large power requirements a large number of planar converters would be required, resulting in a cumbersome design. Cylindrical converters, particularly multi-cell devices can have large emitters and large output power. To simplify the design further, the outer cylinder contains the emitter and the inner cylinder the collector. This is an inverted configuration compared to the traditional thermionic fuel element. Figure 1 shows a schematic of a cylindrical inverted multi-cell (CIM) thermionic converter.

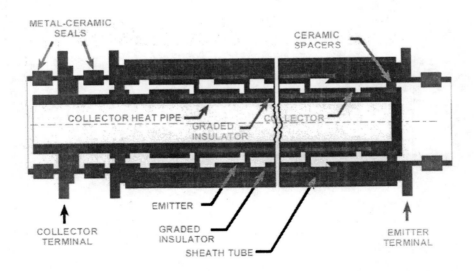

FIGURE 1. Schematic of a cylindrical inverted multi-cell (CIM) thermionic converter.

CP504, *Space Technology and Applications International Forum–2000*, edited by M. S. El-Genk
© 2000 American Institute of Physics 1-56396-919-X/00/$17.00

DESIGN OF CYLINDRICAL INVERTED CONVERTER

Design features and calculated performance data of a cylindrical inverted converter have been published previously (Streckert et al., 1999). Since then, the main experimental components of a single-cell cylindrical inverted converter were designed, fabricated, and tested. Additionally, a functional prototype converter was manufactured.

The single-cell cylindrical inverted converter, shown in Figure 2, is a coaxial system of two electrodes, in which the emitter (–5), is an outer heated clad, and a collector (–6) is set inside the emitter. The emitter is made of single-crystal tungsten with a <111> orientation. Besides tungsten, rhenium is the only other practical metal capable of operating at temperatures up to 2300 K in vacuum. However tungsten is more readily available.

The collector (–6) is made of single crystal molybdenum-niobium alloy. The collector surface is coated with an epitaxial layer of <111> single crystal tungsten. A wick-type heat pipe is used for collector cooling, where sodium is the heat-transfer agent. Radiation of waste heat occurs along the 270 mm collector heat pipe. This portion of the heat pipe is coated with a black coating (–9) with an emissivity coefficient of 0.8 to 0.85.

The emitter is fixed relative to the collector with spacers (–4, 8) made of scandium oxide. Six spacers, positioned radially, are located in slots around the collector body. Previous studies (Kozlov, 1995) show high stability of scandia in a cesium plasma. The gap between the emitter and collector is 0.45 mm at room temperature.

A coaxial metal-ceramic seal (–3) is used for electrical isolation of the emitter and collector. The collars of the seal are made of niobium alloy and the insulator is alumina. Calculations show that the relative movement due to the thermal expansion mismatch between the emitter and collector will be about 1 mm during heat up. To compensate for this expansion mismatch and potential deformation, a bellows (–11) is joined to the metal-ceramic seal.

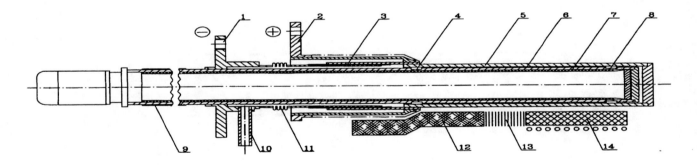

FIGURE 2. Schematic of a cylindrical inverted single-cell converter. 1,2 – collector and emitter electrical terminals. 3 – metal ceramic seal. 4,8 – ceramic spacer. 5 – emitter. 6 – collector. 7 – heat pipe. 9 – high emissivity coating. 10 – cesium supply line. 11 – bellows. 12, 13 – thermal insulation. 14 – electric heater.

Flanges (–1, 2) are used for electric terminals. The collector flange doubles as a mounting bracket, i.e. it is rigidly fixed to the solar receiver body or test stand. When connected in series, the attachment must be done through an electric insulator.

MAIN CONVERTER COMPONENTS DEVELOPMENT AND TEST

An optical photograph of a collector unit attached to a sodium heat pipe is presented in Figure 3. A total of three such units were fabricated. One unit was used for optimizing the methodology of filling the heat pipe with sodium and developing a heat pipe test procedure. The second was used as a practice unit for assembling a mock up converter. The third was used for assembling the operational converter intended for a full scale, cesiated power test. During the test procedure the functionality of the heat pipe was investigated under three separate operational conditions: 1. Inverted, where the evaporation zone of the heat pipe is above the condensation zone (sodium vapor travels against gravity), 2. Horizontal, where the evaporation zone is nearly the same as the condensation zone, 3. Vertical, where the evaporation zone is at the bottom, corresponding to the regular operational conditions of the converter during ground testing.

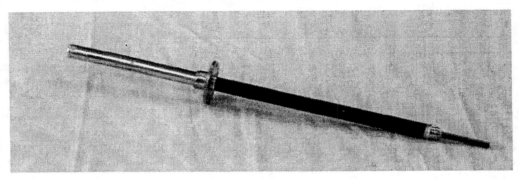

FIGURE 3. Collector unit of cylindrical inverted converter.

The temperature distribution along the length of heat pipe radiation zone for an input power between 7 and 20 W/cm^2 is shown in Figure 4. The variation between thermocouple readings is about ± 10 °C, and represents the accuracy of the measurements. With a thermal input power of 20 W/cm^2 the temperature of the collector heat pipe is about 650 °C.

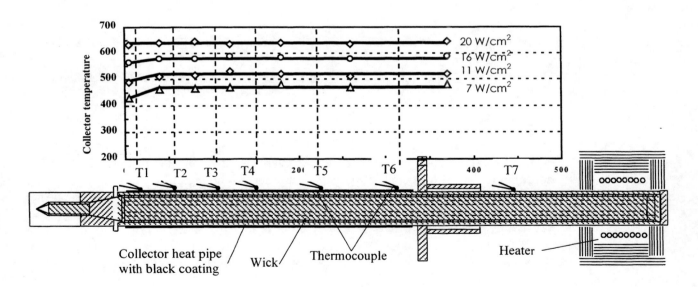

FIGURE 4. Temperature (in kelvin) distribution along collector heat pipe.

An optical photograph of an emitter unit and a metal ceramic seal is shown in Figure 5. The upper part of the emitter (left side of the right part) is coated with a layer of plasma sprayed alumina. In a preliminary test an emitter was thermal cycled between 300 – 1600 °C for 20 cycles. Subsequently it was leak checked and electrical resistance measured. Helium leak rate was less than $5 \cdot 10^{-7}$ l·μm/sec). Total electrical resistance was equal to 1.5 MΩ.

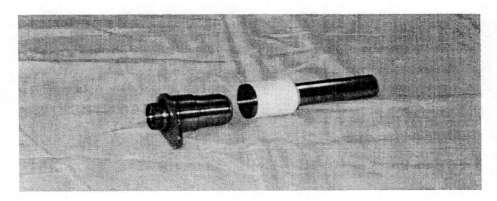

FIGURE 5. Emitter unit and metal ceramic seal of cylindrical inverted converter.

Two types of metal ceramic seals that separate the converter electrodes were developed. One was a coaxial type and the other a conventional "serial" type. The coaxial type may be advantageous for future converter designs, but requires further development. For the present design a conventional serial type metal ceramic seal with Nb collars was chosen. To test for robustness three metal ceramic seals were thermally cycled between 150 and 750 °C, with a heat up rate of 20 °C/min and a cooling rate 20 °C/min. All three units survived the tests. Helium leak rate was less then than $5 \cdot 10^{-7}$ l·μm/sec, The data on electrical resistance versus temperature were similar to those published previously (Izhvanov, 1996).

An optical photograph of the assembled inverted converter is shown in Figure 6. The fully assembled converter was subjected to a hot vacuum test. The objectives of the hot vacuum test were as follows:

- After exposure to operating temperature leak check all converter components.
- Electrical resistance check of electrodes at operating temperature.
- Preliminary converter outgassing.

A specially developed ohmic heater with a heating zone of 40 mm was used for the converter heat up. The test was carried out at an emitter temperature of 1550 °C and a collector temperature of 690 °C. The heat pipe temperature was measured at five different locations along the length of the radiator. The variation in thermocouples readings was consistent with the initial heat pipe test and was no greater than ± 10 °C. In the test process electrical resistance between electrodes was measured by determining vacuum I-V curves. It was equal to 250 Ohm, similar to the resistance measured at 700 °C. After an isothermal hold, the electrodes and structural converter components were outgassed and leak checked. The interelectrode gap was pressurized with helium to $9.8 \cdot 10^4$ Pa. The helium leak rate was less than the detection limit of $5 \cdot 10^{-7}$ l·μm/sec. The preparation for the cesiated power test started after completion of the hot vacuum test.

During the fabrication process of the first prototypical cylindrical inverted converter a number of thermal tests were performed on its main components and ultimately on the completed converter. The tests confirmed the adequacy of the design and demonstrated the readiness of the converter for a full-scale power test.

FIGURE 6. Overall view of the assembled single-cell cylindrical inverted converter.

SUMMARY

A cylindrical inverted thermionic converter design concept has been presented. This type of thermionic converter is intended for use in a solar powered satellite platform. The design and fabrication of a prototype single-cell cylindrical inverted converter have been demonstrated. The materials and design concepts for the main converter components have been studied and verified in thermal cycling tests. Full-scale testing of the completed prototype unit is in progress.

ACKNOWLEDGMENTS

This work was supported by The Boeing Company under letter contract 98055006. The authors wish to thank Dr. P. Andreev and Dr. G. Zaritsky for the heat pipe tests and helpful discussions.

REFERENCES

Streckert, H., Begg, L., Nikolaev, Y., et. al., (1999) "Cylindrical Inverted Thermionic Converter for Solar Power and Propulsion Systems," Proceedings 34th Intersociety Energy Conversion Engineering Conference, ISBN 0-7680-0475, 1999-01-2460.

Kozlov, O., Vybyvanets, V., Olson, D., (1995). "Research of Ceramic Material Stability in Thermionic Converter Interlectrode Gap," Proceedings 12th Symposium on Space Nuclear Power and Propulsion, M. S. El-Genk and M. D. Hoover eds, American Institute for Physics, New York, AIP Conference Proceedings, **2**, 699-704.

Izhvanov, O., Vasilchenko, A., (1996), TOPAZ-2 Single Cell TFE Electric Insulation Properties Study," Proceedings of 13th Symposium on Space Nuclear Power and Propulsion, M. S. El-Genk eds, American Institute for Physics, New York, **3**, 1215-1220.

Development and Testing of Conductively Coupled Multi-Cell TFE Components

Holger H. Streckert[1], Lester L. Begg[1], Yuri V. Nikolaev[2], Valentin S. Kolesov[2], Oleg L. Izhvanov[2], Nikolai L. Lapochkin[2], David L. Tsetskhladse[2]

[1]GENERAL ATOMICS, 3550 GENERAL ATOMICS COURT, San Diego, CA 92121-1194
[2]State Research Institute of SIA LUTCH, Podolsk, Moscow region, 142100 Russian Federation
[1] (858) 455-2911; [2](095) 1379876

Abstract. The technology development and modeling results for conductively coupled multi-cell thermionic fuel element components are presented. Different design versions of the converter structural units are considered. The results of thermal test of the main converter components are presented.

INTRODUCTION

A preliminary design and development of a radiatively coupled multi-cell thermionic fuel element (TFE) was discussed by Wilson (1997). The design had the unique capability of being testable without nuclear fuel. The concept of designing and electrically testing multi-cell TFE's was further developed. Nikolaev et al. (1998) described the design and calculated operating parameters for a conductively coupled in-core TFE. In comparison to the radiatively coupled TFE, the conductively coupled TFE has a number of design and operating advantages, including: 1, the radiatively coupled design inefficiently radiates heat from the fuel clad to the emitter requiring considerably more thermal input; 2, no special radiation clad is needed; 3, significantly lower fuel temperature or electric heater temperature for testing; 4, the absence of leak proof seals between power generating cells of the TFE. The last issue is especially important to ensure TFE reliability.

The most stressed component of a conductively coupled multi-cell thermionic fuel element (CC/MC TFE) is the emitter tri-layer, which consists of the combination of a fuel clad, a ceramic insulating layer, and an emitter tube. Developing a highly reliable TFE depends initially on the development of a reliable emitter tri-layer stack that can operate at temperatures of 1500-1700 °C under thermal cycling conditions. An experimental three-cell TFE was designed (Nikolaev et al., 1998; Streckert et al., 1999) to investigate this problem. The preset efforts are focused on the development, fabrication, and testing of the main components of the CC/MC TFE, comprising the high temperature emitter tri-layer, collector tri-layer, and intercell connectors.

DESIGN OF CONDUCTIVELY COUPLED THREE-CELL TFE

The design schematic of a three-cell conductively coupled TFE is shown in Figure 1. The TFE consists of: emitter tri-layer stack –2, four-layer collector stack –4, intercell connectors –3, end parts, bellows, and seals – 1, 7, 9, that provide needed electrical and mechanical isolation of the TFE electrodes. Interelectrode gap of 0.4 mm is maintained by the intercell connectors –3 and spacers –6, which locate the emitter assembly relative to the collector tube. To simplify the intercell connector system, the last cathode and anode are mounted directly onto the emitter assembly fuel clad and the sheath tube, respectively. Therefore these parts double as current busses and they terminate in current leads –8, 10. The seals –1 and 7 enclose the TFE internal plenum. They simultaneously provide thermal, mechanical and electric isolation of the TFE electrodes. The ceramic to metal seal –9 provides thermal, mechanical and electric isolation of the TFE from the reactor case.

CP504, *Space Technology and Applications International Forum–2000*, edited by M. S. El-Genk

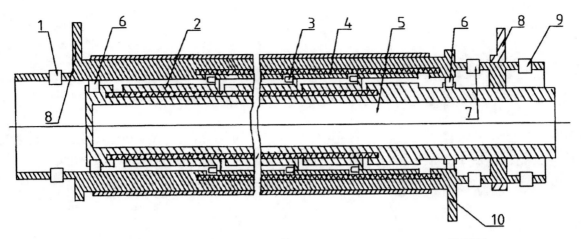

FIGURE 1. General view of conductively coupled three-cell TFE.
1, 9 – metal to ceramic seals; 2 – trilayer emitter stack; 3, – intercell connectors,
6 – spacers;4 – collector stack; 5 – fuel (or electric heater) cavity 7 – seal;
8, 10 – current leads.

The TFE is heated by an electric heater, which is inserted into the emitter cavity –5. Dissipation of waste heat is provided through the helium gap between a sleeve and the wall of a cooling channel where the TFE is located. Cesium vapor is introduced through a seal –1, which is connected via a tube to a cesium reservoir.

EMITTER UNIT

The emitter assembly (Figure 2) consists of a fuel clad (– 1), made of single crystal tungsten/niobium alloy, 0.4 mm thick alumina insulating layer (– 3), and discreet tungsten emitters (– 6). Niobium transition layers (– 2, 4) are applied to obtain better adhesion between the ceramic, fuel clad, and emitters. The emitters are separated from each other by axial gaps that are electrically insulating. The alumina ceramic layers are additionally coated with scandia (– 5) to prevent potential interaction of the alumina with the cesium plasma.

Several fabrication methods for joining metal to ceramic layers were performed and tested, including high temperature brazing and hot isostatic pressing. In principle both technologies provide the necessary integrity of an emitter stack. The hot isostatic pressing was chosen as more promising to achieve long service life. Experimental tri-layer emitter stacks with two or three emitter layers were subjected to thermal tests. During the tests, electrical properties of the ceramic insulation and the emitter stack integrity were examined. Thermal test consisted of thermal cycling (15-30 cycles for 50 to 100 hours) with an isothermal hold at 1600 °C.

Electrical resistance and breakdown voltage of the TFE emitter stack were investigated during the test. A source of direct current and standard electric resistance were introduced into an electric circuit bridging the emitter – ceramic layer – fuel clad. Electric current of the circuit $(I= \Delta Vst/Rst$) was determined by measurement of the voltage drop, ΔVst, across a standard resistance, Rst. Electric resistance, $Rcer$, of the ceramic layer of an emitter stack was determined from the data of electric current of the circuit and by the voltage drop, $\Delta Vcer$, directly between an emitter and the fuel clad, $Rcer =Rst\Delta Vcer./\Delta Vst$. The current leaking through the insulation of the current leads and wires that are attached to the article and set inside the hot zone may influence the measurements. This electrical resistance was determined separately by removing the test article. In addition, the leak resistance was checked during testing by periodic measurements of the total insulation resistance value.

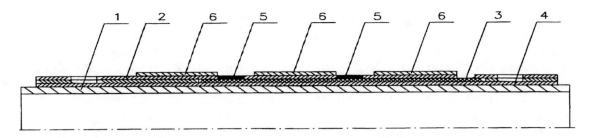

FIGURE 2. Emitter trilayer stack for the three-cell conductively coupled TFE.
1 – fuel clad; - 2, 4 – Nb transition layers; 3 – ceramic layer of Al_2O_3;
5 - ceramic layer of Sc_2O_3; 6 – single crystal tungsten emitter.

Assessment of the ceramic layer breakdown voltage was carried out before heat up and after a thermal excursion with a steady state hold at 800 °C. The assessment was done by examination of the I – V curve in the range of 1 to 50 volts input potential.

Typical results of measurements of electrical resistance are presented in Figures 3 and 4. The data on specific electric resistance of the alumina layer are similar to those published previously (Shtern, 1973). It can be seen, that the parameters for the ceramic layer resistance are stable and reproducible during the isothermal hold and thermal cycling. Breakdown voltage of the ceramic layer at 1600 °C is greater than 10 volts.

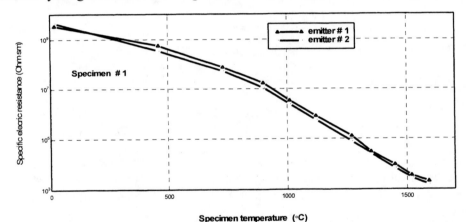

FIGURE 3. Specific electric resistance for ceramic layer of an emitter stack as a function of emitter temperature.

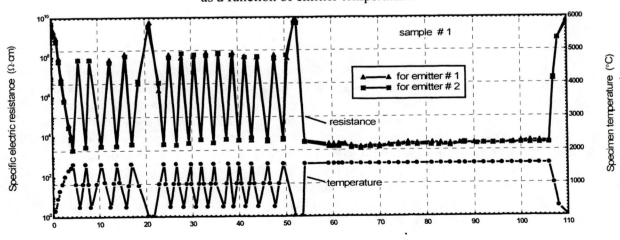

FIGURE 4. Specific electrical resistance of the ceramic for the emitter tri-layer stack under thermal cycling.

COLLECTOR STACK

The fabrication method for the collector stack (Figure 5) was based on previous technology, which had been developed for conventional in-core TFE's (Nikolaev, 1996) and demonstrated to have high reliability. The collector of the TFE consists of two parts: a W/Nb alloy electrode and a four-layer tube, made of Nb + Al$_2$O$_3$ + Nb + Al$_2$O$_3$. A thin layer of tungsten (–1) is needed to improve the collector emission parameters. The basic Nb layer (–5) of the bimetallic electrode is joined with the first Nb layer (–3) of the four-layer tube by brazing. The three-layer (W + Nb + Al$_2$O$_3$) stack is a collector for the power generating cell of the TFE. The collectors of power generating cells are insulated electrically from each other by a ceramic alumina layer (– 4).

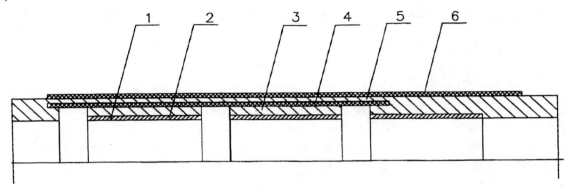

FIGURE 5. Multi-layer collector of the conductively coupled three cell TFE. 1 – tungsten layer; 2, 3 – niobium layers; 4, 6 –ceramic layers of plasma sprayed Al$_2$O$_3$; 5 – niobium sheath tube.

The TFE collectors and the sheath tube (–5) are joined with a ceramic alumina layer (–4) by diffusion welding. The outer surface of the sheath tube is coated with plasma sprayed alumina.

The experimental collector stacks were subjected to a similar test as described above for the emitter tri-layer stacks. The range of thermal cycling was 150 to 750 °C. Cycles numbered 15 to 30. The electrical resistance for the inner and outer ceramic layers at 20 °C is equal to $3.0 \cdot 10^9$ Ohm, the same as at 750 °C.

Break down voltage of individual ceramic layers exceeded 50 Volts at temperatures up to 750 °C. Ultrasonic checks after thermal cycles did not reveal delaminations of the tested collectors

INTERCELL CONNECTORS

The intercell connectors of a conventional multi-cell in-core TFE generally do not encounter the problems associated with the need to compensate for large differences in thermal expansion between the emitter and the collector. The lengths of the emitter and collector usually are equal and the greater thermal expansion of the emitter due to higher temperatures is not large. Expansion can be compensated by relative movement of the emitter unit. In the present design this problem is considerably greater and is a major engineering challenge. The fuel clad is a continuous tube with emitters mechanically coupled. The thermal expansion mismatch relative to the collector sheath tube can be approximately 0.15 mm for a three-cell TFE. Therefore the electrical connectors between the emitters and collectors of the multi-cell device have to compensate this elongation of the fuel clad. Intercell connectors that are shown in Figure 6 were developed to accommodate the thermal expansion mismatch and minimize electrical losses.

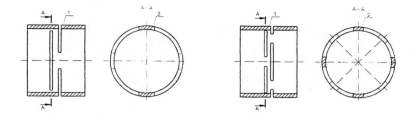

FIGURE 6. Intercell connector design, type a, left; type b, right; 1 – slots; 2 – beams.

Two types of intercell connectors were developed, modeled, manufactured and tested. The cylinder of the first type of connector (type a) is perforated by two rows of slots, 0.5 mm width with a length 5.2 mm. The slots form two equivalent beams, each 6 mm wide. The second type (type b) is also made as a thin walled cylinder of the same thickness and outer diameter. The slots form four equivalent parts with four beams, each 2.6 mm wide located along an axis with a length of 2.3 mm. Calculated electric resistance of the intercell connector made from niobium is $\sim 4 \cdot 10^{-4}$ ohm. Electrical resistance of the intercell connector made from molybdenum is less than $2.5 \cdot 10^{-4}$ ohm.

The results of mechanical tests of the intercell connectors are presented in Table 1 and Figure 7 for one axis compression under isothermal conditions. Deformation to 0.2 mm of the intercell connector from Figure 6a takes place in a quasi ductile region at ambient temperature, while the intercell connector from Figure 6b survives plastic deformation. In the ductile region, the intercell connector of type b is approximately 2-2.5 times more rigid than the intercell connector of type a. The loading that results in a displacement of 0.2 mm at room temperature (20 °C) for the type b intercell connector is approximately 30% higher than for the type a intercell connector. With increasing temperature the indicated difference in loading goes down. Absolute loading values decrease also. At 1200 °C a loading of 130-150 newtons, and at 1500 °C a loading of 45 newtons was measured.

TABLE 1. Load required for 0.2 mm displacement of an intercell connector at three test temperatures.

Type of intercell connector	Load, newtons		
	T = 20 °C	T = 1200 °C	T = 1500 °C
a	450 ± 10	130 ± 10	45 ± 5
b	580 ± 10	150 ± 10	45 ± 5

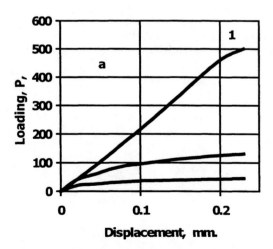
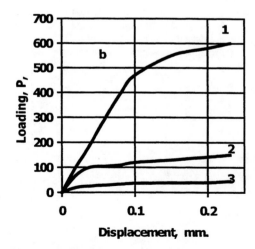

FIGURE 7. Deformation of intercell connectors (type a, left; type b, right) at three test temperatures.
1 –20 °C; 2 – 1200 °C; 3 – 1500 °C

The surfaces of tested intercell connectors were examination by optical microscopy. Testing consisted of 30 thermal cycles between 600 and 1500 °C. Small distortions of the surface texture were observed in the regions where horizontal and vertical beams intersected. This is where the largest deformation of material occurs due to deflection of the beams. The formation of cracks was not observed.

Based on the preliminary tests, a three-cell conductively coupled TFE was fabricated. After assembly, the TFE was checked and tested for a variety of parameters. Preparation for a full power test is in progress.

CONCLUSIONS

Thermal tests of the main components of a multi-cell TFE, which can be tested without nuclear fuel, were accomplished. The temperature dependence of the electrical resistance for the ceramic insulation of an emitter a tri-layer up to 1630 °C was determined. The electrical resistance and break down voltage up to 1630 °C of ceramic insulation made of alumina was measured. These data indicated the feasibility of the tri-layer emitter stack. Experimental development of intercell connectors between power generating cells was performed. In conditions of thermal cycling the intercell connector design demonstrated stability of electrical and mechanical properties. Based on the test results the development and fabrication of a CC/MC TFE power test with electric heating is in progress.

ACKNOWLEDGMENTS

This work was supported by the Defense Threat Reduction Agency under contract number DSWA01-97-C-0088.

REFERENCES

Wilson V.C., "Outline for a Multi-Cell Nuclear Thermionic Fuel element that may be Pretested with Electric Heat," Proceedings 14th Symposium on Space Nuclear Power and Propulsion, M.S. El-Genk and M.D. Hoover eds, American Institute of Physics, Albuquerque, NM, USA, 1, 1567-1572, 1997.

Nikolaev Yu.V., Begg L.L. et. al., "Conductively Coupled Multi-Cell TFE with Electric Heating Pretest Ability," Proceedings 15th Symposium on Space Nuclear Power and Propulsion, M.S. El-Genk and M.D. Hoover eds, American Institute of Physics, Albuquerque, NM, USA, 1, 318-323. 1998.

Nikolaev Yu.V., et. al., "Multi-Cell TFE`s of Thermionic NPP`s," Proceedings 12th Symposium on Space Nuclear Power and Propulsion, M.S. El-Genk and M.D. Hoover eds, American Institute of Physics, Albuquerque, NM, USA, 2, 699-704, 1995.

Streckert, L. L. Begg, D. Pelessone, "Design of Conductively Coupled Multi-Cell Thermionic Fuel Element," Space Technology and Applications International Forum – 1999, ISBN 1-56396-879-7, edited by M. S. El-Genk, E12, 1458 (1999).

Shtern Z.Yu., Krzhizhanovsky R.S., "Thermophysical Properties of Non Metal Materials (Oxides)," Reference Book, Energiya, 1973.

Thermodynamic Characterization of a Diamond-Based Electron Emitter

T.S. Fisher[1a], A.M. Strauss[1b], J.L. Davidson[2], and W.P. Kang[2]

[1a,b]*Department of Mechanical Engineering, and*
[2]*Department of Electrical Engineering and Computer Science*
Vanderbilt University, Nashville, TN 37235

[1a]*Telephone: 615-322-2956; Email: Timothy.S.Fisher@Vanderbilt.edu*

Abstract. This paper contains a thermodynamic analysis of electron emission from a micro-fabricated diamond tip array. The analysis is based on experimental measurements of the current-voltage characteristics of an actual device. Field enhancement, applied field, and electrical current density are shown to influence thermodynamic performance. The idealized thermodynamic analysis predicts cooling rates above 10 W/cm^2 for an existing device under room temperature operation and that 100 W/cm^2 may be possible for future devices.

INTRODUCTION

Small-scale cooling systems for space applications are increasingly in demand as the need to transfer heat from microelectronic components, such as ULSI devices and IR focal plane arrays, intensifies. For digital electronics, surface cooling rates in excess of 30 W/cm^2 will be required for sub-100 nm technology while maximum device junction temperatures will remain at or below 100°C (Semiconductor Industry Association, 1997). Further, the cooling systems must be reliable and manufacturable. The present paper describes a concept for creating high capacity, direct electrical-to-thermal energy conversion for compact cooling based on field emission phenomena.

The required cooling capacity can be achieved by a variety of methods, including single- and two-phase fluid flow, jet impingement, and both traditional and non-traditional refrigeration schemes. For liquid-flow and air-impingement systems, intricate flow networks and flow-generating mechanisms are required which can limit their practicality and long-term reliability.

Alternative devices, such as thermoelectric refrigerators, eliminate moving parts by using electrical current to produce refrigeration. However, thermoelectric refrigerators are typically inefficient and have modest cooling capacities. Contemporary research is seeking to improve the efficiency and utility of thermoelectrics (*e.g.*, Chen, 1997). Recently, several researchers (Mahan, 1994; Shakouri and Bowers, 1997) have suggested that, in theory, high-efficiency room-temperature refrigeration can be achieved via small-scale electron emission devices. The physical principle of this concept involves thermionic electron emission, whereby hot electrons are ejected over a potential barrier. The primary challenges in achieving thermionic refrigeration involve finding new materials with low potential barriers (*i.e.*, work functions) and minimizing reverse heat transfer.

Field emission provides an alternate means of electron emission whereby electrons tunnel through a potential barrier. Field emission is particularly appealing because electrical current densities can be significantly larger than those generated by thermionic emission (Nation et al., 1999). This extremely high electrical current density implies that high thermal current densities are possible, as suggested recently by Miskovsky and Cutler (1999). The present work provides a thermodynamic analysis and cooling predictions based on experimental current-voltage data from diamond emitters.

CP504, *Space Technology and Applications International Forum–2000*, edited by M. S. El-Genk
© 2000 American Institute of Physics 1-56396-919-X/00/$17.00

THEORETICAL BACKGROUND

In this work, we consider diamond emitters that preferentially emit a high-energy distribution of electrons to produce a cooling effect. The origin of the tunneling electrons in diamond can be the valence band or conduction band depending on geometric and material properties (Modinos, 1984). Emission has been shown to originate from the diamond's valence band in studies on single-crystal diamond (Bandis and Pate, 1996) and tetrahedral amorphous carbon (Satyanarayana et al., 1997). However, some forms of polycrystalline diamond grown by chemical vapor deposition emit electrons from the conduction band, aided by a high content of graphite-like (so-called sp^2) bonding. The focus in the present work is on the latter, which produces the emission of high-energy electrons.

Tunneling into the conduction band is shown schematically in the band diagrams of Fig. 1. Figure 1(a) shows the unbiased state, with slight band bending due to space charge effects (Silva et al., 1999). The band gap, $E_g = E_C - E_V$, of diamond is 5.5 eV (Robertson, 1998), and the difference between the base electrode's Fermi energy and the cathode's valence band energy is typically $E_{FB} - E_V = 1.4$ eV, which is the energy associated with grain boundaries. Cathode doping can alter this energy difference. The effective work function ϕ_{eff} is dictated by the vacuum energy level, E_{vac}. This work function represents the electron energy required to exceed all potential barriers. The parameter $\chi = E_{vac} - E_C$ is the electron affinity and represents the energy required to eject an electron from the conduction band into vacuum. Numerous recent studies indicate that χ is negative (although it is shown positive in Fig. 1) when the diamond surface is terminated by hydrogen (e.g., Ristein et al., 1998; Baumann and Nemanich, 1998). This negative electron affinity can significantly enhance emission.

Under a voltage bias (positive on the anode), the electric field causes the bands in the cathode and vacuum to shift, as shown in Fig. 1(b). Near the base-electrode/cathode interface, significant band bending narrows the potential barrier width w. This narrow barrier increases the probability of quantum tunneling, and field emission occurs. A second tunneling process may also occur at the cathode/vacuum barrier for $\chi > 0$. Note that the average energy, ϕ_{ave}, of tunneled electrons can be positive relative to the base electrode's Fermi level due to the triangular-like shape of the barrier. Emitted electrons originate from the high-energy tail of the Fermi-Dirac electron distribution at non-zero absolute temperatures and are transported quasi-ballistically (Cutler et al., 1996) through the cathode into vacuum. In effect, only high-energy electrons are emitted, and the electrical current caused by this preferential emission produces a commensurate transfer of thermal energy.

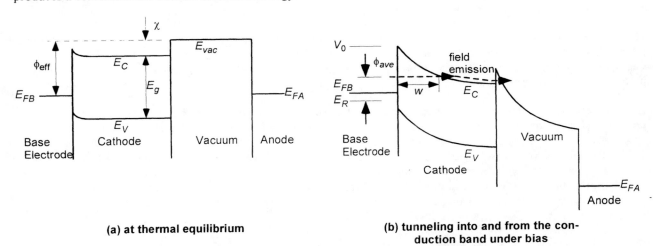

(a) at thermal equilibrium

(b) tunneling into and from the conduction band under bias

FIGURE 1. Band diagrams for field emission from diamond cathodes. (a) Unbiased state. (b) Under bias with tunneling into and from diamond's conduction band.

The emission current density J_E has been described by Fowler and Nordheim (1928) for quantum tunneling through a triangular barrier and takes the form:

$$J_E = \frac{2e}{h^3} \int_0^{V_0} \int_{-\infty}^{\infty} \int_{-\infty}^{\infty} f(E_x, p_y, p_z) D(E_x) dp_y dp_z dE_x , \tag{1}$$

where h is Planck's constant, f is the Fermi-Dirac energy distribution, and D is the quantum tunneling transmission coefficient, given by:

$$D(E_x) = 4 \frac{\sqrt{E_x(V_0 - E_x)}}{V_0} \exp\left\{ \frac{-4}{3F_{local}} \left(\frac{2m}{\hbar^2} \right)^{0.5} (V_0 - E_x)^{1.5} \right\},$$ (2)

where V_0 is the maximum potential at the base electrode-diamond interface and m is the electron mass. The term F_{local} is the local electric field at the interface. This field is greater than the average field, denoted by F_{ave}, due to geometric enhancement, polycrystalline structure, and cathode doping. The local and average fields are often related by a field enhancement factor β (Nation et al., 1999) as:

$$F_{local} = \beta \cdot F_{ave}.$$ (3)

The integral in Eq. (1) was computed numerically for all results presented herein.

To determine thermodynamic quantities, the average energy of emitted electrons must be known. This energy, E_{ave}, can be computed by taking the energy moment of Eq. (1). Prior work (Chung et al., 1994) has shown that the energy, E_R, of electrons that replace emitted electrons is less than the Fermi level. We assume here that the replacement energy is 0.2 eV below the Fermi level. The net thermal energy flux of emitted electrons can be expressed as:

$$J_Q = (\phi_{ave} + 2kT)J_E,$$ (4)

where $\phi_{ave} = E_{ave} - (E_F - E_R)$ and the second term on the right side is a kinetic energy term. The heat transfer rate described by Eq. (4) flows from the base electrode, through the cathode and vacuum, and finally to the anode. For the sake of clarity and brevity, we neglect reverse radiative heat transfer from the anode to the cathode and possible self-heating effects due to current flow in the anode. The latter assumption is consistent with molecular dynamics simulations (Cutler et al, 1996), which predict quasi-ballistic electron transport through the cathode under high local fields. We note that the foregoing assumptions are not conservative, but rather, assume conditions that represent ideal thermodynamic behavior. Future work will incorporate the effects on non-ideal behavior.

RESULTS AND DISCUSSION

The following results are based primarily on experimental current-voltage data reported recently by Wisitsora-at et al. (1999). The experimental study included results from boron-doped and undoped diamond tip emitters with and without heat treatment. Higher emission currents at lower voltages were reported due to heat treatment. Data from the heat treated samples were analyzed according to Eq. (1) to estimate the *local* electric field at the base electrode/cathode interface. The resulting value of F_{local} was then compared to experimental values of F_{ave} according to Eq. (3) to determine approximate values of the field enhancement factor β for both doped and undoped samples. The resulting calculations indicated best-fit values of $\beta = 535$ for the undoped cathode and $\beta = 1300$ for the boron-doped sample. The assumed cathode temperature for all results was 300 K. The following presentation includes electrical and thermodynamic performance predications for each of these samples, as well as a hypothetical sample with a field enhancement factor of $\beta = 2500$. Given the recent advances and trends in diamond emission technology, this value of β appears to be a reasonable estimate of future performance.

Electrical current density depends strongly on field enhancement. Figure 2 shows current density J_E as a function of the average applied field F_{ave} for three field enhancement factors, $\beta = 535$, 1300, and 2500. The results indicate that the field required to produce electrical current decreases from approximately $F_{ave} = 5$ V/μm to 1 V/μm as the field enhancement factor increases from $\beta = 535$ to 2500. As indicated in Eq. (4), the thermal current density is directly proportional to the electrical current, and consequently, the field enhancement factor should have a significant influence on thermal transport.

Thermal transport is a strong function of the average energy carried by emitted electron relative to replacement electrons. This quantity is represented by ϕ_{ave} as described previously (see Fig. 1). The average energy ϕ_{ave} is shown as a function of the average applied field F_{ave} in Fig. 3 for $\beta = 535$, 1300, and 2500. The results for low fields show relatively high average energies. The low-field behavior is due to the relatively wide tunneling distance for low fields across most of the energy distribution. In effect, at low fields only the highest energy electrons are able to tunnel through the narrow region at the top of the triangular barrier shown in Fig. 1. As the average field increases (and causes higher local fields), more low-energy electrons are able to tunnel through the narrower barrier, and the average energy of emitted electrons decreases, as shown for all curves in Fig. 3.

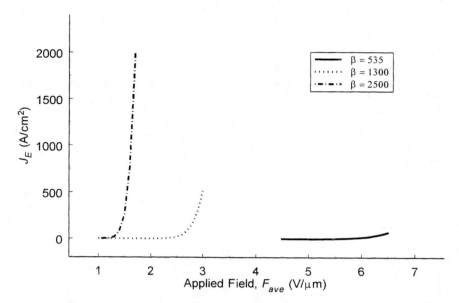

FIGURE 2. Electrical current density J_E as a function of applied field F_{ave} for three field enhancement factors, $\beta = 535$, 1300, and 2500.

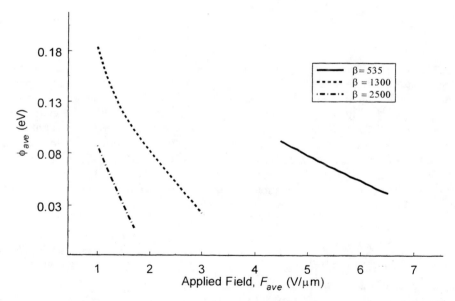

FIGURE 3. Average electron energy ϕ_{ave} due to emission as a function of applied field F_{ave} for three field enhancement factors, $\beta = 535$, 1300, and 2500.

The effects presented in Figs. 2 and 3 combine, through Eq. (4), to produce thermal energy transport from the cathode to the anode. Figure 4 shows the resulting thermal energy flux J_Q as a function of average applied field for $\beta = 535$, 1300, and 2500. For the lowest field enhancement ($\beta = 535$), the thermal energy flux reaches several Watts per square centimeter for fields above 6 V/μm. This level of cooling is roughly equivalent to that provided by thermoelectric and thermionic coolers. For higher field enhancement factors, the results in Fig. 4 predict cooling rates above 10 W/cm^2 for current technology ($\beta = 1300$) and above 100 W/cm^2 for projected field enhancements ($\beta = 2500$). These levels of cooling significantly exceed those of existing technologies.

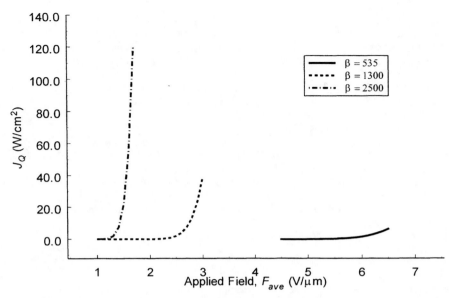

FIGURE 4. Heat transfer rate J_Q as a function of applied field F_{ave} for three field enhancement factors, $\beta = 535$, 1300, and 2500.

CONCLUSIONS

This work has presented theoretical results for the thermal cooling rates possible from diamond field emitters. The results are based primarily on experimental data from diamond field emitters. Cooling rates above 10 W/cm^2 are predicted for existing diamond emitters, and rates above 100 W/cm^2 are projected for future technology. These idealized performance metrics significantly exceed those of existing thermoelectric and thermionic technologies and suggest that field emission cooling could find wide applicability in space technologies.

Future work should focus on refining the theoretical assumptions and obtaining experimental confirmation of this work. This paper assumes that the average energy of replacement electrons is 0.2 eV below the Fermi level. This assumption is central to the reported cooling effect. Further, the work assumes that the emission current produces no self-heating in the emitter tips due to quasi-ballistic electron transport (Cutler et al., 1996). Further modeling of electron transport and scattering should clarify these issues. Finally, experiments will be conducted to characterize thermal transport and to study the influences of electron energy distributions and electron scattering on cooling performance.

ACKNOWLEDGMENTS

This work was supported by the Semiconductor Research Corporation under Contract 99-NJ-720.

REFERENCES

Bandis, C. and Pate, B.B., "Simultaneous Field Emission and Photoemission from Diamond," *Applied Physics Letters*, **69**, 366-368 (1996).

Baumann, P.K., and Nemanich, R.J., "Surface Cleaning, Electronic States and Electron Affinity of Diamond (100), (111) and (110) Surfaces," *Surface Science*, **409**, 320-335 (1998).

Chen, G., "Size and Interface Effects on Thermal Conductivity of Superlattices and Periodic Thin-Film Structures," *ASME Journal of Heat Transfer*, **119**, 220-229 (1997).

Chung, M.S., Cutler, P.H., Miskovsky, N.M., and Sullivan, T.E., "Energy Exchange Processes in Electron Emission at High

Fields and Temperatures," *Journal of Vacuum Science and Technology B*, **12**, 727-736 (1994).

Cutler, P.H., Huang, Z.-H., Miskovsky, N.M., D'Ambrosio, P., and Chung, M., "Monte Carlo Study of Hot Electron and Ballistic Transport in Diamond: Low Electric Field Region," *Journal of Vacuum Science & Technology B*, **14**, 2020-2026 (1996).

Fowler, R.H. and Nordheim, L.W., "Electron Emission in Intense Fields," *Proceedings of the Royal Society A*, **119**, 173-181 (1928).

Mahan, G.D., "Thermionic Refrigeration," *Journal of Applied Physics*, **76**, 4362-4366 (1994).

Miskovsky, N.M. and Cutler, P.H., "A New Technique for Microelectronic Cooling Using a Composite Metal(Semiconductor[S])/Diamond (Wide Band Gap [WBG] Material) Thin Film Device with Internal Field Emission," *1999 Int'l Vacuum Microelectronics Conference*, Darmstadt, Germany, 1999, pp. 276-277.

Modinos, A., *Field, Thermionic, and Secondary Electron Emission Spectroscopy*, Plenum, New York, 1984.

Nation, J.A., Schächter, L., Mako, F.M., Len, L.K., Peter, W., Tang, C.-M., and Srinivasan-Rao, T., "Advances in Cold Cathode Physics and Technology," *Proceedings of the IEEE*, **87**, 865-889 (1999).

Ristein, J., Stein, W. and Ley, L., "Photoelectron Yield Spectroscopy on Negative Electron Affinity Diamond Surfaces: A Contactless Unipolar Transport Experiment," *Diamond and Related Materials*, 7, 626-631 (1998).

Robertson, J., "Theory of Electron Field Emission From Diamond and Diamond-Like Carbon," *Materials Research Society Symposium*, **498**, 197-208 (1998).

Satyanarayana, B.S., Hart, A., Milne, W.I. and Robertson, J., "Field Emission from Tetrahedral Amorphous Carbon," *Applied Physics Letters*, **71**, 1430-1432 (1997).

Semiconductor Industry Association, *National Technology Roadmap for Semiconductors: Technology Needs*, SEMATECH, Austin, TX, 1997, p. 135.

Shakouri, A., and Bowers, J.E., "Heterostructure Integrated Thermionic Coolers," *Applied Physics Letters*, **71**, 1234-1236 (1997).

Silva, S.R.P., Amaratunga, G.A.J., and Okano, K., "Modeling of the Electron Field Emission Process in Polycrystalline Diamond and Diamond-Like Carbon Thin Films," *Journal of Vacuum Science and Technology B*, **17**, 557-561 (1999).

Wisitsora-at, A., Kang, W.P., Davidson, J.L., Gurbuz, Y., and Kerns, D.V., "Field Emission Enhancement of Diamond Tips Utilizing Boron Doping and Surface Treatment," *Diamond and Related Materials*, **8**, 1220-1224 (1999).

An Advanced Thermionic Theory: Recent Developments

Albert C. Marshall

Defense Threat Reduction Agency, 1680 Texas Street SE, Kirtland Air Force Base
Albuquerque, New Mexico 87117-5669
(505) 853-0946; acmarsh@sandia.gov

Abstract. In previous papers I have shown that a revision is required to the basic approach for predicting net currents in thermionic energy conversion diodes. Revised equations were developed to properly account for electron reflection and temperature coefficient effects. Electron spectrum equations for averaging transmission coefficients were also developed. In this paper the spectrum equations are simplified and several new developments are presented that relate to the revised methodology. Recent developments include a demonstration of the general applicability of the new equations, equations for space charge that include the effects of reflection and an emitting collector, and electron cooling equations with electron reflection effects included. In addition, methods are developed to apply the new equations to non-uniform surfaces.

PREVIOUS WORK SUMMARY

Previous work includes empirical net current equations with reflection effects (Nottingham, 1956) and equations derived for the special case of mirror reflection (Fin *et al.*, 1973 and Eryomin *et al.*, 1991). These investigators assumed a cold collector. My previous work (Marshall, 1998) includes a demonstration that the conventional approaches for predicting net currents in thermionic diodes either conflict with the second law of thermodynamics or conflict with the basic physics of thermionic diodes. Revised net current equations were developed that resolve the observed conflicts by correctly including electron reflection and temperature coefficient effects. The new equations also permit a hot collector. These developments are summarized in the following. Although spectra equations for averaging transmission coefficients were also derived, a simpler approach for predicting spectra has been developed. The spectra equations are discussed under the topic of recent developments.

Conflict Identified

The basic equation for thermionic currents has been successfully used in the study and development of thermionic energy conversion diodes. For some conditions, however, predictions are quite inaccurate (Rasor, 1998 and Balestra *et al.*, 1978). Net current density J for these diodes is conventionally predicted by subtracting the collector (back) emission current density J_C^* from the emitter emission current density J_E^*; i.e., $J = J_E^* - J_C^*$. The asterisk is used to indicate that current densities are computed using the apparent emission constant A^*. The subscripts E and C are used throughout this paper to indicate emitter and collector electrode, respectively. The expressions for J_E^* and J_C^* are obtained from the basic form of the Richardson-Dushman (RD) equation (Hatsopoulous and Gyftopoulous, 1973). To obtain J_E^* and J_C^* for vacuum energy conversion diodes, the work function in the RD equation must be replaced by $(\psi_m - \mu)$, where ψ_m is the maximum in the potential energy barrier between electrodes and μ is the Fermi energy level of the electrode (illustrated in Fig. 1). The value of ψ_m depends on the magnitude of the work functions of both electrodes, the output voltage V, and the space charge between electrodes. The Fermi energy for the emitter is taken as the zero of potential energy and the quantity $(\psi_m - \mu_C)$ can be replaced by the equivalent $(\psi_m - eV)$, where e is the electron charge. With these substitutions, the standard equation is written:

CP504, *Space Technology and Applications International Forum–2000*, edited by M. S. El-Genk

$$J = A_E^* T_E^2 \exp\left(\frac{-\psi_m}{kT_E}\right) - A_C^* T_C^2 \exp\left(\frac{eV - \psi_m}{kT_C}\right). \tag{1}$$

Here, T is the surface temperature and k is Boltzmann's constant. Because these devices are heat engines Eq. (1) must be consistent with the second law of thermodynamics (Hatsopoulous and Gyftopoulous, 1973). At thermal equilibrium Eq. (1) balances only when $A_E^* = A_C^*$ and when J and V have opposite signs or J and V both equal zero. In other words power must be supplied to produce a net current, as required by the second law of thermodynamics. In the foregoing discussion the emitter and collector emission constants were assumed to be identical. If unequal values of A_E^* and A_C^* are used in Eq. (1) the equation allows positive values for both J and V for $T_C = T_E$. Consequently, the conventional equation predicts power production at thermal equilibrium, in conflict with the second law of thermodynamics.

Apparent emission constants were examined to resolve the conflict in Eq. (1). The reported A^* are frequently obtained using diode current-voltage (JV) measurements to extract a temperature independent factor (A^*). Measured A^* are in general given by $A^* = A\,(1 - \overline{R})\exp(-\alpha_T/k)$ where A is the theoretical value of the emission constant ($120\ \text{A/cm}^2\text{K}^2$) and the other terms are nearly-constant factors (Herring and Nichols, 1949). The quantities $\overline{R}$ and α_T are the average electron reflection coefficient and the temperature coefficient of the work function ϕ, respectively. The exponential factor in A^* results because ϕ often has the form $\phi_0 + \alpha_T T$. Inserting $\phi_0 + \alpha_T T$ into the RD equation yields the term, $\exp(-\alpha_T/k)$. For the retarding range Fig. 1 shows that the work function with or without the αT term (curves a and b, respectively) give identical values for ψ_m. Consequently, including α_T for this case has no effect on J. If $\exp(-\alpha_T/k)$ is incorrectly included in A^*, rather than using it to determine the value of ψ_m, the net current will be under-predicted and the second law can be violated.

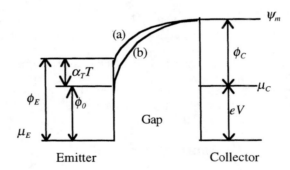

FIGURE 1. Potential energy diagram for a vacuum energy conversion diode (retarding range).

The $(1 - \overline{R})$ term, however, is not part of the work function; thus, an attempt to avoid the second law conflict by using the "effective" work function $\phi' = \phi_0 - kT \ln(1 - \overline{R})$ cannot be justified. The second law conflict demonstrates that the A^* approach (using A^* in place of A) cannot be correct, and the alternative ϕ' approach (including the reflection term in ϕ') is also found to be incorrect. The parameter $\overline{R}$ accounts for reflection at the electrode surface of electrons with energies ε_z greater than ψ_m, where ε_z is the energy associated with electron momentum component p_z normal to the surface. Possible reflection mechanisms include reflection due to an abrupt change in electrical potential near the electrode surface, reflection from patches with different work functions, reflection due to forbidden energy bands, and two-dimensional diffraction effects. Rasor (1998) has stressed the importance of electron reflection and significant electron reflection has been reported for electrode materials in the energy range important for thermionic energy conversion (Balestra et al., 1978 and Zollweg, 1964).

Revised Equations

The preceding discussion applies to internally reflected electrons (from within the electrode); however, quantum symmetry rules require that surfaces reflecting internal electrons must also reflect external electrons from the opposite electrode (Herring and Nichols, 1949). A revised net current equation was derived that includes the effects of external reflection and temperature coefficients. The revised equation is

$$J = \tau\left(J_E - \Gamma J_C\right) . \tag{2}$$

The quantities J_E and J_C are defined as ideal emission current densities, where A^* is always equal to the theoretical emission constant $A = 120$ amperes/cm^{2}K^2 and ψ_m is computed using temperature dependent work functions. The parameters, τ and Γ are defined as the effective transmission coefficient and the symmetry index, respectively, given by

$$\tau \equiv \overline{D}_E \overline{D}_C^I \Big/ \left(\overline{D}_E^I + \overline{D}_C^I - \overline{D}_E^I \overline{D}_C^I\right),$$

$$\tag{3}$$

$$\text{and} \qquad \Gamma \equiv \overline{D}_C \overline{D}_E^I \Big/ \overline{D}_E \overline{D}_C^I.$$

Here, $\overline{D} = \left(1 - \overline{R}\right)$ is the average transmission coefficient. The superscript I is used to indicate that the coefficient has been averaged over the external spectrum from the opposite electrode, rather than the internal spectrum. At thermal equilibrium $\Gamma = 1.0$ and Eq. (2) is always consistent with the second law. The original and revised equations were used to predict JV curves for typical operating conditions, assuming $\overline{D}_E = 0.7$, $\overline{D}_C = 0.3$ and $\alpha_T = 2 \times 10^{-5}$. Calculational results presented in Fig. 2 show that the original equation using A^* can grossly overpredict the net current. If no temperature coefficient is present, agreement between the predictions improves, but the original equation still overpredicts the net current by more than a factor of two. The original equation can be forced to approximately match the revised equation predictions by adjusting A^* or ϕ'; however, these adjustments are large and arbitrary. Furthermore, adjustments provide little insight into the nature of electrode surfaces and can result in very poor predictions when operating conditions are changed.

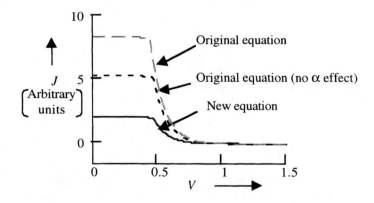

FIGURE 2. Effect of revised equations on JV curves.

RECENT DEVELOPMENTS

My most recent theoretical developments have not been reported previously. New theoretical developments include: a demonstration of the general applicability of the new equations, simplified spectrum equations, space charge equations that account for effects of reflection and an emitting collector, and electron cooling equations with electron reflection effects included. In addition, methods are developed to apply the new equations to non-uniform surfaces. The proof of general applicability provides confidence that the theory can be used for any homogenous electrode material. Though the spectrum equations are essentially equivalent to my previously reported equations (Marshall, 1998) the approach presented in this paper simplifies spectra computation. The revised space charge equation completes the basic framework for the new theory. The space charge equation is presented and discussed in the following subsection, but the very tedious derivation is not provided. Equations for electron cooling with electron reflection effects are obtained by a simple extension of the derivation of the revised net current equations; these revised electron cooling equations are included for completeness. My suggested approach for applying the new equations to patchy electrodes assures that the theory can be used for any electrode material, including non-homogenous surfaces. A discussion of recent developments follows.

Generality of Equations

The RD equation is often regarded as a crude approximation because the simple assumption of a free electron gas in a vacuum is used in its derivation. If the RD equation is a crude approximation, then the general applicability of the revised equations (based on the RD equation) may be questioned. However, the RD equation, J_E, and J_C can be shown to be idealizations that gives the maximum current density. Consider a diode with a real emitter material and an ideal collector at thermal equilibrium with no applied voltage. At thermal equilibrium, the second law requires that

$$\mathcal{J}_{EC} = \mathcal{J}_{CE}, \tag{4}$$

where $\mathcal{J}_{EC}$ and $\mathcal{J}_{CE}$ are the actual emitted current densities from the emitter and collector, respectively, that are transmitted into the opposite electrode. From the definition of the ideal current density, for $T_E = T_C$ and $V = 0$,

$$J_E = J_C = AT^2 \exp(-\psi_m/kT). \tag{5}$$

The current density from the collector is J_C, by definition, but the emitter may not absorb all electrons from the collector because the emitter is not an ideal material. The fraction of the ideal current from the collector absorbed by the emitter can be represented by $\overline{D}_E$, where $\overline{D}_E \leq 1.0$; therefore,

$$\mathcal{J}_{CE} = \overline{D}_E J_C. \tag{6}$$

From Eqs. (4), (5), and (6), we find

$$\mathcal{J}_{EC} = \overline{D}_E J_E. \tag{7}$$

But all electrons leaving the emitter are absorbed by the ideal collector; thus,

$$\mathcal{J}_E = \overline{D}_E J_E, \tag{8}$$

where $\mathcal{J}_E$ is the current density emitted from the real emitter.

The derivation of the equation for an ideal electrode is based on the assumption of a free electron gas in a vacuum with no reflection at the barrier (smooth potential shape). If we postulate a material capable of emitting a current density greater than J_E, then electrons must be reflected by the vacuum. However, a vacuum offers no impedance to the flow of electrons, and $\overline{D}_E \leq 1.0$ must be true for any material. Several other arguments can be used to reach this conclusion.

If we allow the collector to cool, emission from the emitter is virtually unchanged. (For some materials, a very small reduction in emitter current may result; however, this nuance does not affect the revised equations or our conclusions.) Also observe that $\mathcal{J}_E$ is valid for any emitter temperature and any value of ψ_m. Hence, we can drop the device of an ideal collector at thermal equilibrium. We conclude that J_E is the maximum current density for any conditions, and the actual current density from a real electrode can be represented by the product of the ideal current density and a transmission coefficient.

Electron Cooling

The equations for electron cooling with reflection effects included can be easily derived. The equations are

$$Q_{Eec} = \tau \vartheta_{EC} J_E \left(2kT_E + \psi_m \right)/e, \tag{9}$$

and

$$Q_{Cec} = \tau \Gamma \vartheta_{CE} J_C \left[2kT_C + \left(\psi_m - eV \right) \right]/e, \tag{10}$$

where Q_{EC} and Q_{CE} are the electron cooling heat flux for electrons emitted by the emitter and collector, respectively. The parameter ϑ is the inelastic scattering heat transfer factor ≈ 1.0.

Internal Spectrum

Full utilization of the proposed equations requires detailed transmission and reflection data as well as a method for calculating electron spectra (used to average coefficients). The electron spectrum for unreflected electrons depends only on the electron energy relative to ψ. Using this observation and my previously derived spectrum equation (Marshall, 1998) the internal spectrum $f_E(\varepsilon_z, \theta)$ for unreflected electrons reaching ψ_m is given by

$$f_E(\varepsilon_z, \theta) = \left(\frac{2\varepsilon_z \tan\theta \sec^2\theta}{k^2 T_E^2} \right) \exp\left(\frac{-\varepsilon_z \sec^2\theta}{kT_E} \right). \tag{11}$$

The parameter θ is the angle of incidence with respect to the normal and ψ_m is selected as the zero of energy. This approach is simpler than my previous approach in that the spectra, as well as the transport and reflection coefficients, are always referenced to ψ_m rather than using two references (a location inside the electrode and the location of the maximum in the potential energy barrier).

External Spectrum

We found that emission from any real material is given by the product of the ideal current density and a transmission coefficient. Thus, electrons from the emitter that will strike the collector for the first time are described by the fractional spectrum $f_{C1}^E(\varepsilon_z, \theta)$ at ψ_m given by

$$f_{C1}^E(\varepsilon_z, \theta) = \int_0^{\pi/2} \int_0^{\infty} D_E(\varepsilon_z, \theta) f_E(\varepsilon_z, \theta) \, d\varepsilon_z d\theta, \tag{12}$$

where $D_E(\varepsilon_z, \theta)$ is the energy/angle dependent transmission coefficient. Note that $D_E(\varepsilon_z, \theta)$ also depends on $\Delta\psi = \psi_m - \mu_E$. The integral of the spectrum $f_E(\varepsilon_z, \theta)$ over all energies and over angles between 0 and $\pi/2$ is equal to 1.0; however, the integral of the fractional spectrum is less than 1.0 if $\overline{D}_E$ is less than 1.0. For fractional spectra, the superscripts indicate the emitting electrode and subscripts indicate the electrode struck and the number of times the electrode is struck by this group of electrons. The collector reflected fractional spectrum that will return to the emitter $f_{E1}^E(\varepsilon_z', \theta')$ is

$$f_{E1}^E(\varepsilon_z', \theta') = \int_0^{\pi/2} \int_0^\infty S_C(\varepsilon_z, \theta, \varepsilon_z', \theta', \Delta\psi_C) f_{C1}^E(\varepsilon_z, \theta) \, d\varepsilon_z \, d\theta \qquad \text{for} \quad \pi/2 < \theta' < \pi. \qquad (13)$$

Here, $S_C(\varepsilon_z, \theta, \varepsilon_z', \theta', \Delta\psi_C)$ is the scattering kernel for the probability of reflection from ε_z, θ to ε_z', θ' evaluated for $\Delta\psi_C$. Continuing this process we obtain $f_{C2}^E(\varepsilon_z, \theta)$, $f_{E2}^E(\varepsilon_z', \theta')$, $f_{C3}^E(\varepsilon_z, \theta)$, etc. The energy/angle dependent incident spectra for the emitter $f_E^I(\varepsilon_z', \theta')$ is then given by

$$f_E^I(\varepsilon_z', \theta') = \chi_E(\varepsilon_z', \theta') / \chi_E, \qquad (14)$$

where

$$\chi_E(\varepsilon_z', \theta') = \sum_{n=1}^\infty \left[J_E f_{En}^E(\varepsilon_z', \theta') + J_C f_{En}^C(\varepsilon_z', \theta') \right], \qquad (15)$$

and χ_E is the integral of $\chi_E(\varepsilon_z', \theta')$ over all ε_z' and $\pi/2 < \theta' < \pi$. The average transmission coefficients are obtained from

$$\overline{D}_E = \int_0^{\pi/2} \int_0^\infty D_E(\varepsilon_z, \theta) f_E(\varepsilon_z, \theta) \, d\varepsilon_z \, d\theta = \int_0^{\pi/2} \int_0^\infty \int_0^{\pi/2} \int_0^\infty S_E(\varepsilon_z, \theta, \varepsilon_z', \theta', \Delta\psi_E) f_E(\varepsilon_z, \theta) \, d\varepsilon_z' \, d\theta' \, d\varepsilon_z \, d\theta, \qquad (16)$$

and

$$\overline{D}_E^I = \int_{\pi/2}^\pi \int_0^\infty \int_{\pi/2}^\pi \int_0^\infty S_E(\varepsilon_z, \theta, \varepsilon_z', \theta', \Delta\psi_E) f_E^I(\varepsilon_z', \theta') \, d\varepsilon_z' \, d\theta' \, d\varepsilon_z \, d\theta. \qquad (17)$$

Calculations using the spectra equations show that the angle-dependent current density peaks at small angles for high electron energies (about 10 degrees at a few eV). However, the energy-integrated current density is dominated by the low energy contributions because the electron distribution decreases exponentially as energy increases. As a result, the energy-integrated current density peaks at a surprisingly large angle of about 45 degrees. Thus, estimates of electron reflection based on the assumption of normal incidence may be very inaccurate.

Space Charge

The effect of space charge on ψ_m was developed by Langmuir for the special case of no reflection and a cold collector. Kniazzeh (1959) studied the effect of a hot collector on space charge, but his study did not include electron reflection. More recently investigators (Eryomen, *et al.*, 1991) have developed space charge equations for the special case of mirror reflection with a cold collector. I derived a general equation for space charge that includes the effects of both reflection and a hot collector. These equation also allow for energy and angle changes during reflection and are applicable to any diode using homogenous electrodes. The general space charge equation for the emitter side of the barrier is

$$\frac{d\gamma}{d\xi} = -\left[\begin{array}{l} \left\{ \alpha_{E1}\left[\exp(\gamma) - \exp(\gamma)\mathrm{erf}\sqrt{\gamma} + 2\sqrt{\gamma/\pi} - 1\right] - \alpha_{E2}\left[2\sqrt{\gamma/\pi} - \exp(\gamma)\mathrm{erf}\sqrt{\gamma}\right] \right\} \\ + \varsigma/\delta \left\{ \alpha_{C1}\left[\exp(\delta\gamma) - \exp(\delta\gamma)\mathrm{erf}\sqrt{\delta\gamma} + 2\sqrt{\delta\gamma/\pi} - 1\right] - \alpha_{C2}\left[2\sqrt{\delta\gamma/\pi} - \exp(\delta\gamma)\mathrm{erf}\sqrt{\delta\gamma}\right] \right\} \end{array} \right]^{1/2}. \qquad (18)$$

Here $\delta = T_E/T_C$ and γ and ξ are the dimensionless potential and distance, given by

$$\gamma = e\left(\psi_m - \psi\right)/kT_E , \tag{19}$$

and

$$\xi = \sqrt{e/e_0}\left(\frac{2\pi m}{k^3 T_E^3}\right)^{\frac{1}{4}}\exp\left(-\frac{e\psi_m}{2kT_E}\right)(z - z_m). \tag{20}$$

Also,

$$\varsigma = \sqrt{\delta}\,\frac{J_{CS}}{J_{ES}}\exp\left\{\frac{e\left[\Delta\psi_{EC} - (T_E - T_C)/T_E\right]}{kT_C}\right\}. \tag{21}$$

The parameters e_0, z, and z_m are the dielectric constant for a vacuum, the spatial dimension, and the position of the maximum in the potential, respectively. The subscript S refers to saturation conditions and the αs are reflection parameters. For the collector side Eq.(18) also applies, except that the αs are replaced by βs and the minus before the brackets is replaced by a plus. An excellent approximation for the αs and βs gives

$$\alpha_{E1} \approx \beta_{E1} \approx \left[\overline{D}_E\left(2 - \overline{D}_C^I\right)\right]\Big/\left[\overline{D}_E^I + \overline{D}_C^I - \overline{D}_E^I\overline{D}_C^I\right],$$
$$\tag{22}$$

and

$$\alpha_{C1} \approx \beta_{C1} \approx \left[\overline{D}_C\left(2 - \overline{D}_E^I\right)\right]\Big/\left[\overline{D}_E^I + \overline{D}_C^I - \overline{D}_E^I\overline{D}_C^I\right].$$

Also, $\alpha_{E2} \approx \beta_{C2} \approx 2.0$ and $\alpha_{C2} \approx \beta_{E2} \approx 0$. An iterative solution approach to Eq. (18) was developed and calculations were performed to determine the effect of electron reflection on space charge. Predicted JV curves were obtained for typical operating conditions, assuming $\overline{D}_E = 0.7$ and $\overline{D}_C = 0.3$. For this case the revised equations predict a significant increase in space charge relative to predictions using Langmuir's equation and A^*. Using the new formulations, the combined effect of reflection on space charge and net current calculations resulted in predicted net currents only one third of the value obtained using the original equations. However, the errors incurred using the original equations depend strongly on the values of D for the emitter and collector, and adjusting A^* can significantly reduce the error produced in the original equations.

Patchy Surfaces

The preceding equations and methods were developed for homogenous surfaces. The equations for current density can be applied to surfaces containing patches with different ϕs and Ds by using an associated transmission coefficient given by

$$\overline{D}_{Ea} = \frac{\sum_{i-1}^{N} F_{Ei}\overline{D}_{Ei}\exp\left(-\overline{\psi}_{mi}/kT_E\right)}{\exp\left(-\overline{\psi}_{mL}/kT_E\right)} , \tag{23}$$

where F_{Ei}, $\overline{D}_{Ei}$, $\overline{\psi}_{mi}$ and N are the fraction of the surface occupied by patch type i, the transmission coefficient for patch i, the maximum in the potential barrier for patch i, and the number of patches types, respectively. The parameter $\overline{\psi}_{mL}$ is the lowest value of ψ_m among all patch types. The bar over ψ_m indicates that ψ_m is averaged over the patch surface. Given this definition, the emission current density can be expressed by

$$\mathcal{J}_E = \overline{D}_{Ea}AT_E^2\exp\left(-\overline{\psi}_{mL}/kT_E\right). \tag{24}$$

The advantage of this approach is that Eqs. (2), (3), and (22) can be used to correctly compute net current densities, including the effect of reflections between the emitter and collector. Equation (23) can also be used in Eq. (18) to determine the effect of electron reflection on space charge for non-homogenous electrodes; however, this approach is not rigorous and may provide a poor approximation for some conditions. Better methods are under development for including the effect of reflection on space charge in patchy diodes.

WORK IN PROGRESS AND CONCLUSIONS

Several other activities related to this work are in progress. These include theory validation experiments, a method for obtaining energy dependent transmission coefficients from current-voltage measurements, a method for obtaining the effect on D due to multiple reflections between the barrier and the surface, methods for including the effect of reflection on space charge for non-homogenous electrodes, and methods for obtaining the effect of patch fields on the value of ψ_m. Although the new theory was developed specifically for vacuum diodes, the basic principles also apply to plasma diodes. Development of revised equations for plasma diodes is planned.

In conclusion, the effect of electron reflection and temperature dependent work functions must be properly included in net current and space charge calculations. The use of the revised equations may significantly improve predictions of thermionic performance and provide greater understanding of the physics of electrode surfaces. Confirmation of the new thermionic theory by validation experiments is needed to provide confidence in the revised equations.

ACKNOWLEDGMENTS

The advice and support of R. Nelson (Defense Threat reduction Agency) and N. Rasor (Consulting Physicist), and the support and encouragement of T. McKelvey, D. Kristensen, and T. Drake (Defense Threat reduction Agency) and D. Berry (Sandia National Laboratories) is greatly appreciated. Programming and calculations performed by D. Gallup and D. King are also greatly appreciated. This work was funded by the U. S. Defense Threat reduction Agency. The author is employed by Sandia National Laboratories, which is operated by the U.S. Department of Energy under Contract DE-AC04-94AL85000. The author is presently on an Interagency Personnel Act assignment with the Defense Threat reduction Agency.

REFERENCES

Balestra, C. L., Huffman F. N., and Wang, C. C., "Topical Report on Electron Reflection from One-Dimensional Barriers," *Thermo Electron Corp., Report*, TE4237-38-79, Waltham, MA, 1978.

Eryomen, S. A., *et al.*, "Specified calculation of a low–temperature close-spaced thermionic converter," in *Nuclear Power Engineering in Space, 2nd Intersociety Conference*, Conference Proceedings, Sukhumi, 1991, pp. 372-381.

Finn, R. M., Nicholson, D. J. and Trischka, J. W., "Thermionic Constants and Electron Reflection for Ta(100) by the Shelton Retarding Field Method," *Surf. Sci.* **34**, pp. 522-546 (1973).

Hatsopoulous, G. N. and Gyftopoulous, E. P., *Thermionic Energy Conversion* Cambridge, MA, MIT Press, 1973.

Herring, C. and Nichols, M. H., "Thermionic Emission" *Rev. Mod. Phys.* **21**, pp. 185-268 (1949).

Kniazzeh, A. "Potential distribution between two parallel thermionic emitters with applications to the thermoelectron engine," *Massachusetts Institute of Technology Thesis*, Massachusetts, 1959.

Marshall, A. C., "An Equation for thermionic currents in vacuum energy conversion diodes," *Appl. Phys, Letters* **73**, 2971-2973 (1998).

Nottingham, W. B., "Thermionic Emission," *Handbuch der Physik, Bd. XXI*, Berlin, Springer-Verlag, 1956, pp. 1-175.

Rasor, N. S., "The Important Effect of Electron Reflection on Thermionic converter performance," in *33rd Intersociety Energy Conversion Engineering Conference-1998*, ANS Conference Proceedings 700262, Colorado Springs, CO, 1998, p. 211.

Zollweg, R. J., "Electron Reflection from Tungsten Crystals, Clean and With Adsorbed Cs and CO," *Surface Sci.* **2**, pp.409-417 (1964).

The Heatpipe Power System (HPS) for Mars Outpost and Manned Mars Missions

David I. Poston[1], Stewart L. Voit[1], Robert S. Reid[1], and Peter J. Ring[2]

[1]*Los Alamos National Laboratory, MS-K551, Los Alamos, NM 87545*
[2]*Advanced Methods and Materials, 1798 Technology Dr., #251, San Jose, CA 95110*

[1]*(505)667-4336; poston@lanl.gov,* [2]*(408)451-3169; ammsj@aol.com*

Abstract. Interest continues to increase in using fission reactors to power robotic missions to Mars within the next decade, as well as manned missions in the future. This paper evaluates the use of the Heatpipe Power System (HPS) for these missions (the paper focuses on the reactor core, and does not significantly evaluate power conversion and shielding possibilities). The HPS is a safe, simple reactor that is designed to have a low development time and cost. Previously, most HPS designs have used refractory metals to allow high-temperature operation (for increased power conversion efficiency) and higher thermal conductivity. For use on Mars, a stainless-steel or super-alloy system may be required in order to avoid corrosion. As a result, Los Alamos National Laboratory (LANL) has begun a program to evaluate and test a stainless-steel HPS core. This paper will describe some analysis of potential non-refractory HPS designs, as well as work that is currently underway to build a full-scale unfueled HPS core. Resistance heated testing will be used to evaluate the thermal performance of the core under a wide variety of conditions; including the simulated Martian atmosphere. Analytical and experimental results thus far indicate that the HPS is very well suited for Martian applications. A near-term, low-cost, low-risk design is proposed for a Mars Outpost application that could provide 40 kWt, and thus 2 kWe with 5% efficient thermoelectrics. A higher power core, 350 kWt, is proposed as a potential surface power source for a manned Mars mission that could provide up to 100 kWe, depending on the efficiency of the power conversion system.

INTRODUCTION

Space fission power systems can potentially enhance or enable ambitious lunar and Martian surface missions, as well as many solar system and deep space science missions. Research into space fission power systems has been ongoing (at various levels) since the 1950s, but to date the United States (US) has flown only one space fission system, SNAP-10A, in 1965. Cost and development time have been significant reasons why space fission systems have not been used by the US. High cost and long development time are not inherent to the use of space fission power. However, high cost and long development time are inherent to any program that tries to do too much at once. Nearly all US space fission power programs have attempted to field systems capable of high power, even though more modest systems had not yet been flown. All of these programs have failed to fly a space fission system.

Martian surface power is an obvious application for space fission systems. Fission systems are currently being considered for robotic (Mars Outpost) missions proposed for later this decade. This system will probably need to be very compact and light-weight (<500 kg), and provide between 1 – 5 kWe of power. This low-power system will be a precursor to a higher-power system that could be used for surface power on a manned Mars mission. The manned surface power system might provide between 50 – 250 kWe, and will be used for life support, operations, in-situ propellant production, scientific experiments, high intensity lamps for vegetable growth, etc. On Mars, a solar array would require the area of several football fields to produce this average power output.

The Heatpipe Power System (HPS) is a potential, near-term, low-cost space fission power system that could be used for Martian surface power. The proposed HPS designs are composed of independent modules, and all components use existing technology and operate within the existing database. The HPS has relatively few system integration issues; thus, most engineering issues can be solved with module tests. Two hardware programs are ongoing that aim to prove that the HPS is a viable near-term low-cost space fission systems. The results to date have been very encouraging, and highlights from these programs are contained in this paper.

CP504, *Space Technology and Applications International Forum–2000*, edited by M. S. El-Genk
2000 American Institute of Physics 1-56396-919-X

ATTRIBUTES OF THE HPS

The HPS is a safe, simple reactor that is designed to have a low development time and cost. The HPS incorporates lessons learned from previous space fission power development programs, and possesses the following features:

- Safety. The HPS is designed to remain subcritical during all credible launch accidents without using in-core shutdown rods. The core can be designed to achieve passive subcriticality due to high radial reflector worth and the use of resonance absorbers, plus the core is very amenable to fueling in space or at its final destination. The HPS also passively removes decay heat and is virtually nonradioactive at launch (no plutonium).

- Reliability. The HPS has no single-point failures and is capable of delivering rated power even if several modules and/or heatpipes fail.

- Long life. The low power density of the HPS core gives the potential for long life. At 100 kWt, fuel burnup limits will not be reached for several decades.

- Modularity. The HPS consists of independent modules, and most potential engineering issues can be resolved by testing modules with resistance heaters (used to simulate heat from fission).

- Testability. Full HPS tests can be performed using resistance heaters, with very few operations required to replace the heaters with fuel and ready the system for launch. Flight qualification is accomplished with resistance-heated system tests and zero-power criticals. The HPS flight unit can be tested at full power before launch (this would not be possible in systems where the fuel cannot be readily replaced by resistance heaters).

- Versatility. The HPS can use a variety of fuel forms, structural materials, and power converters.

- Scalability. The HPS design approach scales well to >1000 kWt. Very high power (>10 MWt) systems based on the HPS approach are possible but are much more complex than lower power versions (large number of heatpipes) and would most likely suffer a system mass penalty compared with other options.

- Simplicity. There are few system integration issues. There are no in-core shutdown rods, hermetically sealed vessel or flowing loops, electromagnetic pumps, coolant thaw systems, gas separators, or auxiliary flow loops.

- Fabricability. The HPS has no pumped coolant loops and does not require a pressure vessel with hermetic seals. There are no significant bonds between dissimilar metals and thermal stresses are low. There are very few system integration issues, thus making the system easier to fabricate.

- Storability. The HPS is designed so that the fuel can be stored and transported separately from the system until shortly before launch. This capability will reduce storage and transportation costs significantly.

- Low mass. The HPS has a high fuel fraction in the core, which reduces core, reflector, and shield mass. The HPS has no pumped coolant loops or associated components, further reducing mass.

- Early Milestones. Several milestones early in the development of the HPS will prove the viability of the concept. The most significant early milestones were the successful development and testing of an HPS module.

- Near term. An HPS capable of enhancing or enabling missions of interest can be built with existing technology.

- Reduced program expense. The attributes of the HPS should allow for inexpensive (<$100M) development. After development, the unit cost should be <$20M.

UTILIZING THE HPS ON THE SURFACE OF MARS

A space fission power system (SFPS) is defined by the following important attributes: (1) high specific power (power/mass), otherwise it would not be economical to launch the system off the Earth; (2) reliability, because there is little or no hands-on maintenance after the system leaves the Earth; and (3) safety, because the system must remain safe while it operates and while it is transported to its final destination (launch accidents present a unique hazard). These three attributes are also of primary importance for a system designed to operate on the surface of any planet (other than Earth), moon, asteroid, or comet. Such a system might best be classified as a terrestrial SFPS. There are several differences between an SFPS designed for terrestrial and space applications; these differences include: materials compatibility, heat rejection, shielding, reliability, and safety. Materials compatibility and heat rejection issues are affected by the presence of an atmosphere. Shielding issues can differ because of the presence of indigenous material and/or vastly different shielding requirements. Reliability and safety issues can be affected by how the system responds to seismic, meteorological, and astronomical events.

Material compatibility issues will have a major impact on the design and performance of a Martian HPS. An SFPS is usually designed with refractory metals because of their high temperature capability, high thermal conductivity, and in some cases better neutronic characteristics. Previous HPS designs have used niobium 1-zirconium, molybdenum, and/or tungsten depending on the application. One major drawback of refractory metals is that they tend to corrode very easily, thus their use in an atmospheric setting may be limited. Although the Martian atmosphere is relatively benign (mostly low-density CO_2) as compared with Earth, materials compatibility will still be a major issue. Several materials compatibility tests will have to be conducted, but it may be that a super-alloy-based SFPS (possibly stainless-steel, Inconel, or HT-9) will be required on the Martian surface. Preliminary testing of Inconel in high temperature CO_2 at Brookhaven National Laboratory has shown promising performance up to 1000°C (Greene, 1999). If a super-alloy SFPS is required, then the power output of a given system will probably be lower than for a refractory metal counterpart (because of lower temperature limits and thermal conductivity). One advantage of a super-alloy system is that it should have a lower development and unit cost (because of lower material cost, easier manufacturing, larger material database, etc.); however, because of lower performance, the actual cost per watt delivered to space will probably be higher. The power reduction of a specific HPS design may be up to a factor of two (in the worst case), as compared with a similar refractory metal design. However, it should still be possible to economically design and build a super-alloy HPS that can deliver 50–150 kWe on the Martian surface.

There are several factors that could benefit the design of an HPS for use on Mars. The atmosphere could be used to aid in heat transfer and heat rejection. Indigenous material could be moved, or existing craters/hills could be used to significantly reduce the required shield mass sent from Earth. If the system is used on a manned mission, then the astronauts could perform tasks that would allow simplification of the system. For example, the reliability requirements of some components could be relaxed if the potential for maintenance exists. Also, launch safety issues could be simplified if an astronaut could partially fuel the reactor at its destination or remove some attached safety mechanism. Of course there are also disadvantages in using the HPS on Mars. In addition to some of the material issues mentioned above, dust could create a lot of problems for the reactor control, power conversion, and heat rejection subsystems. Also, the system would have to survive potential meteorological and seismic events. However, these disadvantages are common to all systems that are deployed on the Martian surface.

MARTIAN SURFACE HPS CORE DESIGN

The HPS core uses similar (or identical) independent modules to meet the performance requirements for a given mission. A wide variety of core layouts have been evaluated that use 12 to >100 modules. A schematic of a 12-module HPS is shown in Fig. 1. The fuel pins are bonded structurally and thermally to a central heatpipe, which transfers heat to an ex-core power conversion system. The heatpipe also provides structural support for the fuel pins. The rated power of an HPS system is based on a worst case heatpipe failure, and in most cases multiple heatpipes can fail before a significant power decrease occurs.

Two fuel types have been evaluated for use in the Martian HPS: uranium nitride (UN) and uranium dioxide (UO_2). The use of UN results in the most compact, high-power core. UN fuel pins may need to be sealed hermetically, although the peak fuel temperature will be ~1150 K during nominal operation and ~1300 K adjacent to a failed heatpipe. At these relatively low temperatures the pins may not need to be sealed. UO_2 has a lower uranium loading than UN; however, the pins do not have to be sealed hermetically and can be operated at a higher temperature than UN pins (although this is not an issue for a non-refractory core, which is clad

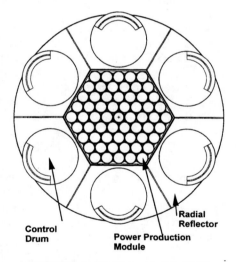

FIGURE 1. Twelve module HPS.

temperature limited). One advantage of UO_2 is that there is a larger experience base and production capability for this fuel. For both fuels, burnup limits (based on experimental results) are not reached for several decades in most designs. Carbide and advanced fuels could also be considered for the Martian HPS.

Several options are available for cladding the Martian HPS. Some of the most promising candidates are stainless-steel, Inconel, and HT-9. Several other super-alloys could be considered, and there is even the potential that a refractory metal could be used (if tests show little corrosion or if a robust coating is developed). Simple corrosion testing in a simulated Martian atmosphere could go a long way in narrowing down the field of candidate materials. For the purposes of this study, two materials were considered – SS and Inconel. The Martian HPS cores are designed so that the clad temperature does not exceed 1250 K for a worst case heatpipe failure (the maximum clad temperature is closer to 1100 K during nominal operation).

The non-refractory HPS core will run at relatively low temperatures (<1300 K), therefore sodium is best suited for the heatpipe working fluid. There is more experience with Na heatpipes than any other type of high-temperature liquid metal pipe. There is also considerable irradiation data on these types of heatpipes. A total of 29 liquid metal heatpipes have been irradiated to significant fast neutron fluences, with no failures. A stainless steel/sodium heatpipe operated at 1100 K to a fast fluence of 2.2×10^{22} n/cm^2 (Merrigan, 1997), which is more than an order of magnitude higher than that required by most potential near-term HPS missions.

There are several potential options for electric power conversion on a Martian HPS. If desired, thermoelectric converters identical to those used by radioisotope thermoelectric generators (RTGs) could be coupled to the HPS. Thermoelectrics of this kind are flight-qualified, but they provide a low efficiency (<5%). This may be acceptable for the Mars Outpost reactor if the power requirement is only a few kWe. Advanced thermoelectrics, close-space thermionic converters, alkali metal thermoelectric converters (AMTECs), Stirling, Rankine, and Brayton power conversion are also options. Dynamic power conversion may be more desirable for the manned application, because of the higher power requirement and because of the potential for hands-on maintenance. The option also exists for using the Martian atmosphere to dump waste heat, A paper by Lipinski et. al. describes an innovative radiator design that rejects heat to the atmosphere (Lipinski, 1999).

Neutron and gamma shielding will probably consist of an optimal mix of material brought from earth and indigenous material. Because of its small size and the lack of activated coolant in its radiator, the HPS can be well shielded with relatively little extra mass needed from earth. For manned missions, it may be desirable to shield the HPS so that a radiation exclusion zone is not needed. A paper by Houts et. al. describes various options for using indigenous materials for shielding, including the use of liquefied or frozen CO_2 from the Martian atmosphere (Houts, 1999).

The HPS is designed to remain subcritical during all credible launch accidents. This can be accomplished by keeping the system radius small, keeping the reflector worth high, and strategically placing neutron absorbers in the core. The positive reactivity effect of core flooding or compaction is offset by (1) the negative reactivity worth of the control drums in the reflector or (2) the negative reactivity effect of losing the reflector and surrounding the reactor with wet sand or water. This effect eliminates the need for in-core safety rods. For deep-space or planetary surface missions where reentry after reactor startup is impossible, passive launch safety can be ensured by fueling the reactor in space (or on site) or using retractable boron wires to provide shutdown. This allows the removal of resonance absorbers from the core and reduces system mass and volume. The HPS is very amenable to fueling after launch – this option may be even more attractive on a manned mission if an astronaut is available for the job. The HPS is virtually nonradioactive at launch (no plutonium in the system).

HPS STAINLESS-STEEL POINT DESIGNS

Neutronic and thermal calculations have been performed to optimize the mass and power level of various HPS core geometries. Criticality and power density calculations were performed with MCNP (Briesmeister, 1997). Each core is designed with a cold clean k-eff of 1.03. These cores will need to contain additional neutron absorbing material in the interstitials at launch, or will have to be launched partially fueled (totally passive safety may be achieved with a mass penalty of approximately 50 kg). Thermal calculations were performed with a 3D finite-difference FORTRAN code. This code discretely models each pin (fuel, gap, and clad), the heatpipes, and the shroud. The most significant limitation of the model is that the heatpipes are simply modeled with a constant temperature boundary condition on the inner wall (which should be reasonable for steady-state calculations). The thermal power is based on a maximum clad temperature of 1250 K, and the heatpipe temperature is 973 K. The calculations assume a surface emissivity of 0.4 (which is probably conservative because the SS will become carbonized) and the presence of the Martian atmosphere in all gaps (~5 mbar CO2). It is also assumed that each pin and heatpipe is brazed to an interstitial SS tricusp on all sides (as is being done for the HPS full-core demo, which is discussed below).

UO_2 and UN designs were evaluated at each geometry. Due to the higher uranium loading and higher thermal conductivity of UN, the UN cores on average produced much higher power (+38%) and much lower mass (-33%). The advantage in using UN is much larger than is seen for refractory metal systems because power output is not limited by the fuel temperature. Because of this higher performance, the nitride system is considered the baseline; although issues such as sealing the pins (if required), fabrication availability, existing nuclear data, or other issues could favor UO_2. Inconel designs were also evaluated for each geometry. In each case the Inconel design provided slightly lower power (-3%) at a slightly higher mass (+2%). This, plus the fact that Inconel parts are more difficult to fabricate and are more expensive, led to the choice of SS for the baseline designs. This is not to say the Inconel, HT-9, or any other material might not be the best choice. There will be several factors other than neutronic and thermal performance that will drive the decision, such as: corrosion in CO_2, radiation damage characteristics, existing nuclear database, etc. A summary of SS clad UN HPS cores is listed in Table 1 (these cores are not fully optimized).

TABLE 1. Summary of stainless-steel clad, uranium-nitride fueled HPS cores.

Design parameter	hpsmars1	hpsmars2	hpsmars3	hpsmars4	hpsmars5
Number of heatpipes	12	19	30	57	121
Number of fuel-pins	48	102	138	152	336
Core flat-flat dimension (cm)	20.2	18.2	18.4	20.0	21.8
Core active length (cm)	34	31	34	37	36
Axial reflector length (cm)	4	4	4	4	4
Pin length (cm)	42	39	42	45	44
Reflector thickness (cm)	6.5	6.2	6.4	6.5	6.0
Reflector outer radius (cm)	17.1	15.8	16.1	17.0	17.4
Reflector height (cm)	38	35	38	40	40
Control drum radius (cm)	3.25	3.10	3.20	3.25	3.00
Pin outer diameter (cm)	2.54	1.60	1.40	1.35	1.00
Clad thickness (mm)	1.27	0.51	0.51	0.51	0.51
Reactor power (kW)	**40**	**50**	**85**	**350**	**640**
Nominal operation					
Power to heatpipes (kW)	38.2	48.8	83.1	347.6	637.4
Max fuel temperature (K)	1077	1150	1069	1191	1198
Max clad temperature (K)	1074	1113	1064	1172	1170
Max heatpipe power (kW)	3.6	3.0	3.0	7.2	6.0
Max radial heat flux (W/cm2)	17.4	21.3	25.6	59.2	68.4
Max axial heat flux (kW/cm2)	0.8	1.8	2.4	6.3	9.4
Worst case failed heatpipe					
Power to heatpipes (kW)	**37.7**	**47.8**	**82.8**	**347.6**	**637.4**
Max fuel temperature (K)	1270	1286	1264	1260	1262
Max clad temperature (K)	1253	1246	1249	1248	1244
Max heatpipe power (kW)	4.8	3.4	4.2	8.6	7.2
Max radial heat flux (W/cm2)	26.7	26.9	41.6	81.2	93.0
Max axial heat flux (kW/cm2)	1.1	2.1	3.3	7.5	11.2
Mass (kg)	**195**	**153**	**173**	**199**	**217**
UN mass (kg)	90	75	83	92	101
Cold/clean k-eff	1.031	1.034	1.032	1.029	1.030
Fuel burnup per year (%)	0.019	0.029	.044	0.164	0.272

The mass listed in Table 1 includes the core, core support, primary heat transport, and reflector (does not include power conversion, heat rejection, shielding, etc.). The clad, heatpipes, tricusps, and structure are SS-304. The fuel is 97.6% enriched UN at 13.56 g/cc (96% TD); which is the spec of existing SP-100 fuel at LANL TA-55. The reflector is BeO and it is clad with 1-mm SS. The core shroud is 2-mm SS and the upper and lower support plates are 1-cm SS. The poison on the control drums consists of a 1-mm layer of Mo-Re sandwiched between two 5-mm layers of B_4C (the use of sliding reflector would lower mass slightly). The axial reflector region contains BeO pellets within the fuel pins. The highest power cores require very high-performance heatpipes, which will probably require some development (the heatpipe requirements could be relaxed with a slight mass penalty).

The power capability of the HPS is highly dependent on the temperature delivered to the power conversion system. If the heatpipes were operated at 700 K instead of 973 k, then the power output of the system would double (provided the heatpipes were designed to carry that much power at 700 K). The lower heatpipe temperature would allow twice the conductive delta-T from the fuel to the heatpipe, which would allow twice the power density (not exactly because of variable thermodynamic properties and radiation non-linearity).

The 12-module design (HPSMARS1) is the one that is currently being manufactured for the HPS full-core demo. This design uses a thicker clad because these are the dimensions of the tubes chosen for the demo. The other designs use a rather thin clad because of the relatively low burnup (the lower power cores would take several decades to reach 1% burnup); if necessary the clad can be made thicker for a small mass penalty. On Table 1, the higher power cores have a higher heatpipe to fuel pin ratio and smaller diameter pins. The higher heatpipe to fuel ratio also causes a slight increase in mass; although there is only a 42% increase in mass between cases 2 and 5 while there is a 1,250% increase in power. At very high powers (>1000 kWt) it may be desirable to design modules that consist of cylindrical fuel pins surrounded by non-cylindrical heatpipes. Limited data is available on non-cylindrical heatpipes, but such data can be obtained inexpensively by testing electrically heated modules.

HPSMARS1 and HPSMARS2 are excellent candidates for a Mars Outpost mission. Although HPSMARS2 has better performance, HPSMARS1 might be a better choice because it is a more conservative design and it is the same core that is being built and tested in the HPS demo program. Given the reactor power of 40 kWt (actually 37.7 kWt through the heatpipes), 5% efficient thermoelectric converters would supply ~2 kWe at a total mass of <500 kg (and much lower mass if minimal shielding needs to be brought from Earth). This performance would make the HPS very attractive for a Mars Outpost mission, especially considering the low development cost and time for the system.

Of the options listed on Table 1, HPSMARS4 is probably the best candidate for manned mission surface power – HPSMARS5 produces considerably more power, but it has a very large number of modules and the heatpipe requirements (especially radial and axial heat flux) are difficult at these temperatures. A power of 350 kWt would produce 52 kWe with a 15% efficient conversion system, and ~100 kWe with a very efficient dynamic system (Brayton, Rankine, Stirling, or other?). The total mass of the system will be highly dependent on the conversion system, heat rejection method, and shielding. Finally, if specific mission requirements arise, then an HPS core could be optimized specifically for that mission.

THE REFRACTORY METAL HPS MODULE TEST

The first hardware produced within the HPS project was a 4-pin Mo module with a Mo/Li heatpipe. This module was designed to be 1 of 12 identical modules required by a 100-kWt HPS, and was optimized to make maximum use of existing hardware and facilities. The module consisted of three electron discharge machined (EDM'd) molybdenum pieces brazed to a central molybdenum/lithium heatpipe. The EDM'd pieces simulated the molybdenum fuel pins that would be used in an actual system. The pin's outer diameter was 2.54 cm, which allowed existing resistance heaters to be used for testing. The heated length of the pins was 0.30 m. Fabrication cost for the first module, including the central heatpipe, was ~$75K. The use of existing resistance heaters and facilities reduced the cost of testing the module; the total cost of the entire project, from design to fabrication to testing, was ~$100K.

Fabrication of the first HPS module was completed in January 1997, and initial tests were completed in February 1997. Initial testing of the HPS module was performed at the New Mexico Engineering Research Institute (NMERI) in Albuquerque, New Mexico, using resistance heaters and test equipment purchased from Russia during the TOPAZ International Program. After the NMERI tests, the module was removed from the test chamber and found to be in excellent condition. Further tests were carried out at LANL using radio frequency (RF) heating instead of resistance heating. The RF heated tests repeatedly demonstrated module operation at 1400 K and 4 kWt. The NMERI and LANL tests demonstrated the following:

- High-power (>4 kWt) and high-temperature (>1400 K) heatpipe operation against gravity.
- Multiple restart capability – 9 module startups (frozen to >1300K and >2.5 kWt) and shutdowns.
- Heatpipe operation with high, non-uniform radial heat fluxes.
- Advanced refractory metal bonding techniques.

Although this testing was very successful, we were unable to demonstrate the very high operating power (8.5 kWt) that was originally set as a goal. The thermal power was limited to 4 kWt by the amount of heat that could be radiated from the heatpipe condenser at 1400 K. Diagnostics show that the module and heatpipe were functioning well. In the future, resistance heaters capable of providing 8.5 kWt and a gas-gap calorimeter should be used to enable the full power operation of the module.

THE HPS FULL-CORE DEMO

This goal of this project is to construct and test a full Heatpipe Power System (HPS) core. The material chosen for this demo core is stainless-steel, with SS/Na heatpipes. Stainless steel was chosen because of its relatively low cost, and because it is a candidate material for a Martian surface reactor. The project will attempt to demonstrate the full thermal-hydraulic performance of the HPS core in FY01, with full core testing beginning in the end of FY00. This would represent the first full thermal-hydraulic demonstration of US space fission core since 1960s. Resistance-heated tests are orders of magnitude less expensive than fission-heated tests and it is much easier to determine the cause of any failures if the system has not been activated (no hot cell required). If the resistance heated system tests are successful, we are very confident that we can convert the core into an operational fission system. From a nuclear standpoint all components within the core operate well within the established database, and no nuclear effects are anticipated. Due to time and budget constraints, the power level of the HPS demo system will be relatively low (~25 kWt). This core consists of 12 modules; Figure 2 shows a schematic of the HPS demo module and Figure 3 shows a cross section of the demo core. The module contains four 2.54-cm diameter SS tubes that are bonded to a central heatpipe; this is accomplished by brazing each part to an interstitial tricusp. Advanced Methods and Materials of San Jose, California has developed a braze technique for the HPS module that provides excellent mechanical and thermal bonding. In all tests thus far, diagnostics (including acoustic microscopy) have verified that a robust braze has been achieved. A full-size prototype module was successfully fabricated in September 1999.

An innovative composite-annular wick structure was devised for the module heatpipe that eliminates the necessity for hot acid etching. Hot acid etching has been used to form the wick structure for most high performance heatpipes previously developed at Los Alamos. Tests were conducted with room temperature fluids that verified the capillary pumping capacity of the composite wick. A prototype heatpipe was completed in July 1999 and was successfully tested to full power in August 1999. As was the case for the refractory metal module test, the power output of these heatpipes will be limited by the thermal coupling to the calorimeter (projected to reject ~2.5 kWt per module). If calorimeter coupling could be increased (gas-gap or higher emissivity), and if the heater capability exists, that the heatpipe should be able to provide close to 10 kWt before reaching any physical limitations.

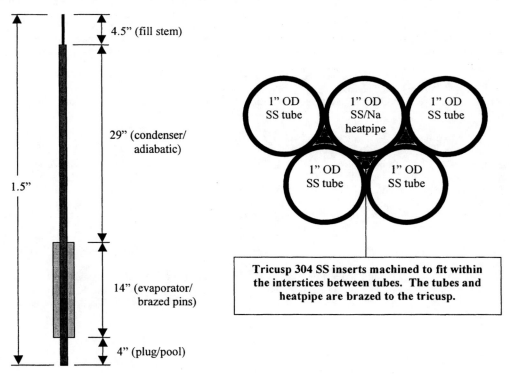

FIGURE 2. Schematic of HPS demo module.

The core assembly will be very simple. Each heatpipe will have a flange brazed just above the evaporator region, which will be bolted to an upper support plate. The bottom of each module will be positioned in a lower support plate, but will be free to expand axially. After assembly is complete, heaters will be inserted in each of the open module tubes and cylindrical calorimeters will be slid over each heatpipe condenser. The calorimeters will measure the total power flow through each heatpipe, and thus for the entire system. Several thermocouples will be attached to each module and heatpipe, so that temperature gradients can be established and thermal models can be benchmarked. These thermocouple readings will be used to determine the level of thermal coupling within and between modules. The major components of the test plan are:

- Nominal and simulated failed heatpipe conditions.
- Mars atmosphere and vacuum conditions.
- Mars gravity and zero-g conditions (with and against gravity).
- Full power, decay power, and transient (startup/shutdown) conditions.

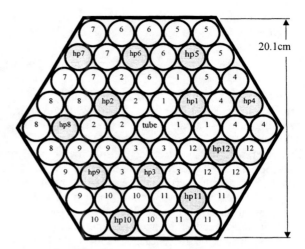

FIGURE 3. Cross Section of HPS demo core.

After the thermal-hydraulic testing has been completed, the next step might be to acquire and attach an electric power conversion system to the demo core, or perhaps begin reflector, control system, and fuel fabrication in order to perform zero-power criticals testing (to confirm neutronic calculations).

SUMMARY

This paper evaluates the of use the Heatpipe Power System (HPS) for Mars surface power applications. Previously, most HPS designs have used refractory metals to allow high-temperature operation and higher thermal conductivity. For use on Mars, a stainless-steel or super-alloy system may be required in order to avoid corrosion. As a result, Los Alamos National Laboratory (LANL) has begun a program to evaluate and test a stainless-steel HPS core. This paper describes the analysis of potential non-refractory HPS designs, as well as work that is currently underway to build a full-scale unfueled HPS core. Resistance heated testing will be used to evaluate the thermal performance of the core under a wide variety of conditions; including the simulated Martian atmosphere. Analytical and experimental results thus far indicate that the HPS is very well suited for Martian applications. A low-risk design is proposed for a Mars Outpost application that could provide 40 kWt, and thus 2 kWe with 5% efficient thermoelectrics. A higher power core, 350 kWt, is proposed as a potential surface power source for manned Mars missions, that could provide up to 100 kWe depending on the efficiency of the power conversion system.

ACKNOWLEDGMENTS

This research was supported by the Los Alamos National Laboratory Center for Space Science and Exploration.

REFERENCES

Briesmeister, J. F., *MCNP—A General Monte Carlo N-Particle Transport Code*, LA-12625-M, Version 4B, Los Alamos National Laboratory, Los Alamos, NM (1997).

Greene, G., A. (1999), personal communication, Brookhaven National Laboratory, Upton, NY, March 1999.

Houts, M. G., Poston, D. I., Trellue, H. R., Baca, J. A., Lipinski, R. J., "Planetary Surface Reactor Shielding Using Indigenous Materials," *Space Technology and Applications International Forum-1999 (STAIF-99)*, edited by Mohamed S. El-Genk, AIP Conference Proceedings 458, 1999, pp. 1476-1479.

Lipinski, R. J., Wright, S. A., Lenard, R. X., Harms, G. A., "A Gas-Cooled Reactor Surface Power System," *Space Technology and Applications International Forum-1999 (STAIF-99)*, edited by Mohamed S. El-Genk, AIP Conference Proceedings 458, 1999, pp. 1470-1475.

Merrigan, M. (1997), personal communication, Los Alamos National Laboratory, Los Alamos, NM, May 1997.

Poston, D. I., Houts, M. G., Emrich, W. J., "Development Status of the Heatpipe Power and Bimodal Systems," *Space Technology and Applications International Forum-1999 (STAIF-99)*, edited by Mohamed S. El-Genk, AIP Conference Proceedings 458, 1999, pp. 1197-1204.

A Deep Space Power System Option Based On Synergistic Power Conversion Technologies

Jeffrey G. Schreiber

NASA John H. Glenn Research Center at Lewis Field, MS 301-2, 21000 Brookpark Road, Cleveland, OH 44135
Jeffrey.Schreiber@grc.nasa.gov; (216) 433-6144

Abstract. Deep space science missions have typically used radioisotope thermaoelectric generator (RTG) power systems. The RTG power system has proven itself to be a rugged and highly reliable power system over many missions, however the thermal-to-electric conversion technology used was approximately 5% efficient. While the relatively low efficiency has some benefits in terms of system integration, there are compelling reasons why a more efficient conversion system should be pursued. The cost savings alone that are available as a result of the reduced isotope inventory are significant. The Advanced Radioisotope Power System (ARPS) project was established to fulfill this goal. Although it was not part of the ARPS project, Stirling conversion technology was being demonstrated with a low level of funding by both NASA and DOE. A power system with Stirling convertors, although intended for use with an isotope heat source, can be combined with other advanced technologies to provide a novel power system for deep space missions. An inflatable primary concentrator would be used in combination with a refractive secondary concentrator (RSC) as the heat source to power the system. The inflatable technology as a structure has made great progress for a variety of potential applications such as communications reflectors, radiators and solar arrays. The RSC has been pursued for use in solar thermal propulsion applications, and it's unique properties allow some advantageous system trades to be made. The power system proposed would completely eliminate the isotope heat source and could potentially provide power for science missions to planets as distant as Uranus. This paper will present the background and developmental status of the technologies and will then describe the power system being proposed.

INTRODUCTION

The Thermo-Mechanical Systems branch at the NASA Glenn Research Center is responsible for developing a wide variety of advanced technologies for spacecraft. Among the technologies that are currently being pursued are 1) advanced free-piston Stirling convertors for potential use with a radioisotope heat source to provide power for deep space exploration, 2) inflatable solar concentrators that may potentially be used for earth orbiting solar thermal systems, and 3) a novel refractive secondary concentrator (RSC) intended primarily for solar thermal propulsion. For each one of the technologies there exists a clear rationale that shows its value at the system or mission level, for the intended application. Being that each is technology is targeted at a different application, these technologies are currently being developed independent of one another.

Each of the advanced technologies have characteristics that make them very attractive when compared to the alternatives, however it has also recognized that they each have limitations. The free-piston Stirling convertor has long been considered a candidate for space power applications, in part because of the demonstrated high conversion efficiency, and also because of the potential for long life. The current status of this technology will be discussed later in this paper. Inflatable primary concentrators are used to concentrate incident solar rays to a focal spot. This is essentially the same function performed by solar concentrators in past applications, however the inflatable technology dramatically reduces the areal mass to levels far below those achieved with rigid or deployable concentrators in the past. The RSC utilizes the principle of total internal reflectance and can perform two important functions. First, it can further concentrate the solar flux from a primary concentrator to achieve a higher concentration ratio or temperature, and secondly it can transport the flux deep into a cavity and deposit it with a tailored profile.

CP504, *Space Technology and Applications International Forum–2000*, edited by M. S. El-Genk
2000 American Institute of Physics 1-56396-919-X

There has long been interest in developing a power system for deep space exploration that reduces the inventory of the isotope fuel (plutonium) necessary, or potentially eliminates the need entirely. The radioisotope thermoelectric generator (RTG) used in the past has proven itself to be very rugged and highly reliable, however the relatively low conversion efficiency must be compensated for by the inventory of plutonium contained in the General Purpose Heat Source (GPHS) blocks. For reasons of environmental concerns and mission cost, a highly efficient conversion system requiring less plutonium, or possibly eliminating the plutonium would be highly desirable. One system investigated in the past was the power antenna (Cassapakis, 1996, Lichodziejewski, 1999) which used an inflatable parabolic reflector that served two purposes. It functioned as a reflector for the communications system, and also as a solar reflector for the power system. The power conversion technology used was a photovoltaic array situated in concentrated sun light, in the general region of the focal spot. Through a NASA SBIR, L'Garde investigated this option and proposed that it had merit for low power missions, potentially to Jupiter and Saturn.

A system was conceived originally at the NASA Glenn Research Center (GRC) that integrates a free-piston Stirling based power conversion system intended to be used with GPHS blocks, an inflatable primary concentrator, and the RSC into a relatively low power level (less than 1 kWe) solar dynamic power system. It is believed that with careful consideration, a system could be configured that exploits the strengths of each of the technologies and provides a viable non-nuclear power system option based on technology essentially being developed for other applications. After conceptualizing the system, a brief study was performed that indicated that the system indeed may be advantageous for some space exploration missions (Mason, 2000). A key attribute of the system proposed is that all of the subsystems are currently being developed under existing projects, and that the system was conceived by recognizing the synergism that existed in these previously unrelated projects.

THE TECHNOLOGIES

A brief description will be given of the technologies that have been incorporated into the proposed power system. The description will include the characteristics of the technology, a description of the intended applications, and an indication of the status of development.

Radioisotope Stirling Conversion System

The Department of Energy (DOE) and NASA GRC are developing a Stirling convertor as a possible high efficiency radioisotope electric power system option for future space exploration. A 1997 study team with joint participation from NASA and DOE determined that the Stirling conversion option was a low risk alternative with dependable background of materials, lifetime, and demonstrated performance (Frazier, 1998, Mondt, 1998). While the Advanced Radioisotope Power System (ARPS) project went forward with the Alkali Metal Thermal to Electric Converter (AMTEC) as the conversion technology, a low level effort supporting the free-piston Stirling convertor

was initiated. This was performed primarily through a DOE contract with the Stirling Technology Company (STC) of Kennewick, WA, and was supported by NASA GRC through Small Business Innovation Research (SBIR) contracts and GRC in-house capabilities. This effort has resulted in operational 55 We convertors which have demonstrated full power operation and predicted efficiency. The convertors are shown in figures 1 and 2. The technologies used in these 55 We convertors have previously been used by STC in other convertors that have demonstrated long life with over 53,000 hours of continuous operation with no degradation (White, 1999).

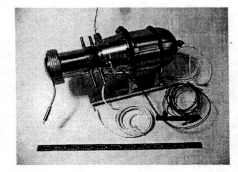

FIGURE 1. A 55 We Free-Piston Stirling Technology Demonstrator

As work has progressed on the 55 We convertor, many of the previously unresolved concerns regarding free-piston Stirling convertors have been addressed. Multi-convertor systems operating in a synchronous mode have been demonstrated. Synchronous operation of pairs of convertors can lead to a dynamically balanced system with low vibration. An Advanced Vibration Reduction System (AVRS) is also being developed to further reduce vibration (Thieme, 1999). With the Stirlng convertor technology for this application progressing, attention is starting to focus on system integration issues. Orbital Science Corporation (OSC) has developed a system concept (Schock, 1999). One of the early system concepts developed by OSC is shown in figure 3. Under contract to DOE, Lockheed Martin Astronautics, Valley Forge, PA, is also developing system concepts.

In the concept shown, an insulated cavity exists in the center of the structure where the GPHS blocks would be secured. The housing acts as a structure to which the Stirling convertors are mounted, however it also contains the multi-layer insulation (MLI) that surrounds the GPHS blocks. Although other configurations are being studied, the highly insulated structure for the GPHS blocks is a common feature.

FIGURE 2. A Pair of 55 We Stirling Convertors Demonstrating Low Vibration

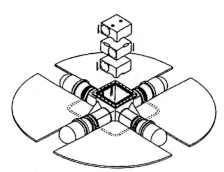

FIGURE 3. Radioisotope Stirling Power System Concept Proposed by Orbital Science Corporation.

Solar Concentrator

The NASA GRC has had a longstanding interest in the development of advanced, lightweight solar concentrators. Efforts at GRC date back to the early 1960's with the development of deployable concentrators with rigid facets. An early effort working with TRW produced a deployable concentrator knows as the Sunflower because of the arrangement of the hinged pedals that deployed to form the concentrator. A subsequent effort at GRC in 1965 produced a 6 meter diameter concentrator with a specific mass of 4.6 kg/m^2. In the early stages of Space Station Freedom (SSF), one of the power system options was a hybrid system including both photovoltaic and solar dynamic power conversion. In 1989, a concentrator was developed for SSF based on a box-beam supporting structure with multiple triangular facets. The 456 facets in this design, without the supporting box beam structure, had a specific mass of 2.4 kg/m^2. In 1994 a system level test was performed with a downsized version of the SSF concentrator, as shown in figure 4.

Thin film inflatable technology offers the potential to significantly reduce the mass of solar concentrators to levels far below those typically quoted for advanced rigid concentrators. There is current interest in inflatable technology for a number of applications, of which the inflatable solar concentrator is one. Companies such as SRS Technologies, ILC Dover, UAT, and L'Garde are developing concepts for large thin film inflatable solar concentrators. The demonstration of the Inflatable Antenna Experiment by L'Garde in 1994, as shown in figure 5, was a major step forward in demonstrating the potential of this technology.

FIGURE 4. Rigid Concentrator Used During The Brayton Solar Dynamic Power System Test at GRC.

Inflatables had previously flown in space, however this was the first demonstration in space of a parabolic reflector supported by a taurus and struts.

There is also interest in using light weight concentrators for solar thermal propulsion systems (Partch, 1999). Solar thermal propulsion provides an option for orbit transfer and in-space maneuvers with a trade off in specific impulse and thrust levels, each being somewhere between those levels achieved with chemical propulsion and electric propulsion. This application requires a temperature well in excess of 2000K in the solar receiver cavity. The optical accuracy requirements levied on the inflatable concentrator are very demanding and are one of the key elements being addressed in current efforts.

Recognizing that the requirements placed on a solar concentrator for either solar thermal propulsion or solar thermal power are similar, GRC began coordinating efforts with the Air Force Research Laboratory (AFRL) in the development of inflatable concentrators. A Phase I SBIR contract with SRS Technologies of Huntsville, AL produced a technology demonstration inflatable concentrator as shown in figure 6. This concentrator was subsequently tested in collaboration with AFRL in the Tank 6 solar thermal vacuum facility at GRC. The concentrator achieved an approximate 2 mrad slope error. The Phase II contract has been awarded and the effort is currently underway. While this technology looks promising, many issues such as inflation control, rigidization, and UV tolerance need to be addressed.

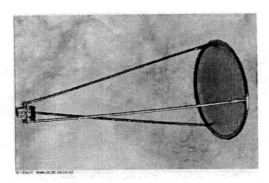

FIGURE 5. Inflatable Antenna Experiment by L'Garde.

FIGURE 6. SRS Solar Concentrator Being Installed in GRC Tank 6 for Testing.

Secondary Concentrators

An effort is underway at NASA GRC to develop a highly efficient secondary solar concentrator. It is based on an innovative, single crystal refractive secondary concentrator (RSC) concept which uses the principle of total internal reflection. The RSC alleviates some of the limitations of reflective secondary concentrators in terms of concentration ratio, efficiency, and distribution of the flux exiting the concentrator. Concentration ratios of more than 25:1 are possible with the RSC, and efficiency at high power is projected to be 90% or greater. When used with a high quality primary concentrator, this allows a total concentration ratio (i.e., the primary and secondary concentrators combined) on the order of 1,000:1. A prototype RSC is shown in figure 7.

FIGURE 7. Refractive Secondary Concentrator at GRC.

The RSC designs analyzed and tested thus far perform two distinct functions. First, the conical segment concentrates the light via total internal reflection. Secondly, the extractor can be used to transport the solar flux into a receiver cavity, and the extractor can then be tailored to deliver the flux with a favorable distribution. The ability of the extractor to transport the energy deep into the receiver differs from the reflective secondary concentrators which tend to distribute the flux near the aperture of the receiver.

The effort at GRC has been directed primarily toward the use of this technology for solar thermal propulsion where high concentration ratios and high temperature in the receiver are necessary. This effort was initiated in support of the Shooting Star solar thermal propulsion experiment proposed by the NASA Marshall Space Flight Center. Other applications would be for solar thermal furnaces and possibly for solar thermal power conversion. While this work is in its early stages, the results are very promising (Wong, 1999). The use of the RSC has a significant impact on the primary concentrator. For example, in an optimized system the accuracy requirements on the primary concentrator may be able to be relaxed, and a mass or cost savings may be realized.

THE POWER SYSTEM

The ARPS program is based on the desire of both NASA and DOE to have an advanced radioisotope power system with higher efficiency than the RTGs, and thus requiring less plutonium. Extending this logic would lead one to try to develop a power system for deep space missions that needs no plutonium at all. Such a system was envisioned at GRC by integrating the technologies that were discusses earlier; that is, a system with a lightweight inflatable primary concentrator, a high efficiency secondary concentrator, and a high efficiency free-piston Stirling conversion system. While the proposed system is a solar dynamic power system, which in general have been studied and tested in the past (Mason, 1999), there are some unique characteristics of the proposed system. Unique traits of the individual technologies allowed system level trades to be performed that are different than experienced in previous systems.

By taking advantage of the high concentration ratio and high efficiency of the RSC, the accuracy requirements levied on the primary concentrator can be relaxed. This feature should result in savings in terms of both mass and cost. System level analysis would be used to determine the final configuration of the optical concentration system. The RSC also incorporates an extractor which can take the solar energy deep into a cavity and deposit it in a predetermined distribution. This solar concentration system would then be integrated with a free-piston Stirling based power conversion system, intended for use with GPHS blocks. The extractor would take the solar flux into a cavity designed to house the GPHS blocks, however this would now be heated with the solar energy. A further extension of the system would use the primary concentrator as proposed in the study of the power antenna, and thus it would perform duties for both the power system and the communication system.

The proposed system leads to a highly integrated spacecraft which might be suitable for a range of deep space missions. Missions to Mars may very well be served with a photovoltaic power system because the solar intensity is still relatively high. Similarly, missions to Neptune or Pluto may require unreasonably large primary concentrators because of the extremely low solar intensity, and thus may be better served with a high efficiency isotope power system. It is believed however, that there may be a range of missions which could benefit from using the proposed non-isotope power system. If these missions are able to be performed using the proposed power system, then there are great benefits incurred because of the elimination of the isotope fuel. These benefits are in terms of both environmental concerns and cost, which would be on the order of 10's of millions of dollars per mission.

CONCLUDING REMARKS

While a detailed study of this concept has not been performed, nor has a system level demonstration been performed, an initial investigation indicates that this solar powered conversion system would be a viable option for missions out to Jupiter and Saturn and possibly Uranus. Advances in each of the technologies are needed; the primary concentrator must demonstrate low mass and life for the candidate missions, the RSC must demonstrate high power throughput, and the integrated system utilizing existing Stirling convertors must be developed. While it is recognized that each of the technologies needs to be advanced, it should also be pointed out that each of the technologies needed in this system are currently being developed under separate programs. The level of performance achieved for each of the technologies will determine which missions might benefit by using this power system. The system is not proposed as a solution for all missions, however if some of the typical missions that had previously used isotope fueled power systems could now be flown without any isotope for power, the benefits are substantial.

REFERENCES

Frazier, T.A.: "Advanced Conversion Technology Review Panel Report," Proceedings of the 33rd Intersociety Energy Conversion Engineering Converence, Paper IECEC-98-39, 1998.

Cassapakis, C. and Thomas, M.: "A Power Antenna for Deep Space Missions," in *Solar Engineering*, edited by J. H. Davidson et al., ASME Converence Proceedings, Book Number H01046, 1996.

Lichodziejewski, D. and Cassapakis, C.: "Inflatable Power Antenna Technology," AIAA Paper 99-1074, American Institute of Aeronautics and Astronautics, 1999.

Mason, L.S.: "Technology Projections for Solar Dynamic Power," NASA TM-1999-208851, 1999.

Mason, L.S.: Solar Stirling for Deep Space Applications, *in Space Technology and Applications International Forum*, edited by M. S. El-Genk, AIP Conference Proceedings, American Institute of Physics, 2000.

Mondt, J.F. and Nesmith, B.J.: "Advanced Convertor Technology Evaluation and Selection for ARPS," in *Space Technology and Applications International Forum*, edited by M. S. El-Genk, AIP Conference Proceedings, American Institute of Physics, 1998.

Partch, R.E.; Holmes, M.R.; and Pearson, J.C.: "Inflatable Concentrator Performance Characterization for Space Flight Applications," in Renewable and Advanced Energy Systems for the 21st Century, American Society of Mechanical Engineering, Maui, 1999.

Schock, Alfred; Or, Chuen; and Kumar, Vasanth: "Radioisotope Power System Based on Improved Derivative of Existing Stirling Engine and Alternator," in *Space Technology and Applications International Forum*, edited by M. S. El-Genk, AIP Conference Proceedings, American Institute of Physics, 1999.

Thieme, L.G., Qui, S. and White, M.A.: "Technology Development for a Stirling Radioisotope Power System for Deep Space Missions," Proceedings of the 34th Intersociety Energy Conversion Engineering Conference, Society of Automotive Engineers, 1999, Paper IECEC-99-01-2454.

White, M.A., Qui, S, Olan, R.W., and Ereznik, R.M.: "Technology Demonstration of a Free-Piston Stirling Advanced Radioisotope Space Power System," in *Space Technology and Applications International Forum*, edited by M. S. El-Genk, AIP Conference Proceedings, American Institute of Physics, 1999.

Wong, Wayne A., and Macosko, Robert P.: "Refractive Secondary Concentrators for Solar Thermal Applications," NASA TM-1999-209379, 1999.

Five Major NASA Health and Safety Issues

Raymond B. Gavert

NASA Headquarters, Office of Health Affairs, 300 E Street, SW, Washington, DC

(202)358-4682; rgavert@hq.nasa.gov

Abstract. The goal has been set to establish NASA as number one in safety in the nation. This includes Systems and Mission Safety as well as Occupational Safety for all NASA employees and contractors on and off the job. There are five major health and safety issues important in the pursuit of being number one and they are: (1) Radiation (2) Hearing (3) Habitability/Toxicology (4) Extravehicular Activity (EVA) (5) Stress. The issues have features of accumulated injury since NASA's future missions involve long time human presence in space i.e., International Space Station operations and Mars missions. The objective of this paper is to discuss these five issues in terms of controlling risks and enhancing health and safety. Safety metrics are discussed in terms of the overall goal of NASA to be number one in safety.

INTRODUCTION

Safety of NASA's manned missions in the past has been carried out over relatively short periods measured in days. In the future, safety of humans on spacecraft will involve changes in design features, operational procedures, and requirements necessary to enable long term human presence in space. The International Space Station (ISS) will be in orbit for fifteen or more years. Going to Mars will involve mission times in years. The overall goal of NASA is to be number one in safety in the nation for its workers and its missions on the ground, through the air and into space. The term health and safety covers those areas where injury or illness may be involved. The objective of this paper is to discuss NASA's five major health and safety issues in terms of controlling risks and enhancing health and safety. The five major health and safety issues are detailed next. Systems and mission safety will then be discussed with examples of the International Space Station and Mars missions. Finally, safety metrics will be discussed.

FIVE MAJOR HEALTH AND SAFETY ISSUES

NASA's Office of Health Affairs has major concern over five health and safety issues. The issues are: (1) Radiation (2) Hearing (3) Habitability/Toxicology (4) EVA and (5) Stress.

Radiation Health and Safety

Radiation health and safety of NASA workers on Earth and in space involves two categories of radiation (1) Ionizing and (2) Non-Ionizing. The higher energy ionizing radiation includes X-rays, Gamma rays and particle type radiation such as Electrons, Protons, Neutrons, Alpha particles and HZE (heavy primary nuclei) particles. Non-ionizing radiation includes lower energy sources such as Lasers, Microwaves, and Radar. Human missions beyond Earth orbit may include nuclear radiation sources such as Radioisotope Thermal Generators. These will present additional radiation problems along with the ionizing radiation from solar disturbances, and galactic cosmic rays which are parts of the total natural radiation environment.

CP504, *Space Technology and Applications International Forum–2000*, edited by M. S. El-Genk
2000 American Institute of Physics 1-56396-919-X

Whereas current standards for non-ionizing radiation appear adequate, the area of ionizing radiation requires more research followed by changes in the standards. Schimmerling (1994) has pointed out that Ionizing radiation is a significant risk to humans living and working in space. He has warned that the biological effect of the particles cannot be extrapolated in a straightforward way from available X-ray and Gamma ray data. He proposed the development of an independent database obtained from studies of radiation biology of heavy charged particles generated at ground-based accelerator laboratories with verification of the predictions made by space experiments. His warnings have encouraged NASA to start a Radiation Protection Initiative. Some of the initiative activities include: development of research needs; instrumentation, telescience, robotics; internal NASA coordination; coordination with other Federal agencies, private institutions and international partners; and development of an implementation plan including predictions and validations.

Hearing Health and Safety

Added concern for hearing health and safety in space missions has emerged from NASA' s participation with the Russians on their Mir Space Station. Although the Mir was designed to 60 decibels (dB)(A) maximum, operational noise has been higher. As a countermeasure, the Mir crew members were advised to wear special earphones to protect their hearing. Some cosmonauts on Mir over six months have had temporary and possibly permanent damage to their hearing. Significant sources of noise comes from compressors, fans, pumps in the 55 to 70 dB(A) range.

NASA and the Russians have collaborated in reducing noise on the Russian Service Module which will provide life support and utilities, thrusters, and habitation functions for the International Space Station. The objectives were: (1) conduct a noise survey (2) Identify and agree on major noise contributors requiring mitigation (3) develop an action plan to quite the equipment. After the noise survey both sides agreed on what items needed to be modified with mufflers, vibration isolators, acoustic wrap. The Russians agreed to redesign to reduce noise of significant noise sources such as fans, pumps, refrigerators. Both continuous and intermittent noise sources were investigated. Preliminary data indicated maximum noise up to 70 to 74 dB(A) vs. 60dB(A) specification. Verification will be done on Earth and on-orbit.

On the ground, NASA has a hearing conservation program that is tighter than the OSHA Standard 29 CFR 1910.95. For example, where the OSHA standard says action must be taken to reduce noise if the worker is exposed to 85 dB(A) on an average exposure over an 8hr period, NASA will take action to reduce noise at 80 dB(A). In space the NASA standards are tight and will likely become tighter for long term missions.

Habitability/Toxicology Health and Safety

There is currently a draft Human-Rating Requirements document underway for the next generation of NASA spacecraft. A human rated system is one that incorporates those designs, features, operational procedures, and requirements necessary to accommodate human participants. The requirements are aimed at establishing a capability to conduct manned operations, including safe recovery from any credible emergency situation. One of the proposed requirements is that the cumulative probability of safe crew return over the life of the program exceeds 0.99. Habitability /Toxicology Health and Safety concerns goes beyond simply returning the crew. Bringing back a crew member with a bad back or damaged kidneys, for example, would not be a successful outcome from a health and safety standpoint.

Kuroda et.al.(1999) in their summary of the International Workshop on Human Factors in Space held in Tokyo, Japan, pointed out that, while almost any inconvenience, awkward habitation or uncomfortable crew mix can be tolerated for a few days in space, with longer missions the contribution of habitability and behavior factors to mission success will rise in importance.

NASA Standard 3000 "Manned Systems Integration Standards" (1995) is the principal reference document for space flight health and safety requirements for crew habitability and life support systems design. That document draws on experiences all the way back to Mercury program. Research and revisions are needed for habitability/Toxicology health and safety over long space mission periods since most of the existing data are for short term missions.

EVA Health and Safety

NASA Standard 3000 (1995) is also the principal reference document for Extravehicular Activity (EVA). EVA is the term for the activity performed by an astronaut outside a spacecraft. The U.S. space suit used in EVAs is the Extravehicular Mobility Unit (EMU). The current EMU is designed for a total maximum duration of 7 hours. It provides environmental protection, mobility, life support and communications. EVA hazards include: excessive touch temperatures; radiation; micrometeoroids and debris; chemical contamination; sharp edges; hazardous equipment; depressurization; untethered drift; ignition in oxygen-rich life support systems; overpressurization; electrical voltage shocks.

Safe and productive work in space requires a basic understanding of suits, airlocks, tools, facilities, vehicle interfaces and operational techniques. Decompression, or the "bends", is a continuing problem. It is essential that sufficient time be taken in allowing the body to adjust to pressure changes and to take care to avoid cuts and other damage to the suit. Airlocks permit suited crew members to transfer from the spacecraft without depressurizing the spacecraft internal areas. Hatch designs must be fault free to open and close. A lesson learned on Shuttle Flight STS-80 was that it is important to validate were specified self-locking inserts are required. On that flight the hatch would not open because screws had backed out and were wedged in the hatch gears.

In general, it is always safer to stay in the spacecraft and minimize EVAs using robotics to perform external operations as well as designing external components for easy assembly and maintenance should EVA be needed.

Stress/Immunology Health and Safety

NASA has been working in collaboration with the Australians on space analogue studies in Anarctica. Lugg and Shepanek (1998) have pointed out that Anarctica is a hostile, dangerous environment establishing a biological and psychological isolation similar to conditions astronauts might experience during long duration missions in space. Research has demonstrated a lowered responsiveness of the immune system under the stress of isolation and confinement. There was almost a 50% reduction in T cell proliferation in the Antarctica "winterover". The T cells are considered a major component of the body's defense system. Tingate et al., (1997) have hypothesized that the lack of challenge by new bacteria to the immune system results in an overall immune system reduction. This work complements research at NASA's Johnson Space Flight Center which has been looking into the effects the stress of spaceflight has on the reductions in the astronauts' immune system. Possible countermeasures might be to have the astronauts take antiviral drugs during the mission or give siliva tests to crew members which will provide warnings to modify their activities.

In addition to the effects of stress on the immune system, there are a range of human factors involved in stress. The summary of the International Workshop on Human Factors in Space (Kuroda et al., 1999) has emphasized that human factors involve consideration of behavioral, psychological, physiological and operational factors that will influence human performance and safety. Mental Stress and Monitoring was one of the workshops. Other workshops included: Personal, Interpersonal and Group Dynamics; Behavior and Performance in Isolation;

Perception and Cognition, Habitability; Human Physical Performance and Performance Shaping Factors. Prioritized research directions were developed such as developing controls that allow the isolation and investigation of unique contributors to stress, the introduction of off-normal stressors, and the controlled introduction of multiple stressors. Suggestions were made for having on-orbit human factors research facilities on the International Space Station.

On the ground, NASA provides an Employee Assistance Program (EAP) for obtaining professional help on problems related to stress. NASA includes the employee's immediate family and is also working to extend the resolution of stress problems to contracted employees.

SYSTEMS AND MISSION HEALTH AND SAFETY

The five major health and safety issues can also be looked at in terms of systems and missions safety and health. The two most prominent are the International Space Station (ISS) and the Mars Mission. The ISS is an example of long term human presence in Low Earth Orbit. The Mars Mission is an example of long term human travel beyond Earth Orbit.

International Space Station

The International Space Station (ISS) will house an international crew of up to seven for stays of 3 to 6 months. The ISS was designed to a total safety design goal. ISS efforts are now shifting to On-Orbit Institutional Safety (NASA, 1998). It is NASA's intent to meet the occupational health and safety and other regulatory requirements similar to those here on Earth such as the standards of the Occupational Safety and Health Administration (OSHA) of the Department of Labor as well as other organizations like the American National Standards Institute (ANSI).

Burrough in his book *Dragonfly* (1998) has made clear what happens in a Space Station type crisis. On June 25, 1997, there were two Russian Cosmonauts and one American Astronaut manning the Mir Space Station. The Mir at that time would not have done well with an OSHA type inspection having cramped dimly lit passages, clutter, loose wires hanging through passageways, and coolant leaks. The Russians were attempting a first time manual docking of the unmanned Progress Supply Vessel with the Mir. The Mir Commander was under a high level of stress and exhausted for multiple reasons. Because of the coolant leaks, it was necessary to stay up late finding the leaks and operating pumps to help fix the problem. He had even collided head on with a large floating toxic balloon of ethylene glycol that had leaked out and was free in the spacecraft. There was no emergency wash available, necessitating the use of wet and dry towels to clean off. There were stressful dialogs with the ground station as if the commander was a "hamster" doing their bidding. Just before the docking experiments the ground had scheduled him for an unusually long 12 day sleep study which gave him insomnia. Needless to say, the stress and exhaustion plus a poorly planned docking procedure resulted in the supply vehicle colliding with the Mir's Spector science module. The Spector module was finally sealed off by the crew from the rest of the Mir but the clutter, wires through the hatch, and lack of easy access to emergency equipment almost caused total failure of the space station.

There are more problems in the example than just meeting current OSHA safety and health standards. Neither OSHA nor current NASA standards deal sufficienctly with stress and human factors. The whole cluster of factors presented in the International Workshop on Human Factors in Space (Kuroda et al., 1999) were involved in the problem namely: Mental Stress and Measurement; Habitability; Personal, Interpersonal Group Dynamics; Behavior and Performance in Isolation Environments; Perception and Cognition; Human Physical performance and Performance Shaping Factors. Clearly there is much work to be done in these areas.

Mars Mission

The NASA study on the "Human Exploration of Mars" (NASA 1997) provides insights into the factors of these five important health and safety issues. Human travel to Mars with a crew size up to 6 will involve 4 to 6 months going, approximately 18 to 20 months on the surface, and other 6 months to return to Earth. Astronaut health and safety will have top priority with special concerns for long term injury or illnesses.

Radiation health and safety will have its highest risk during the interplanetary phases of the mission where there is constant background radiation and periodic exposure from solar flares. When situated on Mars, the risk will be reduced somewhat by protection from the Martian atmosphere. Extensive EVA activity will be needed on the Mars surface and the radiation risks will increase for those using the suits since they are not as radiation potective as being inside a capsule. Nuclear thermal propulsion for getting to Mars would require substantially less propellant than the theoretical best chemical engine but public concerns over radiation safety risk may disallow its use.

Hearing health and safety measures planned to conserve hearing on the International Space Station should help improve the hearing protection for the Mars crew members. On the Martian surface there may be noise from mining activities and transportation endeavors. These risks will have to examined in detail to assure hearing health and safety.

Habitability/Toxicology Health and Safety is related to the types of habitats involved in the interplanetary, surface and transitional (landing/takeoff) phases of the Mars mission. A common habitat design for all phases has been considered including pressure vessel, electrical distribution, hatches and docking mechanisms. Mission critical systems need to be as automated as much as possible in order to minimize crew risks related to failures. Performance of the habitat is expected to be tested by attaching the development unit to the International Space Station.

EVA health and safety will depend primarily on the EVA suit and the airlock system. EVAs will be an essential part of the Mars mission for doing constructing and maintaining surface facilities, geologic field work, deployment, operation, and maintenance of instruments. The suit will have a pressure shell, atmospheric and thermal control, communications, monitoring and display, and nourishment and hygiene elements. The design and locations of the airlock facilities will have a major design impact on the pressurized habitat architecture. Keeping safety first while attempting to achieve high mobility and dexterity in EVAs will be a key design challenge.

Stress health and safety in the Mars mission must be more than designing good habitat features. Six months testing in a Mars development habitat attached to the International Space Station would be a great help in monitoring stress and learning more about human factors as well as improving design and operational procedures. Health and wellness should be promoted by good communications with families back home. The availability of effective telemedicine should also reduce stresses of isolation.

SAFETY METRICS

As an Agency, NASA files an Annual Occupational Safety and Health Report to the Department of Labor. NASA's Lost Time Injury/Ilness Rate, without exceptions, was 0.517 cases /100 employees in 1998, one of the lowest of all Federal Agencies. Accidents are reported in terms of: Slips/Trips/Fall, Lifting/Moving, Bumped into/Struck by, Repetitive Motion, and Other. Injuries and illness related to NASA's major health and safety issues tend to be long term in nature. Safety statistics that look good in the early years might prove much worse when the long term damages begin to be reported. It is important to recognize the problems early and do something about them.

The Department of Labor metrics are based on all of the NASA employees. The actual number of NASA astronauts at risk on the International Space Station or on a mission to Mars might be less than 15 in a given year and not show any sharp changes in overall Department of Labor metrics. Clearly, front page newspaper headlines about an injured or ill astronaut in space will get more public attention than any OSHA statistics. NASA's drive to be the best in health and safety in all areas should further promote health and safety of the astronauts. For complete health and safety coverage, NASA will soon extend its metrics to its contractor work force. Following that, it is expected that the next step will include metrics involving the Internationals, since systems and mission health and safety for the International Space Station and the Mars mission will be strongly dependent on partner health and safety practices.

SUMMARY AND CONCLUSIONS

It is NASA's goal to be the preeminent organization in safety in the nation. The five major health and safety issues are particularly of concern in NASA's long term missions.

Of the five major issues, radiation health and safety improvements are now in progress under NASA's Radiation Safety Initiative. Continuing action is needed in terms of research with requirements changes for hearing, habitability/toxicology, and EVA issues. Stress health and safety including the human factors needs the most attention since NASA and other standards in the past have not given full attention to these subjects.

The International Space Station (ISS) and the Mars mission were provided as examples of systems and missions health and safety. The ISS program is starting to follow health and safety standards similar to the Occupational Safety and Health Administration (OSHA) and other standards used on Earth. The Mir mishap example illuminates the importance of stress and human factors in long time operations in space. A Mars mission will reqire years to complete. There will be significant potential exposure to radiation, possible noise problems affecting hearing, cramped habitat with as many as six crew members, considerable EVA activity, and stress of isolation. Stress problems may be improved by greater attention to human factors, good communications and effective telemedicine. Considerable research and new requirements will be needed.

Health and safety metrics provided by NASA to the Department of Labor are among the lowest of the Federal Agencies and the statistics will go lower under NASA's safety initiative to be number one in safety in the nation. NASA takes a broad look on safety beyond that of its direct employees. New metrics will be developed to include NASA's contractors. Systems and mission safety for the ISS and the Mars mission will have strong dependence on international partner health and safety practices and metrics for them will also need to be included.

ACKNOWLEDGMENTS

Acknowledgment is given to the NASA contributors and workers dedicated to making NASA number one in safety in the nation.

REFERENCES

Burrough, B., *Dragonfly*, Harper Collins,Publisher, 1998, pp 354-391.

Kuroda,I., Young, L.,Fitts, D. "Summary of the International Workshop on Human Factors in Space," July 7-9, 1999 Tokyo, Japan pp 1-6.

Lugg, D., and Shepanek, M., "Space Analogue Studies in Anarctica," 49th International Astronautical Congress, September 28-October 2, 1998, pp. 1-7.

NASA Special Publication 6107, "Human Exploration of Mars: The Reference Mission of the NASA Mars Exploration Study Team," edited by S. Hoffman and D. Kaplan, Lyndon B. Johnson Space Center, Houston, Texas, July 1997, 1-1 to 3-35.

NASA-STD-3000 "Manned Systems Integration Standards," Volume 1 Revision B, Lyndon B. Johnson Space Center, Houston, Texas, July 1995, pp. 5-1 to 6-24, 8-1 to 8-53, 14-1 to 14-6.

NASA Report SSP 50308, "On-Orbit Institutional Safety Report for the ISS," Lyndon B. Johnson Space Center, Houston, Texas, 1998, Appendix F pp. 1-31.

Schimmerling, W., " Space and Radiation Protection: Scientific Requirements for Space Research," International Symposium on Heavy Ion Research: Space, Radiation Protection and Therapy, Sophia-Antipolis, France, 21-24, March 1994, pp. 1-5.

Tingate, R., Lugg,D., Muller, H., Stowe, R., Pierson, D., "Antarctic Isolation: Immune and Viral Studies", Immunology and Cell Biology, Vol 75, 1997, pp. 275-283.

The Safe Use of Nuclear Power Systems in Outer Space: Current Safety Standards and a Suggested Framework for the Future

Michael D. White[1] and Daniel F. Stenger[2]

[1]Of Counsel, Blank Rome Comisky & McCauley LLP, 900 17th Street, N.W., Suite 1000, Washington, D.C. 20006; e-mail: white@blankrome.com.; [2]Partner, Hopkins & Sutter, 888 16th Street, N.W., Washington, D.C. 20006; dstenger@hopsut.com.

Abstract. The use of nuclear power sources in outer space is recognized as being particularly suited and essential to some missions. A risk-informed approach to regulating the use of nuclear power sources in space should be followed to avoid unnecessary impediments to their use. Safety requirements should be based upon realistic estimates of expected risk reduction.

INTRODUCTION

"Recognizing that for some missions in outer space nuclear power sources are particularly suited or even essential due to their compactness, long life and other attributes...," thus begins the Preamble to U.N. Resolution 47/68 entitled "Principles Relevant to the Use of Nuclear Power Sources in Outer Space," adopted December 14, 1992. Nuclear power sources have been used in space missions since the early years of space exploration. The use of such power sources in space has proven to be controversial, just as the use of nuclear power on earth. Nonetheless, the potential benefits and essential nature of such power sources has now been expressly recognized. This paper addresses the express and implied international requirements and principles which have been developed for the use of nuclear power sources in outer space.

Nuclear power sources have played a significant role in the space programs of the United States and the former Soviet Union. In the U.S. space program alone approximately 25 missions have employed decay heat power sources. Among the U.S. missions were the Transit series of navigational satellites, the Nimbus meteorological satellites, the Apollo lunar missions, the Pioneer, Viking, Voyager, Galileo and Ulysses planetary probes and the LES communications satellites. (DOE/NE-0071.) Most recently, decay heat power sources were used in the Cassini planetary mission to Saturn. (Cassini FEIS, 1995.) The former Soviet Union made extensive use of nuclear reactors in its Cosmos series of Radar Ocean Reconnaissance Satellites (RORSATs). (Goldman, 1996, p. 111.) It is contemplated that nuclear power sources will be used in future missions to the outer planets and to distances well beyond the orbit of Pluto.

Three types of nuclear power sources have been employed in the above-mentioned missions: radioisotope thermoelectric generators (RTGs) using primarily non-weapons grade plutonium-238 as a radioactive decay heat source for use in conjunction with a thermoelectric generator; radioisotope heater units (RHUs) using primarily plutonium-238 as a decay heat source to provide heat to critical spacecraft components; and nuclear reactors using primarily highly enriched uranium-235 as the fissile material. The United States has employed only RTGs and RHUs in all but one instance. The SNAPSHOT mission of 1965 used a SNAP-10A nuclear reactor. The former Soviet Union and Russia have reportedly used RTGs and RHUs in space missions. As mentioned above, the most prominent use of nuclear reactors in space was by the Soviet Union in its Cosmos series of RORSATs.

Although the use of nuclear power sources has contributed to the success of a number of space missions, such use has not been without mishap. The first major mishap was the re-entry burn-up of the SNAP-9A RTG on the Transit

CP504, *Space Technology and Applications International Forum–2000*, edited by M. S. El-Genk
2000 American Institute of Physics 1-56396-919-X

5-BN-3 mission during a launch failure on April 21, 1964. This resulted in the release of approximately 17,000 Curies of Pu-238 into the atmosphere. In comparison, approximately 9,000 Curies of Pu-238 were released into the atmosphere due to atmospheric nuclear weapons testing. (Cassini, FEIS, 1995, p. 3-44.)

The most well-known and most serious accident was the reentry of Cosmos 954 in 1978 which contaminated an extensive area of Western Canada. Cosmos 954 was a RORSAT mission employing a reactor fueled with highly enriched uranium (HEU). This incident led to the U.N. Committee on the Peaceful Uses of Outer Space (UNCOPUOS) establishing the Working Group on the Use of Nuclear Power Sources in Outer Space. This ultimately led to the preparation and adoption of U.N. Resolution 47/68, the Principles Relevant to the Use of Nuclear Power Sources in Outer Space. (Goldman, 1996, pp. 111-117.)

Despite these mishaps, research efforts in the development of nuclear power sources have resulted in significant knowledge about the advantages of the use of nuclear power systems for space purposes, due to properties such as their compactness and long life. In recognition of the increasing need for reliable, quality sources of power for deep space missions and space stations, the importance of the proper legal and regulatory standards governing the use of nuclear power in space, particularly in terms of safety, cannot be over-emphasized.

The implementation of robust but reasonable regulatory standards, internationally applicable, is necessary to provide a framework to promote the safe use of nuclear power sources in outer space, and potentially on other planets. The implementation of a sound framework is important to maintain public confidence in the use of nuclear power for space missions. However, unnecessary or overly burdensome standards that are impediments to the safe use of nuclear power systems in space must be avoided.

GOVERNING LAW

The United Nations has developed five treaties directed to activities in outer space: (1) Treaty on Principles Governing the Activities of States in the Exploration and Use of Outer Space, Including the Moon and Other Celestial Bodies (Outer Space Treaty); (2) Agreement on the Rescue of Astronauts, the Return of Astronauts, and the Return of Objects Launched into Outer Space (Rescue Treaty); (3) Convention on International Liability for Damage Caused by Space Objects (Liability Convention); (4) Convention on Registration of Objects Launched into Outer Space (Registration Convention); and (5) Treaty Governing the Activities of States on the Moon and Other Celestial Bodies (Moon Treaty). The United States has ratified all but the Moon Treaty. As a ratifying state under the four instruments identified above, the U.S. is subject to those treaty provisions (to the extent that they are self-executing and once implementing legislation is enacted, to the extent they are not self-executing). In addition, U.N. Resolution 47/68, "Principles Relevant to the Use of Nuclear Power Sources in Outer Space," is a resolution, not a treaty, and hence is not binding on the U.S. Nonetheless, its provisions are important to actions taken by the U.S. in using nuclear power sources in outer space. In fact, the U.S. by its actions has implicitly followed the Resolution by notifying the U.N. of its planned activities regarding the Cassini mission to Saturn. (U.N. Suppl. No. 20 (A/52/20), 1997.)

In addition to being bound by its treaty commitments, the U.S. also is bound by its own domestic laws and regulations. For example, in using RTGs and RHUs on the Cassini mission to Saturn and a gravity assist fly-by of Earth, it was necessary for the mission planners to provide a risk assessment in a Final Environmental Impact Statement ("FEIS") and in a supplement to it. The FEIS was prepared in accordance with the National Environmental Policy Act of 1969 (NEPA) (42 U.S.C. 4321 *et seq.*), as amended; the Council on Environmental Quality Regulations for Implementing the Procedural Provisions of NEPA (40 CFR Parts 1500-1508); and the National Aeronautics and Space Administration (NASA) policy and regulations (14 CFR Subpart 1216.3). Importantly, mission activities such as Cassini are not regulated by an organization such as the Nuclear Regulatory Commission (NRC). This is attributable to the research aspects of the use of nuclear power sources in outer space. Rather, such activities are subject to review by the Interagency Safety Review Board, having representatives from NASA, DOD, and DOE.

The primary focus of this paper will be the principles adopted by the U.N. in U.N. Resolution 47/68. However, it is important to note that each of the five international treaties contain provisions which are subject to being invoked concerning the use of nuclear power sources in outer space. By the express terms of the treaties, only the Moon Treaty (not ratified by the U.S.) expressly deals with nuclear power sources through provisions dealing with radioactivity. Article VII of the Moon Treaty provides:

1.   In exploring and using the moon, States Parties shall take measures to prevent the disruption of the existing balance of its environment whether by introducing adverse changes in that environment, by its harmful contamination through the introduction of extra environmental matter or otherwise. States Parties shall also take measures to avoid harmfully affecting the environment of the earth through the introduction of extraterrestrial matter or otherwise.

2.   States Parties shall inform the Secretary-General of the United Nations of the measures being adopted by them in accordance with paragraph 1 of this article and shall also, to the maximum extent feasible, notify him in advance of all placements by them of radioactive materials on the moon and of the purposes of such placements.

3.   States Parties shall report to other States Parties and to the Secretary-General concerning area of the moon having scientific interest in order that, without prejudice to the rights of other States Parties, consideration may be given to the designation of such areas as international scientific preserves for which special protective arrangements are to be agreed upon in consultation with the competent bodies of the United Nations.

The remaining treaties only inherently deal with the use of nuclear power sources, primarily through article provisions which address responsibility, liability, registration and hazardous space objects. These articles are reproduced in the Appendix and include Articles VI and VII of the Outer Space Treaty, Article 5 of the Rescue Treaty, Articles II and XXI of the Liability Convention and Article IV of the Registration Convention.

"PRINCIPLES RELEVANT TO USE OF NUCLEAR POWER SOURCES IN OUTER SPACE"

Although not a treaty, the most comprehensive international pronouncement concerning the use of nuclear power sources in outer space is found in the U.N. Resolution 47/68 entitled "Principles Relevant to Use of Nuclear Power Sources in Outer Space." As mentioned at the outset of this paper, the Preamble of this document expressly recognizes the potential value and essential nature of these power sources for some missions. Eleven principles are articulated. Principle 1 announces the applicability of international law to the use of nuclear power sources in outer space and identifies in particular the U.N. Charter and the 1967 Outer Space Treaty. Principle 2 defines "launching state," "foreseeable," "all possible," "general concept of defense-in-depth," and "made critical." This principle is important in that the terms "foreseeable" and "all possible" are used to "describe a class of events or circumstances whose overall probability of occurrence is such that it is considered to encompass *only credible possibilities for purposes of safety analysis*." (Emphasis added.) Zero-power testing does not come within the term "made critical." Of greatest significance is the flexibility provided in defining the concept of defense-in-depth:

The term "general concept of defense-in-depth" when applied to nuclear power sources in outer space considers the use of design features and mission operations in place of or in addition to active systems, to prevent or mitigate the consequences of system malfunctions. Redundant safety systems are not necessarily required for each individual component to achieve this purpose. Given the special requirements of space use and of varied missions, no particular set of systems or features can be specified as essential to achieve this objective.

Principle 3 is perhaps the most important as it is directed to the guidelines and criteria for safe use. Although the Preamble to the Principles recognizes the potential value and essential nature of nuclear power sources, Principle 3 restricts their use:

> In order to minimize the quantity of radioactive material in space and the risks involved, the use of nuclear power sources in outer space should be restricted to those space missions which cannot be operated by non-nuclear energy sources in a reasonable way.

Principle 3 further provides general goals for radiation protection and nuclear safety. These goals include no significant radioactive contamination of outer space, as well as the observance of the appropriate radiation protective objective for the public recommended by the International Commission on Radiological Protection (ICRP) during normal operations and reentry. Radiation exposure is to be restricted to a limited geographical region and to individuals to the principal limit of 1 mSv per year, with a 5 mSv dose permitted if the average annual effective dose over a lifetime does not exceed 1 mSv per year. Defense-in-depth is applied with system reliability ensured by "redundancy, physical separation, functional isolation and adequate independence" of components.

Nuclear reactors are to use only highly enriched uranium 235 (HEU) as fuel, not to be made critical before having reached their operating orbit or interplanetary trajectory, and designed to ensure that they will not become critical before reaching the operating orbit even in the event of a launch failure, reentry or submersion in water. Nuclear reactors may be used in interplanetary missions, earth orbits sufficiently high that the orbital lifetime is long enough to allow for sufficient fission product decay to the approximate activity of the actinides. For low earth orbit, a highly reliable operational system to ensure effective and controlled disposal of the reactor is required.

Radioisotope generators may be used in earth orbit or missions "leaving the gravity field of the Earth." If used in earth orbit, they must be disposed of, such as by storage in high earth orbit. Radioisotope generators must be protected by a containment system which will protect against heat and aerodynamic forces of reentry, and also "[u]pon impact, the containment system and the physical form of the isotope shall ensure that no radioactive material is scattered into the environment so that the impact area can be completely cleaned of radioactivity by a recovery operation."

Another key aspect of the guidelines is the requirement for a safety assessment. Principle 4 requires that a "thorough and comprehensive safety assessment is conducted" respecting the guidelines and criteria of Principle 3. The "assessment shall cover all relevant phases of the mission and all systems involved, . . .including the means of launching, the space platform, the nuclear power source and its equipment and the means of control and communications between ground and space." The safety assessment and intended time frame of the launch are to be made publicly available, pursuant to Article XI of the 1967 Outer Space Treaty.

Principles 5-9 relate to the consequences of reentry. Principle 5 requires that a launching State make a notification of the risk of reentry of radioactive materials including the system parameters such as best prediction of orbit lifetime, trajectory and impact region, as well as information on the radiological risk including the power source type and physical form, amount and radiological characteristics of the power source. The information is to be timely updated as reentry approaches.

Principle 6 requires consultations with other States, while Principle 7 requires assistance to other States when necessary. In particular, after reentry, "the launching State shall promptly offer, and if requested by the affected State, provide promptly the necessary assistance to eliminate actual and possible harmful effects, including assistance to identify the location of the area of impact of the nuclear power source on the Earth's surface, to detect the re-entered material and to carry out retrieval or clean-up operations." Principle 8 provides for responsibility in accordance with Article VI of the 1967 Outer Space Treaty, and Principle 9 provides for liability and compensation under Article VII of that treaty. Dispute resolution is governed by Principle 10. Principle 11 provides for the periodic re-opening and review of the Principles, now scheduled for the year 2000. Although further revision of the principles has been deferred to 2000, the Working Group has invited the consideration of principles applicable to the terrestrial use of nuclear power and reasons why use of nuclear power in space differs from use on earth. (U.N. Suppl. 20 (A/53/20), 1998.)

U.S. ADHERENCE TO U.N. RESOLUTION PRINCIPLES

Although not technically bound by the principles established in U.N. Resolution 47/68, the United States has to a significant degree adhered to the intent. In accordance with Principle 4, the U.S. notified the U.N. of its proposed actions in the Cassini mission. Moreover, NASA regulations address the use of nuclear power sources. 14 C.F.R. §1216.305(c)(3) specifically identifies such sources as requiring environmental impact statements:

> Specific NASA actions requiring environmental impact statements, all in the R&D budget category, are as follows:
>
> ...Development and operation of nuclear systems, including reactors and thermal devices used for propulsion and/or power generation....

Moreover, in scoping issues, "the range of environmental categories to be considered in the scoping process shall include, but not be limited to: ...radioactive materials.... 14 C.F.R. §1216.307(g).

A VIEW TOWARD THE FUTURE

As has been seen, the only international treaty which directly addresses the use of nuclear power sources in outer space is the Moon Treaty. In addition, there exist several treaty provisions which are worded so as to capture nuclear power sources. The United Nations has adopted Principles Relevant to the Use of Nuclear Power Sources in Outer Space which do directly address the issue. However, those principles do not have the force of law unless adopted as part of a statute or regulation of a spacefaring nation. The United States has only peripherally (inferentially) adopted such principles in its environmental assessments of space missions.

If nuclear power sources are to be viable alternatives for the future of space exploration and commercialization, the statutory and regulatory response must recognize and take into account the inherent spatial separation and transitory nature of the space-based nuclear power sources as compared to those for use on earth.

Specifically, the risk that the public has come to understand and accept on earth is far different from that resulting from the use of nuclear power sources in outer space. Nuclear power plants, for example, have fixed geographic locations, in which the primary potential radiological exposure of the public is limited substantially to emergency planning zones near the power plants. The risk is from a stationary source. Significant benefits of that risk are usually reaped by those in the vicinity of the plant. The risk is managed not only by the utility, but by ongoing oversight of the management and operation of the facility by the Nuclear Regulatory Commission. The time period of the risk is measured over a period of decades (the license term of the facility). As a result, there is a degree of fixed spatial and temporal risk. An entire statutory and regulatory regime has been developed to manage the risk and to severely limit or minimize it. This is accomplished by the use of costly backup systems and a concept of defense-in-depth.

Nuclear power sources for use in outer space may pose potential radiological risks to the environment and to its inhabitants. However, those risks are to a significant degree different from the risks posed by nuclear power plants. First, the risks to the earth and its inhabitants are transitory in nature. They are limited to the launch and potential atmospheric reentry of objects containing nuclear power sources. The radiological risks of a space nuclear reactor using HEU launched before having gone critical are essentially zero. Such risks are confined to premature criticality caused by a launch failure or the reentry and burn-up or crash of a craft having a nuclear reactor which contained fission products resulting from critical operation during its mission. Similarly, with radioisotope decay heat power sources such as RTGs and RHUs, the radiological risks are essentially limited to launch failures and reentry events. For a nuclear power source which burns up in the atmosphere, the radiological risk to the earth may be spread over a wide area of the planet but any individual exposure would be limited.

As a consequence, the risk management for space-based nuclear power sources should be directed at the two major events which can cause the greatest potential harm – reduction of launch failure risks and reduction of atmospheric reentry risks. These reductions can be achieved at launch by using non-radioactive HEU-fueled nuclear reactors which are not taken critical until they are in confirmed stable earth orbits or on an interplanetary trajectory. By delaying criticality until a stable non-reentry path of motion is achieved relative to the earth, the reentry risk is virtually eliminated.

In general terms, reductions in risk for radioisotope decay heat sources can also be achieved by reducing the risk of launch failure by improving the launch system reliability, reducing the risk of radiological release during a launch failure should one occur by improving the containment of the decay heat material, by using such sources in stable orbits/trajectories which would minimize the likelihood of reentry, and by using containment systems which minimize the likelihood of release or dispersal of radioactive material should a reentry occur.

Within the constraints of reduction in risk due to launch failures and reduction in risk due to reentry, mission objectives must be taken into account. An allocation of resources between launch failure reduction and reentry reduction can be made. For example, an interplanetary mission which uses RTGs and RHUs which will not be in earth orbit for any extended period of time is a candidate for greater risk reduction allocation to launch failure reduction and launch failure survivability than reentry reduction. In contrast, a mission employing an initially non-radioactive HEU-fueled reactor poses a minimal radiological launch risk (none if there is no premature criticality), so the allocation of risk reduction should be made to preventing reentry of the reactor once it becomes radioactive.

With these general principles of risk management in mind, further refinements can be made to help avoid unnecessary impediments to the use of nuclear power sources in outer space.

In this connection, recent experience with risk-based or risk-informed regulation of commercial nuclear power reactors in this country may be instructive. By way of background, under the Atomic Energy Act of 1954, the Nuclear Regulatory Commission is charged with regulating power reactors, as well as other licensees that utilize nuclear materials, in order to provide reasonable assurance of adequate protection of public health and safety. The Atomic Energy Act essentially embodies a two-tier health and safety standard. It directs the Commission to maintain a minimum level of "adequate protection" from the operation of power reactors, but also authorizes the Commission to go further and regulate in a manner so as to "*minimize* danger to life or property." Atomic Energy Act § 161, 42 U.S.C. § 2201(i) (emphasis added). In carrying out its mission under the Atomic Energy Act, the Commission has adopted standards governing the imposition of new safety enhancements or "backfits" for power reactors. The Commission's rule on backfitting is set forth in 10 C.F.R. § 50.109.

Under the NRC's standards on backfitting, any requirements found necessary to ensure the minimum level of public health and safety may be adopted without regard to the economic cost of implementation. However, when considering potential new backfits that go beyond ensuring this minimum level of protection, the NRC is free to employ strict cost-benefit analysis. See 10 C.F.R. §§ 50.109(a)(3) and (a)(4). The NRC's backfitting rule, in fact, goes further and requires that proposed new backfits be shown to provide a "substantial increase" in overall protection as well as be cost-justified. (For a discussion of the two-tier standard under the Atomic Energy Act, *see Union of Concerned Scientists v. Nuclear Regulatory Commission*, 842 F.2d 108 (D.C. Cir. 1987).)

Similar concepts can and should be employed in establishing safety standards for the use of nuclear power sources in space. Certainly any safety requirements or features necessary to ensure a minimum acceptable level of risk should be mandated without regard to economic cost. But once an adequate level of risk has been achieved for a mission, the need for additional safety enhancements should be addressed using cost-benefit analysis. The cost-benefit analysis, moreover, can and should be based on realistic estimates of risk reduction using available probabilistic risk assessment (PRA) tools. PRA techniques have now advanced to the point where *quantitative* estimates of risk can be made with a high degree of precision. The Nuclear Regulatory Commission and the nuclear power industry have been pioneers in the general application of PRA. The NRC and the industry are currently employing PRA in various contexts ranging from day-to-day assessments of the risk of taking safety systems out of service for maintenance (see 10 C.F.R. § 50.65(a)(4)) to broad proposals to "risk-inform" all the NRC's regulations governing the operation of power reactors in 10 C.F.R. Part 50.

CONCLUSIONS

Consonant with the principles set forth in U.N. Resolution 47/68, the technology exists today for making accurate predictions of the risk from nuclear-powered space missions, as well as for assessing the need for specific safety features. A risk-informed approach to regulating the use of nuclear power sources in space should be followed in order to avoid unnecessary impediments to their use. Under a risk-informed regime, the need for certain safety features would be judged based on *realistic* estimates of the expected risk reduction. With today's PRA tools for assessing risk, decisions should no longer be made on the basis of *worst-case* scenarios. Moreover, once the risk associated with a mission has been shown to be insignificant based on probabilistic risk assessment, no further consideration of additional safety enhancements should be undertaken. A reduction in an already insignificant risk is itself insignificant.

In short, a risk-based approach to regulating the safety of nuclear power sources in space would assure that minimum safety standards are met at all times without unnecessarily hindering innovation.

ACKNOWLEDGMENT

The authors gratefully acknowledge the assistance of Susan Yim and Larisa Halfer, who participated in the preparation of this paper.

REFERENCES

Final Environmental Impact Statement for the Cassini Mission, NASA, June 1995 (Cassini FEIS).
Goldman, Nathan, *American Space Law: International and Domestic*, Univelt, Inc., 1996.
> Report of the Committee on the Peaceful Uses of Outer Space, General Assembly Official Records, Fifty-second Session, Supplement No. 20 (A/52/20), 1997, Section II B 4, para. 83.
Nuclear Power In Space, U.S. Dept. of Energy Office of Nuclear Energy, Science & Technology (DOE/NE-0071).
Report of the Committee on the Peaceful Uses of Outer Space, General Assembly Official Records, Fifty-third Session, Supplement No. 20 (A/53/20), 1998, Section 2(e)(4), para. 91-92.

APPENDIX - U.N. TREATY PROVISIONS

TREATY ON PRINCIPLES GOVERNING THE ACTIVITIES OF STATES IN THE EXPLORATION AND USE OF OUTER SPACE , INCLUDING THE MOON AND OTHER CELESTIAL BODIES (THE 1967 OUTER SPACE TREATY)

Articles VI and VII of the 1967 Outer Space Treaty respectively provide the basis for responsibility and liability for activities associated with nuclear power sources. Article VI provides:

States Parties to the Treaty shall bear international responsibility for national activities in outer space, including the moon and other celestial bodies, whether such activities are carried on by governmental agencies or by non-governmental entities, and for assuring that national activities are carried out in conformity with the provisions set forth in the present Treaty.

Article VII provides:

Each State Party to the Treaty that launches or procures the launching of an object into outer space, including the moon and other celestial bodies, and each State Party from whose territory or facility an object is launched, is internationally liable for damage to another State Party to the Treaty or to its natural or juridical persons by such object or its component parts on the Earth, in air space or in outer space, including the moon and other celestial bodies.

AGREEMENT ON THE RESCUE OF ASTRONAUTS, THE RETURN OF ASTRONAUTS, AND THE RETURN OF OBJECTS LAUNCHED INTO OUTER SPACE (THE 1968 RESCUE TREATY)

In addition to the 1967 Outer Space Treaty, the 1968 Rescue Treaty includes Article 5 which is directed to the return of space objects and is expressly directed to hazardous space objects:

1.   Each Contracting Party, which receives information or discovers that a space object or its component parts has returned to Earth in territory under its jurisdiction or on the high seas or in any other place not under the jurisdiction of any State, shall notify the launching authority and the Secretary-General of the United Nations.

2.   Each Contracting Party having jurisdiction over the territory on which a space object or its component parts has been discovered shall, upon the request of the launching authority and with assistance from that authority if requested, take such steps as it finds practicable to recover the object or component parts.

3.   Upon request of the launching authority, objects launched into outer space or their component parts found beyond the territorial limits of the launching authority shall be returned to or held at the disposal of representatives of the launching authority, which shall, upon request, furnish identifying data prior to their return.

4.   Notwithstanding paragraphs 2 and 3 of this Article, a Contracting Party which has reason to believe that a space object or its component parts discovered in territory under its jurisdiction, or recovered by it elsewhere, is of a hazardous or deleterious nature may so notify the launching authority, which shall immediately take effective steps, under the direction and control of the said Contracting Party, to eliminate possible danger of harm.

5.   Expenses incurred in fulfilling obligations to recover and return a space object or its component parts under paragraph 2 and 3 of this article shall be borne by the launching authority.

CONVENTION ON INTERNATIONAL LIABILITY FOR DAMAGE CAUSED BY SPACE OBJECTS (THE 1972 INTERNATIONAL LIABILITY CONVENTION)

The 1972 International Liability convention addresses the liability of launching states for damages caused by space objects. It establishes a principle of absolute liability. The convention provides in Article II:

A launching State shall be absolutely liable to pay compensation for damage caused by its space object on the surface of the earth or to aircraft in flight.

Further, Article XXI of the convention obligates the launching state and other nations to provide assistance to any nation suffering large-scale damage from a space object. Article XXI provides:

If the damage caused by a space object presents a large-scale danger to human life or seriously interferes with the living conditions of the population or the functioning of vital centers, the States Parties, and in particular the launching State, shall examine the possibility of rendering appropriate and rapid assistance to the state which has suffered the damage, when it so requests. However, nothing under this article shall affect the rights or obligations of the states Parties under this Convention.

CONVENTION ON REGISTRATION OF OBJECTS LAUNCHED INTO OUTER SPACE (THE 1975 REGISTRATION CONVENTION)

The 1975 Registration Convention establishes information reporting requirements concerning space objects. It provides in Article IV:

1.   Each State of registry shall furnish to the Secretary-General of the United Nations, as soon as practicable, the following information concerning each space object carried on its registry:

(a)   Name of launching State or States;

(b)   An appropriate designator of the space object or its registration number;

(c)   Date and territory or location of launch;

(d)   Basic orbital parameters, including:

(i)   Nodal period;
(ii)   Inclination;
(iii)   Apogee;
(iv)   Perigee;

(e)   General function of the space object.

2.   Each State of registry may, from time to time, provide the Secretary-General of the United Nations with additional information concerning a space object carried on its registry.

3.   Each State of registry shall notify the Secretary-General of the United Nations, to the greatest extent feasible and as soon as practicable, of space objects concerning which it has previously transmitted information, and which have been but no longer are in earth orbit.

Small AMTEC Systems as Battery Substitutes

Thomas K. Hunt, Robert K. Sievers and Andrew C. Patania

Advanced Modular Power Systems, Inc., 4370 Varsity Drive, Ann Arbor, Michigan 48108

thunt@ampsys.com

Abstract. Alkali Metal Thermal to Electric Converter (AMTEC) technology is highly scalable and converters can be designed to provide fuel based electric power at levels ranging from a few watts to several kilowatts. AMTEC is a static, modular, heat to electricity conversion technology with the potential for high efficiency and compact size using heat source temperatures readily achievable even with small combustors. AMTEC systems are currently under development for a variety of spacecraft and satellite applications related to potential NASA and Air Force missions requiring power levels in the 75 watt to multi-kilowatt power range. While AMTEC converters can operate with any heat source delivering heat at 900 K to 1200 K, the small AMTEC systems considered here, are suitable for system integration with small, combustion heat sources. In this paper we describe concept designs for small combustion fired, self-contained AMTEC systems whose size, operating duration and mass make them superior choices for applications normally served by secondary or primary batteries. The scaling properties of AMTEC converters and comparisons with direct hydrogen PEM fuel cells are also discussed.

INTRODUCTION

This paper describes the basic properties of a small AMTEC power system that can serve a variety of applications in lieu of a battery. A 20 W system was used as the baseline for this study, but the results presented are relevant for AMTEC systems over a range of power levels from 0.5 W to 1 kW. The work provides an understanding of how AMTEC system mass, using butane, propane or diesel fuel, scales with power level. The 20 Watt AMTEC system is compared to battery and fuel cell options at similar power and operating duration levels. For missions lasting more than one day, AMTEC systems can offer a clear mass advantage over direct hydrogen fuel cells (operating at 50% conversion efficiency) or batteries. Beyond a one day mission, the proposed AMTEC system, with fuel, has a substantially lower mass even at AMTEC efficiencies much lower than the 15+% that is now routinely achieved in small converters in this size range. The support equipment needed to generate hydrogen for recharging fuel cell systems is a further handicap to fuel cell use. Table 1 shows projected masses for complete, fueled, AMTEC (Weber, 1974, Cole, 1983) and direct hydrogen PEM fuel cells (Daugherty, 1999) and an advanced battery system. The AMTEC mass advantage is clear.

Design

The 20 Watt AMTEC system concept is shown in Figure 1. The system uses a single AMTEC converter unit producing 25 Watts of electric power. The complete system includes air and fuel handling and the combustion system as well as power processing electronics. Five watts from the AMTEC is used to power the blower and fuel pumps. Fuel bottles of various capacity can be easily attached to one end of the system. As shown, the fuel capacity is sufficient for ~ 20 hours of operation. The system layout will be modified as

TABLE 1. System + Fuel Mass for a 7 Day, 20 We Mission.

Converter	System + Fuel	Fuel Only
AMTEC 10%	5.6 kg	2.9 kg
Fuel Cell – 50% (Hydride storage 1.5%)	15.3 kg	14.6 kg
Advanced Battery (Li Polymer/ Na/S)	19.5 kg	19.5 kg

CP504, *Space Technology and Applications International Forum–2000*, edited by M. S. El-Genk
© 2000 American Institute of Physics 1-56396-919-X/00/$17.00

the air flow design is developed. A diesel fuel burner, or for special cases, butane or propane, is planned. It is expected that butane or propane will be the most easily developed, but operation on diesel fuel will be the longer range goal.

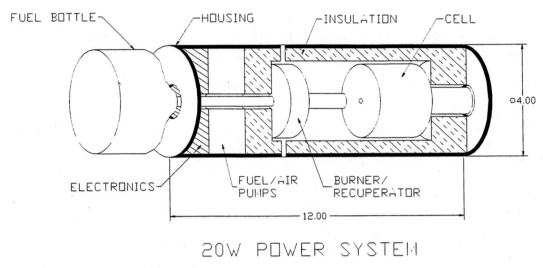

FUEL BOTTLE — HOUSING — INSULATION — CELL

ELECTRONICS — FUEL/AIR PUMPS — BURNER/RECUPERATOR

⌀4.00

12.00

20W POWER SYSTEM

FIGURE 1. 20 W_e Fueled AMTEC System

This concept was used to develop an initial mass estimate as a part of the scaling study. Comparative masses for AMTEC, fuel cell, and battery are shown in Table 1. The mass breakdown for the AMTEC system is shown in Table 2. The converter mass and that of several other AMTEC system designs in various stages of maturity are included in the plot shown in Figure 2. Both of the ARPS (Advanced Radioisotope Power System) data points are radioisotope powered. The OSC design is a candidate for a second generation advanced radioisotope power system option. The 0.5 W data point is from another radioisotope system design, using an advanced vacuum insulation package. The MAPPS I (multi-fueled AMTEC portable power system) value is based on a completed system design. The point for MAPPS II is based on the assumption that the projected design evolution to reduce system weight will be completed in the MAPPS program. A single point for a commercial (Global Thermoelectric's Model 5030) thermoelectric generator is also shown.

Other Technologies

It is useful to consider the characteristics of AMTEC performance in comparison with those of battery and fuel cell technologies for use in a variety of applications for which batteries have historically been the customary

TABLE 2. 20 Watt AMTEC System Mass Breakdown.

Component	Mass
Burner assembly, recuperator, ejector, fan/heat sink, insulation, structural case	1.0kg
AMTEC Converter	.45kg [1]
Electronics	.03kg
Fuel, Propane (7 day mission)	2.02kg [2]
Fuel Canister (7 day mission)	1.2kg
TOTAL	**4.7kg**

[1] Based on power density ~55W/kg
[2] Fuel mass based on propane (LHV=46.39E06 J/kg), 7 day mission, 20% converter efficiency.

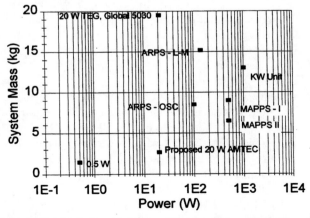

FIGURE 2. Converter System Mass vs. Power, Exclusive of Fuel. for Several AMTEC Systems. A Commercial Thermoelectric Generator Mass is also Shown.

choice. In such service, power level and energy storage density are the key parameters once one has determined that the voltage requirements can be met by a given technology and system design.

The curves shown in Figure 3 show the effective energy density of a fueled 20 Watt AMTEC system as a function of the length of time for which it contains the needed fuel supply. The more fuel is carried, the longer the system can operate and the higher the overall system energy density is expected to be since the 'overhead' mass of the converter, at a given power level, becomes a smaller fraction of the total system mass as the fuel stored increases. At very high fuel loadings, the mass of the converter becomes nearly irrelevant and the energy density saturates at a value dependent only on the system efficiency and the packaging properties of the fuel. The assumptions made for these figures are that the AMTEC converter with its burner, controls and empty fuel

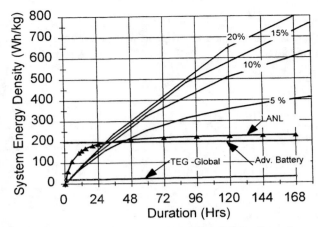

FIGURE 3. Energy Density of AMTEC Cells at Several Efficiencies and of the Los Alamos Direct Hydrogen PEM Fuel Cell with Hydrogen Storage at 1.5 Wt. % (Triangles).

container (capable of holding up to a 7 day supply of fuel) has a mass of 2.7 kg. The fuel is assumed to be a hydrocarbon fuel and while the total chemical energy for such fuels is ~ 13.2 kWh_{th}/kg the lower heating value (the heat one can obtain without condensing the water of combustion) is more appropriate for these purposes and typical lower heating value energy densities of about 11.7 kWh_{th}/kg are used here.

An additional curve (indicated by triangles) in Fig. 3 shows data reported recently by Los Alamos (LANL)[1] for an advanced proton exchange membrane (PEM) fuel cell. The indicated fuel cell mass includes adequate hydrogen storage for the operating durations shown, and shows clearly that it is a relatively smaller issue for very short duration missions than for multi-day operation. With current technology, hydrogen, whether stored as a compressed gas in an adequately strong cylinder or as a metal hydride in a relatively low pressure container as in the LANL case, is substantially heavier than a liquid hydrocarbon fuel storing an equivalent energy. In the case of this particular 20 watt cell concept, the AMTEC system specific energy exceeds that for the PEM fuel cell if the operating period extends past about 30 to 40 hours. The horizontal line for the advanced battery is shown at about 200 Wh/kg, a value near the upper limit for the best secondary batteries that have been proposed. The best primary batteries can reach energy storage densities as high as 300 to 400 Wh/kg, but, for many applications, pose issues of uncertain charge state and disposal. Values for a commercial thermoelectric system (data from Global Thermoelectric, Inc. Model 5030) are also shown. Figure 4 puts the same data in a context giving the total mass for complete, portable systems with both converter and fuel supply (for the given duration) plotted as a function of the operating duration at a constant 20 watts output. The LANL PEM fuel cell values are plotted along with the similar projected values for an advanced battery technology such as sodium sulfur or optimized lithium polymer.

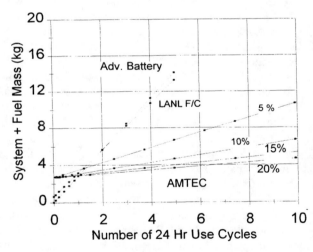

FIGURE 4. Fueled System Mass vs Operational Days for 20 Watt Systems.

It should be noted that other system designs may exhibit different energy density points at which the fuel cell and AMTEC systems have equivalent operating times. It is clear from the lower duration region in Figure 4, that the basic mass of the PEM fuel cell stack whose data is shown, is less than for a comparable power AMTEC system.

Nevertheless, it appears that for missions longer than a few days, the fuel cell system, for which a 50% conversion efficiency has been assumed, appears to be heavier than the values expected for the AMTEC system even if the AMTEC system were to operate at only 5% efficiency, well below efficiencies now routinely achieved. The mass of fuel cell and battery systems is dominated by the fuel contribution as soon as one needs to stock or maintain logistics for 2 days or more. For short missions lasting less than one day only, it appears that primary and/or advanced secondary batteries may be as good or better in basic energy storage density than either AMTEC or fuel cell systems. For such cases, however, the rapid recharge capability of AMTEC or fuel cells can still provide a substantial advantage over batteries since the 'state of charge' is much easier to determine for a fueled system. As noted above, the problem of toxic waste disposal common for some primary batteries also argues against their common use.

While the system concept described here uses propane or butane, for field use it is expected that AMTEC systems can be designed to operate with logistics fuels and that this can provide a further operational advantage over fuel cells relying either on stored hydrogen or on reforming hydrocarbon fuels.

If AMTEC systems are to be considered as battery replacements for field use, it is important to compare their characteristics relative to the current and proposed alternatives. Table 3 shows the power and mass aspects of three alternatives for providing 20 watt service for one day. The complete 20 W AMTEC converter has a mass of 2.7 kg. Note that while the LANL fuel cell data indicates a very low mass for the converter itself, the fuel mass for the one day, with its hydride storage medium or pressure tank, dominates the total. For missions beyond 1 or 2 days, it is the added mass required to store the necessary hydrogen that pushes the combined fuel and converter mass beyond that of AMTEC systems.

For AMTEC systems, there is a trade off between efficiency and mass since optimizing burner efficiency generally requires incorporating a recuperation system which adds mass while it enhances overall system efficiency. The mass cited here for the 20 W_e AMTEC system includes a high effectiveness recuperator. The support equipment needed to produce hydrogen for recharging the portable fuel cell storage containers and the powered generator needed for recharging batteries, raise additional logistics issues for each of those systems.

TABLE 3. Comparison of Mass and Support Needs for Several Portable Electric Power Supply Systems.

Approach	Power Level	Energy Storage	Fuel Mass (for 24 hrs)	Converter Mass	Support Equipment
AMTEC (15%)	20 W	500 Whe	0.28 kg	2.7 kg	None
Fuel Cell	20 W	500 Whe	2.2 kg	0.5 kg	Electric Generator (to power H_2 generator), H_2 generator (from diesel fuel or water)
Conventional Pb-acid Battery	15 W	450 Whe	20 kg	0	Generator, PMAD
NiMH$_x$ Battery	20 W	480 Whe	8 kg	0	Generator, PMAD
Advanced Battery	20 W	480 Whe	2.8 kg	0	Generator, PMAD

Energy Density of Stored Fuel

It is clear from the foregoing that the utility of fueled energy converters as substitutes for conventional batteries for longer operation applications, hinges on the energy density of their respective fuel supplies. Consider storage of hydrogen in metal hydrides for use in fuel cells and liquid hydrocarbon storage for use in AMTEC systems. Energy Conversion Devices (ECD)(Sapru, K. et al 1999) has published data asserting that for their metal hydride hydrogen storage systems, 15 gms of H_2 requires 0.25 L of hydride (2% H_2 by weight) to 0.55 L of hydride (1% H_2 by weight) and provides 250 Wh$_e$ through operation in a 50% conversion efficiency PEM fuel cell. This implies that 15 gms of hydride-stored H_2 (at 2%) occupies ~ 750 gms of material + gas. Liquid hydrocarbons store ~ 11.7 kW$_{th}$/kg at a density of ~ 0.82 gm/cm^3 (diesel). Since both the hydride storage and liquid fuels are essentially at atmospheric pressure and ambient temperatures, the container masses are expected to be similar (for a given volume) and negligible for long operation in each case. Table 4 indicates the approximate relative system energy densities for 50% efficient fuel cells with hydride storage and AMTEC systems at several efficiencies with liquid fuel storage.

The numbers in the last two columns do not reflect the mass of the converter and so represent the energy density to be expected for the situation where the fuel supply is large and the operation duration is long so that the relative contribution of the converter mass is small.

TABLE 4. Fuel Energy Density Comparison (no allowance for converter mass)

Converter/Fuel	Fuel Energy Density (kWh$_{th}$/kg)	Fuel Energy Density (kWh$_{th}$/L)	Efficiency x Energy Density (KWh$_e$/kg)	Efficiency x Energy Density (KWh$_e$/L)
AMTEC (15%) - Liquid Hydrocarbon	11.7	9.6	1.76	1.44
AMTEC (10%) - Liquid Hydrocarbon	11.7	9.6	1.17	0.96
AMTEC (5%) - Liquid Hydrocarbon	11.7	9.6	0.58	0.48
Fuel Cell (50%) – ECD Hydride (2%)	0.67	2	0.34	1.0
Fuel Cell (50%) – ECD Hydride (1%)	0.33	1.0	0.165	0.50
Advanced Battery (Li-polymer, Na/S)	0.2	0.24	0.2	0.24

SUMMARY

For missions extending beyond 1 day, AMTEC systems at current efficiencies (15% - 20%) can have substantially lower total masses (converter + fuel) than either hydride storage based PEM fuel cells (operating at 50% efficiency) or next generation advanced batteries. For a 7 day mission at 20 watts, the mass for an AMTEC system is less than a half to a third that for a PEM fuel cell system described recently.

ACKNOWLEDGMENTS

The authors wish to thank Jerry Martin of Mesoscopic Devices, Inc*., for helpful discussions related to appropriate burner technology for low power AMTEC devices and its requirements and limitations.

REFERENCES

*   Mesoscopic Devices, Inc., 3400 Industrial Lane, Suite 7B, Broomfield, CO 80020.

Cole, T., "Thermoelectric Energy Conversion with Solid Electrolytes," *Science*, 221, 915 (1983).
Daugherty, M. Haberman, D., Salter, C., Lokken, O., Ibrahim, S., Chernieack, M, Wilson, M., and Stetson, N.T., "Simple, Rugged and Reliable fuel Cells for Use in Portable Applications", *Proceedings of the Conference on Commercialization of Small Fuel Cells & the Latest Battery Technologies for Portable Applications,* 29 – 30 April 1999, Bethesda, MD.
Sapru, K, Stetson, N.T., Ramachandran, S., Evens, J., Lu, M., "Metal Hydride Storage for Portable Power," *Proceedings of the Conference on Commercialization of Small Fuel Cells & the Latest Battery Technologies for Portable Applications,* (1999).
Weber, N., "A Thermoelectric Device Based on Beta-Alumina Solid Electrolyte", Energy Conversion *14*, 1-8 (1974).

Conical Evaporator and Liquid-Return Wick Model for Vapor Anode, Multi-Tube AMTEC Cells

Jean-Michel Tournier and Mohamed S. El-Genk

Institute for Space and Nuclear Power Studies, Department of Chemical and Nuclear Engineering
School of Engineering, The University of New Mexico, Albuquerque, NM 87131
(505) 277 – 5442, FAX: – 2814, email: mgenk@unm.edu

Abstract. A detailed, 2-D thermal-hydraulic model for conical and flat evaporators and the liquid sodium return artery in PX-type AMTEC cells was developed, which predicts incipient dryout at the evaporator wick surface. Results obtained at fixed hot and cold side temperatures showed that the flat evaporator provided a slightly lower vapor pressure, but reached the capillary limit at higher temperature. The loss of performance due to partial recondensation over up to 20% of the wick surface of the deep conical evaporators was offset by the larger surface area available for evaporation, providing a slightly higher vapor pressure. Model results matched the PX-3A cell's experimental data of electrical power output, but the predicted temperature of the cell's conical evaporator was consistently ~ 50 K above measurements. A preliminary analysis indicated that sodium vapor leakage in the cell (through microcracks in the BASE tubes' walls or brazes) may explain the difference between predicted and measured evaporator temperatures in PX-3A.

INTRODUCTION

Several vapor anode, multi-tube Alkali-Metal Thermal-to-Electric Conversion (AMTEC) cells of Pluto-Express (PX) type have been tested in vacuum at well controlled conditions at the Space Vehicles Directorate of the Air Force Research Laboratory (AFRL) (Merrill et al., 1998; El-Genk and Tournier, 1998a). These cells consist of 5-7 Beta''-Alumina Solid Electrolyte (BASE) tubes arranged in a circle inside the cell, surrounding the central liquid sodium return artery and evaporator structure (Fig. 1). The permeability and the average pore size of the artery are larger than those of the cell evaporator to minimize the pressure losses. The average pore radius in the evaporator wick is quite small (< 5 μm) to provide sufficient capillary pressure head to circulate the sodium working fluid in the cell and balance the pressure losses (El-Genk and Tournier, 1998b). These include the liquid sodium pressure losses in the return artery and cell condenser and the sodium vapor pressure losses in the high and low pressure cavities, the BASE tube membrane, in the anode and cathode electrodes, and at the vapor/liquid interface in the condenser. Therefore, the cell evaporator maintains the sodium vapor flow through the BASE tubes commensurate with the external load demand (or electric current). In addition to supplying sodium vapor, the cell evaporator temperature should be maintained al least 20 K below that at the cold end of the BASE tube. This requirement ensures no or little sodium vapor condensation inside the BASE tubes, to avoid shorting the cell. The BASE tubes in the vapor anode, PX-series cells are connected electrically in series to increase the output voltage of the cell.

The heat consumed in the liquid sodium evaporation in the cell evaporator is supplied by conduction from the BASE support plate (Fig. 1). A mismatch between the rate of the liquid sodium flow from the cell condenser and the rate of sodium evaporation could cause local dryout of the evaporator wick. Such dryout could be initiated at high load current, by a vapor leakage in the ceramic-metal brazes between the BASE tubes and the metal support plate, or by a partial blockage of the liquid sodium return artery or the condenser porous structure (Fig. 1). On the other hand, a lower evaporation rate, due to insufficient evaporator area or heat supply from the cell's hot plate, would decrease the cell current, electric power output, and conversion efficiency. Therefore, the shape and the pore size of the evaporator wick as well as its separation distance from the BASE tubes' support plate should be selected based on rigorous parametric and optimization analysis. To the best of the authors' knowledge such an analysis has never been done. To ensure sufficient surface area for evaporation in the PX-cells, deep conical evaporators were used in earlier PX-cell designs.

CP504, *Space Technology and Applications International Forum–2000*, edited by M. S. El-Genk
2000 American Institute of Physics 1-56396-919-X

In the PX-type cells (such as PX-3A) with deep conical evaporators (small apex angle), the effective separation distance between the evaporator and the BASE tubes' support plate increased, increasing the conduction resistance to the heat flow from the cell's hot side to the evaporator surface. In an attempt to increase the heat flow to the cell evaporator, SS or nickel rings were stacked along the evaporator standoff structure (Merrill et al., 1998; El-Genk and Tournier, 1998a). This arrangement, however, increased the effective evaporator temperature, T_{ev}, making it difficult to maintain a positive temperature margin in the cell at low cell electric current. To minimize the impact of these conflicting requirements on the cell operation, subsequent cell designs have used either shallow conical (large apex angle) or flat evaporators, and reduced the separation distance between the evaporator and the hot support plate. Test results of these cells showed performance improvement and no evidence that the cell operation was vapor flow limited. However, the effect of the evaporator shape on the cell performance has never been quantified, requiring a detailed two-dimensional thermal-hydraulic model for optimization analysis.

As part of the AFRL test and evaluation program, the University of New Mexico's Institute for Space and Nuclear Power Studies has developed a comprehensive AMTEC Performance and Evaluation Analysis Model (APEAM). This cell model has been validated using the test results of the PX-series cells. APEAM includes detailed models of the sodium vapor pressure losses in the low-pressure cavity of the cell (Tournier and El-Genk, 1999a); a radiation/conduction heat flow model (Tournier and El-Genk, 1999b); an electrochemical model of the BASE tubes and the electrodes (Tournier and El-Genk, 1999c); an electric circuit model of the entire cell (Tournier and El-Genk, 1999c); and a simplified lumped model of the cell evaporator. APEAM has accurately predicted the performance of PX-cells with either shallow cone or flat evaporators (Tournier and El-Genk, 1999d). However, the simple evaporator model in APEAM could neither simulate the operation of the deep conical evaporator in earlier cell designs and the detailed vaporization processes along the evaporator surface, nor predict incipient dryout in the evaporator wick.

The objectives of this paper are to develop a detailed, two-dimensional thermal-hydraulic model of the evaporator and liquid return artery in the PX-series cells, and investigate the effect of the apex angle of the conical evaporator on the cell operation. This model is capable of not only predicting the temperature field in the evaporator wick and the local temperature and evaporation/condensation rate at the evaporator surface, but also the capillary limit (or incipient dryout) in the wick. This paper briefly describes the present conical evaporator and liquid-sodium return artery model and the numerical approach used to solve the highly non-linear governing equations. Analyses are performed to investigate the effect of the evaporator cone angle on the performance of the PX-3A cell. The calculated evaporator temperature is compared with measured

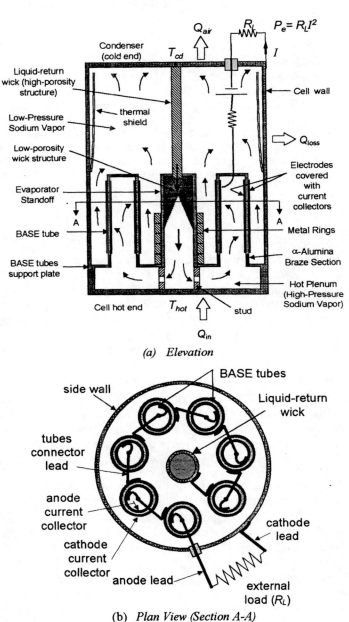

(a) Elevation

(b) Plan View (Section A-A)

FIGURE 1. Cross-section views of PX-series, vapor anode, multi-tube AMTEC cell (not to scale).

values in PX-3A cell. In addition, a preliminary analysis is performed to investigate the effect of sodium vapor leakage in the BASE membranes or brazes on the evaporator temperature and the cell's performance.

VAPOR ANODE, MULTI-TUBE, PX-SERIES AMTEC CELLS

Cross-section views of a PX-type AMTEC cell are shown in Fig. 1. The BASE tubes and the housing of the evaporator assembly (or standoff), which are brazed to a stainless steel (SS) support plate, divide the cell into two separate cavities. The low-pressure (cathode) and high-pressure (anode) sides of the BASE tubes are covered with titanium nitride (TiN) metal electrodes, which also provide a conduction path for the electrons to the molybdenum mesh current collectors. The BASE tubes are electrically connected in series to increase the terminal voltage, hence decreasing the electrical current and the internal losses in the cell.

The heat from the cell's hot plate is transported by conduction and radiation to the BASE tubes and the evaporator structure. The circumferential radiation shield, laid on the inside of the cell wall above the BASE tubes, reduces the parasitic heat losses to the wall, increasing the cell conversion efficiency. Some of the PX-cells have used a metal stud to enhance the heat conduction from the cell's hot plate to the evaporator and the BASE tubes' support plate. The conical evaporator provides a large surface area for evaporating the liquid sodium returning from the condenser (Fig. 1a). PX cells have used a Creare type condenser wick to maintain a continuous thin film of liquid sodium on its surface, effectively reflecting thermal radiation toward the interior, and reducing the parasitic heat losses. The condenser porous structure is hydrodynamically coupled to that in the liquid sodium return artery. The liquid return artery extends from the cell condenser to the evaporator wick and is made of a relatively large pore size, high permeability, metal wick structure. The evaporator wick, however, is made of a very small pore size structure, to provide high capillary head for circulating the sodium working fluid in the cell.

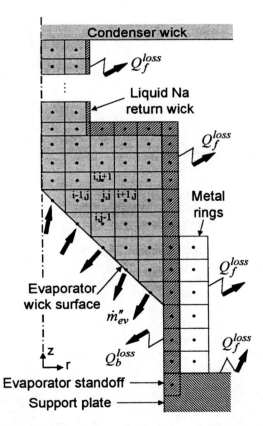

FIGURE 2. A layout of the computation grid in the conical evaporator and the liquid sodium return artery of PX-Type cell.

CONICAL EVAPORATOR AND RETURN ARTERY MODEL

A transient, two-dimensional thermal-hydraulic model of the conical evaporator and the liquid sodium return artery in PX-series cells is developed. It accounts for the liquid flow and the heat transfer by conduction and convection in the porous wicks of the evaporator and liquid return artery, and for the effect of gravity. It also accounts for the radial and axial heat conduction in the evaporator standoff and the conduction rings. The model calculates the rates of sodium vaporization/condensation along the evaporator surface and incorporates the radiation heat transfer between the evaporator and liquid return artery structures and other components in the PX-cell.

Previous analyses have shown that radiation heat transfer between the evaporator structure and other components in the cell was negligible, compared to the heat transfer by conduction for sodium evaporation (Tournier and El-Genk, 1999b). Therefore, the effect of radiation heat exchange was considered only in the upper, annular cavity of the cell, above the BASE tubes (Fig. 1). The present model assumes that both the evaporator and the return artery are fully saturated with liquid sodium and predicts incipient dryout at the surface of the evaporator wick, based on the capillary Pascal relationship. Details on the governing equations and boundary conditions are documented elsewhere (Tournier and El-Genk, 1995 and 1999e).

Governing Equations and Method of Solution

The well-known volume-averaged homogeneous enthalpy method is used to calculate the temperature field and the heat transfer in the porous evaporator wick. A two-dimensional (staggered) cylindrical grid (Fig. 2) represents the structures of the conical evaporator and the liquid sodium return artery. The liquid densities and the local pressures, temperatures, and enthalpies are calculated at the centers of the control volumes, while the mass fluxes are calculated at the interfaces between the control volumes. The enthalpies of the liquid sodium and the solid wick are linearized in terms of temperature, using their respective specific heat capacities. The effective wick thermal conductivities in the radial and axial directions are calculated as functions of the local volume porosity and the conductivities of the liquid sodium and solid wick structure (Tournier and El-Genk, 1995; Chi, 1976).

The present model accounts for the various modes of heat transfer occurring at the boundaries of the control volumes by radiation heat exchange and liquid evaporation. The radiant energy loss terms are linearized as function of the local temperature, using the procedure described in a previous paper (Tournier and El-Genk, 1999b). The local evaporation/condensation rates of sodium at the evaporator wick surface are predicted using the kinetic theory of gases (Tournier and El-Genk, 1996):

$$(\dot{m}_{ev}'')_{i,j} = a_{cc}\left(\frac{M}{2\pi R_g T_{i,j}}\right)^{1/2}\left[P_{sat}(T_{i,j}) - P_a\right] = \chi_{i,j}\left[P_{sat}(T_{i,j}) - P_a\right] \quad , \tag{1}$$

where $T_{i,j}$ is the wick surface node temperature and P_a is the vapor pressure in the evaporator (anode) cavity. The accommodation coefficient, a_{cc} for liquid metals is close to unity in well evacuated, clean systems, which is assumed to be the case in the PX-cells (Tournier and El-Genk, 1996). For computational efficiency, the heat of vaporization appearing in the energy balance equation is also linearized in terms of temperature as:

$$(\dot{m}_{ev}'' h_{LV})_{i,j}^{n+1} = (\dot{m}_{ev}'' h_{LV})_{i,j}^* + (\dot{m}_{ev}'')_{i,j}^*\left(\frac{\partial h_{LV}}{\partial T}\right)_{i,j}^* T_{i,j}' + (h_{LV})_{i,j}^* \times \sum_{\substack{cone \\ surface}} \frac{\partial(\dot{m}_{ev}'')_{i,j}^*}{\partial T_{k,l}} \times T_{k,l}' \quad . \tag{2}$$

At steady state, the total vaporization rate in the conical evaporator is related to the cell's electric current obtained from the electrical circuit model in APEAM, as:

$$\sum_{\substack{cone \\ surface}} (A_{ev}\dot{m}_{ev}'')_{i,j} = \sum_{\substack{cone \\ surface}} A_{ev}^{i,j}\chi_{i,j}\left[P_{sat}(T_{i,j}) - P_a\right] \equiv \dot{m} = N_B\frac{MI}{F} \quad . \tag{3}$$

When rearranged, this equation gives the sodium vapor pressure in the evaporator cavity, P_a as:

$$P_a = \left[\sum_{\substack{cone \\ surface}} A_{ev}^{i,j}\chi_{i,j}P_{sat}(T_{i,j}) - N_B\frac{MI}{F}\right] \times \left[\sum_{\substack{cone \\ surface}} A_{ev}^{i,j}\chi_{i,j}\right]^{-1} \quad . \tag{4}$$

As this equation indicates, P_a is a very complex function of the cell current, I, and the local temperatures, $T_{i,j}$ along the conical evaporator surface. The partial derivatives of P_a with respect to these temperatures, obtained from Equation (4), are used in Equation (2) to express the partial derivatives of the vaporization rates with temperatures. This linearization is a key to obtaining a fast-converging solution.

The phase-change at the evaporator wick surface and the radiation heat transfer boundary conditions caused the governing equations to be highly non-linear. As a result, the discretized energy balance equations extended beyond the classical penta-diagonal linear system, and were solved using a banded, Gauss-elimination algorithm, with row normalization and partial pivoting (Tournier and El-Genk, 1999b).

The liquid sodium flow in the fully saturated isotropic porous wick is modeled using the Forchheimer-extended Darcy's continuity and momentum equations (Tournier and El-Genk, 1995), which include the effect of gravity. Because of the very small sodium flow rates in PX-series AMTEC cells (< 25 g/hr), the Forchheimer extension could be neglected in the momentum equations. The wick permeability is calculated as a function of the volume porosity and the effective pore size (Tournier and El-Genk, 1995; Chi, 1976).

To predict the onset of dryout at the evaporator wick surface, the vapor and liquid in the porous wick are hydrodynamically (also thermally) coupled at the liquid-vapor interface, using the radial momentum jump condition, or the extended Pascal relationship (Tournier and El-Genk, 1995). The maximum capillary pressure head occurs when the local radius of curvature of the liquid meniscus at the liquid-vapor interface equals the wick pore radius. The vapor void fraction in the wick interfacial pores then equals unity, a value that defines the onset of dryout in the evaporator wick. The efficient iterative solution procedure developed to solve the governing and constituent equations has been described elsewhere (Tournier and El-Genk, 1999b and 1999e). Calculation results are presented and discussed next.

RESULTS AND DISCUSSION

The present model was used to predict the performance of the conical evaporator in the PX-3A cell. This stainless-steel cell is 31.75 mm in diameter and 101.6-mm high, and has five, 32 mm-long BASE tubes. These tubes have 25.4 mm-long TiN electrodes covered with Mo mesh current collectors. The cell has a SS thermal radiation shield and a SS conduction stud having a 38 mm^2 cross-section area (El-Genk and Tournier, 1998a). The cell's evaporator wick is 4.9 mm in diameter, 21.5 mm-high, and encased in a 0.71 mm-thick SS tube, surrounded by 3 nickel rings (1.1 mm-thick). The deep evaporator cone had a half angle $\alpha \sim 8°$. The base of the conical evaporator wick is situated 5.2 mm above the BASE tubes' support plate. The felt-metal liquid return artery is 3.2 mm in diameter and ~ 60 mm-long. In the experiment, the cell was tested in the vertical position, with the condenser in the downward position, so that the cell's artery and evaporator wicks were pumping liquid sodium against gravity. The permeabilities of the artery (30% porous) and of the evaporator wick (40% porous) were assumed equal to 5.6×10^{-13} m^2 and 4.5×10^{-13} m^2, respectively (Hunt et al., 1993). The effective pore radius of the evaporator wick is 4 μm.

Effect of Cone Angle on Thermal Performance of Evaporator

The present model was used to investigate the effect of the cone angle on the thermal performance of the evaporator. The length of the evaporator standoff (5.2 mm) was maintained constant in all the cases analyzed. Figure 3 shows the predicted wick surface temperature, vaporization mass flux, and vapor pressure along the evaporator surface. The calculations were performed at fixed hot and cold side temperatures and a cell current of 2.5 A (or a Na mass flow rate = 10.73 g/hr). The wick surface temperature was essentially uniform in all cases, within ± 1 K (Fig. 3a). The vaporization, however, occurred mostly in the outer most region, near the standoff wall (Fig. 3b). The effective vapor pressure of the flat

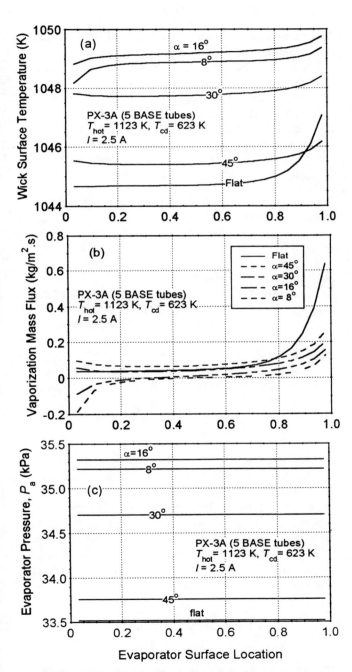

FIGURE 3. Predicted effect of cone angle on evaporator performance in PX-3A cell, at T_{hot} = 1123 K and I = 2.5 A.

evaporator, 33.5 kPa, was the lowest of all cases studied (Fig. 3c). By contrast, the shallow cone evaporators ($\alpha = 45°$, $30°$) exhibited the most uniform wick surface temperature and vaporization mass flux. Their shape provided an optimum conduction path to all points on the surface of the wick. The $30°$ shallow conical evaporator wick, which had twice the surface area of the flat evaporator (Table 1), provided a vapor pressure 1.2 kPa higher that the later.

The deep cone evaporators ($\alpha = 8°$ and $16°$), which had the highest surface area (7.2 and 3.6 times that of the flat evaporator, respectively), exhibited the best thermal performance (Figs. 3a and 3c). However, the long conduction path to the apex of the conical evaporator wick resulted in a cooler surface temperature, causing recondensation of Na vapor (Fig. 3b), as commonly observed in heat pipes. As a result, the performance of a deep conical evaporator does not increase proportionally with its surface area.

Figure 4 shows the effects of the cone angle of the evaporator, the cell current, and the hot side temperature on the location of zero evaporation on the evaporator surface. At the currents commonly encountered in PX-series cells ($I > 1.5$ A), the flat and shallow conical evaporators do not experience recondensation (or *heat pipe effect*). The deep conical evaporators, however, are less effective than the flat of shallow conical evaporators, due to the recondensation occurring in the former at all cell currents. A $16°$ conical evaporator typically exhibits re-condensation over 10% of its surface, while a $8°$ conical evaporator will experience recondensation over 20% of its surface (Fig. 4). Note that this *heat pipe effect* is not strongly affected by the operating temperature of the cell.

The loss in total evaporation rate due to the partial recondensation of sodium vapor in the deep conical evaporators is largely offset by the large surface area available for evaporation, compared to a flat evaporator (a factor 7.2 for $\alpha = 8°$). The deep cone evaporators provide a sodium vapor pressure that is between 1.7 and 1.8 kPa higher than a flat evaporator (Fig. 3c). The performance of the $8°$ and $16°$ conical evaporators is very similar, with that of the later slightly better at cell currents below ~ 3A (Fig. 3c).

Effect of Cone Angle on Capillary Limit of Evaporator

The shape of the evaporator wick also affects its ability to circulate the liquid sodium back from the condenser (or capillary limit). As expected, the flat evaporator

TABLE 1. Evaporator surface area in PX-3A, as a function of half cone angle, α.

α	8°	16°	30°	45°	flat
A_{ev} (mm^2)	135.5	68.4	37.71	26.67	18.86
Area ratio	7.2	3.6	2.0	1.414	1

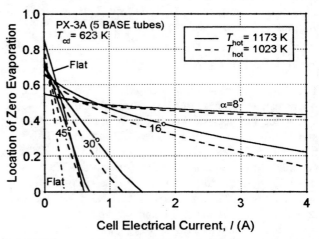

FIGURE 4. Location of zero evaporation point as a function of cone angle, cell current and hot side temperature.

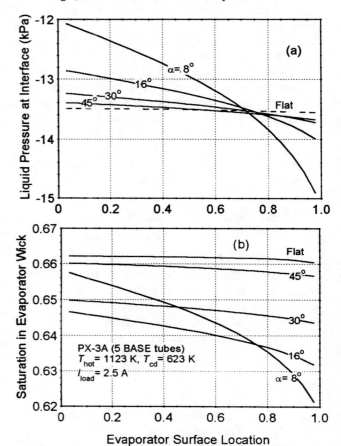

FIGURE 5. Effect of cone angle on liquid pressure and saturation in evaporator wick ($T_{hot} = 1123$ K, $I = 2.5$ A).

exhibits the lowest liquid pressure losses of all evaporators (Fig. 5a). The deeper is the cone (the lower the cone angle), the lower is the liquid pressure at the tip of the evaporator wick near the standoff (Fig. 1). As shown in Fig 5a, the sodium liquid in the evaporator wick is in tension, reaching its highest value next to the standoff wall. A deep, 8° conical evaporator exhibits a 1.4 kPa lower liquid pressure than a flat evaporator. Figure 5b shows the calculated liquid volume fraction in the pores at the evaporator wick surface. Clearly the conical evaporators would exhibit incipient dryout at a lower operating temperature than a flat evaporator. Note that the model results showed that, when operating the cell at a hot side temperature of 1173 K, dryout occurred in all evaporator wicks at all cell currents.

Comparison with Experimental Data

The two-dimensional evaporator model was coupled to APEAM to predict the performance of the PX-3A cell, which had a deep conical evaporator ($\alpha = 8°$). The model assumed a charge-exchange current $B = 80$ $A.K^{1/2}/Pa.m^2$, a contact resistance $R_{cont} = 0.09$ $\Omega.cm^2$, and zero electrical or sodium vapor leakage. Experimental data were generated at AFRL at a fixed cold side temperature ($T_{cd} = 623$ K) and hot side temperatures of $T_{hot} = 1023$ K, 1123 K and 1173 K (Merrill et al., 1998).

The predicted evaporator temperature and the cell's electrical power are shown in Figs. 6a and 6b, as functions of cell current and hot side temperature, along with the measured values. The predicted electric power is in good agreement with the measurements, except perhaps at the highest hot side temperature (1173 K) (Fig. 6b). The model predicted that dryout occurred in the evaporator wick, thus it is possible that the cell operation was capillary limited. The predicted evaporator temperature, however, is ~ 50 K higher than the measured values (Fig. 6a). Assuming that the measuring thermocouple was actually in contact with the evaporator wick surface, the measurements' uncertainty could be ± 15 K (Huang and El-Genk, 1998). The slope of the calculated T_{ev}-I curves, nonetheless, was almost identical with that of the measurements.

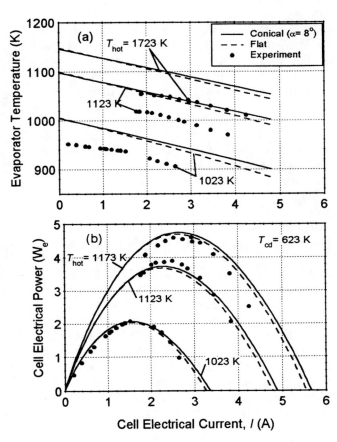

FIGURE 6. Predicted and measured performance of PX-3A cell, using 8° conical and flat evaporators.

As indicated by the dashed lines in Figs. 6a and 6b, changing the evaporator shape from a deep cone to a flat surface resulted in only a 1.5% decrease in the peak electrical power (and conversion efficiency). This is because, although the cell's voltage is a logarithmic function of the evaporator vapor pressure, PX-series cells with TiN electrodes exhibited large electrical internal losses (Tournier and El-Genk, 1999c).

Effect of Sodium Vapor Leakage on Performance of PX-Series Cells

A process that may explain the relatively large difference between the predicted and measured evaporator temperatures in PX-3A is sodium vapor leakage through the BASE tubes' membranes or brazes. In this case, the cell evaporator would be actually circulating a sodium mass flow rate larger than that diffusing through the BASE membranes (which is directly proportional to the cell's electric current). As shown in Fig. 6a, the larger the cell current (or the evaporator mass flow rate), the lower is the evaporator surface temperature. At fixed T_{hot}, the evaporator temperature decreases almost linearly with the generated vapor mass flow rate (or cell current in the absence of leakage). This is because the energy consumed in the vaporization of the liquid Na is proportional to the sodium flow rate, and must be conducted up the standoff structure and the metal rings to the evaporator wick (Fig. 1a).

The evaporator flow rate would be higher than the diffusion rate through the BASE in two possible situations: (a) When the temperature margin in the cell is negative, causing condensation of sodium inside the BASE tubes; (b) When there is a vapor leakage from the cell's hot plenum to the low-pressure cavity. A preliminary analysis of the later is performed next.

The present model assumed a vapor leakage path through a number of circumferential cracks in the ceramic BASE tubes and/ or the ceramic-metal brazes. The BASE tubes were 0.508 mm-thick and had a 7.6 mm outer diameter. The pressure drop through these diametral cracks was calculated using the Dusty-Gas-Model (Tournier and El-Genk, 1996 and 1999a), which accurately simulates all vapor flow regimes (continuum, transition and free-molecular). The model calculated the number of cracks to bring the evaporator temperature down by a fixed ΔT, at T_{hot} = 1123 K and the electric current corresponding to the measured peak power in PX-3A (2.5 A). As shown in Fig. 7, one 7 μm-wide crack, or ten 3 μm-wide cracks per BASE tube would allow a vapor leakage flow rate of 4.4 g/hr, lowering the calculated evaporator temperature by 20 K. This vapor leakage rate is equivalent to an increase in the cell current of 1.03 A, thus the cell evaporator would have been circulating Na at a flow rate of 15.1 g/hr ($I_{leak}+I$ = 1.03+2.5 = 3.53 A).

The circulation of excess sodium, which does not expand through the BASE membrane, negatively impacts the cell's conversion efficiency. The later was estimated at the measured electrical power output of the cell in Fig. 6b, by correcting the heat input predicted by APEAM for the additional heat required to vaporize the excess leaking sodium, as:

$$Q_{in} = Q_{in}^{APEAM} + \dot{m}_{leak}\left[h_{LV}(T_{ev}) + C_p^L(T_{ev} - T_{cd})\right] . \quad (5)$$

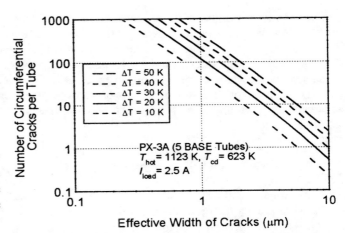

FIGURE 7. Number of circumferential cracks per tube to bring the PX-3A evaporator temperature down by ΔT.

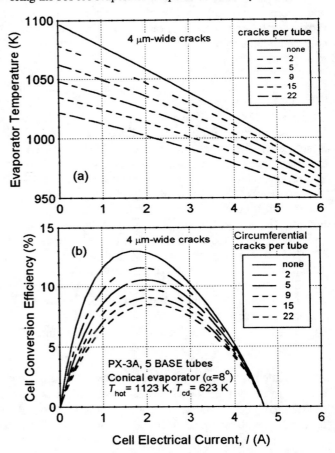

FIGURE 8. Effect of sodium vapor leakage on the conversion efficiency of PX-3A cell (T_{hot} = 1123 K).

Results are presented in Fig. 8 for T_{hot} = 1123 K and 4 μm-wide circumferential cracks. The leakage of sodium vapor affects the evaporator temperature more at low than at high cell current (Fig. 8a). This is because the higher the cell current, the lower the evaporator vapor pressure is and the lower the leakage flow rate. The PX-3A cell had a predicted peak efficiency of 13% at T_{hot} = 1123 K, at a cell current I = 1.75 A, when assuming no Na vapor leakage. However, assuming the existence of a total of ten, 4 μm-wide cracks in the cell's BASE tubes, the cell efficiency could drop by 1.5 points (to 11.5%, with a corresponding sodium leakage rate of 2.5 g/hr). More dramatic drops in the cell's efficiency could result at higher leakage rates.

SUMMARY AND CONCLUSIONS

A detailed, two-dimensional thermal-hydraulic model for conical and flat evaporators and the liquid sodium return artery in the PX-type, AMTEC cells was developed. The model accounts for liquid flow and heat transfer in the saturated porous wicks, sodium vaporization and condensation along the evaporator wick surface; and thermal radiation exchange between the liquid return artery and the other cell components. It predicts the capillary limit (or incipient dryout) at the evaporator wick surface based on the extended Pascal relationship, and uses an advanced methodology for linearizing the governing equations and boundary conditions. The highly nonlinear nature of the governing equations and the boundary conditions is due to the change-of-phase at the evaporator surface, the thermal radiation exchange with the surrounding structure, and to a lesser extent the temperature-dependent thermophysical properties of the working fluid.

Results showed that the wick surface temperature was essentially uniform within $\pm$ 2 K for both flat and conical evaporators. The flat evaporator provided the lowest vapor pressure and lowest liquid pressure losses. The deep cone evaporators ($\alpha = 16°$ and $8°$) with very high surface area (3.6 and 7.2 times that of the flat evaporator, respectively) exhibited good thermal performance, despite the partial recondensation of Na vapor over 10 to 20% of the evaporator surface, near the apex of the conical evaporators. The loss of performance due to Na condensation was largely offset by the large surface area available for evaporation. However, the deep conical evaporators would reach the capillary limit (or incipient dryout) at a lower operating temperature than a flat evaporator.

The present evaporator model was coupled to APEAM to predict the performance of the PX-3A cell, which had a deep conical evaporator ($\alpha = 8°$). The predicted electric power as a function of the cell current was in good agreement with reported measurements. Changing the evaporator shape from a deep cone to a flat surface would have resulted in a ~ 1.5% decrease in the peak electrical power of the cell. The predicted vapor temperature in the PX-3A cell at different hot side temperatures and a condenser temperature of 623 K, however, was consistently ~ 50 K higher than the measured values. The difference between measured and predicted values (considering a measurement uncertainty of $\pm$ 15 K) could have been caused by a Na vapor leakage through micro-cracks in the BASE tube walls or the BASE brazes, or by condensation at the cold end of the BASE tubes. A preliminary analysis of the former suggested that one 7-μm or ten 3 μm circumferential cracks per BASE tube would have caused a vapor leakage flow rate of 4.4 g/hr, reducing the calculated evaporator temperature by 20 K. Such a leakage of sodium vapor negatively impacts the cell efficiency. The PX-3A cell had a predicted peak efficiency of 13% at T_{hot} = 1123 K, at I = 1.75 A, assuming no leakage of Na vapor. However, the presence of two circumferential cracks on average (4 μm wide) per BASE tube would decrease the cell efficiency by 1.5 points (to 11.5%, where the leakage rate is 2.5 g/hr). Future work will further investigate the effect of vapor leakage on the cell performance, and will focus on predicting incipient dryout and the liquid recession in the evaporator wick of PX-type cells.

NOMENCLATURE

English

A	Surface area (m^2)
B	Temperature-independent charge-exchange current (A.K$^{1/2}$/Pa.m^2)
F	Faraday's constant (F = 96,485. C / mole)
I	Cell electrical current (A)
h_{LV}	Sodium latent heat of vaporization (J / kg)
M	Sodium molecular weight (M = 23 g/mole)
$\dot{m}_{ev}''$	Sodium vaporization mass flux (kg / m^2.s)

$\dot{m}_{leak}$	Sodium vapor leakage flow rate (kg / s)
N_B	Number of (series-connected) BASE tubes
P_a	Anode sodium vapor pressure (Pa)
P_{sat}	Sodium saturation pressure (Pa)
R_{cont}	Contact resistance between current collector and electrode (Ω.cm^2)
R_g	Perfect gas constant (R_g = 8.314 J / mol.K)
T	Temperature (K)
T'	Temperature correction vector (K)

Greek

α Half apex angle of conical wick surface

Subscript / Superscript

cd   Condenser

ev   Evaporator wick surface
hot   Cell's hot end
n   Old time step
n+1   New time step
*   Best estimate of new-time variable available at the time of computation.

ACKNOWLEDGMENTS

This research is internally funded by the University of New Mexico's Institute for Space and Nuclear Power Studies.

REFERENCES

Chi, S. W., *Heat Pipe Theory and Practice*, Hemisphere Publishing Co., Washington, D.C., 1976, **2**, 47-51.

El-Genk, M. S., and Tournier, J.-M., "Recent Advances in Vapor Anode, Multi-Tube, Alkali-Metal Thermal-to-Electric Conversion Cells for Space Power," in *Proceedings of the 5th European space Power Conference (ESPS-98)*, held 21-25 September 1998, in Tarragona, Spain, SP-416, European Space Agency Publications Division, 1998a, Paper No. 98-1046, pp. 257-264.

El-Genk, M. S., and Tournier, J.-M., "Optimization of Liquid-Return Artery in a Multi-Tube AMTEC," in *Proc. of the Space Technology and Applications International Forum (STAIF-98)*, CONF-980103, M. S. El-Genk, Ed., American Institute of Physics, New York, NY, 1998b, AIP Conf. Proceedings No. 420, **3**, 1586-1593.

Huang, L., and El-Genk, M. S., "Experimental Uncertainties in Vacuum Tests of PX-Series AMTEC Cells," in *Proceedings of the Space Technology and Applications International Forum (STAIF-98)*, CONF-980103, M. S. El-Genk, Ed., American Institute of Physics, New York, NY, 1998, AIP Conf. Proc. No. 420, **3**, 1471-1478.

Hunt, T. K., Sievers, R. K., Ivanenok, J. F., Pantolin, J. E., and Butkiewicz, D. A., "Capillary Pumped AMTEC Module Performance," in *Proceedings of the 28th Intersociety Energy Conversion Engineering Conference*, American Chemical Society, 1993, Paper No. 93307, 1: 849–854.

Merrill, J., Schuller, M. J., and Huang, L. "Vacuum Testing of High-Efficiency Multitube AMTEC Cells: February 1997-October 1997," in *Proceedings of the Space Technology and Applications International Forum (STAIF-98)*, CONF-980103, M. S. El-Genk, Ed., American Institute of Physics, New York, NY, 1998, AIP Conference Proceedings No. 420, 3: 1613-1620.

Tournier, J.-M., and El-Genk, M. S., "Transient Analysis of the Startup of a Water Heat Pipe from a Frozen State," *Numerical Heat Transfer, Part A* **28**, 461-486 (1995).

Tournier, J.-M., and El-Genk, M. S., "A Vapor Flow Model for Analysis of Liquid-Metal Heat Pipe Startup from the Frozen State," *International Journal of Heat and Mass Transfer* **39**(18), 3767-3780 (1996).

Tournier, J.-M., and El-Genk, M. S., "Sodium Vapor Pressure Losses in a Multitube, Alkali Metal Thermal-to-Electric Converter," *Journal of Thermophysics and Heat Transfer* **13** (1), 117–125 (1999a).

Tournier, J.-M., and El-Genk, M. S., "Radiation Heat Transfer in Multitube, Alkali Metal Thermal-to-Electric Converter," *Transactions of the ASME – Journal of Heat Transfer* **121** (1), 239–245 (1999b).

Tournier, J.-M., and El-Genk, M. S., "An Electric Model of a Vapor-Anode, Multi-Tube Alkali-Metal Thermal-to-Electric Converter," *Journal of Applied Electrochemistry* **29** (11), 1263–1275 (1999c).

Tournier, J.-M., and El-Genk, M. S., "Analysis of Test Results of a Ground Demonstration of a Pluto/Express Power Generator," *Energy Conversion and Management* **40**, 1113–1128 (1999d).

Tournier, J.-M., and El-Genk, M. S., "A Thermal Model of the Conical Evaporator in Pluto/Express, Multi-Tube AMTEC Cells," in *Proceedings of the Space Technology and Applications International Forum (STAIF-99)*, CONF-990103, M. S. El-Genk, Ed., American Institute of Physics, New York, NY, 1999e, AIP Conference Proceedings No. 458, **2**, 1526-1533; CD ROM.

Performance Measurements of Advanced AMTEC Electrodes

Michael Schuller, Brad Fiebig, Patricia Hudson, and Imran Kakwan

Center for Space Power, Texas Engineering Experiment Station, Texas A&M University, College Station, TX, 77843
schullr@acs.tamu.edu, bnf5693@acs.tamu.edu, waste@myriad.net, i-kakwan@tamu.edu; (409) 845-8768

Abstract. These results are from sodium exposure test cell experiments with advanced AMTEC electrodes performed at Texas A&M University. The majority of the results are for metal electrodes; the minority of the results are for ceramic electrodes. Initial results for iridium and titanate electrodes have been good, but degrade with time.

INTRODUCTION

The Alkali Metal Thermal to Electric Converter is a high efficiency device for directly converting heat to electricity, first described by Weber in 1974 (Weber, 1974). AMTEC operates as a thermally regenerative electrochemical cell by expanding sodium through the pressure differential across a sodium beta" alumina solid electrolyte (BASE) membrane (Cole, 1983).

While AMTEC technology is still being developed, laboratory devices have achieved efficiencies as high as 19% and system design studies indicate that efficiencies as high as 30% are achievable in the near term and 35% or more may be possible. AMTEC can provide all the advantages of a static power system (low vibration, redundancy, no wear) at efficiencies normally achieved only in dynamic systems. System designs using AMTEC have shown 27% cell and 23% system efficiencies (Schock, 1997), while laboratory experiments with developmental multi-tube cells have achieved 16% efficiencies (Merrill, 1997).

Further, because AMTEC requires energy input at modest temperatures, and not at a specific wavelength, it is easily adapted to any heat source, including radioisotope, concentrated solar, external combustion, or reactor. This adaptability makes AMTEC very attractive for development because it has so many potential applications, both space and terrestrial, for a single basic design.

A key component of achieving high efficiency in an AMTEC device is the electrode, which serves three main functions. The first function of the electrode is to provide reaction sites at which sodium may be oxidized or reduced, creating the charge separation and recombination, which is the heart of AMTEC. The second function is to provide a means to transport electrons to or from the reaction site. The third function is to provide a means to transport neutral sodium to or from the reaction site.

Historically, AMTEC electrodes have been TiN or Mo. These electrodes have deficiencies, either in performance or lifetime. This paper documents work being done at Texas A&M University to improve the performance and lifetime of AMTEC electrodes, for the purpose of increasing the efficiency and lifetime of AMTEC devices.

PROCEDURE

The following text describes our procedures in general terms. Specifics of the treatment of each experiment varied with experience and equipment available. Our goal was to run each test 1000 hours at 1123-1173 K to develop aging data, but oxygen infiltration curtailed the tests short of that goal.

CP504, *Space Technology and Applications International Forum–2000*, edited by M. S. El-Genk
© 2000 American Institute of Physics 1-56396-919-X/00/$17.00

Electrode Deposition

All of the electrode samples, except for the iron titanate/sodium titanate sample, were deposited by sputtering. The BASE (beta" alumina solid electrolyte) tubes were prepared by firing in air at 973 K for one hour, then masking the tubes with aluminum foil. The typical BASE tube had four test electrodes, each .00626 m wide, separated by three bare regions, each .00376 m wide. Sputtering parameters varied slightly for each material, but each sputtering run was 30 minutes long, at a power level of 50 watts. After sputtering, the aluminum foil was removed and the electrodes were wrapped with 150 mesh Mo screen, which served as the current collector.

The iron titanate/sodium titanate electrodes were deposited by evaporation. Initial trials used tungsten boats to hold the titanate powders. When these electrodes were examined by scannng electron microscopy (SEM), we discovered that a considerable amount of tungsten had also been deposited. The tungsten reacted with the sodium titante to form sodium tungstate, which evaporates at AMTEC operating temperatures, depriving the electrode of the sodium it needs to operate properly. After this problem was discovered, other boats, made of tantalum or alumina coated tungsten were used for evaporating the powders. These approaches were not successful, because tungsten or tantalum still evaporated at too high a rate, poisoning the electrode.

The Sodium Exposure Test Cell

The basic test device used in this work was the SETC (sodium exposure test cell) developed by JPL for measuring AMTEC electrode parameters (Ryan, 1998). Some changes were made to the configuration and materials of the JPL SETC in the present work. Specifically, we used a Nb1Zr liner in the hot zone of the SETC, to better simulate the environment of an ARPS (Advanced Radioisotope Power System) AMTEC cell. In a typical SETC, three electrode material specimens were mounted in the hot zone of the SETC. One of these specimens was usually WRh, which served as the standard, since its properties are relatively well known.

The Electrochemical Measurements

Two electrochemical measurements were made for these experiments: two-point EIS (electrochemical impedance spectroscopy) curves and two-point current-voltage curves. Both data sets were collected using a Solartron 1250 Frequency Analyser and a Solartron 1286 Electrochemical Interface. The EIS curves were run from 64,000 Hz to 0.1 Hz to measure the apparent charge transfer resistance of the electrode pair, from which we can calculate B, the temperature independent exchange current. The current-voltage curves provide the limiting current, from which we can calculate G, the porosity or neutral sodium impedance. This paper focuses on the EIS data.

Post Test Examination

After each SETC was shut down, the samples were sectioned, with one part potted and polished to observe the sample through the thickness of the BASE tube and a second part mounted, without polishing, to observe the sample normal to the electrode. The samples were examined using electron microprobe and SEM techniques, including backscattered electron and secondary electron imaging, energy dispersive and wavelength dispersive spectroscopy, and x-ray mapping.

RESULTS

Figure 1 is a typical EIS curve for these specimens. The curve is a plot of the imaginary impedance versus the real impedance. Sweeps generally run from left to right as the Solartron moves from high frequency to low frequency. The difference between the high frequency intercept of the real axis and the low frequency intercept of the real axis is the apparent charge transfer resistance, R_{ACT} which is related to B, the temperature independent exchange current of the electrode pair, by

$$B = \sqrt{\frac{T_{el}}{P_{el}P_{Na}}} \left[\frac{RT_{el}}{R_{ACT}F} \right] \qquad (1)$$

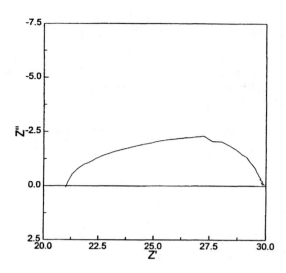

FIGURE 1. EIS curve. The vertical axis is the imaginary impedence and the horizontal axis is the real impedence. Units are ohms.

where Tel and Pel are the temperature and pressure of sodium at the electrode, while PNa is the pressure of sodium at the sodium pool, R is the universal gas constant, and F is the Faraday.

Table 1 contains computed B values for one set of samples. The sodium pool temperatures are those measured by the thermocouple in the sodium pool; the parenthetical values are adjusted for the hottest position of the sodium pool. Because of the physical distribution of sodium found during post test disassembly, we suspect that the actual sodium pool temperature, and hence the actual sodium pressure over the test electrodes, was significantly higher than what was measured. The "B (adjusted)" column of Table 1 contains B values using effective sodium pool temperatures calculated from the width of the sodium pool (measured after disassembly) and the temperature gradient in the SETC, which are much closer to those reported in the literature.

TABLE 1. Temperature independent exchange current.

Sample	Time (hrs)	T_{el} (K)	T_{Na} (K) (adjusted)	B	B (adjusted)
Titanate	1	873	468 (489)	63-69	36-39
	23	973	481 (504)	80-97	45-55
	30	973	486 (511)	157-182	85-98
	50	1023	487 (515)	407-501	207-255
	56	1023	491 (518)	373-460	192-237
	72	1073	485 (516)	934-1100	440-518
WRh	1	898	468 (489)	55	31
	23	998	481 (504)	103-105	58-59
	30	998	486 (511)	301-328	162-177
	50	1048	487 (515)	302-452	153-230
	56	1048	491 (518)	225-302	116-156
	72	1148	485 (516)	278-503	131-237
Ir	1	923	468 (489)	88-99	50-57
	23	1023	481 (504)	191-213	80-120
	30	1023	486 (511)	227-359	122-193
	50	1073	487 (515)	386-518	196-263
	56	1073	491 (518)	297-429	153-221
	72	1123	485 (516)	314-470	148-221

SEM Data

Aside from morphological changes to the electrodes, SEM data provides useful insights into the movement and interaction of materials in the electrodes, current collectors, and BASE tubes. Figures 2 and 3 are typical secondary electron images taken normal to the electrode surface. They show the physical structure of the electrode before (Figure 2) and after (Figure 3) sodium exposure. Figure 4 is a backscattered electron image of a sample after sodium exposure. The figure shows, from left to right, a bright area, which is the electrode (W, mostly), a gray area (Mo), a dark gray area (BASE), a second grey area (Mo), and a second bright area (W). The most interesting feature of the picture is the two grey bands, which are caused by Mo from the current collector depositing on the BASE.

CONCLUSIONS

As shown in Table 1, B values did vary significantly with electrode and pool temperature and showed significant variability between electrode pairs at constant temperature. Two main reasons for this variability have been advanced. First, due to the geometry of the SETC specimens, most of the sodium reactions take place in a small portion of each active electrode. Changes in the electrode area, which are accounted for in the equations for B and G, may affect the calculated values, while having no effect on the real values. Second, due to the geometry of the sodium pool within the SETC, the actual vapor pressure of sodium over the electrodes, which has a large effect on the values of B and G, may be seriously underestimated by using the saturation pressure corresponding to the measured pool temperature. Much of the inconsistency of the B values shown in Table 1 can be explained by relatively small variations in the actual sodium pressure at the test electrodes. Since the vapor pressure of sodium is a sensitive function of the temperature of the liquid sodium pool, small variations in the geometry and/or temperature of the sodium pool could have a significant effect on the results of these experiments.

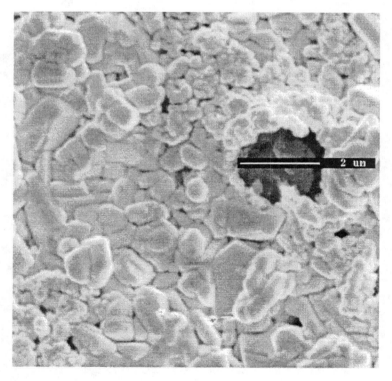

FIGURE 2. Electrode before sodium exposure. Secondary electron image at 5000x magnification.

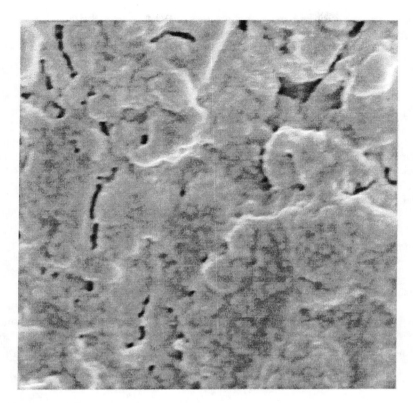

FIGURE 3. Electrode after sodium exposure. Secondary electron image at 5000x magnification.

Another potential source of error in the SETC may be the Mo deposits found in the post-test SEM. Due to the geometry of the SETC specimens, sodium transport between the electrodes is primarily at the surface of the BASE and most of the oxidation/reduction reactions take place in the portion of each electrode nearest the other active

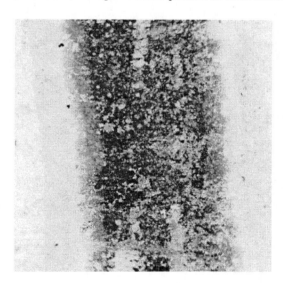

FIGURE 4. Backscattered electron image, at 63x magnification, of sample after sodium exposure. White band is tungsten, gray band is molybdenum, dark region is beta" alumina.

electrode. Since the Mo deposits occurred between the electrodes, it is probable that they formed a secondary electrode that had a significant impact on the measured values of R_{act}.

FUTURE WORK

The results of these experiments, and the conclusions drawn from those results, have led to several SETC design modifications. First, the location of the samples has been changed to reduce the temperature gradients within and between specimens. Second, we have added an antechamber to the cold end of the SETC to contain the sodium pool in a small physical space whose temperature is well known. Third, we are investigating alternatives to Mo as the current collector mesh material. The results of these experiments have also prompted us to examine several variants of the basic SETC experiment. First, we intend to examine the effect of varying the electrode area in a test. Second, we intend to examine alternative geometries that force the sodium to move through the BASE, preferably radially, rather than along its surface.

The initial good results for the titanate electrodes were not sustained over time, due to the effects of tungsten on the sodium titanatae. The promising performance of the titanate electrodes led us to explore other means of depositing the electrode materials, which will be discussed in future papers.

ACKNOWLEDGMENTS

This work was supported by DOE contract DE-AC03-98SF21559. The SEM work was performed on the Cameca SX-650 Electron Microprobe at the Department of Geology and Geophysics at Texas A&M University, with the able assistance of Dr. Ray Guillamet.

REFERENCES

Cole, "Thermoelectric Energy Conversion with Solid Electrolytes," Science, 221, 915 (1983).

Merrill, et al., "Vacuum Testing of High Efficiency Multi-BASE Tube AMTEC Cells," 32nd IECEC, Honolulu, HI, 1997.

Ryan, et al., "The Sodium Exposure Test Cell to Determine Operating Parameters for AMTEC Electrochemical Cells," 33rd IECEC, Colorado Springs, CO, 1998.

Schock, et al., "Design and Performance of Radioisotope Space Power Systems Base on OSC Multi-tube AMTEC Converter Designs," 32nd IECEC, Honolulu, HI, p489, 1997.

Weber, "A Thermoelectric Device Based on Beta-Alumina Solid Electrolyte," Energy Conversion 14, 1(1974)

Lifetimes of AMTEC Electrodes:
Molybdenum, Rhodium-Tungsten, and Titanium Nitride

M. A. Ryan, V. B. Shields, R. H. Cortez, L. Lara, M. L. Homer and R. M. Williams

Jet Propulsion Laboratory, California Institute of Technology, 4800 Oak Grove Drive, Pasadena CA 91109
(818) 354-8028;   mryan@jpl.nasa.gov

Abstract. The lifetime of three types of AMTEC electrodes is predicted from the rate of grain growth in the electrode. Grain size is related to electrode performance, allowing performance to be correlated with grain growth rate. The rate of growth depends on physical characteristics of each material, including the rates of surface self-diffusion and molecule mobility along grain boundaries. Grain growth rates for molybdenum, rhodium-tungsten and titanium nitride electrodes have been determined experimentally and fit to models in order to predict operating lifetimes of AMTEC electrodes. For lifetimes of 10 years or more, Rh_xW electrodes may be used at any operating temperature supportable by the electrolyte. TiN electrodes may be used in AMTEC cells only at operating temperatures under 1150 K, and Mo may be used only below 1100 K.

INTRODUCTION

AMTEC, the Alkali Metal Thermal to Electric Converter, is a direct energy conversion device capable of near-Carnot efficiencies. The device is an alkali metal concentration cell which uses a β"-alumina solid electrolyte (BASE) tube as a separator between a high pressure region containing liquid metal, generally sodium, at 900 - 1300 K and a low pressure region containing a condenser at 400 - 700 K (Weber, 1974; Cole, 1983). Single cells have performed at high power densities, with open circuit voltages up to 1.6 V and current densities up to 2.0 A/cm² (Williams, *et al.*, 1989). Efficiencies as high as 13-15% have been reported (Underwood, *et al.*, 1993; Sievers *et al.*, 1994).

NASA's interest in AMTEC is as the power source for outer planet exploratory spacecraft. For an AMTEC device to be feasible for either space or terrestrial applications, it must have an operating lifetime of 7 - 15 years. Tests done in AMTEC research have shown that, on the time scales studied (up to 8000 hours), the electrode is the component most likely to influence device performance and limit operating lifetime. The ideal AMTEC cathode is one which has three primary characteristics:

1. It is thin enough to allow ready transport of sodium atoms from the electrode-electrolyte interface through the electrodes to the low pressure vapor space, for vaporization and transport to the condenser;
2. It is thick enough to offer good electrical conductivity for the electrons traveling from the external circuit to the interface where recombination of electrons and sodium ions takes place; and
3. It has a lifetime which will allow continuous operation with minimal degradation in performance for seven (mission to Europa) to fifteen years (mission to Pluto).

Electrode materials tested at JPL in AMTEC and Sodium Exposure Test Cell (SETC) experiments have included refractory metal and metal alloy electrodes, including molybdenum, tungsten, rhodium-tungsten and platinum-tungsten, and titanium nitride electrodes (Wheeler, *et al.*, 1988; Williams, *et al.*, 1989; Williams, *et al.*, 1990). This paper considers the characteristics of molybdenum, rhodium-tungsten and titanium nitride electrodes, and predicts operating lifetimes of those materials as AMTEC electrodes based on data taken during several hundreds or thousands of hours of operation under AMTEC conditions.

CP504, *Space Technology and Applications International Forum–2000*, edited by M. S. El-Genk
© 2000 American Institute of Physics 1-56396-919-X/00/$17.00

All electrodes discussed here are considered for use as cathodes, although use as anodes is not precluded, and were sputter-deposited to thicknesses of 0.75 - 1.5 µm.

ELECTRODE LIFE MODEL

The electrode life model developed over the last several years at JPL is based on a model of grain growth and the relationship between grain radius and electrode performance. Grain radius has been correlated to the temperature independent exchange current, B, and to electrode power experimentally. Development of a model of electrode life proceeds through four steps: determination of a model of grain growth for each electrode material, development of an expression for the relationship between B and grain radius, development of an expression for the relationship between B and Power, and finally, calculation of electrode life by setting an arbitrary decline in power and relating it to grain radius through the previous two expressions.

Grain Growth Model

Lifetime characteristics of molybdenum and two preparations of Rh_xW have been previously reported (Ryan, *et al.*, 1994). A model for grain growth based on a model of coalescence of spheres was developed and applied to Mo, RhW and Rh_2W. The grain radius was correlated with electrode performance through the temperature independent exchange current, B (Williams, *et al.*, 1990b), and B was independently correlated with electrode power (Ryan, *et al.*, 1994). Further data have been used to confirm the applicability of that model. The model, which is based on the Herring Law (Herring, 1950; Binh & Uzan, 1987) allows the calculation of the time, t, for two grains to coalesce into one from an initial radius R_0 as

$$ t = \frac{0.89\, R_0^{\,4}}{D_S N_0 \gamma_s \Omega^2 / kT} , \tag{1}$$

where D_S is the surface self-diffusion coefficient of the electrode material, N_0 is the atom density of electrode material on the surface of the electrolyte, γ_s is the surface energy of the material, Ω is the atomic volume of the material, k is the Boltzmann constant and T is the temperature. D_S may be calculated from this expression from measured grain growth over a measured time at a single temperature. N_0 is calculated from the atomic volume as $N_0 = \Omega^{-2/3}$ (Binh & Uzan, 1987). The other factors may be, or have been, measured directly. From this equation, the final radius of a grain may be computed from the number of rounds of coalescence of spheres.

In Mo electrodes, observation of the morphology of an electrode surface led to the use of the coalescence of spheres model. After a few hundreds of hours of operation at T>1050 K, scanning electron micrographs showed a clear pattern of coalescence of Mo material into spheres (Wheeler, *et al.*, 1988). The Herring Law model for grain growth which was used for Mo was applied to Rh_xW electrodes with success, although observation of the surface did not show that spheres were formed in the same way as in Mo. While the Herring Law model for grain growth can be used successfully to predict grain size in Rh_xW after a period of AMTEC operation, it does not fully describe the process of grain growth in that material.

Observed grain growth in titanium nitride did not follow the coalescence of spheres model, nor did application of the Herring Law model predict grain size successfully. A model based on grain boundary mobility was developed from measured grain growth of TiN in low pressure sodium vapor. The development of that model is discussed in detail elsewhere (Shields, *et al.*, 1999). In the grain boundary mobility model, the relationship in time between the initial radius R_0 of a grain to its final radius R_f can be expressed as:

$$ R_f = R_0 \left[1 + \frac{2c\gamma_s M \cos\phi\,(\exp(-E_A/RT))\,t}{R_0^{\,n}} \right]^{1/n} , \tag{2}$$

where c is a proportionality constant, γ_s is the surface energy of TiN, M is the mobility of TiN on itself, ϕ is the angle between grains, E_A is the activation energy for grain growth, R is the gas constant, and n = 3.2. To use this expression, we have used observed grain sizes grown in sodium atmosphere at known temperatures for known times in SETC experiments (Ryan, *et al.*, 1998) and derived a value for $2c\gamma_sM\cos\phi$, which is approximately constant over the temperature range 1050-1250 K.

As neither γ_s nor M can be easily measured in our laboratory and we have not yet found these values for TiN in sodium atmosphere in the literature, we have applied the model using the derived value

$$a = 2c\gamma_sM\cos\phi \quad , \tag{3}$$

making Equation (2)

$$R_f = R_0 \left[1 + \frac{a\,[\exp(-E_A/RT)]\,t}{R_0^{\,n}} \right]^{1/n} \quad . \tag{4}$$

R_0 is measured on electron micrographs; ϕ can be taken from the same micrographs, but with less precision because photo angle is important, so it is included in a. E_A was taken from a literature value of 230 kJ/mol for the formation of TiN films by evaporation of Ti in a nitrogen atmosphere; the activation energy for grain growth may, in fact, be somewhat lower, which would result in faster grain growth. As this E_A is the same order of magnitude as was found for Rh_xW electrodes, 2.4 - 2.7 eV, this energy was used for the first order model.

The function of the proportionality constant, c, is to provide dimensional conversion factors for the dimensions of the radius to be correct. The value of n, 3.2, is taken from literature values where the grain growth characteristics of TiV and TiMn compounds were studied. The source of these values is discussed in detail elsewhere (Shields, *et al.*, 1999).

Relationship of B and Grain Radius

In studies of the grain growth rates of Mo, Pt_xW and Rh_xW electrodes, an empirical correlation between grain size and exchange current expressed as B has been developed (Ryan, *et al.*, 1994). The electrode reaction, reduction of sodium ions (Na^+), occurs at the electrode/electrolyte interface, at the perimeter of the grains. As grain size increases, total material volume is conserved and the number density of grains, and thus total contact between grain and electrolyte, decreases. B as measured in SETC experiments has been plotted as a function of observed radius R_f. The values of B calculated from experiment were fit to the expression

$$B = a - bR_f^{1/2} \quad , \tag{5}$$

where $a = 1.421 \times 10^2$ and $b = 6.218$ and R_f is expressed in nm.

This relationship between exchange current and grain radius is empirical, and is based on observed behavior of Mo, PtW and RhW electrodes. It has not yet been shown to be valid for TiN electrodes. However, it is valid to assume that exchange current will decrease as grain size increases and the length of the reaction zone decreases. The length of the TiN reaction zone may change according to a different relation than does the reaction zone in other electrode materials, but as a first order model, this approach gives an idea of the lifetime of a TiN electrode.

Relationship of Power and B

B, the exchange current, is a measure of the efficiency with which the reaction $Na^+ + e^- \rightarrow Na^0$ is carried out at the electrode/electrolyte interface. It is a sensitive probe of electrode performance, which is reflected in the overall power conversion in an AMTEC cell. A semi-empirical relationship between Power and B has been developed and previously presented (Ryan, *et al.*, 1991; Ryan, *et al.*, 1994). B can fall to ~ 50 from a starting value of 100 with only a 10% drop in power. Power drops to 70% of initial power when B drops to 20; any further decline in B results in rapid decline in power and failure of the device.

Electrode Lifetime

Electrode lifetime is the period over which electrode performance is sufficient for necessary operation device operation; from the perspective of spacecraft operation, it may be defined as the time it takes for produced power to fall to a defined level. That level is an arbitrary one; for the purposes of this discussion, times for power to fall to 90% and 70% of initial power will be presented. Using the relationship of B and Power described above, and the relationship of grain radius and B described in Equation 5 and Figure 1, time to grow to a grain radius of 200 nm results in a power level of 90% of initial power, and a grain radius of 400 nm results in a power level of 70% of initial power.

EXPERIMENTAL

Three SETC experiments using TiN electrodes provided by AMPS, Inc. were run. The experiment has been previously described in detail (Ryan, *et al.*, 1998). All three experiments were contained in stainless steel vacuum chambers within titanium liners to prevent volatile components of the steel from reaching the electrode and electrolyte. The sodium pool temperature was kept at 500 - 600 K. Each experiment was run at a different temperature and for a different length of time: 980 hours at 1120 K, 1850 hours at 1170 K and 500 hours at 1220 K. Current-voltage curves and two probe impedance measurements were made every 24 - 72 hours and the time evolution of the apparent charge transfer resistance R_{ACT} was plotted. The temperature independent exchange current, B, was computed from R_{ACT} (Williams, *et al.*, 1990; Ryan, *et al.*, 1998).

Three SETC experiments using RhW and Rh_2W electrodes were run at 1070 or 1120 K for 1000 to 1500 hours; the coalescence of spheres model was used to determine an expression for D_S in those preparations of Rh_xW. One Rh_xW SETC experiment using two preparations of Rh_xW, $Rh_{1.5}W$ and $Rh_{2.5}W$ was run at 1170 K for 3000 hours. $Rh_{1.5}W$ data were analyzed for this paper, using a D_S which is intermediate between that for RhW and Rh_2W to calculate the grain radius for $Rh_{1.5}W$ operated for 3000 hours at 1170 K.

Mo electrodes were operated in three AMTEC experiments which operated for periods of 100 to 300 hours at 1050 or 1170 K, and in one SETC experiment 1070 K for 500 hours. The coalescence of spheres model was used to determine an expression for D_S for Mo.

RESULTS AND DISCUSSION

Grain Radii

Grain Boundary Mobility Model - TiN The value of *a*, Equation 3 was derived from Equation 4 and the data from TiN experiment 1 (1120 K, 980 hours) and applied to the time and temperature data for TiN experiments 2 and 3. The value of *a* was derived from the model by setting $R_0 = 30$ nm, $E_A = 230$ kJ/mol, n = 3.2, T = 1120 K, and t = 980 hours as described above and finding the value that yielded a final grain radius R_f within 10% of the measured radius. *a* was found to be constant in the temperature range for which we have data (1120 - 1220 K). γ_s and M will be temperature dependent, but relatively weakly compared with the value of $(exp(-E_A/RT))$. The value of ϕ will be weakly time dependent, but it has been kept constant for the present purposes. The value computed is $a = 7.0 \times 10^{13}$. Using the same value of *a*, final grain radius calculated for other conditions correlated well with measured values:

TABLE 1. Calculated and Experimental Grain Radii for Three SETC Experiments with TiN Electrodes.

R_0 (nm)	T (K)	time (hours)	R_f calc (nm)	R_f exp (nm)
30	1120	980	82	88
30	1170	1850	138	127
30	1220	500	124	117

Coalescence of Spheres Model - Mo and $Rh_{1.5}W$ The grain size of $Rh_{1.5}W$ was measured after 3000 hours of operation and compared with grain size computed for this temperature and time according to the model previously presented (Ryan, *et al.*, 1994). New Mo experiments were not run.

TABLE 2. Surface Self Diffusion Coefficients for Mo and Rh_xW in Na Vapor.

	E_A (eV)	A (cm²/sec)	T (K)
Mo	8.44	3.0×10^{26}	1050-1200
RhW	2.70	2.2×10^{-3}	1050-1200
Rh_2W	2.14	1.3×10^{-5}	1050-1200

The surface self diffusion coefficient can be expressed as $D_s = A \exp(-E_A/kT)$. The values derived are shown in Table 2. In the new experiment of $Rh_{1.5}W$, the measured grain radius after 3000 hours at 1170 K is 60 nm; the calculated grain radius is 55 nm, in good agreement with the observed size. According to Equation (5), B at this grain radius should be ~ 95. During the 3000 hour SETC experiment, the measured B for $Rh_{1.5}W$ electrodes was ~ 100.

Electrode Lifetime

Electrode lifetimes for Mo, TiN and $Rh_{1.5}W$ electrodes are shown in Figures 1 and 2 as functions of operating temperature. Starting grain radius, R_0, for Mo is 10 nm, for $Rh_{1.5}W$ is 5 nm and for TiN is 30 nm. All are measured from electron micrographs of the as-sputtered material. Figure 2a shows time for grains to grow to a radius of 200 nm, resulting in a power decline of 10%, $P = .9P_i$. Figure 2b shows time for grains to grow to a radius of 400 nm, resulting in a power decline of 30%, $P = .7P_i$. For all electrodes, the initial value of B has been assumed to be 100. JPL experiments have consistently shown the initial value of B in TiN to be about 60-70% of that of Rh_xW. If the initial value of B for TiN is assumed to be 60, the lifetime will be as shown in Figure 2.

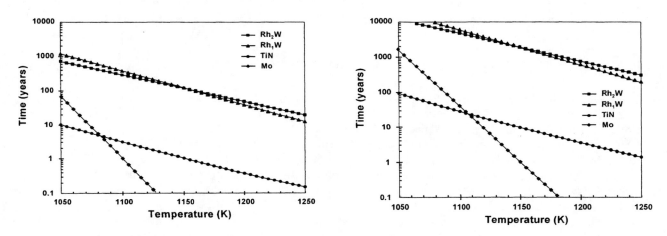

FIGURE 1. (a) Time for Mo, TiN and Rh_xW electrodes to grow to a grain radius of 200 nm, corresponding to B ~ 50 and P = .9P_i; (b) Time for Mo, TiN and Rh_xW electrodes to grow to a grain radius of 400 nm, corresponding to B ~ 20 and P = .7P_i.

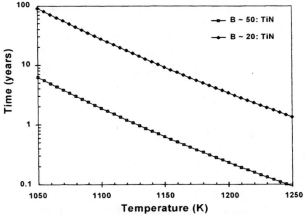

FIGURE 2. Time for TiN grain growth to result in a decline in exchange current to B ~ 50 (squares) and B ~ 20 (diamonds), assuming the initial B is 60.

CONCLUSION

Using models of grain growth developed for TiN and for Rh$_x$W and data taken in SETC experiments, we have predicted the lifetime of AMTEC electrodes made with those materials. With an operating temperature of 1125 K, Rh$_x$W electrodes will not decline appreciably in the 15 year lifetime required for an outer planet mission. As can be seen in Figure 2, Rh$_x$W electrodes are predicted to fall to 90% of initial power in ~100 years if there are no degradation mechanisms other than grain growth in the electrode. In contrast, Mo cannot be operated at more that 70% of initial power at 1125 K for more than a few weeks. The maximum temperature at which Mo can be operated to preserve a power of more than 90% of initial power for ten years is 1075 K. TiN is predicted to decline to 90% of initial power in ~2 years and to 70% of initial power in ~16 years. The predicted lifetime for TiN is based on a grain growth model developed for that material and the relationships among grain radius, exchange current and power density observed in other electrode materials. Experiments to confirm the applicability of the grain boundary mobility model of TiN growth are underway. The predicted lifetime for TiN must be validated with further experiments in which exchange current is correlated with both grain size and power density in the electrode.

ACKNOWLEDGMENTS

The research reported in this paper was carried out by the Jet Propulsion Laboratory, California Institute of Technology, under a contract with the National Aeronautics and Space Administration. Research was supported by NASA Code S.

REFERENCES

Binh, V.T. and R. Uzan, "Tip Shape Evolution: Capillarity-Induced Matter Transport by Surface Diffusion," *Surface Science*, **179**, 540-543 (1987).

Cole, T., "Thermoelectric Energy Conversion with Solid Electrolytes," *Science*, **221**, 915-920 (1983).

Herring, C., "Effect of Change of Scale on Sintering Phenomena," *J. Appl. Physics*, **21**, 301-306 (1950).

Ryan, M.A., B. Jeffries-Nakamura, D. O'Conner, M.L. Underwood, and R.M. Williams, "AMTEC Electrode Morphology Changes as Studied by Electrochemical Impedance Spectroscopy and Other Techniques," *Proc. Symposium on High Temperature Electrode Materials*, D.D. Macdonald and A.C. Khandkar, eds., The Electrochemical Society, **91-6**, 115-120 (1991).

Ryan, M.A., A. Kisor, R.M. Williams, B. Jeffries-Nakamura, and D. O'Connor, "Lifetimes of Thin Film AMTEC Electrodes," *Proc. 29th IECEC*, American Institute of Aeronautics and Astronautics, **2**: 877-880 (1994).

Ryan, M.A, R.M. Williams, L. Lara, R.H. Cortez, M.L. Homer, V.B. Shields, J. Miller, and K.S. Manatt, "The Sodium Exposure Test Cell To Determine Operating Parameters For AMTEC Electrochemical Cells", *Proc. 33rd IECEC*, American Nuclear Society, I335 (1998).

Shields, V.B., R.M. Williams, M.A. Ryan, and M.L. Homer, "Model for Grain Growth in AMTEC Electrodes" *Proceedings of 34th IECEC*, SAE, 2703 (1999).

Sievers R.K., T.K. Hunt, D.A. Butkiewicz *et al.* "Prototype AMTEC Cell Development," *Proceedings of 29th IECEC*, American Institute of Aeronautics and Astronautics, **2**, 894-898 (1994).

Underwood M.L., R.M. Williams, M.A. Ryan, B. Jeffries-Nakamura, D. O'Connor, "Recent Advances in AMTEC Recirculating Test Cell Performance," *Proc. of 10th Symp. on Space Nuclear Power*, M.S. El-Ghenk and M.D. Hoover, eds., American Institute of Physics, 885-889 (1993).

Weber, N., "A Thermoelectric Device Based on β"-Alumina Solid Electrolyte," *Energy Conv.*, **14**, 1-8 (1974).

Wheeler B.L., R.M. Williams, B. Jeffries-Nakamura, J.L. Lamb, M.E. Loveland, C.P. Bankston and T. Cole, "Performance and Impedance Studies of Thin, Porous Molybdenum and Tungsten Electrodes for the Alkali Metal Thermoelectric Converter," *J. Appl. Electrochem.*, **18**, 410-415 (1988).

Williams, R.M., B. Jeffries-Nakamura, M.L. Underwood, B.L. Wheeler, M.E. Loveland, S.J. Kikkert, J.L. Lamb, T. Cole, J.T. Kummer and C.P. Bankston, "High Power Density Performance of WPt and WRh Electrodes in the Alkali Metal Thermoelectric Converter," *J. Electrochem. Soc.*, **136**, 893-894 (1989).

Williams, R.M., B. Jeffries-Nakamura, M.L. Underwood, C.P. Bankston and J.T. Kummer, "Kinetics and Transport at AMTEC Electrodes II. Temperature Dependence of the Interfacial Impedance of Na$_{(g)}$/Porous Mo/Na-β" Alumina," *J. Electrochem. Soc.*, **137**, 1716-1722 (1990).

Analyses of Nb-1Zr/C-103, Vapor Anode, Multi-Tube AMTEC Cells

Jeffrey C. King[1] and Mohamed S. El-Genk[2]

Institute for Space and Nuclear Power Studies and Chemical and Nuclear Engineering Dept.
The University of New Mexico, Albuquerque, NM 87131
[1](505) 277-3321, FAX:-2814, ajc@isnps.unm.edu
[2](505) 277-5442, FAX:-2814, mgenk@unm.edu

Abstract. A high performance, Nb-1Zr/C-103, vapor anode, multi-tube AMTEC cell design is presented. The cell measures 41.27 mm in diameter, is 125.3 mm high, and has eight BASE tubes connected electrically in series. The hot structure of the cell (hot plate, BASE tubes support plate, hot plenum wall, evaporator standoff, evaporator wick, and side wall facing the BASE tubes) is made of Nb-1Zr. The cold structure of the cell (condenser, interior cylindrical thermal radiation shield, the casing and the wick of the liquid sodium return artery, and side wall above the BASE tubes) is made of the stronger, lower thermal conductivity niobium alloy C-103. This cell, which weighs 163.4 g, could deliver 7.0 W_e at 17% efficiency and load voltage of 3.3 V, when using TiN BASE electrodes characterized by B = 75 $A.K^{1/2}/m^2.Pa$ and G = 50 and assuming BASE/electrode contact resistance of 0.06 Ω-cm^2 and leakage resistance of the BASE braze structure of 3 Ω. For these performance parameters and when the interior cylindrical C-103 thermal radiation shield is covered with low emmisivity rhodium, the projected specific mass of the cell is 23.4 g/W_e. The BASE brazes and the evaporator temperatures were below the recommended limits of 1123 K and 1023 K, respectively. In addition, the temperature margin in the cell was at least + 20 K. When electrodes characterized by B=120 $A.K^{1/2}/m^2.Pa$ and G = 10 were used, the cell power increased to 8.38 W_e at 3.5 V and efficiency of 18.8%, for a cell specific mass of 19.7 g/W_e. Issues related to structure strength of the cell and the performance degradation of the BASE and electrodes are not addressed in this paper.

INTRODUCTION

Several stainless steel (SS), PX-Series cells, fabricated by Advanced Modular Power Systems (AMPS), have been tested in vacuum at the Air Force Research Laboratory in Albuquerque, NM. Some of these cells have operated continuously for more than 8,000 hours without failure, but with appreciable degradation in performance (Merrill et al. 1998, 1999), possibly due to mass transport of volatile alloying elements and/or degradation of the TiN electrodes covering the sodium beta"-alumina solid electrolyte (BASE). The release of volatile alloying elements such as manganese and the slow solubility of iron in sodium, as well as their deposition in other parts of the cell, could have contributed to the observed degradation in the performance of the PX-cells (El-Genk et al., 1998, 1999). The test results of the PX-type cells compared favorably with the predictions of the AMTEC Performance and Evaluation Analysis Model (APEAM) (Tournier and El-Genk 1999a,b).

In an attempt to replace the SS structure of the PX-series cells, El-Genk et al. (1999) performed parametric analyses of nickel/Haynes-25 AMTEC cells. One of these cells, designated cell D, was 41.25 mm in diameter and 127 mm high and had eight BASE tubes, 0.4 mm thick and 50.8 mm long, connected electrically in series. The hot structure, including the cell wall facing the BASE tubes (~200 μm thick), the BASE tubes support plate, the evaporator stand off, evaporator wick, and the interior cylindrical thermal radiation shield, was made of nickel. The thin (~100 μm thick) cell wall above the BASE tubes and the casing and wick of the liquid sodium return artery were made of Haynes-25 to take advantage of this alloy's low thermal conductivity and high strength. The cell is insulted on the outside by an inch of Min-K, separated from the cell wall by a small gap to minimize heat losses through the wall. When operated at a condenser temperature of 640 K and heat input of 51.2 W_{th}, the cell efficiency was 19 % at 3.5 V. The calculated temperatures of the BASE braze and of the evaporator were 1171 K and 1058 K, respectively,

CP504, *Space Technology and Applications International Forum–2000*, edited by M. S. El-Genk

and the temperature margin in the cell was +29K. While the latter is quite sufficient to ensure that Na vapor would not condense inside the BASE tubes, the former are slightly higher than the recommended design values of 1123 K and 1023 K, respectively. This cell weighed 192 g for a projected specific mass of 19.7 g/W$_e$ (El-Genk et al., 1999). The cell design is similar to that shown in Figure 1a, except for slight changes in some dimensions (see Table 1).

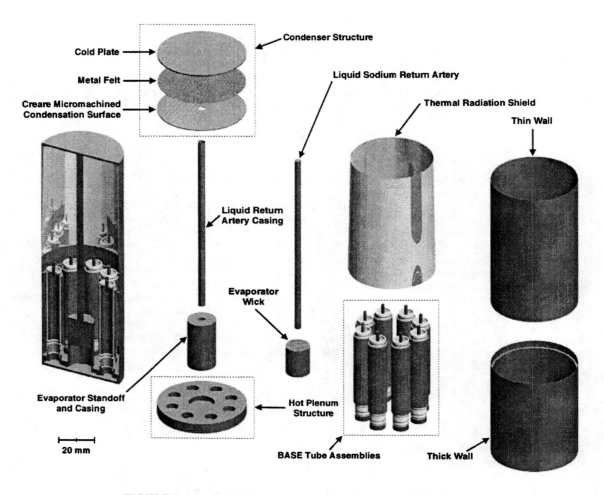

FIGURE 1a. An exploded trimetric view of the present Nb-1Zr/C-103 cell.

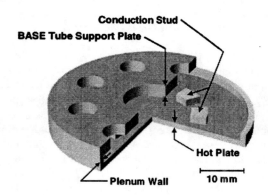

FIGURE 1b. A trimetric view of the hot plenum structure of the present Nb-1Zr/C-103 cell.

However, because of concerns regarding the compatibility of sodium with nickel, particularly in the presence of minute amounts of oxygen, attention was focused on replacing both nickel and Haynes-25 with refractory alloys. Ongoing efforts at AMPS include fabricating and testing as well as optimizing the performance of Nb-1Zr cells (Hendricks, et al. 1999). In a companion paper, Nb-1Zr has been recommended for the cell hot's structure (hot plate, BASE tubes support plate, hot plenum wall, evaporator standoff, evaporator wick, and side wall facing the BASE tubes). The stronger, lower thermal conductivity, niobium alloy C-103, is better suited for the cold structure of the cell (condenser, interior cylindrical thermal radiation shield, the casing and the wick of the liquid sodium return artery, and side wall above the BASE tubes) (King and El-Genk, 2000).

AMPS' Nb-1Zr cell (EPX-1) designed for meeting the power requirement of the Advanced Radioisotope Power System (ARPS) was 50.5 mm in diameter and 101.6 mm high with one cylindrical and 21 conical thermal radiation shields (Hendricks et al., 1999). The cell had eight BASE tubes, 10.16 mm in diameter and 25.4 mm in active length. When modeled using electrodes characterized by $B=120$ A.K$^{1/2}$/m^2.Pa and $G = 10$, the projected cell power was 8.94 W$_e$ at 3.5 V and 16.4% efficiency. The values of the BASE/electrode contact resistance and of the BASE braze leakage resistance were not reported. When electrodes characterized by $B=80$ A.K$^{1/2}$/m^2.Pa and $G = 50$ were considered, the cell power decreased to 8.56 We at 3.5 V with an efficiency of 15.7% (Hendricks et al., 1999). The calculated maximum BASE tube (or braze joint) temperature was well below the 1123 K limit. The evaporator temperature (1061-1063 K), however, was higher than the recommended 1023K (see Table 1). The estimated mass of the cell was ~ 350 g.

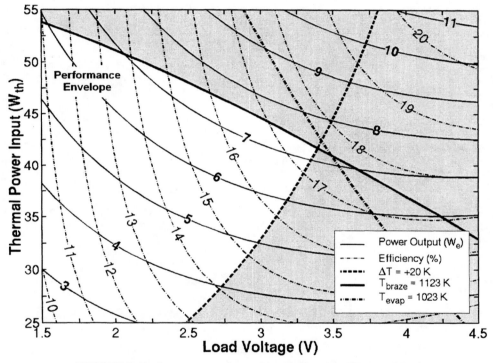

FIGURE 2. Performance envelope of the present nickel/Haynes-25 cell.

The objective of this paper is to design and investigate the performance of AMTEC cells with Nb-1Zr hot structure and C-103 cold structure. The effect on the cell performance of using low emissivity rhodium coating of the inner surface of the thin (~100 µm thick) cell side wall above the BASE tube and of the cylindrical thermal radiation shield (see Figure 1) is also investigated. In addition, parametric analyses are performed to assess the effect of reducing the thickness of the hot plate, BASE tubes support plate, conduction studs, and the hot plenum on the performance and the mass of the cell. The performance of the Nb-1Zr/C103 cell is compared with that of a nickel/Haynes-25 cell of similar design and dimensions. The present analysis is conducted using APEAM, developed at the University of New Mexico's Institute for Space and Nuclear Power Studies and benchmarked extensively using test results of the PX-type cells, generated at the Air Force Research Laboratory in Albuquerque, NM (Tournier and El-Genk, 1999a,b).

TABLE 1. Design-performance parameters of nickel/Haynes-25, and refractory cells.

Design and Performance Parameters	Present nickel and Haynes-25 Cell	AMPS' Nb-1Zr cell (EPX-1)	Present Nb-1Zr/C-103 cell[@]
Design Parameters			
Cell diameter (mm)	41.27	50.8	41.27
Cell height (mm)	127	101.6	125.3
Evaporator (Evap.) shape	Flat	Not reported	Flat
Evap. material	Ni	Mo	Nb-1Zr
Evap. position from support plate (mm)	9.98	Not reported	6.17
Evap. standoff material	Ni	Nb-1Zr	Nb-1Zr
Evap. standoff thickness (mm)	0.71	Not reported	0.71
Standoff outer diameter (mm)	13.29	Not reported	13.29
BASE support plate thickness (mm)	2.54	N/A	0.635
Hot plate thickness (mm)	1.27	N/A	0.318
Hot plenum wall thickness (mm)	1.17	N/A	0.293
Conductive stud area (mm^2)/material	102/nickel	Not reported	10/Nb-1Zr
Number of BASE tubes	8	8	8
BASE tube outer diameter (mm)	7.62	10.16	7.62
Length of BASE tube assembly (mm)	50.1	Not reported	50.1
BASE tube braze segment materials	TiNi, α-Al$_2$O$_3$, Ta, Ni	TiNi, α-Al$_2$O$_3$, Ta, Nb-1Zr	TiNi, α-Al$_2$O$_3$, Ta, Nb-1Zr
Electrode length (mm)	38.1	25.4	38.1
Electrode area per tube (cm^2)	9.1	8.1	9.1
Current collector material	60-mesh Mo	Mo screen	60-mesh Mo
Thermal radiation shield/material	Cylindrical / nickel	Cylindrical & up to 21 conicals	Cylindrical /C-103 with rhodium coating
Condenser type	Creare	Not reported	Creare
Condenser material	Haynes-25	Not reported	C-103
Sodium return artery material	Haynes-25	Mo	C-103
Sodium return artery diameter (mm)	3.18	Not reported	3.18
Hot side structural material	Ni	Nb-1Zr	Nb-1Zr
BASE/electrode contact resistance (Ω-cm^2)	0.06	N/A	0.06
BASE Brazes leakage resistance (Ω)	3.0	N/A	3.0
Cold structural material	Haynes-25	Nb-1Zr	C-103
Performance Parameters			
Hot plate temperature (K)	1145 [+], 1147[#]	N/A	1171[+], 1174[#]
Cold plate temperature (K)	600	N/A	600
Electric power output W$_e$)	7.22 [+], 8.38[#]	8.65[*], 8.94[#]	6.89[+], 7.97[#]
Corresponding Efficiency (%)	17.6[+], 18.8[#]	15.7[*], 16.4[#]	17.3+, 18.4[#]
Load Voltage (V)	3.5	3.5	3.5
Braze temperature (K)	1123	< 1123	1123
Evaporator temperature (k)	1023[+], 1009[#]	1061 - 1063	1014[+], 999[#]
Temperature margin, □T (K)	+18[+], +28[#]	N/A	+ 29[+], +39[#]
Comments		BASE tube assembly includes compression stud	Inner surface of cold wall is also coated with Rh.

Electrodes B = 120 A.K$^{1/2}$/m^2.Pa and G = 10; + Electrodes B = 75 A.K$^{1/2}$/m^2.Pa and G =50; * Electrodes B = 80 A.K$^{1/2}$/m^2.Pa and G =50; and @ low mass cell.

DESCRIPTION OF THE Nb-1Zr/C-103 AMTEC CELL

Figure 1a shows an exploded trimetric view of the present Nb-1Zr/C-103 and nickel/Haynes-25 cells. These cells have eight BASE tubes, 0.4 mm thick, 7.1 mm in outer diameter, and 50.1 mm long, connected electrically in series. The BASE ceramic-metal braze joints are made of TiNi, Ta, and α-Al$_2$O$_3$ (see Table 1 and Figure 1a). Owing to its lower thermal conductivity and higher strength, C-103 is very suitable for the cold structure of the cell. In this part of the cell, the wall (101 μm thick) is much thinner than the hotter portion of wall (~ 200 μm thick), facing the BASE tubes. Joining the Nb-1Zr and C-103 should be quite easy since both alloys have the same primary metal, niobium. The hot structure of the cell is made of Nb-1Zr (see Figures 1a and 1b).

As shown in Figure 1a , the surface of the evaporator wick is flat. Its average pore size of < 5 μm is much smaller than that of the liquid sodium return artery, in order to produce high capillary pressure head for circulating the sodium working fluid through the cell. It has been shown that, for the same electric current, a flat evaporator operates at ~ 10 K lower surface temperature than a conical evaporator, resulting is a higher temperature margin in the cell (Tournier and El-Genk, 2000). Table 1 compares the design and performance parameters of AMPS' EPX-1 cell (Hendricks et al., 1999) and of the present NB-1Zr/C-103 and nickel/Haynes-25 cells.

Performance of the Present Nickel/Haynes-25 Cell

Figure 2 shows the calculated performance envelope and the cell's electric power output and efficiency contours of the present nickel/Haynes-25 AMTEC cell. The white region in the figure represents the performance envelope of the cell in which none of the design temperature limits are exceeded. These limits are $T_{evap} \leq 1023$ K, $T_{braze} \leq 1123$ K, and $\Delta T \geq 20$ K. As indicated in Figure 2, the constraining design limits are ΔT and T_{braze}. A maximum electric power of 7.5 W_e is achieved at a cell efficiency of 16.6% and load voltage of 2.9 V. The corresponding thermal power input is 45.5 W_{th}.

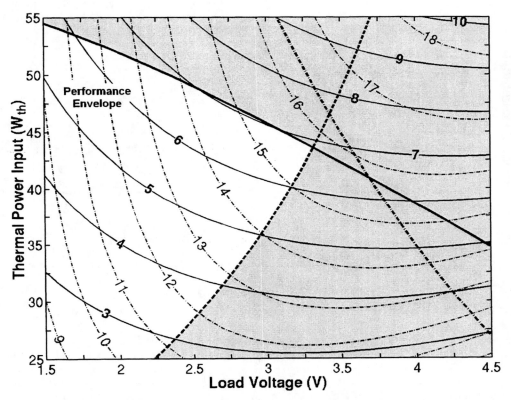

FIGURE 3. Performance envelope of the present NB-1Zr/C-103 cell without Rh coating.

At lower thermal power input, the maximum cell performance follows the right boundary of the performance envelope representing $\Delta T = +20$ K, to lower cell voltage, electric power, and conversion efficiency. At input thermal power beyond 45.5 W_{th}, the performance of the cell follows the top boundary representing $T_{braz} = 1123$ K, decreasing both the cell efficiency and the electric power output. At a load voltage of 3.5 V, not only is the cell electric power (7.1 W_e) lower, but also all the temperature design limits are violated as the operation point falls to the right of the performance envelope.

Performance of Present Nb-1Zr/C-103 Cell

The calculated performance parameters and envelopes for two design options of the present Nb-1Zr/C-103 refractory cell are presented in Figures 3 and 4, respectively. In the first option, the surfaces of the C-103 thermal radiation shield, the casing of the sodium artery, and the cold wall above the BASE tubes are not covered by rhodium. They are coated with rhodium, however, in design option two. The surface emissivity of Nb-1Zr and C-103 are both taken as temperature dependent. The emissivity of Nb-1Zr at 1000 K was 0.15 while that of C-103 at the same temperature was taken equal to that of niobium (0.13). Emissivity data for C-103 could not be found in the literature. The emissivity of rhodium at 1000 K is only 0.06.

For option one, at the highest load voltage within the operation envelope (3.35 V), the electric power of the cell is 6.8 W$_e$ at an efficiency of 15.8 % (see Figure 3). However, the maximum electric power output in the operation envelope is 7.1 W$_e$ and the cell's efficiency and load voltage are 15% and 2.85 V, respectively.

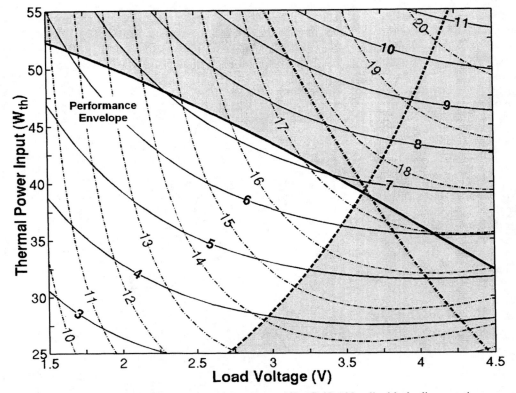

FIGURE 4. Performance envelope of the present NB-1Zr/C-103 cell with rhodium coating.

Covering the inner surface of the cell wall above the BASE tubes, the surface of the casing of the liquid sodium return artery and of the thermal radiation shield with low emissivity rhodium improved the cell performance (see Figure 4). The maximum load voltage within the operation envelope increased to 3.65 V, at which the cell electric power and efficiency are 6.8 W$_e$ and 17.3%, respectively. The maximum electric power of the cell within the performance envelope also increased to 7.3 W$_e$ at an efficiency of 16% and a load voltage of 2.8 V. The results in Figures 3 and 4 clearly demonstrate the superiority of the second design option of the present Nb-1Zr/C-103 cell. The cell mass is 194 g, including the sodium working fluid, the cell electric feed throughs, and the liquid sodium fill port.

Low Mass Nb-1Zr/C-103 Cell

An analysis was performed to reduce the mass of the Nb-1Zr/C-103 cell in design option two, by reducing the thickness the hot plenum wall, the BASE tubes support plate, the hot plate, and the conduction studs (see Figures 1a and 1b). Reducing the thickness of the hot plenum wall (see Table 1) by as much as 75% slightly decreased the cell efficiency and ΔT, and resulted in a mass saving of 1.6 g. When the thickness of the BASE support plate was reduced by 75%, the cell efficiency slightly increased and ΔT decreased slightly. The corresponding mass saving was 16.8 g. Varying the thickness of the hot plate from 25% to 200% insignificantly affected the cell performance (see Table 1). In order to maintain the cell's structural strength, the thickness of the hot plate (see Table 1) was reduced by only 75%, resulting in a mass saving of 10.9 g. When the cross-sectional area of the conduction studs was decreased by as much as 90%, ΔT was insignificantly affected and the cell efficiency decreased only slightly. This is probably because of the high thermal conductivity of Nb-1Zr. The corresponding mass saving is only 1.3 g.

When all the above changes in the cell structure were incorporated, the mass saving was 30.6 g, reducing the cell's total mass from 194 g to 163.4 g. As shown in Figure 5, the maximum load voltage in the performance envelope of the cell increased to 3.75 V, at which point the cell's electric power and efficiency decreased to 6.4 W_e and 17.2%, respectively. The maximum electric power of the cell is 7.2 W_e at 16.2% efficiency and a load voltage of 2.8 V.

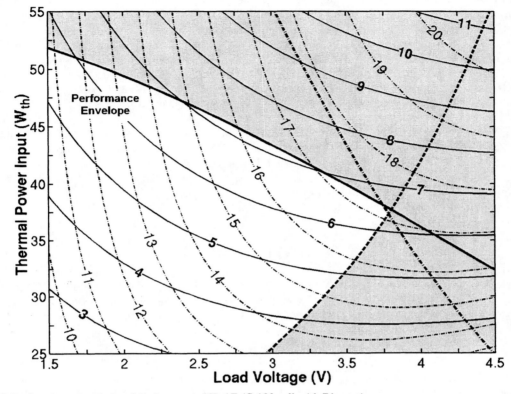

FIGURE 5. Performance envelope of the low-mass NB-1Zr/C-103 cell with Rh coating.

The results presented in Figures 2-5 are for BASE electrodes characteristics of B = 75 A.K$^{1/2}$/m^2.Pa and G = 50. Using high performance electrodes having characteristics of B = 120 A.K$^{1/2}$/m^2.Pa and G = 10 significantly improved the cell performance. The maximum load voltage increased to 4.15 V, at which the cell electric power is 7.1 W_e and efficiency is 18.4%. The maximum cell electric power of 8.3 We occurred at a load voltage of 3.0 V with a conversion efficiency of 17.3 %. At these operation parameters the cell's specific mass is 19.7 g/W_e.

SUMMARY AND CONCLUSIONS

The performance of a number of Nb-1Zr/C103, multi-tube vapor anode AMTEC cells is analyzed. The cell's hot structure (hot plate, BASE tubes support plate, hot plenum wall, evaporator standoff, evaporator wick, and side wall facing the BASE tubes) is made of Nb-1Zr. However, the cold structure of the cell (condenser, interior cylindrical thermal radiation shield, the casing and the wick of the liquid sodium return artery, and side wall above the BASE tubes) is made of the stronger, lower conductivity niobium alloy C-103. The performance of a nickel/Haynes-25 cell of similar design is also evaluated and compared with that of Nb-1Zr/C-103 cells. Details of the design and performance parameters of these cells and of AMPS' EPX-1 cell are listed in Table 1.

The performance contours and envelopes of the present cells are calculated and compared. The cell performance envelope is bounded by the isotherms corresponding to the temperature limits for the BASE braze joints and the evaporator (1123 K and 1023 K, respectively) and to a temperature margin $\Delta T = +20K$.

Results indicated that the performance of the Nb-1Zr/C-103 cell is superior to the nickel/Haynes-25 cell. Covering the surface of the C-103 wall, the casing of the liquid sodium return artery, and the thermal radiation shield with a low emissivity rhodium coating further improved the cell's performance. The maximum load voltage within the performance envelope increased from 3.35 V to 3.65 V, the electric power remained unchanged at 6.8 W_e, but the cell efficiency increased from 15.8% to 17.3%. The mass of the Nb-1Zr/C-103 reference cell (194g) is only 2 grams heavier than the nickel/Haynes-25 cell. These performance parameters are calculated with BASE electrode characteristics of B = 75 A.$K^{1/2}$/m^2.Pa and G = 50. The mass of the Nb-1Zr/C103 cell is reduced by ~15.77% (or 30.6 g), by reducing the thickness of the structure in the cell's hot plenum, to 163.4 g, with little change in performance. The calculated maximum electric power of this cell was 7.2 W_e at a load voltage of 2.8 V and an efficiency of 16.3%; the corresponding specific mass of the cell is 22.7 g/W_e. For a load voltage of 3.5 V, the maximum cell power within its performance envelope is 6.89 W_e at an efficiency of 17.3%.

When BASE electrodes having characteristics of B = 120 A.$K^{1/2}$/m^2.Pa and G = 10 are used, the performance of the low-mass Nb-Zr/C-103 cell further improved. The maximum electric power within the cell's performance envelope increased by 15 % to 8.3 W_e, cell efficiency increased 1.1 percentage points to 17.3%, and the load voltage increased by 7.14% to 3 V. At this performance level, the cell specific mass decreased by 13% to 19.7 g/We. For a load voltage of 3.5 V, the maximum electric power within the performance envelope of the cell increased by 13%, from 6.89 to 7.97 W_e, and the cell efficiency increased 6.35%, from 17.3% to 18.4%.

ACKNOWLEDGMENTS

This research was sponsored by the University of New Mexico's Institute for Space and Nuclear Power Studies (ISNPS) and supported by the New Mexico Space Grant Consortium. The authors wish to thank Dr. Jean-Michel Tournier of ISNPS for his valuable comments as well as help throughout the course of this research.

REFERENCES

El-Genk, M., Tournier, J.-M., James, R., and Mayberry C., "Super-Alloy, AMTEC Cells for the Pluto/Express Mission," *in Proceedings of Space Technology and Applications International Forum*, edited by M.S. El-Genk, AIP Conference Proceedings 458, 2, 1999, pp. 1293-1300.

El-Genk, M. and Tournier, J.-M., "Recent Advances in Vapor-Anode, Multi-tube Alkali Metal Thermal-to-Electric Conversion Cells for Space Power," *in Proceedings of 5th European Space Power Conference (ESPS-98), SP-416, European Space Agency*, 1998, paper No. 1046.

Hendricks, T., Huang, C. and Huang, L., "AMTEC Cell Optimization for Advanced Radioisotope Power System (ARPS) Design," in *Proceedings Intersociety Energy Conversion Engineering Conference*, Society of Automotive Engineers, Inc., Paper No. 1999-1-2655.

King, J., and El-Genk, M., "A Review of Refractory Materials for Vapor-Anode AMTEC Cells," *in this Proceedings Space Technology and Applications International Forum*, edited by M.S. El-Genk, 2000.

Merrill, J., Schuller, M., and Huang, L., "Vacuum Testing of High-Efficiency Multitube AMTEC Cells: February 1997-October 1997," *in Proceedings of the 1998 Space Technology and Applications International Forum*, edited by M.S. El-Genk, AIP Conference Proceedings 420, 3, 1998, pp. 1613-1620.

Merrill, J. and Mayberry, C., "Experimental Investigation of Multi-AMTEC Cell Ground Demonstration Converter Systems Based on PX-3 and PX-5 Series AMTEC Cells," *in Proceedings of Space Technology and Applications International Forum*, edited by M.S. El-Genk, AIP Conference Proceedings 458, 2, 1999, pp. 1369-1377.

Tournier, J.-M., and El-Genk, M., "Analysis of Test Results of a Ground Demonstration of a Pluto/Express Power Generator," *J. Energy Conversion and Management,* 40 (1999a), 1113-1128.

Tournier, J.-M., and El-Genk, M., "Performance Analysis of Pluto/Express, Multitube AMTEC cells," *J. Energy Conversion & Management,* 40 (1999b), 139 – 173.

Tournier, J.-M. and El-Genk, M., "Conical Evaporator and Liquid Return Wick Model for Vapor Anode, Multi-Tube AMTEC Cells," in the current *Proceedings of the Space Technology and Applications International Forum*, edited by M.S. El-Genk, 2000.

A Review of Refractory Materials for Vapor-Anode AMTEC Cells

Jeffrey C. King[1] and M.S. El-Genk[2]

Institute for Space and Nuclear Power Studies and Chemical and Nuclear Engineering Dept.
The University of New Mexico, Albuquerque, NM 87131
[1](505) 277-3321, FAX:-2814, ajc@isnps.unm.edu
[2](505) 277-5442, FAX:-2814, mgenk@unm.edu

Abstract. Recently, refractory alloys have been considered as structural materials for vapor-anode Alkali Metal Thermal-to-Electric Conversion (AMTEC) cells, for extended (7-15 year) space missions. This paper reviewed the existing database for refractory metals and alloys of potential use as structural materials for vapor-anode sodium AMTEC cells. In addition to requiring that the vapor pressure of the material be below 10^{-9} torr (133 nPa) at a typical hot side temperature of 1200 K, other screening considerations were: (a) low thermal conductivity, low thermal radiation emissivity, and low linear thermal expansion coefficient; (b) low ductile-to-brittle transition temperature, high yield and rupture strengths and high strength-to-density ratio; and (c) good compatibility with the sodium AMTEC operating environment, including high corrosion resistance to sodium in both the liquid and vapor phases. Nb-1Zr (niobium-1% zirconium) alloy is recommended for the hot end structures of the cell. The niobium alloy C-103, which contains the oxygen gettering elements zirconium and hafnium as well as titanium, is recommended for the colder cell structure. This alloy is stronger and less thermally conductive than Nb-1Zr, and its use in the cell wall reduces parasitic heat losses by conduction to the condenser. The molybdenum alloy Mo-44.5Re (molybdenum-44.5% rhenium) is also recommended as a possible alternative for both structures if known problems with oxygen pick up and embrittlement of the niobium alloys proves to be intractable.

INTRODUCTION

Alkali-Metal Thermal-to-Electric Conversion (AMTEC) is a promising new technology, currently under development for use in future space exploration missions. This technology, which also has great promise for terrestrial electric power generation in remote areas, offers several benefits over other static energy conversion devices. In addition to a relatively high conversion efficiency and moderate operation temperatures, AMTEC cells can provide much higher voltages than thermoelectric or thermionic devices. Numerous multi-tube, vapor-anode sodium AMTEC cells of the Pluto-Express (PX) series, in extended vacuum tests lasting from a few months to more than two years, have demonstrated conversion efficiencies of up to 15% and electric power output of 3-5 We per cell at a load voltage of 2.5 - 3 V (Merrill et al., 1997, Merrill and Mayberry, 1999).

A typical vapor-anode, multi-tube AMTEC cell consists of two cavities separated by 5 to 9 beta"-alumina solid electrolyte (BASE) tubes and a sodium condenser/return artery/evaporator structure. In the inner or high pressure cavity (the anode side of the BASE tubes), the pressure of the sodium vapor is ~ 60 KPa. In the outer low pressure cavity, the sodium vapor pressure is ~20-60 Pa. The electric potential developed across the BASE tubes, which are electrically connected in series, is directly proportional to the logarithm of the sodium pressure ratio across the BASE tubes. A remote condenser plate opposite to the hot plate and a thin metallic cell wall enclose the outer cavity. To reduce parasitic heat losses by conduction to the condenser, the cell wall and return artery need to be as thin and thermally non-conductive as possible. To further reduce the parasitic heat losses by thermal radiation from the BASE tubes, the inner surface of the cell's side wall should have a low emissivity. To accomplish this, a low emissivity thermal radiation shield is also installed on the inside of the cell wall above the BASE tubes. The high reflectivity of the liquid sodium film covering the condenser surface also reduces heat loses by thermal radiation. In addition, the outside of the cell is thermally insulated using low conductance multi-foil and molded Min-K insulations (Hendricks and Huang, 1998).

CP504, *Space Technology and Applications International Forum–2000*, edited by M. S. El-Genk
© 2000 American Institute of Physics 1-56396-919-X/00/$17.00

Previous PX-series AMTEC cells have used stainless steel as their structural material. Some of these cells have undergone continuous testing for almost 2 years (Merrill et al., 1997; Merrill and Mayberry, 1999). These cells have exhibited a rapid initial decrease in performance, followed by a slower, but continuous, performance fall-off. It has been speculated that this decrease in cell performance was caused by a degradation of the cathode electrode, as reported in single effect electrode tests performed at the Jet Propulsion Laboratory (JPL) (Ryan et al., 1998). Other potential causes of the degradation in the PX-cells performance include potential contamination of the electrodes and the BASE by volatile materials such as manganese and chromium from the stainless steel cell structure. In post-test visual examinations of PX series cells that had failed early in vacuum testing, massive material transport and deposition, in the form of large globules, was detected on the outer surface of the sodium return artery casing (Merrill et al., 1997). Similar material deposition has also been noted on the inside surface of the cell wall near the condenser. So far, no detailed evaluation of the composition of these deposits has been performed.

As a replacement for stainless steel, El-Genk et al. (1999) have investigated the performance of PX-type cells with a nickel hot-end structure and Haynes-25 at the cell cold-end. Although the predicted cell performance was up to 30% higher than the stainless-steel PX series cells, several lifetime issues were raised. One of these issues is that the vapor pressure of nickel at the typical hot side temperature of 1200 K is higher than 10^{-9} torr (133 nPa) (Ryan, 1999). A vapor pressure of 10^{-9} torr has been recommended by JPL as a practical limit to ensure minimal mass loss and transport within the cell during its operation lifetime of 7-15 years. Although this vapor pressure limit may eventually be proven to be quite conservative, it is nonetheless a recommended rule of thumb until actual data on volatilization and transport become available.

Meanwhile, the attention of the AMTEC community began to focus on ensuring long operation lifetime of the PX-type cells by selecting structural materials with known good compatibility with liquid and gaseous sodium, low volatility, good strength and fabricability, and known thermophysical properties. Advanced Modular Power Systems (AMPS) has replaced stainless steel in the PX-series cells with the refractory alloy Nb-1Zr (niobium with 1% zirconium) (Svedberg and Sievers, 1998). Other refractory alloys were also considered for use as AMTEC structural materials, but to date, only cells constructed of Nb-1Zr have been fabricated and are currently being tested. Unfortunately, a concern related to the potential for rapid oxygen pick-up and embrittlement of this alloy has been raised (Kramer, et. al., 1999).

Modeling has shown that parasitic heat losses from conduction along the cell sidewalls and return artery and radiative heat transfer to the cold end of cell greatly effect cell efficiency (El-Genk et al., 1999). To reduce these parasitic heat losses, a structural material with low thermal conductivity and low emissivity is desirable. Since the thermal conductivity of all the candidate refractory metals is higher than that of stainless steel, conduction losses will be more significant in refractory metal cells. Fortunately, the emissivities of all of the refractory metals of interest are lower than that of nickel and stainless steel, which somewhat offsets the higher thermal conductivity of these materials. In addition, to minimize thermal stresses in the cell structure, the chosen material should have a small thermal expansion coefficient.

Although a structural material or alloy that ideally meets the long lifetime operation requirements for vapor-anode AMTEC cells might not exist, there is a need to review the literature on refractory metals and alloys to identify the most suitable candidates. Refractory metals were extensively studied for space power applications during the SNAP and SP-100 programs in the sixties and eighties, respectively, and some were selected for use in liquid-metal cooled space nuclear reactor power systems. PWC-11 (niobium-1% zirconium with 0.5% carbon) was developed during these programs and was selected as the structural material in the SP-100 nuclear reactor power system, which used molten lithium as the coolant (Titran, Stephens and Petrasek, 1988). Coolant materials considered were lithium, sodium, potassium, and Na-K alloy. Na-K and potassium were also considered as working fluids in ground tests of alkali metal Rankine systems (DiStefano, 1989). Unfortunately, these programs were either cancelled or terminated before successful implementation in a space environment.

The objectives of this paper are to review the literature on refractory metals and alloys to identify potential structural materials for vapor anode sodium AMTEC cells that are likely to meet the operation and lifetime requirements. The screening criteria used are based on either recommended rules of thumb or desirable properties. These properties are: (a) low thermal conductivity and emissivity; (b) low ductile-to-brittle transition temperature; (c) high yield and rupture strengths along with high strength to density ratios; (d) low thermal expansion coefficient; and (e) good compatibility with the sodium AMTEC operating environment (including low vapor pressure and low susceptibility to alkali metal corrosion).

REFRACTORY METALS

In common practice, the term "refractory metals" refers to the twelve metallic elements with melting points equal to or greater than that of chromium (Tietz and Wilson, 1965). These elements are shown as a subset of the periodic table in Figures 1a and 1b (Winter, 1999). The disk areas in these figures are proportional to the densities (Figure 1a) or melting points (Figure 1b) of the elements. The melting point of technetium (2430 K) technically qualifies it as a refractory metal; however, technetium is a man-made, radioactive element and has been omitted from this review. Of the twelve refractory metals, only the six Group 5 and 6 elements (vanadium {V}, niobium {Nb}, tantalum {Ta}, chromium {Cr}, molybdenum {Mo}, and tungsten {W}) are usually considered as alloy base metals. However, since chromium is one of the volatile alloying elements of concern when using stainless steel and the superalloys as AMTEC structural materials, it has been excluded. Five of the remaining elements (hafnium {Hf}, ruthenium {Ru}, osmium {Os}, rhodium {Rh} and iridium {Ir}) are sufficiently rare that they are usually only considered as alloying additions, coatings, or catalysts in specialized applications. Although rhenium {Re} has been used for a variety of special applications in recent years (Sherman, Tuffias and Kaplan, 1999), it is extremely heavy and hard to work, and thus is considered only as an alloying additive.

Historically, element number 41 (niobium), along with the naturally occurring niobium/tantalum alloy, has been referred to as columbium {Cb}. In accordance with the current International Union for Pure and Applied Chemistry (IUPAC) standards, element number 41 is designated "niobium" for the remainder of this work. The term "refractory metal" refers to the pure refractory elements and "refractory alloy" indicates an alloy based on one of the refractory metals.

Thermal Conductivity

Figure 2 (Wilkinson, 1969; Benjamin, 1980; Noravian, 1998) presents the temperature dependent thermal conductivities of the

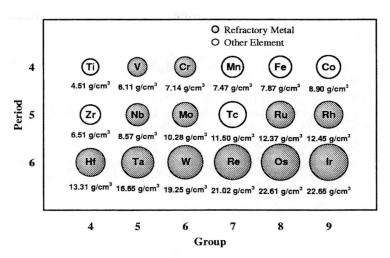

FIGURE 1a. Refractory metal densities (data from Winter, 1999).

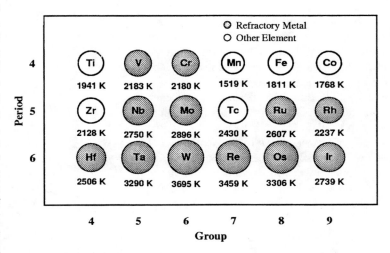

FIGURE 1b. Refractory metal melting points (data from Winter, 1999).

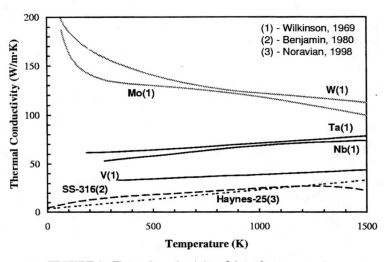

FIGURE 2. Thermal conductivity of the refractory metals.

1393

candidate refractory metals. Molybdenum and tungsten, which have the highest thermal conductivities, pose significant challenges to reducing the parasitic heat losses in the AMTEC cells. The thermal conductivity of niobium and tantalum are also relatively high. Of the refractory metals shown in Figure 2, only vanadium has a thermal conductivity that is only slightly higher than that of the stainless steel used in the laboratory tested PX-type AMTEC cells. Based on thermal conductivity considerations alone, vanadium would be the first choice, followed by niobium and tantalum.

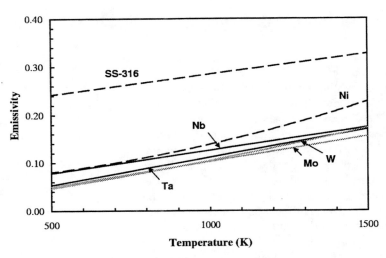

FIGURE 3. Thermal radiative emissivity of the refractory metals (data from Touloukian et al., 1970).

Emissivity

As shown in Figure 3 (Touloukian, et al., 1970), all of the Group 5 and 6 elements have similar thermal radiative emissivities that are lower than stainless steel and nickel. The Group 6 elements have very slightly lower emissivities than the Group 5 elements. Sufficient data on the emissivity of vanadium were not available and it is predicted that vanadium has an emissivity between those of nickel and niobium. Based on the small variations between the candidates, no strong distinctions can be made based on emissivity.

Thermal Expansion

The linear thermal expansion coefficients of the refractory metals of interest are delineated in Figure 4 (Touloukian et al., 1970). Tungsten and molybdenum have the lowest

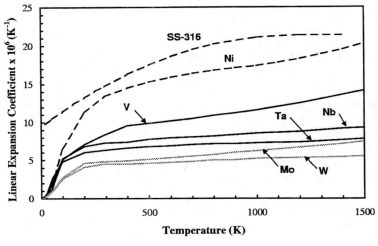

FIGURE 4. Linear thermal expansion coefficients of the refractory metals (data from Touloukian et al., 1970).

thermal expansion, followed by tantalum. Niobium and vanadium have higher thermal expansion, but are still lower than nickel and stainless steel. While a low thermal expansion is desirable, the fact that all refractory metals thermally expand less than stainless steel means that any of these metals are acceptable based on this criteria.

Ductility

Although it is not strictly a structural property, the ductility of the refractory metals, particularly at room temperature, is an important consideration for their use as an AMTEC structural material. Unfortunately, many of the refractory metals (particularly Group 6 elements) are not especially ductile at room temperature (Wilkinson, 1969). While brittle materials could be manufactured and handled safely, a material that is ductile at all temperatures of interest will greatly simplify fabrication and handling of the AMTEC cells. Figure 5 presents two measures of the ductility of the refractory metals, the reduction in area at fracture and the ductile-to-brittle transition temperature. It is important to note that the values of these parameters are very dependent on purity, fabrication techniques and handling of the material. All Group 5 elements (V, Nb, Ta) are ductile at all temperatures of interest. While the ductile-to-brittle transition of molybdenum occurs at or near room temperature, much work has been done to produce ductile molybdenum alloys and satisfactorily ductile alloys exist (such as Mo-TZM and Mo-Re) (Hagel, Shields and Tuominen, 1984). It has also been shown that highly pure, multiply-refined molybdenum is much more ductile than commercial grade molybdenum (Ault, 1966); however, it is not clear that the required purity levels can be obtained economically on a commercial scale. In addition to being very heavy, tungsten remains brittle at AMTEC cold-end operating temperatures (600-650 K). Based on their ductility, vanadium, niobium and tantalum are good candidates; molybdenum may be acceptable; and tungsten should be avoided.

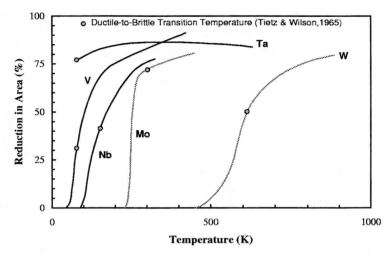

FIGURE 5. Ductility of the refractory metals (adapted from Wilkinson, 1969).

Yield Strength

Yield strength is a good measure of the short-term strength of the refractory metals. Based on the data presented in Figure 6 (Tietz and Wilson, 1965), molybdenum is the strongest refractory metal at AMTEC hot-end temperatures (~1200 K), followed by tantalum and tungsten. Niobium and vanadium both exhibit comparably lower strengths. Since a large fraction of the cell's structural stresses will result from the weight of the structural material, and because AMTEC cells designed for space applications need to be of low mass, consideration of the strength-to-density ratio of the refractory metals is important (Figure 7). Based on this ratio, molybdenum is clearly superior at hot-end temperatures and vanadium emerges as a strong contender at cold-end temperatures and up to 1000 K. For strength, molybdenum appears to be the best candidate and vanadium also warrants consideration. Niobium is the poorest candidate at lower temperatures (<1000 K), but is comparable to vanadium at AMTEC hot-side temperatures (~1200 K) (Figures 6 and 7).

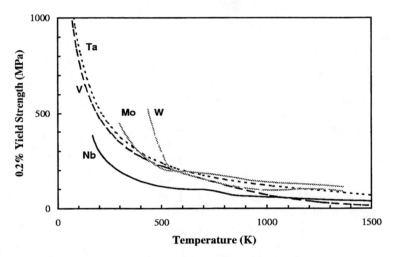

FIGURE 6. Yield strength of the refractory metals (adapted from Tietz and Wilson, 1965).

Creep Resistance

At a temperature above roughly one-third of a material's melting point, creep, or long-term deformation under load, becomes important

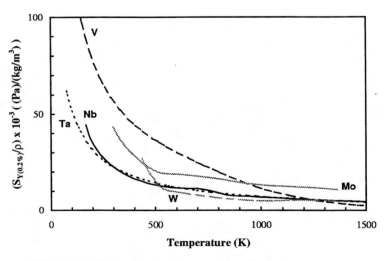

FIGURE 7. Yield strength to density ratio of the refractory metals.

(Evans and Wilshire, 1993). Many parameters have been used to quantify creep resistance and as a result, comparable data is difficult to come by. The 100-hour rupture strength has been chosen as a representative criterion. Based on the data presented in Figure 8 (Tietz and Wilson, 1965), tungsten is clearly the standout material in this case. Molybdenum also demonstrates greater creep resistance than either niobium or tantalum. While comparable data for vanadium were not found, it is expected to be the least creep resistant of the five candidate refractory metals, owing to its lower melting point.

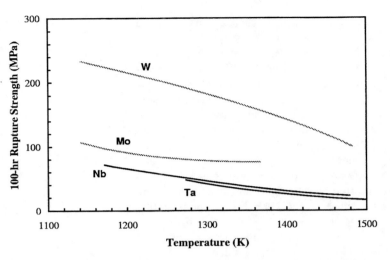

FIGURE 8. 100-hr rupture strength of the refractory metals (adapted from Tietz and Wilson, 1965).

Compatibility with the AMTEC Operating Environment

Three main mechanisms govern corrosion of refractory metals in alkali metal systems: dissolution, mass transfer, and reactions with impurities such as dissolved oxygen (DiStefano, 1989). Mass transfer of volatile metals within the PX-series AMTEC cells is one of the main reasons why refractory metals and alloys are being considered as alternatives to stainless steel and superalloy materials.

Figure 9 (Nesmeyanov, 1963) compares the sublimation vapor pressures of the refractory metals and alloying additions of interest with the major constituents of the stainless steels and superalloys. All of the refractory metals under consideration have vapor pressures below 10^{-9} torr (133 nPa) at all AMTEC operating temperatures (<1200 K). A sublimation vapor pressure of 10^{-9} torr has been recommended as a rule of thumb by the

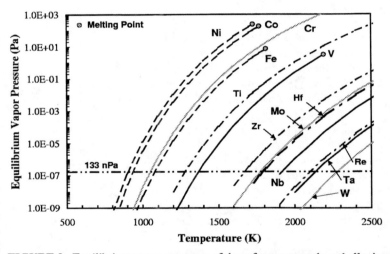

FIGURE 9. Equilibrium vapor pressures of the refractory metals and alloying additions (data from Nesmeyanov, 1963).

Jet Propulsion Laboratory to minimize mass transfer over a 10 to 15 year lifespan (Ryan, 1999). Reliable quantitative data on the corrosion of the refractory metals by liquid and vapor alkali metals is scarce; however, several sources have summarized the results of the numerous capsule and single- and two-phase loops tests conducted during the SNAP and SP-100 programs (DiStefano, 1989 and DeVan, DiStefano and Hoffman, 1984). In general, these tests concluded that compatibility of the refractory metals with sodium was very good providing that the oxygen contents of the sodium and the refractory metal were controlled. Niobium and tantalum were particularly susceptible to oxygen driven attack (even in sodium containing less than 10 ppm oxygen), requiring alloying with an oxygen getter such as zirconium or hafnium (to stabilize dissolved oxygen) in order to withstand long-term alkali-metal exposure. Vanadium has demonstrated behavior similar to niobium and tantalum in other compatibility tests (Greenberg, Ruther and Levin, 1967). Molybdenum and tungsten do not appear to be as susceptible to oxygen assisted alkali-metal attack and non-gettered alloys have demonstrated acceptable performance in capsule tests (Klopp, 1985). Given that the Group 6 elements (molybdenum and tungsten) have low vapor pressures and are better able to withstand sodium attack without alloying additions, they are preferred over the Group 5 metals (vanadium, niobium and tantalum). The later have similarly low vapor pressures (with the exception of vanadium, which is still in the acceptable range), but will require further investigation to identify acceptably performing alloys.

Ranking of the Refractory Metals

A rating of the candidate refractory metals based on the selection criteria detailed earlier is presented in Tables 1 & 2. A metal rated "poor" for a given property either fails to meet an accepted criterion or is significantly worse than other metals considered. An "acceptable" metal appropriately meets a criterion and is not significantly better or worse than other metals. A metal that is significantly better than the others is rated "excellent". The ratings for each category were used to determine a relative ranking for each of the metals in terms of the thermophysical and structural properties. The metals were not ranked for compatibility in the AMTEC operating environment since it is likely that any of them would perform satisfactorily if oxygen contamination can be controlled.

Based on the evaluation of the thermophysical properties of the refractory metals, vanadium would be the first choice as an AMTEC structural material followed by niobium. Looking at structural properties, molybdenum is recommended, followed by vanadium. Since there are some gaps in the data available for vanadium, particularly concerning long term creep-rupture performance, and because of the currently very limited experience with vanadium as a space power material, the authors recommend niobium and molybdenum as the best candidates for near term investigation as AMTEC structural materials. Unfortunately, both niobium and molybdenum possess certain weaknesses (susceptibility to oxygen driven alkali metal corrosion in the case of niobium and low temperature brittleness in the case of molybdenum). Both of these problems have been addressed by the development of alloys of these two refractory metals, which are reviewed next.

TABLE 1. Ranking of the refractory metals based on thermophysical properties.

	V	Nb	Ta	Mo	W
Density	Excellent	Acceptable	Poor	Acceptable	Poor
Thermal Conductivity	Acceptable	Acceptable	Acceptable	Poor	Poor
Emissivity	Acceptable	Acceptable	Acceptable	Acceptable	Acceptable
Linear Expansion	Acceptable	Acceptable	Acceptable	Acceptable	Acceptable
Rank	1	2	3	3	4

TABLE 2. Ranking of the refractory metals based on structural properties.

	V	Nb	Ta	Mo	W
Ductility	Acceptable	Acceptable	Acceptable	Acceptable	Poor
Yield Strength	Acceptable	Acceptable	Acceptable	Acceptable	Acceptable
Yield Strength/ Density Ratio	Excellent	Acceptable	Acceptable	Excellent	Acceptable
Rupture Strength	Likely Acceptable	Acceptable	Acceptable	Acceptable	Excellent
Rank	2	3	3	1	4

REFRACTORY ALLOYS

During SNAP program in the 1960's, a number of companies were involved in the development of many different niobium alloys (Condliff and Marsh, 1987). Unfortunately, only a few of those alloys remained commercially available (Svedberg and Sievers, 1998). Nb-1Zr (niobium-1% zirconium) was an early development and remained in use in the nuclear industry and as a general purpose high-temperature alloy. PWC-11 (niobium-1% zirconium-0.1% carbon) was developed in the 1960's as a higher strength alternative to Nb-1Zr (Titran, Stephens and Petrasek, 1988) but never entered commercial production. Nonetheless, a large amount of research and development was performed on PWC-11 during the SP-100 program (Condliff and Marsh, 1987). Since the fabricability and working characteristics of PWC-11 are very similar to those of the commonly available Nb-1Zr, it will be considered as a candidate AMTEC cell material. The niobium alloy C-103 (niobium-10% hafnium-1% titanium-0.5% ziconium) is a commercially available aerospace alloy that was developed during the Apollo program and has seen significant use in military aircraft. All three niobium alloys contain either zirconium and/or hafnium, which have been demonstrated to dramatically improve the alloy's resistance to oxygen driven alkali metal corrosion (DiStefano, 1989). The density of the candidate niobium alloys are very similar (Nb-1Zr; 8.58 g/cm^3; PWC-11, 8.6 g/cm^3; C-103, 8.86 g/cm^3).

The development of the molybdenum alloys was historically less energetic than the development of niobium alloys, but several promising candidates exist. TZM (molybdenum-0.5% titanium-0.1% zirconium) is readily available and offers improved strength and ductility over basic molybdenum (Klopp, 1985). Recent work on the ductilization of molybdenum by the addition of rhenium has resulted in the availability of a range of molybdenum-rhenium binary alloys. These alloys offer significantly improved ductility up to 44.5% rhenium; however, above 44.5% rhenium, an undesirable sigma phase is formed (Kramer et al., 2000). The authors recommend Mo-14Re (molybdenum-14% rhenium) and Mo-44.5Re (molybdenum-44.5% rhenium) for further consideration as AMTEC structural materials. Mo-44.5Re is commercially available and is currently being tested as an AMTEC candidate (Kramer, et al., 2000). Mo-14Re is only available as a special order, but offers the best ductility improvement for a lower rhenium addition. The density of the molybdenum alloys spans a greater range than the niobium alloys (from 10.16 g/cm^3 for TZM to 13.50 g/cm^3 for Mo-44.5Re).

In ranking the refractory metal alloys, only the properties that led to significant discrimination between the refractory metals (density, thermal conductivity and yield strength) were reconsidered. The thermal conductivity of the candidate refractory alloys is presented in Figure 10. In each case, the alloys have lower thermal conductivities than the parent metals. This is particularly significant for C-103, which has a thermal conductivity that is close to that of stainless steel and Haynes-25, which will result in significantly lower thermal conductive losses in the AMTEC cell. Unfortunately, the thermal conductivity of TZM is very high. Data for Mo-14Re was lacking, but the thermal conductivity of the alloy is expected to be intermediate to Mo-44.5Re and pure molybdenum.

Figure 11 compares the yield strengths of the candidate alloys. Of note are the particularly high strengths of Mo-44.5Re and TZM. While C-103's strength is comparable to that of the other niobium alloys at hot-end temperatures (~1200 K), it is superior to the other niobium alloys at cooler temperatures (<1000 K). Using C-103 would allow for a thinner cold wall, which would reduce heat losses by conduction and increase cell efficiency. PWC-11 is stronger than Nb-1Zr,

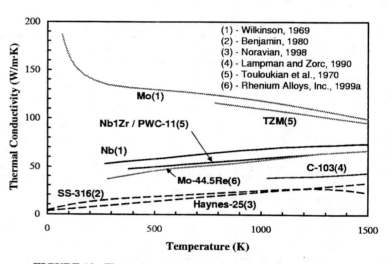

FIGURE 10. Thermal conductivity of selected refractory alloys.

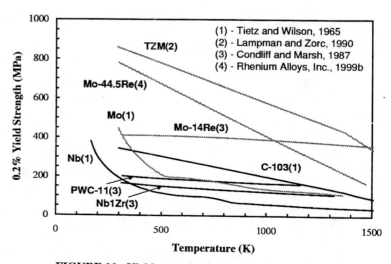

FIGURE 11. Yield strength of selected refractory alloys.

but not dramatically so, considering the readily available C-103 as an alternative.

It is also important that any refractory alloy chosen be compatible with AMTEC operating conditions. This means that the vapor pressure of the constituent metals should be acceptably low ($<10^{-9}$ torr or 133 nPa) and that the alloy demonstrates adequate resistance to alkali-metal corrosion. The vapor pressures of the metal additions in each of the candidate alloys are shown in Figure 9. In every case, the vapor pressure is below the 133 nPa rule of thumb, although titanium approaches this limit at AMTEC hot end temperatures (~1200 K). Capsule testing of Nb-1Zr and TZM has shown good resistance to attack by liquid and vapor sodium (DiStefano, 1989). Such data are not available for PWC-11, molybdenum-rhenium, and C-103. Data for niobium alloys have shown that alloys containing zirconium or hafnium perform satisfactorily, so PWC-11 and C-103 are judged to be acceptable. The only data available for molybdenum-rhenium alloys are for Mo-50Re in lithium and for tungsten-rhenium in several liquid metals (DeVan, DiStefano and Hoffman, 1984 and Klopp, 1985). In both cases, acceptable resistance was noted and molybdenum-rhenium is likely to perform satisfactorily in the AMTEC environment.

Table 3 ranks the candidate niobium and molybdenum alloys relative to the other alloys with same primary metal using the same screening criteria used for the refractory metals. While all of the niobium alloys in Table 3 are acceptable, C-103 is rated excellent with regard to every criterion. From a strength standpoint, all of the molybdenum alloys are rated acceptable to excellent; unfortunately, their high thermal conductivity is problematic and only the addition of large amounts of low thermal conductivity, rhenium makes Mo-44.5Re acceptable at the expense of significantly increasing density. While the preferred niobium alloy is clearly C-103, Nb-1Zr is rated acceptable in every category; and, significant experience with this particular alloy has been developed during the SNAP and SP-100 programs. The higher thermal conductivity of Nb-1Zr will be of benefit at the hotter end of the cell when it is matched with a lower conductance material (such as C-103) for the cooler part of the cell, particularly the thin (~100 μm) cell wall above the BASE tubes.

Recent investigations (Kramer et al., 1999 & 2000) have raised the concern of oxygen pickup and embrittlement of Nb-1Zr with Mo-44.5Re being recommended as an alternative material (Kramer et al., 2000). While the oxygen embrittlement concerns are likely to apply to most of the niobium alloys, including C-103, the higher thermal conductivity of the molybdenum alloys will increase parasitic conduction heat losses in the cell and will likely dramatically reduce cell efficiency. Mo-44.5Re is also ~50% more dense than C-103 and Nb-1Zr. Therefore, the niobium alloys, Nb-1Zr and C-103, are preferred, unless the problem of oxygen embrittlement turns out to be intractable.

TABLE 3. Ranking of the Refractory Metal Alloys

	Niobium Alloys			Molybdenum Alloys		
	Nb-1Zr	**PWC-11**	**C-103**	**Mo-14Re**	**Mo-44.5Re**	**TZM**
Thermal Conductivity	Acceptable	Acceptable	Excellent	Likely Acceptable	Acceptable	Poor
Yield Strength	Acceptable	Acceptable	Excellent	Acceptable	Excellent	Excellent
Fabricability	Excellent	Excellent	Excellent	Acceptable	Acceptable	Acceptable

SUMMARY AND CONCLUSIONS

Vapor anode, multi-tube AMTEC cell technology has been under development for more than four years for consideration for use in future space exploration missions. These cells are currently being considered for use in conjunction with radioisotope heat sources to provide the electric power requirements for the Europa spacecraft in year 2003-2004 and the Pluto-Express (PX) flyby mission in 2005-2006. Early PX-series cells were made with stainless steel structures and some have been tested in vacuum for more that 8000 hours. The test results have raised concerns regarding the use of stainless steel as a structural material for vapor-anode, sodium AMTEC cells. The volatility of iron and other alloying elements such as magnesium, nickel and chromium, along with their solubility in sodium, particularly in the presence of minute amounts of oxygen (10-15 ppm), were considered potential contributors to the observed degradation in the cell performance in the vacuum tests.

Recently, refractory metals and alloys are being considered to replace stainless steel as structural materials in vapor-anode AMTEC cells. Despite their strength and very low vapor pressure at typical AMTEC cell operating temperatures (1150-1200 K and 550-650 K for the hot plate and condenser, respectively), several of the refractory metals are very soluble in sodium at these temperatures in the presence of trace amounts of oxygen (<10 PPM). To enhance the corrosion resistance of these refractory metals, alloying elements with a high affinity for oxygen (gettering elements) are added to sequester the sodium soluble oxygen.

The existing database for refractory metals and alloys for potential use as vapor-anode sodium AMTEC structural materials is reviewed and evaluated. In addition to requiring that the vapor pressure of the selected material be below 10^{-9} torr (133 nPa) at a typical hot side operational temperature of 1200 K, other screening criteria were considered, including: (a) low thermal conductivity, low thermal radiation emissivity, and low linear thermal expansion coefficient; (b) low ductile-to-brittle transition temperature, high yield and rupture strengths and high strength-to-density ratio; and (c) good compatibility with the AMTEC operating environment, including high corrosion resistance to sodium in both the liquid and vapor phases.

None of the refractory metals were perfect candidates and none were ranked first in all screening criteria; however, niobium and molybdenum were ranked the highest. Niobium, unfortunately, is subject to corrosion by sodium even in the presence of less than 10 ppm oxygen. This problem could be alleviated by alloying with gettering elements such as zirconium and hafnium, which sequester oxygen and hence increase the corrosion resistance of the primary refractory metal. The use of molybdenum is limited by low its ductility, particularly at room temperature, and its high thermal conductivity. Fortunately, alloying additions, particularly rhenium, can dramatically improve the ductility of the primary metal while lowering the thermal conductivity.

The alloy Nb-1Zr (niobium-1% zirconium) is recommended for the hot cell structure. This includes the cell's hot plate, the BASE tube support plate, the evaporator standoff and evaporation wick, and the thick (~200μm) wall facing the BASE tubes. The cell wall above the BASE tubes is almost half the thickness of the hotter part of the wall to minimize parasitic conduction heat losses to the cell condenser. For this part of the structure, a strong material with low thermal conductivity is preferred. Therefore, the niobium alloy C-103, containing the oxygen getters zirconium and hafnium along with titanium, is recommended for the cold structure of the AMTEC cell. This includes the side wall above the BASE tubes, the condenser structure, the casing and the wick of the liquid sodium return artery, and the interior thermal radiation shield. C-103 is stronger and less thermally conductive than Nb-1Zr.

ACKNOWLEDGMENTS

This research was sponsored by the University of New Mexico's Institute for Space and Nuclear Power Studies (ISNPS) and supported by funding from the New Mexico Space Grant Consortium. The authors wish to thank Dr. Jean-Michel Tournier of ISNPS for his valuable comments and assistance through out the course of this research.

REFERENCES

Ault, G. M., *A Decade of Progress in Refractory Metals*, Philadelphia, PA, American Society for Testing and Materials, 1966, pp. 5-6.

Benjamin, D., senior editor, *Metals Handbook, Ninth Edition, Vol. 3*, Metals Park, OH, American Society for Metals, 1980.

Condliff, A.F., and Marsh, R.J., "Selection of Structural Materials for Space Nuclear Power Generators," in *Space Nuclear Power Systems 1986*, edited by M.S. El-Genk and M.D. Hoover, Orbit Book Company, Malabar, FL, 1987, pp. 275-281.

DeVan, J.H., DiStefano, J.R., and Hoffman, E.E., "Compatibility of Refractory Alloys with Space Reactor Coolants and Working Fluids," in *Proceedings of the Symposium on Refractory Alloy Technology for Space Nuclear Power Applications*, edited by R.H. Cooper and E.E. Hoffman, Report CONF-8308130, U.S. Department of Energy, 1984, pp. 34-85.

DiStefano, J. R. "Review of Alkali Metal and Refractory Alloy Compatibility for Rankine Cycle Applications," in *Space Nuclear Power Systems 1988*, edited by M.S. El-Genk and M. D. Hoover, Orbit Book Company, Malabar, FL, 1989, pp. 299-310.

El-Genk, M.S., Tournier, J.-M., James, R., and Mayberry, C., "Super-Alloy, AMTEC Cells for the Pluto/Express Mission," in *Proceedings of the Space Technology and Application International Forum (STAIF-99)*, edited by M.S. El-Genk, AIP Conference Proceedings No. 458, American Institute of Physics, New York, 1999, pp. 1293-1300.

Evans, R.W., and Wilshire, B., *Introduction to Creep*, London, The Institute of Materials, 1993, pp. 1-6.

Greenberg, S., Ruther, W.E., and Levin, H.A., "Corrosion of Vanadium Base Alloys in Sodium at 550° to 750°C," in *Alkali Metal Coolants - Proceedings of the Symposium on Alkali Metal Coolants*, International Atomic Energy Agency, Vienna, Austria, 1967, pp. 63-84.

Hagel, W.C., Shields, J.A., and Tuominen, S.A., "Processing and Production of Molybdenum and Tungsten Alloys," in *Proceedings of the Symposium on Refractory Alloy Technology for Space Nuclear Power Applications*, edited by R.H. Cooper and E.E. Hoffman, Report CONF-8308130, NTIS, U.S. Department of Energy, 1984, pp. 98-113.

Hendricks, T.J., and Huang, C., "System Design Impacts of Optimization of the Advanced Radioisotope Power System (ARPS) AMTEC Cell," in *Proceedings of the 33rd Intersociety Energy Conversion Engineering Conference*, American Nuclear Society, LaGrange Park, IL, 1998, Paper No. 98-407.

Klopp, W.D., "Technology Status of Molybdenum and Tungsten Alloys," in *Space Nuclear Power Systems 1984*, edited by M.S. El-Genk and M. D. Hoover, Orbit Book Company, Malabar, FL, 1985, pp. 359-370.

Kramer, D.P., Ruhkamp, J.D., McNeil, D.C., Mintz, G.V., and Howell, E.I., "Mechanical Testing Studies on Niobium-1% zirconium in Association with Its Applications as Cell Wall Material in an AMTEC Based Radioisotope Space Power System," in *Proceedings of the 34th Intersociety Energy Conversion Engineering Conference*, Society of Automotive Engineers, Inc., Warrendale, PA, 1999, Paper No. 1999-01-2608.

Kramer, D.P., Ruhkamp, J.D., McNeil, D.C., Howell, E.I., Williams, M.K., McDougal, J.R., and Booher, R.A., "Investigation of Molybdenum-44.5% Rhenium as Cell Wall Material in an AMTEC Based Space Power System," in *Proceedings of Space Technology and Application International Forum (STAIF-2000)*, edited by M.S. El-Genk, AIP Conference Proceedings, American Institute of Physics, New York, 2000, in these proceedings.

Lampman, S.R., and Zorc, T.B., technical editors, *Metals Handbook, Tenth Edition, Vol. 2*, Metals Park, OH, American Society for Metals, 1990.

Merrill, J., Schuller, M., Sievers, M., Borkowski, C., Huang, L., and El-Genk, M., "Vacuum Testing of High Efficiency Multi-Base Tube AMTEC Cells," in *Proceedings of the 32nd Intersociety Energy Conversion Engineering Conference*, Institute of Electric and Electronics Engineers, Piscataway, NJ, 1997, Paper No. 97-379, pp. 1184-1189.

Merrill, J., and Mayberry, C., "Experimental Investigation of Multi-AMTEC Cell Ground Demonstration Converter Systems Based on PX-3 and PX-5 Series AMTEC Cells," in *Proceedings of the Space Technology and Application International Forum (STAIF-99)*, edited by M.S. El-Genk, AIP Conference Proceedings No. 458, American Institute of Physics, New York, 1999, pp. 1369-1377.

Nesmeyanov, A.N., *Vapor Pressure of the Chemical Elements*, edited by R. Gary, New York, Elsevier Publishing Company, 1963, pp. 150-456.

Noravian, H., Personal Communication with J.M. Tournier, Timonium, MD, Analytix Corporation, 1998.

Rhenium Alloys, Inc., online material property data, 1999a, http://www.rhenium.com/data2.htm.

Rhenium Alloys, Inc., data in the MatWeb online material properties database, 1999b, http://www.matls.com.

Ryan, M.A., Personal Communication, Pasadena, CA, Jet Propulsion Laboratory, March 18, 1999.

Ryan, M.A., Williams, R.M., Lara, L., Cortez, R.H., Homer, M.L., Shields, V.B., Miller, J., and Manatt, M.S., "The Sodium Exposure Test Cell to Determine Operating Parameters for AMTEC Electrochemical Cells," in *Proceedings of the 33rd Intersociety Energy Conversion Engineering Conference*, American Nuclear Society, LaGrange Park, IL, 1998, Paper No. 98-335.

Sherman, A.J., Tuffias, R.H., and Kaplan, R.B., "CVD Processing, Properties, and Applications of Rhenium," electronic document, Pacoima, CA, Ultramet, 1999, http://www.ultramet.com/rhenium2.htm.

Svedberg, R.C., and Sievers, R.K., "Refractory Metals for ARPS AMTEC Cells," in *Proceedings of the 33rd Intersociety Energy Conversion Engineering Conference*, American Nuclear Society, LaGrange Park, IL, 1998, Paper No. 98-397.

Tietz, T.E., and Wilson, J.W., *Behavior and Properties of Refractory Metals*, Stanford, Ca, Stanford University Press, 1965, pp. 28-39, 377-409.

Titran, R.H, Stephens, J.R., and Petrasek, D.W., "Refractory Metal Alloys and Composites for Space Nuclear Power Systems," *NASA Technical Memorandum 101364*, Cleveland, OH, NASA Lewis Research Center, 1988, pp. 5-6.

Touloukian, Y.S., Powell, R.W., Ho, C.Y., Klemens, P.G., *Thermophysical Properties of Matter, the TPRC Data Series, Vol. 1, 7 & 12*, New York, IFI/Plenum, 1970.

Wilkinson, W.D. *Properties of Refractory Metals*, New York, Gordon and Breach Science Publishers, 1969, pp. 29-138.

Winter, M.J., "WebElements," electronic periodic table and property database, Sheffield, England, University of Sheffield, 1999, http://www.shef.ac.uk/chemistry/web-elements/.

Investigation of Molybdenum-44.5%Rhenium as Cell Wall Material in an AMTEC Based Space Power System

Daniel P. Kramer, Joe D. Ruhkamp, Dennis C. McNeil, Edwin I. Howell,
Melvin K. Williams, James R. McDougal and Robert A. Booher

Mound Isotope Power Systems, Babcock & Wilcox of Ohio, Building 102, P.O. Box 3030, Miamisburg, OH 45343
(937)865-3558, kramdp@doe-md.gov

Abstract. A new generation of radioisotope space power systems based on AMTEC (Alkali Metal Thermal to Electrical Conversion) technology is presently being developed. The future application of this technology, as the electrical power system for outer planet deep space missions, is ultimately dependent on it being robust enough to withstand the mission's operational environments (high temperatures, dynamic loadings, long mission durations etc). One of the critical material selections centers on the cell wall whose physical and chemical properties must provide it with sufficient strength and material compatibility to successfully complete the mission. Niobium-1%zirconium has been selected as the baseline cell wall material with a molybdenum/rhenium alloy being the cell wall backup material. While these refractory materials have been commercially available for a number of years, several of their physical and mechanical properties have not been completely characterized especially within the expected operating parameters of an AMTEC based space power system. Additional characterization of the selected refractory alloys was initiated by fabricating mechanical test specimens out of ~0.5mm (0.020") thick sheets of material. Test specimens were heat treated at 1073K and 1198K for up to 150 hours under an argon cover gas containing small concentrations of oxygen. Room temperature and high temperature (1073K/1198K) mechanical tests were performed to determine the effect of time, temperature, and atmosphere on the mechanical properties of the refractory alloys. In addition, since the fabrication of AMTEC cell walls requires the welding of sheet material into a cylindrical shape, preliminary electron beam welding studies were performed. Comparison of the various test results obtained on Mo-44.5%Re and Nb-1%Zr samples are discussed.

INTRODUCTION

Over the last several years there has been a concerted effort to determine the potential usefulness of the AMTEC technology in future deep space power systems. One of the most important material issues in the projected application of this technology is the selection of the appropriate alloy for the AMTEC cell wall. The theoretical design and dimensional aspects of an AMTEC cell wall have been the basis of several previous papers (El-Genk, 1998; Hendricks, 1998). Independent of the selected cell design/dimensions, the material selection for the cell wall is critical since it is the main structural member that encases all of the internal working components of the cell including; the sodium inventory, condenser, artery, beta tubes, etc. It is crucial in selecting the cell wall material that it exhibits the required physical and mechanical properties to insure the successful completion of the mission. Some of the more important characteristics are; good high-temperature mechanical properties, weldability, sodium compatibility, corrosion resistance, and machinability. Several different alloys have been considered for application in fabricating AMTEC cell walls including Nb-1%Zr and Mo/Re alloys.

Previous research on niobium alloys determined that they have excellent sodium resistance at elevated temperatures provided that the total amount of available oxygen is minimal (Romano, 1967). After investigating several refractory niobium alloys, niobium-1%zirconium was selected as the baseline AMTEC cell wall material. However, research has demonstrated that when heated to elevated temperatures under an inert atmosphere containing small concentrations of oxygen the mechanical properties of niobium-1%zirconium are detrimentally effected (Kramer, 1999). This is of importance since it is anticipated that the final assembly of an AMTEC based radioisotope power system will take place in an inert atmosphere chamber that contains small concentrations of oxygen. All of the RTG's (Radioisotope Thermoelectric Generators) which were used to power the Galileo, Ulysses, and Cassini missions were assembled at Mound in the Inert Atmosphere Assembly Chamber or IAAC. The IAAC is ~2 meters wide x ~4.5 meters long x ~3.5 meters high. On one side of the IAAC is a row of mechanical manipulator arms and on the opposite side are two rows of glove ports. The argon atmosphere is closely controlled to contain <5ppm oxygen, <10ppm moisture, and <15ppm nitrogen.

CP504, *Space Technology and Applications International Forum–2000*, edited by M. S. El-Genk
2000 American Institute of Physics 1-56396-919-X

It is anticipated that the final assembly of an AMTEC based power system would likely take place in the IAAC or some similar inert atmosphere chamber. During assembly of the power system, the General Purpose Heat Source (GPHS) modules, containing the plutonium (238) dioxide fuel, would be placed within the AMTEC converter. This operation may take several days to perform with subsequent heating of the AMTEC cell walls. Since it was determined that niobium-1%zirconium may undergo mechanical property changes under the anticipated assembly conditions, the present work was initiated to determine the effect of time at temperature and test atmosphere on the mechanical properties of a molybdenum-rhenium alloy which is the designated backup AMTEC cell wall material.

MOLYBDENUM-RHENIUM TESTING STUDIES

The specific alloy selected for this study was molybdenum-44.5%rhenium. Adding rhenium to molybdenum tends to increase the ductility of the resultant molybdenum alloy. However, higher rhenium additions may result in the formation of an undesired sigma phase in the material. The selected alloy molybdenum-44.5% rhenium was obtained from Rhenium Alloys, Inc. (Elyria, OH) as ~0.5 mm thick (0.020") sheet. The material was produced via powder metallurgy techniques using −250 mesh 99.95% molybdenum and −325 mesh 99.99% rhenium powders as the raw materials. The powders were blended and pressed into sheet bars. Sintering was performed under hydrogen after which the sheet bar was hot-rolled/annealed several times until the final thickness was obtained. Specific details on the preparation of molybdenum-rhenium sheet material via powder metallurgy processing can be found in the following reference (Leonhardt, 1999).

Mo-44.5%Re tensile specimens were fabricated out of the purchased sheet stock using wire (molybdenum) electrode discharge machining (WEDM). The specimens were ~7.6 cm (3") long with a gauge length of ~2.5 cm whose gauge width was ~0.63 cm (0.25"). Each grip area was ~1.6 cm (0.62") long and ~1.6 cm (0.62") wide. After the specimens were fabricated, all of their edges were lightly sanded using 600 grit silicon carbide paper to remove any material that may have been left from the WEDM process. The specimens were cleaned in a liquid vapor degreaser prior to testing in the experimental setup shown schematically in Figure 1.

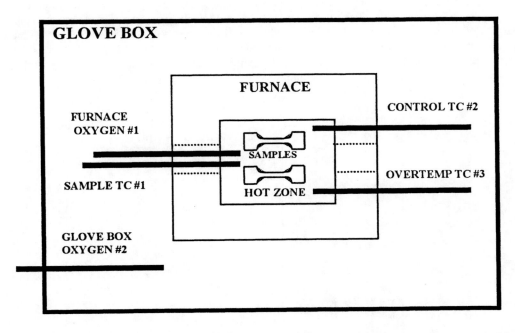

FIGURE 1. Schematic (top view) of the experimental setup that allowed sample temperature and oxygen content of the furnace and the glove box to be monitored.

The initial Mo-44.5%Re tensile specimens were heat treated in an electrically heated box furnace (hot zone ~10 cm wide x ~9.5 cm high x ~11.5 cm deep) that was positioned within an argon atmosphere glove box. The heating experiments were performed at either 1073K(800°C) or 1198K (925°C). The box furnace was placed within a flowing argon atmosphere controlled glove box which is ~1 meter wide x ~1 meter high x ~2 meters long. Oxygen concentrations within the glove box were monitored using a Vacuum/Atmospheres Oxygen Analyzer Model AO-316-C. Calibration of the oxygen monitor was performed using two procedures. The first employed the manufacturers recommended calibration procedure based on atmospheric air. The second procedure utilized purchased argon/oxygen mixture gas standards with a certified parts per million content of oxygen. One argon/oxygen gas mixture contained 43ppm of oxygen while the second certified mixture contained 4ppm of oxygen in argon.

During the experimental runs the oxygen content of the glove box and the oxygen content of the gas within the furnace next to the test specimens were monitored. Gas samples of the glove box were obtained via the use of a small "vacuum" pump that pulled the gas sample from the glove box via the oxygen #2 port through the analyzer. Direct gas sampling of the furnace atmosphere was accomplished by connecting a small length of hose from the oxygen #2 port to the cold end of an alumina tube (furnace oxygen #1) whose open end was placed next to the samples. Control TC #2 (thermocouple) was used to control the furnace temperature, Overtemp TC #3 was used to automatically shut down the furnace if the furnace temperature increased significantly above a selected set point, and Sample TC #1 measured the temperature of the test specimens.

Room Temperature Mechanical Test Results Obtained on Mo-44.5% Test Specimens Heat Treated at 1073K or 1198K Under an Argon Atmosphere (<5ppm Oxygen)

Tensile specimens that were fabricated out of the 0.5mm Mo-44.5% sheet stock were heated at 1073K (800°C) or 1198K (~925°C) for 25 or 150 hours in the glove box furnace. Oxygen concentration in the argon cover gas was determined to be less than 5ppm during the course of the experiments. After heating, the specimens were removed from the glove box and several of their mechanical properties were determined. Room temperature mechanical test results (ultimate tensile strength, yield strength, and percent elongation) were obtained on two tensile specimens for each time/temperature parameter. Figure 2 (left) presents the ultimate tensile strength and yield strength results and Figure 2 (right) shows the percent elongation results obtained on the specimen bars heat treated at 1073K or 1198K for 25 or 150 hours. The results show that for the heat treatment conditions tested there were no significant variations in the mechanical properties of the Mo-44.5%Re samples as a function of temperature or time.

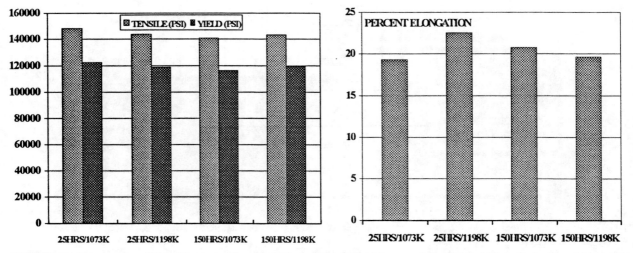

FIGURE 2. Comparison of the ultimate tensile and yield strengths (left) and percent elongation (right) obtained on Mo-44.5%Re specimens heat treated at 1073K or 1198K for 25 or 150 hours under argon with <5ppm of oxygen.

Comparison of Room Temperature Mechanical Test Results on Mo-44.5%Re and Nb-1%Zr Tensile Specimens Heat Treated at 1073K and 1198K

Figure 3 shows a comparison of the room temperature mechanical properties (ultimate tensile strength, yield strength and percent elongation) obtained on Mo-44.5%Re and Nb-1%Zr test specimens. Niobium-1%zirconium has been selected as the primary AMTEC cell wall material for space power systems. The Nb-1%Zr mechanical property presented formed the basis of a recent report.[5] Test specimens were prepared from 0.5mm Nb-1%Zr sheet stock in a similar procedure as described earlier in this paper. The specimens were heat treated in the same glove box furnace for the identical time/temperature parameters. Three Nb-1%Zr test specimens were heat treated at each experimental condition.

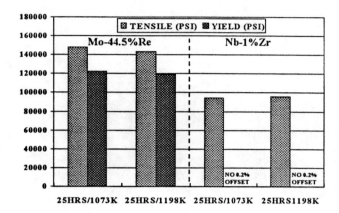 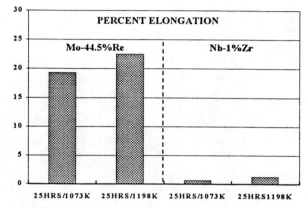

FIGURE 3. Comparison of the room temperature mechanical properties of Mo–44.5%Re and Nb-1%Zr test specimens heat treated at 1098K or 1198K for 25 hours under argon with ~5ppm oxygen.

The data presented in Figure 3 shows marked differences in the room temperature mechanical properties of heat treated Mo-44.5%Re and Nb-1%Zr. Heat treated Mo-44.5%Re has a higher ultimate tensile strength compared to heat treated Nb-1%Zr. A yield strength is not reported for the heat treated Nb-1%Zr test specimens since there was no 0.2% offset of the stress-strain curve obtained during the tests. The percent elongation results shows that the Mo-44.5%Re had a much greater percent elongation than did the Nb-1%Zr test specimens. Low percent elongation values such as those measured for the Nb-1%Zr specimens reflects the lack of ductility or "brittleness" of these specimens. While the percent elongation values measured on the Mo-44.5%Re specimens shows that those specimens even after the heat treatment were ductile.

High Temperature (1073K and 1198K) Mechanical Test Results Obtained on Mo-44.5%Re and Nb-1%Zr Test Specimens

All of the previous mechanical tests performed on the heat treated Mo-44.5%Re and Nb-1%Zr test specimens were performed at room temperature. However, many of the mechanical properties of materials may be a function of temperature. In order to determine some of the high temperature mechanical properties of these two materials a series of high temperature mechanical tests were performed on unheat treated Mo-44.5%Re and Nb-1%Zr test specimens. Due to limited resources only one test specimen of each material was tested at each temperature (1073K and 1198K). Figure 4 is a picture of the high temperature tantalum element vacuum furnace used showing a test specimen prior to testing. Prior to heating, ~0.47cm TZM (Mo-0.5Ti-0.1Zr-0.02W) dowel pins were inserted in ~0.5cm holes that had been previously drilled into the grip ends of the test specimens to help apply the appropriate load for the tests. Three thermocouples were positioned along the gauge length of the specimen to measure temperature during the tests. The tests were performed by first pumping the test chamber down to a vacuum in the 10^{-7} torr range. Ramp time to temperature was typically ~1.5 hours and the vacuum was continuously monitored to maintain the high vacuum conditions (10^{-7} torr) during the test. Soak times at temperature to obtain equilibrium temperature conditions was less than 15 minutes. The high temperature mechanical test results obtained on Mo-44.5%Re and Nb-1%Zr test specimens at 1073K and 1198K are presented in Figure 5.

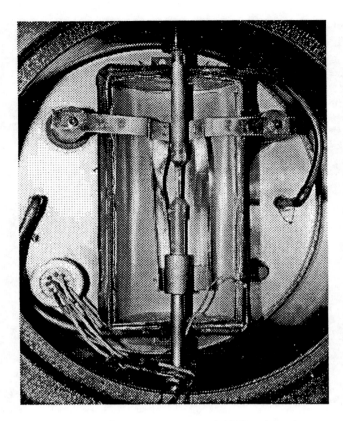

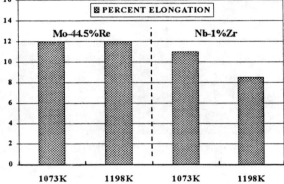

FIGURE 4. High temperature mechanical test setup.

FIGURE 5. High temperature mechanical test results.

Electron Beam Welding (EBW) Tests of Mo-44.5%Re Coupons

In the fabrication of AMTEC cells it is anticipated that to some degree welding of the selected cell wall material will be required. This is necessary to obtain cells that are hermetically sealed to contain the sodium inventory within the cell and to keep the outside environment out. While in some scenarios the cell wall could be manufactured via brazing techniques, it is more likely that the cell wall will require at least some welding. A study was performed to determine the feasibility of electron beam welding of Mo-44.5%Re coupons. The ~1cm wide x ~3cm coupons were fabricated out of the same ~0.5mm thick sheet stock used in fabricating the mechanical test specimens. Coupons were butt welded together using an electron beam welder while under vacuum (~1 x 10^{-5} torr). Polished cross-sections of the welds were prepared which showed complete weld penetration. As expected, the grains in the heat effected zone were many times larger than the parent material. Small amounts of porosity was also observed in the weld area. Simple longitudinal and transverse face bend tests (not guided bend tests) (Figure 6) were performed on several of the weld samples which showed that the welded samples exhibited ductility.

DISCUSSION

Tensile bars fabricated out of Mo-44.5%Re were heat treated for various times and temperatures under an argon cover gas containing <5ppm of oxygen. Room temperature mechanical tests on the heat treated test specimens showed only minor variations in the tensile strength, yield strength, and percent elongation measurements as a function of time and temperature. The room temperature Mo-44.5%Re results were compared to earlier reported room temperature mechanical test data obtained on heat treated Nb-1%Zr showing that the strength and ductility of the Mo-44.5%Re specimens were superior to Nb-1%Zr specimens.

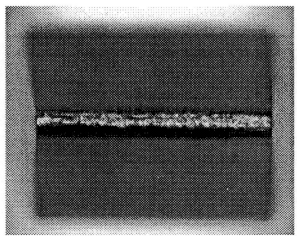 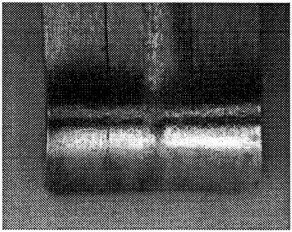

FIGURE 6. Top views of a (left) simple transverse face bend test (sample width ~1.5cm) and (right) longitudinal face bent test (sample width ~1cm) performed on electron beam welded Mo-44.5%Re coupons.

High temperature mechanical tests were performed under vacuum on both Mo-44.5%Re and Nb-1%Zr specimens at 1073K and 1198K. These results showed that the Mo-44.5%Re specimens exhibited ultimate tensile and yield strengths that were approximately twice the values obtained on the Nb-1%Zr samples. Initial electron beam welding experiments were also performed showing the feasibility of obtaining ductile Mo-44.5%Re butt welds.

ACKNOWLEDGMENTS

The authors would like to thank Mr. Sam Farler, Mr. Al Hoddap, and Mr. David Burden of the Mound Manufacturing Center, Inc., Miamisburg, OH, whose expertise in the WEDM of Mo-44.5%Re and Nb-1%Zr made this work possible. Also Dr. C. William Merten of Mound Engineering & Analysis Group, Inc., Miamisburg, OH, and Mr. Gregory Kasten, METCUT Research Inc., Cincinnati, OH, for their mechanical testing expertise.

REFERENCES

El-Genk, M.S., and Tournier, J.M., "Parametric Analyses of Vapor-Anode, Multitube AMTEC Cells for Pluto/Express Mission," in *Space Technology and Applications International Forum - 15th Symposium on Space Nuclear Power and Propulsion,* edited by M. El-Genk, AIP Conference Proceedings 420, New York, 1998, pp.1461-1470.

Hendricks, T.J., and Huang, C., "System Design Impacts on Optimization of the Advanced Radioisotope Power System (ARPS) AMTEC Cell," in *Proceedings of the 33rd Intersociety Engineering Conference on Energy Conversion*, edited by S. Anghaie, American Nuclear Society, 1998, paper No. IECEC-98-407.

Kramer, D.P., Ruhkamp, J., McNeil, D., Mintz, G., and Howell, E., "Mechanical Testing Studies on Niobium-1%Zirconium in Association with its Application as Cell Wall Material in an AMTEC Based Radioisotope Space Power System," in *Proceedings of the 34rd Intersociety Engineering Conference on Energy Conversion,* Society of Automotive Engineers, Inc., 1999, paper No. IECEC-1999-01-2608.

Leonhardt, T., Carlen, J., Buck, M., Brinkman, C., Ren, W., and Stevens, C., "Investigation of Mechanical Properties and Microstructure of Various Molybdenum-Rhenium Alloys," in *Space Technology and Applications International Forum - 16th Symposium on Space Nuclear Power and Propulsion,* edited by M. El-Genk, AIP Conference Proceedings 458, New York, 1999, pp. 685-690.

Romano, A.J., Fleitman, A.H., and Klamut, C.J., "Evaluation of Li, Na, K, Rb, and Cs Boiling and Condensing in Nb-1%Zirconium Capsules," *Nuclear Applications*, **vol. 3**, p. 110-116, February 1967.

THERMAL STABILITY OF BETA"-ALUMINA SOLID ELECTROLYTE UNDER AMTEC OPERATING CONDITIONS

Roger M. Williams, Margie L. Homer, James Kulleck, Liana Lara, Adam K. Kisor, Roger H. Cortez, Virgil B. Shields, and Margaret A. Ryan

Jet Propulsion Laboratory, California Institute of Technology, 4800 Oak Grove Drive, Pasadena, CA 91109

Abstract. A critical component of alkali metal thermal-to electric converter (AMTEC) devices for long duration space missions is the sodium beta"-alumina solid electrolyte ceramic (BASE), for which there exists no substitute. The major phase in this ceramic, sodium beta"- alumina shows no evidence of thermal decomposition in AMTEC environments including clean liquid sodium and low pressure sodium gas, at temperatures below 1173K, or in vacuum below 1273K. This paper presents additional results of ionic conductivity and exchange current studies in sodium exposure test cells (SETCs) to characterize the changes occurring in BASE below 1273K in low pressure sodium vapor. Also presented are additional annealing studies to characterize the kinetics of processes occurring in in the BASE ceramic in the AMTEC operating regime.

INTRODUCTION

The Jet Propulsion Laboratory has been investigating the degradation of critical AMTEC operating components under accelerated conditions for several years. The thermochemical data available and the phase relations of beta"- alumina, beta-alumina, and alpha alumina suggest that slow sodium oxide loss from beta"-alumina might be possible at AMTEC operating conditions, above 973K, especially at the higher temperatures up to 1223K. (Rog, 1992; Duncan, 1988; Williams, 1997) Experimental studies have shown that BASE is converted to alpha alumina at the surface when heated to 1273K or higher in vacuum for 100-500 hours. (Williams, 1998; Williams 1999) However, experiments have not revealed any decomposition of the sodium beta" alumina phase in BASE at temperatures of 1173K and below in vacuum or in low pressure sodium gas or in liquid sodium, except under conditions where reactive metals such as Ca, Cr, or Mn come into contact with the BASE. (Ryan, 1994; Williams 1995; Williams, 1998; Williams, 1999) Several earlier studies have demonstrated that interconversion of the beta, and beta"-alumina, and alpha alumina and sodium aluminate are kinetically slow at temperatures below 1600K, and apparently negligible below 1200K for exposure times of hours. (Kummer, 1972; Youngblood, 1977; Hodge, 1983; Hodge,1984; Mosely 1985) We cannot rule out minor loss of sodium oxide from the beta"alumina phase, if it occurs without a phase change. Nor can we rule out extremely slow decomposition of sodium beta" alumina which has been too slow to detect in our tests. However, in the event of very slow, highly activated loss of soda from the beta" alumina phase we would expect surface formation of degradation products, resulting in a decrease in the exchange current of stable electrodes on BASE. This would also be the most critical failure mode for BASE other than complete phase decomposition. If a few percent of the beta"alumina phase in the solid electrolyte ceramic were converted to a non-conducting form, distributed homogeneously throughout the ceramic, the solid electrolyte would still function, with slightly higher resistance. However, formation of an insulating layer on the order of 1.0nm at the surface would reduce the electrode/electrolyte interface's exchange current substantially.

CP504, *Space Technology and Applications International Forum–2000*, edited by M. S. El-Genk

Our present experimentally derived model for changes in BASE under ordinary and accelerated AMTEC conditions includes no crystalline phase transformation of the beta" alumina phase, but allows for slight soda loss from that phase. The model ascribes most changes in the performance of BASE to loss of excess soda present as $NaAlO_2$ at grain boundaries, with a maximum mass loss of about 0.5% or the BASE mass. The BASE ceramic usually contains about 1% of the $NaAlO_2$ phase, due to the requirements of the liquid sintering process used to fabricate dense BASE ceramic.

We have recently reported that BASE with RhW electrodes in an SETC experiment in which the hot BASE is protected from metal evaporating from the stainless steel vacuum chamber by a niobium 1%zirconium shield shows no evidence of a sustained decrease in conductivity at 1173K over 2500 hours. (Williams, 1999) Furthermore, in the same experiment, there was no sustained drop in the exchange current of the RhW electrode/ BASE interface, indicating that the surface of the BASE undergoes no decomposition under these conditions. The exchange current in AMTEC cells has been discussed in previous work. (Williams, 1990a, Williams, 1990b) Earlier work in SETC experiments showed that, with shielding, little degradation in either the ionic conductivity of the BASE or the exchange current of the TiN/BASE interface is observed at 1123 and 1173K. Break-in behavior of TiN on BASE was observed in an SETC experiment at 1123K without a titanium shield, and during a test, at 1223 K in which the shield ruptured. Post-mortem results of a test with broken shield showed that Mn vapor can lead to degradation of BASE with loss of sodium, uptake of Mn by the solid electrolyte ceramic, and formation of $MnAl_2O_4$ and Al_2O_3.

EXPERIMENT

BASE stability in low pressure (0.1-10 Pa) sodium atmosphere was tested by measuring the electrolyte resistance of BASE in a four probe configuration during sodium exposure test cell (SETC) operating at temperatures from 1100K and 1223K for 500 to 3000 hours, and by characterizing the time dependence of the exchange current at the TiN/BASE or RhW interface with two-probe impedance measurements. (Williams 1995). SETC experiments were run with titanium, niobium 1%zirconium, and molybdenum/tantalum shields between hot stainless steel components and RhW or TiN electrodes on BASE in low pressure sodium vapor. Our investigation utilized lithium stabilized beta"-alumina ceramic from Ionotec (Britain) and Cerametec (Utah) as well as ceramics from those manufacturers modified by pretreatment.

Sodium meta-aluminate, $NaAlO_2$, was prepared as the anhydrous phase by dehydration of the hydrated reagent, and was characterized by XRD . All of our tests on BASE were carried out with Ionotec or Cerametec lithia stabilized BASE ceramics, which have near identical nominal compositions. Some of these samples were modified by annealing processes prior to experiments. The nominal composition of Cerametec BASE is Na_2O $(Li_2O)_{0.176}$ $(Al_2O_3)_{6.21}$ and Ionotec BASE is 99.25 weight % Na_2O $(Li_2O)_{0.176}$ $(Al_2O_3)_{6.16}$ but includes 0.75 weight % ZrO_2. The Cerametec BASE had a varying content of $NaAlO_2$ near 1% but sometimes higher, while Ionotec BASE always is close to 1 weight % $NaAlO_2$. Mass loss, annealing, and phase change studies of BASE, modified BASE samples, and pressed pellets of sodium aluminate were carried out at several temperatures for periods of hundreds to thousands of hours either in low pressure sodium or in vacuum with zirconium getters as previously described.

RESULTS AND DISCUSSION

After a small initial decrease in BASE conductivity, there was no systematic variation in conductivity for up to 2500 hours, for cells with RhW electrodes. However, the four probe conductivity sometimes showed abrupt decreases, followed by slow relaxation to the conductivity before the decrease. Similar transient perturbations were shown, more often, as increases in the ohmic series resistance R_{ser}, measured between pairs of electrodes, and in the width of the arcs between high and low frequency intercepts, designated as R_{act}, obtained from impedance measurements, even though there was also no systematic and continuous decrease in these parameters. Perturbations in measurements in SETC cells are occasionally caused by instrument problems or by electrical contacts developing between one or more of the leads and the vacuum chamber, which is usually at ground. These conditions were ruled out in the cases described here, because instrumental problems effect all measurements done at the same time, and electrical contacts to ground can be ruled out with measurements with an ohmmeter.

The perturbations in the conductivity from the four probe measurement is interpreted as a resistance increase in the section of the BASE cylinder between the two innermost electrodes. Similarly, the series resistance measured in the two probe experiments has a significant but not exclusive contribution form the electrolyte resistance in the gap between the two electrodes. The separate occurrence of all three resistance increases is only consistent with localized resistance increases in the BASE itself, although other phenomena could account for each of the two probe resistance changes. We have found that BASE samples exposed to low pressure sodium gas for hundreds to thousands of hours at 1023K to 1123K appear somewhat more susceptible to fracture when stressed, compared to as-received ceramic, but we have not quantitatively measured this effect. However, microscopic examination shows no evident cracks following the tests.

Mass loss and phase change investigations showed that sodium aluminate, $NaAlO_2$, undergoes thermal decomposition at temperatures well below 1173K. At 1123K for 500 hours, the beta" alumina phase appears to be the major new phase formed, but some beta alumina is also formed. At 1073K for 500 hours, $NaAlO_2$ pellets lose about 10% of their mass, but diffraction due to sodium beta" or beta alumina is not seen. It is possible that $NaAlO_2$ may lose sodium oxide, but that the kinetics of crystallization of the beta" and beta phases are very slow at this temperature. This lends support for our model of the resistance fluctuations and, potentially, mechanical strength changes in BASE at the low pressure side under AMTEC operating conditions, at temperatures as low as 1023K to 1073K. X-ray diffraction (XRD) shows several weak peaks and shoulders in the diffraction patterns of the BASE ceramic which do not fit the beta"-alumina pattern. Some of these peaks can be assigned to very small amounts of $NaAlO_2$, Na_2CO_3, alpha Al_2O_3, and ZrO, with perhaps in preferential orientation at grain boundaries, or they might be due to an unidentified phase. Both $NaAlO_2$ and Na_2CO_3 present in the ceramic may be expected to thermally decompose under AMTEC operating condition.

Our model for the abrupt local resistance increases followed by a slow return to the original resistance is that thermal decomposition of the $NaAlO_2$ phase releases sodium and oxygen vapor resulting in void formation at grain boundaries between the crystals or the dominant beta" alumina phase. This decomposition and attendant volume decrease of the intergranular phase is driven by its thermodynamic instability at the low pressure of oxygen and sodium in our SETC cells. The same process might occur more slowly at the low pressure side of the BASE electrolyte in an AMTEC cell, and might to some degree be suppressed by maintaining a significant sodium vapor flux at all times. Thermal decomposition of Na_2CO_3 would be still faster and harder to suppress than $NaAlO_2$ decomposition, but Na_2CO_3 is not specified by the manufacturers as an impurity phase and is not expected to survive the ceramic firing process; it should only be present at or near the BASE surface as a consequence of handling in air after fabrication. Recovery of the BASE sample resistance suggests that the voids formed are extremely small and close due to sintering or creep of the ceramic, which must be very slow at these temperatures. Closure of macroscopic cracks or voids is not observed in these ceramics at these temperatures. The phenomenon of local resistance increases may be related to, or accelerated by mechanical stresses on the sample, but is probably primarily driven by the thermal instability of high sodium oxide activity phases.

CONCLUSIONS

Loss of a small amount of sodium oxide from BASE at temperatures below 1173K occurs without detectable conversion of the major phase, sodium beta" alumina to alpha alumina, but instead from decomposition of sodium aluminate, NaAlO2, at grain boundaries. The decomposition of NaAlO2 results in transient, reversible, local changes in the solid electrolyte's ionic conductivity when the electrolyte is under stress. We have not seen evidence of a continuous conductivity or exchange current decrease, in tests at 1173K in several sodium exposure test cell experiments of several thousand hours with rhodium-tungsten electrodes.

ACKNOWLEDGMENTS

The research described in this paper was performed by the Jet Propulsion Laboratory, California Institute of Technology, under a contract with the National Aeronautics and Space Administration, and was supported by NASA. The authors would like to thank Neill Weber for access to unpublished results and calculations, and to Tom Hunt and Neill Weber for access to an unpublished report on beta"alumina, and to Jim Rasmussen and Tom Hunt for helpful discussions.

REFERENCES

Duncan, G. K., and West, A. R. "The Stoichiometry of Beta"-Alumina Phase Diagram Studies in the System Na_2O-MgO-Li_2O-Al_2O_3," Solid State Ionics, 28, 338-343 (1988).

Hodge, J. D., "Kinetics of the Beta"-to-Beta-Transformation in the System Na_2O-Al_2O_3," J.Am. Ceram. Soc., 66, 166-169, (1983).

Hodge, J. D. , "Phase-Relations in the System Na_2O-Li_2O-Al_2O_3," J. Am. Ceram. Soc., 67, 183-185 (1984).

Kummer, J. T. , "Beta-Alumina Electrolytes," Prog. Solid State Chem., 7, 141 (1972).

Moseley, P. T. , The Solid Electrolyte, Properties and Characteristics, in "The Sodium- Sulfur Battery," edited by Sudworth, J. L., and Tilley, A. R. , Chapman and Hall, New York, (1985).

Rog,G., Kozlowska-Rog, A., and Zakula,K., "Determination of the Standard Free Energies of Formation of the Sodium Aluminates by e.m.f. Measurements," J. Chem. Thermodynamics, 24, 41-44 (1992)

Ryan, M.A., Williams, R. M., Allevato, C. E. , Vining, C. B. , Lowe-Ma, C. K. , and Robie, S. B. , "Thermophysical Properties of Polycrystalline Sodium Beta"-Alumina Ceramic," J. of Phys. Chem. of Solids, 55, 1255-1260 (1994).

Ryan, M. A., Williams, R. M., Lara, L., Cortez, R. H., Homer, M. L., Shields, V. B., Miller, J., and Manatt, K. S., The Sodium Exposure Test Cell to Determine Operating Parameters for AMTEC Electrochemical Cells Proceedings of the 33rd Intersociety Engineering Conference on Energy Conversion, Colorado Springs, CO, published by the Amer. Nucl. Soc., paper #335 (1998)

Williams,R. M., Loveland, M. E., Jeffries-Nakamura, B., Underwood, M. L., Bankston, C. P., Leduc, H. and Kummer, J. T., "Kinetics and Transport At AMTEC Electrodes, I. The Interfacial Impedance Model," J. Electrochem. Soc., 137, 1709-1716 (1990a)

Williams, R. M., Jeffries-Nakamura, B., Underwood, M. L., Bankston, C. P., and Kummer, J. T., "Kinetics and Transport at AMTEC Electrodes. II. Temperature Dependence of the Interfacial Impedance of $Na(g)$/Porous Mo/Na-Beta"- Alumina," J. Electrochem. Soc., 137, 1716-1723 (1990b)

Williams, R.M. , Ryan, M. A., Homer, M. L., Lara, L. , Manatt, K., Shields, V., Cortez, R. H., Kulleck , J., "The Thermal Decomposition of Sodium Beta"-Alumina Solid Electrolyte Ceramic," Proceedings of the 33rd Intersociety Engineering Conference on Energy Conversion, Colorado Springs, CO, published by the Amer. Nucl. Soc., paper #333 (1998)

Williams, R. M. , Ryan, M. A., and Phillips, W. M., "Loss of Alkali Oxide from Beta" Alumina and its Importance to AMTEC Life Issues," , in Proceedings of the 32nd Intersociety Engineering Conference on Energy Conversion, Honolulu HA, 1997, published by the American Institute of Chemical Engineers, pp 1220 -1223 (1997).

Williams, R. M., Kisor, A., and Ryan, M. A., "Time Dependence of the High Temperature Conductivity of Sodium and Potassium Beta"-Alumina in Alkali Metal Vapor," J. Electrochem. Soc., 142, 4246 (1995).

Williams,R.M., Ryan, M.A., Homer,M. L., Lara,L., Manatt,K., Shields,V., Cortez,R.H., and Kulleck, J., "The Thermal Stability Of Sodium Beta"-Alumina Solid Electrolyte Ceramic In AMTEC Cells, "in Proceedings of the Space Nuclear Power Symposium, STAIF, Albuquerque, NM , p. 1306 (1999)

Williams, R. M., Ryan, M. A., Homer, M. L., Lara, L., Manatt, K. Shields,V., Cortez, R. H., and Kulleck, J. "The Kinetics Of Thermal Interconversion Of Sodium Beta"- Alumina With Sodium Beta Alumina Or Alpha Alumina At Temperatures Below 1600K In Gaseous And Liquid Sodium," {PRIVATE } to be published in "Proceedings of the Symposium on Solid State Ionic Devices" Electrochemical Society Softbound Proceedings Volume Series , The Electrochemical Society, Pennington, New Jersey (1999)

Youngblood,G., Virkar,A., Cannon,W., and Gordon, R. W., "Sintering processes and Heat Treatment Schedules for Conductive, Lithia Stabilized Beta"-Alumina," Am. Ceram. Soc. Bull., 56, 206 (1977).

AIMStar: Antimatter Initiated Microfusion For Pre-cursor Interstellar Missions

Kevin J. Kramer,[1] Raymond A. Lewis,[1] Kirby J. Meyer,[1]
Gerald A. Smith[1] and Steven D. Howe[2]

[1]*Department of Physics, The Pennsylvania State University, University Park, PA 16802*
(814) 863-3076; smith@leps3.phys.psu.edu
[2]*Synergistic Technologies, Inc., Los Alamos, NM 87544*

Abstract. We address the challenge of delivering a scientific payload to 10,000 A.U. in 50 years. This mission may be viewed as a pre-cursor to later missions to Alpha Centauri and beyond. We consider a small, nuclear fusion engine sparked by clouds of antiprotons, and describe the principle and operation of the engine and mission parameters. An R&D program currently in progress is discussed.

INTRODUCTION

The interstellar medium provides several clues to the evolution of the solar system, yet it cannot be fully exploited. Mewaldt (Mewaldt, 1995) states that a "termination shock" of the interstellar wind at approximately 80 AU limits earth-based scientific investigation of high-energy particles and magnetic fields beyond this point. Moreover, speculation about the existence of Brown Dwarves in the Oort Cloud at 10,000 AU, also difficult to detect from ground-based sources, has piqued interest concerning the sun's formation as well as the mass of the solar system and ultimately the universe.

An interstellar scientific payload can answer three unresolved issues: (1) the precise location of the shock and the magnitude of the magnetic fields beyond; (2) composition of high-energy particles and plasma in the interstellar medium; and (3) the presence of Brown Dwarves and other clumps of matter not previously detected. To accomplish its goal, a space probe must not only carry the desired scientific equipment, but must also reach the Oort Cloud within a certain period of time to enrich our knowledge. Until now, failure of meeting the latter of these conditions has held interstellar exploration to a dream.

The challenge of developing a means to travel to the stars in the life span of a human being is long-standing. In their book Mirror Matter , Forward and Davis (Forward, 1988) review several proposed concepts, including rockets utilizing nuclear fusion reactions or antimatter annihilation, as well as rocketless systems such as laser-pushed lightsails. More recently, particle-pushed magsails (Guterl, 1995) and solar sails (Frisbee, 1999) have been proposed as another example of rocketless propulsion. A comprehensive review of fusion rocket concepts has been presented recently by Frisbee (Frisbee, 1999)

Each concept faces fundamental technological problems. For example, fusion rockets require massive (several hundred tonnes) high powered "drivers", being either a laser or particle beam array (ICF systems) or a magnetic torus or mirror (MCF systems). Antimatter systems generally call for quantities of antimatter far in excess of current or conceivable future production capabilities. Rocketless systems require intense high power laser or particle beams with exceptional focusing requirements, etc.

We have developed a new concept, called Antimatter Initiated Microfusion (AIM), which is a hybrid of nuclear fusion and antimatter technologies. Any system designed for deep space missions must meet the following criteria: (1) high

CP504, *Space Technology and Applications International Forum–2000*, edited by M. S. El-Genk
© 2000 American Institute of Physics 1-56396-919-X/00/$17.00

specific power: $\alpha > 1$ kW/kg; (2) high exhaust velocities: $v_{ex}(max) = 10^4$ km/s; and (3) continuous power with near zero maintenance for several years. The following sections describe the AIMStar concept and expected performance for stellar missions.

SPARKING THE FUSION BURN

Physicists have been working for fifty years in an attempt to "spark" fusion fuel into a significant burn. ICF experiments use short pulses of intense laser or particle beams to compress and heat targets by ablation of surrounding materials. Using intense magnetic fields, MCF experiments attempt to continuously compress and heat fusion fuel into a burn. In general, results to date have been measurable, but unfortunately incomplete, burns of the target material. Inefficient coupling of beams to the target and plasma instabilities have been largely responsible for the failure for full ignition.

These experiments have generally attempted fusion of hydrogen isotopes deuterium (D) and tritium (T), which has a low ignition temperature. The fusion reaction:

$$D + T \rightarrow n(14.1 \text{ MeV}) + \alpha(3.5 \text{ MeV}), \tag{1}$$

presents serious problems for space applications: (1) large amounts of radioactive tritium are required. From space transportation safety considerations, this requires special shielding; (2) the neutrons require absorbers if their energy is to be used to heat propellant, and (3) if not fully shielded, the neutrons will cause severe radiation damage to the engine and payload. In all, the additional weight required may be at least 1 tonne, which would be intolerable for a small, fast stellar probe. Therefore, for this study we have considered DHe^3 fusion:

$$D + He^3 \rightarrow p(14.7 \text{ MeV}) + \alpha(3.6 \text{ MeV}), \tag{2}$$

which is aneutronic provided the fuel is burned at a sufficiently high temperature so that the competing $D + D \rightarrow n + He^3$ fusion rate is insignificant. We therefore hereafter illustrate AIMStar performance characteristics using DHe^3 as the fusion fuel.

We propose to inject small fusion fuel droplets into a cloud of antiprotons confined in a very small volume within a reaction Penning trap. The reaction trap (Figure 1) is roughly the size of a shoebox, weighing perhaps 10 kg. It is fed 10^{11} antiprotons on a periodic basis from a portable trap (not shown) positioned about 1 meter away on axis, safe from fusion debris. Radial confinement within a 0.8 cm maximum diameter orbit is provided by a 20T axial magnetic field. Axial trapping of the 2 cm long cloud of antiprotons within a 10 keV space charge electric potential on the electrodes is shown in Figure 2.

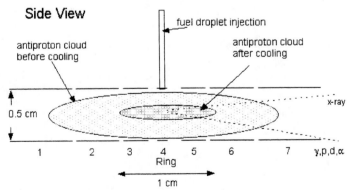

FIGURE 1. Expanded side view of the AIMStar reaction trap.

The key ingredient for heating of the 42 ng DHe^3 liquid droplet (Figure 2) is antiproton-induced fission fragments which have a range of 45 μm in the droplet. In order to spark the microfusion process, 5×10^8 antiprotons are annihilated in a 2% molar admixture of a pre- or actinide metal, such as Pb^{208} or U^{238}, with the DHe^3. Annihilation takes place on the surface of the antiproton cloud, pealing back 0.5% of the cloud. The power density released by the fission fragments into the DHe^3 is about 5×10^{13} W/cm^3, fully ionizing the D and He^3 atoms. This is roughly comparable to a 1 kJ, 1 ns laser depositing its energy over a 200 μm ICF target, a system much too massive for driving a small space probe.

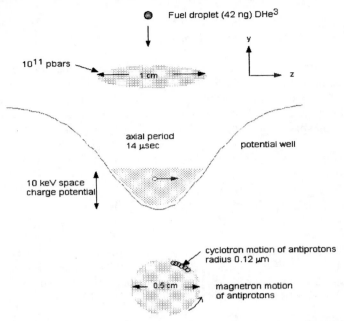

FIGURE 2. Fuel injection cycle, showing fuel droplet and confined antiproton cloud (side and end view).

We note that it has been shown recently that the fission fragments from antiproton-induced fission are not radioactive (Smith, 1997), so there is no concern of accumulative radioactive contamination of the engine and spacecraft as the engine burns. The heating of the plasma takes place in 1 ns, and is confined in the center of the trap by application of a weak nested well potential (Figure 3). Antiprotons, as well as ionized electrons, are stored off trap center for use in

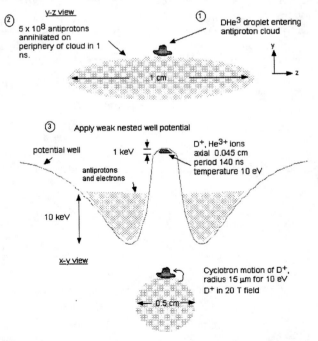

FIGURE 3. Fuel heating cycle, showing antiproton annihilation and confinement of D$^+$,H$^+$ ions, antiporton and electrons in a weak nested potential (side and end views).

subsequent cycles of the engine.

Transverse profiles of the antiprotons are shown in Figures 2 and 3, before and after injection of the fuel droplet, respectively. The antiprotons and D^+, He^{++} ions are confined by magnetic cyclotron and magnetron forces, which along with the axial electric force comprise a set of three fundamental harmonic forces in a Penning trap.

COMPRESSING THE FUEL TO A FULL BURN

In order to compress the fully ionized DHe^3 droplet to high density and temperature sufficient to start a fusion burn, we apply a strong nested well potential as shown in Figure 4. The application of a 600 kV potential, which presents a new and important challenge to Penning trap operation, results in a 100 keV ion plasma with density $n = 6 \times 10^{17}$ ions/cm^3, which when combined with a $\tau = 20$ ms lifetime satisfies Lawson's criterion ($n\tau > 5 \times 10^{15}$ s/cm^3) for a full fusion burn. Because the kinetic pressure of the plasma under these conditions exceeds the magnetic pressure, a question which must be best answered experimentally arises as to the lifetime of the plasma against this instability.

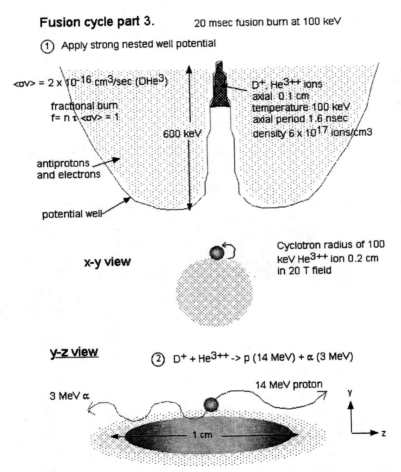

FIGURE 4. Fuel burn cycle, showing D^+,He^{++} ion compression to high density conditions over a 20 msec period, by application of a strong nested potential (side and end view)

Antiprotons not consumed in the original ionization of the droplet await later use, trapped in the wings of the potential well. Upon completion of the burn, the potential is returned to its original configuration (Figure 1), minus 0.5% of the original load of 10^{11} antiprotons. The four cycle process is repeated 50 times, followed by one cycle used to load another 10^{11} antiprotons from the storage trap into the reaction trap. The duty factor of the reaction trap is 99.5%, and delivers

0.75 MW (15 kJ/20 ms cycle) of continuous power in the form of protons and alpha particles.

SPECIAL CHALLENGES

The techniques described above are required to establish conditions for a complete fusion burn. They present two identifiable challenges in technology. First, the requisite density (6×10^{17} ions/cm^3) shown in Figure 4 exceeds the Brillouin space charge limit by several orders of magnitude. This we propose to mitigate by dynamic injection and manipulation of electrons in the ion cloud, following the prescription outlined by Ordonez. (Ordonez, 1996). This technique has been generally applied in other electromagnetic fusion ion confinement schemes, for example by Bussard (Bussard, 1991), Krall (Krall, 1992) and Barnes et al. (Barnes, 1993).

Second, injection of 600 kV of high voltage onto electrode gaps in the trap presents special problems associated with breakdown and stability. However, the resultant fields (1200 kV/cm) are not far beyond proposed fields (600 kV/cm) in other current and similar applications of traps, such as that of Barnes et al. (Barnes, 1993). This work will bear watching as we approach experimental tests of our proposed system.

SPACECRAFT DESIGN

A preliminary design of the AIMStar spacecraft is shown in Figure 5. The reaction traps, antiproton storage, and engine are located to the aft of the vessel in a special "booster rocket" used only to accelerate the payload (shown at right) to a velocity of ~ 0.003c. At time of burnout the booster engine separates, leaving only the payload that is fully expanded into the form seen in Figure 6. Separation of the booster is important to permit communications with Earth.

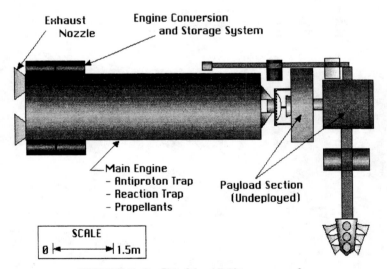

FIGURE 5. Profile of the AIMStar spacecraft.

Figure 6 describes the general locations of the various subsystems, as well as the scientific apparatus used for the mission. The magnetometer, found farthest from the central hub, is used to examine magnetic fields of the interstellar medium and determine the location of the termination shock discussed previously. A near-infrared spectrometer is used to examine Brown Dwarves in the Oort Cloud, and the optical imager serves a dual-purpose to detect large clumps of cold matter and to tell the spectrometer where to point. The astrophysics package, containing an ion-mass spectrometer, investigates high-energy matter and plasma contained within the interstellar medium.

Assuming that Ka-band will be employed on the Deep Space Network by 2030, a 100 bps data rate at 10,000 AU can be achieved by use of an 8m parabolic antenna. This requires 780W of power, which can be acquired through the use of RTG's envisioned for the future (AMTEC's).

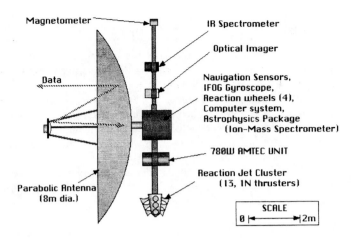

FIGURE 6. Profile of the AIMStar payload sections.

MISSION ANALYSIS

We have developed a 50 year, pre-cursor mission to 10,000 A.U. (Table 1). For comparison, we include parameters for both DHe³ and DT fusion-driven systems. The DT system has lower ignition temperatures, but presents problems associated with containment of 14 MeV neutrons and tritium. We are presently designing a chamber in which the energy of protons and alpha particles from the fusion reactions is transferred to hydrogen propellant. The numbers in Table 1 assume 100% energy transfer efficiency, which is of course not realistic, and will be updated as our work progresses.

We see that the parameters under the assumptions stated above meet the minimum expectations for deep space missions outlined earlier. Further, the antiproton requirements are modest. Therefore, we believe that further work and defining experiments should be carried out on this very promising technology.

TABLE 1. AIMStar 50 year Mission to 10,000 AU.

	DT	DHe³
ΔV (km/s)	956	956
V_e (m/s)	5.98×10^5	5.98×10^5
I_{sp} (s)	61,000	61,000
Power (MW)	33	0.75
Thrust (N)	55.2	1.25
dm/dt (kg/s)	9.22×10^{-5}	2.09×10^{-6}
t_b (yr)	0.50	22
Distance @ burnout (AU)	37	1635
α_{ave} (kW/kg)	30.5	0.69
N_{pbar} (μg)	130	28.5

1) $\Delta V/V_{ex} = 1.6$ for energy optimization (from rocket eq. this fixes payload to total mass ratio = 0.2);
2) 361 kg dry mass, 1444 kg propellant.

RESEARCH AND DEVELOPMENT PROGRAM

In collaboration with Penn State, Synergistic Technologies, Inc. (STI) is developing an Antimatter Plasma Gun (APG) under a NASA STTR Phase I award. We can foresee that the APG has potentially important near term applications as an ion thruster, and in a scaled up version as a spark or ignitor of nuclear fusion reactions to drive fast, interstellar probes. The technical objectives of this activity are to design and confirm computationally the specifications and applicability of the APG to important plasma applications technologies, and especially ion thrusters for near-term space propulsion.

As discussed earlier in this paper, the APG would utilize simultaneous double-well, nested confinement of negatively and positively charged particles in a Penning trap of cylindrical geometry. The negatively charged species is antiprotons, which are an extraordinarily high energy density and compact source for heating and ionization of neutral atoms when confined in a trap. Partially ionized positive ions created by antiproton annihilation with these atoms are captured and stored in the trap, along with the remaining antiprotons (and electrons resulting from the ionization process). Questions to be addressed in this research include: Can the positively charged ions be captured, along with the remaining antiprotons and ionization electrons, in separate wells by application of a double-well, nested electric potential? What vacuum is required to sustain storage for up to about one second, before injection of more atoms?

Radial confinement of the particles would be provided by an axial magnetic field, perhaps as much as 2-5 T. Can this field be provided by permanent magnets? If not, would a hot or cold electromagnet be the next best choice?

The positive ions would be extracted from the trap in the form of a bright, low emittance beam ready for applications. What configuration of extraction potentials on the trap electrodes are required to provide a well-defined beam of 1 keV energy, for example? Can the beam be extracted in very short pulses, and with what duty factor? Whata ion current is expected?

The HiPAT trap (Smith 1998, Smith 1999) presently under development at MSFC will serve as a storage facility for 10^{12} antiprotons. Initially, up to 10% of these antiprotons would be transferred from HiPAT to the APG. This number defines the approximate space charge limit for a non-neutral plasma cloud of antiprotons in an APG magnetic field. This process would be repeated with a periodicity which depends on the desired duty factor of beam delivery by the APG. Can the transfer be done efficiently, and what electrostatic lens configurations are required?

As indicated above, a small droplet of neutral atomic matter, e.g. LiH/U, would be injected into the antiproton cloud. What physical mechanism is required for a stable and repeatable injection of droplets of about 100 ng mass at a rate of about 1 Hz?

The high density cloud of positively charged ions and electrons created by the antiproton annihilation form a neutral plasma of temperature 1-10 eV. When the double-well, nested electrostatic potential configuration is placed on the trap electrodes, confining the positive ions in the center of the trap, the remaining antiprotons and electrons are pushed to confining wells at the ends of the trap. In order to achieve the highest possible density of ions, it is desirable to recirculate electrons from the outer wells back through the central well. What are the electric fields and frequencies required for dynamic manipulation of the mobile electrons form the outer wells to the central well and back, such that a load of positive ions many orders of magnitude above the space charge limit for a non-neutral plasma remains trapped?

This process of injection of neutral atoms is repeated, perhaps every one second, accompanied by injection at a slower rate of more antiprotons from HiPAT into the APG. Each injection of atoms will consume about 0.1% of the antiprotons in the APG. Allowing for a storage time of order one second per injection, the full 10^{12} fill of antiprotons in HiPAT will support plasma formation for approximately 3-4 hours with 100% duty factor, 30-40 hours with 10% duty factor, etc., depending on specific beam requirements.

Because of the extremely high power density imparted to the plasma by antiprotons, i.e. about 1.9×10^{14} W/cc, the APG can be compact and light of weight, making it an ideal portable plasma gun. Unlike other devices currently used for low temperature plasma research, its compactness and cost could make the APG widely available to universities, small-to-medium sized businesses, and government research laboratories for a myriad of applications.

ACKNOWLEDGMENTS

The authors gratefully acknowledge support from NASA and the Boeing Company.

REFERENCES

Barnes, D.C. et al., "Alternate Fusion: Continuous Inertial Confinement," *Plasma Phys. Control. Fusion* **35**, 929-940 (1993).

Bussard, R.W., Some Physics Considerations of Magnetic inertial-Electrostatic Confinement: A New concept for Spherical Converging-Flow Fusion," *Fus. Tech.* **19**, 273-293 (1991).

Forward, R.L. and Davis, J., *Mirror Matter*, New York, Wiley Science Editions, 1988.

Frisbee, R.H., "Interstellar Mission Propulsion Studies - Status Update," Tenth Annual NASA/JPL/MSFC/AIAA Advanced Propulsion Workshop, University of Alabama, Huntsville, AL, April 5-5, 1999.

Guterl, F., "A Small Problem of Propulsion," *Discover Magazine*, October, 1995, pp. 100-105.

Krall, N.A., The Polywell™: A Spherically Convergent Ion Focus Concept," *Fus. Tech.* **22**, 42-49 (1992).

Mewaldt, R.A. et al, A Small Interstellar Probe to the Heliospheric Boundary and Interstellar Space," *Acta Astronautica* **35** (suppl.) 35, 267 (1995).

Ordonez, C.A., "Time-Dependent Nested-Well Plasma Trap," *IEEE Trans. Plasma Sci.* **24**, 1378-1382 (1996).

Smith, G.A., "Antiproton-Catalyzed Microfission/fusion Propulsion Systems for Exploration of the Outer Solar System and Beyond, *JPL Workshop on Advanced Propulsion*, Pasadena, CA, May 20-23, 1997.

Smith, G.A. and Meyer, K.J., "Preliminary Design for the High Performance Antimatter Trap." NASA/ASEE Summer Faculty Fellowship Program report, MSFC, Huntsville, Alabama, August 1998.

Smith, G.A. and Kramer, K.J., "Enabling Exploration of Deep Space: High Density Storage of Antimatter," NASA/ASEE Summer Faculty Fellowship Program report, MSFC, Huntsville, Alabama, July 1999.

Performance Optimization of the Gasdynamic Mirror Propulsion System

William J. Emrich, Jr.[1] and Terry Kammash[2]

[1]NASA - Marshall Space Flight Center, Bldg. 4666, Room 370,Huntsville, AL. 35812
[2]Dept. of Nuclear Engineering and Radiological Sciences, University of Michigan, Ann Arbor, MI 48109
(256) 544-7504; bill.emrich@msfc.nasa.gov
(313) 764-0205; tkammash@engin.umich.edu

Abstract. Nuclear fusion appears to be a most promising concept for producing extremely high specific impulse rocket engines. Engines such as these would effectively open up the solar system to human exploration and would virtually eliminate launch window restrictions. A preliminary vehicle sizing and mission study was performed based on the conceptual design of a Gasdynamic Mirror (GDM) fusion propulsion system. This study indicated that the potential specific impulse for this engine is approximately 142,000 sec. with about 22,100 N of thrust using a deuterium-tritium fuel cycle. The engine weight inclusive of the power conversion system was optimized around an allowable engine mass of 1500 Mg assuming advanced superconducting magnets and a Field Reversed Configuration (FRC) end plug at the mirrors. The vehicle habitat, lander, and structural weights are based on a NASA Mars mission study which assumes the use of nuclear thermal propulsion[1] Several manned missions to various planets were analyzed to determine fuel requirements and launch windows. For all fusion propulsion cases studied, the fuel weight remained a minor component of the total system weight regardless of when the missions commenced. In other words, the use of fusion propulsion virtually eliminates all mission window constraints and effectively allows unlimited manned exploration of the entire solar system. It also mitigates the need to have a large space infrastructure which would be required to support the transfer of massive amounts of fuel and supplies to lower a performing spacecraft.

INTRODUCTION

One of the great deterrents to the large scale exploration of the solar system has been and continues to be the tremendous cost associated with putting the massive amounts equipment and infrastructure into space to support such endeavors. Consider, for example, the case for a manned Mars mission which by almost any account would be the easiest interplanetary mission to accomplish. If one were to use a strictly chemical system having a specific impulse (I_{sp}) of 450 s to perform this mission a tremendous amount of infrastructure in the form of refueling stations and multistage expendable vehicles would be required. Nuclear thermal systems with an I_{sp} of about 950 s while requiring far less infrastructure than chemical systems would, nevertheless, still require massive amounts of fuel (in the order of several times the dry weight of the transfer vehicle) to accomplish the mission. Both of these systems, while they are able to successfully accomplish a Mars mission, do so only at great cost due primarily to the massive amounts of fuel which must be transported into orbit to compensate for the inherent efficiency limitations of the engines. For more ambitious manned missions, such as to the outer planets, it is unlikely that either chemical systems or probably even nuclear thermal systems will be able to accomplish the voyage.

Ideally, for solar system exploration, one would want a vehicle with specific impulses in the range of 10,000 to 200,000 s and at least moderate levels of thrust. Fusion engines, if they can be built in reasonable sizes, match these requirements quite closely and would be most suitable as the primary propulsion system for an interplanetary vehicle. Such a vehicle would be quite capable accomplishing manned missions to any planet in the solar system. In the present study, one particular type of fusion engine, the Gasdynamic Mirror, was chosen as the basis for a vehicle design upon which various planetary mission analyses were performed[2].

ENGINE DESCRIPTION

The Gasdynamic Mirror or GDM upon which this study is based is a modification of the simple mirror design first proposed many years ago. This particular reactor configuration for a variety of reasons seems to be particularly well suited for fusion propulsion applications. The device would operate at much higher plasma densities and with much

CP504, *Space Technology and Applications International Forum–2000*, edited by M. S. El-Genk
© 2000 American Institute of Physics 1-56396-919-X/00/$17.00

larger length to diameter ratios than previous mirror machines. Several advantages accrue from such a configuration. First, the high length to diameter ratio minimizes to a large extent certain magnetic curvature effects which lead to plasma instabilities causing a loss of plasma confinement. In particular, this feature should make it possible to suppress the magnetohydrodynamic (MHD) "flute" instability which is known to plague classical mirror fusion designs. Second, the high plasma density should result in the plasma behaving much more like a conventional fluid with a mean free path shorter than the length of the device. The short mean free path implies that a majority of the fuel ions will undergo fusion reactions prior to being reflected at the magnetic mirror reflectors at the ends of the device. This characteristic helps reduce problems associated with "loss cone" microinstabilities which result from a depletion of the particle velocity distribution function in that these instabilities will not cause the longitudinal confinement time to be reduced below a certain level.

For the present study, a number of parameters have been optimized so as to maximize the engine thrust to weight ratio subject to various constraints such as maximum permissible vehicle mass, maximum allowable magnetic field strengths, etc. The vehicle, in particular, was constrained to a maximum mass of 1500 Mg and the mirror magnets were constrained to maximum centerline field strengths of 15 tesla. The resulting optimization yielded in the set of parameters presented in Table 1.

TABLE 1. GDM Reference Conditions.

Plasma Density (ion/cm^3)	2.2×10^{16}
Plasma Temperature (keV)	10
Fuel Mixture	½D + ½T
Vacuum Beta	0.95
Gain Factor (Q)	2.935
Plasma Length (m)	72
Plasma Diameter (cm)	8
Magnetic Field at Center (tesla)	13.8
Magnetic Field at Mirror (tesla)	15
Specific Impulse (s)	142200
Thrust (N)	22100
Thrust Power (MW)	11890
Fusion Power (MW)	14860
Injector Power (MW)	5064
Bremsstralung Power (MW)	317
Synchrotron Power (MW)	103
Neutron Power (MW)	11890
Neutron Wall Load (MW/m^2)	187
Specific Power (kw/kg)	133

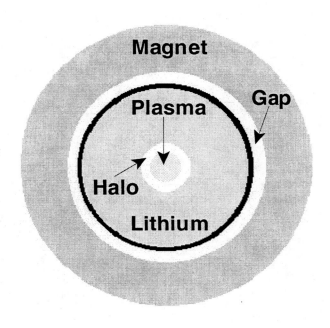

FIGURE 1. GDM Cross-sectional View.

The use of the deuterium/tritium (DT) fuel cycle for the GDM vehicle may seem a bit odd since deuterium and helium-3 are generally regarded as being preferable due to their essentially aneutronic nature.

$$D + T \rightarrow {}^4He + n\ (14.1\ MeV) \tag{1}$$

Unfortunately, the optimization program converged on a D to He^3 ratio of zero. The D-He^3 fuel combination apparently results in vehicle lengths which are impractically long and heavy. To cope with the high flux of 14 MeV neutrons which is the unavoidable consequence of using the DT fuel combination, therefore, it was decided to implement a design in which the first wall of the reactor is made entirely of lithium. During engine operation, the high neutron wall load resulting from the intense fusion reactions will cause the lithium to liquify. A slow rotation induced in the vehicle, will then produce centrifugal forces which will force the lithium to remain in place creating in essence a liquid first wall. A number of advantages accrue from this design. First, the material problems associated with radiation damage to the first wall are greatly reduced due to its being a liquid state. Second, the lithium can act as a coolant which when pumped through an appropriate heat exchanger can be used to heat a gaseous working fluid to power a Brayton cycle thermal electric conversion system. This energy would be used to provide electrical power for the fusion neutral beam injector system and for general station keeping. Third, as shown below, the lithium can act as a breeding material to produce the tritium fuel required by the engine.

$$n + {}^6Li \Rightarrow T + {}^4He \tag{2}$$

$$n + {}^7Li \Rightarrow T + {}^4He + n' \tag{3}$$

The use of bred tritium thus eliminates the need to transport large quantities of the radioactive element into orbit. Figure 1 illustrates a cross-section of the GDM engine concept used in this study.

Perhaps the most critical component in the GDM engine is the system of magnets used to confine the plasma in a stable configuration. Magnetic fields must be generated which are of sufficient strength to contain the hot fusing plasma, while at the same time not being so strong so as to overstress the engine's structural members. Calculations were performed to determine magnet sizes and current densities required to produce the necessary fields. These calculations show that with today's technology, it is possible to attain the required field strengths for the midsection magnets. Plasma physics considerations require centerline field strengths along the engine midsection to be approximately 14 tesla. The mirror magnets are a different story, however, in that for the present design, mirror centerline field strengths of 100's of tesla would normally be required. Magnetic fields having this intensity have never been produced and represent an obstacle to constructing GDM fusion engines of reasonable size. The use of Field Reversed Configurations (FRC) are proposed as a solution to this problem by creating in effect a magnetic plug which will have the effect of retarding plasma loss rates by creating high effective mirror ratios. The mirror magnets need be only strong enough to prevent the ejection of the FRC plug. These mirror fields are estimated to be only about 10% higher than the central magnets[3]. For this present study, optimizations yielded the following relationships between the magnetic field strength and the overall engine weight and thrust to weight ratio. The optimizations independently varied the mirror ratios at each end of the engine, the ratio of Larmor radius to plasma radius and the plasma temperature to yield either minimum engine mass or maximum engine thrust to weight ratio. Figures 2 through 5 illustrate the results of these studies.

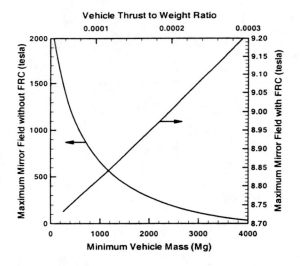

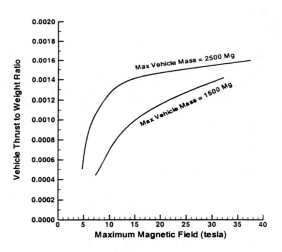

FIGURE 2. GDM Vehicle Thrust to Weight Ratio   **FIGURE 3.** GDM Performance Characteristics

In order to maintain the fusion reactions within the engine, it is proposed that neutral beam injectors be employed both to fuel and heat the plasma. The power to run the injectors would be derived from the fusing plasma itself through a combination of direct and thermal electric energy convertors. The critical "Q" value (Energy Out / Energy In) required for the engine system to be self sustaining will, therefore, be heavily dependent on the conversion system efficiencies.

For the present study conversion efficiencies of 30% were assumed for the Brayton cycle thermal electric conversion system, 90% for the direct electric conversion system, and 100% for the neutral beam injector system. Optimization studies found that for a minimum mass engine, charged particle power should generally be split evenly between that used for thrust and that used for direct electrical energy generation.

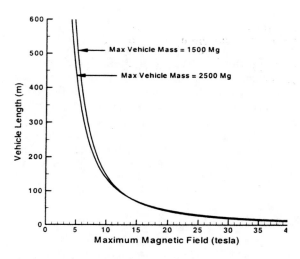

FIGURE 4. GDM Length Parametrics   **FIGURE 5.** GDM Asymmetry Effects

The energy required to start the system initially comes from a capacitor bank which is charged by a set of fuel cells. The capacitor bank was sized to produce a 1000 MW-sec pulse of electricity to start the fusion reactions. Advanced capacitors currently under development are anticipated to have charge storage densities of 36 kJ/kg. The fuel cells used in the system also provide station keeping power for the vehicle when the fusion engine system is shutdown. Recharging of the fuel cells is accomplished by tapping off some of the power generated by the fusion electrical conversion systems during engine operation.

TABLE 2. GDM Vehicle Weights

Magnets (Mg)	30
Radiator (Mg)	1077
Thermal Electric Convertor (Mg)	55
Direct Electric Convertor (Mg)	53
Neutral Beam Injector (Mg)	23
Fuel Cell/Capacitor System (Mg)	35
Tritium Breeding System (Mg)	10
Lithium Shield (Mg)	37
Magnet Cooling System (Mg)	19
Structure (Mg)	36
Habitat (Mg)	65
Lander (Mg)	60
Total (Mg)	1500

The habitat portion of the vehicle consists of a set of eight modified space station modules and is sized to provide long term accommodations for a crew of six. Radiation shielding is provided through the use of an internal "storm shelter" composed of foodstuffs, water, and waste products. It is estimated that approximately 100 kw of electricity will be required to operate the vehicle, exclusive of the fusion engine. The particular design considered for this study does not include a completely closed life support system. Only the oxygen/carbon dioxide portion of the system is closed, the foodstuffs are considered consummables and must be brought along for the missions.

The lander used for this study is a modified Reusable Launch Vehicle (RLV) having a vertical takeoff, vertical landing design. This vehicle could serve as a means not only of transferring a crew and equipment to and from earth orbit but also to and from the orbits of other planetary bodies. It is anticipated that this vehicle will include a small nuclear powered in-situ fuel processing system such that it would be possible to make multiple trips to the main vehicle and to other locations on the planetary surface using locally available resources.

The system weights for all the major systems for the vehicle, including the habitat, lander, engine system, etc. were estimated and included for all the mission studies. A summary of the weights is given in Table 2.

CONCLUSIONS

The results of this study indicate that the fusion rocket is capable of opening the entire solar system to continuous extensive human exploration with little of the in-space infrastructure required to support the refueling operations associated with NTR or chemical systems. Voyages beyond Mars become feasible due to the ultra high efficiencies of these engines. From an economic standpoint, the benefits associated with the low mission fuel requirements characteristic of fusion systems should quickly outweigh the development cost differential perceived to be in favor of fission or chemical systems. Indeed, it may be that the development costs associated with fusion systems may be less than corresponding fission systems because of the greatly reduced test facility requirements needed to clean up the NTR hydrogen exhaust effluent. Fusion engines, contrary to popular thought, epitomize the "cheaper, faster, better" concept to space exploration. It will only be through vehicles of this type that truly practical solar system exploration and colonization will be possible on a large scale at an affordable cost.

REFERENCES

Emrich, W. and Young, A., "Nuclear Propulsion System Options for Mars Missions," *AIAA Space Programs and Technologies Conference*, March 24-27, 1992, Paper # AIAA 92-1496.

Kammash, T. and Lee, M., "Gasdynamic Fusion Propulsion System for Space Exploration," *Journal of Propulsion and Power*, Vol. 11, No. 3, May-June 1995, Page 544-553.

Kammash, T., Emrich, W., and Godfroy, T., *"Promising Approaches to Mass Reduction of the GDM Fusion Propulsion System,"* 10th NASA/JPL/MSFC/AIAA Advanced Space Propulsion Workshop, April 5-8, 1999.

Colliding Beam Fusion Reactor Space Propulsion System

Frank J. Wessel,[1] Michl W. Binderbauer,[1] Norman Rostoker,[1]
Hafiz Ur Rahman,[2] Joseph O'Toole[3]

[1] *University of California*, Irvine, CA 92697
[2] *University of California, IGPP*, Riverside CA 92521
[3] *Los Alamos National Laboratories*, Los Alamos, NM 87545

Abstract. We describe a space propulsion system based on the Colliding Beam Fusion Reactor (CBFR). The CBFR is a high-beta, field-reversed, magnetic configuration with ion energies in the range of hundreds of keV. Repetitively-pulsed ion beams sustain the plasma distribution and provide current drive. The confinement physics is based on the Vlasov-Maxwell equation, including a Fokker Planck collision operator and all sources and sinks for energy and particle flow. The mean azimuthal velocities and temperatures of the fuel ion species are equal and the plasma current is unneutralized by the electrons. The resulting distribution functions are thermal in a moving frame of reference. The ion gyro-orbit radius is comparable to the dimensions of the confinement system, hence classical transport of the particles and energy is expected and the device is scaleable. We have analyzed the design over a range of $10^6 - 10^9$ Watts of output power (0.15-150 Newtons thrust) with a specific impulse of, $I_{sp} \sim 10^6$ sec. A 50 MW propulsion system might involve the following parameters: 4-meters diameter x 10-meters length, magnetic field $\sim$ 7 Tesla, ion beam current $\sim$ 10 A, and fuels of either D-He^3, P-B^{11}, P-Li^6, D-Li^6, etc.

INTRODUCTION

Fusion space propulsion is attractive due to the high-energy-density of its fuel, the high-specific-impulse of its exhaust, and the abundance of fusion fuels throughout the solar system. However, present fusion space propulsion concepts involve large, complex, and costly systems with perhaps insurmountable engineering obstacles. These drawbacks are primarily the result of energy and particle transport processes that are non-classical and a community bias toward the use of D-T fuels.

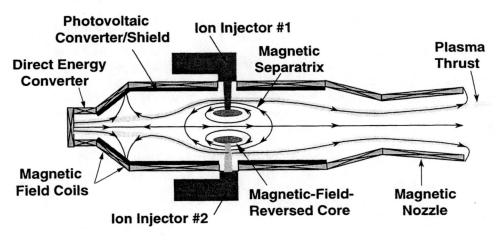

FIGURE 1. Colliding Beam Fusion Space Propulsion concept.

The Colliding Beam Fusion Reactor (Rostoker, et. al., 1993 and 1997) is based on near-term technologies and the use of advanced (aneutronic) fusion fuels. Repetitively-pulsed (ion) beams inject high energy particles into the magnetic-field reversed, plasma-core of the CBFR to sustain momentum and energy losses that

CP504, *Space Technology and Applications International Forum–2000*, edited by M. S. El-Genk

result from Ohmic dissipation and fusion burn-up. A large-bore superconducting magnet provides a constant external magnetic field. A schematic illustration of the propulsion system is provided in Figure 1.

The fuel core in the CBFR is a field-reversed configuration (FRC). FRCs have been studied for decades in the international plasma physics community. (Tuszewski, 1988; Steinhauer, et. al., 1996) The main advantages of the FRC are its high beta (hence, reduced synchrotron power losses), cylindrical symmetry (ease of access and serviceability), and a "magnetic sepratrix" that surrounds the plasma core that acts as a natural divertor of non-resonant particle energies. Typical laboratory FRC parameters are: radial dimension, $r \sim 2\rho_{ion}$, temperatures, $T_e \leq 100$ eV, $T_i \leq 1$ keV, and density, $n \sim 5 \times 10^{15}$ cm^{-3}.

Laboratory FRCs are typically created by a single-pulse, magnetic-induction system with a current decay lifetime of the order of the L/R timescale, $\sim 10^{-3}$ sec. The CBFR concept extends the lifetime to the steady state using tangential injection and trapping of intense, repetitively-pulsed ion beams. The pulsed ion beam enters the magnetic field as a current- and charge-neutralized plasma stream. On entry the Lorentz force causes the lateral edges of the beam to undergo a charge separation. If the beam current density is sufficient to sustain the required electric field, the beam propagation continues undeflected by the $\vec{E} \times \vec{B}$ drift.(Wessel, et. al., 1990) A dense (background) plasma shorts the $\vec{E}$ field causing the beam to be trapped inside the FRC core.

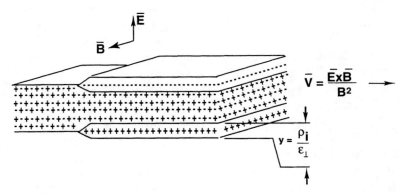

FIGURE 2. Dense plasma beam propagation into a transverse magnetic field.

Intense ion beam current drive in the CBFR has an additional important advantage: classical confinement and transport has been observed in experiments involving the injection of 100 keV particles into the (1 keV) thermal plasma of a tokamak.(Heidbrink, 1994) Moreover, for some experimental conditions FRCs have shown similar behavior. If confirmed classical transport would reduce substantially the size of a fusion reactor compared to other fusion-power concepts. Moreover, the needed ion beam injector and superconducting magnet technologies largely exist and are highly developed so that a multi-megawatt CBFR could be assembled and tested in the near term. When one considers the prospect for future decades of continuing fusion studies along present lines this is a compelling argument in favor of the CBFR.

EQUILIBRIUM

A one-dimensional plasma equilibrium model for the CBFR is employed, based on rigid rotor solutions of the Vlasov-Maxwell equations and drifted Maxwellian distributions. For electrons and one type of ion we obtain the following exact solution:[1]

$$n_e(r) = n_e(r) = \frac{n_o}{\cosh^2\left[(r^2 - r_o^2)/r_o\Delta r\right]} \, , \tag{1}$$

$$B_z(r) = -B_o\left[1 + \sqrt{\beta}\tanh\left(\frac{r^2 - r_o^2}{r_o\Delta r}\right)\right] \, , \tag{2}$$

$$E_r(r) = \frac{m_e}{e}r\omega_e^2 - \frac{r\omega_e}{c}B_z - \frac{T_e}{en_e}\frac{dn_e}{dr} \, . \tag{3}$$

where r_o is the radius at which the density is a maximum and $-B_o$ is the externally applied magnetic field,

$$\Delta r = \frac{2\sqrt{2}}{r_o}\left(\frac{T_e + T_i}{4\pi n_o e^2}\right)^{1/2}\frac{c}{|\omega_i - \omega_e|} , \tag{4}$$

$$\beta = 8\pi n_o \frac{T_e + T_i}{B_o^2} , \tag{5}$$

and Δr is defined by the equation,

$$N_i = \int^{r_B} 2\pi r\, dr n_i = 2\pi r_o \Delta r n_o. \tag{6}$$

The initial ion velocity, $V_i = r\omega_i$, is determined by the injection energy. The maximum particle density is also determined by design. The temperatures cannot be determined from the Vlasov-Maxwell equations. Higher order processes must be included – Coulomb collisions as well as particle sources and sinks. Other features of the higher order processes have already been included such as the assumption of drifted Maxwell distributions for electrons and ions which must be verified a posteriori. The external magnetic field B_o and V_e can be identified by considering the conservation of momentum of a single fluid description,

$$\rho\frac{V_\theta^2}{r} = \frac{\partial}{\partial r}\left(P + \frac{B_z^2}{8\pi}\right) , \tag{7}$$

where $\rho V_\theta = \sum_i m_i \int f_i \mathbf{v}\, d\mathbf{v} = \sum_j n_j m_j r\omega_j$ and $\rho = \sum_j n_j m_j \simeq n_i m_i$ since $m_i \gg m$. Eq. (7) can be integrated from $r = 0$ to $r = r_B = \sqrt{2}r_o$. At these limits $P = n(r)[T_e + T_i] \simeq 0$ so that,

$$\int_0^{r_B} \rho\frac{V_\theta^2}{r} dr = m_i\omega_i^2 \int_0^{r_B} n_i r\, dr = r_o\Delta r n_{io} m_i \omega_i^2 .$$

Additionally assuming $r_o \gg \Delta r$,

$$\int_0^{r_B} \frac{\partial}{\partial r}\frac{B_z^2}{8\pi} dr = \frac{1}{8\pi}\left[(B_o + B_m)^2 - (B_o - B_m)^2\right] = \frac{1}{2\pi}B_o B_m ,$$

where $B_m = \sqrt{8\pi n_o(T_e + T_i)}$. Substituting Eq. (4) for $r_o\Delta r$ the result is,

$$\omega_e = \omega_i\left[1 - \frac{\omega_i}{\Omega_o}\right] , \tag{8}$$

where $\Omega_o = eB_o/m_i c$ is the ion cyclotron frequency in the externally applied field B_o. Eq. (8) determines $V_e = r\omega_e$. If $\omega_i = \Omega_o$, $V_e = 0$. By increasing the externally applied field B_o, the value of ω_e can be controlled and therefore the value of the plasma width according to Eq. (4).

With minor approximations these solutions are relevant to a two-component plasma with ions rotating at the same velocity. The only change to the above is to use the average ion mass in the cyclotron frequency, $\Omega_o = <Z>eB_o/<m>c$. A numerical solution to the Vlasov/Maxwell equation for D-T confirms the accuracy of Eqs. (1)-(3); the results are displayed in Figure 3. Notice in particular the narrow radial extent of the plasma core in addition to, electrostatic confinement of the electrons.

DESIGN OF A COLLIDING BEAM FUSION REACTOR SPACE PROPULSION SYSTEM

The present analysis considers a 100 MW Colliding Beam Fusion Reactor system; Table 1 summarizes the fusion core design parameters. The ion species have equal density and velocity. The temperatures and line density are determined as described above. The circulating current is, $I_\theta = \frac{eN_e}{2\pi}(\omega_i - \omega_e)$, where ω_i is the ion gyro-frequency. The inductance per unit length is, $L = 2\pi^2 r_o^2/c^2$ and the resistance is, $R = \frac{2\pi r_o m}{N_e e^2}\left(\frac{1}{t_{e1}} + \frac{1}{t_{e2}}\right)$.

The intense ion beam injection timescales are illustrated in Figure 4. An increment of current, ΔI, is injected into the system with a beam pulse duration, $\tau_{beam} \leq \Delta t$ and an inter-pulse time T. The number of ions per

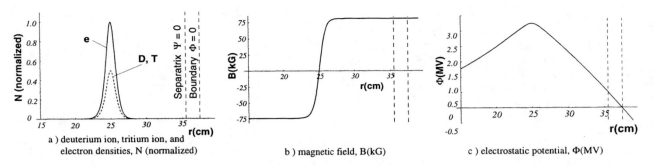

a) deuterium ion, tritium ion, and
electron densities, N (normalized)

b) magnetic field, B(kG)

c) electrostatic potential, Φ(MV)

FIGURE 3. Equilibrium solutions for a D-T Colliding Beam Fusion Reactor.

pulse balances the fuel ions that disappear due to fusion reactions in time T, $\Delta N_i = N_i T / < t_{fi} >$, where $t_{fi} = \frac{1}{<\sigma v > n_i}$. The L/R time is longer than the momentum exchange times so that heating does not take place. The electron-ion momentum exchange time is,

$$t_{ei} = \frac{3}{8}\sqrt{\frac{\pi}{2}}\frac{T_e^{3/2}m^{1/2}}{n_i Z_i^2 e^4 \ln \Lambda} \simeq 10^{-3} \text{ sec },$$

assuming $n_i \sim 10^{15}$ cm^{-3} and $T_e \simeq 64$ keV. The ion-electron momentum exchange time is

$$t_{ie} = \frac{n_i m_i}{n_e m}t_{ei} \simeq 1 \text{ sec }.$$

During time T the ion velocities change very little. The current decay is almost entirely due to the change in electron velocity which involves very little energy change. The dissipated energy during the period $(t_n + \Delta t, t_n + T)$ comes from the stored magnetic energy,

$$\frac{1}{2}LI^2|_{t_n+\Delta t} - \frac{1}{2}LI^2|_{t_n+T} = I^2 RT .$$

The loss of magnetic energy is replaced by the injected/trapped beam during the period $(t_{n+1}, t_{n+1} + \Delta t)$.

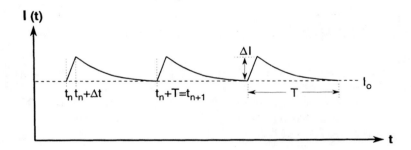

FIGURE 4. Current and energy replacement in a CBFR using intense, pulsed beams.

The typical pulse duration for intense ion beams is, $< 1\mu$sec with pulse-repetition-rates ~ 100 Hz and an ion energy in the range, $E_{ion} = 0.1 - 1$ MeV. The beam current/pulse depends on the design of the accelerator and the ion diode. It is typically in the range, 1-1,000 kA. A commercial ion beam system is advertised to produce 40 kA, 400 keV, 0.5 μsec beams of protons with a pulse-repetition rate of 10 Hz.(Quantum Manufacturing, 1999) Improvements in the rep-rate and ion energy would be needed for a 100 MW CBFR. Concerns related to plasma clearing in the ion diode, at high repetition rate, might be averted through the use of multiple-beam injectors.

The beam parameters are summarized in Table 1. The current per unit length is,

$$I_{input} = \frac{2}{3}\sum_{i=1,2}\frac{N_i}{t_{fi}}e \quad (A/cm).$$

TABLE 1. Core Parameters for a 100 MW CBFR Space Propulsion System.

	D-T	D-He^3	p-B^{11}
Densities $\times 10^{15}$ (cm^{-3}) ($n_e = 1$)			
n_1	0.5	0.33	0.5
n_2	0.5	0.33	0.1
Velocities $\times 10^9$ (cm/s)			
V_e	0.23	0.44	0.66
$V_1 = V_2$	0.54	0.66	0.76
Fuel Ion Energy (keV)			
$\frac{1}{2} m_1 V_1^2$	300	450	300
$\frac{1}{2} m_2 V_2^2$	450	675	3300
Temperatures (keV)			
T_i	96	217	235
T_e	100	170	85
$r_o \Delta r$ (cm^2)	114	203	310
Current $I \times 10^5$ (A/cm)	1.42	1.75	1.27
Magnetic Field (kG)			
B_o	5.88	8.25	15.3
$B_o + B_m$	94.7	121	96.3
Current Decay Time L/R (sec)	42	195	36
Reactivity $\langle \sigma v \rangle \times 10^{16}$ (cm^3/s)	12	3.0	6.4

For the CBFR the average injected beam current is, $< I_{beam} >= I_{input} \times l$ A. The peak current pulse is,

$$\hat{I}_{beam} = \frac{I_{input} T}{\tau_{beam}} \quad (kA).$$

The beam power is,

$$P_{input} = \frac{2}{3} \sum_{i=1,2} \frac{N_i}{t_{fi}} \frac{1}{2} m_i v_i^2 + I^2 R \quad (MW).$$

The ion diode voltage is, $\Delta\phi_{ion\ diode} = P_{in}/ < I_{beam} >$ (kV). The fusion reactivity is, $< \sigma v >$ (cm^3/sec), which gives a fusion power, $P_f = 1.6 \times 10^{-19} n_1 n_2 < \sigma v > \epsilon_f$ (W/cm^3), where ϵ_f (MeV) is the fusion energy/reaction. The peak output power per unit length is, $\hat{P}_n$ (kW/cm).

The results for a 100 MW CBFR propulsion system are summarized in Table 2 for three fuel cycles. The system masses and specific impulses for each are nearly equivalent. All systems involve a first wall radius of the order of, 42 cm, and a length of the order of, l= 1.1 m for D-T, 4.6 m for D-He^3, and 6.9 m for P-B^{11}. Higher, or lower, output powers would be possible by simply adjusting the length of the reactor.

D-T produces over 80% of its fusion energy in the form of energetic neutrons that cannot be easily vectored as thrust. The conversion of neutron energy also involves a thermal cycle that is at most 30 % efficient. Moreover, a heavy shield must be used to limit biological hazards to operations personnel and to extend the lifetime of the material structures. These aspects largely negate the advantages for D-T fuel. The remaining fuels, D-He^3 and B^{11}, are primarily aneutronic and the dominant fraction of their fusion energy is in the form of energetic-charged particles and Bremsstrahlung radiation. Methods exist to capture and convert the radiation with an efficiency as high as 60% .(Monkhorst, 1999) Normally, the charged particle flow ("exhaust") emerges from both ends of the system, althoutgh it may be possible to alter the magnetic configuration to enhance assymetric flow. In a space propulsion system the flow from one end of the reactor would be used to provide thrust through a magnetic nozzle. Flow through the opposite end of the system could be used to sustain power to the reactor in addition to providing power for on-board ship operations. In such a system direct energy conversion is expected to be as high as 90%. Excess power might also be used to drive an auxiliary thruster system, perhaps at lower I$_{sp}$, should that be beneficial. Taking all these factors into account the P-B^{11} system is the most attractive, characterized by the highest thrust-to-power and thrust-to-mass ratios of the three systems considered here.

CONCLUSION

The scaleable CBFR is well suited to fusion space propulsion applications. Individual 100 MW modules could be arranged to provide output powers at any level: for example, a multi-module system of 100 MW units could provide a GW thrust/power source, or a single large unit could be developed to obtain a desired output power. Compared to toroidal systems that burn D-T fuels the cylindrical nature of a D-He^3 or P-B^{11} CBFR provides dramatically enhanced spacecraft performance metrics, in addition to reduced costs, higher reliability, and enhanced lifetime of the reactor.

TABLE 2. Parameters for a 100 MW Colliding Beam
Space Propulsion System.

	D-T	D-He^3	p-B^{11}
Total Output Power, P_o (MW)	100	100	100
Fusion Energy/Reaction, E_F (MeV)	17.4	18.2	8.68
Nuclear (Particle) Power, P_N (MW)	99	84	77
Radiation Power, P_B (MW)	0.9	15.9	22.8
Recirculated Power, P_C (MW)	6.9	11.8	38
Thrust Power, P_T (MW)	29.9	67.8	50.8
Nuclear Power/Radiation Power, P_N/P_R	112	5.28	3.84
Recirculated Power/Nuclear Power, P_C/P_N	.070	.141	.493
Thrust Power/Total Power, P_T/P_o	0.3	0.68	0.51
Specific Impulse, $I_{sp} \times 10^6$ (s)	1.3	1.4	1.4
Plasma Radius, r (cm)	30	30	30
Plasma Width, Δr (cm)	3.8	6.8	10.3
First Wall Radius, R (m)	.42	0.42	0.42
Chamber Wall Thickness, ΔR (m)	3	0.6	0.6
Chamber Length, L (m)	1.1	4.6	6.9
Average Material Density, $\delta \times 10^3$ (kg/m^3)	4.0	4.0	4.0
Total Mass, $M \times 10^4$ (kg)	3.5	2.9	4.4
Mass/Total Power, $M/P_o \times 10^{-3}$ (kg/W)	0.35	0.29	0.44
Thrust/Total Power, T/P_o (mN/MW)	37.8	95.5	281
Thrust/Mass, $T/M \times 10^{-4}$ (N/kg)	1.1	3.2	6.4

REFERENCES

Heidbrink, W. W. and Sadler, G. J., "The Behavior of Fast Ions in Tokamak Experiments," *Nucl. Fusion* **34**, 535(1994).

Monkhorst, H., private communication, University of Florida, Gainsville, 1999.

Quantum Manufacturing, Inc., Albuquerque, NM, 1999.

Rostoker, N. et.al., "Magnetic Fusion with High Energy Self-Colliding Ion Beams," *Phys. Rev. Lett.* **70**, 1818(1993).

Rostoker, et. al., "Colliding Beam Fusion Reactor," *Science* **21**, 1419(1997).

Steinhauer, L., et. al, "FRC 2001: A White Paper on FRC Development in the Next Five Years," *Fusion Technology* **30**, 116(1996).

Tuszewski, M., "Field Reversed Configurations," *Nuclear Fusion* **28**, 2033(1988).

Wessel, F. J., et. al., "Propagation of Neutralized Plasma Beams," *Phys. Fluids* **B2**, 1467(1990).

Evaluation of a Fusion-Driven Thruster for Interplanetary and Earth-to-Orbit Flight

H.D. Froning Jr.[1] John J. Watrus,[2] Michael H. Frese[2], Richard A Gerwin[2]

[1]Flight Unlimited 5450 Country Club Drive, Flagstaff, AZ 86004
[2]NumerEX, 2309 Renard Place S.E. Albuquerque, NM 8716-4259

Abstract. Investigations performed under a NASA Small Business Independent Research (SBIR) Phase I Award have shown that hydrogen propellant heating from fusion energy on the order of a few 10^{14} W/kg, corresponding to power absorption rates of a few GW, can produce on the order of 100 kN of thrust with a 0. 5 kg/s propellant mass flow rate.

INTRODUCTION

The results of this investigation have applicability to the thrust chamber element of any fusion propulsion system that heats hydrogen propellant to plasma form. However, they are probably the most applicable to the thruster subsystem of the "Quiet Energy Discharge" (QED) electrostatic confinement fusion propulsion system described by Dr. R.W. Bussard in previous STAIF Proceedings (Bussard 1993a, 1993b, 1993c and 1994). As shown in Figure 1, aneutronic fusion reactions (that emit no neutrons and cause no radioactivity) are accomplished within a reactor core as protons and Boron 11 ions are driven into sufficiently close proximity for fusion by the electrostatic repulsion of surrounding electrons that, themselves, are confined within the central portion of the reactor by quasi-spherical polyhedral magnetic fields.

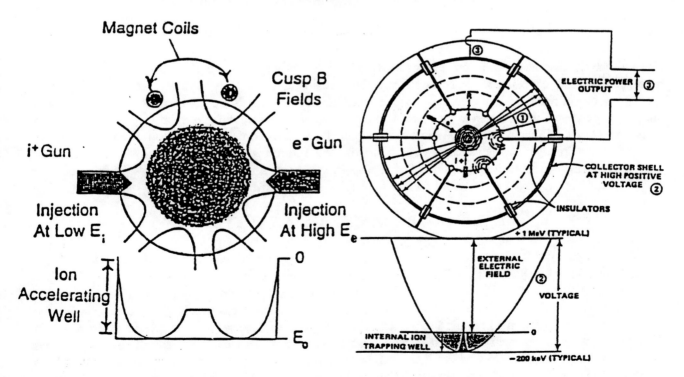

FIGURE 1. Aneutronic Fusion by Electrostatic Confinement.

CP504, *Space Technology and Applications International Forum–2000*, edited by M. S. El-Genk
© 2000 American Institute of Physics 1-56396-919-X/00/$17.00

Outward expanding fusion products are Helium 4 ions whose energies and charges are such that they can be effectively slowed and collected as electricity by spherical grid structures that are maintained at appropriate voltage gradients. The electricity collected from the expanding fusion products is transformed into electron beams that deposit their energy into flowing propellent within a thrust chamber, producing a very high temperature gas/plasma which expands within a magnetically insulated exhaust nozzle to produce thrust at very high specific impulse (Isp). A typical QED engine concept is shown in Figure 2.

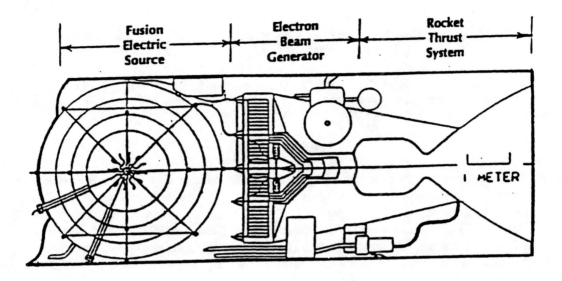

FIGURE 2. QED Fusion Rocket Propulsion System.

Because the emitted fusion energy contains no neutrons or radioactivity, and because most of it is transformed into electricity before it can reach reactor walls and cause erosion and heat, only modest reactor shielding mass (for absorption of relatively low-energy x-rays) is needed, and cooling needs are minimal. And because fusion energy is generated and converted into electrical energy within thin structures of relatively modest size and mass, predicted QED engine thrust-to-weight ratios for any Isp level are orders of magnitude greater than those of other fusion propulsion systems that have been proposed for future flight. Furthermore, the "clean" exhaust and relatively high thrust-to-weight of QED fusion propulsion systems are attractive, not only for flight beyond earth orbit, but flight from earth to orbit as well (Bussard 1993a and Froning 1993).

INITIAL PARAMETRIC ANALYSIS

QED fusion energy is transformed directly into electrical energy, and the electrical energy is deposited into flowing hydrogen propellant by means of electron beams. And electron energies from such beams - which are of the order of 10 KeV - transforms flowing liquid hydrogen propellant into a highly ionized high-speed plasma flow. A detailed analysis of the electron/plasma interaction associated with electron beams was not performed. But the strong magnetic field required to guide the strongly ionized plasma flow through the thrust chamber was sufficient for confinement of injected electrons to spiral orbits of sufficient length for all electron energies to be deposited within the flow before its exit from the combustion region. One-dimensional flow analysis explored the feasibility of efficient fusion energy deposition into flowing hydrogen propellant for generation of thrust; determined the influence of propellant heating rate upon thrust generation and upon the radiant heat that reaches thrust chamber walls; and defined the thrust chamber geometry and established initial conditions for more detailed two-dimensional analysis.

Preliminary and detailed analysis indicated that wall heating from resistive diffusion of plasma across field lines would be several orders of magnitude less than radiant heating. Thus, the critical heating factor was the radiant heat load that is imposed upon thrust chamber walls - and which must be dissipated by space radiators and/or regenerative cooling. Radiant heating was calculated using a model for optically thin radiative cooling that included the effects of bound-bound, free-bound, and free-free transitions. Here, preliminary one-dimensional flow analysis revealed the influence of propellant heating rates up to 6×10^{14} W/kg on achieved thrust and plasma cooling rate - which, for conservatism, was assumed to be the radiant heat reaching thrust chamber walls. These analyses are summarized in Figures 3 and 4 for a low thrust (20-30KN) and a high thrust (80-90KN)family of thruster designs.

Although the lower thrust family is associated with lower flow temperature (1 to 2 eV) and heating, its higher flow density resulted in greater radiative cooling - as indicated in Figure 3. As a consequence, large amounts of "line" radiation reached thrust chamber walls because of excited state neutrals going to the ground state. By contrast, higher heating rate and flow temperature (6 to 10 eV) is associated with the higher thrust family. But orders of magnitude less radiation from such flow reaches chamber walls (despite increased bremsstrahlung radiation) because line radiation diminishes towards zero. It is seen that the lower and higher thrust families are separated by a region - a region where propellant heating rates are comparable to radiative cooling rates and unsteady flow phenomenon appear.

DETAILED TWO-DIMENSIONAL ANALYSIS

The one-dimensional analysis identified the heating rate and propellant flow density to be used in a more detailed two-dimensional analysis, together with the dimensions of the axi-symmetric thrust chamber. These dimensions, which were found to be comparable to those estimated by Bussard for the thrust chamber portion of a QED power and propulsion system, were: a throat diameter of 1.2m, a maximum chamber diameter of 3.5m; and a total length of 3.0m. Detailed two-dimensional flow analysis were performed over the entire thrust chamber region for a heating rate of 3×10^{14} W/kg and a propellant flow density of 6×10^{-5} g/m³. Thrust levels in the 80-100kN range were achieved with applied magnetic fields of the order of 3 tesla - resulting in magnetic field pressure in the range of a few 10^5J/m³ in the central region thrustor. Maximum pressures, temperatures and ion number densities in the central region were of the order of: 9×10^4 J/m³, 37eV, and 6×10^{16} cm⁻³. Figure 5 through 8 shows contours of: magnetic field intensity, temperature, pressure and ion number density within the thrust chamber. The radiative output to the thrust chamber walls, including both bremsstrahlung and line radiation, was approximately 20MW. It is not yet known if regenerative cooling by the hydrogen propellant is sufficient to handle such a heating load, or if additional cooling by means of space radiators is required. Mission analysis indicated that only about 20 percent of the available hydrogen propellant would be needed for thrust generation. Thus 80 percent would be available for regenerative cooling.

EARTH-TO-ORBIT MISSION ANALYSIS

Mission analysis for regeneratively cooled QED fusion propulsion systems indicate that thrust levels in the 75 to 100 kN and specific impulse levels in the 4500 to 5000 sec range could enable: (a) fast interplanetary trips (less than 3 months between earth and Mars) with more payload and less total vehicle mass than currently envisioned nuclear fission systems. Mission analysis also indicated that such thrust levels could enable economical earth-to-orbit trips by reusable launch vehicles (RLVs) that would have only one-fourth to one-sixth the takeoff mass of currently envisioned RLVs. For the all-rocket fusion powered RLV, the chemical rocket system provided an initial boost of about 4,100 m/s over a 5 minute interval and the fusion rocket system

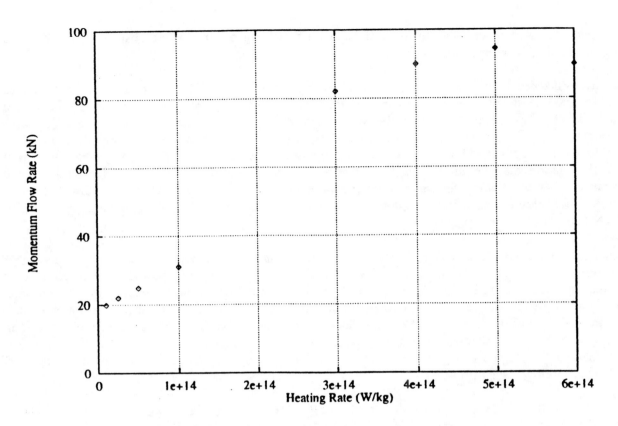

FIGURE 3. Variation of thrust with heating rate.

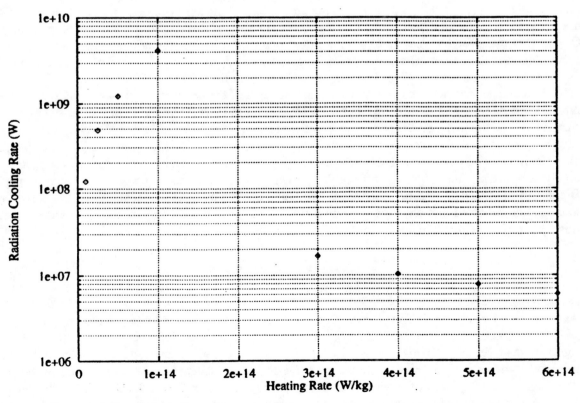

FIGURE 4. Variation of radiative cooling rate with heating rate.

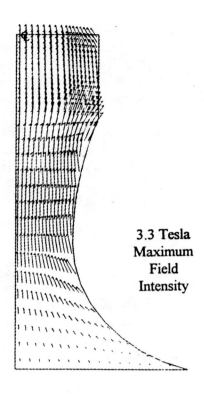

FIGURE 5. Magnetic field vector distribution.

3.3 Tesla
Maximum
Field
Intensity

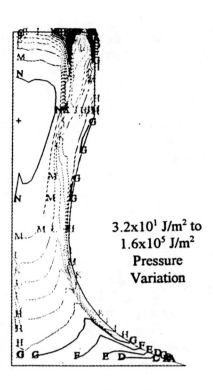

FIGURE 6. Steady state temperature distribution.

3.2×10^1 J/m^2 to
1.6×10^5 J/m^2
Pressure
Variation

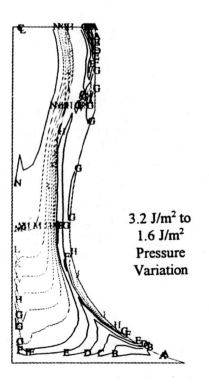

FIGURE 7. Steady state pressure distribution.

3.2 J/m^2 to
1.6 J/m^2
Pressure
Variation

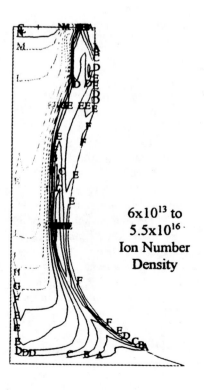

FIGURE 8. Steady state Ion density distribution.

6×10^{13} to
5.5×10^{16}
Ion Number
Density

providing the final 6,540 m/s with consumption of about 16,000 kg of hydrogen propellent for fusion propulsion and cooling to reach orbital speed over a 26 minute period. The lighter chemical rocket/airbreathing system of the lighter airbreathing RLV, provided an initial boost of about 4,560 m/s over an 8 minute interval while its fusion rocket system provided the final 6,080 m/s over a 28 minute interval to reach orbit with consumption of about 14,000 kg of hydrogen propellant for fusion propulsion and cooling.

The takeoff mass of the all-rocket fusion powered RLV was approximately 280 Mg while that of the airbreathing one was approximately 190 Mg - with both capable of carrying 19 Mg of payload to low earth orbit and 11 Mg of payload to the International space station. By contrast, the takeoff mass of all-rocket and airbreathing RLVs with the same payload and all-chemical propulsion that were configured in the NASA Access-to-Space Study in 1993 (Volume 3: Design Data), were 1127 Mg and 417 Mg respectively.

SUMMARY AND CONCLUSIONS

This work represents only a first iteration of a fusion powered thruster design. Additional work is needed to optimize flow parameters to reduce, if possible, radiant heat that bathes thrust chamber walls. But this work is encouraging, indicating in that approximately 99 percent of the flow enthalpy from fusion heating can be utilized for propulsion - with only about 1.0 percent going into waste heat. Additional work would determine the possibility of reducing waste heat further and the possibility of regenerative cooling (circulation of hydrogen propellant through thrust chamber walls before being used for propulsion) handling the radiant heat load without recourse to the added bulk and weight of space radiators.

ACKNOWLEDGMENTS

This work was performed under a Small Business Independent Research (SBIR) Phase I Contract awarded to NumerEx. This work was sponsored and directed by the George C. Marshall Space Flight Center under NASA Contract Number NAS 8-98106.

REFERENCES

Bussard, R.W., Jameson, L.W., (1993a) "The QED Engine Spectrum: Fusion-Electric Propulsion for Air-Breathing to Interstellar Flight," 29th Joint Propulsion Conference, Monterey, CA, Paper #AIAA.4.1-93-2005.

Bussard, R.W., Jameson, L.W., Froning, H.D., (1993b) "The QED Engine: Fusion-Electric Propulsion for CIS Oort/Quasi-Interstellar (QIS) Flight," 7th Interstellar Space Exploration Symp. Proc.

Bussard, R.W., Jameson, L.W., (1993c) "The QED Engine System: Direct-Electric Fusion-Powered Rocket Propulsion Systems," in Proc. 10th Symposium of Space Nuclear Power and Propulsion, CONF-930103, M.S. El-Genk and M.D. Hoover, eds., American Institute of Physics, NY, AIP Conf. Proc. 271, vol 3: p 1601-p 1611.

Bussard, R.W., Jameson, L.W., (1994) "Design Considerations for Clean QED Fusion Propulsion Systems," in Proc. 11th Symposium on Space Nuclear Power and Propulsion, CONF 940101, M.S. El-Genk and M.D. Hoover, eds, American Institute of Physics, NY, AIP Conf. Proc. 301, vol 3: p 1289-p 1296.

Froning, H.D., Bussard, R.W., (1993) "Fusion-Electric Propulsion Technology for Hypersonic Flight," 44th Congress of the International Astronautical Federation, Graz, Austria, Paper #IAF-93-5.3.473.

Watrous, J.J., Gerwin, R.A., Frese, M.H., (1998) "A Fusion-Driven Earth-to-Orbit Thruster (Initial Design Considerations for an Electron Beam Heated, Magnetic Nozzle Thruster)," NASA Contract Number Nas8-98106, Numerex Report 98-02.

Helium Release from ^{238}PuO$_2$ Fuel Particles

Mohamed S. El-Genk and Jean-Michel Tournier

Institute for Space and Nuclear Power Studies and Chemical and Nuclear Engineering Dept.
The University of New Mexico, Albuquerque, NM 87131, USA
(505) 277–5442, email: mgenk@unm.edu

Abstract. Coated plutonia fuel particles have recently been proposed for potential use in future space exploration missions that employ radioisotope power systems and/or radioisotope heater units (RHUs). The design of this fuel form calls for full retention of the helium generated by the natural radioactive decay of ^{238}Pu, with the aid of a strong zirconium carbide coating. This paper reviews the potential release mechanisms of helium in small-grain (7-40 μm) plutonia pellets currently being used in the General Purpose Heat Source (GPHS) modules and RHUs, during both steady-state and transient heating conditions. The applicability of these mechanisms to large-grain and polycrystalline ^{238}PuO$_2$ fuel kernels is examined and estimates of helium release during a re-entry heating pulse up to 1723 K are presented. These estimates are based on the reported data for fission gas release from granular and monocrystal UO$_2$ fuel particles irradiated at isothermal conditions up to 6.4 at.% burnup and 2030 K. It is concluded that the helium release fraction from large-grain ($\geq$ 300 μm) plutonia fuel kernels heated up to 1723 K could be less than 7%, compared to ~ 80% from small-grain (7-40 μm) fuel. The helium release fraction from polycrystalline plutonia kernels fabricated using Sol-Gel techniques could be even lower. Sol-Gel fabrication processes are favored over powder metallurgy, because of their high precision and excellent reproducibility and the absence of a radioactive dust waste stream, significantly reducing the fabrication and post-fabrication clean-up costs.

INTRODUCTION

The early technology of coated fuel particles came as a spin off from the ROVER and NERVA Nuclear Rocket Programs carried out in the sixties and early seventies at Los Alamos National Laboratory, Atomic International, and Westinghouse Electric Company. The coated plutonia particles design also draws on a vast and proven fabrication technology of UO$_2$ and (U,Pu)O$_2$ fuels for High-Temperature Gas-Cooled Reactors (HTGRs) (Minato et al., 1997). The attractive features of coated particles fuel in fission reactors have been their ability to operate at high temperatures (up to 1700 K in HTGRs and 3000 K in nuclear thermal propulsion reactors) and achieve high fuel burnup (> 30 at.%), with little concern about fuel swelling and fission products release. The primary coating of the fuel particles (SiC at < 1300 K or ZrC for operation at higher temperatures) serves as a pressure vessel for containing fission products and constraining fuel swelling during irradiation. The fuel kernel could be fabricated using Sol-Gel techniques (Haas et al., 1967; Huschka and Vygen, 1977; Förthmann and Blass, 1977; Lahr, 1977) or powder metallurgy processes (Burnett et al., 1963; Ford and Shennan, 1972; Allen et al., 1977). Unlike powder metallurgy, Sol-Gel techniques do not involve any milling and grinding and hence, do not generate a post-fabrication radioactive solid waste stream. In addition, all used chemicals are recycled and reused. The literature on the fabrication, gas release, and material properties of coated particles fuel has recently been reviewed and documented (El-Genk and Tournier, 1999).

Coated plutonia (^{238}PuO$_2$) fuel particles are being investigated for potential use in Radioisotope Heater Units (RHUs) and Radioisotope Power Systems (RPSs) (Scholtis et al., 1999; Tournier and El-Genk, 2000). The proposed fuel particles consist of a plutonia kernel (~ 300–1200 μm in diameter) coated with a thin layer (5 μm) of Pyrolytic Graphite (PyC) and a strong outer coating of zirconium carbide (see Figure 1). The thickness of the ZrC coating depends on the fuel kernel size and internal pressure of the released helium gas (Tournier and El-Genk, 2000). The amount and the rate of helium release depend on the fuel temperature and microstructure (granular or polycrystalline), and the helium inventory in the fuel matrix (or storage time).

CP504, *Space Technology and Applications International Forum–2000*, edited by M. S. El-Genk
© 2000 American Institute of Physics 1-56396-919-X/00/$17.00

The ^{238}PuO$_2$ coated particles fuel offers promise for enhanced safety and unique design flexibility, as the heat sources can be fabricated in a variety of shapes, sizes, and power densities. The coated fuel particles could be dispersed in a solid graphite matrix made into different shapes or used in the form of a heating tape or a paint, for a variety of applications. The coated particles fuel graphite-compact (CPFC) in a pellet form can potentially be used to develop high specific power RHUs using the current Fine Weave Pierced Fabric (FWPF) aeroshell structure. When the fuel pellet, Pt-alloy cladding, and inner graphite sleeve in the Light Weight RHU (LWRHU) were replaced with CPFC, the thermal power

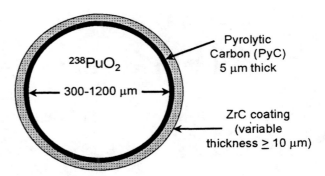

FIGURE 1. Coated plutonia fuel particle.

output increased from ~ 1.0 W$_{th}$ to 1.54, 2.4 and 2.52 W$_{th}$, for 100%, 10% and 5% helium release from the fuel kernels, respectively (Tournier and El-Genk, 2000). Such higher thermal powers are possible at essentially the same LWRHU mass (~ 40 g), and conservatively assuming a pre-launch storage time of 10 years and a fuel temperature of 1723 K. This temperature is the predicted maximum during an accidental re-entry heating pulse (Schock, 1981; Tournier and El-Genk, 2000). The higher the release fraction of He, the thicker is the ZrC coating needed to accommodate the internal gas pressure in the coated fuel particles, reducing the loading of the plutonia fuel in the CPFC. Therefore, there is a need to investigate He gas release in coated plutonia fuel kernels as a function of temperature and storage time before launch in order to maximize the fuel loading in the CPCF. In addition to the storage time and temperature, the He release depends on the grain size and the fuel microstructure.

The objectives of this paper are to: (a) review the release mechanisms of helium in small grain (7-40 μm) plutonia pellets currently being used in the General Purpose Heat Source (GPHS) modules and RHUs; (b) examine the applicability of these mechanisms to the helium release from large-grain ($\geq$ 300 μm) and polycrystalline ^{238}PuO$_2$ fuel kernels; (c) provide best estimates of the He release fraction from coated plutonia particles as a function of temperature, up to 1723 K. These estimates are based on the reported fission gas release data for granular and monocrystal, spherical UO$_2$ fuel particles irradiated at isothermal conditions up to 6.4 at. % burnup (Turnbull, 1974; Friskney et al., 1977; Turnbull et al., 1977; Friskney and Turnbull, 1979).

BASIC MICROSTRUCTURES OF FUEL MICROSPHERES

There are two basic microstructures for the fuel microspheres or kernels (Figure 2): (a) the *granular microstructure* obtained by powder metallurgy processes; and (b) the *polycrystalline microstructure* obtained by the Sol-Gel processes. A granular fuel kernel consists of a number of polycrystalline grains of almost the same size (typically 7–40 μm). These grains are separated by common grain boundaries that develop during sintering at high pressure and temperature. This process also controls the as-fabricated porosity in the fuel grains and at the triple interface of the grains to accommodate fission gases during operation in nuclear reactors. Typical values of the as-fabricated fuel kernel porosity range between 5 and 15% (see Figure 2a).

Granular fuel kernels are fabricated using the binderless agglomeration process (Burnett et al., 1963; Ford and Shennan, 1972; Allen et al., 1977). This process could be used to fabricate highly spherical and virtually monosized oxide fuel kernels in the range from 200 to 1000 μm in diameter, with a porosity between 5 and 20% (see Figure 2a). Green spheroids are initially prepared from ceramic-grade, plutonium dioxide powder, having a specific surface area of 2-4 m^2/g. The fuel powder is mixed (using ball milling) with proprietary carbon blacks, having a specific surface area of 15-30 m^2/g. The resulting powder is granulated to form seeds which are then grown to the desired kernel size in a vibrating pan fed with fresh fuel powder. The fuel green spheres, or "seeds" are formed by the simple action of gyrating the fine submicron fuel powder into a bowl. These seeds, which have a size distribution in the range between 150 and 200 μm in diameter, are spherical in shape. Fine fuel powder is added at a controlled rate to the charge of seeds while being continuously gyrated into a bowl. The PuO$_2$/C "green" particles are then spheroidized in a rotary sieve or planetary mill.

Porous PuO$_2$ kernels are fabricated from fuel green microspheres using a two-stage heat treatment process to control the as-fabricated porosity and the stoichiometry of the oxide fuel. First, the green fuel microspheres are

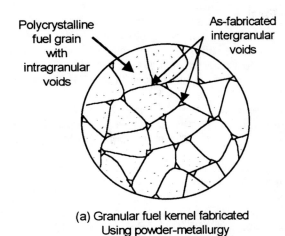

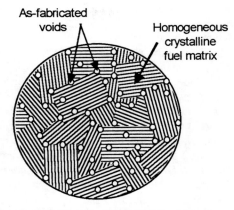

(a) Granular fuel kernel fabricated
Using powder-metallurgy

(b) Polycrystalline fuel kernel fabricated
using Sol-Gel techniques

FIGURE 2. Basic fuel microstructures for coated fuel particles.

consolidated by sintering in carbon monoxide at 1923 K to prevent carbide formation. The sintered particles are then decarbonized in a flowing CO/CO_2 at 1573 K in a continuous furnace, where the removal of the carbon leaves the desired porosity within the fuel microspheres (Figure 2a) (Burnett et al., 1963; Ford and Shennan, 1972; Allen et al., 1977).

Polycrystalline fuel kernels with diameters as large as 1200 μm can be fabricated using Sol-Gel techniques (Haas et al., 1967; Huschka and Vygen, 1977; Förthmann and Blass, 1977; Lahr, 1977). These techniques do not require milling or grinding (as during the fabrication of granular fuel by powder metallurgy), thus do not generate any radioactive dust or aerosols. These all wet chemical processes use solutions or sols (or suspensions) of fissile or fertile materials dispersed into uniform liquid (gel like) droplets. The spherical shape of these droplets is fixed by the gelation (by water or ammonia extraction) achieved by either precipitation or dehydration reactions. The gelled fuel microspheres with an almost perfect sphericity are then washed, dried, and fired at high temperature to remove water and volatile additives and sinter the spheres to the desired density and fuel stoichiometry. The sintered fuel kernels have a polycrystalline fuel structure with tiny, intragranular voids (< 1μm in diameter) (see Figure 2b).

The amount of inter-granular and intra-granular porosity in the fuel kernels (5-20%) is controlled during the sintering process. The former tends to be very tiny, of sub-micron size, while the latter could be a few microns in diameter. Thus, for helium release, the polycrystalline fuel particles (Figure 2b) may be regarded as a single grain of the kernel diameter, with as-fabricated intragranular porosity made of tiny pores.

The release mechanisms of fission gases in granular, mixed-oxide and plutonia fuel are summarized next. In addition, the similarities and differences pertaining to helium release in plutonia fuel pellets and microspheres are indicated and discussed.

MECHANISMS OF GAS RELEASE IN $(U,Pu)O_2$ AND PuO_2 FUELS

There are strong similarities between the release mechanisms of fission gases and volatile fission products in granular oxide fuel and those of helium in granular $^{238}PuO_2$. The major difference is the restructuring and cracking of oxide fuel occurring during irradiation in fission reactors. In fission reactors, the power density in the fuel pellets is orders of magnitude higher than that by natural radioactive decay of ^{238}Pu in plutonia pellets. When combined with the inherently low thermal conductivity of oxide and mixed-oxide fuels, fuel pellets in fission reactors develop a steep, parabolic radial temperature distribution and significantly higher fuel temperatures, particularly at the pellet centerline. The combination of high temperature and radial temperature gradient causes significant restructuring of the oxide fuel pellets in fission reactors (Olander, 1976). However, owing to the similar ceramic nature of both oxide and plutonia fuels, for the same operating temperature and as-fabricated porosity and grain size, the gas release mechanisms in both fuel types are identical.

The processes by which gas bubbles nucleate within the fuel grains, grow, diffuse, and eventually coalesce at the grain boundaries are extremely complex (Olander, 1976). The fuel temperature, however, appears to provide a useful index for determining the release mechanisms and rates of fission gases in oxide and mixed-oxide fuels in fission reactors and of helium in plutonia fuel in RHUs. In summary, the prevailing release mechanisms of the gases and volatile products generated in ceramic fuel by either fission in nuclear reactors or radioactive decay of $^{238}PuO_2$ (Lyons et al., 1972; Mueller et al., 1974; Olander, 1976; Peterson et al., 1984; Scaffidi-Argentina et al., 1997) can be divided into four sequential mechanisms:

(a) *Atomic diffusion* near the surface of the fuel pellet or particles when the fuel temperature is < 900 K. The release fraction in this case is typically very low since it is limited by the available area for release at the fuel surface and the relatively low diffusion coefficient in the fuel matrix. At such low fuel temperatures, most of the generated gases are accumulated within the fuel grains.

(b) *Accumulation in and diffusion to newly forming intergranular gas bubbles*, due to the enhanced mobility of tiny intragranular gas bubbles and the growth of fuel grains at temperatures ~ 1150 K or higher. At such temperatures, some open porosity may form at the grain boundaries, resulting in a higher gas release. The gas release fraction, however, remains well below 10%, due to the limited surface area for release.

(c) *Separation and formation of open porosity at grain boundaries*, which occur at fuel temperatures of 1150-1500 K. At these high temperatures, most accumulated gases in the grain boundary bubbles are released through open porosity at the grain boundaries.

(d) *Atomic and volume diffusion* in fuel grains, to the open porosity at the grain boundaries. This mechanism, which is dominant at fuel temperatures in excess of 1600-1700 K, solely depends on the fuel temperature, since gas release is not limited by release surface area. At these temperatures, the open porosity at the grain boundaries are fully accessible for gas release from the fuel grains.

These release mechanisms have been confirmed by experimental data on helium gas release from GPHS and RHU plutonia fuel pellets as a function of temperature, storage time, and mode of heating (steady-state and transient re-entry heating) (Angelini et al., 1970; Mulford and Mueller, 1973; Mueller et al., 1974; Land, 1980; Peterson and Starzynski, 1982; Peterson et al., 1984). Results described later in this paper also indicate the strong dependence of the release fraction of He on the as-fabricated grain size of the fuel.

Results of the helium gas release experiments from granular plutonia pellets (7-40 μm grain size) have indicated that transient heating increases the helium release rate, but the release fraction of helium depends on the temperature reached during the transient (Peterson and Starzynski, 1982; Peterson et al., 1984). Similar release fractions were reported at the same temperatures in steady-state heating experiments of granular plutonia pellets and smaller samples (Mulford and Mueller, 1973; Mueller et al., 1974). Therefore, it is evident that increasing the fuel grain size and/or the elimination of grain boundaries could significantly reduce the release fraction of He gas in $^{238}PuO_2$ fuel.

The grain boundaries are absent in polycrystalline fuel particles fabricated using Sol-Gel techniques (Figure 2b). Therefore, in such polycrystalline plutonia fuel particles, helium gas release is expected to be much lower than in granular fuel particles of the same diameter. The only contributing mechanisms for He gas release in polycrystalline plutonia particles will be atomic and volume diffusion, which are limited by the effective mass diffusion coefficient of the gas in the fuel matrix and the availability of surface area for gas release. The absence of grain boundaries in polycrystalline fuel particles limits the gas release to the geometrical surface area of the fuel kernel, even at temperatures in excess of 1600 K.

The next section presents a summary of the equivalent-sphere diffusion model for the release of fission gases and volatile fission products and discusses the effects of the grain size and the decay constant (or half-life) on the release fraction of various radioactive species. The applicability of the results to predicting the release fraction of noble, non-radioactive gases, such as helium in plutonia fuel, is also discussed.

EQUIVALENT-SPHERE DIFFUSION MODEL FOR FISSION GAS RELEASE

The equivalent-sphere model, originally proposed by Booth and Rymer (Olander, 1976), has been used to describe fission gas and helium release from granular, oxide fuel in fission reactors and granular plutonia pellets in GPHS and RHU units (Angelini et al., 1970; Mulford and Mueller, 1973; Mueller et al., 1974; Land, 1980; Peterson and Starzynski, 1982; Peterson et al., 1984). The model assumes that gases diffuse to the free surface of so-called equivalent spheres, from which it is released. The equivalent-sphere model has successfully been used to correlate gas-release experiments. The model treats granular fuel as a collection of equivalent spheres of uniform size, which have the same surface-to-volume ratio as the average release unit in the fuel matrix. Thus, the diameter of the equivalent sphere for gas release is $d = 2a = 6V/S_R$, where a in the sphere radius and V and S_R are the equivalent sphere's volume and surface area, respectively.

The total fuel surface area for the gas release, however, can be measured using gas-adsorption techniques, thus the effective value of d (or a) is experimentally determined. When the average grain size of as-fabricated fuel is known, it may be substituted as an approximate value for d. The actual value of d, however, could be different, particularly as intragranular fuel cracking occurs, typically at temperatures < 1400 K. Once the effective geometry of the fuel specimen for gas release has been characterized, the results of gas-release experiments can be used to determine the effective mass diffusion coefficient of the gases in the fuel matrix, D.

In order to account for the effect of radioactive decay, the following diffusion equation:

$$\frac{\partial C}{\partial t} = \dot{B} + \frac{D}{r^2}\frac{\partial}{\partial r}\left(r^2\frac{\partial C}{\partial t}\right) - \lambda C \tag{1}$$

is used to determine the release fraction, F, assuming an initial gas concentration $C(a,0) = 0$. In this equation, $\dot{B}$ is the production or birth rate of the gaseous isotope of interest and λ is its radioactive decay constant. The approximate solution of Equation (1) for stable gases (i.e. when the third term on the right hand side of Equation (1) is zero) gives the release fraction, F, as (Olander, 1976):

$$F \approx 6\sqrt{\tau/\pi} \ , \tag{2}$$

where, $\tau = \int\limits_{o}^{t} D'dt$ is a dimensionless time, and

$$D'(T) = D(T)/a^2(T) \propto \left(S_R/V\right)^2 \times D \ . \tag{3}$$

In this Equation, D', the effective diffusion coefficient (s^{-1}), depends on the release surface-to-volume ratio of the fuel and the mass diffusion coefficient of the gaseous species, D (m^2/s), which are both functions of the fuel temperature, T. The release fraction, F, can be expressed in terms of the effective radius of the average release unit, a, fuel temperature, T, and release time, t, as:

$$F \propto \left(\frac{S_R}{V}\right) \times \sqrt{Dt} \quad \propto \quad \frac{1}{a(T)}T^\gamma\sqrt{t} \ . \tag{4}$$

As this Equation indicates, F is inversely proportional to the effective radius of the average release unit in the fuel, and increases with the square root of time. Note that the fractional gas release, F, in Equation (3) is not the release-to-birth rate ratio, $\dot{R}/\dot{B}$.

For radioactive fission gas species that have attained a steady-state concentration in the fuel, the fractional release, F^* equals in this case the release-to-birth rate ratio, and is given by (Olander, 1976):

$$F^* \equiv \dot{R}/\dot{B} \approx \left(\frac{S_R}{V}\right) \times \sqrt{\frac{D}{\lambda}} \quad \propto \quad \sqrt{\frac{D'(T)}{\lambda}} \quad \propto \quad \frac{1}{a(T)}T^\gamma\sqrt{T_{1/2}} \ . \tag{5}$$

This equation shows F^* to be inversely proportional to the average fuel grain size, a, and to increase with the square root of the half-life (or inversely proportional to the square root of the radioactive decay constant). As the half-life increases, F^* approaches the release fraction of non-radioactive species, given by Equation (2). Thus, for the same fuel material, temperature, and grain size, F^* increases exponentially with the half-life of the gas species, approaching an asymptote. This conclusion is critical to the applicability of the release data of fission gases to the prediction of helium release in plutonia fuel particles, as detailed later.

The effective diffusion coefficient, D', increases exponentially with temperature and accounts for the increases in both the mass diffusion coefficient, D, and the effective release area, S_R, (including fuel cracks and open grain boundaries) with temperature. Neglecting the as-fabricated open porosity, the surface-to-volume ratio of plutonia fuel microspheres for gas release can be expressed as:

$$\left(\frac{S_R}{V}\right) = (1-\alpha)\frac{6}{d_p} + \alpha\frac{6}{d_g} \quad , \tag{6}$$

where,

$$\begin{aligned}
\alpha &= 0, \quad \text{for} \quad && T < 900\,\text{K}, \\
0 &< \alpha < 1, \quad \text{for} \quad && 900\,\text{K} < T < 1450\,\text{K}, \\
\alpha &= 1, \quad \text{for} \quad && T > 1450\,\text{K}.
\end{aligned} \tag{7}$$

The normalized surface area for helium gas release from plutonia fuel particles can thus be written as:

$$\left(\frac{S_R}{S_p}\right) = (1-\alpha) + \alpha\frac{d_p}{d_g} \quad , \tag{8}$$

where S_p is the geometrical surface area of the as-fabricated fuel particle.

The effect of fuel temperature on the effective area for gas release in granular and polycrystalline fuels, S_R, is illustrated in Figure 3. At temperatures below ~ 900 K, gas release occurs by atomic diffusion from the fuel matrix, and S_R is equal to the geometrical surface area of the as-fabricated fuel sample (i.e. $\alpha = 0$). Between 900

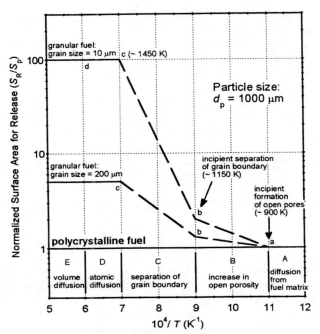

FIGURE 3. Effective release area for gases in granular and polycrystalline fuels.

and 1150 K, $0 < \alpha < 1$ and S_R increases with temperature in granular fuel due to the formation and the coalescence of grain boundary bubbles. Above ~1150 K, S_R increases rapidly with temperature as α approaches unity, due to the formation of open porosity caused by the separation of the grain boundaries. At ~1450 K, S_R reaches its maximum value and $\alpha = 1.0$, as the separation at the grain boundaries is complete. Above this temperature, the release fraction in granular fuel is no longer limited by the surface area available for release, but rather by the atomic and volume diffusion of the gases in the fuel grains. *It should be noted that the separation of the grain boundaries does not necessarily cause powdering of the fuel or breakup of its structure, but rather a full release of the gas at the grain boundaries through the formation of open porosity or tunnels.*

Figure 3 illustrates the effect of the fuel grain size on S_R and potentially on the helium gas release in plutonia pellets or particles. For the same fuel temperature and particle geometry, increasing the grain size from a typical 10 μm to 200 μm could reduce the release fraction at high fuel temperatures (>1450 K) by more than an order of magnitude. As indicated earlier, a polycrystalline fuel kernel could be regarded as a single grain with a size equal to the kernel diameter. Thus, the effective surface area for gas release in polycrystalline particles would remain essentially unchanged and equal to the geometrical surface area of the particle. Therefore, the helium gas release in polycrystalline PuO_2 fuel would be expected to be significantly lower than in granular fuel pellets or microspheres.

To illustrate this point, consider a plutonia fuel microsphere, 1-mm in diameter, at high temperatures > 1450 K. A granular fuel made of 200 μm grains has a surface area for release that is 5 times that of a polycrystalline particle, while a 10 μ-grain fuel particle would have 100 times the surface area for helium release of a polycrystalline particle of the same diameter (Figure 3). These estimates are based on the experimental data showing complete release of the gas at the grain boundaries at fuel temperatures > ~ 1450 K (Angelini et al., 1970; Mulford and Mueller, 1973; Mueller et al., 1974; Land, 1980; Peterson and Starzynski, 1982; Peterson et al., 1984).

Most gas release data reported in the literature for coated fuel particles were obtained for oxide fuels irradiated in fission reactors. In order to apply the reported data to the helium gas release in plutonia fuel kernels, fission gas release measurements must be made at isothermal conditions, which is the case in plutonia fuel particles. Fortunately, such measurements were obtained at the Berkeley Nuclear Laboratories (U.K.). The release fractions of radioactive fission gases (Xe and Kr isotopes) and volatile fission products (Cs, I and Te isotopes) were measured for small-grain, large-grain and monocrystal UO_2 particles. The applicability of the reported data to the He release in spherical plutonia particles operating at the same temperatures and having same microstructure is discussed next.

ISOTHERMAL RELEASE IN GRANULAR AND MONOCRYSTAL FUEL PARTICLES

The only detailed studies of the isothermal release of fission gases and volatile fission products from granular and monocrystal fuel particles (Figure 4) were those conducted at Harwell and at Berkeley Nuclear Laboratories in the U.K. (Turnbull, 1974; Friskney et al., 1977; Turnbull et al., 1977; Friskney and Turnbull, 1979). These studies investigated the effects of grain size, radioactive decay, fuel burnup (up to 6.4 at.%) and temperature (up to 2023 K) on the release-to-birth rate ratio of the various gaseous and volatile isotopes. The temperature of the fuel particles during irradiation was kept nearly uniform, as would be expected during actual operation in fission reactors at low power density and in plutonia fuel particles in RHUs. UO_2 fuel microspheres consisting of small grains (average grain size of 10 μm) (Figure 4a) and large grains (an effective grain size between 300 and 600 μm) (Figure 4b) were irradiated up to a 6.4 at.% burnup. Monocrystal right cylinders of natural (0.72 wt.% ^{235}U) stoichiometric UO_2 were also irradiated (Figure 4c). In addition, granular fuel specimens of 2.0% enriched UO_2 of near theoretical density, in the form of small cylinders, 10 mm long and 3 mm in diameter, having 7-μm and 40-μm grain sizes were irradiated (Turnbull, 1974).

During irradiation, the individual fuel particles were wrapped in tungsten mesh to prevent them from touching each other or the walls of the molybdenum container. The particles were irradiated isothermally in an electrically heated rig in the UKAEA reactor DIDO. Fission heating (thermal neutron flux of ~ 2.6 x 10^{17} m^2/s) produced a temperature drop between centerline and outer surface of the specimens of less than 100 K. The fuel particles were continuously swept with a He–2%H$_2$ gas mixture to carry the released fission gases and volatile fission products. The volatile fission products were deposited using a cold finger while fission gases were collected using charcoal traps cooled with liquid nitrogen. The amounts of the various gaseous species released were measured by γ-spectroscopy, after correcting for their radioactive decay. The release rates of fission gases and volatile fission products were calculated based on these measurements, while their birth rates were calculated using computer codes (Friskney and Turnbull, 1979).

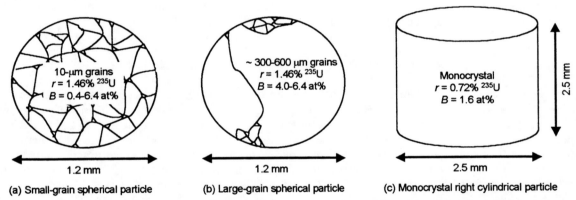

FIGURE 4. UO_2 fuel samples used in isothermal fission gas release studies at Harwell and Berkeley Nuclear Laboratories (Turnbull, 1974; Friskney et al., 1977; Turnbull et al., 1977; Friskney and Turnbull, 1979).

Measured Release Data

Figure 5 presents the measured release-to-birth rate ratios of ^{133}Xe, which has a half-life of 5.2 days, for the different fuel particles, as a function of irradiation temperature. The small-grain (10 μm) UO_2 microspheres released about 20 times more gas that the large-grain (~300–600 μm) ones. This ratio is comparable to that of the grain size ratio, consistent with Equation (5), which predicts the release-to-birth rate ratio to be inversely proportional to the grain size. As expected, the monocrystal, natural uranium particles released less gas than the large-grain microspheres, due to the absence of grain boundaries in the former. However, the difference was not as large as might be expected, because the rate of fission was about half that in the large-grain fuel particles, due to the lower ^{235}U enrichment.

Figures 5 and 6 show a change in the fission gas release-to-birth rate ratio in granular fuel particles at ~ 1350–1450 K, indicating a change in the release mechanism. Below ~ 1100 K, fission gases and volatile fission products

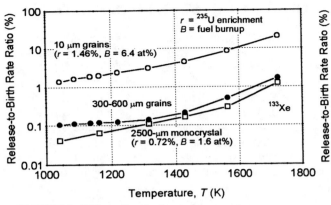

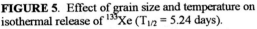

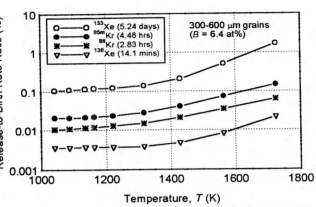

FIGURE 5. Effect of grain size and temperature on isothermal release of ^{133}Xe ($T_{1/2}$ = 5.24 days).

FIGURE 6. Effect of half-life and temperature on isothermal release of noble fission gases from large-grain ($\sim$ 300-600 μm) UO$_2$ spheres irradiated to 6.4 at.%.

diffuse to the grain boundaries, where they are trapped in the grain boundary bubbles or released by diffusion along the grain boundaries. Above $\sim$ 1100 K, the growth and coalescence of intergranular bubbles cause separation of the grains, forming open porosity and effectively increasing the surface area available for release. The increase in fission gas release due to grain boundary separation does not occur in the monocrystal fuel particle due to the absence of grain boundaries. This explains why release-to-birth rate ratio data for the monocrystal particles did not exhibit a change in slope with increasing fuel temperature below 1600 K (Figure 6). The increase in the release-to-birth rate ratios for granular and monocrystal particles at temperatures > 1600 K is caused by the higher mobility of gas atoms and tiny intragranular bubbles in the fuel grains. The annealing of fission defects in the fuel matrix at such high temperatures could also have contributed to the increase in the mobility and the diffusion coefficient of fission gases, hence, increasing the gas release rate.

Effect of Half-Life on Release-to-Birth Rate Ratio

The database showing the effect of half-life on the release of noble gases and volatile fission products from the large-grain ($\sim$ 300–600 μm) UO$_2$ microspheres is presented in Figures 7-8 (Friskney and Turnbull, 1974; Turnbull, 1974; Turnbull et al., 1977). Since gaseous radioisotopes decay as they diffuse through the fuel grains and are being trapped at the grain boundaries for a period of time, those having longer half-lives would exhibit higher release-to-birth rate ratios. According to Equation (5), the release-to-birth rate ratio is proportional to the square root of the half-life, for the relatively short-lived isotopes that reach equilibrium early in time. Such dependence of the release-to-birth rate ratio on half-life is evident in Figures 7a and 8 for the species having half-lives < $\sim$ 10 days.

At 1723 K and 4.0 at.% burnup, the small-grain fuel particles (7-40 μm) released essentially all gases and volatile fission products ($\sim$ 80% release), whereas only $\sim$ 7% was released from the large-grain ($\sim$ 300-600 μm) fuel particles (Figures 7 and 8). It is worth noting that the release fraction from the small-grain UO$_2$ particles is almost the same as that reported for granular plutonia pellets, of the same average grain size (7-12 μm), during re-entry heating tests to 1723 K (Peterson and Starzynski, 1984). When the data in Figure 7a was plotted versus the square root of half-life, it exhibited an exponential increase with half-life, reaching asymptotic values at large half-lives (Figure 7b). These asymptotic values are representative of the release fractions of stable gases in granular UO$_2$ fuel during fission and of helium in plutonia fuel particles of the same grain size.

Based on these experimental data, helium gas release from small-grain (7-40 μm) plutonia fuel at 1723 K would be $\sim$ 80%, which is in agreement with the experimental data generated at LANL for GPHS and LWRHU granular plutonia pellets (Angelini et al., 1970; Mulford and Mueller, 1973; Mueller et al., 1974; Land, 1980; Peterson and Starzynski, 1982; Peterson et al., 1984). For large-grain ($\geq$ 300 μm) and polycrystalline fuel microspheres, when heated up to 1723 K, the helium gas release could be more than an order of magnitude lower ($\sim$ 7%). The data presented in Figure 7b is in excellent agreement with the theory (Equation (5)), showing the strong effect of the grain size on $\dot{R}/\dot{B}$ of fission noble gases and volatile fission products, particularly evident for $T_{1/2}$ > 1 year.

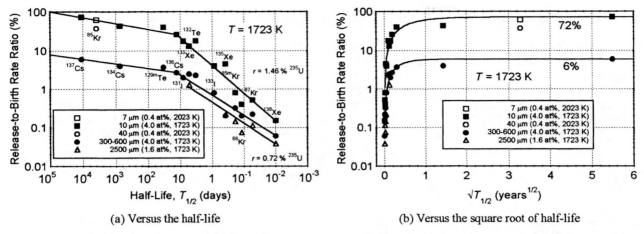

(a) Versus the half-life

(b) Versus the square root of half-life

FIGURE 7. Effects of half-life and fuel microstructure on isothermal release of fission gases and volatile fission products from granular and monocrystal UO_2 fuel samples irradiated at high temperature.

Effect of Fuel Temperature

The data presented in Figure 8 illustrates the effect of fuel temperature on the isothermal release-to-birth rate ratios of fission gases and volatile fission products from the large-grain (~ 300-600 μm) UO_2 particles (Turnbull, 1974; Turnbull et al., 1977; Friskney and Turnbull, 1979). At 1042 K, only the data for the short-lived noble gases and volatile fission products was reported. This data was extrapolated to higher half-lives using a factor of three, which is the same as that for the high-temperature data between 10 and 10^5 days half-life (Figure 8). This extrapolation is appropriate since the difference between the release-to-birth rate ratios of the radioactive and the stable gases equals the decay rate of the former, which is independent of temperature.

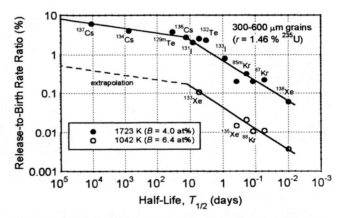

FIGURE 8. Effects of half-life and temperature on isothermal release of fission gases and volatiles from large-grain (~ 300-600 μm) UO_2 spheres enriched to 1.46%.

The data delineated in Figure 8 indicate that at 1042 K, less than 1% of the fission gases and volatile fission products were released. Similar release fraction would be expected for helium gas in plutonia fuel particles having 300 to 600 μm grains. Also, since the nominal operating temperature in LWRHUs (~ 800 K) is several hundred degrees lower than 1042 K, the helium gas release at the operating temperature in coated particles of large grain size (≥ 300 μm) plutonia fuel would be practically nil.

In summary, the isothermal release data of the fission gases and volatile fission products presented in this section clearly demonstrated the strong effects of the fuel grain size and the half-life on the steady-state release-to-birth rate ratios of the species. The data is consistent with the theory, Equations (4)-(8), particularly the dependence of $\dot{R}/\dot{B}$ on the half-life for the radioactive species that have attained equilibrium ($T_{1/2}$ < 10 days). For the small-grain (7-10 μm) UO_2 fuel, the release-to-birth rate ratio of the long-lived gases and volatile fission products at 1723 K was nearly 80%. For the large-grain (~ 300-600 μm) fuel particles, the measured release-to-birth rate ratios at same burnup of 4.0 at % and 1723 K were about an order of magnitude lower, ~ 7%, decreasing to < 1% at a fuel temperature of 1042 K and higher burnup of 6.4 at.%. The release-to-birth rate ratios for both small- and large-grain fuel particles increased exponentially with the half-life of the released species, as indicated by the theory (Figure 7b).

Application to Helium Gas Release in Plutonia Fuel Kernels

The reported data for the isothermal release of fission gases and volatile fission products from the granular fuel particles have provided a solid foundation for predicting the helium gas release in $^{238}PuO_2$ kernels, having similar grain size and as-fabricated porosity. The application of the reported data to helium release in plutonia particles includes a certain degree of conservatism. For example, the weakening of the grain boundaries by the bombardment of fission products, which increases fission gas release, does not occur in the α-emitter plutonia fuel. In addition, the constraint imposed by the ZrC coating could decrease the release of helium gas from the $^{238}PuO_2$ fuel kernels. In addition, due to the absence of grain boundaries, the helium release in polycrystalline $^{238}PuO_2$ fuel kernels, fabricated using Sol-Gel techniques, is expected to be significantly lower than in large-grain fuel kernels.

Based on the reported data for the isothermal release of fission gases in UO_2 fuel particles (Turnbull, 1974; Friskney et al., 1977; Turnbull et al., 1977; Friskney and Turnbull, 1979), the following conclusions relative to the helium release in plutonia fuel can be drawn:

(a) The helium release from small-grain (7-40 μm) plutonia fuel at 1723 K would be ~ 80%, decreasing to less than 10% at 1042 K. This conclusion is in excellent agreement with the actual helium release data obtained at Los Alamos National Laboratory for GPHS and LWRHU plutonia pellets.

(b) The release from large-grain ($\geq$ 300 μm) plutonia fuel at 1723 K could be ~ 7%, decreasing to ~ 0.8% at 1042 K.

(c) In polycrystalline plutonia fuel kernels fabricated using Sol-Gel processes, the helium release fraction could be even lower than that reported for the large-grain $^{238}PuO_2$ kernels (i.e. < 7% at 1723 K and < 0.8% at 1042 K).

SUMMARY AND CONCLUSIONS

The coated plutonia fuel particles have recently been proposed for potential use in future space exploration missions that employ radioisotope power systems and/or RHUs. The particles vary in size from 300 to 1000 μm and consist of a $^{238}PuO_2$ kernel with a thin (5 μm) PyC inner coating and a strong ZrC outer coating. The thickness of the ZrC is selected to ensure full retention of the helium gas generated by the radioactive decay of ^{238}Pu. The thickness of the ZrC coating, therefore, depends on the actual release fraction of the gas during a simulated re-entry heat pulse, following a storage time of as much as 10 years. During such transient heating, the fuel temperature could reach 1723 K, compared to only ~ 800 K during nominal operation. Reducing the thickness of the ZrC coating increases the fuel loading of the coated particles, and hence the specific thermal power of the coated particles fuel compacts.

This paper reviewed the potential release mechanisms of helium in small-grain (7-40 μm) plutonia pellets, currently being used in GPHS modules and LWRHUs, and in large-grain ($\geq$ 300 μm) and polycrystalline plutonia kernels of the coated particles. The helium release mechanisms are similar to those of fission gases and volatile fission products in oxide and mixed-oxide fuels in fission reactors. The applicability of these mechanisms to small-grain, large-grain and polycrystalline $^{238}PuO_2$ fuel particles is examined and estimates of the helium release during a re-entry heating pulse up to 1723 K are presented. These estimates, based on the reported data of fission gas release from granular and monocrystal UO_2 fuel particles irradiated at isothermal conditions up to 6.4 at.% burnup and 2030 K, performed at Harwell and Berkeley Nuclear Laboratories (U.K.), are in good agreement with the helium release tests performed at LANL for small-grain (7-40 μm) plutonia GPHS and LWRHU pellets.

It is concluded that the helium release fraction from large-grain ($\geq$ 300 μm) plutonia fuel kernels heated up to 1723 K could be less than 7%, compared to ~ 80% in small-grain (7-40 μm) fuel. The release fraction of helium in large-grain plutonia fuel kernels could be less than 1% at 1000 K and nil at the nominal operation temperature in RHUs of ~ 800 K. Due to the absence of grain boundaries, the fraction of helium gas released in polycrystalline plutonia kernels fabricated using Sol-Gel techniques could be significantly lower than those in large-grain fuel kernels, for the same storage time and operation temperature.

Helium gas release experiments involving large-grain ($\geq$ 300 μm) and polycrystalline, coated and un-coated plutonia fuel kernels are recommended. In these experiments, the release rate and release fraction of helium gas would be measured as functions of storage time and fuel temperature up to 1800 K, during both steady-state and transient heating conditions. The fuel kernels fabricated using both the Sol-Gel and the powder-metallurgy agglomeration techniques could be used. The PyC undercoating, the ZrC coating, and the PyC overcoating of the kernels would be applied using the state-of-the-art Chemical-Vapor-Deposition techniques. The proposed tests, which may run continuously for up to two years, could be performed at an approved and qualified DOE facility.

ACKNOWLEDGMENTS

This research was funded by Sandia National Laboratories (SNL), Kirtland Air Force Base, Albuquerque, NM, under Contract No. BE-2543, to the University of New Mexico's Institute for Space and Nuclear Power Studies. The opinions expressed in this paper are solely those of the authors. We are grateful to Dr. Ronald J. Lipinski, SNL, and Mr. Joseph A. Sholtis, Jr., Sholtis Engineering & Safety Consulting, for their continuous technical support.

REFERENCES

Allen, P. L., Ford, L. H., and Shennan, J. V., "Nuclear Fuel Coated Particle Development in the Reactor Fuel Element Laboratories of the U. K. Atomic Energy Authority," *Nuclear Technology* **35**, 246–253 (1977).

Angelini, P., McHenry, R. E., Scott, J. L., Ernst, W. S., Jr., and Prados, J. W., *Helium Release from $^{238}PuO_2$ Microspheres*, Report No. ORNL-4507, Oak Ridge National Laboratory, Oak Ridge, Tennessee, March 1970.

Burnett, R. C., Bisdorff, L., and Gough, J. R. C., *Development of Coated Particle Fuel – Part I: the Fissile Fertile Particle*, Dragon Project Report N°. 151, also in *Proceedings of the Dragon Project Fuel Element Symposium*, held January 28-29, 1963, in Bournemouth, England, CONF-630103-8 (1963).

El-Genk, M. S., and Tournier, J.-M., *Study of a Coated-Particle Fuel Form for Advanced Radioisotope Heat Sources – Part I: A Review of Coated-Particle Fuel Performance, Fabrication Techniques and Materials Properties*, Report UNM–ISNPS–5–1999, Sandia National Laboratories, Contract BE–2543, (1999).

Ford, L. H., and Shennan, J. V., "the Mechanism of Binderless Granulation and Growth of Ceramic Spheres," *Journal of Nuclear Materials* **43**, 143 (1972).

Förthmann, R., and Blass, G., "Fabrication of Uranium-Plutonium Oxide Microspheres by the Hydrolysis Process," *Journal of Nuclear Materials* **64**, 275–280 (1977).

Friskney, C. A., Turnbull, J. A., Johnson, F. A., Walter, A. J. and Findlay, J. R., "The Characteristics of Fission Gas Release from Monocrystalline Uranium Dioxide during Irradiation," *Journal of Nuclear Materials* **68**, 186–192 (1977).

Friskney, C. A., and Turnbull, J. A., "The Characteristics of Fission Gas Release from Uranium Dioxide during Irradiation," *Journal of Nuclear Materials* **79**, 184–198 (1979).

Haas, P. A., Kitts, F. G., and Beutler, H., "Preparation of Reactor fuels by Sol-Gel Processes," *Chemical Engineering Progress Symposium Series* **63** (80), 16–27 (1967).

Huschka, H., and Vygen, P., "Coated fuel Particles: Status of Fabrication Technology," *Nuclear Technology* **35**, 238–245 (1977).

Land, C. C., *Microstructural Damage Produced by Helium in Aged $^{238}PuO_2$ Fuels*, Report N°. LA-8083, Los Alamos Scientific Laboratory, Los Alamos, New Mexico, January 1980.

Lahr, H., "Fabrication and Properties of Particle Fuels," *Kerntechnik* **19**, 159 (1977).

Lyons, M. F., Boyle, R. F., Davies, J. H., Hazel, V. E., and Rowland, T. C., "UO$_2$ Properties Affecting Performance," *Nuclear Engineering and Design* **21**, 167–199 (1972).

Minato, K., Ogawa, T., Fukuda, K., Sekino, H., Kitagawa, I. and Mita, N., "Fission Product Release from ZrC-Coated Fuel Particles during Post-Irradiation Heating at 1800 and 2000 °C," *Journal of Nuclear Materials* **249**, 142–149 (1997).

Mulford, R. N. R., and Mueller, B. A., *Measurements of Helium Release from Materials Containing $^{238}PuO_2$*, Report N°. LA-5215, Los Alamos Scientific Laboratory, Los Alamos, New Mexico, July 1973.

Mueller, B. A., Rohr, D. D., and Mulford, R. N. R., *Helium Release and Microstructural Changes in $^{238}PuO_2$*, Report N°. LA-5524, Los Alamos Scientific Laboratory, Los Alamos, New Mexico, April 1974.

Olander, D. R., *Fundamental Aspects of Nuclear Reactor Fuel Elements*, U.S. ERDA Report No. TID-26711-P1, Technical Information Center, Energy Research and Development Administration, Oak Ridge, Tennessee, 1976, Chapters 13–15.

Peterson, D. E., and Starzynski, J. S., *Re-entry Thermal Testing of Light-Weight Radioisotope Heater Units*, Report N°. LA-9226, Los Alamos National Laboratory, Los Alamos, New Mexico, March 1982.

Peterson, D. E., Early, J. W., Starzynski, J. S., and Land, C. C., *Helium Release from Radioisotopic Heat Sources*, Report N°. LA-10023, Los Alamos National Laboratory, Los Alamos, New Mexico, 1984.

Scaffidi-Argentina, F., Donne, M. D., Ronchi, C., and Ferrero, C., "ANFIBE: A Comprehensive Model for Swelling and Tritium Release from Neutron-Irradiated Beryllium – I: Theory and Model Capabilities," *Fusion Technology* **32**, 179–195 (1997).

Schock, A., "Light-Weight Radioisotope Heater Unit," in *Proceedings of the 16th Intersociety Energy Conversion Engineering Conference*, held in Atlanta, GA, 9–14 August 1981, Paper N°. 819175, AIAA, 1981, pp. 343–354.

Sholtis, J. A., Jr., Lipinski, R. J., and El-Genk, M. S., "Coated Particle Fuel for Radioisotope Power Systems (RPSs) and Radioisotope Heater Units (RHUs)," in *Proceedings of the Space Technology and Applications International Forum – 1999*, M. S. El-Genk, Ed., American Institute of Physics Conf. Proceedings N°. 458, Woodbury, New York, 1999, **2**, 1378–1384.

Tournier, J.-M., and El-Genk, M. S., "Performance Analysis of Coated $^{238}PuO_2$ Fuel Particles Compact for Radioisotope Heater Units," in these Proceedings, M. S. El-Genk, Ed., American Institute of Physics, Woodbury, New York, 2000.

Turnbull, J. A., "The Effect of Grain Size on the Swelling and Gas Release Properties of UO$_2$ during Irradiation," *Journal of Nuclear Materials* **50**, 62–68 (1974).

Turnbull, J. A., Friskney, C. A., Johnson, F. A., Walter, A. J., and Findlay, J. R., "The Release of Radioactive Gases from Uranium Dioxide During Irradiation," *Journal of Nuclear Materials* **67**, 301–306 (1977).

PROCESSING OF MIXED URANIUM/REFRACTORY METAL CARBIDE FUELS FOR HIGH TEMPERATURE SPACE NUCLEAR REACTORS

Travis Knight and Samim Anghaie

Innovative Nuclear Space Power and Propulsion Institute (INSPI)
PO Box 116502, University of Florida, Gainesville, FL 32611-6502
Phone: 352-392-1427, email: Knight@inspi.ufl.edu, Anghaie@ufl.edu

Abstract. Single phase, solid-solution mixed uranium/refractory metal carbides have been proposed as an advanced nuclear fuel for high performance, next generation space power and propulsion systems. These mixed carbides such as the pseudo-ternary, (U, Zr, Nb)C, hold significant promise because of their high melting points (typically greater than 3200 K), thermochemical stability in a hot hydrogen environment, and high thermal conductivity. However, insufficient test data exist under nuclear thermal propulsion conditions of temperature and hot hydrogen environment to fully evaluate their performance. Various compositions of (U, Zr, Nb)C were processed with 5% and 10% metal mole fraction of uranium. Stoichiometric samples were processed from the constituent carbide powders while hypostoichiometric samples with carbon-to-metal (C/M) ratios of 0.95 were processed from uranium hydride, graphite, and constituent refractory carbide powders. Processing techniques of cold pressing, sintering, and hot pressing were investigated to optimize the processing parameters necessary to produce dense (low porosity), homogenous, single phase, solid-solution mixed carbide nuclear fuels for testing. This investigation was undertaken to evaluate and characterize the performance of these mixed uranium/refractory metal carbides for space power and propulsion applications.

INTRODUCTION

Because of its high performance potential, nuclear thermal propulsion could be utilized for manned missions and cargo transport to the moon or mars, unmanned explorations of the outer planets, and earth orbit transfers of satellites or other space-based assets. Observed high melting point and high thermal conductivity of single phase, solid-solution mixed uranium/refractory metal carbides such as the pseudo-ternary carbide, (U, Zr, Nb)C, portend their usefulness as an advanced fuel for high performance, next generation space power and propulsion systems. This study was undertaken to optimize the processing parameters and measure melting points and hot hydrogen corrosion rates for compositions of (U, Zr, X)C, where X=Hf, Nb, Ta, or W, with uranium metal mole fractions of 5% to 10% to give uranium densities of 0.8 to 1.8 $g \cdot cm^{-3}$.

The Innovative Nuclear Space Power and Propulsion Institute (INSPI) has been conducting research on ultrahigh temperature nuclear fuels and materials for space power and propulsion systems for more than 14 years. Research into mixed uranium/refractory metal carbide fuels has been carried out at the INSPI high temperature laboratories located at the University of Florida as well as in Russia through a collaborative research effort between INSPI and the Russian Scientific Research Institute, LUTCH, from 1993 to 1997. Results of this work has been summarized previously (Dyakov, 1994; Knight, 1999). Tosdale reported higher melting points and greater oxidation resistance for (U, Zr, Nb)C over the corresponding binary carbide mixtures (Tosdale, 1967). Work done on pseudo-binary carbides during the Rover/NERVA program and by Czechowicz et al. showed the importance of composition particularly carbon-to-metal (C/M) ratio to the microstructure and melting point of mixed carbides (Lyon, 1973; Czechowicz, 1991). This paper describes some of the recent achievements in optimizing the processing parameters and fabricating pseudo-ternary carbide nuclear fuels--namely (U, Zr, Nb)C--with varying C/M ratios with the goal of this work being the evaluation and characterization of advanced nuclear fuels for space power and propulsion applications.

CP504, *Space Technology and Applications International Forum–2000*, edited by M. S. El-Genk
© 2000 American Institute of Physics 1-56396-919-X/00/$17.00

METHODS

Cold Pressing and Sintering

Earlier efforts had shown that processing homogenous, single phase, solid-solution pseudo-ternary carbides was difficult requiring high temperatures and/or long times for sintering (Knight, 1999). Because the sintering of carbides should take place at temperatures about three fourths of their melting points, sintering temperatures in the neighborhood of 2600 to 2800 K are required for refractory carbides containing low uranium compositions of 5% to 10% metal mole fraction as called for in this study. Previous samples required sintering at temperatures as high as 2600 K and times greater than two hours in order to produce homogenous samples of low porosity. Such long times at high temperatures are often difficult to maintain and samples experience a large degree of grain growth. In order to decrease the amount of time required to process samples, higher processing temperatures were attempted. Changes were made to the induction furnace to allow temperatures in the range of 2800 to 2900 K. A smaller graphite susceptor was designed along with a smaller coil. This smaller susceptor has less surface area and therefore radiates less heat to the coil, which must be removed by the heat exchanger. Previously this had been the limiting factor in achieving higher temperatures. Also, the smaller coil provides better coupling of the electromagnetic field with the susceptor. A 20 kW, 450 kHz power supply was used for all induction heating experiments.

Previously samples had been cold pressed in the same graphite susceptors that were used for sintering. The low strength of graphite for containing the lateral pressure during pressing caused the graphite die/susceptor to crack. This had largely been overcome by using a tapered graphite die/susceptor and biaxial pressing of the sample and graphite die separately in a large graphite block with a complementary taper. However, this method was abandoned in favor of the new smaller graphite susceptor of uniform diameter designed for higher temperatures in the induction furnace. To avoid the problem of cracking the graphite die/susceptor, stainless steel dies were used to press the powders followed by transfer to the new, smaller graphite susceptor for sintering. A binder of three weight percent stearic acid was added to give the pre-sintered samples adequate green strength to be transferred to the graphite susceptor without breaking apart. These changes to the experimental setup permits cold pressing samples at 120 MPa and sintering at temperatures as high as 2800 K for more than an hour.

This method was used to process samples from mixed carbide powders--namely UC, ZrC, and NbC--with a uranium metal mole fraction of 10%. Additionally, samples processed from mixtures of metal hydrides and carbides were processed containing 5% and 10% metal mole fraction of uranium as described in the next section.

Composition and Carbon-to-Metal-Ratio

It has been shown that changes in the carbon-to-metal (C/M) ratio can have a dramatic effect on the melting point of refractory carbides as well as mixed refractory carbides (Lyon, 1973; Czechowicz, 1991). Their highest melting point usually occurs for C/M ratios less than one. Carmack used metal hydrides of ZrH_2 and HfH_2 to produce hypostoichiometric, mixed carbides (Carmack, 1991). Accary et al. produced UC by the decomposition of UH_3 mixed with graphite. Also noted was the very small size and flake-like shape of UH_3 particles (Accary, 1961). In order to produce samples with varying carbon-to-metal-ratios, alternate starting mixtures were used substituting varying amounts of uranium hydride (UH_3) and graphite for the UC used in previous experiments. Based on calculations of Gibb's free energies for the decomposition of UH_3 using the FACT computer code, the hydrogen is evolved at temperatures above 676 K (Bale, 1996). During sintering at temperatures of 2500 K, all the hydrogen is predicted to be evolved from mixtures containing UH_3, graphite, and refractory metal carbides leaving behind uranium metal to form mixed uranium/refractory metal carbides. Compositions processed in this study had target C/M ratios of 0.95 and uranium metal mole fractions of 5% and 10%. Uranium hydride for these samples was produced from uranium metal rod of 5mm in diameter heated to 473 K in an atmosphere of flowing Ar-7%H. As the hydriding reaction takes place, the rod appears to swell and crack and UH_3 particles flake off exposing more uranium metal. These flakes of UH_3 were then mixed with the desired compositions of graphite and carbides of zirconium and niobium in a ball mill for 24 hours. Finally, similar to the previous all carbide mixtures, cold pressing was done at 120 MPa followed by sintering at 2500 K for various lengths of time from 3 to 30 minutes.

Hot Pressing

Hot pressing as an alternate method for processing mixed carbides was attempted. This method has been noted to provide good results with low porosity (Fischer, 1964; Butt, 1993). A hot press was constructed employing water cooled copper electrodes bolted to larger graphite electrodes that attach to the 1.27 cm (0.5 inch) graphite punch (attached to the upper electrode) and the matching graphite die (attached to the lower electrode) containing the mixed carbide powders for hot pressing. A total of 5 MPa was applied to the samples by placing large lead bricks on top of the press. Compaction was monitored by use of a depth gauge contacting the weight atop the hot press. Hot pressing was first used to process binary carbide samples of ZrC and NbC without uranium to compare with samples processed by cold pressing and sintering with induction heating. This method was investigated to determine whether or not there was any advantage to be realized by hot pressing samples instead of the two step process of cold pressing and sintering that was used initially.

RESULTS

Samples of the pseudo-ternary carbide, $(U_{0.1}, Zr_{0.45}, Nb_{0.45})C$, were produced by cold pressing and sintering by induction heating for varying processing times and temperatures in order to study the effects these parameters have on the quality of samples produced. As stated earlier, the desired sintering temperature is usually three-fourths the melting point of the material placing it between 2600 and 2800 K for the mixed carbides of interest in this study. Scanning electron micrographs of five representative samples in Figure 1 illustrate the effects of cold pressing and sintering time, as well as sintering temperature. A theoretical density of 8.11 $g \cdot cm^{-3}$ was calculated for this composition from the crystallographic densities of the component carbides. Based on this density, samples were processed with relative densities ranging from 73% to 94% of theoretical density. The lowest densities were found for those samples which received no cold pressing prior to sintering (sample A in Figure 1), while the most dense samples were those that were cold pressed at 120 MPa and sintered for either long times or high temperatures such as samples D and E respectively in Figure 1.

Sample A was sintered at the lower end of the desired sintering temperature, 2600 K, for a moderate amount of time, 20 min., but without first being cold pressed. Similar to sample A, but having been cold pressed at 120 MPa prior to sintering, sample C was also processed for 20 min. at 2600 K. Both samples A and C are representative of the intermediate stage of sintering with smooth pores. The absence of cold pressing in sample A means the initial density was lower with less contact area between particles and therefore less area for sintering to occur. Accordingly, the pores of sample A are more numerous, where those of sample C are fewer and elongated. These samples exhibit some grain growth (grain size on the order to 20 to 30 μm) which occurs in the latter part of the intermediate sintering stage. Sample B sintered at the lower sintering temperature of 2600 K for only five minutes represents the initial stage of sintering which is very porous with a large amount of open space. The final stage of sintering is illustrated by samples D and E with a closed structure and isolated, spherical pores. Sample D processed at the lower sintering temperature, 2600 K, for a long time, 142 min. exhibits excessive grain growth with grains on the order of 100 μm. Only a small degree of grain growth is exhibited by sample E with grains on the order of 10 to 20 μm. This sample was sintered at the upper sintering temperature of 2800 K but for only a short time, 5 min.

Samples of the pseudo-binary carbide $(Zr_{0.7}, Nb_{0.3})C$ were processed by cold pressing and sintering by induction heating and by hot pressing. Figure 2 illustrates the similar results obtained from the two methods. Sample F was hot pressed for 100 min., while sample G was cold pressed at 120 MPa and sintered for 20 min. at 2800 K. These methods produced similar results with densities between 86% and 91% of theoretical density. Analysis of samples processed from the metal hydrides is incomplete and will be reported on at a later time.

CONCLUSIONS

Cold pressing and sintering of mixed carbide fuels to produce homogenous, dense (low porosity) samples is possible for high sintering temperatures and/or long time periods. Based on these results a processing temperature of 2800 K for sintering times on the order of ten minutes should be sufficient to provide dense, homogenous samples without excessive grain growth. Lower temperatures of 2500 to 2600 K can be used but require longer times (greater than

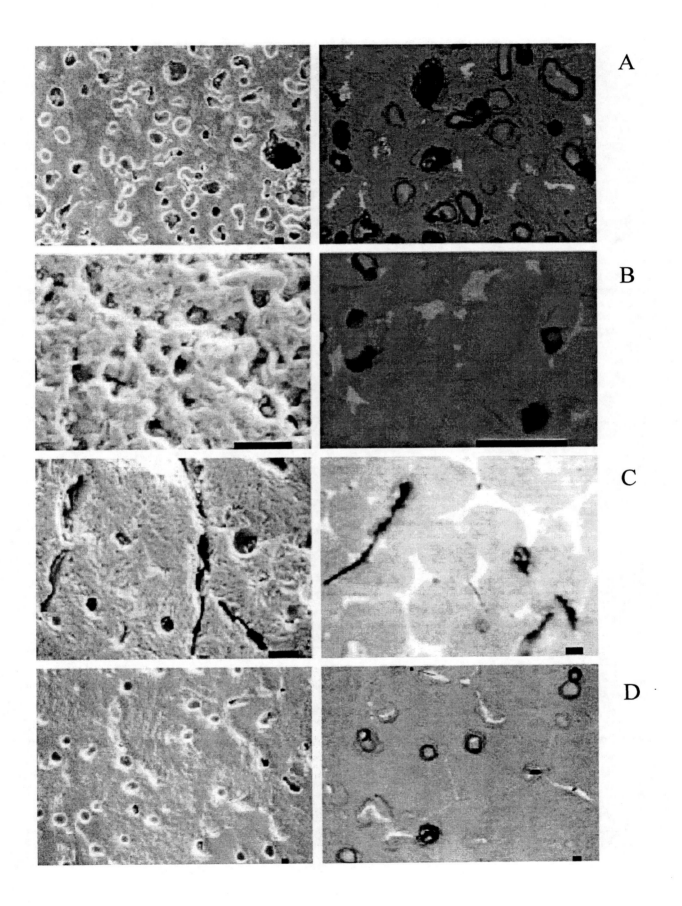

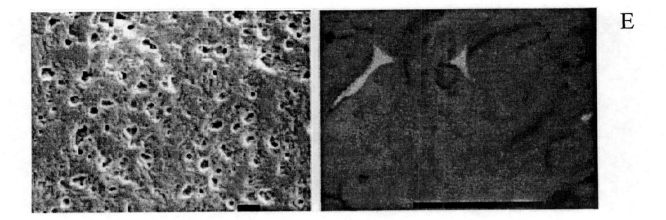

FIGURE 1. Five samples of $(U_{0.1}, Zr_{0.45}, Nb_{0.45})C$ processed from the constituent carbide powders and sintered for various times and temperatures by induction heating. All samples except A were cold pressed at 120 MPa prior to sintering. The black bar along the bottom right hand corner of each image represents 10 μm. The right image for each sample shows the compositional contrast with light areas representing areas of mostly uranium carbide. Sample A processed for 20 min. at 2600 K, Sample B processed for 5 min. at 2600 K, Sample C processed for 20 min. at 2600 K, Sample D processed for 142 min. at 2600 K, and Sample E processed for 5 min. at 2800 K.

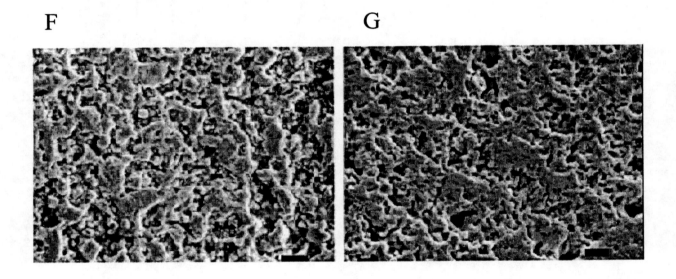

FIGURE 2. Two samples of $(Zr_{0.7}, Nb_{0.3})C$. Sample F was processed by hot pressing for 100 min., while sample G was processed by cold pressing at 120 MPa and sintering for 20 min. by induction heating at 2800 K. The black bar along the bottom right hand corner of each image represents 10 μm.

two hours) and samples will experience large grain growth (on the order of 100 μm). Cold pressing of the powders with a binder is necessary to achieve high densities (low porosity).

Alternate processing routes and methods still require some evaluation in order to determine their overall effectiveness. Metal hydrides can be used to alter the C/M ratio of the final mixed carbide in order to determine the composition with the best performance characteristics. Preliminary results show that fine UH_3 powders can be produced from the metal without great effort. Analysis of the processed samples is required to determine its overall effect and the necessary conditions for processing high quality samples from the metal hydride. Hot pressing of mixed carbides can be used to produce high quality samples but requires fairly long processing times of more than an hour. Further study is necessary to determine the optimum processing parameters for hot pressing of mixed uranium/refractory metal carbides.

Finally, these results show that high quality samples of mixed uranium/refractory metal carbides can be produced under the specified conditions of processing pressure, temperature, and time. These samples will be necessary for the full evaluation of their performance at high temperatures and exposure to flowing hot hydrogen as required for high performance, next generation space power and propulsion systems.

ACKNOWLEDGMENTS

This work was performed for NASA Marshall Space Flight Center under grant NAG8-1251 and for the Department of Defense, Ballistic Missile Defense Organization (formerly SDIO), Innovative Science and Technology Office under contract NAS-26314, managed by NASA Lewis Research Center through INSPI. The authors also wish to thank Dr. Robert Hanrahan of Los Alamos National Laboratory for his many valuable contributions and assistance on this study.

REFERENCES

Accary, A. and R. Caillat, "Sintering with a Chemical Reaction as Applied to Uranium Monocarbide," Powder Metallurgy, edited by W. Leszynski, Interscience Publishers, New York, NY, 1961, pp. 209-220

Bale, C. W., A. D. Pelton, and W. T. Thompson, "FACT 2.1 - User Manual," Ecole Polytechnique de Montreal/Royal Military College, Canada, July 1996

Butt, Daryl, Edmund Storms, Terry Wallace, *Knowledge Status Report on Mixed Uranium/Refractory Metal Carbides Useful for Nuclear Therman Propulsion*, LA-CP-93-41, Nuclear Fuels Technology Group, Los Alamos National Laboratory, 1993

Carmack, W. J., *Melting Studies on Refractory Carbides*, M. S. Thesis, University of Washington, 1991

Czechowicz, D. G., F. G. Hampel, and E. K. Storms, "High Temperature Mixed Carbide Fuels for Space Propulsion Reactors," in *8th Symposium On Space Nuclear Power and Propulsion*, edited by M. S. El-Genk et al., AIP Conference Proceedings #910116, 1991, pp. 1059-1063

Dyakov, E. and M. Tishchenko, "Manufacture and Tests of the Fuel Elements in Hydrogen," *INSPI Contract Report*, Research Institute of the Scientific and Industrial Association LUTCH, May 1994

Fischer, J. J., "Hot-Pressing Mixed Carbides of Ta, Hf, and Zr," Ceramic Bulletin, **43**, 3, 1964, pp. 183-185

Knight, T. W. and S. Anghaie, "Processing Of Pseudo-Ternary Carbide Fuels For High Temperature Space Nuclear Reactors," in *16th Symposium on Space Nuclear Power and Propulsion*, edited by M. S. El-Genk et al., New York, AIP Conference Proceedings, 1999

Lyon, L. L., *Performance of (U, Zr)C-Graphite (Composite) and of (U, Zr)C (Carbide) Fuel Elements in the Nuclear Furnace 1 Test Reactor*, Los Alamos National Laboratory, LA-5398-MS, September 1973

Tosdale, J. P., *Refractory Metal-Carbide Systems*, M. S. Thesis, Ames Laboratory, Iowa State University, Ames, IA, 1967

Processing Tungsten Single Crystal by Chemical Vapor Deposition

Zhigang Xiao[1], Ralph H. Zee[1], and Lester L. Begg[2]

[1]Materials Research and Education Center, Auburn University, AL 36849
[2]General Atomics, P.O. Box 85608, San Diego, CA 92186-5608
[1](334) 844-0368, xiaozhi@eng.auburn.edu; (334) 844-3320, rzee@eng.auburn.edu;
[2](858) 455-2482, les.begg@gat.com

Abstract. A tungsten single crystal layer has been fabricated on molybdenum single crystal substrate through the hydrogen (H_2) reduction of the tungsten hexafluoride (WF_6) in low pressure. Substrate temperature, reaction chamber pressure, and flow rate of WF_6 and H_2, are critical process parameters during deposition. A comprehensive analysis for the effects of these parameters on single crystal layer growth has been processed and optimized growth conditions have been achieved. The different orientation of the substrate shows the different deposition rate for tungsten. Low index plane has higher deposition rate than high index plane. The kinetics of the deposition process has also been investigated. SEM surface analysis indicates that the single crystal layer is smooth in macro-scale and rough and step-growth format in micro-scale.

INTRODUCTION

Fabrication of refractory metal single crystals, especially for tungsten, is always the need for high temperature applications and ballistic impact. There are several techniques existed for growing single crystals of tungsten, such as electron beam floating zone melting, plasma-arc melting, strain annealing, and chemical vapor deposition. The mechanism of growing tungsten single crystals from melt and solid states has been extensively studied (Glebovvsky, and Semenov, 1999; Savitsky, etc., 1982; Katoh, etc., 1991; Sell and Grimes, 1964) and the corresponding techniques for that have also been well developed. Polycrystalline tungsten coating product made by chemical vapor deposition has already been applied in thermionic fuel element (TFE) cladding (Yang and Hudson, 1967). However, the high creep rate and low performance at high temperatures of polycrystalline materials comparing with single crystalline materials drive a demand for fabricating single crystal of tungsten and its alloys in order to meet highly requirements in applications. Smirnov et al. (1996) have fabricated single crystal tungsten coating by chemical transport reactions in tungsten-bromine system, which allowed tungsten bromine (WBr_2) decomposed at high temperature and then tungsten was transported and deposited on to a molybdenum single crystal substrate. In their work, they also cited another group who had made tungsten single crystalline TFE cylindrical cladding by chemical transport reaction in tungsten-chlorine system. What they have accomplished have no microstructures present regarding to the quality of the crystals. In this research work, tungsten single crystals were processed by hydrogen (H_2) reduction of tungsten hexafluoride (WF_6) in flow chemical vapor deposition system. The microstructures corresponding with processing parameters are analyzed with scanning electron microscopy (SEM) as well as x-ray diffraction.

EXPERIMENTAL

The chemical vapor deposition experiments were accomplished using the existed facility (Fig. 1) at General Atomics located in San Diego, California. Molybdenum single crystal mandrel, which was used as a substrate for deposition, was chemically cleaned before loaded in a quartz tube reaction chamber. An induction coil is outfit outside of quartz tube as a heating source. A K-type thermocouple was used to monitor the substrate temperature of the mandrel during deposition. A uniform temperature profile on the mandrel was obtained by adjusting the gap of the

CP504, *Space Technology and Applications International Forum–2000*, edited by M. S. El-Genk
© 2000 American Institute of Physics 1-56396-919-X/00/$17.00

induction coil. The mandrel surface temperature was read by pyrometer before the deposition occurring. The residual WF_6, H_2 and reaction product HF in reaction chamber were evacuated with mechanical pump. The microstructure of tungsten coating was revealed by etching with 10% $K_3Fe(CN)_6$ solution. The observation of microstructure was undertaken in optical microscope and SEM. The surface single crystal orientation was detected by Laue x-ray back reflection technique.

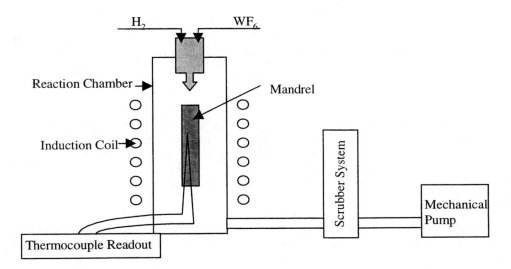

FIGURE 1. Schematic diagram of chemical vapor deposition process at General Atomics.

RESULTS AND DISSCUSION

Three processing parameters, WF_6 and H_2 flow rate, substrate temperature, and reaction chamber pressure, were selected as factors on the crystal growth of tungsten. Since H_2 flow rate is a critical factor comparing WF_6 flow rate during the deposition process, H_2 flow rate was kept at certain level in our experiment according to the reaction formula (1) which indicates the ratio of H_2 to WF_6 as 3:1.

$$WF_6 + 3 H_2 = W + 6HF. \tag{1}$$

As the ratio of H_2 to WF_6 increases, the probability of multigrain growth is higher, in other words, it is unlikely to get single crystal film. Gretz and Hirth (1967) stated that CVD deposition under larger supersaturations which relate to equilibrium constant for the dissociation or reduction reaction involved in the deposition, is controlled by growth, producing a fine-grained, roughly random, deposit, and with growth rates tending to approach the maximum possible rate. Under lower supersaturations, nucleation can become rate controlling, preferred-orientation deposits can form, and growth rates can be less than the maximum possible rate. This is also true for the effect of reaction chamber pressure during deposition occurring, which is discussed later.

The substrate temperature was varied up to 1973 K in our experiment. Lower substrate temperature will have smooth surface layer but very low growth rate. Higher temperature will give high growth rate but unlikely single crystal film. The optimum substrate temperature range was determined for optimal growth condition to be within the temperature range investigated. The effect of substrate temperature on nucleation of process is to affect the excitation of various degree of freedom in the system, such as the rotation degree of freedom for WF_6 in the vapor, and then further to affect the nucleation rate on the substrate. For crystal growth, the effect of substrate temperature is to affect the mean-free-path of particles in the vapor and diffusivity of particles on the surface. High substrate temperature enhances the diffusion of particles on the surface, which can benefit the formation of single crystal film. On the other hand, high substrate temperature increases nucleation rate, which should be prohibited during deposition. Therefore, a compromise substrate temperature should be reached for obtaining single crystal film. Our experiment results also proved this.

The reaction chamber pressure is another critical parameter during single crystal film deposition. Two pressure conditions were selected during the experiment. Lower pressure showed a smooth crystal growth layer. Higher reaction pressure conducted an abnormal multigrain growth on some surface area, which interrupted a single crystal layer growth even it is not from the substrate at beginning. This can be explained that high reaction pressures imply that the particles compete with each other to deposit onto the surface and also it increases supersaturation in the vapor. Both of them facilitate abnormal multigrain growth.

Figure 2 shows tungsten single crystal layer with one extra-grain growing out from it. It is clearly indicates that the extra-grain has different growth orientation from its matrix. Figure 3 shows that different orientation of the substrate results in the different growth rate. The results of Laue x-ray back reflection analysis pointed out that the plane with low growth rate and high growth rate are in (110) and (100) orientation, respectively. Usually, low-index planes have lower surface energy than high-index planes, therefore it is ready for atoms adsorbed on low-index plane surface, and results in a higher growth rate.

The surface microstructures of tungsten single crystal film as well as polycrystal under SEM are shown in Figures 4 and 5. The surface of single crystal layer (Fig. 4) is in step-growth format and not smooth in micro-scale but is pretty flat comparing abnormal multigrain growth on the surface at low magnification (Fig. 5). The abnormal multigrain microstructure can appear under either at high substrate temperatures or larger ratio of H_2 to WF_6 or higher reaction chamber pressure. The location of multigrain growth, generally, occurred at transition region between low growth rate and high growth rate regions if the substrate is in cylinder form.

In our experiment, we also found that abnormal grain growth normally does not start from the substrate initially, as Figure 1 illustrated. This might be due to the intrinsic impurity, which comes from residual reactant WF_6 and product HF, causing heterogeneous nucleation on a substrate, and then induced abnormal crystal growth. Green et al. (1968) pointed out that a number of drastic film orientation changes caused by impurities, which influence both the nucleation and the coalescence process.

FIGURE 2. Abnormal grain growth from single crystal matrix.

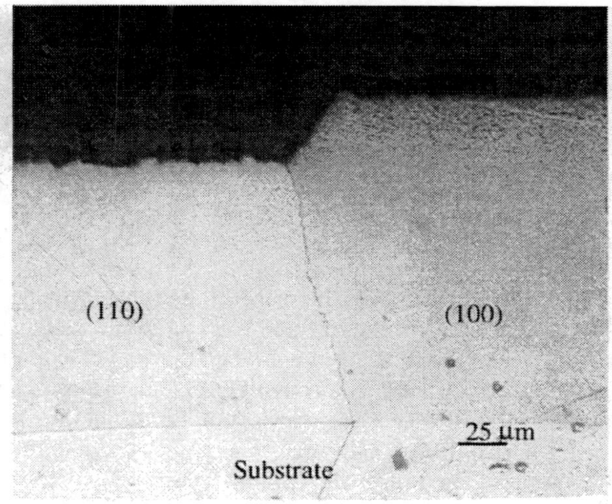

FIGURE 3. Single crystal layer growing from different orientation of substrate.

CONCLUSIONS

For processing tungsten single crystal film in chemical vapor deposition process, the ratio of H_2 to WF_6, substrate temperature, and reaction chamber pressure must be selected properly and corresponded each other appropriately. Low ratio of H_2 to WF_6, appropriate substrate temperatures, and reaction chamber pressure will produce better tungsten single crystalline film. Low index plane has a faster growth rate than high index plane during deposition.

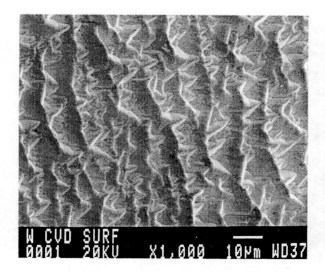 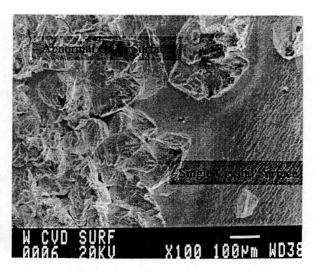

FIGURE 4. Single crystal surface morphology at high magnification.

FIGURE 5. Abnormal grain surface and single crystal surface morphology at low magnification.

ACKNOWLEDGMENTS

This work has been financially supported by the U. S. Defense Threat Reduction Agency through a contract with General Atomics. Mr. Richard J. Precourt, Mr. Jack Knipping, and Mr. Tim Heffernan at General Atomics are gratefully acknowledged from Zhigang Xiao for their technical assistances.

REFERENCES

Glebovvsky, V. G. and Semenov, V. N., " The perfection of tungsten single crystals grown from the melt and solid state", *Vacuum*, **52**, 71-74 (1999).

Green, A., Bauer, E., and Dancy, J., "The influence of impurities on the formation of single-crystal films", in *Molecular Processes on Solid Surfaces*, edited by E. Drauglis, R. D. Gretz, and R. I. Jaffee, New York, McGraw-Hill, 1968, pp. 479-497.

Gretz, R. D. and Hirth, J. P., "Nucleation and Growth Processes in CVD", *Proceedings of the Conference on Chemical Vapor Deposition of Refractory Metals, Alloys, and Compounds*, edited by A. C. Schaffhauser, Gatlinburg, Tennessee, 1967, pp. 73-97

Katoh, M., Iida, S., Sugita, Y., "X-ray characterization of tungsten single crystals grown by secondary recrystallization method", *Journal of Crystal Growth*, **112**, 368-372 (1991).

Savitsky, E. M., Burkhanov, G. S., and Kirillova, V. M., "Single crystals refractory and rare metals, alloys, and compounds", in *Crystals, Growth, Propeties and Applications*, Berlin Heidelberg, Springer-verlag, 1982, pp107-148.

Sell, H. G. and Grimes, W. M., "Modes of operation of an electron beam floating zone melting furnace: growing of single crystals of metals and alloys", *The Review of Scientific Instruments*, Vol. **35**, No. 1, 64-68 (1964).

Smirnov, V. P., Klinkov, A. Ye., Yastrebkov, A. A., Sidorov, Yu. I., Koshkin, L. Ye., and Afanasiev, N. G., "Description of . Tungsten Vapor Deposition Process and Equipment in Use; Result of Measuring Work Function of Emitter Samples", Final Report. Contract DNA 001-96-C-0020(item 0003), 1996.

Yang, L. and Hudson, R. G., "Evaluation of Chemically Vapor Deposited Tungsten as Eletron Emitters for Nuclear Thermionic Application", *Proceedings of the Conference on Chemical Vapor Deposition of Refractory Metals, Alloys, and Compounds*, edited by A. C. Schaffhauser, Gatlinburg, Tennessee, 1967, pp. 329-348.

Performance Analysis of Coated ^{238}PuO$_2$ Fuel Particles Compact for Radioisotope Heater Units

Jean-Michel Tournier and Mohamed S. El-Genk

Institute for Space and Nuclear Power Studies / Department of Chemical and Nuclear Engineering
School of Engineering, The University of New Mexico, Albuquerque, NM 87131, USA
(505) 277–5442, mgenk@unm.edu

Abstract. A fuel form consisting of coated plutonia fuel particles dispersed in a graphite matrix is being investigated for use in Radioisotope Heater Units (RHUs). The fuel particles consist of a ^{238}PuO$_2$ kernel (300–1200 μm in diameter), a 5-μm PyC inner coating and a ZrC outer coating ($\geq$ 10 μm). The latter, an extremely strong material at high temperatures, serves as a pressure vessel for maintaining the integrity of the fuel particle and containing the helium generated by radioactive decay. Parametric analyses compared the thermal powers of the coated particle fuel compact (CPFC) RHU and LWRHU. Both utilize Fine-Weave Pierced Fabric (FWPF) aeroshell and PyC insulation sleeves. During normal operation, the fuel temperature is ~ 800 K, but could reach as much as 1723 K during an accidental re-entry heating. Assuming full helium release, a single-size particle (500 μm) fuel compact would maintain its integrity at a temperature of 1723 K, after 10 years storage time before launch. When replacing the LWRHU fuel pellet, Pt-alloy clad and inner PyC insulation sleeve with CPFC, the calculated thermal power of the CPFC-RHU is 1.5, 2.3 and 2.4 times that of LWRHU, for 100%, 10%, and 5% helium release, respectively, with little change in total mass. A fuel compact using binary-size particles (300 and 1200 μm diameters) would deliver 15% more thermal power. A one-dimensional, transient thermal analysis of the CPFC-RHU showed that during accidental re-entry the maximum fuel temperature in the CPFC would be 1734 K.

INTRODUCTION

Coated plutonia particles are a promising fuel form for advanced radioisotopic power systems (RPS) and heater units (RHU) (Sholtis et al., 1999). Potential benefits include enhanced safety, design flexibility, and higher specific thermal power. Coated particles fuels have been implemented successfully in the Nuclear Rocket Program in the late sixties and early seventies (Altseimer et al., 1971) and in the High-Temperature Gas Cooled Reactors (HTGRs) (Gulden and Nickel, 1977). More recently, this fuel form has been proposed for use in innovative nuclear thermal propulsion concepts such as the Particle Bed Reactor (PBR) (Powell and Horn, 1986, Dobranich and El-Genk, 1991) and the Pellet Bed Reactor (PeBR) (El-Genk et al., 1992, Morley and El-Genk, 1995). In addition, ZrC-coated UO$_2$ fuel particles are being developed at the Japan Atomic Energy Research Institute for use in advanced HTGRs (Minato et al., 1997) up to 1600 K. The ZrC coating is much stronger and chemically more stable at high temperatures (it has a large heat of formation = −207.1 kJ/mole) than SiC used in earlier generations of HTGR fuel. It is one of the most ductile refractory materials at high temperatures, eutectically melts at 3123 K (El-Genk and Tournier, 1999), and has higher yield strength.

To capitalize on the extensive database on the fabrication of and fission gas release in coated oxide fuel particles, coated ^{238}PuO$_2$ fuel particles were proposed for use in RHUs (Sholtis et al., 1999). The diameter of these coated plutonia particles is selected to eliminate any potential radiological health concerns in case of release. The literature on the fabrication, gas release, and material properties of this fuel form has been reviewed and documented (El-Genk and Tournier, 1999). This review indicated that plutonia fuel kernels fabricated by sol-gel processes are attractive, due to their potential high retention of helium gas. Also, unlike powder metallurgy currently in use, sol-gel processes do not generate dust nor a solid waste stream. Liquid waste stream is also minimized since all chemicals are recycled and re-used.

The objective of this paper is to analyze the performance of ZrC-coated plutonia fuel particles dispersed in a graphite matrix, and quantify the effects of the following parameters on the thermal power of the CPFC-RHU:

CP504, *Space Technology and Applications International Forum–2000*, edited by M. S. El-Genk

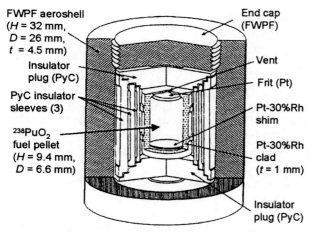

FIGURE 1a. Current LWRHU (Johnson 1997).

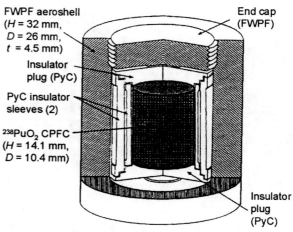

FIGURE 1b. Proposed CPFC-RHU.

(a) fuel temperature; (b) storage time before launch; and (c) helium release fraction from the fuel kernel. Both single-size particles and binary-size particles compacts are considered. Current LWRHUs use a granular $^{238}PuO_2$ fuel pellet encapsulated in a vented, Pt-alloy clad, and surrounded by 3 successive insulation layers of PyC (Figure 1a). The CPFC-RHU uses the same Fine-Weave Pierced Fabric (FWPF) aeroshell and PyC insulation sleeves, but replaces the fuel pellet, Pt-alloy clad and inner PyC insulation sleeve with CPFC (Figure 1b).

COATED FUEL PARTICLE DESIGN

The plutonia fuel particle (Figure 2) consists of a $^{238}PuO_2$ fuel kernel, 300–1200 μm in diameter, having an as-fabricated porosity of 5–35%; an inner coating of pyrolytic carbon (PyC) of low or intermediate density (1.5–1.9 g/cm^3) that is 5 μm thick; and a zirconium carbide (ZrC) outer coating. The ZrC outer coating is the primary containment, designed to withstand the pressure of the helium gas released from the fuel kernel. The thickness of the ZrC coating, t_{ZrC}, is determined in the present analysis for different kernel size, helium inventory and fractional release, and maximum fuel temperature. Both PyC and ZrC coatings are deposited by Chemical Vapor Deposition (CVD) techniques. The plutonia fuel kernels are fabricated using powder metallurgy or sol-gel processes. A PyC over-coating could be deposited to provide a microscopically rough surface to facilitate bonding of the coated particles in the graphite matrix, and/or to provide graphite material for filling the interstitials in the particles compact. The PyC inner-coating prevents corrosion of the fuel kernel during CVD deposition of ZrC, and provides some open porosity to accommodate the helium gas released from the fuel kernel. It also protects the ZrC from recoil by alpha particles.

The ZrC is an extremely strong material at high temperatures which serves as a pressure vessel for maintaining the integrity of the fuel particle and containing the helium generated by radioactive decay. The sol-gel plutonia kernels ($\geq$ 300 μm in diameter) would most likely function as a single large grain for the release of He gas. Helium release data suggest that at temperatures < 900 K, all helium would be contained within the fuel matrix (El-Genk and Tournier, 2000). Even at temperatures in excess of 1700 K, He gas release from such large-grain fuel kernels could be < 10%. This stipulation is supported by irradiation test results of large-grain ($\geq$ 300 μm), uncoated UO_2 fuel kernels at isothermal conditions, up to 6.4 at.% burnup.

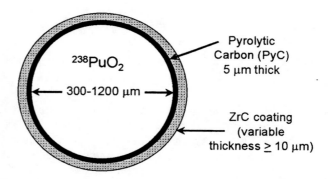

FIGURE 2. Reference design of the coated fuel particle.

Experimental results showed that at a fuel temperature of 1723 K, the fission gas release was ~ 8%, compared to ~ 100% for small-grain (7–40 μm) fuel kernels (El-Genk and Tournier, 2000).

Radiological Effect of Coated Plutonia Fuel Particles

For a typical plutonia fuel density of $\rho_p \approx 10{,}000$ kg/m^3, the actual kernel diameter is about a third of its aerodynamic diameter, $D_{aero} = D_p \times \sqrt{\rho_p/1{,}000}$ (Hoover, 1999). For $D_{aero} > 25$ μm ($D_p > 8$ μm), kernels could not be inhaled into the lungs nor the thoracic cavity, and present no radiological risk. Kernels having $D_{aero} = 100$ μm ($D_p = 32$ μm) are not respirable, but there is a 50% chance they might be deposited in the nasal passages. In this case, the radiological risk to the nasal tissues is negligible since the ZrC coating effectively attenuates alpha particles emitted near the surface of the plutonia kernel. Very large particles ($D_{aero} > 1.5$ mm or $D_p > 474$ μm) cannot be taken into the nasal passages, hence present no radiological risk whatsoever, actual or perceived.

If coated fuel particles were accidentally released and dispersed in the biosphere and some were ingested, they would simply pass through the digestive track, since plutonia is essentially non-soluble and thus would not be absorbed in the body. Fuel kernels with $D_p = 300\text{--}1200$ μm ($D_{aero} = 950\text{--}1580$ μm) used in the following analyses cannot be inhaled, thus present no radiological risk whatsoever. The thermal and stress analysis model developed to calculate the ZrC coating thickness of the coated plutonia particles is described next.

THERMAL AND STRESS ANALYSIS MODEL OF COATED PLUTONIA PARTICLES

The inner PyC layer deposited by CVD at 1723 K has a density ≤ 1.7 g/cm^3, or volume porosity $\geq 25\%$. The thickness of the ZrC coating deposited also at 1723 K is ≥ 10 μm. The ZrC coating is under compression below 1723 K, free of any thermal stresses at 1723 K, but under tension above 1723 K. Thus, at 1723 K, the fuel kernel and the PyC and ZrC coatings would be in contact and free of thermal stresses. Upon cooling down, the fuel kernel would shrink faster than the PyC and ZrC coatings, since PuO$_2$ has higher thermal expansion coefficient, providing an additional void space to accommodate released He gas. The present model does not account for this additional void volume, and is therefore conservative below 1723 K. The released He is assumed to occupy only the as-fabricated open porosity in the fuel kernel and the PyC layer (which is very small compared to the former). The amount of open porosity in the plutonia kernel is assumed to behave similarly to that in commercial UO$_2$ fuel pellets (Gontar et al., 1997):

$$\varepsilon_f^{open} = 1.067 \times \varepsilon_f - 0.03, \quad T \leq 1723 \text{ K}. \tag{1}$$

Above 1723 K, the model assumes conservatively that the fuel kernel is constrained by the PyC and ZrC coatings, and the open pores volume is reduced due to the fuel thermal expansion. No credit is taken for the strength of the PyC layer; the ZrC alone must withstand the released He pressure. The maximum tangential stress in the ZrC coating is calculated using the spherical shell model (Kaae, 1969), assuming zero pressure at the coating outer surface. Also, no credit is taken for the strength and constraint imposed by the graphite matrix of the compact. From a design standpoint, the ZrC thickness is calculated assuming the maximum tangential stress in the ZrC coating equals 80% of the yield strength of ZrC (Dobranich and El-Genk, 1991), which decreases exponentially with temperature (Figure 3).

POWER ESTIMATES OF COATED PARTICLE FUEL COMPACTS (CPFC)

The CPFCs analyzed in this work are made of single-size and binary-size coated ^{238}PuO$_2$ particles dispersed in a graphite matrix (Figure 4). In LWRHUs (Figure 1a), since He is vented (Tate, 1982), the primary role of the Pt-30%Rh clad is to resist impact following an accidental re-entry. The CPFC design relies on the ZrC coating to fully contain the released He. In a potential impact following accidental re-entry, most of the shock's mechanical energy will be absorbed by the weaker graphite structure of the CPFC, resulting in cracks and/or compaction of the graphite matrix, and leaving the ZrC-coated fuel particles intact. The ZrC coating is strong enough to withstand the pressure of He released following a long storage time; its yield strength is an order of magnitude higher than Pt-30%Rh (770 MPa versus 60 MPa at 1447 K, 430 MPa versus 30 MPa at 1773 K).

The CPFC fuel loading is a function of the as-fabricated fuel porosity, kernel size, coating thickness and packing fraction of particles. Figure 5 shows that the packing density of binary-size particles in CPFC increases linearly

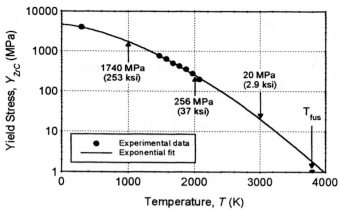

FIGURE 3. Yield stress of zirconium carbide as a function of temperature.

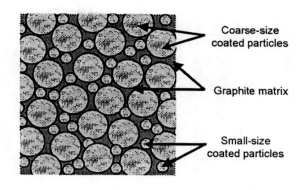

FIGURE 4. Binary-size coated particle fuel compact (CPFC).

with D_{p1}/D_{p2} up to ~ 6.5, beyond which the packing density increases very little (McGeary, 1961). At this value, the intersticial pore size, $(2/\sqrt{3}-1)D_{p1}$, between the coarse closed-packed spheres would accommodate the smaller particles (see Figure 4) such that $D_{p1}/D_{p2} > (2/\sqrt{3}-1)^{-1} = 6.464$. A binary mixture with $D_{p1}/D_{p2} = 1, 2, 3, 4$ and 5 could have a packing fraction of 0.625, 0.656, 0.688, 0.719 and 0.75, respectively.

Specific Thermal Power of CPFC

In the present study, the plutonia fuel contains 80.05 at.% of ^{238}Pu, has $\rho_f^{TD} = 11.46$ g/cm^3 (Johnson, 1997), a Beginning-Of-Life (BOL) activity of 448 GBq/g (12.1 Ci/g), and a specific thermal power of 0.40 W/g (the ^{238}Pu decay constant $= 2.5048 \times 10^{-10}\, s^{-1}$, molar weight = 238.24 g/mole, and the as-fabricated ^{238}Pu atom density = 1.784 x 10^{24} atoms/kg). The calculated thermal power density of the single-size particles CPFC (of 500 μm fuel kernels, $t_{PyC} = 5$ μm and maximum packing fraction of 0.625) is compared in Figure 6 with that of LWRHU fuel pellet. The specific thermal power of the CPFC decreases rapidly with increasing ZrC coating thickness and as-fabricated fuel porosity. A Cassini fuel pellet with an as-fabricated porosity, $\varepsilon_f = 15\%$ has a thermal power density, $q'''_{LWRHU} = 3.9$ W/cm^3, while a single-size CPFC with $\varepsilon_f = 15\%$ for which $t_{ZrC} = 30$ μm has a BOL power density of 1.63 W/cm^3, or 42% of q'''_{LWRHU}, due to the lower fuel loading. However, as will be shown later, this disadvantage can be more than compensated for by increasing the volume of the fuel compact in the aeroshell of the current LWRHU, at no mass penalty.

High Thermal Power CPFC-RHU

The CPFC-RHUs performance results presented in this section are for a storage time of up to 20 years, assuming a hypothetical re-entry accident during which the fuel temperature reaches 1723 K (Schock, 1981). At such temperature, the fractional release of He gas from large-grain (≥ 300 μm) and polycrystalline sol-gel fuel kernels could be less that 10% (El-Genk and Tournier, 2000). Nonetheless, the calculations assumed different He gas release fractions: $F = 1.0, 0.10$ and 0.05.

Figure 7a shows the calculated ZrC coating thickness to withstand the helium gas pressure at 1723 K, assuming full release ($F = 1.0$). Figure 7b presents the ratio of the thermal power of the CPFC-RHU to that of LWRHU, when replacing the fuel pellet (0.287 cm^3) and the Pt-alloy clad (0.370 cm^3) in the latter with 500-μm single-size CPFC (Design I). Note that the upper envelop of the curves in Figure 7b corresponds to the maximum thermal power of the CPFC-RHU. The plateaux at small storage times to launch correspond to the minimum ZrC coating thickness of 10 μm specified in the CPFC design.

In a CPFC-RHU with 25% as-fabricated fuel porosity, a 36-μm thick ZrC coating could withstand the internal pressure of all He gas generated after 10 years storage, but the fuel loading of the CPFC is only 31% by volume. In

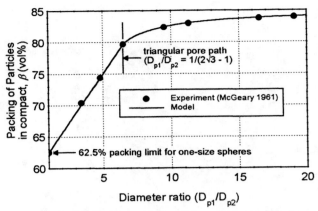

FIGURE 5. Maximum packing of two-size coated fuel particles (data from McGeary, 1961).

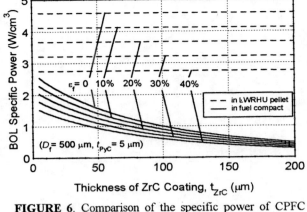

FIGURE 6. Comparison of the specific power of CPFC and LWRHU Fuel Pellet.

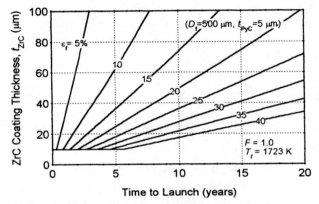

FIGURE 7a. Required ZrC coating thickness in CPFC for full helium release and T_f = 1723 K.

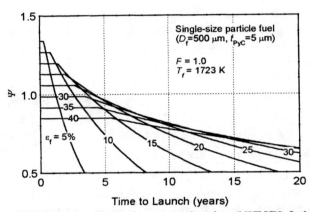

FIGURE 7b. Thermal power ratio when LWRHU fuel pellet and Pt-alloy clad are replaced with single-size CPFC.

the current LWRHU, the Pt-30%Rh cladding occupies a 29% larger volume than the fuel pellet itself (Tate, 1982). Thus, replacing also the cladding with CPFC provides more than a two-fold increase (a factor of 2.29) in RHU thermal power. As shown in Figure 7b, for design option I, the CPFC-RHU thermal power is 80% of that of LWRHU, assuming full He release after 10 years storage.

The three PyC insulation sleeves in the current LWRHU occupy a relatively large volume (Figure 1a). They were designed to maintain the Pt-alloy temperature below its eutectic point (Sholtis et al., 1999) during an accidental re-entry heat pulse (Pt-30%Rh forms an eutectic with carbon at 2033 K). A portion of this insulation could be replaced with CPFC (Figure 1b), without risking overheating the coated plutonia particles, since their ZrC coating has a much higher melting point (ZrC forms an eutectic with carbon at 3123 K). To verify this statement, a thermal analysis of the CPFC-RHU was performed next to calculate the fuel compact temperature during a re-entry heating pulse.

Thermal Analysis of CPFC-RHU

To predict the CPFC maximum temperature during a re-entry heating pulse, a one-dimensional, transient thermal model of the RHU was developed. The model was setup initially such that its predictions of the thermal response of the LWRHU during the worst-case re-entry accident (inertial angle of −4.85°) agreed with those reported by Schock (1981). He has used a comprehensive 3-D transient model to simulate the LWRHU thermal response during re-entry. The present 1-D transient model reproduced the 3-D results to within ± 30 K over the full 300 s re-entry transient, and the Pt-30%Rh clad maximum temperature (1726 K) within 3 K and 5 s (at 270 s).

The present analysis showed that in the CPFC-RHU design I, the fuel compact maximum temperature was 1688 K after 278 s into the re-entry transient. The initial (operating) fuel temperature in the CPFC-RHU was 709 K, or 100 K lower than in the LWRHU. When the inner PyC insulation was also removed and replaced with CPFC (Design II), the initial fuel temperature of the CPFC-RHU was slightly higher (735 K), and the CPFC temperature reached a maximum of 1734 K after 255 s into the re-entry transient.

EFFECT OF HELIUM ON CPFC-RHU PERFORMANCE

In the present study, when the inner PyC insulation sleeve in the LWRHU was also replaced with CPFC (Design II), the thermal power of the CPFC-RHU was 1.83 times that of Design I, in which only the fuel pellet and the Pt-30%Rh clad were replaced with CPFC (the total volume enclosed by the inner insulation sleeve is 1.20 cm^3; this sleeve accounts for 12 vol% of the cavity between the Pt-30%Rh clad and the aeroshell (Tate, 1982)). As shown in Figure 8, the CPFC-RHU designed for fuel temperature up to 1723 K, and storage time to launch up to 10 years, assuming full He release, provides 50% more thermal power than the LWRHU, at only 95% of the total mass. Using a binary-size CPFC with 300 and 1200 μm diameter particles increases the thermal power of the CPFC-RHU by an additional 15% to 1.7 times that of the LWRHU (with the CPFC-RHU mass essentially equal to that of the LWRHU, 40 g).

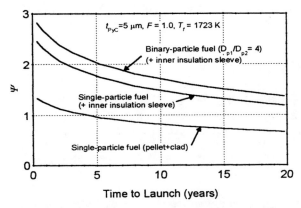

FIGURE 8. Comparison of CPFC-RHU thermal power with that of LWRHU assuming 100% helium gas release.

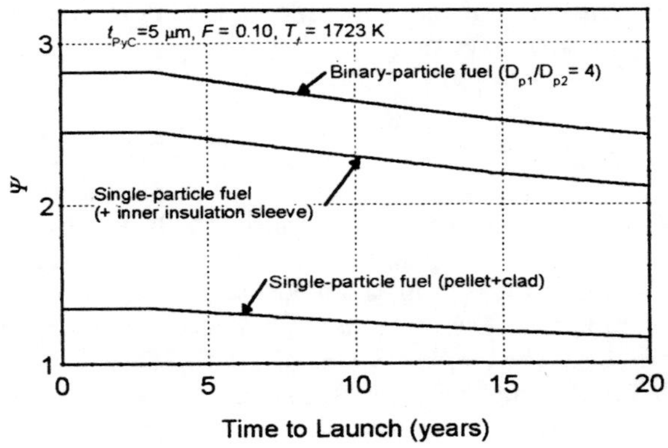

FIGURE 9. Comparison of CPFC-RHU thermal power with that of LWRHU assuming 10% helium gas release.

To investigate the effect of He gas release on the performance of the CPFC-RHU, results were also obtained for $F = 0.1$ and $F = 0.05$. The predicted improvements in the CPFC-RHU's thermal power are delineated in Figure 9 for $F = 0.1$ and a maximum fuel temperature of 1723 K. A compact with 11% as-fabricated fuel porosity, the 11-μm ZrC coating could withstand the released He pressure after a 10 years storage time. When the fuel pellet and the Pt-30%Rh clad in the LWRHU are replaced with single-size CPFC, the RHU provides 25% more thermal power than the LWRHU (Figure 9), which represents a 55% increase over the CPFC-RHU assuming full He release ($F = 1.0$).

When the inner insulation sleeve in the LWRHU is also replaced with single-size CPFC (Design II), an RHU stored for 10 years would provide 2.3 times the thermal power of LWRHU. Using binary-size (300 and 1200 μm particles diameters) CPFC, the thermal power of the CPFC-RHU increases by an additional 15%, to 2.6 times that of LWRHU, with essentially the same mass (~ 40 g).

TABLE 1. Thermal power increase in CPFC-RHU designed to resist 1723 K heating after 10 years.

| CPFC-Type (Design Option) | Helium gas release fraction, F | | |
	1.0	0.10	0.05
Single-size (D_f=500 μm)[a]	$\psi = 0.80$	$\psi = 1.25$	$\psi = 1.31$
	(ε_f = 25%, t_{ZrC} = 36 μm)	(ε_f = 11%, t_{ZrC} = 11 μm)	(ε_f = 7%, t_{ZrC} = 10 μm)
Single-size (D_f=500 μm)[b]	$\psi = 1.46$	$\psi = 2.29$	$\psi = 2.40$
Binary-size	$\psi = 1.68$	$\psi = 2.63$	$\psi = 2.76$
(D_{f1} = 1200 μm, D_{f2} = 300 μm)[b]	(t_{ZrC1}=86 μm, t_{ZrC2}= 22 μm)	(t_{ZrC1}=24 μm, t_{ZrC2}= 10 μm)	(t_{ZrC1}=19 μm, t_{ZrC2}= 10 μm)

[a]: Replace LWRHU pellet and Pt-alloy clad with CPFC. [b]: Replace LWRHU pellet, clad and inner insulation sleeve with CPFC.

The effect of He release fraction is best illustrated in Figure 10, when the fuel pellet, Pt-30%Rh clad and inner PyC insulation sleeve in the LWRHU were replaced with a single-size CPFC. After a storage time of 10 years, a decrease in He release from 100% to 10% increases the fuel loading and hence the thermal power of the CPFC-RHU by 55%. However, decreasing He release from 10% to 5% results only in a ~ 6% increase in the CPFC-RHU's thermal power. These results suggest that He release in excess of 10% strongly impacts the power density of the CPFC-RHU, but not below 10%.

The results summarized in Table 1 demonstrate the potential of the CPFC-RHU for achieving specific thermal powers as high as 65 W_{th}/kg, or 2.5 times that of LWRHU. Such good performance, together

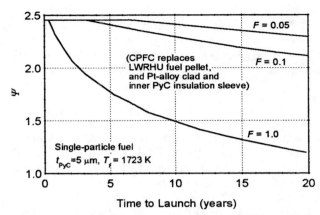

FIGURE 10. Effect of helium gas release on thermal power of CPFC-RHU for T_f = 1723 K (design II).

with the potential for enhanced safety and design flexibility, make CPFC an attractive fuel form. Remaining issues for future investigations include measuring the He gas release from large-grain and sol-gel $^{238}PuO_2$ fuel kernels at temperatures up to 1900 K.

SUMMARY AND CONCLUSIONS

A parametric analysis is performed which investigated the potential of CPFC-RHUs. The CPFC consists of ZrC-coated, single- or binary-size $^{238}PuO_2$ kernels dispersed in a graphite matrix. The thickness of the ZrC coating depends on the operating temperature, storage time before launch, size of plutonia kernel, and the assumed release fraction of He gas. Unlike the LWRHU, the proposed CPFC-RHU retains all He generated in the coated plutonia particles. The thermal power of the CPFC-RHU is compared with that of LWRHU, assuming that both utilize the FWPF aeroshell and the PyC insulation sleeves of the latter.

Results showed that for a storage time of 10 years before launch, a maximum fuel temperature of 1723 K during a re-entry heating, and assuming full He release from the fuel kernel, replacing both the fuel pellet and the Pt-30%Rh clad in the LWRHU with single-size (500 μm) CPFC (Design I) provided only 80% of the LWRHU thermal power (~ 1 W_{th}).

However, when the inner graphite insulation sleeve is also replaced with CPFC (Design II), the thermal power of the CPFC-RHU increases to 1.54 W_{th}, 2.40 W_{th} and 2.52 W_{th}, when assuming 100%, 10% and 5% He release, respectively. Replacing the single-size CPFC with binary-size (300 μm and 1200 μm) particles compact increases the thermal power of the CPFC-RHU by an additional 15% to 1.77 W_{th}, 2.76 W_{th} and 2.90 W_{th}, for 100%, 10% and 5% He release, respectively.

Additional investigations of the impact response of CPFC and of the coated plutonia particles are recommended to better quantify any radiological uncertainties. Additional issues for further investigation include the sol-gel fabrication of the plutonia fuel kernels, and He gas release from coated and uncoated plutonia fuel kernels as a function of temperature and storage time.

ACKNOWLEDGMENTS

This research was funded by Sandia National Laboratories (SNL), Kirtland Air Force Base, Albuquerque, New Mexico, under Contract No. BE-2543, to the University of New Mexico's Institute for Space and Nuclear Power Studies. The opinions expressed in this paper are solely those of the authors. The authors are thankful to Dr. Ronald J. Lipinski, SNL, and Mr. Joseph A. Sholtis, Jr., Sholtis Engineering & Safety Consulting, for their continuous technical support throughout this work effort.

NOMENCLATURE

English

D Diameter (m)

D_{aero} Aerodynamic diameter of particle

F Fraction of He released from the fuel kernel

q RHU thermal power (W_{th})

q''' Volumetric thermal power density (W_{th} / m^3)

T Temperature (K)

t_{PyC} Thickness of PyC layer (m)

t_{ZrC} Thickness of ZrC coating (m)

Greek

e Porosity

ρ Density (kg / m^3)

Ψ Thermal power ratio, $q_{CPFC-RHU} / q_{LWRHU}$

Subscript/Superscript

f PuO_2 fuel kernel

open   Open

TD   Theoretical density

1   Coarse particles in a binary-size CPFC

2   Fine particles in a binary-size CPFC

REFERENCES

Altseimer, J. H., et al., "Operating Characteristics and Requirements for the NERVA Flight Engine," *J. Spacecraft* **8** (7), 766 (1971).

Dobranich, D., and El-Genk, M. S., "Thermal Stress Analyses of the Multi-Layered fuel Particles of a Pellet-Bed Reactor," *Nuclear Technology* **94**, 372 (1991).

El-Genk, M. S., Buden, D., and Mims, J. "Nuclear Reactor Refuelable in Space," *United States Patent* No. 5,106,574, April 1992.

El-Genk, M. S., and Tournier, J.-M., *Study of a Coated-Particle Fuel Form for Advanced Radioisotope Heat Sources – Part I: A Review of Coated-Particle Fuel Performance, Fabrication Techniques and Materials Properties*, Report UNM–ISNPS–5–1999, Sandia National Laboratories, Contract BE–2543, 1999.

El-Genk, M. S., and Tournier, J.-M., "Helium Release from $^{238}PuO_2$ Fuel Particles," in these Proceedings, 2000.

Gontar, A. S., Nelidov, M. V., Nikolaev, Yu. V., and Schulepov, L. N., *Fuel Elements of Thermionic Converters*, a Special Issue of *Technology: Journal of the Franklin Institute*, Volume 333A Numbers 2–6 (1996), R. L. Hunter, Ed., Cognizant Communication Corporation, New York, 1997, pp. 175–176.

Gulden, T. D., and Nickel, H., "Preface: Coated Fuel Particles," *Nuclear Technology* **35**, 206–213 (1977).

Hoover, M., *Personal Communication*, Lovelace Inhalation Toxicology Research Institute, Albuquerque, New Mexico, January 15, 1999.

Johnson, E. W., *Light Weight Radioisotope Heater Unit (LWRHU) – Final Safety Analysis Report (FSAR) for the Cassini Mission*, EG&G Mound Applied Technologies, Report No. MLM-3826, UC-713, February 1997.

Kaae, J. L., "A Mathematical Model for Calculating Stresses in a Pyrocarbon- and Silicon Carbide-Coated Fuel Particle," *Journal of Nuclear Materials* **29**, 249–266 (1969).

McGeary, R. K., "Mechanical Packing of Spherical Particles," *Journal of the American Ceramic Society* **44** (10), 513-522 (1961).

Minato, K., Ogawa, T., Fukuda, K., Sekino, H., Kitagawa, I., and Mita, N., "Fission Product Release from ZrC-Coated Fuel Particles during Post-Irradiation Heating at 1800 and 2000 °C," *Journal of Nuclear Materials* **249**, 142–149 (1997).

Morley, N. J., and El-Genk, M. S., "Neutronics and Thermal-Hydraulics Analyses of the Pellet Bed Reactor for Nuclear Thermal Propulsion," *Nuclear Technology* **109**, 87–107 (1995).

Powell, J. R., and Horn, F. L., "High Power Density Reactors Based on Direct Cooled Particle Beds," in *Space nuclear Power systems 1985*, M. S. El-Genk and M. D. Hoover, Eds., Orbit Book Company, Malabar, FL, Chapter 39, 1986, pp. 319–329.

Schock, A., "Light-Weight Radioisotope Heater Unit," in *Proceedings of the 16th Intersociety Energy Conversion Engineering Conference*, held in Atlanta, GA, 9–14 August 1981, Paper N°. 819175, AIAA, 1981, pp. 343–354.

Sholtis, J. A., Jr., Lipinski, R. J., and El-Genk, M. S., "Coated Particle Fuel for Radioisotope Power Systems (RPSs) and Radioisotope Heater Units (RHUs)," in *Proceedings of the Space Technology and Applications International Forum*, M. S. El-Genk, Ed., AIP Conference Proceedings N°. 458, Woodbury, NY, 1999, **2**, pp. 1378–1384.

Tate, R. E., *The Light-Weight Radioisotope Heater Unit (LWRHU): A Technical Description of the Reference Design*, Los Alamos National Laboratory, Report No. LA-9078-MS, UC-33a, January 1982.

Coated Particle Fuel for Radioisotope Power Systems and Heater Units: Status and Future Research Needs

Mohamed S. El-Genk[1], Jean-Michel Tournier[1], Joseph A. Sholtis, Jr.[2], and Ronald J. Lipinski[3]

[1]Institute for Space and Nuclear Power Studies and Department of Chemical and Nuclear Engineering
The University of New Mexico, Albuquerque, NM, (505) 277-5442, mgenk@unm.edu

[2]Sholtis Engineering & Safety Consulting, P.O. Box 910, Tijeras, NM, (505) 281-4358, Sholtis@aol.com

[3]Sandia National Laboratories, P.O. Box 5800, Albuquerque, NM, (505) 284-3651, rjlipin@sandia.gov

Abstract. Coated particle fuel has been proposed recently for use in Radioisotope Power Systems (RPSs) and Radioisotope Heater Units (RHUs) for a variety of space missions requiring power levels from mWs to 10's or even hundreds of Watts. It can be made into different shapes and sizes of solid compacts, heating tapes, or paints. Using a conservative design approach, this fuel form could increase by 2.3-2.4 times the thermal power output of a LWRHU, while offering promise of enhanced safety. These performance figures are based on using single-size (500 μm) compacts of ZrC coated $^{238}PuO_2$ kernels and assuming 10% and 5% He release, respectively, at 1723 K, following 10 years of storage. Using binary-size (300 and 1200 μm) fuel kernels in the compact increases the thermal power output by an additional 15%. $^{238}PuO_2$ fuel kernels are intentionally sized (≥ 300 μm in diameter) to prevent any adverse radiological effects. They are non-respirable and non-inhalable and, if ingested, would simply be excreted with no radiological effects. The $^{238}PuO_2$ fuel kernels are contained within a strong ZrC coating, which is designed to fully retain the fuel and the helium gas. Helium retention in large grain (≥ 300 μm) granular and polycrystalline fuel kernels is possible even at high temperatures (> 1700 K). The former could be fabricated using binderless agglomeration or similar processes, while the latter could be fabricated using Sol-Gel or thermal plasma processes, with potentially less radioactive waste and fabrication contamination. In addition to summarizing the results of a recent effort investigating the performance of coated fuel particle compact (CPFC) and helium gas release, this paper identifies and discusses future research and testing needs.

INTRODUCTION

Recently, coated plutonia particles fuel was proposed for potential use in advanced radioisotope heater units (RHUs) and radioisotope power systems (RPSs) (Sholtis et al. 1999). The fuel compact consists of coated $^{238}PuO_2$ fuel kernels dispersed into a graphite matrix with a packing density for single size particles of up to 62.5% by volume. This packing density could be increased to ~ 73%, using two particle sizes with a diameter ratio of 4.0 (Tournier and El-Genk, 1999, 2000). The fuel kernels are covered with a 5-μm thick, Pyrolytic Graphite (PyC) inner coating for protection during the application of the outer ZrC coating by Chemical Vapor Deposition (CVD). The CVD processes are also used to apply the PyC coating. The ZrC, a very strong material that is not only strong but ductile at temperature in excess of 2000 K, serves as the containment vessel of the fuel kernel and the He gas generated by the radioactive decay of ^{238}Pu. The thickness of the ZrC coating depends on the fuel temperature, the helium release fraction, and the storage time and grain size of the granular plutonia fuel. Figure 1 shows a cross-sectional view of a coated plutonia fuel particle and Figure 2 shows an illustration of a binary-size coated particle fuel compact (CPFC).

Since the graphite matrix is designed to be structurally weaker than the ZrC coating, cracking of the coated particles fuel compact (CPFC) upon impact on solid surfaces is likely to occur through the graphite, leaving the coated fuel particles intact. The design concept of the CPFC is analogous to that of the "*pomegranate*" fruit (Figure 3), in which the holding structure is spongy as well as weak under applied tensile stress to protect the fruit seeds. Similarly, the graphite matrix in the CPFC would be spongy and structurally weaker than the ZrC coating of the fuel kernels. It

CP504, Space Technology and Applications International Forum–2000, edited by M. S. El-Genk
© 2000 American Institute of Physics 1-56396-919-X/00/$17.00

would accommodate the thermal expansion of the fuel particles and protect them during impact on a solid surface, following a launch or reentry accident.

The coated fuel particles offer a promise for enhanced safety and higher specific power than current state-of-the-art Light Weight Radioisotope Heater Units (LWRHUs). The fuel kernels are intentionally sized (≥ 300 μm) to prevent any adverse radiological effects. They are non-respirable, non-inhalable, and if ingested, would simply be excreted with no radiological effects (Tournier and El-Genk, 2000; Hoover, 1999). In addition, this coated fuel form offers excellent design flexibility as the CPFC could be made into different shapes and sizes to provide thermal power from milli-watts to tens or even hundreds of watts.

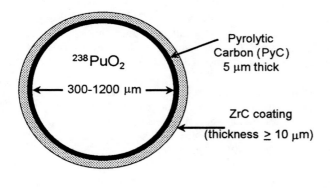

FIGURE 1. Cross-section of a coated fuel kernel showing the PyC and ZrC coatings.

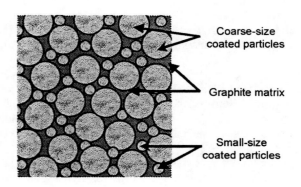

FIGURE 2. Binary-size CPFC.

FIGURE 3. A photograph of a split "*Pomegranate fruit*" illustrating the design concept of the CPFC.

This fuel form can also be made into heating tapes, buttons, or paint for a variety of new applications as RHUs or miniaturized RPSs for satellites, spacecraft, and planetary exploration probes. For example, a button-like CPFC heat source could be used with miniaturized thermoelectrics to provide electrical power in the mWe range for more than 10 years. The excess heat could be used for thermal management, thus achieving 100% energy utilization. In addition, larger size CPFC in the form of pellets or disks could be used in higher power RHUs as well as in RPSs to produce power in the range from 1.0 to tens or even hundreds of We.

In current General Purpose Heat Sources (GPHSs) and LWRHUs, the refractory cladding of the plutonia fuel pellets is kept at relatively high temperatures (> 1173 K) to ensure sufficient ductility when impacting solid surfaces. In addition to being very heavy, these claddings (platinum-30%-rhodium in LWRHUs and iridium in GPHSs) must be kept well below their melting temperatures and those of any eutectics that could form in a solid propellant fire. These temperatures are maintained with the aid of multi-layered, low conductivity, PyC insulation sleeves. Conversely, the CPFC has no temperature constraint. Thus, most of the internal thermal insulation sleeves and the refractory cladding in the GPHS and LWRHU could be replaced with CPFC. This would result in higher thermal power, at lower or same mass, or in smaller size and lower weight, for the same thermal power.

This fuel form is also ideal for scientific probes requiring both thermal and electric power and in which the appropriate type and shape of the RHU can be fabricated to optimize the design, operation, and the functionality of the probe. Some of the space missions that employ planetary probes for in-situ analysis of surface materials require that the He gas be fully retained in order to avoid contaminating the environment and skewing the sensitive measurements. Such an option is not attainable using the current RHU and GPHS designs.

During FY 99, an exploratory effort sponsored by the Department of Energy was initiated to investigate the potential of the coated particle fuel form for RHUs and address fabrication and performance issues. This effort was performed jointly by Sandia National Laboratories, Sholtis Engineering and Safety Consulting, and the University of New Mexico's Institute for Space and Nuclear Power Studies. The specific tasks investigated were:

(a) Review the fabrication technology of coated plutonia fuel particles.
(a) Review the release mechanisms of helium gas in small grain (7-40 μm) granular plutonia pellets in GPHSs and LWRHUs, and examine the applicability of these mechanisms to the He release from large grain (≥ 300 μm) and polycrystalline fuel kernels.
(c) Review the spectrum of credible launch and reentry accident environments that the coated particle fuel could potentially experience. Based on this review, design and functional requirements for coated particle fuel were established.
(d) Develop a design and performance model of coated fuel particles to investigate the impact of using single-size and binary-size CPFC on the thermal power level of the RHU. Also quantify the effects on the RHU thermal power of the helium gas release, fuel temperature, and storage time before launch.
(e) Identify future research and testing needs to confirm the coated particle fuel's potential operation and safety promise.

This paper provides a summary of the work done under this joint exploratory research effort and presents excerpts of the results obtained. In addition, future research needs are identified and discussed. A more detailed account of the work done and of the results obtained can be found elsewhere (Tournier and El-Genk, 1999, 2000; El-Genk and Tournier, 1999, 2000; and Sholtis, 1999).

FABRICATION TECHNIQUES

There are two basic microstructures of fuel kernels (Figure 4): granular and polycrystalline. Granular UO_2 and mixed-oxide fuel kernels with small grain sizes (7-40 μm) have been successfully fabricated on a large scale using the binderless agglomeration processes (Burnett et al., 1963; Ford and Shennan, 1972; Allen et al., 1977). This fabrication process, however, has not been used for the fabrication of large grain (100-300 μm) fuel kernels. Polycrystalline UO_2 and mixed-oxide fuel kernels have been successfully fabricated on a bench scale using the Sol-Gel plasma processes (Haas et al., 1967; Huschka and Vygen, 1977; Förthmann and Blass, 1977; Lahr, 1977). Granular fuel kernels fabricated using binderless agglomeration or similar processes consist of a number of polycrystalline grains, typically 7-40 μm in average diameter. The common grain boundaries develop during the sintering of the oxide fuel kernels at high pressure and temperature. The as-fabricated porosity in the fuel matrix and at the triple interface of the grains is controlled during sintering. Typical as-fabricated porosity in oxide fuel kernels ranges from 5 to 15% (see Figure 4a). The produced fuel kernels are highly spherical, and virtually monosized particles with diameters ranging from 200 to 1000 μm.

Polycrystalline, UO_2 and mixed-oxide fuel kernels, having diameters as large as 1200 μm, have been fabricated using the Sol-Gel techniques (Haas et al., 1967; Huschka and Vygen, 1977; Förthmann and Blass, 1977; Lahr, 1977). These techniques do not require milling or grinding (as during the fabrication of granular fuel by powder metallurgy), thus generating very little, if any, radioactive dust or aerosols. Liquid waste, however, may be produced in the Sol-Gel processes, depending on the efficiency of recycling the chemicals used and of the fabrication procedures. The gelled fuel microspheres, with an almost perfect sphericity, are sintered to the desired density and stoichiometry by heating to high temperature in air or in an oxidizing atmosphere. The sintered fuel kernels typically have a polycrystalline structure with tiny, intragranular voids (< 1 μm in diameter) (see Figure 4b). As with granular fuel, the as-fabricated porosity in the Sol-Gel, oxide fuel kernels (5-20%) is controlled during sintering.

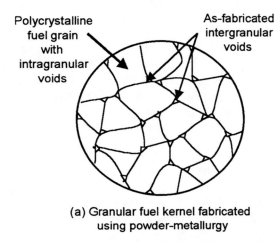

(a) Granular fuel kernel fabricated using powder-metallurgy

(b) Polycrystalline fuel kernel fabricated using Sol-Gel techniques

FIGURE 4. Microstructures of plutonia fuel kernels.

HELIUM GAS RELEASE

With respect to the helium release from $^{238}PuO_2$ fuel kernels, the polycrystalline structure (Figure 4b) may be regarded as a single grain of the kernel diameter, with as-fabricated intragranular porosity made of tiny pores. In granular fuel kernels, the helium gas release unit has an average size equal to that of the fuel grain, significantly increasing the helium gas release. No data have been found on the fabrication of granular or polycrystalline $^{238}PuO_2$ fuel kernels using the binderless agglomeration or the Sol-Gel techniques. The data on the fabrication of monocrystalline fuel kernels using thermal plasma processes are also nil (Silver, 1999).

Results of helium release from granular $^{238}PuO_2$ pellets (7-40 μm grain size) have indicated that transient heating increased the release rate. However, the He release fraction was the same as that measured during isothermal steady state experiments performed at fuel temperatures of up to 2023 K (Peterson and Starzynski, 1982; Peterson et al., 1984; Mulford and Mueller, 1973; Mueller et al., 1974). The reported data on helium gas release from granular $^{238}PuO_2$ fuel pellets covered grain sizes from 7-50 μm and storage time from a few months up to ~ 8 years. The results showed that He release below 900 K was nil. However, at higher temperatures, increasing the grain size decreased the He release fraction. At the peak fuel temperature of 1723 K during a simulated re-entry heating pulse, the measured release fraction from granular RHU $^{238}PuO_2$ fuel pellets was as much as 88%. The identified helium release mechanisms for these small grain, RHU $^{238}PuO_2$ fuel pellets and test samples were in agreement with the release mechanisms discussed later and with the equivalent-sphere model for fission gas release. According to the equivalent-sphere model (Olander, 1976), for radioactive fission gases which attained steady-state concentrations in the fuel, the fractional release, F^*, which equals the release-to-birth rate ratio, $\dot{R}/\dot{B}$, can be expressed as (El-Genk and Tournier, 2000):

$$F^* \equiv \dot{R}/\dot{B} \approx \left(\frac{S_R}{V}\right) \times \sqrt{\frac{D}{\lambda}} \propto \sqrt{\frac{D'(T)}{\lambda}} \propto \frac{1}{a(T)} T^\gamma \sqrt{T_{1/2}} \; , \tag{1}$$

where V is the total volume of the fuel kernel and λ is the radioactive decay constant of the released isotope. This equation shows F^* to increase inversely proportional to the average fuel grain radius, a, but directly proportional to the square root of the half-life, $T_{1/2}$. As the half-life increases, F^* approaches the release fraction of non-radioactive species (Figure 5). Thus, for the same fuel material, temperature, T and grain size, F^* increases exponentially with half-life of the gas species, approaching an asymptote, which corresponds to the release fraction of non-radioactive gases, such as helium in granular $^{238}\text{PuO}_2$ fuel. The effective diffusion coefficient, D' (s^{-1}) = D/a^2, which increases exponentially with the fuel temperature, accounts for the increases in both the gas mass diffusion coefficient, D (m^2/s), and the effective release area, S_R (m^2). This release area is a function of temperature and includes any developed cracks in the fuel matrix and open porosity at the grain boundaries.

Neglecting the as-fabricated open porosity in the fuel, the normalized surface area for the helium gas release from plutonia fuel kernels can be expressed as (El-Genk and Tournier, 1999 and 2000):

$$\left(\frac{S_R}{S_p}\right) = (1-\alpha) + \alpha\frac{d_p}{d_g} \; . \tag{2}$$

In this equation, S_p is the geometrical surface area of the as-fabricated fuel particle, and d_p and d_g are the diameter of the fuel kernel and the average diameter of the fuel grain, respectively. The value of coefficient α ($0 \leq \alpha \leq 1.0$), depends on the effective surface area available for He release from the fuel matrix, S_R.

Helium Release Mechanisms

Below ~ 900 K, He release occurs by atomic diffusion from the fuel matrix, and S_R is equal to the geometrical surface area of the as-fabricated fuel sample (i.e. $\alpha = 0$). Between 900 K and 1150 K, $0 < \alpha < 1$ and S_R increases with temperature in granular fuel due to the formation and the coalescence of grain boundary bubbles. Above ~1150 K, S_R increases rapidly with temperature as α approaches unity, due to the formation of open porosity at the grain boundaries. At ~1450 K, S_R reaches its maximum value ($\alpha = 1.0$), as the formation of the open porosity at the grain boundaries is complete. Above this temperature, He gas release in granular fuel is no longer limited by the surface area available for release, but rather by the atomic and volume diffusion of the gas in the fuel grains.

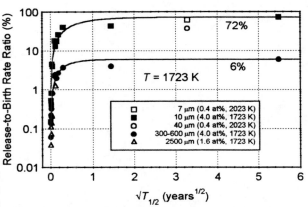

FIGURE 5. Release fraction of noble gases and volatile fission products in UO$_2$ fuel particles, versus half-life.

Therefore, for the same fuel temperature and kernel diameter, increasing the grain size from 10 μm to 200 μm could reduce He release at high fuel temperatures (>1450 K) by more than an order of magnitude (El-Genk and Tournier, 2000). As indicated earlier, a polycrystalline fuel kernel could be regarded as a single grain with a size equal to the diameter of the kernel. Thus, the effective surface area for the He gas release in large grain ($\geq$ 200 μm) or polycrystalline particles would essentially be equal to the geometrical surface area of the kernel. This would significantly reduce the effective area available for the He gas release, relative to that in small-grain fuel.

Estimates of Helium Release

Gas release data have been reported for UO$_2$ fuel particles irradiated in fission reactors (Turnbull, 1974; Friskney et al., 1977; Turnbull et al., 1977; Friskney and Turnbull, 1979) at isothermal conditions. These authors investigated the effects of grain size, radioactive decay, fuel burnup (up to 6.4 at. %) and fuel temperature (up to 2023 K) on the release-to-birth rate ratio of the various noble fission gases and volatile isotopes. The UO_2 fuel particles used in these experimental investigations included small-grain (average grain size of 10 μm) and large-grain (an effective grain size of 300-600 μm) particles irradiated up to a 6.4 at. % burnup. Monocrystal right cylinders of natural (0.72 wt. % ^{235}U) stoichiometric UO_2 were also irradiated in the experiments. The amounts of the various gaseous species released were measured by γ-spectroscopy, after correcting for their radioactive decay. The release rates of fission gases and volatile fission products were calculated based on these measurements, while their birth rates were calculated using computer codes (Friskney and Turnbull, 1979).

Results showed that at 1723 K and 4.0 at.% burnup, the small-grain fuel particles (7-40 μm) released essentially all gases and volatile fission products (~ 80% release), whereas only ~ 7% was released from the large-grain (~ 300-600 μm) fuel particles. It is worth noting that the release fraction from the small-grain UO_2 particles was almost the same as that reported for He in granular plutonia pellets of the same average grain size (7-12 μm), at the same temperature (Peterson and Starzynski, 1984). The data showing the effect of half-life on the release fraction of noble gases and volatile fission products from the various UO_2 fuel samples at 1723 K is presented in Figure 5 (El-Genk and Tournier, 2000). According to Equation (1), the release-to-birth rate ratio is proportional to the square root of the half-life, for the relatively short-lived isotopes that reach equilibrium early in time. Such dependence of the release-to-birth rate ratio on the half-life is evident in Figure 5; the gas release data exhibited an exponential increase with half-life, reaching asymptotic values at large half-lives. These asymptotic values are representative of the release fractions of stable gases in granular UO_2 fuel during fission and helium in plutonia fuel particles of the same average grain sizes.

Based on these experimental data, helium gas release from small-grain (7-40 μm) plutonia fuel at 1723 K would be ~ 80%. This is in agreement with the experimental data generated at LANL for GPHS and LWRHU granular plutonia pellets (Angelini et al., 1970; Mulford and Mueller, 1973; Mueller et al., 1974; Land, 1980; Peterson and Starzynski, 1982; Peterson et al., 1984). For large-grain ($\geq$ 300 μm) and polycrystalline fuel microspheres, when heated up to 1723 K, the helium gas release could be more than an order of magnitude lower ($\leq$ 7%). The experimental fission gas release data also showed that, at 1042 K, less than 1% of the fission gases and volatile fission products were released from the large-grain fuel particles. Similar release fraction would be expected for helium in plutonia fuel particles having similar grain sizes, and even lower release fractions are expected in polycrystalline fuel kernels. Therefore, the helium gas release at the operating temperature in coated, large-grain and polycrystalline $^{238}PuO_2$ fuel kernels at the nominal operating temperature in a RHU is practically nil. The operating temperature of coated particle fuel in a RHU (~ 800 K) is more than two hundred degrees lower than 1042 K (Tournier and El-Genk, 2000).

In summary, the helium release from small-grain (7-40 μm) $^{238}PuO_2$ fuel kernels at temperatures $\geq$1723 K is expected to be ~ 80-90 %, decreasing to less than 10% at 1042 K. This conclusion is in excellent agreement with the actual helium release data obtained at Los Alamos National Laboratory for GPHS and LWRHU plutonia pellets. The release of He from large-grain ($\geq$ 300 μm) $^{238}PuO_2$ fuel kernels at $\geq$ 1723 K, however, could be ~ 7%, decreasing to ~ 0.8% at 1042 K. In polycrystalline plutonia fuel kernels fabricated using Sol-Gel processes, the helium release fraction could be even lower than the estimates for the large-grain $^{238}PuO_2$ kernels (< 7% at 1723 K and < 0.8% at 1042 K).

PERFORMANCE OF COATED PARTICLE FUEL FOR RHUs

A performance and stress analysis model of coated $^{238}PuO_2$ fuel kernels was developed to calculate the required thickness of ZrC coating to fully retain the He gas released as a function of the fuel temperature, Helium gas release fraction, and storage time of CPFC before launch (Tournier and El-Genk, 2000). At 1723 K, the model assumes conservatively that the fuel kernel is constrained by the PyC and ZrC coatings. No credit was taken in the model for the strength of PyC inner coating or the constraint by the graphite matrix in the CPFC, assuming zero pressure on the outside of the ZrC coating. The induced tangential stress in the ZrC coating is calculated using the spherical shell model (Kaae, 1966) since its thickness is much smaller than the diameter of the fuel kernel. Calculations showed that when replacing the fuel pellet, the refractory alloy cladding, and the inner PyC insulation sleeve in a LWRHU (Johnson, 1997) with a CPFC (Figures 6a and 6b), the CPFC-RHU could provide 2.3 times the thermal power (~ 2.4 W) of the LWRHU. These calculations are conservative, assuming storage of 10 years, fuel temperature of 1723 K (the same as the peak temperature expected during a re-entry heating pulse), replacing only the inner insulation sleeve in LWRHU with CPFC, and 10% He gas release. This He release is higher than estimated earlier for large-grain and polycrystalline $^{238}PuO_2$ fuel kernels. The CPFC was made of single-size 500 μm coated fuel kernels packed in graphite at a density of 62.5% by volume.

When a binary-size CPFC of 300 and 1200 μm particles was used, the thermal power of the CPFC-RHU increased by an additional 15% to ~ 2.6 W, at essentially the same mass as the LWRHU (~ 40 g). At this performance level, the CPFC-RHU's specific thermal power is ~ 66 W/kg. These calculations, which are also conservative, were conducted at CPFC temperature of 1723 K and 11% as-fabricated fuel porosity, and assuming a 10% release of the He gas generated in the fuel kernels. The calculated thickness of the ZrC coating to withstand the internal He gas pressure at these conditions was 10, 11, and 24 μm for the 300, 500, and 1200 μm diameter fuel kernels in the CPFC, respectively. As indicated earlier, using large grain (≥ 300 μm) or polycrystalline $^{238}PuO_2$ fuel kernels could reduce the He release fraction at 1723 K to below 7%. When the assumed He release fraction was decreased to 5%, the calculated thermal power of the single-size CPFC-RHU was 2.4 W, and increased to 2.76 W for the binary-size CPFC-RHU (Tournier and El-Genk, 2000). The corresponding specific thermal power is 60 and 69 W/kg, respectively.

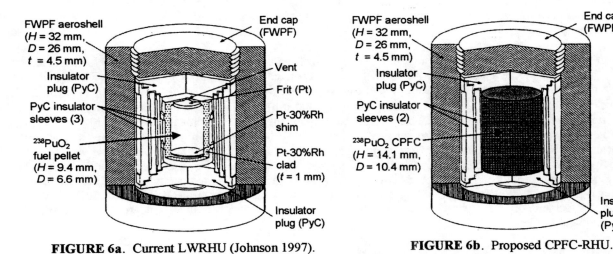

FIGURE 6a. Current LWRHU (Johnson 1997).   **FIGURE 6b**. Proposed CPFC-RHU.

FUTURE RESEARCH NEEDS

The results of the present investigations of potential performance, helium gas release, and the status of the fabrication technology of coated $^{238}PuO_2$ fuel particles have confirmed the promise of the CPFC. Results showed that CPFC could increase the thermal power of RHUs and potentially that of the GHPSs, with no increase in their total mass. Several technical issues, however, have been raised that are worthy of further investigations. The first issue is related to acquiring and demonstrating the fabrication technology of large grain (≥ 200 μm) and polycrystalline $^{238}PuO_2$ fuel kernels (300-1200 μm in diameter) using either state-of-the art Sol-Gel and/or thermal plasma processes. The costs of production, decontamination, and of mastering each of these technologies need to be investigated and compared. The quality of the fuel kernels fabricated by the different techniques is also an important factor. This includes the $^{238}PuO_2$ fuel kernels structure strength and ability to retain the He gas within the fuel matrix at high temperatures, typical of those expected during a re-entry heating pulse (>1700 K).

As indicated in this research, the retention of the He gas generated in the $^{238}PuO_2$ fuel kernels depends strongly on the fuel microstructure. Almost full He release is expected in small grain (7-40 μm) fuel kernels at 1723 K, making them a poor selection for the CPFC. The large ZrC coating thickness required to retain the released He gas in these small-grain fuel kernels would significantly reduce the fuel loading, and hence the thermal power of CPFC-RHU or

CPFC-GPHS below that of the current state-of-the-art (Tournier and El-Genk, 2000). Conversely, large-grain or polycrystalline $^{238}PuO_2$ fuel kernels, which are the primary choice for CPFC, could retain as much as 93-95% of the He gas generated in the fuel matrix at temperatures in excess of 1700 K. As a result, the smaller thickness of the ZrC coating needed increases the fuel loading and, hence, the thermal output of the CPFC-RHU and potentially of CPFC-GPHS, well in excess of those of the current state-of-the-art LWRHU and GPHS designs.

The next issue worthy of future investigation is related to confirming the He gas release estimates from coated and uncoated, large-grain ($\geq$ 200 μm) and polycrystalline, $^{238}PuO_2$ fuel kernels. Helium gas release experiments for these fuel particles are recommended. The measured He release fractions could be compared with those reported for the small grain (7-40 μm) RHU fuel pellets. These experiments should investigate the effects of storage time, fuel temperature, and heating mode (steady-state and ramped heating transient) on both the release rate and release fraction of He. For the coated fuel particles, theses experiments may also investigate the potential failure modes and threshold temperature for ZrC coating failure, if any.

The coating of UO_2 and mixed-oxide fuel kernels with PyC and ZrC has been demonstrated successfully for many years for commercial, high temperature gas cooled reactors (HTGRs) with SiC outer coating and recently with ZrC coating for operation at higher temperature and fuel burnup (Minato et al., 1997). The PyC and ZrC coatings for UO_2 and mixed oxide fuels are typically applied using CVD processes. Owing to the relatively shorter half-life of ^{238}Pu (86 years), however, the alpha particles released by the radioactive decay of ^{238}Pu could affect the quality of the coating in two ways. The heat dissipated in the decay process and the bombardment of the deposited coating by the emitted alpha particles could affect the quality of the PyC and ZrC coatings during the CVD processes. This is an important issue worthy of future investigation.

An additional issue for future research is related to the fracture strength of the CPFC upon impacting solid surfaces. Both fracture and impact tests involving CPFC and detailed stress analysis are recommended. The results could guide the future development of the graphite matrix of the CPFC, which can protect the coated fuel particles during handling, re-entry heating, as well as during impact on solid surfaces. As indicated earlier, the graphite matrix should be structurally strong and of low density to accommodate thermal expansion of the coated fuel particles during a re-entry heating pulse. It has to fracture upon impact with solid surfaces in order to protect the coated fuel kernels, "*the pomegranate concept.*" Ultimately, mechanical, thermal, and aeroablation testing of coated particle fuel in simulated accident environments will also be needed.

SUMMARY AND CONCLUSIONS

An exploratory research effort was initiated to investigate the potential of coated $^{238}PuO_2$ fuel particles for future use in advanced RHUs and RPSs. The fabrication technology of UO_2 fuel particles and the release mechanisms and measured release fractions of fission gas in small-grain (7-40 μm), large-grain ($\geq$ 300 μm), and polycrystalline fuel kernels at isothermal irradiation conditions up to 2023 K and up to 6.4 at. % burnup were reviewed. Based on the reported data, estimates of the He gas release were made for small-grain, large-grain, and polycrystalline $^{238}PuO_2$ fuel kernels. In addition, a design model of the coated particle fuel was developed and used to investigate the performance of CPFC-RHUs. The performance of both single-size and binary-size CPFC-RHUs was investigated as a function of the He release fraction, for 10 years storage before launch and fuel temperature up to 1723 K. This temperature corresponds to the peak value expected during a re-entry heating pulse.

Results indicated the need to fabricate $^{238}PuO_2$ fuel kernels that could retain most of the He generated by the ^{238}Pu radioactive decay, even at fuel temperatures as high as 1723 K. This is possible using large grain ($\geq$ 200 μm) or polycrystalline $^{238}PuO_2$ fuel kernels. The former could be fabricated using binderless agglomeration or similar processes while the latter could be fabricated using Sol-Gel or thermal plasma processes. Although these processes have successfully been used in the fabrication of UO_2 and mixed-oxide fuel kernels, they have not been demonstrated for the fabrication of $^{238}PuO_2$ fuel kernels. In addition, CVD that is being used to apply the PyC and ZrC coatings for the former is yet to be demonstrated for the latter.

Estimates of the He gas release from large-grain and polycrystalline $^{238}PuO_2$ fuel kernels were made, based on isothermal measurements of the released noble gases and volatile fission products from UO_2 fuel particles during irradiation up to 6.4 at. % burnup. Based on these measurements, it was concluded that He release in large-grain ($\geq$ 300 μm) $^{238}PuO_2$ fuel kernels at 1723 K could be less than 7% and even lower in polycrystalline fuel kernels. At fuel temperatures $\leq$ 1000 K, the He release will be nil.

Performance analysis of conservatively designed CPFC-RHU indicated that the thermal power could be 2.3 and 2.4 times that of the LWRHU, at essentially the same total mass and using the same Fine-Weave Pierced Fabric (FWPF) aeroshell, and two outer PyC insulation sleeves. These performance figures of the CPFC-RHU were calculated assuming a single-size (500 μm) coated $^{238}PuO_2$ fuel particles compact, 11% as-fabricated fuel porosity, and 10% and 5% helium gas release, respectively. In the CPFC-RHU, the fuel pellet and its refractory cladding and the inner PyC insulation sleeve in the LWRHU were replaced with CPFC. When using binary-size (300 and 1200 μm) CPFC at the same conditions, the thermal power of the CPFC-RHUs increased by an additional 15%.

In conclusion, the CPFC is indeed a promising fuel form for use in advanced RHUs and RPSs. In addition to enhancing the thermal power output, it offers enhanced safety and unique design flexibility as it could be fabricated in different sizes and shapes. The CPFC-RHUs and RPSs could meet the thermal and electric power needs for future spacecraft and planetary exploration in the range from a few milli-watts to tens and even hundreds of watts, for more than ten years. Several remaining issues, however, are worthy of future investigations. These issues are:

(a) Investigate and demonstrate the fabrication techniques of large-grain (> 200 μm) and polycrystalline plutonia fuel kernels and the technical issues related to the application of the PyC and ZrC coatings using CVD processes.

(b) Perform fracture impact tests and detailed analysis of CPFCs to provide data to benchmark models. These data and the analysis results could also be used to guide the development and the selection of the appropriate graphite matrix material. Ultimately, mechanical, thermal, and aeroablation testing of coated particle fuel in simulated accident environments will also be needed.

(c) Perform He gas release tests from large grain and polycrystalline fuel kernels, both coated and uncoated, to confirm the current estimates and the release mechanisms of the He gas.

ACKNOWLEDGMENTS

This research was funded by Sandia National Laboratories (SNL), Kirtland Air Force Base, Albuquerque, NM, under contracts No. BE-2543, to the University of New Mexico's Institute for Space and Nuclear Power Studies, and BE-2544, to Sholtis Engineering & Safety Consulting, Tijeras, NM. The opinions expressed in this paper are solely those of the authors.

REFERENCES

Allen, P. L., Ford, L. H., and Shennan, J. V., "Nuclear Fuel Coated Particle Development in the Reactor Fuel Element Laboratories of the U. K. Atomic Energy Authority," *J. Nuclear Technology* **35**, 246–253 (1977).

Angelini, P., McHenry, R. E., Scott, J. L., Ernst, W. S., Jr., and Prados, J. W., *Helium Release from $^{238}PuO_2$ Microspheres*, Report No. ORNL-4507, Oak Ridge National Laboratory, Oak Ridge, Tennessee, March 1970.

Burnett, R. C., Bisdorff, L., and Gough, J. R. C., *Development of Coated Particle Fuel – Part I: the Fissile Fertile Particle*, Dragon Project Report No. 51, also in *Proceedings of the Dragon Project Fuel Element Symposium*, held January 28-29, 1963, Bournemouth, England, CONF-630103-8 (1963).

El-Genk, M. S., and Tournier, J.-M., *Study of a Coated-Particle Fuel Form for Advanced Radioisotope Heat Sources – Part I: A Review of Coated-Particle Fuel Performance, Fabrication Techniques and Materials Properties*, Report No. UNM–ISNPS–5–1999, Sandia National Laboratories, Contract BE–2543, Institute for Space and Nuclear Power Studies, The University of New Mexico (1999).

El-Genk, M. S., and Tournier, J.-M., "Helium Release From $^{238}PuO_2$ Fuel Particles," *in these Proceedings*, M. S. El-Genk, Ed., American Institute of Physics, Woodbury, New York, 2000.

Ford, L. H., and Shennan, J. V., "The Mechanism of Binderless Granulation and Growth of Ceramic Spheres," *Journal of Nuclear Materials* **43**, 143 (1972).

Förthmann, R., and Blass, G., "Fabrication of Uranium-Plutonium Oxide Microspheres by the Hydrolysis Process," *J. Nuclear Materials* **64**, 275–280 (1977).

Friskney, C. A., Turnbull, J. A., Johnson, F. A., Walter, A. J. and Findlay, J. R., "The Characteristics of Fission Gas Release from Monocrystalline Uranium Dioxide during Irradiation," *J. Nuclear Materials* **68**, 186–192 (1977).

Friskney, C. A., and Turnbull, J. A., "The Characteristics of Fission Gas Release from Uranium Dioxide during Irradiation," *J. Nuclear Materials* **79**, 184–198 (1979).

Haas, P. A., Kitts, F. G., and Beutler, H., "Preparation of Reactor Fuels by Sol-Gel Processes," *Chemical Engineering Progress Symposium Series* **63** (80), 16–27 (1967).

Hoover, M., *Personal Communications*, Lovelace Inhalation Toxicology Research Institute, Albuquerque, NM, January 1999.

Huschka, H., and Vygen, P., "Coated fuel Particles: Status of Fabrication Technology," *J. Nuclear Technology* **35**, 238–245 (1977).

Johnson, E. W., *Light Weight Radioisotope Heater Unit (LWRHU) – Final Safety Analysis Report (FSAR) for the Cassini Mission*, EG&G Mound Applied Technologies, Report No. MLM-3826, UC-713, 1997.

Kaae, J. L., " A mathematical Model for calculating Stresses in a Polyocarbon- and Silicon carbide-Coated Fuel Particle," *J. Nuclear Materials* **29**, 249–266 (1966).

Land, C. C., *Microstructural Damage Produced by Helium in Aged $^{238}PuO_2$ Fuels*, Report No. LA-8083, Los Alamos Scientific Laboratory, Los Alamos, New Mexico, 1980.

Lahr, H., "Fabrication and Properties of Particle Fuels," *Kerntechnik* **19**, 159 (1977).

Minato, K., Ogawa, T., Fukuda, K., Sekino, H., Kitagawa, I. and Mita, N., "Fission Product Release from ZrC-Coated Fuel Particles during Post-Irradiation Heating at 1800 and 2000 °C," *J. Nuclear Materials* **249**, 142–149 (1997).

Mueller, B. A., Rohr, D. D., and Mulford, R. N. R., *Helium Release and Microstructural Changes in $^{238}PuO_2$*, Report No. LA-5524, Los Alamos Scientific Laboratory, Los Alamos, New Mexico, 1974.

Mulford, R. N. R., and Mueller, B. A., *Measurements of Helium Release from Materials Containing $^{238}PuO_2$*, Report No. LA-5215, Los Alamos Scientific Laboratory, Los Alamos, New Mexico, 1973.

Olander, D. R., *Fundamental Aspects of Nuclear Reactor Fuel Elements*, U.S. ERDA Report No. TID-26711-P1, Technical Information Center, Energy Research and Development Administration, Oak Ridge, Tennessee, 1976, Chapters 13–15.

Peterson, D. E., and Starzynski, J. S., *Re-entry Thermal Testing of Light-Weight Radioisotope Heater Units*, Report No. LA-9226, Los Alamos National Laboratory, Los Alamos, New Mexico, 1982.

Peterson, D. E., Early, J. W., Starzynski, J. S., and Land, C. C., *Helium Release from Radioisotopic Heat Sources*, Report No. LA-10023, Los Alamos National Laboratory, Los Alamos, New Mexico, 1984.

Sholtis, J. A., Jr., Lipinski, R. J., and El-Genk, M. S., "Coated Particle Fuel for Radioisotope Power Systems (RPSs) and Radioisotope Heater Units (RHUs)," *in Proceedings of the Space Technology and Applications International Forum – 1999*, M. S. El-Genk, Ed., American Institute of Physics Conf. Proceedings No. 458, Woodbury, New York, 1999, **2**, 1378–1384.

Sholtis, J. A., Jr., *Test plan for Pu-238 Oxide Coated Particle Fuel and Fuel Forms for Space Radioisotope Heater Units (RHUs) and Space Radioisotope Power Systems (RPSs)*, SESC/SNL#BE-2544/1099, Sholtis Engineering and Safety Consulting, Tijeras, NM (1999).

Silver, G., *Personal Communications*, Los Alamos National Laboratory, Los Alamos, NM, 1999.

Tournier, J.-M. and El-Genk, M. S., *Study of a Coated-Particle Fuel Form for Advanced Radioisotope Heat Sources – Part II: Coated-Particle Fuel Design Model and Potential of Advanced Radioisotope Heat Sources*, Report No. UNM–ISNPS–5–1999, Sandia National Laboratories, Contract BE-2543, Institute for Space and Nuclear Power Studies, The University of New Mexico (1999).

Tournier, J.-M., and El-Genk, M. S., "Performance Analysis of Coated $^{238}PuO_2$ Fuel Particles Compact for Radioisotope Heater Units," *in these Proceedings*, M. S. El-Genk, Ed., American Institute of Physics, Woodbury, New York, 2000.

Turnbull, J. A., "The Effect of Grain Size on the Swelling and Gas Release Properties of UO_2 during Irradiation," *J. Nuclear Materials* **50**, 62–68 (1974).

Turnbull, J. A., Friskney, C. A., Johnson, F. A., Walter, A. J., and Findlay, J. R., "The Release of Radioactive Gases from Uranium Dioxide During Irradiation," *J. Nuclear Materials* **67**, 301–306 (1977).

Milliwatt Thermoelectric Generator for Space Applications

Daniel T. Allen, John C. Bass, Norbert B. Elsner,
Saeid Ghamaty and Charles C. Morris

Hi-Z Technology, Inc., 7606 Miramar Road, San Diego CA 92126, 858 695 6660, e-mail@hi-z.com

Abstract. A small thermoelectric generator is being developed for general use in space, and in particular for any of several proposed Mars atmospheric probes and surface landers that may be launched in the 2003 to 2006 time period. The design is based on using an existing 1 watt radioisotope heater unit as the generator heat source. That is the Light-Weight Radioisotope Heater Unit (RHU) which has already been used to provide heating alone on numerous spacecraft, including the 1997 Pathfinder/Sojourner Mars lander. Important technical issues that need to be addressed in the detailed design are the mechanical integrity of the overall power supply in consideration of the impact of landing on Mars and the subsequent performance of the thermal insulation around the heat source, which is critical to delivering the output power. The power supply is intended to meet a 20-year operational lifetime. Hi-Z is developing milliwatt modules that make use of micro fabrication techniques. For this generator modules are being fabricated that produce approximately 40 milliwatts at a T-hot of 250°C and a T-cold of 25°C. The module is composed of an 18 x 18 array of 0.38 mm (0.015") square x 22.9 mm (0.900") long N and P elements. The modules use bismuth-telluride based alloys that are fine grain metallurgy prepared materials that can endure the demanding fabrication techniques. The paper describes the design status to date, and it presents the analytical approach, the testing program plan and a manufacturing schedule that is needed to meet the launch dates being considered. Electrical performance and life test data for the modules is also presented.

BACKGROUND

The generator being developed was described in its preliminary design stage previously (Bass, 1998). The U.S. Department of Energy is now furthering this technology. While the DOE is developing this generator for general use in space and there is no specific mission as yet, NASA has identified a class of applications for this size of radioisotope powered generator (Chemielewski, 1994). These are the "micro missions", which are physically small-scale spacecrafts and/or probes and which are planned to be executed for relatively low cost. Solar photovoltaic and even primary battery power sources suit some such micro missions, but for those with the objective of long operational lifetime through hostile environments a radioisotope power supply is the only option. There are several proposed Mars atmospheric and surface probes which are comprised of multiple landers that will do such things as report weather or seismic data. These missions need only tens of milliwatt of continuous power to accomplish their objective, since energy can be stored in ultra capacitors to power burst mode communication to Mars orbiters that will relay data back to Earth science teams.

The starting point for the generator design is to utilize the existing 1 watt Light Weight Radioisotope Heater Unit (RHU) which is used on U.S. spacecraft for localized heating of components (Tate, 1992). The RHU was designed and developed almost twenty years ago. It consists of a pellet of $^{238}PuO_2$ clad in 1mm (0.039 inch) thick Pt-30%Rh nested in a three layer pyrolitic graphite insulating assembly enclosed in a carbon-carbon composite aeroshell. For example, three RHUs are in place of the Pathfinder/Sojourner unit that explored Mars in 1997, and so the RHU has already met qualification standards for launch, travel and landing on Mars. The only difference in application of the RHU to a generator is that the potential accidental Earth re-entry conditions are complicated by the presence of the generator structure around the RHU capsule and the higher initial RHU temperature that results. The generator structure is essentially all aluminum, however, and it should disintegrate early in the sequence, and this is not expected to significantly change the re-entry scenario. In addition, analyses of the fuel and clad temperature on re-entry have shown that initial temperatures have very little effect on their peak values because of the overpowering re-entry heat pulse (Shock, 1981). However, a more recent study indicates margin on clad integrity may be inadequate (Shock, 1995). These differences need to be addressed with further analyses and possibly with experimental verification. In any event, an extensive launch approval process is required for each application of the RHU.

CP504, *Space Technology and Applications International Forum–2000*, edited by M. S. El-Genk
© 2000 American Institute of Physics 1-56396-919-X/00/$17.00

THERMOELECTRIC MODULE

The thermoelectric converter module is an array of 18 x 18 elements each 0.38 mm (0.015 inch) square and 22.9 mm (0.90 inch) long. Figure 1 shows the relative size of one of the modules. The elements alternate P- and N-type $(Bi, Sb)_2 (Se, Te)_3$ oriented semiconductor material. Separating the element is 25 μm (0.001 inch) Kapton® insulation. Even though the bismuth-telluride material has a fine-grained metallurgy, individual thermoelectric elements of such dimensions would be extremely fragile to handle. In fact, in manufacture the elements are never unsupported.

The fabrication technique is shown in Figure 2. Starting point for fabrication of a module are vacuum hot pressed pucks of the N and P alloy. The puck is sliced with a precision saw into 0.38 mm (0.015 inch) plates. The plates are alternated into

FIGURE 1. 40 mW Thermoelectric Module.

stacks with self-adhesive Kapton® insulation film. The stack is bonded under pressure, and then the assembly can be cross-cut again into slices. These slices are alternated again with the insulation and bonded into the final matrix. Output of the module is 4 volts.

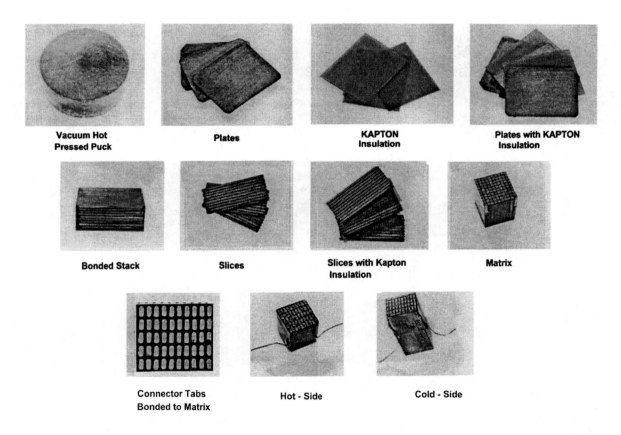

Vacuum Hot Pressed Puck   Plates   KAPTON Insulation   Plates with KAPTON Insulation

Bonded Stack   Slices   Slices with Kapton Insulation   Matrix

Connector Tabs Bonded to Matrix   Hot - Side   Cold - Side

FIGURE 2. Assembly for Fabrication of 40mW Modules.

Performance of the first module built and tested is shown in Figure 3 for reference thermal conditions on each end. The first unit was held on test at Hi-Z for 2750 hours, and Figure 4 is the plot of optimum output power

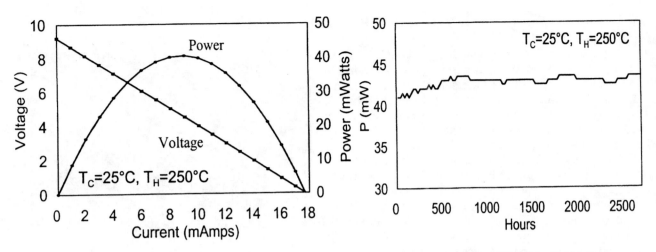

FIGURE 3. 40mW Module I vs. V Curve.

FIGURE 4. 40 mW Module #1 Life Test.

GENERATOR DESIGN

One watt of heat is not a large amount of power to begin with, and so a design challenge is to thermally insulate the RHU to minimize parasitic heat losses. Figure 5 illustrates the generator in crossection and graphically shows the heat flow "budget". Hot and cold side design basis temperatures for the thermoelectric module were selected 250°C and 25°C. At these temperatures 0.82 watts will flow through the module if heat loss down the capsule holder tie wires is 0.01 watt each and heat flow through the remaining capsule insulation is limited to 0.14 watts. The insulation will be multiple alternate layers of 13μm (0.0005 inch) thick aluminized Kapton® and 79μm (0.0031 inch) Cryotherm® glass fiber paper. The volume between the RHU capsule holder and the generator pressure shell will be evacuated at assembly. Gas generated by the isotope fuel decay will mostly be retained in the fuel body. However, the RHU is vented to the capsule container volume which is sealed, and so gas that is released from the RHU accumulates there and does not reach the insulation. This type of multifoil vacuum insulation has been utilized in space radioisotope generators in

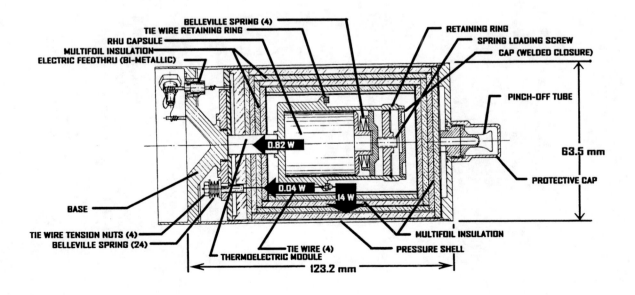

FIGURE 5. Milliwatt Space Generator

the past. However, these specific materials need experimental qualification, which is an objective of the ongoing generator development effort.

At the beginning-of-life with 1 watt of heat from the RHU, the generator is expected to provide 40 milliwatt of electric output at 5 volts. The generator is proposed for 15 to 20-year mission lifetimes. This is consistent with the known decay of the isotope heat source of approximately 1% per year. Overall degradation due to that and extrapolated thermoelectric performance decline will be about 25% over 20 years, and so 30 milliwatt is the reference end-of-life output. Design of the generator is similar to a series of 75 Milliwatt units built by DOE for the US Navy and deployed in the 1970's and 80's. Twenty-seven of these generators, which were designed for 15-year lifetimes, were made. The operating hour history of sixteen of these was documented, and this is summarized in Figure 6. Figure 7 shows representative life test data for these generators. The generators were operating normally past the times represented in Figure 6, but they were removed from service, defueled and scrapped before final performance data points could be obtained. This operating record of over 100 years of generator operation provides verification of the expected lifetime and performance of this generator.

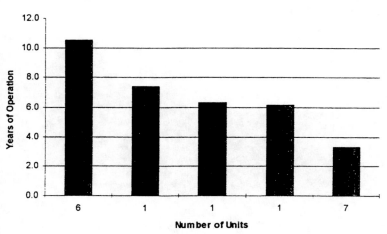

FIGURE 6. Heritage of Operating History.

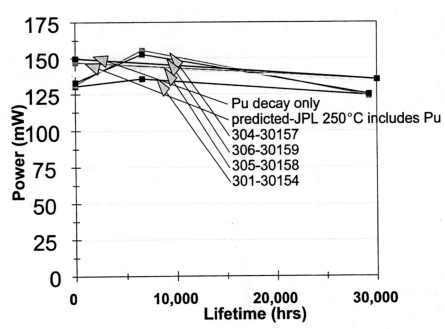

FIGURE 7. Representative Life Test Data, 75 mW Generator.

Because the potential applications of the generators include the Mars surface landers, another design issue of significance is the mechanical shock load on the unit in the landing. The typical trajectory to Mars involves an entry velocity there at about 5 km/s. Although Mars does have an atmosphere to dissipate this kinetic energy and the landing scenario includes further parachute deceleration, the lander still could have a surface impact velocity at 30 to 40 m/s (65 to 90 miles per hour). The ultimate shock load depends on design of the probe, especially the design of a "crushable impact attenuation feature". It is estimated that the deceleration along the axis parallel to the possible landing trajectories will be less than 2000 g's and probably less than 1000 g's. A representative mock-up of the generator was in early 1998 subjected to a sequence of high-g shock tests at NASA/Ames Research Center, and the assembly survived 3600 to 3800 g's along the longitudinal axis of the generator but failed at this load when inclined at 45°. Further mechanical testing is an early objective of the current development program.

The development timetable up to September 2000 calls for fabrication and testing the generator components critical to the design, with attention to the performance, stability and longevity of the thermoelectric module, efficiency of the thermal insulation and endurance of all of the generator components under moderate g loads. First articles of all of the generator components have already been fabricated. Figure 8 is a collage of generator component photos. These have been assembled into test articles for mechanical shock load testing. Two complete units are to be assembled but without the RHU. One will be mechanically shock tested as a unit proof test. The other unit will be tested for thermal and electrical performance and then placed on life test.

FIGURE 8. Model of Generator.

Non-nuclear testing presents the problem of how to assuredly apply 1 watt to a surrogate RHU. In previous radioisotope generator development an electric heater has been used, but for this generator the power desired is on the same order as the uncertainty in losses along heater electrical leads. Evaluation of generator efficiency is important, and so for this reason, a solid-state adjustable continuous wave IR laser will be used to beam the 1 watt into the evacuated generator through a window and small holes in the insulation, as is shown schematically Figure 9.

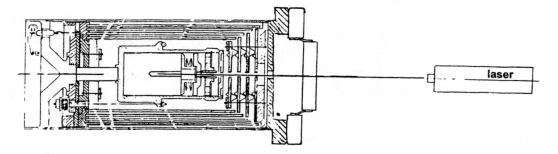

FIGURE 9. Use of Laser for Non-RHU Heating.

CONCLUSIONS

At the end of the present work effort a year from now, at least one of the two complete units will be made available for RHU loading and actual radioisotope-fueled proof testing.

ACKNOWLEDGMENT

The work reported herein was supported by the U. S. Department of Energy, Office of Space and Defense Power Systems, under contract no. DE-AC03-98SF21556.

REFERENCES

Bass, John C., & Norbert B. Elsner "Small Universal Thermopile Power Source For Space and Terrestrial Applications", *Proceedings, Fifteenth Symposium on Space Nuclear Power and Propulsion*, Albuquerque, 25-29 Jan 1998, AIP Conference Proceedings 420.

Chemielewski, Arthur B., and Richard Ewell "The Powerstick", AIAA-94-3816-CP, 29th IECEC, 1994, Monterey, California.

Shock, A. "Light-Weight Radioisotope Heater Unit", *Proceedings, Sixteenth IECEC*, 1981, ASME H00179-A, pp.343-354.

Shock, A., and Chuen T. Or "Parametric Design Study of 'Power Stick' and Its Derivatives", *Proceedings, Twelfth Symposium on Space Nuclear Power and Propulsion*, edited by M. S. El-Genk, American Institute of Physics, New York,1995, AIP Conference Proceedings 324, pp.461-470.

Tate, R.E. "The Light Weight Radioisotope Heater Unit (LWRHU): A Technical Description of the Reference Design", LA-9078-MS, January 1992, Los Alamos National Laboratory (avail. "http://lib-www.lanl.gov").

Advanced Power for Planetary Missions

Robert D. Cockfield and E. Wayne Tobery

Lockheed Martin Astronautics, 230 Mall boulevard, King of Prussia, Pennsylvania 19403
(610-354-3190, 610-354-6396)
robert.d.cockfield@lmco.com, edward.w.tobery@lmco.com

Abstract. Radioisotope power systems are an enabling and essential technology for missions to the outer planets. Previous planetary missions such as Voyagers I and II, Galileo, Ulysses, and Cassini have relied on Radioisotope Thermoelectric Generators (RTGs). These RTGs, while demonstrating outstanding reliability and extremely long life, have relatively low power conversion efficiency. Advanced power systems now being developed offer the prospect of significant improvement in efficiency with corresponding reduction in the plutonium fuel required. The Advanced Radioisotope Power System (ARPS) being developed by Lockheed Martin under contract from DOE, utilizes cells that rely on Alkali Metal Thermal to Electric Conversion (AMTEC) as a means of energy conversion. AMTEC cells currently under development are fabricated from refractory alloys, capable of matching the elevated temperature of the General Purpose Heat Source previously used in RTGs. Preliminary design of the power system is based on the requirement to provide two generators for each mission, with each generator capable of providing a minimum of 105 watts of electrical power, six years after launch. The selected generator configuration includes a stack of four GPHS modules, with 16 AMTEC cells arranged on four sides surrounding the stack. This paper will describe the preliminary design of the generator, and will provide estimates of the mass, performance, and thermal characteristics.

GENERATOR DESCRIPTION

Figure 1 shows the general arrangement of the ARPS generator. The heat source is composed of a stack of four General Purpose Heat Source (GPHS) modules, the same radioisotope heat source that was used in RTGs for the Galileo, Ulysses, and Cassini missions. Each module consists of a graphite aeroshell containing two Graphite Impact Shells (GIS). Within each GIS are two fuel capsules, composed of plutonia clad with iridium. The stack of fuel modules is supported by a series of components that carry loads to the end domes. This heat source support system incorporates a stack of spring washers that provide an axial preload on the modules in order to withstand launch dynamic loads. Arranged radially around the stack of fuel modules are sixteen AMTEC cells, four cells on each of the four sides. AMTEC cells convert the thermal energy of the heat source to electricity, using a principle that relies on the transfer of sodium ions through beta phase alumina. Cells are supported by attachment to the housing through a bracket that also serves as the thermal path for waste heat to the housing. This bracket incorporates an electrical insulator that provides electrical isolation of the cell from the housing and the electrical circuit. Positive and negative electrodes on the cells are joined with copper straps to provide a 28-volt output at the electrical connector interface with the spacecraft. Straps are connected to provide redundancy in a series/parallel circuit. Multifoil thermal insulation surrounding the cells consists of sixty layers of molybdenum foil with quartz cloth spacers, and in local areas, thicker alumina felt spacers. Other features of the generator design include a gas management valve to permit the generator to be filled with inert gas, and a pressure relief device to vent the gas for vacuum operation after launch. Bathtub fittings on the generator housing provide a means of mechanical attachment to the spacecraft.

AMTEC Cells

AMTEC cells for the ARPS Program are currently being developed by AMPS (Advance Modular Systems, Inc.). The status of development has been extensively reported in a number of papers (Sievers, 1997; Hendricks, 1998, 1999).

CP504, *Space Technology and Applications International Forum–2000*, edited by M. S. El-Genk
© 2000 American Institute of Physics 1-56396-919-X/00/$17.00

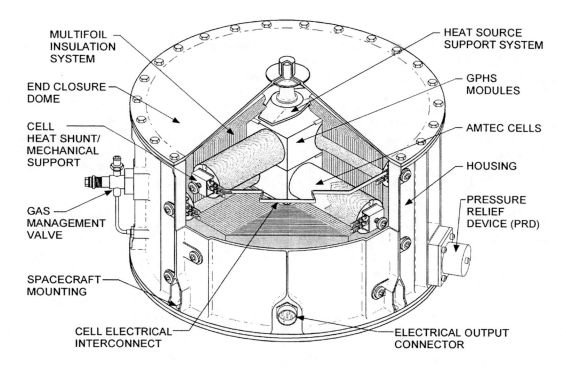

FIGURE 1. ARPS Generator Concept (Not to Scale).

Structural Analyses

Structural analyses of the generator were performed using an ABAQUS finite element model, to determine both detailed stresses and dynamic modes and frequencies. The quasi-static loading conditions used in the analysis were based on mass-acceleration curves provide by JPL and intended to envelope the dynamic response to launch environments for several candidate expendable launch vehicles. A 28-g limit load was applied independently in axial and lateral directions, combined with a heat source preload of 600 lb. and a 5-psi differential pressure due to the inert cover gas in the generator. Factors of Safety of 1.25 on yield and 1.40 on ultimate were used, with the exception that a Factor of Safety of 1.00 was permitted where a dedicated test article was planned to verify structural integrity. Positive Margins of Safety were calculated for all loading conditions and all components.

Thermal/Performance System Model

The primary functions of the thermal system are to maintain component temperatures within allowable limits, maximize thermal system efficiency to the extent possible, and to transfer some of the converter waste heat into the spacecraft propulsion module cavity. Specific objectives are to maintain maximum shell temperature less than 300°C during all mission phases and maintain the cell condensers as close as practical to 370°C to optimize cell performance.

The systems model consists of the thermal model, which couples all of the components, including the cell model, and an electrical circuit model, which couples the cells electrically. The systems model makes use of the Variables Routine within SINDA to simultaneously calculate temperatures, incorporate the cell model results, and solve the electrical equations. Each of these aspects of the systems model will now be discussed.

The thermal model consists of a geometric math model using the TRASYS code and a thermal math model using the SINDA code. The TRASYS model consists of an external and internal models. The external model calculates radiation interchange factors between the converter nodes to each other and to the external environment. It also calculates solar, albedo, and earth radiation absorbed fluxes as determined by spacecraft orientation and proximity to the planets. The internal TRASYS model is used to calculate radiation interchange factors between the hot faces of the cells and the GPHS modules, including reflections from the insulation.

The AMPS cell model is incorporated into the systems model through a series of curves which yield heat into the cell and electrical power output as a function of cell hot face temperature, condenser temperature and output voltage. To do this the cell model is run over a range of temperatures covering the expected operating range. These results are tabulated and subroutines are used to perform three-dimensional interpolation within the systems code. The cell is represented into the systems model with nodes representing the cell hot face and condenser temperature. Convergence is obtained when the cell data obtained from the curves matches the temperatures, heat flow and voltages of the system.

The electrical circuit model is based on the design of the cell interconnect wiring which has two parallel, crossed strapped, strings of eight cells, with each string constrained to produce 28 volts. Kirchoff's equations are solved prior to each thermal update and iteration continues until convergence is achieved between cell and system heat flows, temperatures, and voltages.

The systems model can determine the effect of a cell failure on the performance of the other cells providing; Thermal and electrical characteristics of the failed cell are known or can be estimated, and, the tabulated cell data is expanded to include the operating range of the working cells.

The purpose of the multifoil insulation design is to minimize heat losses between the modules and generator walls while at the same time creating an adiabatic cell wall. The concept to achieve an adiabatic cell wall is to space the insulation so that the foil temperatures match the cell wall temperatures along the length. At the hot end the first foil layer is close to the module temperature and significantly above the cell hot face temperature. The solution is to space the foils with minimum separation using quartz cloth only until a foil temperature matches the cell temperature at the base tube support plate. This concept is illustrated in Figure 2. Along the cell wall from hot end to cold end the foil layers are spaced using alumina paper and quartz cloth to match foil temperature to the cell wall temperature. At the cold end again the foil layers are closely spaced until a layer matches the cell condenser temperature. The final design will be consistent with effects of cell hot face emissivity on the module, foil, and cell hot face temperatures and the cell wall distribution.

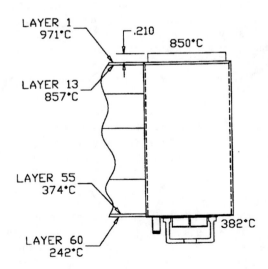

FIGURE 2. Insulation Design.

The systems model was used to determine generator temperatures, heat balance, and performance at Beginning of Mission (BOM) conditions. The assumptions used at BOM are as follows:

1) Heat dissipation from four GPHS modules = 973.2 watts
2) Shell and dome emissivity = 0.91
3) Average cell hot face emissivity = 0.67
4) Adiabatic cell walls
5) The ARPS generator is shaded by the high gain antenna of the spacecraft, resulting in an average sink temperature of –40°C.
6) The generator inboard dome radiates to the spacecraft propulsion module cavity having an average temperature of 30°C.

The results of the analyses at BOM are shown in Figures 3 and 4. Generator temperatures are shown in Figure 3. The shell temperatures on the flats where the cells attach range from 217 to 241°C. The mounting bracket which attaches the cells to the shell is sized to produce the desired condenser average condenser temperature of 370°C. This was achieved by producing a condenser temperature range of 361 to 382°C. Cell hot face temperatures only vary between 845 and 850°C as they are influence by the modules which only vary by 3°C. The generator heat balance is summarized in Figure 4. The variation in heat flow to the cells is small with a total of 916.9 watts into the cells out of total of 973.2 watts. This results in a thermal efficiency of 94.2%. The heat losses from the modules to the generator housing are summarized in the figure. Based on the inboard dome temperatures shown in the previous figure, the heat transferred into the spacecraft propulsion cavity was calculated to be 118 watts.

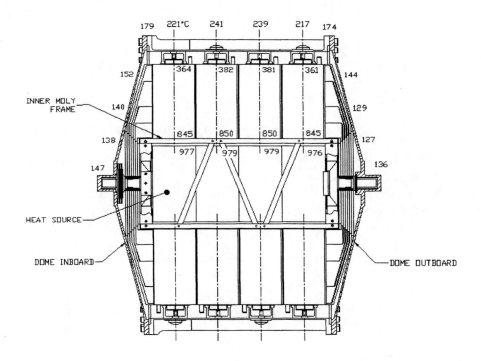

FIGURE 3. Generator Temperatures at Beginning of Mission (BOM).

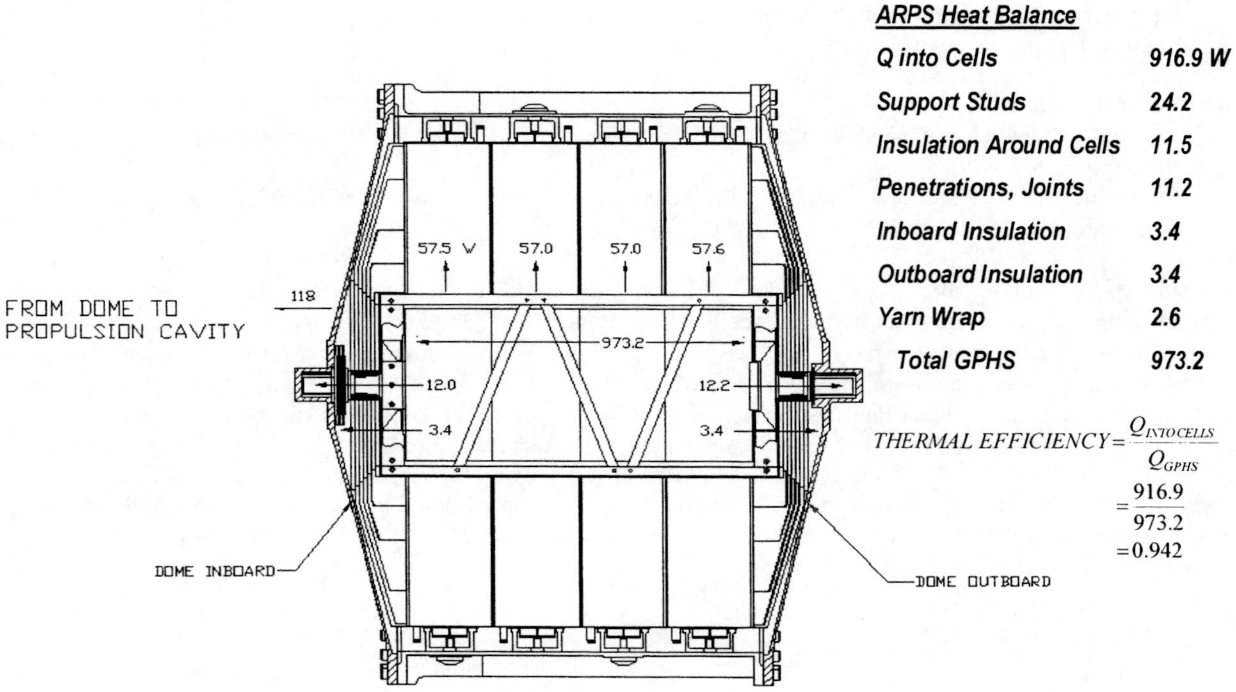
ARPS Heat Balance
Q into Cells 916.9 W
Support Studs 24.2
Insulation Around Cells 11.5
Penetrations, Joints 11.2
Inboard Insulation 3.4
Outboard Insulation 3.4
Yarn Wrap 2.6
Total GPHS 973.2
THERMAL EFFICIENCY = Q INTO CELLS / Q GPHS
= 916.9 / 973.2
= 0.942
FROM DOME TO PROPULSION CAVITY
118
57.5 W
57.0
57.0
57.6
973.2
12.0
12.2
3.4
3.4
DOME INBOARD
DOME OUTBOARD

FIGURE 4. Generator Heat Balance at Beginning of Mission (BOM).

Performance Summary

Table 1 lists characteristics that summarize the electrical and thermal performance of the generator.

TABLE 1. ARPS Generator Characteristics.

Number of GPHS Modules	4
Thermal Power Input	973 watts (Beginning of Mission)
Number of AMTEC Cells	16
Number of Cell in Series	8
Electrical Power Output	
Beginning of Mission	141 watts
Six Years After Launch	112 watts
Design Life	16 years
Voltage Output	28 VDC
Temperatures (Peak)	
Heat Source Surface	979°C
Cell Hot Side	850°C
Cell Condenser	382°C
Housing	241°C (Peak)

Mass Summary

An estimate of the generator mass and its major components is summarized in Table 2, based on the design as defined at the Preliminary Design Review in June 1999. The 20% contingency included is in recognition of the lack of maturity of the cell design, as compared with the known mass of the heat source and the estimates based on mature design components that have been used on RTGs.

TABLE 2. Mass Estimate for ARPS Generator.

Heat Source Modules	5.73
AMTEC Cells	6.66
Housing	2.14
End Closures	1.46
Multifoil Insulation	2.66
Inner Molybdenum Frame	0.32
Heat Source Supports	0.34
Pressure Relief Device	0.43
Gas Management Valve	0.16
Electrical Connector	0.13
Cell Interconnects	0.32
C-Seals, Fasteners	0.30
Total	20.67
Contingency (20%)	4.13
"Not to Exceed" Mass	24 .70 Kg

ACKNOWLEDGMENTS

The authors wish to acknowledge the following Lockheed Martin personnel who contributed to the work reported in this paper: R. A. Kull, F. A. Greenwood, C. E. Kelly, and D. Vacek.

REFERENCES

Hendricks, T. J., Huang, C., and Huang, L., "AMTEC Cell Optimization for Advance Radioisotope Power System (ARPS) Design," *Proceedings of 34th Intersociety Energy Conversion Engineering Conference,* Vancouver, British Columbia, Canada, Paper 1999-0102655.

Hendricks, T. J., and Huang, C., "System Design Impacts on Optimization of the Advanced Radioisotope Power System AMTEC Cell," *Proceedings of the 33rd Intersociety Energy conversion Engineering Conference,* edited by Samin Anghaie, La Grange Park, IL, American Nuclear Society, 1998, Paper #98-407.

Sievers, R. K., Pantolin, J. E., Svedberg, R. C., Butkiewizc, D. A., Borkowski, C. A., Huang, C., Hendricks, T. J., and Hunt, T. K., "AMTEC Cell Design and Development," *Proceedings of the 32nd Intersociety Energy Conversion Engineering Conference,* New York, American Institute of Chemical Engineers, 1997, pp. 1125-1129.

A Filament Wound Carbon-Carbon Composite for Impact Shell Application

Ralph Zee[1] and Glenn Romanoski[2]

[1] Materials Research and Education Center, Auburn University, Auburn, AL 36849
[2] Carbon and Insulation Materials Technology Group, Oak Ridge National Laboratory, Oak Ridge, TN 37831

[1] (334) 844-3320, rzee@eng.auburn.edu
[2] (423) 574-4838, romanoskigr@ornl.gov

Abstract. The performance and safety of the radioisotope power source depend in part on the thermal and impact properties of the materials used in the general purpose heat source (GPHS) through the use of an impact shell, thermal insulation and an aeroshell. Within the aeroshell are two graphite impact shells, made of fine-weave pierced-fabric (FWPF) that encapsulate four iridium alloy clad isotopic fuel pellets and provides impact protection for the clad. Impact studies conducted at Los Alamos National Laboratory showed that impact shells typically fractured parallel to their longitudinal axis. The objective of this effort is to develop new impact shell concepts with improved performance. An effort to develop alternative carbon-carbon composites for the graphite impact shell was conducted. Eight braided architectures were examined in this study. The effects of the number of graphitization cycles on both the density and circumferential strength of these braided structures were determined. Results show that a filament wound carbon-carbon composite possesses the desired density and circumferential strength important to GPHS.

INTRODUCTION

The performance and safety of radioisotope power source depends in part on the thermal and impact properties of the materials used in the general purpose heat source (GPHS) through the use of an impact shell, thermal insulation and an aeroshell. Each heat source is comprised of 18 modules providing a nominal 300W electric power. Each module is made of fine-weave pierced-fabric (FWPF) (fabricated by Textron Specialty Materials, Lowell, MA) carbon-carbon composite aeroshell to protect it from the thermal requirements of reentry. Within the aeroshell are two graphite impact shells, also made of FWPF, that encapsulate four iridium alloy clad fuel pellets and provides further impact protection for the clads. The basic design of the aeroshell is shown in Figure 1. Experimental evaluation of the ability of the modules to contain the fuel while sustaining impact was undertaken by Los Alamos National Laboratory. The safety test program included a series of tests designed to simulate reentry of the general purpose heat source modules into earth atmosphere (Schonfeld, 1984(a), 1984(b), 1984(c)) (George, 1984(a), 1984(b), 1984(c)) Results have been reported by Cull (Cull, 1989). Postmortem analysis showed that impact shells typically fractured parallel to their longitudinal axis. It was recognized that the composite fibers were not suitably oriented to carry impact loads at these locations. Post impact analysis revealed that the pellet and cladding deformed during impact. Although the present GPHS satisfies the operational and safety requirements, any improvement in enhanced performance and system simplicity would be highly desirable.

An effort to develop alternative carbon-carbon composites for the graphite impact shell of the radioisotope thermoelectric generator (RTG) was recently completed. This paper will be devoted to the results on consolidation and circumferential strength of eight configurations considered as candidate materials for impact shell. Results showed that the best material is a filament wound carbon-carbon composite, which possessed the best combination

CP504, *Space Technology and Applications International Forum–2000*, edited by M. S. El-Genk
© 2000 American Institute of Physics 1-56396-919-X/00/$17.00

of thermal and mechanical properties important to GPHS. With the correct winding architecture, this filament wound material could have a circumferential strength nearly six times that of fine-weave pierced-fabric (FWPF) presently used for the impact shell.

EXPERIMENTAL PROCEDURES

The aeroshell and impact shell are currently machined from a block of Textron 3D Fineweave™ carbon-carbon composite (FWPF). The impact shell is machined with its longitudinal axis parallel to the z-axis of the composite. Intermediate densities around 300 to 400 kg/m³ above the preform density gave the greatest impact protection. In this study, eight candidate materials with fiber volume contents ranging from 33% to 60% and densities ranging form 751 to 1324 kg/m³ were fabricated by Fiber Materials Inc. (FMI). The basic parameters of these materials are shown in Table 1. To achieve the winding configuration with circumferential strength, the fiber density in the circumferential direction is a major parameter examined in this study. This parameter is given in terms of the number of plates per meter which corresponds to the number of circumferential carbon fiber bundles per axial meter. Braid H is a hybrid of Braids A and F. The outside and insider layers of this architecture were comprised of Braid F (1180 plates/m) whereas the middle layers were made of Braid A (200 plates/m). This yields an effective fiber volume of 49% and an effective circumferential fiber density of 853 plates/m. Amoco T-300 carbon fibers were used in these architectures. The currently specified fineweave pierced-fabric (FWPF) material was also evaluated to serve as a comparative baseline.

The eight preforms were processed at FMI (Fiber Materials Inc) each with up to five cycles of infiltration and graphitization. The densities of the components were measured after each cycles (except the first cycle). The circumferential strengths of the composites were determined after successive cycles using a split mandrel technique.

TABLE 1. C-C Composites for the Impact Shell.

Sample Designation	Fiber Volume	Circumferential Fiber Density
Braid A	33%	200 plates/m
Braid B	37%	390 plates/m
Braid C	42%	590 plates/m
Braid D	47%	790 plates/m
Braid E	53%	980 plates/m
Braid F	57%	1,180 plates/m
Braid G	60%	1,770 plates/m
Braid H	49%	853 plates/m
FWPF*	49%	N/A

*FWPF is included for comparison purposes.

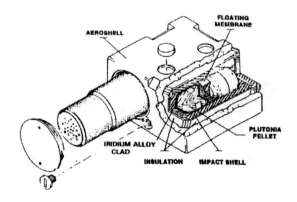

FIGURE 1. Schematic drawing showing an aeroshell and impact shell.

RESULTS AND DISCUSSION

The densities of the candidate composites are given in Table 2 and Figure 2. The density of all the structures increases with the number of graphitization cycles. Furthermore, the braid structures with higher fiber volumes and higher densities of circumferential fibers also possess higher densities. This indicates that the tightly wound structure does not prohibit the infiltration of the resin and the subsequent graphitization process. This in turn suggests that it is potentially possible to control the density of all the architectures examined using different numbers of graphitization cycles. The cross sectional structures of the eight carbon-carbon composites fabricated in this investigation are shown in Figure 3. Braids A to H contain progressively increasing fractions of fiber content. The amount of fiber in the circumferential direction also progressively increases from Braids A to G as expected. Braid H is a combination of Braids A and F. The microstructure of this hybrid configuration shows the changing architecture through the thickness.

TABLE 2. Density of Braided Architecture Fabricated by FMI.

Architecture	Density (kg/m^3)				
	Preform	Preform + 2 cycles	Preform + 3 cycles	Preform + 4 cycles	Preform + 5 cycles
FMI Braid #A	581	751	748	791	804
FMI Braid #B	660	794	893	920	1091
FMI Braid #C	748	859	1007	1012	1048
FMI Braid #D	836	971	1057	1068	1159
FMI Braid #E	924	1024	1090	1196	1245
FMI Braid #F	1003	1039	1086	1150	1192
FMI Braid #G	1056	1194	1298	1299	1324
FMI Braid #H		1066	1089	1167	1195

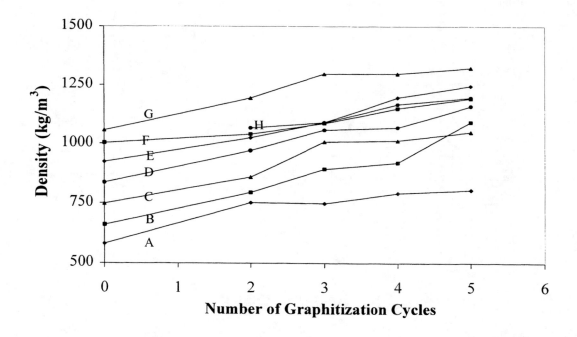

FIGURE 2. Change in composite density with number of carbonization cycles.

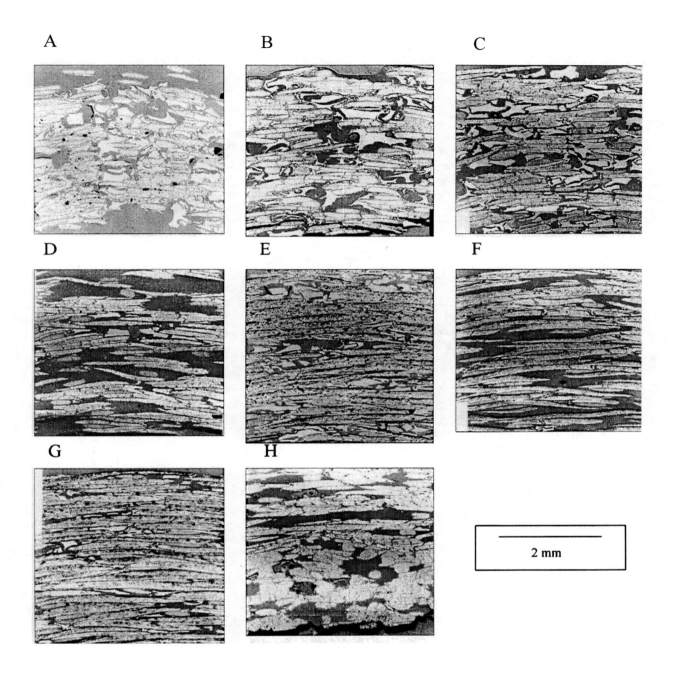

FIGURE 3. Cross sectional structure of the eight carbon-carbon composites fabricated in this investigation. Braids A to G contain progressively increasing fraction of fiber content. The amount of fiber in the circumferential direction also increases from Braid A to G. Braid H is a hybrid between Braids A and F.

Split mandrel tensile tests were performed on the eight candidate materials. The test results are summarized in Table 3 and Figure 4. Three of the eight braid structures examined (Braids E, F and G) possess circumferential strengths that exceed the weak orientation (45°) of the fine-weave pierced-fabric standard. These braids all had high volume fraction of carbon fibers and these fiber were oriented preferentially in the circumferential direction. The architecture G (with the highest oriented fiber) when processed after five cycles exhibited a strength which is double

TABLE 3. Circumferential Strength of Impact Shell Candidates.

Architecture	Circumferential Strength (MPa)			
	Preform +2 cycles	Preform +3 cycles	Preform +4 cycles	Preform +5 cycles
FMI Braid #A	0.7	1.5	0.2	0.8
FMI Braid #B	5.4	9.7	15.9	13.7
FMI Braid #C	12.7	15.3	14.4	22.7
FMI Braid #D	15.1	20.3	26.5	18.3
FMI Braid #E	18.3	24.7	25.4	29.1
FMI Braid #F	19.2	29.1	27.4	30.4
FMI Braid #G	37.6	43.4	--	56.4
FMI Braid #H	13.4	12.9	13.9	12.7
Textron FWPF	25.2 (@ 45°)[*]		29.2 (@ 0°)[*]	
Filament Wound with 80% fiber in circumferential direction	150[#]			

[*] Represents direction of loading with respect to the X and Y fiber orientations.

[#] Expected value based on the rule-of-mixture principle.

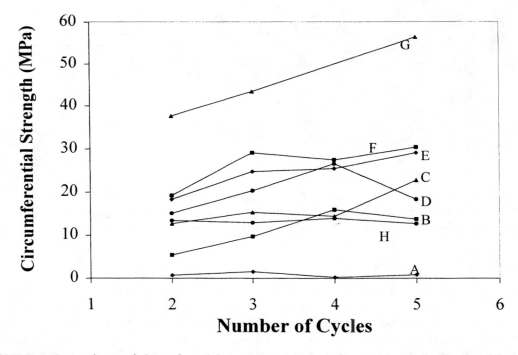

FIGURE 4. Dependence of circumferential strength on number of processing cycles for the eight specimens.

that observed in the FWPF standard. Results from this study clearly indicate the importance of fiber density, in particular the effect of fiber orientation in the circumferential direction. A calculation was made to determine the expected circumferential strength of a filament wound architecture with 80% purely circumferential fibers and 20% at ±45° for axial stiffness. This filament wound material would have circumferential strength approaching 150 MPa, nearly six times that of fine-weave pierced-fabric. Such an architecture was not fabricated because ORNL was trying to achieve high impact energy absorption in addition to high circumferential strength. Consequently, ±45° lay-up was used to achieve an open architecture with intermediate density.

CONCLUSION

Eight braided carbon-carbon composite architectures with controlled content of carbon fiber and orientation were examined in this study. These structures were given up to five cycles of graphitization. The dependence of density and circumferential structure on processing (number of cycles) and architecture was determined. Results show that the density of the composite increases with number of densification cycles. Furthermore, circumferential fibers are beneficial to increasing the circumferential strength of the composites.

ACKNOWLEDGMENTS

This work was sponsored by the Department of energy Office of Space and Defense Power Systems, under contracts DE-AC03-98SF21558 and DE-AC05-84OR21400.

REFERENCES

Cull, T.A., "General-Purpose Heat Source Development: Extended Series Test Program, Large Fragment Tests," LA-11597-MS, UC-713, Los Alamos National Laboratory, Los Alamos, N. M., August 1989.

George, T.G., and Schofeld, F.W., "General-Purpose Heat Source Development: Safety Test Program, Postimpact Evaluation, Design Iteration Test 4," LA-10217-SR, Los Alamos National Laboratory, Los Alamos, N. M., 1984(a).

George, T.G., and Schofeld, F.W., "General-Purpose Heat Source Development: Safety Test Program, Postimpact Evaluation, Design Iteration Test 5," LA-10232-SR, Los Alamos National Laboratory, Los Alamos, N. M., 1984(b).

George, T.G. and Schofeld, F.W., "General-Purpose Heat Source Development: Safety Test Program, Postimpact Evaluation, Design Iteration Test 6," Los Alamos National Laboratory, Los Alamos, N. M., 1984(c).

Schonfeld, F.W., "General-Purpose Heat Source Development: Safety Test Program, Postimpact Evaluation, Design Iteration Test 1," LA-9680-SR, Los Alamos National Laboratory, Los Alamos, N. M., 1984(a).

Schonfeld, F.W., and George, T.G, "General-Purpose Heat Source Development: Safety Test Program, Postimpact Evaluation, Design Iteration Test 2," LA-10012-SR, Los Alamos National Laboratory, Los Alamos, N. M., 1984(b).

Schonfeld, F.W. and George, T.G., "General-Purpose Heat Source Development: Safety Test Program, Postimpact Evaluation, Design Iteration Test 3," LA-10034-SR, Los Alamos National Laboratory, Los Alamos, N. M., 1984(c).

Thermoelectric Properties of $Bi_{2(1-x)}In_{2x}Te_3$ Ternary Samples

Jiro Nagao and Marhoun Ferhat

Materials Division, Hokkaido National Industrial Research Institute, AIST, MITI, Sapporo 062-8517, Japan

[+81-11-857-8948; nagao@hniri.go.jp]

Abstract. Low temperature thermoelectric properties of $Bi_{2(1-x)}In_{2x}Te_3$ ternary samples have been measured in the 50-300 K temperature range. At room temperature, the thermal conductivity decreases with increasing In_2Te_3 content, reaches a minimum at x=0.2 and then slightly increases with increasing In_2Te_3 concentration. For x=0.1 and x=0.2 samples, the thermal conductivity varies approximately as $1/T$, which may indicate that the defect scattering with phonon is predominant to thermal conduction. Seebeck coefficient and electrical resistivity have been measured and the thermoelectric figure merit calculated. The maximum value obtained is $ZT=0.5$ around 300 K, a value comparable to those of state-of-the-art thermoelectric semiconductors. From x-ray diffraction measurements, it is found that these samples are composites with In_2Te_3 particles in Bi_2Te_3. In this case, the phonons would be scattered by the inclusion particles, depending on their size which is critical for achieving an increase in ZT compare to Bi_2Te_3. An increase in ZT could be achieved for an optimal particle size of In_2Te_3.

INTRODUCTION

Thermoelectric devices are attractive to waste heat recovery and small refrigeration applications. However, more efficient materials (*i.e.* materials with high thermoelectric figure of merit) are required to expand to commercial users. A low thermal conductivity is one of the conditions to achieve a high thermoelectric figure of merit, $Z = \alpha^2/\rho(\kappa_c + \kappa_l)$, where α is the Seebeck coefficient, ρ is the electrical resistivity and κ_c, κ_l are carrier and lattice contributions to the thermal conductivity, respectively (Goldsmid, 1982). For a good thermoelectric semiconductor, a high Seebeck coefficient, low electrical resistivity and low thermal conductivity are naturally required. Here the electrical resistivity, Seebeck coefficient and carrier contribution to the thermal conductivity are closely related to the each other. A decrease in the electrical resistivity simultaneously leads to the decrease in the Seebeck coefficient and the increase in the carrier contribution to the thermal conductivity. This shows a difficulty for improving Z by controlling these electronic parameters in conventional semiconductors. On the other hand, a lattice contribution to thermal conductivity is only related to the lattice properties such as phonon energy and phonon dispersion spectrum (Slack, 1979). A phonon behavior can be controllable without changing the electrical properties. Therefore reducing thermal conductivity may be a way to achieve higher Z.

To reduce the thermal conductivity of thermoelectric materials, several ways have been performed and tested. One of the efficient methods may be the introduction of small particles as phonon scattering centers. For example, the reduction of thermal conductivity of $CoSb_3$ by adding PbS was reported (Anno *et al.*, 1998). These results explain that small particles become efficient phonon scattering centers. It has been known that the introduction of point

CP504, *Space Technology and Applications International Forum–2000*, edited by M. S. El-Genk
© 2000 American Institute of Physics 1-56396-919-X/00/$17.00

defects is also an effective way to reduce the thermal conductivity (Stary *et al.*, 1995). Therefore we considered that the introduction of small particles with defect structures into the thermoelectric materials can strongly reduce the thermal conductivity.

Bi_2Te_3 and its related alloys have attractive thermoelectric properties and possess relatively high thermoelectric figures of merit around room temperature. For Bi_2Te_3 r, an investigation of the lattice thermal conductivity in the $PbTe$-Bi_2Te_3 system was performed and a decrease of lattice thermal conductivity was observed (Christakudi *et al.*, 1992). On the other hand, In_2Te_3 is a semiconducting compound with a large number of vacancies in the In lattice sites (Zaslavski and Sergeeva, 1961). Therefore it is likely that the In_2Te_3 may become phonon scattering centers due to defects when Bi_2Te_3-In_2Te_3 ternary composites could be formed. Furthermore, quite recently, the enhancement of thermoelectric power factor in composite materials was reported (Bergman and Fel, 1999). They pointed out that the power factor of composites can be greater than those of both the pure components, with the greatest enhancement always achieve in a parallel slabs microstructure with definite volume fractions for the two components. From the phase diagram of the Bi_2Te_3-In_2Te_3 system, the existence of a eutectic region for small In_2Te_3 concentrations (Belotskii *et al.*, 1970) can be observed. Thus it is expected to form a composite structure in $Bi_{2(1-x)}In_{2x}Te_3$ system. In this paper we report the low temperature thermoelectric properties of $Bi_{2(1-x)}In_{2x}Te_3$ ternary composites. The structure was investigated by the XRD measurements. From the electrical resistivity, Seebeck coefficient and thermal conductivity data, thermoelectric figures of merit have been calculated.

EXPERIMENTAL DETAILS

Polycrystalline $Bi_{2(1-x)}In_{2x}Te_3$ alloys were synthesized by a reaction of Bi_2Te_3 and In_2Te_3 alloys. Bi_2Te_3 and In_2Te_3 binary alloys were prepared from melting of high purity Bi (99.999%) or In (99.9999%) and Te (99.9999%) powders. Stoichiometric mixtures were sealed in evacuated quartz ampoules (2×10^{-5} Torr) and heated at 1100 K for 4 h for Bi_2Te_3 and 1200 K for 4h for In_2Te_3, respectively. The ampoules were then slowly cooled through the solidus temperatures to room temperature. The samples were then crushed into powders. Desired amounts of Bi_2Te_3 and In_2Te_3 powders were then weighted, loaded and sealed in evacuated quartz tubes (2×10^{-5} Torr). These mixtures were heated at 1300 K for 1 h and rapidly cooled by quenching them into ice bath. X-ray diffraction (XRD) and electron probe micro-analysis (EPMA) were used to analyze the structure and composition of the $Bi_{2(1-x)}In_{2x}Te_3$ samples.

Disk-shaped samples were cut into the size of 8 mm in diameter and 0.5 mm in thickness. Electrical conductivity and Seebeck coefficient were measured in the temperature range 50-300 K by a van der Pauw and a temperature gradient methods, respectively. The Seebeck coefficient was measured with a temperature difference of 10 K. The Au film of 50 nm thickness was deposited on the samples to obtain an ohmic contact. The ohmic nature was confirmed by the linearity of current-voltage. Thermal conductivity was measured in the temperature range 80-300 K by a laser flash method using a ruby laser with 6 J/pulse power and $\lambda=694.3$ nm wavelength.

RESULTS AND DISCUSSION

Figures 1 (a) and (b) show the XRD results for two $Bi_{2(1-x)}In_{2x}Te_3$ samples with x=0.1 and x=0.2. The peaks corresponding to Bi_2Te_3 and In_2Te_3 compounds are clearly observed and the intensity of peaks for In_2Te_3 becomes relatively strong when the In_2Te_3 concentration increases. From the phase diagram, the $Bi_{2(1-x)}In_{2x}Te_3$ system presents a eutectic in small In_2Te_3 concentration regions. Therefore the $Bi_{2(1-x)}In_{2x}Te_3$ samples with x<0.2 seem to consist of the Bi_2Te_3-In_2Te_3 composite structure.

Figure 2 shows the room temperature thermal conductivity of $Bi_{2(1-x)}In_{2x}Te_3$ ternary composites. The room temperature thermal conductivity decreases, and reaches minimum at x=0.2, then slightly increases with increasing the concentration of In_2Te_3. We can see the strong reduction of thermal conductivity in small In_2Te_3 concentrations. This reduction of thermal conductivity may be due to phonon scattering with eutectic regions of In_2Te_3 in Bi_2Te_3 matrix.

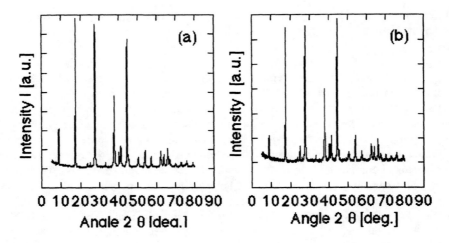

FIGURE 1. X-ray diffraction patterns for $Bi_{2(1-x)}In_{2x}Te_3$ samples with (a) x=0.1 and (b) x=0.2.

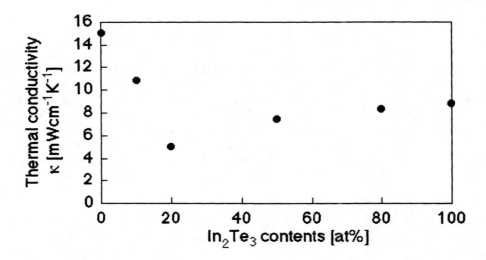

FIGURE 2. Room temperature thermal conductivity of $Bi_{2(1-x)}In_{2x}Te_3$ samples.

Figure 3 shows the temperature dependence of thermal conductivity for two $Bi_{2(1-x)}In_{2x}Te_3$. The thermal conductivity depends on the relation proportional to inverse temperature T^{-1}. This shows that defect scattering with phonons is dominant in the heat conduction process (Dey and Chaudhuri, 1975). This suggests that the eutectic region of In_2Te_3 in Bi_2Te_3 play the similar role as that of defects due to the defective structure of In_2Te_3.

Figures 4 and 5 show the temperature dependence of the electrical resistivity and Seebeck coefficient for the $Bi_{2(1-x)}In_{2x}Te_3$. In the x≤0.2 region, the electrical resistivity slightly increases with increasing In_2Te_3 concentration. Seebeck coefficient decreases with increasing In_2Te_3 concentration. We consider that the decrease in Seebeck coefficient is due to a small increase in the carrier concentration. Thus the small increase in electrical resistivity may be related to the carrier scattering with the In_2Te_3 particles.

From the measurement results of electrical resistivity, Seebeck coefficient and thermal conductivity, thermoelectric figures of merit *ZT* were calculated. Figure 6 shows the temperature dependence of the figure of merit *ZT* for the x=0.1 and x=0.2 samples. A maximum of about *ZT*=0.5 at 300 K is achieved for x =0.1 sample. For x=0.2 sample, the maximum of *ZT* is achieved at higher temperature than 300 K. We believe that the composites may lead to materials with higher *ZT*.

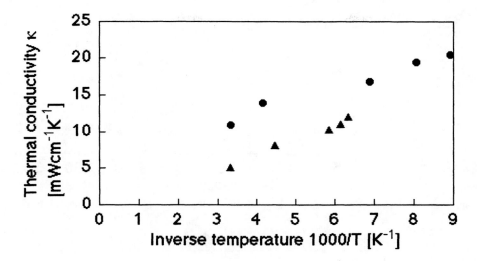

FIGURE 3. Temperature dependence of thermal conductivity for two $Bi_{2(1-x)}In_{2x}Te_3$ samples with ●: x=0.1 and ▲: x=0.2.

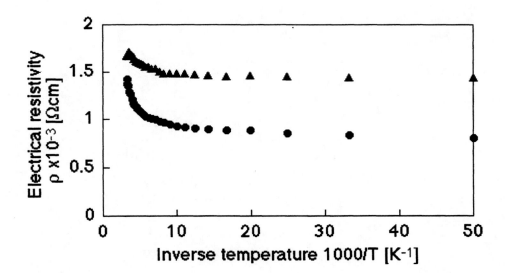

FIGURE 4. Temperature dependence of electrical resistivity for $Bi_{2(1-x)}In_{2x}Te_3$ samples with ●: x=0.1 and ▲: x=0.2.

CONCLUSION

Structure and low temperature thermoelectric properties of $Bi_{2(1-x)}In_{2x}Te_3$ ternary samples were investigated. From the XRD measurement and thermal conductivity behavior, it was demonstrated that the $Bi_{2(1-x)}In_{2x}Te_3$ samples with x≤0.2 are of composite type. From the Seebeck coefficient and electrical resistivity measurements, the thermoelectric figure of merit for x=0.2 sample would reach a *ZT*=1.0 at slightly above 300 K. This value is comparable to those of state-of-the-art thermoelectric semiconductors. This suggests that the composite materials involving defect structure may lead to materials with higher *ZT*..

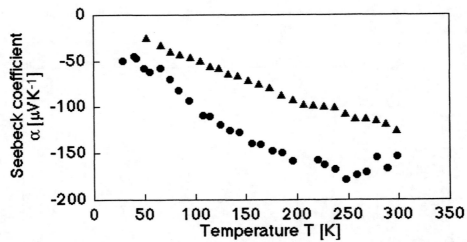

FIGURE 5. Temperature dependence of Seebeck coefficient for $Bi_{2(1-x)}In_{2x}Te_3$ samples with ●: x=0.1 and ▲: x=0.2.

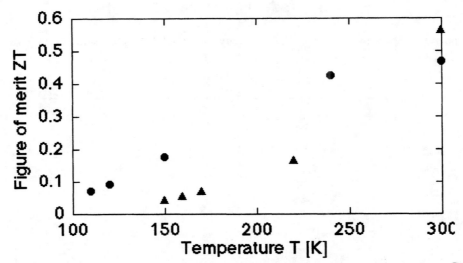

FIGURE 6. Temperature dependence of dimensionless figures of merit for $Bi_{2(1-x)}In_{2x}Te_3$ with ●: x=0.1 and ▲: x=0.2.

ACKNOWLEDGMENTS

The authors express their thanks to Dr. T. Okutani of the HNIRI and Prof. K. Mukasa and Dr. E. Hatta of Hokkaido University for their useful discussions. The authors also thank Mr. M. Sasamori and Miss. S. Nagahara for their experimental assistance. One of the authors (M.F.) is also grateful to the Science and Technology Association of Japan for a fellowship research grant at HNIRI.

REFERENCES

Anno., H., Tashiro, H., Kaneko, H., and Matsubara, K., in *Proceeding of 17ᵗʰ International Conference on Thermoelectrics-1998*, edited by Koumoto, K., IEEE catalog No. 98TH8365, IEEE, Piscataway, 1998, pp.326-329.

Belotskii, D. P., Bankina, V. F., and Babyuk, P. F., *Izv. Akad. Nauk SSSR, Neorg. Mater.* **6**, 988 (1970).

Bergman, D. J., and Fel, L. G., *J. Appl. Phys.* **85**, 8205 (1999).

Christakudi, T. A., Christakudis, G. Ch, and Borissova, L. D., *Phys. Stat. Sol. B* **171**, K67 (1992).

Dey, T. K., and Chaudhuri, K. D., *J. Low Temp. Phys.* **23**, 419 (1976).

Goldsmid, H. J., *Electronic Refrigeration*, Pion, London, 1982, pp.82.

Slack, G. A., *Solid State Phys.* **34**, 1 (1979).

Stary, Z., Navrátil, J., Novotn´y, R., and Plechácek, in *Proceeding of 14ᵗʰ International Conference on Thermoelectrics-1995*, edited by, Vedernikov, M. V., A. F. Ioffe Physical-Technical Institute, St Petersburg, Russia, 1995, pp.92-95.

Zaslavskii, A. I. and Sergeeva, V. M., *Sov. Phys. Solid State* **2**, 2556 (1961); [*Fiz. Tverd. Tela* **2**, 2872 (1960)].

Miniaturized Radioisotope Solid State Power Sources

J.-P. Fleurial, G.J. Snyder, J. Patel, J.A. Herman,
T. Caillat, B. Nesmith and E.A. Kolawa

Jet Propulsion Laboratory/California Institute of Technology
MS 277-207, 4800 Oak Grove drive, Pasadena, California 91109, USA

(818) 354-4144; jean-pierre.fleurial@jpl.nasa.gov

Abstract. Electrical power requirements for the next generation of deep space missions cover a wide range from the kilowatt to the milliwatt. Several of these missions call for the development of compact, low weight, long life, rugged power sources capable of delivering a few milliwatts up to a couple of watts while operating in harsh environments. Advanced solid state thermoelectric microdevices combined with radioisotope heat sources and energy storage devices such as capacitors are ideally suited for these applications. By making use of macroscopic film technology, microgenerators operating across relatively small temperature differences can be conceptualized for a variety of high heat flux or low heat flux heat source configurations. Moreover, by shrinking the size of the thermoelements and increasing their number to several thousands in a single structure, these devices can generate high voltages even at low power outputs that are more compatible with electronic components. Because the miniaturization of state-of-the-art thermoelectric module technology based on Bi_2Te_3 alloys is limited due to mechanical and manufacturing constraints, we are developing novel microdevices using integrated-circuit type fabrication processes, electrochemical deposition techniques and high thermal conductivity substrate materials. One power source concept is based on several thermoelectric microgenerator modules that are tightly integrated with a 1.1W Radioisotope Heater Unit. Such a system could deliver up to 50mW of electrical power in a small lightweight package of approximately 50 to 60g and 30cm^3. An even higher degree of miniaturization and high specific power values (mW/mm^3) can be obtained when considering the potential use of radioisotope materials for an alpha-voltaic or a hybrid thermoelectric/alpha-voltaic power source. Some of the technical challenges associated with these concepts are discussed in this paper.

INTRODUCTION

Deep space missions have a strong need for compact, high power density, reliable and long life electrical power generation and storage under extreme temperature conditions. Conventional power generating devices become inefficient at very low temperatures (temperatures lower than 200K encountered during Mars missions for example) and rechargeable energy storage devices cannot be operated thereby limiting mission duration. At elevated temperatures (for example temperatures of 600K and higher for solar probe or Venus lander) thin film interdiffusion destroy electronic devices used for generating and storing power. Solar power generation strongly depends upon the light intensity, which falls rapidly in deep interplanetary missions (beyond 5 a.u.) or in planetary missions in the sun shadow. Moreover, it has been observed during the Mars pathfinder mission that significant performance degradation occurred when solar cells get covered with dust particles. Radioisotope thermoelectric generators (RTGs) have been successfully used for a number of deep space missions RTGs. However, their energy conversion efficiency and specific power characteristics are quite low, and this technology has been limited to relatively large systems (more than 100W).

The National Aeronautics and Space Administration (NASA) and the Jet Propulsion Laboratory (JPL) have been planning the use of much smaller spacecrafts that will incorporate a variety of microdevices and miniature vehicles such as microdetectors, microsensors and microrovers. Except for electrochemical batteries and solar cells, there are currently no available miniaturized power sources. Novel technologies that will function reliably over a long duration mission (ten years and over), in harsh environments (temperature, pressure, and atmosphere) must be developed to enable the success of future space missions. It is also expected that such micro power sources could have a wide range of terrestrial applications, in particular when the limited lifetime and environmental limitations of batteries are key factors. The first step towards such compact power sources consists of a miniaturized, versatile milliwatt power source (MWPS) device that is currently being developed. The original MWPS concept was advanced by engineers at JPL (Chmielewski and Ewell, 1994) and consists of a 1.1 W Radioisotope Heating Unit (RHU) to provide the high temperature source for a thermoelectric converter which generates sufficient electrical power (~40 mW) to trickle-charge a rechargeable battery pack. The battery power can then be used in low duty cycle, low power applications. The MWPS approach is based on several technologies developed earlier by space power programs. The RHU was developed by the Department of Energy for the Galileo and Ulysses missions, the thermoelectric converter, or thermopile, is a combination of the early RTG technology, pace maker technology with new packaging techniques, and a lithium-ion rechargeable space battery being developed at JPL, under a separate ongoing program. The prototype MHPS, illustrated in Figure 1, is 67 mm in diameter, 81 mm long, and weighs about 0.41 kg.

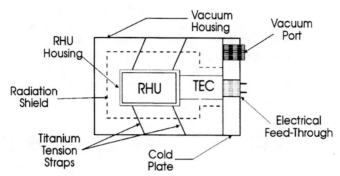

FIGURE 1. Milliwatt Power Source (MHPS) concept incorporating a radioisotope heater unit (RHU), a thermoelectric converter (TEC) and rechargeable batteries (not shown).

In collaboration with industry, JPL is now pursuing the fabrication and testing of MHPS. This effort is described in more details elsewhere (Bass, 1998; Borshchevsky et al., 1997). However, the MHPS will still be too bulky for several next generation miniature "sciencecrafts" and a couple of new approaches to ultra-compact solid state power sources is presented here.

THERMOELECTRIC GENERATORS

Solid state thermoelectric generators covering a wide range of power outputs, from nanowatts to kilowatts, have demonstrated attractive characteristics such as long life, the absence of moving parts or emissions, low maintenance and high reliability. In spite of a large number of potential civilian and military applications, their use has been severely limited due to their relatively low energy conversion efficiency and high development costs. As a consequence more efficient advanced radioisotope power systems (ARPS) are now being developed, based on competing thermal-to-electric technologies. One approach to higher performance thermoelectric devices and systems is the discovery and infusion of novel thermoelectric materials more efficient above room temperature than the current state-of-the-art Bi_2Te_3, PbTe or SiGe alloys. Recent results in several laboratories have successfully identified superior materials in several temperature ranges (Fleurial et al, 1996; Sales et al., 1996; Caillat et al., 1997). There is currently an effort to introduce some of these new compounds into simple unicouple configurations to demonstrate increased conversion efficiency of up to 20% (Caillat et al, 1999).

A second approach is to significantly improve the design, specific power (watts per unit area or volume) and lower the costs of generator devices even when using state-of-the-art thermoelectric materials. This is of great interest when considering large-scale applications using waste heat recovery schemes, or low power devices integrated with electronics and optoelectronics components. For both aerospace and terrestrial applications, there is a growing need for developing miniaturized on-chip low power batteries with long life, high voltage, resistance to extreme temperatures and low environmental impact characteristics (Rowe, 1995; Fleurial et al., 1997). Figures 2.a and 2.b illustrate the potential of miniaturized thermoelectric power sources based on state-of-the-art Bi_2Te_3 alloys. The calculations show that in spite of the low conversion efficiency, high specific power values (in W/cm^3) can be achieved even for relatively small temperature differentials, in the 20 to 200K. This is due to the fact that thermoelectric devices are scalable, provided that the aspect ratio of the thermoelements (or legs) is kept constant and that electrical and thermal resistance loss can be kept low. Thus miniaturized thermoelectric devices with leg size in the tens of microns range (instead of tens of millimeters) are very attractive.

However, current thermoelectric module technology is ill suited to such development due to mechanical and manufacturing constraints for thermoelement dimensions (100-200µm thick minimum) and number (100-200 legs maximum). In addition to the widespread use of semi-manual assembly techniques that results in high costs for more compact configurations, these devices have typically undesirable high current and low voltage characteristics. Much smaller devices capable of high voltage (up to 5V) power output in the nW to tens of µW range have also been developed: monolithic structures and more recently thin film devices. Most of the monolithic module configurations have been used in nuclear battery type devices, operating across large temperature differences (100-200K), with a small amount of radioisotope material (usually PuO_2) as the heat source (Rowe, 1995; Bass, 1998). The specific power density of the monolithic thermopiles is typically measured in tens of mW/cm^3, but falls to about 60 $µW/cm^3$ when taking into account the complete power source package. Thin film devices producing 20 mW at 4V under load with a temperature difference of 20K have been recently described (Stordeur and Stark, 1997). The 0.22 cm^3 device is comprised of 2250 thermocouples deposited on Kapton thin foils packed together and was fabricated using integrated circuit-type techniques. However, in spite of this remarkable achievement that could allow for batch fabrication of these devices, the specific power density still remains quite low, close to 90 $µW/cm^3$ (heat source not included). This is mainly due to the fact that the length of the thermoelectric legs is supported by the Kapton substrate, thus introducing a very significant thermal shunt and dramatically degrading conversion efficiency.

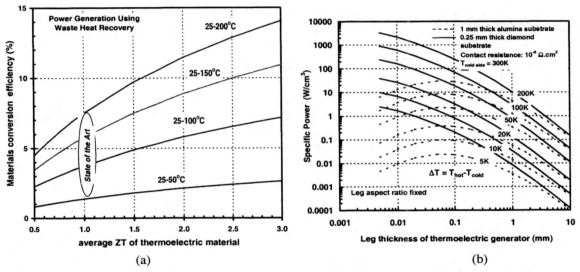

(a)   (b)

FIGURE 2. (a) Improvement of thermoelectric generator performance when using more efficient materials. The temperature differences reported here are relevant to state-of-the-art Bi_2Te_3 alloys. (b) Calculated increase in specific power (per unit volume) with increasing leg miniaturization (constant cross section to length leg aspect ratio) and increasing temperature differences of operation. Data are shown for both low (alumina, blue dots) and high (diamond, red lines) thermal conductivity substrates.

MINIATURIZED THERMOELECTRIC GENERATOR DEVICES

To circumvent key shortcomings described in the preceding section, the Jet Propulsion Laboratory (JPL) is pursuing the development of vertically integrated thermoelectric microdevices that can be fabricated using a combination of thick film electrochemical (ECD) and integrated circuit (IC) processing techniques (Fleurial et al., 1998). Indeed, current prototype devices leave much room for performance improvement, as illustrated in Figure 2b. Even for relatively small temperature differences, such as 10 to 20K, high specific power outputs in the 1 to 10 W/cm^3 are potentially achievable provided that the legs be no thicker than 50 to100 µm.

Microdevice configuration

The term "vertically integrated" here refers to the conventional thermoelectric module configuration shown in Figure 3. This design eliminates the large heat losses observed in planar thin film thermoelectric devices where the legs are deposited onto a supporting substrate. However, planar configurations do offer a very convenient way of fabricating electrical interconnects between the thin film legs by using traditional masking techniques. Thermal resistances due to heat transfer through the metallizations and substrates, as well as electrical resistances due to the interconnects between n-type and p-type thermoelectric legs, rapidly become important issues when increasing device miniaturization. High thermal conductivity substrates, thin metallizations and intimate contact with the heat source and heat sink media are key to minimizing thermal issues when the microgenerators operate in particular under low temperature differences and high heat flux conditions. Since high voltage power outputs are highly desirable from a power conditioning aspect, this means that the microdevices will typically possess several thousands of very short thermocouples. Electrical contact resistances can thus easily become a very large fraction of the total internal device resistance. However, low values are routinely obtained in the electronic semiconductor industry and similar processing techniques have been developed here. Finally thermally stable diffusion barriers are needed to maintain the integrity of the multilayered stack of substrates, metallic interconnects and thermocouples. The effectiveness of amorphous transition metal nitride diffusion barriers for metallizations on diamond, AlN and thermally oxidized silicon substrates has been recently demonstrated (Kacsich et al., 1998).

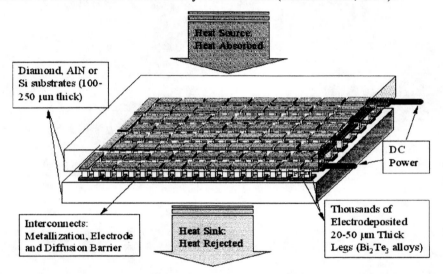

FIGURE 3. Schematic representation of a vertically integrated thick film thermoelectric generator using thin high thermal conductivity substrates.

Microdevice fabrication

Hot side temperatures for microdevice applications that we are currently considering are 200 to 500K. $Bi_{2-x}Sb_xTe_{3-y}Se_y$ alloys are the state-of-the-art materials best suited to these temperatures of operation. Since the thickness of the legs selected in our various device concepts ranges from 10 to 60μm, we have actively pursued the development of an electrochemical thick film deposition process. ECD constitutes an inexpensive way to synthesize semiconducting films (Pandey et al., 1996) and, depending on the current density used in deposition, the deposition rate can be varied widely, up to several tens of microns per hour. In addition, slight variations in the deposition potential or solution concentration may possibly be used to induce off-stoichiometric films, thus providing p- or n-type doping through stoichiometric deviation. The electrodeposition of thermoelectric materials has not been widely investigated (Muraki and Rowe, 1991; Takahashi et al., 1994) and new experimental methods have been developed to obtain p-type and n-type $Bi_{2-x}Sb_xTe_{3-y}Se_y$ compositions which are optimal for thermoelectric power generation in the temperature range of interest. An additional advantage of ECD is that some of the interconnect layers necessary to the fabrication of these devices, such as Cu for the electrical path or Ni for the Cu diffusion barrier can also be deposited by using different aqueous solutions. More details have been reported elsewhere (Fleurial et al., 1998).

Building on the availability of new thick photoresist commercial products, we have developed templates suitable to the electrochemical deposition of legs as thick as 70μm and as small as 6μm in diameter. Actually, it has been determined that to be able to tightly control the geometry of the legs and prevent "mushrooming" growth, electrodeposition must be conducted in equally thick photoresist templates. The thick positive photoresist template is patterned with deep square or round shaped holes that must be pre-aligned on top of metallic interconnects.

Figures 4.a and 4.b illustrate the result of IC-type processing. More processing steps are required to successively deposit n- and p-type legs on top of the bottom substrate interconnects, and then ensure proper joining to a top substrate with similarly patterned interconnects. Based on commercial electrolytes, we have used ECD techniques to deposit high quality Cu, Ni and Pb-Sn solder layers as well. The Pb-Sn layer can be used to form solder bumps on top of the legs, as done for flip-chip bonding (Annala et al., 1997). These processing steps are illustrated in Figure 5. The combination of ECD and IC-type techniques offers a degree of flexibility in designing and fabricating microdevices. Typically, a single photolithography mask can combine all of the necessary patterns to completely fabricate one generator configuration.

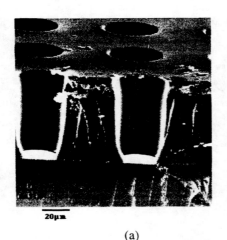

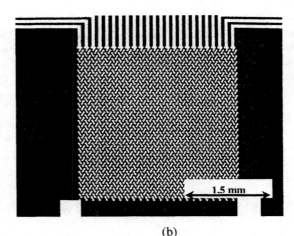

(a)   (b)

FIGURE 4. (a) Thick positive photoresist template on top of a metallized Si/SiO$_2$ substrate. Deep cylindrical holes where the thermoelectric leg will be deposited can be seen. (b) Cu metallization on top of a Si/SiO2 substrate where interconnects have been patterned for subsequent deposition of the thick photoresist template and thermoelectric legs. The fully metallized square pads are for providing electrical contact tests.

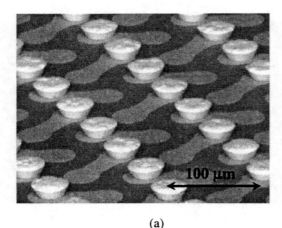

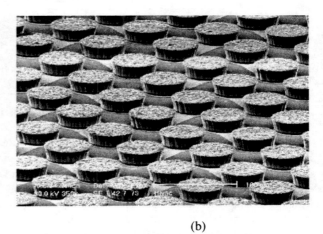

(a)   (b)

FIGURE 5. (a) Bi$_2$Te$_3$ legs electrodeposited on top of Cu interconnect metallizations (using a Si/SiO$_2$ substrate) and (b) Ni/Bi$_2$Te$_3$/Ni/Pb-Sn legs electrodeposited on top of Cu interconnects.

Advanced Milliwatt Power Source

This concept proposes to integrate highly miniaturized thin film thermoelectric converters with a single RHU into a lightweight, rugged, long life, compact and modular solid-state power source capable of delivering tens of milliwatts. A schematic of the complete advanced milliwatt power source (AMWPS) is shown in Figure 6.a. Figure 6.b illustrates how the spherical power source could be integrated with next generation microrovers designed for planetary and asteroid missions.

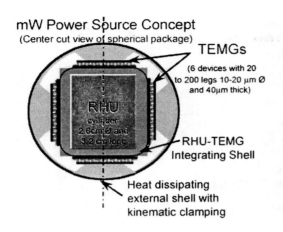

mW Power Source Concept
(Center cut view of spherical package)

TEMGs

(6 devices with 20 to 200 legs 10-20 μm Ø and 40μm thick)

RHU cylinder 2.6cm Ø and 3.2 cm long

RHU-TEMG Integrating Shell

Heat dissipating external shell with kinematic clamping

(a)

AMWPS

(b)

FIGURE 6. (a) Center cut view of an advanced milliwatt power source (AMWPS) concept based on a single RHU thermal source and six thermoelectric thin film microgenerators (TEMGs). The schematic shows both the external shell and internal jacket for the RHU to allow for the bonding of planar thermoelectric microdevices. (b) Illustration of an AMWPS device integrated with a next generation microrover.

The RHU unit is a cylinder 32mm high and 26mm in diameter and weighs about 40g. The palletized fuel is surrounded by a platinum-alloy capsule, pyrolitic graphite thermal insulation and a graphite ablation shell. Based on the 1.1W RHU thermal output and thermoelectric microgenerators fabricated with thick films of Bi_2Te_3 alloys and operating across a 250K to 300K temperature difference, the novel power sources could produce up to 50mW. The total weight of the MWPS device is predicted to be close to 60g, compared to the 350-400g weight of the MWPS based on bulk thermopiles. Energy storage devices such as capacitors could also be used in combination with this power output to deliver much higher energy levels at brief time intervals (for example 0.5W at 10% duty cycle). Technical issues associated with this concept include in particular the development of an external jacket for the RHU to allow for the bonding of planar thermoelectric microdevices and that is compatible with RHU atmospheric re-entry requirements. In addition thermal and mechanical studies must be conducted to minimize heat losses and maximize mechanical ruggedness.

ALPHA-VOLTAICS

Another attractive solid state power generator device investigated at JPL is based on the direct conversion of the kinetic energy of alpha particles into electricity (Patel, 1999). This device is expected to exhibit a high conversion efficiency (over 14%) and to function continuously over a long period of time in the temperature range of 20 to 800K without any recharging needs or the presence of any sunlight thanks to a unique long life design. The use of alpha particle kinetic energy for conversion into electricity using an existing SiC photodiode was reported earlier (Rybicki et al., 1996). This particular device was found to have degraded in a rapid manner because it was not specifically designed to avoid the crystal damage from implanted alpha particles. Results on similar devices using beta and gamma rays were also published (Olsen, 1974), but these power sources exhibited extremely low conversion efficiency (0.1 to 4%) and required substantial shielding to reduce the dose from its radiation to adjacent electronics. Such devices also had a relatively short lifetime (2-6 years) and delivered very low power levels (a few milliwatts).

The device design pursued by JPL is aiming at minimizing lattice damage from alpha particles in the active semiconductor p-n junction, as illustrated in Figure 7. The key design feature of the technology lies in the determination of the diode dimensions so that alpha particles with energy of 5.8 to 6.1 MeV do not stop in the active device volume but in the inactive substrate layers. Alpha particles cause severe lattice damage when they stop since they lose a large fraction of their kinetic energy just before stopping in materials. Nevertheless, some lattice damage is expected in the active p-n junction, but it has been observed that such damage is continuously annealed during the ionization process in semiconductors. Semiconductors such as GaAs or SiC that are stable at high temperatures will be used in the fabrication of the alpha particle-based power source. Curium-244, a nearly pure alpha particle source with negligible soft gamma emission and an 18-year half-life has been initially selected as the radioisotope material. Initial studies indicate that optimally configured miniaturized alpha voltaic power sources could offer high specific power values close to a milliwatt per cubic millimeter (mW/mm³).

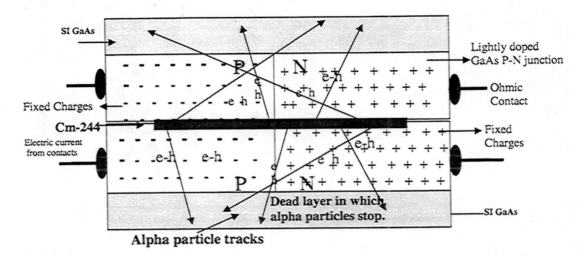

FIGURE 7. Schematic representation of an alpha-voltaic power generator device. The high-energy alpha particles emitted by the radioisotope thin film power source cross the thin depleted GaAs p-n junction, creating electron-hole pairs that are collected at the metallic contacts. Most of the lattice damage from the alpha particle kinetic energy occurs in the inactive substrate layers.

CONCLUSION

Some new NASA missions, such as asteroid rovers, Europa cryobots (link modules), or the Mars Network, are designed for deep space, cold planetary environments and call for milliwatt to watt power source requirements. Current electrical power source and storage technology is either too bulky, limited to a narrow temperature range of operation, or dependent on sunlight intensity. Radioisotope solid state energy conversion devices offer attractive possibilities for configuring rugged, lightweight, long life, modular power sources that could potentially achieve high specific power values in the mW/mm³ range. Thermoelectric microgenerators based on state-of-the-art materials could be combined with RHUs, or even other more compact advanced fuel packages, to deliver tens of milliwatts. However, that technology to date is limited to bulky configurations based on monolithic thermopiles or to inefficient planar thin film devices. To fabricate high performance microdevices in a "classic" vertically integrated module configuration, a combination of electrochemical deposition techniques and integrated circuit technology is now under development. Thick photoresist templates up to 70 μm have been successfully developed and patterned using conventional UV photolithography, resulting in the reproducible fabrication of highly packed arrays of thousands of legs as small as 6μm in diameter. We are now focusing on the fabrication of operational prototype devices with high specific power, high voltage characteristics that in particular could operate in a 200 to 500K temperature range. Direct conversion of the kinetic energy of alpha particles into electricity is also a very attractive technology. JPL is currently focusing on the study of simple p-n junction devices based on wide band gap materials such as GaAs that combine high conversion efficiency and long lifetime. The key challenge is designing the alpha-voltaic device to minimize lattice damage from the alpha particles. Because of their complementary energy conversion processes, it might even be possible to fabricate hybrid radioisotope alpha-voltaic/thermoelectric microgenerators, provided than the alpha-voltaic device can operate at elevated temperatures. If such technologies can be successfully developed, it is likely that they will be introduced in many terrestrial applications

ACKNOWLEDGMENTS

The work described in this paper was performed at the Jet Propulsion Laboratory, California Institute of Technology under contract with the National Aeronautics and Space Administration. Part of this work was supported by the U.S. Office of Naval Research, award No. N00014-96-F-0043 and the U.S. Defense Advanced Research Projects Agency, award No. 99-G557.

REFERENCES

Annala, P., Kaitila, J. and Salonen, J., 1997, "Electroplated Solder Alloys for Flip-Chip Inetrconnections", Phys. Scripta, T69, 115.

Bass, J.C., 1998, "Preliminary Development of a Milliwatt Generator for Space", *Proc. XVII Int. Conf. Thermoelectrics*, Nagoya, Japan, May 24-28, IEEE Catalog No. 98TH8365, p. 433.

Borshchevsky, A. Chmielewski, C. K. Huang, J. Bass, D. Bugby, A. Pustovalov, V. Gusev, 1997, *Proceedings of the 32nd Intersociety Energy Conversion Engineering Conference*, July 27-August 1, 1997, Honolulu, Hawaii, p. 465.

Caillat, T., Fleurial, J.-P. and Borshchevsky, A., 1997, J. Phys. Chem. Solids, 7 1119.

Caillat, T., Fleurial, J.-P., Snyder, G. J., Zoltan, A., Zoltan, D., and Borshchevsky, A., 1999, "A New High Efficiency Segmented Thermoelectric Unicouple", *Proceedings of the 34th Intersociety Energy Conversion Engineering Conference*, August 1-5, 1999, Vancouver, BC, Canada.

Chmielewski A. B., and Ewell R., 1994, *Proceedings 29th Intersociety Energy Conversion Engineering Conference*, American Institute of Aeronautics and Astronautics, Inc., Vol. 1, 311.

Fleurial, J.-P., Borshchevsky, A., Caillat, T., Morelli, D. T., and Meisner, G. P., 1996, , "High Figure of Merit in Ce-Filled Skutterudites", *Proc.15th International Conference on Thermoelectrics*, ed. T. Caillat (IEEE Catalog 96TH8169), p. 91.

Fleurial, J.-P., Borshchevsky, A., Caillat, T., and Ewell, R., 1997, "New Materials and Devices for Thermoelectric Applications", *Proceedings of the 32nd Intersociety Energy Conversion Engineering Conference*, July 27-August 1, Honolulu, Hawai (2), p. 1080.

Fleurial, J.-P., et al., 1998, "Development of Thick-Film Thermoelectric Microcoolers Using Electrochemical Deposition" in: *Thermoelectric Materials 1998* eds. T.M. Tritt, M.G. Kanatzidis, H.B. Lyon, and G.D. Mahan, MRS Volume 545, *MRS 1998 Fall Meeting Symp. Proc.*.

Kacsich, T., Kolawa, E., Fleurial, J.-P., Caillat, T., and Nicolet, M.-A., 1998, "Films of Ni-7%V, Pd, Pt, and Ta-Si-N as Diffusion Barriers for Copper on Bi_2Te_3", J. Phys. D, **31**, 1.

Muraki, M. and Rowe, D.M., 1991, "Structure and Thermoelectric Properties of Thin Film Lead Telluride Prepared by Electrolytic Deposition", *Proc. Xth Int. Conf. on Thermoelectrics*, Cardiff, Wales, UK, p. 174.

Olsen, L.C., 1974, "Advanced beta-voltaic power sources", in Energy conversion, pp.754-762.

Pandey, R.K., Sahu ,S.N. and Chandra, S., 1996 in *Handbook of Semiconductor Deposition*, Ed. M. Dekker, New York .

Patel, J., 1999, "Novel long-life alpha particles based power sources for highly integrated microelectronics and MEMS", New Technology Report: NPO-20654, to be published in *NASA Tech. Briefs*.

Rowe, D.M., 1995, "Miniature Semiconductor Thermoelectric Devices" in *Thermoelectric Handbook*, ed. by M. Rowe (Chemical Rubber, Boca Raton, FL), p. 441.

Rybicki,, G., et. al, 1996, "Silicon Carbide Alpha-voltaic Battery", *Proceedings of the 25th IEEE Photo-voltaic Specialists Conference* May, 1996, Washington D.C.

Sales, B.C., Mandrus, D. and Williams, R.K., 1996, Science, Vol. 22, 1325-1328.

Stordeur, M. and Stark, I., 1997, "Low Power Thermoelectric Generator – Self-Sufficient Energy Supply for Microsystems", *Proc. XVI Int. Conf. Thermoelectrics*, Dresden, Germany, August 26-29, IEEE Catalog No. 97TH8291, p. 575.

Takahashi, M., Katou, Y., Nagata, K. and Furuta, S., 1994, Thin Solid Films, 240 (1-2), 70.

High Efficiency Segmented Thermoelectric Unicouples

Thierry Caillat, Alex Borshchevsky, Jeff Snyder, and Jean-Pierre Fleurial

Jet Propulsion laboratory/California Institute of Technology, Pasadena, CA 91109

818-354-0407; thierry.caillat@jpl.nasa.gov

Abstract. A new version of a segmented thermoelectric unicouple incorporating advanced thermoelectric materials with superior thermoelectric figures of merit has been recently proposed and is currently being developed at the Jet Propulsion Laboratory. The advanced segmented unicouple currently being developed would operate over a 300 to 975K temperature difference and the predicted efficiency is about 15%. There has been recently a growing interest for thermoelectric power generation using various waste heat sources such as the combustion of solid waste, geothermal energy, power plants, automobile, and other industrial heat-generating processes. Hot-side temperatures ranging from 370 to 1000K have been reported in the literature for some of these potential applications. Although the segmented unicouple currently being developed is expected to operate at a hot-side temperature of 975K, the segmentation can be adjusted to accommodate various hot-side temperatures depending on the specific application. This paper illustrates various segmentation options corresponding to hot-side temperatures ranging from 675 to 975K. The predicted efficiency and characteristics of the corresponding unicouples are described.

INTRODUCTION

A new version of a segmented thermoelectric unicouple incorporating advanced thermoelectric materials with superior thermoelectric figures of merit has been recently proposed and described (Caillat, 1998 and 1999a) (Fleurial, 1997a and 1997b). This segmented unicouple, under development at the Jet Propulsion Laboratory (JPL), contains a combination of state-of-the-art thermoelectric materials based on Bi_2Te_3 and novel p-type Zn_4Sb_3, p-type $CeFe_4Sb_{12}$-based alloys and n-type $CoSb_3$-based alloys developed at JPL. To achieve high thermal to electrical efficiency, it is desirable to operate thermoelectric generator devices over large temperature differences and also to maximize the thermoelectric performance of the materials used to build the devices. The advanced segmented unicouple currently being developed would operate over a 300 to 975K temperature difference and the segmenting of the n- and p-legs into several sections made of different materials would increase the average thermoelectric figure of merit of the legs compared to using a single material per leg or/and state-of-the-art thermoelectric materials. In the current version, the predicted efficiency would be about 15%. Despite their relatively low efficiency, thermoelectric generator devices are used in various industrial applications because of their high reliability, low maintenance and long life, in particular when considering harsh environments. The most common applications are for cathodic protection, data acquisition and telecommunications. More recently, there is a growing interest for waste heat recovery power generation, using various heat sources such as the combustion of solid waste, geothermal energy, power plants, and other industrial heat-generating processes. Hot-side temperatures ranging from 370 to 1000K have been reported in the literature. Because of the need for cleaner, more efficient cars, car manufacturers worldwide have also expressed some interest for using waste heat generated by the vehicle exhaust to replace or supplement the alternator. If successful, more power would become available to the wheels and the fuel consumption would decrease. According to some car manufacturers, the available temperature range would be from 475 to 675K. There is therefore a variety of potential applications for advanced thermoelectric generators with a diversity of hot-side temperatures. Although the segmented unicouple currently being developed is expected to operate at a hot-side temperature of 975K, the segmentation can be adjusted to accommodate various hot-side

CP504, *Space Technology and Applications International Forum–2000*, edited by M. S. El-Genk
© 2000 American Institute of Physics 1-56396-919-X/00/$17.00

temperatures depending on the specific application. This paper illustrates various segmentation options and the predicted efficiency and characteristics of the corresponding unicouples.

SEGMENTED UNICOUPLE VERSION UNDER DEVELOPEMENT

The segmented unicouple under development incorporates a combination of state-of-the-art thermoelectric materials and novel p-type Zn_4Sb_3, p-type $CeFe_4Sb_{12}$-based alloys and n-type $CoSb_3$-based alloys developed at JPL. In a segmented unicouple as depicted in Figure 1, each section has the same current and heat flow as the other segments in the same leg. Thus in order to maintain the desired temperature profile (i.e. keeping the interface temperatures at their desired level) the geometry of the legs must be optimized. Specifically, the relative lengths of each segment in a leg must be adjusted, primarily due to differences in thermal conductivity, to achieve the desired temperature gradient across each material. The ratio of the cross sectional area between the n-type and p-type legs must also be optimized to account for any difference in electrical and thermal conductivity of the two legs. A semi-analytical approach that includes smaller effects such as the Peltier and Thompson contributions and contact resistance in order to optimize and calculate the expected properties of the device has been used to solve the problem (Swanson, 1961). For each segment, the thermoelectric properties are averaged for the temperature range it is used. At each junction (cold, hot, or interface between two segments), the relative lengths of the segments are adjusted to ensure heat energy balance at the interface. Without any contact resistance between segments, the efficiency is not affected by the overall length of the device; only the relative length of each segment needs to be optimized. The total resistance and power output, however, does depend on the overall length and cross sectional area of the device. The calculated optimized thermoelectric efficiency is about 15% with the hot junction at 975K and the cold junction near room temperature. The optimal geometry is illustrated in Figure 1.

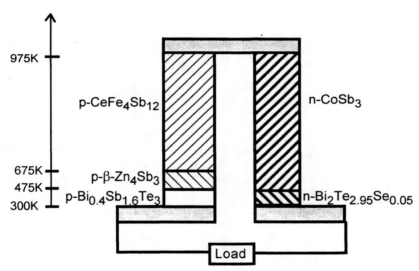

FIGURE 1. Illustration of the advanced unicouple incorporating new high performance thermoelectric materials. The relative lengths of each segment and the cross-sectional areas for the p- and n-legs are drawn to scale. The calculated thermoelectric efficiency is 15%.

High contact resistance between the thermoelectric segments can dramatically reduce the efficiency of a generator. Calculations show that a low contact resistance, less than about 20 $\mu\Omega cm^2$, is required to keep the efficiency from being significantly degraded by the contact resistance. Techniques and materials have been developed to bond the different segments of the unicouple together and also the lower and upper segments to the interconnects (Caillat, 1999b). Electrical contact resistance lower than 5 $\mu\Omega cm^2$ have been obtained for each of the junctions at its projected operating temperature. The unicouple illustrated in Figure 1 has been built and thermal and electrical tests are in progress (Caillat, 1999b). As previously mentioned, the hot-side temperature may vary depending on the specific application and this would require a different optimization of the geometry of the legs. However, the bonding techniques an materials developed for the version of the unicouple operating at a hot-side temperature of 975K could be apply for the fabrication of unicouples operating at lower hot-side temperatures.

POSSIBLE SEGMENTATION VARIATIONS

The maximum hot-side operating temperature of the segmented unicouple is limited to 975K because of the limited temperature stability of the thermoelectric materials used for the upper segments above about 1075K. Efforts are underway to develop skutterudite based materials which could operate at significantly higher temperatures. Preliminary results obtained on arsenide and phosphide skutterudites show that these materials are more refractory than their antimonide analogs and therefore could potentially used at higher temperatures. Efforts are underway to fully assess the potential of some of these materials for thermoelectric applications in particular for use tin segmented unicouples. Increasing the hot-side temperature would result in an increase of the thermoelectric efficiency of the unicouple as illustrated in Figure 2 which shows the variations of the thermoelectric efficiency as a function of the hot-side temperature of the unicouple for two different cold-side temperatures: 300 and 500K. Even if the hot-side temperature is lowered to 700K while the cold-side temperature is maintained at 300K, efficiency about 10% can still be achieved.

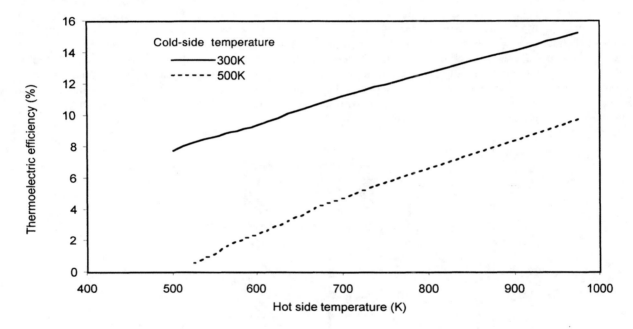

FIGURE 2. Thermal to electrical conversion efficiency as a function of temperature for two different cold-side temperatures : 300 and 500K.

To illustrate the impact of varying the hot-side temperature on the segmented unicouple geometry, some of the thermal, electrical and geometrical properties of the unicouple have been calculated for three different hot-side temperatures : 975, 775, and 675K and are listed in Table 1. Decreasing the hot-side temperature from 975 to 675K results in a decrease of the efficiency from 15 to about 10%. The relative lengths of the segments and cross-sectional areas of unicouples with hot-side temperature of 775 and 675K are illustrated in Figure 3. Decreasing the hot-side temperature from 975 to 775K results in an increase of the lengths of the two lower segments of the p-legs and all three segments have a much closer length than for the 975K version. This could be an important factor if one needs to fabricate very small legs for example for an application for which a relatively high voltage is required. If the lengths of the segments are too disproportionate, it would be more difficult to scale down the size of the legs. The thermoelectric efficiency decreases only from 15% to about 12% going from a hot-side temperature of 975K to a hot-side temperature of 775K. While both version 1 and 2 unicouples have three segments for the p-leg, the unicouple with a hot-side temperature of 675K would only use two p-type materials, *i.e.* Zn_4Sb_3 and $Bi_{0.4}Sb_{1.6}Te_3$. In this version, the two p-type segments would be about equal in length but the p-leg would possess a much lower average thermal conductivity than the n-leg which would result in a much smaller cross sectional area, about 50% smaller than that of the p-type. Much of the difference of the thermal conductivity for the n-leg is attributed to the relative large thermal conductivity of n-CoSb$_3$ (~40mW/cmK) compared to the other materials (~12 mW/cmK). Efforts are currently underway to develop alloys based on CoSb$_3$ which would have similar thermoelectric figure of merit in the temperature range of interest but much lower thermal conductivity. Such materials could be used to replace n-CoSb$_3$ in the segmented unicouple, resulting in a more proportionate geometry for the p- and n-legs with respect to their relative cross-sectional areas.

TABLE 1. Some properties of three segmented unicouples with a hot-side temperature of 975, 775, and 675K.

Property	Version 1	Version 2	Version 3
Hot side temperature (K)	975	775	675
Cold-side temperature (K)	300	300	300
n/p cross sectional area ratio (%)	84	67	48
Power output (Watts)	6.3	2.2	0.8
Current (Amps)	37	19.6	10.24
Efficiency (%)	15.2	12.3	9.8

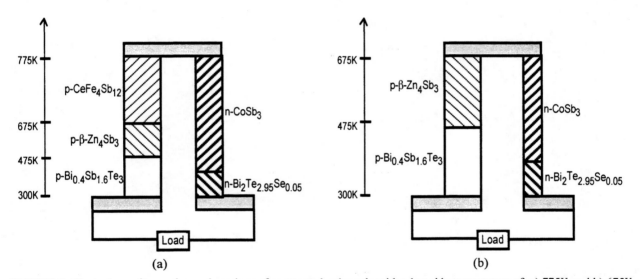

FIGURE 3. Illustrations of two advanced versions of segmented unicouple with a hot-side temperature of a) 775K and b) 675K. The relative lengths of each segment and the cross-sectional areas for the p- and n-legs are drawn to scale.

CONCLUSION

A new version of a segmented thermoelectric unicouple incorporating advanced thermoelectric materials with superior thermoelectric figures of merit is currently under development at JPL. The advanced segmented unicouple would operate over a 300 to 975K temperature difference and the predicted efficiency is about 15%. Techniques and materials have been developed to achieve low electric resistance bonds between the different segments of the unicouple and also between the lower and upper segments to the interconnects Thermal and electrical tests are in progress. Considering the growing interest for waste heat recovery thermoelectric power generation worldwide and the variety of potential heat sources with hot-side temperatures ranging from 370 to 1000K, possible variations of the segmentation to adjust the geometry and optimize the efficiency for a specific hot-side temperature have been described in this paper. Even if the hot-side temperature is decreased to 675K, efficiency values of about 10% can still be achieved. Some adjustments of the segments geometry would be necessary but the techniques and materials developed to date for the version which would operate at a temperature of 975K can be directly used for modified versions of segmented unicouples operating between 675 and 975K.

ACKNOWLEDGMENTS

The work described in this paper was carried out at the Jet Propulsion Laboratory/California Institute of Technology, under contract with the National Aeronautics and Space Administration. This work is supported by the U. S. Defense Advanced Research Projects Agency, Grant No. E748.

REFERENCES

Caillat, T., Borshchevsky, A., and Fleurial, J. -P., in *Proceedings of the 15th Symposium on Space Nuclear Power and Propulsion*, edited by M. S. El-Genk editor, AIP Conference Proceedings 420, New York, USA, 1998, pp. 1647-1651.

Caillat, T., Fleurial, J. -P. , Snyder, G. J., Zoltan, A., Zoltan, D., and Borshchevsky, A., in *Proceedings of the 16th Symposium on Space Nuclear Power and Propulsion*, AIP Conference Proceedings 458, edited by M. S. El-Genk, New York, 1999a, pp. 1403-1408.

Caillat, T., Fleurial, J. -P. , Snyder, G. J., Zoltan, A., Zoltan, D., and Borshchevsky, A., in *Proceeding of 18th International Conference on Thermoelectrics*, edited by Ehrlich, A., IEEE, Piscataway, in press, 1999.

Fleurial, J. -P., A. Borshchevsky, A., and Caillat, T., in *Proceedings of the 1st Conference on Synergistic Power and Propulsion Systems Technology*, edited by M. S. El-Genk, AIP Conference Proceedings 387, American Institute of Physics, New York, 1997a, pp. 293-298.

Fleurial, J. -P., A. Borshchevsky, A., Caillat, T., and Ewell, R., in *Proceedings of the 32nd Intersociety Energy conversion Engineering Conference*, American Institute of Chemical Engineers, New York, 1997b, pp. 1080-1085.

Swanson, B. W., Somers, E. V., and Heike, R. R., *Journal of Heat Transfer*, 77 (1961).

Thermoelectric properties of $Ag_2Te_xSe_{1-x}$ Ternary Compounds

Marhoun Ferhat and Jiro Nagao

Materials Division, Hokkaido National Industrial Research Institute, AIST, MITI, Sapporo 062-8517, Japan

+81-11-857-8948; ferhat@hniri.go.jp

Abstract. A low lattice thermal conductivity is one of the conditions to achieve a high thermoelectric figure of merit, $Z=\alpha^2\sigma/(\kappa_l+\kappa_e)$, where α is the Seebeck coefficient, σ is the electrical conductivity and κ_l, κ_e are the lattice and electronic contributions to thermal conductivity respectively. Silver chalcogenides binary compounds exhibit very low lattice thermal conductivity coupled with high electrical characteristics leading to relatively high thermoelectric figure of merit (e.g. for β-Ag_2Se we obtained $Z\approx2.6\ 10^{-3}\ K^{-1}$ at 300 K). The effect of alloying can normally lead to further decrease the lattice thermal conductivity. In this paper we report on the results of low temperature transport properties of $Ag_2Te_xSe_{1-x}$ ($0\leq x\leq0.5$) ternary compounds in the temperature range from 70K to 300K. The effect of substitutional alloying on their electrical properties is discussed. The thermoelectric figures of merit at 300K are estimated.

INTRODUCTION

Semiconductor chalcogenide compounds of the column I_b of the periodic table are of interest as thermoelectric materials in the low temperature range (<300K). The reason for this is that the I_2-VI binary compounds such as silver telluride (β-Ag_2Te) and silver selenide (β-Ag_2Se), the low temperature polymorphs of silver chalcogenide, exhibit very low thermal lattice conductivity (<8mW/cm K) and very high electronic mobility (10000 cm^2/V.s for β-Ag_2Te and 3000 cm^2/V.s for β-Ag_2Se at 300K).

β-Ag_2Se crystallizes in the orthorhombic system with space group $P2_12_12_1$. The lattice parameters are a=4.333Å, b=7.062Å and c=7.764Å. However, β-Ag_2Te has a monoclinic crystalline structure with P2/n space group. The lattice parameters are a=8.1698 Å, b=8.940 Å, c=8.0653 Å and $\beta=112°793'$.

These compounds undergo a phase transition from the low temperature semiconducting β-phase to the high temperature superionic β-phase. This phase transition occurs at 133°C and 140°C depending on the initial stoechiometry for β-Ag_2Se and β-Ag_2Te, respectively. The thermoelectric properties of these compounds have been first investigated by (Taylor, 1961) and (Wood, 1960). (Dalven, 1967) investigated the electrical properties in the range 4.2K-300K to clarify their conduction modes.

It is well established that the use of solid solution between isomorphic compounds lead to the improvement of their thermoelectric properties. This behavior has been observed experimentally for number of compounds, which are well established in the thermoelectric field such as Bi_2Te_3 with its isomorphic Sb_2Te_3 and Bi_2Se_3. This is essentially attributed to the reduction in the thermal conductivity due to the additional scattering of phonons by the substitutional atoms with different masses and volumes.

The comparison of the crystalline structure of β-Ag_2Te and β-Ag_2Se and the atomic size factor between tellurium and selenium atoms, one could expect the possibility of the formation of solid solution based on anionic substitution on these compounds. This assumption was in fact, experimentally confirmed and the phase diagram of Ag_2Te-Ag_2Se system was first investigated by (Aramov, 1977). The latter authors revealed the existence of an Ag_2Se-based solution with an Ag_2Te content from 10 to 50 mol%. It therefore appeared of interest to investigate the effect of alloying on the electrical properties of silver chalcogenides.

CP504, *Space Technology and Applications International Forum–2000*, edited by M. S. El-Genk
© 2000 American Institute of Physics 1-56396-919-X/00/$17.00

In this paper we present the temperature variation of electrical conductivity, Hall mobility and the Hall coefficient of $Ag_2Te_xSe_{1-x}$ ($0 \leq x \leq 0.5$) from 70K to 300K. The results obtained are discussed and the value of the dimensionless figure of merit is estimate at 300K.

EXPERIMENTAL

Polycrystalline $Ag_2Te_xSe_{1-x}$ (x=0, 0.2, 0.4, 0.5, 1) samples were prepared by slow cooling of the molten mixtures. Stoichiometric mixtures of high purity (6N) of Ag, Se, and Te were sealed in evacuated quartz tube (10^{-3} Pa) and was melted at 1000°C for 3 hours. Continuous stirring was used to ensure homogeneity. The molten mixtures were then slowly cooled to the ambient temperature by natural convection. The samples obtained were then cut into cylindrical shaped ingots of about 1mm thickness for electrical measurements. An X-ray diffraction measurement was carried out to confirm the formation of solid solution. Electron probe microanalysis (EPMA) showed that the synthesized samples have the following composition: $Ag_{2.09}Se_{0.91}$, $Ag_{2.04}Te_{0.21}Se_{0.75}$, $Ag_2Te_{0.42}Se_{0.58}$, $Ag_2Te_{0.51}Se_{0.49}$ and $Ag_{1.98}Te_{0.91}$.

The electrical conductivity and Hall coefficient were measured from 70K to 300K using a standard four-probe technique. Ohmic contacts were made by the evaporation of Au films. The ohmic nature of the contacts was confirmed throughout the above temperature range by the linear current-voltage characteristics. Measurements of the Hall coefficient were performed in a magnetic field of 0.35T. The Seebeck coefficients were measured by the small gradient method with $\Delta T = 10K$ at 300K. Measurements of the thermal conductivity of the binary compounds (β-Ag_2Te and β-Ag_2Se) were carried out by a laser–flash method at 300K.

RESULTS AND DISCUSSION:

The X-ray powder diffraction of $Ag_2Te_xSe_{1-x}$ samples and their corresponding binary compounds (β-Ag_2Te and β-Ag_2Se) are presented in Figure. 1. The interplanar spacing and lattice constants of the binary compounds determined from the diffraction patterns agree well with the ASTM charts.

For the solid solutions with x=0.21, 0.42 and 0.5 one can notice that the diffraction lines are exactly similar to that of β-Ag_2Se indicating clearly the formation of solid solution based on this compounds. However one can see from (Fig. 1). that the angular position of these lines are shifted to smaller Bragg angles as the content of tellurium is increased. In fact, these shifts indicate an increase in the unit cell dimensions as consequence of the substitution of Se^{-2} ions by Te^{-2} ions, which possess larger ionic radius.

In Figure. 2. we present the temperature dependence of the electrical conductivity of $Ag_2Te_xSe_{1-x}$ for x=0, 0.21, 0.42, 0.52 and 1. The Hall coefficients of $Ag_{2.04}Te_{0.21}Se_{0.75}$, $Ag_2Te_{0.42}Se_{0.58}$ and $Ag_2Te_{0.51}Se_{0.49}$ are nearly independent of temperature having a room temperature carrier concentration of 1.07×10^{18} cm^{-3}, 8.85×10^{17} cm^{-3} and 1.33×10^{18} cm^{-3} respectively. This behavior implies a very small impurity or defect–ionization energy and high degeneracy.

The variation of the electrical conductivity with temperature for the same samples is shown in Figure.3. The behavior of the electrical conductivity with temperature is similar to that of Hall coefficient i.e. no variation with temperature in the

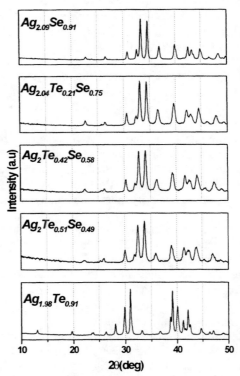

FIGURE 1. X-ray diffraction plots for $Ag_2Te_xSe_{1-x}$ samples (x=0, 0.2, 0.4, 0.5 and 1).

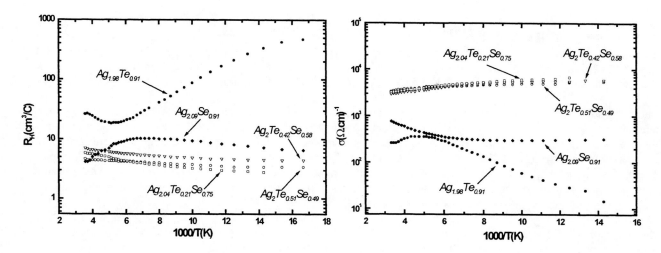

FIGURE 2. Evolution of Hall coefficient with temperature for samples $Ag_2Te_xSe_{1-x}$ (x=0, 0.2, 0.4, 0.5 and 1).

FIGURE 3. Evolution of electrical conductivity with temperature for $Ag_2Te_xSe_{1-x}$ samples(x=0, 0.2, 0.4, 0.5 and 1).

case of the ternary compounds. In the whole temperature range studied the Hall mobility is practically constant (Fig. 4.) with very high value (~ $20000 cm^2/V.s$ at 300K) which suggests again high degree of degeneracy in these compounds. This degeneracy makes the analysis of the Hall and the electrical data complicated for the ternary compounds. For example the intrinsic range which permit the calculation of the energy gap have not been observed even at room temperature. This can be attributed to the high defect concentration, which pushes the transition to the intrinsic range to higher temperatures.

Let us now discuss qualitatively the variation of the electrical parameters with temperature in the case of β-Ag_2Se and β-Ag_2Te. The behavior of β-Ag_2Se is consistent with usual narrow band-gap semiconductors. The Hall coefficient (Fig.2). is approximately constant from 70K to 160K the carrier concentration in this range was calculated to be about 6×10^{17} cm^{-3}. From about 170K the Hall coefficient increase regularly as the temperature is increased up to 300K indicating the onset of the intrinsic conduction. The carrier concentration at 300K reach a value of $1.5 \times 10^{18} cm^{-3}$. In the intrinsic temperature range (>170K) the increase of the conductivity is exponential with an activation energy of 30 meV indicating an energy gap of 60 meV in agreement with the values obtained by

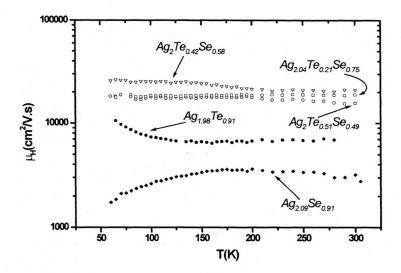

FIGURE 4. Evolution of the Hall mobility with temperature for $Ag_2Te_xSe_{1-x}$ samples (x=0, 0.2, 0.4, 0.5 and 1).

(Dalven, 1960). The Hall mobility in the range 70K–300K present a flat maximum which can be explained by the presence of mixture of lattice and impurity scattering (Fig. 4). For β-Ag$_2$Te the situation is quite different, the Hall coefficient decreases steeply from 70K to 210K where the carrier concentration increases from 1.32×10^{16}cm^{-3} to 3.41×10^{17}cm^{-3} and then decreases slightly to give a value of 2.32×10^{17}cm^{-3} at 300K.

The variation of electrical conductivity (Fig.3.). corresponds well to the variation of the Hall coefficient with temperature. The increase of the carrier concentration leads naturally to an increase of the electrical conductivity and vice-versa if the evolution of the Hall mobility is consistent with the usual scattering mechanism. The activation energy is obtained from the slope of the conductivity between 70K and 210K and gives a value of 60 meV for the energy gap if we suppose that the onset of intrinsic conduction takes place at 70K. However, what is quite anomalous in the (Fig. 2). for β-Ag$_2$Te is the fact that above 210K the carrier concentration decreases again and the Hall mobility in this range is constant with temperature.

In the other hand, the fit of the electrical conductivity by the Mott's equation for variable range hopping (Mott, 1990) is in very good agreement with experimental data. Variable range hopping in this compound is quit hard to explain as the small effective mass in this compounds pushes the metal-insulator transition to very low carrier concentration range. However, if the degree of compensation is very high, it possible to observe this transition in higher concentration range and at higher temperature. Moreover, in the present state, we can not confirm which process is connected to the conduction mechanism in β-Ag$_2$Te. A systematic study over a large concentration range is required to understand the behavior of silver chalcogenides binary compounds.

Finally, in order to clarify weather or not silver chalcogenides and their solid solution are possible candidates for thermoelectricity in the low temperature range, we have performed measurements of their Seebeck coefficient at room temperature.

The thermal conductivity was only measured for β-Ag$_2$Se and β-Ag$_2$Te. We found that for both compounds the total thermal conductivity is 1.1W/m K at room temperature, which is very low if compared to other crystalline semiconductors. In the table 1 we present the value at the room temperature of the electrical conductivity, the Seebeck coefficient and the estimated dimensionless figure of merit ZT for each composition.

In the case of the solid solution we estimated the ZT value in the hypothesis that the thermal conductivity of the ternary compounds is similar to that of the binary compounds. As seen from the table1, values of ZT are very close to the best value obtained for that the state of the art Bi$_2$Te$_3$ based compounds. However, we expect higher value if we consider first, the fact that the value reported here are for materials which are not yet optimized and second the effect of alloying normally leads to decrease in thermal conductivity.

TABLE 1. Thermoelectric parameters of Ag$_2$Te$_x$Se$_{1-x}$ solid solution and their estimated figures of merit (the thermal conductivity at room temperature has been measured only for the binary compounds).

	σ (Ω.cm)$^{-1}$	S (μV/K)	K=K$_l$+K$_e$ (W/m.K)	ZT$_{@300K}$
Ag$_{2.09}$Se$_{0.91}$	800	-200	1.1	0.87
Ag$_{2.04}$Te$_{0.21}$Se$_{0.75}$	3225	-89	1.1	0.69
Ag$_2$Te$_{0.42}$Se$_{0.58}$	2958	-65	1.1	0.33
Ag$_2$Te$_{0.51}$Se$_{0.49}$	3333	-76	1.1	0.52
Ag$_{1.98}$Te$_{0.91}$	263	-164	1.1	0.19

CONCLUSION

In this paper we presented the preliminary results concerning the investigation of the electrical properties of $Ag_2Te_xSe_x$ (x<0.5) solid solution in the low temperature range. The results show that the effect of alloying improves the electrical properties but lead to relatively small decrease in the Seebeck coefficient at room temperature. The estimation of the dimensionless figure of merit at room temperature gives a value, which is very close to the well-established thermoelectric materials. More detailed studies of both the electronic conduction processes and the heat conduction mechanisms are needed to clarify the specific properties of this class of materials.

ACKNOWLEDGMENTS

The authors would like to thank Dr. T. OKUTANI of the HNIRI for useful discussion and Mr. M. SASAMORI and Miss S. NAGAHARA for experimental assistance. One of the author (M.F.) is grateful to STA of Japan for award of fellowship research.

REFERENCES

Aramov. N, Odin. I, and Boncheva-Mladenova. Z, *Thermochimica. Acta*. 20, 107 (1977).
Dalven. D, *Phys. Rev. Lett.* 16 (8), 311(1960).
Dalven. R and Gill. R, *J. Appl. Phys.* 38 (2), 753 (1967).
Mott. N. F, *"Metal-Insulator Transition"* 2nd Edition (Taylor & Francis, London, 1990).
Taylor. P. F, and Wood. C, *J. Appl. Phys.* 32 (1), 1 (1961).
Wood. C, Harrap. V and Kane. W. M, *Phys. Rev.* 121 (4), 978 (1960).

SQUARE LATTICE HONEYCOMB REACTOR FOR SPACE POWER AND PROPULSION

Reza Gouw and Samim Anghaie

*Innovative Nuclear Space Power and Propulsion Institute, University of Florida, 202 NSC,
Gainesville, FL 32611-8300
(352-392-1427, widargo@inspi.ufl.edu)*

Abstract. The most recent nuclear design study at the Innovative Nuclear Space Power and Propulsion Institute (INSPI) is the Moderated Square-Lattice Honeycomb (M-SLHC) reactor design utilizing the solid solution of ternary carbide fuels. The reactor is fueled with solid solution of 93% enriched (U,Zr,Nb)C. The square-lattice honeycomb design provides high strength and is amenable to the processing complexities of these ultrahigh temperature fuels. The optimum core configuration requires a balance between high specific impulse and thrust level performance, and maintaining the temperature and strength limits of the fuel. The M-SLHC design is based on a cylindrical core that has critical radius and length of 37 cm and 50 cm, respectively. This design utilized zirconium hydrate to act as moderator. The fuel sub-assemblies are designed as cylindrical tubes with 12 cm in diameter and 10 cm in length. Five fuel sub-assemblies are stacked up axially to form one complete fuel assembly. These fuel assemblies are then arranged in the circular arrangement to form two fuel regions. The first fuel region consists of six fuel assemblies, and 18 fuel assemblies for the second fuel region. A 10-cm radial beryllium reflector in addition to 10-cm top axial beryllium reflector is used to reduce neutron leakage from the system. To perform nuclear design analysis of the M-SLHC design, a series of neutron transport and diffusion codes are used. To optimize the system design, five axial regions are specified. In each axial region, temperature and fuel density are varied. The axial and radial power distributions for the system are calculated, as well as the axial and radial flux distributions. Temperature coefficients of the system are also calculated. A water submersion accident scenario is also analyzed for these systems. Results of the nuclear design analysis indicate that a compact core can be designed based on ternary uranium carbide square-lattice honeycomb fuel, which provides a relatively high thrust to weight ratio.

INTRODUCTION

In 1955 nuclear propulsion research was initiated in the United States. The technology emphasis occurred in the 1960's and was primarily associated with the Rover/NERVA Program. Los Alamos National Laboratory (LANL) conducted the initial tests for a series of nuclear rockets, and the program was known as the KIWI program (Robbins, 1991). During the Rover/NERVA program, two major problems were identified. One of the problems was core damage due to vibration that was accompanied by cracking of the fuel matrix and loss of material into the propellant flow. The second problem was the loss of fuel matrix uranium and carbon due to coating erosion and cracking, and through diffusion of propellant through the coating (Horman, 1991). On July 20, 1989, President George Bush outlined the Space Exploration Initiative (SEI) during the 20th anniversary of Apollo 11. President Bush called for a return to the Moon "to stay" early in the next century, followed by a journey to Mars using systems "space tested" in the lunar environment. The NERVA Derivative Reactor concept is the current "baseline" concept for the Space Exploration Initiative (SEI) missions (Borowski, 1992).

The Square-Lattice Honeycomb Space Nuclear Rocket Engine is a NERVA Derivative Reactor core with a new nuclear fuel design. It is an attempt to reduce the weight of the nuclear rocket engine and simplify the core design without scarifying the thrust level. There are two designs analyzed, Intermediate-Spectrum Square-Lattice Honeycomb and Moderated Square-Lattice Honeycomb. The main difference between these two designs is the incorporation of zirconium hydride (ZrH$_2$) in the Moderated Square-Lattice Honeycomb. With the utilization of

CP504, Space Technology and Applications International Forum–2000, edited by M. S. El-Genk
© 2000 American Institute of Physics 1-56396-919-X/00/$17.00

ZrH$_2$, the Moderated Square-Lattice Honeycomb reactor is found relatively thermal in contrast to the Intermediate-Spectrum Square-Lattice Honeycomb. A nuclear thermal propulsion system has some advantages over the conventional chemical system for a manned mission to a distant planet, such as Mars. A nuclear thermal rocket produces an enormous energy per unit mass of fuel, and the energy-producing medium is separate from the thrust-producing propellant. From these differences, a nuclear thermal rocket with hydrogen propellant could produce a greater specific impulse (I$_{sp}$) than chemical propulsion systems. The higher specific impulse permits the nuclear rocket to accomplish its mission in a shorter time and to carry a larger payload. The separation between the energy producing medium and the thrust-producing propellant allows the nuclear rocket to utilize propellants of low molecular weight. With the low molecular weight propellant, the rocket will also have an increased thrust to weight ratio.

METHOD OF CALCULATIONS

Two methods of calculation are used to analyze the square-lattice honeycomb, the diffusion theory method and the Monte Carlo method. For the diffusion theory method, the computer code used is VENTURE. Group constants for VENTURE were obtained from the COMBINE code by using a B-1 approximation to the neutron transport equation. The MCNP4B code is used to perform the Monte Carlo analysis. COMBINE is a FORTRAN 77 computer code for the generation of spectrum-averaged multigroup neutron cross-section data suitable for use in diffusion and transport theory reactor design analysis (Grimesey, 1990). VENTURE is an IBM-PC or compatible microcomputer version of the BOLD VENTURE system of connected codes or modules used to analyze the core of a nuclear reactor by applying multigroup diffusion theory. The code system can analyze one, two or three dimensions in various geometries. Variable dimensioning is used throughout the codes, which allows for any number of energy groups and mesh points, with the limitation that the problem fits into the core memory. An important feature of this code system is each code module receives input from, and writes output to, standard interface files (Shapiro, 1990). The Monte Carlo method is another method for solving the transport equations. The Monte Carlo methods are very different from deterministic transport methods. The Monte Carlo method simulates particles interaction and keeps records of their behaviors instead of solving an explicit equation. The average behavior of particles in the physical system is then inferred from the average behavior of the simulated particles. The other difference between Monte Carlo and deterministic methods is the information in the solution. Deterministic methods gives complete information throughout the phase space of the problem, while Monte Carlo supplies information only about specific tallies requested by user (Hendricks, 1997).

CORE DESCRIPTIONS

The fuel of the Moderated Square-Lattice Honeycomb (M-SLHC) designs is constructed from uranium-zirconium-niobium carbide (U,Zr,Nb)C, which is one of several ternary uranium carbides that are under consideration for this concept. The core is fueled with a solid solution of 93% enriched (U,Zr,Nb)C. The fuel is to be fabricated as 1 mm grooved (U,Zr,Nb)C wafers. The fuel wafers are used to form square-lattice honeycomb fuel sub-assemblies containing 30% cross-sectional flow area, as shown Figure 1 (Furman, 1999). The hydrogen propellant is passed through these flow channels and removes the heat from the reactor core.

The core design of the M-SLHC is based on a set of five small disk-shaped square-lattice honeycomb fuel sub-assemblies, which are configured into cylindrical fuel assemblies. The fuel sub-assemblies have a diameter and height of 0.05 m and 0.10 m, respectively. These fuel sub-assemblies are surrounded by 0.005 m thick graphite and followed by 0.005 m thick zirconium oxide (ZrO), as shown in Figure 2. These cylindrical fuel assemblies are then inserted into the zirconium hydride (ZrH$_2$) matrix. There are 18 fuel assemblies in the core, as shown in Figure 3. These fuel assemblies are arranged into two set of circular configurations. The smaller configuration has six fuel assemblies, and the larger configuration has 12. In the middle of the core, there is a cylindrical tube made of Inconel to allow hydrogen to flow through the core. The reflected core has a critical diameter and height of 0.368 m and 0.50 m, respectively. Table 1 presents the M-SLHC reactor specifications.

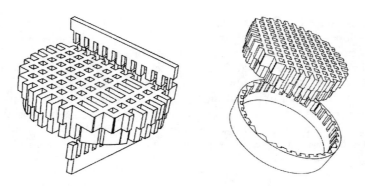

FIGURE 1. Square-Lattice Honeycomb fuel sub-assembly.

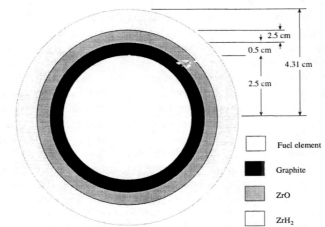

2.5 cm

0.5 cm

4.31 cm

2.5 cm

Fuel element

Graphite

ZrO

ZrH$_2$

FIGURE 2. Moderated Square-Lattice Honeycomb fuel unit cell dimension in region 2.

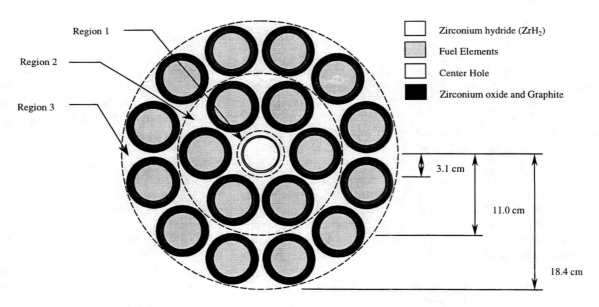

Region 1

Region 2

Region 3

Zirconium hydride (ZrH$_2$)

Fuel Elements

Center Hole

Zirconium oxide and Graphite

3.1 cm

11.0 cm

18.4 cm

FIGURE 3. Moderated Square-Lattice Honeycomb fuel configuration.

Table 1. Moderated Square-Lattice Honeycomb reactor specifications

Properties	M-SLHC
Diameter (m)	0.368
Height (m)	0.50
Radial Reflector thickness (m)	0.10
Top Axial Reflector thickness (m)	0.10
Thickness of Fuel Assembly (m)	0.10
Fuel Type	Solid Solution of (U,Zr,Nb)C
Fuel Enrichment (%)	93
Uranium Density (g/cm^3)	0.4 – 1.0
^{235}U amount (g)	7.66
Propellant	H$_2$

TABLE 2. Moderated Square-Lattice Honeycomb axial temperature zones and uranium density variations

Region	Axial Temperature (K)	Axial Uranium Density (g/cm^3)
1	300	0.4
2	800	0.6
3	1300	1.0
4	1800	1.0
5	2300	1.0

TABLE 3. Moderated Square-Lattice Honeycomb reflector specifications

Parameters	Value
Top axial reflector thickness (m)	0.10
Radial reflector thickness (m)	0.10
Material	Beryllium metal

Five different unit cells are used to represent the materials in the M-SLHC. These unit cells correspond to the five different axial temperature zones and five axial uranium density variations in the core. Three radial regions are defined to perform the analysis of the M-SLHC, as shown in Figure 3. Table 2 presents the temperature zones and axial uranium density variations of the M-SLHC. For each unit cell, a COMBINE run was performed to obtain the average macroscopic and microscopic cross-section of the core. These cross-sections were used to perform criticality calculations using VENTURE. The buckling used in the unit cell is calculated based on the geometric buckling. The unit cell is treated as homogeneous cell. The number densities of the materials in the core are calculated. The volume fractions of the materials are also calculated. Finally homogenized number densities of the materials are calculated by multiplying the pure number densities with their corresponding volume fractions. A homogeneous approximation is used in the resonance calculation. The temperature variations in the core are accommodated by the treatments of the doppler broadening in the code. The method used to calculate the resonance region is the Nordheim method, one of the options in COMBINE. For the critical systems, the reflector configurations are obtained and are shown in Table 3. The reflector material for the M-SLHC cores is beryllium metal. The average cross-section of the materials in the reflector is calculated by incorporating a very small concentration of this material in every regions in the unit cell and requesting the microscopic cross-section to be printed in the output.

RESULTS AND DISCUSSION

Figure 4 presents the neutron energy spectrums for the M-SLHC. This figure is used to determine the neutron spectrum used to classify whether the systems are thermal or fast. Fraction neutrons produced in fission from different neutron energies are also useful for classifying the systems. These fractions are shown in Table 4. The M-SLHC can be classified as a thermal system because almost 78% of the system power is produced by neutron fission in the energy less than 0.2 eV.

Again, the size reduction of the M-SLHC can be contributed to the presence of zirconium hydride; as well as the axial and radial beryllium reflectors. Zirconium hydride has a high neutron scattering peak in the thermal energy range. There is a limitation in COMBINE to account the behavior of zirconium hydride since there is only one temperature in the COMBINE cross-section library. An analysis is performed to determine the effect of the zirconium hydride peak by using MCNP4B. The analysis is performed by varying the temperatures of zirconium hydride while keeping the other parameters the same. The result obtained from this analysis is that the effect of zirconium hydride peak is determined to be insignificant.

Figure 5 show the axial and radial power distributions for the M-SLHC. The value of axial power peaking in Figure 5 is 2.10. The maximum axial power peaking in Figure 5 is encountered at the first fuel region. The high peaking power at the top of the core can be contributed to the top axial beryllium reflector. Figure 5 also shows the neutron leakage from the bottom of the core. The radial power peaking is found in the first fuel region of the core. The value of radial power peaking is 1.35.

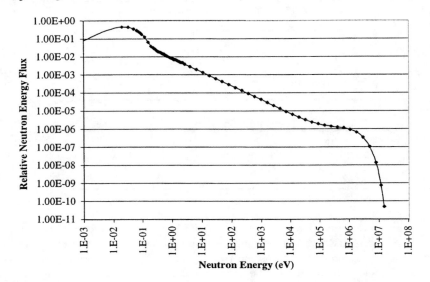

FIGURE 4. Moderated Square-Lattice Honeycomb neutron energy spectrums obtained from COMBINE

TABLE 4. Power fraction of Intermediate-Spectrum and Moderated Square-Lattice Honeycomb produced from fission at sixteen different energy groups

Energy Group	Upper Limit (eV)	Lower Limit (eV)	Fraction of Power
1	1.69E+07	3.68E+06	0.0017
2	3.68E+06	8.21E+05	0.0100
3	8.21E+05	1.11E+05	0.0094
4	1.11E+05	3.18E+04	0.0036
5	3.18E+04	9.12E+03	0.0039
6	9.12E+03	5.53E+03	0.0019
7	5.53E+03	2.04E+03	0.0054
8	2.04E+03	4.54E+02	0.0152
9	4.54E+02	1.01E+02	0.0273
10	1.01E+02	22.6	0.0115
11	22.6	8.32	0.0011
12	8.32	1.86	0.0038
13	1.86	0.7	0.0301
14	0.7	0.2	0.0990
15	0.2	0.015	0.7088
16	0.015	0	0.0673

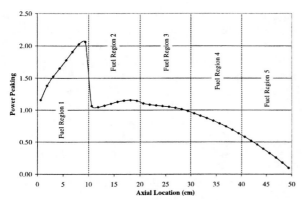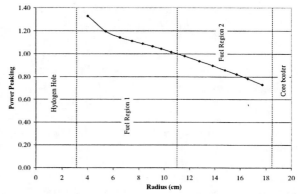

FIGURE 5. Moderated Square-Lattice Honeycomb 16-group model axial and radial power distributions obtained from VENTURE.

Water submersion accident scenario is analyzed for the M-SLHC cores. For this analysis, the hydrogen holes are filled with water, and the reactor is surrounded by water. The analysis is performed at room temperature because water has the highest density at this temperature. The eigenvalue for this water submersion accident for the M-SLHC is 1.3513. Therefore, the M-SLHC will be supercritical in case of the water submersion accident. This performance is not desirable in the reactor system. However, the solution to this problem is to put more absorber materials during launch and remove them after it reaches the earth orbit.

ACKNOWLEDGMENTS

This work was performed for NASA Marshall Space Flight Center under grant NAG8-1251 and for the Department of Defense, Ballistic Missile Defense Organization (formerly SDIO), Innovative Science and Technology Office under contract NAS-26314, managed by NASA Lewis Research Center through INSPI. The authors also wish to thank Dr. Robert Hanrahan of Los Alamos National Laboratory for his many contributions and assistance on this study. This work also was performed with the assistance from Department of Energy Nuclear Engineering/Health Physics Fellowship.

CONCLUSIONS

A detailed neutronic analysis of the square-lattice honeycomb nuclear rocket core was performed. Results indicated that the core, which utilizes uranium ternary carbide fuels inside the zirconium hydride matrix reduces the amount of ^{235}U required to make the core critical. A key feature of the core design is the use of the square-lattice honeycomb fuel, which is more amenable to simple and less expensive manufacturing. Further analysis needs to be performed to establish the optimum design of the core and fuel materials.

REFERENCES

Borowski, S.K. and Clark, J.S., "Nuclear Thermal Propulsion Transportation System for Lunar/Mars Exploration," in proceeding of *Nuclear Power Engineering in Space Nuclear Rocket Engines Conference, by Research and Production Association "LUCH"*, Lewis Research Center, Cleveland, Ohio, September 1992.

Furman, E. "Thermal Hydraulic Design Analysis of Ternary Carbide Fueled Square-Lattice Honeycomb Nuclear Rocket Engine," in proceeding of *16th Symposium on Space Nuclear Power and Propulsion*, edited by M.S. El-Genk et al., American Institute of Physics, New York, 1999.

Grimesey, R.A., Nigg, D.W., and Curtis, R.L. *COMBINE/PC-A Portable ENDF/B Version 5 Neutron Spectrum and Cross-Section Generation Program.* *RSICC Peripheral Shielding Routine Collection*, EG&G Idaho, Inc., Idaho Falls, Idaho, 1990.

Haloulakos, V.E., et al, "Nuclear Propulsion: Past, Present, and Future," in proceeding of *Fifth Symposium on Space Nuclear Power Systems*, edited by M.S. El-Genk et al., American Institute of Physics, New York, 1988.

Hendricks, J.S., *MCNP4B, Monte Carlo N-Particle Tranport Code System Manual.* *RSICC Computer Code Collection*, Los Alamos National Laboratories, Los Alamos, New Mexico et al. 1997.

Horman, F.J., et al, "Particle Fuels Technology for Nuclear Thermal Propulsion," *AIAA/NASA/OAI Conference on Advanced SEI Technologies*, Cleveland, Ohio, September 1991, Paper AIAA 91-3457.

Pelaccie, Dennis G., "A Review of Nuclear Thermal Propulsion Carbide Fuel Corrossion and Key Issues," *Final Report* et. Al. New Mexico University, Albequerque, NM, November 1994.

Robbins, W. H., "An Historical Perspective of the NERVA Nuclear Rocket Engine Technology Program," Lewis Research Center, Brook Park, Ohio, July 1991.

Shapiro, A., Huria, H.C., and Cho, K.W. VENTURE/PC Manual A Multidimensional Multigroup Neutron Diffusion Code System Version 2. RSICC Computer Code Collection, EG&G Idaho, Inc., Idaho Falls, Idaho, 1990.

Technology Needs for Asteroid and Comet Trajectory Deflection Of a Tunguska-Sized Object Using Fission Propulsion

Roger X. Lenard[1] and Michael Houts[2]

[1]Nuclear Technology Research, Organization 6442, Sandia National Laboratories
Albuquerque, NM 87185, (505) 845-3143, rxlenar@sandia.gov
[2]Department TD-40, Advanced Propulsion, NASA Marshall Spaceflight Center,
Huntsville AL 35812, (256) 544-7143, michael_houts@msfc.nasa.gov

Abstract. Recent studies of Near Earth Object Interceptions (DOC, 1993) have shown that impact of the Earth by a civilization-killing sized asteroid are rare. However, some have publicly stated that impact of the Earth by a smaller asteroid, ~100m diameter, such as the one impacting near Tunguska, Siberia, in 1908, occur approximately twice per century (Young, 1999). While such objects will not necessarily result in widespread societal dislocations, such objects are sufficiently energetic to destroy a very large city, such as Los Angeles or New York. Consequential earthquakes and fault disruptions can result in further damage and loss of life. Displacing the trajectory of a Tunguska-sized asteroid, estimated to be <100m in diameter so that it will convincingly miss the Earth is not a trivial venture. If the asteroid is stony in nature, it composition, it will weigh 20-30 million kg. Depending upon when and where the asteroid is discovered, a velocity increment of ~10cm/s is necessary to impart to the asteroid in order for it to convincingly miss the cis-lunar system. The technology requirements for system a system, based on fission propulsion are examined, and a strawman concept is developed.

ASTEROID IMPACT: STATEMENT OF PROBLEM

Recent motion pictures (Zanuck/Brown, 1998) and other publications have identified the potential for civilization-ending events based on impacts by large asteroids. The Near-Earth Asteroid Interception Workshop identified the size threshold for such events as 1-2 km diameter objects impacting the Earth. While the report appeared to down-play the effect of smaller objects, ~100 meters in diameter, the damage associated with these objects can be devastating to a major metropolitan area. The 1908 Tunguska asteroid was estimated to be ~70 meters in diameter and leveled an area about 25 km in radius (Hill and Goda, 1992). Impact frequency appears to be inversely related to asteroid impact energy, i.e., inversely with the cube of their diameter out to a diameter of a few kilometers (Canavan and Solem, 1992), assuming the velocity is roughly constant. A review of some data, therefore indicates that while collision by an object of 10-20 km size, (thought to be the size responsible for the extinction of the dinosaurs), is rare, collisions of an object capable of destroying a major metropolitan area, for example, Los Angeles is much more frequent, perhaps occurring as often as twice per century (Morrison, 1992). Figure 1 shows the approximate impact frequency as a function of impactor size. Figures 2 and 3 show the effect of impactor size, and damage radius as a function of asteroid type and impact velocity. Clearly, based on the potential for severe damage, it is important to mitigate the effects of a Tunguska-sized object when ever feasible. While the initial asteroid detection studies indicated low probability for damage of a city, we believe this is due to an improper scaling of future population trends, consequently, the analysis erred on the low side of the damage distribution. We here discuss the technology requirements to move a Tunguska-size object employing known detection and propulsion technologies. The most important feature of any asteroid or comet deflection scheme is early detection and reliable orbit analysis.

CP504, *Space Technology and Applications International Forum–2000*, edited by M. S. El-Genk
© 2000 American Institute of Physics 1-56396-919-X/00/$17.00

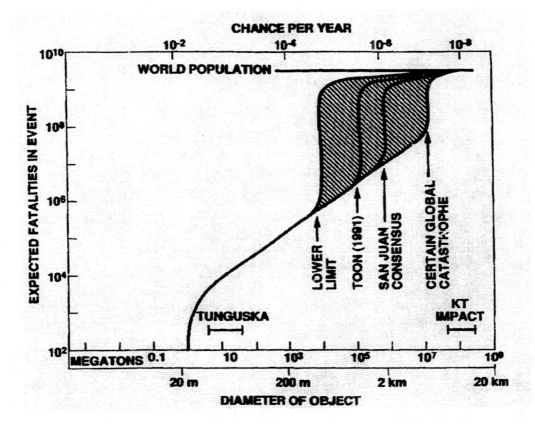

FIGURE 1. Risk versus object size

The technology requirements to move a Tunguska-size object employing known detection and propulsion technologies. As shown in Figure 4, employing solar illumination, detection distance for a 100m class object is less than an AU. For a Tunguska sized object, using existing observation assets, it is about .4 AU, ~60,000,000 km. Unless this detection range is associated with a near-earth-object crossing range where the object's trajectory is such that it may not strike earth for a number of years, we will show that it is almost impossible to locate, track rendezvous with and move such an object in so short a time period. There exist circumstance under just such a scenario can occur, however we will also show that the system requirements are quite extreme by today's technology standards. We assume that the asteroid's average velocity is 25 km/s enroute to earth, and we further assume that any encounter requires us to send the rendezvous spacecraft on an equal intercept velocity.

If one assumes that detection will improve by a factor of four through increasing aperture size and improvements in detector technology, then given an average crossing velocity of 25 km/s, the available action time is ~ 10^7 s, or about four months. During this interval, the interceptor must accelerate and coast to the asteroid, perform a delta-V correction to its orbit and miss the earth. We presume that the delta-V added must be commensurate with a miss distance of greater than the earth's diameter.

If one assumes that the intercepting device is already in space, awaiting for intercept authorization, and is already in a high energy, e.g. geostationary orbit the energy to escape earth is minimal. However, during the available time, the interceptor must accelerate to at least the crossing velocity, and, if it is rendezvous with the target, it must decelerate to the impactor's final velocity.

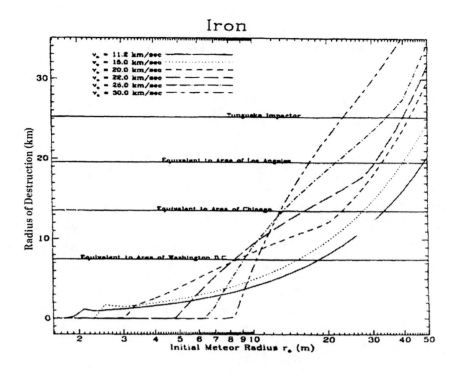

FIGURE 2. Damage Effects from Iron Asteroid.

We first developed hypothetical system concepts, although some of these have a historical and analytical legacy. The low-end system, 100 kWe is based on a joint NASA-Sandia-JPL Kuiper Belt Object (KBO) explorer mission. The high end system at 1 GWe is based, in part on an interstellar rendezvous mission. The low end system specific mass (kg/kWe) or alpha of 35 which is a conservative estimate of a near term technology system to generate this level of electrical power. The opposite extreme is represented by the 1Gwe Interstellar Rendezvous class system with a system alpha of 1 kg/kWe. The 500 kWe class system is characteristic of a joint Sandia National Laboratories/industry commercial space tug concept. This system has a spacecraft alpha of 20 kg/kWe that appears to be feasible at this power level. The 1 MWe system possess an alpha of 15 which is actually close to the typical square root of power ratio nostrum that has been used as a general scaling relationship in the past. As such, the technology inherent in these two systems merely shows the differences in power. The systems proposed are shown in Table 1.

TABLE 1. Asteroid Deflection Spacecraft Parameters.

Low Thrust	Case 1	Case 2	Case 3	Case 4	Case 5
Power Level kWe	100	500	1000	10000	1000000
System Alpha	35	20	15	10	1
Spacecraft Total Mass (kg)	54839	168131	158891	294149	4079170
Electric/Jet Eff	72	75	80	85	90
Specific Impulse (s)	3000	4000	5000	10000	15000
Thrust Level (Nt)	4.8	18.75	32	170	12000

We relate these differences because the differences in overall power relate to the final force-to-mass ratio of the asteroid/spacecraft combination once thrusting commences because the spacecraft mass is insignificant with respect

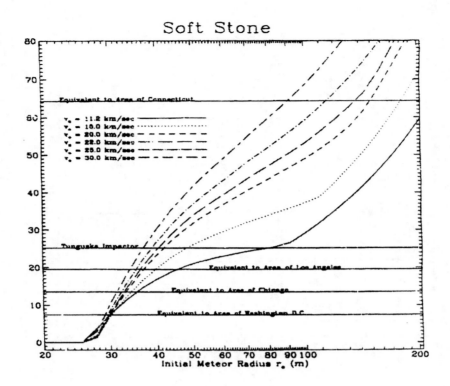

FIGURE 3. Damage Effects from Stony Asteroid.

to the asteroid mass. Lower system specific mass values relates to how rapidly the spacecraft can accelerate to and decelerate from 25 km/s, the proposed intercept velocity. In this regard, a lower alpha assists dramatically because it enables higher accelerations hence longer time for the spacecraft to perform the trajectory deviation maneuver. Further, reductions in spacecraft alpha minimize the total fuel required to reach the asteroid.

To elucidate these differences, we calculate the mass of fuel required to accelerate to and decelerate at the asteroid as shown in Table 2. The spacecraft will be accelerating to 25 km/s and decelerating from this velocity at the same specific impulse as used for moving the asteroid. This may not be the optimum scenario, however it may simplify the system concept somewhat. There is a considerable difference in fuel used in the various concepts. This is because the approach employed was to minimize the overall mass to travel to and move the asteroid the minimum distance of 10,000 km. Consequently, the higher power systems, are able to move the asteroid in a shorter period of time, however, this requires that the asteroid achieve a higher final delta-V. This requires more fuel. The end result of this mechanism is that as the systems become larger, smaller and smaller reaction times and distances are necessary to move the asteroid. As a result, for the small systems, we need many years of reaction time, while for the larger system, it is feasible that a hyperbolic trajectory deep space object could be diverted even if it were heading directly toward the earth.

TABLE 2. Comparison of Asteroid Divert Options.

Low Thrust	Case 1	Case 2	Case 3	Case 4	Case 5
Power Level kWe	100	500	1000	10000	1000000
Travel time (yr)	8.96	8.71	5.63	3.34	0.937
Acceleration fuel (kg)	51339.5	137828.8	123176.9	189982.3	2412503.6
Divert Fuel (kg)	6196.8	9185.6	9600	11063.4	61967.74
Asteroid divert delta V	3.10E+00	6.12E+00	8.00E+00	1.84E+01	1.55E+02

Technology Assessment for Asteroid Divert Mission

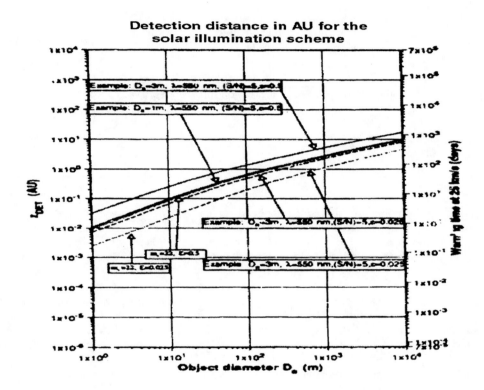

FIGURE 4. Detection ranges for solar illuminated objects.

Technologies required for the various systems shown are exceptionally diverse, ranging from technologies that now exist and could be employed to those which require extensive development. The low-end system employs technologies that are available and ready for implementation. We show, in Table 3, possible technology combinations that could be used to meet the mission objectives. This may not represent the complete envelope, however, it does represent the fact that several options may exist for each category.

REACTOR TECHNOLOGIES

The technology for the SNAP reactor or SNAP-derived reactor essentially exists today and can, with some difficult be recovered. The primary changes that would have to be made with this system is a different conversion system that was used with the original SNAP program. The SNAP program employed two different conversion systems. One system (SNAP-10A) used a thermoelectric converter with very low (1.2%) conversion efficiency. The SNAP – 8 systems used a mercury rankine system at about 8% efficiency. We find that either of these options are unsuitable for our requirements. The thermoelectrics are simply too inefficient at possible SNAP operating temperatures to be useful for this mission. The mercury rankine system has obvious toxicity concerns, however, the SNAP systems were very heavy because tantulum liners had to be used on all surfaces to prevent corrosion.

TABLE 3. Potential Technology Options for Divert Spacecraft Types.

Tech Area	Case 1	Case 2	Case 3	Case 4	Case 5
Reactor	Heat Pipe SNAP Gas Cooled	Heat Pipe SNAP Gas cooled	Gas Cooled Potassium Cooled	Gas Cooled Potassium Cooled	Gas Cooled Liquid U Potassium
Power Conversion	Brayton Steam Rankine Thermoelectric Stirling	Brayton Steam Rankine Stirling	Brayton Potassium Rankine Stirling	Brayton Potassium Rankine Stirling?	Brayton Potassium Rankine
Radiator	Tube fin Heat Pipe Deployable	Tube Fin Deployable	Deployable Liquid Droplet	Deployable Liquid Droplet	Liquid Droplet
Thruster	Xenon Ion Argon Ion Hall VASIMR	Argon Ion Hall VASIMR	Argon Ion Hall MPD VASIMR	Hall MPD VASIMR	Hall MPD New VASIMR
Power Level	100 kW	500 kW	1000 kW	100 MW	1 GW

The mercury rankine system has obvious toxicity concerns, however, the SNAP systems were very heavy because tantulum liners had to be used on all surfaces to prevent corrosion. It is not entirely clear what prompted the migration to that form of rankine system in the original SNAP program. For this reason, we would select a steam rankine cycle because it appears to be optimally suited to the operating temperature range of the SNAP cores.

A considerable amount of recent work has been applied to the Los Alamos Heat Pipe system. The Heat Pipe system has a number of advantages including compact size, modularity, and the potential for being fueled in space, although that represents some logistical and complexity issues. The heat pipe system is scaleable to approximately 1 MWth, and therefore with a suitably efficient conversion system could address the 500 kWe system. While the heat pipe concept is useful for these two lower end options, it is not a likely candidate for the higher power options.

The gas cooled system features moderately compact design at lower power levels, but does not scale as well in power system specific mass as the SNAP or the Heat Pipe system at power levels ~100 kWe. This is due to gas coolant cavities in the core through which the helium-xenon working fluid passes as it obtains energy from the reactor. This system scales well with increasing power and operating temperature. Near term availability of fuel, either uranium nitride (UN) or uranium oxide (UO_2) enable the system to operate with maximum gas exit temperatures ~1300K for UN and ~1600-1700K for UO_2. Limitations on gas exit temperature become a function of case materials or high temperature gas ducting in the reactor to turbine portion of the system.

ENERGY CONVERSION TECHNOLOGIES

Energy conversion technologies are somewhat more problematic because the only energy conversion schemes that have been used with regularity in a space system employing nuclear power system are thermo-electric converters. Unfortunately, these conversion devices only scale in mass linearly with increased power, and at this time are limited to about 5 – 6% thermal to electric efficiencies. However, they are well-understood devices and could be utilized at the low power end of the operating spectrum.

Brayton cycle systems are highly mature and space concepts for a closed cycle brayton system have been demonstrate in the laboratory. Operation testing times for the highly successful Glenn Research Center BRU project accumulated about 55,000 hours of testing. Later tests on the mini-brayton system accumulated about 8,000 hours of testing. Brayton systems scale very well with increased power due to a variety of favorable conditions. Therefore these systems result in lower specific mass values as power levels increase. Due to their inherent simplicity, and potential for high efficiency, brayton systems are always a serious contender at power levels above 100 kWe. An important liability of brayton systems, however, is the comparative low reject temperature for a given cycle efficiency. This implies a large radiator which engenders a substantial penalty at high power levels unless high turbine inlet temperatures or very light deployable radiator concepts are used, or a combination of the two. At the very high power end, a liquid droplet radiator, if this technology can be matured, would provide substantial theoretical reductions in system alpha and potentially reduce deployment issues associated with very large radiators. A concern with liquid droplet radiators is the potential for fluid loss during operation.

Steam rankine systems are a mature technology. They have been operated on earth for over a hundred years and have an enormous operational database. Unfortunately, due to the nature of water systems, they have a relatively low radiator temperature and the water itself can be subject to freezing. Since water expands upon freezing, ducting can rupture if the system is allowed to cool during non-operating conditions. However, total cycle efficiencies of >30% are possible.

Potassium rankine cycles have long been postulated for space power conversion systems where very high power levels ~ MWe are required. This technology is still immature and has not been used in space experiments of any kind. Therefore, the technology remains a system of potential promise with little actual data to justify this approach. However, it is possible that a single pass system could be used with a potassium liquid droplet radiator. For a potentially low ultimate specific mass this approach may be the optimum choice, although substantial technology development is required.

Stirling cycles are a maturing technology and have been demonstrated in non-space flight weight units at the Glenn research Center. These devices operate using (typically) an inert gas loop system employing a free-piston mechanical to electric conversion method. While vibration and life issues have been frequently cited as issues, the developmental program at Glenn Research Center appears to have addressed most of these problems. The scaling relations of stirling engines do not appear to be as favorable as brayton systems, although at the lower power end are probably comparable in mass to brayton cycle systems.

RADIATOR TECHNOLOGIES

As high power level systems are employed, the radiator becomes the dominant item. For a system at 100 kWe, a conventional tube-fin panel approach with a specific mass of <5 kg/m2 appears very low risk and available technology. This type of technology is used in the International Space Station and is high mature. At higher power levels, the ability to compactly store and deploy these devices is increasingly called into question. As a result, several on-going programs to develop light weight, compact, high area radiators are available. Most of these programs are seeking concepts to enable radiators with a double-sided radiating area possessing a specific mass of ~1-2 kg/m2 radiating area. This is of particular importance for brayton systems where system efficiencies are directly related to compressor inlet temperatures. The lower the compressor inlet temperature, the higher the temperature ratio for a fixed turbine inlet temperature; efficiencies of < 50% are possible. Radiator specific mass values of 1 kg/m2 for deployable concepts work well for systems of up to 100 MWe.

Ultimately, a liquid droplet radiator appears to enable specific mass values of <0.1 kg.m2 radiating area. Some technology advances have been made with this concept, although little or no work is presently being conducted. These concepts also appear to alleviate potential micro-meteoroid an debris damage issues since the major radiating area is a droplet sheet. This concept, if feasible would enable the 100 MW and 1 GW systems to achieve the low specific mass values.

ELECTRIC THRUSTER TECHNOLOGIES

There are a number of mature and maturing technologies for this mission, particularly at the low-end range. Ion thrusters for primary propulsion have proven themselves in the Deep Space 1 – NSTAR program. The specific impulse levels required for the asteroid moving mission are identical to those demonstrated in the DS-1 experiment. We note, however, that the total impulse requirements for the system are about an order of magnitude greater than have been demonstrated, although recent technology advances appear to enable this greater total impulse demand. Ion thruster performance improvements to the 500 kW and 1 MW requirements are very close to the demonstrated NSTAR performance. The increase in efficiency is an important improvement need, and is of particular importance in reducing total system mass. Increasing the specific impulse while increasing efficiency for the 100 MW and 1GW system is a major technology improvement, and it is not known if this is possible with present state of the art ion engine technology. An important liability with ion thrusters is their relative poor power densities. Specific mass values of < 2kg/kWe will be difficult to achieve. This limitation appears to preclude them from the 1 GW electric system where the total power system specific mass is 1 kg/kWe.

Hall thrusters are a rapidly maturing technology. In particular, the Hall TAL (Thruster with Anode Layer) and the two-stage Hall thruster have the available performance and also appear to possess high specific power levels. If the Hall thruster can approach 10,000 second specific impulse with a specific mass of < 0.5 kg.kWe it is a good candidate.

The MPD thruster excels at producing high specific power levels, in fact, extant technology approaches 0.35 kg/kWe. However, poor efficiency and less than adequate lifetime are issues that must be addressed.

The VASIMR concept developed at Johnson Spaceflight Center is an emerging technology. It features the ability to vary the specific impulse so situations requiring lowered Isp with a corresponding increase in thrust level can be accommodated. While rapid progress is being made, end-to-end efficiencies are not presently known. Further, this thruster uses liquid hydrogen as a propellant. While hydrogen is readily available, space storage of such large quantities of LH_2 is an issue that must be addressed before this concept is the primary candidate.

SUMMARY AND CONCLUSION

While substantial effort has been associated with the diversion of a large asteroid with the capability to cause destruction on a global basis and perhaps the extinction of the human species, significantly less effort has been associated with addressing the problems associated with smaller asteroids. A Tunguska-sized asteroid could result in damage far exceeding that wrought by Hurricane Andrew whose damage estimate exceeded $26B. The authors contend that developing and demonstrating the ability to move a city-destroying asteroid is or could be soon within our technical abilities. We believe that the cost of developing the technologies required for such a technical capability are less than a several hundred million dollars depending on the level of capability required. Employing existing technologies could result in a near-term technology demonstrator for about the same cost. A mission to the outer asteroid belt could easily demonstrate moving a 70-meter diameter object with the requisite delta – V to force it to miss the Earth if a close encounter were detected. If a more capable system is desired, then depending upon the desired performance requirements, a system could be demonstrated for less than a billion dollars, including technology development. In due consideration of the costs involved from an impact of a Tunguska-sized asteroid with a major metropolitan complex, we believe the costs of developing such a system are a bargain in comparison with potential damage avoided.

We have reviewed the technical requirements for moving an asteroid weighing approximately 70 million kg. Technologies exist and have been demonstrated in part that could allow a near term capability to move such an object given adequate warning time. As technology improves, less warning time is necessary, until at the far end of the spectrum, it may be feasible to divert an asteroid with as little as a year's warning.

REFERENCES

Canavan, G. and Solem, J. 1992, Department of Commerce, "Near Earth Object Interception Workshop Summary", Los Alamos National Laboratory, Sandia Natonal Laboratories, February 1993.

U.S. Department of Commerce report DE93011040, 1993. "Report of the Near-Earth-Object Interception Workshop" Sandia National Laboratories, Feb 1993.

Hills, J.G., and Goda, M. Patrick, 1992, Department of Commerce, "Near Earth Object Interception Workshop Summary", Los Alamos National Laboratory, Sandia Natonal Laboratories, February 1993.

Morrison, D. Chairman, "The Spaceguard Survey: Report of the NASA International Near-Earth Object Detection Workshop (NASA: Jet Propulsion Laboratory), California Institute of Technology.

Young, J. 1999. Public Statements to the Space Technology International Applications Forum, Albuquerque, New Mexico, 1999.

Zanuck/Brown Productions, 1998, "Deep Impact", Directed by Mimi Leder, Paramount Pictures, 1998.

The MagOrion – A Propulsion System for Human Exploration of the Outer Planets

Jason Andrews[1] and Dr. Dana Andrews[2]

[1]Andrews Space & Technology, 214 Main Street, PMB 195, El Segundo, CA 90245,
(310) 779-3646, jandrews@spaceandtech.com
[2]Boeing Advanced Space & Communications Group, 2600 Westminster Blvd. MC SJ-52, Seal Beach, CA 90740,
(562) 797-1242, dana.g.Andrews@boeing.com

Abstract. Manned exploration beyond Mars requires very high specific energy. The only potential solution under discussion is fusion propulsion. However, fusion has been ten years away for forty years. We have an available solution that combines new technology with an old concept – "Project Orion". The proposed "MagOrion" Propulsion System combines a magnetic sail (MagSail) with conventional small yield (0.5 to 1.0 kiloton) shaped nuclear fission devices. At detonation, roughly eighty percent of the yield appears as a highly-ionized plasma, and when detonated two kilometers behind a robust MagSail, approximately half of this plasma can be stopped and turned into thrust. A MagOrion can provide a system acceleration of one or more gravities with effective specific impulses ranging from 15,000 to 45,000 seconds. Dana Andrews and Robert Zubrin published a paper in 1997 that described the operating principles of the MagOrion. We have taken that concept through conceptual design to identify the major operational features and risks. The risks are considerable, but the potential payoff is staggering. Our proposed MagOrion will enable affordable exploration of the solar system.

INTRODUCTION

Near term human exploration of the outer planets, specifically the moons of Jupiter, requires a spacecraft with a new "breakthrough" propulsion system that is capable of generating tens of kilometers per second of delta velocity carrying large 100,000 kg payloads. The MagOrion, with specific impulses ranging from 15,000 to 45,000 seconds, far outperforms both current and near-term propulsion systems like Nuclear Thermal and Ion. Furthermore, the MagOrion, with accelerations of one gravity or greater, is a high thrust device which enables short transfer times and allows ΔV-multiplying thrust maneuvers deep in Jupiter's gravity well.

MagOrion System Description

The MagOrion consists of a central core, which houses the MagOrion propellants, power-plants, and crew systems, attached to a MagSail, which consists of a super-conducting current loop to create a magnetic dipole in free space. This loop of superconducting reinforced wire is attached to the central propulsion and payload core via a network of guidelines and stays. Previous studies (Andrews and Zubrin, 1988, 1989, 1990) have envisioned building large 100 km or 150 km diameter current loops to "sail" in the Solar Wind. We have focussed on reducing the overall size and "ruggedizing" the magnetic loop to make it both more practical and more affordable. As a result, the superconducting loop of our MagOrion is only two kilometers in diameter. The stays and guidelines that connect the core to the current loop are designed to handle the axial thrust loads as well as the radial forces created by the magnetic field and amplified by the fission plasma.

The MagOrion that we are proposing has three propulsion system modes. The MagSail can be used as a low thrust device (100 N) when it interacts with the solar wind, a medium thrust device (50,000 N) when it interacts with a dense ionosphere like the one found around Jupiter, and a high thrust device (1,000,000 N) when it interacts with the fission plasma from a nuclear device. With the MagOrion, we have reoptimized the original MagSail concept to allow one configuration to meet all three design conditions, and at the same time have greatly improved its usefulness for outer planet exploration. A detailed description of the three propulsion modes is listed below:

CP504, *Space Technology and Applications International Forum–2000*, edited by M. S. El-Genk
© 2000 American Institute of Physics 1-56396-919-X/00/$17.00

- **Low ("supercruise") and Medium Thrust:** The MagSail is a device that uses a magnetic field, created by a superconducting current loop, to interact with the indigenous space plasma to generate thrust or drag. While in interplanetary space the MagSail interacts with the Solar Wind, which flows radially away from the sun and has a density of approximately 8 ions per cubic centimeter at 1 AU (Astronomical Unit). Near a planet, the primary interaction is between the MagSail and the planet's ionosphere or plasmasphere. The Earth's plasmasphere extends ten Earth radii (63,000 km) and varies in density depending on altitude. The Earth's ionosphere is located closer to the surface between 100 and 1000 km altitude and is as dense as 10,000 ions per cubic centimeter at 300 km altitude. Jupiter's magnetosphere, however, extends eighty Jupiter radii (5.6 million kilometers) and at Io can be as dense as 4500 ions per cubic centimeter. In both the low and medium thrust cases, because the MagSail interacts with existing mediums, it does not need fuel and it's specific impulse is infinite.

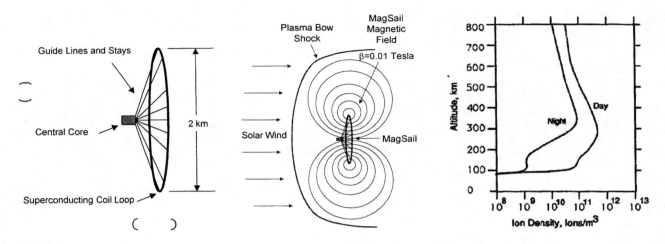

FIGURE 1. Three figures depicting: 1) the relative size of our proposed MagOrion vehicle; 2) how the MagSail interacts with a space plasma; and 3) the ion density in the Earth's ionosphere.

FIGURE 2. Graphic depicting the solar wind and its interaction with the Earth's magnetic field. This is, in principle, exactly how a MagSail works.

- **High Thrust ("MagOrion"):** The low and medium thrust applications of the MagSail are not conducive to missions that require high velocity changes and short trip times. In these situations, the MagSail plasma interaction is augmented using nuclear propellant to artificially create a dense plasma. Small fission nuclear "devices", which weigh 6 kg each and have a yield equivalent to 500 tons to 1000 tons of Trinitrotoluene (TNT), are detonated approximately 2 km behind the MagSail. Approximately 20% of the bomb's energy is released in the form of radiation (X-Rays, Gamma Rays and Neutrons) while the other 80% of the energy is absorbed by the bomb products to form an expanding cloud of extremely hot plasma (>300 kev). This "propellant" is captured by the MagSail and, literally, stopped in its tracks, creating a boundary shock where the dynamic pressure of the plasma equals the magnetic pressure of the compressed magnetic field. This interaction converts the kinetic energy of the bomb plasma into energy in the current loop and the momentum of the plasma into MagOrion thrust. From our preliminary mathematical simulations, a small 6 kg device can generate thrust pulses in excess of one million Newtons, resulting in equivalent Isp's around 20,000 seconds. However, this does not include the thrust from the rebound of the plasma driven by the magnetic field expanding using the stored energy from the pulse. More analysis is required to fully characterize this phenomena. Accordingly, we feel confident that further analysis and test will place the specific impulse in the range from 15,000 to 45,000 seconds. To error on the side of conservatism, we have been using 20,000 seconds of Isp for all on-going mission analysis work.

Recent work has focussed on understanding the specific interaction between the bomb plasma and the magnetic field. Figure 3 depicts the boundary shock location for a two kilometer diameter MagSail. The graph and drawing are approximately to scale. All measurements have been made from the bomb detonation point located two kilometers behind the MagSail at (0,0). Magnetic field lines have been included to represent the resulting field compression and are not to scale.

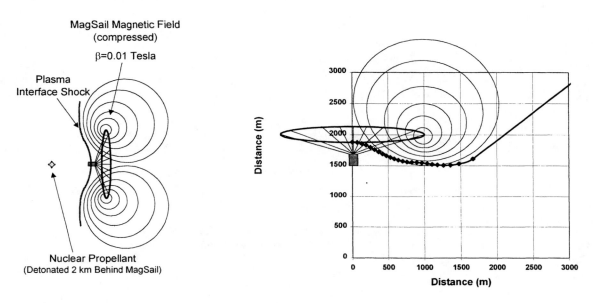

FIGURE 3. Graph depicting the size and location of the maximum penetration of the plasma shock boundary for a bomb detonated two kilometers behind a two kilometer diameter MagSail.

The second part of our analysis of the bomb plasma and magnetic field interaction is understanding and designing for the different forces involved. As the bomb plasma expands and reacts with the magnetic field, the magnetic field compresses, increasing the load forces on the current loop. Figure 4 plots the radial and axial forces on the current loop as a function of time after bomb detonation.

At first blush, it may appear that these are incredibly high loads. In fact, they are. However, we need to remember that they are acting on an equally large system. As an example, the MagSail axial load acts upon a 100 metric ton reinforced HTS current loop. At the peak, the MagSail sees 240 g's of acceleration. However, this large force only acts for one millisecond. Furthermore, because the MagSail is so large (6.3 km in circumference), the local running loads are small (100 kgf/m or 150 lbf/ft). Integrating the total force imparted and the resulting MagSail acceleration relative to the central core, which is suspended by elastic tethers designed to limit its acceleration to one gravity, we can track the axial separation distance imparted by the plasma impulse as a function of time (Figure 5). The maximum differential distance between the two bodies, in this case 2.5 meters, establishes the design requirements for the tether and hydraulic dampers.

As you can see, the MagOrion is actually two vehicles, a MagSail that accelerates with the bomb plasma, and a Central Core, which is suspended by elastic tethers to maintain a relatively constant acceleration. Work has been initiated on the system design details but is not included in this report.

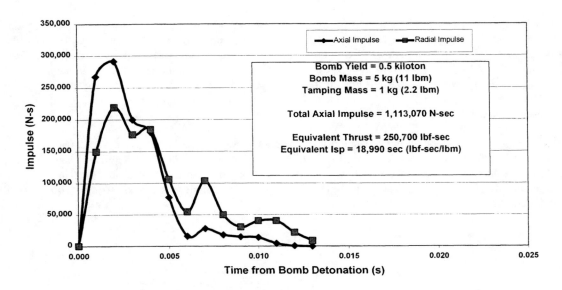

FIGURE 4. Loading conditions on the HTS current loop as a function of time from bomb detonation.

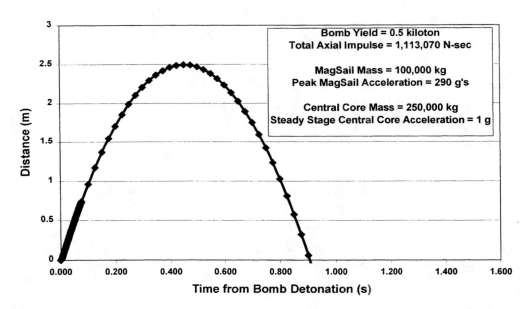

FIGURE 5. Plot of the linear distance change from the Central Core (accelerating at one gravity) for the MagSail due to plasma acceleration as a function of time.

Alternatively, it may be feasible to "tailor" the detonation in such a fashion to manage the plasma temperature profile. Such a profiling would allow the plasma pulse to interact with the MagSail over a longer period of time. Consequently, the impulsive loads are reduced by the dilation factor of the pulse. For example, if we can increase the pulse width by a factor of 100, we should be able to minimize the overall acceleration loads. As a practical matter, we may not be able to avoid some profiling of the pulse width. Another alternative is to decrease the diameter of the MagSail. Preliminary results indicate that this significantly reduces peak loading, especially in the radial direction, but that overall efficiency (specific impulse) suffers.

The radial loads on the MagSail create large hoop stresses, a factor that has proved to be the Achilles heel of similar designs. We feel we have developed a solution that will mitigate this problem. We postulate that the use of tamping material allows us to focus the bomb plasma to maximize performance. Figure 6 depicts graphically how a focused cone of tamping material can tailor the density of the plasma field to control MagSail loading. Increasing the tamping half angle (widening the debris cone) results in a portion of the dense plasma interacting outside the radius of the MagSail, creating a restoring force. By optimizing the tamping material configuration, we can tune the radial loading on the MagSail to all but eliminate the hoop stress. This topic is part of an on-going analysis that is being performed as part of the SBIR Phase 1 in conjunction with the MagSail structural design activity.

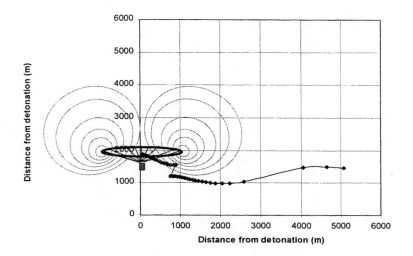

FIGURE 6. The use of tamping material creates a cone of denser plasma (in this case we used a 30 degree half angle) that can be used to tailor the loads on the MagSail.

CONCLUSIONS

Although we feel we understand the basic principles behind the MagOrion propulsion system, more work is required to mature the concept and flush out any showstoppers. However, we are excited by the fact that such a propulsion system appears both technically feasible and physically possible using current technology. The only propulsion systems currently envisioned which have comparable performance to the MagOrion are Nuclear Fusion and Antimatter drive devices. Both of these require several technology breakthroughs to be realized. As a result, the MagOrion represents a critical opportunity to develop a near term, heavy lift, interplanetary propulsion system at a relatively low cost.

REFERENCES

Andrews, D.G. and Zubrin, R.M., "Magnetic Sails and Interstellar Travel", *AIAA Paper 88-553,* 1988.
Andrews, D.G. and Zubrin, R.M., "Use of Magnetic Sails for Mars Exploration Missions", *AIAA Paper 89-2861,* 1989.
Andrews, D.G. and Zubrin, R.M., "Progress in Magnetic Sails", *AIAA Paper 90-2367,* 1990.
Andrews, D.G. and Love, S.G., "Applications of Magnetic Sails", *IAF Paper 91-245,* Montreal, Canada, 1992.
Andrews, D.G. and Zubrin, R.M., "Nuclear Device-Pushed Magnetic Sails (MagOrion)", *AIAA Paper 97-3072,* Seattle, WA, 1997.
Martin, A.R. and Bond, A., "Nuclear Pulse Propulsion: A Historical Review of an Advanced Propulsion Concept", *Journal of the British Interplanetary Society, Vol. 32,* pg. 283-310; 1979.
Taylor, Theodore B., "Third-Generation Nuclear Weapons", *Scientific American, Vol. 256, No. 4,* pg. 30, 1987.

High Power Electromagnetic Propulsion Research
At the NASA Glenn Research Center

Michael R. LaPointe[1] and John M. Sankovic[2]

[1]*Ohio Aerospace Institute, Cleveland, OH 44135*
Phone: (216) 433-6192, E-mail: HorizonTDG@aol.com
[2]*NASA Glenn Research Center, Cleveland, OH 44135*
Phone: (216) 977-7429; E-mail: john.sankovic@grc.nasa.gov

Abstract. Interest in megawatt-class electromagnetic propulsion has been rekindled to support newly proposed high power orbit transfer and deep space mission applications. Electromagnetic thrusters can effectively process megawatts of power to provide a range of specific impulse values to meet diverse in-space propulsion requirements. Potential applications include orbit raising for the proposed multi-megawatt Space Solar Power Satellite and other large commercial and military space platforms, lunar and interplanetary cargo missions in support of the NASA Human Exploration and Development of Space strategic enterprise, robotic deep space exploration missions, and near-term interstellar precursor missions. As NASA's lead center for electric propulsion, the Glenn Research Center is developing a number of high power electromagnetic propulsion technologies to support these future mission applications. Program activities include research on MW-class magnetoplasmadynamic thrusters, high power pulsed inductive thrusters, and innovative electrodeless plasma thruster concepts. Program goals are highlighted, the status of each research area is discussed, and plans are outlined for the continued development of efficient, robust high power electromagnetic thrusters.

INTRODUCTION

High power electric propulsion can provide substantial benefits for Department of Defense (DOD) platform maneuvering and for the orbital, interplanetary, and deep space exploration missions envisioned by the National Aeronautics and Space Administration (NASA). For orbit applications, high power electric propulsion can either lower the required propellant mass or extend the operational lifetime of a spacecraft (Filliben, 1996). On interplanetary and deep space missions, high power electric propulsion can significantly reduce the required propellant mass and reduce the total trip time in comparison to purely chemical propulsion systems (Gilland 1991; Gilland and Oleson, 1992).

High power electromagnetic thrusters have been proposed as primary in-space propulsion options for the NASA Space Solar Power Satellite (Oleson 1999) and for the bold new lunar and interplanetary exploration missions envisioned by the NASA Human Exploration and Development of Space (HEDS) Strategic Enterprise (NASA, 1998). In addition, high power electromagnetic propulsion remains an attractive option for the NASA Interstellar Precursor Mission, an ambitious program to send a small probe to the edge of the heliopause within 10 years (Liefer, 1998). As the lead NASA center for electric propulsion, the Glenn Research Center (GRC) has renewed its program to develop high power electromagnetic thruster technologies to meet these demanding propulsion requirements. The following sections discuss the current GRC research program to develop MW-class magnetoplasmadynamic (MPD) thrusters, high-power pulsed inductive thrusters (PITs), and other advanced electromagnetic thruster concepts.

CP504, *Space Technology and Applications International Forum–2000*, edited by M. S. El-Genk
© 2000 American Institute of Physics 1-56396-919-X/00/$17.00

MAGNETOPLASMADYNAMIC THRUSTER TECHNOLOGY

In its basic form, the MPD thruster consists of a central cathode surrounded by a concentric cylindrical anode (Fig. 1). A high-current arc is struck between the anode and cathode, which ionizes and accelerates a gas propellant. In the self-field version of the MPD thruster, an azimuthal magnetic field produced by the return current flowing through the cathode interacts with the radial discharge current flowing through the plasma to produce an axial body force. In applied-field versions of the thruster, a solenoid magnet surrounding the anode is used to provide additional radial and axial magnetic fields that can help stabilize and accelerate the discharge plasma.

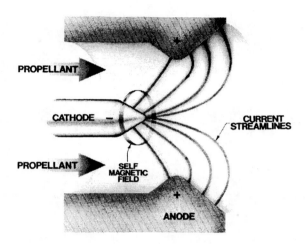

FIGURE 1. MPD thruster diagram.

The MPD thruster has had a long if episodic history of development, and a number of review papers have appeared over the past several decades that chronicle the progress and challenges of this still-promising propulsion technology (Nerheim and Kelly, 1968; Sovey and Mantenieks, 1988; Myers et al., 1991). Although the MPD thruster can be operated with a variety of propellants (Choueiri and Ziemer, 1998), the most efficient performance to date has been obtained with vaporized lithium. Recent MPD thruster experiments at the Moscow Aviation Institute in Russia have demonstrated applied-field thruster efficiencies of around 45% using lithium propellant at power levels approaching 150-kW and specific impulse values above 3500 s (Tichonov et al., 1997).

The goal of the NASA GRC high power MPD program is to extend the lithium performance measurements to MW power levels with domestically produced quasi-steady self-field and applied-field MPD thruster designs. A synergistic approach that combines numerical simulations and quasi-steady MPD thruster testing has been adopted to achieve this goal in a timely manner. The numerical model is a 2-D (cylindrically symmetric), two-temperature, single fluid magnetohydrodynamic code that incorporates classical plasma transport coefficients and Hall effects to predict steady state, self-field MPD thruster performance (LaPointe 1991, 1992). The code has recently been modified to incorporate total voltage calculations, which in turn are used to predict total thruster efficiencies. The simulation has been benchmarked against experimental measurements for a variety of self-field thruster geometries, propellants, and operating conditions, and appears capable of predicting thrust, voltage, and the onset of thruster instabilities with fairly high accuracy. Figure 2a shows experimental thrust measurements and numerical code predictions for the Princeton University full-scale benchmark thruster operated with argon propellant at a mass flow rate of 6 g/s, and Figure 6b shows measured and predicted voltages for the thruster over the same operating regime (Gilland, 1988). Predicted thrust and voltage values match the experimental data to within a few percent, and the predicted onset of instabilities is within 10% of the measured onset of severe voltage oscillations in the thruster. Similar accuracy is achieved with other thruster geometries and propellants, providing some confidence that the numerical model is an accurate predictor of self-field MPD thruster performance. The code is currently being used to design an efficient self-field MPD thruster for operation with lithium propellant.

In tandem with the numerical modeling effort, facility modifications are currently underway at the Glenn Research Center to support a high power MPD thruster test program. A 40-kJ capacitor bank is being installed to provide up to 20-MW of pulsed power over a 2-ms discharge period to an MPD thruster mounted in GRC Vacuum Facility 1 (VF1). The cylindrical VF1 chamber is approximately 1.5 m in diameter by 5 meters long, and is pumped by two

16-inch diffusion pumps that are sufficient to maintain low background pressures during the pulsed discharge experiments. A new vent stack has been installed on VF1 to facilitate operation with lithium, and a new flange and support structure have been installed to provide unhindered access for thruster hardware. A thrust stand and a lithium condensation target are currently being designed for use with both self-field and applied-field MPD thrusters.

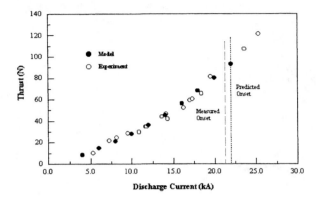

FIGURE 2a. Experimental thrust measurements and numerical model predictions for the Princeton Full Scale Benchmark Thruster (Argon, 6 g/s).

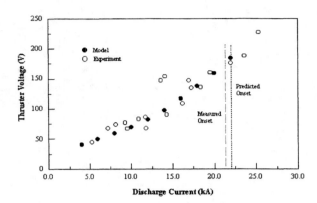

FIGURE 2b. Experimental voltage measurements and numerical model predictions for the Princeton Full Scale Benchmark Thruster (Argon, 6 g/s).

Self-field and applied-field thruster tests will be performed with inert gas propellants to validate the numerical model predictions and to demonstrate pulsed power operation in VF1. Once these preliminary tests have been completed, the facility will be dedicated to the experimental demonstration of an efficient MW-class lithium MPD thruster to support near-earth and deep space exploration missions.

PULSED INDUCTIVE THRUSTER TECHNOLOGY

An alternative high power electric thruster that appears capable of supporting a broad range of mission applications is the TRW Pulsed Inductive Thruster (PIT). In its basic form, the PIT consists of a flat spiral coil covered by a thin dielectric plate (Fig. 3). A pulsed gas injection nozzle distributes a thin layer of gas propellant across the plate surface at the same time that a pulsed high current discharge is sent through the coil. The rising current creates a time varying magnetic field, which in turn induces a strong azimuthal electric field above the coil. The electric field ionizes the gas propellant and generates an azimuthal current flow in the resulting plasma. The current in the plasma and the current in the coil flow in opposite directions, providing a mutual repulsion that rapidly blows the ionized propellant away from the plate to provide thrust. Power, thrust, and specific impulse can be tailored by adjusting the discharge voltage, pulse repetition rate, and injected propellant mass, and there is minimal if any erosion due to the electrodeless nature of the discharge. Because the performance can be varied as a function of power, repetition rate and propellant mass, the thruster can be used for precision orbit maneuvering to support NASA and DOD earth and space science missions, or as a versatile primary propulsion system for planetary and deep space exploration.

The Pulsed Inductive Thruster was conceived and developed by TRW from the mid-1960s through the late 1980s with intermittent DOD and in-house funding (Dailey, 1965; Dailey and Lovberg, 1987). Preliminary thruster performance was less than optimal due to parametric losses in the power delivery circuit (Dailey and Lovberg 1988), and DOD funding for the program terminated in 1988. TRW continued in-house development of the PIT, and in 1991 the NASA Glenn Research Center funded TRW to test a 1-meter diameter PIT with various gas propellants (Dailey and Lovberg, 1993). The highest thruster efficiencies measured during single-pulse operation were obtained with space-storable ammonia and hydrazine propellants. With ammonia propellant and a 15-kV pulsed discharge, the thruster efficiency ranged from 45% at an I_{sp} of 2,000 seconds to above 50% for I_{sp} values approaching 8,000 seconds (Fig. 4a). The same device operated with decomposed hydrazine and a 14-kV pulsed discharge achieved a maximum single-shot thruster efficiency of around 45% at an Isp of 4,000 seconds (Fig. 4b). Although these results showed significant promise for primary propulsion applications, NASA funding for the PIT was terminated in 1992 corresponding to the demise of the NASA Space Exploration Initiative. Interest in the Pulsed Inductive Thruster has recently been revived to support the in-space propulsion requirements of the NASA HEDS Strategic Enterprise missions, as well as other potential applications such as DOD orbital platform maneuvering, NASA deep space exploration missions, and interstellar precursor demonstration flights.

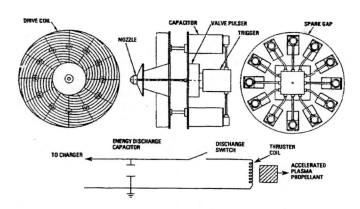

FIGURE 3. Schematic of the TRW Pulsed Inductive Thruster (courtesy TRW, Inc. Redondo Beach, CA)

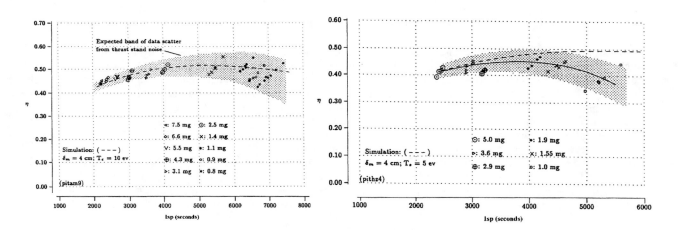

FIGURE 4a. PIT efficiency vs. specific impulse, ammonia propellant (Dailey and Lovberg, 1993).

FIGURE 4b. PIT efficiency vs. specific impulse, hydrazine propellant (Dailey and Lovberg, 1993).

In collaboration with TRW, Inc., and the NASA Marshall Space Flight Center (MSFC), the NASA Glenn Research Center is spearheading a new effort to develop a multiple-repetition rate version of the PIT. During the past fiscal year, GRC funded TRW to initiate design efforts for a multiple rep-rate version of the thruster, and has funded MSFC to initiate a study of pulsed power requirements for limited high frequency thruster operation.

The near term objectives of the joint GRC/TRW/MSFC high power PIT program are: (i) to develop a physics-based numerical model of the PIT ionization and acceleration mechanisms, (ii) to verify prior single-shot performance measurements in a larger vacuum test facility, (iii) to investigate current thruster performance using sequential pulses to determine potential high rep-rate performance issues, and (iv) to design new energy storage and power delivery systems for rapidly pulsed thruster operation. Farther term objectives are to integrate the high rep-rate power system with a thermally cooled version of the PIT in order to demonstrate multiple rep-rate thruster performance at high average power levels. The results of this collaborative research and development program will lead to the near-term deployment of an efficient, high rep-rate thruster capable of meeting a variety of DOD and NASA in-space propulsion requirements.

ADVANCED ELECTROMAGNETIC PROPULSION CONCEPTS

The NASA Glenn Research Center is also pursuing research and development opportunities in other areas of advanced high power electric propulsion to meet future mission requirements. Two such concepts currently under evaluation are the Variable Specific Impulse Magnetoplasma Rocket (VASIMR) and the high power theta-pinch plasma thruster.

The VASIMR thruster is being developed by the Advanced Space Propulsion Laboratory at the NASA Johnson Space Center (JSC) as a high power engine for interplanetary and deep space mission applications (Chang-Diaz et al., 1995). The cylindrical thrust chamber consists of a tandem mirror arrangement of magnetic field coils. A pre-ionized plasma propellant provided by a helicon source is heated through ion-cyclotron resonance to high temperatures and exhausted through a magnetic nozzle to provide thrust. A 10-kW version of the thruster is currently being developed to provide primary in-space propulsion for the NASA Radiation Technology Demonstration (RTD) mission, scheduled for launch circa 2003, and higher power versions are being designed for future interplanetary mission applications. The advantage of the MW-class VASIMR for interplanetary missions is a potential ability to vary the thrust and specific impulse through the mixing of neutral gas with the high temperature exhaust to provide high thrust for planetary orbit maneuvering and escape and high specific impulse for interplanetary cruising. The efficient mixing of neutral gas with a high temperature, high velocity plasma exhaust has not been demonstrated to date. NASA GRC and NASA JSC are thus planning a collaborative effort to experimentally investigate neutral gas mixing with the plasma plume of a high power MPD thruster mounted on a thrust stand at GRC Vacuum Facility 5. The MPD thruster will be operated to match the expected plume characteristics of the VASIMR thruster, and the effectiveness of mixing a neutral gas with the plasma plume will be evaluated to provide design information for the JSC team. Additional discussions are underway between JSC and GRC to install a 100-kW VASIMR engine test bed in an existing GRC vacuum facility, which will be used by the JSC team to evaluate high power VASIMR thruster performance and design issues.

A second advanced electric propulsion concept of potential interest for deep space applications is the pulsed theta-pinch thruster. This device consists of a hollow cylindrical chamber surrounded by a single-turn discharge coil, which in turn is connected to a high-energy capacitor bank. Neutral propellant gas is radially injected into the chamber, and a brief current pulse is sent through the discharge coil to provide partial, uniform pre-ionization of the outer edges of the gas fill. The main storage bank is then fired, and the rapidly rising current pulse generates a time varying axial magnetic field that compresses and heats the ionized propellant. A magnetic mirror located at the upstream end of the discharge chamber provides single-pass reflection of the compressed plasma column, and the high temperature plasma is exhausted from the open downstream end of the chamber to produce thrust. Additional details of the concept are provided in a companion paper in this volume (LaPointe, 2000). A preliminary scaling analysis indicates the thruster can provide comparable specific impulse at significantly higher thrust than current electric propulsion systems. A detailed feasibility study is currently being supported through the NASA Institute for Advanced Concepts; once feasibility is demonstrated, it is anticipated that the thruster will transition to the NASA Glenn Research Center for additional testing and development.

In summary, the NASA Glenn Research Center remains actively involved in the design, development and testing of high power electromagnetic propulsion concepts. As the lead center for electric propulsion, GRC looks forward to participating in the bold new missions of the 21st century.

ACKNOWLEDGMENTS

The Pulsed Inductive Thruster research described in this paper is a collaborative effort between the NASA Glenn Research Center, TRW, Inc., and the NASA Marshall Space Flight Center. Dr. Robert Vondra serves as program manager and principle investigator for the PIT research effort at TRW, Inc. The PIT was conceived and built by Dr. C. Lee Dailey and Dr. Ralph Lovberg, who, although retired, graciously continue to serve as consultants for the resurrected PIT development program. Dr. Ivana Hrbud of the NASA Marshall Space Flight Center serves as the principle investigator for the MSFC effort to design and build advanced high power drivers for the repetitively pulsed version of the PIT. Dr. Franklin Chang-Diaz and Dr. Jared Squire are the principle investigators for the VASIMR thruster program at the NASA Johnson Space Center. Dr. Michael LaPointe is supported at the NASA Glenn Research Center through the Ohio Aerospace Institute under NASA Grant NAG3-2230, whose funding is gratefully acknowledged.

REFERENCES

Chang-Diaz, F., Braden, E., Johnson, I., Hsu, M., Yang, T. F., "Rapid Mars Transits with Exhaust-Modulated Plasma Propulsion," NASA Technical Paper 3539, March 1995.

Choueiri, E. Y., and Zeimer, J. K., "Quasi-Steady Magnetoplasmadynamic Thruster Measured Performance Database," AIAA-98-3472, 34th AIAA Joint Propulsion Conference, Cleveland, OH, July 1998.

Dailey, C. L., "Plasma Properties in an Inductive Pulsed Plasma Accelerator," AIAA Paper 65-637, 1965.

Dailey, C. L., and Lovberg, R. H., "Pulsed Inductive Thruster Comp/Tech," TRW Final Report for AFRPL Contract No. F04611-82-C-0058, TRW Space and Technology Group, Applied Technology Division, One Space Park, Redondo Beach, CA, January 1987.

Dailey, C. L., and Lovberg, R. H., "Pulsed Inductive Thruster Clamped Discharge Evaluation," TRW Final Report for AFOSR Contract No. F49620-87-C-0059, TRW Space and Technology Group, Applied Technology Division, One Space Park, Redondo Beach, CA, December 1988.

Dailey, C. L., and Lovberg, R. H., "The PIT MkV Pulsed Inductive Thruster," NASA CR-191155, July 1993.

Filliben, J. D., "Electric Propulsion for Spacecraft Applications," CPTR 96-64, Chemical Propulsion Information Agency, Columbia, MD, December 1996.

Gilland, J. H., "The Effect of Geometrical Scale Upon MPD Thruster Behavior," M. S. Thesis No. 1811-T, Department of Mechanical and Aerospace Engineering, Princeton University, Princeton, NJ, March 1988.

Gilland, J. H., "Mission and System Optimization of Nuclear Electric Propulsion Vehicles for Lunar and Mars Missions," NASA CR-189058, December 1991.

Gilland, J. H., and Oleson, S., "Combined High and Low Thrust Propulsion for Fast Piloted Mars Missions," NASA CR-190788, November 1992.

LaPointe, M. R., "Numerical Simulation of Self-field MPD Thrusters," NASA CR-187168, August 1991.

LaPointe, M. R., "Numerical Simulation of Geometrical Scale Effects in Cylindrical Self-Field MPD Thrusters," NASA CR-189224, August 1992.

LaPointe, M. R., "High Power Theta-Pinch Propulsion for Piloted Deep Space Exploration," to be published in the *Proceedings of the Space Technology and Applications Forum (STAIF) 2000*, Albuquerque, NM Jan. 2000.

Liefer, S., "Overview of NASA's Advanced Propulsion Concepts Activities," AIAA-98-3183, 34th Joint Propulsion Conference, Cleveland, OH, July 1998.

Myers, R. M., Mantenieks, M. A., and LaPointe, M. R., "MPD Thruster Technology," NASA TM-105242, 1991.

NASA Office of Policy and Plans, "Human Exploration and Development of Space Enterprise," in NASA Strategic Plan, NASA Policy Directive (NPD)-1000.1, Headquarters, National Aeronautics and Space Administration, Washington, D.C., 1998.

Nerheim, N. M., and Kelly, A. J., "A Critical Review of the Magnetoplasmadynamic (MPD) Thruster for Space Applications," NASA CR-92139, February 1968.

Oleson, S., "Advanced Propulsion for Space Solar Power Satellites," AIAA-99-2872, 35th Joint Propulsion Conference, Los Angeles, CA, June 1999.

Sovey, J. S. and Mantenieks, M. A., "Performance and Lifetime Assessment of Magnetoplasmadynamic Arc Thruster Technology," NASA TM-101293, 1988.

Tikhonov, V. B., Semenikhin, S. A., Brophy, J. R., and Polk, J. E., "Performance of 130-kW MPD Thruster with an External Applied Field and Li as a Propellant," IEPC-97-120, 25th International Electric Propulsion Conference, Cleveland, OH, August 1997.

Interstellar Rendezvous Missions Employing Fission Propulsion Systems

Roger X. Lenard[1] and Ronald J. Lipinski[2]

[1]Advanced Propulsion Technology Directorate, TD-40, Marshall Space Flight Center
Huntsville, AL 35812 [(256) 544-2337; Roger.X.Lenard@msfc.nasa.gov
[2]Nuclear Technology Research, Organization 6442, Sandia National Laboratories
Albuquerque, NM 87185 (505) 845-3177; rjlipin@sandia.gov

Abstract. There has been a conventionally held nostrum that fission system specific power and energy content is insufficient to provide the requisite high accelerations and velocities to enable interstellar rendezvous missions within a reasonable fraction of a human lifetime. As a consequence, all forms of alternative mechanisms that are not yet, and may never be technologically feasible, have been proposed, including laser light sails, fusion and antimatter propulsion systems. In previous efforts, [Lenard and Lipinski, 1999] the authors developed an architecture that employs fission power to propel two different concepts: one, an unmanned probe, the other a crewed vehicle to Alpha Centauri within mission times of 47 to 60 years. The first portion of this paper discusses employing a variant of the "Forward Resupply Runway" utilizing fission systems to enable both high accelerations and high final velocities necessary for this type of travel. The authors argue that such an architecture, while expensive, is considerably less expensive and technologically risky than other technologically advanced concepts, and, further, provides the ability to explore near-Earth stellar systems out to distances of 8 light years or so. This enables the ability to establish independent human societies which can later expand the domain of human exploration in roughly eight light-year increments even presuming that no further physics or technology breakthroughs or advances occur. In the second portion of the paper, a technology requirement assessment is performed. The authors argue that reasonable to extensive extensions to known technology could enable this revolutionary capability.

INTERSTELLAR RENDEZVOUS: STATEMENT OF PROBLEM

Many interstellar exploration studies have been conducted under the presumption that fission propulsion is unsuitable for interstellar travel [Frisbee, 1999]. The reasons generally specified are a consequence of the following reasoning: 1. Fission energy production releases about 1 MW-day of energy per gram of fissioned material. This equates to about 0.1% of the mass fraction of fissionable material into energy. The reasoning goes that since $E=10^{-3} mc^2$, $v=\sqrt{(10^{-3})}$ or about 0.03c if all we do is accelerate the fission products. 2. Power densities wherein the fission products are directly released are to low for a practical propulsion system. 3. The combination of low power density (implying low acceleration levels) and low energy content, (implying low final velocities) make fission systems unattractive for the interstellar rendezvous mission.

As a direct result, people have gravitated toward concepts with greater perceived energy density such as fusion, anti matter or concepts that employ beamed energy. There are a number of problems with each of these categories not all of which will be enumerated here. However, it is well known that controlled fusion has been a particularly elusive objective with actual experiments operating well short of ignition and certainly below specific power and total energy values needed for interstellar missions. Terrestrial fusion concepts weigh tens of thousands of tons; it is difficult to ascertain how space concepts will become seriously lower in weight. Yet more advanced concepts such as antimatter may well provide the necessary energy density, however, antimatter propulsion is a very futuristic concept, with exorbitant production costs. For example, present generation efficiencies are approximately 4×10^{-8}/unit energy. Consequently, with electrical power costs of $0.10/kW-hr, each gram of antimatter costs about $62 trillion. If we make reasonable conversion efficiency assumptions that future generation concepts can increase production efficiencies by three orders of magnitude, the costs are now $62T/kg. Even were it feasible to accelerate a spacecraft at 100% energy transfer efficiency, a 1000T crewed spacecraft

CP504, *Space Technology and Applications International Forum–2000*, edited by M. S. El-Genk
© 2000 American Institute of Physics 1-56396-919-X/00/$17.00

moving at 0.1c contains ~10^{22} joules. This would presume the launch of 1000kg antimatter, or $62 quadrillion ($62Q) for the fuel alone.

Use of beamed energy has been extensively studied, with a concept produced by Forward and others of a two-stage laser lightsail. The lightsail itself is 1000 km in diameter and requires a beam director of 160 km diameter. The laser is a 27 TeraWatt device. Even with billions of dollars invested, present laser powers are stymied at ~2 MW beam power. Even given that such a laser could be developed, at present laser costs of $100/watt of beam power, such a laser would cost $2.7Q. The beam director would be similarly expensive, resulting in a rough conservation of cost axiom for futuristic concepts. Additionally, these concepts would have to be space-based, further increasing costs by roughly an order of magnitude.

The shear audacity of the requirements for such futuristic concepts led the authors to investigate a hybrid concept involving fission power driven systems. Numerous authors have studied the necessary power densities for adequate accelerations for modest flight times. The results of such studies are shown in Figure 1. The authors assumed an architecture that involved the following provisos:

1. High temperature, refuelable fission reactors operating at power levels of ~100+ GW could be developed.
2. It would be feasible to beam macro-particles of uranium from near-earth space to the accelerating spacecraft to distances of ~0.2 LY
3. Power system specific mass values ~10^{-3} could be developed when the total power requirements exceed 100 GW.
4. Highly efficient thermal to electric propulsion jet power conversion systems ~0.8 that were light enough for space applications could be developed

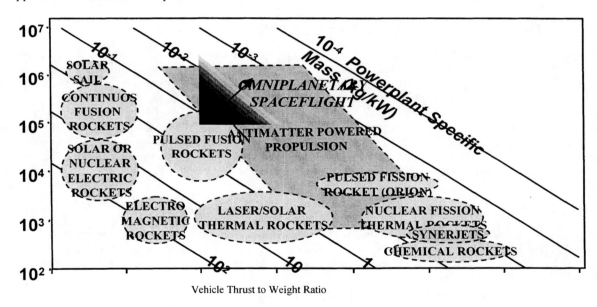

FIGURE 1. Possible Concepts for Various Mission Architectures.

Based on the above assumptions, the following architectural approach was adopted:

1. A fully equipped spacecraft with a refuelable reactor is assembled in space.
2. A system of macro-particle accelerators beams U-235 fuel and electric propulsion propellant to the spacecraft with near-zero interception velocity
3. The spacecraft uses the U-235 for power generation and propellant for acceleration fuel
4. The spacecraft accelerates to ~0.1c
5. The Earth-system based accelerators provide the spacecraft with sufficient uranium fuel for the deceleration leg of the journey.
6. After uranium fuel is supplied by the Earth system, the spacecraft is on its own devices.

A graphic rendition of the proposed architecture is shown in Figure 2.

The physics of this architecture is very favorable to the spacecraft system. Since the uranium fuel and the electric propulsion propellant arrives at the spacecraft at nearly zero velocity, the fuel contains a large quantity of equivalent energy. This energy is equivalent to the energy the spacecraft would have had to provide to it in order to accelerate it to the spacecraft's velocity. When the spacecraft uses that increment of fuel for acceleration purposes, the spacecraft obeys

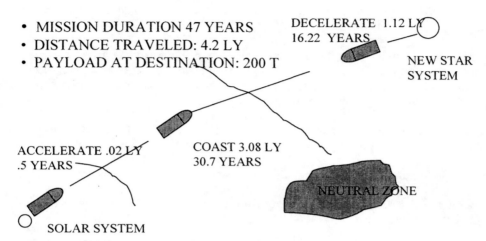

- MISSION DURATION 47 YEARS
- DISTANCE TRAVELED: 4.2 LY
- PAYLOAD AT DESTINATION: 200 T

DECELERATE 1.12 LY
16.22 YEARS

NEW STAR SYSTEM

ACCELERATE .02 LY
.5 YEARS

COAST 3.08 LY
30.7 YEARS

NEUTRAL ZONE

SOLAR SYSTEM

FIGURE 2. A Candidate Fission-Powered Interstellar Rendezvous Architecture.

first law acceleration principles. Energy in the Earth-system accelerator/spacecraft frame of reference is conserved, and the spacecraft acts like it is receiving free energy. Because the Earth system has effectively unlimited power capability, the spacecraft is free to select a lower specific impulse for higher acceleration levels. This can be used to substantially reduce the acceleration times for the spacecraft to achieve 0.1c.

After the spacecraft has reached 0.1c, it then continues to receive uranium fuel in sufficient quantities to decelerate at the target star system. The deceleration leg presumes that the continuously refuelable reactor will use the supplied fuel, fission products will be extracted and the fission products themselves will be used as the propellant. The following equations may be used to evaluate the architecture.

Spacecraft acceleration leg

The acceleration leg manifests substantial flexibility because the Earth system supplies the bulk of the power (in the form of supplied fuel and propellant) for the accelerating spacecraft. As a result, the specific impulse and power level of the spacecraft can be tailored to meet the mission requirements in the most effective manner. The acceleration leg portion of the journey has some competing demands. For example, the spacecraft can either operate in a mode of constant final acceleration level, stabilizing the final distance at end of acceleration trading spacecraft power and specific impulse for macroparticle accelerator power level at the Earth system. Alternatively, the spacecraft can operate at a high power level and higher specific impulse resulting in reduced demands on the Earth system accelerator.

It must be stated that there is "no free lunch" in any interstellar concept employing known technologies or concepts absent breakthroughs in physics. A 1000T spacecraft traveling at 0.1c contains 5×10^{20} joules of energy. The present generating capacity of the US is about 5×10^{15} watts. The spacecraft has the same energy as the entire generating capacity of the US operating for more than a day. Thus while an interstellar mission is not by any means completely outlandish, there is a substantial societal cost to underwrite. However, we note that once a society commits to this enterprise, presently known technologies can achieve the goal.

The spacecraft acceleration is proportional to its jet power and specific impulse. The lower the specific impulse for a given jet power, the higher the acceleration, however at the cost of larger propellant flow rates at the Earth system and consequentially higher macro accelerator power demands.

Jet power = Reactor power * thermal to electric efficiency * Electric to jet efficiency

For purposes of this discussion we assume thermal to electric is near carnot where Th = 3000K, Tc = 200 K thus,

Thermal to electric carnot efficiency = 3000-200/3000 = 2800/3000 = .93333.

If we assume a 90% of Carnot, the thermal to electric conversion efficiency is .84. If we further assume an advanced electric thruster can be built with an electric to jet efficiency of .95, the overall thermal to jet efficiency is 80%. The following table lists some of the specifics for this concept description.

TABLE 1. Interstellar Rendezvous Operating Parameters.

Spacecraft Power (W)	Prop flow rate Flow Rate	thrust (Nt)	accel time (s)	accel dist (m)	final microparticle power level (W)	Isp (s)
1.00E+11	4.759E+00	9.756E+05	3.075E+07	4.613E+14	2.677E+15	2.050E+04
2.00E+11	9.518E+00	1.951E+06	1.538E+07	2.306E+14	5.354E+15	
1.00E+11	2.222E+00	6.667E+05	4.500E+07	6.750E+14	1.250E+15	3.000E+04
2.00E+11	4.444E+00	1.333E+06	2.250E+07	3.375E+14	2.500E+15	

High accelerations are important to reduce pointing accuracy requirements. For example, if the fuel/propellant collector at the spacecraft is 100km in diameter, then the macro particle accelerator pointing accuracy requirements range between 0.07 and 0.2 nano radians. While these are very precise pointing accuracies, they are within about a factor of twenty of those required by the Space Based Laser as envisioned for missile defense. It is reasonable to expect that more precise pointing will be available for such a mission. As may be noted, the acceleration times under these conditions range from about ½ of a year to about a year. This is an acceptably short time frame and it provides for minimal trip times. Also to be noted are the relatively short acceleration distances, ranging from 0.02 to 0.07 LY. As mentioned above, this reduces pointing accuracy requirement for the beamed fuel and propellant. The peak power required for the accelerator system is very large, between 120 and 535 Terawatts. This power must be supplied from some source if the spacecraft is to accelerate rapidly to interstellar velocities. It is likely more easily accomplished from the Earth system than attempting to build such a capability into the spacecraft itself.

Pre-Coast, Coast and Deceleration Legs

Immediately after the acceleration leg is completed, the macro particle accelerator continues to supply the spacecraft with uranium fuel. This fuel is used for the deceleration leg into the target star system. A substantively different arrangement must be made for the deceleration leg, since supplying propellant from Earth is practically impossible. Thus, the spacecraft must decelerate using only on-board power and fuel. To minimize mass it can do this by extracting fission products from the fissioned fuel and using the fission products for electric propulsion propellant.

The amount of uranium propellant is very closely coupled to the spacecraft overall efficiency of converting uranium energy into jet power and thrust. To determine the amount of propellant required, we must determine the minimum specific impulse associated with spacecraft parameters. To do this we determine the energy available to the uranium fission fragments based on thermal energy available from uranium and the conversion efficiency from thermal to jet power. Thus, the lower the end-to-end system efficiency, the lower the feasible specific impulse, and consequently, the more mass required to decelerate the vehicle.

8.64×10^{13} = approximate fission energy available per kg-U

conversion efficiency thermal to jet (given) 0.8

$0.8 * 8.64 \times 10^{13} = 6.912 \times 10^{13}$ joules

The above is the energy per unit mass of fission products available for deceleration. From this we find the exhaust stream velocity:

$V_{ex} = \sqrt{(2*6.912 \times 10^{13})} = 1.176 \times 10^7$ m/s, or $I_{sp} = 1.176 \times 10^6$ s

In order to decelerate, the power supply will have to operate at considerably higher power than for acceleration in order to decelerate rapidly at the star system, or we can operate at the identical reactor power and take longer to decelerate. Selecting 1000 GWth as the desired power level for deceleration. The propellant flow rate can be determine be dividing the energy per unit mass by the power available, 800 GWjet = (1000 GWth times the system end-to-end efficiency of 0.8)

Mass flow rate = $8.0 \times 10^{11}/(\frac{1}{2} * Vex^2) = 3.2 \times 10^{11}/(1.176 \times 10^7)^2 = 1.16 \times 10^{-2}$ kg/s = Mdot, or the mass flow rate consistent with the power system providing precisely the power and energy necessary to accelerate the uranium fissioned and used as propellant.

From the rocket equation, we can determine the total propellant required to decelerate from 0.1c

Mtotal = Mspacecraft * exp(0.1c/Vex) = 12700 T – 1000T = 11700T uranium

Dividing the above by the mass flow rate generates the total deceleration time:

Time = $11700T/1.16 \times 10^{-2}$ kg/s = 2.4×10^9 s = 31.9 years

The above allows us to determine the pre-coast fueling portion of the trip. If we launch 1 kg/s of uranium, it will require 1.270×10^7 sec to provide the deceleration uranium. The power required at the Earth system is reduced to approximately 45 Terawatts for an additional 147 days. The spacecraft is now on its own and free from support from the Earth system.

Total Trip Time

To determine the total trip time, we must determine the total distance to the next star system, 4.2 LY. The distance covered in the acceleration phase and time required is:

Distance = 0.0024 LY   Time = 0.048 years

To find the cost time we must determine the average velocity, distance and time covered during deceleration. Since we use the rocket equation for deceleration, we must find the average velocity during deceleration. This is found by multiplying the initial velocity by ln 2 = 0.693, resulting in the average speed during deceleration of 2.08 X 107 m/s. This results in a deceleration distance of

Dist = 2.08×10^7 m/s * 1.00×10^9 s = 2.08 LY

Thus, the total coast distance is 4.100 – 2.08 = ~2.00 LY, which at 0.1c requires 20 years. Therefore, the total trip time is

20 + 0.048 + 32 = 52 years

Significantly, we find that a trip to a star system 8.2 light years distant would require only an additional forty years. Given advances in gerontology, we should expect life spans at the time of these missions to be over 120 years. Under these conditions, we should expect that crewed exploration missions to star systems within an 8 light year radius of the Earth could be feasible by the middle of the next century.

INTERSTELLAR RENDEZVOUS SYSTEM TECNOLOGY ASSESSMENT

Travel to another star system is an exceedingly difficult task. Implementation approaches are frequently awkward and refractory. Technology solutions result in anfractuous aggregates with exceptional demands. For example, neither fission nor fusion reactions contain sufficient energy to easily carry all the necessary fuel for a spacecraft to accelerate to and decelerate from 0.1c, considered by many to be the minimum practical velocity. Consequently, some form of beamed matter, or interaction with the available matter in free space is necessary to close the energy balance. This has resulted in the advocacy of what appear to be completely impractical concepts employing beamed energy to interact with light sails or microwave sails, wherein the power levels, transmitter optics or antennae and sails appear to be well into the next century if ever feasible from an engineering standpoint. Further, at present costs, these concepts have no practical solution from a fiscal point of view.

Finally, and perhaps most fundamental, is the fact that accelerating to 0.1c appears to be relatively practical among a number of alternatives. The major problem is decelerating and stopping at the desired destination. This is particularly the case with complete beamed energy approaches. Fission and fusion appear to require some external source of supply unless long acceleration times are postulated. Either energy source has sufficient energy to decelerate at the destination provided the spacecraft does not have to carry both the acceleration and deceleration fuel with it. At present, it appears that only antimatter has the ability to provide a source of energy for both acceleration and deceleration.

The exceptional difficulty of the requirements for such futuristic concepts led the authors to investigate a hybrid concept involving fission power driven systems. Numerous authors have studied the necessary power densities for adequate accelerations for modest flight times. In another paper, the authors developed an architecture that involved the following provisos:

The issue addressed in this paper is one that assesses available or potentially available technologies to achieve such mission parameters. We divide this assessment into three areas: Power subsystem, Propulsion Subsystem, Microparticle accelerator and Catcher subsystem.

Power Subsystem

The architecture proposed by the authors employed a spacecraft generating 200 GW electric weighing 800 T. This power system provided all power to the spacecraft for acceleration and deceleration as well as housekeeping functions for the trip. The housekeeping functions consume comparatively little power. The crewed portion of the spacecraft weighs 200 T which was judged to be adequate to house a crew of >30for the trip based on extensions of Moon and Mars exploration studies involving inflatable habitations. All life support functions are considered to be closed including food production. Based on former studies, this requires about 11 kWe per individual; an additional 4 kW for work would bring the total housekeeping to about 15 kWe per person or about 0.5 MWe for the entire crew. This is a trivial amount compared to the propulsion requirements.

Refuelable Reactor Component

The above shows that the power system specific mass is 8 X 10^5 kg/8 X 10^8 kWe = 1 X 10^{-3} kg/kWe, or 1000 kWe/kg. If we divide the power system into roughly two parts: 1. The refuelable reactor, and 2. The power conversion and thermal management system, and we allocate roughly ½ of the mass for the reactor plus shield and ½ for the conversion system, we require a reactor that produces, on a continuous basis, 2000 kWe/kg. Since we noted in our other work that the high temperature reactor operates at a 93% thermal to electric efficiency, the reactor must operate at 3000K and it must produce ~2100 kWth/kg. As a point of comparison, in general terms, we note that the SNAP-10A reactor as a thermal power plant generated about 40 kWth and weighed about 160 kg for that system. Consequently, its power system specific power was .25 kWth/kg. This system was designed to run for one year or about 3 X 10^7 seconds. Based on burn up data it could have easily run for several years at that power level, so a conservative estimate would be for roughly 10^8 seconds of operation at .25 kWth/kg. We also note that the SNAP systems were uncharacteristically conservative in power density. The SNAP 10-A operated at roughly 6 watts/cc, well below pressurized water reactors, and well below the demonstrated 500 watts/cc of the FFTF. If we assume that this kind of liquid metal device (the FFTF) could be extrapolated to space operations, we would have a final power density of roughly 25 kWth/kg. While this is a dramatic step, we are still about two orders of magnitude low. On the other end of the chart, we have Nuclear Thermal Rocket systems. These devices operate typically at high power densities, but for short periods of time. The NERVA program demonstrated 2-3 MW/l power densities, or about 600 kWth/kg. The SNTP program, although never completed, did some preliminary fuel element testing, and based on completed tests, one could extrapolate power densities of ~60 MW/l or about 2000 kWth/kg. As a point of comparison, we note some historic technology comparisons as shown in Figure 3, below.

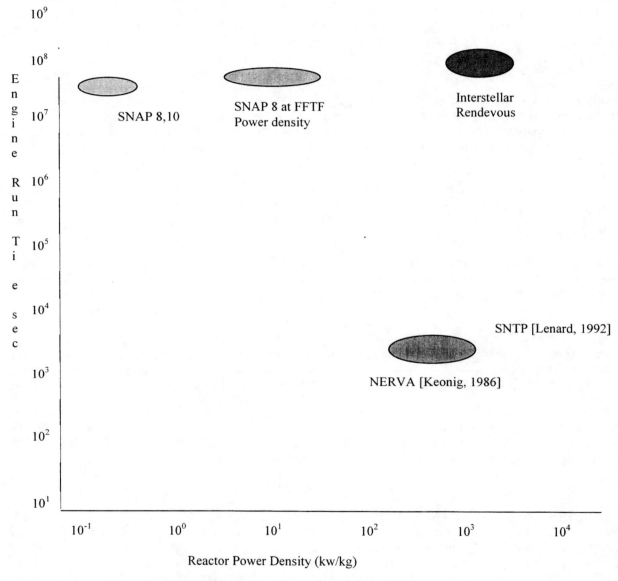

FIGURE 3. Reactor Technology Comparison.

As is clearly denoted above, aspects of the requirements have been demonstrated in either deployed space systems (SNAP-10A), or ground demonstration programs (NERVA and SNTP). Of course, the requirement not shown is to incorporate continuous refueling into a reactor concept that employs the power density of a nuclear thermal rocket with the lifetime of a lower power density space power reactor while also being refuelable. While this technology has never been demonstrated, it appears that it could be achieved.

This portion of the system is comprised of (as a reference only) a high temperature brayton cycle system of high efficiency. It may be that a multiple phase system, such as brayton to rankine to another rankine system to maximize efficiency may be necessary to achieve the required thermal to electric efficiency. However, if the brayton portion can achieve the required power densities, then since rankine systems are of higher theoretical power density, the cascaded systems should also be able to achieve the necessary specific power. Enormous progress has typified the turbine industry, particularly in the areas of increased temperature, efficiency and reliability. As a guide, we require our turbomachinery to operate at the specific mass level of 10 kg/kw shaft power. This is unprecedented when compared to typical turbine components. There are several cases that can be documented, unfortunately few actual systems of a relevant power level have been built and tested. Some of these systems have been identified on the chart below for reference and comparison. Some of the systems are needlessly lower in specific power because there is a potentially misguided notion that the rotating machinery represents the lowest reliability factor of any of the power system components simply because it comprises moving parts. While this may possibly be true in concept, it has not necessarily proved true in practice. Reviewing the chart below shows some historic cases in point.

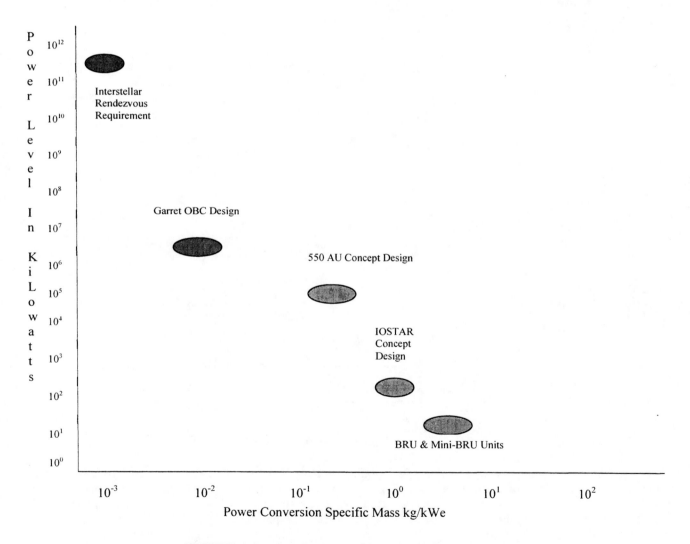

FIGURE 4. Brayton Cycle Design and Operating Points.

A similar problem exists for the conversion of mechanical energy to electrical energy for the interstellar rendezvous mission. Again exceptionally low specific mass values ~10^{-3} kg/kWe are necessary. There has been considerable progress recently in permanent magnet alternators. In recent conversations with Ashman Technologies [Bhargava, 1999] in Santa Barbara California, very high power density units are presently being built for other applications. Many of these units are coupled with very high rotational speed rankine cycle turbine units. Typical operating speeds for some of these devices are approximately 90,000 rpm. These units produce alternating current at very high frequencies. Although alternating current is not viable for direct coupling to the ion propulsion system, at such high frequencies, rectification losses and masses are very low. As with brayton cycle turbines, scaling to higher power levels is very favorable, although with the existing developments have a small data base established. Based on the research by Ashman technologies, the scaling is expected to be less than linear, possibly close to the square root of power. This would imply a projection as shown in the chart in Figure 5, below.

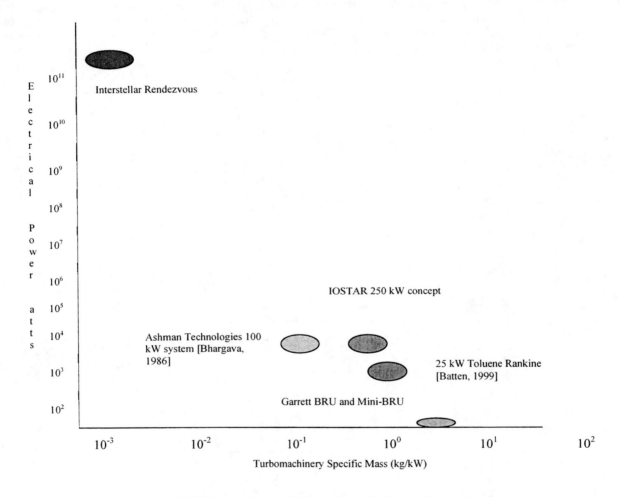

FIGURE 5. Turbomachinery Specific Power Comparisons.

The alternator, not in a redundant configuration, would not achieve the specific power objectives without some modifications. In terms of reducing the weight yet further, the replacement of the permanent magnets and the conventional alternator windings with high temperature superconductors appears to be a good option. The superconducting rotor winding should be able to generate a field approximately 2-3 the flux of the permanent magnets, reducing alternator size and weight. In fact, the mass should be reduced by about of 4-9 excluding structure mass changes that would likely be less profound. Therefore, it appears feasible to meet the specific power objectives of the generation system. The ability to reach an alternator efficiency of 99.95% (an ambitious goal for a superconducting alternator) is extremely important at these power levels. The requirement to radiate several 100 MW of thermal power is difficult when attempting to minimize mass. Wiring weight is also a primary consideration at these power levels. Typical power plant generators operate at 24.5 kV [Lenard, 1999]. Unless the transmission distances are very short, or unless superconducting leads are employed, wire weight will be exceptionally heavy. For this reason, we baseline superconducting leads with a power density of 10,000 amp/cm^2. This will still result in leads over 30 cm in diameter that must be twisted in order to minimize magnetic field effects. The current carried by this conductor is ten million amperes, an unprecedented current carrying capacity for any conductor on a continuous basis.

Power conversion for space nuclear power systems have characteristically been of low efficiency, driven by the nostrum that since the radiator is a high mass entity, its size must be reduced, thus it must operate at high temperature in order to reduce physical size and mass. We have defied this approach by employing two important modifications. First we operate the system at very high ~93% thermal to electric efficiency. This reduces the amount of power that needs to be radiated substantially. Second we envision the use of liquid droplet radiators which use a droplet sheet to radiate power. The droplet sheet requires that it maintain only optical thickness, which at the radiating temperature should be no more than about 30 microns thick. If we employ an organic high vapor pressure sheet, density of 0.8 gm/cm^3, a radiator of this type will have a lower mass limit of 2.4 X 10^{-2} gm/m^2 radiator area, for a two-sided radiator, 1.2 X10^{-2} gm/m^2 radiating area as an extreme lower limit. If the radiator radiates at 200K, the upper limit of radiated power is 81 W/m^2 presuming an emissivity of 0.9. In order to radiate the waste heat, 7% * 2X10^{11} = 1.4 X10^{10} W. Since the radiator radiates at so low power density, a very large radiator is necessary. The total radiating area required is 8.4 X 10^7 m^2, consequently, the radiator is a flat plat roughly 9 km on a side. However, due to the extremely large entrained mass of coolant in a conventionally manifolded flat radiator, we envision a sphere, ~2.5 km radius surrounding the spacecraft with a magnetic collector. The total entrained mass is roughly 50 times the radiating mass, or about 50-100 tons. A superconducting magnetic collector should weigh about 10 tons and eliminate the need for all but the most rudimentary manifolds. The radiator system would have a mass flow rate of approximately 2 X 10^6 kg/s providing a suitable fluid with a heat capacity of .4 cal/gm-degree could be developed.

Electric Propulsion subsystem

The electric propulsion subsystem will certainly prove to be problematic since it must operate in a highly efficient way for very long times at high power densities. There are several propulsion technologies that can possibly be employed, however, no single concept appears to embody all the essential features. For example, if we allocate approximately 300 T to the electric propulsion system , (assuming that power for the electric propulsion system can be directly generated without large power conditioning systems), we require a propulsion system specific power of about 333 kWe/kg. The Magneto Plasma Dynamic thruster has a very high power density, ~tens of Kw/kg, however it is not particularly efficient, and has lifetime issues. The ion engine and two-stage Hall thrusters are efficient ~70+%, but are of low (ion ~0.5 – 1.0 kW/kg) to moderate (Hall, 5-10 kW/kg in larger sizes) power densities. Both possess some lifetime issues although these are being addressed through continuing technology programs. In short, major advances in power density, improvements of a factor of 50-150 are necessary over what has been demonstrated.

Ion engines have demonstrated lifetimes exceeding three years in vacuum chamber, however, lifetimes of 20 years at high power density and high specific impulse have not been even contemplated as yet. What is known is that propellants will have to be changed from Xenon to some other lighter propellant in order to gain optimal efficiency and minimize accelerator grid voltage. There has been no work accomplished on a high efficiency, fission-product ion engine that would have to be used for the deceleration leg.

It may be that a new form of electric propulsion engine will have to be designed, fabricated and tested to meet these requirements if none of the available candidates can meet the diverse requirements. As a consequence of these major requirements, the electric propulsion area is ranked as very high risk and should receive substantial effort if interstellar travel using this conceptual approach is to be contemplated.

Fuel and Propellant Receiver

This architecture requires that fuel (uranium) and propellant for the acceleration portion of the journey (Xenon, Argon or Lithium, for example) be accelerated by a Earth system based accelerator and be collected by a large ~100 km diameter particle collector. The technology for this type of capability should be well-developed by the solar and light sail community. However, care will have to be exercised in velocity control since the light sail cannot withstand large velocity differences.

An alternative architecture could be a variant of the magnetic liquid droplet radiator collector that uses a large magnetic field to collector radiator droplets. Such a collector could possibly be used for the propellant and fuel collection system, although magnetic field strengths must be large and possess a large volume. It is not clear if thrust impingement will result in a problem with incoming fuel and propellant, these issues need to be studied and addressed.

Fuel and Propellant Accelerator

The fuel/propellant accelerator is the highest power consumer in this system. The accelerator must accelerate a cluster of $\sim10^6 - 10^{12}$ atoms of uranium to velocities of 0.1c. For uranium, (the heaviest material accelerated), the particle energies are enormous, $\sim4 \times 10^{14}$ eV $- 4 \times 10^{20}$ eV. The former figure is at the limit of extant accelerators, and the latter figure is completely beyond realistic development at this time. The beam current multiplied by the particle voltage yields the beam power. The Accelerator Production of Tritium program envisioned a 400 MW beam current device. Since the ultimate beam power as mentioned above is ~30 TW, this is about five orders of magnitude greater in total beam power than anything contemplated today.

In our original rendition of this concept, we postulated longer acceleration times, in fact, we presumed that micro particles would have to travel at least two light years, possibly four light years. This resulted in not only extreme pointing requirements, but also in undefined effects due to the interaction of the galactic magnetic field with inevitable charge exchange processes that were difficult to predict. The consequence of these unknowns was to require larger accumulations of atoms in the cluster, resulting in very high single-cluster energies as shown above. Reducing the number of particles from 10^6 to 10^4 would reduce the beam voltage to more reasonable levels, $\sim10^{12}$ eV. The integrated beam power will not change, however, multiple accelerators can be used to meet the total beam current requirements. Unlike lasers, the coherence of this beam is a result of the clusters themselves and does not require large apertures. It is possible that several hundred high power accelerators could be employed in lieu of a single large accelerator. This would introduce needed redundancy into the system as well.

The accelerator is a prodigious undertaking. There does not appear to be a show-stopper in this technology area, although it will be an extremely difficult and expensive development task. It is likely that this will be the most expensive part of the mission, however, it is also likely that these assets can be used for other missions, so the investment can be amortized over multiple missions. At present prices, and introducing quantities of scale, one could roughly calculate the cost of this portion of the technology.

Cost of APT accelerator ~$1B * 100000 (number of APT accelerator equivalents) * LCF 0.81

=   ~$11 Trillion, or roughly $0.40 per watt of beam power

While this is not a trivial investment, it is inexpensive compared to any of the more technically advanced, yet highly speculative options. Similar to the laser for the laser powered lightsail option, the system could, in principal, be used for a variety of different trips to different star systems for a single up-front investment. However, the cost is about three orders of magnitude less than the costs for the laser and beam director. However, unlike the laser-driven lightsail option, the accelerator is in use for only a few months, as opposed to years for the laser system. This could mean that the fission based concept could send a crewed mission to a star system every few years if required.

SUMMARY AND CONCLUSIONS

We have constructed an architecture that allows trips at 0.1c to rendezvous with nearby star systems. It should be noted that the first trips will be one-way, and unless there is some breakthrough in communications technology occurs, it is likely that a completely new society will result once the trip has reached its destination simply due to the time delay and enormous power demands for communications over interstellar distances.

The technology base for these missions is mid- to far-term, available within the next ten to twenty-five years for mission employment given funding and programmatic commitment. We have determined that fission electric propulsion is a very promising concept for these missions requiring no breakthrough in physics and no major engineering advance in either fusion or antimatter. As a matter of analytical review, we see no advantage for fusion systems since while they enjoy an energy advantage of at most a factor of ten in mass; this is insufficient to alleviate the need for some form of logistics supply from Earth during the acceleration portion of the journey in order to provide energy for the deceleration leg of the journey.

Fission systems maintain a distinct advantage in power density over known non-explosive fusion systems. Power density, (alpha) is a critical parameter for these missions and must be minimized at almost all costs. Coupled with their near term availability and known development risks, we find substantial advantages to the fission powered electric propulsion systems to nearby star systems with exploration and settlement crews.

REFERENCES

Batten Bill, 1999. Personal conversations with author on the 25 kWe Solar Thermal Toluene Rankine Cycle engine, Barber Nichols Corporation, Denver CO. May 1999.

Frisbee, R.H. "Interstellar Mission Propulsion Studies – Status Update". Presentation to tenth Annual NASA/JPL/MSFC/AIAA Advanced Propulsion Research Workshop, April 5-8, 1999. Pg 125-153.

Koenig, D.R. "Experience Gained from the Space Nuclear Rocket Program (ROVER), Los Alamos National Laboratory, LA-10062-H, May 1986.

Lenard, R.X. 1999. Brown Boveri steam turbine generator specifications for Bellefonte Nuclear Power plant, site visit by author, 1999.

Lenard, R.X. 1999 "Space Nuclear Thermal Propulsion Briefing", Air Force Phillips Laboratory, May 1992.

Lenard, R.X. and Lipinski, R.J. "Architectures for Fission-Powered Propulsion to Nearby Stars". Presentation to tenth Annual NASA/JPL/MSFC/AIAA Advanced Propulsion Research Workshop, April 5-8, 1999. Pg 218-222i.

Cathode Temperature Reduction By Addition Of Barium In High Power Lithium Plasma Thrusters

James Polk[1], Viktor Tikhonov[2], Sergei Semenikhin[2] and Vladimir Kim[2]

[1] *Jet Propulsion Laboratory, California Institute of Technology, 4800 Oak Grove Dr., Pasadena, CA 91109*
[2] *Research Institute of Applied Mechanics and Electrodynamics, Moscow Aviation Institute, 4 Volokolamskoe shosse, Moscow, 125810, Russia*
[1] *(818) 354-9275, james.e.polk@jpl.nasa.gov;* [2] *(095) 158-0020, riame@sokol.ru*

Abstract. Lithium Lorentz Force Accelerators (LFA's) are capable of processing very high power levels and are therefore applicable to a wide range of challenging missions. The cathode in these coaxial discharge devices operates at a very high temperature to supply the required current and appears to be the primary life-limiting component. One potential method for lowering the cathode temperature is to add a small amount of barium to the lithium propellant. An analytical model of the surface kinetics of this system shows that a relatively small partial pressure of barium can dramatically reduce the cathode operating temperature. Preliminary experiments with a lithium-fuelled thruster demonstrated temperature reductions of 350–400 K with barium addition.

INTRODUCTION

Lorentz force accelerators are the only type of electric thruster with a demonstrated capability to process steady state power levels up to several MWe in a relatively compact device. In these engines a very high current is driven between coaxial electrodes through an alkali metal vapor or gaseous propellant. The current interacts with a self-induced or externally-generated magnetic field to produce an electromagnetic body force on the gas. LFA's can operate efficiently at power levels from 150 kWe up to tens of MWe and are therefore ideally suited for a variety of future missions requiring high power levels. The cathode, which must operate at very high temperatures and is subject to evaporative mass loss, is the primary life-limiting component. This paper describes an analytical model and preliminary feasibility experiments which show that addition of a small amount of barium to the propellant flow can dramatically reduce cathode operating temperatures.

THE CURRENT STATUS OF LITHIUM-FED LFA TECHNOLOGY

The current focus of LFA technology feasibility assessment is on applied-field, lithium-fuelled engines operating at 150–200 kWe. At these power levels the discharge current is not high enough to generate significant self magnetic fields, so an external field generated by a solenoid is used. This field induces azimuthal currents which interact with the radial and axial magnetic field components to accelerate the plasma. Lithium propellant yields very high engine efficiency because it has low frozen flow losses. Because it has a very low first ionization potential and a high second ionization potential, very little power is expended in creating the plasma. Figure 1 shows a schematic of an engine being developed by the Moscow Aviation Institute (MAI) under Jet Propulsion Laboratory sponsorship.

The electrode geometry is designed to balance engine performance considerations with lifetime concerns. The tungsten anode is designed to be radiatively-cooled for this range of power levels, and operates at a temperature of 2000 K or less. The cathode is composed of a bundle of tungsten rods enclosed in a tungsten

CP504, *Space Technology and Applications International Forum–2000*, edited by M. S. El-Genk
© 2000 American Institute of Physics 1-56396-919-X/00/$17.00

tube. An integral heater vaporizes the lithium propellant which then flows through the channels between rods into the discharge. These channels act as small hollow cathodes which, for the proper choice of mass flow rate and current, will very efficiently ionize the lithium. Multichannel hollow cathodes such as these offer more emitting area than comparably sized rod or single channel hollow cathodes. In addition, the attachment is more stable, the operating voltage is lower than large single channel cathodes and there is a higher probability of recapturing material evaporated from the emitting surfaces. The cathode is sized so the current density does not exceed 200 A/cm^2 of cross-sectional area (a total current of 3200 A) and operates at a temperature of about 3000 K. The main insulator between the cathode and anode in the laboratory-model

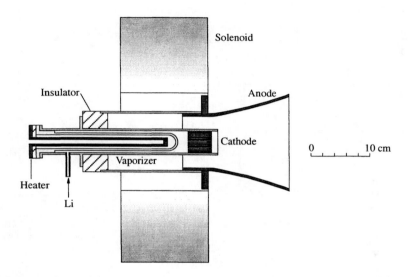

FIGURE 1. 150-200 kWe LFA Developed at the Moscow Aviation Institute.

thruster is aluminum-boron nitride, which provides sufficient life for short-term tests. Hot insulators can react with the lithium vapor, so the engine is designed to maintain a relatively low insulator temperature and isolate it from the lithium plasma. The solenoid in this engine is a water-cooled copper coil.

This engine has been operated at up to 188 kWe and has demonstrated 49 percent efficiency at an Isp of 4500 s. In relatively short duration tests performed so far, the primary life-limiting component appears to be the cathode. With proper design and operating conditions the anode and insulator are not subject to significant wear. The cathode, however, must operate at high temperatures to emit electrons and erodes primarily by evaporation, if the propellant is free of oxidizing contaminants. Erosion rates in terms of mass loss per unit charge transfer as low as 0.1–1 ng/C appear to be achievable if the current density is less than 200 A/cm^2 and the cathode is operating in the hollow cathode mode. At 3200 A, this yields a mass loss rate of 1–10 g/khr, so operation for several hundreds or perhaps a few thousand hours appears feasible. Achieving greater service life capability requires reducing the cathode operating temperature.

AN ANALYTICAL MODEL OF A CATHODE IMMERSED IN AN ACTIVATING VAPOR

The cathode operating temperature is very strongly affected by the cathode work function. Composite surfaces composed of refractory metals activated with submonolayer films of electronegative adatoms often have a work function which is lower than that of the adsorbate or the substrate. The desorption energy of submonolayer films is also generally lower than the bulk sublimation energy of the adsorbate. The work function is strongly dependent on substrate crystal orientation and coverage, and may have a deep minimum at a coverage less than one monolayer because of adatom interactions. Figure 2 shows this behavior for the work function of lithium and barium films on the (110) face of tungsten as a function of the coverage f, defined as the ratio of surface density to the density at the work function minimum, N/N_{min} (Medvedev, 1974,

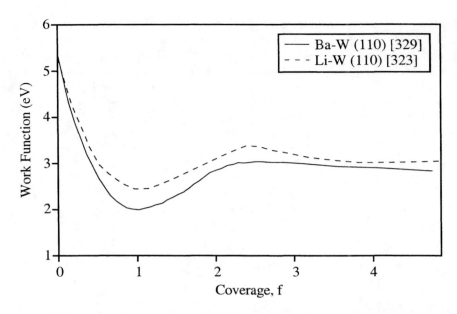

FIGURE 2. Work function of barium-tungsten and lithium-tungsten composite surfaces.

Fedorus, 1972). Because the atoms in the film desorb at cathode operating temperatures, the film material must be replenished. If a sufficient partial pressure of the activator is maintained over the emitting surface, condensation can balance the desorption rate at a given coverage level. To determine the surface emission characteristics of a tungsten cathode immersed in a lithium or barium vapor, the adsorption isotherms for these systems were modeled.

The equilibrium surface coverage of an activator j supplied to the cathode from the gas phase is given by the relationship

$$k_a^j n_{j,s} = k_d^j N_j, \tag{1}$$

where k_a^j is the condensation rate coefficient, $n_{j,s}$ is the activator gas phase number density near the surface, k_d^j is the desorption rate coefficient and N_j is the surface density. Assuming that the adsorption process is non-activated, the adsorption sites are non-localized, there are no competing adsorbate species and that the adsorbate flux to the surface is determined by the random thermal flux of vapor at a temperature equal to the surface temperature T_s, the adsorption isotherm is given by

$$\frac{P}{(2\pi m_j k T_s)^{1/2}} = \omega_j \exp(-E_d^j/k T_s) N_{min}^j f_j, \tag{2}$$

where P is the vapor pressure, m_j is the molecular weight of the activator species, k is the Boltzmann constant, ω_j is a pre-exponential constant, E_d^j is the desorption energy, N_{min}^j is the surface density at the work function minimum and f_j is the fractional surface coverage, N_j/N_{min}^j.

The adsorption of barium and lithium on tungsten appears to satisfy the first two assumptions, so this approach was used to model these systems. A tungsten surface composed of crystal faces preferentially oriented in the (110) direction was assumed because this yields the lowest work function when covered with lithium or barium. This type of surface can be achieved by chemical vapor deposition of tungsten (Hartenstine, 1994) or may result naturally from operation at high temperature because it is the lowest surface energy state (Hatsopoulos, 1979). Experimental measurements of the desorption energy and the pre-exponential factor ω for barium (Medvedev, 1969) and lithium (Medvedev, 1974) on the (110) face of tungsten as functions of coverage and temperature, respectively, are shown in Figures 3 and 4. Lithium has a lower desorption energy and a higher ω than barium, giving it a higher desorption rate for a given coverage and temperature. This is reflected in the calculated adsorption isotherms displayed in Figures 5

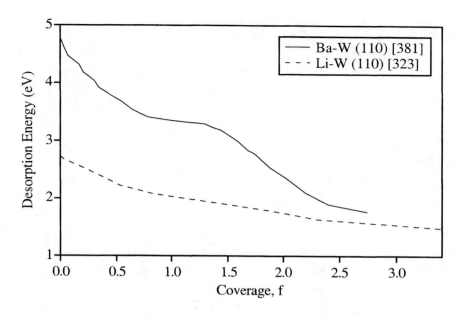

FIGURE 3. Desorption energy of barium and lithium from the tungsten (110) face.

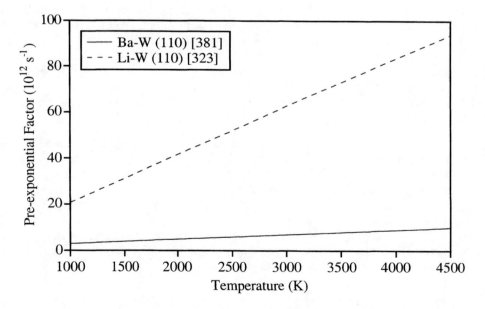

FIGURE 4. Pre-exponential factor for desorption of barium and lithium from the tungsten (110) face.

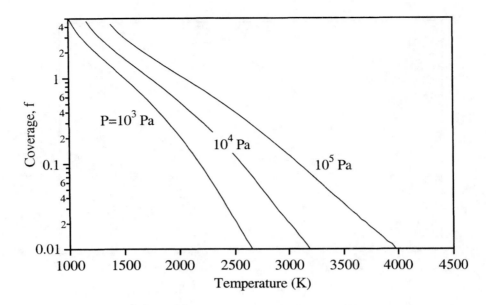

FIGURE 5. Adsorption isotherms for lithium on the tungsten (110) face.

and 6. To maintain the optimum surface coverage of lithium $f_{Li} = 1$ at a given temperature requires a vapor pressure about four orders of magnitude larger than that of barium at $f_{Ba} = 1$. These isotherms were used with the data shown in Figure 2 to calculate the work function for Ba-W and Li-W as a function of surface temperature. The minimum work function occurs at the temperature which gives the optimum coverage. At lower temperatures the work function approaches that of the bulk activator on tungsten and at high temperatures approaches that of pure tungsten. The current density calculated from the Richardson equation is plotted in Figures 7 and 8. For barium a Richardson coefficient $A = 1.5$ A/cm^{2}K^2 was used (Fomenko, 1966), and for lithium $A = 3$ was assumed. This value is equal to that of cesium on tungsten, a more well-characterized alkali metal-tungsten system. As the coverage approaches zero or many monolayers, the value of A should approach 120 A/cm^{2}K^2, the theoretical value for pure metals. The current density for pure tungsten, lithium and barium are plotted in these figures as well, showing that the calculated S-curves for the composite surfaces underestimate the current density at low and high temperatures because of the assumed Richardson coefficient.

These calculations show that to extract high current densities from a cathode immersed in a lithium vapor requires unreasonably high lithium partial pressures. Evidently decreased cathode operating temperatures measured with lithium propellant are due to simultaneous adsorption of lithium and oxygen present as an impurity in the lithium (Babkin, 1979). Such surfaces have a lower work function minimum than lithium alone on tungsten and may also elevate the lithium desorption energy. Unfortunately, the oxygen impurities also cause chemical erosion of the tungsten, apparently negating the lifetime gains associated with the lower temperature. However, barium can be used as a propellant additive to maintain a low work function with quite reasonable partial pressures. Comparing the pure tungsten curve with the activated surface curves suggests that temperature reductions of over 1000 K should be possible.

PRELIMINARY EXPERIMENTS WITH BARIUM ADDITION

In tests of this approach using barium vapor added to the lithium propellant vapor in a high power LFA a temperature reduction of over 1000 K was reported (Ageyev, 1993). Similar experiments were performed at MAI with a 30 kW lithium plasma thruster to verify these findings. This thruster is similar in design to that shown in Figure 1 and operates at current levels of 600–700 A with an efficiency of up to 32 percent at Isp's as high as 3600 s (Tikhonov, 1993). A small tungsten cup filled with barium metal and capped with

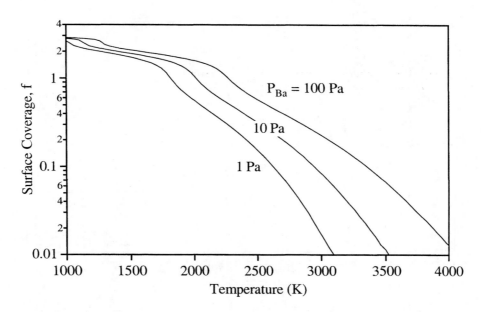

FIGURE 6. Adsorption isotherms for barium from the tungsten (110) face.

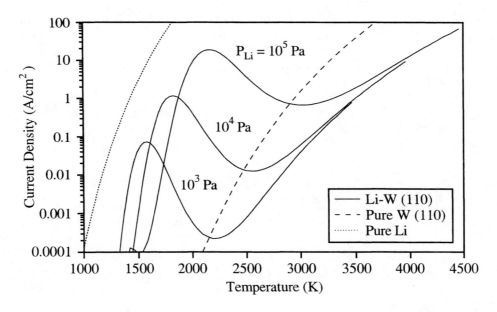

FIGURE 7. S-curves of emission current density for the lithium-tungsten (110) system.

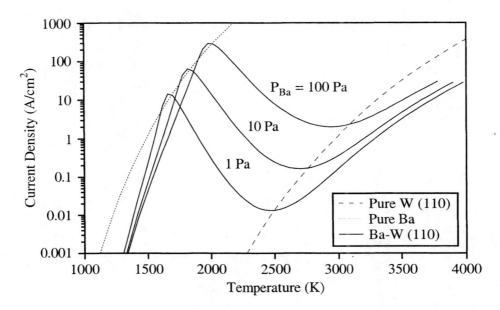

FIGURE 8. S-curves of emission current density for the barium-tungsten (110) system.

a porous tungsten plug was placed inside the cathode tube upstream of the tungsten rods. As this cup was heated by operation of the discharge, barium vapor was released into the lithium propellant flow.

In these preliminary experiments there was no way to control or measure the barium flow rate, and the only possible location for the barium source in this thruster design was near the tip of the cathode where the temperatures were 1500–1700 K. This resulted in very high initial partial pressures of barium during the first few minutes of operation. During this phase of operation the discharge was characterized by low discharge currents I_d and increased values of the discharge voltage V_d. All attempts to raise the discharge current resulted in further increases in discharge voltage with an almost constant discharge current. In each experiment the thruster transitioned to a normal operating mode which most likely corresponded to a lower, more stable barium flow as the supply in the source was depleted. During this phase the thrust T was measured using the apparatus described in (Tikhonov, 1995) and cathode and anode temperatures T_c and T_a were measured with an EOP-66 optical pyrometer. Table 1 compares results obtained with and without barium addition at similar lithium flow rates $\dot{m}$, applied magnetic field strength measured at the cathode tip B_c and discharge current.

TABLE 1. Comparison of thruster operation with and without barium addition.

Mode	$\dot{m}$ (mg/s)	B_c (G)	I_d (A)	U_d (V)	T (N)	η	T_c (K)	T_a (K)
Li + Ba	14.0	1000	700	38.0	0.499	0.335	2513	2153
Li	13.5	1000	700	41.5	0.504	0.325	2903	2273

The experimental results show unambiguously that the addition of barium results in a cathode temperature reduction of 350–400 K and an anode temperature reduction of about 100 K with no significant change in thruster performance.

CONCLUSIONS

The analytical model of activator adsorption and work function lowering on tungsten suggests that substantial cathode temperature reductions are possible with barium addition. Preliminary experiments with an uncontrolled lithium vapor flow qualitatively confirm these predictions. A cathode with a geometry, thermal design and barium vapor flow rate optimized to operate at the minimum work function should be capable of running at a temperature of 2000–2500 K for many thousands of hours. Subsequent work will focus on modeling the effect of the electric field on the transport of lithium and barium ions in the cathode concentration boundary layer and performing experiments with controlled flow rates of barium to determine the optimum operating conditions.

ACKNOWLEDGMENTS

This research was conducted at the Jet Propulsion Laboratory, California Institute of Technology, and the Research Institute of Applied Mechanics and Electrodynamics of the Moscow Aviation Institute. The work was sponsored by the National Aeronautics and Space Administration.

REFERENCES

Ageyev, V. P., V. G. Ostrovsky and V. A. Petrosov (1993) "High-Current Stationary Plasma Accelerator of High Power," in *Proc. 23rd International Electric Propulsion Conference*, IEPC-93-117, held in Seattle, WA, 13-16 September 1993.

Babkin, G.V. and Potapov, A.V., "An Experimental Investigation of the Effect of Oxygen on the Erosion of a Multichannel Tungsten Cathode," *J. Appl. Mech. and Tech. Phys.*, **20**, 312-316 (1979).

Fedorus, A.G., Naumovets A.G. and Vedula, Yu. S. "Adsorbed Barium Films on Tungsten and Molybdenum (011) Face," *Phys. Stat. Sol. (A)*, **13**, 445-456 (1972).

Fomenko, V.S. *Handbook of Thermionic Properties*, New York, Plenum Press Data Division, 1966.

Hartenstine, J.R., Horner-Richardson, K.D., Bosch, D.R. and Jacobson, D.L., "Evaluation of the Thermionic Work Function of the Mandrel Side of Chemically Vapor Deposited and Vacuum Plasma Sprayed Refractory Metals," in *Proc. 29th IECEC*, IECEC 94-3975, held in Monterey, Calif., 1994.

Hatsopoulos, G.N. and Gyftopoulos, E.P., *Thermionic Energy Conversion. Volume II: Theory, Technology and Application*, Cambridge, MA, The MIT Press, 1979.

Medvedev, V.K., "Adsorption of Barium on the (110) Face of a Tungsten Single Crystal," *Soviet Physics–Solid State*, **10**, 2752-2753 (1969).

Medvedev, V.K. and Smereka, T.P. "Lithium Adsorption onto the Fundamental Faces of a Tungsten Single Crystal," *Sov. Phys. Solid State*, **16**, 1046-1049 (1974).

Tikhonov, V. B., S. A. Semenikhin, V. A. Alexandrov, G. A. Dyakonov and G. A. Popov (1993) "Research of Plasma Acceleration Processes in Self-Field and Applied Magnetic Field Thrusters," in *Proc. 23rd International Electric Propulsion Conference*, IEPC-93-076, held in Seattle, WA, 13-16 September 1993.

Tikhonov, V. B., S. A. Semenikhin, J. R. Brophy and J. E. Polk (1995) "The Experimental Performance of the 100 kW Li Thruster with External Magnetic Field," in *Proc. 24th International Electric Propulsion Conference*, IEPC-95-105, held in Moscow, Russia, 19-23 September 1995.

High Power Theta-Pinch Propulsion
For Piloted Deep Space Exploration

Michael R. LaPointe

Horizon Technologies Development Group, Cleveland, OH 44135
Phone: (216) 476-1170; E-mail: HorizonTDG@aol.com

Abstract. The piloted deep space exploration missions envisioned by the NASA Human Exploration and Development of Space initiative will require the development of advanced electric propulsion systems capable of providing high specific impulse for extended periods of operation. Current electric propulsion thrusters are well suited for orbit maneuvering and robotic exploration, but at present they cannot provide the combination of specific impulse, lifetime, and efficiency required for piloted deep space missions. The theta-pinch thruster concept is a high power plasma rocket that can potentially meet these future deep space propulsion requirements. Efficient, partial preionization of a gas propellant followed by rapid adiabatic magnetic compression is used to generate, heat, and expel a high velocity, high density plasma to provide thrust. The concept is electrodeless, and radial compression of the plasma by the magnetic field of the discharge coil mitigates material erosion to ensure long thruster life. Because the heated plasma is free to flow along axial magnetic field lines during compression, a magnetic mirror located at the entrance to the discharge chamber is used to direct the plasma flow out of the thruster. The thrust and specific impulse of the engine can be tailored for a given mission scenario through the selection of propellant species, mass flow rate, compression coil discharge current, and/or the compression coil repetition rate, making this a unique and versatile electric propulsion system.

INTRODUCTION

Chemical rockets and low-thrust electric propulsion systems are well suited for near-Earth orbit applications and long-duration robotic space flight, but advanced propulsion technologies must be developed to enable more complex orbital missions and to promote the human exploration and development of space. To maximize the useful mission payload, the propellant exhaust velocity, v_e, must be comparable to or greater than the required mission velocity, Δv, as described by the rocket equation:

$$(m_f / m_0) = \exp[-\Delta v / v_e] \ , \tag{1}$$

where m_f is the rocket mass after the propellant is expended and m_0 is the initial rocket mass. If the exhaust velocity is much smaller than the required mission velocity, then most of the rocket mass will be in the form of propellant. Less propellant is required if the exhaust velocity is comparable to or larger than the required mission velocity, allowing a larger payload mass fraction. The mission velocities required by the future piloted deep space exploration scenarios envisioned in NASA's Human Exploration and Development of Space (HEDS) initiative fall in the range of 10^4 m/s $< v_e < 10^5$ m/s, corresponding to optimum thruster specific impulse (I_{sp}) values of 1000 s to 10,000 s (Liefer, 1998; NASA 1998). The highest specific impulse provided by chemical combustion is around 500 s, and a significant amount of chemical propellant would have to be expended to perform piloted deep space exploration missions. Electric propulsion thrusters, although capable of achieving the required I_{sp} values, are currently low thrust devices used for primary propulsion on small robotic spacecraft (Polk, 1999; Sovey, 1999) or for auxiliary applications such as satellite stationkeeping and orbit maneuvering (Curran, 1998; Dunning, 1999).

A number of advanced concepts have been investigated over the past several decades in an effort to develop a robust propulsion system for piloted deep space exploration. While each concept has merit, each also suffers from apparent performance or feasibility issues. Current high power electric propulsion systems can provide the necessary specific impulse values, but electrode erosion may limit thruster lifetimes and place unwarranted constraints on future mission options. Electrodeless plasma thrusters would be good candidates for long-duration space missions, but present electrodeless systems cannot produce the proper combination of thrust and specific impulse required for piloted mission applications. Fission rockets offer high thrust, but their maximum specific impulse of around 1000 s is too low for some of the more advanced HEDS mission scenarios. Fusion engines and antimatter-based propulsion systems potentially offer appropriate combinations of high thrust and high specific impulse, but both technologies are still decades away from practical development and deployment. The theta-pinch thruster concept discussed in this paper attempts to bridge the gap between current high power electric propulsion systems and the more exotic fusion and antimatter-based thrusters of the future to provide new, near-term propulsion capabilities in support of piloted deep space exploration.

PINCHED PLASMA PROPULSION CONCEPT

The proposed thruster concept combines efficient preionization with pulsed adiabatic compression to generate a high density plasma exhaust whose temperature and hence specific impulse can be tailored to meet a variety of mission requirements. The body of the thruster consists of a hollow cylindrical chamber surrounded by a single turn discharge coil, which in turn is connected to a capacitor bank or solid-state current driver. Neutral propellant gas is radially injected into the chamber, and a brief current pulse is sent through the single-turn coil to partially preionize the outer edges of the gas fill. The main energy storage bank is then immediately discharged through the single-turn coil. The rapidly rising primary current creates a time-varying axial magnetic field within the chamber, which induces an azimuthal current in the preionized plasma sheath via Faraday's law of induction. The sheath current interacts with the axial magnetic field produced by the coil to generate a radially inward Lorentz force on the plasma ions and electrons. These particles move toward the axis of the chamber, and through collisions sweep the other gas particles along in a "snowplow" effect. During a fast discharge, the partially ionized gas is radially compressed at a speed greater than the local speed of sound, creating a radial shock wave characteristic of irreversible heating processes. Compression continues until the kinetic pressure of the shock-heated plasma balances the magnetic pressure of the axial magnetic field. Balance is generally attained with a fraction of a microsecond, at which point the shock heating is complete. The pulsed current sent through the coil has a rise time much longer than a microsecond, however, and as the current in the coil continues to increase so does the axial magnetic field. The increasing magnetic pressure produced by the increasing field continues to compress and heat the plasma until the peak current is reached, typically within a few microseconds. If the primary compression is slower than the local speed of sound in the plasma, then a shock wave is not formed and the plasma is heated by simple adiabatic compression. In either case, the axial magnetic field that heats the plasma also serves to confine it radially, which prevents the plasma from contacting the walls of the discharge chamber. A magnetic mirror is used to provide single-pass reflection of the axial plasma flow from the upstream end of the discharge chamber. The plasma is free to exit the downstream end of the discharge chamber to provide thrust, and the compression cycle is repeated at a repetition rate prescribed by the plasma exhaust time.

A preliminary scaling analysis has been performed to estimate the size and potential performance of a high power theta-pinch thruster suitable for deep space mission applications. The following section outlines the results of the scaling analysis and presents a preliminary design for the pulsed high power thruster. The final section of the paper discusses preliminary steps toward a numerical model of the thruster plasma physics and outlines a modest research program to experimentally evaluate the concept.

THETA-PINCH THRUSTER SCALING ANALYSIS

A simple scaling analysis was undertaken to estimate potential thruster performance and to provide some guidance for planned MHD numerical simulations. The theta-pinch thruster uses strong, time-varying magnetic fields to radially compress and heat a plasma propellant, which then expands axially out of the chamber to provide thrust. The ideal exhaust velocity of the plasma propellant as a function of temperature is given by (Sutton, 1958):

$$v_e = \sqrt{\frac{2\gamma}{(\gamma - 1)} \frac{R_u T}{M} \eta_i} , \qquad (2)$$

where v_e is the propellant exhaust velocity (m/s), T is the propellant temperature (K), M is the propellant molecular weight (kg/kmol), R_u is the universal gas constant (8314.4 J/K-kmol), γ is the adiabatic index for the propellant gas, and η_i is ideal cycle efficiency. The adiabatic index is given by:

$$\gamma = \frac{N+2}{N} , \tag{3}$$

where N is equal to the number of degrees of freedom of the gas. For ionized propellants $N = 3$, yielding a value of $\gamma = 5/3$ for the adiabatic index. The ideal cycle efficiency is given by:

$$\eta_i = 1 - \left(\frac{P_{ex}}{P_{ch}} \right)^{\frac{(\gamma-1)}{\gamma}} , \tag{4}$$

where p_{ex} is the exhaust pressure and p_{ch} is the chamber pressure. During operation p_{ch} greatly exceeds p_{ex}, hence the value of η_i is generally close to unity. Substituting the values $\gamma = 5/3$ and $\eta_i = 1$ in Equation 2 yields the following expression for the (ideal) propellant exhaust velocity, expressed in m/s:

$$v_e = 2.04 \times 10^2 \sqrt{\frac{T(K)}{M}} . \tag{5}$$

The specific impulse (Isp) of the rocket is related to the exhaust velocity by the equation:

$$I_{sp} = \frac{v_e}{g} , \tag{6}$$

where g is the acceleration of gravity, 9.8 m/s^2, and the specific impulse is expressed in seconds. Equations 5 and 6 can be combined to find the propellant temperature required to achieve a given engine specific impulse:

$$T(K) = M \left(\frac{g\, I_{sp}}{2.04 \times 10^2} \right)^2 . \tag{7}$$

For a given specific impulse, the required plasma temperature can be used to determine the amount of compressional heating that must be provided by the theta-pinch coil. The compression is assumed to be adiabatic, i.e. slow enough that shock waves do not develop. In this case the final plasma temperature is related to the initial plasma temperature through the ratio of initial to final plasma radius:

$$\frac{T_f}{T_0} = \left(\frac{r_0}{r_f} \right)^{4/3} , \tag{8}$$

where T_0 (T_f) is the initial (final) plasma temperature and r_0 (r_f) is the initial (final) plasma radius. The value (r_0/r_f) is the amount of compression that must be supplied by the axial magnetic field during the pulsed discharge. Using Equations 7 and 8 and assuming an initial plasma preionization temperature, the compression ratio required to provide a given specific impulse can be determined.

The pressure of the compressed plasma column, p_f, is related to the initial pressure, p_0, by the adiabatic pressure law:

$$\frac{P_f}{P_0} = \left(\frac{n_f}{n_0} \right)^{5/3} , \tag{9}$$

where n_0 is the initial plasma number density and n_f is the final plasma density after compression. Plasma pressures, densities, and temperatures are assumed to be related by the ideal gas law:

$$p_0 = \sum (n_0 k T_0)_j ,$$ (10a)

$$p_f = \sum (n_f k T_f)_j ,$$ (10b)

where k is the Boltzmann constant (1.3807×10^{-23} J/K), and the summation is over all particle species present in the plasma. Equations 8 through 10 can be combined to find the following expression for the ratio of final to initial pressure:

$$\frac{p_f}{p_0} = \left(\frac{r_0}{r_f} \right)^{10/3} ,$$ (11)

where (r_0/r_f) is just the plasma compression ratio. The plasma pressure is balanced by the pressure of the axial magnetic field created by the time-varying current in the discharge coil. As the current in the coil increases, the magnetic field compresses the plasma column through the "snowplow" effect. The magnetic field strength required to compress the plasma column is given by:

$$\sum p = \frac{B_0^2}{2\mu_0} = \sum (n k T) ,$$ (12)

where B_0 is the axial magnetic field strength (Tesla) and μ_0 is the permittivity of free space ($4\pi \times 10^{-7}$ H/m). The axial magnetic field provides radial compression, but the plasma is still free to move along the axial field lines through the discharge chamber. To provide directed thrust, a strong magnetic mirror field has to be provided at the upstream end of the discharge chamber. The mirror field does not provide perfect refection of the ionized plasma, and some fraction of the particles will be lost through the magnetic mirror. The probability that a particle will be lost through the mirror is given by:

$$P_L = 1 - \left(\frac{R_M - 1}{R_M} \right)^{1/2} ,$$ (13)

where P_L is the probability that the particle will be lost, and R_M is the ratio of the mirror magnetic field strength (B_M) to the strength of the axial magnetic field in the chamber (B_0):

$$R_M = \frac{B_M}{B_0} .$$ (14)

A small loss probability, P_L, is required to ensure that most of the plasma is directed axially downstream and out of the engine. A mirror ratio of $R_M = 2$ yields a loss probability of around 30%, while a mirror ratio of $R_M = 5$ yields a loss probability of around 10%. The limit on the achievable mirror ratio is governed by the strength of the required axial field needed for plasma compression and the constant magnetic mirror field located at the upstream end of the chamber. A reasonable value for the mirror magnetic field is around 10 T, which can be readily achieved using current superconducting magnet technology. It will be assumed in the scaling analysis that the maximum mirror field is 10 T, and that a mirror ratio of at least 5 is required for satisfactory engine performance.

The axial magnetic field that compresses and heats the plasma is produced by a time-varying current pulsed through a single turn coil that runs the length of the discharge chamber. The current required to produce an axial magnetic field of strength B_0 over a discharge chamber of length L is given by:

$$I = \frac{B_0 L}{\mu_0} .$$ (15)

The strength required of the compression magnetic field varies over the discharge period, rising during plasma compression and falling as the plasma exits the thruster. To improve thruster efficiency, a pulse forming network will be used to tailor the current pulse to fit the required magnetic field profile during plasma compression and exhaust.

Assuming the particle losses through the mirror field can be neglected, the thrust, F, produced by the pulsed theta-pinch engine is given by the following expression:

$$F = \dot{m} v_e , \tag{16}$$

where $\dot{m}$ is the propellant mass flow rate (kg/s), v_e is the propellant exhaust velocity (m/s), and the thrust is expressed in Newtons (1 N = 0.2248 lbf). The mass flow rate depends on the propellant mass injected during each pulse, Δm (kg/pulse), and the engine pulse repetition rate, f (number of pulses per second, in Hz):

$$\dot{m} = \Delta m \cdot f . \tag{17}$$

The pulse repetition frequency, f, will be limited by how quickly the plasma evacuates the discharge chamber. The time to evacuate the chamber can be approximated as the time it takes for a particle moving along an axial field line at velocity v_e to travel from the downstream end of the chamber, reflect from the upstream magnetic mirror, and travel back the length of the chamber. The corresponding single pulse time, τ_p, is $2L/v_e$, yielding a maximum pulse repetition frequency of:

$$f = \frac{1}{\tau_p} = \frac{v_e}{2L} . \tag{18}$$

The frequency, f, corresponds to the maximum pulse repetition frequency at a 100% engine duty cycle. The amount of mass injected during each pulse can be expressed in terms of the initial propellant mass density, ρ_0, and the discharge chamber volume initially filled by the plasma, $\pi r_0^2 L$. The propellant mass density is related to the plasma number density by:

$$\rho_0 = m_H M' n_0 , \tag{19}$$

where m_H is atomic mass of hydrogen, M' is the propellant atomic weight expressed in amu, and the plasma number density n_0 is expressed in m^{-3}. The mass injected during each pulse is then given by:

$$\Delta m = \rho_0 \pi r_0^2 L = m_H M' n_0 \pi r_0^2 L . \tag{20}$$

Substituting Equations 17 through 20 into Equation 16 yields the following expression for the thrust, F, at 100% engine duty cycle:

$$F = (\rho_0 \pi r_0^2 L) \left(\frac{v_e}{2L} \right) v_e = (1/2) \rho_0 \pi r_0^2 v_e^2 , \tag{21}$$

or, in terms of initial number density, n_0:

$$F = (8.36 \times 10^{-28}) n_0 M \pi r_0^2 v_e^2 . \tag{22}$$

Note that the length of the discharge chamber does not enter the thrust calculation, due to the approximation made in Equation 18 for the pulse repetition frequency.

PERFORMANCE ANALYSIS FOR HYDROGEN PROPELLANT

The previous set of equations can be used to evaluate the potential performance of the pulsed theta-pinch thruster operated with various propellants. In this section, sample calculations are provided to demonstrate how the scaling analysis can be used to determine the axial magnetic field requirements, pulse repetition frequencies, and thrust values for a theta-pinch thruster operated with hydrogen. An initial plasma temperature of 1160 K (0.1-eV) was chosen for the plasma preionization temperature, and a maximum field of 10 T was assumed for the magnetic mirror at the upstream end of the chamber. The uncompressed hydrogen propellant is assumed to have an initial radius of 1 meter, equal to the radius of the discharge chamber, and an initial number density of 10^{21} m^{-3}. Two thruster cases are analyzed, corresponding to specific impulse values of 5,000 s and 10,000 s, respectively.

Specific Impulse of 5000 s

Equation 7 predicts that a hydrogen plasma temperature of 5.4×10^3 K (4.9-eV) is required to achieve a specific impulse of 5000 s with hydrogen propellant. For the assumed preionization propellant temperature of 1160 K (0.1-eV), the adiabatic compression ratio given by Equation 8 is around 18.5. Assuming an initial plasma (chamber) radius of 1 meter, the final compressed plasma radius will be approximately 0.054 m (5.4 cm). The small plasma radius achieved during radial compression has some beneficial consequences for the design of the magnetic mirror field coil. Since the plasma does not significantly expand along the axial direction during compression, only the compressed plasma column will be affected by the mirror field. The small radius of the plasma column allows the use of a small radius magnetic field coil, which mitigates the fabrication and mass constraints imposed by large-diameter field coils.

For an initial hydrogen number density of 10^{21} m^{-3} and an initial temperature of 1160 K, Equation 10a predicts an initial (uncompressed) plasma pressure of around 16 Pa (0.12 torr). Using an adiabatic compression ratio of 18.5, the maximum pressure of the radially compressed plasma column (Equation 11) is around 2.7×10^5 Pa (2.0×10^3 torr). This is the amount of pressure that must be supplied by the axial magnetic field to radially confine the plasma column. From Equation 12, the magnitude of the confining field is approximately 0.82 T. The magnetic mirror is assumed to have a constant value of 10-T, providing a mirror ratio of 12. From Equation 13, the fraction of particles lost from the mirror field is around 4%, indicating that most of the plasma will be exhausted from the downstream end of the discharge chamber.

The amount of thrust produced by the engine operating at 100% duty cycle is calculated using Equation 22. A specific impulse of 5000 s corresponds to a propellant exhaust velocity of 4.9×10^5 m/s. For an initial hydrogen number density of 10^{21} m^{-3} and an uncompressed plasma (chamber) radius of 1 m, the thrust at 100% duty cycle is approximately 6,300 N. The propellant mass flow rate, given by Equation 16, is around 0.13 kg/s of hydrogen. The maximum pulse repetition rate, corresponding to 100% engine duty cycle, is given by Equation 18. Assuming a chamber length of around 20 m, the maximum repetition rate is approximately 1225 Hz. Given a mass flow rate of 0.13 kg/s and a repetition rate of 1225 Hz, the propellant mass injected into the thrust chamber during each pulse is approximately 1×10^{-4} kg, in agreement with the predictions of Equation 20. Note that a longer chamber length would reduce the maximum pulse repetition frequency, while a shorter chamber length would require a higher repetition rate to sustain the same level of thrust. The engine can be operated at less than 100% duty cycle by decreasing the repetition rate below the maximum rate predicted by Equation 18, which allows operation at the same value of specific impulse but lowers the average thrust.

Specific Impulse of 10,000 s

As predicted by Equation 7, to operate the theta-pinch thruster with hydrogen at a specific impulse of 10,000 s will require a bulk plasma temperature of around 2.3×10^5 K (19.9 eV). Assuming an initial plasma temperature of 1160 K (0.1-eV), the adiabatic compression ratio required to provide this high propellant temperature is approximately 52.4. Assuming the uncompressed plasma number density is again equal to 10^{21} m^{-3}, the initial plasma pressure is 16 Pa (0.12 torr). Using a compression ratio of 52.4 and an initial pressure of 16 Pa, Equation 11 predicts a fully compressed plasma pressure of 8.6×10^6 Pa (6.5×10^4 torr). The axial magnetic field strength required to create this radially confining pressure on the plasma column is approximately 4.6 T. Again assuming a magnetic mirror field strength of 10 T yields a mirror ratio of around 2.1. The fraction of particles lost through the mirror (Equation 13) is approximately 26%, which would significantly impair the operation of the thruster. To reduce the particle loss we can either increase the magnetic mirror field strength or reduce the initial plasma number density. Reducing the initial plasma number density to 10^{20} m^{-3} yields an initial plasma pressure of 1.6 Pa (0.012 torr) and a compressed plasma column pressure of 8.6×10^5 Pa (6.5×10^3 torr). The maximum magnetic field strength required to compress the plasma column to this pressure is approximately 1.5 T, which for a 10-T mirror field corresponds to a mirror ratio of 6.8. The fraction of plasma particles lost from the upstream end of the chamber is then reduced to around 8%, and most of the plasma will be exhausted to provide useful thrust. For an initial plasma number density of 10^{20} m^{-3}, an initial plasma (chamber) radius of 1 m, and an exhaust velocity of 9.8×10^4 m/s (Isp = 10^4 s), the thrust at 100% duty cycle predicted by Equation 22 is approximately 2500 N. The mass flow rate for the thruster operating at 10,000 s and 2500 N is 2.6×10^{-2} kg/s, which is a factor of 5 less than the mass flow rate required for operating the thruster at 5000 s specific impulse and a 100% duty cycle.

The pulse repetition rate required to achieve a 100% engine duty cycle depends on the length of the discharge chamber and the propellant exhaust velocity (Equation 18). For a 20-m long chamber and an exhaust velocity of 9.8×10^4 m/s, the required pulse repetition rate is around 2450 Hz. Reducing the chamber length would require a higher repetition rate to sustain the same level of thrust, while lengthening the chamber length would reduce the required repetition rate. As before, the engine could be operated at less than 100% duty cycle by reducing the repetition rate, which would provide the same value of specific impulse but reduce the average thrust.

PROGRAM PLANS

The preceding analysis indicates that the theta-pinch thruster could be a viable propulsion technology for future deep space exploration missions. Based on this preliminary analysis, a time-dependent MHD model is being developed to more fully examine potential thruster performance. The initial version of the numerical simulation incorporates a single-fluid continuity equation, single-fluid radial and axial momentum equations, a single-temperature energy equation, the full complement of Maxwell's equations, a generalized Ohm's law, and an adiabatic equation of state. The simulation will be used to investigate radial plasma compression and heating, axial exhaust velocity distributions, the potential appearance and growth of plasma instabilities, and the time-dependent characteristics of the pulsed discharge current required for efficient compression and heating of the plasma propellant.

Based on the results of the theta-pinch thruster model, a small prototype thruster will be designed and fabricated for experimental evaluation. The scaled down version of the pinched plasma thruster will be used to verify model predictions and to experimentally demonstrate the feasibility of a pulsed theta-pinch device for propulsion applications. Discussions are currently underway with the NASA Glenn Research Center to secure a vacuum facility for testing a small-scale version of the theta-pinch thruster during fiscal year 2001.

If the numerical modeling and experimental test programs are successful, it is anticipated that the theta-pinch thruster will transition into a high power electric propulsion technology development program, leading to a new plasma thruster capable of supporting the future human exploration and development of space.

CONCLUDING REMARKS

A preliminary scaling analysis has been used to evaluate the potential performance of a high power theta-pinch plasma thruster for piloted deep space mission applications. Two performance regimes were analyzed, corresponding to thruster operation with hydrogen propellant at specific impulse values of 5000 s and 10,000 s. The thrust produced by the engine operating at 5000 s and a 100% duty cycle is predicted to be 6300 N, significantly larger the thrust produced by current high power electric propulsion systems at comparable specific impulse. The required peak compression magnetic field was approximately 0.8 T, and the pulse repetition frequency for 100% duty cycle was calculated to be 1225 Hz for a 20-m long discharge chamber. A 10-T superconducting magnet located at the upstream end of the discharge chamber was assumed to provide magnetic mirroring of the compressed plasma column. The corresponding mirror ratio for the system is thus around 12, and roughly 96% of the plasma is exhausted from the open downstream end of the chamber to provide thrust. The thrust produced by the engine operating with hydrogen propellant at 10,000 s specific impulse and a 100% duty cycle was calculated to be around 2500 N, again significantly exceeding the thrust produced by current high power electric propulsion systems at comparable specific impulse. The peak compression magnetic field for this case was 1.5 T, and the pulse repetition frequency for 100% duty cycle was calculated to be 2450 Hz for a 20-m long chamber. In each case, shortening the chamber length would require a higher repetition rate to sustain the same level of thrust, while lengthening the chamber would reduce the required pulse repetition frequency for engine operation at 100% duty cycle. Operating the engine at less than 100% duty cycle would provide the same values of specific impulse but would lower the average thrust.

Based on the results provided by this preliminary analysis, a time-dependent MHD model is being developed to more fully investigate the fundamental physics and potential performance of the theta-pinch thruster. Concurrent with the numerical modeling effort, a modest experimental program has been proposed to evaluate the performance of a small theta-pinch thruster at the NASA Glenn Research Center. If successful, the high power theta-pinch thruster will evolve into a truly unique and versatile propulsion system that will enable and support future piloted deep space exploration missions.

ACKNOWLEDGMENTS

This research was funded through a Phase I grant by the NASA Institute for Advanced Concepts, whose support is gratefully acknowledged. NIAC Grant Number 07600-022, period of performance 1 June 1999 - 30 November 1999.

REFERENCES

Curran, F., "Overview of NASA's Program in Electric Propulsion," AIAA-98-3180, presented at the 34th Joint Propulsion Conference, Cleveland, OH, July 12-15, 1998.

Dunning, J. and Sankovic, J., "An Overview of the NASA Electric Propulsion Program," AIAA-99-2161, presented at the 35th Joint Propulsion Conference, Los Angeles, CA, June 20-23, 1999.

Liefer, S., "Overview of NASA's Advanced Propulsion Concepts Activities," AIAA-98-3183, presented at the 34th Joint Propulsion Conference, Cleveland, Oh, July 12-15, 1998.

NASA Office of Policy and Plans, "Human Exploration and Development of Space Enterprise," in *NASA Strategic Plan*, NASA Policy Directive (NPD)-1000.1, Headquarters, National Aeronautics and Space Administration, Washington D.C., 1998.

Polk, J., Anderson, J., Goodfellow, K., Garner, C., Sovey, J., Rawlin, V., Patterson, M., and Hamley, J., "In-Flight validation of the NSTAR Ion Thruster Technology on the Deep Space One Mission, AIAA-99-2274, presented at the 35th Joint Propulsion Conference, Los Angeles, CA, June 20-23, 1999.

Sovey, J., Rawlin, V., and Patterson, M., "A Synopsis of Ion Propulsion Development Projects in the United States: SERT1 to Deep Space One," AIAA-99-2270, presented at the 35th Joint Propulsion Conference, Los Angeles, CA, June 20-23, 1999.

Sutton, G. P., *Rocket Propulsion Elements*, 2nd ed., New York, J. Wiley & Sons, 1958, pp 45-53.

IEC Fusion: The Future Power and Propulsion System for Space

Walter E. Hammond[1], Matt Coventry[2], John Hanson[3], Ivana Hrbud[4], George H. Miley[2], Jon Nadler[2]

[1]*HQ Air Force/SB, 1180 Air Force Pentagon, Washington, D.C 20330-1180*
[2]*Fusion Studies Lab, University of Illinois at Urbana-Champaign, Urbana, IL 61801*
[3]*Space Transportation Directorate, Mail Code TD-54, NASA Marshall Space Flight Center, Huntsville, AL 35812*
[4]*Propulsion Research Center, Mail Code TD-40, NASA Marshall Space Flight Center, Huntsville, AL 35812*

[1]*256-544-0584; Walter.Hammond@msfc.nasa.gov*

Abstract. Rapid access to any point in the solar system requires advanced propulsion concepts that will provide extremely high specific impulse, low specific power, and a high thrust-to-power ratio. Inertial Electrostatic Confinement (IEC) fusion is one of many exciting concepts emerging through propulsion and power research in laboratories across the nation which will determine the future direction of space exploration. This is part of a series of papers that discuss different applications of the Inertial Electrostatic Confinement (IEC) fusion concept for both in-space and terrestrial use. IEC will enable tremendous advances in faster travel times within the solar system. The technology is currently under investigation for proof of concept and transitioning into the first prototype units for commercial applications. In addition to use in propulsion for space applications, terrestrial applications include desalinization plants, high energy neutron sources for radioisotope generation, high flux sources for medical applications, proton sources for specialized medical applications, and tritium production.

INTRODUCTION

A preliminary conceptual design for an IEC space power unit was recently described by Miley (1999 a). The use of the IEC as a power unit admittedly has to rely on a still-inadequate experimental database. Thus, while the experimental database is growing, extrapolation to an actual power unit design contains many uncertainties. For space power applications, employing D-^{3}He as the IEC fuel is attractive because of reduced radioactivity and neutron emissions, and its relatively large reactivity (Miley, 1997 a). The design of an IEC reactor includes four main components: the IEC itself, a direct energy converter, the step-down electronics, and an energy storage/pulse-forming unit. In addition, cooling systems and waste heat radiation play a major role. A block diagram of a 1-MWe IEC power plant concept is shown in Figure 1 (Miley, 1999 b). Studies (Miley 1994-1999) show that small-scale IEC reactor designs are conceiveable, with very attractive operational characteristics. If higher power levels are desired, a somewhat higher power unit appears possible or, alternately, multiple units can be employed. The IEC is one of the few fusion concepts that permits small size and a modest power level. In addition to the compact size and lightweight design of the IEC, it is ideally suited for coupling energy out using a charged particle electrostatic direct convertor. This in turn leads to a high plant efficiency.

The commercialization of IEC technology has already begun with the Daimler-Chrysler development of relatively low intensity, portable neutron source for activation analysis (Sved, 1997) This is the first commercial application in the world using confined, fusing plasma. Spin-off applications for IEC technology include:
1) Intense Neutron Source for Radioisotope Generation
2) High Flux Source for Medical Radioisotope Oncology
3) High Flux Source for Isotope Production for Medical Applications
4) D-^{3}He Proton Source for Specialized Medical Applications
5) Moderate Flux Neutron Source for Advanced Activation Analysis/Nondestructive Evaluation
6) Tritium Production Using D-D IEC Reactor

CP504, *Space Technology and Applications International Forum–2000*, edited by M. S. El-Genk

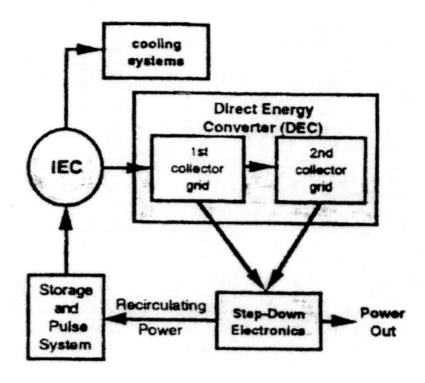

FIGURE 1. Power flow diagram for a conceptual D-He^3 IEC power plant (Miley, et. al. 1999b).

Comparison with other concepts

An assessment was recently made of thirteen previously published fusion space propulsion concepts and how well their operational parameters enabled fast solar system travel (Williams and Borowski, 1997). The process identified requirements (missions, payload mass ratios, and trip times) for fusion-class systems, established requisite ranges of operational parameters (specific power, specific impulse, and nozzle jet efficiency), and assessed system characteristics (mass properties, power output, jet power, and thrust). Preliminary mission analysis was conducted to establish candidate mission requirements for fast solar system travel, from which operational parameters were defined. Thirteen previously published fusion space propulsion concepts, spanning a variety of reactor types, fuels, and fusion processes were surveyed and system characteristics were assessed and compared. An IEC concept by R.W. Bussard (Bussard, 1993, 1994 and Froning, 1997) was an electrostatic confinement fusion reactor fueled with p-^{11}B and is schematically shown in Figure 2. This was the sole concept not to use fusion energy directly to increase the energy of the exhaust mass. Instead, the fusion energy was used to generate electricity by direct conversion, which in turn powered electron beams to heat high pressure plasma for expansion and acceleration. Only conceptual-level design was provided on the fusion reactor and energy conversion equipment. Mass property breakdowns were provided for the fusion reactor, thrust, and electricity systems only. Deterministic mission analysis was based on a continuous high thrust, straight-line transfer method. Performance data was provided for a Mars transfer. The Bussard IEC concept resulted in a total system thrust-to-weight ratio of some 20 milli-g's and a 68 metric ton dry weight. This system thrust-to-weight parameter was only exceeded by a much heavier (486 metric ton) inertial confinement reactor by Hyde (Hyde, 1983) fueled with D-D. The Hyde concept requires heavier and more complex KrF laser driver technology to implode fuel pellets, and has not been shown to be capable of reaching break-even.

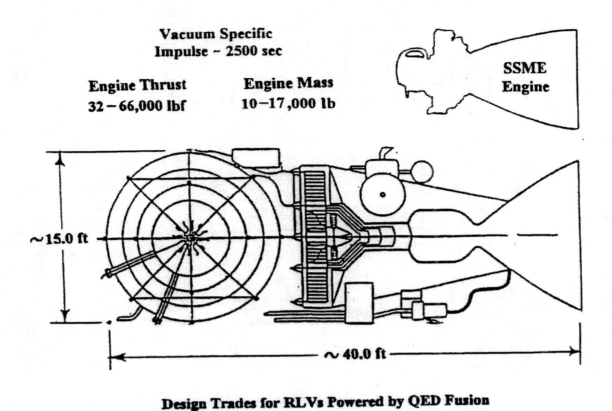

FIGURE 2. Conceptual schematic of Quiet Electric Discharge (QED) fusion engine with electric power converter (Bussard, 1994).

SYSTEM CONCEPT FOR FUSION PROPULSION

The bulk of the energy output from D-^{3}He IEC fusion reactions is in the form of high-energy charged particles, a 14.7-MeV proton and a 3.54-MeV alpha particle. In addition, "side" reactions produce some neutrons, but the energy fraction carried by neutrons can be held to $<<10\%$, so their effect is minimized. To collect the energy of the charged particles and convert it to useful electrical power, a "venetian blind" type direct energy converter is employed (Moir and Barr, 1973). This type of direct energy converter offers a high efficiency of about 80% for the fairly monoenergetic fusion product ions in the IEC. Details are discussed in (Miley, 1999b). The power flow analysis of the IEC reactor components requires knowledge of how the fusion rate scales with injected ion current. Experiments have been done at low currents and these results combined with theoretical studies allow an estimate of the power scaling law at high-currents of injected ions. A minimum scaling of the fusion reaction rate with I^3 (where I = ion current) is predicted theoretically for beam-beam reactions in a deep double well (Miley et al., 1997d). There are, however, several theories that support even steeper scaling rates with current due to greater compression associated with standing wave formation (Bussard, 1991; Bussard et al., 1992). Pulsed operation requires state-of-the-art pulsed power technology to achieve high peak currents, thus taking advantage of the strong scaling with current. Details of a 1-MWe IEC power plant, weighing about 1000 kg. exclusive of cooling radiators and shielding, are given in (Miley, 1999b). Such a pilot unit would provide a most important start in the development of D-^{3}He IEC power. Its predicted performance is impressive, but later generation units would have the possibility of even much better performance. This would come from two improvements: increased Q values (Q = or >1) and improved energy conversion using direct energy conversion.

The physics issues surrounding a successful IEC fusion energy reactor involve resolving the following power drains: 1) electron confinement, 2) ion losses to the grid, 3) charge exchange, 4) ion source profile, and 5) ion collisional time scale. Approaches being undertaken now by the IEC experimental community to solve these problems are discussed in (Miley, 1999b). Numerous studies, spanning almost four decades, have reported exceedingly high neutron yields, as a function of plasma density and energy, and direct measurements of potential well formation. Additional study is also needed to determine if the interior grid used to initiate the IEC fusion reaction can be dispensed with once the fusion process begins, because it is unlikely that present-day cathode-grid technology is sufficient to make a grid that can contain a Q>1 IEC plasma.

IEC Mission Analysis for Mars, Saturn, and Pluto Missions

It was decided to simulate one-way manned missions from Earth to Mars, Saturn, and Pluto. This was intended to give a perspective of the utility of IEC-driven propulsion systems for interplanetary travel to near, mid, and far locations in the solar system. It was further postulated that departures and arrivals were from/to locations in orbits near the respective planets, not to/from their surfaces, as the IEC as envisioned in this paper would be suitable as a "fast freighter" or for fast human transport, but other propulsion systems would be necessary to actually perform the ascent from and descent to the planetary surfaces. A crew of four was assumed for all missions. Table 1 below shows the parameters used and derived in this study, and some final parameters are displayed in the following graphs. Two basic cases were considered: a baseline case and a far-term case. The baseline case used a propulsion module of six deuterium-driven IEC's operating at under breakeven conditions, coupled with a moderate-size IEC reactor operating past the breakeven point to power the six modules. The baseline case assumed 25% deuterium and 75% hydrogen gas injected into the output deuterium plasma from a "jet mode" IEC. The far-term case assumed all-deuterium output plasma flow from the six jet mode propulsion IECs, which would themselves be operating past the breakeven point, so that an additional power-producing IEC to drive them would not be needed. Assumptions for all cases include continuous thrusting and a desire for minimum transfer time. Transfer time was essentially minimized by assuming direct flight out to the planet, with radial thrusting relative to the Sun, as opposed to a lower-energy transfer that makes use of the already-existing orbital motion. This assumes that planetary transfers occur only when the planetary geometries are correct, which occurs roughly every two years for Mars and every year for Saturn and Pluto. Judging from the results, travel from Earth to any planet in the solar system should be possible at any time, using this propulsion and the assumptions made here. The maximum travel time should not exceed the travel time to Pluto plus a small amount. Design of trajectories for the more general case was not attempted as part of this study.

TABLE 1. Baseline and Far Term IEC Four-person Fast Spaceship Mission Analysis Results.

Parameter	Baseline			Far Term		
	Mars	Saturn	Pluto	Mars	Saturn	Pluto
IEC modules and power conditioning (kg)	6,000	6,000	6,000	6,000	6,000	6,000
Six nozzles (kg)	1,200	1,200	1,200	1,200	1,200	1,200
Driving IEC to power thrust modules (kg)	3,000	3,575	4,150	0.0	0.0	0.0
Crew and consumables (kg)	21,601	35,300	70,623	19,993	25,033	35,932
Structure (kg)	10,000	11,500	13,000	10,000	11,500	13,000
Contingency and reserves (kg)	10,000	10,000	10,000	10,000	10,000	10,000
Propellant tanks (kg)	1,962	11,238	35,158	873	4,285	11,666
Propellant (kg)	19,617	112,377	351,575	8,726	42,850	116,657
Final planetary orbit radius (units of planetary radii)	1.5	2.0	1.5	1.5	2.0	1.5
Total Initial Mass in LEO (kg)	73,379	191,188	491,704	56,791	100,867	194,454
Final mass in destination orbit (kg)	53,765	78,803	140,121	48,064	58,011	77,789
Total travel time (days)	17.5	101.0	324.5	8.1	39.6	107.7
Time to Earth escape speed (sec)	46,536	130,791	356,282	7,051	14,291	29,712
Max speed achieved (km/s)	107	298	430	229	758	1,264
Max thrust angle from vertical (deg)	3.0	5.1	8.5	1.37	1.94	2.7

Thrust and specific impulse were assumed constant. The baseline case used a thrust for the six IEC propulsion modules of 8,670 N (1951 lbf), which assumes ¾ hydrogen and ¼ deuterium. The far-term case, with all deuterium, had a thrust of 34,680 N (7,803 lbf) for the six propulsion IEC's. Specific impulse was 70,500 sec for the baseline scenario and 282,000 sec in the far-term case. A crew of four was assumed and was allocated consumables of 40 kg/day per person.

Trajectories were generated using numerical integration. A number of simplifying assumptions were made for this study. All planets were assumed to be in the same plane, in circular orbits about the Sun at their mean radius. Earth orbit departure was from a 470-km altitude circular orbit. Each concept vehicle was simulated as thrusting in the direction of its velocity until it reached escape velocity from the Earth. Then, its position and velocity were re-initialized as being at the Earth's distance and speed relative to the Sun. Thrusting was assumed to be radially outward from the Sun, with a slight offset from that direction to maintain circular speed about the Sun at each distance (the maximum angular offset required is shown in Table 1). The offset assures that the horizontal speed upon arrival at the target planet matches the speed of that planet. The radial speed at planetary arrival was nulled out by turning the vehicle around and thrusting in the opposite direction at the appropriate time. Thus rendezvous with the target planet was assured. Arrival in the final planetary orbit was simulated by beginning in that orbit and

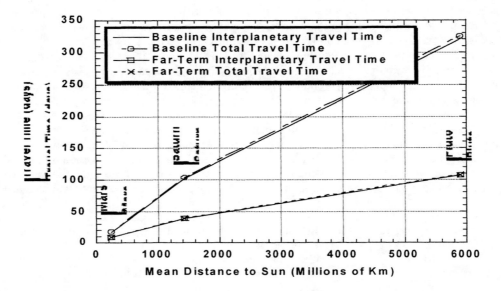

FIGURE 3. Travel Time versus Distance.

thrusting in the direction of the velocity vector until reaching escape velocity. Figure 3 shows the travel time versus distance plot for the baseline and far-term cases, and Figure 4 shows the initial and final masses for the IEC-driven spaceship, again for both baseline and far-term cases.

An interesting result is that the propellant masses and trip times for the baseline case are closer to a factor of two above the far term case rather than a factor of four, even though the acceleration is only one fourth. This is understandable since the distance traveled with a given acceleration goes up as the square of the travel time (ignoring perturbing terms such as solar gravity), so that the time to achieve the same distance with ¼ the acceleration goes up by a factor of two. Since the flow rate is the same in both the baseline and far-term cases, the propellant consumed also goes up by a factor of two. The time to escape the Earth goes up by a much higher factor as orbital mechanics plays a much more important role here. Other methods of Earth escape and planetary arrival were not investigated as part of this analysis. The masses for Saturn are slightly above a straight-line curve, in part because Saturn's large mass requires more energy expenditure for planetary arrival.

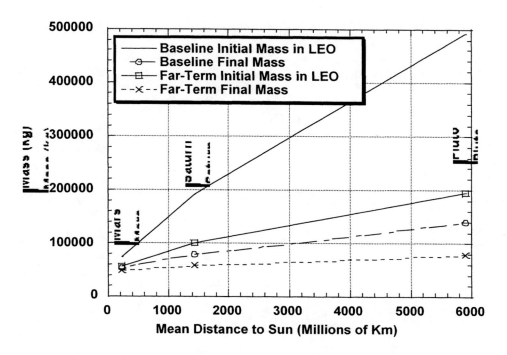

FIGURE 4. Initial and Final Masses.

Need for Continued IEC Funded Research

Further study is required to understand the multiple potential well formation in IEC devices and the scaling of such formation with ion current, energy, etc. This is the central key to developing IEC technology for advanced applications beyond that of small scale neutron activation analysis. The creation of two double wells will increase the IEC efficiency as the device rises in power. Operation with present IEC devices is in the mode of linear scaling of neutron output with cathode-grid current. It is believed that too high of a background operating pressure is truncating ion confinement, preventing the formation of deep, multiple potential wells. Construction of next-generation IEC devices is needed to reduce background neutral gas pressure and increase ion injection currents, so as to measure reaction rates as a function of cathode-grid current.

CONCLUSIONS

Inertial-Electrostatic Confinement offers several possible advantages for the confinement of a Q>1 fusion plasma. Numerous past experiments have produced data that warrants further investigation by the fusion community. Due to the simplistic nature and relatively small physical scale of an IEC device, it is possible to continue exploring the potential of IEC concepts with modest experimental programs in universities and federal laboratories. A series of IEC fusion experiments is needed to resolve the relevant technical issues, so that private industry can aggressively pursue the engineering and manufacturing development of Q>1 IEC reactor systems. The study performed herein was only intended to introduce the concept to the interplanetary travel community. Further, more detailed studies and refinements will be necessary in the future, particularly as the true flight weight of operational IEC units is better quantified. It is shown here that fast travel times throughout the solar system are possible with this technology.

ACKNOWLEDGMENTS

The authors wish to acknowledge the generous contributions from the Space Transportation Directorate at NASA Marshall Space Flight Center for the use of interplanetary flight simulation codes and computer time.

REFERENCES

Bussard, R.W., *Fusion Technology*, Vol. 19, No. 2, pp. 273-293 (1991).Bussard, R.W., Jameson, L.W., and King, K.E., "Ion-Acoustic Waves and Ion-Wave Trapping in IEC Systems," *Bulletin of the American Physical Society*, Vol. 37, No. 6, p. 1582 (1992).

Bussard, R.W., "The QED (Quiet Electric Discharge) Engine System: Direct-Electric Fusion-Powered Propulsion Systems," American Institute of Physics, *Proceedings of the 10th Sysposium on Space Nuclear Power and Propulsion*, CONF-930103, M.S. El-Genk and M.D. Hoover, eds. American Institute of Physics, New York, AIP Conf. Proc. No. 271, Vol. 3, pp. 1601-1611 (Jan. 1993).

Bussard, R.W. and Jameson, L.W., "Design Considerations for Clean QED Fusion Propulsion Systems," in *Proc. 11th Symposium on Space Nuclear Power and Propulsion*, CONF-940101, M.S. El-Genk and M.D. Hoover, eds. American Institute of Physics, New York, AIP Conf. Proc. (Jan. 1994).

Froning, H.D. Jr., "Roadmap for QED (Quiet Electric Discharge) Engine Research and Development,"work performed under NASA Purchase Order No. H28027D for the NASA Marshall Space Flight Center, AL (July 1997).

Hirsch, R.R., "Inertial-Electrostatic Confinement of Ionized Fusion Gases," *J. Appl. Phys.*, v. 38, no. 11, p. 4522 (1967).

Hyde, R.A., "A Laser-Fusion Rocket for Interplanetary Propulsion," 34th International Astronautical Federation, Budapest, Hungary (Oct. 1983).

Miley, G.H., Javedani J., Nebel R., Nadler, J, Gu, Y., Satsangi, A. and Heck, P., "An Inertial Electrostatic Confinement Neutron/Proton Source," *Third International Conference on Dense Z Pinches*, eds. Haines, Malcolm, and Knight, Andrew, American Institute of Physics (AIP) Conference Proceedings 299, AIP Press, New York, pp. 675-689 (1994).

Miley, G.H., Gu, Y., Ohnishi, M., Yamamoto, M., Hasegawa, M., Yoshikawa, K., "Potential Well Structure and Scaling Studies for the IEC," International Sherwood Fusion Theory Conference, Madison, Wisconsin, April 23-26, (1997 a).

Miley, G.H., "A Novel 2.5 MeV D-D Neutron Source," *Journal of Brachytherapy International*, Vol. 1, No. 1, pp. 111-121, (1997 b).

Miley, G.H., Bromley, B., Jurczyk, B., Gu, Y., "Progress in IEC Research for Near-Term Thrusters and Future Fusion Propulsion," *JPL Propulsion Workshop*, Pasadena, CA., May 21, (1997 c).

Miley, G.H., "The Inertial Electrostatic Confinement Approach to Fusion Power," *Current Trends in International Fusion Research*, edited by Emilio Panarella, Plenum Press, New York, NY, pp. 135-148 (1997 d).

Miley, G.H., Bromley, B., Jurczyk, B., Stubbers, R., DeMora, J., Chacon, L., and Gu, Y., "Scaling of the Intertial Electrostatic Confinement (IEC) for Near-Term Thrusters and Future Fusion Propulsion," *Proceedings of the 15th Symposium on Space Nuclear Power and Propulsion* (Space Technology & Applications International Forum STAIF 98), Part 3, pp. 1373-1375, Albuquerque, NM (Jan. 1998).

Miley, G.H., "A D-3He IEC Power Unit for Space Applications," *Proceedings, Space Tech. & Applications International Forum* (STAIF 99), U. of New Mexico, Albuquerque, NM (Jan. 1999 a)

Miley, G.H., J. Nadler, B. Jurczyk, R. Stubbers, J. DeMora, L. Chacon, and M. Nieto, "Issues for Development of Intertial Electrostatic Confinement (IEC) for Future Fusion Propulsion," AIAA Paper No. 99-2140, *35th AIAA/ASME/SAE/ASEE Joint Propulsion Conf. & Exhibit*, Los Angeles, CA. (June, 1999 b).

Moir, R.W. and Barr, W.L., "Venetian-Blind Direct Energy Converter for Fusion Reactors," *Nuclear Fusion*, Vol. 13, No. 35 (1973).

Sved, J., "The Commercial IEC Portable Neutron Source," *Transactions of the American Nuclear Society*, Vol. 77, p. 504 (1997).

Williams, C.H., and Borowski, S.K., "An Assessment of Fusion Space Propulsion Concepts and Desired Operating Parameters for Fast Solar System Travel," AIAA technical paper no. 97-3074, *33rd AIAA/ASME/SAE/ASEE Joint Propulsion Conf. & Exhibit*, Seattle, WA. (July, 1997).

MULTIMEDIA DISPLAY SESSION

Commercial Combustion Research:
ISS Hardware and Commercial Products

F.D. Schowengerdt

Center for Commercial Applications of Combustion in Space,
Colorado School of Mines, Golden, CO 80401
(303) 384-2091, fschowen@mines.edu

Abstract. This display features some of the commercial combustion hardware planned for the International Space Station, as well as one of the early commercial products to come out of the Center for Commercial Applications of Combustion in Space (CCACS) at the Colorado School of Mines. Backing up the display is a board showing all of the combustion research within CCACS. Each of the projects is described, along with the hardware planned for the Space Shuttle and the International Space Station. The industrial partners in CCACS are listed, along with the project interests of each. Finally, summaries of the investments made to date by all the parties are shown.

The primary ISS hardware to be used for commercial combustion research, and the first commercial hardware to be sent to ISS, is SpaceDRUMS™. The centerpiece of the display is a full-scale mockup of the SpaceDRUMS™ facility, which is scheduled to be launched to ISS aboard UF-1 in January, 2001. This facility was developed by CCACS industrial member Guigné International Ltd. (GIL), in partnership with SpaceHab™. It is capable of high-precision positioning, manipulation and shaping of both solid and liquid samples. The positioning system operates by application of pulsed acoustic restoring forces. The main advantage of this approach over other containerless processing technologies (i.e., electromagnetic, electrostatic, optical, aerodynamic or conventional acoustic techniques) is the ability to process large samples with complete elimination of any forces that may have a detrimental effect on the material being processed. The facility can position samples up to 5 cm in diameter inside a large spherical chamber. No forces act on the object until it floats outside a pre-defined volume, at which time short acoustic pulses are applied by an array of 20 acoustic projectors arranged at the vertices of a dodecahedron, which are digitally controlled through a feedback loop by a set of three orthogonal infrared cameras. The sample can be allowed to drift away from the assigned position (center or any point inside the chamber) by a few millimeters or by as much as 10 cm. The shape of the restoring force pulse is designed to induce minimum disturbances, even in a liquid sample. In the process, the speed and direction of rotation of solid samples is controlled and they are effectively isolated from all external sources of vibration. The ISS facility will make possible containerless combustion synthesis of exotic material samples, handling of large glass samples and fiber-pulling investigations. The latter experiments will be done in cooperation with researchers at the Marshall Space Flight Center (MSFC). The same facility specifically designed for the high-temperature needs of glass and ceramic sample processing will also be used for the processing of porous ceramic for bone replacement and other applications using combustion synthesis. The facility can also be used for room temperature processing of many types of materials, including protein crystals.

The display also contains a new commercial product, a de-modulating digital camera, which was developed in the course of the combustion research within CCACS and in conjunction with member Princeton Instruments (now a division of Roper Scientific). The camera operates through lock-in detection and amplification applied to the signals associated with each pixel in the field of view. By employing a modulation/de-modulation technique, weak images can be extracted from the strong backgrounds that characterize combustion experiments. Particular portions of flames and solid matter within flames can be observed using this camera.

CP504, *Space Technology and Applications International Forum–2000*, edited by M. S. El-Genk

The Soft Stowage® Catalog: A New Approach to Procuring Space Qualified Hardware

David A. Smith

Boeing, Spacelab Program, 499 Boeing Blvd., JC-70 Huntsville, AL 35824
(256) 461-3533; davidalan.smith@boeing.com

EXTENDED ABSTRACT

The patented Soft Stowage® Human Space Logistics System had already proven itself within the Shuttle system of reusable carriers where it has been used extensively to transport cargo both up to and down from the Russian Mir Space Station. For the International Space Station (ISS) however, Boeing wanted to offer a seamless product line that offered launch/landing and orbital stowage hardware, as well as associated integration services that reduce the time, documentation, and cost of transporting goods between earth and earth orbit. To meet that objective Boeing developed a comprehensive Soft Stowage® commercial catalog that offers both fixed pricing and delivery of standard items six weeks from order. The ability to obtain modular stowage accommodation elements through a standardized catalog promises to significantly reduce the cost and time to get payload to orbit. To date, Boeing's Soft Stowage® Catalog has supported delivery of over 600 elements to Spacelab, SPACEHAB, ISS and other payload customers.

While other soft bags have been used in space, Soft Stowage® fabric carriers provide the only integrated family of standard equipment capable of routinely transporting payload to orbit without using much heavier "hard" lockers or containers. The standardization of its rugged design and related integration services directly contribute to its low operating cost and its ability to be marketed in a catalog format. While not all payload is suited to transportation in a soft container, the system's mass and volume savings has allowed payloads ranging from Russian Space Agency space suits and Mir replacement oxygen generators to science instruments like NASA's Optical Properties Monitor to be cost effectively transported to Mir and back. The adaptability/conformity offer by the Soft Stowage® system is particularly useful when articles to be returned from orbit are of different shapes and sizes from those carried up.

Benefits of ordering this system from a published catalog include:
- Modular hardware elements
- Standardized interfaces
- Customization by addition of available accessories
- Fixed prices
- Six week delivery from standard item order
- Optional analytical integration support (depending on spacecraft used)
- Special order of unique hardware elements

Benefits of using this system include:
- More usable payload mass to orbit: less support structure needed
- More usable payload volume: optimizes use of existing volume
- Simple payload-to-vehicle interface: adapter plates and straps
- Test verified analytical models
- Decreased payload development effort and cost to you the user by procuring "off the shelf" hardware
- Established certification documentation

CP504, *Space Technology and Applications International Forum–2000*, edited by M. S. El-Genk
2000 American Institute of Physics 1-56396-919-X

- Conformal packaging: supports "last minute" payload packing, delivery and installation into vehicle
- Safe, easy payload transfer: soft materials prevent damage during intervehicular transport
- Reusability: foam pillows facilitate on-orbit repacking of bag and contents for return
- Non-combustible: all materials defined in NASA's MAPTIS database

The Soft Stowage® Catalog is functionally divided into two categories of equipment: Launch/landing logistics delivery and On-orbit logistics. By providing unique parts to perform each function in an optimal manner, Soft Stowage® produces efficiencies many times greater than a "one size fits all" solution. In contrast, a standard International Space Station Stowage Rack must provide a generic interface and launch/landing capability to any item it transports to an orbiting pressurized module. This generic capability demands a complex structure that once on orbit (where it spends most of its useful life), is consequently overbuilt for its mission.

LAUNCH/LANDING TRANSPORT ELEMENTS

These catalog items provide the means to transport payload during Shuttle launch and landing phases. They are further divided into two classes: Non-rack and Rack delivery accommodations.

Non-rack Delivery Accommodations

Accommodations consist of Adapters, Restraints, Middeck Locker Volume Equivalent (MLVE) Flight Bags of different sizes and Rack Volume Equivalent (RVE) Inserts which are used in combination to transport payloads in any location not occupied by a rack (i.e. center aisle, endcone, bulkhead, or unoccupied rack location). Structurally efficient, and of simple design, these components offer mass savings up to a factor of 4 over current metallic stowage containers. For SPACEHAB, Flight Bags and Adapters replaced most Middeck Lockers for an average mass saving of 15 lb/locker (70% reduction). Because of their inherent mass and volume savings, Flight Bags have been on SPACEHAB logistics missions to Mir, ISS Node 1 launch and SPACEHAB resupply mission to ISS. While Flight Bags and Inserts can be used on any vehicle, specific Adapter designs may be needed to meet each vehicle's unique interface. For example, flying a large Adapter in place of an ISS rack would require different interface hardware from that required for SPACEHAB Middeck Locker Adapter.

Rack Delivery Accommodations

Accommodations consist of the Softrack® system of Adapters and Liners that are installed into an International Standard Payload Rack. They were initially developed for Spacelab logistics missions to the Mir Space Station to replace the heavy rack secondary structure required to accommodate standard, aluminum stowage containers. For SPACEHAB stowage racks, the Liners (with Adapters) reduce secondary support structural mass by 39% while at the same time increasing usable rack volume by 70%. Since Liners come in different lengths (relative to the depth of the rack), the full depth of the rack can be used by payload using a combination of Liners whose lengths vary according to their local vertical location in a rack.

ON ORBIT LOGISTICS ACCOMMODATIONS

Because of the space station's high inclination orbit, only 20% of a typical ISS module's rack complement can be installed at launch. The remaining "empty" rack locations (up to 16 rack positions/module) are covered with Rack Volume Closeouts (RVCO's) for launch. These 25 lb RVCOs are simple partitions installed to control cabin airflow between modules and must be disposed of once a replacement ISS rack is delivered by NASA's Multi-Purpose Logistics Module (MPLM), ESA's Autonomous Transfer Vehicle, or NASDA's H-II Transfer Vehicle. It may require many Earth To Orbit Vehicle flights over several years to bring up the remaining module racks, so accommodations for stowage, trash management and powered payloads will be extremely limited during initial station operations.

The Zero-g Softrack® (ZSR), essentially a lightweight, fabric ISS rack, can be used to replace these RVCOs and provide rack-like utility years before a standard ISS rack can be delivered. It consists of a fabric, Rack Volume Equivalent (RVE) Insert of different standard configurations capable of carrying various payload containers, an aluminum and fabric Shell that provides the Insert with a standardized interface to the ISS rack standoff structure. The Shell incorporates a "seat track" interface to support crew IVA and includes an optional rack center post that can be used to provide a rigid, on-orbit interface to the Frame for front panel mounted payloads. It can be folded for stowage or transportation and, like an ISS rack, supports rack pivoting for access to the pressure shell. The fabric nature of ZSR promotes both noise reduction and passive microgravity isolation. Empty (no payload), the Shell combination nominally weighs 45 lb. Over time the function of a particular ZSR can easily be changed by replacing the Insert (empty weight of 10-25 lb) with another one tailored to offer new accommodations. The utility offered by the ZSR is such that ISS is currently planning to fly 31 ZSRs of different catalog configurations located in US ISPR locations scattered throughout NASA, ESA and NASDA modules.

BENEFITS OF A CATALOG APPROACH

A published catalog can much more broadly inform the user community on the advantages of a product or service than a comparable technical presentation. Basing a catalog on the web further leverages the reach of a particular product by supporting remote access, convenient updating and reduced production costs.

However, not all products lend themselves for sale in a catalog medium. Generally, the developer/provider must invest their own capital into development of the system so they have the freedom to market as they please. Fitting a product line into a catalog format requires a streamlining of the product fabrication, certification, delivery and quoting processes-all of which requires investment on top of that needed to develop the basic product. Obviously, many of the products and services developed in support of NASA's Human Space program do not lend themselves to catalog sales. There are three criteria that should be evaluated before developing a catalog:

- Does it make technical sense to offer a product via catalog? The only way to effectively market the Soft Stowage® system with all of its various components, sizes and options was through a catalog medium.
- Can the total projected sales justify the additional investment required of creating and maintaining a catalog and a catalog process? The ISS offers the Soft Stowage® system at least 10 years of projected sales with customers ranging from government to private.
- And specific to NASA, can you develop and provide the necessary certification data for all anticipated applications cost effectively? The Soft Stowage® system was qualified to envelop the requirements of all human space vehicles and the complementary certification data was formatted for compliance with both NASA STS and ISS specifications.

NASA's projected ISS utilization budget demands new approaches that decrease cost and risk while providing new capabilities. Solutions can not be limited to new technology, but must embrace sustainable technology. Boeing continues its catalog experiment with the Soft Stowage® system in order to increase the transport and orbital accommodation options available to the Human Space customer while evaluating whether the catalog business case will be ultimately profitable to Boeing.

Spaceborne and Ground Thermal Infrared Observations Provide a Firm Basis for the Space Station FireMapper

James W. Hoffman[1], Philip J. Riggan[2], James A. Brass[3]

[1]*Space Instruments, Inc., 4403 Mancester Ave., Encinitas, CA 92024*
(760) 944-7001, jhoffsi@aol.
[2] *USDA Forest Service, Pacific Southwest Research Station, Riverside, CA 92057*
(909) 315-0182
[3]*NASA Ames Research Center, Moffett Field, CA 94035-1000*
(650) 604-5232

Abstract. A series of successful spaceborne and ground based experiments have provided a firm basis for the design of the Space Station based FireMapper instrument. The prototype of the FireMapper Instrument was called ISIR (Infrared Spectral Imaging Radiometer) and was fabricated by Space Instruments, Inc. for NASA under an SBIR (Small Business Innovation Research) contract. ISIR was flown successfully on shuttle mission STS-85 in August, 1997 and collected over 60 hours of thermal infrared data in 4 spectral bands over a variety of land, ocean, and cloud scenes. This was the first spaceborne test of a thermal imager using an uncooled microbolometer detector array. This proof of concept space experiment demonstrated that the FireMapper will have sufficient sensitivity to observe both fires and tropical deforestation from the Space Staion. Samples of the multispectral imagery obtained on mission STS-85 will be shown.

Although a significant amount of data was obtained on STS-85, several problems with the calibration system were brought to light. These problems have now been resolved with the design of a completely new onboard calibration system. The new calibration system for FireMapper eliminates the need for a complex motor driven iris and provides a direct view blackbody system that allows absolute calibration of the entire instrument in all spectral bands.

In addition to the space experiments, a series of ground experiments have been performed to test the newly derived calibration techniques and to verify dynamic range performance when viewing brush fires and high temperature flame fronts. These ground experiments were performed with the TIR (Thermal Imaging Radiometer) which was developed and fabricated by Space Instruments. As part of a joint research project with the USDA Forest Service, field trials were performed from 1997 to 1999 in which terrestrial scenes and controlled burns were observed in the 8.5 micron, 10.8 micron and 8 to 12.5 micron spectral regions. Time sequences of flame dynamics were also recorded. Calibration algorithms were developed and tested as part of this research project. These techniques and algorithms are continuously being tested and improved by means of ongoing ground based field and laboratory experiments. The most accurate of these calibration techniques and algorithms are now being incorporated Into the FireMapper program. Samples of calibrated flame and terrestrial images will be presented.

In addition to the thermal infrared sensor, the Space Station FireMapper will include 2 high resolution visible/near infrared imaging systems. These will have a spatial resolution approximately 5 times better than the infrared system. They will provide precision monitoring of deforestation due to illegal logging operations. Narrow band spectral filters will additionally allow continuous vegetation index mapping of agricultural and forest areas. An airborne prototype of the visible/near infrared mapping system is now being fabricated at Space Instruments for the USDA Forest Service for flight demonstrations on their research Piper Navajo aircraft.

CP504, *Space Technology and Applications International Forum–2000*, edited by M. S. El-Genk
© 2000 American Institute of Physics 1-56396-919-X/00/$17.00

ISS Qualified Thermal Carrier Equipment

Mark S. Deuser, John C. Vellinger, Wm. M. Jennings IV

Space Hardware Optimization Technology (SHOT), Inc., 5605 Featherengill Road
Floyd Knobs, Indiana 47119
(812) 923-9591; Fax: 923-9598
Mdeuser@shot.com, Jvellinger@shot.com, Mjennings@shot.com

Abstract. Biotechnology is undergoing a period of rapid and sustained growth, a trend which is expected to continue as the general population ages and as new medical treatments and products are conceived. As pharmaceutical and biomedical companies continue to search for improved methods of production and, for answers to basic research questions, they will seek out new avenues of research. Space processing on the International Space Station (ISS) offers such an opportunity! Space is rapidly becoming an industrial laboratory for biotechnology research and processing. Space bioprocessing offers exciting possibilities for developing new pharmaceuticals and medical treatments, which can be used to benefit mankind on Earth. It also represents a new economic frontier for the private sector.

SHOT is the first commercial vendor to successfully design, develop and verify the next generation of space-qualified thermally controlled single locker equipment, which is capable of transporting and processing microbiology experiments while utilizing the power, data, command, and air resources aboard the ISS (in particular the Express Rack). SHOT has continued to stay on the cutting edge of technology associated with the development of space-qualified thermal controllers. Modular components allow SHOT to remain current in the following technologies: insulation, maximized experiment volume, maximized experiment weight, modular experiment attachment, power efficiency, command and data via remote location, compliance with ISS resource requirements, etc.

For over eight years, the thermal carrier development team at SHOT has been working with government and commercial sector scientists who are conducting microgravity experiments that require thermal control. SHOT realized several years ago that the hardware currently being used for microgravity thermal control was becoming obsolete. It is likely that the government, academic, and industrial bioscience community members could utilize SHOT's hardware as a replacement to their current microgravity thermal carrier equipment. Moreover, SHOT is aware of several international scientists interested in utilizing our space qualified thermal carrier. SHOT's economic financing concept could be extremely beneficial to the international participant, while providing a source of geographic return for their particular region. Beginning in 2000, flight qualified thermal carriers are expected to be available to both the private and government sectors.

SHOT is committed to maintaining its leadership role in the commercial marketing of space-qualified thermal carriers by offering high-value flight hardware that allows economical access to microgravity for biological research. SHOT's mission is to provide thermal carriers that can be used in developing products that further the advancement of biological sciences and promote the development of new treatments for human dysfunction.

CP504, *Space Technology and Applications International Forum–2000*, edited by M. S. El-Genk
© 2000 American Institute of Physics 1-56396-919-X/00/$17.00

Texas A&M Vortex Type Phase Separator

Frederick Best

Texas A&M University (TAMU), Department of Nuclear Engineering, NASA Commercial Center for Space Power
College Station, TX. 77843-3133, (409) 845-4108, frb449a@acs.tamu.edu

Abstract. Phase separation is required for regenerative biological and chemical process systems as well as thermal transport and rejection systems. Liquid and gas management requirements for future spacecraft will demand small, passive systems able to operate over wide ranges of inlet qualities. Conservation and recycling of air and water is a necessary part of the construction and operation of the International Space Station as well as future long duration space missions. Space systems are sensitive to volume, mass, and power. Therefore, it is necessary to develop a method to recycle wastewater with minimal power consumption. Regenerative life support systems currently being investigated require phase separation to separate the liquid from the gas produced.

The microgravity phase separator designed and fabricated at Texas A&M University relies on centripetal driven buoyancy forces to form a gas-liquid vortex within a fixed, right-circular cylinder. Two-phase flow is injected tangentially along the inner wall of this cylinder producing a radial acceleration gradient. The gradient produced from the intrinsic momentum of the injected mixture results in a rotating flow that drives the buoyancy process by the production of a hydrostatic pressure gradient. Texas A&M has flown several KC-135 flights with separator. These flights have included scaling studies, stability and transient investigations, and tests for inventory instrumentation. Among the hardware tested have been passive devices for separating mixed vapor/liquid streams into single-phase streams of vapor only and liquid only.

Johnson Space Center (JSC) identified a fixed container, passive vortex phase separator system as being applicable to their proposed Immobilized Microbe Microgravity Water Processing System (IMMWPS). The device is used to separate liquid from gas produced in the biologic bed. The Texas A&M phase separator was selected and the flight hardware fabricated and tested. IMMWPS is in the process of being manifested on STS-107. The phase separator has many inherent features that make it attractive for weightless applications. One feature is the scalable swing volume. This volume allows the separator to act as a liquid accumulator. This feature was a primary design consideration for the IMMWPS experiment. The wing volume allows for varying inlet quality as well as inlet-outlet volumetric flow mismatch. Other features that have been proposed include the device acting as a direct contact heat exchanger, settling tank, or a 0-g wash system.

CP504, *Space Technology and Applications International Forum–2000*, edited by M. S. El-Genk
© 2000 American Institute of Physics 1-56396-919-X/00/$17.00

THE MAGNETIC PARTICLE PLUME SOLAR SAIL CONCEPT

William H. Knuth

Orbital Technologies Corporation (ORBITEC), 1212 Fourier Drive, Madison, Wisconsin 53717

(608) 827-5000, knuthw@orbitec.com

Abstract. A magnetic particle space radiator was proposed in the late 1950s as a means to dissipate waste heat from space nuclear systems. The concept was a plume of hot magnetic particles confined to and traversing a magnetic field produced by super conducting magnets in the space vehicle. The large surface area of the hot particles was expected to effectively radiate away the heat. The cooling particles followed along the lines of the magnetic field and eventually returned to the vehicle where they again picked up a fresh charge of waste heat for return out to the plume. This paper presents a new concept for consideration

The same basic magnetic particle plume idea is proposed in this paper, except the purpose of the plume would be to receive momentum (and possibly electric power) from the solar wind in the manner of a solar sail. Recent nano-technologies allow the magnetic particles to be 2-3 orders of magnitude smaller than envisioned for the heat radiator, and the magnetic field would be stronger than we envisioned in the '50s. The application of the magnetic solar sail would be for propelling space-faring vehicles on long duration exploration of the solar system and possibly beyond.

A plume of about 1000 meters thick, and approximately 50 square kilometers in projected area is envisioned. Such a plume might have an average nano-particle concentration of 10^5 particles per cubic meter, where the nano-particles may have a density of say 10^9 particles per lb. The 50 square kilometer plume would contain about 5000lb of nano-particles. A 1 kilometer thick plume of such magnetic particle concentration is expected be fairly opaque magnetically because each particle has its own magnetic field surrounding it, so the plume is able to effectively deflect the ionized stream of the solar wind. In so doing the overall electromagnetic interactions should include momentum transfer to the plume as well as generation of an electric potential by cutting the magnetic field of the plume. Since the plume is magnetically attached to the space vehicle, it pulls the vehicle along.

A first look is provided at the elements of the system, together with an estimate of the thrust potential and the approximate weights of the system. The system appears have the potential to develop on the order of 50lb and 100lb of thrust and weigh on the order of 15,000lb

AUTHOR INDEX

DATE DUE

OCT 2 200L			
NOV 0 3 2003			